Stem Cell Therapy for Autoimmune Disease

Stem Cell Therapy for Autoimmune Disease

Richard K. Burt
Chief, Division of Immunotherapy
Department of Medicine
Feinberg School of Medicine
Northwestern University
Chicago, Illinois, U.S.A.

Alberto M. Marmont
Professor Emeritus
Division of Hematology and Stem Cell Transplantation
San Martino's Hospital
Genoa, Italy

LANDES BIOSCIENCE
GEORGETOWN, TEXAS
U.S.A.

STEM CELL THERAPY FOR AUTOIMMUNE DISEASE

LANDES BIOSCIENCE / EUREKAH.COM
Georgetown, Texas, U.S.A.

Copyright ©2004 Landes Bioscience / Eurekah.com
All rights reserved.
No part of this book may be reproduced or transmitted in any form or by any means, electronic or mechanical, including photocopy, recording, or any information storage and retrieval system, without permission in writing from the publisher.
Printed in the U.S.A.

Please address all inquiries to the Publishers:
Landes Bioscience / Eurekah.com, 810 South Church Street, Georgetown, Texas, U.S.A. 78626
Phone: 512/ 863 7762; FAX: 512/ 863 0081
www.landesbioscience.com
www.eurekah.com

ISBN: 1-58706-031-0

While the authors, editors and publisher believe that drug selection and dosage and the specifications and usage of equipment and devices, as set forth in this book, are in accord with current recommendations and practice at the time of publication, they make no warranty, expressed or implied, with respect to material described in this book. In view of the ongoing research, equipment development, changes in governmental regulations and the rapid accumulation of information relating to the biomedical sciences, the reader is urged to carefully review and evaluate the information provided herein.

Library of Congress Cataloging-in-Publication Data

Stem cell therapy for autoimmune disease / [edited by] Richard K. Burt,
Alberto M. Marmont.
    p. ; cm.
Includes index.
Includes bibliographical references and index.
  ISBN 1-58706-031-0
  1. Autoimmune diseases--Treatment. 2. Stem cells--Transplantation.
  [DNLM: 1. Autoimmune Diseases--therapy. 2. Stem Cell
Transplantation. WD 305 S824 2004] I. Burt, Richard K., 1956- II.
Marmont, A. M. (Alberto M.)
  RC600.S746 2004
  616.97'806--dc22
                    2003026938

This book is in appreciation of the support, encouragement,
and patience of my wife, Shalina, children, Michael, Rajan, Reena, Shantha,
and my patients who have allowed me the priviledge to be their doctor.
All of whom have been the inspiration to continue.
Richard K. Burt

This book is dedicated to Paul Ehrlich, who created the
fail-safe dictum of "horror autotoxicus", but also appreciated that
violation of this concept was conducive to disease.
Alberto M. Marmont

CONTENTS

1. When is a Stem Cell Really a Stem Cell? ... 1
   Gerald J. Spangrude

2. Embryonic Stem Cells: Unique Potential to Treat Autoimmune Diseases ... 5
   Dan S. Kaufman and James A. Thomson

3. Neural Stem Cells and Oligodendrocyte Progenitors in the Central Nervous System ... 11
   Jennifer A. Jackson and Diana L. Clarke

4. Turning Blood into Liver ... 18
   Bryon E. Petersen and Neil D. Theise

5. Adipose Tissue-Derived Adult Stem Cells: Potential for Cell Therapy ... 24
   Laura Aust, Lyndon Cooper, Blythe Devlin, Tracey du Laney, Sandra Foster, Jeffrey M. Gimble, Farshid Guilak, Yuan Di C. Halvorsen, Kevin Hicok, Amy Kloster, Henry E. Rice, Anindita Sen, Robert W. Storms and William O. Wilkison

6. Hematopoietic Stem Cell Biology: Relevance to Autoimmunity ... 31
   Richard J. Jones

7. Properties and Therapeutic Potentials of Adult Stem Cells from Bone Marrow Stroma (MSCs) ... 35
   Darwin J. Prockop

8. Regeneration of Cardiomyocytes from Bone Marrow Stem Cells and Application to Cell Transplantation Therapy ... 39
   Keiichi Fukuda

9. Clinical Trials of Hematopoietic Stem Cells for Cardiac and Peripheral Vascular Diseases ... 49
   Hiroaki Matsubara

10. The Stem Cell Continuum: A Plastic Plasticity ... 55
    Peter J. Quesenberry, Jean-Francois Lambert, Gerald A. Colvin, Mark Dooner, Christine I. McAuliffe, Mehrdad Abedi, Deborah Greer, Delia Demers, Jan Cerny, Brian E. Moore, Evangelos Badiavas and Vincent Falanga

11. Adult Stem Cell Plasticity ... 59
    Sean Lee and Diane S. Krause

12. Collection and Expansion of Stem Cells ... 73
    Linda Kelley and Ian McNiece

13. The Extracellular Matrix as a Substrate for Stem Cell Growth and Development and Tissue Repair ... 87
    Stephen F. Badylak and Mervin C. Yoder

14. Gene Transfer into Human Hematopoietic Stem Cells: Problems and Perspectives ... 92
    Serguei Kisselev, Tatiana Seregina, Richard K. Burt and Charles J. Link

15. The Etiopathogenesis of Autoimmunity ... 106
    Howard Amital and Yehuda Shoenfeld

16. Overview of Immune Tolerance Strategies ... 113
    Charles J. Hackett and Helen Quill

17. Death Receptor-Mediated Apoptosis and Lymphocyte Homeostasis ... 119
    Lixin Zheng, Richard M. Siegel, Jagan R. Muppidi, Felicita Hornung and Michael J. Lenardo

18. Shifting Paradigms in Peripheral Tolerance ... 132
    Jonathan D. Powell and Ronald H. Schwartz

19. Dendritic Cells Control the Balance between Tolerance and Autoimmunity ... 139
    Simon W. F. Milling and G. Gordon MacPherson

20. $CD4^+$ T Regulatory Cells and Modulation of Undesired Immune Responses ... 148
    Rosa Bacchetta, Megan K. Levings and Maria-Grazia Roncarolo

21. Major Histocompatibility Complex and Autoimmune Disease ... 155
    Ursula Holzer and Gerald T. Nepom

22. Analyzing Complex Polygenic Traits: The Role of Non-HLA Genes in the Susceptibility to Autoimmune Disorders ... 164
    Bernard R. Lauwerys and Edward K. Wakeland

23. Drug-Induced Autoimmunity ... 173
    Robert L. Rubin and Anke Kretz-Rommel

24. Evidence for a Role of Infections in the Activation of Autoreactive T Cells and the Pathogenesis of Autoimmunity ... 182
    J. Ludovic Croxford and Stephen D. Miller

25. Molecular Analysis of Immunity ... 194
    Daniel Douek

26. Immune Reconstitution after Hematopoietic Stem Cell Transplantation — 206
Andreas Thiel, Tobias Alexander, Christian A. Schmidt, Falk Hiepe, Renate Arnold, Andreas Radbruch, Larissa Verda and Richard K. Burt

27. Historical Perspective and Rationale of HSCT for Autoimmune Diseases — 223
Alberto M. Marmont

28. High-Dose Immune Suppression without Hematopoietic Stem Cells for Autoimmune Diseases — 232
Robert A. Brodsky

29. Autologous Stem Cell Transplantation in Animal Models of Autoimmune Diseases — 237
D.W. van Bekkum

30. Allogeneic Hemopoietic Stem Cell Transplantation in Animal Models of Autoimmune Disease — 245
Susumu Ikehara

31. Mobilization and Conditioning Regimens in Stem Cell Transplant for Autoimmune Diseases — 253
Ewa Carrier and Richard K. Burt

32. Infection in the Hematopoeitic Stem Cell Transplant Recipient with Autoimmune Disease — 262
Valentina Stosor and Teresa R. Zembower

33. Immunological Aspects of Multiple Sclerosis with Emphasis on the Potential Use of Autologous Hemopoietic Stem Cell Transplantation — 277
Paolo A. Muraro, Henry F. McFarland and Roland Martin

34. Axonal Injury and Disease Progression in Multiple Sclerosis — 284
Carl Bjartmar and Bruce D. Trapp

35. Monitoring Disease Activity in Multiple Sclerosis — 290
Lorri Lobeck

36. Intense Immunosuppression Followed by Autologous Stem Cell Transplantation in Severe Multiple Sclerosis Cases: MRI and Clinical Data — 303
G.L. Mancardi, R. Saccardi, A. Murialdo, F. Pagliai, F. Gualandi, A. Marmont, M. Inglese, P. Bruzzi, M.P. Sormani, M.G. Marrosu, G. Meucci, L. Massacesi, A. Bertolotto, A. Lugaresi, E. Merelli, M. Filippi and the Italian Gitmo-Neuro Intergroup on ASCT for Multiple Sclerosis

37. Hematopoietic Stem Cell Transplantation for Multiple Sclerosis: Finding Equipoise — 309
Athanasios Fassas and Richard K. Burt

38. Molecular and Cellular Pathogenesis of Systemic Lupus Erythematosus — 320
George C. Tsokos, Yuang-Taung Juang, Christos G. Tsokos and Madhusoodana P. Nambiar

39. Definition, Classification, Activity and Damage Indices in Systemic Lupus Erythematosus — 328
Jennifer M. Grossman and Kenneth C. Kalunian

40. Lupus Nephritis — 339
Annie Y. Suh and Robert M. Rosa

41. Hematopoietic Stem Cell Transplantation for Systemic Lupus Erythematosus — 347
Ann E. Traynor, Richard K. Burt and Alberto Marmont

42. Treatment of Rheumatoid Arthritis — 358
Stuart Weisman and Arthur Kavanaugh

43. Haemopoietic Stem Cell Transplantation for Rheumatoid Arthritis—World Experience and Future Trials — 367
John A. Snowden, John J. Moore, Sarah J. Bingham, Steve Z. Pavletic and Richard K. Burt

44. Autologous Stem Cell Transplantation for Refractory Juvenile Idiopathic Arthritis (JIA) — 378
Nico Wulffraat

45. Immunology of Scleroderma — 388
Carol M. Artlett

46. Hematopoietic Stem Cell Transplantation for Systemic Sclerosis — 398
Andrew M. Yeager, Diane BuchBarker, Thomas A. Medsger, Jr. and Albert D. Donnenberg

47. High-Dose Immunosuppressive Chemotherapy with Autologous Stem Cell Support for Chronic Autoimmune Thrombocytopenia — 404
Richard D. Huhn, Patrick F. Fogarty, Ryotaro Nakamura and Cynthia E. Dunbar

48. High-Dose Chemotherapy with Haematopoietic Stem Cell Transplantation in Primary Systemic Vasculitis, Behçet's Disease and Sjögren's Syndrome — 411
Christoph Fiehn and Manfred Hensel

49. Hematopoietic Stem Cell Transplantation in the Treatment of Chronic Inflammatory Demyelinating Polyradiculoneuropathy 419
George Hutton, Yu Oyama, Richard K. Burt and Uday Popat

50. Hematopoietic Stem Cell Therapy for Patients with Refractory Myasthenia Gravis 429
Richard K. Burt

51. Hematopoietic Stem Cell Transplantation in Patients with Autoimmune Bullous Skin Disorders 434
Joan Guitart and Richard K. Burt

52. Idiopathic Inflammatory Myositis 437
Yu Oyama, Walter G. Barr and Richard K. Burt

53. Hematopoietic Stem Cell Transplantation as Treatment for Type 1 Diabetes 442
Júlio C. Voltarelli, Richard K. Burt, Norma Kenyon, Dixon B. Kaufman and Elizabeth C. Squiers

54. Autologous Hematopoietic Stem Cell Transplantation for Crohn's Disease 448
Robert M. Craig and Richard K. Burt

55. Bronchial Asthma and Idiopathic Pulmonary Fibrosis as Potential Targets for Hematopoietic Stem Cell Transplantation 457
Júlio C. Voltarelli, Eduardo A. Donadi, José A. B. Martinez, Elcio O. Vianna and Willy Sarti

56. Autologous Stem Cell Transplantation in Relapsing Polychondritis 468
Falk Hiepe, Andreas Thiel, Oliver Rosen, Gero Massenkeil, Gerd-Rüdiger Burmester, Andreas Radbruch and Renate Arnold

57. Allogeneic Hematopoietic Stem Cell Transplantation for Autoimmune Diseases 474
Shimon Slavin, Alberto Marmont and Richard K. Burt

Index 481

EDITORS

Richard K. Burt
Chief, Division of Immunotherapy
Department of Medicine
Feinberg School of Medicine
Northwestern University
Chicago, Illinois, U.S.A
Chapter 14, 26, 31, 37, 41, 43, 49, 50, 51, 52, 53, 54, 57

Alberto M. Marmont
Professor Emeritus
Division of Hematology and Stem Cell Transplantation
San Martino's Hospital
Genoa, Italy
Chapter 27, 36, 41, 57

CONTRIBUTORS

Mehrdad Abedi
Research Department
Roger Williams Medical Center
Providence, Rhode Island, U.S.A.
Chapter 10

Tobias Alexander
Clinical Immunology
German Rheumatism Research Centre Berlin
Berlin, Germany
Chapter 26

Howard Amital
Center for Autoimmune Diseases
Department of Medicine 'B'
Sheba Medical Center
Tel-Hashomer
Tel-Aviv University
Tel Aviv, Israel
Chapter 15

Renate Arnold
Department of Haematology and Oncology
University Hospital Charité
Berlin, Germany
Chapter 26, 56

Carol M. Artlett
Division of Rheumatology
Jefferson Medical College
Thomas Jefferson University
Philadelphia, Pennsylvania, U.S.A.
Chapter 45

Laura Aust
Artecel Sciences, Inc.
Durham, North Carolina, U.S.A.
Chapter 5

Rosa Bacchetta
San Raffaele Telethon Institute for Gene Therapy
Université Vita-Salute San Raffaele
Milan, Italy
Chapter 20

Evangelos Badiavas
Research Department
Roger Williams Medical Center
Providence, Rhode Island, U.S.A.
Chapter 10

Stephen F. Badylak
Department of Biomedical Engineering
Purdue University
West Lafayette, Indiana, U.S.A.
Chapter 13

Walter G. Barr
Division of Rheumatology
Feinberg School of Medicine
Northwestern University
Chicago, Illinois, U.S.A.
Chapter 52

Antonio Bertolotto
Department of Neurology
San Luigi Gonzaga Hospital
Orbassano, Italy
Chapter 36

Sarah J. Bingham
Rheumatology Research Unit
University of Leeds
Leeds, U.K.
Chapter 43

Carl Bjartmar
Department of Neurosciences
Lerner Research Institute
Cleveland Clinic Foundation
Cleveland, Ohio, U.S.A.
Chapter 34

Robert A. Brodsky
Division of Hematologic Malignancies
Sidney Kimmel Comprehensive Cancer Center
Johns Hopkins School of Medicine
Baltimore, Maryland, U.S.A.
Chapter 28

Paolo Bruzzi
Unit of Clinical Epidemiology and Trials
National Cancer Institute
Genova, Italy
Chapter 36

Diane BuchBarker
Stem Cell Transplantation Program
Division of Hematology/Oncology
Department of Medicine
University of Pittsburgh School of Medicine
Pittsburgh, Pennsylvania, U.S.A.
Chapter 46

Gerd-Rüdiger Burmester
Department of Internal Medicine
Rheumatology and Clinical Immunology
University Hospital Charité
Berlin, Germany
Chapter 56

Ewa Carrier
Blood and Marrow Transplant Program
University of California San Diego
San Diego, California, U.S.A.
Chapter 31

Jan Cerny
Research Department
Roger Williams Medical Center
Providence, Rhode Island, U.S.A.
Chapter 10

Diana L. Clarke
ES Cell International
Cambridge, Massachusetts, U.S.A.
Chapter 3

Gerald A. Colvin
Research Department
Roger Williams Medical Center
Providence, Rhode Island, U.S.A.
Chapter 10

Lyndon Cooper
Artecel Sciences, Inc.
Durham, North Carolina, U.S.A.
Chapter 5

Robert Craig
Division of Immunotherapy and Gastroenterology
Feinberg School of Medicine
Northwestern University
Chicago, Illinois, U.S.A.
Chapter 54

J. Ludovic Croxford
Department of Microbiology-Immunology
Interdepartmental Immunobiology Center
Feinberg School of Medicine
Northwestern University
Chicago, Illinois, U.S.A.
Chapter 24

Delia Demers
Research Department
Roger Williams Medical Center
Providence, Rhode Island, U.S.A.
Chapter 10

Blythe Devlin
Artecel Sciences, Inc.
Durham, North Carolina, U.S.A.
Chapter 5

Eduardo A. Donadi
Division of Clinical Immunology
University Hospital
School of Medicine of Ribeirão Preto
University of São Paulo
Ribeirão Preto, Brazil
Chapter 55

Albert D. Donnenberg
Stem Cell Transplantation Program
Division of Hematology/Oncology
Department of Medicine
University of Pittsburgh School of Medicine
Pittsburgh, Pennsylvania, U.S.A.
Chapter 46

Mark Dooner
Research Department
Roger Williams Medical Center
Providence, Rhode Island, U.S.A.
Chapter 10

Daniel Douek
National Institute of Allergy and Infectious Disease
National Institutes of Health
Bethesda, Maryland, U.S.A.
Chapter 25

Tracey du Laney
Artecel Sciences, Inc.
Durham, North Carolina, U.S.A.
Chapter 5

Cynthia E. Dunbar
National Heart, Lung and Blood Institute
National Institutes of Health
Bethesda, Maryland, U.S.A.
Chapter 47

Vincent Falanga
Research Department
Roger Williams Medical Center
Providence, Rhode Island, U.S.A.
Chapter 10

Athanasios Fassas
Department of Hematology
The George Papanicolaou General Hospital
Thessaloniki, Greece
Chapter 37

Christoph Fiehn
University of Heidelberg
Department of Hematology, Oncology
  and Rheumatology
Heidelberg, Germany
Chapter 48

Massimo Filippi
Neuroimaging Research Unit
Department of Neuroscience
Scientific Institute
Ospedale San Raffaele
Milan, Italy
Chapter 36

Patrick F. Fogarty
National Heart, Lung and Blood Institute
National Institutes of Health
Bethesda, Maryland, U.S.A.
Chapter 47

Sandra Foster
Artecel Sciences, Inc.
Durham, North Carolina, U.S.A.
Chapter 5

Keiichi Fukuda
Institute for Advanced Cardiac Therapeutics
Keio University School of Medicine
Tokyo, Japan
Chapter 8

Jeffrey Gimble
Artecel Sciences, Inc.
Durham, North Carolina, U.S.A.
Chapter 5

Deborah Greer
Research Department
Roger Williams Medical Center
Providence, Rhode Island, U.S.A.
Chapter 10

Jennifer M. Grossman
Department of Medicine/Rheumatology
University of California at Los Angeles
Los Angeles, California, U.S.A.
Chapter 39

Francesca Gualandi
Division of Hematology and Stem Cell
  Transplantation
San Martino's Hospital
Genova, Italy
Chapter 36

Farshid Guilak
Artecel Sciences, Inc.
Durham, North Carolina, U.S.A.
Chapter 5

Joan Guitart
Department of Dermatology
Feinberg School of Medicine
Northwestern University and Northwestern
  Memorial Hospital
Chicago, Illinois, U.S.A.
Chapter 51

Charles J. Hackett
Division of Allergy, Immunology and Transplantation
National Institute of Allergy and Infectious Diseases
National Institutes of Health
Bethesda, Maryland, U.S.A.
Chapter 16

Yuan Di C. Halvorsen
Artecel Sciences, Inc.
Durham, North Carolina, U.S.A.
Chapter 5

Manfred Hensel
University of Heidelberg
Department of Hematology, Oncology
  and Rheumatology
Heidelberg, Germany
Chapter 48

Kevin Hicok
Artecel Sciences, Inc.
Durham, North Carolina, U.S.A.
Chapter 5

Falk Hiepe
Department of Rheumatology
University Hospital Charité
Berlin, Germany
Chapter 26, 56

Ursula Holzer
Benaroya Research Institute
Virginia Mason Research Center
Seattle, Washington, U.S.A.
Chapter 21

Felicita Hornung
Cell Biology Section
Laboratory of Viral Diseases
National Institute of Allergy and Infectious Diseases
National Institutes of Health
Bethesda, Maryland, U.S.A.
Chapter 17

Richard D. Huhn
Coriell Institute
Camden, New Jersey, U.S.A.
Chapter 47

George J. Hutton
Department of Neurology
Baylor International MS Center
Baylor College of Medicine
Houston, Texas, U.S.A.
Chapter 49

Susumu Ikehara
First Department of Pathology
Transplantation Center
Regeneration Research Center for Intractable Diseases
Kansai Medical University
Moriguchi City, Osaka, Japan
Chapter 30

Matilde Inglese
Neuroimaging Research Unit
Department of Neuroscience
Scientific Institute
Ospedale San Raffaele
Milan, Italy
Chapter 36

Jennifer A. Jackson
Curis Inc.
Cambridge, Massachusetts, U.S.A.
Chapter 3

Richard J. Jones
Sidney Kimmel Comprehensive Cancer Center
Johns Hopkins School of Medicine
Baltimore, Maryland, U.S.A.
Chapter 6

Yuang-Taung Juang
Department of Cellular Injury
Walter Reed Army Institute of Research
Silver Spring, Maryland, U.S.A.
and
Department of Medicine
Uniformed Services University of the Health Sciences
Bethesda, Maryland, U.S.A.
Chapter 38

Kenneth C. Kalunian
Division of Rheumatology, Allergy and Immunology
UCSD Division of Rheumatology
University of California, San Diego
La Jolla, California, U.S.A.
Chapter 39

Dan S. Kaufman
University of Wisconsin
Madison, Wisconsin, U.S.A.
Chapter 2

Dixon B. Kaufman
Division of Transplant Surgery
Department of Surgery
Feinberg School of Medicine
Northwestern University
Chicago, Illinois, U.S.A.
Chapter 53

Arthur Kavanaugh
Center for Innovative Therapy
Division of Rheumatology, Allergy and Immunology
University of California, San Diego
La Jolla, California, U.S.A.
Chapter 42

Linda Kelley
University of Utah
Salt Lake City, Utah, U.S.A.
Chapter 12

Norma Kenyon
Diabetes Research Institute
Department of Surgery
University of Miami School of Medicine
Miami, Florida, U.S.A.
Chapter 53

Serguei Kisselev
Stoddard Cancer Research Institute
Iowa Methodist Medical Center
Des Moines, Iowa, U.S.A.
Chapter 14

Amy Kloster
Artecel Sciences, Inc.
Durham, North Carolina, U.S.A.
Chapter 5

Diane S. Krause
Department of Laboratory Medicine
Yale University School of Medicine
New Haven, Connecticut, U.S.A.
Chapter 11

Anke Kretz-Rommel
Alexon Antibody Technologies
San Diego, California, U.S.A.
Chapter 23

Jean-Francois Lambert
Research Department
Roger Williams Medical Center
Providence, Rhode Island, U.S.A.
Chapter 10

Bernard R. Lauwerys
Service de Rhumatologie
Cliniques Universitaires Saint-Luc
Brussels, Belgium
Chapter 22

Sean Lee
Department of Laboratory Medicine
Yale University School of Medicine
New Haven, Connecticut, U.S.A.
Chapter 11

Michael J. Lenardo
Laboratory of Immunology
National Institute of Allergy and Infectious Diseases
National Institutes of Health
Bethesda, Maryland, U.S.A.
Chapter 17

Megan K. Levings
San Raffaele Telethon Institute for Gene Therapy
Université Vita-Salute San Raffaele
Milan, Italy
Chapter 20

Charles Link
Stoddard Cancer Research Institute
Iowa Methodist Medical Center
Des Moines, Iowa, U.S.A.
Chapter 14

Lorri Lobeck
Department of Neurology
Medical College of Wisconsin
Milwaukee, Wisconsin, U.S.A.
Chapter 35

Alessandra Lugaresi
Department of Oncology and Neurosciences
University of Chieti
Chieti, Italy
Chapter 36

G. Gordon MacPherson
Sir William Dunn School of Pathology
University of Oxford
Oxford, U.K.
Chapter 19

Giovanni Luigi Mancardi
Department of Neurological Sciences, Opthalmology
  and Genetics
University of Genova
Genova, Italy
Chapter 36

Maria Giovanna Marrosu
Department of Neurosciences
University of Cagliari
Cagliari, Italy
Chapter 36

Roland Martin
National Institute of Neurologic Disease and Stroke
National Institutes of Health
Bethesda, Maryland, U.S.A.
Chapter 33

José A. B. Martinez
Pulmonary Division
University Hospital
School of Medicine of Ribeirão Preto
University of São Paulo
Ribeirão Preto, Brazil
Chapter 55

Luca Massacesi
Department of Neurological and Psychiatric Sciences
University of Firenze
Firenze, Italy
Chapter 36

Gero Massenkeil
Departments of Internal Medicine
Rheumatology and Clinical Immunology
Hematology and Oncology
Berlin, Germany
Chapter 56

Hiroaki Matsubara
Department of Medicine and Cardiovascular Division
Kyoto Prefecture University School of Medicine
Kamigyo-ku, Kyoto, Japan
Chapter 9

Christine I. McAuliffe
Research Department
Roger Williams Medical Center
Providence, Rhode Island, U.S.A.
Chapter 10

Henry F. McFarland
National Institute of Neurologic Disease and Stroke
National Institutes of Health
Bethesda, Maryland, U.S.A.
Chapter 33

Ian McNiece
Johns Hopkins Oncology Center
Baltimore, Maryland, U.S.A.
Chapter 12

Thomas A. Medsger, Jr.
Stem Cell Transplantation Program
Division of Rheumatology
Department of Medicine
University of Pittsburgh School of Medicine
Pittsburgh, Pennsylvania, U.S.A
Chapter 46

Elisa Merelli
Department of Neurology
University of Modena
Modena, Italy
Chapter 36

Giuseppe Meucci
Department of Neurology
Hospital of Livorno
Livorno, Italy
Chapter 36

Stephen D. Miller
Department of Microbiology-Immunology
Interdepartmental Immunobiology Center
Feinberg School of Medicine
Northwestern University
Chicago, Illinois, U.S.A.
Chapter 24

Simon W. F. Milling
Sir William Dunn School of Pathology
University of Oxford
Oxford, U.K.
Chapter 19

Brian E. Moore
Research Department
Roger Williams Medical Center
Providence, Rhode Island, U.S.A.
Chapter 10

John J. Moore
Department of Haematology
St. Vincent's Hospital
Sydney, Australia
Chapter 43

Jagan R. Muppidi
Laboratory of Immunology
Autoimmunity Branch
National Institute of Arthritis and Musculoskeletal
 and Skin Diseases
National Institutes of Health
Bethesda, Maryland, U.S.A.
Chapter 17

Paolo A. Muraro
National Institute of Neurologic Disease and Stroke
National Institutes of Health
Bethesda, Maryland, U.S.A.
Chapter 33

Alessandra Murialdo
Department of Neurological Sciences, Opthalmology
 and Genetics
University of Genova
Genova, Italy
Chapter 36

Ryotaro Nakamura
City of Hope
Duarte, California, U.S.A.
Chapter 47

Madhusoodana P. Nambiar
Department of Cellular Injury
Walter Reed Army Institute of Research
Silver Spring, Maryland, U.S.A.
and
Department of Medicine
Uniformed Services University of the Health Sciences
Bethesda, Maryland, U.S.A.
Chapter 38

Gerald T. Nepom
Benaroya Research Institute
Virginia Mason Research Center
Seattle, Washington, U.S.A.
Chapter 21

Yu Oyama
Division of Immunotherapy
Feinberg School of Medicine
Northwestern University
Chicago, Illinois, U.S.A.
Chapter 49, 52

Francesca Pagliai
Careggi Hospital
Bone Marrow Transplantation Unit
Firenze, Italy
Chapter 36

Steve Z. Pavletic
Department of Internal Medicine
University of Nebraska Medical Center
Omaha, Nebraska, U.S.A.
Chapter 43

Bryon E. Petersen
Department of Pathology, Immunology
 and Laboratory Medicine
University of Florida
Gainesville, Florida, U.S.A.
Chapter 4

Uday Popat
Department of Medicine
Center for Cell and Gene Therapy
Baylor College of Medicine
Houston, Texas, U.S.A.
Chapter 49

Jonathan D. Powell
Departments of Oncology and Pharmacology
Johns Hopkins School of Medicine
Baltimore, Maryland, U.S.A.
Chapter 18

Darwin J. Prockop
Center for Gene Therapy
Tulane University Health Sciences Center
New Orleans, Louisiana, U.S.A.
Chapter 7

Peter J. Quesenberry
Research Department
Roger Williams Medical Center
Providence, Rhode Island, U.S.A.
Chapter 10

Helen Quill
Division of Allergy, Immunology and Transplantation
National Institute of Allergy and Infectious Diseases
National Institutes of Health
Bethesda, Maryland, U.S.A.
Chapter 16

Andreas Radbruch
Cell Biology Group
German Rheumatism Research Centre
Berlin, Germany
Chapter 26, 56

Henry E. Rice
Artecel Sciences, Inc.
Durham, North Carolina, U.S.A.
Chapter 5

Maria-Grazia Roncarolo
San Raffaele Telethon Institute for Gene Therapy
Université Vita-Salute San Raffaele
Milan, Italy
Chapter 20

Robert Rosa
Division of Nephrology and Hypertension
Feinberg School of Medicine
Northwestern University
Chicago, Illinois, U.S.A
Chapter 40

Oliver Rosen
Department of Internal Medicine
Rheumatology and Clinical Immunology
Hematology and Oncology
University Hospital Charité
Berlin, Germany
Chapter 56

Robert L. Rubin
Department of Molecular Genetics and Microbiology
University of New Mexico School of Medicine
Albuquerque, New Mexico, U.S.A.
Chapter 23

Riccardo Saccardi
Careggi Hospital
Bone Marrow Transplantation Unit
Firenze, Italy
Chapter 36

Willy Sarti
Division of Clinical Immunology
University Hospital
School of Medicine of Ribeirão Preto
University of São Paulo
Ribeirão Preto, Brazil
Chapter 55

Christian A. Schmidt
Department of Haematology and Oncology
University Hospital Greifswald
Greifswald, Germany
Chapter 26

Ronald H. Schwartz
National Institute of Allergy and Infectious Disease
National Institutes of Health
Bethesda, Maryland, U.S.A.
Chapter 18

Anindita Sen
Artecel Sciences, Inc.
Durham, North Carolina, U.S.A.
Chapter 5

Tatiana Seregina
Stoddard Cancer Research Institute
Iowa Methodist Medical Center
Des Moines, Iowa, U.S.A.
Chapter 14

Yehuda Shoenfeld
Center for Autoimmune Diseases
Department of Medicine 'B'
Sheba Medical Center
Tel-Hashomer
Tel-Aviv University
Tel Aviv, Israel
Chapter 15

Richard M. Siegel
Autoimmunity Branch
National Institute of Arthritis and Musculoskeletal
    and Skin Diseases
National Institutes of Health
Bethesda, Maryland, U.S.A.
Chapter 17

Shimon Slavin
Department of Bone Marrow Transplantation
    and Cancer Immunotherapy
Hadassah University Hospital
Jerusalem, Israel
Chapter 57

John A. Snowden
Department of Haematology and Division
    of Genomic Medicine
Royal Hallamshire Hospital
Sheffield, U.K.
Chapter 43

Maria Pia Sormani
Unit of Clinical Epidemiology and Trials
National Cancer Institute
Genova, Italy
Chapter 36

Gerald J. Spangrude
Departments of Oncological Sciences, Pathology
 and Medicine
Division of Hematology
University of Utah
Salt Lake City, Utah, U.S.A.
Chapter 1

Elizabeth C. Squiers
Genzyme/SangStat
Fremont, California, U.S.A.
Chapter 53

Robert W. Storms
Artecel Sciences, Inc.
Durham, North Carolina, U.S.A.
Chapter 5

Valentina Stosor
Division of Infectious Diseases
Feinberg School of Medicine
Northwestern University
Chicago, Illinois, U.S.A.
Chapter 31

Annie Y. Suh
Division of Nephrology and Hypertension
Feinberg School of Medicine
Northwestern University
Chicago, Illinois, U.S.A.
Chapter 40

Neil D. Theise
Department of Pathology
New York University School of Medicine
New York, New York, U.S.A.
Chapter 4

Andreas Thiel
Clinical Immunology
German Rheumatism Research Centre Berlin
Berlin, Germany
Chapter 26, 56

James A. Thomson
University of Wisconsin
Madison, Wisconsin, U.S.A.
Chapter 2

Bruce D. Trapp
Department of Neurosciences
Lerner Research Institute
Cleveland Clinic Foundation
Cleveland, Ohio, U.S.A.
Chapter 34

Ann E. Traynor
Division of Immunotherapy
Feinberg School of Medicine
Northwestern University
Chicago, Illinois, U.S.A.
Chapter 41

Christos G. Tsokos
Department of Cellular Injury
Walter Reed Army Institute of Research
Silver Spring, Maryland, U.S.A.
and
Department of Medicine
Uniformed Services University of the Health Sciences
Bethesda, Maryland, U.S.A.
Chapter 38

George C. Tsokos
Department of Cellular Injury
Walter Reed Army Institute of Research
Silver Spring, Maryland, U.S.A.
and
Department of Medicine
Uniformed Services University of the Health Sciences
Bethesda, Maryland, U.S.A.
Chapter 38

D.W. van Bekkum
Leiden, The Netherlands
Chapter 29

Larissa Verda
Division of Immunotherapy for Autoimmune Diseases
Feinberg School of Medicine
Northwestern University
Chicago, Illinois, U.S.A.
Chapter 26

Elcio O. Vianna
Pulmonary Division
University Hospital
School of Medicine of Ribeirão Preto
University of São Paulo
Ribeirão Preto, Brazil
Chapter 55

Júlio C. Voltarelli
Division of Clinical Immunology
Bone Marrow Transplantation Unit
University Hospital
School of Medicine of Ribeirão Preto
University of São Paulo
Ribeirão Preto, Brazil
Chapter 53, 55

Edward K. Wakeland
Center for Immunology
Southwestern Medical Center
University of Texas
Dallas, Texas, U.S.A.
Chapter 22

Stuart Weisman
Center for Innovative Therapy
Division of Rheumatology, Allergy and Immunology
University of California, San Diego
La Jolla, California, U.S.A.
Chapter 42

William O. Wilkison
Artecel Sciences, Inc.
Durham, North Carolina, U.S.A.
Chapter 5

Nico Wulffraat
Pediatric BMT Unit
University Medical Center Utrecht
Utrecht, The Netherlands
Chapter 44

Andrew M. Yeager
Stem Cell Transplantation Program
Division of Hematology/Oncology
Department of Medicine
University of Pittsburgh School of Medicine
Pittsburgh, Pennsylvania, U.S.A.
Chapter 46

Mervin C. Yoder
Cancer Research Institute
Indiana University School of Medicine
Indianapolis, Indiana, U.S.A.
Chapter 13

Teresa R. Zembower
Division of Infectious Diseases
Feinberg School of Medicine
Northwestern University
Chicago, Illinois, U.S.A.
Chapter 32

Lixin Zheng
Laboratory of Immunology
National Institute of Allergy and Infectious Diseases
National Institutes of Health
Bethesda, Maryland, U.S.A.
Chapter 17

Representation of Shiva. ©2001 Indra Sharma/Mandala Publishing. Used with permission.

PREFACE

The Immune System and Its Treatment with Stem Cell Therapy: Creator, Preserver, and Destroyer

The creation of something new means loss of the old, while destruction means, for better or worse, a new beginning. This duality of existence is an element of Eastern religions and philosophies. For example, Shiva, the Hindu Deity of creation and preservation, is also the God of destruction. The paradox and complexity of the immune system is that it embodies these same attributes. It is the protector and preserver without which human life is impossible, as well as in the case of autoimmune disease, the destroyer of that which it protects.

Shiva represents contradictions that coexist in conscious reality. The many appendages are images for contradictory and varied functions within one consciousness. The hands hold objects representing these attributes, a flame for destruction, the trident for creation, and a snake as imagery for mastery over nature. Shiva's duality is sometimes also manifest in gender, pictured not as male or female but as a "Half Woman Lord". So it is with the immune system, autoimmune diseases more common in women, and tasks that are interrelated with numerous arms and opposing functions. Attempts at treating one component invariably affects, not uncommonly in an adverse way, other circuits or components of immune surveillance. The immune system interacts with nature to hold mastery over environmental pathogens, destroying or coaxing them to live in peaceful coexistence, but other times influenced by pathogens, xenobiotics, gender hormones, or other environmental triggers to destroy the host it also protects.

Central tolerance, that is negative selection within the thymus, eliminates T cells with opposite avidities. T cells expressing either too strong or having no affinity for self-epitopes undergo apoptotic destruction, while those with mild/moderate affinity for self-peptides are positively selected. These T cells may in turn mediate autoimmunity and/or down regulate autoimmune responses through regulatory T cells. The process of thymic education that selects for repertoires with mild to moderate self-recognition results in an autoimmune duality. Autoimmune cells are physiologic while autoimmune disease is pathologic. Autoimmunity is normal, but autoimmune disease is abnormal. Unlike a malignancy, in which a tumor cell is always viewed as pathologic, for autoimmune disease, treatment must also preserve that which the therapy is designed to destroy. It is for these reasons that this textbook on stem cell therapy for autoimmune disease has numerous chapters devoted to basic immunology. The clinical art of balance in treating an intertwined and at times contradictory immune system begins with the science of immunology.

Stem cell transplantation may be complicated by treatment-related mortality and like the immune system that it regenerates has equal potential to either create and preserve or destroy. The dual nature that defines stem cells is differentiation that ultimately leads to death and self-renewal which leads to immortality. What types of stem cells are there? How are they collected? What are their attributes and characteristics? This textbook devotes many chapters to familiarize the reader with the basic science, clinical aspects, and new questions being raised in the field of stem cell biology. Blood stem cells for tolerance and tissue regeneration are a rapidly developing research and clinical field that is being applied to autoimmune diseases.

In clinical trials, autologous hematopoietic (blood) stem cells are being used to reduce the cytopenic interval following intense immune suppressive transplant regimens. While as yet not delineated, some possible mechanisms and pathways leading to tolerance after hematopoietic stem cell transplantation are suggested in these chapters. Tissue regeneration from blood stem cells is also suggested by animal experiments on stem cell plasticity or metamoirosis (i.e., change in fate) as described within this textbook. Ongoing early clinical trials on tissue regeneration from blood stem cells are described in the chapter on stem cell therapy for cardiac and peripheral vascular disease. Whether autologous hematopoietic stem cells, through the process of mobilization and reinfusion, may be manipulated to contribute to tissue repair in autoimmune diseases is a future area for translational research.

Allogeneic stem cell sources include siblings or other related donors, unrelated donors, cord blood, and finally although not yet in clinical trial, embryonic stem cells. Allogeneic stem cells offer the advantage of changing genetic susceptibility to autoimmune disease. In addition they possess the ability of eradicating the autoimmune clones which survive the conditioning regimens, which in these patients are preferably of the nonmyeloablative (NST) modality. Thus, a Graft-versus-Autoimmunity effect is elicited, as discussed in the chapters by Marmont and Slavin et al.

Embryonic stem cells are being applied in vitro and in animal models to both tolerance and regenerative medicine. In animal models, embryonic stem cells have been used as an alternate marrow donor source resulting in engraftment occurring across major histocompatibility barriers without graft versus host disease (Burt et al, unpublished). When working with embryonic stem cells, simultaneous creation, preservation, and destruction are not just metaphysical discussions but real issues yet to be resolved by society.

Embryonic stem cell lines have been developed from blastocysts that were otherwise destined to be destroyed following in vitro fertilization. In vitro fertilization involves the collection of numerous oocytes that are fertilized not within the fallopian tubes but rather in a Petri dish containing even more numerous donor spermatocytes. While numerous blastocysts (pre-implantation embryos) form, only one will be implanted in the uterus, the others are cryopreserved. When the couple no longer desires children, what becomes of the remaining blastocysts? Paradoxically, in the philosophical spirit of attempting to preserve the life of the unborn, blastocysts are currently destroyed rather than being allowed to live on as embryonic stem cell lines that contribute to the tissue and existence of another individual. Does the blastocyst have a soul and if so, is it better to allow its existence to continue in another individual or to destroy it entirely?

Hematopoietic stem cells as a therapeutic tool to induce tolerance were first applied to human autoimmune disorders in 1996. Since then, the door has opened on trials involving numerous autoimmune diseases and allogeneic as well as autologous hematopoietic stem cells. Stem cells are becoming a new therapeutic tool to supplement or replace traditional approaches of surgery, pharmacy, and radiotherapy. As the embryonic stem cell-related ethical issues involving creation and destruction are resolved, their clinical application may follow the clinical paths, trails, and trials already being explored with hematopoietic stem cells.

Richard K Burt, M.D.
Alberto M. Marmont, M.D.

CHAPTER 1

When is a Stem Cell Really a Stem Cell?

Gerald J. Spangrude

Introduction

In recent years, data from numerous experimental studies has suggested that the potential uses of stem cells in medicine may reach far beyond bone marrow transplantation. How applicable is recent research to modern medicine, and how soon might we expect to see stem cells applied to tissue engineering problems? These and other questions are explored in this introductory chapter. It is altogether fitting that a discussion of the therapeutic potentials of stem cell therapy be grounded in our field, being the first to apply stem cell therapy to the clinical management of acquired and inherited diseases. But what is a stem cell? In the context of bone marrow transplantation, we understand the answer to this question in a concrete and functional sense due to decades of research and clinical applications that grew out of the need to understand the effects of ionizing radiation on biological systems. In the years following the Second World War, a considerable amount of scientific effort was focused on the prevention and treatment of radiation sickness. From these studies came the observation that transplants of spleen or bone marrow cells contribute to cellular recovery following lethal radiation.[1] Almost 50 years after this dramatic insight, we now understand that the ability of such transplants to reconstitute hematopoiesis following radiation depends upon the presence of extremely rare stem cells found predominantly in the bone marrow but capable of mobilization into peripheral tissues via the blood vascular system.[2]

After many years shrouded in mystery and controversy, the characteristics of blood stem cells were gradually revealed through novel assays[3-5] and methods for isolation of these rare cells.[6,7] We now understand that the definition of a stem cell must include the two essential characteristics of self-renewal (cellular division maintains stem cell potential) and multipotency (differentiation into functionally distinct lineages). To complicate matters, it is clear that progenitor cells, which are multipotent but lack self-renewal potential, are often difficult to distinguish from true stem cells.[8] Finally, at least some confusion persists in the tissue stem cell field, where unipotent precursor cells which maintain a tissue through a self-renewing process are often considered stem cells.

The general field of stem cell biology has been the subject of intense public interest in recent years for several reasons. First, the demonstration that recipients of bone marrow transplants harbor donor-derived cells in a variety of tissues has radically changed our expectations for the applications of this type of therapy,[9] even though many questions have been raised by these interesting findings.[10] Second, the derivation of totipotent human stem cells from both embryonic and fetal sources has introduced a potential new source of tissue for engineering applications. Equally important, this new technology marks the genesis of a new level of conflict between science and religion that surpasses that raised by older questions of creationism versus evolution. The potential use of stem cells derived from adult tissues introduces yet another side to this complex story. How are we to define when a stem cell is a stem cell? It is in this vein that I examine a few of the historical aspects of stem cell biology in order to better understand where we have come from at this stage in the development of the stem cell field.

Embryonic Stem Cells: A Timeline

Lewis[11] has correctly identified the origins of the stem cell biology field in the work of Leroy Stevens, a developmental biologist who identified frequent testicular tumors arising spontaneously in strain 129 mice at the Jackson Laboratories. This work was published to little fanfare beginning in 1958.[12] However, the curiosity of Mintz and Illmensee led to a startling observation. When malignant teratocarcinoma cells were mixed into developing mouse embryos, the environment of the embryo harnessed the unregulated growth of the tumor and directed these cells to proper channels of proliferation and differentiation.[13] The result was chimeric mice in which a significant portion of the body mass was derived from the teratocarcinoma. This startling discovery was viewed at the time as evidence for environmental regulation of malignant growth, but the potential of these cells was certainly not overlooked by developmental biologists. Embryonic stem cell lines were derived from the inner cell mass of mouse blastocysts in 1981,[14,15] as shown in Figure 1. These cells were adapted for growth in culture without differentiation, but could differentiate into mesoderm, endoderm, and ectoderm in vitro and in vivo. The derivation of embryonic stem cell lines was rapidly exploited to give birth to the field of targeted mutagenesis,[16,17] an entirely new approach to the investigation of complex mammalian biological systems. Today, it is difficult for scientists to imagine a world in which the genome could not be mutated in a specific manner. The true power of stem cell biology was revealed to the world at large with the announcement that the transfer of

---

Adapted with permission from Bone Marrow Transplantation, Vol. 32, Supplement 1, Aug 2003.

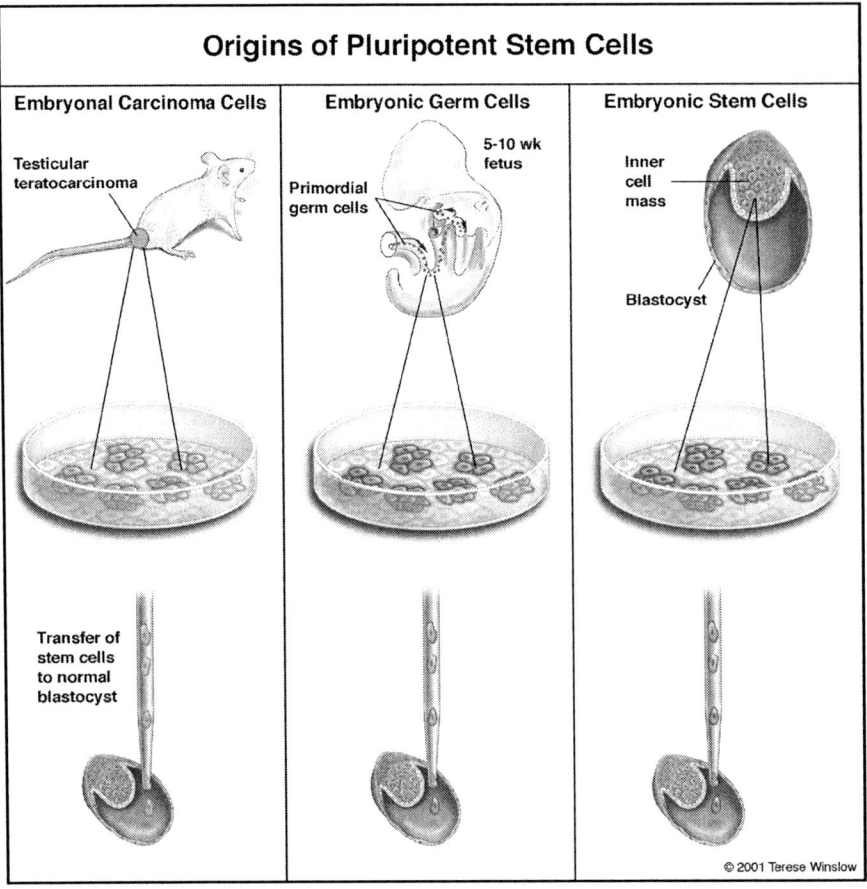

Figure 1. From the initial descriptions of the ability of testicular carcinoma cells to produce pluripotent embryonic stem cells, these cells have since been derived from blastocysts as well as the primordial germ cells in the developing genital ridge in both mouse and man. Figure courtesy of Terese Winslow, used with permission of the artist.

nuclei derived from adult somatic cells into enucleated oocytes produced, at a low frequency, viable offspring clonally derived from the donor of the nuclei.[18]

Mouse to Man

The successful application of targeted mutagenesis in the mouse was not the only useful application of embryonic stem cell lines. A variety of investigators utilized these cell lines to model the development of the early embryo in culture systems, and successfully recapitulated several aspects of embryogenesis. When the application of in vitro fertilization to the clinical problem of infertility resulted in the birth of the first test-tube baby in 1978, the stage was set for the eventual derivation of human embryonic stem cells from embryos fertilized in vitro but not implanted into a womb.[19] Since these early embryos are frozen in quantities that exceed clinical need, large banks of fertilized embryos destined for destruction now exist around the world as a consequence of the widespread application of in vitro fertilization. Some of these embryos have been cultured to derive embryonic stem cell lines, however the derivation of cell lines in addition to those already in existence has been deemed unnecessary and will not be supported by federal funding agencies in the United States.

A second approach to the application of stem cell technology in humans utilizes tissue derived from the genital ridge of aborted first trimester fetuses (Fig. 1).[20] These cells, which normally develop into mature gametes, can be cultured under specific conditions to produce cell lines with all known characteristics of blastocyst-derived embryonic stem cells but lacking apparent tumorigenic potential. This combination of multipotential differentiation in the absence of tumor formation has lead to the proposed use of these cells in clinical trials to treat spinal cord injury, Parkinson's disease, and other cell-based therapies. With the specter of the cloning of human beings looming before us, the National Academy of Sciences initiated a comprehensive analysis of this brave new world.[21] The current state of federal funding will support the utilization of fetal-derived embryonic germ cells in clinical applications, most likely because the derivation and use of these cells avoids some of the concerns raised by the concept of frozen embryos as sources of embryonic stem cells. Embryonic germ cells are unable to be implanted into a surrogate mother to produce a genetically normal human, unlike the embryos formed during in vitro fertilization. As such, the only embryos that might be formed by embryonic germ cells would be genetic mosaics of the germ cell and a blastocyst in which such cells might be introduced, or would be the product of somatic cell nuclear transfer. Since the latter process can be performed using a wide variety of cell types, the embryonic germ cell provides no special advantage in this sense.

Adult Stem Cells

Undifferentiated cells that are found in a differentiated adult tissue are considered adult stem cells, particularly when these cells contribute to ongoing tissue maintenance or repair. These cells may be capable of self-renewal, but do not replicate indefinitely in culture. Adult stem cells may differentiate to produce progenitor, precursor, and mature cells, but these activities are usually limited to the cells contained in the tissue of origin. Adult stem cells usually comprise a small minority of the total tissue mass, and as such are usually quite difficult to identify and isolate. Adult stem cells have been described in regenerating tissues such as the liver, epithelium and muscle, as well as in tissues like the brain, which previously was thought not to possess extensive regenerative properties. By far, the most well-characterized example of adult stem cells is that of the hematopoietic system.

Hematopoietic Stem Cells: Paradigms for Stem Cell Biology

The limited life-span of most blood cells demands that a continual source of these cells be assured throughout life. It is likely for this reason that the hematopoietic system has so readily lent itself to applications involving clinical transplantation. Indeed, the challenges faced by cells utilized in bone marrow transplantation are not so different than the normal physiologic role played by these cells during maintenance of hematopoiesis over the lifetime of the normal mammal. Compared to the hematopoietic system, other tissues of the adult mammal display relatively limited potential for replacement from endogenous stem cells in response to tissue injury. Furthermore, no other tissue is characterized by such a wide variety of different cell lineages which all arise from a common stem cell in a developmental process that continues throughout life. The ability to model many of these differentiation pathways under controlled conditions in vitro, and the availability of recombinant proteins that select or direct differentiation along specific lineages makes the hematopoietic system the premier paradigm for the field of stem cell biology.

Plasticity

The concept of stem cell plasticity refers to the phenomenon of trans-differentiation, which is the ability of an adult stem cell from one tissue to differentiate as a specialized cell type of another tissue. A recent study showed that neural stem cells were capable of regenerating blood lineages in transplant recipients,[22] and the field rapidly advanced as examples of muscle, epithelium, liver, and other tissues derived from heterologous stem cells (usually bone marrow-derived) were reported.[23] With few exceptions, these studies involved transplantation of large numbers of cells, leaving open the possibility that distinct classes of stem cells were responsible for regeneration of the different tissues. Furthermore, the magnitude of tissue replacement has often been minor, suggesting that this approach to tissue engineering will require extensive optimization in order to be clinically useful. Finally, technical artifacts[24] and difficulty in reproducing some of the reported findings[25] suggest that caution is indicated in interpreting many of the experiments. The concept that stem cells derived from adult tissues will provide a viable alternative to the embryo as a source of material for tissue engineering is far from validated.

Stem Cells as Therapeutic Agents

The ability of stem cells to provide a self-renewing source of normal differentiated cells has been extensively exploited in the bone marrow transplantation field. Applications include the treatment of marrow failure syndromes, leukemia and lymphoma, and certain inherited blood disorders, and autoimmune diseases to which this book is devoted.

Recent advances in vector design have made gene correction a feasible approach to the treatment of a number of genetic diseases, including X-linked severe combined immunodeficiency disease, hemophilia, and a number of autoimmune disorders. Clinical application of stem cell therapy depends on robust self-renewal and differentiation of the transplanted cells, and in this sense trans-differentiation must be sufficiently frequent and robust to achieve enough tissue replacement to be clinically useful. However, if harnessed and properly regulated, it is not difficult to imagine the application of stem cell therapy in diverse settings such as diabetes (generation of islet cells from stem cells), repair of damaged heart muscle, and rebuilding the nervous system after injury or age-related decline.[26]

Regulatory and Funding Issues

The United States government now provides some funding for human stem cell research. While NIH funds cannot be used for derivation of human ES lines, this type of research can be performed using private sources of funding if proper informed consent is obtained under a protocol approved by an institutional review board according to NIH guidelines. NIH funds can be used for research that utilizes existing embryonic stem cell lines, as well as for derivation and use of embryonic germ cells from fetal tissues. This apparent discrepancy in policy arises due to the consideration that established stem cell lines and aborted fetal tissues are not embryos and can not, by themselves, develop into human beings. The NIH guidelines and the FDA regulate experimental and clinical use of human pluripotent stem cells and fetal tissues. As of September 25, 2002, 5 NIH grants have been approved and funded for a total of $4.2 million, and administrative supplements for embryonic stem cell research have been awarded to 30 additional investigators.[27]

The Future of Stem Cell Therapy

A number of challenges remain before the promise of stem cell therapy can be translated into application. First and foremost, the political and ethical conflicts that surround the use of human embryonic and fetal tissue for medical applications must be resolved. The concept that stem cells derived from adult tissues will substitute for those obtained from fetal or embryonic sources is simply too premature to be used as a basis for legislation and regulation. While the combination of gene therapy with stem cell therapy has proven to be effective for certain diseases, methods of gene delivery must be improved to prevent unpredictable adverse events. Animal models must be refined to allow comprehensive analysis of potential risks and benefits prior to clinical application. These and other barriers stand before us, marking the path toward new applications in clinical medicine.

References

1. Ford CE, Hamerton JL, Barnes DWH et al. Cytological identification of radiation-chimaeras. Nature 1956; 177:452-454.
2. Wright DE, Wagers AJ, Gulati AP et al. Physiological migration of hematopoietic stem and progenitor cells. Science 2001; 294:1933-1936.
3. Till JE, McCulloch EA. A direct measurement of the radiation sensitivity of normal mouse bone marrow cells. Radiat Res 1961; 14:213-222.
4. Bradley TR, Metcalf D. The growth of mouse bone marrow cells in vitro. Aust J Exp Biol Med Sci 1966; 44:287-299.
5. Thean LE, Hodgson GS, Bertoncello I et al. Characterization of megakaryocyte spleen colony-forming units by response to 5-fluorouracil and by unit gravity sedimentation. Blood 1983; 62:896-901.

6. Visser JWM, Bauman JGJ, Mulder AH et al. Isolation of murine pluripotent hemopoietic stem cells. J Exp Med 1984; 159:1576-1590.
7. Spangrude GJ, Heimfeld S, Weissman IL. Purification and characterization of mouse hematopoietic stem cells. Science 1988; 241:58-62.
8. Orlic D, Bodine DM. What defines a pluripotent hematopoietic stem cell (PHSC): Will the real PHSC please stand up! Blood 1994; 84:3991-3994.
9. Korbling M, Katz RL, Khanna A et al. Hepatocytes and epithelial cells of donor origin in recipients of peripheral-blood stem cells. N Engl J Med 2002; 346:738-746.
10. Abkowitz JL. Can human hematopoietic stem cells become skin, gut, or liver cells? N Engl J Med 2002; 346:770-772.
11. Lewis R. A stem cell legacy: Leroy Stevens. The Scientist 2000; 14:19.
12. Stevens LC. Studies on transplantable testicular teratomas of strain 129 mice. J Natl Cancer Inst 1958; 20:1257-1270.
13. Mintz B, Illmensee K. Normal genetically mosaic mice produced from malignant teratocarcinoma cells. Proc Natl Acad Sci USA 1975; 72:3585-3589.
14. Evans MJ, Kaufman MH. Establishment in culture of pluripotential cells from mouse embryos. Nature 1981; 292:154-156.
15. Martin GR. Isolation of a pluripotent cell line from early mouse embryos cultured in medium conditioned by teratocarcinoma stem cells. Proc Natl Acad Sci USA 1981; 78:7634-7638.
16. Doetschman T, Gregg RG, Maeda N et al. Targeted correction of a mutant HPRT gene in mouse embryonic stem cells. Nature 1987; 330:576-578.
17. Thomas KR, Capecchi MR. Site-directed mutagenesis by gene targeting in mouse embryo-derived stem cells. Cell 1987; 51:503-512.
18. Wilmut I, Schnieke AE, McWhir J et al. Viable offspring derived from fetal and adult mammalian cells. Nature 1997; 385:810-813.
19. Thomson JA, Itskovitz-Eldor J, Shapiro SS et al. Embryonic stem cell lines derived from human blastocysts. Science 1998; 282:1145-1147.
20. Shamblott MJ, Axelman J, Wang S et al. Derivation of pluripotent stem cells from cultured human primordial germ cells. Proc Natl Acad Sci USA 1998; 95:13726-13731.
21. Committee on science, engineering, and public policy. Scientific and medical aspects of human reproductive cloning. Washington, DC: National Academy Press, 2002. http://www.nap.edu/catalog/10285.html.
22. Bjornson CR, Rietze RL, Reynolds BA et al. Turning brain into blood: A hematopoietic fate adopted by adult neural stem cells in vivo. Science 1999; 283:534-537.
23. Anderson DJ, Gage FH, Weissman IL. Can stem cells cross lineage boundaries? Nature Med 2001; 7:393-395.
24. Wurmser AE, Gage FH. Stem cells: Cell fusion causes confusion. Nature 2002; 416:485-487.
25. Morshead CM, Benveniste P, Iscove NN et al. Hematopoietic competence is a rare property of neural stem cells that may depend on genetic and epigenetic alterations. Nat Med 2002; 8:268-273.
26. Department of Health and Human Services. Stem cells: Scientific progress and future research directions. National Institutes of Health 2001. http://www.nih.gov/news/stemcell/scireport.htm.
27. Zerhouni E. Stem cell research. Senate Appropriations Subcommittee on Labor HHS Education. Washington, DC: 2002. http://www.nih.gov/about/director/092502sctestimony.htm.

CHAPTER 2

Embryonic Stem Cells: Unique Potential to Treat Autoimmune Diseases

Dan S. Kaufman and James A. Thomson

Introduction

Embryonic stem (ES) cells are pluripotent cells that can be maintained indefinitely in culture as undifferentiated cell lines, yet retain their ability to form any cell type. The derivation of human embryonic stem cells provides a new model system to learn about human developmental biology. These cells may also be a suitable starting point to produce novel cell-based therapies to treat a variety of diseases. This chapter describes the characteristics of human embryonic cells and the potential of these cells to be used for hematopoietic cell transplantation, tolerance induction, and treatment of autoimmune diseases.

This book, "Stem Cell Therapy for Autoimmune Disease," reflects the recently growing interest in stem cells and cell-based therapies. Among the various medical specialties, hematologists are probably the most familiar with the concept of stem cells. Many of the diseases seen in the clinical fields of hematology and hematopoietic stem cell transplantation (HSCT) are clonal in nature and demonstrate how a defect in a single early precursor cell can lead to a systemic disease. For example, in chronic myelogenous leukemia (CML), the transformation of a single hematopoietic progenitor leads to an overwhelming production of mature and immature hematopoietic cells. If untreated, this "stem cell malignancy" will lead to death of the afflicted individual within a few years. However, HSCT can cure CML and other hematological malignancies by replacement of the abnormal clone with normal hematopoietic stem cells (HSCs). Additionally, it is also possible to treat many nonmalignant hematologic conditions such as congenital immunodeficiencies and sickle cell anemia with HSCT.[1,2] The ability to use HSCT to treat severe autoimmune disease (as described in this book) is one of the newest applications of stem cell biology.[3] Indeed, the autoimmune process can be considered another form of stem cell abnormality. In many of these cases, a clone of self-reacting lymphocytes become abnormally activated, leading to the destruction of normal cells and tissues. Autologous or allogeneic HSCT aims to alleviate this process by the elimination or suppression of these rogue lymphocytes. As described in this chapter, the derivation of human ES cells may eventually lead to novel methods of stem cell therapies for autoimmune diseases.

Stem cells are defined as specific cell types that have two important properties: self-renewal and differentiation. Self-renewal refers to the ability of these cells to undergo cell division for prolonged periods as cells that maintain multipotent potential without evidence of differentiation down a particular developmental lineage. However, in the proper environment or with the proper stimuli, a stem cell retains the ability to form more specialized cells such as blood, muscle, liver, or skin. Broadly speaking, there are two main categories of stem cells: "adult" stem cells and embryonic stem cells. Adult stem cells are derived from post-natal tissue and are typically thought to have a limited developmental potential. Hematopoietic stem cells (HSCs) found in the bone marrow produce blood cells, neural stem cells (NSCs) found in the central nervous system give rise to neurons and glial cells, hepatic stem cells found in the liver, produce hepatocytes and biliary cells are all examples of adult stem cells. In contrast, ES cells are derived from preimplantation blastocysts and have the potential to form any cell type in the body.

Research Prior to Derivation of Human ES Cells

While the derivation of human ES cells has recently sparked considerable enthusiasm (and some debate) over the potential of these cells to be used to treat human disease, it is important to recognize that decades of basic scientific research preceded the isolation and characterization of human ES cells. Studies on mouse and human embryonal carcinoma (EC) cell lines were a crucial precedent to work on ES cells. Differences and similarities between human EC cells, murine ES cells, and human ES cells are listed in Table 1. EC cells are multipotent malignant precursor cells derived from teratocarcinomas.[4] Mouse and human EC cells can be isolated from these tumors and be maintained in vitro as undifferentiated cells. When treated with agents to induce differentiation (such as retinoic acid) or reimplanted into immunodeficient animals, multiple differentiated cell lineages can be derived. EC cells have provided important insights regarding the mechanisms of normal mammalian development and abnormal tumorigenesis. However, the use and interpretation of work on these cells is limited because EC cells typically have genetic abnormalities, and their developmental potential is often restricted to a few cell lineages. For these reasons, it became important to derive and characterize embryonic stem cells, the normal nonmalignant counterpart to EC cells. This breakthrough was first reported by two groups in 1981.[5,6]

Table 1. Comparison of primordial cell lines

	Murine ES Cells	Human EC Cells	Human ES Cells
Source	Peri-Implantation Embryo	Teratocarcinomas	Pre-Implantation Blastocysts
Karyotype	Normal	Aneuploid	Normal
Surface antigens	SSEA1$^+$	SSEA1$^-$	SSEA1$^-$
	SSEA3$^-$	SSEA3$^+$	SSEA3$^+$
	SSEA4$^-$	SSEA4$^+$	SSEA4$^+$
	TRA-1-60$^-$	TRA-1-60$^+$	TRA-1-60$^+$
	TRA-1-81$^-$	TRA-1-81$^+$	TRA-1-81$^+$
			CD133$^+$
Culture conditions to maintain self-renewal	Requires LIF or other gp130/stat3 agonist	LIF independent, most feeder-independent	Require MEF feeder cells or MEF conditioned media. Does not require LIF
In vivo potential	Totipotent	Limited lineages	Unknown

Abbreviations: ES= embryonic stem; EC= embryonal carcinoma; LIF= leukemia inhibitory factor; MEF= mouse embryonic fibroblast; SSEA= stage specific embryonal antigens; TRA= transcription factor.

Mouse ES cells are typically derived from the inner cell mass (ICM) of early, preimplantation stage blastocysts. This preimplantation stage of development occurs after fertilization of the oocyte and before attachment to the uterus. During this time, the zygote undergoes several rounds of cell division. The cells produced at this early stage are not committed to become any particular part of the body. Indeed, it is possible to split the developing embryo in half at this stage, and each half has the ability to develop normally. The first cell differentiation is evident after 5-6 days of development with the appearance of the outer cell mass (trophectoderm) and the inner cell mass (ICM) (Fig. 1). The trophectoderm develops into the outer layers of the placenta; whereas the ICM, a small cluster of about a dozen cells, will eventually derive all the cells of the fetal and adult body, and some extraembryonic structures. ES cells are derived by careful isolation of the ICM and culturing these cells under conditions that permit them to divide without undergoing differentiation down specific developmental lineages. Under proper conditions, ES cells can be maintained for months or years as undifferentiated cells without evidence of senescence. ES cells naturally express high levels of telomerase and maintain a normal karyotype (unlike EC cells).[7] However, under specific conditions, the ES cells can be induced to differentiate into specific cell types of interest. This potential is best demonstrated by inducing mouse ES cells to form chimeras with an intact embryo. Careful analysis and breeding of the resulting chimeras show that all cells within an adult organism can be formed from a single ES cell.[8]

Work with mouse ES cells has unraveled some of the basics of mammalian developmental biology and genetics. Not surprisingly, research using mouse ES cells has been particularly amenable to defining mechanisms of hematopoietic development. Work by many groups has demonstrated that mouse ES cells can be induced to differentiate into hematopoietic lineages in vitro.[9] Intricate time-course experiments show the regulation of specific genes during specific developmental stages.[10,11] Other studies carefully alter external stimuli (cytokines, adhesion molecules) to determine how these factors affect hematopoiesis.[12] Perhaps the most precise experiments have used genetic homologous recombination to delete specific genes within ES cells. For example, deletion of the genes for the vascular endothelial growth factor (VEGF) receptors flk-1 and flt-1 leads to death at approximately day 9 of mouse embryogenesis. Analysis of these mice demonstrates lack of normal hematopoietic and endothelial cell development and leads to the presumption that these receptors (and VEGF) are required for normal hematopoiesis.[13,14] These results correspond to other studies that find flk-1 is expressed on HSCs.[15]

While research on mouse ES cells has been fruitful to elucidate mechanisms of mammalian development, mouse and human embryogenesis are distinctly different. For example, the relative size and structure of the placenta and other fetal structures are quite dissimilar.[16] These discrepancies lead to the very likely possibility that mouse ES cells may not always closely model normal human developmental biology. Indeed, the yolk sac, an important organ of early hematopoiesis, is quite different between mouse and man.[17] Therefore, it is valuable to produce a model system that more closely resembles human development. Toward this goal, our group (Thomson) derived ES cells from nonhuman primates: rhesus monkeys and common marmosets.[18,19] Importantly, the earliest stages of embryogenesis is very analogous between humans and nonhuman primates.[16]

Characteristics of Human ES Cells

Using lessons learned from derivation of nonhuman primate ES cells, it became feasible to generate ES cells from human preimplantation blastocysts. With informed consent and protocols approved by the local institutional review board, fertilized oocytes no longer desired by couples undergoing in vitro fertilization (and destined to be discarded) were donated to this research endeavor. The technique used to derive human ES cells was similar to that used for nonhuman primate ES cells. The oocytes were cultured to blastocyst stage and immunosurgery used to isolate the ICM, which is cultured on mitotically-inactivated mouse embryonic fibroblast feeder cells. The ICM-derived cells divided and were serial passed without evidence of differentiation—thereby established into lines of human ES cells. The multipotent nature of these cells was initially demonstrated by intramuscular injec-

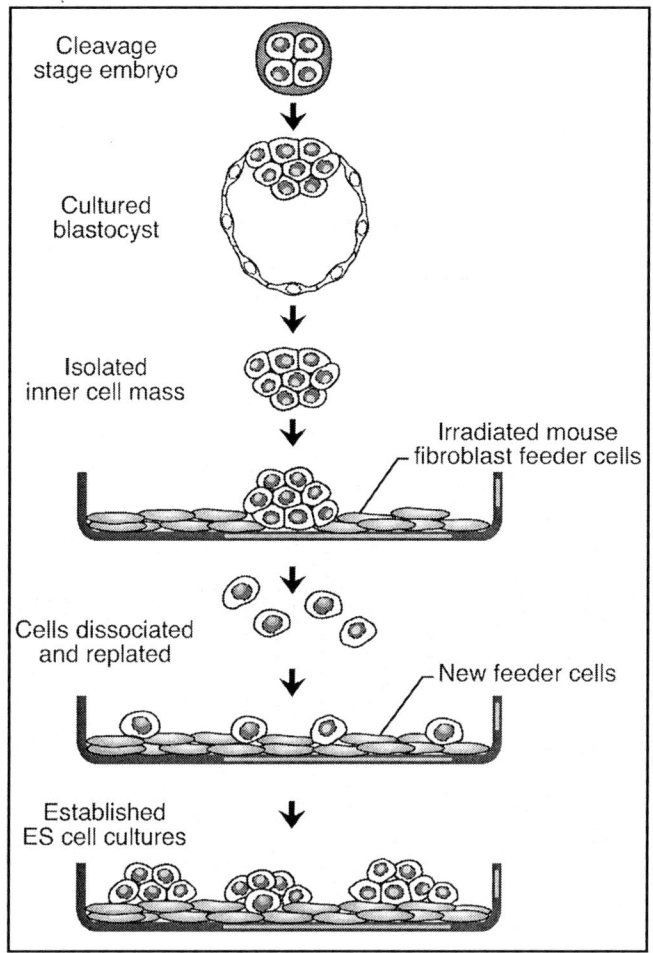

Figure 1. Derivation of human ES cell lines. Human blastocysts were grown from cleavage-stage embryos produced by in vitro fertilization. Inter cell mass (ICM) cells were separated from trophectoderm by immunosurgery and plated onto irradiated mouse fibroblast feeder cells. Colonies were then expanded and cloned with maintenance of undifferentiated state (reproduced with permission from Stem Cells).

tion into immunodeficient mice. Derivatives of all three embryonic lineages (endoderm, ectoderm, and mesoderm) could be easily identified within these teratomas.[20]

Since the initial description of human ES cells, several other groups have established similar lines.[21,22] By convention, all human ES cell lines should meet the following criteria: (1) derivation from preimplantation embryo; (2) prolonged undifferentiated proliferation; (3) ability to differentiate into cells representative of all three embryonic germ layers, and (4) maintenance of normal karyotype. Importantly, clonally derived lines of human ES cells that meet the above criteria have also been established. This provides unambiguous evidence that a single human ES cell can produce all cell lineages.[23]

The characterization of human ES cells immediately demonstrated fundamental differences between these cells and mouse ES cells. Leukemia inhibitory factor (LIF) or other agonists of the gp130-STAT3 signaling cascade are required for maintenance of mouse ES cells as undifferentiated cells. Removal of LIF rapidly induces differentiation of these cells. In contrast, LIF is not sufficient to sustain human (and nonhuman primate) ES cells.[18,20] Recent studies suggest this species difference may reflect the ability of the mouse reproductive cycle to undergo diapause.[24] Diapause, a state of suspended embryo growth prior to uterine implantation, requires LIF or other gp130 agonists. Humans do not have such a capacity and presumably do not require LIF or similar molecules to regulate embryonic development.

Mouse and human ES cells also differ in the expression of particular glycolipids on the cell surface. Undifferentiated human ES and EC cells express the stage-specific embryonal antigens (SSEA-3 and SSEA-4), as well as TRA-1-60 and TRA-1-81, but do not express SSEA-1.[20] In contrast, mouse ES cells express SSEA-1, but not SSEA-3, SSEA-4, TRA-1-60, or TRA-1-81.[16] These differences again reflect a fundamental embryological difference between mouse and human development. The enzymes that produce the glycosylation events which lead to production of these glycolipids are regulated differently between species, though the reason for this difference remains unclear.

Recently, the culture conditions required for growth of undifferentiated human ES cells have been refined. Whereas the original derivation of these cells used irradiated mouse embryonic fibroblast (MEF) feeder cells and media containing fetal bovine serum, it is now possible to grow these cells in serum-free media and on plates coated with either Matrigel® or laminin.[25] Growth without feeder cells still requires "conditioned media" taken from cultures of MEFs. The requirement for "conditioned media" suggests that soluble factor(s) produced by the MEFs are essential for maintenance of undifferentiated ES cell growth. These refinements in the culture methods are the start of a process to identify the conditions that are essential for maintenance of human ES cell self-renewal. Eventually, use of a chemically defined media will aide in large-scale growth of human ES cells as a prelude to better cell-based therapies.

The excitement surrounding the derivation and characterization of human ES cells stems from the potential of these cells to be induced to produce any cell type in the body. These cells may become an important source to replace diseased or damaged cells and tissues. For example, insulin producing pancreatic beta-cells may be derived to treat diabetics, dopaminergic neurons may be produced to treat patients with Parkinson's disease, and hematopoietic cells may be developed for HSCT and transfusion medicine. However, before any of these treatments become a reality, several barriers must be over come (more extensively reviewed elsewhere, ref. 26). Foremost among these barriers is to derive efficient means to induce differentiation of human ES cells to the cell types of interest. The fundamental differences between mouse and human ES cells strongly suggests that mechanisms to induce differentiation of mouse ES cells may not work equally well with human ES cells. Indeed, one main strategy to induce differentiation of mouse ES cells is withdrawal of LIF from the culture media. Since human and rhesus ES cells do not require LIF for undifferentiated growth, elimination of LIF is not a suitable means to induce differentiation.

Despite the still relatively recent isolation of human ES cells and the potential difficulties in inducing differentiation of these cells down defined pathways, it is remarkable that several types of differentiated cell lineages already have been described. To varying degrees, hematopoietic[27] (Fig. 2), neural,[28,29] insulin-producing pancreatic beta cells,[26,30] cardiac muscle,[31] and hepatocytes[32] have all been characterized. Next, it will be crucial to demonstrate normal physiologic function of these cells in vitro and in vivo. This in vivo testing may be done initially using transplantation into immunodeficient mice, or neonatal mice. However, the availability of rhesus ES cells presents the opportunity to derive nonhuman primate models for preclinical testing. In-

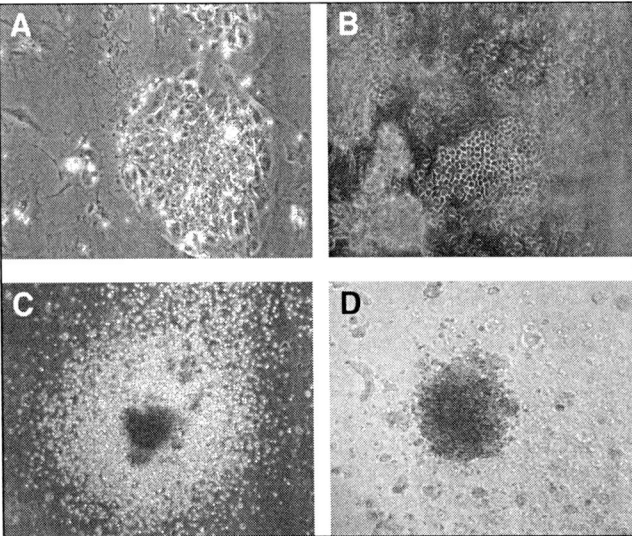

Figure 2. Undifferentiated and differentiated human ES cells. A) Undifferentiated Human embryonic stem (ES) cells maintained on irradiated mouse fibroblast feeder cells. Colony of cells with uniform morphology and large nuclei characteristic of human ES cells, original magnification 200X. B) Cobblestone area-forming cells (CAFC) derived from human ES cells. Human ES cells were allowed to differentiate on S17 mouse bone marrow stromal feeder cells with appearance of CAFC typical of hematopoietic precursor cells, original magnification 200X. C) Colony-forming unit-granulocyte, macrophage (CFU-GM) and D) Burst-forming unit-erythroid (BFU-E) derived colonies from human ES cells. Undifferentiated human ES cells allowed to differentiate on S17 cells were harvested and plated in methylceloulose based media supplemented with hematopoietic cytokines resulting the appearance of typical CFU-GM and BFU-E (red, hemoglobinized cells) derived colonies. Original magnification of C is 50X and D is 100X. More complete description of derivation of hematopoietic colonies from human ES cells in ref. 27.

deed, rhesus monkey models for hematopoietic cell transplantation, diabetes mellitus and Parkinson's disease already exist.[26,33]

Human ES Cells, Hematopoietic Cell Transplantation and Tolerance

The development of hematopoietic cells from human ES cells is a natural extension of attempts over the past thirty years to produce new and/or improved sources of cells for hematopoietic cell transplantation. HSCT is typically used to treat hematologic malignancies such as leukemia, lymphoma, or multiple myeloma. While transplantation of HSCs from a donor (allogeneic HSCT) often affords the best chance to cure these otherwise fatal diseases, too frequently a suitable HLA-matched sibling or unrelated donor cannot be found. Indeed, it has been estimated that only approximately one-third of patients who would benefit from an allogeneic HSCT actually receive a transplant.[34-36] The remaining patients either undergo autologous HSCT (using their own HSCs) or further chemotherapy. Often, these alternatives do not produce as high a cure rate as allogeneic HSCT,[37,38] leading to efforts to find alternative sources of HSCs. These alternatives include umbilical cord blood cells,[39,40] activated autologous lymphocytes,[41] or other ex vivo expanded HSCs.[42,43] Moreover, the growing use of nonmyeloablative allogeneic HSCT is expanding the potential pool of patients who may benefit from these treatments.[44]

Table 2. Strategies to prevent immune-mediated rejection of human ES cell-derived cells and tissues

- Immunosuppressive medications
- HLA-defined ES cell "banks"
- Deletion/insertion of HLA genes
- Insertion of genes for immune-modifying proteins
- Deletion of genes for co-stimulatory immune response proteins
- Nuclear reprogramming to produce HLA-matched ES cells
- Hematopoietic chimerism

The development of transplantable HSCs from human ES cells may be quite difficult. However, the enormous benefits that may be gained makes the potential for this development extremely exciting. There are thousands of patients a year who would benefit if such a breakthrough is accomplished. Even if it does not become readily feasible to produce these transplantable blood cells from human ES cells, the basic research to understand the mechanisms of the earliest stages of human hematopoietic cell development may be applied toward better growth of other alternative sources of HSCs. For example, cord blood expansion may become more feasible if we more clearly understand the particular genes and proteins that maintain the long-term engraftment potential of HSCs.

If (perhaps when) human ES cell-derived HSCs can be used for HSCT, these cells may have important clinical implications beyond the current indications of HSCT. There is good reason to conjecture that these ES cell-derived HSCs can be used to create tolerance for transplantation of other ES cell-derived cells and tissues. Avoidance of immune-mediated rejection of ES cell-derived cells will be one of the important barriers to overcome prior to routine clinical use of these cells. Various strategies to prevent rejection have been outlined elsewhere[26,45] (and Table 2). One intriguing strategy takes advantage of the multipotent nature of ES cells to derive HSCs and a second cell line of interest (beta cell, for example) from the same parental ES cell line. Transplantation of the ES cell-derived HSCs would be used to induce a state of hematopoietic chimerism in a patient. Then, this patient should be immunologically tolerant to the second HLA-matched cell lineage derived from the same parental ES cell line.

This hypothesis has precedent in many clinical cases where patients who have undergone allogeneic HSCT subsequently receive a second organ (typically a kidney) from the same donor as the hematopoietic cells. Due to immune reconstitution from HSCs that are a perfect tissue match for kidney, often no additional immune suppression is needed, and in many cases, immunosuppressive medications can be discontinued without loss of the bone marrow or solid organ graft.[46] Animal models more clearly demonstrate that transplantation of highly purified HSCs will induce tolerance to HLA-matched organs, but these animals will still reject third-party, non HLA matched grafts.[47,48] Studies using large animals (nonhuman primates, dogs, miniature swine) show nonmyeloablative conditions followed by allogeneic HSCT can create a state of hematopoietic mixed chimerism. These chimeric animals are tolerant to solid organ grafts from the hematopoietic cell donor.[49] Prospective studies are now being done in human trials that combine nonmyeloablative allogeneic HSCT with transplantation of a kidney from the same donor. Initial reports of these studies are promising.[50]

This potential to use human ES cell-derived HSCs to induce hematopoietic chimerism and tolerance has particular relevance to treatment of autoimmune disease. One of the major goals of human ES cell-based research is to develop replacement cells that can be transplanted to replace diseased or damaged tissues. Therefore, devising methods to induce human ES cells to produce pancreatic islet cells to treat diabetics or oligodendrocytes to benefit patients with multiple sclerosis seems quite promising. However, immune mechanisms of graft rejection and autoimmunity again become a potentially significant impediment to this type of cellular therapy. If the ES cell-derived cells (beta cells for example) are not HLA-matched to the host, these cells will likely be rejected if immunosuppressive drugs are not used. Of course drugs such as cyclosporin, tacrolimus, and corticosteroids have significant complications such as infections, renal failure, diabetes, and lymphoproliferative disorders. However, the unique characteristics of ES cells may allow methods to permit transplantation of these cells without need for immunosuppressive medications.[26,45] For example, genetic recombination may be used to delete or substitute HLA genes within the ES cells. In this manner, the engineered cells would not be regarded as foreign to the host. Derviation of ES cells from the products of nuclear reprogramming may one day also be able to accomplish this feat.[51] However, even if these perfect HLA-matched cells can be produced, the underlying autoimmune process may continue to pose a problem. Indeed, one would fully expect that without immunosuppression, transplantation of perfect HLA-matched pancreatic islet cells would be recognized and destroyed by the same clone of autoimmune lymphocytes that produced the type 1 diabetes in the first place. This has been unfortunately seen in diabetic patients who receive a pancreas transplant from an identical twin.[52] However, as described above, the plasticity of the ES cells may be used to create both HSCs and a second cell type of interest, such as a pancreatic beta cell. Cotransplantation of the two cell types derived from the same parental ES cell should permit tolerance to the new pancreatic cells due to the production of tolerant lymphoctyes from the HLA-matched ES cell-derived HSCs.

Finally, basic studies of immune system development using human ES cells should lead to better understanding of the mechanisms of immunologic education and tolerance. Derivation of lymphocytes from ES cells will lead to a population of cells that are immunlogically naïve. Altering the conditions of the growth of these cells (for example, exposure to thymic tissue) may permit dissection of the stimuli that induce tolerance or reactivity to particular antigens. Using this system to evaluate basic immune mechanisms may eventually lead to understanding the initiation of the autoimmune process and new measures to alleviate these diseases.

References

1. Buckley RH, Schiff SE, Schiff RI et al. Hematopoietic stem-cell transplantation for the treatment of severe combined immunodeficiency. N Engl J Med 1999; 340:508-516.
2. Walters MC, Storb R, Patience M et al. Impact of bone marrow transplantation for symptomatic sickle cell disease: An interim report. Multicenter investigation of bone marrow transplantation for sickle cell disease. Blood 2000; 95:1918-1924.
3. Burt RK, Traynor AE. Hematopoietic stem cell transplantation: A new therapy for autoimmune disease. Stem Cells 1999; 17:366-372.
4. Andrews PW, Przyborski SA, Thomson JA. Embryonal carcinoma cells as embryonic stem cells. In: Marshak DR, Gardner R, Gottlieb D, eds. Stem cell biology. Cold Spring Harbor, NY: Cold Spring Harbor Laboratory Press, 2001:231-265.
5. Evans MJ, Kaufman MH. Establishment in culture of pluripotential cells from mouse embryos. Nature 1981; 292:154-156.
6. Martin GR. Isolation of a pluripotent cell line from early mouse embryos cultured in medium conditioned by teratocarcinoma stem cells. Proc Nat Acad Sci USA 1981; 78:7634-7638.
7. Smith A. Embryonic stem cells. In: Marshak DR, Gardner R, Gottlieb D, eds. Stem cell biology. Cold Spring Harbor, NY: Cold Spring Harbor Laboratory Press, 2001:205-230.
8. Nagy A, Rossant J, Nagy R et al. Derivation of completely cell culture derived mice from early-passage embryonic stem cells. Proc Natl Acad Sci USA 1993; 90:8424-8428.
9. Keller GM. In vitro differentiation of embryonic stem cells. Curr Opin Cell Biol 1995; 7:862-869.
10. Keller G, Kennedy M, Papayannopoulou T et al. Hematopoietic commitment during embryonic stem cell differentiation in culture. Mol Cell Biol 1993; 13:473-486.
11. Robertson SM, Kennedy M, Shannon JM et al. A transitional stage in the commitment of mesoderm to hematopoiesis requiring the transcription factor scl/tal-1. Development 2000; 127:2447-2459.
12. Nishikawa SI, Nishikawa S, Hirashima M et al. Progressive lineage analysis by cell sorting and culture identifies flk1+ve-cadherin+ cells at a diverging point of endothelial and hemopoietic lineages. Development 1998; 125:1747-1757.
13. Shalaby F, Rossant J, Yamaguchi TP et al. Failure of blood-island formation and vasculogenesis in flk-1-deficient mice. Nature 1995; 376:62-66.
14. Fong GH, Rossant J, Gertsenstein M et al. Role of the flt-1 receptor tyrosine kinase in regulating the assembly of vascular endothelium. Nature 1995; 376:66-70.
15. Ziegler BL, Valtieri M, Porada GA et al. KDR receptor: A key marker defining hematopoietic stem cells. Science 1999; 285:1553-1558.
16. Thomson JA, Marshall VS. Primate embryonic stem cells. Curr Top Dev Biol 1998; 38:133-165.
17. Palis J, Yoder MC. Yolk-sac hematopoiesis: The first blood cells of mouse and man. Exp Hematol 2001; 29:927-936.
18. Thomson JA, Kalishman J, Golos TG et al. Isolation of a primate embryonic stem cell lines. Proc Natl Acad Sci USA 1995; 92:7844-7848.
19. Thomson JA, Kalishman J, Golos TG et al. Pluripotent cell lines derived from common marmoset (callithrix jacchus) blastocysts. Biol Reprod 1996; 55:254-259.
20. Thomson JA, Itskovitz-Eldor J, Shapiro SS et al. Embryonic stem cell lines derived from human blastocysts. Science 1998; 282:1145-1147.
21. Reubinoff BE, Pera MF, Fong CY et al. Embryonic stem cell lines from human blastocysts: Somatic differentiation in vitro. Nat Biotechnology 2000; 18:399-404.
22. NIH human embryonic stem cell registry. 2001. http://escr.nih.gov/
23. Amit M, Carpenter MK, Inokuma MS et al. Clonally derived human embryonic stem cell lines maintain pluripotency and proliferative potential for prolonged periods of in vitro culture. Dev Biology 2000; 227:271-278.
24. Nichols J, Chambers I, Taga T et al. Physiological rationale for responsiveness of mouse embryonic stem cells to gp130 cytokines. Development 2001; 128:2333-2339.
25. Xu C, Inokuma MS, Denham J et al. Feeder-free growth of undifferentiated human embryonic stem cells. Nat Biotechnol 2001; 19:971-974.
26. Odorico JA, Kaufman DS, Thomson JA. Multilineage differentiation from human embryonic stem cell lines. Stem Cells 2001; 19:193-204.
27. Kaufman DS, Lewis RL, Hanson ET et al. Hematopoietic colony-forming cells derived from human embryonic stem cells. Proc Natl Acad Sci USA 2001; 98:10716-10721.
28. Zhang S-C, Wernig M, Duncan ID et al. In vitro differentiation of transplantable neural precursors from human embryonic stem cells. Nat Biotechnology 2001; 19:1129-1133.
29. Reubinoff BE, Itsykson P, Turetsky T et al. Neural progenitors from human embryonic stem cells. Nat Biotechnol 2001; 19:1134-1140.
30. Assady S, Maor G, Amit M et al. Insulin production by human embryonic stem cells. Diabetes 2001; 50:1691-1697.

31. Kehat I, Kenyagin-Karsenti D, Snir M et al. Human embryonic stem cells can differentiate into myocytes with structural and functional properties of cardiomyocytes. J Clin Invest 2001; 108:407-414.
32. Lebkowski JS, Gold JD, Chiu C-P et al. Differentiation of human embryonic stem cells into hepatocytes, cardiomyocytes, and neurons: Transplantation applications. Blood 2001; 98:548a.
33. Kim HJ, Tisdale JF, Wu T et al. Many multipotential gene-marked progenitor or stem cell clones contribute to hematopoiesis in nonhuman primates. Blood 2000; 96:1-8.
34. Oudshoorn M, Cornelissen JJ, Fibbe WE et al. Problems and possible solutions in finding an unrelated bone marrow donor. Results of consecutive searches for 240 Dutch patients. Bone Marrow Transplant 1997; 20:1011-1017.
35. Howe CWS, Radde-Stepanick T. Hematopoietic cell donor registries. In: Thomas ED, Blume KG, Forman SJ, eds. Hematopoietic cell transplantation. Malden, MA: Blackwell Sciences, 1999:2:503-512.
36. Patients proceeding to stem cell transplant: National Marrow Donor Program 2001. http://www.marrow.org/NMDP/SLIDESET/sld025.htm.
37. Silver RT, Woolf SH, Hehlmann R et al. An evidence-based analysis of the effect of busulfan, hydroxyurea, interferon, and allogeneic bone marrow transplantation in treating the chronic phase of chronic myeloid leukemia: Developed for the American Society of Hematology. Blood 1999; 94:1517-1536.
38. Stockerl-Goldstein KE, Blume KG. Allogeneic hematopoietic cell transplanation for adult patients with acute myeloid leukemia. In: Thomas ED, Blume KG, Forman SJ, eds. Hematopoietic Cell Transplantation. Malden, MA: Blackwell Sciences, 1999:2:823-834.
39. Laughlin MJ, Barker J, Bambach B et al. Hematopoietic engraftment and survival in adult recipients of umbilical-cord blood from unrelated donors. N Engl J Med 2001; 344:1815-1822.
40. Gluckman E, Rocha V, Boyer-Chammard A et al. Outcome of cord-blood transplantation from related and unrelated donors. Eurocord Transplant Group and the European Blood and Marrow Transplantation Group. N Engl J Med 1997; 337:373-381.
41. Lum LG, LeFever AV, Treisman JS et al. Immune modulation in cancer patients after adoptive transfer of anti-CD3/anti-CD28-costimulated t cells-phase I clinical trial. J Immunother 2001; 24:408-419.
42. Lundell BI, Mandalam RK, Smith AK. Clinical scale expansion of cryopreserved small volume whole bone marrow aspirates produces sufficient cells for clinical use. J Hematother 1999; 8:115-127.
43. Emerson SG. Ex vivo expansion of hematopoietic precursors, progenitors, and stem cells: The next generation of cellular therapeutics. Blood 1996; 87:3082-3088.
44. McSweeney PA, Niederwieser D, Shizuru JA et al. Hematopoietic cell transplantation in older patients with hematologic malignancies: Replacing high-dose cytotoxic therapy with graft-versus-tumor effects. Blood 2001; 97:3390-3400.
45. Kaufman DS, Odorico JS, Thomson JA. Transplantation therapies from human embryonic stem cells- circumventing immune rejection. e-biomed: J Regenerative Medw 2000; 1:11-15.
46. Dey B, Sykes M, Spitzer TR. Outcomes of recipients of both bone marrow and solid organ transplants. A Review Medicine (Baltimore) 1998; 77:355-369.
47. Gandy KL, Weissman IL. Tolerance of allogeneic heart grafts in mice simultaneously reconstituted with purified allogeneic hematopoietic stem cells. Transplantation 1998; 65:295-304.
48. Shizuru JA, Weissman IL, Kernoff R et al. Purified hematopoietic stem cell grafts induce tolerance to alloantigens and can mediate positive and negative t cell selection. Proc Natl Acad Sci USA 2000; 97:9555-9560.
49. Wekerle T, Sykes M. Mixed chimerism as an approach for the induction of transplantation tolerance. Transplantation 1999; 68:459-467.
50. Spitzer TR, Delmonico F, Tolkoff-Rubin N et al. Combined histocompatibility leukocyte antigen-matched donor bone marrow and renal transplantation for multiple myeloma with end stage renal disease: The induction of allograft tolerance through mixed lymphohematopoietic chimerism. Transplantation 1999; 68:480-484.
51. Lanza RP, Cibelli JB, West MD. Prospects for the use of nuclear transfer in human transplantation. Nat Biotechnol 1999; 17:1171-1174.
52. Sutherland DE, Goetz FC, Sibley RK. Recurrence of disease in pancreas transplants. Diabetes 1989; 38:85-87.

CHAPTER 3

Neural Stem Cells and Oligodendrocyte Progenitors in the Central Nervous System

Jennifer A. Jackson and Diana L. Clarke

Introduction

The adult vertebrate central nervous system (CNS) consists of four major differentiated cell types: neurons, astrocytes, oligodendrocytes and ependymal cells. Historically, there has been disagreement on how these differentiated cell types are generated in the CNS. Progress remains hindered by the complexity of cell structure in this system, the lack of specific cell surface markers to identify distinct cell types and the presence of numerous transit amplifying cell populations that rapidly generate early progenitors. At present it is clear that some cells, termed neural stem cells, can generate neurons as well as astrocytes and oligodendrocytes of the glial lineage both in vitro and in vivo. However, controversy still exists over whether the majority of glia in the CNS are generated by multipotential stem cells or progenitor cells that were born as committed glioblasts. Nevertheless, the existence of stem cells in the CNS has important implications for understanding both the mechanisms that generate neural diversity during embryonic development and the recruitment and differentiation of neural stem cells present in the adult. This review summarizes our knowledge of stem cells that comprise the CNS and examines the broad plasticity reported for adult CNS stem cell populations.

Interdependence of Neurons and Glia in the CNS

Glial cells constitute the majority of the cells in the CNS. These cells provide the structural scaffolding that is important for migration of early neuroblasts, are a major source of adhesion molecules that participate in the formation of neural networks, form limiting membranes that separate the CNS from other tissues and aid in the rapid conduction of electrical impulses down axons of mature neurons. There is now growing recognition that glia, possibly through their immense glial networks, may possess communication skills that complement those of neurons themselves.[1,2] Astroglia are stellate, branched supporting cells that contact the soma, dendrites and axons of neurons, the soma and processes of oligodendrocytes and associate intricately with other astrocytes. Their close association with the surface of various neurons is thought to mediate the exchange of substances between these two cell types.[3] Their multiple processes also contact, induce and maintain the tight junctions in endothelial cells that effectively form the blood-brain barrier. Oligodendroglia, also a supporting glial cell type present in the CNS, extend many processes, each of which contacts and repeatedly envelopes a stretch of axon with subsequent condensation of its spiraling plasma membrane. This myelination of axons imparts insulating properties that allow the rapid propagation of action potentials throughout the CNS without continuous regeneration of the action potential along each segment of an axon.

Origin of Neural and Glial Progenitors in the CNS

All cells comprising the CNS are originally derived from the early neuroepithelium that forms as the neural plate along the midline of the developing embryo. As development proceeds, this single layer of pseudostratified epithelium folds to form the neural tube (Fig. 1). The differentiation of the neuroepithelial stem cells into neurons and glia then proceeds in a temporal specific manner that is specific for each region of the developing neural tube.[4,5] Generally neurogenesis occurs first, followed by gliogenesis. This patterning of the neural tube is thought to begin at the neural plate stage of development through inductive interactions that create organizing centers at the dorsal and ventral poles.[6-8] These specialized neuroepithelial cells generate signals that induce, often in a concentration dependent manner, the expression of patterning genes in adjacent neuroepithelial cells. These patterning genes generally encode homeodomain transcription factors, and their expression patterns divide the cells in the neuroepithelium into different domains along the rostral-caudal and dorso-ventral axes of the neural tube.[9,10] These patterning genes are thought to specify neuronal subtype identity and control the duration of specific types of neurogenesis occurring in the brain and spinal cord during each developmental stage.

The exact mechanisms underlying the developmental changes in the stem cell and precursor population in any given region of the neural tube are not understood. However, by midgestation in the developing cortex of the brain, for example, young neurons have migrated beyond the germinal ventricular zone (VZ) of the neuroepithelium with the aid of newly formed glial cells in this region (Fig. 2). These radial glia contact the inner ventricular surface and the outer pial surface of the neural tube, guiding neuronal migration away from the VZ and forming the second germinal zone, the subventricular zone (SVZ). When early neuroblast formation has ceased, the neuroepithelial stem cells begin to differentiate into glioblasts. Clonal studies suggest that most glia

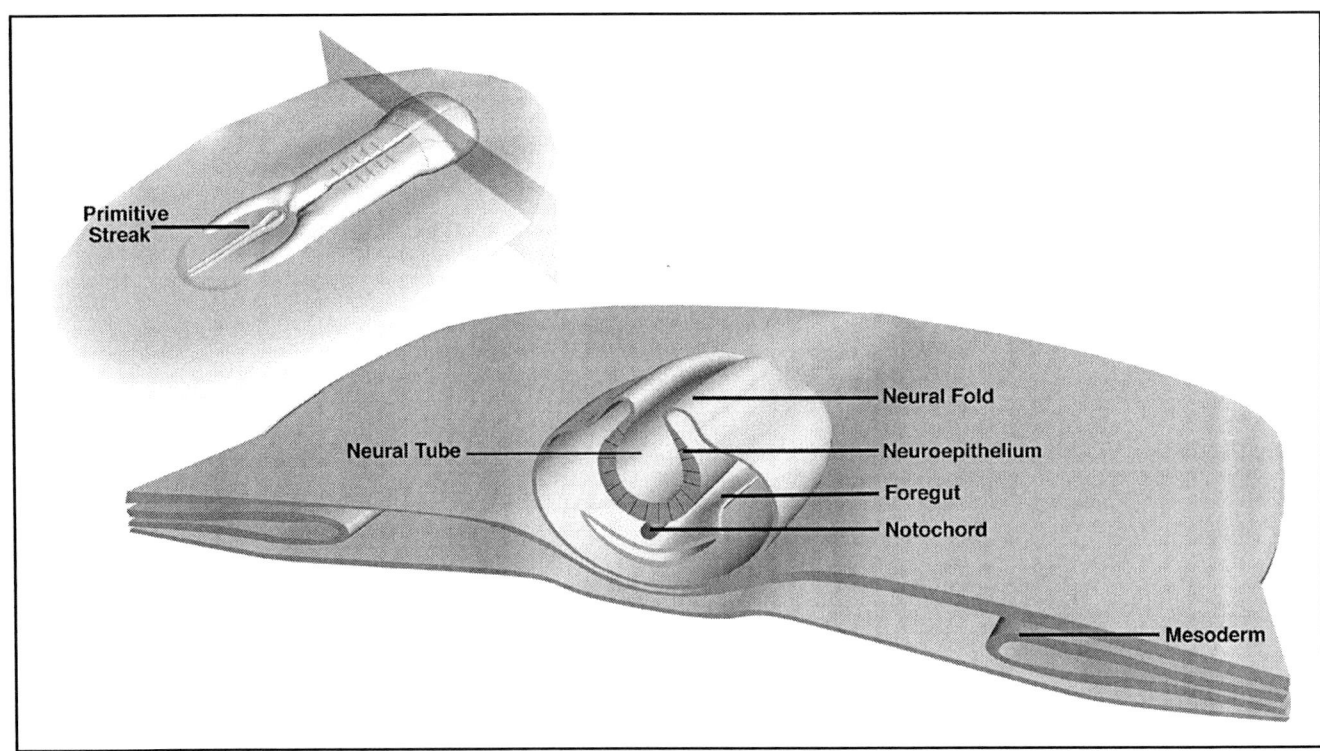

Figure 1. Early development of the central nervous system (CNS). Cross-section through the head anlage of a developing chick embryo: Initially, cells forming the neural plate lie on both sides of the primitive streak. As the primitive streak migrates caudally, the mesoderm invaginates, and the neural plate forms the neural groove in the midline of the embryo. Changes in cell shape cause the neural groove to expand upward and fold inward to form the neural tube. Cells lining the neural tube are the early neuroepithelial stem cells.

originate from stem cells in the neuroepithelium.[11] These cells migrate out into the adjacent SVZ where they proliferate and become astrocytes and oligodendrocytes. Lineage tracing studies using stereotactically injected retrovirus support the view that the majority of progenitors within this germinal matrix are glial precursors that generate either astrocytes or oligodendrocytes.[12,13] Some SVZ cells give rise to both oligodendrocytes and astrocytes, and a rare cell will develop into both neurons and glia,[14] athough this remains a controversial issue.[15]

When glioblast formation ceases shortly after birth, the germinal VZ disappears throughout the neuroaxis and many of the remaining neuroepithelial cells become ependymal cells. The ependymal cells persist throughout adulthood lining the luminal surface of the ventricular system of the brain and the central canal of the spinal cord. These cells possess multiple cilia on their apical surface that effectively move the cerebral spinal fluid throughout these regions. Similarly, the SVZ, decreases in size and persists immediately adjacent to the ependymal cell layer throughout most of the ventricular regions of the brain. However, a SVZ region is not present in the developing or mature regions of the spinal cord.

As compartmentalization of the CNS becomes apparent, neural stem cells in the early mammalian CNS are considered to be concentrated in seven major areas: the olfactory bulb, VZ and SVZ of the forebrain; the hippocampus, cerebellum, cerebral cortex and the spinal cord. Their number and pattern of development vary in different species.[4,16,17] However, once the patterning of the different CNS compartments is in place, it is believed that stem cells located in these different regions of the developing CNS are developmentally distinct and are not a single population of stem cells that are dispersed over multiple sites.[18] Stem cells isolated from the spinal cord generate spinal cord progeny.[19] Stem cells isolated from the basal forebrain generate more GABA containing neurons than stem cells derived from dorsal regions.[20]

Adult Neural Stem Cells

Proliferative stem cell compartments are not exclusive to the developing CNS. However, after embryonic development, the exact characteristics of the neural stem cell in the CNS have remained elusive and controversial. Numerous pioneering experiments have demonstrated that specific regions of the mammalian CNS undergo a moderate, yet continuous level of neurogenesis postnatally and throughout adult life[21-24] (Fig. 3). To date, neurogenesis in the adult mammalian CNS is known to utilize at least one dividing progenitor cell population[25] and two different multipotent stem cell populations.[26-28] The putative progenitor cell population resides in the subgranular zone of the dentate gyrus located in the hippocampus, the region of the brain involved in learning and memory. The two remaining stem cell populations have been reported to exist in and near the anterior lateral ventricular wall of the cerebral cortex, both of which exist in the adult as highly differentiated glial cell types; SVZ astrocytes and ventricular ependymal cells. While there is only a limited amount of neurogenesis that occurs in the adult hippocampus, stem cells located in the SVZ are thought to continuously replace interneurons in the olfactory bulb. It remains controversial whether the rapidly dividing multipotent stem cells in the SVZ are a distinct stem cell population that contributes to the generation of olfactory interneurons. It has been suggested that the adjacent ependymal cell layer, which has been shown to divide at a comparatively slower rate in vivo, may give rise to the SVZ cells.[26,29] Although glial cells in the SVZ are derived from the VZ during early

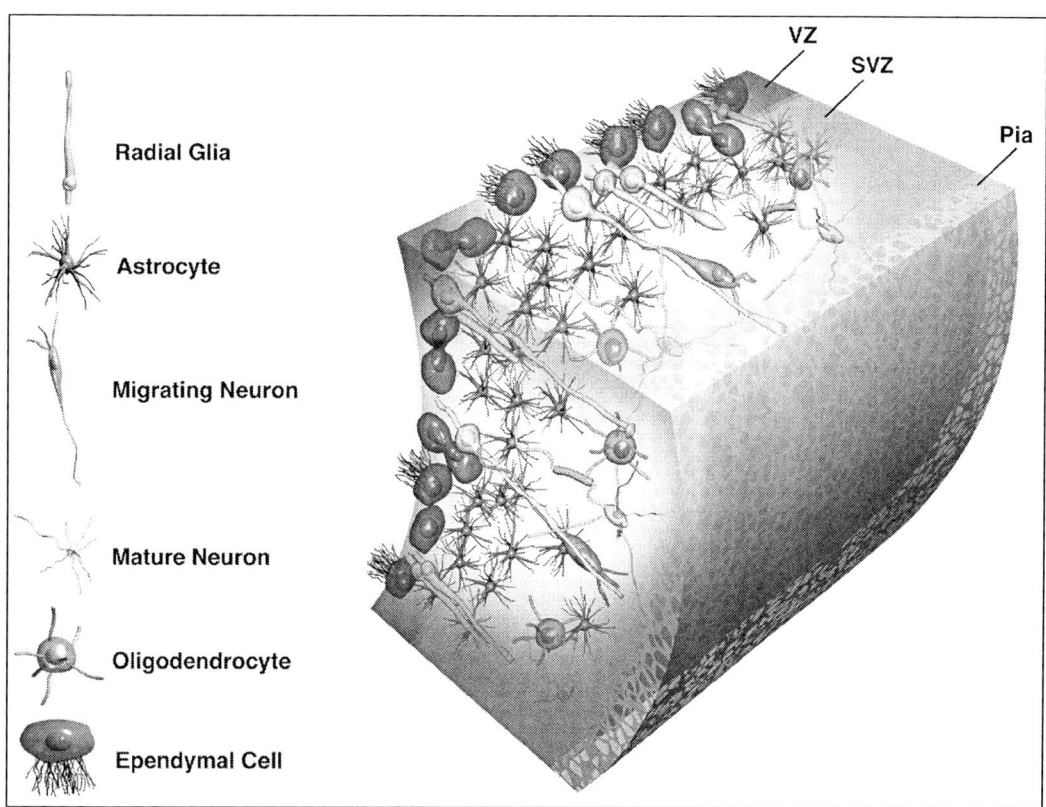

Figure 2. Early differentiation of the neuroepithelium in the neocortex of the brain. Early in development, neuroepithelial stem cells reside in the luminal cellular layer of the neural tube in an area generally termed the ventricular zone (VZ). These cells begin to divide rapidly, initially generating radial glia and early restricted neuroblasts. The neuroblasts exit from the cell cycle, delaminate from the neuroepithelium, accumulate within the ventricular zone and activate a number of genes indicative of generic neuronal differentiation. Activation of these genes is thought to put into motion the expression of specific cascades of factors that specify both neuronal determination and differentiation in different regions of the cortex. As the neuroblasts mature, radial glia provide a primary substrate for the migration of post-mitotic neurons from the germinal ventricular zone to the emerging layers of the neocortex. After neuroblast formation subsides, neuroepithelial stem cells in the ventricular zone generate glioblasts. These glioblasts migrate out into the adjacent subventricular zone (SVZ) where they proliferate and become primarily astrocytes and oligodendrocytes. These cells then migrate away from this region to populate other regions of the developing cortex.

embryonic development, a lineage relationship between these two multipotent adult cell types has not been firmly established. However, the location of these two adult neural stem cell populations has some surprising parallels to early neurogenic development of the ventricular regions in the embryonic brain. The ependymal cells, within the lateral ventricular wall, occupy a position analogous to the embryonic VZ cells and are thought to be derived directly from a subset of embryonic VZ cells. While the ependymal cells are highly differentiated glial cells that line the luminal surface of the adult ventricular system, these seemingly differentiated cells express several proteins expressed by neural stem cells during normal development including nestin, musashi, and Notch 1 receptors.

While the identification and existence of adult neural stem cells in the ventricular and SVZ was a surprising finding, even more surprising is the suggestion that they are highly differentiated glial cell types and not remnants of nascent undifferentiated stem cells that are specified early during embryonic development. Numerous trophic factors have been shown to influence the actual developmental fate of a progenitor or multipotent stem cell from these regions, which may differ from its developmental potential.[30,31] In culture, cells from these regions can differentiate into neurons in response to instructive extracellular signals that promote the production or activity of proneural transcription factors. Similarly, they can differentiate into glia in response to a variety of extracellular signals such as ciliary neurotrophic factor (CNTF), bone morphogenic proteins (BMPs), transforming growth factor-α (TGFα), and neuregulin-1 (Nrg-1)/glial growth factor-2 (GGF2). However it is not known whether these same factors directly induce glial differentiation in vivo. Because environmental conditions can be provided in vitro by adding specific trophic factors to the culture medium, neural stem cells differentiating in cultures have been shown to exhibit considerable plasticity not normally observed in vivo. Thus, understanding the environmental conditions necessary to promote specific types of differentiation from a stem cell may show that the limitations that these cells perceive in vivo are controlled by environmental cues and not necessarily by intrinsic commitment of the stem cell itself.

Oligodendroglial Progenitors

The multipotent neural stem cells or early neural progenitor cells present in the central nervous system of embryonic, neonatal and adult animals can also differentiate into intriguing lineage-restricted progenitors termed oligodendroglial progenitors. Most oligodendrocyte precursor cells in the developing central

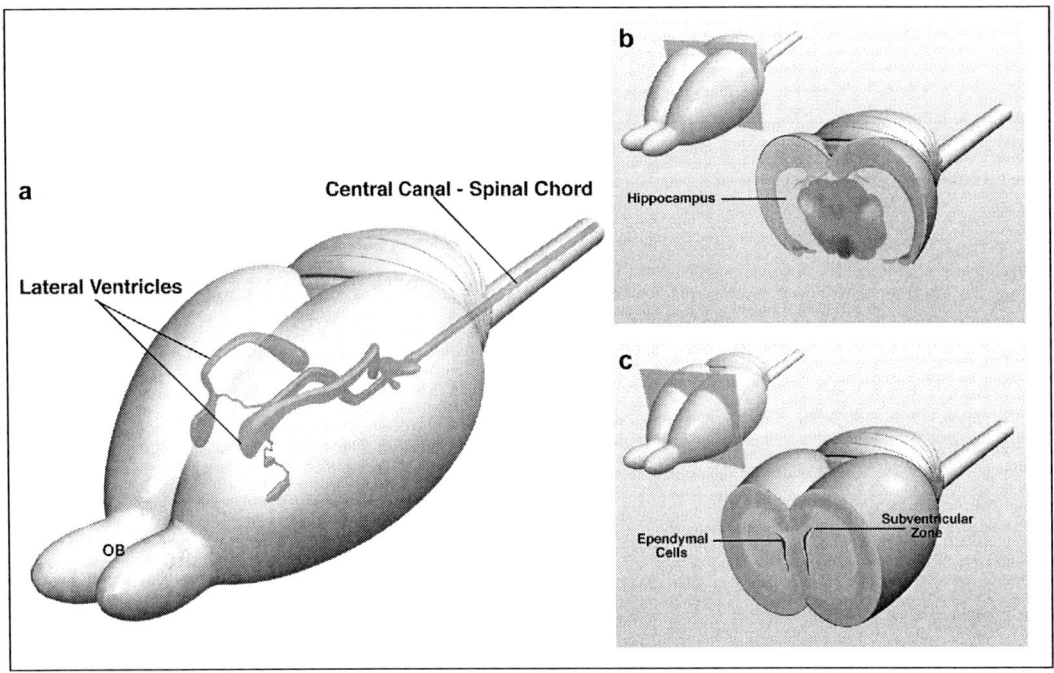

Figure 3. Neurogenic regions of the mammalian CNS. Neural stem cells in the adult CNS have been identified in four different regions: the ependymal cell layer lining the lateral ventricular regions of the brain, the adjacent subventricular zone, the hippocampus and the spinal cord. The ventricular system of the brain is continuous with the central canal of the spinal cord (A). Coronal sections through different regions of the brain show the location of the hippocampus near the posterior region of the cerebral hemispheres (B). The ependymal cell layer lining the luminal surface of the lateral ventricles and the adjacent subventricular zone is shown in C.

nervous system terminally differentiate into oligodendrocytes that myelinate axons. Terminally differentiated oligodendrocytes do not divide, dedifferentiate, or reenter the cell cycle. However, precursors to oligodendrocytes do exist and their division persists, albeit at a slow rate, throughout life, with a certain potentiality bias for myelin repair.

Oligodendrocyte progenitors have been characterized in rodent species by their bipolar morphology and by the presence of specific markers. In vitro studies as well as in vivo experiments have shown that these cells are actively proliferating and posses migratory properties. Oligodendrocyte progenitors arise from multipotential cells in spatially restricted germinal zones thoughout the brain and spinal cord. These progenitors migrate long distances away from these zones in the CNS before they settle along fiber tracts of the future white matter and then transform into preoligodendrocytes. The pre-oligodendrocyte is a multiprocessed, post-migratory cell that retains the ability to divide but is not responsive to the mitogen, platelet-derived growth factor (PDGF).[32-34] These cells can be identified by their acquisition of the marker, O4.[35] The preoligodendrocyte can further differentiate into an immature oligodendrocyte, characterized in the rat by the appearance of the marker GalC, and the loss of expression of GD3 and A2B5 antigens on the cell surface. CNP (2′,3′-cyclic nucleotide 3′-phosphodiesterase), the earliest known myelin-specific protein to be synthesized by developing oligodendrocytes also appears at this time.

The mature oligodendrocyte is characterized by the expression of myelin basic protein (MBP), myelin associated-glycoprotein (MAG) and myelin proteolipid protein (PLP). In vitro analyses suggest that maturation of oligodendrocytes from the precursor stage to the mature cell is identical in culture, even in the absence of neurons, as in intact tissue. Thus the capacity of oligodendrocyte progenitors to differentiate into oligodendrocytes is intrinsic to the lineage.[36] However, co-culture with neurons increases myelin gene expression, such as PLP, MBP and MAG.

Plasticity of the Oligodendrocyte Lineage

Recently, mouse oligodendroglial progenitors derived from totipotent embryonic stem cells were transplanted into a myelin-deficient rat model.[37] These cells were able to interact with the host neurons and efficiently myelinate axons in the brain and spinal cord. Similarly, oligodendrocyte progenitors have been isolated from the adult brain. These cells have been propagated extensively in vitro as neurospheres—clonal spheroid cell aggregates that detach from the tissue culture dish and grow in suspension[21]—to generate a large number of multipotent stem cell progeny that maintain their myelinating potential.[38] Phenotypic characterization of these cells has indicated that these oligodendrocyte progenitors resemble neonatal rather than adult progenitors. The ability to generate such cells from both embryonic stem cells and the adult brain opens many possibilities to explore the potential of these cells for repairing myelin disorders. Moreover, the ability of the adult-derived oligodendrocyte progenitors to apparently dedifferentiate to a more primative state, suggests that these cells may also represent a unique progenitor cell population that may be permissive to manipulation both in vitro and in vivo.

Are Neural Stem Cells Derived from the Adult CNS Irreversibly Determined?

Adult spinal-cord–derived cells, which normally generate only glia[4,26] can make interneurons if injected into the adult hippocampus.[39] Similarly, adult hippocampal-derived stem cells can make olfactory interneurons after transplantation into the SVZ.[40] However, the ability of adult-derived cells to produce complex projection neurons that span long distances in the mature CNS, has not been demonstrated.

Only recently have people investigating the differentiation potential of adult neural stem cells discovered that their differentiation repertoire extends well beyond the boundaries of the cell types identified in the CNS. This revelation has caused quite a stir among traditional developmental biologists and many valid arguments have been raised and remain to be answered.[20,41] If this newly discovered potentiality is true, it may cause us to re-evaluate or define the methods that are used to determine whether a neural stem cell is truly irreversibly specified or not. It is possible that embryonic or adult neural stem cells, when cultured in vitro, exhibit a renewed potentiality that has not been tested or observed in direct transplantation assays. This renewed potential may involve complex inter- and intracellular mechanisms that may have been unwittingly imparted to these cells in culture. Alternatively, the need to restrict the potential of a stem cell may decrease as an organism matures.[42] Thus, during embryonic development, cells may be exposed to overlapping sets of extracellular signals. This may initially necessitate their use of cell-autonomous mechanisms to restrict their differentiation. Therefore, transplanting restricted cells from one location to another does not affect their differentiation potential. However, once an organism matures, stem cells in different tissues may be spatially segregated into specific niches where they no longer actively encounter signals that restrict their differentiation potential. These cells may be released from their cell autonomous programs making them more responsive to environmental cues that can influence their differentiation.

Differentiation to Hematopoietic Cells

Intriguingly, in the past few years several reports have indicated that both hematopoietic and neural stem cells both demonstrate surprising plasticity.[43] Bone marrow derived cells have been shown to generate cells expressing neuronal markers in the brain.[44,45] Similarly, embryonic or adult-derived neural stem cells from the brain, when injected intravenously into sublethally irradiated mice, have been shown to generate hematopoietic derivatives.[46] In this study, in vitro clonogenic assays, immunocytochemistry and flow cytometric analysis were used to test whether these cells had adopted a hematopoietic identity. In the clonogenic assays, cells from the bone marrow of transplanted animals were plated in the presence of defined cytokines. Colonies founded by single hematopoietic precursors of neural stem cell origin were 13% pure granulocyte, 30% granulocyte-macrophage, 22% pure macrophage and 19% mixed colonies. A few colonies did not originate from the donor stem cell population confirming that some endogenous hematopoietic progenitors had survived the irradiation. Importantly, none of the neural stem cell cultures or progeny of clones used in the transplantation proliferated or formed colonies in the clonogenic assays. The neural stem cells used in this study also generated neurons, astrocytes and oligodendrocytes in vitro, arguing against the possibility that the blood forming cells from the brain were stray hematopoietic stem cells that contributed to the repopulation of the hematopoietic lineages.

Differentiation to Other Somatic Cell Lineages

The myogenic potential of adult neural stem cells has also been recently demonstrated. Co-culture of primary mouse or human neural stem cells with myoblasts or injection of the neural stem cells into skeletal muscle of adult mice resulted in the differentiation of these cells to myocytes, a process which required direct contact between the neural stem cell and the muscle cells.[47] A conceptually different approach has also demonstrated the myogenic differentiation potential of neural stem cells.[48] This approach took advantage of the large number of inductive signals present in embryonic stem cells undergoing differentiation in vitro, a process that ultimately results in the differentiation of the ES cells to various cell lineages. In order to expose neural stem cells to such a milieu, the cells were cultured in close proximity to differentiating embryoid bodies. In these experiments, differentiated neural stem cells gave rise primarily to myocytes, identified by the expression of the markers desmin and myosin heavy chain, and exhibited morphological features such as syncytium formation.

Two separate methods have been used to simultaneously demonstrate the broad differentiation potential of neural stem cells.[48] These experiments have relied heavily on the abundance of factors present in the early embryonic environment to demonstrate such potential. In a separate set of experiments, adult neural stem cells were injected into early developing chick or mouse embryos. In the developing chick, neurospheres or clusters of clonally-derived neural stem cells were placed directly on or near the primitive streak of a gastrulating embryo. This allowed the neural stem cells to integrate into the epiblast, move with the primitive ectoderm cells through the primitive streak during gastrulation and be distributed throughout the three definitive germ layers. In the mouse, dissociated neural stem cells, or small neurospheres, were aggregated with morulae or injected into blastocysts. In both of these early embryo models, neural stem cell progeny were found in numerous tissues of the host, in environments that inevitably had exposed them to a large number of different inductive signals. The neural stem cells integrated in a mosaic pattern into many tissues and were morphologically indistinguishable from neighboring host cells. Contributions to tissues were large and occasionally comprised as much as 30% of the entire organ. Cell type specific antibodies were used to verify whether the neural stem cell derived cells had remained neural or if they had taken on a phenotype appropriate for the tissue into which they had integrated. In both the chick and mouse assays, neural stem cell progeny had indeed differentiated to many embryonic cell types including hepatocytes, cardiomyocytes and epidermal cells, thus representing cells that originate from all three germ layers. Importantly, in experiments where multipotent secondary clonal cultures were established, it was shown that a single neural stem cell, had the potential to generate progeny for all three germ layers.

While many of the transplantation and embryonic differentiation studies have demonstrated the ability of adult neural stem cell populations to populate and differentiate into cell types specific for a particular tissue in vivo, their longevity and functional contribution to a living organism have not been demonstrated. Many of the obstacles hampering the definitive detection of donor cells in these early experiments can now be addressed as new genetically marked strains of mice and methods increasing the efficiency of cellular contribution have been identified. It will be of particular interest to determine whether the contribution of a donor stem cell to a particular adult stem cell niche will allow the original donor stem cell to retain its ability to self-renew, continuously contributing to the neogenesis of cell types particular to that tissue. It will also be of interest to determine whether stem cells introduced into the early embryonic environment exhibit germline contribution.

Neural Stem Cell Potency

The culture of many neural stem cell populations as heterogenous nonadherent clusters of cells, termed neurospheres, may be necessary to prevent the irreversible determination and differentiation of these cells. Clonal analysis of neural stem cell populations typically results in the clonal growth of 4 to 6% of the total starting population, provided the cells are exposed to conditioned medium isolated from densely proliferating neural stem cell cultures.[26] However, if single cells are seeded densely in culture medium immediately following dissociation and allowed to physically aggregate with other similar cells, the majority of the cell population survives and continues to proliferate in culture. Even more striking, is the observation that contribution of adult-derived neural stem cells to embryonic tissues required that small clusters of clonal or heterogenous neural stem cell populations be introduced into the early blastocyst or gastrulating chick embryo in order to reproducibly observe contribution of these cells to various germ layers.[48] Together, these observations have important implications for both the existence of short-range factors that may help maintain proliferation and prevent the differentiation of these cells[49] and cell surface molecules that may be important in specifying the number of cells that are allowed to initiate a differentiation program at any given time.[50-52]

Developmental biologists have long made the distinction between the determination of myogenic cells, as a state that is inherently plastic and easily reversible in vitro, and terminal differentiation, a more committed state that leads to terminal muscle differentiation. In a similar manner, proneural genes that promote the initiation of the neuronal program, are thought to activate a second wave of proteins that are expressed in neural stem cells, the determination genes. When their activity reaches a threshold, these proteins activate the expression of downstream differentiation genes to promote exit from the cell cycle and terminal neuronal differentiation. If, for example, it was possible to block the expression of the determination and proneural genes, neural stem cells may revert back to a more primative ground state where they would have the option to proliferate and become a neuron, glial cell or any number of different cell types at a later time.[53] Evidence exists for the multilineage expression of genes in stem cells prior to lineage commitment.[54,55] Thus regulation of gene expression in a population of closely associated cells may prove to be critical in maintaining neuroepithelial cells or stem cells in a plastic or more primitive state.

References

1. Barres BA, Barde Y. Neuronal and glial cell biology. Curr Opin Neurobiol 2000; 10(5):642-648.
2. Ullian EM, Sapperstein SK, Christopherson KS et al. Control of synapse number by glia. Science 2001; 291(5504):657-661.
3. Yudkoff M, Nissim I, Daikhin Y et al. Brain glutamate metabolism: Neuronal-astroglial relationships. Dev Neurosci 1993; 15:343-350.
4. Rao MS. Multipotent and restricted precursors in the central nervous system. Anat Rec 1999; 257:137-148.
5. McConnell SK. Constructing the cerebral cortex: Neurogenesis and fate determination. Neuron 1995; 15:761-768.
6. Altmann CR, Brivanlou AH. Neural patterning in the vertebrate embryo. Int Rev Cytol 2001; 203:447-482.
7. Weinstein DC, Hemmati-Brivanlou A. Neural induction. Annu Rev Dev Biol 1999; 15:411-433.
8. Wolpert L. Positional information and pattern formation in development. Dev Genet 1994; 15:485-490.
9. Kobayashi D, Kobayashi M, Matsumoto K et al. Early subdivisions in the neural plate define distinct competence for inductive signals. Development 2001; 129:83-93.
10. Lumsden A, Krumlauf R. Patterning the vertebrate neuraxis. Science 1996; 274:1109-1115.
11. Rao MS, Mayer-Proschel M. Glial-restricted precursors are derived from multipotent neuroepithelial stem cells. Dev Biol 1997; 188:48-63.
12. Levison SW, Goldman JE. Multipotential and lineage restricted precursors coexist in the mammalian perinatal subventricular zone. J Neurosci Res 1997; 48:83-94.
13. Price J, Thurlow L. Cell lineage in the rat cerebral cortex: A study using retroviral-mediated gene transfer. Development 1988; 104:173-182.
14. Levison SW, Goldman JE. Both oligodendrocytes and astrocytes develop from progenitors in the subventricular zone of postnatal rat brain. Neuron 1993; 10:201-212.
15. Price J. Glial cell lineage and development. Curr Opin Neurobiol 1994; 4:680-686.
16. Frisén J, Johansson CB, Lothian C et al. Central nervous system stem cells in the embryo and adult. Cell Mol Life Sci 1998; 54:935-45.
17. Gage, FH. Mammalian Neural Stem Cells. Science 2000; 287:1433-1438.
18. Gaiano N, Fishell, G. Transplantation as a tool to study progenitors within the vertebrate nervous system. J Neurobiol 1999; 36:152-161.
19. Kalyani AJ, Piper D, Mujtaba T et al. Spinal cord neuronal precursors generate multiple neuronal phenotypes in culture. J Neurosci 1998; 18:7856-7868.
20. Temple S. The Development of neural stem cells. Nature 2001; 414:112-117.
21. Altman J, Das GD. Autoradiographic and histological studies of postnatal neurogenesis. J Comp Neurol 1966; 126:337-90.
22. Bayer SA, Yackel JW, Puri PS. Neurons in the rat dentate gyrus granular layer substantially increase during juvenile and adult life. Science 1982; 216:890-892.
23. Reynolds BA, Weiss S. Generation of neurons and astrocytes from isolated cells of the adult mammalian nervous system. Science 1992; 255:1707-1710.
24. Lois C, Alvarez-Buylla A. Proliferating subventricular zone cells in the adult mammalian forebrain can differentiate into neurons and glia. Proc Natl Acad Sci USA 1993; 90:2074-2077.
25. Eriksson PS, Perfilieva E, Bjork-Eriksson T et al. Neurogenesis in the adult human hippocampus. Nat Med 1998; 4:1313-1317.
26. Johansson CB, Momma S, Clarke DL et al. Identification of a neural stem cell in the adult mammalian central nervous system. Cell 1999; 96:25-34.
27. Doetsch F, Caille I, Lim DA et al. Subventricular zone astrocytes are neural stem cells in the adult mammalian brain. Cell 1999; 97:703-716.
28. Rietze RL, Valcanis H, Brooker GF et al. Purification of a pluripotent neural stem cell from the adult mouse brain. Nature 2001; 412:736-739.
29. Barres BA. A new role for glia: Generation of neurons! Cell 1999; 97:667-670.
30. Johe KK, Hazel TG, Muller T et al. Single factors direct the differentiation of stem cells from the fetal and adult central nervous system. Genes Dev 1996; 10:3129-3140.
31. Kuhn HG, Winkler J, Kempermann G et al. Epidermal growth factor and fibroblast growth factor-2 have different effects on neural progenitors in the adult rat brain. J Neurosci 1997; 17:5820-5829.
32. Gao FB, Apperly J, Raff M. Cell-intrinsic timers and thyroid hormone regulate the probability of cell-cycle withdrawl and differentiation of oligodendrocyte precursor cells. Dev Biol 1998; 197:54-66.
33. Hart IK, Richardson WD, Bolsover SR et al. PDGF and intracellular signaling in the timing of oligodendrocyte differentiation. J Cell Biol 1989; 109:3411-3417.
34. Pringle NP, Richardson WD. A singularity of PDGF alpha receptor expression in the dorso-ventral axis of the neural tube may define the origin of the oligodendrocyte lineage. Development 1993; 117:525-533.
35. Sommer I, Schachner M. Monoclonal antibodies to oligodendrocyte cell surfaces: An immunocytological study in the central nervous system. Dev Biol 1981; 83:311-327.

36. Temple S, Raff M. Clonal analysis of oligodendrocyte development in culture: Evidence for a developmental clock that counts cell divisions. Cell 1986; 44(5):773-9.
37. Brustle O, Jones KN, Learish RD et al. Embryonic stem cell-derived glial precursors: A source of myelinating transplants. Science 1999; 285:754-756.
38. Zhang S, Ge B, Duncan ID. Adult brain retains the potential to generate oligodendroglial progenitors with extensive myelination capacity. Proc Natl Acad Sci USA 1999; 96:4089-4094.
39. Shihabuddin LS, Horner PJ, Ray J et al. Adult spinal cord stem cells regenerate neurons after transplantation in the adult dentate gyrus. J Neurosci 2000; 20:8727-8735.
40. Suhonen JO, Peterson DA, Ray J et al. Differentiation of adult hippocampus-derived progenitors into olfactory neurons in vivo. Nature 1996; 383:624-627.
41. Weissman IL, Anderson DJ, Gage F. Stem and progenitor cells: Origins, phenotypes, lineage commitments, and transdifferentiations. Annu Rev Cell Biol 2001; 17:387-403.
42. Blau HM, Baltimore D. Differentiation requires continuous regulation. J Cell Biol 1991; 112:781-783.
43. Terskikh AV, Easterday MC, Li L et al. From hematopoiesis to neuropoiesis: Evidence of overlapping genetic programs. Proc Natl Acad Sci USA 2001; 98:7934-7939.
44. Brazelton TR, Rossi FMV, Keshet GI et al. From marrow to brain: Expression of neuronal phenotypes in adult mice. Science 2000; 290:1775-1779.
45. Mezey E, Chandross KJ, Harta G et al. Turning Blood into Brain: Cells bearing neuronal antigens generated in vivo from bone marrow. Science 2000; 290:1779-1782.
46. Bjornson CR, Rietze RL, Reynolds BA et al. Turning brain into blood: a hematopoietic fate adopted by adult neural stem cells in vivo. Science 1999; 283:534-537.
47. Galli R, Borello U, Gritti A et al. Skeletal myogenic potential of human and mouse neural stem cells. Nature Neurosci 2000; 3:986-991.
48. Clarke DL, Johansson CB, Wilbertz J et al. Generalized potential of adult neural stem cells. Science 2000; 288:1660-1663.
49. Taupin P, Ray J, Fischer WH et al. FGF-2-responsive neural stem cell proliferation requires CCg, a novel autocrine/paracrine cofactor. Neuron 2000; 28(2):385-397.
50. Wang S, Barres BA. Up a notch: Instructing gliogenesis. Neuron 2000; 27:197-200.
51. Kondo T, Raff M. Oligodendrocyte precursor cells reprogrammed to become multipotential CNS stem cells. Science 2000; 289:1754-1757.
52. Tsai RYL, McKay RDG. Cell contact regulates fate choice by cortical stem cells. J Neurosci 2000; 20:3725-3735.
53. Kintner, C. Neurogenesis in embryos and in adult neural stem cells. J Neurosci 2002; 22(3):639-643.
54. Hu M, Krause D, Greaves M et al. Multilineage gene expression precedes commitment in the hematopoietic system. Genes Dev 1997; 11:774-785.
55. Busslinger M, Nutt SL, Rolink AG. Lineage commitment in lymphopoiesis. Curr Opin Immunol 2000; 12:151-158.

CHAPTER 4

Turning Blood into Liver

Bryon E. Petersen and Neil D. Theise

Introduction

Hepatic oval "stem" cells have been studied in rodents since the mid-1950s when they were first described by E. Farber.[1] Hepatic oval "stem" cells have been shown to be a cell type capable of developing into hepatocytes or bile duct epithelial cells.[2,3] Numerous in vivo and in vitro studies have shown a central role of oval cells in liver biology and carcinogenesis.[4-7] In addition to the effects seen in the liver, oval cells (or cells very similar to oval cells) have been seen in the architecture of the regenerating pancreas after injury.[8,9] While there is a specific function for this cell population in the liver, the mechanism(s) by which oval cells activate, proliferate and differentiate are poorly understood. In spite of four decades of research and over 300 publications on hepatic oval "stem" cells, their ontogeny remains unidentified.

The existence of "oval-like" cells in humans has also been debated. Ductular reactions in different acute and chronic liver diseases have been observed.[10] Though these cells are morphologically different from oval cells in rodent models (they are not usually oval!), they share many phenotypic features of oval cells in rodents. This has led many investigators to speculate that they serve a similar regenerative function.[11-17]

In this chapter, we seek to briefly summarize the history of investigations into these cell populations and to highlight the most recent investigations, which have finally established a consensus regarding the role of these cells in liver regeneration, and raised new questions about the replenishment of the liver "oval" stem cell compartment from hematopoietic stem cells.[3,18,19]

Animal Models

"Oval cell proliferation" is an orchestrated set of events that describes the occurrence of nonparenchymal cells detected in large numbers at early stages of carcinogenesis. The population of cells is very heterogeneous and contains cells that differ in both their differentiation-state and potential in their lineage commitment. The "oval cell compartment" is also used to describe the cells invading the parenchyma after the administration of certain carcinogens.[20,21] It is the view of some investigators that cell lineage analyses cannot be accomplished in a culture system(s) and that the markers used in studies of hepatocarcinogenesis can be considered reliable lineage markers for work done in vivo.[6] Farber also reported that in each oval cell compartment, the oval cells will express different isoenzyme profiles, however, certain hepatic markers such as alpha-fetoprotein (AFP) and biliary cytokeratin-19 (CK-19) will always be expressed.[6] From the dimethylaminoazbenzene (DMAB) model presented by Farber (1956), the single most important conclusion provided by this data is that hepatocytes could be generated from nonparenchymal progenitor cells.[1] This model establishes the key principle that some type of progenitor cell exists, even if it turns out that the differentiation of this precursor cell occurs only infrequently.

Evarts et al (1987 and 1989), show that the oval cell compartment is activated extensively in a 2-Acetylaminoflourene/two thirds partial-hepatectomy (2-AAF/PHx) model. This model is a variation of the Solt-Farber protocol and has become the trade benchmark.[22,23] Figure 1 represents a generic timeline for this model of liver regeneration from oval "stem" cells. These conditions markedly suppress mature hepatocyte proliferation and, when followed by partial hepatectomy activate, oval cell proliferation. Further studies, presented by the same investigators, showed that proliferation of the oval cells are associated with the activation of the hepatic stellate cells, which may regulate the developmental fate of the progenitor cells either directly, by secreting growth factors such as hepatocyte growth factor (HGF), transforming growth factor-alpha (TGF-α) and TGF-β, or indirectly, through the effects of stellate cell produced extra-cellular matrix components that can also be induced by TGF-β.[24] It has been shown that progenitor cell proliferation and differentiation could also be regulated by autocrine production of TGF-α, acid fibroblast growth factor (aFGF) and IGF-II (insulin-like growth factor II), since it has been shown that oval cells can make all of these factors.[7]

Over the past 40 years several laboratories have performed extensive studies concerning oval cell characterization with regards to developing techniques to produce new models for oval cell activation in the rat.[25-28] These models basically follow the protocol established by Thorgiersson and followed by others[25,26,28,29] with the exception that a chemical injury was used to induce liver injury, while Thorgiersson et al used a partial hepatectomy (PHx) model with surgical removal of 2/3 of the liver. In our models, oval cell activation occurs with expression of the same markers as in other models. It has also been shown by the group of Sell et al, that when the periportal region of the liver was damaged by allyl alcohol (AA), the oval cell response was quite different as to the actual number and location of oval cells seen in either the

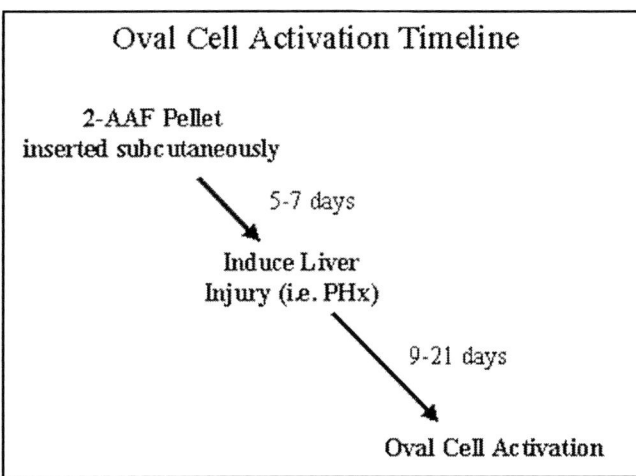

Figure 1. Time line of events for activation of oval cell proliferation. The presence of 2- acetylaminofluorene (2-A AF) is necessary to suppress hepatocyte proliferation and allow extended proliferation of oval cells by partial hepatectomy (pHx). The diagram represents the different stages of oval cell proliferation

2-Acetylaminoflourene/carbon tetrachloride (2-AAF/CCl$_4$) or 2-AAF/PHx-model and that, at the earliest stages of proliferation, these cells were "null" cells, i.e., negative for all usual phenotypic markers of oval cells.[30]

The analysis of liver development and the studies of cell populations during hepatocarcinogenesis provide a framework for predicting the location of putative liver stem cells in adult normal liver and the types of markers they may express. Given the pattern of proliferation of the oval cell in carcinogenesis and the histogenesis of primitive intra-hepatic bile ducts in development, it would be logical to expect that the stem cells should be evident in the smallest unit of the biliary tree. These units are called canals of Hering. These units line the hepatocytes and ductular cells and extend through the limiting plate to form a connection between the parenchyma and interlobular units.[20,31,32] The canals of Hering have also been called cholangioles, or terminal ducts or ductules, although these terms are imprecise and should be avoided.[20] In a study published in the *American Journal of Pathology* (1997), our laboratory showed that when the bile ductual epithelial is destroyed by exposure to methylene dianaline (DAPM) 24 hrs prior to hepatic damage (2-AAF/hepatic injury), oval cell proliferation is inhibited.[33] This study was the first to show that there is a direct association between an intact bile ductular epithelia and oval cell activation. However, this does not prove that oval cells arise from bile ductular cells. DAPM exposure could have a toxic effect upon the oval cells either in a direct or indirect manor. Until recently, amongst those researchers who believed that oval cells represented a liver stem/progenitor compartment, the concept that oval cells could only be derived from the biliary tree was considered to be dogma. Recent studies have shown that this may not be the case.[3,18,19,34,35]

While characterizing the cells from our model, we discovered that the oval cells highly express the hematopoietic stem cell marker Thy-1.1.[36] In addition, it has been shown that oval cells also express CD34 and Flt-3 as well as c-Kit, all known to be hematopoietic stem cell markers.[37-39] Cells were labeled with Thy1.1-FITC antibody and sorted by flow cytometry. From this type of isolation technique we are able to obtain a 95-97% pure population of Thy-1.1+ cells, which have been shown to also

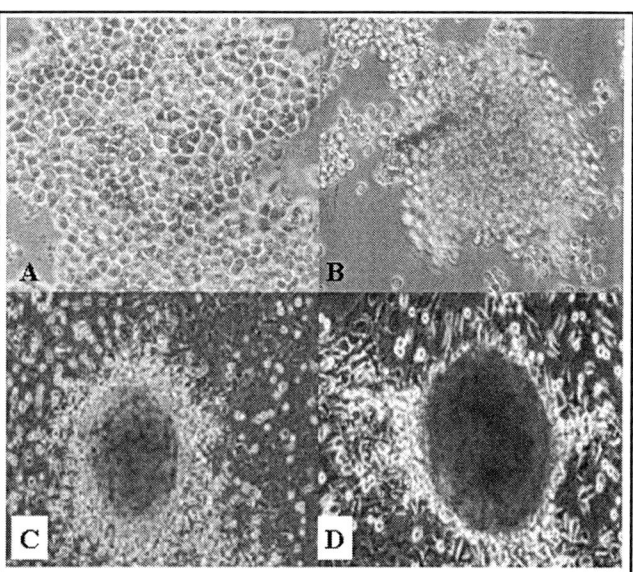

Figure 2. Thy-1 + sorted oval cells in culture. A) Cells in culture for three days. B) 12 days in culture, notice the cells are beginning to form clusters. C) Oval cells in continuous culture for 6 months. The clusters are now tightly formed spheroids. D) Oval cells frozen and thawed and grown in culture for over 12 months.

express the traditional oval cell markers of AFP, CK-19, gamma glutamyl-transferase (GGT) and OV-6 (oval cell marker 6) and were Desmin negative.[36] This method is to date the highest reported purity of oval cells. Using this method we have also been able to place these cells into culture on fibronectin-coated plates in a serum free Iscove's medium supplemented with various growth factors. After 3 days in culture they begin to form colonies (Fig. 2A). This type of colony formation closely resembles colonies described by other investigators.[40,41] As the colonies continue to grow, they form tightly-compact colonies (Fig. 2B and 2C), which appear to retain what would be considered characteristics for oval cells, staining positive for both AFP and GGT (data presented as poster discussion at EB/ASIP meeting April 1997, San Francisco, CA).[42] It also appears that oval cells are able to change in morphology depending upon culture conditions. If oval cells are cocultured with porcine microvascular endothelial cells (PMEC), they give a strong epithelial morphology. Whereas, if oval cells 3 day in culture colonies are overlaid with Matrigel they appear to become stellate cells 7 days later.[2] To date, we have been able to grow primary hepatic oval cells in a continuous culture for over twenty months with no end in sight. We are in the process of investigating the signals involved in self-proliferation and differentiation. During this period, we have been able to cryopreserve these cells and place them back into culture seemingly without any ill effects to their proliferative capacity (Fig. 2D).

Recently, we have published results that long-term hepatic oval cells could be induced to trans-differentiate into a hormone insulin-producing cell type. In this study, we showed that oval cells cultured for over 6 months in a growth stimulating media and switched to a high glucose nicotinamide not only increased the production of pro-insulin, but also released the active two chain form of insulin into the media. This shows that the cells were able to adapt to fit their environment. This, however, does not show function. In addition to the in vitro experiments we performed, we also transplanted the trans-differentiated oval cells into chemically induced diabetic mice. In this study it was shown that 10 days post transplant, glucose levels in mice that did not receive cells (controls) remain extremely high (approximately 400 mg/dl). Whereas, the mouse that received the largest number of trans-differentiated cells had glucose levels drop back to normal levels (about 87 mg/dl). These data show that not only did the cells make the transition over to a pancreatic cell type, but were also capable of functioning in a normal fashion and correct a diabetic condition.[42] The current study represents a new potential approach in cell therapy-based treatment of insulin-dependent diabetes by generating insulin-producing cells from stem cells derived from nonpancreatic tissue. As such, hepatic oval "stem" cells and possibly bone marrow-derived stem cells may provide a realistic hope that adult stem cells from one tissue lineage can trans-differentiate along the lineage of other tissues, thus circumventing the need for embryonic stem cells. It is critical that this phenomenon be further explored to determine how this trans-differentiation of hepatic stem cells, and possibly bone marrow stem cells, into endocrine pancreas can be best exploited.

Two lines of investigation have led to the recognition that oval cells can originate in extra-hepatic locations, including the bone marrow, and can enter the liver through the circulation. Based upon previous findings that oval cells express hematopoietic markers, we examined the possibility that another site of oval cell origins may exist outside the liver. We have recently demonstrated that bone marrow progenitor cells are capable of engrafting into the recipient liver. We demonstrated that adult rat male bone marrow contained a sub-population of cells capable of engraftment with differentiation to hepatocyte following 2 AAF/hepatic injury of adult female rat liver.[2,3] Another line of investigation in human livers showed that progenitor cell proliferations were derived from cytokeratin 19 positive cells lining the canals of Hering (see below).[20] However, as mentioned above, experimental models involving periportal injury demonstrated that, when the parenchymal zones containing the canals of Hering were obliterated, the proliferating oval cells were negative for cytokeratin 19.[26,30] This discrepancy in data led investigators to explore the possibility that hepatocytes may be derived from an extra-hepatic source in mice, even in the absence of overt hepatic injury.[18] Subsequent confirmations of these studies have been reported in other animal models as well as in humans.[18,34,35,43]

To address the above issue, we have reported a possible mechanism for oval cell activation in the injured liver and a possible mechanism for recruitment of bone marrow derived stem cells to the liver as a second wave of stem cell aided regeneration.[44] During embryogenesis, the liver develops from a diverticulum of the floor of the foregut,[20,45,46] where the founder cells invade the mesenchyme of the septum transversum. Endodermal cells eventually give rise to hepatocytes and the bile duct epithelial cells, while the mesenchyme gives rise to cells that make up the sinusoidal lining. During fetal development the liver functions as the hematopoietic organ.[47,48] The hematopoietic cells found in the developing liver are of extra-hepatic origin, being derived from stem cells of the yolk sac[32,49] and the aorta-gonad-mesonephros (AGM).[50] Recruitment of the extra-hepatic cells to the embryonic liver is required for proper development, but the signals required by which HSCs respond and the mechanism of their movement within the fetus is not totally understood. It has been suggested that this movement could be controlled through the stromal cell-derived factor-1 alpha (SDF1α)/CXC chemokine receptor (CXCR)4 homing interaction between the hematopoietic and stromal cells.[51,52] The importance of SDF1α and CXCR4 in hematopoiesis is supported by observations of embryonic lethality in knockout mice with targeted disruption of the genes for either SDF1α[53] or its receptor CXCR4.[54-56] The expression of CXCR4 on the majority of CD34 positive and negative cells, and a demonstrated role of SDF1α in inducing chemotaxis in these cells, strongly suggests that the most primitive hematopoietic population including stem cells are responsive to a SDF1α chemotactic gradient.[57-59] As stated earlier, several reports have shown hepatic oval cells and HSCs share a similar immunohistochemical profile, being positive for Thy-1, CD34 and c-Kit.[36-39] In addition, the findings of Lagasse et al (2000) demonstrated that it was indeed the Sca$^+$/Thy^{lo}/kit$^+$/lin$^-$ sub-population of the bone marrow cells that was capable of becoming hepatic tissue.[35] It would seem logical that some type of signal would be needed to bring cells from an extra-hepatic source to aid in the regenerative process.

Using two different types of rat liver regeneration models, normal (nonoval cell aided, (i.e., allyl alcohol (AA) and carbon tetrachloride (CCl_4)) and oval cell aided models (i.e., 2-acetylaminofluorene (2-AAF)/CCl_4, 2-AAF/PHx and 2-AAF/AA),[36,37] we tested whether or not SDF1α protein regulation was affected in a positive or negative manner. The results described in this study show that when oval cells are involved in the regenerative process, SDF1α is up regulated. However, under normal, nonoval cells aided regeneration, SDF1α protein expression was not detected. In the former condition, it was found that the oval cells express the SDF1α receptor, CXCR4, while SDF1α was expressed by the liver parenchyma. These data indicate a possible mechanism by which oval cells are activated in the injured liver.

The SDF1α/CXCR4 interaction is unique, meaning that CXCR4 is the only known receptor for SDF1α and because of its specificity and its location throughout the body it makes this protein-receptor interaction a good candidate for homing bone marrow derived cells to various sites of injury. This interaction could also be viewed as a signal to initiate the oval cell compartment in certain forms of liver regeneration. These experiments begin to shed light on the age-old problem of what exactly is the mechanism in oval cell activation, and may bring into focus the reason why cells migrate to the liver under certain injury models. This may some day lead to a better understanding of the hepatic and hematopoietic interaction in oval cell activation and proliferation. This may in turn lead to better clinical relevance in treatment of patients through stem cell therapies, but clearly, further experiments are needed.

Human Studies

The study of hepatic regeneration from a progenitor or stem cell compartment has been as controversial in humans as it has been in animal models. However, consensus that such cells exist, both in intrahepatic and extrahepatic locations, has emerged in the last few years.[60-63] In humans, investigations have proliferated in the last two decades due to the rapid expansion of liver transplantation as a widespread treatment for acute and chronic liver failure.[11-17,64-68] This development not only resulted in more extensive investigations into the morphologic changes in diverse forms of hepatic regeneration by the examination of whole explant organs, but also because serial biopsies for graft dysfunction following transplantation made human time course studies possible.

The types of studies into the human equivalent of the oval cell can be grouped into three loose categories: morphologic/phenotypic studies, usually based on immunohistochemical analysis of clinical specimens in two and 3 dimensions; time course studies in serial biopsies following organ transplantation; and isolation and in vivo culture of progenitor cell populations from pediatric and adult livers.

Morphological phenotype studies by Michael Gerber and colleagues, as well as other groups, focused primarily on ductular reactions (also referred to as "ductular proliferations") in the setting of acute or chronic liver disease secondary to toxins or viral infection or vascular insufficiency, usually post-transplantation. Immunohistochemical staining for biliary markers (e.g., cytokeratins 7 and 19) and hepatocyte markers (e.g., alpha-fetoprotein, HepPar1) routinely demonstrate duct-like reactive structures, often referred to as "ductular hepatocytes,[13,14,17,64-66,69] which showed networks of parenchymal cells with biliary phenotypes, hepatocyte phenotypes, and intermediate cell populations located between these compartments with mixed phenotypes. The presence of these intermediate cells suggested a transition between the cholangiocyte-like populations and the hepatocytes. Moreover, cells of such combined phenotypes echoed phenotypes of early hepatoblasts in human fetal liver.[14] However, an unresolvable debate focused on the inability of such studies to show in which direction the transformation was going: were these cholestatic, damaged hepatocytes undergoing biliary metaplasia[70] or a progenitor cell compartment undergoing proliferation and differentiation into new hepatocytes?[64,65]

The question of directionality was convincingly settled by three dimensional analysis. First, in the same study that demonstrated the combined phenotype of early human hepatoblasts, the presence of isolated oval cells, positive for biliary-type cytokeratins, in the hepatic parenchyma were identified in normal human livers.[14] When these cells were studied in serial sections, they were found to be cross-sections of the canals of Hering.[20] The canals had traditionally been thought to extend from the terminal branches of the bile ducts within the portal tracts just to points of connection to the hepatocyte canalicular system in the limiting plate of hepatocytes, just at the margins of the portal tract stroma. This 3-dimensional analysis, however, showed that they extended beyond the limiting plate, sometimes extending as much as a third of the way into the hepatic lobule. Three dimensional examination of a case of massive hepatic necrosis secondary to acetaminophen injury then showed that the ductular reaction actually comprised a highly proliferative,[71] greatly complex, arborizing expansion of the canals of Hering, thus confirming that the canal of Hering "is comprised of, or at least harbors bipotent hepatic progenitor cells in humans."

These phenotypic studies go beyond clarification of the existence of hepatic progenitor cells, highlighting important signalling pathways in the normal development of the liver as well as in pathologic processes. For example, it has been demonstrated that Jagged1/Notch2 signalling is an important pathway for morphogenesis in the developing liver and probably is important in restitution in acute and chronic liver injury.[72] The cells expressing these factors are the same ductular reactions, which are now identified as the hepatic progenitor cell compartment.

The second type of study is exemplified by the work of Roskams et al (1998) in which serial biopsy specimens from patients with an acute onset of liver injury and regeneration demonstrated the early appearance of a ductular reaction positive for OV-6, biliary type cytokeratins (7 and 19) and Chromogranin A.[11] With time, these ductular reactions extended topographically into the hepatic lobule, and the more distal cells of the reaction took on ever more distinctive hepatocyte morphology and immunophenotypes. Thus, with a time course study in humans, the ductular reaction could be watched as it behaved like an activated stem cell compartment.

Thirdly, attempts to isolate stem/progenitor cells from human livers have been successful in isolating cells which resemble oval cells isolated from rodent models.[16,67,73] These studies have relied on HEA positive cells, which coexpressed CD34 or c-kit, echoing the bone marrow stem cell experiments which have been reported in parallel with these efforts, as well as markers such as neural cell adhesion molecule (NCAM) and bcl-2. In culture, such cells have been successfully differentiated into biliary-like populations and structures. The possibility that these cells can be manipulated into hepatocyte phenotypes and morphologies is being explored, with some encouraging preliminary data.

In humans, the canal of Hering data, as mentioned above, was one of the routes of investigation which led to demonstration that some hepatic progenitors arrived from extra-hepatic sources, certainly in part or entirely, from the bone marrow.[18-20] Confirmation of this process in humans has been made.[34] In fact, in the study in which quantification of this process was attempted,[18] in cases in which there was a marked ductular reaction in response to post-transplant strictures from biliary anastamosis or recurrence of primary disease (primary sclerosing cholangitis, hepatitis C), it was shown that up to 40% of hepatocytes and cholangiocytes were deriving from the circulation. Thus, the contribution of extra-hepatic progenitors to organ reconstitution in humans was not only nontrivial, but obviously of clinical significance.

In conclusion, the possibilities for stem cell based therapies seem limitless in the treatment of hepatic disorders. However, there is still a long road ahead and setbacks that need to be worked out. The German philosopher Nietzsche once said "Many a man fails as an original thinker simply because his memory is too good." It is this type of mentality that has kept the blinders on many very prominent scientists, and may have kept stem cell research at a slower pace. This may or may not have been a bad thing. It now appears that the blinders have been taken off, and the pace of stem cell research is now proceeding at a very rapid pace, perhaps at too rapid of a pace. Mother nature had billions of years through the process of evolution to refine these cells, which carry the awesome power to develop independently, through a very precise set of instructions enabling them to differentiate into a variety of cells types[74-77] and to a fully grown organism.[78] The stem cell dogmas of yesterday are not withstanding the research findings of today, and many researchers are redefining cell biology at its most basic level. This places the stem cell researcher at the crossroads: do we try and harness this power for the good of mankind (knowing full well the Pandora's box that comes with it) or do we awkwardly back away from it. We, as scientists, can not and must not go forging recklessly ahead without fully understanding the basic nature of these stem cells. Further research and understanding will eventually define the stem cell's usefulness in both the basic and clinical sciences.

References

1. Farber E. Similarities in the sequence of early histologic changes induced in the liver of the rat by ethionine, 2-acetylaminoflouorene, and 3'-methyl-4-dimethylaminoazbenzene. Cancer Res 1956; 16.
2. Petersen BE. Hepatic "stem" cells: Coming full circle. Blood Cells Mol Dis 2001; 27:590-600.
3. Petersen BE, Bowen WC, Patrene KD et al. Bone marrow as a potential source of hepatic oval cells. Science 1999; 284:1168-1170.
4. Farber E. Hepatocyte proliferation in stepwise development of experimental liver cell cancer. Dig Dis Sci 1991; 36:973-978.

5. Farber E, Rubin H. Cellular adaptation in the origin and development of cancer. Cancer Res 1991; 51:2751-2761.
6. Fausto N, Lemire JM, Shiojiri N. Oval cells in liver carcinogenesis; cell lineages in hepatic development and identification of stem cells in normal liver. In: Sirica AE, ed. The role of cell types in hepatocarcinogenesis. Boca Raton: CRC Press, 1992:89.
7. Grisham JW, Thorgiersson. Liver stem cells. In: Potten CS, ed. Stem Cells. San Diego: CA, Academic Press, 1997:233-282.
8. Rao MS, Reddy JK. Hepatic transdifferentiation in the pancreas. Semin Cell Biol 1995; 6:151-156.
9. Reddy JK, Rao MS, Yeldandi AV et al. Pancreatic hepatocytes. An in vivo model for cell lineage in pancreas of adult rat. Dig Dis Sci 1991; 36:502-509.
10. Roskams T, Desmet V. Ductular reaction and its diagnostic significance. Semin Diagn Pathol 1998; 15:259-269.
11. Roskams T, De Vos R, Van Eyken P et al. Hepatic OV-6 expression in human liver disease and rat experiments: Evidence for hepatic progenitor cells in man. J Hepatol 1998; 29:455-463.
12. Crosby HA, Hubscher SG, Joplin RE et al. Immunolocalization of OV-6, a putative progenitor cell marker in human fetal and diseased pediatric liver. Hepatology 1998; 28:980-985.
13. Hsia CC, Evarts RP, Nakatsukasa H et al. Occurrence of oval-type cells in hepatitis B virus-associated human hepatocarcinogenesis. Hepatology 1992; 16:1327-1333.
14. Haruna Y, Hayashi N, Yuki N et al. Serum preS1 and preS2 antigens as prognostic markers in interferon therapy for chronic hepatitis B. Scand J Gastroenterol 1992; 27:615-619.
15. Sell S. Comparison of liver progenitor cells in human atypical ductular reactions with those seen in experimental models of liver injury. Hepatology 1998; 27:317-331.
16. Baumann U, Crosby HA, Ramani P et al. Expression of the stem cell factor receptor c-kit in normal and diseased pediatric liver: Identification of a human hepatic progenitor cell? Hepatology 1999; 30:112-117.
17. Thung SN. The development of proliferating ductular structures in liver disease. An immunohistochemical study. Arch Pathol Lab Med 1990; 114:407-411.
18. Theise ND, Nimmakayalu M, Gardner R et al. Liver from bone marrow in humans. Hepatology 2000; 32:11-16.
19. Theise ND, Badve S, Saxena R et al. Derivation of hepatocytes from bone marrow cells in mice after radiation-induced myeloablation. Hepatology 2000; 31:235-240.
20. Sell S, Ilic Z. Liver development in liver stem cells. Austin: R.G. Landes Company, 1997; 30.
21. Fausto N. Liver stem cells. In: Arias IM, ed. The Liver Biology and Pathobiology. New York: Raven Press, 1994:1501.
22. Evarts RP, Nagy P, Nakatsukasa H et al. In vivo differentiation of rat liver oval cells into hepatocytes. Cancer Res 1989; 49:1541-1547.
23. Evarts RP, Nagy P, Marsden E et al. A precursor-product relationship exists between oval cells and hepatocytes in rat liver. Carcinogenesis 1987; 8:1737-1740.
24. Evarts RP, Hu Z, Fujio K et al. Activation of hepatic stem cell compartment in the rat: Role of transforming growth factor alpha, hepatocyte growth factor, and acidic fibroblast growth factor in early proliferation. Cell Growth Differ 1993; 4:555-561.
25. Petersen BE, Zajac VF, Michalopoulos GK. Hepatic oval cell activation in response to injury following chemically induced periportal or pericentral damage in rats. Hepatology 1998; 27:1030-1038.
26. Yin L, Lynch D, Ilic Z et al. Proliferation and differentiation of ductular progenitor cells and littoral cells during the regeneration of the rat liver to CCl4/2-AAF injury. Histol Histopathol 2002; 17:65-81.
27. Hixson DC, Chapman L, McBride A et al. Antigenic phenotypes common to rat oval cells, primary hepatocellular carcinomas and developing bile ducts. Carcinogenesis 1997; 18:1169-1175.
28. Park DY, Suh KS. Transforming growth factor-beta1 protein, proliferation and apoptosis of oval cells in acetylaminofluorene-induced rat liver regeneration. J Korean Med Sci 1999; 14:531-538.
29. Thorgeirsson SS, Evarts RP, Bisgaard HC et al. Hepatic stem cell compartment: Activation and lineage commitment. Proc Soc Exp Biol Med 1993; 204:253-260.
30. Yavorkovsky L, Lai E, Ilic Z et al. Participation of small intraportal stem cells in the restitutive response of the liver to periportal necrosis induced by allyl alcohol. Hepatology 1995; 21:1702-1712.
31. Alpini G, Phillips JO, LaRusso N. The biology of the biliary epithelia. In: Arias IM, ed. The Liver: Biology and Pathobiology. New York: Raven Press, 1994:623.
32. Houssaint E. Differentiation of the mouse hepatic primordium. I. An analysis of tissue interactions in hepatocyte differentiation. Cell Differ 1980; 9:269-279.
33. Petersen BE, Zajac VF, Michalopoulos GK. Bile ductular damage induced by methylene dianiline inhibits oval cell activation. Am J Pathol 1997; 151:905-909.
34. Alison MR, Poulsom R, Jeffery R et al. Hepatocytes from nonhepatic adult stem cells. Nature 2000; 406:257.
35. Lagasse E, Connors H, Al Dhalimy M et al. Purified hematopoietic stem cells can differentiate into hepatocytes in vivo. Nat Med 2000; 6:1229-1234.
36. Petersen BE, Goff JP, Greenberger JS et al. Hepatic oval cells express the hematopoietic stem cell marker Thy-1 in the rat. Hepatology 1998; 27:433-445.
37. Omori N, Omori M, Evarts RP et al. Partial cloning of rat CD34 cDNA and expression during stem cell- dependent liver regeneration in the adult rat. Hepatology 1997; 26:720-727.
38. Omori M, Omori N, Evarts RP et al. Coexpression of flt-3 ligand/flt-3 and SCF/c-kit signal transduction system in bile-duct-ligated SI and W mice. Am J Pathol 1997; 150:1179-1187.
39. Fujio K, Evarts RP, Hu Z et al. Expression of stem cell factor and its receptor, c-kit, during liver regeneration from putative stem cells in adult rat. Lab Invest 1994; 70:511-516.
40. Germain L, Noel M, Gourdeau H et al. Promotion of growth and differentiation of rat ductular oval cells in primary culture. Cancer Res 1988; 48:368-378.
41. Goff JP, Shields DS, Boggs SS et al. Effects of recombinant cytokines on colony formation by irradiated human cord blood CD34+ hematopoietic progenitor cells. Radiat Res 1997; 147:61-69.
42. Yang L, Li S, Hatch H et al. In vitro trans-differentiation of adult hepatic stem cells into pancreatic endocrine hormone-producing cells. Proc Natl Acad Sci USA 2002; in press.
43. Korbling M, Katz RL, Khanna A et al. Hepatocytes and epithelial cells of donor origin in recipients of peripheral-blood stem cells. N Engl J Med 2002; 346:738-746.
44. Hatch HM, Zheng D, Jorgensen MJ et al. SDFα1/CXCR4: A mechanism for hepatic oval cell activation and bone marrow stem cell recruitment to the liver of rats. Cloning and Stem Cells 2002; in press.
45. DuBois AM. The embryonic liver. In: Rouiller C, ed. The liver, morpholgy, biochemistry, physiology. New York: Academic Press, 1963:1.
46. Carlson BM. Digestive and respiratory systems and body cavities. In: Carlson BM, ed. Human Embryology and Developmental Biology. St. Louis: Mosby Publishing, Inc., 1999:320.
47. Watanabe Y, Aiba Y, Katsura Y. T cell progenitors in the murine fetal liver: Differences from those in the adult bone marrow. Cell Immunol 1997; 177:18-25.
48. Gallacher L, Murdoch B, Wu D et al. Identification of novel circulating human embryonic blood stem cells. Blood 2000; 96:1740-1747.
49. Filipe A, Li Q, Deveaux S et al. Regulation of embryonic/fetal globin genes by nuclear hormone receptors: A novel perspective on hemoglobin switching. EMBO J 1999; 18:687-697.
50. Marshall CJ, Kinnon C, Thrasher AJ. Polarized expression of bone morphogenetic protein-4 in the human aorta- gonad-mesonephros region. Blood 2000; 96:1591-1593.
51. Medvinsky AL, Dzierzak EA. Development of the definitive hematopoietic hierarchy in the mouse. Dev Comp Immunol 1998; 22:289-301.
52. Maekawa T, Ishii T. Chemokine/receptor dynamics in the regulation of hematopoiesis. Intern Med 2000; 39:90-100.
53. Nagasawa T, Hirota S, Tachibana K et al. Defects of B-cell lymphopoiesis and bone-marrow myelopoiesis in mice lacking the CXC chemokine PBSF/SDF1. Nature 1996; 382:635-638.

54. Kawabata K, Ujikawa M, Egawa T et al. A cell-autonomous requirement for CXCR4 in long-term lymphoid and myeloid reconstitution. Proc Natl Acad Sci USA 1999; 96:5663-5667.
55. Tachibana K, Hirota S, Iizasa H et al. The chemokine receptor CXCR4 is essential for vascularization of the gastrointestinal tract. Nature 1998; 393:591-594.
56. Zou YR, Kottmann AH, Kuroda M et al. Function of the chemokine receptor CXCR4 in haematopoiesis and in cerebellar development. Nature 1998; 393:595-599.
57. Jo DY, Rafii S, Hamada T et al. Chemotaxis of primitive hematopoietic cells in response to stromal cell- derived factor-1. J Clin Invest 2000; 105:101-111.
58. Broxmeyer HE, Kim CH. Regulation of hematopoiesis in a sea of chemokine family members with a plethora of redundant activities. Exp Hematol 1999; 27:1113-1123.
59. Youn BS, Mantel C, Broxmeyer HE. Chemokines, chemokine receptors and hematopoiesis. Immunol Rev 2000; 177:150-174.
60. Sell S. Heterogeneity and plasticity of hepatocyte lineage cells. Hepatology 2001; 33:738-750.
61. Petersen BE, Terada N. Stem cells: A journey into a new frontier. J Am Soc Nephrol 2001; 12:1773-1780.
62. Crosby HA, Strain AJ. Adult liver stem cells: Bone marrow, blood, or liver derived? Gut 2001; 48:153-154.
63. Strain AJ, Crosby HA. Hepatic stem cells. Gut 2000; 46:743-745.
64. Fiel MI, Antonio LB, Nalesnik MA et al. Characterization of ductular hepatocytes in primary liver allograft failure. Mod Pathol 1997; 10:348-353.
65. Haque S, Chandra B, Gerber MA et al. Iron overload in patients with chronic hepatitis C: A clinicopathologic study. Hum Pathol 1996; 27:1277-1281.
66. Rubin EM, Martin AA, Thung SN et al. Morphometric and immunohistochemical characterization of human liver regeneration. Am J Pathol 1995; 147:397-404.
67. Fabris L, Strazzabosco M, Crosby HA et al. Characterization and isolation of ductular cells coexpressing neural cell adhesion molecule and Bcl-2 from primary cholangiopathies and ductal plate malformations. Am J Pathol 2000; 156:1599-1612.
68. Crosby HA, Hubscher S, Fabris L et al. Immunolocalization of putative human liver progenitor cells in livers from patients with end-stage primary biliary cirrhosis and sclerosing cholangitis using the monoclonal antibody OV-6. Am J Pathol 1998; 152:771-779.
69. Gerber MA, Thung SN, Shen S et al. Phenotypic characterization of hepatic proliferation. Antigenic expression by proliferating epithelial cells in fetal liver, massive hepatic necrosis, and nodular transformation of the liver. Am J Pathol 1983; 110:70-74.
70. Van Eyken P, Sciot R, Desmet VJ. A cytokeratin immunohistochemical study of alcoholic liver disease: Evidence that hepatocytes can express 'bile duct-type' cytokeratins. Histopathology 1988; 13:605-617.
71. Koukoulis G, Rayner A, Tan KC et al. Immunolocalization of regenerating cells after submassive liver necrosis using PCNA staining. J Pathol 1992; 166:359-368.
72. Nijjar SS, Crosby HA, Wallace L et al. Notch receptor expression in adult human liver: A possible role in bile duct formation and hepatic neovascularization. Hepatology 2001; 34:1184-1192.
73. Crosby HA, Kelly DA, Strain AJ. Human hepatic stem-like cells isolated using c-kit or CD34 can differentiate into biliary epithelium. Gastroenterology 2001; 120:534-544.
74. Krause DS, Theise ND, Collector MI et al. Multi-organ, multi-lineage engraftment by a single bone marrow-derived stem cell. Cell 2001; 105:369-377.
75. Malouf NN, Coleman WB, Grisham JW et al. Adult-derived stem cells from the liver become myocytes in the heart in vivo. Am J Pathol 2001; 158:1929-1935.
76. Asahara T, Murohara T, Sullivan A et al. Isolation of putative progenitor endothelial cells for angiogenesis. Science 1997; 275:964-967.
77. Bjornson CR, Rietze RL, Reynolds BA et al. Turning brain into blood: A hematopoietic fate adopted by adult neural stem cells in vivo. Science 1999; 283:534-537.
78. Clarke DL, Johansson CB, Wilbertz J et al. Generalized potential of adult neural stem cells. Science 2000; 288:1660-1663.

CHAPTER 5

Adipose Tissue-Derived Adult Stem Cells: Potential for Cell Therapy

Laura Aust, Lyndon Cooper, Blythe Devlin, Tracey du Laney, Sandra Foster, Jeffrey M. Gimble, Farshid Guilak, Yuan Di C. Halvorsen, Kevin Hicok, Amy Kloster, Henry E. Rice, Anindita Sen, Robert W. Storms and William O. Wilkison

Introduction

The term "adult stem cell" has traditionally been used to refer to the hematopoietic progenitors isolated from bone marrow and transplanted into patients after high dose chemotherapy. Until recently, cell differentiation was conceptualized in a step-wise, unidirectional process. Unique progenitors gave rise to tissues composed of specific or "terminally differentiated" cell types (adipocytes, osteoblasts, chondrocytes, etc).

New evidence has forced a paradigm shift in the concept of an "adult stem cell".[1] First, recent evidence suggests that transplanted bone marrow-derived hematopoietic stem cells not only differentiate along the hematopoietic pathway, but are also found as mature cells in the skin, brain, muscle, intestinal epithelium and liver of recipients.[2-9] Second, it has been observed that bone marrow-derived stromal cells are multipotent.[10-11] Bone marrow stromal cells differentiate along the osteoblast, adipocyte, cardiac myocyte, neuronal, and other pathways.[10-13] Similar multipotent stem cells have been derived from other adult tissues, including the dermis,[9] skeletal muscle,[14a,b] and adipose tissue.[15-18] It is now accepted that these "adult stem cells" display a wide range of plasticity that extends across traditional embryologic dermal boundaries.[1]

Technology has encouraged individuals worldwide to adopt a more sedentary life style, and the accessibility of "fast foods" has led to an increase in their caloric intake. Consequently, an increasing percentage of the population is over-weight or obese. Adipose tissue presents an abundant, accessible, and replenishable source of adult stem cells. Subcutaneous adipose tissue can be harvested by liposuction in an outpatient setting; over 385,000 lipoplasties were performed in 2001, according to the American Society for Aesthetic Plastic Surgery.

Rheumatologic diseases are an important target for cell based therapies. In combination with biomaterials and cytokines, adult stem cells can be used to regenerate tissues such as bone and cartilage in patients with acute or chronic disorders. Bone marrow transplantation is now being explored as a potential therapeutic modality for rheumatologic diseases and adult stem cells may improve engraftment in these procedures. This review focuses on adipose tissue as a novel source of adult stem cells.

Cell Isolation and Characterization

Adipose tissue is composed of a heterogeneous cell population that includes mature adipocytes (fat cells), the un-differentiated stromal cells, vascular cells, and neuronal cells.[19] Liposuction waste material consists of small fragments (<10 mm^3) of human subcutaneous adipose tissue suspended in neutral buffered saline solution. Like other soft tissues, the adipose cells are organized within and connected by extracellular matrices. To dissociate the tissue into single cells and propagate the adult stem cells, the tissue is mixed with an equal volume of Krebs Ringers Buffer containing type I collagenase and digested at 37° C with gentle agitation for 60 to 80 minutes (outlined in Table 1). The floating mature adipocytes are separated from the pelleted stromal vascular fraction by centrifugation.[15,20] The pellet is washed and the cells allowed to adhere to a plastic culture vessel in the presence of a growth medium containing 10% fetal bovine serum. These steps remove contaminating blood cells because they do not adhere to plastic. Following a 24 to 48 hour incubation, the cells are fed with media supplemented with cytokines. The cells typically reach confluence in 4 to 6 days. The final population of adipose-derived cells can then be harvested in their undifferentiated state by trypsin digestion and either used immediately or stored by cryopreservation for later use. These stromal cells, which normally differentiate to form mature adipocytes, can be reprogrammed to differentiate into a variety of cell types.[15,16,18] Because of their multipotent phenotype, the stromal cells are referred to as Adipose Derived Adult Stem Cells or ADAS cells. A gram of subcutaneous adipose tissue yields between 20,000 and 100,000 ADAS cells; these can be further expanded in vitro.[17]

The human ADAS cells have a fibroblast like morphology and lack the intracellular oil droplets typically seen in mature adipocytes (Fig. 1). The adherent cells can then be cultured in medium that selectively promotes differentiation into mature adipocytes.

After expansion in culture, the primary isolated human ADAS cells display a distinct phenotype based on cell surface protein expression (characterized by flow cytometric analysis) and cytokine expression (characterized by polymerase chain reaction and ELISA) (Table 2). The profile is similar, but not identical, to that described for human bone marrow stromal cells.[10] For example,

Table 1. Isolation and culture of adipose-derived adult stem (ADAS) cells

Steps	Procedure
Isolation	Obtain liposuction waste material Digest with type I collagenase Centrifuge Isolate and wash stromal vascular pellet
Expansion	Plate on plastic ware in presence of DMEM/F10 medium supplemented with 10% fetal bovine serum Feed after 1-3 days with medium supplemented with TGFβ, EGF, bFGF
Harvest	At confluence, harvest cells for cryopreservation, immediate use, or differentiate along lineage specific pathway
Differentiation	Adipogenic Factors - dexamethasone, indomethacin, isobutylmethylxanthine, insulin Osteogenic Factors - dexamethasone, 1, 25 dihydroxyvitamin D_3, ascorbate 2 phosphate, β glycerophosphate Chondrogenic Factors - dexamethasone, ascorbate, TGFβ Myogenic Factors - 2-5% horse serum Neurogenic Factors - brain derived neurogenic factor, bFGF, EGF, antioxidant, valproic acid, forskolin, insulin

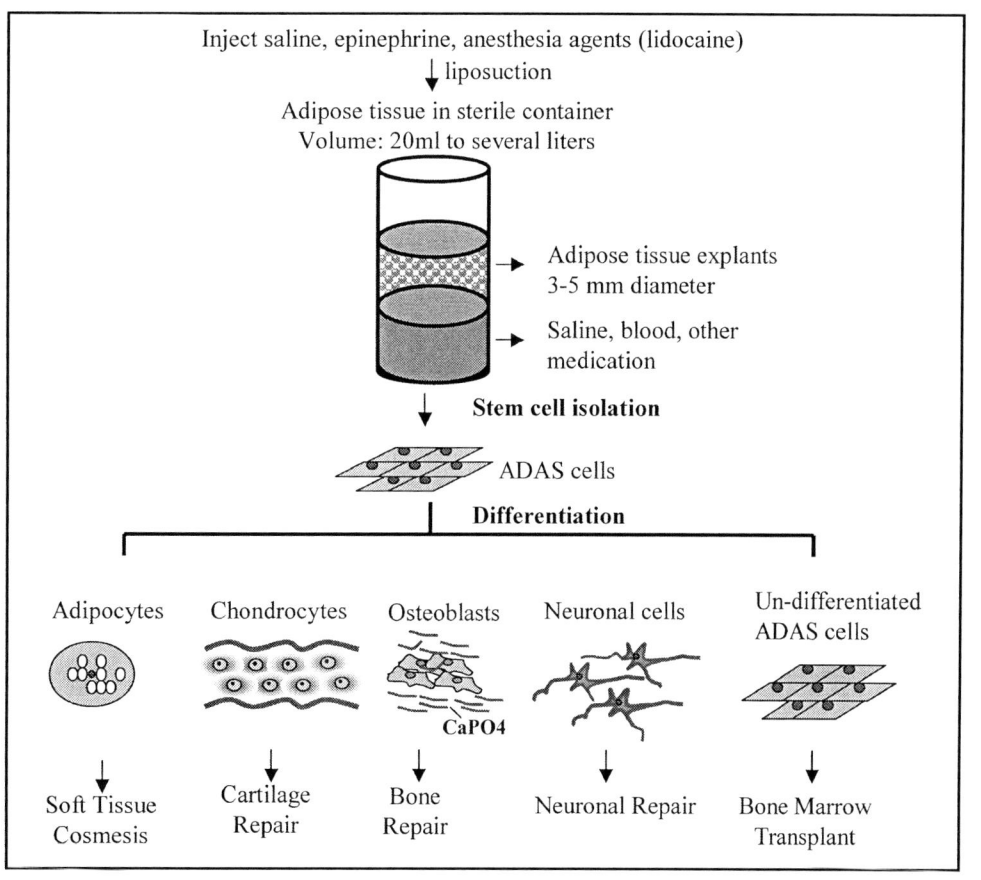

Figure 1. Adipose-derived adult stem (ADAS) cell expansion in culture.

Table 2. ADAS cell protein expression phenotype

Abundance	Surface Proteins	Cytokines
High	CD10 CALLA (common acute lymphoblastic leukemia antigen) CD13 aminopeptidase CD29 integrin b_1 CD55 Decay Accelerating Factor CD59 Complement Protectin CD90 Thy 1 CD146 Muc 18 HLA Class I	Interleukin 6,8 Macrophage CSF
Low to Moderate	CD9 tetraspan protein CD34 CD44 hyaluronate receptor $CD49_{c,d,e}$ integrins a_{3-5} CD 54 ICAM-1 CD105 endoglin CD106 VCAM-1 CD166 ALCAM	Interleukin 7,11 Granulocyte CSF Leukemia Inhibitory Factor Stem Cell Factor GM-CSF Bone Morphogenetic Proteins Flt3 Ligand
Absent	CD11 CD 14 CD45 CD56 N-CAM HLA Class II	Interleukin 1α and β

unlike human bone marrow stromal cells, ADAS cells do not express the STRO-1 (a marker of osteoprogenitor stem cells) surface antigen or the interleukin 1 cytokines.[10,21] The ADAS surface protein phenotype is also similar to that described for skeletal muscle derived adult stem cells.[14a,b] Both populations express the common acute lymphoblastic leukemia antigen (CD10), aminopeptidase (CD13), and the major histocompatibility antigen class I; however, unlike muscle derived adult stem cells, ADAS cells do not express the neural cell adhesion molecule or N-CAM (CD56).[14a,22]

Adipose Differentiation

The undifferentiated ADAS cells differentiate to adipocytes upon exposure to the adipogenic factors outlined in Table 1. Three to 4 days post induction, multiple small oil droplets appear in the perinuclear region. These oil droplets progressively increase in size as the cells continue to be cultured under adipogenic conditions. Within 10 to 14 days, the differentiated adipocytes display many of the characteristics of adipocytes in vivo. The ADAS adipocytes (1) express differentiation selective genes including aP2, PPARγ2, C/EBPα, stearoyl-CoA desaturase, and ob; (2) increase insulin stimulated glucose uptake; (3) increase the rate of lipolysis in the presence of a general beta-receptor agonist like isoproterenol; (4) secrete leptin protein.[15,17,23,24] Human ADAS cells isolated from individual donors have similar characteristics, though some heterogeneity in their responsiveness to the adipogenic induction regimen is observed.[17] The differentiation process can occur in a three dimensional lattice. Human ADAS cells continued to undergo adipocyte differentiation when cultured on the synthetic polymer, polytetrafluoroethylene.[25]

Cultured adipocytes can potentially be used in soft tissue cosmesis to reduce wrinkles and acne or surgical scars when implanted subcutaneously in small volumes. The ability of implanted cultured cells to establish adipose tissue in vivo was first demonstrated in a rodent model.[26] Green and Kehinde injected the murine preadipocyte cell line 3T3-F442A subcutaneously over the sternum of athymic (nude) mice. These cells formed fat pads, which were indistinguishable from normal adipose tissue. Mandrup et al[27] performed similar transplantation experiments using the 3T3-F442A preadipocytes labeled with β-galactosidase by retroviral transduction. Histology performed on the injection site demonstrated that the preadipocytes differentiated to mature adipocytes in vivo with typical intracellular monolocular lipid formation that occupied most of the cell volume. The subcutaneous adipose observed at the injection site stained positive for β-galactosidase, indicating that it was derived from the transplanted cells. Neovascularization was evident two weeks after implantation.

The formation of adipose tissue has been accomplished by embedding human preadipocytes on bovine collagen scaffolds and implanting the constructs subcutaneously in nude mice for up to 8 weeks.[28] Mature monolocular adipocytes were identified from the implants. The adipocytes, however, remained clustered at the surface of the scaffolds; this may be a consequence of the relatively small pore size of the scaffolds and restrictions on cell motility and migration. In similar experiments, preadipocytes isolated from rat adipose tissue were seeded onto synthetic poly (lactic-coglycolic) acid polymers and implanted subcutaneously in host animals.[29] This resulted in the formation of adipose tissue within a 2 to 5 week period.

In addition to the evidence that cultured preadipocytes can form fat pads, transplanted adipose tissue has been shown to retain its function in a murine model for the disease "lipodystrophy". Patients with the rare Berardinelli-Seip syndrome exhibit a loss of subcutaneous adipose tissue (lipodystrophy) and severe type II diabetes. A transgenic mouse was developed that expressed a dominant negative protein that heterodimerizes with and inactivates members of the C/EBP and Jun families of B-ZIP transcription factors. These animals have no adipose tissue and exhibit elevated serum glucose levels. Transplantation of normal adipose tissue to these animals resulted in partial or complete reversal of all aspects of the diabetic phenotype.[30] These studies demonstrate that transplanted fat pads continue to display the normal morphology and functions of adipose tissue in the recipient. Plastic surgeons have injected adipose explants for cosmetic surgeries (microlipoinjection).[31] However, the transplantation of intact adipose tissue faces limitations; the transplanted tissue may be damaged during harvest or injection, reducing the longevity of the cells and resulting in a fibrotic response rather than the formation of a new adipose depot.[32]

Bone Differentiation

It is well established that bone marrow derived stromal cells or "mesenchymal stem cells" (MSC) are multipotent. Bone marrow stromal cells not only differentiate to form osteoblasts both in vitro[10,33,34] and in vivo,[35-37] but are also capable of adipocyte differentiation.[11] This observation led to a corollary hypothesis that adipose derived adult stem cells would not only form fat but also differentiate into bone forming osteoblasts. This was confirmed in recent in vitro experiments.[16,18,38] The osteoblast phenotype is induced in ADAS cells when they are grown in medium that contains fetal bovine serum supplemented with calcitriol (1,25-vitamin D3), dexamethasone, beta glycerol phosphate and ascorbic acid. In the presence of osteogenic media, ADAS cells express osteoblast associated genes including bone alkaline phosphatase, type I collagen, bone morphogenetic protein-2 (BMP-2), parathyroid hormone (PTH) receptor, osteopontin, and osteocalcin within a time frame similar to that seen in bone marrow derived mesenchymal stem cells.[16] Alkaline phosphatase and osteocalcin protein levels are induced by 6 days of osteo-induction.[16,18] After 14 days of induction, ADAS cells produce mineralized matrix confirmed by von Kossa and alizarin red staining analysis. Fourier transform infrared analysis confirmed that the mineralized matrix produced by ADAS cells is hydroxy-apatite in nature.[16]

Krebsbach et al[36] found that the synthesis of bone marker proteins in vitro by bone marrow stromal-derived osteoblasts does not necessarily correlate with the ability to synthesize bone in vivo. Nevertheless, recent studies indicate that ADAS cells not only display bone markers in vitro but also synthesize osteoid in vivo[39] (unpublished observations). Human ADAS cells loaded onto a hydroxyapaptite/tricalcium phosphate biomatrix form new bone when implanted subcutaneously in immunodeficient mice. These data indicate that ADAS cells have potential clinical utility in the treatment of nonhealing fractures and other bone disorders.

Cartilage Differentiation

Articular cartilage is the thin layer of deformable, load-bearing material that lines the bony ends of all diarthrodial joints. The primary functions of cartilage are to support and distribute forces generated during joint loading and to provide lubricating surfaces to prevent wear or degradation of the joint. Cartilage is a metabolically active tissue with relatively slow state of turnover by a sparse population of specialized cells, chondrocytes. Cartilage has limited capacity for intrinsic repair, and even minor lesions or injuries may lead to progressive damage and joint degeneration. Chondral or osteochondral lesions may be a significant source of pain and loss of function and rarely heal spontaneously. The poor repair capability of cartilage is often attributed to the lack of blood supply to the affected area or due to the lack of a source of un-differentiated cells that can promote repair.[40,41] Recently, a cell based cartilage repair product became available for clinical application. The Carticel™ (Genzyme, Cambridge MA) procedure involves the isolation and amplification of autologous chondrocytes and subsequent reimplantation into the defect, which is covered by a flap of autologous periosteal tissue.[42] Other potential sources of cell therapy include chondrocytes isolated from elastic cartilage,[43] bone marrow derived mesenchymal stem cells (MSC),[44,45] and ADAS cells.[18,46,47]

The conditions that favor chondrogenesis include growth factors of the FGF and TGF family as well as physical conditions such as lowered oxygen tension and appropriate matrices that maintain cells in a rounded shape, as found in cartilage in vivo.[44,48-50] When cultured in media containing fetal bovine serum, ascorbate 2-phosphate, dexamethasone, and transforming growth factor beta 1 (TGF-β1), and embedded in alginate matrix or grown in pellet culture, ADAS cells expressed molecules characteristic of the chondrocyte phenotype. The molecules expressed by ADAS cells included cartilage matrix molecules including collagen type II, VI, and chondroitin 4-sulfate, a primary component of aggrecan.[18,38,46,47] Unlike mesenchymal stem cells, however, ADAS cells under chondrogenic conditions do not appear to express type X collagen, a marker of hypertrophic chondrocytes.[44] The cultured adipose derived chondrocytes maintained their chondrocyte phenotype after being implanted subcutaneously in nude mice for 4 to 12 weeks.[46] These findings indicated the potential of adipose tissue derived cells as a source of cells for cartilage tissue engineering.

Neuronal Differentiation

Cellular therapeutics are exciting and potentially powerful tools for the treatment of neurological disorders.[51] The generation of neuronal stem cells would revolutionize the therapy of neurologic diseases.[51] Brain derived stem cells and embryonic stem cells have been successfully isolated and differentiated in vitro and in vivo toward a range of neuronal phenotypes.[51] Brain cells are relatively inaccessible and moral and ethical considerations limit the use of embryonic stem cells for therapeutic application. Recently, bone marrow stromal derived adult stem cells were differentiated toward a neuro/glial phenotype both in vitro and in vivo.[52,53] Woodbury et al developed a differentiation cocktail containing an antioxidant (either butylated hydroxyanisole or 2-mercaptoethanol) to induce neuronal differentiation.[53] Upon exposure to the neuronal differentiation conditions in the presence of antioxidant, valproic acid, forskolin, and insulin, human ADAS cells are able to undergo rapid morphological reorganization toward a neuronal/glial phenotype.[54] Within several hours following neuronal induction, ADAS cells begin to exhibit morphologic changes consistent with neuronal differentiation. Immunohistochemical characterization of induced ADAS cells demonstrated that they express protein markers associated with a neuronal cell fate including the preneuron marker protein nestin, intermediate filament M, and the differentiated neuron nuclear protein.[54] Examination by Western blot analysis has confirmed the expression of several neuronal proteins following neuronal induction of human ADAS cells, including nestin, NeuN, and glial cell marker protein glial fibrillary acidic protein (GFAP) by 24 hours of induction. Exposure of the ADAS cells to bFGF and EGF augmented the neuronal induction of ADAS cells. Finally, when the ADAS cells are cultured on neuro-supportive substrates the cells rapidly reorganize to form neurospherelike structures. Neurosphere formation is one of the classic characteristics attributed to neuronal stem cells. The neurospherelike structures formed by ADAS cells contain nestin and NeuN positive cells (unpublished observations). Clonal analysis of cells derived from these structures will help determine ADAS cells' capacity to generate functional neuronal cells.

Skeletal Myocyte Differentiation

Studies have begun to examine the ability of ADAS cells to differentiate along the myocyte lineage.[18,55,56] When incubated in the presence of hydrocortisone and horse serum for one week, ADAS cells expressed the protein MyoD1, the transcriptional "master regulatory gene" for myogenesis.[55,56] Over the next five weeks, the number of cells expressing MyoD1 increased by 2.4-fold, accounting for 15% of the total cell population.[56] The expression of MyoD1 was accompanied by the appearance of myosin heavy chain protein and the morphologic appearance of myotubes.[56] These findings suggest that ADAS cell differentiation along the myocyte lineage is inducible and that the cells may have potential application for the repair and regeneration of damaged skeletal muscle.

Hematopoietic Support

Hematopoietic stem cells (HSC) are maintained in the bone marrow endosteum and are surrounded by stromal cells that provide a nurturing microenvironment by providing distinct adhesion molecules and secreted cytokines. The ADAS cells express a variety of cell surface antigens that suggest that they may also provide this function. These include antigens such as the hyaluronate receptor (CD44) as well as the heterodimeric integrin α/β collagen (VLA2, VLA5) and fibronectin (VLA4) receptors (Table 2). In addition to cell surface antigens that potentially mediate cell-to-cell communication, the ADAS cells also express at least two proteases (CD10 and CD13) on their cell surface.[22] These extracellular proteases may provide a capacity to modify or reconstruct the cellular microenvironment, but may also down-regulate the transfer of signals to other cells in intimate contact. The ADAS cells express hematopoietic-specific cytokines as well (Table 2). When confluent, quiescent ADAS cell cultures were induced with lipopolysaccharide (LPS, 100 ng/ml), the level of mRNAs increased for the interleukins 6, 7, 8, and 11 (IL-6, -7, -8, -11) as well as for the cytokines leukemia inhibitory factor (LIF), macrophage-colony stimulating factor (M-CSF), granulocyte-macrophage-colony stimulating factor (GM-CSF), granulocyte-colony stimulating factor (G-CSF), flt-3 ligand (Flt3L), kit ligand (KL, also known as stem cell factor), tumor necrosis factor α (TNFα) and the bone morphogenetic proteins 2 and 4 (BMP-2, -4) (unpublished observations). Both murine and human bone marrow-derived stromal cells also express many of these same cytokines.[21,57] Thus, the ADAS cells may fulfill a fundamentally important niche for the maintenance of both myeloid and lymphoid progenitors.

To directly test whether ADAS cells can support hematopoietic cells, CD34+ hematopoietic stem cells isolated from umbilical cord blood (UCB) were placed in coculture on irradiated confluent monolayers of ADAS cells under conditions that support myeloid progenitors. In these assays, the cocultures were established using a commercially available medium for the long-term culture of myeloid cells. After 3 weeks in coculture, the entire culture was harvested and the cells were used to establish secondary cultures in a methylcellulose-based medium that contains cytokines that support myeloid colony formation. The cocultures contained myeloid progenitors for up to 6 weeks (unpublished observations). In similar experiments, ADAS cells have supported lymphoid and myeloid progenitors in 12 day cocultures (unpublished observations). This work documents the ability of human adipose-derived stromal cells to support the in vitro proliferation and expansion of committed and primitive hematopoietic progenitors. The ADAS cells have potential clinical applications for patients receiving hematopoietic stem cell transplants. ADAS cells may be coinfused at the time of hematopoietic stem cell transplantation to accelerate and enhance engraftment. Alternatively, ADAS cells may be used to expand a population of short term progenitors ex vivo, prior to transplantation.

Allogeneic Transplantation

The ability to transplant ADAS cells across major histocompatibility barriers would expand the opportunities for their use in a clinical setting. As summarized in Table 2, ADAS cells do not display major histocompatibility (MHC) Class II proteins on their surface. Recent studies using primate bone marrow stromal cells in mixed lymphocyte reactions (MLR) suggest that bone marrow stromal cells from an unrelated individual suppress lymphocyte proliferation in vitro.[58] Although it remains to be determined if transplantation of HLA-mismatched ADAS cells will be tolerated in vivo, these findings in the literature indicate that further experiments are warranted to evaluate the feasibility of this approach.

Gene Delivery

Gene therapy is a promising approach to introduce DNA-encoding therapeutic products to correct or modulate inborn genetic defects.[59] Technical hurdles and safety concerns have interfered with or delayed clinical trials, including the risks presented by direct exposure of the patient to high doses of adenovirus and other vectors.[60,61] Using ADAS cells as a gene delivery vehicle could address this and other issues for the following reasons: First, the nonimmortalized ADAS cells can be genetically manipulated ex vivo by viral infection (unpublished data) or by gene gun-mediated introduction of genes.[62] These ex vivo procedures allow selection of gene product-producing cells as well as removal of excess vectors or viral particles that could result in potentially fatal inflammatory responses.[63] Second, ADAS cells can be differentiated to several cell lineages. Genetically manipulated cells with specific phenotypes can be implanted at the most favorable microenvironment for maximal effects. Finally, adipose tissue is present in most parts of the body and in proximity to most vital organs. It is likely that the genetically manipulated ADAS cells can be implanted at the most relevant location in their native state for maximal effects. Clearly, more work is required in this field before it is of clinical utility, but it deserves continued attention in the future.

Conclusions and Future Directions

In summary, human adipose tissue is a reservoir for adult stem cells that can be easily harvested without causing significant morbidity to the patient. The ADAS cells can be induced to follow multiple differentiation pathways. These include mesodermally derived cells, such as adipocytes, chondrocytes, osteoblasts, hematopoietic support cells, and other epidermal lineages (neuronal). Further research and induction methods may demonstrate that additional lineages, such as pancreatic islet cells, cardiomyocytes, hepatocytes, and smooth muscle, are also possible. Current approaches focus on the use of ADAS cells for autologous transplantation, however allogeneic transplantation may also be possible and experiments are underway. Human ADAS cells can be easily obtained in large numbers that could be developed for a variety of clinical applications. It is likely that this cell therapy will have direct application to rheumatologic conditions and other autoimmune disorders, especially in the areas of cartilage and bone repair.

References

1. Anderson DJ, Gage FH, Weissman IL. Can stem cells cross lineage boundaries? Nat Med 2001; 7:393-395.
2. Ferrari G, Cusella-De Angelis G, Coletta M et al. Muscle regeneration by bone marrow-derived myogenic progenitors. Science 1998; 279:1528-1530.
3. Gussoni E, Soneoka Y, Strickland CD et al. Dystrophin expression in the mdx mouse restored by stem cell transplantation. Nature 1999; 401:390-4.
4. Petersen BE, Bowen WC, Patrene KD et al. Bone marrow as a potential source of hepatic oval cells. Science 1999; 284:1168-1170.
5. Alison MR, Poulsom R, Jeffrey R et al. Hepatocytes from nonhepatic adult stem cells. Nature 2000; 406:257-267.
6. Theise ND, Badve S, Saxena R et al. Derivation of hepatocytes from bone marrow cells in mice after radiation-induced myeloablation. Hepatology 2000; 31:235-240.
7. Krause DS, Theise ND, Collector MI et al. Multi-organ, multi-lineage engraftment by a single bone marrow-derived stem cell. Cell 2001; 105:369-77.
8. Körbling M, Katz RL, Khanna A et al. Hepatocytes and epithelial cells of donor origin in recipients of peripheral-blood stem cells. N Engl J Medicine 2002; 346:738-746.
9. Toma JG, Akhavan M, Fernandes KJ et al. Isolation of multipotent adult stem cells from the dermis of mammalian skin. Nat Cell Biol 2001; 3:778-84.
10. Pittenger MF, Mackay AM, Beck SC et al. Multilineage potential of adult human mesenchymal stem cells. Science 1999; 284:143-147.
11. Nuttall ME, Gimble JM. Is there a therapeutic opportunity to either prevent or treat osteopenic disorders by inhibiting marrow adipogenesis? Bone 2000; 27:177-84.
12. Li Y, Chen J, Wang L et al. Treatment of stroke in rat with intracarotid administration of marrow stromal cells. Neurology 2001; 56:1666-1672.
13. Toma C, Pittenger MF, Cahill KS et al. Human mesenchymal stem cells differentiate to a cardiomyocyte phenotype in the adult murine heart. Circulation 2002; 105:93-8.
14a. Young HE, Steele TA, Bray RA et al. Human pluripotent and progenitor cells display cell surface cluster differentiation markers CD10, CD13, CD56 and MHC Class I. Proc Soc Exp Biol Med 1999; 221:63-71.
14b. Young HE, Steele TA, Bray RA et al. Human reserve pluripotent mesenchymal stem cells are present in the connective tissues of skeletal muscle and dermis derived from fetal, adult, and geriatric donors. Anat Rec 2001; 264:51-62.
15. Halvorsen YDC, Bond A, Sen A et al. Thiazolidinediones and glucocorticoids synergistically induce differentiation of human adipose tissue stromal cells: Biochemical, cellular and molecular analysis. Metabolism 2001; 50:407-413.
16. Halvorsen YDC, Franklin D, Bond AL et al. Extracellular matrix mineralization and osteoblast gene expression by human adipose tissue-derived stromal cells. Tissue Eng 2001; 6:729-741.
17. Sen A, Lea-Currie YR, Sujkowska D et al. Adipogenic potential of human adipose derived stromal cells from multiple donors is heterogeneous. J Cellular Biochem 2001; 81:312-319.
18. Zuk PA, Zhu M, Mizuno H et al. Multilineage cells from human adipose tissue: Implications for cell-based therapies. Tissue Eng 2001; 7:211-28.
19. Ailhaud G, Grimaldi P, Negrel R. Cellular and molecular aspects of adipose tissue development. Annu Rev Nutr 1992; 12:207-33.
20. Hauner H, Entenmann G, Wabitsch M et al. Promoting effects of glucocorticoids on the differentiation of human adipocyte precursor cells cultured in a chemically defined medium. J Clin Invest 1989; 84:1663-1670.
21. Majumdar MK, Thiede MA, Mosca JD et al. Phenotypic and functional comparison of cultures of marrow-derived messenchymal stem cells (MSCs) and stromal cells. J Cell Physiol 1998; 176:57-66.
22. Gronthos S, Franklin DM, Leddy HA et al. Surface protein characterization of human adipose tissue-derived stromal cells. J Cell Physiol 2001; 189:54-63.
23. Zhou L, Halvorsen Y-D, Cryan EV et al. Analysis of the pattern of gene expression during human adipogenesis by DNA microarray. Biotechnol Techniques 1999; 13:513-517.
24. Pederson T, Rondinone CM. Regulation of proteins involved in insulin signaling pathways in differentiating human adipocytes. Biochem Biophys Res Commun 2000; 276:162-8.
25. Kral JG, Crandall DL. Development of a human adipocyte synthetic polymer scaffold. Plast Reconstr Surg 1999; 104:1732-1728.
26. Green H, Kehinde O. Formation of normally differentiated subcutaneous fat pads by an established preadipose cell line. J Cell Physiol 1979; 101:169-171.
27. Mandrup S, Loftus TM, MacDougald OA et al. Obese gene expression at in vivo levels by fat pads derived from s.c. implanted 3T3-F442A preadipocytes. Proc Nat Acad Sci USA 1997; 94:4300-4305.
28. Von Heimburg D, Zachariah S, Heschel I et al. Human preadipocytes seeded on freeze-dried collagen scaffolds investigated in vitro and in vivo. Biomaterials 2001; 22:429-438.
29. Patrick Jr CW, Chauvin PB, Hobley J et al. Preadipocyte seeded PLGA scaffolds for adipose tissue engineering. Tissue Eng 1999; 5:139-151.
30. Gavrilova O, Marcus-Samuels B, Graham D et al. Surgical implantation of adipose tissue reverses diabetes in lipoatrophic mice. J Clin Invest 2000; 105:271-278.
31. Fournier PF. Facial recontouring with fat grafting. Dermatol Clin 1990; 8:523-537.
32. Sommer B, Sattler G. Current concepts of fat graft survival: Histology of aspirated adipose tissue and review of the literature. Dermatol Surg 2000; 26:1159-1166.
33. Owen M. Marrow stromal stem cells. J Cell Sci Suppl 1988; 10:63-76.
34. Nuttall ME, Patton AJ, Olivera DL et al. Human trabecular bone cells are able to express both osteoblastic and adipocytic phenotype: Implications for osteopenic disorders. J Bone Miner Res 1998; 13:371-382.
35. Kuznetsov S, Krebsbach P, Satomura K et al. Single-colony derived strains of human marrow stromal fibroblasts form bone after transplantation in vivo. J Bone Miner Res 1997; 12:1335-1347.
36. Krebsbach P, Kuznetsov S, Satomura K et al. Bone formation in vivo: Comparison of osteogenesis by transplanted mouse and human marrow stromal fibroblasts. Transplantation 1997; 63:1059-1069.
37. Horwitz EM, Prockop DJ, Fitzpatrick LA et al. Transplantability and therapeutic effects of bone marrow-derived mesenchymal cells in children with Osteogenesis imperfecta. Nat Med 1999; 5:309-313.
38. Huang JI, Beanes SR, Zhu M et al. Rat extramedullary adipose tissue as a source of osteochondrogenic progenitor cells. Plast Reconstr Surg 2002; 109:1033-1041.
39. Justesen J, Stenderup K, Pedersen SB et al. Evidence for differentiation ability of human preadipocytes into osteoblasts. J Bone Miner Res 2001;16 Suppl 1:Abstract SU273.

40. Guilak F, Butler D, Goldstein S. Functional tissue engineering - The role of biomechanics in articular cartilage repair. Clinical Orthop 2001; 391S:S295-S305.
41. Minas T, Nehrer S. Current concepts in the treatment of articular cartilage defects. Orthopedics 1997; 20: 525-538.
42. Gillogly SD, Voight M, Blackburn T. Treatment of articular cartilage defects of the knee with autologous chondrocyte implantation. J Orthop Sports Phys Ther 1998; 28:241-251.
43. Arevalo-Silva CA, Cao Y, Vacanti M et al. Influence of growth factors on tissue-engineered pediatric elastic cartilage. Archives of Otolaryngol Head Neck Surg 2000; 126:1234-8.
44. Johnstone B, Hering TM, Caplan AI et al. In vitro chondrogenesis of bone marrow-derived mesenchymal progenitor cells. Exper Cell Res 1998; 238:265-272.
45. Gao J, Dennis JE, Solchaga LA et al. Tissue-engineered fabrication of an osteochondral composite graft using rat bone marrow-derived mesenchymal stem cells. Tissue Engineering 2001; 7:363-371.
46. Erickson GR, Franklin D, Gimble JM et al. Chondrogenic potential of adipose tissue-derived stromal cells in vitro and in vivo. Biochem Biophys Res Commun 2002; 290:763-9.
47. Wickham MQ, Erickson GR, Gimble JM et al. Multipotent human adult stem cells derived from infrapatellar fat pad of the knee. Clin Orthop. In press.
48. Iwasaki M, Nakata K, Nakahara H. Transforming growth factor-beta 1 stimulates chondrogenesis and inhibits osteogenesis in high density culture of periosteum-derived cells. Endocrinology 1993; 132:1603-8.
49. Hickok NJ, Haas AR, Tuan RS. Regulation of chondrocyte differentiation and maturation. Microsc Res Tech 1998; 43:174-190.
50. Murphy CL, Sambanis A. Effect of oxygen tension and alginate encapsulation on restoration of the differentiated phenotype of passaged chondrocytes. Tissue Engineering 2001; 7:791-803.
51. Gage FH. Mammalian neural stem cells. Science 2000; 287:1433-8.
52. Sanchez-Ramos J, Song S, Cardozo-Pelaez F et al. Adult bone marrow stromal cells differentiate into neural cells in vitro. Experimental Neurology 2000; 164:247-256.
53. Woodbury D, Schwarz EJ, Prockop DJ et al. Adult rat and human bone marrow stromal cells differentiate into neurons. J Neurosci Res 2000; 61:364-370.
54. Safford KM, Hicok KC, Safford SD et al. Neurogenic differentiation of murine and human adipose-derived stromal cells. Biochem Biophys Res Commun 2002; 294:371-379.
55. Mizuno H. The myogenic potential of human processed lipoaspirates – Part I: Morphological, Immunohistochemical analysis and gene expression. J Jpn Soc Plast Recontr Surg 2001; 21:427-436.
56. Mizuno H, Zuk PA, Zhu M et al. Myogenic differentiation by human processed lipoaspirate cells. Plast Reconstr Surg 2002; 109:199-209.
57. Gimble JM, Pietrangeli C, Henley A et al. Characterization of murine bone marrow and spleen derived stromal cells: Analysis of leukocyte marker and growth factor mRNA transcript levels. Blood 1989; 74:303-311.
58. Bartholomew A, Sturgeon C, Siatskas M et al. Mesenchymal stem cells suppress lymphocyte proliferation in vitro and prolong skin graft survival in vivo. Exp Hematol 2002; 30: 42-48.
59. Rice MC, Czymmek K, Kmiec EB. The potential of nucleic acid repair in functional genomics. Nature Biotechnol 2001; 19:321-326.
60. Kay MA, Glorioso JC, Naldini L. Viral vectors for gene therapy: The art of turning infectious agents into vehicles of therapeutics. Nat Med 2001; 7:33-40.
61. Nishikawa M, Huang L. Nonviral vectors in the new millennium: Delivery barriers in gene transfer. Hum Gene Ther 2001; 12:861-870.
62. Becker DJ, MacDougald OA. Transfection of adipocytes by gene gun-mediated transfer. Biotechniques 1999; 26(2):660-668.
63. Marshall E. Gene Therapy death prompts review of adenovirus vector. Science 1999; 286:2244-2245.

CHAPTER 6

Hematopoietic Stem Cell Biology: Relevance to Autoimmunity

Richard J. Jones

Introduction

Animal models have been helpful in better understanding the biology of both hematopoietic stem cells (HSC) and autoimmunity. Autoimmunity is caused by a complex interplay of genetic and environmental factors. Spontaneous animal models of autoimmunity result from germ-line mutations; these include the lpr/lpr mouse as a model for arthritis and vasculitis,[1] the NOD (nonobese diabetic) mouse for diabetes, and the NZB/W mouse for systemic lupus erythematosus.[2] Although studies of murine susceptibility genes are likely to provide insights into the genetic mechanisms involved in the human predisposition to autoimmunity, these models differ significantly from most cases of human autoimmune disease. Spontaneous animal models with germ-line mutations virtually always develop autoimmunity, while only a fraction of individuals who harbor susceptibility genes develop autoimmunity; there are rare exceptions to this, such as patients with the autoimmune lymphoproliferative syndrome (ALPS), who, like lpr/lpr mice, inherit loss-of-function mutations of the Fas gene (refer to Chapter 17).[3,4] Induced animal models of autoimmunity focus on the role of environmental factors in the development of autoimmunity.[2] These animal models more closely approximate most cases of human autoimmunity, and have the advantage over spontaneous models in that the onset and progression of the disease can be controlled (refer to Chapters 29 and 30).

Immunoablation, followed by autologous or syngeneic blood or marrow transplantation (BMT), can eliminate autoimmunity in induced animal models. Conversely, syngeneic or autologous BMT cannot cure spontaneous (i.e., genetic) animal models of autoimmunity. Allogeneic BMT, however, can cure these animals by replacing the host HSC, that harbor the genetic abnormality, with normal HSC.

Hematopoietic Stem Cells: Background

Many groups[5-7] have now been able to isolate rare populations of blood cells, small numbers of which can generate all lympho-hematopoietic lineages for the lifetime of a mouse. Recently, multiple groups have also suggested that HSC may be able to produce non-hematologic tissues.[8] HSC are small-sized cells that have lymphoid morphology,[5,6,9] do not express antigens specific for differentiated blood lineages (i.e., are lin^{neg}), are mostly in the G_0 phase of the cell cycle,[6,7,9] and express high levels of aldehyde dehydrogenase (ALDH).[5,10,11] However, other characteristics that define HSC have been a matter of debate. For example, it now appears that there are CD34^{neg}, as well as CD34$^+$, HSC (Fig. 1).[5,12-15]

One explanation for the discrepancies regarding HSC characteristics may be that there are several classes of HSC, which represent different stages of development. These classes or stages of HSC have varying capacity for long-term production of the different hematologic lineages; primitive or "high quality"[16,17] HSC are capable of life-long production of all hematologic lineages while "low quality"[16,17] HSC generate only limited hematopoiesis. However, different classes or stages of HSC are no doubt a continuum, and thus specific definitions for distinct HSC classes should be viewed as somewhat subjective. Nevertheless, our data,[5,9,18] now confirmed by many groups,[16,19] demonstrate that certain characteristics can be used to separate different classes of HSC (Fig. 1). I will (maybe arbitrarily) define high quality HSC as small CD34^{low}lin^{neg} ALDH$^+$ HLA-DR^{low} cells.[5] In contrast, low quality HSC are operationally defined as small CD34$^+$ lin^{neg} ALDH$^+$ DR$^+$ cells. Large CD34$^+$ lin$^+$ ALDH^{low} DR$^+$ cells are generally considered the phenotype of committed progenitors, such as granulocyte-macrophage colony-forming units (CFU-GM). We found that high quality HSC do not generate assayable progenitors, including day 12 spleen colony-forming units (CFU-S), upon direct culture and are unable to produce rapid engraftment after transplantation.[5,9] In addition to being responsible for long-term engraftment after BMT,[5,9] high quality HSC are capable of marked cellular expansion in vitro, producing large numbers of assayable progenitors after liquid culture.[20] Conversely, low quality HSC produce early, but not long-term, engraftment after BMT.[5] In mice, low quality HSC appear to be responsible for day 12 CFU-S,[5] and in humans they may be the cells responsible for engraftment of immunodeficient mice (SCID-repopulating cell or SRC).[21]

High Versus Low Quality HSC: Clinical Relevance

High quality HSC appear to be relatively resistant to disease, with low quality HSC probably being the target of most acquired stem cell disorders. Although bone marrow failure in most cases

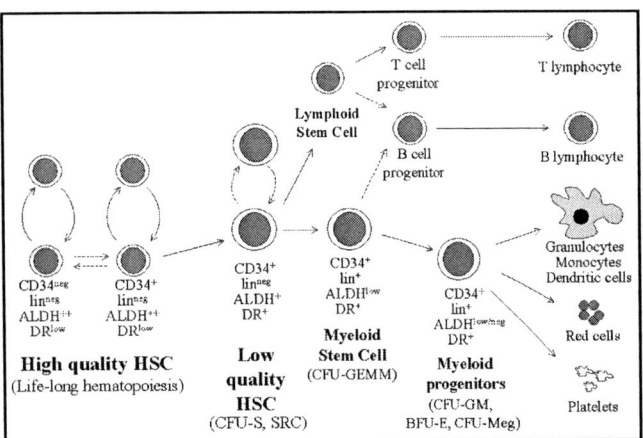

Figure 1. Hematopoietic stem cells: high versus low quality. High quality stem cells are able to self-renew and generate life-long hematopoiesis, while low quality HSC produce only limited hematopoiesis but are probably the cell origin of most leukemias. Dotted lines represent areas that have not been definitively established. HSC= hematopoietic stem cell; neg= negative; lin= lineage-specific cell surface antigens; ALDH= aldehyde dehydrogenase; DR= HLA-DR; CFU-S= spleen colony-forming unit; SRC= NOD/SCID mouse repopulating cell; CFU-GEMM= granulocyte, erythrocyte, monocyte, megakaryocyte colony-forming unit; CFU-GM= granulocyte-macrophage colony-forming unit.

of severe aplastic anemia (SAA) has generally been regarded to be the result of an autoimmune attack (regardless of the inciting event) against high quality HSC,[22,23] immunologic deficiencies are uncommon. Lymphocyte counts are usually normal in SAA, as is B and T cell function.[22,24] In addition, complete recovery of normal host hematopoiesis is a well-recognized outcome of allogeneic BMT for SAA.[25-28] In fact, recovery of normal host hematopoiesis appears to be a fairly common event in SAA, reported in up to 20% of matched sibling transplants[27] and in the majority of successful matched unrelated transplants.[28] We also found that durable, long-term complete remissions can be induced with high (transplant) dose cyclophosphamide (without allogeneic BMT) in most patients with SAA.[29,30]

Thus, clinical data suggest that high quality HSC persist in many patients with SAA. In fact, we have been able to isolate, from most patients with SAA, relatively normal numbers of cells that phenotypically and functionally resemble high quality HSC (unpublished data). Based on these data, we hypothesize that the immune attack in SAA is against low quality HSC, while high quality HSC elude, or are resistant to, immune attack. There are a number of potential mechanisms by which high quality HSC could evade immune attack. Low/absent HLA-DR expression by high quality HSC,[16,31] may account for their relative invisibility to autoreactive effectors. In addition, Fas-Fas ligand may be involved in the pathophysiology of SAA,[32] and Fas has been reported to be absent on high quality HSC, with its expression increasing as HSC differentiate.[33,34]

We hypothesize that the success of high (transplant) dose cyclophosphamide without bone marrow transplantation (BMT) in SAA[29,30] is likely the result of the persistence of high quality HSC, which are resistant to cyclophosphamide. HSC express high levels of ALDH,[5,10,11] an enzyme that inactivates cyclophosphamide,[35] while lymphocytes express little or no ALDH.[10] Although high-dose cyclophosphamide appears to cure the majority of newly-diagnosed SAA patients, it is associated with a long duration of aplasia; this may be at least partly because the high-quality HSC are quiescent and full differentiation into mature blood cells is a slow process.

Chronic myeloid leukemia (CML) and myelodysplastic syndromes (MDS) have long been considered stem cell disorders. The leukemic stem cells in these diseases appear to be $CD34^+$ and $lin^{neg}/CD38^{neg}$.[36-38] However, although B cells are sometimes derived from the leukemic clone in CML[39,40] and MDS,[41-43] T cells are virtually never involved. Moreover, normal HSC can be separated from leukemic stem cells in these diseases using characteristics (such as small size[36] and low HLA-DR expression[37,38]) of high-quality HSC. Thus, CML and MDS stem cells appear to have characteristics of low-quality, rather than high-quality, stem cells. Many groups have now shown that most cases of AML also arise from $CD34^+$ $lin^{neg}/CD38^{neg}$ stem cells,[44,45] that have the characteristics of low quality HSC.[46] Recent data suggest that B-lineage acute lymphocytic leukemia (ALL) stem cells also often exhibit a similar phenotype.[47] Since the initial neoplastic event in most leukemias (CML, MDS, AML, and even ALL) appears to originate in low quality HSC, it should not be surprising that some acute leukemias share bi-lineage potential with CML and MDS. The phenotype of a particular leukemia is probably a function of the degree of differentiation allowed by the disease's leukemogenic events.

Thus, high-quality HSC generally appear to be relatively resistant to leukemogenic changes.[17] Low-quality HSC being the cell of origin for leukemia may explain why the risk for therapy-related leukemia lasts only 10 years, while the risk for therapy-related solid tumors can be life-long. The duration of risk for therapy-related leukemia (10 years) may represent the life-span of low-quality HSC.

Rapid engraftment that will prevent the acute complications of marrow aplasia induced by the high dose conditioning regimen is the primary objective of the graft in BMT for most diseases. A graft that contains high numbers of low quality HSC, such as mobilized peripheral blood grafts,[48] will generate early, rapid engraftment. Moreover, as treatment for malignancy, these grafts need not even contain high quality HSC. No BMT conditioning regimen is truly myeloablative as demonstrated by the frequent occurrence of mixed chimerism after T cell-depleted allogeneic BMT. Thus, absence of high quality HSC in the graft should not lead to late graft failure after BMT, if residual normal host hematopoiesis reappears. BMT for SAA may also not require high quality HSC, since high quality HSC persist in many SAA patients and complete recovery of normal host hematopoiesis is a well-recognized outcome of allogeneic BMT. However, high quality HSC with "unlimited" proliferative potential are required for treating genetic diseases with either allogeneic, or genetically-modified autologous BMT. In these diseases, eventual loss of donor engraftment, as may occur with the transplantation of low quality HSC (even if it occurred late after BMT), would be associated with resurgence of abnormal host hematopoiesis and the genetic disease.

Stem Cell Therapy in Autoimmune Diseases

High-dose immunoablative therapy followed by autologous BMT is being used with increased frequency for severe autoimmune diseases.[49,50] The stimulus to explore this approach emanates from autoimmune animal models demonstrating marked improvement or complete eradication of autoimmune disease following syngeneic BMT.[51,52] There also have been case reports of allogeneic BMT (performed chiefly for a malignancy) in which a concurrent autoimmune disease was eradicated.[53] Allogeneic BMT is not currently utilized for the routine treatment of autoimmune disorders because of the significant morbidity and mortality currently associated with the procedure.

Relapses have been common in most reported series of autologous BMT for autoimmune diseases.[49,54] However, it is not clear whether the reappearance of disease is attributable to:

(1) failure to eradicate autoaggressive lymphocyte clones by the high-dose BMT conditioning regimen,

(2) relapse of autoimmune disease through auto-antigen re-challenge, or

(3) reinfusion of autoaggressive lymphocyte clones with the autologous graft.

The success of syngeneic BMT in animal models[51,52] and allogeneic BMT in patients[53] against autoimmune diseases suggests that the high-dose immunoablative regimen is frequently able to eradicate autoreactive lymphocytes. If genetic predisposition were the major determinant of autoimmunity, allogeneic BMT would be required to replace the host HSC that harbor the genetic abnormality. However, only a fraction of individuals who harbor susceptibility genes develop autoimmunity, suggesting that environmental factors are the predominant determinants for the development of autoimmunity.

Thus, the major cause of relapse after autologous BMT for autoimmune diseases may be the re-infusion of autoreactive effector cells with the autologous graft. Autoreactive lymphocytes in the autologous graft could be removed or "purged" prior to reinfusion of the graft; however, at present, it may not be possible to deplete all autoaggressive lymphocytes from an autologous graft. Mobilized peripheral blood grafts usually contain about 10^9 T cells, and CD34 isolation, for example, will remove only about 3-4 logs of T cells. The backbone of the immunoablative regimen used in most autologous BMT trials for autoimmune disease is high (transplant) dose cyclophosphamide. Since HSC are resistant to cyclophosphamide through their high expression of ALDH, transplant doses of cyclophosphamide do not require BMT (as there is endogenous stem cell recovery), thus eliminating the potential for re-infusion of autoreactive effector cells.

Based on the success of transplant dose cyclophosphamide (50 mg/kg/day x 4 days) without BMT in one autoimmune disease, SAA, we have applied this approach to patients with a variety of severe, refractory autoimmune disorders,[55] including systemic lupus erythematosus,[56] autoimmune hematologic disorders,[57] pemphigus,[58] chronic inflammatory demyelinating polyneuropathy, myasthenia gravis, and rheumatoid arthritis with Felty's syndrome. The majority of the patients who have received transplant dose cyclophosphamide with "endogenous stem cell recovery" have achieved a durable complete remission, despite having severe, life-threatening disease that had failed a median of 3 treatments (including lower-dose cyclophosphamide in many patients). In contrast to SAA where the marrow reserve is limited, hematologic recovery after transplant dose cyclophosphamide in other autoimmune diseases is rapid with a median time of 15 (range 11-22) days to an absolute neutrophil count of 500/μL and the median number of platelet and red cell transfusions being 2 (range 0-10) and 4 (range 0-8), respectively.

Conclusions

HSC, the progenitors of all blood cells, represent a continuum of cells at different stages of development. High quality[16,17] HSC are capable of life-long production of all hematologic lineages, while low quality[16,17] HSC generate rapid, but limited, hematopoiesis. Low quality HSC appear to be the target of the immune attack in SAA, as well as the cells of origin of most leukemias, including B-lineage ALL. Low quality HSC also appear to be the most important component of most BMT grafts, as they produce rapid engraftment and prevent the acute complications of marrow aplasia induced by the high dose BMT conditioning regimen. High quality HSC may only be required when using BMT to treat genetic diseases.

References

1. Suda T, Nagata S. Why do defects in the Fas-Fas ligand system cause autoimmunity? J Allergy Clin Immunol 1997; 100(6 Pt 2):S97-101.
2. Taneja V, David CS. Lessons from animal models for human autoimmune diseases. Nat Immunol 2001; 2(9):781-784.
3. Drappa J, Vaishnaw AK, Sullivan KE et al. Fas gene mutations in the Canale-Smith syndrome, an inherited lymphoproliferative disorder associated with autoimmunity. N Engl J Med 1996; 335(22):1643-1649.
4. Ramenghi U, Bonissoni S, Migliaretti G et al. Deficiency of the Fas apoptosis pathway without Fas gene mutations is a familial trait predisposing to development of autoimmune diseases and cancer. Blood 2000; 95(10):3176-3182.
5. Jones RJ, Collector MI, Barber JP et al. Characterization of mouse lymphohematopoietic stem cells lacking colony-forming activity. Blood 1996; 88(2):487-491.
6. Spangrude GJ, Heimfeld S, Weissman IL. Purification and characterization of mouse hematopoietic stem cells. Science 1988; 241:58-62.
7. Li CL, Johnson GR. Long-term hemopoietic repopulation by Thy-1lo, Lin-, Ly6A/E+ cells. Exp Hematol 1992; 20:1309-1315.
8. Krause DS, Theise ND, Collector MI et al. Multi-organ, multi-lineage engraftment by a single bone marrow-derived stem cell. Cell 2001; 105(3):369-377.
9. Jones RJ, Wagner JE, Celano P et al. Separation of pluripotent hematopoietic stem cells from multipotent progenitors (CFU-S). Nature 1990; 347:188-189.
10. Jones RJ, Barber JP, Vala MS et al. Assessment of aldehyde dehydrogenase in viable cells. Blood 1995; 85:2742-2746.
11. Storms RW, Trujillo AP, Springer JB et al. Isolation of primitive human hematopoietic progenitors on the basis of aldehyde dehydrogenase activity. Proc Natl Acad Sci USA 1999; 96(16):9118-9123.
12. Osawa M, Hanada K, Hamada H et al. Long-term lymphohematopoietic reconstitution by a single CD34-low/negative hematopoietic stem cell. Science 1996; 273(5272):242-245.
13. Goodell MA, Rosenzweig M, Kim H et al. Dye efflux studies suggest that hematopoietic stem cells expressing low or undetectable levels of CD34 antigen exist in multiple species. Nat Med 1997; 3(12):1337-1345.
14. Sato T, Laver JH, Ogawa M. Reversible expression of CD34 by murine hematopoietic stem cells [see comments]. Blood 1999; 94(8):2548-2554.
15. Nakamura Y, Ando K, Chargui J et al. Ex vivo generation of CD34(+) cells from CD34(-) hematopoietic cells. Blood 1999; 94(12):4053-4059.
16. Van Zant G, de Haan G, Rich IN. Alternatives to stem cell renewal from a developmental viewpoint. Exp Hematol 1997; 25(3):187-192.
17. Jones RJ. Leukemic stem cells—where have they gone wrong? Blood 2001; 97(12):3681-3682.
18. Jones RJ, Celano P, Sharkis SJ et al. Two phases of engraftment established by serial bone marrow transplantation in mice. Blood 1989; 73:397-401.

19. Orlic D, Bodine DM. What defines a pluripotent hematopoietic stem cell (PHSC): Will the real PHSC please stand up! Blood 1994; 84:3991-3994.
20. Jones RJ, Collector MI, Barber JP et al. Is there more than one class of lymphohematopoietic stem cells (LHSC)? Blood 1995; 86:590a, Abstract.
21. Bhatia M, Bonnet D, Murdoch B et al. A newly discovered class of human hematopoietic cells with SCID- repopulating activity. Nat Med 1998; 4(9):1038-1045.
22. Camitta BM, Storb R, Thomas ED. Aplastic anemia (second of two parts): Pathogenesis, diagnosis, treatment, and prognosis. N Engl J Med 1982; 306:712-718.
23. Young NS, Maciejewsi J. The pathology of acquired aplastic anemia. New England Journal of Medicine 1997; 336(19):1365-1372.
24. Gale RP, Champlin RE, Feig SA et al. Aplastic anemia: Biology and treatment. Ann Intern Med 1981; 95:477-494.
25. Sensenbrenner LL, Steele AA, Santos GW. Recovery of hematologic competence without engraftment following attempted bone marrow transplantation for aplastic anemia: Report of a case with diffusion chamber studies. Exp Hematol 1977; 77(1):51-58.
26. May WS, Sensenbrenner LL, Burns WH et al. BMT for severe aplastic anemia using cyclosporin. Bone Marrow Transplant 1993; 11:459-464.
27. Weitzel JN, Hows JM, Jeffreys AJ et al. Use of a hypervariable minisatellite DNA probe (33.15) for evaluating engraftment two or more years after bone marrow transplantation for aplastic anaemia. Br J Haematol 1988; 70(1):91-97.
28. Socie G, Landman J, Gluckman E et al. Short-term study of chimaerism after bone marrow transplantation for severe aplastic anaemia. Br J Haematol 1992; 80(3):391-398.
29. Brodsky RA, Sensenbrenner LL, Jones RJ. Complete remission in acquired severe aplastic anemia following high-dose cyclophosphamide. Blood 1996; 87:491-494.
30. Brodsky RA, Sensenbrenner LL, Smith BD et al. Durable treatment-free remission following high-dose cyclophosphamide for previously untreated severe aplastic anemia. Ann Intern Med 2001; 135:477-483.
31. Rusten LS, Jacobsen SE, Kaalhus O et al. Functional differences between CD34- and DR- subfractions of CD34+ bone marrow cells. Blood 1994; 84(5):1473-1481.
32. Maciejewski JP, Selleri C, Sato T et al. Increased expression of Fas antigen on bone marrow CD34+ cells of patients with aplastic anaemia. Br J Haematol 1995; 91(1):245-252.
33. Nagafuji K, Shibuya T, Harada M et al. Functional expression of Fas antigen (CD95) on hematopoietic progenitor cells. Blood 1995; 86(3):883-889.
34. Stahnke K, Hecker S, Kohne E et al. CD95 (APO-1/FAS)-mediated apoptosis in cytokine-activated hematopoietic cells. Exp Hematol 1998; 26(9):844-850.
35. Hilton J. Role of aldehyde dehydrogenase in cyclophosphamide-resistant L1210 leukemia. Cancer Res 1984; 44:5156-5160.
36. Bedi A, Zehnbauer BA, Collector MI et al. BCR-ABL gene rearrangement and expression of primitive hematopoietic progenitors in chronic myeloid leukemia. Blood 1993; 81:2898-2902.
37. Verfaillie CM, Miller WJ, Boylan K et al. Selection of benign primitive hematopoietic progenitors in chronic myelogenous leukemia on the basis of HLA-DR antigen expression. Blood 1992; 79:1003-1010.
38. Leemhuis T, Leibowitz D, Cox G et al. Identification of BCR/ABL-negative primitive hematopoietic progenitor cells within chronic myeloid leukemia marrow. Blood 1993; 81:801-807.
39. Torlakovic E, Litz CE, McClure JS et al. Direct detection of the Philadelphia chromosome in CD20-positive lymphocytes in chronic myeloid leukemia by tri-color immunophenotyping/FISH. Leukemia 1994; 8(11):1940-1943.
40. Nitta M, Kato Y, Strife A et al. Incidence of involvement of the B and T lymphocyte lineages in chronic myelogenous leukemia. Blood 1985; 66(5):1053-1061.
41. Boultwood J, Wainscoat JS. Clonality in the myelodysplastic syndromes. Int J Hematol 2001; 73(4):411-415.
42. Okada M, Okamoto T, Takemoto Y et al. Function and X chromosome inactivation analysis of B lymphocytes in myelodysplastic syndromes with immunological abnormalities. Acta Haematol 2000; 102(3):124-130.
43. White NJ, Nacheva E, Asimakopoulos FA et al. Deletion of chromosome 20q in myelodysplasia can occur in a multipotent precursor of both myeloid cells and B cells. Blood 1994; 83(10):2809-2816.
44. Bonnet D, Dick JE. Human acute myeloid leukemia is organized as a hierarchy that originates from a primitive hematopoietic cell. Nat Med 1997; 3(7):730-737.
45. Lapidot T, Sirard C, Vormoor J et al. A cell initiating human acute myeloid leukaemia after transplantation into SCID mice. Nature 1994; 367(6464):645-648.
46. Feuring-Buske M, Hogge DE. Hoechst 33342 efflux identifies a subpopulation of cytogenetically normal CD34(+)CD38(-) progenitor cells from patients with acute myeloid leukemia. Blood 2001; 97(12):3882-3889.
47. George AA, Franklin J, Kerkof K et al. Detection of leukemic cells in the CD34(+)CD38(-) bone marrow progenitor population in children with acute lymphoblastic leukemia. Blood 2001; 97(12):3925-3930.
48. Gianni AM, Siena A, Bregni M et al. Granulocyte-macrophage colony-stimulating factor to harvest circulating hemopoietic stem cells for autotransplantation. Lancet 1989; 2:580-585.
49. Burt RK, Traynor AE, Pope R et al. Treatment of autoimmune disease by intense immunosuppressive conditioning and autologous hematopoietic stem cell transplantation. Blood 1998; 92(10):3505-3514.
50. Brodsky RA, Smith BD. Bone marrow transplantation for autoimmune diseases. Curr Opin Oncol 1999; 11(2):83-86.
51. Levite M, Zinger H, Zisman E et al. Beneficial effects of bone marrow transplantation on the serological manifestations and kidney pathology of experimental systemic lupus erythematosus. Cell Immunol 1995; 162:138-145.
52. Adachi Y, Inaba M, Amoh Y et al. Effect of bone marrow transplantation on antiphospholipid antibody syndrome in murine lupus mice. Immunobiology 1995; 192:218-230.
53. Lowenthal RM, Cohen ML, Atkinson K et al. Apparent cure of rheumatoid arthritis by bone marrow transplantation. J Rheumatol 1993; 20(1):137-140.
54. Euler HH, Marmont AM, Bacigalupo A et al. Early recurrence or persistence of autoimmune diseases after unmanipulated autologous stem cell transplantation. Blood 1996; 88(9):3621-3625.
55. Brodsky RA, Petri M, Smith BD et al. Immunoablative high-dose cyclophosphamide without stem cell rescue for refractory severe autoimmune disease. Ann Intern Med 1998; 129(12):1031-1035.
56. Brodsky RA, Petri M, Smith BD et al. Durable complete remissions following high-dose cyclophosphamide for refractory systemic lupus erythematosus (SLE). Blood 2000: 96(11): 422a, Abstract.
57. Moyo VM, Smith BD, Brodsky I et al. High-dose cyclophosphamide for refractory autoimmune hemolytic anemia. Blood; in press.
58. Nousari HC, Brodsky RA, Jones RJ et al. Immunoablative high-dose cyclophosphamide without stem cell rescue in paraneoplastic pemphigus: Report of a case and review of this new therapy for severe autoimmune disease. J Am Acad Dermatol 1999; 40(5 Pt 1):750-754.

CHAPTER 7

Properties and Therapeutic Potentials of Adult Stem Cells from Bone Marrow Stroma (MSCs)

Darwin J. Prockop

Introduction

In addition to hematopoietic stem cells, bone marrow contains a second class of adult stem cells that have been referred to as fibroblastoid colony-forming units, nonhematopoietic mesenchymal stem cells or marrow stromal cells (MSCs). MSCs have attracted increasing attention for their potential use in cell and gene therapy because they have several appealing features:[1,2]

(a) They are readily isolated from a patient by simple bone marrow aspiration under local anesthesia.

(b) They can be rapidly expanded in culture a billion-fold or more.[3,4] Therefore, adequate numbers of autologous cells can be generated for treatment of most conditions. The use of a patient's own MSCs avoids the immune rejection encountered with cells from unrelated individuals.

(c) Although the cells can be expanded rapidly, they are not immortal. Therefore, they do not pose a danger of producing tumors as is seen with embryonic stem cells and with most immortal cell lines.[5]

(d) MSCs can differentiate in vitro and in vivo into multiple cell lineages including osteoblasts, chondrocytes, adipocytes, myocytes,[6] pneumocytes,[7] epithelial cells[8] and early neural precursors.[9-13]

(e) They can be readily transduced with genes with or without the use of viral vectors.[14-17]

(f) As increasing evidence has now demonstrated, the cells have the remarkable property that they home to sites of tissue injury and repair the tissue either by differentiating into tissue-specific cell phenotypes[7,18-22] or by creating a milieu that increases the capacity of the endogenous cells to repair the tissue.[23-25]

These features of MSCs probably account for the fact that promising results have been reported with the use of the cells in a number of animal models for human diseases, including models for osteogenesis imperfecta,[18] spinal cord injury,[12,24-27] stroke,[23,28] Parkinsonism,[29,30] myelin defects,[31] and cardiomyopathies.[32,33] They also probably account for the reports that promising results have been obtained in one clinical trial[34] of children with severe brittle bone disease (osteogenesis imperfecta). The results obtained with MSCs may also be related, in a manner still not defined, to promising results obtained with other sub-fractions of cells from bone marrow, including results obtained with cells defined as multipotent adult progenitor cells (MAPCs)[35] in a rat model for stroke[36] and with Lin⁻c-kit⁺ bone marrow cells,[37] with side population cells (SP cells) from marrow,[38] and with bone marrow cells mobilized by stem cell factor or granulocyte-colony stimulating factor[39] in cardiac ischemia.

Early Studies with MSCs

MSCs were first discovered about 30 years ago by Friedenstein et al,[40,41] who found that they were easily separated from whole bone marrow by their tight adherence to tissue culture plasticware. They also discovered that the cells can be easily differentiated into osteoblasts, chondrocytes and adipocytes both in vitro and in vivo. A large number of subsequent investigators confirmed and extended Friedenstein's observations.[1,2,6,14,42-48] Pereira and his associates[18,49] were the first to demonstrate that after MSCs were infused systemically into marrow-ablated mice, progeny of the MSCs appeared in a variety of tissue including bone, cartilage, lung, skin and liver. Also, when the progeny of the MSCs appeared in several tissues, they seemed to acquire the phenotype of the resident cells. These observations were subsequently confirmed and extended by other investigators.[14,20,50] When MSCs are infused directly into the brains of rats or mice, they engraft and migrate in a manner similar to neural stem cells.[13,51] Some of the cells infused into the ventricles of newborn mice differentiated into astrocytes and others probably became neurons.[13] These observations have now been confirmed and extended by other investigators.[12,23,31,36,52-54]

MSCs as Part of a Natural Repair System

The repair of wounds and other tissue injury has long been of interest to investigators but is still a poorly understood process. For example, Cohnheim in 1867[55] published an extensive thesis in which he injected dyes into the arteries feeding wounds in a number of different tissues. He observed that all the cells in the repairing wound were labeled by dyes, including both inflammatory cells and cells that today would be defined as fibroblasts. He concluded that all the cells for wound repair are derived from the blood stream, and by implication, from the bone marrow. Cohnheim's thesis was pursued by a large number of investigators over the years but the results were inconsistent.[56-57] The results were inconsistent, primarily because it is difficult to design experiments that distinguish invasion of stem-like cells through the bloodstream from proliferation of local stem-like cells such as pericytes at the site of wound repair. Most recent evidence, however, supports Cohnheim's thesis.

Some of the recent evidence to support Cohnheim's thesis consists of identification of cells called fibrocytes from the bloodstream that apparently become the fibroblasts found in repairing wounds.[58-61] Another part of the evidence consists of observations that demonstrate homing of MSCs to sites of tissue damage. One of the first reports demonstrating that MSCs home to sites of tissue damage came from the work of Ferrari et al[19] who demonstrated that after muscle in a mouse is damaged with infusion of a cardiotoxin, prelabeled MSCs appear at the site of injury and differentiate into muscle cells. In contrast, Pereira and his associates[49] did not see intravenously infused MSCs appearing in uninjured muscle of normal mice. Subsequently, Mackenzie and Flake[21] found human MSCs appearing in increased numbers in wounded and regenerating tissues after the cells were engrafted into fetal sheep. Lu et al,[62] found cells home to sites of traumatic brain injury in a rat model after intra-arterial infusion, and Chen et al,[23,52] saw similar effects in a rat model for cerebral ischemia after either intravenous or intracerebral infusion. Wang et al[33] found that MSCs engrafted into heart muscle and differentiated into fibroblasts and cardiomyocytes after myocardial infarction. In one of the most dramatic observations, Kotton et al[7] demonstrated the cells homed to, engrafted, and differentiated into type I pneumocytes in a mouse model in which the lungs were injured with intra-tracheal bleomycin. We have recently obtained similar results with engraftment into rat lungs that were injured with monocrotaline (preliminary results).

Problems with Expanding MSCs ex Vivo

Although MSCs are readily expanded in culture under the conditions employed by Friedenstein et al[40] and others,[2,63] the cells undergo subtle changes as they proliferate. As the cells are expanded through several passages in culture, they gradually cease to proliferate and they lose the ability to differentiate into adipocytes.[64] Late passage human MSCs (hMSCs) retain the ability to differentiate into osteoblasts, apparently as a default pathway. We[3,65] recently confirmed much earlier observations[43] that the cells in early passage were heterogeneous in that they contained two morphologically distinct cell types: small, spindle-shaped cells and large flat cells. The small spindle-shaped cells in the cultures (rapidly, self-renewing, or RS cells) were multipotential for differentiation and underwent rapid, symmetrical division. The large, flat cells (mature MSCs or mMSCs) in the cultures had a limited potential to differentiate and replicated slowly. To obtain cultures enriched for RS cells, it was essential to plate the cells at extremely low densities of 3 to 50 cells per cm^2 and lift the cultures during the late log phase of growth.[64-66] After low density plating, the doubling times of the cells were as short as 10 hours.[65,66] More densely plated cultures were morphologically homogeneous but they grew more slowly and largely consisted of mMSCs. These and related observations indicate that each preparation of MSCs must be carefully assayed for content of early progenitor RS cells by morphology, propagation rate, clonogenicity, FACS analysis, and multilineage differentiation.[65,66] The difficulties in expanding hMSCs in culture are compounded by the fact that there is no consensus as to characteristic surface epitopes that can be used to identify the cells. A series of antibodies to surface epitopes have been employed by several investigators,[67-69] but none have been shown to identify subpopulations of MSCs. We ourselves have now screened over 200 antibodies but have not found any that efficiently distinguish RS cells from mMSCs.[65]

Limitations of in Vitro and in Vivo Assays for MSCs

In vitro assays that have been employed for MSCs for the last several decades include: (a) the rate of proliferation of the cells; (b) differentiation of the cells into osteoblasts, adipocytes and chondrocytes; and (c) clonogenicity assayed by plating the cells at very low densities.[1,2,63] Each of these assays is time consuming and difficult to standardize and quantitate. In vivo assays for MSCs have also been problematic. For example, hMSCs have been assayed by impregnating ceramic matrices with cell suspensions and then embedding the ceramic matrices under the skin of immunodeficient mice.[59,71] The cells in the matrices have been shown to differentiate into both bone and cartilage. However, the in vivo assays are difficult to quantitate and time consuming.

Relationship of MSCs to Other Stem Cells

Stem cells or stem-like cells from multiple sources are now under investigation but much of the information on the similarities and differences among the cells is incomplete. A few distinctions, however, can be made. hMSCs are not immortal,[2,64] do not express telomerase,[70] and have not formed tumors after infusion into animals in extensive experiments carried out over the last 30 years.[1,2] Therefore, they do not have the tumorogenic propensities of embryonic stem cells (ES),[5] and other immortal cell lines. With the exception of one report suggesting rare engraftment as thymocytes in utero,[20] standard preparations of MSCs have not been observed to give rise to hematopoietic cells. Therefore, they differ from CD34+ hematopoietic cells, side population (SP) hematopoietic cells (SP cells are a small homogeneous population of hematopoietic stem cells able to efflux Hoechst dye)[38,72] or embryonic stem (ES) cells[73] and they are not as pluripotent as ES cells. However, some of the data suggesting broad pluripotentiality of some stem cells such as neural stem cells[74] must be reevaluated. As indicated by recent reports,[75] the apparent plasticity observed in some experiments may be explained by cell fusion and formation of tetraploid cells. The cell fusion events are rare but can mimic cell differentiation in experiments in which stem cells are injected into embryos or blastocytes and the fused cells are then extensively amplified during development.

Multipotent adult progenitor cells (MAPCs) studied by Verfaillie and associates[35] have greater plasticity for differentiation than MSCs in that they can differentiate into hematopoietic cells. Also, they express telomerase and are immortal. However, MAPCs are similar to MSCs in that they are also prepared from bone marrow by their adherence to tissue culture surfaces. The major differences from the protocols for preparation of hMSCs appear to be: (a) MAPCs are depleted for Lin and glycophorin A, plated on fibronectin-coated dishes, and grown in 1% or 2% fetal calf serum plus a cocktail of cytokines. In contrast, most protocols for hMSCs consist of plating all nucleated cells from bone marrow on uncoated tissue culture plastic in 10 to 20% fetal calf serum (FCS) and then passing the cells in medium containing 10 to 20% FCS. (b) Perhaps most importantly, the cultures for MAPCs grow slowly for 25 or more population doublings before the more rapidly dividing, multipotential, and immortal cells defined as MAPCs appear in the cultures. In contrast, cultures of hMSCs expand rapidly as initially isolated, particularly if plated at low densities,[64,65] but they gradually lose their proliferative capacity and multipotentiality as they approach about 50 population doublings. Therefore, it is not clear whether the protocols for preparation of MAPCs: (a) select for a rare cell in marrow not found in cultures of MSCs; (b) select for a rare cell found in standard cultures of MSCs; or (c) select for a rare cell in cultures of MSCs that undergo some subtle change that makes the cell immortal and gives it greater multipotentiality than MSCs.

The relationship between MSCs and SP cells[38,72] has also not been clearly defined. The rapidly proliferating and multipotential RS cells in cultures of hMSCs are similar in size to SP cells. However, we have not been able to identify SP cells in hMSC cultures because the Hoescht dyes used to identify SP cells produced rapid necrosis of hMSCs under the conditions used by Goodell et al.[72] The presence of the necrotic cells and cell fragments makes it difficult to determine whether hMSCs cultures include a sub-population of SP cells.

Cells with many similarities to hMSCs have recently been isolated from fat tissue obtained by liposuction (refer to Chapter 5).[76] The cells show a similar potential to differentiate in vitro into osteoblasts, adipocytes, chondrocytes, myocytes and neural-like cells. They have not yet been extensively characterized as hMSCs. Also, it is not clear whether: (a) the cells arise from pericytes in the abundant vessels of adipose tissue; (b) they represent progeny of hMSCs from the bloodstream; or (c) they are a unique lineage of adult stem cells.

References

1. Prockop DJ. Marrow stromal cells as stem cells for nonhematopoietic tissues. Science 1997; 276:711-74.
2. Pittenger MF, Mackay AM, Beck SC et al. Multilineage potential of adult human mesenchymal stem cells. Science 1999; 284:143-147.
3. Colter DC, Class R, DiGirolamo CM et al. Rapid expansion of recycling stem cells in cultures of plastic-adherent cells from human bone marrow. Proc Natl Acad Sci USA 2000; 96:7294-7299.
4. Hung SC, Chen NJ, Hseih SL et al. Isolation and characterization of size-seived stem cells from human bone marrow. Stem Cells 2002; 20(3):249-58.
5. Bjorklund LM, Sanchez-Pernatue R, Chung S et al. Embryonic stem cells develop into functional dopaminergic neurons after transplantation in a Parkinson rat model. Proc Natl Acad Sci USA 2002; 99(4):1755-7.
6. Wakitani S, Saito T, Caplan AI. Myogenic cells derived from rat bone marrow mesenchymal stem cells exposed to 5-azacytidine. Muscle Nerve 1995; 18:1417-1426.
7. Kotton DN, Ma BY, Cardoso WV et al. Bone-marrow derived cells as progenitors of lung alveolar epithelium. Development 2001; 128(24):5181-88.
8. Spees JL, Wang MY, Prockop DJ. In vitro differentiation of adult stem cells to an epithelial-like phenotype. Abstracts of the 8th Annual Meeting of the International Society for Cellular Therapy. Barcelona, Spain: May 25-28 2002.
9. Woodbury D, Schwarz EJ, Prockop DJ et al. Adult rat and human bone marrow stromal cells differentiate into neurons. J Res 2000; 61:364-370.
10. Sanchez-Ramos J, Song S, Cardozo-Pelaez F et al. Adult bone marrow stromal cells differentiate into neural cells in vitro. Exp Neurol 2000; 164:247-256.
11. Deng W, Obrocka M, Fischer I et al. In vitro differentiation of human marrow stromal cells into early progenitors of neural cells by conditions that increase intracellular cyclic AMP. Biochem Biophys Res Comm 2001; 282:148-152.
12. Chopp M, Zhang XH, Li Y et al. Spinal cord injury in rat: Treatment with bone marrow stromal cell transplantation. Neuroreport 2000; 11(13):3001-3005.
13. Kopen GC, Prockop DJ, Phinney DG. Marrow stromal cells migrate throughout forebrain and cerebellum and they differentiate into astrocytes following injection into neonatal mouse brains. Proc Natl Acad Sci USA 1999; 96:10711-10716.
14. Keating A, Guinn B, Larava P et al. Human marrow stromal cells electrotransfected with human factor IX (FIX) cDNA engraft in SCID mouse marrow and transcribe human FIX. Exp Hematol 1996; 24:180 (abstr).
15. Schwarz EJ, Reger RL, Alexander GM et al. Rat marrow stromal cells rapidly transduced with a self-inactivating retrovirus synthesize L-DOPA in vitro. Gene Therapy 2001; 8(16):1214-23.
16. Zhang XY, La Russa VF, Bao L et al. Lentiviral vectors for sustained transgene expression in human bone marrow-derived stromal cells. Mol Ther 2002; 5(5Pt1):555-65.
17. Peister A, Mellad JA, Wang MY et al. Stable transfection of adult stem cells from bone marrow stroma by electroporation. Abstracts of the 8th Annual Meeting of the International Society for Cellular Therapy. Barcelona, Spain: May 25-28 2002.
18. Pereira RF, O'Hara MD, Laptev AV et al. Marrow stromal cells as a source of progenitor cells for nonhematopoietic tissues in transgenic mice with a phenotype of osteogenesis imperfecta. Proc Natl Acad Sci 1998; 95:1142-1147.
19. Ferrari G, Cusella-De Angelis G, Coletta M et al. Muscle regeneration by bone marrow-derived myogenic progenitors. Science 1998; 279:1528-1530.
20. Liechty KW, MacKenzie TC, Shaaban AF et al. Human mesenchymal stem cells engraft and demonstrate site-specific differentiation after in utero transplantation in sheep. Nat Med 2000; 6(11):1282-1286.
21. Mackenzie TC, Flake AW. Human mesenchymal stem cells persist, demonstrate site-specific multipotential differentiation, and are present in sites of wound healing and tissue regeneration after transplantation into fetal sheep. Blood Cells Mol Dis 2001; 27(3):601-4.
22. Dezawa M, Takahashi I, Esaki M et al. Sciatic nerve regeneration in rats induced by transplantation of in vitro differentiated bone-marrow stromal cells. Eur J Neurosci 2001; 14(11):1771-6.
23. Chen J, Li Y, Wang L et al. Therapeutic benefit of intracerebral transplantation of bone marrow stromal cells after cerebral ischemia in rats. J Neurol Sci 2001b; 189(1-2):49-57.
24. Hofstetter CP, Schwarz EJ, Hess D et al. Marrow stromal cells form guiding strands in the injured spinal cord and promote recovery. Proc Natl Acad Sci USA 2002; 99(4):2199-2204.
25. Ankeny DP, McTigue DM, Wei P et al. Transplanted bone marrow stromal cells provide a scaffold for axonal regrowth and lead to activation of hindlimb airstepping. Abstracts from the 31st Annual Meeting of the Society for Neurosciences. San Diego, CA: November 2001; 10-15, 2001.
26. Honmou O, Sasaki M, Oka S et al. Mesenchymal stem cells derived from bone marrow remyelinate demyelinated axons in the adult rat spinal cord. Abstracts of the 31st Annual Meeting of the Society of Neurosciences. San Diego, CA: November 10-15 2001.
27. Sasaki M, Honmou O, Akiyama Y et al. Transplantation of an acutely isolated bone marrow fraction repairs demyelinated adult rat spinal cord axons. Glia 2001; 35(1):26-34.
28. Li Y, Chopp M, Chen J et al. Intrastriatal transplantation of bone marrow nonhematopoietic cells improves functional recovery after stroke in adult mice. J Cereb Blood Flow Metab 2000; 20(9):1311-1319.
29. Schwarz EJ, Alexander GM, Prockop DJ et al. Multipotential marrow stromal cells transduced to produce L-DOPA: Engraftment in a rat model of Parkinson's disease. Human Gene Therapy 1999; 10:2539-2549.
30. Li Y, Chen J, Wang L et al. Intracerebral transplantation of bone marrow stromal cells in a 1-methyl-4-phenyl-1,2,3,6-tetra-hydropyridine mouse model of Parkinsons' disease. Neurosci Lett 2001; 316(2):67-70.
31. Jin HK, Carter JE, Huntley GW et al. Intracerebral transplantation of mesenchymal stem cells into acid sphingomyelinase-deficient mice delays the onset of neurological abnormalities and extends their life span. J Clin Invest 2002; 109(9):1183-1191.
32. Toma C, Pittenger MF, Cahill KS et al. Human mesenchymal stem cells differentiate ito a cardiomyocyte phenocyte in the adult murine heart. Circulation 2002; 105(1):93-8.

33. Wang JS, Shum-Tim D, Chedrawy E et al. The coronary delivery of marrow stromal cells for myocardial regeneration: Pathophysiologic and therapeutic implications. J Thorac Cardiovasc Surg 2001; 122(4):699-705.
34. Horwitz EM, Prockop DJ, Gordon PL et al. Clinical responses to bone marrow transplantation in children with severe osteogenesis imperfecta. Blood 2001; 97(5):1227-31.
35. Reyes M, Lund T, Lenvik T et al. Purification and ex vivo expansion of postnatal human marrow mesodermal progenitor cells. Blood 2001; 98(9):2615-25.
36. Zhao LR, Duan WM, Reyes M et al. Human bone marrow stem cells exhibit neural phenotypes and ameliorate neurological deficits after grafting into the ischemic brains of rats. (2002) Exp Neurol 2002; 174(1):11-20.
37. Orlic D, Kajstura J, Chimenti S et al. Bone marrow cells regenerate infarcted myocardium. Nature 2001a; 410(6829):701-5
38. McKinney-Freeman SL, Jackson KA, Camargo FD et al. Muscle-derived hematopoietic stem cells are hematopoietic in origin. Proc Natl Acad Sci USA 2002; 99(3):1341-6.
39. Orlic D, Kajstura J, Chimenti S et al. Mobilized bone marrow cells repair the infarcted heart, improving function and survival. Proc Natl Acad Sci USA 2001b; 98(18):10344-9.
40. Friedenstein AJ, Gorskaja U, Kalugina NN. Fibroblast precursors in normal and irradiated mouse hematopoietic organs. Experimental Hematology 1976; 4:267-274.
41. Friedenstein AJ, Chailakhyan RK, Gerasimov UV. Bone marrow osteogenic stem cells: In vitro cultivation and transplantation in diffusion chambers. Cell and Tissue Kinetics 1987; 20:263-272.
42. Piersma AH, Brockbank KGM, Ploemacher RE et al. Characterization of fibroblastic stromal cells from murine bone marrow. Exp Hematol 1985; 13:237.
43. Mets T, Verdonk G. In vitro aging of human bone marrow-derived stromal cells. Mech Ageing Dev 1981; 16:81-89.
44. Owen ME, Friedenstein AJ. Stromal stem cells: Marrow-derived osteogenic precursors. In: Cell and Molecular Biology of Vertebrate Hard Tissues, Ciba Foundation Symp. Chichester, UK: 1988:42-60.
45. Caplan AI. Mesenchymal stem cells. J Ortho Res 1991; 9:641-650.
46. Beresford JN, Bennett JH, Devlin C et al. Evidence for an inverse relationship between the differentiation of adipocytic and osteogenic cells in rat marrow stromal cell cultures. J Cell Sci 1992; 102:341-351.
47. Castro-Malaspina H, Gay RE, Resnick G et al. Characterization of human bone marrow fibroblast colony-forming cells (CFU-F) and their progeny. Blood 1980; 56:289-301.
48. Kuznetsov SA, Krebsbach PH, Satomura K et al. Single-colony derived strains of human marrow stromal fibroblasts form bone after transplantation in vivo. J Bone Miner Res 1997; 12(9):1335-1347.
49. Pereira RF, Halford KW, O'Hara MD et al. Cultures adherent cells from marrow can serve as long-lasting precursor cells for bone, cartilage and lung in irradiated mice. Proc Natl Acad Sci 1995; 92:4857-4861.
50. Hou Z, Nguyen Q, Frenkel B et al. Osteoblast-specific gene expression after transplantation of marrow cells: Implications for skeletal gene therapy. Proc Natl Acad Sci USA 1999; 96(13):7294-9.
51. Azizi SA, Stokes DG, Augelli BJ et al. Engraftment and migration of human bone marrow stromal cells implanted in the brains of albino rats – similarities to astrocyte grafts. Proc Natl Acad Sci USA 1998; 95:3908-3913.
52. Chen J, Li Y, Wang L et al. Therapeutic benefit of intravenous administration of bone marrow stromal cells after cerebral ischemia in rats. Stroke 2001a; 32(4):1005-11.
53. Fukuda K. Development of regenerative cardiomyocytes from mesenchymal stem cells for cardiovascular tissue engineering. Artif Organs 2001; 25(3):187-193.
54. Qu T, Kim HM, Sugaya K. In vivo differentiation and migration properties of mesenchymal stem cells. Abstracts from the 31st Annual Meeting of American Society for Neurosciences. San Diego, CA: November 2001.
55. Cohnheim J. Ueber entzündung und eiterung. Path Anat Physiol Klin 1867; Med 40:1.
56. Ross R, Everett NB, Tyler R. Wound healing and collagen formation. VI. The origin of the wound fibroblast studied in parabiosis. J Cell Biol 1970; 44:645-54.
57. Davidson JM. Inflammation: Basic principles and clinical correlates. In: Gallin JI, Goldstein IM, Snyderman R, eds. 2nd ed. New York: Raven, 1992:809.
58. Chesney J, Bucala R. Peripheral blood fibrocytes: Mesenchymal precursor cells and the pathogenesis of fibrosis. Curr Rheumatol Rep 2000; 2(6):501-505.
59. Kuznetsov SA, Mankani MH, Gronthos S et al. Circulating skeletal stem cells. J Cell Biol 2001; 153(5):1133-1140.
60. Zvaifler JJ, Marinova-Mutafchieva L, Adams G et al. Mesenchymal precursor cells in the blood of normal individuals. Arthritis Res 2000; 2(6):477-88.
61. Hartlapp I, Abe R, Saeed RW et al. Fibrocytes induce an angiogenic phenotype in cultured endothelial cells and promote angiogenesis in vivo. FASEB J 2001; 15(12):2215-24.
62. Lu D, Li Y, Wang L et al. Intraarterial administration of marrow stromal cells in a rat model of traumatic brain injury. J Neurotrauma 2001; 18(8):813-9.
63. Bruder SP, Jaiswal N, Haynesworth SE. Growth kinetics, self-renewal, and the osteogenic potential of purified human mesenchymal stem cells during extensive subcultivation and following cryopreservation. J Cell Biochem 1997; 64(2):278-94.
64. DiGirolamo CM, Stokes D, Colter D et al. Propagation and senescence of human marrow stromal cells in culture. A simple colony-forming assay identifies samples with the greatest potential to propagate and differentiate. Br J of Haem 1999; 107:275-281.
65. Colter DC, Sekiya I, Prockop DJ. Identification of a sub-population of rapidly self-renewing and multipotential adult stem cells in colonies of human marrow stromal cells. Proc Natl Acad Sci USA 2001; 98(14):8741-8745.
66. Sekiya I, Larson BL, Smith JR et al. Expansion of human adult stem cells from bone marrow stroma: Conditions that maximize the yield of early progenitors and for evaluating their quality. Stem Cells 2002; 20:530-541.
67. Gronthos S, Simmons PJ, Graves SE et al. Integrin-mediated interactions between human bone marrow stromal precursor cells and the extracellular matrix. Bone 2001; 28(2):174-81
68. Barry FP, Boynton RE, Haynesworth S et al. The monoclonal antibody SH-2, raised against human mesenchymal stem cells, recognizes an epitope on endoglin (CD105). Biochem Biophys Res Commun 1999; 265(1):134-9
69. Barry F, Boynton R, Murphy M et al. The SH-3 and SH-4 antibodies recognize distinct epitopes on CD73 from human mesenchymal stem cells. Biochem Biophys Res Commun 2001; 289(2):519-24.
70. Sekiya I, Vuoristo JT, Larson BL et al. In vitro cartilage formation by human adult stem cells from bone marrow stroma defines the sequence of cellular and molecular events during chondrogenesis. Proc Natl Acad Sci USA 2002b; 99(7):4397-402.
71. Mankani MH, Kuznetsov SA, Fowler B et al. In vivo bone formation by human bone marrow stromal cells: Effect of carrier particle size and shape. Biotechnol Bioeng 2001; 72(1):96-107.
72. Goodell MA, Rosenzweig M, Kim H et al. Dye efflux studies suggest that hematopoietic stem cells expressing low or undetectable levels of CD34 antigen exist in multiple species. Nat Med 1997; 3(12):1337-45.
73. Lumelsky N, Blondel O, Laeng P et al. Differentiation of embryonic stem cells to insulin-secreting structures similar to pancreatic islets. Science 2001; 292(5520):1389-94.
74. Clarke DL, Johansson CB, Wilbertz J et al. Generalized potential of adult neural stem cells. Science 2000; 288(5471)1559-61.
75. Ying QL, Nichols J, Evans EP et al. Changing potency by spontaneous fusion. Nature 2002; 416(6880):545-8.
76. Zuk PA, Zhu M, Mizuno H et al. Multilineage cells from human adipose tissue: Implications for cell-based therapies. Tissue Eng 2001; 7(2):211-28.

CHAPTER 8

Regeneration of Cardiomyocytes from Bone Marrow Stem Cells and Application to Cell Transplantation Therapy

Keiichi Fukuda

Introduction

Although heart transplantation is the ultimate therapy for the treatment of severe heart failure, it has not been widely examined, because it requires donor hearts, and the inadequate supply of donor hearts is often a major problem everywhere in the world. As a result, the current challenge in cardiology is how to reserve pump failure by cell transplantation or regenerative medicine. Recent studies have shown that transplanted fetal cardiomyocytes can survive in heart scar tissue and that the transplanted cells limit scar expansion and prevent post-infarction heart failure. Transplantation of cultured cardiomyocytes into damaged myocardium has been proposed as a future method of treating heart failure,[1,2] but this revolutionary concept remains unfeasible in clinical settings because of the difficulty of obtaining donor fetal hearts. A cardiomyogenic cell line has long been awaited, and such a line might be capable of substituting for fetal cardiomyocytes in this therapy.

A number of studies have demonstrated that cardiomyocytes can differentiate from various multipotent stem cells, including embryonic stem (ES) cells[3] and embryonic carcinoma (EC) cells.[4] ES cells are an attractive cell source in regenerative medicine, but because the transplanted ES cells are allogeneic, the recipients must take immunosuppressant drugs throughout their lives. Use of these reagents impairs the quality of life of the recipients, and transplantation of undifferentiated ES cells often causes teratocarcinoma. In addition, the establishment of human ES cells involves ethical problems and is not allowed in every country. Because of these circumstances, the regeneration of cardiomyocytes from adult autologous stem cells has long been awaited.

Recent reports have demonstrated the existence of pluripotent stem cells in adult tissues. Roy et al reported the existence of neural stem cells in the brain that can differentiate into neurons, oligodendrocytes, and astrocytes in vitro.[5] Marrow stromal cells have been shown to possess many characteristics of mesenchymal stem cells,[6] and pluripotent progenitor marrow stromal cells can differentiate into various cell types, including osteoblasts,[7,8] myocytes,[9] adipocytes, tenocytes, and chondroblasts.[10] We recently reported the differentiation of mesenchymal stem cells into cardiomyocytes after exposure to 5-azacytidine and the establishment of cell line CMG (cardiomyogenic) that differentiates into cardiomyocytes in vitro.[11] CMG cells exhibit spontaneous beating and express atrial natriuretic peptide (ANP) and brain natriuretic peptide (BNP), and they may provide a useful and powerful tool for cardiomyocyte transplantation after further characterization of their cardio-myocyte phenotype.

This chapter describes the characteristics of bone marrow-derived regenerated cardiomyocytes and discusses the possibility of using them for cardiovascular tissue engineering. The expression and function of α_1, β_1, and β_2-adrenergic receptors and muscarinic M_1 and M_2 receptors in CMG cells is also described, because these receptors play a critical role in modulating cardiac function.[12]

Does the Heart Have its Own Stem Cell Compartment?

It is well known that skeletal muscle cells contain stem cells, called "satellite cells". Satellite cells can both proliferate by cell division and differentiate into skeletal muscle cells, and the differentiated skeletal muscle cells can fuse to form myotubes. By contrast, fetal cardiomyocytes can proliferate by cell division, but they undergo terminal differentiation and stop dividing after birth. A number of studies have reported that cardiomyocytes increase in size by cell hypertrophy, not by cell hyperplasia. To our knowledge there have been no reports of the presence of satellite-cell-like cardiac stem cells in the heart. Beitrani et al recently reported that human cardiomyocytes express Ki67, a marker of cell division, and the M phase of the nucleus of the cardiomyocytes was observed in the border zone area of recent myocardial infarction in autopsied hearts.[13] These findings suggested that only a very few adult cardiomyocytes can divide after the terminal differentiation. Although their findings were very interesting, these cells were insufficient to improve cardiac function, since the population of these cells was very small.

Mesenchymal Marrow Stem Cells as a Possible Source of Cardiomyocytes: The Cardiomyogenic (CMG) Cell?

Figure 1 shows the classification of the stem cell system of adults.[14] Bone marrow stromal cells were previously used as a feeder layer to culture hematopoietic stem cells, and are known to be of mesodermal origin and produce various cytokines and

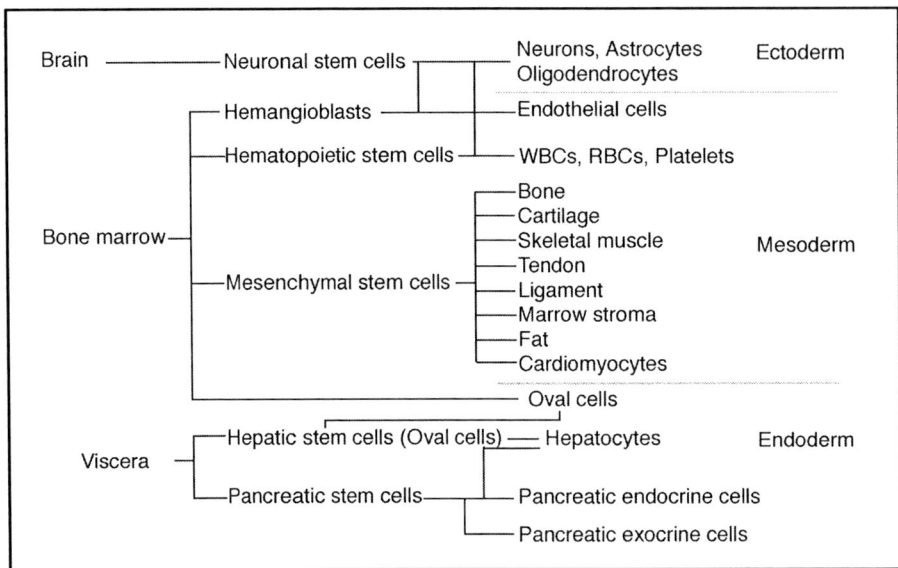

Figure 1. Classification of pluripotent stem cells in adult tissues. Bone marrow contains several kinds of stem cells. Mesenchymal stem cells may differentiate into various mesoderm-derived cells, such as osteoblasts, chondroblasts, adipocytes, skeletal muscle cells and cardiomyocytes.

growth factors. In the late 1990s, a number of papers reported that bone marrow stromal cells contain multipotent stem cells for non-hematopoietic tissues, called "marrow mesenchymal stem cells", that could differentiate into osteoblasts, chondroblasts, and adipocytes. All of these cells were known to be of mesodermal origin. If marrow mesenchymal stem cells are multipotent, we hypothesized that they might have the ability to differentiate into cardiomyocytes and instituted this study. We also recognized that bone marrow cells could be obtained from patients themselves and that autologous cells would not be rejected after cell transplantation.

Method of Establishing Bone-Marrow Derived Cardiomyocytes

Female C3H/He mice were anesthetized with ether, their femora were excised, and bone marrow cells were collected. The procedures were performed in accordance with the guidelines for animal experimentation of Keio University. Primary culture of the marrow cells was performed according to Dexter's method. Cells were cultured in Iscove's modified Dulbecco's medium (IMDM) supplemented with 20% fetal bovine serum and penicillin (100 μg/ml)/streptomycin (250 ng/ml)/ amphotericin B at 33°C in humid air containing 5% CO2. After a series of passages, the attached marrow stromal cells became homogeneous and were devoid of hematopoietic cells. The marrow stromal cells basically did not require co-culture with blood stem cells. Immortalized cells were obtained by frequent subculture for more than 4 months. Cell lines from different dishes were subcloned by limiting dilution. To induce cell differentiation, cells were treated with 3 μmol/L of 5-azacytidine for 24 hours. Subclones that included spontaneously beating cells were screened by microscopic observation (first screening), and cells surrounding spontaneous beating cells were subcloned with cloning syringes. Subcloned cells were maintained, exposed to 5-azacytidine again for 24 hours, and clones that showed spontaneous beating most frequently were screened (second screening). The clonal cell line thus obtained was named the CMG cell.

As a result of repeated rounds of limiting dilution, we succeeded in isolating 192 single clones, several of which differentiated into cardiomyocytes and showed spontaneous beating. The experiments were reproducible, but the percentage of cells that differentiated into cardiomyocyte differentiation was specific to each clone. Phase-contrast photography and/or immunostaining with anti-sarcomeric myosin antibodies were used to identify the morphological changes in the CMG cells. CMG cells showed a fibroblast-like morphology before 5-azacytidine treatment (0 week), and this phenotype was retained through repeated subculturing under non-stimulating conditions. After 5-azacytidine treatment, however, the morphology of the cells gradually changed (Fig. 2). Approximately 30% of the CMG cells gradually increased in size at 1 week, and they formed a ball-like appearance, or had lengthened in one direction to exhibit a stick-like morphology. They connected with adjoining cells after 2 weeks and had formed myotube-like structures at 3 weeks. The differentiated CMG myotubes maintained the cardiomyocyte phenotype and beat vigorously for at least 8 weeks after the final 5-azacytidine treatment and did not de-differentiate.[11] Most of the other non-myocytes had an adipocyte-like appearance.

Regenerated Cardiomyocytes Display a Fetal Ventricular Phenotype

Various cardiac contractile protein isoforms are differentially expressed in cardiomyocytes at different developmental stages and in different chambers. At around the time of birth there is a developmental switch in the ventricular muscle of small mammals from expression of β-myosin heavy chain (MHC), which is the predominant fetal form, to expression of α-MHC. There is also a developmental switch from expression of α-skeletal actin, which is the predominant fetal and neonatal form, to that of α-cardiac actin, the predominant adult form. We investigated the contractile protein isoforms of bone marrow-derived CMG cells to characterize their phenotype as cardiomyocytes. Table 1 summarizes the results. Fetal, neonatal, and adult ventricle and atrium were used as controls.[14] Expression of both α- and β-*MHC* was detected in differentiated CMG cells by RT-PCR, but β-*MHC* expression was overwhelmingly greater than that of α-*MHC*. CMG cells expressed both α-*cardiac* and α-*skeletal actin*, but the

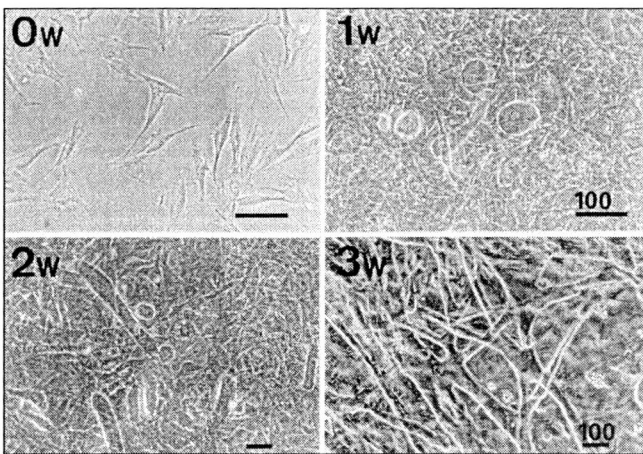

Figure 2. Phase-contrast photographs of marrow-derived cardiomyogenic (CMG) cells before and after 5-azacytidine treatment. Upper left: Marrow mesenchymal cells have a fibroblast-like morphology before 5-azacytidine treatment (0 week). Upper right: One week after treatment, some cells gradually increased in size, and developed a ball-like or stick-like appearance. These cells began spontaneous beating thereafter. Lower left: Two weeks after treatment, the ball-like or stick-like cells connected to adjoining cells, and began to form myotube-like structures. Lower right: Three weeks after treatment. Bars indicate 100 μm.

α–skeletal actin gene was expressed at markedly higher levels than the α-cardiac actin gene. Interestingly, CMG cells expressed the myosin light chain (MLC)-2v gene, but not the MLC-2a gene. MLC-2v is specifically expressed in ventricular cells, while MLC-2a is specifically expressed in atrial cells. Skeletal muscle cells do not express either α-MHC or MLC-2v. These results indicated that differentiated CMG cells possess the specific phenotype of the fetal ventricular cardiomyocytes.[11]

CMG Cells Have a Cardiomyocyte-Like Ultrastructure

Representative transmission electron micrographs are shown in Figure 3. A longitudinal section of the differentiated CMG myotubes clearly revealed the typical striation and pale-staining pattern of the sarcomeres.[11] CMG myotube nuclei were positioned in the center of the cell, not beneath the sarcolemma. The most conspicuous feature of the differentiated CMG myotubes was the presence of membrane-bound dense secretory granules measuring 70-130 nm in diameter. They were thought to be atrial granules, and were especially concentrated in the juxtanuclear cytoplasm, but some were also located near the sarcolemma. These findings indicated that the CMG cells possessed cardiomyocyte-like rather than skeletal muscle ultrastructure.

Developmental Stage of Undifferentiated and Differentiated CMG Cells

Various cardiac specific transcription factors have been cloned, and their genes are serially expressed in the developing heart during myogenesis and morphogenesis. Figure 4 shows the time course of the expression of cardiomyocyte-specific transcription factors in fetal developing heart and CMG cells. The genes coding Nkx2.5[15] (homeobox type transcription factor specifically expressed beginning in the early developing heart), GATA4[16] (GATA-motif-binding Zinc finger type transcription factor expressed beginning in the early stage developing heart), HAND1/2 (basic helix-loop-helix type transcription factor expressed in the heart and autonomic nervous system), and MEF2-B/C[17] (muscle enhancement factor: a MADS box family transcription factor expressed in the myocytes) were expressed in the early stage of heart development, and MEF2A and MEF2-D in the middle stage. The CMG cells already expressed GATA4, TEF-1[18] (transcription enhancement factor 2), Nkx2.5, HAND, and MEF2-C before exposure to 5-azacytidine, and they expressed MEF2-A and MEF2-D after exposure to 5-azacytidine. This pattern of gene expression in CMG cells was similar to that of developing cardiomyocytes in vivo,[11] and indicated that the developmental stage of the undifferentiated CMG cells is close to that of cardiomyoblasts or the early stages of heart development. We estimated that the stage of differentiation of the CMG cells lies between the cardiomyocyte-progenitor stage and the differentiated cardiomyocyte stage.

Serial Changes in Action Potential Shape in CMG Cells Simulate Those of Fetal Ventricular Cardiomyocytes in Vivo

CMG cells exhibit at least two types of distinguishable morphological action potentials: sinus-node-like potentials (Fig. 5A) and ventricular myocyte-like potentials (Fig. 5B).[11] The cardiomyocyte-like action potential recorded from these spontaneous beating cells is characterized by: (1) a relatively long action potential duration or plateau; (2) a relatively shallow resting membrane potential; and (3) a pacemaker-like late diastolic slow depolarization. Peak-and-dome-like morphology was observed in ventricular-myocyte-like cells. Figure 5C shows the time course of the percentages of the sinus node-like and ventricular-myocyte-like action potentials. All action potentials recorded from CMG cells until 3 weeks were sinus-node-like action potential. The ventricular-myocyte-like action potentials were first recorded after 4 weeks, and their percentage gradually increased thereafter.

Table 1. Isoforms of the contractile proteins in differentiated CMG cells

	Atrium			Ventricle		
Developmental Stage	Fetus	Adult	Fetus	Neonate	Adult	CMG
α-actin	skeletal	cardiac	skeletal>cardiac	skeletal	cardiac	skeletal>cardiac
myosin heavy chain	α>β	α	β>α	α>β	α	β>α
myosin light chain	2a	2a	2v	2v	2v	2v

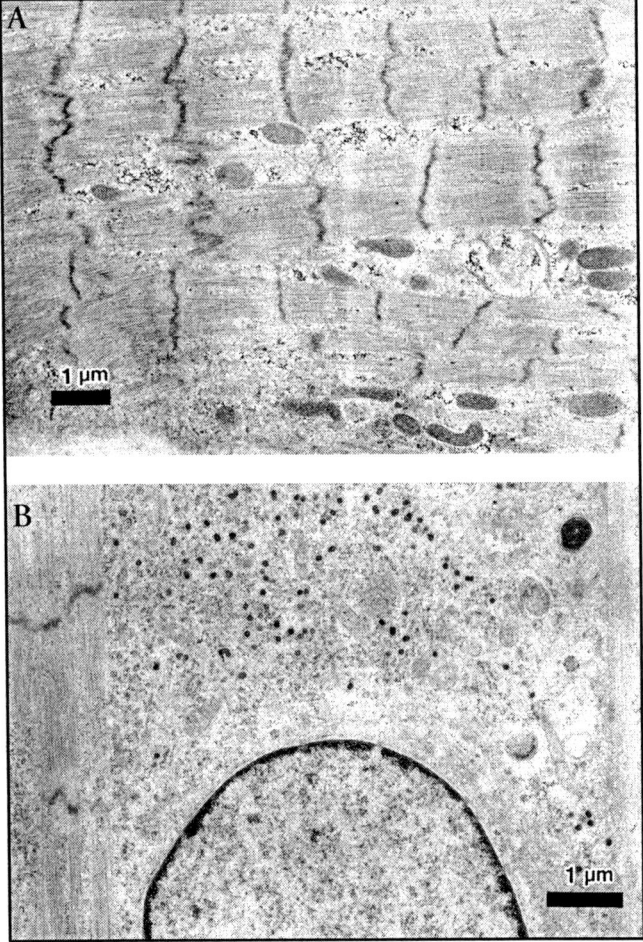

Figure 3. Transmission electron micrograph of CMG cells. A) Differentiated CMG cells had well-organized sarcomeres. Abundant glycogen granules and a number of mitochondria were observed. B) Ultrastructural analysis revealed atrial granules measuring 70-130 nm in diameter in the sarcoplasm that were especially concentrated in the juxtanuclear cytoplasm. Bar indicates 1 μm.

The observation of several distinct patterns of action potential in CMG cells may reflect different developmental stages. Yasui et al studied action potentials and the occurrence of one of the pacemaker currents, I(f), by the whole-cell voltage and current-clamp technique at the stage when a regular heartbeat is first established (9.5 days post coitum) and at 1 day before birth.[19] They showed a prominent I(f) in mouse embryonic ventricles in the early stage, and that it decreased by 82% before birth in tandem with the loss of regular spontaneous activity by the ventricular cells. They concluded that the I(f) current of the sinus node type is present in early embryonic mouse ventricular cells. Loss of the I(f) current during the second half of embryonic development is associated with a tendency for the ventricle to lose pacemaker potency. Our findings in CMG cells may reflect the developmental changes in the action potentials that occur in embryonic ventricular cardiomyocytes.

Expression of α_1- and β-Adrenergic Receptor mRNA in CMG Cells

In the heart in vivo, α and β adrenergic receptors play a key role in modulating cardiac hypertrophy and cardiac function, such as heart rate, contractility, and conduction velocity. CMG cells expressed all the α_1 receptor subtypes (α_{1A}, α_{1B}, and α_{1D}) before 5-azacytidine exposure (Fig. 6A),[12] and their expression in undifferentiated CMG cells may be explained by their ubiquitous or wide expression in vivo.[20] A low level of expression of α_{1A} was observed before 5-azacytidine exposure, and it increased markedly after exposure. Expression of α_{1B} was unaffected by 5-azacytidine. A high level of expression of α_{1D} was detected before 5-azacytidine exposure, but it decreased considerably after exposure. This transcriptional switch may be attributable to the CMG cells having acquired the cardiomyocyte phenotype. The ventricular cardiomyocytes in vivo mainly expressed α_{1A} and α_{1B}, and expressed a low level of α_{1D} receptor. The temporal changes in expression of α_1-adrenergic receptor subtypes in CMG cells are very similar to the postnatal changes observed in neonatal rat heart.[21,22]

The cardiomyocytes of the mammalian hearts express both β_1 and β_2-adrenergic receptors, the β_1 receptor being the predomi-

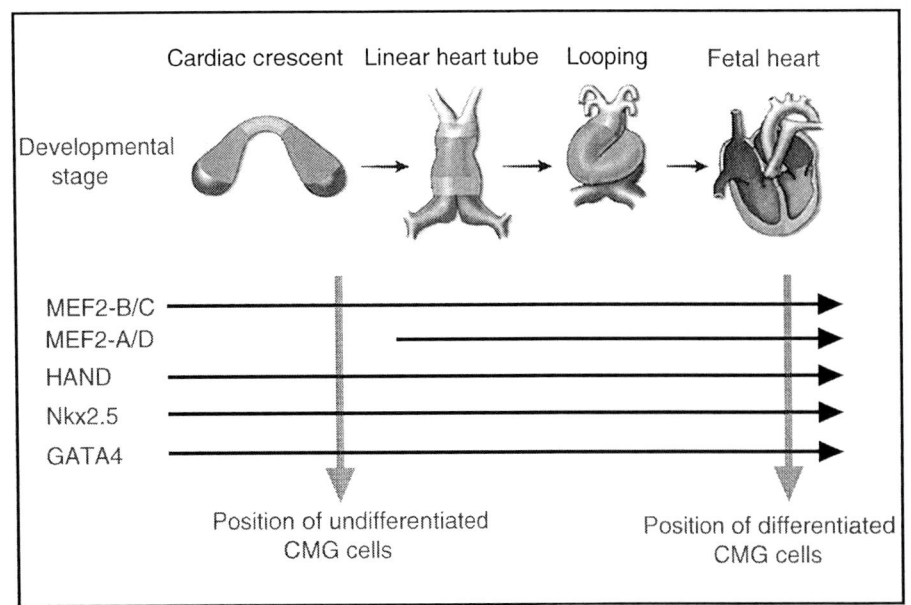

Figure 4. Expression of cardiac-specific transcription factors in the developing heart and in CMG cells. The horizontal arrows indicate the time course of the expression of cardiac-specific transcription factors in the developing fetal heart. The dotted vertical arrows indicate the expression of these factors in undifferentiated and differentiated CMG cells. CMG cells expressed MEF2-A and MEF2-D after 5-azacytidine exposure, when they acquired a cardiomyocyte phenotype.

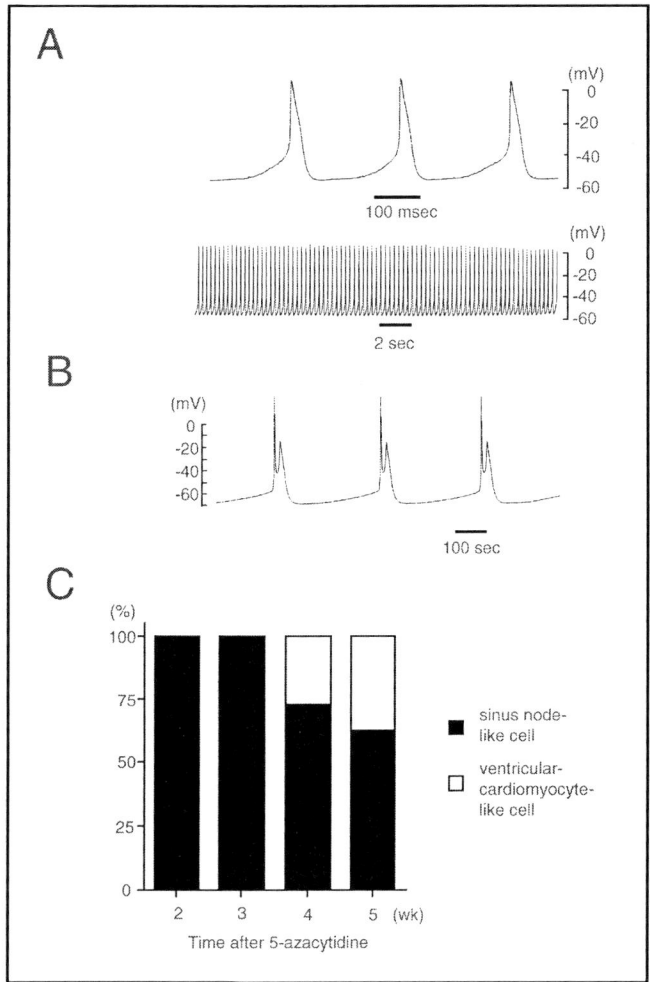

Figure 5. Representative tracing of the action potentials of CMG myotubes. Action potential recordings from spontaneous-beating cells were obtained with a conventional microelectrode at day 28 after 5-azacytidine exposure. The action potentials were classified into two groups: A) sinus-node-like action potentials and B) ventricular-cardiomyocyte-like action potentials. C) Percentages of CMG cells exhibiting sinus-node-like and ventricular-cardiomyocyte-like action potentials after 5-azacytidine exposure. A ventricular cardiomyocyte-like action potential was first recorded 4 weeks after 5-azacytidine exposure, and it rapidly became more prevalent thereafter.

nant subtype (approximately 75–80% of total β receptors).[23] CMG cells did not express β_1 and β_2 receptor transcripts before 5-azacytidine exposure, but RT-PCR showed expression of their mRNAs forward day 1 after exposure onward, and exposure, and expression was stable after 1 week (Fig. 6B).[12] CMG cells expressed β_1 and β_2 mRNA after acquiring the cardiomyocyte phenotype. The temporal pattern of expression of these receptors differed from that of α_1.

Phenylephrine Induces Activation of ERK1/2 and Hypertrophy in CMG Cells Via α_1 Receptors

ERK1/2 was activated by phenylephrine, an α_1 stimulant, within as little as 5 minutes, and the activation peaked at 10 minutes (Fig. 7A,B). The phenylephrine-induced phosphorylation was completely inhibited by prazosin (Fig. 7C), and phenylephrine

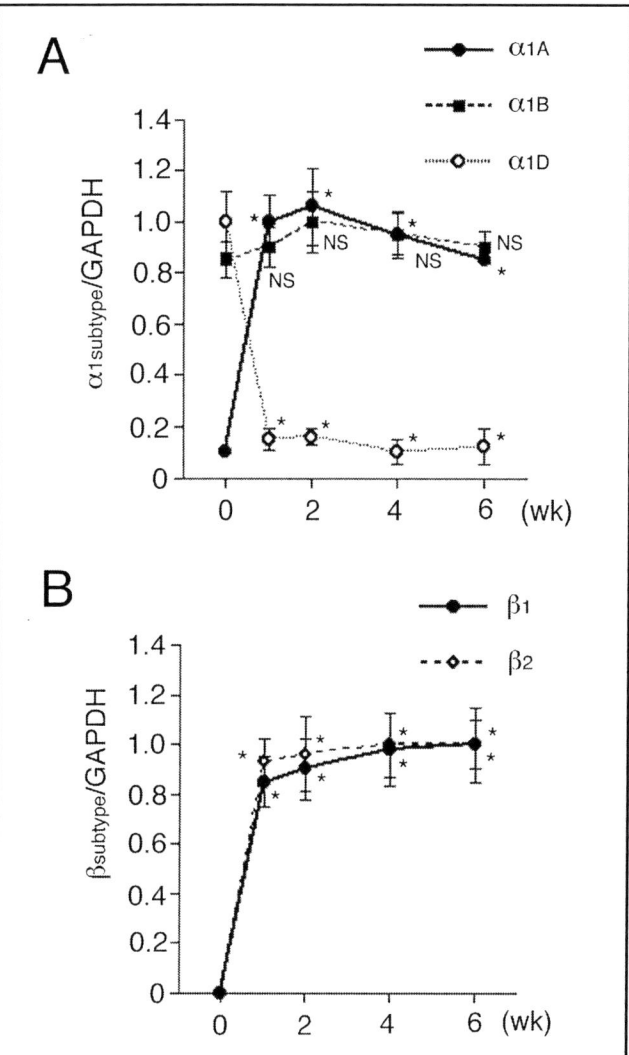

Figure 6. Temporal expression of α_1- and β-adrenergic receptor subtype messenger ribonucleic acid (mRNA) in CMG cells. A) Densitometric analysis was performed, and the ratio of the reverse transcriptase polymerase chain reaction (RT-PCR) product of α_1 subtype (α_{1A}, α_{1B}, α_{1D}) receptors to that of glyceraldehyde-3-phosphate dehydrogenase (GAPDH) is shown. Data were obtained from 5 separate experiments and are shown in arbitrary units compared to the controls. Values are mean ± SE. *=$p<0.01$ vs. controls (before 5-azacytidine exposure). NS= not significant. B) Densitometric analysis was performed, and the ratio of the RT-PCR product of β subtype (β_1 and β_2) receptors to that of GAPDH is shown.

increased the cell area and perimeter of the CMG cardiomyocytes (Fig. 7D,E). These findings indicated that CMG cells express functionally active α_1-adrenergic receptors.[12]

Isoproterenol Increases the cAMP Content, Spontaneous Beating Rate, and Contractility of CMG Cells

Isoproterenol, a β stimulant, increased the cAMP content of CMG cells, and propranolol completely inhibited the isoproterenol-induced cAMP accumulation (Fig. 8A,B). Isoproterenol was applied to the cells to determine whether it would increase the spontaneous beating rate (Table 2), and the results showed

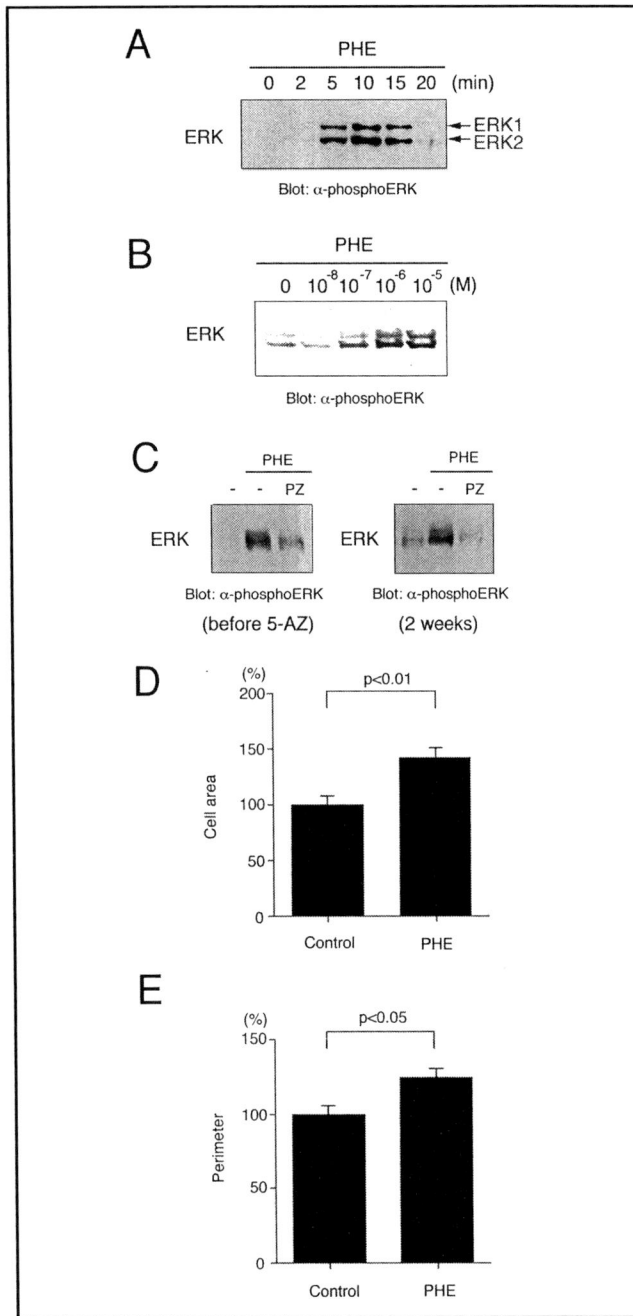

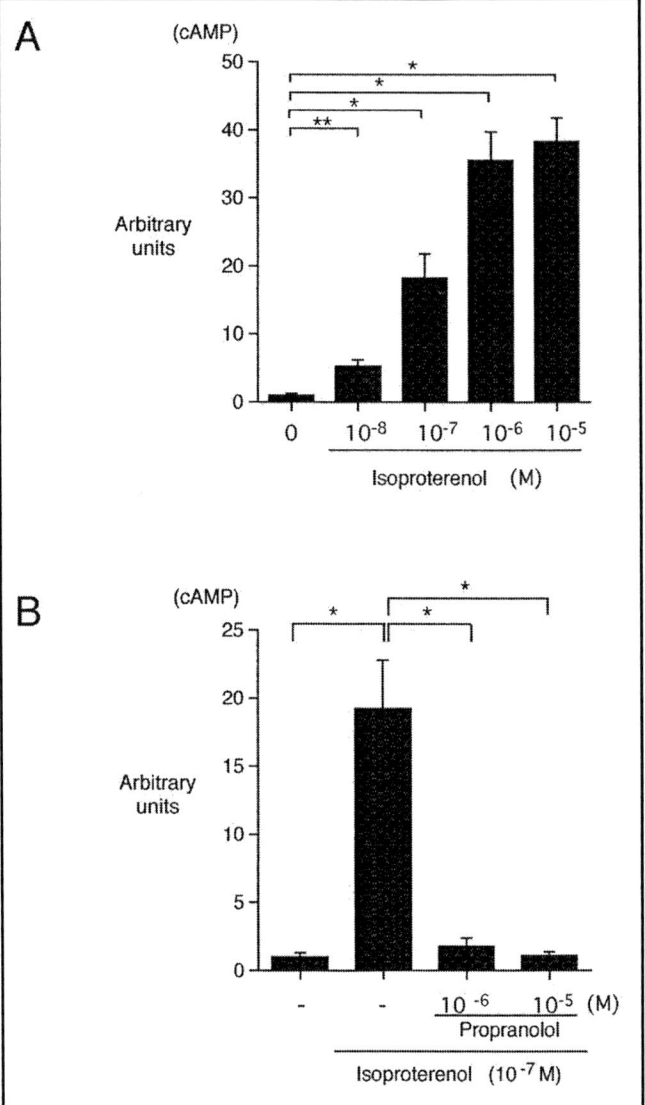

Figure 7. Effect of phenylephrine on phosphorylation of ERK1/2 and cell size in CMG cells. A-C. A) Cells at 2 weeks after 5-azacytidine exposure were stimulated with phenylephrine (PHE) (10^{-4} mol/L), and Western blot analysis was performed to detect phosphorylation of ERK1/2. B) Cells were stimulated with phenylephrine (10^{-7}-10^{-5} mol/L) for 10 minutes, and phosphorylation of ERK was detected. C) Prazosin (10^{-6} mol/L) was added to cells 20 minutes before stimulation with phenylephrine (10^{-6} mol/L). PHE= phenylephrine; PZ= prazosin. D-E. Cells were serum depleted for 24 h, stimulated with phenylephrine for 24 h, and stained with anti-sarcomeric myosin antibody. Cell area (D) and perimeter (E) were quantitated with NIH Image software. (n= 100) *= p<0.01 vs. control.

Figure 8. β receptor-mediated cyclic 3'5' adenosine monophosphate (cAMP) accumulation in CMG cells. A) Effect of isoproterenol on cAMP accumulation in CMG cells at 2 weeks after 5-azacytidine exposure. B) Cells were preincubated with propranolol (10^{-6} or 10^{-5} mol/L) for 20 minutes and stimulated with isoproterenol (10^{-7} mol/L) for 10 minutes. Data were obtained from 5 separate experiments and are shown as arbitrary units compared with the controls. *= p<0.01, **= p<0.05 vs. controls.

that it increased it significantly to 48% over the rate in the control cells.[12] Preincubation with propranolol (non-selective β blocker), or CGP20712A (β$_1$-selective blocker) strongly reduced the isoproterenol-induced increase in beating rate, and preincubation with ICI118551 (β$_2$-selective blocker) only slightly decreased the beating rate. The increase in beating rate was similar to that of adult murine cardiomyocytes and ES cell-derived cardiomyocytes.

We also investigated the effect of isoproterenol on the contractile function of CMG cells and found that it increased cell motion distance, % shortening, and contractile velocity. The isoproterenol-induced increase in contractility was almost completely inhibited by both propranolol and CGP20712A. Collectively, these results indicated that the β$_1$ and β$_2$-adenergic receptors expressed in CMG cells are functional, and that the

Table 2. Isoproterenol increased the spontaneous beating rate and contractility of CMG cells, mainly via β_1 receptors

	Control	Vehicle	Isoproterenol (10^{-7} mol/L)		
			Propranolol (10^{-7} mol/L)	CGP20712A (10^{-7} mol/L)	ICI118551 (10^{-7} mol/L)
% increase in beating rate	-	47.6±8.4*	10.0±1.9†	13.8±2.4†	37.6±1.9‡
cell motion (μm)	5.0±0.3	6.8±0.7*	5.6±0.8‡	5.3±0.6‡	ND
% shortening (%)	6.9±0.5	8.5±1.2*	7.2±0.8‡	5.6±0.6‡	ND
contractile velocity (μm/s)	71.1±5.2	100.9±11.0*	71.3±8.8‡	70.6±6.6‡	ND

CMG cells at 4 weeks after 5-azacytidine exposure were initially exposed to prazosin (10^{-6} mol/L) for 30 minutes to block α_1-adrenergic receptors. Cells were then preincubated for 20 minutes with vehicle (PBS), propranolol, CGP20712A, or ICI118551, and then stimulated with isoproterenol. The beating rate was counted 3 minutes after stimulation. Contractile parameters were analyzed 90 seconds after stimulation. Each contractile parameter value was calculated as the mean of 3 randomly selected beats in one cell. PBS was added to the control. Values are means ± SE (n= 100, each). *= p<0.05 vs. control; †= p<0.01 vs. vehicle (isoproterenol only); ‡= p<0.05 vs. vehicle; ND= not determined.

isoproterenol-induced increase in spontaneous beating rate and contractility is mainly mediated by β_1 receptors. The β_1 receptor was the predominant subtype that mediated changes in the beating rate in CMG cells, and the beating rate and the contractility were significantly increased by isoproterenol, and completely inhibited by propranolol and CGP20712A. β_1-Receptors played a critical role in mediating the isoproterenol-induced signaling in differentiated CMG cells. This expression pattern was consistent with that of cardiomyocytes in vivo.

Phenylephrine and Isoproterenol Induce Atrial Natriuretic Peptide (ANP) and Brain Natriuretic Peptide (BNP) mRNA Expression

Hypertrophic stimuli are well known to induce reprogramming of gene expression in cardiomyocytes. Phenylephrine and isoproterenol significantly induced expression of the ANP (24 hour) gene, and they also induced the BNP (1 hour) gene (Fig. 9). These findings demonstrated that α and β adrenergic signal transduction systems in CMG cells are linked to the gene expression that induces cardiac hypertrophy.

CMG Cells Express Muscarinic Receptor mRNA after 5-Azacytidine Exposure

Heart rate, conduction velocity, and contractility were negatively regulated by the parasympathetic nervous system in cardiomyocytes, and muscarinic (cholinergic) receptors play an important role in mediating this function. To date, 5 subtypes (M_1-M_5) of muscarinic receptors have been cloned. The expression of the muscarinic receptors is tissue-specific, and cardiomyocytes mainly express M_2 receptors in the mouse and human.[22] The M_1 receptor subtype is also expressed in murine neonatal and adult cardiomyocytes. Figure 10A shows the temporal expression pattern of M_1 and M_2 receptor mRNA. Neither receptor was detected prior to 5-azacytidine exposure. CMG cells began to express these receptors when they acquired the cardiomyocyte phenotype.

M_1 receptors coupled to Gq/G_{11} and activated phospholipase Cβ via Gqα, leading to inositol triphosphate (IP_3) production, and M_2 receptors coupled to Gi/G_0/Gz and activated phospholipase Cβ via Gi$\beta\gamma$, leading to IP_3 production.[25,26] Carbachol, an acetylcholine homologue, increased the content of a second messenger, IP_3 (inositol triphosphate), in CMG cells (Fig. 10B), and preincubation with atropine (non-selective muscarinic blocker) and AFDX116 (M_2-selective blocker) inhibited the carbachol-induced IP_3 production (Fig. 10C). These findings indicated that muscarinic receptors can transduce their signals, and that M_2 receptors play a critical role in this carbachol-induced IP_3 production in CMG cells. This expression pattern is similar to that of cardiomyocytes in vivo.

Significance of Expression of Adrenergic and Muscarinic Receptors in CMG Cells

Cardiomyocytes in vivo respond to stimulation by both sympathetic and parasympathetic nerves, and such stimulation alters the heart rate, conduction velocity, and contractility, enabling the cells to adapt to rapid changes in systemic oxygen demand. To date, and to our knowledge, ES cells and mesenchymal-stem-cell-derived CMG cells are the only possible candidates for regeneration of cardiomyocytes. We have already transplanted these cells into normal adult mouse hearts, and have observed that transplanted cells survived in recipient hearts for at least several weeks. Regenerated cardiomyocytes must express functional adrenergic and muscarinic receptors to be useful for transplantation, and although we did not investigate all signaling pathways and their functions, CMG cells are potential candidates for cardiomyocyte cell transplantation, because they possess such receptors.

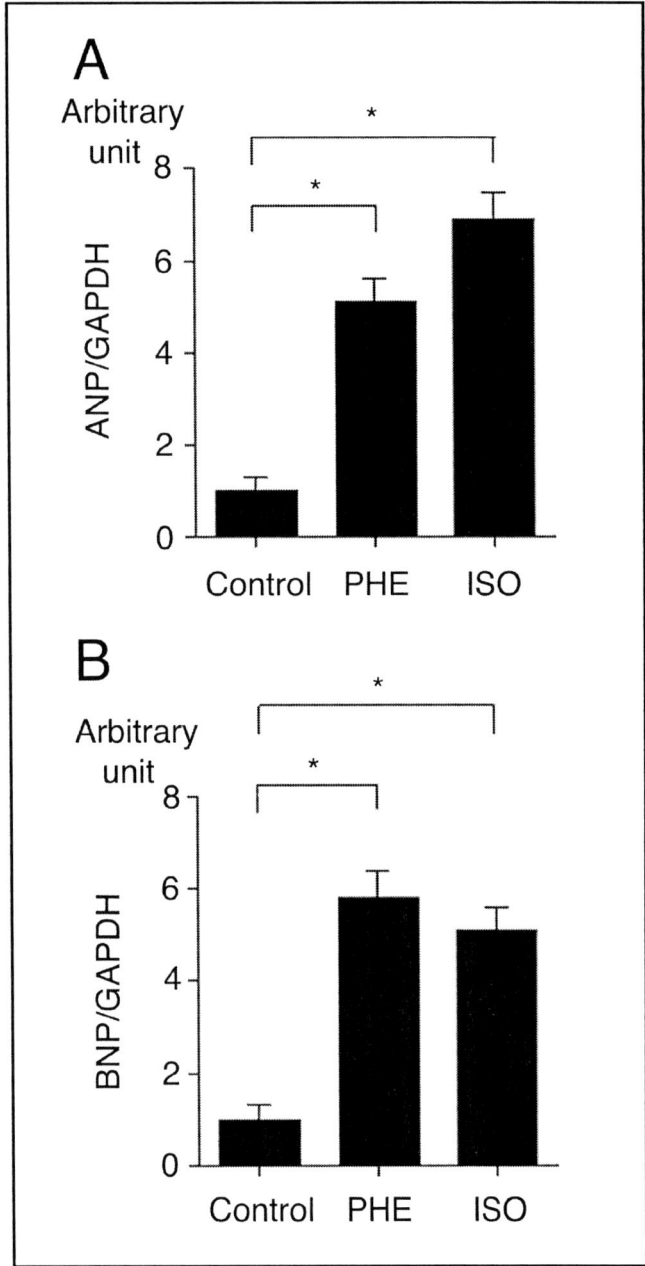

Figure 9. Both α and β stimulation induced mRNA expression of atrial natriuretic peptide (ANP) and brain natriuretic peptide (BNP) genes. CMG cells were serum depleted for 24 hours and pretreated with propranolol and stimulated with phenylephrine (PHE) (50 µM) or isoproterenol (ISO) (100 µM). RNA was extracted for 1 hour (BNP) and 24 hours (ANP), and reverse transcriptase polymerase chain reaction (RT-PCR) was examined. All values were normalized to glyceraldehyde-3-phosphate dehydrogenase (GAPDH). *= p<0.01 vs. control.

Cell Transplantation Therapy for the Treatment of Heart Failure

We have already transplanted CMG cells into normal adult mouse hearts, and observed that the transplanted cells could survive in the recipient heart for at least several months. Fibroblasts, smooth muscle cells, and skeletal muscle cells were the first cells used for transplantation into scar tissue secondary to experimental myocardial infarction in the heart in vivo. While transplantation

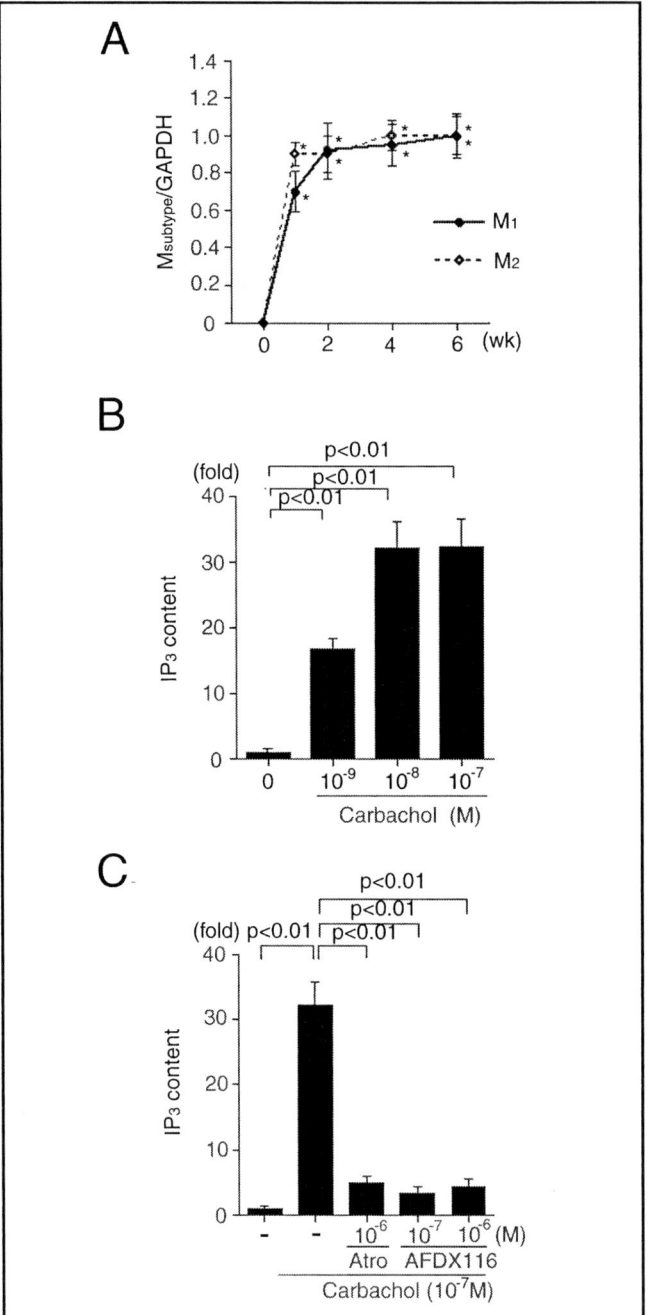

Figure 10. Expression and function of M_1- and M_2-muscarinic receptors in CMG cells. A) The ratio of the reverse transcriptase polymerase chain reaction (RT-PCR) product of muscarinic subtype to that of glyceraldehyde-3-phosphate dehydrogenase (GAPDH) is shown. Data were obtained from 5 separate experiments and are shown as arbitrary units over controls. *= p<0.01 vs. controls. B) Effect of carbachol on inositol triphosphate (IP_3) production in CMG cells at 2 weeks after 5-azacytidine exposure. C) Effect of atropine (10^{-6} mol/L) and AFDX116 (M_2-selective blocker) (10^{-7} or 10^{-6} mol/L) on carbachol-induced IP_3 production. Data were obtained from 5 separate experiments and are shown as arbitrary units compared with the controls. *=p<0.01 vs. controls. Atro= atropine.

of these cells into scar tissue might improve cardiac remodeling or diastolic function, it is unlikely to improve systolic function. Transplantation of cardiomyocytes, however, might rescue systolic function. The only potential sources of regenerated cardiomyocytes available to date are embryonic stem (ES) cells and mesenchymal stem cells. ES cells differentiate into cardiomyocytes in vitro and have both advantages and disadvantages for cardiomyocyte regeneration. Transplanted ES cells may form teratomas if some undifferentiated totipotent cells are still present, and recipients must take immunosuppressants, because ES cells are allogeneic. By contrast, since mesenchymal stem cells do not carry any inherent risks of tumor formation and are syngeneic, it is reasonable to use autologous mesenchymal stem cells to treat heart disease. Nevertheless, there is a need to improve both the current methods for identification and culture of mesenchymal stem cells, and for induction of CMG cell differentiation, which are still inefficient and slow. Identification of specific growth factors, cytokines, or extracellular matrix factors that regulate cardiomyocyte differentiation may help to accelerate this process faster and make it more efficient.

In Vivo Evidence that Marrow Cells Can Generate Functional Cardiac Tissues

Recent studies have revealed that bone-marrow-derived cells differentiate into various types of cells in vivo. Shimizu et al reported that smooth-muscle-like cells (SMCs) in graft-vs-host arterial lesions could arise from circulating bone-marrow-derived precursors. They used murine aortic transplants to formally identify the source of SMCs in lesions in grafted arteries.[27] Allografts in beta-galactosidase transgenic recipients showed that intimal SMCs arose almost exclusively from host cells, and bone-marrow transplantation of beta-galactosidase-expressing cells into aortic allograft recipients demonstrated that the intimal cells included those of marrow origin.

Kocher et al showed that bone marrow from adult humans contains endothelial precursors with phenotypic and functional characteristics of embryonic hemangioblasts and that they can be used to directly induce new blood vessel formation in the infarct-bed (vasculogenesis) and proliferation of preexisting vasculature (angiogenesis) after experimental myocardial infarction.[28] The neoangiogenesis resulted in decreased apoptosis of hypertrophied myocytes in the peri-infarct region, long-term salvage and survival of viable myocardium, reduction in collagen deposition, and sustained improvement in cardiac function.

We also observed that transplanted bone marrow cells differentiated into cardiomyocytes in the recipient heart in vivo (unpublished observation). These findings provided direct evidence that bone marrow cells can regenerate various types of cells in cardiac tissue. We expect cardiac tissues damaged by myocardial infarction or other diseases to be repaired by bone-marrow-derived stem cells in the near future, and the precise mechanism should be investigated to achieve this goal.

References

1. Soonpaa MH, Koh GY, Klug MG et al. Formation of nascent intercalated disks between grafted fetal cardiomyocytes and host myocardium. Science 1997; 264:98-101.
2. Delcarpio JB, Claycomb WC. Cardiomyocyte transfer into the mammalian heart. Cell-to-cell interactions in vivo and in vitro. Ann NY Acad Sci 1997; 52:267-285.
3. Wobus AM, Wallukat G, Hescheler J. Pluripotent mouse embryonic stem cells are able to differentiate into cardiomyocytes expressing chronotropic responses to adrenergic and cholinergic agents and Ca^{2+} channel blockers. Differentiation 1991; 48:173-182.
4. Wobus AM, Kleppisch T, Maltsev V et al. In: Cardiomyocyte-like cells differentiated in vitro from embryonic carcinoma cells P19 are characterized by functional expression of adrenoceptors and Ca^{2+} channels. Vitro Cell Dev Biol Anim 1994; 30:425-434.
5. Roy NS, Wang S, Jiang L et al. In vitro neurogenesis by progenitor cells isolated from the adult human hippocampus. Nat Med 2000; 6:271-277.
6. Prockop DJ. Marrow stromal cells as stem cells for nonhematopoietic tissues. Science 1997; 276:71-74.
7. Rickard DJ, Sullivan TA, Shenker BJ et al. Induction of rapid osteoblast differentiation in rat bone marrow stromal cell cultures by dexamethasone and BMP-2. Dev Bio 1994; 161:218-228.
8. Friedenstein AJ, Chailakhyan R, Gerasimov UV. Bone marrow osteogenic stem cells: In vitro cultivation and transplantation in diffusion chambers. Cell Tissue Kinet 1987; 20:263-272.
9. Ferrari G, Angelis GC, Colleta M et al. Muscle regeneration by bone marrow-derived myogenic progenitors. Science 1998; 279:1528-1530.
10. Ashton BA, Allen TD, Howlett CR et al. Formation of bone and cartilage by marrow stromal cells in diffusion chambers in vivo. Clin Orthop 1980; 151:294-307.
11. Makino S, Fukuda K, Miyoshi S et al. Cardiomyocytes can be generated from marrow stromal cells in vitro. J Clin Invest 1999; 103:697-705.
12. Hakuno D, Fukuda K, Makino S et al. Bone marrow-derived cardiomyocytes (CMG cell) expressed functionally active adrenergic and muscarinic receptors. Circulation 2002; 15:380-386.
13. Beltrami AP, Urbanek K, Kajstura J et al. Evidence that human cardiac myocytes divide after myocardial infarction. N Engl J Med 2001; 344:1750-7.
14. Fukuda K. Development of regenerative cardiomyocytes from mesenchymal stem cells for cardiovascular tissue engineering. Artificial Organs 2001; 25:183-193.
15. Linnets TJ, Parsons LM, Harley L et al. Nkx-2.5: A novel murine homeboy gene expressed in early heart progenitor cells and their myogenic descendants. Development 1993; 119:419-431.
16. Arceci RJ, King AA, Simon MC et al. Mouse GATA-4: A retinoic acid-inducible GATA-binding transcription factor expressed in endodermally derived tissues and heart. Mol Cell Biol 1993; 13:2235-2246.
17. Edmondson DG, Lyons GE, Martin JF et al. Mef2 gene expression marks the cardiac and skeletal muscle lineages during mouse embryogenesis. Development 1994; 120:1251-1263.
18. Chen Z, Friedrich GA, Soriano P. Transcriptional enhancer factor 1 disruption by a retroviral gene trap leads to heart defects and embryonic lethality in mice. Genes Dev 1994; 8:2293-2301.
19. Yasui K, Liu W, Opthof T et al. I(f) current and spontaneous activity in mouse embryonic ventricular myocytes. Circ Res 2001; 88: 536-42
20. Alonso-Llamazares A, Zamanillo D, Casanova E et al. Molecular cloning of alpha 1d-adrenergic receptor and tissue distribution of three alpha 1-adrenergic receptor subtypes in mouse. J Neurochem 2001; 65:2387-2392.
21. Stewart AF, Rokosh DG, Bailey BA et al. Cloning of the rat alpha 1C-adrenergic receptor from cardiac myocytes. alpha 1C, alpha 1B, and alpha 1D mRNAs are present in cardiac myocytes but not in cardiac fibroblasts. Circ Res 2001; 75:796-802.
22. Rokosh DG, Stewart AF, Chang KC et al. Alpha1-adrenergic receptor subtype mRNAs are differentially regulated by alpha1-adrenergic and other hypertrophic stimuli in cardiac myocytes in culture and in vivo. Repression of alpha1B and alpha1D but induction of alpha1C. J Biol Chem 1996; 271:5839-5843.
23. Rockman HA, Koch WJ, Lefkowitz RJ. Cardiac function in genetically engineered mice with altered adrenergic receptor signaling. Am J Physio 1997; 272:H1553-H1559.

24. Sharma VK, Colecraft HM, Rubin LE et al. Does mammalian heart contain only the M2 muscarinic receptor subtype? Life Sci 1997; 60:1023-1029.
25. Nakamura F, Kato M, Kameyama K et al. Characterization of Gq family G proteins GL1 alpha (G14 alpha), GL2 alpha (G11 alpha), and Gq alpha expressed in the baculovirus-insect cell system. J Biol Chem 1995; 270:6246-6253.
26. Berstein G, Blank JL, Smrcka AV et al. Reconstitution of agonist-stimulated phosphatidylinositol 4,5-bisphosphate hydrolysis using purified m1 muscarinic receptor, Gq/11, and phospholipase C-beta 1. J Biol Chem 1995; 267:8081-8088.
27. Shimizu K, Sugiyama S, Aikawa M et al. Host bone-marrow cells are a source of donor intimal smooth- muscle-like cells in murine aortic transplant arteriopathy. Nat Med 2001; 7:738-741.
28. Kocher AA, Schuster MD, Szabolcs MJ et al. Neovascularization of ischemic myocardium by human bone-marrow-derived angioblasts prevents cardiomyocyte apoptosis, reduces remodeling and improves cardiac function. Nat Med 2001; 7:430-436.

CHAPTER 9

Clinical Trials of Hematopoietic Stem Cells for Cardiac and Peripheral Vascular Diseases

Hiroaki Matsubara

Introduction

Blood vessels are primarily composed of two cell types: endothelial cells, lining the inside and smooth muscle cells, covering the outside. While angiogenesis research has generally been focused on these two vascular cell types, recent evidence indicates that the bone marrow may also contribute to this process, both in the embryo and the adult. Following commitment to the endothelial lineage, marrow-derived angioblasts assemble into a primitive vascular plexus of veins and arteries, a process called vasculogenesis. This primitive vasculature is subsequently refined into a functional network by angiogenesis (vascular sprouting from preexisting vessels, vascular fusion and intussusception) and by remodeling and muscularization (arteriogenesis) of newly formed vessels.[1]

Preclinical studies have indicated that angiogenic growth factors promote the development of collateral arteries, a concept called "therapeutic angiogenesis".[2] Angiogenesis can be achieved either by the use of growth factor proteins or genes encoding these proteins. Limited clinical data from protein- and gene-delivery trials suggested that both approaches are safe. However, a great deal more clinical experience will be necessary to resolve the safety concerns such as potentiation of pathological angiogenesis (e.g., malignancy) and "bystander" effects of the delivered factors (e.g., effects on kidney or atheroma).[3] Given the investment of the formed mature vessels with periendothelial matrix and pericyte/smooth muscle cells, combinations of various angiogenic growth factors may be preferable in future therapies.[4]

We discovered that endothelial progenitor cells (EPC) in the CD34+ stem cell fraction of adult human peripheral blood participate in postnatal neovascularization after mobilization from the bone marrow.[5-7] We[7] and Kalka et al[8] reported that mononuclear cells from adult human peripheral or cord blood improved capillary density in hindlimb ischemia. Marrow stromal cells have many of the characteristics of stem cells for mesenchymal tissues, and also secrete a broad spectrum of angiogenic cytokines,[9] raising the possibility that marrow implantation into ischemic limbs effectively enhances angiogenesis by supplying EPC as well as angiogenic cytokines or factors. We have demonstrated in animal studies that bone marrow mononuclear cell (BM-MNC) implantation into ischemic limbs[7] or myocardium[10,11] promotes collateral vessel formation with incorporation of EPC into new capillaries, and that local concentrations of angiogenic factors such as basic fibroblast growth factor (bFGF), vascular endothelial growth factor (VEGF), and angiopoietin-1 or angiogenic cytokines interleukin-1 (IL-1) and tumor necrosis factor alpha (TNFα) were increased in implanted tissues. Neither tissue injury by inflammatory cytokines released from injected cells nor differentiation into other lineage cells such as osteoblasts or fibroblasts was observed in implanted ischemic tissues. Side effects, such as an increase in cardiac enzymes, malignant arrhythmia or differentiation into other lineage cells, were not observed in the BM-MNC-implanted myocardium. On the basis of these animal studies, we have started a clinical trial to test the therapeutic availability by cell therapy using autologous BM-MNC in patients with ischemic limbs or ischemic myocardium. This review summarizes the results of a clinical trial of angiogenic cell therapy by implantation of bone marrow cells or peripheral blood-derived cells.

Therapeutic Angiogenesis for Ischemic Limbs by Autologous Transplantation of Bone Marrow Cells

CD34+ hematopoietic stem cells include EPC and play a key role in neocapillary formation.[4,5] Although CD34- cells enhance CD34+ cells-mediated angiogenesis,[4] the underlying mechanism remains undefined. As BM-MNC contained both CD34+ and CD34- cells, we hypothesized that these cell fractions may release angiogenic factors to enhance angiogenesis in addition to the supply of EPC. Interestingly, we found that CD34- rather than CD34+ cells mainly expressed basic fibroblast growth factor (bFGF) >> vascular endothelial growth factor (VEGF) > angiopoietin-1 mRNAs but not angiopoietin-2, while CD34+ cells predominantly expressed their receptors (Fig. 1), suggesting that BM-MNC have the natural ability of the marrow cells to supply EPC as well as to secrete various angiogenic factors or cytokines. We have reported that autologous marrow implantation in rabbit or rat ischemic limbs augmented collateral vessel formation, and the implanted cells were incorporated into neocapillaries.[7,12] Neither tissue injury by inflammatory cytokines released from injected cells nor differentiation into other lineage cells such as osteoblasts or fibroblasts was observed in implanted ischemic tissues.[7,10] On the

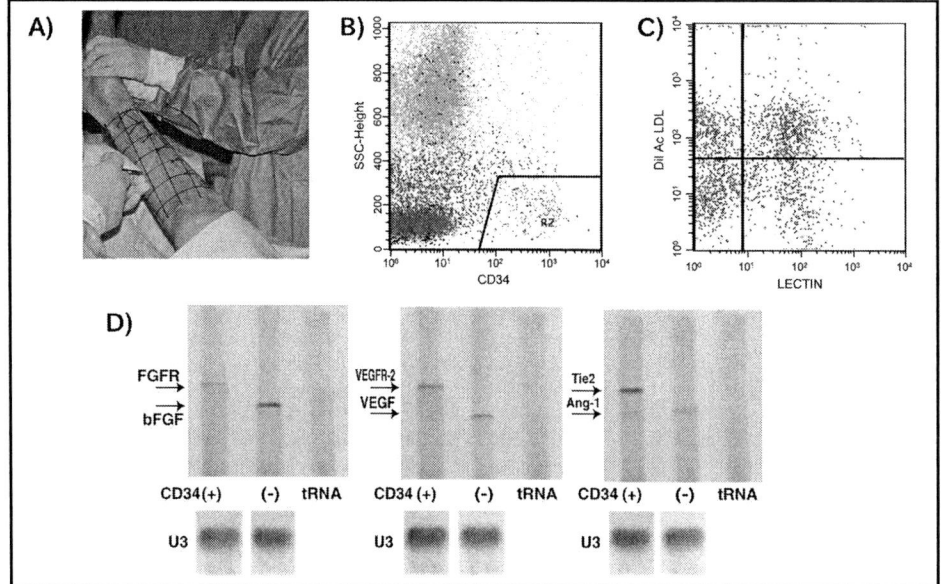

Figure 1. Identification of endothelial lineage cells and expression of angiogenic factors in autologous bone marrow mononuclear cells (BM-MNC). A) Injection of autologous bone marrow mononuclear cells (BM-MNCs) into ischemic muscle; B) FACS analysis showing CD34 population in BM-MNCs; C) Endothelial-lineage cells (assessd by Lectin-binding and low density lipoprotein (LDL) uptake) in CD34 positive fraction; D) RNase protection assay in BM-MNCs-derived CD34 positive and negative fractions.

basis of these animal studies, we have started a clinical trial to test the therapeutic safety and efficiency of cell therapy using autologous BM-MNC in patients with ischemic limbs.[13]

Patients qualified for marrow implantation if they had chronic critical limb ischemia including rest pain and/or nonhealing ischemic ulcers and were not candidates for nonsurgical or surgical revascularization. Requisite hemodynamic deficits included a resting ankle-brachial pressure index (ABI) <0.6 in the affected limb on two consecutive examinations performed at least 1 week apart. Patients with poorly controlled diabetes mellitus (HbA1c>6.5 and proliferative retinopathy) or with evidence of malignant neoplasm (during last 5 years) were excluded. Patients (n=45) were treated on a multicenter, randomised, double-blind trial; BM-MNC (active treatment) or PB-MNC (control cells) were randomly, double-blindly implanted into right or left ischemic limbs. We used peripheral blood mononuclear cells (PB-MNC) as a more appropriate cell control than saline, since peripheral blood was partially contaminated during marrow aspiration procedure (~10% of marrow cells) and the number of CD34$^+$ cells including EPC was ~500-fold less in PB-MNC than that in BM-MNC.

We obtained ~500 ml of marrow cells from each patient and isolated 2.8×10^9 to 0.7×10^9 MNC (1.6×10^9 cells [SD 0.6]), which included 9.6×10^7 to 0.84×10^7 CD34$^+$ cells (3.7×10^7 [SD 1.8]). The BM-MNC were implanted by intramuscular injection into the gastrocnemius muscle of ischemic lower limbs (x40 sites) using a 26-gauge needle. Ankle-brachial index (ABI) values in BM-MNC-implanted limbs were increased by 0.1 (95% CI 0.07 to 0.12) from 0.37 [0.31 to 0.42] at baseline to 0.46 [0.40 to 0.52] at week 4 (P<0.0001). In contrast, PB-MNC-injected limbs showed much smaller increases in ABI (0.02 increase, 95% CI 0.01 to 0.024). In the follow-up study, improvement of ischemic status (ABI, tissue oxygen concentration, rest pain scale, pain-free walking time, improvement of ischemic ulcers) was maintained during 24-weeks of observation (Fig. 2) (see detail data in ref. 13). Angiography revealed a marked increase in visible collateral vessel numbers in 60% of patients (Fig. 3). Vessel numbers assessed by capillary/muscle fiber ratio (2.3 [SD 0.6]) were markedly increased compared with the contralateral saline-injected muscle (0.74 [0.31]) (Fig. 4A). CD31-positive endothelial cells express Ki-67 in the marrow-implanted limb (Fig. 4B). Ki-67 is a nuclear protein that is expressed in proliferating cells and nearly absent in normal vessels. No Ki-67 expression was detected in the saline-injected limb, suggesting the presence of proliferating endothelial cells in newly formed vessels.

In considering the clinical potential of therapeutic angiogenesis, it is important to determine whether growth of new capillaries (vasculogenesis) or of pre-existing collateral (angiogenesis) is the therapeutic goal. For newly formed vessels to survive, they must be remodeled and acquire a smooth muscle coat.[14] Given the complexity of vessel formation, therapies using a single angiogenic factor may produce incompletely functioning endothelial channels.[3] We found that CD34$^-$ fraction in BM-MNC

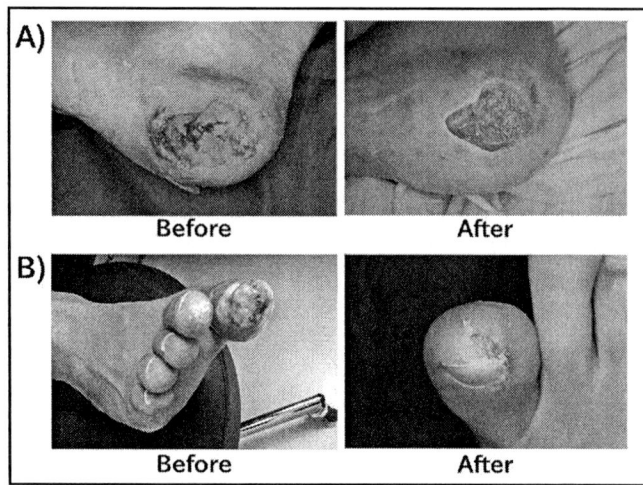

Figure 2. Limb salvage after marrow implantation.

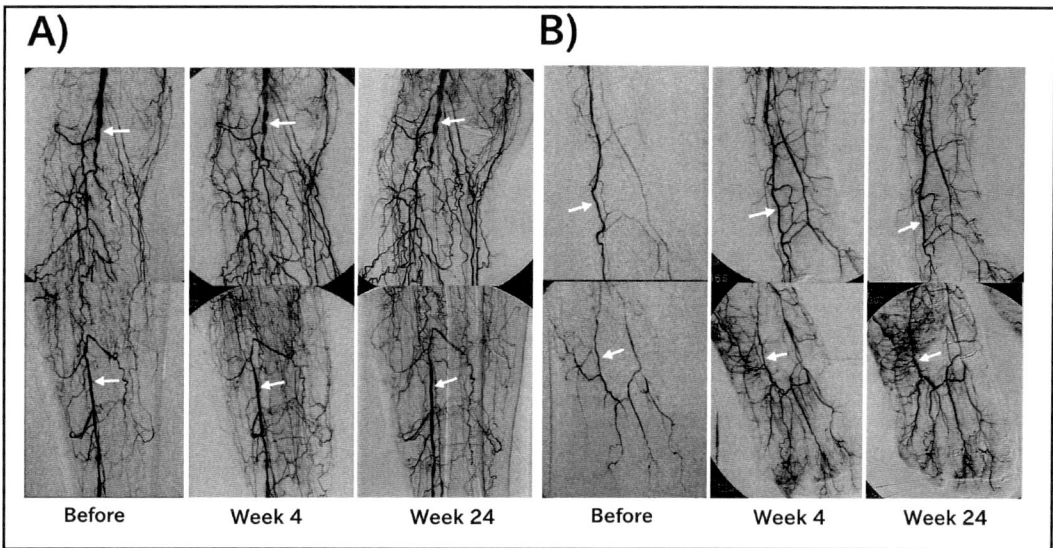

Figure 3. Angiographic analysis of collateral vessel formation. Angiographically visible collateral branches were markedly increased at the knee and upper-tibia levels (panel A), lower-tibia, ankle and foot levels (panel B) 4 and 24 weeks after bone marrow mononuclear cell (BM-MNC) implantation. Note that contrast densities in supra-femoral, posterior-tibial and dorsal pedal arteries (indicated by arrows) are similar in before and after figures.

synthesized not only angiogenic growth factors (VEGF and bFGF), but also angiopoietin-1 known to have important functions in maturation and maintenance of the vascular system. Recently, infusion of EPC or $CD34^+$ cells was shown to induce angiogenesis in ischemic limbs.[6-7] Thus, it is concluded that the efficacy of BM-MNC implantation therapy is due to the supply of EPC (included in $CD34^+$ fraction) as well as multiple angiogenic factors (released from $CD34^-$ fraction). This combined

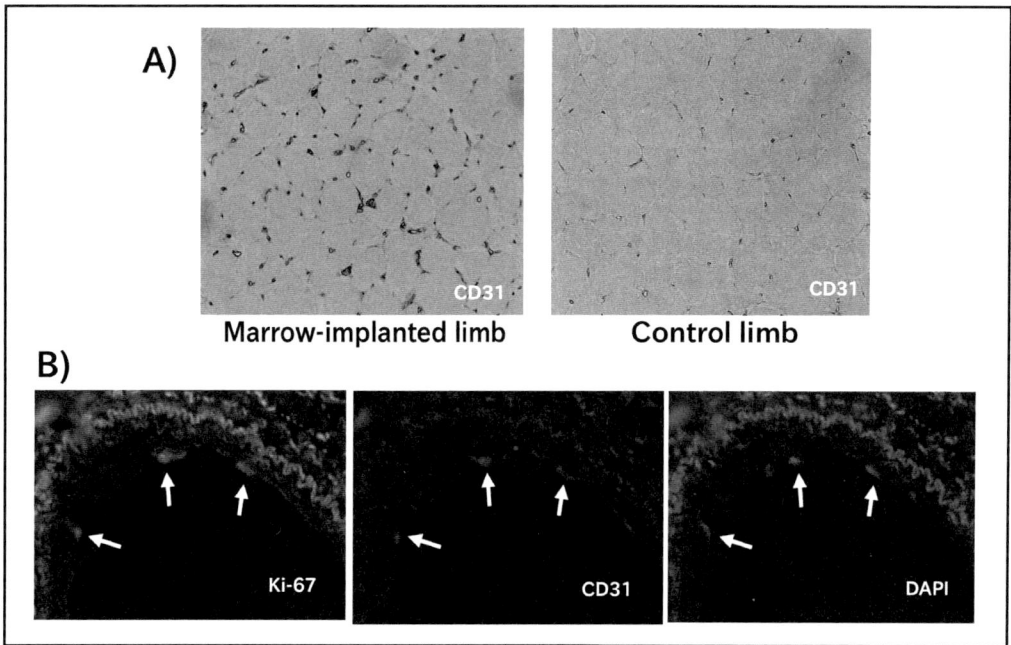

Figure 4. Immunohistochemical analysis of new proliferating vessel formation. Panel A) Vessel formation in specimens of bone marrow mononuclear cell (BM-MNC)-implanted limb (gastrocnemius) was analyzed with anti-CD31 antibody (DAKO) specific to vascular endothelial cells. Staining was viewed by immunoperoxidase reaction using diaminobenzidine. CD31-positive vessels (dark staining) were markedly increased in BM-MNC-implanted limb compared with those in saline-injected control limb (200x). Panel B) Sections from BM-MNC-implanted limb were double-immunostained with mouse monoclonal anti-CD31 antibody (DAKO) and rabbit polyclonal anti-Ki-67 antibody (DAKO), and counterstained with DNA binding dye 4', 6-diamidino-2-phenylindole (DAPI) (Molecular Probe), followed by incubation with Fluorescein-5-isothiocyanate (FITC)-conjugated anti-rabbit antibody or tetramethyl rhodamine isothiocyanate (TRITC)-conjugated anti-mouse secondary antibody (400x). Arrows indicate Ki-67- and CD31-positive endothelial cells.

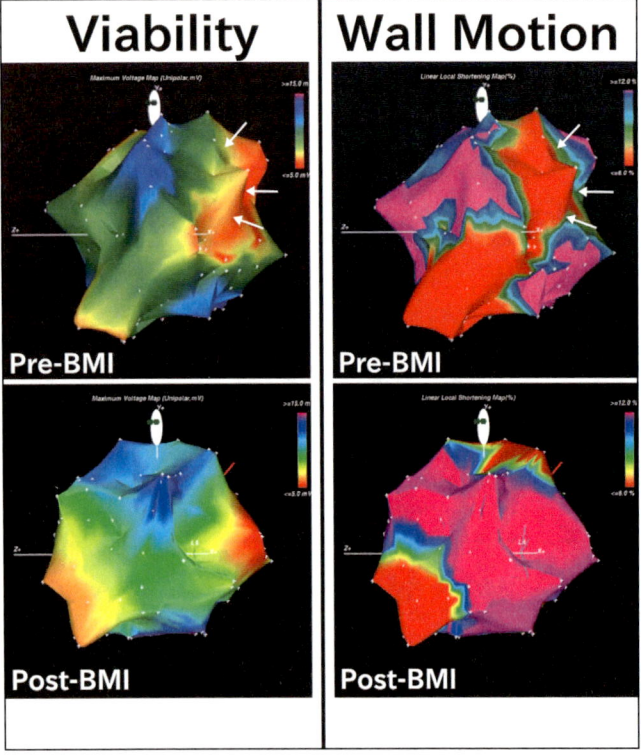

Figure 5. NOGA electromechanical mapping performed in 65-year old man. NOGA images in left lateral projection before BM-MNC implantation (pre-BMI) show myocardial viability and wall motion; red zone (which depicts abnormal wall motion) on linear local shortening map (Wall Motion, top right), together with observed viability (green/yellow in Viability, top left) on unipolar voltage map, constitute focus of electromechanical uncoupling that suggests ischemic hibernating myocardium of posterolateral region. BM-MNC were injected into hibernating area indicated by arrows. Same projection 4 months after BMI (post-BMI) discloses almost complete resolution of wall motion in hibernating region (shift from red zone to purple zone), corresponding to changes observed on SPECT scan (Fig. 2A). Vertical and horizontal axes are presented by red and white lines, respectively.

therapy may lead to the formation of stable capillary vessels, as suggested in the findings that improvement of limb ischemia was sustained during 1 year follow-up.

Therapeutic Angiogenesis for Ischemic Limbs by Peripheral Blood (PB)-Derived Mononuclear Cells

PBMNC,[15] polymorphonuclear leukocytes (PMN)[16] and platelets[17] synthesize and release high levels of VEGF as well as platelet derived growth factor (PDGF) and transforming growth factor beta (TGF-β) known to be prerequisites for investment of stable vessels with pericytes.[18] We have used PB-derived mononuclear cells as a control injection cell and compared its angiogenic activity with that of BM-MNC in patients with ischemic limbs.[13] Although increase in limb blood pressure was significantly lower than BM-MNC, we found moderate improvement in ischemic status, implying the potential angiogenic effect by PB-MNC implantation. Our study using immunodeficient nude rats demonstrated that intramuscular implantation of human PB-MNC enhanced neocapillary formation in a PB-MNC-

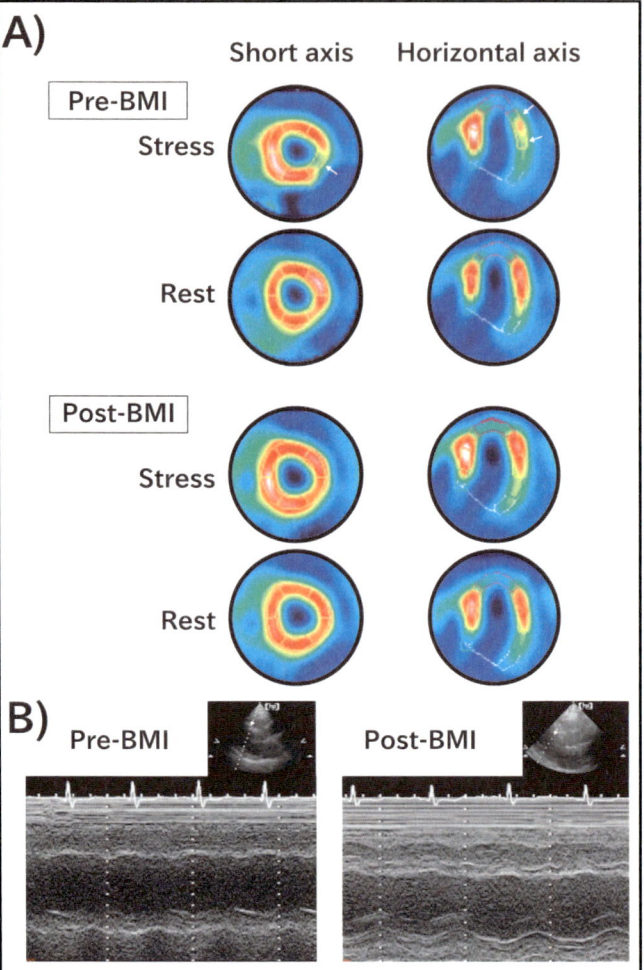

Figure 6. SPECT-sestamibi scan analysis. Panel A) Persantine SPECT-sestamibi scans recorded before and after cell implantation. White color depicts maximal uptake of radionuclide uptake. Scans before cell implantation pre BM-MNC implantation (pre-BMI) show reversible posterolateral wall defect (indicated by arrows). Scans 4 months after cell implantation (post-BMI) show complete normalization of posterolateral wall perfusion after pharmacological stress. Panel B) M-mode echocardiograms before BM-MNC implantation (pre-BMI) and after BM-MNC implantation (post-BMI). An increase in left ventricular (LV) posterior wall contractility occurred after cell implantation. Echocardiography showed the improvement of systolic function in posterior wall (fractional shortening: 19 to 33%) and decrease in LV end-diastolic dimension (from 61 to 43 mm).

derived VEGF-dependent mechanism, while PMN reduced PB-MNC-mediated angiogenesis by PMN-derived neutrophil elastase.[7] In contrast, Minamino et al[19] reported that intramuscular implantation of PB-MNC induces more increase in limb blood pressure and better improvement of ischemic status compared with our previous observation,[13] suggesting that efficacy of angiogenic therapy by PB-MNC may depend on the degree of limb ischemia of patients. Inaba et al[20] showed that intramuscular implantation of CD34+ cells isolated from GCSF-mobilized peripheral blood mononuclear cells also enhanced blood perfusion in ischemic limbs; however, the angiogenic action appeared to be weaker than that of PB-MNC. Taken together with our data,[13] these clinical studies suggest the feasibility of angiogenic cell therapy using peripheral blood-derived cells for the treatment of ischemic tissues. As ischemic status of limbs or numbers of injected cells are different between patients in these pilot studies, further controlled studies using large numbers of patients will be needed to define the efficacy between PB-MNC and CD34+ cells.

Therapeutic Angiogenesis for Ischemic Myocardium by Bone Marrow Cells

Myogenic cell grafting in a damaged myocardium is a promising approach to the treatment of heart failure. We and others have previously shown the efficacy of intramyocardial transplantation of BM-MNC in animal models of ischemic heart failure.[10,11] Side effects, such as an increase in cardiac enzymes, malignant arrhythmia or differentiation into other lineage cells, were not observed in the BM-MNC-implanted myocardium.[10,11] A marked increase in cardiac function may be due to not only neovascularization but also partly cardiomyogenesis derived from marrow-derived hematopoietic cells[21] and/or marrow-derived mesenchymal cells.[22] On the basis of the results of these pre-clinical experiments using porcine models,[10,11] we performed the first clinical trial of a sole cell therapy using BM-MNC implantation targeted into an ischemic hibernating myocardium.

A 65-year-old man suffered from refractory angina (CCS class IV) due to posterolateral ischemia in spite of repeated bypass surgery (2 times) and angioplasty (5 times). Echocardiography showed severe posterolateral hypokinesia. Viability in the posterolateral wall was confirmed by a biphasic response using low-dose and high-dose dobutamine. The hibernating myocardium focus was identified on NOGA electromechanical mapping (Biosense-Webster) by preserved viability (voltage mapping) associated with an impairment in linear local shortening (mechanical mapping) (Fig. 5).[5] The average number of nitroglycerin (NTG) spray was 10-15 per day. Based on the severity of refractory angina as well as the presence of hibernating myocardium, we decided to perform BM-MNC implantation to improve regional blood perfusion. After aspiration of the marrow cells, an intercostal space was opened to visualize the left ventricular lateral wall. 0.1 ml of BM-MNC was injected into 30 different sites (total 3×10^8 cells) on the left ventricular (LV) posterolateral wall by a 27 gauge needle.

Angina occurrence decreased dramatically as early as 7 days after BM-MNC implantation and at 1-year follow-up, the patient's clinical status was markedly improved and he had stable CCS class 1 disease (no NTG spray use). There was no substantial arrhythmia on 24 h Holter recordings performed monthly for 1 year. NOGA nonfluoroscopic electromechanical mapping system (Biosense-Webster) analysis (Fig. 5) indicated that the reduced motion of the posterolateral hibernating area (shown in red in mechanical mapping), identified by preserved viability associated with impaired linear local shortening, markedly improved toward normal wall motion level (shown in purple) 4 months after BM-MNC implantation. The area of ischemic myocardium decreased from 36% before BM-MNC implantation to 11% after implantation. The result of single photon emission computed tomography (SPECT)-sestamibi myocardial perfusion scans corresponded to improved perfusion in the posterolateral myocardium observed in NOGA electromechanical analysis (Fig. 6), and reduced blood perfusion in the posterolateral region after pharmacological stress, was not observed 3 months after BM-MNC implantation. Echocardiography showed systolic function improvement in the posterior wall (fractional shortening: 19 to 33%) and a decrease in the LV end-diastolic dimensions (from 61 to 43 mm). In angiographical analysis, the numbers of detectable collateral vessels did not appear to change, while ejection fraction (EF) calculated by left ventriography increased from 42% to 53%.

Together, these results suggest that improved systolic function of the hibernating myocardium is related to the implanted BM-MNC. We[10,11] demonstrated revascularization of an ischemic myocardium by BM-MNC-derived neocapillary formation, while BM-MNC were also shown to include cardiomyocyte-committed hematopoietic stem cells[21] as well as cardiomyocyte-committed mesencymal stem cells.[22] Thus, our findings show the feasibility of autologous BM-MNC implantation in a patient with an ischemic myocardium. Very recently, two research teams have reported catheter-mediated implantation of BM-MNC as a sole therapy in 8 patients with stable angina.[23] and injection of AC133+ hematopoietic stem cells in 6 patients with recent myocardial infarction at the time of coronary artery bypass.[24] Although the number of patients is too small to derive any meaningful efficacy and definitive safety data, most patients had significant improvement in symptoms and many had improvement in physiological variables such as regional wall motion and target area perfusion, consistent with our present observation. Thus, cell-based investigational therapies for ischemic heart diseases were shown to be safe and feasible. Additional larger randomized placebo-controlled double-blind studies with mechanistic and clinical endpoints are in progress.

References

1. Carmeli P, Luttun A. The Emerging role of the bone marrow-derived stem cells in (therapeutic) angiogenesis. Thromb Haemost 2001; 86:289-297.
2. Isner M, Asahara T. Angiogenesis and vasculogenesis as therapeutic strategies for postnatal neovascularization, J Clin Invest 1999; 103:1231-1236.
3. Simons M, Bonow RO, Chronos NA et al. Clinical trial in coronary angiogenesis: Issues, problems, consensus. Circulation 2000; 102:e73-89.
4. Ferrara N, Alitalo K. Clinical application of angiogenic growth factors and their inhibitors. Nat Med 1999; 5:1359-1364.
5. Asahara T, Murohara T, Sullivan A et al. Isolation of putative progenitor endothelial cells for angiogenesis. Science 1997; 275:964-967.
6. Asahara T, Masuda H, Takahashi T et al. Bone marrow origin of endothelial progenitor cells responsible for postnatal vasculogenesis in physiological and pathological neovascularization. Circ Res 1999; 85:221-228.
7. Iba O, Matsubara H, Nozawa Y et al. Angiogenesis by implantation of peripheral blood mononuclear cells and platelets into ischemic limbs. Circulation 2002; 106: 2019-2025.
8. Kalka C, Masuda H, Takahashi T et al. Transplantation of ex vivo expanded endothelial progenitor cells for therapeutic neovascularization. Proc Natl Acad Sci 2000; 97:3422-3427.
9. Prockop DJ. Marrow stromal cells as stem cells for nonhematopoietic tissues. Science 1997; 276:71-74.
10. Kamihata H, Matsubara H, Nishiue T et al. Implantation of autologous bone marrow cells into ischemic myocardium enhances collateral perfusion and regional function via side-supply of angioblasts, angiogenic ligands and cytokines. Circulation 2001; 104:1046-1052.
11. Kamihata H, Matsubara H, Nishiue N et al. Improvement of collateral perfusion and regional function by catheter-based implantation of peripheral blood mononuclear cells into ischemic hibernating myocardium. Arterioscler Thromb Vasc Biol 2002; 22:1804-1810.
12. Shintani S, Murohara T, Ikeda H et al. Augmentation of postnatal neovascularization with autologous bone marrow transplantation. Circulation 2001; 103:897-895.

13. Tateishi-Yuyama E, Matsubara H, Murohara T et al. Therapeutic angiogenesis for patients with limb ischemia by autologous transplantation of bone marrow cells: A pilot study and a randomised controlled trial. Lancet 2002; 360:427-435.
14. Carmeliet P. VEGF gene therapy: Stimulating angiogenesis or angioma genes? Nature Med 2000; 6:1102-1103.
15. Salven P, Orpana A, Joensuu H. Leukocytes and platelets of patients with cancer contain high levels of vascular endothelial growth factor. Clin Cancer Res 1999; 5:487-491.
16. Gaudry M, Bregerie O, Andrieu V et al. Intracellular pool of vascular endothelial growth factor in human neutrophils. Blood 1997; 90:4153-4161.
17. Wartiovaara U, Salven P, Mikkola H et al. Peripheral blood platelets express VEGF-C and VEGF which are released during platelet activation. Thromb Haemost 1998; 80:171-175.
18. Folkman J, D'Amore PA. Blood vessel formation: What is its molecular basis? Cell 1996; 87:1153-1155.
19. Minamino T, Toko H, Tateno K et al. Peripheral-blood or bone-marrow mononuclear cells for therapeutic angiogenesis. Lancet 2002; 360:2083-2084.
20. Inaba S, Egashira K, Komori K. Peripheral-blood or bone-marrow mononuclear cells for therapeutic angiogenesis. Lancet 2002; 360:2083-2084.
21. Anversa P, Nadal-Ginard B. Myocyte renewal and ventricular remodeling. Nature 2002; 415:240-243.
22. Makino S, Fukuda K, Miyoshi S et al. Cardiomyoctes can be generated from marrow stromal cells in vitro. J Clin Invest 1999; 103:697-705.
23. Tse HF, Kwong YL, Chan JKF et al. Angiogenesis in ischaemic myocardium by intramyocardial autologous bone marrow mononuclear cell implantation. Lancet 2003; 361:47-49.
24. Stamm C, Westphal B, Kleine HD et al. Autologous bone-marrow stem-cell transplantation of myocardial regeneration. Lancet 2003; 361:45-46.

CHAPTER 10

The Stem Cell Continuum: A Plastic Plasticity

Peter J. Quesenberry, Jean-Francois Lambert, Gerald A. Colvin, Mark Dooner, Christine I. McAuliffe, Mehrdad Abedi, Deborah Greer, Delia Demers, Jan Cerny, Brian E. Moore, Evangelos Badiavas and Vincent Falanga

Introduction – Hierarchical Model

Conventional models of hematopoietic regulation propose that a primitive undifferentiated stem cell produces progeny with progressively more lineage restriction and lineage specific function and less renewal, proliferative and differentiation potential. The stem cell itself has tremendous proliferative and differentiative potential, which is diminished with differentiation to different progenitor classes, and is markedly decreased or abolished with differentiation to end-stage functioning cells.[1] This is a classic hierarchical model, and is illustrated in Figure 1.

In this model, it is assumed that stem cell proliferation must proceed as a population by asymmetric divisions. This means that while some stem cells may produce two cells destined for differentiation (extinction) and others produce two stem cells (a hyperproliferative model), most produce one cell destined for differentiation and one which remains a stem cell (an asymmetric division). As a population, the probability of producing a stem cell or a cell destined for differentiation is 0.5 and, thus, the stem population as a whole remains balanced. A model of stem cell division is shown in Figure 2. Apoptotic death presumably plays a role here and would provide another means of modulating the system.

Continuum Model

In early studies, we observed that treatment of unseparated BALB/c marrow cells in vitro with the cytokines interleukin-3 (IL-3), interleukin-6 (IL-6), interleukin-11 (IL-11), and steel factor (SF) led to an expansion of progenitors by 48 hours of culture, but with loss of engraftment capacity.[2,3] The loss of engraftment was seen at 4 weeks or longer after transplantation, and with longer periods of study, the engraftment defect was seen to get progressively more severe; it was not seen at 1 to 3 weeks post-cell infusion.[4] Further studies evaluated engraftment capacity of unseparated marrow cells cultured under set conditions (IL-3, IL-6, IL-11, and SF in non-adherent teflon bottles) and found dramatic shifts in engraftment occurring at 2 to 4 hour intervals prior to cell division.[5] Engraftment in these latter experiments was determined at 8 weeks and 6 months with similar results. In virtually every experiment, engraftment capacity returned to input values. Thus, these studies indicated that as marrow stem cells progressed through a cytokine driven cell cycle their engraftment capacity was reversibly modulated.

These experiments were with cytokine stimulated marrow cells. They were actually prompted by previous gene therapy experiments with the multi-drug resistant (MDR-1) gene in which we found high rates of in vitro transduction with very low rates of engraftment of the transduced cells.[6] This was selective because the non-transduced cells engrafted well. In this experiment, we used IL-3, IL-11, IL-6, and SF to drive the cells into cell cycle and promote transduction. We speculated that engraftment capacity might be linked to cell cycle transit and, thus, we evaluated highly purified lineage negative / Rhodamine low / Hoechst low (Lin$^-$ Rho^{lo} Ho^{lo}) stem cells for cycle progression under the same culture conditions with IL-3, IL-6, IL-11, and SF. We found that these stem cells entered S phase between 16 to 18 hours and completed mitosis by 36 to 38 hours.[7] When these data were juxtaposed in our experiments with unseparated marrow, an engraftment nadir mapped to late S/early G_2.

We have more recently mapped highly purified Lin$^-$ Rho^{lo} Ho^{lo} cells through cell cycle and found 1) similar engraftment nadirs with 2) similar kinetics. These studies indicate that the engraftment phenotype of the marrow hematopoietic stem cell changes with phase of cell cycle and independent of cell division. These observations are schematically portrayed in Figure 3.

These changes suggested that we would see changes in cell surface proteins and gene expression at times when engraftment was altered, and this has, in fact, been the case. We determined expression of c-kit, c-fms, and receptors for IL-2, IL-3, IL-11, and IL-6 at 24 hours of cell cycle transit of Lin$^-$ Rho^{lo} Ho^{lo} cells and saw up regulation of these receptors in S phase.[8] We also determined expression of alfa-4 adhesion protein and found depression of this adhesion protein at 48 hours of culture.[9,10] Lastly, we determined, in collaboration with Dr. Sherman Weissman, gene expression using a gene display technique.[11] Here we showed dramatic changes with a virtual inversion of the expression of stem cell specific genes at between 0 and 48 hours of culture. A total of stem cell specific genes were identified. The adhesion protein fluctuation suggested that homing to bone marrow might underlay the alterations in bone marrow engraftment.

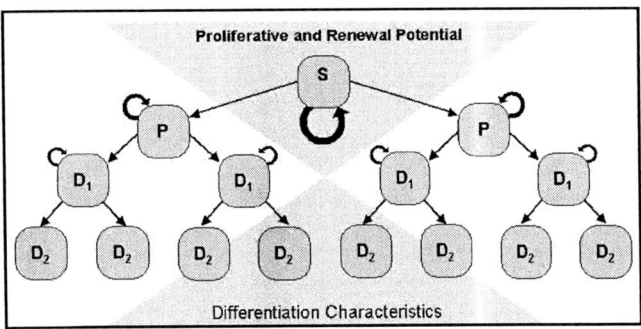

Figure 1. Hierarchical model of Hematopoiesis. S= stem cell; P= progenitor; D= differentiated cells.

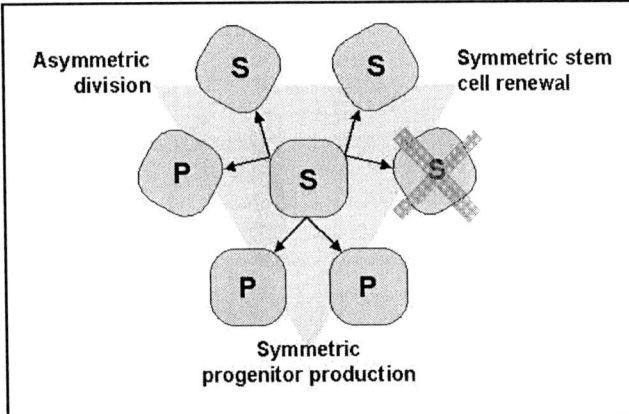

Figure 2. Balanced stem cell division. S= stem cell; P= progenitor (differentiation); Symmetric progenitor production (extinction); Symmetric stem cell renewal (hyperproliferative- leukemia); X= apoptosis.

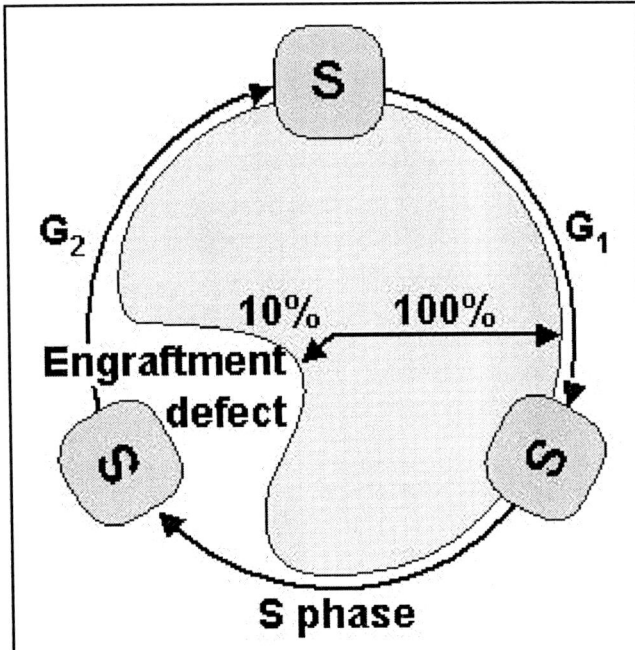

Figure 3. Stem cell engraftment modulation with cytokine induced cell cycle transit. 100%= full engraftment; 10%= defective engraftment; S= stem cell.

We established a homing assay in which Lin$^-$ Sca$^+$ marrow cells are labeled with the supra-vital dye CFDA-SE (carboxyfluorescein diacetate, succinimidyl ester), infused, and then recovered after 3 hours.[12] A large event sorter analysis was then carried out to determine the number of fluorescent positive events per total marrow cells. Using this direct homing assay, we determined that normal BALB/c murine marrow cells show a plateau of homing by 1 hour, and show linear homing between 50,000 and one million cells. A standard assay was established infusing 250,000 CFDA-SE labeled cells and recovery of cells for analysis at 3 hours after cell infusion. Using this assay, we determined that 7.45 to 9.32% of infused cells homed to normal marrow. Marrow was assessed for homing after 48 hours in IL-3, IL-6, IL-11, and SF, a time at which alfa-4 integrin levels were consistently depressed and a point at which engraftment is consistently depressed. Homing of these cells was consistently depressed after 48 hours of culture in these cytokines.

One possibility is that when primitive stem cells (as represented by Lin$^-$ Rho^{lo} Ho^{lo} and engraftment capacity) are induced by cytokine exposure to enter and progress through cell cycle, gene expression is altered including adhesion protein and cytokine receptors leading, in turn, to altered homing, which in turn leads to altered engraftment (Fig. 4).

Cell Cycle Status of Lymphohematopoietic Stem Cell

The primitive hematopoietic stem cell has been considered as a dormant non-cycling cell. This was shown by Bradford and colleagues[13] to be a fallacious concept. They fed mice with oral BrdU chronically and then harvested marrow and purified primitive hematopoietic stem cells (Lin$^-$ Rho^{lo} Ho^{lo}) and by 12 weeks 89% were labeled. These investigators calculated a cycle turnover rate for these marrow stem cells of 4.3 weeks. Cheshire,[14] using a different marrow strain and stem cell separation, confirmed this data but with different kinetics. We have also confirmed these data showing a progressive increase in Lin$^-$ Rho^{lo} Ho^{lo} labeling over time.[15] These data relate to the significance of our observations on the changing stem cell engraftment phenotype with cell cycle transit. If stem cells were truly dormant in G$_0$, then our observations might represent in vitro changes without physiologic significance. However, this is not the case. Stem cells are either slowly, continuously cycling or intermittently

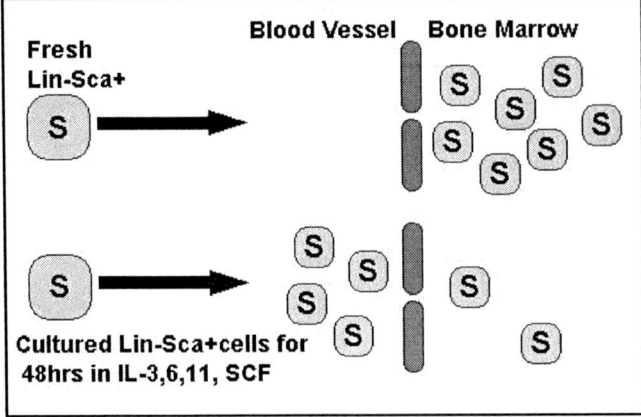

Figure 4. Cytokine treatment induces a homing defect in murine hematopoietic stem cells. Lin$^-$ Sca$^+$= purified murine marrow stem cells.

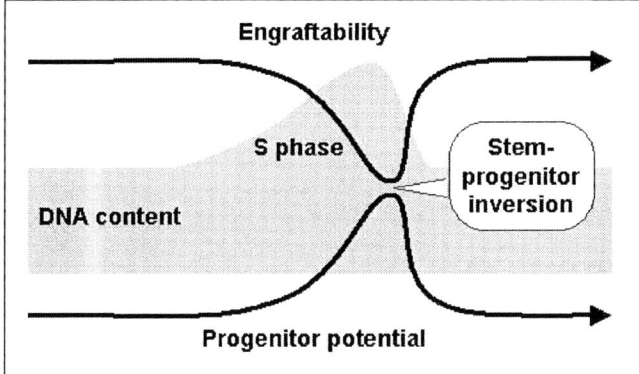

Figure 5. Progenitor/stem cell inversion. During cell cycle passage, progenitors increase and engraftment decreases before cell division and in a reversible fashion.

cycling; but, in any case, virtually all cycle over 12 weeks time. Thus, one might anticipate that the stem cell phenotype would also be continually changing.

Stem/Progenitor Cell Inversions

In recent studies, we have cultured BALB/c or C57BL/6 marrow cells in either teflon flasks or microgravity simulating rotating wall vessels in the presence of an alternative cytokine cocktail, FLT-3 ligand, thrombopoietin (TPO) and steel factor. These studies have been very informative showing large shifts in progenitor numbers, high proliferative potential-colony forming cell (CHPP-CFC) and colony forming unit-culture (CFU-C), during the first cell cycle transit of marrow cells stimulated with TPO, FLT-3, and steel factor.[16] When colony numbers were increased, engraftment was decreased and, in most cases, colony numbers returned to baseline as did engraftment levels. These events took place during one cell cycle, before division. Thus, the increase in progenitors and the decrease in engraftment occurred in a reversible fashion before cell division. We term these events progenitor/stem cell inversions, and interpret these data as indicating a stem cell/progenitor continuum rather than a hierarchy (Fig. 5). This does not, of course, exclude a hierarchical structure at the more differentiative levels of hematopoiesis.

Marrow to Nonhematopoietic Tissues

There have been a number of studies indicating the potential of marrow cells to produce nonhematopoietic tissues.[17-27] These studies have shown marrow cells giving rise to bone, cartilage, fat, skeletal muscle, cardiac myocytes, lung cells, hepatocytes, keratinocytes, hair follicle cells, and neural cells. Some of these studies have shown quantitatively significant production of nonhematopoietic cells by purified marrow cells, which, in turn, improved organ function or reversed disease manifestations. This was true of the studies of Lagasse et al[25] on hepatocyte repopulation in the fumarylacetoacetate hydrolase (FAH) deficient mouse and Orlic & colleagues[26,27] with studies on cardiac myocyte production in mice with ischemic cardiac lesions. Others have reported clonal origin of many tissue cells from single marrow stem cells.[28] We initially studied bone cell production from whole male BALB/c marrow which had been engrafted into female BALB/c nonablated mice.[17] In this setting we estimated that there were up to 3,200 donor male osteocytes per femur. Donor male osteocytes were present at 6 weeks after cell infusion and continued to be produced out to 6 months (the furthest time point evaluated).

Table 1. Potential factors in marrow cell to nonhematopoietic transdifferentiation

1. Tissue injury
2. Time from cell infusion
3. Method of infusion, intravenous, intramuscular, etc.
4. Mobilization from marrow versus directly infused at time of injury
5. Number of cells administered – there may be different threshold levels for different tissues

We found that radiation of the host had no effect, and that there was a relative cell threshold with significant bone cell production only seen at infused marrow cell levels of over 80 million cells.

More recent data examining tissue transdifferentiation shortly after marrow cell infusion or in previously established green fluorescent protein (GFP) marrow chimeric mice, who were then subjected to injury, with or without G-CSF mobilization, has shown marrow cell production of hepatocytes, skin keratinocytes, hair follicle cells, and cardiac myocytes. Several general themes have developed with this work.[29] Tissue injury seems important, although that was not the case with the bone studies. Longer times from cell infusion appear to give better results, and based on our bone studies, high levels of infused cells may be critical. This is summarized in Table 1.

In these studies, donor cell labels have included male DNA beta galactosidase and greenfluorescent protein. Each has its problems especially with double labeling, and specific approaches have to be carefully worked out. Recent reports have suggested that cell fusion and tetraploidy might be occurring in some studies.[30,31] This would, of course, not negate studies such as those of Lagasse[25] and Orlic.[26,27] Studies addressing this interesting possibility have been initiated in a number of laboratories.

Summary

Marrow cells appear to have striking functional and hierarchical plasticity. Our data indicates that the previously defined hematopoietic stem cell to progenitor cell hierarchy may, in fact, be a reversible phenotype continuum. Whether such a continuum extends to nonhematopoietic cells and tissues is a critical question (Fig. 6).

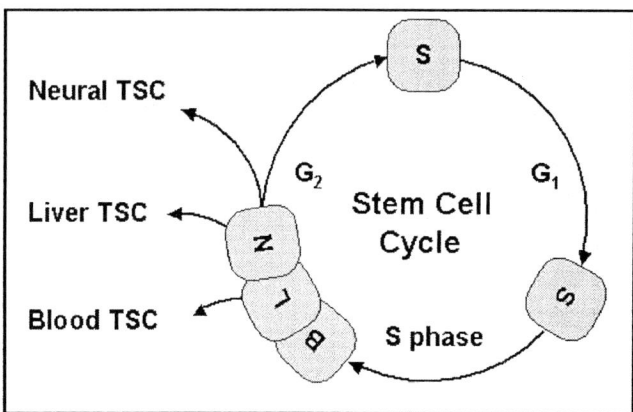

Figure 6. Theoretical cell cycle associated marrow cell conversions. TSC= tissue stem cell.

Acknowledgements

Special thanks to Sandra Bibby for excellent work in preparing this manuscript.

References

1. Quesenberry PJ, Colvin GA. Hematopoietic stem cells, progenitor cells and cytokines. In: Beutler E, Lichtman MA, Coller BS, eds. Williams Hematology. 6th ed. New York: McGraw Hill Text Co. 2001:153-174.
2. Peters SO, Kittler EL, Ramshaw HS et al. Murine marrow cells expanded in culture with IL-3, IL-6, IL-11, and SCF acquire an engraftment defect in normal hosts. Exp Hematol 1995; 23:461-469.
3. Peters SO, Kittler EL, Ramshaw HS et al. Ex vivo expansion of murine marrow cells with interleukin-3, interleukin-6, interleukin-11, and stem cell factor leads to impaired engraftment in irradiated hosts. Blood 1996; 87:30-37.
4. Peters SO, Habibian H, Quesenberry PJ. Cytokine modulation of murine stem cell engraftment: The role of adherence to plastic surfaces. Intl J Hematol 2002; 76:84-90.
5. Habibian HK, Peters SO, Hsieh CC et al. The fluctuating phenotype of the lympho-hematopoietic stem cell with cell cycle transit. J Exp Med 1998; 188:393-398.
6. Kittler EL, Peters SO, Crittenden RB et al. Cytokine-facilitated transduction leads to low-level engraftment in non-ablated hosts. Blood 1997; 90:865-872.
7. Reddy GP, Tiarks CY, Pang L et al. Cell cycle analysis and synchronization of pluripotent hematopoietic progenitor stem cells. Blood 1997; 90:2293-2299.
8. Reddy PV, McAuliffe CI, Pang L et al. Cytokine receptor repertoire and cytokine responsiveness of Ho^{dull}/Rh^{dull} stem cells with differing potentials for G1/S phase progression. Exp Hematol 2002; 30:792-800.
9. Berrios VM, Dooner GJ, Nowakowski G et al. Cytokine and extracellular matrix regulation of adhesion receptor expression and function by primitive murine hematopoietic stem cells. Exp Hematol 2001; 29:1326-35.
10. Becker PS, Nilsson SK, Li Z et al. Adhesion receptor expression by hematopoietic cell lines and murine progenitors: Modulation by cytokines and cell cycle status. Exp Hematol 1999; 27:533-541.
11. Lambert JF, Liu M, Colvin GA et al. Marrow stem cells shift gene expression and engraftment phenotype with cell cycle transit. J Exp Med 2003; in press.
12. Cerny J, Dooner MS, Quesenberry PJ. Cytokine treatment of hematopoietic stem/progenitor cells induces a homing defect in nonmyeloablated hosts. Exp Hematol 2000; 28:51(abstract).
13. Bradford GB, Williams B, Rossi R et al. Quiesence, cycling, and turnover in the primitive hematopoietic stem cell compartment. Exp Hematol 1997; 25:445-453.
14. Cheshire SH, Morrison SJ, Liao X et al. In vivo proliferation and cell cycle kinetics of long-term self-renewing hematopoietic stem cells. Proc Natl Acad Sci USA 1999; 96:3120-3125.
15. Pang L, Reddy PV, McAuliffe CI et al. Hemotopoietic cells: Stem cells and cell lines J Cell Physiol 2003; in press.
16. Colvin G, Lambert JF, McAuliffe C et al. Studies on BrdU labeling of hematopoietic cells: stem cells and cell lines. J Cell Physiol 2003; in press.
17. Nilsson SK, Dooner MS, Weier HU et al. Cells capable of bone production engraft from whole bone marrow transplants in nonablated mice. J Exp Med 1999; 189:729-734.
18. Pereira R, Halford K, O'Hara M et al. Cultured adherent cells from marrow can serve as long-lasting precursor cells for bone, cartilage, and lung in irradiated mice. Proc Natl Acad Sci 1995; 92:4857-4861.
19. Ferrari G, Cusella-DeAngelis G, Coletta M et al. Muscle regeneration by bone marrow-derived myogenic progenitors. Science 1998; 279:1528-1530.
20. Eglitis MA, Mezey E. Hematopoietic cells differentiate into both microglia and macroglia in the brains of adult mice. Proc Natl Acad Sci USA 1997; 94:4080-4085.
21. Brazelton TR, Rossi FM, Keshet GI et al. From marrow to brain: Expression of neuronal phenotypes in adult mice. Science 2000; 290:1775-1779.
22. Mezey E, Chandross KJ, Harta G et al. Turning blood into brain: Cells bearing neuronal antigens generated in vivo from bone marrow. Science 2000; 290:1779-1782.
23. Theise ND, Badve S, Saxena R et al. Derivation of hepatocytes from bone marrow cells in mice after radiation-induced myeloablation. Hepatology 2000; 31:235-40.
24. Theise ND, Nimmakayalu M, Gardner R et al. Liver from bone marrow in humans. Hepatology 2000; 32:11-16.
25. Lagasse E, Connors H, Al-Dhalimy M et al. Purified hematopoietic stem cells can differentiate into hepatocytes in vivo. Nat Med 2000; 6:1229-1234.
26. Orlic D, Kajstura J, Chimenti S et al. Transplanted adult bone marrow cells repair myocardial infarcts in mice. Ann NY Acad Sci 2001; 938:221-229.
27. Orlic D, Kajstura J, Chimenti S et al. Mobilized bone marrow cells repair the infracted heart, improving function and survival. Proc Natl Acad Sci 2001; 98:20344-20349.
28. Krause DS, Theise ND, Collector MI et al. Multi-organ, multi-lineage engraftment by a single bone marrow-derived stem cell. Cell 2001; 105:369-377.
29. Abedi M, Badiavas E et al. Trafficking and transdifferentiation of bone marrow cells in a skin injury model. Exp Hematol 2002; 30:47(abstract).
30. Terada N, Hamazaki T, Oka M et al. Bone marrow cells adopt the phenotype of other cells by spontaneous cell fusion. Nature 2002; 416:542-5.
31. Ying QL, Nichols J, Evans EP et al. Changing potency by spontaneous fusion. Nature 2002; 416:545-8.

CHAPTER 11

Adult Stem Cell Plasticity

Sean Lee and Diane S. Krause

Introduction: Stem Cell Plasticity—A New Discovery and a New Debate

The concept that bone marrow derived stem cells (BMSC) can give rise to cells of disparate lineages began to gain widespread support in 1998, as articles first surfaced describing the presence of marrow-derived myocytes and hepatocytes in bone marrow transplant recipient mice. Subsequent reports have expanded the nonhematopoietic, nonmesenchymal cell types that can differentiate from BMSC to include mature cells of the lung, heart, GI tract, kidney, skin, and the CNS (including neurons). In addition to BMSC, neuronal stem cells are also capable of differentiating into multiple nonneural cell types.

These initial reports challenged the existing dogma of tissue specific stem cell differentiation in adults, and sparked significant debate. Many in the scientific arena have suggested alternative interpretations for these results, although some of these have not yet been strictly ruled out, the plasticity theory continues to be the best supported. The debate regarding bone marrow and other adult stem cells has spilled over into the political and public arenas as well. Opponents of human embryonic stem cell research see the plasticity of these adult cells as a means of avoiding the destruction of human blastocysts that is required to obtain pluripotent embryonic stem (ES) cells for research. The vigor with which both sides of the ES cell issue debate the worthiness of adult stem cells has contributed to the confusion over this new scientific discovery by bringing it into the mass media before science could illuminate its true nature and potential. Despite daily improvements in our knowledge, the mechanisms of cell plasticity are not yet clear, and the extent of usefulness of these cells as research tools or medical therapies remains equally enigmatic. Yet, despite the controversy and the early stage in our understanding of stem cell plasticity, hopes abound that if we can control this plasticity we may be able to use adult stem cells to produce new tissues for transplant and/or repair diseased organs. It is also hoped that these cells will provide new insights into the mechanisms of early development, cancer, and other pathologies.

In this chapter, we discuss the traditional dogma of stem cells and how recent findings have called this dogma into question. We review the current data that purport to show adult stem cell plasticity with an emphasis on bone marrow-derived stem cells. Also discussed are the political impact and potential therapeutic applications of these data.

Background

Strictly defined, a stem cell is a cell capable of extensive self-renewal, enabling it to persist long term, and of differentiation into at least one mature cell type. However, this definition alone is imprecise. There are many different types of stem cells that can be defined by: (1) the stage of differentiation at which they are present; (2) where they are found in the organism; and (3) their differentiative potential. In the 2-4 cell stage of embryogenesis, each of the cells in the blastula is totipotent in that each is capable of differentiating into all embryonic and extraembryonic cell types necessary for development of the fetus. In contrast, stem cells in the adult are traditionally considered to have limited potential. For example, intrahepatic liver stem cells are only thought capable of differentiation into the primary cell types in the liver – hepatocytes and cholangiocytes. At the heart of the debate surrounding adult stem cell plasticity is whether predominantly tissue-specific adult stem cells are capable of differentiating into cell types outside of their primary lineage if in an appropriate microenvironment. To truly understand the debate surrounding BMSC plasticity, one must first understand the role stem cells play in ontogeny and adult physiology.

Old Paradigm: Lineage Restriction

In traditional descriptions of ontogeny, the differentiation capacities of stem cells become more and more limited as the organism completes its development. Strictly defined, totipotent stem cells are only known to exist between the single-celled zygote and morula stages of embryogenesis, as only the cells present in these stages are capable of producing an entire organism. Once the morula begins to hollow out into the blastocyst, two pluripotent populations of cells are formed. The 30 or so cells in the inner cell mass form every cell type found in the embryonic and adult organism, and the cells of the trophoblast create the extraembryonic tissues (e.g., the placenta) necessary for the development of the embryo in utero. Although embryonic stem cells can form trophoblast-like cells in vitro, the inner cell mass cells (also called embryonic stem cells) produce only the embryo in vivo, and cannot in a strict sense be termed totipotent. Once the blastocyst has formed, gastrulation ushers in the incredibly complex program that will lead a clump of 30 homogenous cells towards the production of a mature organism with billions of highly ordered cells of hundreds of types. In order to carry out this program, every cell in the blastocyst must respond appropriately to

---

Stem Cell Therapy for Autoimmune Disease, edited by Richard K. Burt and Alberto M. Marmont. ©2004 Landes Bioscience/Eurekah.com.

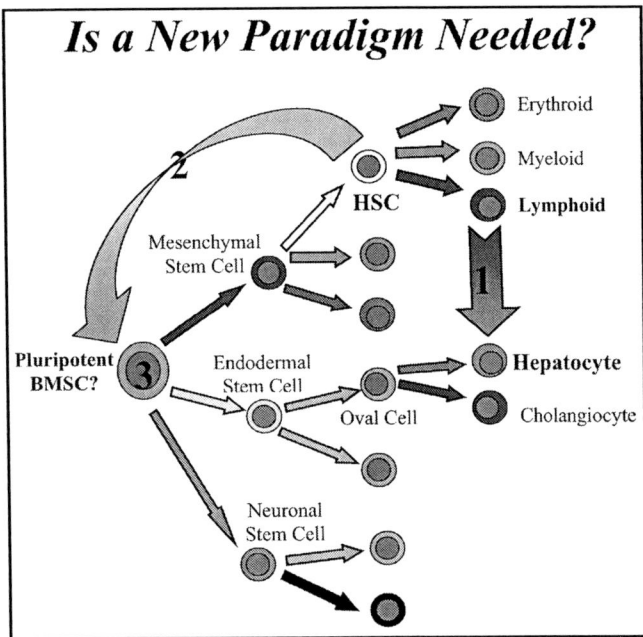

Figure 1. Stem cells within the bone marrow (BM) may contribute to tissues outside the mesenchymal lineage in three ways. 1) Transdifferentiation of a partially committed cell to a different lineage. 2) Dedifferentiation of a committed cell to an uncommitted precursor cell that can differentiate down alternative lineages. 3) A pluripotent cell exists in the BM that can differentiate into multiple lineages.

its environment so that it may correctly balance its choice between replicating itself and differentiating. Replication will provide the number of cells needed for an adult organism, whereas differentiation will allow for the formation of new tissue types, from the formation of the three germ layers, to completion of organogenesis. Indeed, for some tissues this process will continue well after birth. After all organs have finished their formation, adult stem cells remain in each tissue in order to replace cells lost to injury, disease, or normal cell turnover. Unlike their embryonic counterparts, however, adult stem cells have traditionally been ascribed specific, limited cell types into which they are capable of differentiating. For the most part, the adage "like begets like" holds true for adult stem cell tissue or lineage restriction, as stem cells in the liver produce hepatocytes and cholangiocytes, stem cells in the bone marrow (BM) produce blood, bone, cartilage, and other mesenchymal tissues, etc. Such restriction may permit an improved efficiency of stem cell differentiation, as despite their relative scarcity (for example HSCs represent only one in 1×10^5 cells in the BM[1]), adult stem cells can maintain both their tissues and their own numbers in all but the most severe pathological circumstances.

From the viewpoint of researchers who want to study development or to produce cell-based therapies, the limited capability of adult stem cells is disappointing. Although they can be relatively easily isolated from small biopsy specimens, their immutability and commitment make individual adult stem cell types uninformative regarding early developmental programs and difficult to apply to most diseases. Embryonic stem cell lines are therefore of unique importance to researchers, as only they can be used to produce any cell type in vitro. In fact, since the first isolation of human ES cells from blastocysts in 1998,[2] scientists have learned how to coax these cells in vitro to become many different mature cell types including cardiomyocytes, neurons, and pancreatic islet cells.[3] Unfortunately, the use of human ES cells is highly controversial in the United States because such research requires the destruction of early human embryos. Current federal law restricts the number of ES cell lines that federally funded labs can use to approximately 60, only 10 or so of which are available and adequately characterized. Many researchers think that this number is woefully inadequate considering the limited genetic variability of existing cell lines and the vast number of genetic disorders that could be addressed with ES cells.[4]

New Paradigm: Plasticity, Metamoirosis and the HSC

The tissue-restricted view of adult stem cell differentiation has been challenged since 1998 when work was published showing that bone marrow-derived stem cells could differentiate into skeletal muscle myocytes.[5] Since that time, numerous studies have yielded similar results for other tissues, including liver, lung, GI tract, heart, and the CNS. These new findings suggest that adult stem cells are plastic, that is they have the ability to switch from the lineage of the tissue in which they reside to engender specialized, functional cells of a different tissue. Some have termed this process 'transdifferentiation.' This word may not accurately describe the findings, however, as no one has reported the exact mechanism for how adult stem cells switch lineages. Transdifferentiation represents only one of several possible mechanisms of plasticity, as it describes the process of a 'committed' cell differentiating from one lineage to another (see Fig. 1). It therefore implies that, for example, a BMSC can begin to differentiate down the hematopoietic lineage into a monocyte, and then, under appropriate stimuli, directly differentiate into a hepatocyte. Another possible mechanism of plasticity is that an adult stem cell dedifferentiates from a cell committed to a certain lineage to an uncommitted progenitor. This less committed cell is then capable of choosing among various lineage pathways down which to differentiate. Finally, there may be uncommitted stem cells present in very low numbers in various tissues, including the BM, capable of directly differentiating down multiple lineages. Whatever the mechanism, findings of plasticity suggest that adult stem cells retain at least some of the developmental programming used in embryogenesis. Indeed, the pluripotency of plastic adult stem cells makes them similar to ES cells. This prospect is exciting to many researchers, because if pluripotent stem cells can be isolated from adults, they would provide an accessible alternative to ES cells for some studies of development, and for the production of cell-based therapies. Also, adult pluripotent cells could be used for autologous purposes.

An unfortunate byproduct of the plasticity field's youth, is its lack of nomenclature that accurately represents the processes of adult stem cell differentiation. Given the inaccuracy of the present terminology, we will use the term metamoirosis (from the Greek 'change of fate') in our discussion to describe the process by which a stem cell can give rise to cell types outside its tissue of origin's lineage, without favoring any specific mechanism. This term is appropriate because under traditional descriptions of stem cell tissue restriction, the fate of any cell endogenous to the BM, whether committed to the hematopoietic or mesenchymal lineages or not, is to produce blood or mesenchymal tissues. As for other terminology, the words 'transdifferentiation' and 'dedifferentiation' will only be used to refer to those specific mechanisms. The term plasticity is more complicated. Strictly used, it implies that a committed stem cell has switched lineages by either trans or dedifferentiation. However, perhaps because no uncommitted cell has been described in the adult to date, it is often used more loosely to describe any change of fate, and should not be taken to imply a specific mechanism when used in this review. The term metamoirosis is more accurate, however, and should remind the reader of the three mechanisms that could underlie a change of cell fate.

Although metamoirosis has been described in neuronal stem cells (NSCs) and mesenchymal stem cells (MSCs), discussions of adult stem cell plasticity most often focus on the hematopoietic stem cell (HSC). The HSC is the most extensively studied adult stem cell, and to date has shown the greatest differentiation potential of these cells. Studies are currently underway to find the extent of HSC potential, to explore the mechanisms underlying HSC metamoirosis, and to discover ways by which their plasticity can be utilized for scientific and therapeutic use. Unfortunately, a number of factors have confounded this work. First, confusion is sometimes created by the improper use of the term HSC in studies using unpurified cell populations from the BM or peripheral blood. Such cell populations, more properly termed BMSC, can include HSC, MSC, and possibly other undefined stem cell types. In addition, even when the term HSC describes a purified population of cells, different definitions of HSC are used as these cells have yet to be purified to homogeneity from the BM. Perhaps the best definition of the HSC offered to date was produced by Irving Weissman, who reports that all HSC activity is included in the c-kit$^+$ Lin$^-$ Sca-1$^+$ BM cell population,[6] and that individual HSCs in the c-kit$^+$ Thy1.1^{lo} Lin$^-$ Sca-1$^+$ subset of BM cells (the so-called KTLS population) can self-renew and long-term repopulate the BM, but cannot significantly repopulate nonhematopoietic tissues.[7,8] Work in our laboratory used a more functional definition to show that a single BMSC (defined as Lin$^-$, elutriation first fraction cells that could home to the BM within 48 hours of primary transplant and long term repopulate the BM of secondary transplant recipients) was able to reconstitute multiple tissues of multiple developmental lineages.[9] The disparate differentiation capabilities of these two highly HSC-enriched BM cell populations may reflect that the presence of different HSC subpopulations, or the presence of nonHSC BMSC in one of these populations. In fact, there are known subpopulations of HSC that vary in their self-renewal and differentiation capabilities.[10] A final confounder is that there may be more than one cell type in the BM capable of metamoirosis. Until the plastic cell populations can be exhaustively defined, readers of the plasticity literature must be careful to determine the exact nature of the cell populations used in these studies.

Evidence for BMSC Plasticity

Over the past five years, the number of studies of BMSC plasticity and the number of tissues BMSC have been found to supplement have increased greatly. Each study has in turn inspired a great deal of healthy scientific skepticism, such that experimental designs have become increasingly more rigorous. Critical assessment of the articles on both sides of the plasticity debate requires that readers be attentive to several issues: (1) Are BM-derived nonhematopoietic cells present, and if so, how many? (2) What population(s) of cells (e.g., whole BM vs. highly selected subpopulations) was used in the experiment? (3) How were BMSC-derived cells identified and shown to be part of the tissue of interest? For example, fluorescent in situ hybridization (FISH) can be used to detect Y-chromosomes in sex mismatched BM transplant studies and tissue-specific markers can be used to determine what cell types BMSC have become. (4) What is the pathophysiological state of the tissue of interest and the BM? Many studies use injury models to enhance BM engraftment and one must also take into account the potential radiation injury due to whole body irradiation prior to BM transplant. (5) How do BMSC interact with other stem cells present in the target tissue? The interplay between BMSC and the tissue's endogenous adult stem cell compartment may have important implications for the process of BMSC engraftment. (6) Are there corroborating or conflicting studies in the literature?

As we proceed to discuss the evidence for and against stem cell metamoirosis in the next section, we will attempt to address each of these questions. Although not every paper will speak to all of them, together, the papers published on each tissue should provide the reader of this review with our current knowledge regarding each of these questions.

Liver

The liver originates as a collection of proliferating endodermal cells that form a bud on the ventral floor of the foregut in the early mouse embryo.[11] These undifferentiated endodermal cells soon give rise to liver precursor cells called hepatoblasts, which are thought to be bipotent as they differentiate into both hepatocytes and cholangiocytes in culture.[12,13] As hepatoblasts gradually divide and differentiate into hepatocytes, they migrate as chords into the septum transversum and begin to associate with primitive endothelial cells.[14] Although hepatoblasts do seem to form transient intercellular connections between each other, true epithelialization is limited in the early liver by the abundance (approximately 80%) of hematopoietic cells.[14-16] Liver development therefore extends well into the neonatal period.[11] In the adult, the liver is unique in its extensive regenerative capacity. Under most circumstances, hepatocytes themselves are responsible for this regeneration. Even after a two-thirds partial hepatectomy, each hepatocyte can rapidly enter the cell cycle and replicate two or three times to regenerate the lost cell mass.[17] Only when the liver suffers severe damage does a more primitive, bipotent, stem cell become activated.[18,19] These adult liver stem cells, termed oval cells, are thought to exist periportally in the terminal hepatic ductules (canals of Hering).[20] In response to injury, these cells begin to proliferate, leading to an initial increase in the numbers of bile ductule cells, and then differentiate to produce hepatocytes.[18-20] This provides the liver with multiple mechanisms for the regeneration of hepatocytes; most mild or moderate forms of injury are healed by hepatocyte growth, and severe injuries are supplemented by the activation of oval cells.

The first demonstration that bone marrow could supply cells to the liver was reported by Petersen and colleagues in 1999.[21] In three different experimental rat models, bone marrow cells gave rise to up to 0.16% of hepatocytes (defined by their morphology and the expression of the hepatocyte-specific markers H4 and C-Cam) and 0.1% of oval cells (defined by morphology and Thy-1 expression). Bone marrow-derived oval cells were present 9 days after liver injury and hepatocytes appeared by day 13 post-transplant, suggesting that the BM might first seed the liver's endogenous stem cell compartment, which then differentiates into BM-derived hepatocytes (see Chapter 4). In each of the models used, the recipient animals were treated with 2-acetylaminofluorene and carbon tetrachloride to induce oval cell proliferation. A study published a year later by Theise and colleagues[22] found that such acute liver injury was not necessary for BM engraftment in the liver. Also in contrast to the Petersen study, in which only whole BM was used, Theise used both male whole BM and a subpopulation of CD34$^+$Lin$^-$ male BM cells to engraft lethally irradiated female recipients. Six months after whole bone marrow transplant, liver engraftment was 1.5%, and eight months after transplant of CD34$^+$Lin$^-$ cells, the average liver engraftment was 0.66%. Confocal microscopy was also performed as a further step to ensure that the Y-chromosomes being counted were indeed within the nuclei of hepatocytes, and not in the nuclei of abutting blood cells. This study therefore showed that hepatotoxin-induced injury is not necessary for BMSC to produce hepatocytes, and that engraftment of bone marrow derived hepatocytes is maintained in the long term. In addition, the use of CD34$^+$Lin$^-$ BM cells was a first step toward defining the bone marrow cell type responsible for hepatocyte production.

Human BMSC are also capable of differentiating into hepatocytes. The first study to show this used archival pathology specimens from six patients, two of whom were women who had received male bone marrow transplants and the other four were men who had been transplanted with female livers.[23] Again, FISH probes for the Y-chromosome were used to identify donor-derived cells, and immunohistochemistry for CAM 5.2 and Cytokeratins 8, 18, and 19 identified hepatocytes, cholangiocytes, and bile ducts. The bone marrow derived hepatocyte and cholangiocyte engraftment ranged between 5 and 40%, and 4 and 38%, respectively. The wide range in engraftment is likely due to several factors, but the pathological process affecting each patient is likely the most significant. The patient with the highest level of engraftment had histological features consistent with fibrosing cholestatic recurrent hepatitis C, which is associated with high degree of cell turnover, and had expired four months post transplant. None of the allograft patients had signs of acute or chronic rejection, and neither of the bone marrow transplant patients had graft versus host disease (GVHD) or recurrent neoplasia. The authors also make note that, similar to animal experiments, there tends to be clustering of Y-positive liver cells. This finding is consistent with Petersen's hypothesis that BM engraftment of the liver occurs through the oval cell compartment.

Corroborating data were published in which the authors analyzed hepatocytes from nine female patients who had received bone marrow from male donors, and from eleven female livers that had been transplanted into male patients and then removed again due to recurrent disease.[24] This study identified donor-derived cells using Y-chromosome in-situ hybridization, and defined hepatocytes by morphology and cytokeratin 8 staining. Between 0.5% and 1.5% of hepatocytes in these patients' livers were Y-positive. As in the previous study, the authors note that many of the Y$^+$ hepatocytes are clustered. This paper also addresses a unique question. It had been proposed that transplacental passage of fetal cells could lead to microchimerism for the Y-chromosome in women who had become pregnant with a male child. This could then account for the Y-positive hepatocytes seen in female livers in these studies. Livers from such women had no detectable Y-positive leukocytes or hepatocytes using Y-chromosome FISH. Furthermore, one of the patients in the study who showed Y-chromosome positive hepatocytes had never been pregnant with a male child. The specific subpopulation of BM cells responsible for hepatocyte and cholangiocyte engraftment can not be determined by these studies.

An important discovery that highlights the therapeutic potential of HSC was made by Lagasse and colleagues in a study in which they used both whole bone marrow and purified BM subpopulations enriched for HSC to rescue mice with a fatal error in liver metabolism.[25] The fumarylacetoacetate hydrolase (FAH$^{-/-}$) mouse suffers progressive renal tubule damage and liver failure unless treated with the drug 2-(2-nitro-4-trifluoro-methylbenzoyl)-1,3-cyclohexanedione (NTBC).[26] Up to 50% of hepatocytes (defined by morphology and expression of albumin, FAH, CD26, or E-cadherin) are BM-derived when FAH mice are transplanted with either whole BM or c-kit^{high}Thy^{lo}Lin$^-$Sca-1$^+$ BM cells. The putative BM-derived hepatocytes did not express hematopoietic markers, and were often found in clusters, or nodules. This engraftment of BMSC as FAH$^{+/+}$ cells was sufficient to rescue the recipient FAH$^{-/-}$ animals. Lagasse also determined that, separately, c-kit$^+$, Sca-1$^+$, and Lin$^-$ BM fractions, but not c-kit$^-$, Sca-1$^-$, or Lin$^+$ fractions could reconstitute the BM and livers of irradiated recipients. This study thus showed that a highly selected population of BMSC can produce sufficient numbers of healthy hepatocytes to rescue animals with a fatal metabolic liver disease. In addition, only the BM subpopulations that can engraft the hematopoietic system show the ability to become functional hepatocytes.

The engraftment seen in FAH$^{-/-}$ mice is impressive, but such high levels of liver repopulation may not occur in human diseases where the extent of liver damage and the state of endogenous liver progenitors differ from that of FAH$^{-/-}$ mice. It is likely that therapeutic uses of BMSC will require techniques for enhancing liver engraftment. A study addressing this issue was published by Mallet and colleagues in 2002.[27] Transgenic mice expressing the antiapoptotic gene Bcl-2 under a liver-specific promoter were used as BM donors for lethally irradiated female animals. After transplant, some of the recipient mice were treated with the anti-Fas antibody Jo2 which selects for Bcl-2 expressing donor cells in mouse livers.[26] In mice that received Jo2, bone marrow derived hepatocytes comprised 0.05 to 0.08% of total liver cells by quantitative reverse transcriptase polymerase chain reaction (RT-PCR) for Bcl-2 mRNA. Most of the BM-derived hepatocytes were found in clusters, indicating clonal growth. Such clustering was not found in recipients that were not treated with Jo2. Donor-derived (Y-chromosome$^+$ by FISH) cells were identified as being epithelial cells by immunohistochemistry for cytokeratins 8, 18, and 19. Despite the low engraftment rates seen in this study, a protocol that enhanced survival of donor-derived cells was shown to increase BM engraftment into the liver.

Another question integral to furthering our understanding of metamoirosis in the liver and how stem cells might be used in therapeutic situations is the precise kinetics of liver cell engraftment following BM transplant. Petersen[21] and Theise[22] reported finding BM-derived hepatocytes within 13 and 7 days, respectively. Wang and colleagues, published work specifically aimed at better defining the time course and frequency of liver repopulation by BMSC in FAH$^{-/-}$ mice.[28] Individual BM-derived hepatocytes first appeared 7 weeks after transplant, clusters of these hepatocytes appeared after 11 weeks, and donor-derived cells comprised more than 30% of all hepatocytes by 22 weeks. Taken together, it appears that very low levels of engraftment occur within a few weeks of BM transplant, but that clinically relevant hepatocyte engraftment could take several months.

BM subpopulations enriched for HSC are not the only cells capable of giving rise to hepatocytes. Plastic adherent cells from postnatal bone marrow, termed multipotent adult progenitor cells (MAPCs), can differentiate into cells with morphological, phenotypic, and functional characteristics of hepatocytes in vitro.[29] Pancreatic tissue can also give rise to hepatocytes[30,31] and injection of whole pancreas cell suspensions can rescue FAH$^{-/-}$ mice by engrafting as FAH$^{+/+}$ hepatocytes.[30] In contrast, purified pancreatic duct cells can not engraft into the liver. Unfortunately, because only suspensions of unpurified whole pancreas were therapeutic, one cannot be sure whether the liver repopulating cells were indeed of pancreatic origin and not BMSC or another exogenous stem cell population residing in the pancreas. Consistent with these in vivo data, the pancreatic cell line AR42J-B13 can be induced in vitro to differentiate into cells with hepatocyte morphology and marker expression.[31]

Brain

Traditional descriptions of neurogenesis have restricted the time frame of neuron production in the CNS to early ontogeny.[32,33] More recently, however, it has become widely recognized that in the olfactory bulb and the subgranular zone of the dentate gyrus (part of the hippocampus), neuron development and migration persists into adulthood.[34-39] The cells that give rise to fully differentiated neurons are called neuronal stem cells (NSC), and these cells also differentiate into the two subtypes of macroglia, the astroglia and oligodendroglia.[34] In addition, as stem cells, NSC are capable of self-renewal.[34] The current view of adult production of neurons is that NSC, located mainly in the subventricular zone (SVZ),[38,40,41] but also in the hippocampus[42] and the spinal cord,[36] can migrate to regions of the CNS where new neurons are needed. Migration is mainly to the olfactory bulb in the case of SVZ cells and locally in the case of hippocampal NSC, and it has been postulated that as many as 1/2000 CNS neurons is replaced each day.[43] Questions remain, however, as to the true extent of NSC migration through the brain, and whether these cells will prove successful in various proposed therapeutic applications. Excitement over these cells has been intensified by the suggestion that, in addition to their capacity to renew neural tissues, NSC can differentiate into cells of other lineages. NSC contribute to tissues of all germ layers in chimeric chick and mouse embryos,[44] and when cocultured in vitro with skeletal muscle myocytes, purified NSC differentiate to myocytes.[45] Work by Bjornson and colleagues demonstrated that NSC could engraft as blood cells (including monocytes, granulocytes, B-cells, and T-cells) when transplanted into sublethally irradiated mice.[46] Although these exciting findings have not yet been reproduced, they suggest that NSC can produce blood cells. But can BMSC return the favor? This question is significant because NSC cannot be isolated in large numbers,[47] and have not been autografted. In contrast, BMSC are relatively easy to isolate, and are used in autograft protocols. Current data indicate that BMSC can give rise to neurons and glia under certain conditions, but these data are highly controversial.[32,33,48] It is therefore important to read this body of literature critically, but with an open-mind.

In one of the earliest reports of BMSC plasticity in any nonhematopoietic tissue, Eglitis and Mezey described the derivation of microglia and astroglia from transplanted bone marrow.[49] Although in 1997 hematopoietic cells were not thought to cross the blood-brain barrier to generate new microglia,[50-55] under current knowledge the derivation of microglia from BM is not surprising as these CNS macrophage-type cells are known to be of hematopoietic origin.[56] The production of BM-derived astrocytes, on the other hand, suggests metamoirosis, as these macroglial cells are usually of neuroectodermal origin.[57,58] These authors did not observe BM-derived neurons in the CNS. As discussed above, such a finding would be significant since any production of new neurons in the adult CNS is thought to be exceedingly rare, let alone the production of new neurons descended from a different cell lineage.

The first two attempts to demonstrate BM-derived neurons were published in the journal *Science* in December of 2000. Brazelton and colleagues[59] used a GFP transgenic mouse strain as a donor to recipients lacking the transgene. Analysis focused on the olfactory bulb, but GFP$^+$ (donor) cells were also found in the hippocampus, cortex, and cerebellum. Using laser scanning confocal microscopy to visualize very thin (~0.3μm) sections of the brain, GFP$^+$ cells that coexpressed neuron specific antigens (NeuN, neurofilament NF-H, and class III β-tubulin) comprised 0.2-0.3% of all olfactory bulb neurons after BM transplantation. The highest frequency of BM-derived neurons was in the superficial axonal layer of the olfactory bulb, a region known to exhibit high rates of neurogenesis.[60,61] The majority of the GFP$^+$ neuron-like cells also had a triangular shape, indicative of neurons, and a few also possessed long cellular extensions, which could represent axonal outgrowths. The phosphorylation state of cyclic AMP response element-binding (CREB) transcription factor in the putative GFP$^+$ neurons was similar to that in surrounding neurons, suggesting a functional similarity in how these cells and neighboring neurons interact with the extracellular environment.[62] In contrast to other reports, no BM-derived astrocytes were detected in this study.

Mezey and colleagues[47] transplanted immunodeficient PU.1$^{-/-}$ mice (which lack macrophages, neutrophils, mast cells, osteoclasts, B-cells, and T-cells) with wild-type male whole BM within 24 hours of birth. One to four months after transplant, donor-derived cells (Y-chromosome positive) engrafted as both glia and neurons (identified by the expression of NeuN). NeuN stained cells also expressed neuron-specific enolase (NSE). Most donor-derived neurons were found in the cortex, but fewer numbers of cells were detected in the hypothalamus, hippocampus, amygdala, periaqueductal gray, and striatum. Interestingly, large numbers of Y-chromosome$^+$ cells were found in areas in direct contact with the cerebral spinal fluid, indicating that BM-derived cells infiltrate the brain largely via the CSF, first seeding the ependymal and subependymal regions. From there, these cells migrate deeper into the brain and differentiate to produce a variety of neuronal cells. However, this study did not determine whether the BM-derived cells pass through areas traditionally associated with high NSC concentrations (such as the SVZ, olfactory bulb migratory tract, and hippocampus). In separate experiments, these authors were able to confirm the ability of BM cells to undergo metamoirosis towards the neuronal lineage in vitro.

Despite the description of bone marrow-derived neurons in various regions of the brain, Mezey and colleagues did not see Y chromosome positive large motor neurons in the brainstem or spinal cord.[47] A more recent study confirmed a lack of BM-derived motor neurons, but was able to show the presence of bone marrow derived immature neurons in mouse spinal cord and dorsal root ganglia.[63] In summary, these data indicate that BM can produce a small, but detectable, number of immature neurons in various regions of the CNS. The distribution pattern of these neurons indicates that they enter the CNS via the CSF, and follow traditional migration pathways through the brain to reside, most commonly, in regions known to have high neuronal proliferation rates.

Not all studies on BM-derived neurons have supported this picture, however. Priller and colleagues[64] found BM-derived cells (which expressed GFP) in the brain within two weeks of transplant, but these cells did not have neuronal morphology. Not until 12 to 15 months post-transplant were BM-derived neurons found, making up approximately 0.1% of all cerebellar neurons.

These were identified as Purkinje cells by their numerous synaptic contacts and expression of calbindin-D28K and glutamic acid synthase, but not the glial markers glial fibrillary acidic protein (GFAP), ionized calcium-binding adapter molecule 1 (Iba1), or S100β (a calcium binding protein). In contrast to other reports,[47,59] BM-derived neurons were not found outside the cerebellum. The discrepancy may reflect differences in the experimental systems used by these groups, in techniques used, or in how they defined neurons both morphologically and using protein markers.

One of the major concerns in this new field of BMSC plasticity in the CNS is the apparent irreproducibility of the results. In two recent reports, no BMSC-derived neurons could be found in the CNS. Castro and colleagues[65] transplanted lethally irradiated C57Bl/6 mice with LacZ-labeled whole bone marrow or BM SP cells (side population in the marrow that are enriched for HSC[66,67]). Some of the recipient mice were injured by cortical stab or contusion in an attempt to increase bone marrow engraftment in the CNS. No LacZ neurons were found in any of the animals. The only LacZ stained cells were perivascular and had globular morphology, indicating they were most likely hematopoietic (though no lineage specific staining was performed to confirm this). In the same recipients, BMSC derived muscle cells were detected.

When a single GFP$^+$ highly purified BMSC (lin$^-$ Sca$^+$Kit$^+$Thy1^{lo}, or KTLS) was transplanted into wild type recipient mice, 18% of the recipients showed significant hematopoietic engraftment, with GFP$^+$ cells comprising an average of 20.2% of peripheral blood leukocytes at 14 weeks.[8] Various tissues in these mice, including brain, kidney, GI tract, liver, and muscle were analyzed for the presence of GFP$^+$ cells 4 to 9 months post transplant. Only one donor derived neuron, defined as a GFP$^+$, MAP-2$^+$ Perkinje cell, was found in over 13 million brain cells analyzed. In a parabiotic mouse model (in which a mouse carrying a genetic marker was surgically connected to a wild-type mouse such that their circulatory systems were contiguous), no neurons were found to have derived from the transgenic partner after analysis of more than 3 million brain cells. Low level engraftment with 7 donor-derived hepatocytes in 470,000 counted, occurred in the liver. No BMSC-derived tissue-specific cells were found in the kidney, gut, skeletal or cardiac muscle, or lung. With such low levels of engraftment, the researchers concluded that any bone marrow contribution to nonhematopoietic tissues is a rare and physiologically insignificant phenomenon.

The most important difference between this study and the others discussed above, and the most likely explanation for why less metamoirosis was observed, is the use of a highly purified population of HSC (KTLS). The authors emphasize this point, writing that there may be other populations of cells in the BM that have a higher capacity to engraft as nonhematopoietic cells than does the KTLS population. At present, many authors ascribe the pluripotency of BMSC to HSC, but a strong case has been made that MSC can give rise to neurons. Kopen and colleagues[68] found that by twelve days after BrdU labeled MSC were stereotactically injected into the lateral ventricles of mice, the MSC had migrated throughout the brain. Some of the BrdU$^+$ cells coexpressed GFAP, indicating that they may have differentiated into astrocytes. Cells were also located in the reticular formation of the brainstem that coexpressed BrdU and neurofilament indicative of a neuronal differentiation and did not express the astrocyte protein GFAP.

In vitro growth of mouse and human bone marrow stromal cells with epidermal growth factor or brain derived neurotrophic factor induces expression of nestin (a marker of neural progenitors), GFAP (a glial marker), and neurofilament (a neuronal marker).[69] In vitro MSC metamoirosis also occurs when rat or human stromal cells are cultured in DMEM with β-mercaptoethanol.[70] Up to 80% of the cells so cultured expressed the neuronal antigens neuron-specific enolase and neurofilament-M.

In addition to the MSCs used in these studies, multipotent adult progenitor cells (MAPCs) that can be purified from BM, muscle, and brain, can give rise to cells resembling glia, neurons, endothelium, and hepatocytes under certain in vitro conditions.[71] MAPCs do not express blood markers such as CD44, CD45, MHC I and II, and c-kit. The true identity of the multipotential cell, or indeed cells, in the bone marrow is one of the most important unanswered questions in the field.

What conclusions can be drawn from the present state of this work? Although most groups that have looked for bone marrow-derived neurons have found at least a small number, others have found none despite careful analysis. Most explanations of such discrepancies are technical, including differences in the mouse strains, neuronal markers, or BM tagging systems used, and in which brain regions were analyzed.[72] Still, since the majority of researchers doing these experiments find BM-derived neurons using a variety of experimental systems, it is probable that BM can produce neurons. Although the differentiation of BMSC to brain is rare, it is possible that BMSC may be induced to produce neurons more rapidly and in significant numbers. For example, Corti et al[73] found that a three fold increased engraftment of donor-derived, NeuN-expressing cells in the temporal cortex could be achieved by treating transplant recipients with granulocyte-colony stimulating factor (G-CSF) and stem cell factor (SCF). Better optimization of BMSC engraftment in the CNS and their application to therapeutics will remain difficult until the true identity of the multipotential cell, or cells, in the bone marrow is determined. Clarification of this issue remains integral to the further development of all aspects of BMSC science, and will require further study in the coming years.

Skeletal Muscle

Skeletal muscle derives embryologically from somites.[74] Signals from adjacent embryonic structures induce myogenesis in dorsal regions of the somites, giving rise to the dermomyotome, which in turn is the major source of muscle precursor cells, termed satellite cells.[75,76] Satellite cells continue to be the major source of muscle cells in the adult, despite a significant drop in their numbers during postnatal development.[77] In aged mice, satellite cells are reduced even farther.[78] These precursor cells have unique morphology and express a unique set of proteins including M-cadherin, Myf-5, CD34, and Pax-7.[79,80] Satellite cells are normally quiescent, but can be activated to both self-renew and produce more committed myocyte precursors upon muscle injury or weight bearing exercise.[81,82] After activation, myocyte precursors undergo multiple cycles of cell division before fusing with existing myofibers or with each other to make new myofibers.[77,81,83]

Two puzzling findings suggested the possibility that extramuscular cells contribute to the satellite cell and myocyte populations. First, it was noted that the number of myocyte progenitors formed after tissue injury was much greater than the number of satellite cells present in the tissue[84] suggesting that cells from outside of the muscle can populate the satellite cell pool and can contribute to muscle fibers as they do to bone, cartilage, and lung.[5,85] Second, satellite cells express molecules also known to be expressed on cells of the hematopoietic lineage, such as CD34, and a variety of endothelial markers.[79,76]

The first to clearly demonstrate that BM could contribute to muscle fibers were Ferrari and colleagues.[5] When LacZ-expressing whole BM cells were either injected into injured skeletal muscle or used to transplant recipient mice that were later subjected to muscle injury, LacZ$^+$ muscle fibers appeared within two weeks of the injury. Donor-derived nuclei were present in both immature centrally nucleated myocytes and mature peripherally nucleated myocytes, indicating that BMSC contribute first to some form of myocyte precursor which develops into mature myofibers. Although there was no LacZ expression in any nonmuscle tissues, rare LacZ$^+$ monocytes were found in the muscle. This study could therefore not rule out the possibility that LacZ$^+$ myocytes were in fact derived from the fusion of monocytes with myocyte precursors.

BMSC can also differentiate into skeletal and cardiac muscle cells in different mouse models of muscle injury.[86] Wild type male whole bone marrow was transplanted into female C57BL/10ScSnHim *mdx* mice, which lack the cytoskeletal protein dystrophin and are a murine model of Duchenne's muscular dystrophy[87] (reviewed in ref. 88). In this disease, muscle fibers continuously degenerate and regenerate, requiring increased satellite cell activity. In control transplant experiments using wild type BM recipients, only rare Y-chromosome positive skeletal myocytes and no Y$^+$ cardiac myocytes formed. In contrast, in the *mdx* mice, numerous Y$^+$ myocytes formed in skeletal and cardiac muscle. Many of the donor-derived myocytes expressed the myoregulatory proteins myf-5 and myogenin, and some also expressed dystrophin. Although this study did not quantify donor-derived dystrophin$^+$ myofibers, in a similar study up to 4% of myofibers in recipient mice expressed dystrophin.[89] These data hold promise for the future of BMSC-based therapy for Duchenne's and other muscular dystrophies.

Enhancement of BMSC to skeletal muscle myocyte differentiation also occurs in response to exercise-induced muscle damage.[90] After transplantation, BMSC first replaced satellite cells destroyed by the whole body irradiation. These BMSC-derived satellite cells then differentiated to produce up to 3.5% of myofibers in mice with exercise-induced muscle damage. The BMSC-derived myoblasts also expressed the muscle-specific proteins Myf-5, c-MetR, α-7-integrin, and could self renew in culture.

Other data indicate that the plasticity between the BM and muscle may be reciprocal. Gussoni showed that, as in the BM, there is a side population in muscle that stains faintly with Hoechst dye.[89] These cells are Sca-1$^+$ Lin$^-$, but do not express the hematopoietic markers CD45, c-kit, or markers of committed muscle cells. When transplanted into lethally irradiated mice, muscle SP cells repopulated both the bone marrow and skeletal muscle. These SP cells thus appeared to be a population of early stem cells capable of multi-organ reconstitution. Others have also isolated bone marrow repopulating cells from muscle.[91] The data become conflicting here, however, because despite originally defining the BM-reconstituting cell population in muscle as Sca-1$^+$Lin$^-$ c-Kit$^+$CD45$^-$, the same group more recently published that only CD45$^+$ muscle cells can repopulate the bone marrow.[92] Although these data send conflicting messages of the true source of the BM-repopulating cells in the muscle, they all agree that the interplay of stem cells between the muscle and the BM is more detailed than previously thought.

Heart

The various cell types found in the heart, including cardiac myocytes, endocardial endothelium, vascular endothelium, vascular smooth muscle, and fibroblasts derive from different mesodermal populations in the embryo.[93-96] By embryonic day 23 the primitive human heart begins beating, and soon after, looping and cardiac cushion formation produce the chambers and valves necessary to couple the electrical activity of the heart to its mechanical purpose of pumping blood through the lungs and body.[97]

In contrast to the complex cellular migration and differentiation events in the embryonic heart, adult cardiac myocytes are traditionally considered to be fully differentiated and incapable of cycling.[98] This dogma has been challenged in recent years, however, by the discovery of a possible cardiac-specific stem cell,[99] and by reports of proliferating cardiomyocytes in transplanted or diseased human and animal hearts.[100-102] Still, the proliferation seen in these studies is relatively modest and has not been found to produce any clinically significant benefits to patients with heart disease. Studies attempting to restore functional cardiomyocytes to injured myocardium have therefore focused on cell types with higher proliferation capacity, including satellite cells from skeletal muscle and ES cells.[103-105] Although many of these studies have shown promising results, the inability of these cell types to accurately recapitulate the structural and functional properties of adult cardiomyocytes, and the difficulty of using them autologously, make them imperfect candidates for the treatment of heart disease.[106] BMSC have also been shown capable of engrafting in the heart as cardiomyocytes, and if such metamoirosis can be optimized, these cells may offer a more clinically useful alternative for cellular cardiac therapies.

Tomita and colleagues published a report showing that MSC could engraft into and improve the function of injured myocardium.[107] Treatment of MSC with 5-azacytidine induced formation of myotubes that express cardiac specific troponin I and myosin heavy chain. These cells were labeled with BrdU and injected into the myocardial scar in rat that had undergone cardiac cryoinjury three weeks before. Five weeks after injection, BrdU cells with cardiomyocyte morphology and cardiac troponin I expression had entered the scar area. Although located in the center of the scar and uncoupled from the heart's conduction system, these MSC-derived cardiomyocytes increased angiogenesis, reduced scar size, and improved post-mortem cardiac function measurements.

Others confirmed these results using untreated MSC in different animal models. MSC expanded and chemically labeled in culture integrated into cardiac muscle when injected into healthy myocardium.[108] Labeled cells were aligned in intercalated disks, exhibited myocyte morphology, and stained positively for connexin 43. The presence of this component of gap junctions indicates that the donor-derived cells were electrically coupled to the endogenous myocardium. When human MSC, transfected with the LacZ gene, were injected into the uninjured myocardium of immunodeficient mice, LacZ$^+$ cells were found within four days.[109] Although these did not initially express protein markers of cardiac muscle, after fourteen days, donor cells were rod shaped and aligned with other cardiomyocytes, had increased in size, and expressed desmin, cardiac troponin T, α-actinin, phospholamban, and β-myosin heavy chain. Unfortunately, engraftment rates were low in both these studies; only about 1 LacZ$^+$ cell per tissue section was found by Toma et al.

Another group of studies assessed the capability of a HSC enriched BM population to produce cardiomyocytes. In the first of these studies, Lin⁻ c-kit⁺ BM cells from male enhanced green fluorescent protein (eGFP) mice were injected into the anterior and posterior boarders of infracted myocardium 3 to 5 hours after inducing infarction in female mice by ligation of the left main coronary artery.[110] Nine days after injection, myocardial repair occurred in 12 of 30 mice. Repaired regions contained new myocytes, 53% of which were donor-derived, and 93% of which had incorporated BrdU (indicating they were proliferating). All myocytes in the reconstituting region expressed GATA-4 and MEF2, proteins that are specific for cardiac gene activation, and 40% of these cells expressed the transcription factor Csx/Nkx2.5 that is specific to the early stages of myocyte differentiation (refer to Chapter 8). Importantly, cardiac function tests indicated that mice injected with Lin⁻ c-kit⁺ cells had better contractile function than did control mice injected with either no cells or Lin⁻ c-kit⁻ cells. The cardiac regenerative capacity of BMSC was confirmed by Jackson and colleagues, who transplanted male LacZ⁺ BM side population (SP) cells (SP cells were isolated by the marker profile CD34$^{-/lo}$ c-kit⁺ Sca-1⁺) into lethally irradiated female mice, and ten weeks later occluded the left anterior descending artery in the recipient mice.[111] Surviving mice were sacrificed two to four weeks later, and up to 0.06% of cardiac myocytes were LacZ⁺ and CD45⁻. These data provide strong evidence for the regeneration of injured cardiac myocytes by BM cells. Further studies are currently underway looking into how BM metamoirosis can be optimized to provide therapeutically useful effects in patients suffering from myocardial infarction or cardiomyopathy.[112] Currently available data suggest that the implantation of BMSC into patients is not harmful, and may even restore some function of infarcted myocardium,[112] but controlled phase II trials have not yet been completed.

Lung

The development of the lungs begins with the formation of the laryngo-tracheal groove in the endodermal epithelium lining the embryonic pharynx.[113] Although lung development has progressed far enough to permit gas exchange by birth, alveolarization continues until age seven in humans.[114] In contrast to other organs, the adult lung is traditionally divided into multiple compartments, each with a separate type of stem cell.[115] Thus, the precursor cells in the trachea and bronchi are mucous secretory cells in the basal layer of the epithelium. In the bronchioles, Clara cells are thought to be the major precursor cell type, and in the alveoli, type II pneumocytes act as stem cells. Such a division makes sense anatomically and functionally, as more specialized stem cells may be more efficient in replacing the specialized cell types in the lung than a single, more primitive cell. Whether or not there also exists a less differentiated stem cell population in the lungs, capable of contributing to all these compartments, remains unknown.[116] Either way, these separate stem cell compartments add an extra dimension of complexity to the interpretation of data from experiments on BMSC plasticity in the lung, as ideally, one should compare the BM contributions to each compartment. Unfortunately, existing data on BMSC-derived lung cells focuses only on the ability of BMSC to differentiate into alveolar and bronchial epithelium.

A number of studies from our lab have provided evidence that HSC-enriched BMSC populations can give rise to lung epithelial cells. In one study, a single male HSC was found capable of producing epithelial cells in a variety of tissues in transplanted female mice.[9] Donor-derived cells, detected by Y-chromosome FISH, that expressed epithelial-specific cytokeratins and not hematopoietic lineage markers were present in both bronchial and alveolar epithelium. Another report from our lab showed that either whole BM or CD34⁺ Lin⁻ BM cells could contribute to type I and type II pneumocytes in lethally irradiated female recipient mice.[117] Male donor cells were detected by Y-chromosome FISH and epithelial cells were stained by either immunohistochemistry for cytokeratins 7 and 8 (type I and II) or by FISH for the surfactant B promoter (type II only).

A third study in our laboratory showed that BM may also be capable of delivering specific transgenes to the lung epithelium.[118] Male mouse bone marrow transduced with a retrovirus encoding eGFP was transplanted into lethally irradiated female mice. At 2, 5, and 11 months after transplant, eGFP was expressed by 1.0 - 7.0% of lung epithelial cells. Thus, stable transgene expression by lung epithelial cells was achieved by the transplantation of retrovirally transduced BM cells.

Others have reported that MSC can contribute to the lung. After adherent cultured BM cells were injected into lethally irradiated recipients, the genetically marked donor cells made up between 2 and 8% of the lung.[85] When LacZ-labeled BM was intravenously injected into wild-type mice with bleomycin-induced lung injury, LacZ⁺ donor-derived cells formed aggregates in the lung epithelium.[119] The LacZ⁺ cells had a type I pnuemocyte morphology and location, expressed the type I pnuemocyte marker T1α, and bound to the *Lycopersicon esculentum* lectin (thought to only bind to type I pneumocytes in the lung). No donor-derived type II pneumocytes were identified in this study, however.

Gastrointestinal Tract

The gastrointestinal tract is derived from the endoderm. This embryonic layer forms as a single-cell thick cup covering the mesoderm and ectoderm at the end of gastrulation, but it rapidly elongates and expands into a tubular structure. Buds sprout off of the primitive gut tube to form endodermal organs such as the pancreas, liver, and lungs, and the tube itself gives rise to the gastrointestinal tract.[120] In the adult animal, complex migration and differentiation programs persist in this organ system, facilitating the rapid turn over of epithelial cells. Throughout the adult GI tract, new epithelial cells are produced by putative stem cell compartments at sites where cellular migration originates.[121] In the small bowel, these sites are located at the base of the crypts, from which cells migrate upwards toward the villi tips where they will eventually be shed into the bowel lumen. In the gastric glands of the stomach, the migration pattern is bidirectional, as the stem cell compartment exists in the middle portion of the tubule and maturing cells migrate both upwards and downwards.[121] During their migration, these cells differentiate into all of the different cell types found in the gut epithelium, including the columnar enterocytes, mucin-secreting cells, endocrine cells, and Paneth cells. Despite the cellular diversity in the gut, each crypt in the bowel or gastric gland in the stomach is derived from a single stem cell.[122-125] There are only a few instances when crypts and glands have a polyclonal origin, including in the early stages of normal development,[126] in certain regions of the adult GI tract,[127] and in certain neoplasias.[128]

These data provide a context for assessing the actions of bone marrow derived cells in this organ system. For example, if bone marrow derived cells enter into the GI tract through the resident stem cell compartment, then entire crypts should be donor-derived. This would be the ideal situation for the development of BMSC-based therapies, as therapeutic cells could produce fully functional crypts. Even if BMSC are capable of bypassing the resident GI stem cell compartments to directly enter the GI tract epithelium, if BMSC can repetitively seed the gut, BMSC-based therapies may still have long term efficacy.

Unfortunately, current data for BMSC engraftment of epithelial cells of the GI tract are few and inconclusive. Our lab showed BMSC-derived epithelial cells in the esophagus, stomach, small bowel, and colon in female mice each transplanted with a single male HSC.[9] Although up to 2.5% of cells in these epithelia were donor-derived, these cells were found singly, implying that they were able to infiltrate existing crypts and glands without transiting through the tissue-specific stem cell compartment. The tissues were not rigorously examined to determine if any completely donor-derived crypts existed, however. A population of pluripotent BMSC that copurify with mesenchymal stem cells, termed MAPC, also can engraft into the gut epithelium when transplanted into nonirradiated or minimally irradiated host mice.[129] When MAPC expressing β-galactosidase were transplanted into immunodeficient NOD/SCID nonirradiated recipients, up to 11.9% of GI epithelial cells were β-Gal+ cytokeratin+ and CD45−. Furthermore, entire crypts were MAPC-derived, indicating that these cells may enter the GI epithelium via the endogenous crypt stem cell compartment. BM from primary recipients was also used to transplant secondary irradiated NOD/SCID recipient mice, leading to a GI epithelial engraftment of up to 8%. Other data show that bone marrow can also differentiate to pericryptal myofibroblasts in the GI tracts of both mice and humans.[130] Still, others have been unable to find BM-derived cells in the GI tract. Wagers and colleagues were unable to find any nonhematopoietic cells after examining 360,000 cells in the GI tract of mice transplanted with a BMSC population highly enriched for HSC.[8] Although the mechanism, extent, and even the possibility of bone marrow contribution to the GI tract remain unknown, future studies will undoubtedly address these issues further. These studies must be designed and carried out with the developmental and environmental context of the GI tract in mind to ensure that the data are as informative as possible.

Kidney

The kidney develops from the ureteric bud and the metanephrogenic mesenchyme. Through a series of complex reciprocal interactions, these two mesenchymal structures differentiate to produce the nephrons, glomeruli, and supporting structures of the kidney.[131] Cells in the metanephrogenic mesenchyme contribute to all epithelial regions of the nephron, except the collecting duct,[132] and similar cells can produce myofibroblasts, smooth muscle, endothelial cells, and epithelial cells in vitro.[133] In the adult kidney, the identity of the tissue-specific stem cell has evaded attempts at description to date, but accumulating evidence is providing clues regarding the location of the intrarenal stem cell. Tubular epithelial cells of the nephron are capable of regeneration after injury, and in certain pathologic conditions tubular epithelial cells are capable of a phenotypic conversion to fibroblast morphology and protein expression patterns.[134,135] Glomerular epithelial cells have also been found to metamoirose into myofibroblasts in disease models.[136] Together, these data suggest that there is a population of stem cells capable of regeneration and differentiation residing in the nephron epithelium.

A number of reports have described the capability of extrarenal stem cells to contribute to cell populations in the kidney. Poulsom and colleagues found that in both female mice transplanted with male BM and in male humans transplanted with female kidneys, nonhematopoietic Y-chromosome+ cells engrafted in the kidneys.[137] In the mouse model up to 7.9% of the tubular epithelial cells contained Y-chromosomes, and in humans up to 20% of tubular epithelial cells were Y-chromosome+ in the transplanted female kidneys. The Y-chromosome+ cells lacked hematopoietic markers, but had morphological characteristics of epithelial cells and expressed a variety of epithelial cell markers. The authors also reported seeing what appeared to be Y-chromosome containing myofibroblasts in the mouse kidneys and Y-chromosome+ endothelial cells in the human transplanted kidneys. In other studies, BMSC repopulated glomerular mesangial cells in a mouse BM transplant model[138] and male recipient cells repopulated the endothelium of female renal allografts in humans.[139]

It is important to note here that in these human renal allograft studies it is not known whether the recipient-derived cells in allografted kidneys were derived from the bone marrow or some other recipient tissue. Still, given the similar results seen in mouse BM transplant models and the fact that the major source of MSC and HSC is the BM, it is a reasonable hypothesis that many of the recipient cells in these allografts originated in the BM.

As is the case in other tissues, not all researchers who have looked for BM-derived cells in the kidney have found them. Wagers and colleagues detected no donor HSC-derived kidney cells despite examining almost 1 million cells in the kidney.[8] This may indicate that this HSC population is not the BM population responsible for engraftment into the kidney. Other studies have failed to see recipient cell engraftment in the glomeruli or tubules of allografted human kidneys.[140] Studies using transplantation of specific purified BM cell populations followed by careful analysis of kidney engraftment would help to resolve the issue regarding which cells can and cannot reconstitute the kidney.

Skin

The development of mature skin, with its multiple layers and appendages, is dependent upon an intricate cross communication between the embryonic ectoderm and mesoderm. The tri-layered embryonic epidermis must proliferate rapidly in order to cover the entire surface of the developing organism, and during this period of rapid cell growth, signals from the underlying mesenchymal dermis induce an alteration in the differentiation program of some epidermal cells to produce hair, glands, and nails.[141,142] Only after this rapid growth phase has ended near the end of fetal life does the rigid, water-tight stratum corneum form as the outermost layer of the epidermis.[141,143]

Although the growth potential of adult skin is greatly reduced from that of embryonic epidermis, anyone who has suffered even a minor scratch or burn knows that adult skin retains a remarkable regenerative capacity. This capacity is dependent upon two general classes of replicative cells in the skin, the true stem cells, and their progenitor daughter cells, termed transit amplifying (TA) cells.[144] The TA cells divide as needed before differentiating into keratinocytes,[144] whereas epidermal stem cells have a relatively low replicative rate under steady state conditions.[145] Epidermal stem cells are broken into three populations, the keratinocyte stem cells in the basal layer of the interfollicular epidermis (in both the tips and bases of dermal papillae), those in the germinal matrix at the base of hair follicles, and those in the upper segment of the outer root sheath of the hair follicle (the bulge region cells).[146] In addition, the sebaceous gland precursors may represent a fourth stem cell population in the skin.[144] Although some have proposed that the bulge region cells are ultimately responsible for the production of a majority of cells in all skin lineages,[147,148] others believe that under normal conditions the stem cell compartments are each restricted to their own subset of the skin.[144] In either case, it is accepted that there is significant multipotency and cross differentiation amongst these populations, as any of these stem cells can produce fully differentiated keratinocytes, hair follicles, and sebaceous glands.[144] This multipotency may extend to other cell types in the skin, as corneal TA cells (which reside at the center of the cornea and are widely considered terminally differentiated) from rabbits can form pilosebaceous units (sweat glands) after being associated with murine embryonic dermis.[149]

The list of plastic stem cells capable of producing skin is not limited to cells of the integumentary system. Our group reported that up to 4% of epidermal cells were donor-derived when a single BMSC was transplanted into lethally irradiated mice.[9] Körbling and colleagues reported that up to 7% of keratinocytes were Y-chromosome+ in a sample of women who had received BM transplants from male donors.[150] In similar experiments, Y-chromosome+ donor-derived cells expressing cytokeratins comprised 14% of the cells in the epidermis.[151] However, no donor-derived cells could be found in primary keratinocyte cultures from these patients. The authors therefore hypothesized that unlike keratinocytes, these BM-derived epidermal cells may not be capable of self-renewal in vitro. As for other tissues discussed in this review, the exact nature of the BM-derived cells in skin remains uncertain and will require further study.

Arguments Against Plasticity

Many scientists are skeptical of stem cell plasticity reports because there is no solid theory describing the mechanism lineage switching. Indeed, there is little evidence for transdifferentiation or dedifferentiation in mammals, mechanisms that would be necessary if the precursor cells responsible were already committed to an alternate lineage. However, there is evidence for these mechanisms in certain vertebrates. Urodele amphibians (e.g., newts) can regenerate their limbs, tails, jaws, retinas, lenses, and regions of their ventricles and atria after heart damage.[152,153] The mechanism of tissue regeneration is thought to involve local reprogramming of differentiated cells to a less differentiated state, which then can begin dividing to produce new tissues.[152] Interestingly, it appears that the signals driving this process in the newt can be used to cause dedifferentiation of mammalian cells.[154] Although these data from newts provide some biological precedent for dedifferentiation, the exact mechanism of mammalian plasticity will need to be determined before it becomes a universally accepted theory.

Critics of mammalian stem cell plasticity also point to the large number of studies that have failed to find BM-derived cells in a variety of tissues. At least some of the discrepancies between reports may be due to the use of different techniques to detect BM-derived cells,[72] but others may reflect real differences in the subpopulations of BM cells used. Unfortunately, confusion over which BM cell populations are pluripotent is often due to the imprecise use of the term HSC. When used by Wagers and colleagues, HSC refers to a highly purified population of BM cells that are c-kit+ Thy^{lo} Lin- Sca-1+,[7] but others attribute instances of plasticity to HSC despite using impure BM cell populations. The fact that the HSC-enriched population used by Wagers did not produce significant numbers of BM-derived cells in brain, liver, kidney, gut, muscle, or lung,[8] therefore, does not directly refute findings that cells from whole BM can engraft in these tissues. Furthermore, Wagers findings do not directly refute data obtained by our group using another highly purified BMSC population to reconstitute lethally irradiated recipient mice. In these mice, between 0.19 and 20% of the cells in different tissues were BM-derived.[9] Thus, two different HSC-enriched BM populations that are both capable of hematopoietic engraftment appear to differ greatly in their pluripotency. The resolution of this issue awaits the direct comparison of multiple cell populations in otherwise identical conditions.

In the absence of a proven mechanism for stem cell plasticity, there are at least two alternative explanations for how cells with the cytological and protein expression patterns of one tissue could derive from a different tissue. First, there might be a cell in the BM and elsewhere that retains an ES cell-like pluripotency, which can follow a standard differentiation pattern to produce cells of various lineages. Under the nomenclature established at the beginning of this paper, a stem cell residing in a certain adult tissue producing cell types not found in that tissue constitutes a change of fate, whether or not that stem cell is a committed adult stem cell or an uncommitted ES-like cell. Current data from studies of BMSC plasticity do not assess whether the BM-derived cells seen in different tissues are derived from committed or uncommitted stem cells, and doing so will be difficult unless markers capable of discriminating between these stem cell types are described. In the end, this argument may be purely semantic. For example, if dedifferentiation is responsible for allowing HSC to exit the blood lineage, then an uncommitted stem cell exists at least transiently.

The second theory that has generated a great deal of discussion in the plasticity community is the notion of cell fusion. This idea describes the possibility that hematopoietic lineage cells, especially macrophages, fuse with epithelial cells to produce a single cell that expresses markers of both recipient-derived epithelial cells and donor-derived BM cells. Two recently published studies support this idea.[155,156] In each study, primary cells (either neuronal stem cells or BM-derived cells) from female mice were cocultured with male murine embryonic stem cells and fused cells formed. In both studies, the in vitro culture conditions were designed to allow for selective survival and growth of fused cells. These cells had the tetraploid sex chromosome complement XXXY, but were otherwise morphologically and functionally similar to other ES cells. Although these studies clearly demonstrate that adult mammalian cells can fuse with ES cells under specific culture conditions, these conditions do not necessarily reflect the in vivo environments in which stem cell plasticity is thought to occur. The frequency of hybrid cell formation in these studies was at best 1 in 10,000, whereas studies of BMSC plasticity have found as many as 1 in 50 nonhematopoietic cells to be BM-derived without strong selection, and as high as 1 in 2-3 when strong selective pressures were present.

Despite the low frequency of cell fusion and the artificial conditions used to induce fusion in these studies, a number of authors have incorporated attempts to rule out fusion as the source of apparent BM-derived cells in studies of BMSC plasticity. In a study that found GFP+ Purkinje cells in the cerebellum of mice transplanted with GFP-labeled BM cells, the DNA complement of the Purkinje cells was not consistent with the possibility that they were formed by GFP+ macrophages fusing with existing neurons.[64] In addition, there was no correlation between the number of BMSC-derived Purkinje cells and macrophage engraftment in the cerebellum, and the GFP+ neurons did not stain with hematopoietic markers. In a similar study, the authors investigated the nuclear content of putative BMSC-derived GFP+ neurons in mice transplanted with GFP transgenic BM cells to rule out the possibility of fusion.[63] Only one nucleus was found in each of these neurons and the DNA ploidy was the same as in neighboring neurons. Thus, fusion could not explain the presence of the GFP+ neurons in either of these studies.

Other attempts to determine whether fusion can account for observations of stem cell plasticity in vivo have used FISH for X and Y-chromosomes after male to female transplantation. However, quantitative FISH for X and Y chromosomes in tissue sections is fraught with technical quandaries, especially in tissues (e.g., liver) known to undergo cellular fusion events. For example, if a rare XXXY cell is found by FISH in a 4 μm liver section from a female animal transplanted with male BM, it cannot be known whether this is the result of fusion, or if it is simply a polyploid cell with two Y chromosomes outside the plane of section. Careful controls using male and female tissues to normalize the numbers of cells with different chromosome complements will be needed in any such experiments.

Although the arguments against plasticity have not been substantiated, neither can they be dismissed. Data in support of the fusion hypothesis remain exceedingly sparse, and a clear answer to the question of whether fusion can explain instances of apparent metamoirosis awaits a carefully controlled and well-executed in vivo study. Current data also cannot confirm or rule out the existence of early, uncommitted stem cells in the BM and other tissues. If such a cell does exist, true cases of plasticity involving transdifferentiation or dedifferentiation will likely prove less common than many stem cell scientists currently believe them to be. However, if such pluripotent stem cells could be isolated from adults, expanded in vitro, and used autologously, their therapeutic potential would rival or exceed that of plastic adult stem cells. Again, only careful experimentation will reveal the true nature of the stem cells responsible for metamoirosis.

Conclusion

Taken together, the data provide sufficient evidence to conclude that BMSC, and other types of adult stem cells thought to be tissue-specific, are capable of metamoirosis. Although speculation on the potential of adult stem cells to treat or cure various diseases abounds, a great deal of work remains before we can confirm the scientific and medical usefulness of stem cell plasticity. One of the most pressing questions facing the field is what cell (or cells) in the BM is (are) responsible for lineage switching? It remains unclear whether there is an uncommitted, prehematopoietic BMSC that can give rise to HSC and MSC as well as to cell types outside the mesenchymal lineage. If such a cell does exist it would likely mean that true instances of plasticity are more rare than currently believed, but an ES-like cell in the adult could prove to be an ideal vector for cellular therapeutics. If such a cell does not exist, the cell(s) in the marrow (and those in other tissues that exhibit plasticity) responsible will need to be isolated and the mechanism, whether it be transdifferentiation, dedifferentiation, or some other process, governing lineage switching defined before metamoirosis can become a universally accepted theory. Finally, with a better understanding of the process, we can begin to test methods for controlling metamoirosis towards the ultimate goal of enhancing BMSC engraftment in nonhematopoietic tissues to therapeutic levels. Although in most studies published to date engraftment rates are less than one percent, too low to be effective for most therapeutic purposes, a few studies have reported promisingly high engraftment rates. Most notably, a study by Wang and colleagues reported BMSC engraftment rates over 30% in the livers of FAH-/- mice.[30]

The field of stem cell plasticity remains in an early stage of understanding despite an ever-growing literature. Still, it is hoped that further in vitro and in vivo analysis of stem cell differentiation into multiple cell types will provide the knowledge needed to develop and apply effective stem cell-based therapies in human patients.

References

1. Harrison DE, Stone M, Astle CM. Effects of transplantation on the primitive immunohematopoietic stem cell. J Exp Med 1990; 172(2):431-7.
2. Thomson JA et al. Embryonic stem cell lines derived from human blastocysts. Science 1998; 282(5391):1145-7.
3. Bishop AE, Buttery LD, Polak JM. Embryonic stem cells. J Pathol 2002; 197(4):424-9.
4. Weissman IL. Stem cells—scientific, medical, and political issues. N Engl J Med 2002; 346(20):1576-9.
5. Ferrari G et al. Muscle regeneration by bone marrow-derived myogenic progenitors. Science 1998; 279(5356):1528-30.
6. Uchida N, Weissman IL. Searching for hematopoietic stem cells: Evidence that Thy-1.1lo Lin- Sca-1+ cells are the only stem cells in C57BL/Ka-Thy-1.1 bone marrow. J Exp Med 1992; 175(1):175-84.
7. Smith LG, Weissman IL, Heimfeld S. Clonal analysis of hematopoietic stem-cell differentiation in vivo. Proc Natl Acad Sci USA 1991; 88(7):2788-92.
8. Wagers AJ et al. Little evidence for developmental plasticity of adult hematopoietic stem cells. Science 2002; 297(5590):2256-9.
9. Krause DS et al. Multi-organ, multi-lineage engraftment by a single bone marrow-derived stem cell. Cell 2001; 105(3):369-77.
10. Morrison SJ et al. Identification of a lineage of multipotent hematopoietic progenitors. Development 1997; 124(10):1929-39.
11. Duncan SA. Mechanisms controlling early development of the liver. Mech Dev 2003; 120(1):19-33.
12. Rogler LE. Selective bipotential differentiation of mouse embryonic hepatoblasts in vitro. Am J Pathol 1997; 150(2):591-602.
13. Spagnoli FM et al. Identification of a bipotential precursor cell in hepatic cell lines derived from transgenic mice expressing cyto-Met in the liver. J Cell Biol 1998; 143(4):1101-12.
14. Medlock ES, Haar JL. The liver hemopoietic environment: I. Developing hepatocytes and their role in fetal hemopoiesis. Anat Rec 1983; 207(1):31-41.
15. Feracci H et al. The establishment of hepatocyte cell surface polarity during fetal liver development. Dev Biol 1987; 123(1):73-84.
16. Montesano R et al. In vivo assembly of tight junctions in fetal rat liver. J Cell Biol 1975; 67(2PT.1):310-9.
17. Forbes S et al. Hepatic stem cells. J Pathol 2002; 197(4):510-8.
18. Alison MR et al. Liver damage in the rat induces hepatocyte stem cells from biliary epithelial cells. Gastroenterology 1996; 110(4):1182-90.
19. Alison MR, Golding M, Sarraf CE. Liver stem cells: When the going gets tough they get going. Int J Exp Pathol 1997; 78(6):365-81.
20. Paku S et al. Origin and structural evolution of the early proliferating oval cells in rat liver. Am J Pathol 2001; 158(4):1313-23.
21. Petersen BE et al. Bone marrow as a potential source of hepatic oval cells. Science 1999; 284(5417):1168-70.
22. Theise ND et al. Derivation of hepatocytes from bone marrow cells in mice after radiation-induced myeloablation. Hepatology 2000; 31(1):235-40.
23. Theise ND et al. Liver from bone marrow in humans. Hepatology 2000; 32(1):11-6.
24. Alison MR et al. Hepatocytes from nonhepatic adult stem cells. Nature 2000; 406(6793):257.
25. Lagasse E et al. Purified hematopoietic stem cells can differentiate into hepatocytes in vivo. Nat Med 2000; 6(11):1229-34.
26. Grompe M et al. Pharmacological correction of neonatal lethal hepatic dysfunction in a murine model of hereditary tyrosinaemia type I. Nat Genet 1995; 10(4):453-60.

27. Mallet VO et al. Bone marrow transplantation in mice leads to a minor population of hepatocytes that can be selectively amplified in vivo. Hepatology 2002; 35(4):799-804.
28. Wang X et al. Kinetics of liver repopulation after bone marrow transplantation. Am J Pathol 2002; 161(2):565-74.
29. Schwartz RE et al. Multipotent adult progenitor cells from bone marrow differentiate into functional hepatocyte-like cells. J Clin Invest 2002; 109(10):1291-302.
30. Wang X et al. Liver repopulation and correction of metabolic liver disease by transplanted adult mouse pancreatic cells. Am J Pathol 2001; 158(2):571-9.
31. Shen CN, Slack JM, Tosh D. Molecular basis of transdifferentiation of pancreas to liver. Nat Cell Biol 2000; 2(12):879-87.
32. Leblond CP. Classification of cell populations on the basis of their proliferative behavior. National Cancer Institute Monographs 1964; 14:119-150.
33. Jacobson M. Developmental Neurobiology. New York: Plenum Press, 1991:776.
34. Gage FH. Mammalian neural stem cells. Science 2000; 287(5457):1433-8.
35. Rakic P. Young neurons for old brains? Nat Neurosci 1998; 1(8):645-7.
36. Weiss S et al. Multipotent CNS stem cells are present in the adult mammalian spinal cord and ventricular neuroaxis. J Neurosci 1996; 16(23):7599-609.
37. Reynolds BA, Weiss S. Generation of neurons and astrocytes from isolated cells of the adult mammalian central nervous system. Science 1992; 255(5052):1707-10.
38. Morshead CM et al. Neural stem cells in the adult mammalian forebrain: A relatively quiescent subpopulation of subependymal cells. Neuron 1994; 13(5):1071-82.
39. Lois C, Alvarez-Buylla A. Proliferating subventricular zone cells in the adult mammalian forebrain can differentiate into neurons and glia. Proc Natl Acad Sci USA 1993; 90(5):2074-7.
40. Doetsch F et al. Subventricular zone astrocytes are neural stem cells in the adult mammalian brain. Cell 1999; 97(6):703-16.
41. Johansson CB et al. Identification of a neural stem cell in the adult mammalian central nervous system. Cell 1999; 96(1):25-34.
42. Palmer TD, Takahashi J, Gage FH. The adult rat hippocampus contains primordial neural stem cells. Mol Cell Neurosci 1997; 8(6):389-404.
43. Kempermann G, Kuhn HG, Gage FH. Genetic influence on neurogenesis in the dentate gyrus of adult mice. Proc Natl Acad Sci USA 1997; 94(19):10409-14.
44. Clarke DL et al. Generalized potential of adult neural stem cells. Science 2000; 288(5471):1660-3.
45. Rietze RL et al. Purification of a pluripotent neural stem cell from the adult mouse brain. Nature 2001; 412(6848):736-9.
46. Bjornson CR et al. Turning brain into blood: A hematopoietic fate adopted by adult neural stem cells in vivo. Science 1999; 283(5401):534-7.
47. Mezey E et al. Turning blood into brain: Cells bearing neuronal antigens generated in vivo from bone marrow. Science 2000; 290(5497):1779-82.
48. Rakic P. Neurogenesis in adult primate neocortex: An evaluation of the evidence. Nat Rev Neurosci 2002; 3(1):65-71.
49. Eglitis MA, Mezey E. Hematopoietic cells differentiate into both microglia and macroglia in the brains of adult mice. Proc Natl Acad Sci USA 1997; 94(8):4080-5.
50. Perry VH, Gordon S. Macrophages and microglia in the nervous system. Trends Neurosci 1988; 11(6):273-7.
51. Hickey WF, Kimura H. Perivascular microglial cells of the CNS are bone marrow-derived and present antigen in vivo. Science 1988; 239(4837):290-2.
52. Ling EA, Wong WC. The origin and nature of ramified and amoeboid microglia: A historical review and current concepts. Glia 1993; 7(1):9-18.
53. Kitamura T, Miyake T, Fujita S. Genesis of resting microglia in the gray matter of mouse hippocampus. J Comp Neurol 1984; 226(3):421-33.
54. Krall WJ et al. Cells expressing human glucocerebrosidase from a retroviral vector repopulate macrophages and central nervous system microglia after murine bone marrow transplantation. Blood 1994. 83(9):2737-48.
55. Neuhaus J, Fedoroff S. Development of microglia in mouse neopallial cell cultures. Glia 1994; 11(1):11-7.
56. Kaur C et al. Origin of microglia. Microsc Res Tech 2001; 54(1):2-9.
57. Baumann N, Pham-Dinh D. Biology of oligodendrocyte and myelin in the mammalian central nervous system. Physiol Rev 2001; 81(2):871-927.
58. Skoff RP, Knapp PE. The origins and lineages of macroglial cells. In: Neuroglia H, Kettenmann, Ransom BR, ed. Oxford, New York: Oxford University Press, 1995:135-148.
59. Brazelton TR et al. From marrow to brain: Expression of neuronal phenotypes in adult mice. Science 2000; 290(5497):1775-9.
60. Luskin MB. Restricted proliferation and migration of postnatally generated neurons derived from the forebrain subventricular zone. Neuron 1993; 11(1):173-89.
61. Lois C, Alvarez-Buylla A. Long-distance neuronal migration in the adult mammalian brain. Science 1994; 264(5162):1145-8.
62. Deisseroth K, Bito H, Tsien RW. Signaling from synapse to nucleus: Postsynaptic CREB phosphorylation during multiple forms of hippocampal synaptic plasticity. Neuron 1996; 16(1):89-101.
63. Corti S et al. Neuroectodermal and microglial differentiation of bone marrow cells in the mouse spinal cord and sensory ganglia. J Neurosci Res 2002; 70(6):721-33.
64. Priller J et al. Neogenesis of cerebellar Purkinje neurons from gene-marked bone marrow cells in vivo. J Cell Biol 2001; 155(5):733-8.
65. Castro RF et al. Failure of bone marrow cells to transdifferentiate into neural cells in vivo. Science 2002; 297(5585):1299.
66. Goodell MA et al. Isolation and functional properties of murine hematopoietic stem cells that are replicating in vivo. J Exp Med 1996; 183(4):1797-806.
67. Goodell MA et al. Dye efflux studies suggest that hematopoietic stem cells expressing low or undetectable levels of CD34 antigen exist in multiple species. Nat Med 1997; 3(12):1337-45.
68. Kopen GC, Prockop DJ, Phinney DG. Marrow stromal cells migrate throughout forebrain and cerebellum, and they differentiate into astrocytes after injection into neonatal mouse brains. Proc Natl Acad Sci USA 1999; 96(19):10711-6.
69. Sanchez-Ramos J et al. Adult bone marrow stromal cells differentiate into neural cells in vitro. Exp Neurol 2000; 164(2):247-56.
70. Woodbury D et al. Adult rat and human bone marrow stromal cells differentiate into neurons. J Neurosci Res 2000; 61(4):364-70.
71. Jiang Y et al. Multipotent progenitor cells can be isolated from postnatal murine bone marrow, muscle, and brain. Exp Hematol 2002; 30(8):896-904.
72. Blau H et al. Something in the eye of the beholder. Science 2002; 298(5592):361-2. author reply 362-3.
73. Corti S et al. Modulated generation of neuronal cells from bone marrow by expansion and mobilization of circulating stem cells with in vivo cytokine treatment. Exp Neurol 2002; 177(2):443-52.
74. Christ B, Ordahl CP. Early stages of chick somite development. Anat Embryol (Berl) 1995; 191(5):381-96.
75. Cossu G, Tajbakhsh S, Buckingham M. How is myogenesis initiated in the embryo? Trends Genet 1996; 12(6):218-23.
76. De Angelis L et al. Skeletal myogenic progenitors originating from embryonic dorsal aorta coexpress endothelial and myogenic markers and contribute to postnatal muscle growth and regeneration. J Cell Biol 1999; 147(4):869-78.
77. Birschoff. The satellite cell and muscle regeneration. In: Engel AG, Franszini-Armstrong C, eds. Myology. New York: McGraw-Hill, 1994:97-118.
78. Renault V et al. Human skeletal muscle satellite cells: Aging, oxidative stress and the mitotic clock. Exp Gerontol 2002; 37(10-11):1229-36.
79. Beauchamp JR et al. Expression of CD34 and Myf5 defines the majority of quiescent adult skeletal muscle satellite cells. J Cell Biol 2000; 151(6):1221-34.
80. Seale P et al. Pax7 is required for the specification of myogenic satellite cells. Cell 2000; 102(6):777-86.

81. Appell HJ, Forsberg S, Hollmann W. Satellite cell activation in human skeletal muscle after training: Evidence for muscle fiber neoformation. Int J Sports Med 1988; 9(4):297-9.
82. Schultz E, Jaryszak DL, Valliere CR. Response of satellite cells to focal skeletal muscle injury. Muscle Nerve 1985; 8(3):217-22.
83. Grounds MD, Yablonka-Reuveni Z. Molecular and cell biology of skeletal muscle regeneration. Mol Cell Biol Hum Dis Ser 1993; 3:210-56.
84. Grounds MD et al. Identification of skeletal muscle precursor cells in vivo by use of MyoD1 and myogenin probes. Cell Tissue Res 1992; 267(1):99-104.
85. Pereira RF et al. Cultured adherent cells from marrow can serve as long-lasting precursor cells for bone, cartilage, and lung in irradiated mice. Proc Natl Acad Sci USA 1995; 92(11):4857-61.
86. Bittner RE et al. Recruitment of bone-marrow-derived cells by skeletal and cardiac muscle in adult dystrophic mdx mice. Anat Embryol (Berl) 1999; 199(5):391-6.
87. Sicinski P et al. The molecular basis of muscular dystrophy in the mdx mouse: A point mutation. Science 1989; 244(4912):1578-80.
88. Gillis JM. Understanding dystrophinopathies: An inventory of the structural and functional consequences of the absence of dystrophin in muscles of the mdx mouse. J Muscle Res Cell Motil 1999; 20(7):605-25.
89. Gussoni E et al. *Dystrophin* expression in the *mdx* mouse restored by stem cell transplantation. Nature 1999; 401(6751):390-4.
90. LaBarge MA, Blau HM. Biological progression from adult bone marrow to mononucleate muscle stem cell to multinucleate muscle fiber in response to injury. Cell 2002; 111(4):589-601.
91. Jackson KA, Mi T, Goodell MA. Hematopoietic potential of stem cells isolated from murine skeletal muscle. Proc Natl Acad Sci USA 1999; 96(25):14482-6.
92. McKinney-Freeman SL et al. Muscle-derived hematopoietic stem cells are hematopoietic in origin. Proc Natl Acad Sci USA 2002; 99(3):1341-6.
93. Reese DE, Mikawa T, Bader DM. Development of the coronary vessel system. Circ Res 2002; 91(9):761-8.
94. Tevosian SG et al. FOG-2, a cofactor for GATA transcription factors, is essential for heart morphogenesis and development of coronary vessels from epicardium. Cell 2000; 101(7):729-39.
95. Dettman RW et al. Common epicardial origin of coronary vascular smooth muscle, perivascular fibroblasts, and intermyocardial fibroblasts in the avian heart. Dev Biol 1998; 193(2):169-81.
96. Mikawa T, Gourdie RG. Pericardial mesoderm generates a population of coronary smooth muscle cells migrating into the heart along with ingrowth of the epicardial organ. Dev Biol 1996; 174(2):221-32.
97. Larsen WJ. Human embryology. 2nd ed. New York: Churchill Livingston, 1997:512.
98. Soonpaa MH, Field LJ. Survey of studies examining mammalian cardiomyocyte DNA synthesis. Circ Res 1998; 83(1):15-26.
99. Quaini F et al. Chimerism of the transplanted heart. N Engl J Med 2002; 346(1):5-15.
100. Beltrami CA et al. Proliferating cell nuclear antigen (PCNA), DNA synthesis and mitosis in myocytes following cardiac transplantation in man. J Mol Cell Cardiol 1997; 29(10):2789-802.
101. Kajstura J et al. Myocyte proliferation in end-stage cardiac failure in humans. Proc Natl Acad Sci USA 1998; 95(15):8801-5.
102. Leri A et al. Telomerase activity in rat cardiac myocytes is age and gender dependent. J Mol Cell Cardiol 2000; 32(3):385-90.
103. Kessler PD, Byrne BJ. Myoblast cell grafting into heart muscle: Cellular biology and potential applications. Annu Rev Physiol 1999; 61:219-42.
104. Taylor DA et al. Regenerating functional myocardium: Improved performance after skeletal myoblast transplantation. Nat Med 1998; 4(8):929-33.
105. Klug MG et al. Genetically selected cardiomyocytes from differentiating embryonic stem cells form stable intracardiac grafts. J Clin Invest 1996; 98(1):216-24.
106. Hughes S. Cardiac stem cells. J Pathol 2002; 197(4):468-78.
107. Tomita S et al. Autologous transplantation of bone marrow cells improves damaged heart function. Circulation 1999; 100(19Suppl):II247-56.
108. Wang JS et al. Marrow stromal cells for cellular cardiomyoplasty: Feasibility and potential clinical advantages. J Thorac Cardiovasc Surg 2000; 120(5):999-1005.
109. Toma C et al. Human mesenchymal stem cells differentiate to a cardiomyocyte phenotype in the adult murine heart. Circulation 2002; 105(1):93-8.
110. Orlic D et al. Bone marrow cells regenerate infarcted myocardium. Nature 2001; 410(6829):701-5.
111. Jackson KA et al. Regeneration of ischemic cardiac muscle and vascular endothelium by adult stem cells. J Clin Invest 2001; 107(11):1395-402.
112. Stamm C et al. Autologous bone-marrow stem-cell transplantation for myocardial regeneration. Lancet 2003; 361(9351):45-6.
113. Kauffman MH. The Atlas of Mouse Development. London, San Diego: Academic Press, 1992:445-448. 512.
114. Warburton D et al. The molecular basis of lung morphogenesis. Mech Dev 2000; 92(1):55-81.
115. Otto WR. Lung epithelial stem cells. J Pathol 2002; 197(4):527-35.
116. Emura M. Stem cells of the respiratory epithelium and their in vitro cultivation. In Vitro Cell Dev Biol Anim 1997; 33(1):3-14.
117. Theise ND et al. Radiation pneumonitis in mice: A severe injury model for pneumocyte engraftment from bone marrow. Exp Hematol 2002; 30(11):1333-8.
118. Grove JE et al. Marrow-derived cells as vehicles for delivery of gene therapy to pulmonary epithelium. Am J Respir Cell Mol Biol 2002; 27(6):645-51.
119. Kotton DN et al. Bone marrow-derived cells as progenitors of lung alveolar epithelium. Development 2001; 128(24):5181-8.
120. Wells JM, Melton DA. Vertebrate endoderm development. Annu Rev Cell Dev Biol 1999; 15:393-410.
121. Wright NA. Epithelial stem cell repertoire in the gut: Clues to the origin of cell lineages, proliferative units and cancer. Int J Exp Pathol 2000; 81(2):117-43.
122. Ponder BA et al. Derivation of mouse intestinal crypts from single progenitor cells. Nature 1985; 313(6004):689-91.
123. Griffiths DF et al. Demonstration of somatic mutation and colonic crypt clonality by X-linked enzyme histochemistry. Nature 1988; 333(6172):461-3.
124. Thompson M et al. Gastric endocrine cells share a clonal origin with other gut cell lineages. Development 1990; 110(2):477-81.
125. Tatematsu M et al. Clonal analysis of glandular stomach carcinogenesis in C3H/HeN<==>BALB/c chimeric mice treated with N-methyl-N-nitrosourea. Cancer Lett 1994; 83(1-2):37-42.
126. Schmidt GH, Winton DJ, Ponder BA. Development of the pattern of cell renewal in the crypt-villus unit of chimaeric mouse small intestine. Development 1988; 103(4):785-90.
127. Nomura S et al. Clonal analysis of isolated single fundic and pyloric gland of stomach using X-linked polymorphism. Biochem Biophys Res Commun 1996; 226(2):385-90.
128. Novelli MR et al. Polyclonal origin of colonic adenomas in an XO/XY patient with FAP. Science 1996; 272(5265):1187-90.
129. Jiang Y et al. Pluripotency of mesenchymal stem cells derived from adult marrow. Nature 2002; 418(6893):41-9.
130. Brittan M et al. Bone marrow derivation of pericryptal myofibroblasts in the mouse and human small intestine and colon. Gut 2002; 50(6):752-7.
131. Al-Awqati Q, Oliver JA. Stem cells in the kidney. Kidney Int 2002; 61(2):387-95.
132. Herzlinger D et al. Metanephric mesenchyme contains multipotent stem cells whose fate is restricted after induction. Development 1992; 114(3): 565-72.
133. Oliver JA et al. Metanephric mesenchyme contains embryonic renal stem cells. Am J Physiol Renal Physiol 2002; 283(4):F799-809.
134. Okada H et al. Epithelial-mesenchymal transformation of renal tubular epithelial cells in vitro and in vivo. Nephrol Dial Transplant 2000; 15 Suppl 6:44-6.
135. Strutz F, Muller GA. Transdifferentiation comes of age. Nephrol Dial Transplant 2000; 15(11):1729-31.
136. Ng YY et al. Glomerular epithelial-myofibroblast transdifferentiation in the evolution of glomerular crescent formation. Nephrol Dial Transplant 1999; 14(12):2860-72.

137. Poulsom R et al. Bone marrow contributes to renal parenchymal turnover and regeneration. J Pathol 2001; 195(2): 229-35.
138. Imasawa T et al. The potential of bone marrow-derived cells to differentiate to glomerular mesangial cells. J Am Soc Nephrol 2001; 12(7):1401-9.
139. Lagaaij EL et al. Endothelial cell chimerism after renal transplantation and vascular rejection. Lancet 2001; 357(9249):33-7.
140. Andersen CB, Ladefoged SD, Larsen S. Cellular inflammatory infiltrates and renal cell turnover in kidney allografts: A study using in situ hybridization and combined in situ hybridization and immunohistochemistry with a Y-chromosome-specific DNA probe and monoclonal antibodies. Apmis 1991; 99(7):645-52.
141. Byrne C, Hardman M, Nield K. Covering the limb—formation of the integument. J Anat 2003; 202(1):113-23.
142. Sengel P. Pattern formation in skin development. Int J Dev Biol 1990; 34(1):33-50.
143. Hardman M.J et al. Barrier formation in the human fetus is patterned. J Invest Dermatol 1999; 113(6):1106-13.
144. Niemann C, Watt FM. Designer skin: Lineage commitment in postnatal epidermis. Trends Cell Biol 2002; 12(4):185-92.
145. Lavker RM, Sun TT. Epidermal stem cells: Properties, markers, and location. Proc Natl Acad Sci USA 2000; 97(25):13473-5.
146. Potten CS, Booth C. Keratinocyte stem cells: A commentary. J Invest Dermatol 2002; 119(4):888-99.
147. Taylor G et al. Involvement of follicular stem cells in forming not only the follicle but also the epidermis. Cell 2000; 102(4):451-61.
148. Oshima H et al. Morphogenesis and renewal of hair follicles from adult multipotent stem cells. Cell 2001; 104(2):233-45.
149. Ferraris C et al. Adult corneal epithelium basal cells possess the capacity to activate epidermal, pilosebaceous and sweat gland genetic programs in response to embryonic dermal stimuli. Development 2000; 127(24):5487-95.
150. Korbling M et al. Hepatocytes and epithelial cells of donor origin in recipients of peripheral-blood stem cells. N Engl J Med 2002; 346(10):738-46.
151. Hematti P et al. Absence of donor-derived keratinocyte stem cells in skin tissues cultured from patients after mobilized peripheral blood hematopoietic stem cell transplantation. Exp Hematol 2002; 30(8):943-9.
152. Brockes JP. Amphibian limb regeneration: Rebuilding a complex structure. Science 1997; 276(5309):81-7.
153. Oberpriller JO et al. Stimulation of proliferative events in the adult amphibian cardiac myocyte. Ann NY Acad Sci 1995; 752:30-46.
154. McGann CJ, Odelberg SJ, Keating MT. Mammalian myotube dedifferentiation induced by newt regeneration extract. Proc Natl Acad Sci USA 2001; 98(24):13699-704.
155. Terada N et al. Bone marrow cells adopt the phenotype of other cells by spontaneous cell fusion. Nature 2002; 416(6880):542-5.
156. Ying QL et al. Changing potency by spontaneous fusion. Nature 2002; 416(6880):545-8.

CHAPTER 12

Collection and Expansion of Stem Cells

Linda Kelley and Ian McNiece

Introduction

Stem cells are defined as undifferentiated cells that can proliferate and have the capacity of both self-renewal and differentiation. Most adult tissues have multipotential stem cells that are capable of producing a limited range of differentiated cell lineages appropriate to their location. Likewise, recent evidence suggests that stem cells of some tissues can give rise to differentiated cells of distinct tissues under certain circumstances. Therefore, stem cells provide a theoretically inexhaustible supply of cells with potential to replace tissues under conditions where age, disease or trauma has led to tissue damage or dysfunction. Specifically, stem cells offer great promise for tissue regeneration, cell-based transplantation therapies and gene therapy protocols. However, since the stem cell frequency in most adult tissues is rare, potential clinical applications will require the ability to expand stem cells ex vivo or in vivo. Most of the research regarding stem cell expansion has been done on hematopoietic stem cells. To date, no studies have demonstrated convincing data that the true multipotential hematopoietic stem cell can be expanded ex vivo without coincident induction of differentiation and generation of committed progenitor cells. Once differentiated, the progenitor cells acquire tissue-specific functional properties and the self-renewal properties of the stem cell population are lost. Since hematopoietic stem cells appear to have unlimited self-renewal properties in vivo, this suggests that the ex vivo culture conditions used to date are not representative of the in vivo conditions that support stem cell self-renewal. Recent studies described later in this chapter, have made great strides toward defining the genes, culture conditions and cell to cell interactions responsible for self-renewal. Through continued investigation it is anticipated that stem cell expansion ex vivo will be achievable.

However, depending on the clinical application, expansion and differentiation of progenitor cells rather than stem cells, may be acceptable to achieve the desired therapeutic effect. In regards to autoimmune disease, expansion of hematopoietic progenitor cells would be appropriate for supplementing hematopoietic stem cell grafts to shorten chemotherapy-induced pancytopenia post-transplantation. In regards to other diseases, mature functional cells will be required for therapeutic purposes. An example would be diabetes, in which transplantation of islet cells, rather than stem cells, will be required for treatment. Therefore it is important to consider the types of cells that will be needed for future regenerative therapies, understanding that there will be a role for stem, progenitor or mature cells, depending on the specific disease and purpose of the treatment.

Potential applications for stem cell expansion include: (i) procuring enough stem cells, from bone marrow, peripheral blood or cord blood, to achieve and sustain rapid engraftment of a transplant recipient following myeloablative conditioning regimens; (ii) increasing the pool of stem cells to be used as targets for delivery of gene therapy; and (iii) generation of highly purified cell populations free of contaminating tumor cells or other undesirable cells, such as mature lymphocytes, that are capable of causing graft-versus-host-disease (GVHD). This chapter will review the current methods for procuring and expanding stem and progenitor cells, with an emphasis on hematopoietic cells, including the anticipated clinical use and outcomes, where known.

Hematopoietic Stem Cells

Collection and Expansion

Historically, hematopoietic stem cells were collected exclusively from bone marrow. Bone marrow harvest requires that the donor be admitted to the hospital, undergo general anesthesia, and have the procedure performed in the operating room by a physician specifically trained in bone marrow harvest. In most instances the marrow is recovered by multiple needle sticks from the posterior iliac crests. On rare occasions, the bone marrow is recovered from the sternum. Although adverse events associated with bone marrow harvest are infrequent, significant risks associated with anesthesia, bleeding, cardiac puncture, and infection exist. With the introduction of hematopoietic growth factors as agents to mobilize stem cells from bone marrow into the peripheral blood,[1-3] apheresis procedures have become commonplace to collect peripheral blood stem cells. The primary advantage of peripheral blood stem cell collection to the donor is the reduced morbidity associated with bone marrow harvest. However, significant advantages also exist for the recipient. Hematopoietic recovery is accelerated with peripheral blood progenitor cells compared to bone marrow, without a significant increase in the incidence or severity of acute or chronic GVHD, despite an increased number of T lymphocytes in mobilized peripheral blood versus bone marrow.[4-6] More recently, umbilical cord blood has been used as an alternative to bone marrow as a source of transplantable cells.[7-9] However, due to the limited volume of cord blood and lower absolute number of stem cells, most cord blood transplants have been performed in pediatric patients. Several studies have been performed to determine if a variety of obstetric factors (e.g., long gestation, long labor, high infant and placenta weights, short interval between delivery of the infant and cord clamping),[10] the mode of birth (e.g., vaginal versus cesarean section)[11-14] or the timing of collection (e.g., before or after the placenta is delivered)[15,16] affect the volume of cord blood collected. In general, there is not a consensus drawn from these studies as to the best method of collecting cord blood, suggesting that the most important factor may be the experience and speed of the individual collecting the cord blood.

Assays for Measurement of Expanded Stem Cell Pools

Although the Holy Grail of stem cell biology is to expand stem cells while maintaining their ability to self-renew, this has been difficult to achieve ex vivo. Since it is currently not possible to purify human stem cells to a homogeneous population, several functional stem cell assays have been developed to quantify the stem cell pool. In vitro assays used as surrogates to measure stem cells include the long term culture initiating cell (LTCIC) assay, which detects primitive cells capable of giving rise to colony-forming cells (CFCs) after 5 weeks of culture on competent feeder layers;[17,18] the cobblestone area-forming cell (CAFC) assay, which detects primitive cells at limiting dilution on irradiated stromal layers after 28 days of culture;[19,20] the high proliferative potential colony forming cell (HPP-CFC) assay;[21] and the extended-LTC-IC (E-LTC-IC), which defines a small subpopulation of LTC-IC with more extensive proliferative capacity and the ability to be maintained for up to 10 weeks in culture.[22] These assays measure a heterogeneous population of primitive myeloid progenitors, but not cells with multi-lineage differentiation or self-renewal potential. More recently, studies have described culture systems that allow the differentiation of single human Lin⁻CD34⁺ cells into cells with myeloid, natural killer, B-lymphoid, dendritic, and/or T-lymphoid phenotypes, showing that a single cell can differentiate ex vivo into multiple lineages.[23-25] However, none of these assays is able to generate secondary primitive progenitors that again have multi-lineage differentiation potential. One exception is the myeloid-lymphoid initiating cell (ML-IC) assay described by Punzel et al[26] ML-IC are grown on a stromal cell feeder layer in the presence of early acting cytokines and give rise to multiple secondary progenitors that can reinitiate myeloid and lymphoid hematopoiesis ex vivo. Although the ML-IC represents a more primitive cell than previously described using an in vitro assay, the authors did not show that the ML-IC was capable of self-renewal. Taken together these results suggest that the self-renewal capacity of stem cells is most likely regulated by a variety of complicated mechanisms that currently cannot be reproduced in vitro. These mechanisms most likely include a role of the local microenvironment, cell-cell interactions and/or the presence of secreted cytokine(s) that have yet to be identified.

In mice, stem cells are quantified by use of the competitive in vivo repopulation assay, which measures the repopulating potential of stem cells to all lineages in animals that have been myeloablated and transplanted with donor cells of two different phenotypes or genotypes for purposes of identification.[27-30] The competitive repopulation assay is considered to be the most stringent measure of hematopoietic stem cells.[31] Since repopulation assays are not possible in humans, surrogate assays for transplantation of human progenitors in xenogeneic transplant recipients have been developed. These include severe combined immunodeficient (SCID)-mice,[32-34] beige-nude-SCID (BNX)-mice,[35,36] nonobese diabetic (NOD)-SCID-mice[37-40] and fetal sheep.[41,42] Using these transplant models, human cells of myeloid, natural killer (NK), T-lymphoid, and B-lymphoid lineages are produced for several months or years after transplantation. Analysis of the long-term engraftment of multi-lineage human cells or the ability to serially transfer human cells from primary to secondary and tertiary recipients provides an estimate of the initial stem cell pool.

Using the beige-nude-SCID (BNX) mouse, Nolta et al demonstrated that ex vivo expansion of CD34⁺ cells from adult bone marrow interferes with multilineage differentiation following transplantation as few, if any, B cells were generated.[36] In the fetal sheep model, exhaustion of human hematopoiesis was observed with ex vivo expanded, but not unmanipulated, adult bone marrow.[43] It is unknown how the requirement of human stem cells to home to and engraft in a xenogeneic microenvironment impacts on the absolute stem cell measurement, however compared with controls (unexpanded cells), both studies indicated that expansion compromised long-term engraftment. Likewise, in a syngeneic nonhuman primate model of retrovirally-marked stem cell transplantation, engraftment of ex vivo expanded cells was significantly poorer than had been observed with unexpanded cells that had been retrovirally transduced.[44,45] Although the in vivo transplant models indicate that multipotential stem cells are not effectively expanded with the approaches previously utilized, it is commonly believed that continued research into the mechanisms that govern the proliferation and differentiation of hematopoietic stem cells will lead to the development of culture systems that are capable of stem cell expansion.

Molecular Mechanisms Associated with Stem Cell Expansion

In order to expand stem cells, the cells must undergo repeated symmetrical cell divisions, in which both daughter cells retain the characteristics of the original cell. However, in single cell experiments investigating the proliferative potential of human fetal liver stem cells, cell cycle properties and replating potential were unevenly distributed among the daughter cells, suggesting an inherent asymmetric distribution of functional properties.[46,47] Likewise, others have demonstrated that asymmetric divisions occur in hematopoietic progenitors, which seemed to be related to more primitive function.[48-52] Using an assay to detect primitive in vitro function (myeloid lymphoid-initiating cell assay, ML-IC), Punzel et al[53,54] demonstrated that single human CD34⁺CD38⁻ cord blood cells gave rise to daughter cells with unequivocal ML-IC capacities, which were independent of the cytokine combinations used in the cultures. Together, these data suggest that the stem cell pool cannot be significantly expanded in vitro, however, again this may reflect differences in stem cell regulation in tissue culture versus in vivo.

Another explanation for the inability to expand stem cells may be that they are unable to divide in the culture conditions used to date. Most data suggest that ex vivo expansion recruits stem cells from a G_0 to a G_1 state, which may be responsible for the defective repopulation ability of expanded stem cell grafts.[55,56] Most likely this results from the permissive role that the growth factors used in expansion cultures have on survival and proliferation, rather than directing self-renewal.

Stem cell expansion may necessitate a requirement for stem cell homing to the BM microenvironment. Studies in which hematopoietic cells were cultured short- or long-term with high concentrations of cytokines have shown that migration diminishes depending on the duration of cytokine exposure and on the type of cytokine used. Ex vivo culture of stem cells is associated with down regulation of expression of members of the β_1 integrin family[57-59] as well as the chemokine receptor, CXCR4,[60] both of which play key roles in homing and engraftment. Additional studies of the role of homing receptors on stem cell engraftment are needed in order to optimize expansion of engrafting stem cells.

An increase in cell death, or apoptosis, of stem cell progeny may be associated with the inability to expand stem cells. Ex vivo culture is associated with increased expression of Fas ligand CD95[61] and down-regulation of the anti-apoptosis *Bcl2* gene[62] on CD34+ cells. Further evidence that apoptosis is a normal mechanism of regulation of the stem cell pool is provided by the observation that transgenic mice that overexpress Bcl-2 have an increased stem cell pool.[62] The same authors demonstrated that Bcl-2-overexpressing stem cells survive better in vitro and have a competitive engraftment advantage over wild-type stem cells in vivo.

Recent data indicate that a variety of regulatory molecules that are active in early development, homeobox (HOX) transcription factors, may also play a role in the maintenance of stem cell self-renewal. Multiple HOX family members are expressed in the most primitive hematopoietic stem cell-enriched populations and their expression is down regulated in terminally differentiating CD34- cells.[63] Overexpression of the homeobox gene, Hoxb4, in murine[64-66] and human cord blood[67] stem cells leads to enhanced proliferation of clonogenic progenitors in vitro and an enhanced ability to regenerate the most primitive stem cell compartment following serial transplantation in mice. Importantly, none of the mice in these experiments demonstrated hematological abnormalities following transplantation. HOXB4-transduced murine bone marrow cells led to rapid, extensive and highly polyclonal stem cell expansion in vitro resulting in a 1000-fold net increase of stem cells that retained full lymph-myeloid repopulating potential.[68] Together these data suggest that HOXB4 overexpression enhances the rate of stem cell expansion without impairing normal differentiation or causing transformation. Other members of the HOX family, including HOXC4,[69] HOXa9,[70] LIM-homeobox 2 (LH2)[71] and HOX11[72] have been shown to have similar effects as HOX4B on stem and early progenitor cell expansion, but in some cases led to immortalization[71,72] or transformation[70] after prolonged expression. Recent review articles provide an extensive description of the role of HOX genes in normal hematopoiesis and leukemogenesis.[73,74] Together these findings suggest exciting new areas for genetic manipulation of HOX genes in hematopoietic stem cell regulation, and in the case of HOXB4, perhaps for therapeutic stem cell expansion.

Additional genetic regulators of a variety of signaling pathways involved in embryonic development may also regulate stem cell self-renewal in hematopoietic, as well as other tissue types. Stimulation of Notch activation in hematopoietic stem cells leads to increased primitive progenitor activity both in vitro and in vivo and maintenance of multipotentiality.[75-79] In addition to hematopoietic cells, both neuroepithelial progenitor[80] and germ cell[81] self-renewal appears to be regulated by the Notch signaling pathway. Sonic hedgehog (Shh) signaling has been implicated in regulation of self-renewal of highly enriched human CD34+CD38-Lin- stem cells in vitro,[82] neuronal precursor cells[83] and ovarian stem cells.[84] Finally, recent studies suggest that the Wnt (secreted signaling factors which influence cell fate and cell behavior in developing embryos) signaling pathway is involved with stem cell self-renewal in mouse fetal liver[85] and human bone marrow.[86] Retroviral transduction of activated β-catenin, a downstream activator of the Wnt signaling pathway, in cultured human keratinocytes with high proliferative potential resulted in increased epidermal stem cell self-renewal and decreased differentiation.[87] Mice lacking TCF-4, one of the transcriptional regulators of the Wnt signaling pathway, had a decrease in the number of undifferentiated progenitors in the crypts of the gut epithelium during fetal development, suggesting a role for the Wnt pathway in self-renewal of gut epithelial stem cells.[88] Further evidence that the Notch, Shh and Wnt signaling pathways are involved in stem cell self-renewal is provided by the observation that they are also frequently disrupted in cancer cells that have acquired unlimited self-renewal properties.[89] The molecular mechanisms for stem cell self-renewal and whether the Notch, Shh and Wnt pathways cooperate in some way to regulate self-renewal remain to be determined.

The Road to Clinical Trials

Expansion of Hematopoietic Progenitor Cells by Growth Factors

The first demonstration of the use of growth factors to generate increased numbers of specific cell populations was performed by Bradley et al in the early 1980's using crude conditioned media (CM) as a source of hematopoietic growth factors.[90] In these studies it was shown that incubation of post fluorouracil (FU) mouse bone marrow cells in WEHI-3 CM for 7 days, resulted in a 60-fold increase of primitive progenitor cells (CFU-S), and a 53-fold increase in committed progenitor cells (GM-CFC). In subsequent studies from the same group, it was shown that preincubation with CM could expand primitive murine progenitor cells (HPP-CFC) and cells with in vivo marrow repopulating ability.[91,92] Using a similar culture system of human bone marrow cells in teflon bottles, it was shown that the combination of recombinant human granulocyte macrophage colony-stimulating factor (rhGM-CSF) plus recombinant human interleukin 3 (rhIL-3) could generate a 7-fold increase in committed progenitor cells (GM-CFC).[93] In 1991, Bernstein et al[94] demonstrated that incubation of single CD34+Lin- cells in the combination of IL-3, granulocyte colony stimulating factor (G-CSF) and stem cell factor (SCF) gave rise to a 10-fold increase of colonies in vitro.

The use of ex vivo expansion to generate mature neutrophil precursors was proposed in 1992 by Haylock et al[95] These authors demonstrated that the combination of IL-1, IL-3, IL-6, GM-CSF and SCF could generate a 1,324-fold increase in nucleated cells and a 66-fold increase in GM-CFC. The static cultures utilized CD34+ cells as the starting population. Several investigators have demonstrated a requirement for CD34 selection of the starting cells for optimal expansion.[95-98] Subsequent studies were performed using optimal culture conditions in teflon bags and with fully defined media appropriate for clinical applications.[99] This work utilized a growth factor cocktail consisting of SCF, G-CSF and megakaryocyte growth and development factor (MGDF).[99] Other cocktails of growth factors are effective in expanding CD34⁺ cells; however the availability of clinical grade growth factors has been limited due to commercial considerations.

The in vivo potential of ex vivo expanded cells was first reported in murine studies by Muench et al[100] This study demonstrated that bone marrow cells expanded in SCF plus IL-1 engrafted lethally irradiated mice and were capable of sustaining hematopoiesis long term in these animals. In addition, the bone marrow from engrafted mice could repopulate secondary recipients. The authors concluded that the expansion of mouse bone marrow cells did not adversely affect the proliferative capacity and lineage potential of the stem cell compartment.[100]

Preclinical Studies of ex Vivo Expanded Cells in Baboons

Recent studies in normal baboons,[101] have demonstrated the potential clinical benefit of ex vivo expanded cells. Andrews et al harvested peripheral blood progenitor cells (PBPC) from G-CSF mobilized normal baboons and expanded the CD34⁺ cells for 10 days using human growth factors SCF, G-CSF and MGDF. After culture the cells were washed and infused into baboons that had undergone lethal irradiation. The fold-expansion obtained was low compared to human CD34⁺ cells and most likely resulted from species variations of the growth factors and culture conditions developed for expansion of human cells. However, GM-CFC were expanded 7- to 8-fold. Animals transplanted with expanded CD34⁺ cells and given post transplant G-CSF and MGDF, had a significantly shorter duration of neutropenia and significantly higher WBC and PMN nadirs compared to animals in the control groups. In fact, two of 3 animals had no days with neutrophil counts below 500/µl, a clinical endpoint used for documentation of neutrophil engraftment. In these studies, in vitro expansion did not influence platelet recovery despite the use of MGDF in cultures and after transplantation. Further studies will be needed to determine culture conditions that will enhance platelet recovery of PBPC. An important point highlighted by these studies is the requirement for growth factors after the transplantation of the ex vivo expanded cells, as animals receiving expanded cells without growth factors post transplant had no significant improvement in engraftment compared to control animals, whereas animals who received expanded cells and growth factors post transplant had faster recovery of neutrophils.[101] It is well known that culture of hematopoietic cells in vitro requires growth factors for survival and proliferation and in the absence of growth factors cells go into an apoptotic state and die. Therefore, it is possible that transplantation of expanded cells in the absence of growth factor treatment of the recipient results in apoptosis of the cells after infusion.

Clinical Studies of the Role of ex Vivo Expanded Hematopoietic Progenitor Cells

One of the first clinical studies using expanded stem cells in humans was conducted by Naparstek et al[102] who performed a short culture of BM in the presence of GM-CSF and IL-3. Patients undergoing allogeneic BMT for malignant hematological diseases were transplanted with two-thirds of the BM on the scheduled day of transplant and one-third of the BM following incubation in GM-CSF plus IL-3 (100 ng/ml) on day +4. The BM was cultured in RPMI with pooled inactivated human AB serum (7-10%) for 4 days. No significant acceleration of neutrophil recovery was observed, however, the median time to reach an unsupported platelet count of 25,000/µl was 17 days for patients receiving expanded cells versus 23 days for the control group.[102] No data was provided in the report on the fold expansion of nucleated cells or GM-CFC, although the increase in the number of GM-CFC correlated with rapid platelet recovery in 12 patients with durable engraftment (p=0.02). The authors concluded from the study that in vitro culture of BM cells with IL-3 plus GM-CSF may provide a cell product capable of enhancing marrow recovery.

Brugger et al[103] expanded 10% of a CD34-selected autologous PBPC product using a combination of SCF, IL-1β, IL-3, IL-6 and erythropoietin (Epo) for 12 days in culture with RPMI plus 2% autologous plasma. The median number of cells generated ex vivo was 2.4 to 23.1 million, which corresponded to a median increase of 62.4-fold (range 33.4 to 115.5) in the number of total nucleated cells. A median increase in CFC of 50.3-fold (range 14.4 to 92.5) was observed. The patients had rapid and sustained engraftment however, the time to recovery of hematopoiesis was not significantly different from historical controls. No toxic effects were observed with the infusion of the expanded cells. The contribution of the expanded cells was difficult to assess in this study, as the chemotherapy used may not have been totally myeloablative, so that the contribution of endogenous recovery was unknown.

In a second study, Williams et al[104] infused cells expanded in PIXY321 (a GM-CSF/IL-3 fusion protein) for 12 days. Nine patients with metastatic breast cancer were mobilized with chemotherapy plus G-CSF and one apheresis product from each patient was CD34-selected using the Isolex 300i (Baxter) CD34 selection device. The CD34⁺ cells were cultured in gas permeable bags containing serum-free X-VIVO 10 medium (BioWhittaker) supplemented with 1% human serum albumin and 100 ng/ml PIXY321. At day 12 of culture the median expansion was 26-fold (range 6 to 64). The final product contained an average of 29.3% CD15⁺ neutrophil precursors with a range of 18.5% to 48.1%. The patients received the cryopreserved unmanipulated PBPC products on day 0 and the expanded cells on day +1. The median number of expanded cells reinfused was 44.6 million/kg with a range of 0.8 to 156.6 million cells/kg. Although the infusion of expanded cells was well tolerated, no clinical effects of the expanded cells were observed.

Alcorn et al[105] expanded CD34⁺ cells from cryopreserved PBPC. This study is difficult to interpret as there were significant losses of total cells and progenitors after thawing and selection. The recovery of cells post cryopreservation was 58% for total cells and only 21% for GM-CFC. The CD34⁺ cells were selected using the Isolex 300i and cultured for 8 days in Progenitor-34 media (Life Technologies) supplemented with 5-10% autologous serum and SCF (10 ng/ml), IL-3 (10 ng/ml), IL-6 (20 ng/ml), IL-1β (10 ng/ml) and Epo (2 U/ml). The mean increase in total cell number and GM-CFC was 21- and 139-fold, respectively. Ten patients with nonmyeloid malignancy were reinfused with the expanded cells without any adverse effect. The time to neutrophil and platelet recovery was identical to that of historical controls.

In more recent studies, three groups using similar culture conditions have reported decreased time to neutrophil recovery with ex vivo expanded PBPC.[106-108] Two groups cultured CD34⁺ cells in Defined Media (Amgen Inc) supplemented with the growth factors rhSCF, rhG-CSF and rhMGDF at 100 ng/ml for 10 days in teflon bags (American Fluoroceal),[106,107] while the third group cultured a mononuclear fraction in the same media and growth factors.[108]

Reiffers et al[106] recently reported the results of a phase I/II study in myeloma patients (N=14) using ex vivo expanded PBPC. A median of 4.1×10^6 CD34$^+$ cells were expanded for 10 days after which time the cells were washed and reinfused into patients. Unexpanded PBPC were also transplanted into patients on day +1 and the patients received rhG-CSF until neutrophil engraftment was achieved. The post-transplant neutropenia was markedly reduced in these patients compared to historical controls. The median number of neutropenic days (ANC < 500/μl) was 1.5 with a range of 0 to 7 days. Nine of the 14 patients never had a neutrophil count below 500/μl. In a comparable group of myeloma patients (n=242) receiving unexpanded PBPC, the median duration of severe neutropenia was 6 days (range 2 to 24).

Paquette et al[108] cultured unselected PBPCs in teflon bags in SCF, G-CSF and MGDF for 9 days. Twenty-four breast cancer patients received between 2 and 24×10^9 PBPCs cultured at 1, 2 or 3×10^6 cells/ml following high-dose chemotherapy and infusion of at least 5×10^6 CD34$^+$ unexpanded PBPC/kg. The fold expansion of CD34$^+$ cells in culture was reported to be inversely proportional to the initial cell density. No toxicity resulted from infusion of the expanded cells. Eleven patients (46%) recovered neutrophils to > 500/μl by days 5 or 6 post transplant. None of 78 historical control breast cancer patients had neutrophil recovery by the sixth day. Eight patients (33%) had 3 or fewer days of neutropenia and 11 patients did not experience neutropenic fevers or require broad-spectrum antibiotics.

In a study performed at the University of Colorado, 21 patients with high risk stage II, III, or IV breast cancer were treated with ex vivo expanded PBPC.[107] All patients were mobilized with rhG-CSF (10 μg/kg/day) for 9 days and PBPC were harvested on days 5 through 9, with CD34$^+$ cell selection preformed on the first four collections. The fifth collection was frozen unselected as a backup product. CD34$^+$ selection was performed using the Isolex 300i. After selection, each product was frozen in liquid nitrogen. On day -10 of treatment, two PBPC products were thawed and placed into ex vivo expansion culture. The cells were diluted in Defined Media supplemented with 100 ng/ml each of rhSCF, rhMGDF and rhG-CSF to 20,000 cells per ml in 800 ml of media and transferred into teflon bags (American Fluoroceal). The bags were incubated at 37°C for 10 days in a 5% CO$_2$ incubator. On day 0 of treatment the cultures were harvested using a cell washer (Cobe) and the media and growth factors removed with washing. Following ex vivo expansion of the CD34 selected cells, patients in cohort 1 (N=10) were reinfused with expanded cells on day 0 followed by unexpanded CD34$^+$ cells. Patients in cohort 2 (N=10) received only ex vivo expanded cells and the unexpanded CD34$^+$ cells were maintained frozen in liquid nitrogen as a backup source of hematopoietic cells. Transplantation of ex vivo expanded PBPC resulted in rapid engraftment of neutrophils (ANC>500/μl) with one patient engrafting on day 4 and a number of patients engrafting on days 5 and 6. The median time of neutrophil engraftment was day 6 in the first 15 patients. Historical controls had a median time to neutrophil engraftment of 9 days, with a range of 7 to 30 days. The patients in cohort 2 were transplanted with only expanded cells and all patients are alive and have maintained a durable graft at two years post-transplantation. No significant effect on platelet recovery was observed in any of the cohorts studied. These patients will be monitored long term to determine if expansion of the cells compromises long term engraftment. However, since the expanded cells were not marked by retroviral transduction, the contribution of endogenous hematopoietic recovery will be difficult to assess. Failure of the expanded cells to have an effect on platelet recovery may result from culture conditions that do not support platelet precursors. Alternatively, the expanded cells may require thrombopoietin (TPO) following transplantation to drive platelet production. These results demonstrate that infusion of expanded cells improves neutrophil engraftment, and suggest that this approach may be beneficial to reduce complications associated with neutropenia.

Ex Vivo Expansion of Cord Blood Cells

Bone marrow transplantation (BMT) from HLA-matched related and unrelated donors has been successfully used to treat patients with hematological malignancies. The major limitation to BMT is the availability of a suitable donor. The National Marrow Donor Program (NDMP) has identified a pool of 5,000,000 potential donors and as of May, 2003 had facilitated over 15,000 unrelated donor BM transplants. However, the availability of an unrelated donor BM is still limited due to: (i) the length of time for the donor search process (range 1 month to 6 years);[109] (ii) donor availability at the time of request and; (iii) limited availability of donors in certain racial and ethnic populations. Consequently, less than 40% of patients who could benefit from BMT have a suitable donor identified and of those who have a donor identified, less than 40% receive a transplant. Over the past 10 years, cord blood (CB) has been investigated as an alternative source of hematopoietic tissue for allogeneic transplantation of patients lacking an HLA-matched marrow or PBPC donor.[110] The first evidence that CB could be used to engraft an allogeneic recipient was provided by Gluckman et al in 1989 following successful transplantation of a child with Fanconi's anemia.[111] Since that time over 1,000 CB transplants have been performed worldwide and the current status of CB banking and transplantation have been reviewed in several recent reports.[112-114]

The ease of collection, potential availability to ethnic groups who are underrepresented in the NMDP Registry and the decrease in search time are some advantages of CB compared to BM or PBPC.[115] In addition, CB contains fewer and/or more naive T cells than BM or PBPC and thus permits a greater degree of mismatch with less GVHD. However, the total number of nucleated cells in CB is low compared to BM or PBPC, which largely restricts the majority of CB recipients to children with an average weight of 20 kg. The progression-free survival rates reported thus far are comparable to the results achieved with allogeneic BMT, with a suggestion of decreased GVHD; however the time to neutrophil and platelet engraftment in CB recipients is delayed compared to BM or PBPC.[9,116-118]

A common finding among all studies using allogeneic CB for transplantation is that higher nucleated cell doses have a positive impact on neutrophil engraftment and is one of the factors which predicts for better survival.[9,116-118] The study by Gluckman et al suggested a minimum total nucleated cell dose of 3.7×10^7 per kg of recipient's weight.[116] Others have evaluated the CD34 cell content in the graft and recommended a minimum of 1.7×10^5 (see ref. 118) and $2-3 \times 10^5$ [119] CD34$^+$ cells per kg. Since the volume of cord blood is limited (approximately 70-100 ml per collection), so are the numbers of total nucleated and CD34$^+$ cells, thus the size of the CB recipient is usually limited to pediatric patients in order to infuse the minimum cell dose. For CB to become a routine stem cell source for adult patients, expansion of the CB stem and progenitor cell pool will be necessary.

Preclinical Studies of ex Vivo Expansion of CB

Preclinical studies with CB products have demonstrated the potential to expand the numbers of GM-CFC in vitro and potentially broaden the use of CB transplantation to adult patients. CB expanded for 10 days in a growth factor cocktail of SCF, G-CSF and MGDF had a mean 73-fold expansion of GM-CFC.[120] An average total of 490 x 10^4 GM-CFC was harvested after the 10-day culture, representing a potential graft product of 23 x 10^4 GM-CFC/kg for a 20 kg patient.[120] Compared to a PBPC graft this would be approximately equivalent to the minimal GM-CFC harvest for optimal engraftment. Total nucleated cells were expanded approximately 100-fold. In other studies, use of early acting cytokines, Flt3L and TPO, were effective in promoting proliferation of early progenitor cells (CD34$^+$CD38$^-$) and increased cell expansion in long term cultures.[121-124] Yamaguchi et al[125] achieved a 250-fold expansion of CD34$^+$CD38$^-$ CB cells after three weeks culture on a monolayer of human BM-derived primary stromal cells in the presence of SCF, TPO, Flt2/3L and human AB serum. Together these studies demonstrate that certain early progenitor cell populations can be successfully expanded in vitro but do not assess the ability of the expanded cells to reconstitute long-term hematopoiesis in a myeloablated recipient.

To determine the effect of cell expansion on long-term engraftment potential, several investigators have utilized the nonobese diabetic severe combined immunodeficient (NOD/SCID) mouse model. Tanavde et al[126] compared the expansion and engraftment potential of CD34$^+$ CB and mobilized PBPC cultured ex vivo for 1-4 weeks in serum-free medium containing Flt-3L, SCF and TPO. Expanded cells were then infused into NOD/SCID mice. The ex vivo expanded CB cells generated greater numbers of total nucleated cells, CD34$^+$ cells and colony-forming cells in culture than did PBPC. In addition, the expanded cord blood cells demonstrated NOD/SCID engrafting potential, whereas expanded PBPC did not. In an interesting study by Su et al[127] platelet-derived growth factor (PDGF) significantly enhanced ex vivo expansion of LT-CIC and engraftment of human CD45$^+$ cells in NOD/SCID mice using CD34$^+$ CB cells cultured with TPO, IL-6 and Flt-3L for 12-14 days. The mechanism of PDGF-enhancement of CD34$^+$ CB cells remains to be determined, but may result from a direct effect of PDGF on committed CD34$^+$ progenitors or indirectly, via stimulation of the stromal feeder cells. To determine the role of serum in ex vivo expanded CB cells, Berger et al,[128] compared serum-free media with human or fetal calf serum. In this study, CD34$^+$ CB cells were cultured with Flt-3L, TPO and SCF for up to 8 weeks. Following 2 weeks of culture the expanded cells were transplanted into NOD/SCID mice. Although the presence of either serum was required for expansion of CFC and LT-CIC, cells expanded in serum-free media were able to engraft both the bone marrow and spleen in all of the mice examined. In a second study to evaluate serum-free media, Lam et al[129] compared X-Vivo-10, QBSF-60 (Quality Biological) and Stem Span SFEM (Stem Cell Technologies) with Iscove's modified Dulbecco's (IMDM) medium with fetal calf serum(FCS). The culture conditions included TPO, SCF, and Flt-3L with and without G-CSF, and/or IL-6 for 8-10 days. The study demonstrated that both QBSF-60 and StemSpan SFEM supported high yields of early progenitors (CD34$^+$CD38$^-$cells, CFU-granulocyte erythroid macrophage megakaryocyte [GEMM], CFU-GM, burst-forming unit-erythroid [BFU-E], and CFU-E). Likewise, QBSF-60- expanded cells (others were not tested) engrafted and differentiated into both myeloid and lymphoid lineages in NOD/SCID mice.

Although encouraging results have been obtained from all of these studies, the long-term engraftment potential could not be evaluated due to the relatively short lifespan of the NOD/SCID mice. However, in a more recent study Piacibello et al[130] utilized lentiviral vectors to transfer Green Fluorescent Protein (GFP) into CD34$^+$ CB cells that were cultured for up to 4 weeks in Flt-3L, TPO, SCF, and IL-6 prior to serial transplantation into NOD/SCID mice. Mice receiving the transduced expanded cells had higher levels of human engraftment than those transplanted with transduced, nonexpanded cells. Engraftment was multilineage and multiclonal (as determined by Southern Blot and Inverse PCR analysis of marked cells).

To determine the effect of ex vivo expansion on long-term engrafting cells, McNiece et al[131] used a fetal sheep xenogeneic transplant model. CD34$^+$ CB cells were cultured with SCF, MGDF and G-CSF for 14 days in Defined Media in the absence of serum. Starting cells (uncultured) and cultured cells were transplanted in 60-day-old fetal sheep and evaluated at various time points post transplant for the presence of human cells. Long-term engrafting cells were assessed by serial passage into secondary and tertiary recipients. The study demonstrated that although the expanded CB cells provided more rapid engraftment, no human cells were detectable at 16 months post transplant, indicating that the initial engraftment was transient. On the contrary, unexpanded cells persisted at 16 months and beyond. Secondary animals failed to engraft with expanded cells suggesting that although ex vivo expanded cells may be able to provide rapid short-term engraftment, the long-term potential of the cells may be compromised. The authors concluded that clinical protocols may require the use of two fractions of cells, an expanded CB graft to provide rapid short-term engraftment and an unmanipulated fraction of CB to provide long-term engraftment.

Clinical Trials Utilizing ex Vivo Expanded CB Cells

Shpall et al have recently reported the results of 37 patients (25 adults, 12 children) with hematological malignancies (34) or breast cancer (3), who were appropriate candidates for high dose chemotherapy requiring cellular support, and who were transplanted with expanded cord blood cells.[132] A fraction of each CB graft was CD34-selected and cultured ex vivo for 10 days in Defined Medium with SCF, G-CSF and MGDF. The remainder of the CB was infused without further manipulation. One cohort of patients had 40%, and a second cohort 60%, of their graft expanded. All patients received a median of 0.99 x 10^7 total nucleated cells per kg. The median time to engraftment of neutrophils was 28 days (range, 15-49 days) and of platelets was 106 days (range, 38-345 days). All evaluable patients who were followed for 28 days or longer achieved engraftment of neutrophils and, at a median follow-up of 30 months, 35% of the patients were alive. No significant differences were observed between the two cohorts of patients. The data suggest that a potential benefit of expanded CB cells may be the lack of graft failure compared to studies using unexpanded CB. Gluckman et al[116] reported that only 48% of patients greater than 45 kg body weight had engrafted neutrophils by day 60 post transplant. Although this study demonstrates that the use of expanded CB for transplantation is feasible and may provide a means of increasing cell numbers to provide more suitable grafts for adults, it does not specifically address the primary concern of loss of long-term engraftment potential of ex vivo expanded cells. The contribution, if any, of ex vivo expanded cells to long-term multilineage engraftment will require a marking study with long follow-up in a population of patients with a relatively good prognosis.

In a second study Kurtzberg et al,[133] transplanted 21 patients with unmanipulated CB cells on day 0 and expanded CB cells on day +12. No significant effects on engraftment kinetics were observed in these patients. Stiff et al[134] also reported on expansion of CB cells in the Aastrom Bioscience system for 9 patients. The median weight of the patients was 74 kg with a range of 47 to 117 kg. The median time to neutrophil engraftment was 26 days with a range of 14 to 36 days. Engraftment of platelets was delayed in these patients. The authors concluded that ex vivo expanded CB cells may be useful in adults with otherwise incurable hematologic disorders. The Aastrom Bioscience system has also been used for expansion of bone marrow. Studies have demonstrated that expansion of a small aliquot of bone marrow cells can provide short and long term engraftment following myeloablative chemotherapy.[135]

Embryonic Stem Cells

Collection and Expansion

The recent derivation of human embryonic stem (ES)[136,137] and embryonic germ (EG)[138] cell lines has implicated their potential use in regenerative medicine. Undifferentiated ES and EG cells are genetically normal, and in principle, are capable of differentiating into every cell type in the human body. This renders them ideal cell lines for the derivation of purified tissue types suitable for transplantation. Current research centers on promoting stem cell differentiation to the desired lineage, derivation of highly purified cell populations, confirmation of lack of oncogenic potential, and implantation in a form that will replace or augment the function of diseased or degenerating tissues in the absence of immune rejection.

To derive human ES cells, cleavage-stage embryos are grown into blastocysts and the inner cell mass is removed.[136,137] Another approach to deriving human ES cells is via somatic nuclear transfer or cloning. This involves the transfer of a somatic cell nucleus into an enucleated oocyte,[139-141] and subsequent embryonic development to the blastocyst stage prior to the isolation of ES cells from the inner cell mass. Human EG cells are derived from the genital ridges of early-stage embryos.[138]

Although emphasis has been placed on differentiation of ES cells into specific lineages, it is important to understand the processes that maintain ES cells in the undifferentiated state, so as to prevent spontaneous differentiation and increase cell yields. This will have economic implications when protocols for deriving specific cell lineages are scaled up for clinical applications. Protocols for maintaining ES cells in an undifferentiated state have undergone some optimization;[142] however even in well maintained cultures some colonies undergo spontaneous differentiation. Human ES cells require attachment to a substrate to retain an undifferentiated state. Current lines have been derived by culture on mouse embryonic fibroblasts (MEFs) and are sustained in media with fetal calf serum.[136,137] The use of animal-derived extracts raises concern of transmission of unknown xenogeneic viruses and pathogens to patients who have undergone transplantation with ES cell-derived tissues.

Ideally, human ES cell lines should be established in the absence of MEF layers, perhaps on human embryonic fibroblasts and in defined synthetic media. Xu et al demonstrated that human ES cells do not require physical contact with MEFs and can be maintained on a substrate of Matrigel (Becton Dickinson, Franklin Lakes, and NJ).[143] Matrigel is a basement membrane matrix prepared from a mouse sarcoma cell line using medium previously conditioned by MEFs. Matrigel consists of collagen V and laminin, although other components are not fully defined. A suitable fully defined synthetic medium for the culture of human ES cells has not been described. Leukemia inhibitory factor (LIF) is sufficient to support the growth of mouse ES cells in the absence of feeders but does not appear to play a similar role in human ES cells.[137]

Clearly there are still significant safety issues that must be addressed before cells and tissues derived from human ES cells can be used for transplantation into patients. Even highly enriched cultures of specific cell types derived from human ES cells are sometimes contaminated with other cells types and undifferentiated ES cells.[144] Transplantation of mitotically active, undifferentiated ES cells into a patient could result in the formation of teratomas. Inactivation of proliferating cells with mitotic inactivators, such a mitomyocin-c or cytosine arabinoside may be useful in preventing teratoma formation in this setting.

Mesenchymal Stem Cells

Collection and Expansion

Mesenchymal stem cells (MSCs) are derived from bone marrow plated in fetal calf serum-containing medium to generate colonies of adherent fibroblast-like cells that can differentiate into bone, adipocytes, chondrocytes, and skeletal myocytes.[145] MSCs can be purified on the basis of their ability to adhere to plastic or with monoclonal antibodies (SH2, SH4, or Stro1).[146-148] Reyes et al have recently described the isolation and ex vivo expansion of cells from postnatal bone marrow that can differentiate at the single cell level into MSCs, as well as cells of visceral mesodermal origin, such as endothelium.[149] To differentiate mesodermal from hematopoietic progenitors, cells were first selected from bone marrow that did not express hematopoietic markers (CD45 and GlyA). CD45$^-$GlyA$^-$ cells were plated on plastic coated with fibronectin in tissue culture medium containing epidermal growth factor (EGF), platelet-derived growth factor (PDGF-BB) and 2% fetal calf serum. Small clusters of adherent cells developed after 7 to 10 days of culture, which were replated every 3 to 5 days under the same culture conditions, for up to 60 cell doublings without undergoing senescence or genetic instability. Expanded cells failed to express hematopoietic cell markers and differentiated in vitro into osteoblasts, chondrocytes, adipocytes, stroma cells and skeletal myoblasts. In a second study, Jiang et al reported a population of adult bone marrow-derived cells that copurified with MSCs and had characteristics of mesenchymal, visceral mesoderm, neuroectoderm and endoderm cells, and were termed multipotent adult progenitor cells or MAPCs.[150] To further determine the extent of differentiation in vivo, β-gal/Neo-containing MAPCs were introduced into early blastocysts that were subsequently transferred to foster mothers and allowed to develop and be born. Analysis of 6-20 week old mice demonstrated that MAPCs contributed to cells of the hematopoietic lineage, epithelium of the liver, lung and gut. In murine transplantation studies using unirradiated NOD/SCID recipients, β-gal/Neo-containing MAPCs engrafted rapidly and demonstrated persistent contribution to bone marrow, spleen, peripheral blood, epithelium of the lung, liver and intestine of all recipient animals.[150] These results demonstrate that MAPCs can contribute functionally to most somatic tissues, even in undamaged recipients. As MAPCs proliferate extensively without obvious senescence or loss of differentiation potential, they represent an ideal source of cells for therapy of inherited enzyme deficiencies or degenerative diseases such as muscular dystrophy. In contrast to ES cells, MAPCs can be selected from autologous bone marrow and used undifferentiated or after genetic manipulation in local or systemic therapies. Likewise, since MAPCs differentiate into hematopoietic cells in vivo, they can be used to establish hematopoietic chimerism, allowing for an allogeneic cell therapy approach.

Neural Stem Cells

Collection and Expansion

Neural stem cells (NSCs) are operationally defined as mitotically competent, self-renewing, and multipotent cells, able to differentiate along the three main cell lineages in the nervous system (neurons, astro-, and oligodendro-glia). Fulfilling the expectation of being able to control and tailor proliferation, migration and differentiation capabilities represents a challenging goal since very little of the in vivo biology of NSCs is presently known. In recent years, NSCs have been identified, isolated and methods developed for their long term propagation in vitro.[130,151,152] NSCs have been isolated from both embryonic and adult human nervous system. In most cases cells have been obtained from cadaveric, post-mortem tissues; biopsied material; and embryo-derived cells.

Based on current experimental evidence, it is most likely that neural cell lines, rather than stem cells themselves will be most useful for therapeutic purposes. Immortality, in the absence of transformation, represents a valuable property of neural cell lines that make them an attractive therapeutic tool. From the biosafety perspective, any long-term cultured, ex vivo expanded NSC preparation may be seen as a potential source of transformed cells, and bearing some risk for tumor generation. Extensive validation tests will be needed to exclude these possibilities before use in human therapies. Molecular switches designed to regulate or silence transgenes and/or to eradicate the cells may also be needed.

In vitro expansion of NSCs can be accomplished by epigenetic or genetic modifications. Cells can be stimulated to grow until they approach their natural senescence limit in culture using mitogens (epigenetic). Alternatively, cells can be genetically modified to override natural senescence limits and become established. Sah et al described the generation of multiple cell lines of neuronal or neuronal plus glial potential, derived from human embryonic brain.[153] Initially the cells were grown as adherent cultures using basic fibroblast growth factor (bFGF) as a mitogen. The cells were transfected with a retroviral vector containing v-myc driven by a modified cytomegalovirus (CMV) promoter to make it responsive to a tetracycline-regulated transactivator (tet-off version). The resulting cell lines were determined to be clonal by retroviral integration sites and when tetracycline was added and the mitogen turned off, the cells stopped dividing and differentiated. Differentiation was promoted by retinoic acid or by the supply of neurotrophic factors. Functionally, the cells showed voltage-dependent outward sodium currents and ligand-gated currents characteristic of mature neurons, demonstrating that v-myc did not subvert the capacity of the cells to become post-mitotic and terminally differentiated. Subsequently, similar technology has been used for generation of immortal human dorsal root ganglion neurons,[154] human motoneurons from the spinal cord,[155] and multipotent cells with the capacity to differentiate into neurons, astrocytes and oligodendrocytes[156] from human embryonic brain.

An alternative source of neural cell lines is based on the existence of NSCs that divide continuously when cultured in serum-free medium. The original paper that defined this approach reported that the striatum of adult mice could be dissociated enzymatically and cultured as a cell monolayer in defined, serum-free medium.[157] In cultures with EGF, a small percentage of cells detach and form multicellular spheres, neurospheres that quickly increase in size and divide. Upon plating on an adhesive substrate, neurospheres differentiate into mixed cultures containing cells with the morphology and antigen-expression profile of neurons and astrocytes. The methodology for generating rodent neurospheres has recently been extended to the human.[158] Human neurospheres were derived from first trimester fetal forebrain and dissociated by mechanical shearing to yield living cells. These were cultured in N2-defined medium with bFGF, EGF and LIF. When left in suspension, cells in neurospheres proliferated vigorously and could be serially passage to a 1000-fold increase in cell number. Upon plating on an adhesive substrate, neurospheres give rise to oligodendrocytes, astrocytes, and neurons. Neurospheres from adult human brains have been cultured and characterized and are described in a recent review article by David Gottlieb.[159]

Mouse ES cells cultured in the absence of LIF and the presence of retinoic acid differentiate into neural lineage cells that can be further differentiated into neurons, astrocytes, and oligodendrocytes after transient culture in bFGF.[160,161] The resting potential, sodium action potentials, potassium and calcium currents in mouse ES cell-derived neurons closely resembled those in primary neuronal cultures. Human ES cells, cultured as embryoid bodies (EB) for four days, followed by plating onto fibronectin-coated surfaces in defined proliferation medium supplemented with growth factors, yield 50-60% cells positive for neural markers poly-sialyl-NCAM or A2B5.[162] When retinoic acid was included at the EB stage, 90% of the cells were positive for neural markers. Calcium imaging assays demonstrated that the human ES cell-derived neurons could respond to neurotransmitters, demonstrating functional competence.

The expanded neural cell lines described above have been successfully transplanted into rodents with a variety of neurological traumas and degenerative diseases with promising results (reviewed in refs. 144,163). Transplanted cells are capable of migrating to affected regions, differentiating into astrocytes, neurons and oligodendrocytes with no signs of uncontrolled growth, and providing some normal neuronal functions. An important aspect regarding transplanted cell lines expressing v-myc was the profound natural (not pharmacologic) downregulation in vivo following differentiation,[152,156] illustrating the safety of the cells and the conditional nature of the immortalization. If long-term cultured NSCs are to be used in the clinical setting in the future, additional research will be necessary to assure their safety and efficacy; however preliminary data in rodents are very encouraging and lend optimism for the development of relevant therapies to alleviate human neurodegenerative diseases.

Muscle Stem Cells

Collection and Expansion

Normal repair and regeneration of skeletal muscle fibers occurs following injury from muscle satellite cell activation, proliferation and migration to the site of injury.[164] Satellite cells are a unique population of usually quiescent cells located outside of the myofiber, between the sarcolemma and the covering basement membrane, and account for approximately 5% of the nuclei present in muscle fibers. When activated, satellite cells have the capacity to divide extensively in order to produce sufficient myoblasts to replace damaged muscle fiber. Recovery of satellite cells via standard tissue dissociation techniques, fail to extract more that 99% of the known myogenic population of mature muscle,[165] bringing into question the number of cell types that contribute to muscle development in vivo. Satellite cells attached to muscle fibers are capable of rapid proliferation and myogenesis in tissue culture,[165] but when transplanted in vivo into recipient murine muscle, the majority of cultured cells undergo rapid necrotic cell death.[166] Only a few cells survive to give rise to regenerated myofibers and satellite cells. More stringent techniques for identifying myogenic precursors within muscle and tracing their fate are required to determine the spectrum of cell types responsible for skeletal muscle regeneration in vivo.

Derivation of muscle-derived cells with stem cell characteristics (MDSC) has been isolated from the skeletal muscle of neonatal mice.[167-169] One feature responsible for the isolation of MDSC is the cells' slow adherence behavior. However, cell therapy applications are currently constrained by the insufficient number of cells that can be obtained from a biopsy and also the cells' limited proliferative capacity. Efforts to expand cultures have been based on cytokine-induced expansion.[170-173] Deasy et al[170] tested a variety of growth factors, including EGF, FGF-2, IGF-1, FLT-3 ligand, HGF and SCF, on the proliferation of primary MDSCs and a clone of MDSC, MC13. Their experiments demonstrated that EGF, FGF-2, IGF-1 and SCF could be used to expand MDSC and MC13 in vitro. In particular EGF stimulation led to the greatest overall increase in cell number in the primary cultures, approximately 2-fold over control cultures lacking EGF. The authors did not evaluate phenotypic expression of MDSC markers or the functional status of the growth factor-expanded cells. Future studies will need to be performed to determine the potential for ex vivo expansion of human muscle cell precursors with potential for therapeutic use.

Cardiac Stem Cells

Collection and Expansion

In contrast to other organ systems, e.g., bone marrow, liver, skin, intestine, that have the capacity for regeneration, the conventional view is that the heart lacks a pool of stem cells capable of self-renewal and differentiation and that terminally differentiated cardiomyocytes are unable to divide and undergo repair. Augmentation of myocardial performance in experimental models of therapeutic infarction and heart failure has been achieved by the transplantation of exogenous cells into damaged myocardium.[174] Historically a wide range of cell types has been used including rat and human fetal ventricular myocytes, ES cell-derived cardiomyocytes and bone marrow-derived adult somatic stem cells; however the optimal source of donor cell type is currently unknown. Limitations that hinder the use of primary human fetal cardiomyocytes include ethical issues and the inability to obtain sufficient quantities of cells to repair large areas of infarcted myocardium.

Although conflicting data regarding the presence in vivo of mitotically active cardiomyocytes have led to controversy over the existence of a cardiomyocyte stem cell population, recent data provide new evidence of their existence. Beltrami et al have shown that mitotic cardiomyocytes can be detected by fluorescent labeling in combination with high-resolution confocal microscopy in regions adjacent to and distant from infarcts in the adult human heart.[175] Additional evidence suggests that primitive stem cells play a pivotal role in the remodeling process following heart transplantation. Quani et al detected Y chromosome-bearing undifferentiated cells of donor and recipient origin in transplanted female hearts in cadaveric male heart transplant recipients.[176] Although the majority of cells bearing the Y chromosome were mature, a subpopulation of immature cells were detected in the atria and left ventricle of the donor heart. The Y chromosome-positive replicating cardiomyocytes expressed transcription factors active in early cardiac development (MEF2, GATA-4) and the intermediate filament, nestin. However, the majority of cardiomyocytes with these cardiac-specific markers were Y chromosome negative. These results prompted the authors to speculate that a population of endogenous primitive stem cells resides in the donor heart and together with primitive cells translocated from the recipient, can proliferate and differentiate into cardiac lineages in response to altered physiological conditions. The question remains whether the primitive cells identified in the donor hearts arose from a resident cardiac stem cell population present in the remnant of the recipient's heart or if they were circulating stem cells which homed to the heart from some other organ such as the bone marrow. Adult somatic stem cells within the bone marrow exhibit remarkable functional plasticity,[177] although the primitive cell populations in the donor hearts were negative for bone marrow-specific markers.

The failure of significant levels of regeneration in the adult human heart following myocardial infarction, or other forms of injury, may not necessarily be due to the absence of cardiac stem cells. Rather it may be the consequence of the inability of damaged myocardium to provide the appropriate molecular signals to activate proliferation and differentiation of resident cells. Future studies will be required to determine the origin of the primitive stem cell population in the adult human heart. In the meantime, advances in stem cell therapy provide alternative sources for the treatment of cardiovascular disease. These include skeletal myoblasts,[178,179] human embryonic stem cells,[180] bone marrow-derived mesenchymal stem cells,[181-184] hematopoietic stem cells[185-190] and blood and bone marrow-derived endothelial progenitor cells.[191-193] These cell types unequivocally improve myocardial function in a variety of animal models of experimental cardiac injury.

Hepatic Stem Cells

Collection and Expansion

The existence of stem cells in the adult liver is controversial, however it is readily accepted that three phenotypically distinct cell lineages contribute to regenerative growth after liver damage, including hepatocytes, cholangiocytes and bone marrow-derived cells.[194] Hepatocytes are capable of reentering the cell cycle and generating replacement hepatocytes following partial hepatectomy. When massive damage is inflicted on the liver, a potential stem cell pool located within the smallest branches of the intrahepatic biliary tree is activated. These so-called "oval cells" amplify the biliary population before the cells differentiate into hepatocytes. Oval cell numbers in human liver rise with increasing severity of liver disease[195] and is widely accepted to be a stem cell response. Adult male hematopoietic stem cells transplanted into lethally irradiated female mice, whose livers were subsequently injured with pharmacological agents, had evidence of Y chromosome-positive ovals cells and hepatocytes at 9 days after injury.[196] Hepatocytes were also derived from bone marrow cell populations in humans.[197,198] Two model systems were investigated: (i) the livers of female patients who had previously received bone marrow transplantation from male donors were examined for the presence of Y chromosome-positive hepatocytes; and (ii) Y positive hepatocytes were investigated in female livers engrafted into male patients. In both sets of patients, Y chromosome-positive hepatocytes were readily identified, with up to 40% of hepatocytes and cholangiocytes derived from bone marrow.[198]

Isolation of primary oval cells has not been possible, however permanent lines of liver epithelial cells have been established that have characteristics similar to oval cells.[199] When these cells were transduced with a β-gal gene and transplanted into the liver of syngeneic rats, the cells integrated into the hepatic plates and acquired the morphologic appearance of mature hepatocytes and expressed liver-specific genes.[200,201] Other less well-characterized cell lines have been established and used in transplantation studies.[202] In general the transplantation studies using oval cell lines have been disappointing due to a slow proliferation rate in vivo.[202]

Fetal liver epithelial cells are rapidly proliferating between embryonic day (ED) 12 and 16, suggesting that liver stem cells capable of repopulating normal liver after transplantation may be present. Dabeva et al[203] isolated cells from ED12-14 fetal rat liver and studied their expression of hepatocyte markers, AFP and albumin, and the biliary epithelial marker CK-19. They identified three distinct subpopulations of epithelial progenitors expressing various combinations of hepatic and biliary markers. Transplantation studies demonstrated that the epithelial cells proliferated in the host liver and new hepatic cords and bile ducts were formed. Up to nine to ten cell divisions of very large clusters of transplanted cells were observed at 6 months following transplantation, which were still proliferating (as determined by in situ hybridization) and achieved up to 10% liver repopulation.[204]

In the case of a diseased human liver, there may not be substantial selective growth advantage for transplanted cells such as occurs in most rodent models. Therefore, it becomes of interest to determine if it is possible to enrich for true liver stem cells that would continue to expand in a recipient liver in the absence of a major growth stimulus. Kubota and Reid[205] have described a population of progenitors from ED13 fetal liver that lack expression of MHC class I, a feature that may allow hepatocytes to escape immunologic surveillance when transplanted into an MHC-incompatible host. Several other cell types expressing hepatic-specific markers have been isolated from embryonic fetal rat livers.[206] Identification of the specific markers allows a potential mechanism for isolation and purification. Although these studies are encouraging, more research is needed to determine if counterparts exist in human liver and, more importantly if the cells can be expanded for use in clinical applications.

Conclusion

The studies outlined above demonstrate the potential clinical utility of ex vivo expanded cells for a diverse group of human diseases. Although in many cases it is still not clear the exact nature of tissue-specific stem cells or if such cells can indeed be expanded in the absence of differentiation, the data are compelling and indicate the momentum of the field. Future investigation into the molecular mechanisms of stem cell renewal, continued study in animal models, translation to human clinical trials and ongoing dialogue regarding ethical and safety issues promise to promote the evolution of this very exciting area of medicine.

References

1. Bensinger W, Price T, Dale D et al. The effects of daily recombinant human granulocyte colony-stimulating factor administration on normal granulocyte donors undergoing leukapheresis. Blood 1993; 81:1883-1888.
2. Sheridan W, Begley C, To L. Phase II study of autologous filgastrim (G-CSF)-mobilized peripheral blood progenitor cells to restore hemopoiesis after high-dose chemotherapy for lymphoid malignancies. Bone Marrow Transplant 1994; 14:103-108.
3. Matsunaga T, Sakamaki S, Kohgo Y et al. Recombinant human granulocyte colony-stimulating factor can mobilize sufficient amounts of peripheral blood stem cells in healthy volunteers for allogeneic transplantation. Bone Marrow Transplant 1993; 11:103-108.
4. Bensinger W, Weaver C, Applebaum F et al. Transplantation of allogeneic peripheral blood stem cells mobilized by recombinant human granulocyte colony-stimulating factor. Blood 1995; 85:1658.
5. Korbling M, Prezepiorka D, Huh Y et al. Allogeneic blood stem cell transplantation for refractory leukemia and lymphoma. Blood 1995; 85:1659-1665.
6. Schmitz N, Dreger P, Suttorp M et al. Primary transplantation of allogeneic peripheral blood progenitor cells mobilized by filgastrim (granulocyte colony-stimulating factor). Blood 1995; 85:1666-1672.
7. Barker J, Davies S, DeFor T et al. Survival after transplantation of unrelated donor umbilical cord blood is comparable to that of human leukocyte antigen-matched unrelated donor bone marrow: Results of a matched pair analysis. Blood 2001; 97:2957-2961.
8. Rocha V, Cornish J, Sievers E et al. Comparison of outcomes of unrelated bone marrow and umbilical cord blood transplants in children with acute leukemia. Blood 2001; 97:2962-2971.
9. Laughlin M, Barker J, Bambach B et al. Hematopoietic engraftment and survival in adult recipients of umbilical-cord blood from unrelated donors. New Engl J Med 2001; 344:1815-1822.
10. Donaldson C, Armitage W, Laundy V et al. Impact of obstetric factors on cord blood donation for transplantation. Br J Haematol 1999; 106:128-132.
11. Sparrow R, Cauchi J, Ramadi L et al. Influence of mode of birth and collection on WBC yields of umbilical cord blood units. Transfusion 2002; 42:210-215.
12. Yamada T, Okamoto Y, Kasamatsu H et al. Factors affecting the volume of umbilical cord blood collections. Acta Obstet Gynecol Scand 2000; 79:830-833.
13. Lim F, van Winsen L, Willemze R et al. Influence of delivery on number of leukocytes, leukocyte subpopulations, and hematopoietic progenitor cells in human umbilical cord blood. Blood Cells 1994; 20:547-559.
14. Vettenranta K, Piirto I, Saarinen-Pihkala U. The effects of the mode of delivery on the lymphocyte composition of a placental/cord blood graft. J Hematotherapy 1997; 6:491-493.
15. Surbek D, Schonfeld B, Tichelli A et al. Optimizing cord blood mononuclear cell yield: A randomized comparison of collection before vs after placenta delivery. Bone Marrow Transplant 1998; 22:311-312.
16. Surbek D, Visca E, Steinmann C et al. Umbilical cord blood collection before placental delivery during cesarean delivery increases cord blood volume and nucleated cell number available for transplantation. Am J Obstet Gynecol 2000; 183:218-221.
17. Sutherland H, Eaves C, Eaves A. Characterization and partial purification of human marrow cells capable of initiating long-term hematopoiesis in vitro. Blood 1989; 74:2755-2764.
18. Sutherland H, Landsdorp P, Eaves A et al. Functional characterization of individual human hematopoietic stem cells cultured at limiting dilution on supportive marrow stromal layers. Proc Natl Acad Sci USA 1990; 87:3584-3588.
19. Ploemacher R, van der Sluijs J, Brons N. An in vitro limiting dilution assay of long-term repopulating hematopoietic stem cells in the mouse. Blood 1989; 74:2755-2764.
20. Breems D, Blokland E, Neben S et al. Frequency analysis of human primitive hematopoietic stem cell subsets using a cobblestone area forming cell assay. Leukemia 1994; 8:1095-1104.
21. McNiece I, Bertoncello I, Kriegler A et al. Colony forming cells with high proliferative potential (HPP-CFC). In: Murphy MJ, ed. Concise reviews in clinical and experimental hematology. OH Miamisburg: Alpha Med Press, 1992:267.

22. Hao Q, Thiemann F, Petersen D et al. Extended long term culture reveals a highly quiescent and primitive human hematopoietic progenitor population. Blood 1996; 88:3306-3310.
23. Hao Q, Smogorzewska E, Barsky L et al. In vitro identification of single CD34+/CD38- cells with both lymphoid and myeloid potential. Blood 1998; 91:4145-4152.
24. Miller J, McCullar V, Punzel M et al. Single adult human CD34+/Lin-/CD38- progenitor cells give rise to NK cells, B-lineage cells, dendritic cells and myeloid cells. Blood 1999; 93:96-105.
25. Baume C, Weissman I, Tsukamoto A et al. Isolation of a candidate human hematopoietic stem cell population. Proc Natl Acad Sci USA 1992; 89:2804-2810.
26. Punzel M, Wissink S, Miller J et al. The myeloid-lymphoid initiating cell (ML-IC) assay assesses the fate of multipotent human progenitors in vitro. Blood 1999; 93:3750-3756.
27. Hodgson G, Bradley T. Properties of hematopoietic stem cells surviving 5-flourouracil treatment. Nature 1979; 281:381.
28. Spangrude G, Heimfeld S, Weissman I. Purification and characterization of mouse hematopoietic stem cells. Science 1988; 241:58.
29. Spangrude G, Johnson J. Resting and activated subsets of mouse multipotent hematopoietic stem cells. Proc Natl Acad Sci USA 1990; 87:7433.
30. Morrison S, Weissman I. The long-term repopulating subset of hematopoietic stem cells is deterministic and isolatable by phenotype. Immunity 1994; 1:661.
31. Harrison D, Zhong R, Jordan C et al. Relative to adult marrow fetal liver repopulates nearly five times more effectively long-term than short-term. Exp Hematol 1997; 25:293.
32. McCune J, Namikawa R, Kaneshima H et al. The SCID-hu mouse: Murine model for the analysis of human hematolymphoid differentiation and function. Science 1988; 24:1632.
33. Fraser C, Kaneshima H, Hansteen G et al. Human allogeneic stem cell maintenance and differentiation in a long-term multilineage SCID-hu graft. Blood 1995; 86:1680.
34. Kollman T, Kim A, Zhuang X et al. Reconstitution of SCID mice with human lymphoid and myeloid cells after transplantation with human fetal bone marrow without the requirement for exogenous human cytokines. Proc Natl Acad Sci USA 1994; 91:8032.
35. Nolta J, Smogorzewska E, Kohn D. Analysis of optimal conditions for retroviral mediated transduction of primitive human hematopoietic cells. Blood 1995; 86:101.
36. Nolta J, Dao M, Wells S et al. Transduction of pluripotent human hematopoietic stem cells demonstrated by clonal analysis after engraftment in immune-deficient mice. Proc Natl Acad Sci USA 1996; 93:2414.
37. Larochelle A, Vormoor J, Hanenberg H et al. Identification of primitive human hematopoietic cells capable of repopulating NOD/SCID mouse bone marrow: Implications for gene therapy. Nat Med 1996; 2:1329.
38. Gan O, Murdoch B, Larochelle A et al. Differential maintenance of primitive human SCID-repopulating cells, clonogenic progenitors, and long-term culture initiating cells after incubation on human bone marrow stromal cells. Blood 1997; 90:641.
39. Bhatia M, Wang J, Knapp U et al. Purification of primitive human hematopoietic cells capable of repopulating immune-deficient mice. Proc Natl Acad Sci USA 1997; 94:5320.
40. Cashman J, Bockhold K, Hogge D et al. Sustained proliferation, multi-lineage differentiation and maintenance of primitive human hematopoietic cells in NOD/SCID mice transplanted with human cord blood. Br J Hematol 1997; 98:1026.
41. Srour E, Zanjani E, Cornetta K et al. Persistence of human multi-lineage, self-renewing lymphohematopoietic stem cells in chimeric sheep. Blood 1993; 82:3333.
42. Zanjani E, Almeida-Porada G, Ascensao J et al. Transplantation of hematopoietic stem cells in utero. Stem Cells 1997; 15:79.
43. Shimizu Y, Ogawa M, Kobayashi M et al. Engraftment of cultured human hematopoietic cells in sheep. Blood 1998; 91:3688.
44. Dunbar C, Seidel N, Doren S et al. Improved retroviral gene transfer into murine and Rhesus peripheral blood or bone marrow repopulating cells primed in vivo with stem cell factor and granulocyte colony-stimulating factor. Proc Natl Acad Sci USA 1996; 93:11871.
45. Dunbar C, Takatoku M, Donahue R et al. The impact of ex vivo cytokine stimulation on engraftment of primitive hematopoietic cells in a nonhuman primate model. Ann NY Acad Sci 2001; 938:236-245.
46. Brummendorf T, Dragowska W, Zijlmans J et al. Asymmetric cell divisions sustain long-term hematopoiesis from single-sorted human fetal liver cells. J Exp Med 1998; 188:1117-1124.
47. Huang S, Law P, Francis K et al. Symmetry of initial cell divisions among primitive hematopoietic progenitors is independent of ontogenic age and regulatory molecules. Blood 1999; 94:2595.
48. Morrison S, Shah N, Anderson D. Regulatory mechanisms in stem cell biology. Cell 1998; 88:287.
49. Lansdorp P. Self-renewal of stem cells. Biol Blood Marrow Transplant 1997; 3:171.
50. Suda T, Suda J, Ogawa M. Disparate differentiation in mouse hematopoietic colonies derived from paired progenitors. Proc Natl Acad Sci USA 1984; 81:2520.
51. Mayani H, Dragowska W, Lansdorp P. Lineage commitment in human hemopoiesis involves asymmetric cell division of multipotent progenitors and does not appear to be influenced by cytokines. J Cell Physiol 1993; 157:579.
52. Punzel M, Ho A. Divisional history and pluripotency of human hematopoietic stem cells. Ann NY Acad Sci 2001; 938:72.
53. Punzel M, Zhang T, Liu D et al. Functional analysis of initial cell divisions defines the subsequent fate of individual human CD34+CD38- cells. Exp Hematol 2002; 30:464-472.
54. Punzel M, Zhang T, Eckstein V et al. Individual human CD34+/CD38- cells with distinct functional fate distribute their functional properties unequally to two 1st generation daughter cells. Blood 2002; 100:26a.
55. Habibian H, Peters S, Hsieh C et al. The fluctuating phenotype of the lymphohematopoietic stem cell with cell cycle transit. J Exp Med 1998; 188:393-398.
56. Glimm H, Oh I-H, Eaves C. Human hematopoietic stem cells stimulated to proliferate in vitro loose engraftment potential during their S/G2/M transit and do not reenter G0. Blood 2000; 96:4185-4193.
57. Giet O, Huygen S, Beguin Y et al. Cell cycle activation of hematopoietic progenitor cells increases very late antigen 5-mediated adhesion to fibronectin. Exp Hematol 2000; 29:515-524.
58. Orschell-Traycoff C, Hiatt K, Dagher R et al. Homing and engraftment potential of Sca1+Lin- cells fractionated on the basis of adhesion molecule expression and position in cell cycle. Blood 2000; 96:1380-1387.
59. Becker P, Nilsson S, Li Z et al. Adhesion receptor expression by hematopoietic cell lines and murine progenitors: Modulation by cytokines and cell cycle status. Exp Hematol 1999; 27:533.
60. Mohle R, Bautz F, Rafii S et al. The chemokine receptor CXCR-4 is expressed on CD34+ hematopoietic progenitors and leukemic cells and mediates transendothelial migration induced by stromal cell-derived factor-1. Blood 1998; 91:4523.
61. Takenaka K, Nagafuji K, Harada M et al. In vitro expansion of hematopoietic stem cells induces functional expression of Fas antigen. Blood 1996; 88:2871-2877.
62. Domen J, Cheshier S, Weissman I. The role of apoptosis in the regulation of hematopoietic stem cells: Overexpression of Bcl-2 increases both their number and repopulation potential. J Exp Med 2000; 191:253-264.
63. Sauvageau G, Lansdorp P, Eaves C. Differential expression of homeobox genes in functionally distinct CD34+ subpopulations of human bone marrow cells. Proc Natl Acad Sci USA 1994; 91:12223-12227.
64. Sauvageau G, Thorsteinsdottir U, Eaves C et al. Overexpression of HOXB4 in hematopoietic cells causes the selective expansion of more primitive populations in vitro and in vivo. Genes Dev 1995; 9:1753-1765.
65. Antonchuk J, Sauvageau G, Humphries R. HOXB4 overexpression mediates very rapid stem cell regeneration and competitive repopulation. Exp Hematol 2001; 29:1125-1134.
66. Thorsteinsdottir U, Sauvageau G, Humphries R. Enhanced in vivo regenerative potential of HOXB4-transduced hematopoietic stem cells with regulation of their pool size. Blood 1999; 94:2605-2612.

67. Buske C, Feuring-Buske M, Abramovich C et al. Deregulated expression of HOXB4 enhances the primitive growth activity of human hematopoietic cells. Blood 2002; 100:862-868.
68. Antonchuk J, Sauvageau G, Humphries R et al. HOXB4-Induced expansion of adult hematopoietic stem cells ex vivo. Cell 2002; 109:39-45.
69. Daga A, Podesta M, Capra M et al. The retroviral transduction of HOXC4 into human CD34+ cells induces an in vitro expansion of clonogenic and early progenitors. Exp Hematol 2000; 28:569-574.
70. Thorsteinsdottir U, Mamo A, Kroon E et al. Overexpression of the myeloid leukemia-associated Hoxa9. Blood 2002; 99:121-129.
71. Pinto do Oa P, Kolterud A, Carlsson et al. Expression of the LIM-homeobox gene LH2 generates immortalized Steel factor-dependent multipotent hematopoietic precursors. EMBO 1998; 17:5744-5756.
72. Keller G, Wall C, Fong A et al. Overexpression of HOX11 leads to the immortalization of embryonic precursors with both primitive and definitive hematopoietic potential. Blood 1998; 92:877-887.
73. Thorsteinsdottir U, Sauvageau G, Humphries R. Hox homeobox genes as regulators of normal and leukemic hematopoiesis. Hematol Oncol Clin North Am 1997; 11:1221-1237.
74. Buske C, Humphries R. Homeobox genes in leukemogenesis. Int J Hematol 2000; 71:301-308.
75. Varnum-Finney B, Xu L, Brashem-Stein C et al. Pluripotent, cytokine-dependent, hematopoietic stem cells are immortalized by constitutive Notch 1 signaling. Nat Med 2000; 6:1278-1281.
76. Carlesso N, Aster J, Sklar J et al. Notch 1-induced delay of human hematopoietic progenitor cell differentiation is associated with altered cell cycle kinetics. Blood 1999; 93:838-848.
77. Han W, Ye Q, Moore M. A soluble form of human Delta-like-1 inhibits differentiation of hematopoietic progenitor cells. Blood 2000; 95:1616-1625.
78. Walker L, Lynch M, Silverman S et al. The Notch/Jagged pathway inhibits proliferation of human hematopoietic progenitors in vitro. Stem Cells 1999; 17:162-171.
79. Ohishi K, Varnum-Finney B, Bernstein I. Delta-1 enhances marrow and thymus repopulating ability of human CD34+CD38- cord blood cells. J Clin Invest 2002; 110:1165-1174.
80. Henrique D, Hirsinger E, Adam J et al. Maintenance of neuroepithelial progenitor cells by Delta-Notch signaling in the embryonic chick retina. Curr Biol 1997; 7:661-670.
81. Austin J, Kimble J. Glp-1 is required in the germ line for regulation of the decision between mitosis and meiosis in C elegans. Cell 1987; 51:589-599.
82. Bhardwaj G, Murdoch B, Wu D et al. Sonic hedgehog induces the proliferation of primitive human hematopoietic cells via BMP regulation. Nat Immunol 2001; 2:172-108.
83. Wechsler-Reya R, Scott M. Control of neuronal precursor proliferation in the cerebellum by sonic hedgehog. Neuron 1999; 22:103-114.
84. Zhang Y, Kalderon D. Hedgehog acts as a somatic stem cell factor in the *Drosophilia* ovary. Nature 2001; 410:599-604.
85. Austin T, Solar G, Ziegler F et al. A role for the Wnt gene family in hematopoiesis: Expansion of multilineage progenitor cells. Blood 1997; 89:3624-3635.
86. van Den Berg D, Sharma A, Bruno E et al. Role of members of the Wnt gene family in human hematopoiesis. Blood 1998; 92:3189-3202.
87. Zhu A, Watt F. β-catenin signaling modulates proliferative potential of human epidermal keratinocytes independently of intercellular adhesion. Development 1999; 126:2285-2298.
88. Korinek V, Barker N, Moerer P et al. Depletion of epithelia stem cell compartments in the small intestine of mice lacking *Tcf-4*. Nat Genet 1998; 19:1-5.
89. Reya T, Morrison S, Clarke M et al. Stem cells, cancer, and cancer stem cells. Nature 2001; 414:105-111.
90. Bradley T, Hodgson G, Kriegler A et al. Generation of $CFU-S_{13}$ in vitro. Alan R Liss Inc 1985; 39-56.
91. McNiece I, Bradley T, Kriegler A et al. Subpopulations of mouse bone marrow high-proliferative-potential colony-forming cells. Exp Hematol 1986; 14:856.
92. McNiece I, Williams N, Johnson G et al. Generation of murine hematopoietic precursor cells from macrophage high-proliferative-potential colony-forming cells. Exp Hematol 1987; 15:972.
93. McNiece I, Andrews R, Stewart F et al. Synergistic interactions of human growth factors in in vitro cultures of human bone marrow cells. Blood 1987; 72:125a.
94. Bernstein I, Andrews R, Zsebo K. Recombinant human stem cell factor enhances the formation of colonies by CD34+ and CD34+lin-cells, and the generation of colony-forming cell progeny from CD34+lin-cells cultured with interleukin-3, granulocyte colony-stimulating factor, or granulocyte-macrophage colony-stimulating factor. Blood 1991; 77:2316.
95. Haylock D, To L, Dowse T et al. Ex vivo expansion and maturation of peripheral blood CD34+ cells into the myeloid lineage. Blood 1992; 80:1405.
96. Purdy M, Hogan C, Hami L et al. Large scale ex vivo expansion of CD34-positive hematopoietic progenitor cells for transplantation. J Hematotherapy 1995; 4:515-525.
97. Briddell R, Kern B, Zilm K et al. Purification of CD34+ cells is essential for optimal ex vivo expansion of umbilical cord blood cells. Exp Hematol 1996; 24:1055.
98. Shieh J-H, Chen Y-F, Briddell R et al. High purity of blast cells in CD34 selected populations are essential for optimal ex vivo expansion of human GM-CFC. Exp Hematol 1994; 22:756a.
99. Stoney G, Briddell R, Kern B et al. Clinical scale ex vivo expansion of myeloid progenitor cells and megakaryocytes under GMP conditions. Exp Hematol 1996; 24:1043a.
100. Muench M, Firpo M, Moore M. Bone marrow transplantation with interleukin-1 plus kit-ligand ex vivo expanded bone marrow accelerates hematopoietic reconstitution in mice without the loss of stem cell lineage and proliferative potential. Blood 1993; 81:3463.
101. Andrews R, Briddell R, Gough M et al. Expansion of G-CSF mobilized CD34+ peripheral blood cells (PBC) for 10 days in G-CSF, MGDF and SCF prior to transplantation decreased post-transplant neutropenia in baboons. Blood 1997; 90:92a.
102. Naparstek E, Hardan Y, Ben-Shahar M et al. Enhanced marrow recovery by short preincubation of marrow allografts with human recombinant interleukin-3 and granulocyte-macrophage colony-stimulating factor. Blood 1992; 80:1673.
103. Brugger W, Heimfeld S, Berenson R et al. Reconstitution of hematopoiesis after high-dose chemotherapy by autologous progenitor cells generated ex vivo. New Engl J Med 1995; 333:283.
104. Williams S, Lee W, Bender J et al. Selection and expansion of peripheral blood CD34+ cells in autologous stem cell transplantation for breast cancer. Blood 1996; 87:1687.
105. Alcorn M, Holyoake T, Richmond L et al. CD34-positive cells isolated from cryopreserved peripheral-blood progenitor cells can be expanded ex vivo and used for transplantation with little or no toxicity. J Clin Oncol 1996; 14:1839.
106. Reiffers J, Cailliot C, Dazey B et al. Abrogation of post-myeloablative chemotherapy neutropenia by ex-vivo expanded autologous CD34-positive cells. Lancet 1999; 354:1092.
107. McNiece I, Jones R, Bearman S et al. Ex vivo expanded peripheral blood progenitor cells provide rapid neutrophil recovery in breast cancer patients following high dose chemotherapy. Blood 2000; 96:3001-3007.
108. Paquette R, Dergham S, Karpf E et al. Ex vivo expanded unselected peripheral blood: Progenitor cells reduce posttransplantation neutropenia, thrombocytopenia, and anemia in patients with breast cancer. Blood 2001; 96:2385-2390.
109. Williams W, Beutler E, Erslev A et al. Hematology. 4th ed. New York: McGraw-Hills Inc, 2002.
110. Cairo M, Wagner J. Placental and/or umbilical cord blood: An alternative source of hematopoietic stem cells for transplantation. Blood 1997; 90:4665-4678.
111. Gluckman E, Broxmeyer H, Auerbach A et al. Hematopoietic reconstitution in a patient with Fanconi's anemia by means of umbilical-cord blood from an HLA-identical sibling. New Engl J Med 1989; 321:1174-1178.

112. Ballen K, Broxmeyer H, McCullough J et al. Current status of cord blood banking and transplantation in the United States and Europe. Biol Blood Marrow Transplant 2001; 7:635-645.
113. Barker J, Wagner J. Umbilical cord blood transplantation: Current state of the art. Current Opinion in Oncology 2002; 14:160-164.
114. Hows J. Status of umbilical cord blood transplantation in the year 2001. J Clin Pathol 2001; 54:428-434.
115. Barker J, Krepski T, DeFor T et al. Searching for unrelated donor hematopoietic stem cells: Availability and speed of umbilical cord blood versus bone marrow. Biol Blood Marrow Transplant 2002; 8:257-260.
116. Gluckman E, Rocha V, Boyer-Chammard A et al. Outcome of cord-blood transplantation from related and unrelated donors. New Engl J Med 1997; 337:373.
117. Rubinstein P, Carrier C, Scaradavou A et al. Outcomes among 562 recipients of placental-blood transplants from unrelated donors. New Engl J Med 1998; 339:1565-1577.
118. Wagner J, Barker J, DeFor T et al. Transplantation of unrelated donor umbilical cord blood in 102 patients with malignant and nonmalignant diseases: Influence of CD34 cell dose and HLA disparity on treatment-related mortality and survival. Blood 2002; 100:1611-1617.
119. Kurtzberg J, Martin P, Chao N et al. Unrelated placental blood in marrow transplantation. Stem Cells 2000; 18:150-156.
120. Briddell R, Kern B, Zilm K et al. Purification of CD34+ cells is essential for optimal ex vivo expansion of umbilical cord blood cells. J Hematother 1997; 6:145-150.
121. Astori G, Malangone W, Adami V et al. A novel protocol that allows short-term stem cell expansion of both committed and pluripotent hematopoietic progenitor cells suitable for clinical use. Blood Cells Mol Dis 2001; 27:715-724.
122. Piacibello W, Sanavio F, Garetto L et al. Extensive amplification and self-renewal of human primitive hematopoietic stem cells from cord blood. Blood 1997; 89:2644-2653.
123. Piacibello W, Sanavio F, Garetto L et al. Differential growth factor requirement of primitive cord blood hematopoietic stem cell for self-renewal and amplification vs proliferation and differentiation. Leukemia 1998; 12:718-727.
124. Gilmore G, De Pasquale D, Lister J et al. Ex vivo expansion of human umbilical cord blood and peripheral blood CD34+ hematopoietic stem cells. Exp Hematol 2000; 28:1297-1305.
125. Yamaguchi M, Hirayama F, Murahashi H et al. Ex vivo expansion of human UC blood primitive hematopoietic progenitors and transplantable stem cells using human primary BM stromal cells and human AB serum. Cytotherapy 2002; 4:109-118.
126. Tanavde V, Malehorn M, Lumkul R et al. Human stem-progenitor cells from neonatal cord blood have greater hematopoietic expansion capacity than those from mobilized adult blood. Exp Hematol 2002; 30:816-823.
127. Su R, Zhang X, Li K et al. Platelet-derived growth factor promotes ex vivo expansion of CD34+ cells from human cord blood and enhances long-term culture initiating cells, nonobese diabetic/severe combined immunodeficient repopulating cells and formation of adherent cells. Br J Haematol 2001; 117:735-746.
128. Berger M, Fagioli F, Piacibello W et al. Role of different medium and growth factors on placental blood stem cell expansion: An in vitro and in vivo study. Bone Marrow Transplant 2002; 29:443-448.
129. Lam A, Li K, Zhang X et al. Preclinical ex vivo expansion of cord blood hematopoietic stem and progenitor cells: Duration of culture; the media, serum supplements, and growth factors used; and engraftment in NOD/SCID mice. Transfusion 2001; 41:1567-1576.
130. Martinez-Serrano A, Rubio F, Navarro B et al. Human neural stem and progenitor cells: In vitro and in vivo properties, and potential for gene therapy and cell replacement in the CNS. Curr Gene Ther 2001; 1:279-299.
131. McNiece I, Almeida-Porada G, Shpall E et al. Ex vivo expanded cord blood cells provide rapid engraftment in fetal sheep but lack long-term engrafting potential. Exp Hematol 2002; 30:612-616.
132. Shpall E, Quinones R, Giller R et al. Transplantation of ex vivo expanded cord blood. Biol Blood Marrow Transplant 2002; 8:368-376.
133. Jaroscak J, Martin P, Waters-Pick B et al. A phase I trial of augmentation of unrelated umbilical cord blood transplantation with ex-vivo expanded cells. Blood 1998; 92:646a.
134. Stiff P, Pecora A, Parthasarathy M et al. Umbilical cord blood transplants in adults using a combination of unexpanded and ex vivo expanded cells: Preliminary clinical observations. Blood 1998; 92:646a.
135. Engelhardt M, Douville J, deReys S et al. Transplantation of ex vivo perfusion culture expanded bone marrow cells produces durable hematopoietic reconstitution after myeloablative chemotherapy. Blood 1998; 92:126a.
136. Thomson J, Itskovitz-Eldor J, Shapiro S et al. Embryonic stem cell lines derived from human blastocysts. Science 1998; 282:1147.
137. Reubinoff B, Pera M, Fong C et al. Embryonic stem cell lines from human blastocysts: Somatic differentiation in vitro. Nat Biotech 2000; 18:399-404.
138. Shambolt M, Axelman J, Wang S. Derivation of pluripotent stem cells from cultured human primordial germ cells. Proc Natl Acad Sci USA 1998; 95:13726-13731.
139. McGrath J, Solter D. Nuclear transplantation in mouse embryos by microsurgery and cell fusion. Science 1983; 220:1300-1302.
140. Campbell K, McWhir J, Ritchie W et al. Sheep cloned by nuclear transfer from cultured cell line. Nature 1996; 380:64-66.
141. Wilmut I, Schneike A, McWhir J et al. Viable offspring derived from fetal and adult mammalian cells. Nature 1997; 385:810-813.
142. Amit M, Carpenter M, Inokuma M et al. Clonally derived human embryonic stem cell lines maintain pluripotency and proliferative potential for prolonged periods of culture. Dev Biol 2000; 227:271-278.
143. Xu C, Inokuma M, Denham J et al. Feeder-free growth of undifferentiated human embryonic stem cells. Nat Biotech 2001; 19:971-974.
144. Draper J, Andrews P. Embryonic stem cells: Advances toward potential therapeutic use. Curr Opin Obstet Gynecol 2002; 14:309-315.
145. Fridenshtein A. Stromal bone marrow cells and the hematopoietic microenvironment. Arkh Patol 1982; 44:3-11.
146. Gronthos S, Graves S, Ohta S et al. The STRO-1+ fraction of adult human bone marrow contains the osteogenic precursors. Blood 1994; 84:4164-4173.
147. Pittenger M, Mackay A, Beck S et al. Multilineage potential of adult human mesenchymal stem cells. Science 1999; 284:147.
148. Wakitani S, Saito T, Caplan A. Myogenic cells derived from rat bone marrow mesenchymal stem cells exposed to 5-azacytidine. Muscle Nerve 1995; 18:1417-1426.
149. Reyes M, Lund T, Lenvik T et al. Purification and ex vivo expansion of postnatal human marrow mesodermal progenitor cells. Blood 2001; 98:2615-2625.
150. Jiang Y, Jahagirdar B, Reinhardt R et al. Pluripotency of mesenchymal stem cells derived from adult marrow. Nature 2002; 418:41-49.
151. Svendsen C, Smith A. New prospects for human stem-cell therapy in the nervous system. Trends Neurosci 1999; 22:357-364.
152. Villa A, Rubio F, Navarro B et al. Human neural stem cells in vitro: A focus on their isolation and perpetuation. Biomed Pharmacother 2001; 55:91-95.
153. Svendsen C, Smith A. Biopotent progenitor cell lines from human CNS. Nat Biotech 1997; 15:574-580.
154. Raymon H, Thode S, Zhou J et al. Immortalized human dorsal root ganglion cells differentiate into neurons with nonciceptive properties. J Neurosci 1999; 19:5420-5428.
155. Li R, Thode S, Zhou J et al. Motoneuron differentiation of immortalized human spinal cord cell lines. J Neurosci Res 2000; 59:342-352.
156. Flax J, Aurora S, Yang C et al. Engraftable human neural stem cells respond to developmental cues, replace neurons, and express foreign genes. Nat Biotech 1998; 16:1033-1039.
157. Reynolds B, Weiss S. Generation of neurons and astrocytes from isolated cells of the adult mammalian central nervous system. Science 1992; 255:1707-1710.
158. Carpenter M, Cui X, Hu Z et al. In vitro expansion of a multipotent population of human neural progenitor cells. Exp Neurol 1999; 158:265-278.

159. Gottlieb D Large scale sources of neural stem cells. Annu Rev Neurosci 2002; 25:381-407.
160. Liu S, Qu Y, Stewart T et al. Embryonic stem cells differentiate into oligodendrocytes and myelinate in culture and after spinal cord transplantation. Proc Natl Acad Sci USA 2000; 97:6126-6131.
161. Bain G, Kitchens D, Yao M et al. Embryonic stem cells express neuronal properties in vitro. Dev Biol 1995; 168:342-357.
162. Carpenter M, Inokuma M, Denham J et al. Enrichment of neurons and neural precursors from human embryonic stem cells. Exp Neurol 2001; 172:393-397.
163. Villa A, Navarro B, Martinez-Serrano A. Genetic perpetuation of in vitro expanded human neural stem cells: Cellular properties and therapeutic potential. Brain Res Bull 2002; 57:789-794.
164. Goldring K, Partridge T, Watt D. Muscle stem cells. J Pathol 2002; 197:457-467.
165. Rosenblatt J, Lunt A, Parry D et al. Culturing satellite cells from single muscle fiber explants. In vitro Cell Dev Biol 1995; 31:773-779.
166. Beauchamp J, Pagel C, Partridge T. A dual-marker system for quantitative studies of myoblast transplantation in the mouse. Transplantation 1997; 63:1794-1797.
167. Qu Z, Balkir L, van Deutkom J et al. Development of approaches to improve cell survival in myoblast transfer therapy. J Cell Biol 1998; 142:1257-1267.
168. Lee J, Qu-Petersen Z, Cao B et al. Clonal isolation of muscle-derived cells capable of enhancing muscle regeneration and bone healing. J Cell Biol 2000; 150:1085-1100.
169. Jankowski R, Halusszczak C, Trucco M et al. Flow cytometric characterization of myogenic cell populations obtained via the preplate technique: Potential for rapid isolation of muscle-derived stem cells. Hum Gene Ther 2001; 12:619-628.
170. Deasy B, Qu-Petersen Z, Greenberger J et al. Mechanisms of muscle stem cell expansion with cytokines. Stem Cells 2002; 20:50-60.
171. Sheehan S, Tatsumi R, Temm-Grove C et al. HGF is an autocrine growth factor for skeletal muscle satellite cells in vitro. Muscle Nerve 2000; 23:239-245.
172. Sheehan S, Allen R. Skeletal muscle satellite cell proliferation in response to members of the fibroblast growth factor family and hepatocyte growth factor. J Cell Physiol 1999; 181:499-506.
173. Haugk K, Wilson H-M, Swisshelm K et al. Insulin-like growth factor (IGF)-binding protein-related protein-1: An autocrine/paracrine factor that inhibits skeletal myoblast differentiation but permits proliferation in response to IGF. Endocrinology 2000; 141:100-110.
174. Hughes S. Cardiac Stem Cells. J Pathol 2002; 197:468-478.
175. Beltrami A, Urbanek K, Kajstura J et al. Evidence that human cardiac myocytes divide after myocardial infarction. New Engl J Med 2001; 344:1750-1757.
176. Quaini F, Urbanek K, Beltrami A et al. Chimerism of the transplanted heart. New Engl J Med 2002; 346:5-15.
177. Fuchs E, Segre J. Stem cells: A new lease on life. Cell 100:143-155.
178. Taylor D, Atkins B, Hungsreugs P et al. Regenerating functional myocardium: Improved performance after skeletal myoblast transplantation. Nat Med 1998; 4:929-933.
179. Kessler P, Byrne B. Myoblast cell grafting into heart muscle: Cellular biology and potential applications. Annu Rev Physiol 1999; 61:219-242.
180. Kehat I, Kanyagin-Karsenti D, Snir M et al. Human embryonic stem cells can differentiate into myocytes with structural and functional properties of cardiomyocytes. J Clin Invest 2001; 108:407-414.
181. Wang J, Shum-Tim D, Galipeau J et al. Marrow stromal cells for cellular cardiomyoplasty: Feasibility and potential clinical advantages. J Thorac Cardiovasc Surg 2000; 120:999-1005.
182. Wang J, Shum-Tim D, Chedrawy E et al. The coronary delivery of marrow stromal cells for myocardial regeneration: Pathophysiologic and therapeutic implications. J Thorac Cardiovasc Surg 2001; 122:699-705.
183. Tomita S, Li R, Weisel R et al. Autologous transplantation of bone marrow cells improves damaged heart function. Circulation 1999; 100:11247-11256.
184. Toma C, Pittenger M, Cahill K et al. Human mesenchymal stem cells differentiate to a cardiomyocyte phenotype in adult murine heart. Circulation 2002; 105:93-98.
185. Orlic D, Kajstura J, Chimenti S et al. Transplanted adult bone marrow cells repair myocardial infarcts in mice. Ann NY Acad Sci 2001; 938:221-229.
186. Orlic D, Kajstura J, Chimenti S et al. Bone marrow cells regenerate infarcted myocardium. Nature 2001; 410:701-705.
187. Orlic D, Kajstura J, Chimenti S et al. Mobilized bone marrow cells repair the infarcted heart, improving function and survival. Proc Natl Acad Sci USA 2001; 98:10344-10349.
188. Goodell M, Jackson J, Majka S et al. Stem cell plasticity in muscle and bone marrow. Ann NY Acad Sci 2001; 938:208-218.
189. Jackson K, Majka S, Wang H et al. Regeneration of ischemic cardiac muscle and vascular endothelium by adult stem cells. J Clin Invest 2001; 107:1395-1402.
190. Kocher A, Schuster M, Szabolcs M et al. Neovascularization of ischemic myocardium by human bone marrow-derived angioblasts prevents cardiomyocyte apoptosis, reduces remodeling and improves cardiac function. Nat Med 2001; 7:430-436.
191. Asahara T, Murohara TSA. Isolation of putative progenitor endothelial cells for angiogenesis. Science 1997; 275:964-967.
192. Shi Q, Rafii S, Hong-De Wu M et al. Evidence for circulating bone marrow-derived endothelial cells. Blood 1998; 92:362-367.
193. Takahashi T, Kalka C, Masuda H et al. Ischemia- and cytokine-induced mobilization of bone marrow-derived endothelial progenitor cells for neovascularization. Nat Med 1999; 5:434-438.
194. Alison M, Poulsom R, Forbes S. Update on hepatic stem cells. Liver 2001; 21:367-373.
195. Lowes K, Brennan B, Yeoh G et al. Oval cell numbers in human chronic liver diseases are directly related to disease severity. Am J Pathol 1999; 154:537-541.
196. Petersen B, Bowen W, Patrene K et al. Bone marrow as a potential source of hepatic oval cells. Science 1999; 284:1168-1170.
197. Alison M, Poulsom R, Jeffery R et al. Hepatocytes from nonhepatic adult stem cells. Nature 2000; 406:257.
198. Theise N, Nimmakayalu M, Gardner R et al. Liver from bone marrow in humans. Hepatology 2000; 32:11-16.
199. Tsao M-S, Smith J, Nelson K et al. A diploid epithelial cell line from normal adult rat liver with phenotypic properties of oval cells. Exp Cell Research 1984; 154:38-52.
200. Coleman W, Wennerberg A, Smith G et al. Regulation of the differentiation of diploid and some aneuploid rat liver epithelium (stem-like) cells by the hepatic micoenvironment. Am J Pathol 1993; 142:1373-1382.
201. Grisham J, Coleman W, Smith G et al Isolation, culture and transplantation of rat hepatocytic precursor (stem-like) cells. Proc Soc Exp Biol Med 1993; 204:270-279.
202. Shafritz D, Dabeva M. Liver stem cells and model systems for liver repopulation. J Hepatology 2002; 36:552-564.
203. Dabeva M, Petkov P, Sandhu J et al. Proliferation and differentiation of fetal liver epithelial progenitor cells after transplantation into adult rat liver. Am J Pathol 2000; 156:2017-2031.
204. Sandhu J, Petkov P, Dabeva M et al. Stem cell properties and repopulation of the rat liver by fetal epithelial progenitor cells. Am J Pathol 2001; 159:1323-1334.
205. Kubota H, Reid L. Clonogenic hepatoblasts, common precursors for hepatic and biliary lineages, are lacking classic major histocompatability complex class I antigen. Proc Natl Acad Sci USA 2000; 97:12132-12137.
206. Forbes S, Vig P, Poulsom R et al. Hepatic stem cells. J Pathol 2002; 510-518.

CHAPTER 13

The Extracellular Matrix as a Substrate for Stem Cell Growth and Development and Tissue Repair

Stephen F. Badylak and Mervin C. Yoder

Introduction

The extracellular matrix (ECM) is a complex mixture of structural and functional proteins, glycoproteins, and proteoglycans arranged in a unique, tissue specific three-dimensional ultrastructure. The structure and the composition of the ECM are both a product of the cells that populate the matrix and a determining factor in the phenotype of these cells. A "dynamic reciprocity" exists between the cells and the ECM that is in part dependent upon the local environment of each tissue.[1-3] Age, mechanical loading and microenvironment are all factors that can affect the ligands that reside within the ECM and in turn affect behavior of the resident cell population including gene expression. The resultant three dimensional ultrastructure of each tissue ECM is, therefore, likely distinctive and specific. The ECM that surrounds undifferentiated progenitor cell populations is poorly understood and existing knowledge is largely an extension of what has been learned from in vitro studies and the reported effects of selected growth substrates upon embryonic stem cell (ESC) differentiation patterns. The present chapter will briefly review the composition and function of the ECM that surrounds mature cells and organs, developing embryonic cells, and the relationship of selected ECM components to the growth and differentiation of such embryonic stem cells.

Composition of Extracellular Matrix

The function of the ECM in supporting cell adhesion, proliferation, migration and differentiation cannot be understood without knowledge of its composition; i.e., function is related to composition in a reciprocal relationship. Collagen, glycosaminoglycans, fibronectin, laminin, entactin and a variety of growth factors constitute the majority (by weight) of recognized ECM components. The logical division of these components into structural and functional components is not possible because many, if not all, of these molecules have both structural and functional roles.

Collagen is the most abundant protein within the ECM. More than 90% of the dry weight of the ECM from most organs is collagen, and most of that collagen is Type I. More than twenty distinct types of collagen have been identified. As already stated, the primary structural collagen in the ECM of mammalian tissues is type I collagen. This protein has been well characterized and is ubiquitous within the animal and plant kingdom.[4] Collagen has maintained a highly conserved amino acid sequence through the course of evolution and has served as a suitable substrate for in vitro and in vivo cell growth across species lines.

Collagen types other than type I collagen exist in the ECM of virtually every tissue and organ, albeit in much lower quantities. These alternative collagen types each provide distinct mechanical and physical properties to the ECM and contribute to the population of ligands that interact with the resident cell populations. By way of example, Type III collagen exists within selected submucosal ECMs, such as the submucosal ECM of the urinary bladder; a location in which less rigid structure is required for appropriate function than is required in a tendinous or ligamentous location. Type IV collagen is present within the basement membrane of all vascular structures and is an important ligand for endothelial cells. Type VI collagen functions as a "connector" of glycosaminoglycan and functional proteins to larger structural proteins such as type I collagen, thus helping to provide a gel like consistency to the ECM. Type VII collagen is an essential component of the anchoring fibrils of keratinocytes to the underlying basement membrane of the epidermis. Each of these collagen types is of course the result of specific gene expression patterns as cells differentiate and tissues and organs develop and spatially organize.

The ubiquitous nature of collagen is one reason that it serves as a favorable substrate for in vitro attachment and growth of differentiated cell types. Most collagen types cause differentiation of progenitor cell populations; i.e., do not support maintenance of the undifferentiated state. Thus, very little if any collagen is found surrounding ESC populations.

The diversity of collagen types is partially responsible for the distinctive biologic activity of ECM when it is used as a scaffold for tissue repair and reconstruction in adult mammals. This diversity and the associated tissue specific ultrastructure are reflective of the difficulty in recreating such a composite scaffold in vitro. Stated differently, scaffolds/substrates for in vitro cell growth or in vivo tissue repair can be readily constructed from individual components of the ECM (e.g., purified Type I collagen) but, it is not yet possible to artificially recreate the entire ECM.

Fibronectin is second only to collagen in quantity within the ECM. Fibronectin exists both in soluble and tissue isoforms and possesses ligands for adhesion of many cell types[5,6] including endothelial cells.[7] The ECM of both submucosal structures and basement membrane structures contain abundant fibronectin[8,9] and the cell friendly characteristics of this protein have made it an attractive ligand for in vitro cell culture and for use as a coating protein for synthetic scaffold materials to promote host biocompatibility and the attachment of selected cell types. Fibronectin has been shown to play a central role in vascular development and is found at an early stage within the ECM of developing embryos. Targeted deletion of fibronectin in the murine embryo results in embryonic lethality with disruption of normal vessel and cardiovascular development. Fibronectin has also been found to be essential for the maintenance of hematopoietic stem cell activity ex vivo, another indication of its important role in the support of the stem cell phenotype. The crucial role of the beta-1 integrin chain in mediating hematopoietic stem cell interactions with fibronectin has been firmly established.[10-12] Loss of the beta-1 integrin receptors in mice results in intrapartum mortality.

Laminin is a complex adhesion protein found in the ECM; especially within basement membrane ECMs.[8] This protein plays an important role in early embryonic development and is perhaps the best studied of the ECM proteins found in ESC-derived embryoid bodies.[13] This trimeric cross-linked polypeptide exists in numerous forms dependent upon the particular mixture of peptide chains.[14,15] The prominent role of laminin in the formation and maintenance of vascular structures is particularly noteworthy when considering the ECM as a scaffold for tissue reconstruction.[16,17] This protein appears to be among the first and most critical ECM factors in the process of cell and tissue differentiation.

The ECM contains various mixtures of glycosaminoglycans (GAGs) depending upon the location of the ECM in the host, the age of the host, and the microenvironment. The GAGs bind growth factors and cytokines, promote water retention, and contribute to the gel properties of the ECM. The heparin binding properties of numerous cell surface receptors and of many growth factors (e.g., fibroblast growth factor family, vascular endothelial cell growth factor) make the heparin-rich GAGs useful components of naturally occurring substrates for cell growth. The glycosaminoglycans present in ECM include chondroitin sulfates A and B, heparin, heparan sulfate, and hyaluronic acid.[18] Hyaluronic acid has been most extensively investigated as a scaffold for tissue reconstruction and as a carrier for selected cell populations in therapeutic tissue engineering applications. The concentration of hyaluronic acid within ECM is highest in fetal and newborn tissues and tends therefore to be associated with desirable healing properties. Hyaluronic acid and its receptors (CD44 and receptor for hyaluronate mediated motility {RHAMM}) play important roles in those processes that require cell migration such as morphogenesis, wound repair, inflammation, and cancer.[19]

The characteristic of the intact ECM that distinguishes it from other substrates for cell attachment, growth and differentiation, such as purified collagens, fibronectin or laminin, is its diversity of structural and functional proteins. In addition, the associated bioactive molecules that reside within the ECM and their unique spatial distribution provides a reservoir of biologic signals. Although cytokines and growth factors are present within ECM in very small quantities, they act as potent modulators of cell behavior. The list of growth factors is extensive and includes vascular endothelial cell growth factor (VEGF), the fibroblast growth factor (FGF) family, epithelial cell growth factor (EGF), transforming growth factor beta (TGF-beta), keratinocyte growth factor (KGF), hepatocyte growth factor (HGF) and platelet derived growth factor (PDGF), among others. These factors tend to exist in multiple isoforms, each with its specific biologic activity. Purified forms of growth factors and biologic peptides have been investigated in recent years as therapeutic means of encouraging blood vessel formation (e.g., VEGF), inhibiting blood vessel formation (angiostatin), stimulating deposition of granulation tissue (PDGF), and encouraging epithelialization of wounds (KGF). However, this therapeutic approach has struggled with determination of optimal dose, the ability to sustain and localize the release at the desired site, and the inability to turn the factor "on" and "off" as needed during the course of tissue repair.

An advantage of utilizing the ECM in its native state as a substrate for cell growth and differentiation is the presence of all the attendant growth factors (and their inhibitors) in the relative amounts that exist in nature and perhaps most importantly, in their native three-dimensional ultrastructure. The ECMs present these factors efficiently to resident cell surface receptors, protect the growth factors from degradation, and modulate their synthesis.[20-23] If one considers the ECM to be a substrate for in vitro cell growth, it is reasonable to think of it in terms of a temporary controlled release vehicle for naturally derived growth factors. As the understanding of the differences in ECM as a function of tissue type, species, and maturity increases, the ability to modulate the differentiation (as lack thereof) of progenitor cells within and ECM environment will improve.

ECM During Embryonic Growth and as a Substrate for Stem Cell Culture

The concept of using three dimensional culture systems for the maintenance and manipulation of stem cells has received considerable attention in recent years.[24,25] Gene expression patterns observed in three-dimensional cultures have been shown to be different and distinct from those observed in two-dimensional monolayer culture systems; thus the ultrastructure of the microenvironment has a profound influence upon cell differentiation.[13] Human and murine embryonic stem cells (ESC) have been grown on gelatin (i.e., polymerized collagen), Matrigel™, and laminin. To maintain the undifferentiated state in the murine system, media conditioned by mouse embryonic fibroblasts or supplemented with leukemia inhibitory factor (LIF) is required. LIF alone is not sufficient to maintain human ESC, but can be maintained without feeder cells when the cells are plated onto Matrigel™ or laminin (but not fibronectin or collagen IV) and then supplemented with conditioned media from the murine fibroblast feeder cells.[26] In the absence of these factors, ESCs form embryoid bodies (EB) that develop an outer endodermal layer, a subendodermal basement membrane (Reichert's membrane), and further differentiate to form an epiblast and central proamniotic-like cavity.[13,27] Although primitive endodermal cell differentiation precedes the assembly of basement membrane ECM, subsequent cell differentiation to form epiblast and proamniotic cavities will not occur without this ECM in place.[13] As indicated earlier, laminin is an essential component of these earliest cell differentiation steps.[10,11,13]

Undifferentiated human ESC have been cultured for more than 100 passages while maintaining a normal karyotype. The doubling time of ESC in vitro appears to be approximately 28-50 hours.[28] These cells express the nuclear transcription factor OCT-4 and telomerase, an enzyme that replaces telomeric repeats lost with each cell division. OCT-4 and telomerase expression are associated with murine and human ESC immortality and pluripotentiality.[29] As mentioned above, the maintenance of the undifferentiated state requires certain culture conditions including the use of LIF (in the murine system), mouse embryonic fibroblast feeder cells, or growth on a laminin or Matrigel™ in the presence of feeder cell conditioned medium. Murine and human hematopoietic stem cells also require interaction with stromal cells or their secreted bioproducts such as growth factors and matrix molecules. Murine hematopoietic stem cells have been shown to increase in number in vitro when selected stromal elements are present.[30] The use of ESC for therapeutic applications requires the ability to control the state of differentiation of ESC in culture; i.e., the ability to either maintain the undifferentiated state or direct the differentiation toward specific cell lineages.

The requirement for feeder cells during culture of ESC demonstrates the obvious supply by the cells of essential factors/signals that modulates the state of cellular differentiation. The ability to culture undifferentiated ESC without the use of feeder cells provides several advantages over the use of the feeder cell technique. Reduced cell contamination of ESC, qualitative and quantitative control of media components, and the ability to manipulate selected media ingredients to investigate the effects of these ingredients upon ESC phenotype are all desirable properties of the feeder cell free technique.

One mechanism for monitoring the differentiation of ESC in vitro is the determination of the expression of selected genes that precede or accompany differentiation. For example, it has been shown that growth factors such as fibroblast growth factor (FGF), connective tissue growth factor (CTGF), and cysteine rich proteins such as osteopontin, AGF-BP5, and Cyr61 are reduced during differentiation of ESC.[28] In addition, expression of both OCT-4 and telomerase are reduced during differentiation.

Similarly, gene expression of alpha fetal protein (AFP) and apolipoprotein A-II are increased during differentiation of the endodermal cell lineage. ESC can be stimulated to form embryoid bodies with cells of all three germ cell layers by exposure to selected ECM factors and conditioned media. Stated differently, the ECM can be used as an indicator of progenitor cell differentiation events.

A novel stem cell population isolated from human, rat, and murine tissues has recently been reported. These cells called multipotent adult progenitor cells (MAPC) can be isolated from murine bone marrow, skeletal muscle, and brain tissue. MAPC display many features of human and murine ESC including indefinite growth potential in vitro, ability to differentiate into most cell types, requirement for LIF to maintain pluripotentiality in vitro, expression of OCT-4 and telomerase, and stability of chromosomal integrity in vitro.[31] Maintenance of rat, human, and murine MAPC requires growth on fibronectin coated tissue culture labware. Addition of certain growth factors such as VEGF or FGF to MAPC cultured on fibronectin led to differentiation of the cells into endothelial or neuronal cell types, respectively. Changing the ECM to Matrigel (not collagen) and adding FGF-4 and hepatocyte growth factor to MAPC cultures resulted in differentiation to hepatocytes in vitro. Thus, ECM proteins participate in the maintenance of pluripotent MAPC in vitro and are involved in the induction of MAPC commitment to various differentiated cell lineages.

Extracellular Matrix Scaffolds for Tissue Repair

The ECM obviously plays an important role in ESC development, tissue differentiation, and organogenesis. The ability of the ECM ligands to modulate cell phenotype and cell behavior likely also has important implications for other biologic processes such as inflammation, immunity and wound healing. It is well accepted that developing fetuses and newborns heal wounds differently (generally better) than fully mature individual and certainly aged individuals. It is plausible that the composition and structure of the ECM is at least in part responsible for the differences noted in the wound healing process as a function of age. Recently, scaffolds derived from naturally occurring (usually porcine derived) ECM have been used for therapeutic applications. For example, it has recently been shown that ECM scaffolds derived from porcine small intestinal submucosa selectively recruit circulating progenitor cells to sites of wound repair; a phenomenon not found in "normal" adult mammalian wound healing.[32]

There is abundant literature on the use of modified ECM scaffolds, especially chemically cross-linked biologic scaffolds, for tissue repair and replacement. Porcine heart valves, decellularized and cross-linked human dermis (Alloderm™), and chemically cross-linked purified bovine type I collagen (Contigen™) are examples of such products currently available for use in humans. Similarly modified ECM scaffolds have been used for the reconstitution of the cornea, skin, cartilage and bones, and nerve regeneration, among others.[33-36]

Porcine derived ECM scaffolds that have not been modified, except for the decellurization process and terminal sterilization, have been successfully used for the repair of numerous body tissues including musculotendinous structures,[37-39] lower urinary tract reconstruction,[40-42] dura mater replacement,[32,43] vascular reconstruction,[44-46] and the repair of full and partial thickness skin wounds.[47] The remodeling process in all of these applications has been remarkably similar. Immediately following implantation in vivo, there is an intense cellular infiltrate consisting of equal numbers of polymorphonuclear leukocytes and mononuclear cells. By 72 hours post implantation, the infiltrate is almost entirely mononuclear cell in appearance with early evidence for neovascularization. These mononuclear cells have not been definitively characterized to date. Between days 3 and 14, the number of mononuclear cells increases, vascularization becomes intense, and there is a progressive degradation of the xenogeneic scaffold with associated deposition of host derived neomatrix. Following day 14, the mononuclear cell infiltrate diminishes and there is the appearance of site specific parenchymal cells that orient along lines of stress (i.e., a microenvironmental factor).[48] These parenchymal cells consist of fibroblasts, smooth muscle cells, skeletal muscle cells, and epithelial cells depending upon the site in which the scaffold has been placed. It has been shown that circulating, marrow derived progenitor cells participate in this remodeling process when ECM scaffolds are used.[49] Environmental factors such as mechanical loading, have also been shown to be important in the remodeling of ECM scaffolds.[48] Of note, there is an absence of tissue necrosis and scar tissue formation during the remodeling of these xenogeneic ECM scaffolds.

ECM scaffolds derived from the urinary bladder submucosa (UBS) have been used for reconstruction of the lower urinary tract and have shown remodeling and healing that is atypical for adult mammals.[50-62] The UBS scaffolds have been either allogeneic or xenogeneic in origin and have been used both alone and with cultured autologous cells. Sections of urethra, ureter, and urinary bladder have shown near normal reconstitution with formation or organized and innervated smooth muscle. There is a substantial body of literature developing that supports the use of intact ECM as a scaffold for tissue repair. More than 100,000 human patients have now been implanted with xenogeneic ECM scaffold derived from the porcine small intestinal submucosa for a variety of applications.

Summary

The extracellular matrix is a complex structure that exists in dynamic reciprocity with the resident cell population. The ECM plays a critical role in cell, tissue and organ development; including the development and differentiation of embryonic stem cells. The understanding of the cause-effect relationship between the ECM and cell behavior and the ECM and biologic events such as organogenesis, inflammation and wound healing is still cursory at best.

References

1. Bissell MJ, Hall HG, Parry G. How does the extracellular matrix direct gene expression? J Theor Biol 1982; 99:31-68.
2. Boudreau N, Myers C, Bissell MJ. From lamini to lamin: Regulation of tissue-specific gene expression by the ECM. Trends Cell Biol 1995; 5:1-4.
3. Ingber D. Extracellular matrix and cell shape: Potential control points for inhibition of angiogenesis. J Cell Biochem 1991; 47:236-241.
4. Vanderrest M, Garrone R. Collagen family of proteins. FASEB J 5:2814-2823.
5. Schwarzbauer JE. Fibronectin: From gene to protein. Curr Opin Cell Biol 1991; 3:786-791.
6. Miyamoto S, Katz BZ, Lafrenie RM et al. Fibronectin and integrins in cell adhesion, signaling, and morphogenesis. Ann NY Acad Sci 1998; 857:119-129.
7. Francis SE, Goh KL, Hodivala-Dilke K at el. Central roles of alpha-5 beta 1 integrin and fibronectin in vascular development in mouse embryos and embryoid bodies. Arterioscler Thromb Basc Biol 2002; 22:927-933.
8. Schwarzbauer JE. Basement membranes: Putting up the barriers. Curr Biol 1999; 9:R242-R244.
9. McPherson TB, Badylak SF. Characterization of fibronectin derived from porcine small intestinal submucosa. Tissue Engineering 1998; 4:75-83.
10. Fassler R, Meyer M. Consequences of lack of beta-1 integrin gene expression in mice. Genes Development 1995; 9:1896-1908.
11. Hirsch E, Iglesias A, Potocnik AJ et al. Impaired migration but not differentiation of haematopoietic stem cells in the absence of beta-1 integrins. Nature 1996; 380:171-175.
12. Potocnik AJ, Brakebusch C, Fassler R. Fetal and adult hematopoietic stem cells require beta-1 integrin function for colonizing fetal liver, spleen, and bone marrow. Immunity 2000; 12:653-663.
13. Li S, Harrison D, Carbonetto S et al. Matrix assembly, regulation, and survival functions of laminin and its receptors in embryonic stem cell differentiation. J Cell Biol 2002; 157:1279-1290.
14. Timpl R. Macromolecular organization of basement membranes. Curr Opin Cell Biol 1996; 8:618-624.
15. Timpl R, Brown J. Supramolecular assembly of basement membranes. BioEssays 1996; 18:123-132.
16. Ponce M, Nomizu M, Delgado MC et al. Identification of endothelial cell binding sites on the laminin gamma-1 chain. Circ Res 1999; 84:688-694.
17. Werb Z, Cu TH, Rinkenberger JL et al. Matrix-degrading proteases and angiogenesis during development and tumor formation. Apmis 1999; 107:11-18.
18. Hodde JP, Badylak SF, Brightman AO et al. Glycosaminoglycan content of small intestinal submucosa: A bioscaffold for tissue replacement. Tissue Engineering 1996; 2:209-217.
19. Entwistle J, Zhang S, Yang B et al. Characterization of the murine gene encoding the hyaluronan receptor RHAMM. Gene. Oct 3 1995; 163(2):233-8.
20. Bonewald LF. Regulation and regulatory activities of transforming growth factor beta. Crit Rev Eukaryotic Gene Expression 1999; 9:33-44.
21. Kagami S, Kondo S, Loster K et al. Collagen type I modulates the platelet-derived growth factor (PDGF) regulation of the growth and expression of beta 1 integrins by rat mesangial cells Biochem Biophys Res Commun 1998; 252:728-732.
22. Roberts R, Gallagher J, Spooncer E. Heparan sulphate bound growth factors: A mechanism for stromal cell mediated haemopoiesis. Nature (London) 1988; 332:376-378.
23. Sjaastad MD, Nelson WJ. Integrin-mediated calcium signaling and regulation of cell adhesion by intracellular calcium. BioEssays 1997; 19:47-55.
24. Rosenzweig M, Pikett M, Marks DF et al. Enhanced maintenance and retroviral transduction of primitive hematopoietic progenitor cells using a novel three-dimensional culture system. Gene Ther 1997; 4:928-936.
25. Tomimori Y, Takagi M, Yoshida T. The construction of an in vitro three-dimensional hematopoietic microenvironment for mouse bone marrow cells emplying porous carriers. Cytotechnology 2000; 34:121-130.
26. Xu C, Inokuma MS, Denham J et al. Feeder-free growth of undifferentiated human embryonic stem cells. Nature Biotech 2001; 19:971-974.
27. Coucouvanis E, Martin GR. Signals for death and survival: A two-step mechanism for cavitation in the vertebrate embryo. Cell 1995; 83:279-287.
28. Lebkowski JS, Gold J, Xu C et al. Human embryonic stem cells: Culture, differentiation, and genetic modification for regenerative medicine applications. Cancer J 2001; 7(Suppl 2):S83-S94.
29. Thomson JA, Itskovitz-Eldor J, Shapiro SS et al. Embryonic stem cell lines derived from human blastocysts. Science 1998; 282:1145-1147.
30. Verfaillie C, et al. Role of bone marrow matrix in normal and abnormal hematopoiesis. Crit Rev Oncol Haematol 1994; 16:201.
31. Jiang Y, Jahagirdar BN, Reinhardt RL et al. Pluripotency of mesenchymal stem cells derived from adult marrow. Nature 2002; 418(6893):41-9.
32. Badylak SF, Park K, McCabe G et al. Marrow-derived cells populate scaffolds composed of xenogeneic extracellular matrix. Exptl Hematol 2001; 29:1310-1318.
33. Bouhadir KH, Mooney DJ. In vitro and in vivo models for the reconstruction of intracellular signaling. Ann NY Acad Sci 1998; 842:188-194.
34. Kim BS, Mooney DJ. Engineering smooth muscle tissue with a predefined structure. J Biomed Mater Res 1998; 41:322-332.
35. Aiken SW, Badylak SF, Toombs JP et al. Small intestinal submucosa as an intra-articular ligamentous repair material: A pilot study in dogs. Vet Comp Orthopaedics Traumatology 1994; 7:124-128.
36. Badylak SF, Arnoczky S, Plouhar P et al. Naturally-occurring ECMs as scaffolds for musculoskeletal repair clinical orthopaedics and related research 1999; 367S:S333-S343.
37. Kropp BP, Sawyer BD, Shannon HE et al. Characterization of small intestinal submucosa-regenerated canine detrusor: Assessment of reinnervation, in-vitro compliance and contractility. J Urol 1996; 156:599-607.
38. Kropp BP, Rippy MK, Badylak SF et al. Regenerative urinary bladder augmentation using small intestinal submucosa: Urodynamic and histopathologic assessment in long term canine bladder augmentations. J Urol 1996; 155:2098-2104.
39. Cobb MA, Badylak SF, Janas W et al. Histology after dural grafting with small intestinal submucosa. Surg Neurol 1996; 46:389-394.
40. Cobb MA, Badylak SF, Janas W et al. Porcine small intestinal submucosa as a dural substitute. Surg Neurol 1999; 51(1):99-104.
41. Sandusky GE, Lantz GC, Badylak SF. Healing comparison of small intestine submucosa and ePTFE grafts in the canine carotid artery. J Surg Res 1995; 58:415-420.

42. Prevel CD, Eppley BL, McCarty M et al. Experimental evaluation of small intestine submucosa as a microvascular graft material. J Microsurg 1994; 15:588-591.
43. Badylak SF, Lantz G, Coffey A et al. Small intestinal submucosa as a large diameter vascular graft in the dog. J Surg Res 1989; 47:74-80.
44. Prevel CD, Eppley BL, Summerlin DJ et al. Small intestinal submucosa (SIS): Utilization as a wound dressing in full-thickness rodent wounds. Ann Plast Surg 1995; 35:381-388.
45. Hodde JP, Badylak SF, Shelbourne KD. The effect of range of motion on remodeling of small intestinal submucosa (SIS) when used as an Achilles' tendon repair material in the rabbit. Tissue Eng 1997; 3:27-37.
46. Vaught JD, Kropp BP, Sawyer BD et al. Detrusor regeneration in the rat using porcine small intestinal submucosal grafts: Functional innervation and receptor expression. J Urol 1996; 155:374-378.
47. Badylak SF, Meurling S, Chen M at al. Resorbable bioscaffold for esophageal repair in a dog model. J Ped Surg 2000; 35 (7):1097-1103.
48. Chen F, Yoo JJ, Atala A. A cellular collagen matrix as a possible "off the shelf" biomaterial for urethral repair. Urology 1999; 54:407-410.
49. Atala A, Guzman L, Retik AB. A novel inert collagen matrix for hypospadias repair. J Urol 1999; 162:1148-1151.
50. Dahms SE, Piechota HJ, Dahiya R et al. Composition and biomechanical properties of the bladder acellular matrix graft: Comparative analysis in rat, pig and human. Br J Urol 1998; 82:411-419.
51. Dahms SE, Piechota HJ, Nunes L et al. Free ureteral replacement in rats: Regeneration of ureteral wall components in the acellular matrix graft. Urology 1997; 50:818-825.
52. Merguerian PA, Reddy PP, Barrieras DJ et al. A cellular bladder matrix allografts in the regeneration of functional bladders: Evaluation of large-segment (> 24 cm) substitution in a porcine model. BJU Int 2000; 85(7):894-8.
53. Piechota HJ, Dahms SE, Nunes LS et al. In vitro functional properties of the rat bladder regenerated by the bladder acellular matrix graft. J Urol 1998; 159:1717-1724.
54. Piechota HJ, Dahms SE, Probst M et al. Functional rat bladder regeneration through xenotransplantation of the bladder acellular matrix graft. Br J Urol 1998; 81:548-559.
55. Piechota HJ, Gleason CA, Dahms SE et al. Bladder acellular matrix graft: In vivo functional properties of the regenerated rat bladder. Urol Res 1999; 27:206-213.
56. Probst M, Dahiya R, Carrier S et al. Reproduction of functional smooth muscle tissue and partial bladder replacement. Br J Urol 1997; 79:505-515.
57. Probst M, Piechota HJ, Dahiya R et al. Homologous bladder augmentation in dog with the bladder acellular matrix graft. BJU Int 2000; 85:362-371.
58. Reddy PP, Barrieras DJ, Wilson G et al. Regeneration of functional bladder substitutes using large segment acellular matrix allografts in a porcine model. J Urol 2000; 164(3Pt2):936-41. 20414890.
59. Sutherland RS, Baskin LS, Hayward SW et al. Regeneration of bladder urothelium, smooth muscle, blood vessels and nerves into an acellular tissue matrix. J Urol 1996; 156:571-577.
60. Wu HY, Baskin LS, Liu W et al. Understanding bladder regeneration: Smooth muscle ontogeny. J Urol 1999; 162:1101-1105.
61. Yoo JJ, Meng J, Oberpenning F et al. Bladder augmentation using allogenic bladder submucosa seeded with cells. Urology 1998; 51:221-225.
62. Badylak SF. Modification of natural polymers: Collagen. In: Atala A and Lanza RP, eds. Methods of Tissue Engineering. 2002:505-514.

CHAPTER 14

Gene Transfer into Human Hematopoietic Stem Cells: Problems and Perspectives

Serguei Kisselev, Tatiana Seregina, Richard K. Burt and Charles J. Link

Introduction

Manipulations with hematopoietic stem cells (HSC) in the context of bone marrow or peripheral blood stem cell transplantation have become a routine procedure.[1,2] Preparative quantities of HSC are readily accessibile. Either marrow or peripheral blood can provide greater than 2 million CD34⁺ HSC per kilogram of donor weight per harvest or apheresis, respectively. Using human HSC surface markers such as CD34 or AC133, it is possible to separate and enrich for HSC ex vivo. The intrinsic properties of HSC make them attractive targets for gene transfer. Gene modified HSC should achieve long-term repopulation in the recipient with equally long-term transgene expression.[3] Nowadays the gene transfer into hematopoietic cells comprises 7.4% (44 protocols) of the total number of United States Food and Drug Administration (FDA) approved gene therapy clinical trials.

Recent publications have demonstrated the ability of HSC to introduce tolerance in autoimmune disorders (refer to chapters on autoimmune diseases and HSC transplantation) and possibly to regenerate non-hematopoietic tissues (see Chapter 3). Application of HSC to induce tolerance or tissue regeneration imply potential broad therapeutic applications for gene modified HSC.

Currently the interplay of genes, environment, and hormones in autoimmune diseases is difficult to separate. Gene-marking techniques allow the fate of engrafted stem cells as well as their progeny to be tracked. The genetic marking of hematopoietic stem cells supplemented with appropriate techniques to detect their progeny may allow evaluation of the efficiency of immunoablative regimens. For example, recurrence of disease in absence of detectable transgene would suggest (but not prove) relapse from memory lymphocytes that survived the conditioning regimen and indicates an inadequate conditioning regimen. In a patient with a clinical relapse, the presence of the genetic marker in cells with an autoreactive phenotype would establish de novo autoimmune disease from engrafted HSC.

In some autoimmune diseases, organ function improves and even normalizes following HSCT. While controversial, it appears that HSC may be able to change fate and generate non-hematopoietic tissue including skeletal myocytes,[4] neuroglial cells and neurons,[5-7] cardiac muscle,[4,8,9] gut,[10] skin,[10] and hepatocytes.[11,12] Gene marking of HSC followed by biopsy of the affected organ systems may help determine if epithelial or tissue parenchymal cells have regenerated from the circulating HSC pool. Identification of disease-resistant and disease-susceptible autoimmune genes may allow for therapeutic gene therapy strategies in which autologous HSC could be genetically altered prior to infusion.

Technical Aspects of the Gene Transfer into Hematopoietic Stem Cells

Identification of True Stem Cells

Overall, the precise phenotype of human HSC remains elusive or at least not as well defined as previously thought.[13,14] Modern techniques of enrichment of HSC are based on their surface phenotype properties, but it has become evident that highly heterogeneous populations of cells characterized by distinct stages of differentiation are currently used for stem cell transplantation.

The main marker useful for isolation and/or enrichment of human hematopoietic progenitor cells for transplantation is a surface antigen termed CD34⁺ whose function may be the regulation, localization and differentiation of hematopoietic stem cells within the hematopoietic microenvironment.[15,16] Despite a long history of successful long term hematopoietic engraftment, reports by Ogawa and colleagues indicate that the CD34⁺ surface antigen expression is a marker of progenitor cell activation, while the main population of quiescent HSC are CD34 negative.[17-19] Other research groups have confirmed this data.[20-23]

There are three main sources of CD34⁺ HSC used for transplantation. Cells originating from all these sources are employed as targets for gene transfer both in experimental studies and clinical trials. CD34⁺ hematopoietic cells can be obtained from bone marrow, umbilical cord blood (CB) and peripheral blood (after mobilization by G-CSF, GM-CSF administration, or chemotherapy).[24] Use of peripheral blood as a source of cells allows a 10-fold greater collection with HSC compared to bone marrow. Cord blood cells have been used for HSC transplantations since 1988[25] mainly in pediatric practice because of the limited numbers of progenitor cells available from the placenta and umbilical cord.

Comparison of efficiency of in vitro gene transfer into HSC derived from the sources listed above reveals that retrovirus (RV)

Table 1. Stem cell phenotypes and functional properties of HSC, described by different researcher groups

Phenotype	Clonogenic Cells	LTC-IC (or CAFC)	Marrow Repopulating Ability
CD34$^+$	Enriched with CFU-GEMM[37,15]	Enriched (2.2%)	Enriched[16]
CD34$^+$CD117$^{+\ (or\ low)}$	20-50-fold enrichment of CFU-GEMM[38,39]	Enriched in CD117^{low} [38]	Enriched[23]
CD34$^+$CD90$^+$	Enriched with HPP-CFC[40]	Enriched[44]	Highly enriched[44,48]
CD34$^+$CD38^{low}	Highly enriched[41]	Highly enriched[45]	Highly enriched[45]
CD34$^+$CD45RA^{low}	Enriched with CFU-GEMM[40]	Enriched[40]	NA
CD34$^+$HLA-Dr^{-low}	Enriched with HPP-CFC[42]	Enriched	Enriched[42]
CD34$^+$HLA-Dr$^-$ CD33$^-$	Enriched with HPP-CFC[43]	Enriched (4.5%)	NA
CD34$^+$Rhd-123^{dim}	Enriched with HPP-CFC	1000-fold enrichment[46]	NA
Hoechst^{Bright}Rhd123^{dim}	Highly enriched	Highly enriched (7.2%)[47]	NA
CD34$^+$CD38^{low} CXCR4$^+$VLA4$^+$	NA	NA	Highly enriched[49]

NA indicates that data is not available; LTC-IC= long term culture initiating cells; CAFC= cobble-stone area forming cells

mediated transduction of cord blood (CB) CD34$^+$ cells is more efficient than with cells collected from peripheral blood (PB) and bone marrow (BM).[26,27] Furthermore, some reports demonstrate that whole populations of CD34$^+$ cord blood mononuclear cells contain more primitive progenitor cells (CD34$^+$CD38^{low} LTC-IC and CD34$^+$ NOD-SCID mice repopulating cells) than PB (after mobilization) or bone marrow.[28-31] These data suggest that gene transfer into the HSC derived from CB could be targeted at immature progenitor cells more easily.

By traditional definition, HSC possess two properties: ability to differentiate into all blood cell lineages and ability to self-renew and repopulate the bone marrow. The multilineage clonogenic potential of the cell population is traditionally verified by colony assay.[14,32] The ability of HSC to repopulate hematopoietic niches is evaluated in vitro by long-term cultures assay (LTC-IC).[33,34] An in vivo animal model of human HSC ability to initiate long-term hematopoietic activity is demonstrated by transplantation of HSC into immunodeficient nonobese diabetic severe combined immunodeficiency (NOD-SCID) mice.[35,36] Table 1 shows some of the cell surface phenotypes described by different research groups and associated functional cell properties.

Gene transfer into more primitive cells rather than into their differentiated progeny is generally preferential since most strategies require the maintenance of the transgene in long term repopulating cells. In fact, inefficient gene transfer into the earliest hematopoietic progenitors has imposed the major limitation on successful application of HSC gene therapy. The main reason for this less than optimal efficiency of gene transfer is inherent in the properties of specific vector systems and the physiology of HSC.

Vectors and Requirements for HSC Transduction

In most cases, the approach to gene therapy, in the context of hematopoietic stem cell transplantation, requires stable integration and long term expression of genes of interest into the stem cell genome. The most useful vectors to accomplish this are retrovirus-based vectors. Retroviral vectors have advantages including availability, efficiency, integration of the transgene into the genome, and well-developed production techniques. However, retroviral-based vectors also have some significant limitations.

Obtaining High Titer Retroviral Vectors

The choice of vector to be used for HSC transduction in clinical trials is defined by the goals of the trial and the gene to be transferred. The vector construct must accomodate the transgene size and should be non-toxic for target cells. Ideally the vector should not generate immune responses against viral transgenes and must be replication incompetent. Development of effective and standard vector production techniques is an area of active investigation.

The development of gene therapy retroviral vectors for clinical trials must be in accordance with FDA mandated Good Manufacturing Practice (GMP) requirements. It is accomplished by genetic alteration of retrovirus, development of appropriate packaging cell lines, establishment of high titer vector producing cells (VPC), collection and storage of VPC supernatant for transduction of the target HSC (see Fig. 1). First of all, clinical grade retrovirus must be certified as replication incompetent, i.e., non-infectious. This is achieved by splitting of genetic elements responsible for retrovirus replication between independent recombinant constructs inside VPC. In brief, retroviruses contain a sequence of DNA known as ψ (psi), the packaging signal, that allows insertion of retrovirus RNA into the developing capsid particle. Without the packaging sequence, the retrovirus particle is an empty shell, i.e., a capsid without RNA. Plasmid DNA containing a selectable marker and the retrovirus that has the packaging signal deleted may be integrated into the nucleus of a cell line by conventional techniques such as calcium phosphate precipitation. Using the selectable marker, the transfected cell (usually a murine fibroblast cell line) may then be isolated and cloned. These cells are called packaging cell lines because they contain a retrovirus (termed a helper retrovirus) that contains retroviral structural genes (capsid (gag), reverse transcriptase (pol), and envelope (env)) necessary to form retroviral particles. However, packaging cell lines cannot incorporate retrovirus RNA into the particle because the helper virus lacks a packaging (ψ) sequence and is, therefore, non-infectious. Another retrovirus that contains the packaging sequence but in which the capsid, RNA reverse transcriptase, and envelope genes are excised and replaced with a therapeutic or identifying gene (i.e., the clinical trial gene) may then be transfected into the packaging cell line. The result is a vector producing cell line (VPC) in which the helper retrovirus provides the genes (gag, pol, env) to manufacture a retroviral particle into which the vector containing the trial gene of interest is incorporated. VPC produce an intact retroviral particle that is infective only once and may be used to infect the target cell with the trial gene.

Despite the precaution of separating helper virus and vector, replication competent retrovirus (RCR) may occur from recombination events within VPC between the helper viral genes and

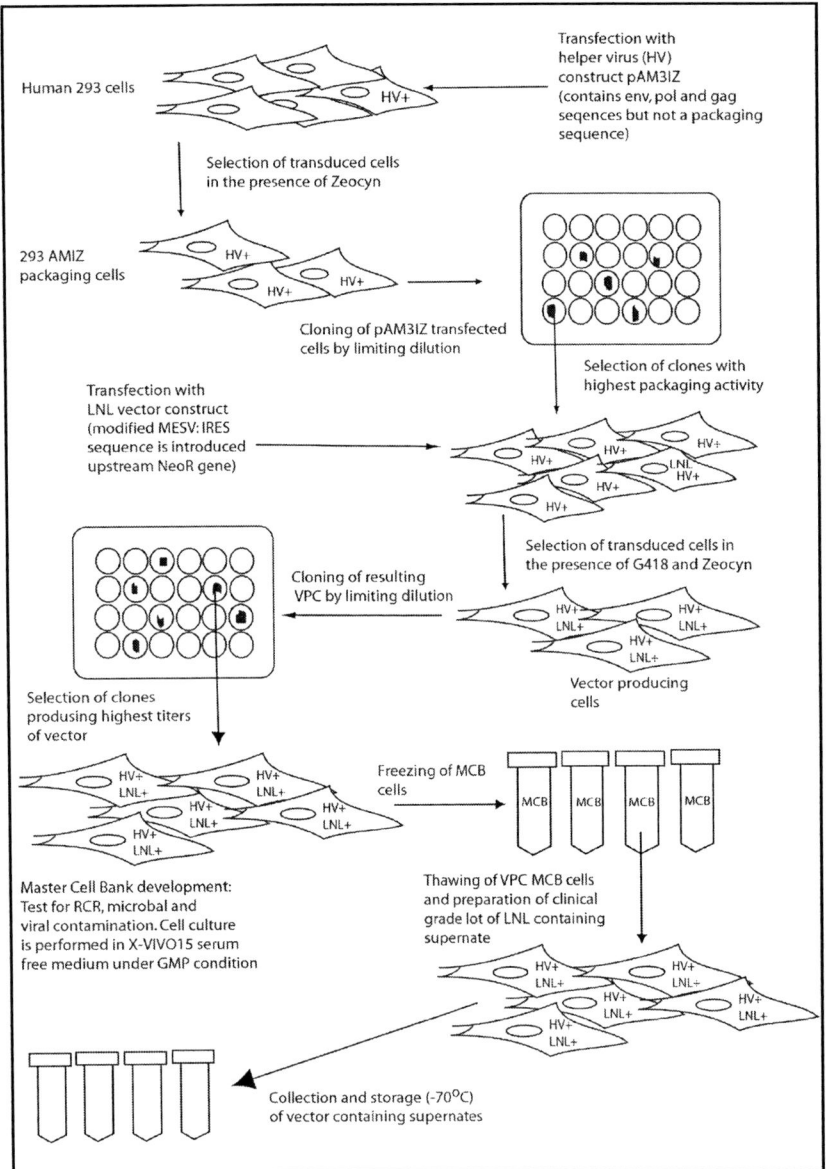

Figure 1. The development of clinical grade retroviral vector producing cells.

high titers of retroviral vectors in the supernates. Physical enrichment of retroviral vector particles is limited because of the fragility of vector envelopes. In contrast to animal experiments, GMP requirements do not allow highly effective methods of transduction of the target cells by co-culturing with virus producing cells (VPC). Instead, for clinical trials, the retrovirus particles released into the media (termed supernates or supernatant) is collected and used to transduce target cells. According to GMP requirements, the vector producing cells are cultured in defined serum-free media or media supplemented with certified human serum. As a rule, the vector titers in the supernates from VPC cultured in serum-free media are relatively low. A second issue is that retroviral particles developed in murine VPC express α galactosyl epitopes that bind human antibodies and can lead to retrovirus destruction by complement.[52] Another problem for the development of virus packaging cells is related to methylation of helper virus genomic elements followed by impairment of vector production performance.

Various laboratories have designed constructs and methods to address the aforementioned problems of clinical grade retroviral vectors. The vector opsonization problem during culturing in human serum-containing media may be avoided by using human instead of murine VPC. The introduction of picornavirus-derived internal ribosome entry site (IRES) followed by the Zeocin resistance gene sequence into the genome of helper virus simplifies the selection of transfected packaging cells and prevents the silencing of helper virus related to 5'methylated cap formation in that the *env* gene derived from amphotropic virus is inserted downstream from the IRES sequence.[53] Transfection of the packaging cell lines with plasmid coding for the vector of interest for the clinical trial may also contain an IRES followed by a different selectable marker such as the Neomycin resistance gene sequence. Cloning of mixed populations of VPC in the presence of Zeocin and Neomycin allows for monoclonal human VPC with reasonably high titers ranging from 2.0 to 3.0 x 10^6 CFU/ml. Cloning of individual VPC is a desirable step because it permits selection of clones with favorable integration sites within the genome. This allows for stable long-term expression of both helper virus and vector.

the vector. Therefore, for use in clinical trials and per FDA guidelines, RCR assays must be performed on the Master Cell Bank, production lots of VPC, and supernates from the production lots to monitor for an outbreak of RCR. Identity testing is needed to demonstrate that the full length vector sequence corresponds to the expected sequence and that genomic DNA of the clonal VPC line contains one full-length vector integrant.

Along with the mandatory test for replication competent retrovirus, all tests for presence of microbial contaminants including bacteria, fungi, mycoplasma, bacterial endotoxin, and adventitious viruses must be negative confirming that clinical grade supernates are free of infectious contaminants. Any packaging or production cell line that is of human origin must show test negative for human pathogens including HIV-I, HIV-II, HBV, HCV, HSV-I, HTLV-I, HTLV-II and CMV.

Potency of production lots of clinical grade supernates is defined as titers (colony forming units (cfu)/ml) of uniform aliquots of frozen supernatant stored at -70°C. Compared to other gene therapy vectors, it is relatively difficult to obtain and concentrate

The Properties of Retroviral Vectors in the Context of the Gene Transfer into Hematopoietic Progenitor Cells

The majority of vector systems used in prior clinical trials for gene transfer into HSC are derivatives of Moloney Murine Leukemia Virus (MoMuLV) (see Fig. 2). These vectors possess three major disadvantages: intracellular transportation of retrovirus into the target cells, transnuclear transport of viral cDNA required for integration into host genome, and transgene silencing.

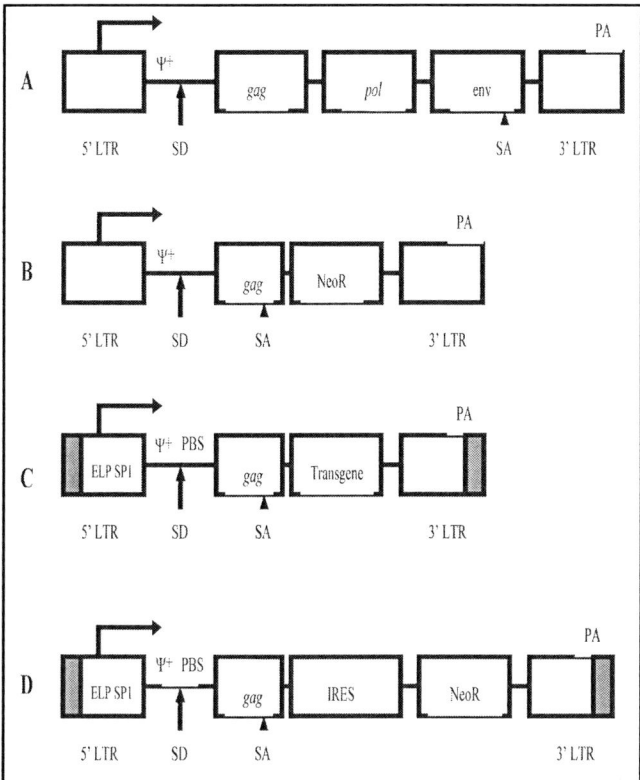

Figure 2. The structures of Moloney murine leukemia virus (MoMuLV) derived oncoretroviral vectors. A) Moloney murine leukemia virus wild type. LTR= long terminal repeat; ψ= packaging signal sequence; SD= splice donor; SA= splice aceptor; PA= polyadenylation sequence. B) LNL6 vector. LTR= long terminal repeat originated from MoMuLV; ψ+ enhanced packaging signal sequence; deleted gag sequence; NeoR= neomycin resistance gene sequence. C) Mouse Embryonic Stem Cell Virus. LTR derived from MPSV; DELP= disrupted binding site for embryonic carcinoma cell factor; SP1= high affinity binding site for the specific protein 1 transcription factor (SP1); PBS= gln t-RNA primer-binding site. D) LNL vector. IRES= Internal ribosome entry site; 3'LTR= long terminal repeat derived from mouse myeloproliferative sarcoma virus.

Intracellular Transportation

The relatively low efficiency of transduction of whole CD34+ populations and lack of long-term persistence of transgenes using MoMuLV-based vectors may be in part a result of low target cells purity and low vector expression. The absorption and intracellular transport of the first generation of MoMuLV-derived vectors are mediated by surface phosphate transporter (RAM-1)[54] and highly dependent on cell activation accompanied by RAM-1 expression. Cell populations enriched with CD34+ cells are heterogeneous and consist mainly of progenitor cells expressing such activation and differentiation markers as CD38, HLA-DR and others. It has been shown that most immature hematopoietic progenitor cells comprise the smallest portion of total CD34+ pool being CD90+, CD38^{low}, HLA-DR^{low} and lineage-negative.[40] It is established that CD34+CD38^{low} as well as CD34+HLA-DR^{low} hematopoietic progenitor cell populations contain most of NOD-SCID repopulating cells.[20,45] Expression of RAM-1 by these cells is quite low.[56] Therefore, gene transfer into the specific repopulating portion of the total CD34+ population could be inefficient because of a competition for RAM-1. In complex cell mixtures, more differentiated short-living cells acquire an advantage in the competition due to their pre-activation in vivo accompanied by higher expression of RAM-1.

Some potential approaches may override this problem. One is to use highly purified primitive progenitor cells in gene transfer clinical trials. Another approach is to use pseudotyped and/or targeted vectors that possess an intracellular transport ability independent of RAM-1 expression. There were three main types of vector systems developed based on the pseudotyped MoMuLV platform. Each vector allows more effective gene transfer into various types of human cells compared with vectors packaged in the amphotropic envelopes. Investigators demonstrated that gibbon ape leukemia virus (GALV) envelope can be successfully used to pseudotype murine retroviruses allowing improved gene transfer into higher primate and human cells.[57] The cellular receptor for the GALV envelope is an inducible sodium phosphate importer (GALVR-1) distinct from RAM-1.[58] GALV-pseudotyped retroviruses significantly improve gene transfer into human HSC capable of repopulating NOD-SCID mice suggesting that GALV-pseudotyped retroviruses can successfully transduce populations of very primitive progenitor cells.[59]

Another viral envelope is derived from vesicular stomatitis virus glycoprotein (VSV-G). VSV-G pseudotyping has been employed as another family of pseudotyped retroviral vector systems. VSV-G pseudotyped murine retroviral vectors possess a number of advantages. The VSV-G envelope is more stable and permits vector concentration without significant loss of activity.[60] Furthermore, VSV-G envelope possesses a tropism to some phospholipids (phosphadidylserine, phosphatidylinositol and GM3 ganglioside) and enters into target cells via endocytic pathway.[61] Therefore, HSC do not require expression of a receptor protein for absorption and intracellular transportation of VSV-G pseudotyped vectors and could be transduced easily.[62] The main disadvantage of VSV-G envelope is its direct toxicity when expressed during production of pseudotyped vectors in VPC.[63]

The use of pseudotyped vector systems based on the feline endogenous retrovirus envelope RD114 appears to be even more efficient than amphotropic or GALV envelopes. These vectors use the neutral amino acid transporter as a receptor.[64] Kelly and colleagues reported efficient stable transduction of human NOD-SCID mouse repopulating cells (90%) using RD114 pseudotyped MoMuLV derived vector as demonstrated by enhanced green fluorescent protein (EGFP) reporter gene fluorescence.[65]

Other approaches aimed at enhancing the efficiency of gene transfer into HSC are based on modulations of the physical interaction of viral particles and the surface of target cells. Some elements of the natural extracellular matrix can absorb and concentrate viral particles due to their charge. Recently, oligopeptides such as CH-296 fragment of fibronectin (RetroNectin™) absorbed on solid surfaces of tissue culture devices were shown to bind viruses via their H-domains.[66] This binding diminishes the negative impact of Brownian movement on vector/cell interactions. Data also suggest that the C-domain and CS-1 sequence of CH-296 are responsible for the interaction of cells with extracellular matrix mediated by integrin molecules. Therefore, VLA4 and VLA5 expressing HSC are held in close contact with vector particles being co-localized on Retronectin™ coated surfaces. These reports indicate that the interactions between HSC and CH-296 allow preservation of the immature phenotype of HSC during their activation in vitro.[67-70] Thus, a variety of approaches are being employed to enhance viral vector entry into cells.

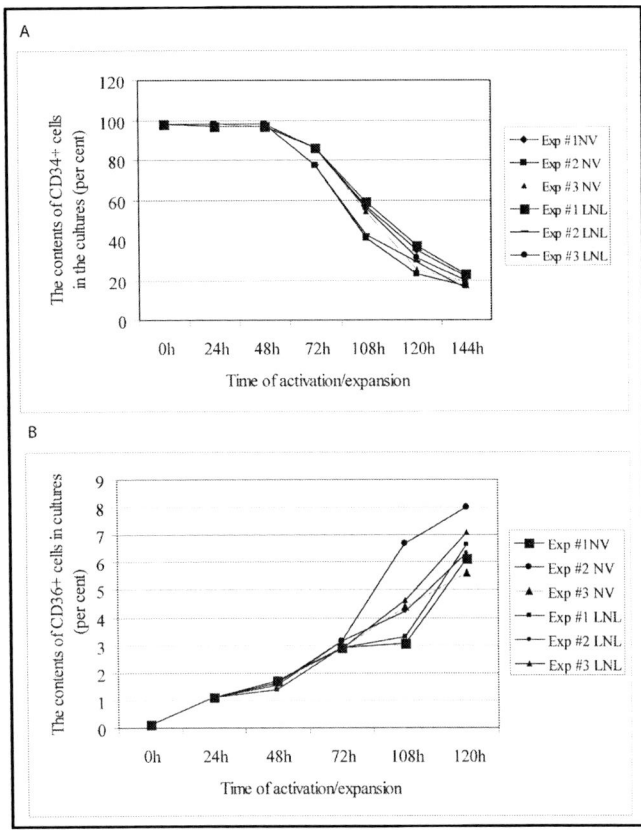

Figure 3. The changes of phenotype of peripheral blood (PB)-derived hematopoietic progenitor cells during their expansion. Mobilized CD34+ cells were separated from human peripheral blood using high speed FACS. Aliquots of the cell suspension (1×10^4 cells/ml) were plated in 1ml of X-VIVO15 medium supplemented with IL-6, IL-11, SCF, FLT3-L, LIF and anti TGF-beta 1 McAb for 24 – 120 hours. The phenotype of the cells was evaluated every 24 hours using the FACS. A) The kinetics of CD34 surface antigen expression during the activation/expansion of the cells and their transduction in vitro using LNL vector. NV= control cells (not undergoing the transduction procedure). LNL= cells have been treated by LNL vector containing supernate, supplemented with cytokines (three rounds of transduction). The results show that the expression of CD34 surface marker starts to decrease after 72h of cell activation/expansion. No significant difference of CD34 expression between transduced and control cells has been detected in three independent experiments. B) The expression of CD36 surface antigen by hematopoietic progenitor cells during their expansion/transduction. The CD36 expression has been estimated using the FACS. The results prove that activation of PB CD34+ cells with IL-6, IL-11, SCF, FLT3-L, LIF and anti TGF-beta1 monoclonal antibodies leads to their differentiation.

Nuclear Transport Problems and Approaches

Another remarkable challenge of the gene transfer into HSC using MoMuLV-derived vectors originates from the inability of murine retroviruses to actively transport the integration complex through the cell's nuclear membranes. Therefore, integration of murine retroviral genomes into a host's chromosomes requires target cells to enter the cell cycle.[71] Once the nuclear membrane is disrupted during mitosis the transduction complex can direct vector integration into the accessible chromosomal DNA. Therefore, murine retrovirus-mediated gene transfer into HSC is dependent upon stem cell activation and division.

Various methods have been used to induce HSC to cycle in vitro using cytokines alone or their various combinations. The challenge is to provide the conditions that maintain the self-renewal proliferation of HSC while minimizing their differentiation that is accompanied by loss of repopulating ability. While it has been shown in some studies that cytokine-mediated activation of HSC allows their expansion without irreversible differentiation, most attempts have not been successful.[44,72-74] Until recently, cytokine combinations used in the trials to expand immature hematopoietic progenitor cells consisted of interleukin-3 (IL-3), stem cell factor (SCF), IL-6, Flt-3 ligand (Flt3-L), thrombopoietin (Tpo) and granulocyte colony stimulating factor (G-CSF). More recent studies show that interleukin-3-containing cytokine combinations promote HSC proliferation but cause the loss of primitive markers on HSC such as MDR-1.[72] This differentiation reduced the frequency of cobblestone area-forming cells (CAFC) and impaired NOD-SCID repopulating ability of expanded cells.[72,75] By contrast, another group demonstrated that a combination of Flt-3-L, IL-6 and SCF preserved an immature phenotype of CD34+Rhd123^{low} cells for 5 to 6 days and stimulated cell divisions in serum free media.[76] The results were further improved by the addition of leukemia inhibitory factor (LIF) and interleukin-11 (IL-11) with the frequency of CAFC gradually increasing up to 6 days in the culture.

Differentiation of mobilized peripheral blood CD34+ cells obtained from adult donors has been evaluated to determine the optimal timing of cytokine exposure. After 12 hours of pre-activation in serum free (X-VIVO 15™) or serum containing media (IMDM with 5% FBS) supplemented with Flt3-L, SCF, IL-6, IL-11 and LIF, expression of activation markers (CD71) by cultured cells became detectable. The rate of expansion of hematopoietic cells under these conditions varied from donor to donor with an average four-fold increase by 72 hours in culture. However, the immature phenotype of cells cultured under these conditions was not preserved for more than three days. From 72 to 120 hours of culture, the number of CD34 positive cells dropped dramatically from 100% to 30% or less (see Fig. 3A). In contrast, the number of cells expressing such differentiation markers as CD13, CD33, CD36, CD41a, CD45^{high} increased progressively in the cultures (see Fig. 3B). Only 8-10% of the cell population expressed CD34+ surface marker by day 6 of culture. The relative number of erythroid burst forming units (BFU-e) per 2×10^3 of CD34+ plated in colony assay are constant up to 72 hours after their activation and expansion and then decrease by 120 hours (see Fig. 4). A clear shift of cell differentiation has been detected after 120 hours of HSC expansion. After 120 hours, myeloid progenitor cells (CFU-G, CFU-GM or CFU-M) became highly prevalent compared to erythroid cells. A more primitive subset of CD34+ cells, i.e., CD34+CD38^{low} cells were detectable in the expansion/activation cultures for up to 10 days in culture (Fig. 5), and 3% of these cells retained their ability to form CFU-GEMM colonies (Fig. 6). Overall these results suggest that an optimal time window for in vitro transduction of activated mobilized CD34+ cells derived from patients' peripheral blood would be between 12 to 72 hours after cytokine addition.

The optimal cytokine stimulation window for transduction of more primitive HSC populations may be different than from unfractionated CD34+ HSC. A number of recent reports describe more primitive subpopulations of CD34+CD38^{low} cells consisting of primitive, highly quiescent cells which cannot be induced into cycling by conventional sets of cytokines. It is possible that their response to the proliferative stimuli is significantly delayed.

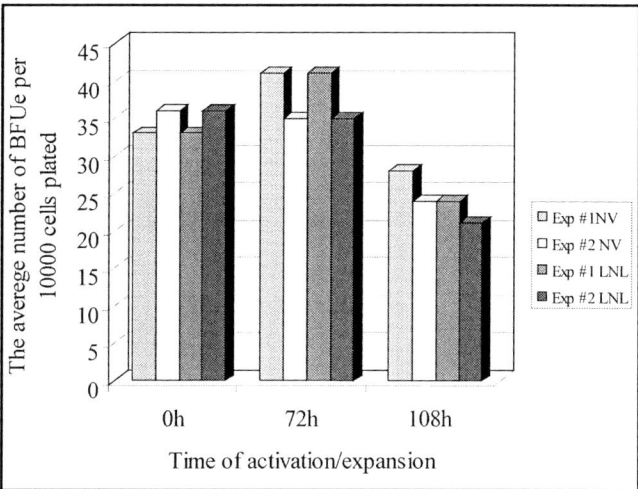

Figure 4. The kinetics of immature erythroid progenitor cells observed in whole CD34$^+$ HSC population during their expansion/retroviral LNL transduction. Mobilized peripheral blood CD34$^+$ were separated from human peripheral blood. Resulting cells were activated by cytokine mixture containing recombinant IL-6, IL-11, SCF, Flt-3L, LIF and anti TGF-beta1 monoclonal antibody and transduced with LNL vector (3 transduction cycles 12 hours each). Aliquots of cell suspension (20 x 10^3 cells each) were collected at 0, 72 and 108 hours of culture. The clonogenic potential of the cells was estimated by colony assay Burst forming unit-e (BFU-e) have been accounted at day 14 of culture. NV= control cells. LNL= cells transduced with retroviral LNL vector. The results show that the number of BFU-e in CD34$^+$ population is decreased significantly after 108 h of cell expansion.

The steady state of some of these primitive cells are maintained in an autocrine manner and mediated by TGF-β1.[41] An abrogation of TGF-β1 activity in vitro by a blocking antibody increases the frequency of CFU-GEMM in colony assays and significantly improves the efficiency of the retrovirus mediated gene transfer into HSC.[77,78] Using anti-TGFβ1 monoclonal antibody with cytokines for HSC activation, genes of interest can be transferred into CD34$^+$ HSC with an efficiency of approximately 50% of BFU-e.

Another way to overcome problems stemming from a necessity to activate HCS for gene transfer mediated by Oncoviridae MoMuLV-derived vectors points to a principal distinction between Oncovidiae and lentiviridae. Lentiviridae are retroviral vectors based on human (HIV) as well as non-human (feline immunodeficiency virus, equine infectious anemia virus, simian immunodeficiency virus) lentiviruses. These viruses can actively transport the pre-integration complex from the cytoplasm into nucleus of target cells with minimal cell activation from G0 to G1 of the cell cycle without division.[79] The effective transduction of CD34$^+$ CD38^{low} as well as NOD-SCID repopulating cells and CD34$^+$ cell population using the HIV-based vectors is well established by in vitro experiments.[80-82]

The main challenges with using HIV-1-based vectors are difficulties in production of high titer supernates, although the technology is progressing. Additional concerns have been expressed with the safety of HIV-derived vectors and the potential risk of homologous recombination between integrated HIV-based vectors with replication-competent retrovirus. Naldini and colleagues recently described a self-inactivating lentiviral vector and packaging systems that may reduce the risk of generating replication competent vector by recombination.[83] In brief, the novel HIV-based vectors are packaged using three non-overlapping expression constructs. Two of them code for HIV proteins and third

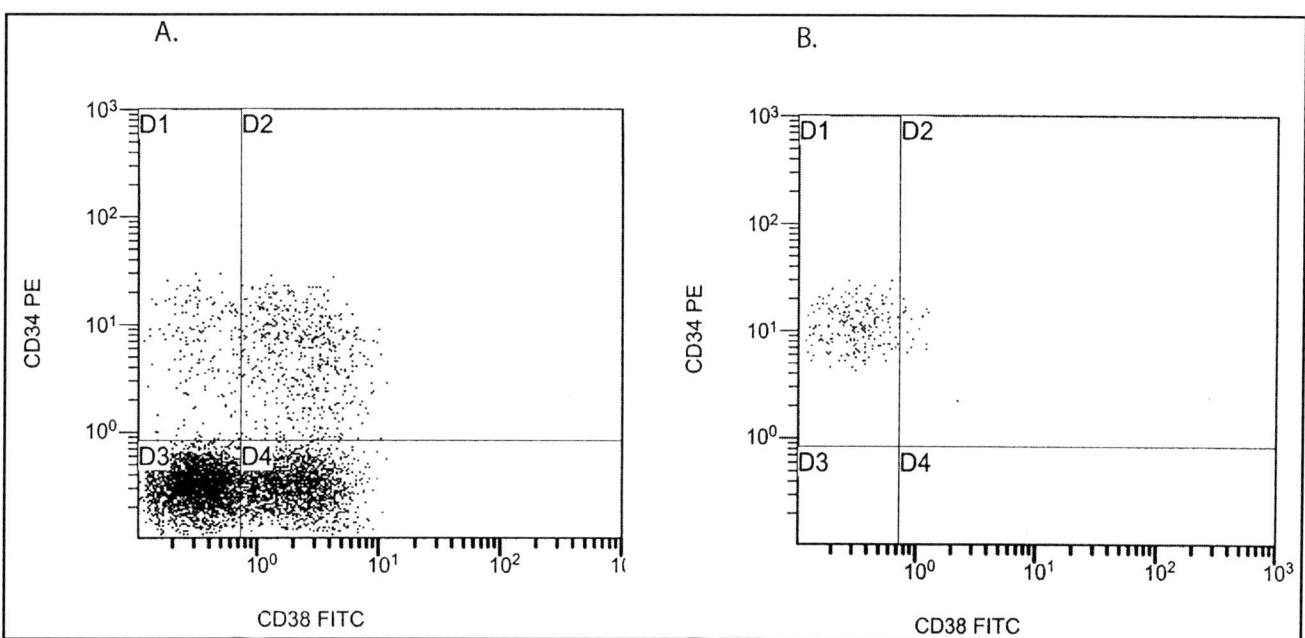

Figure 5. The preserving of immature phenotype by some of expanded HSC. Human peripheral blood CD34$^+$CD38^{low} cells separated by FACS (98% of purity) were subjected to 3 cycles of retroviral LNL transduction followed by expansion in X-VIVO15 media supplemented with recombinant IL-6, IL-11, SCF, Flt3-L, LIF and anti-TGF-beta1 monoclonal antibody for 10 days. At the end of culture resulting cells have been collected, stained with anti-CD34 and anti-CD38 monoclonal antibody followed by FACS analysis. A) CD34$^+$CD38^{low} cells in total expanded population (day 10 of expansion). B) CD34$^+$CD38^{low} cells originated from expansion culture and separated by FACS.

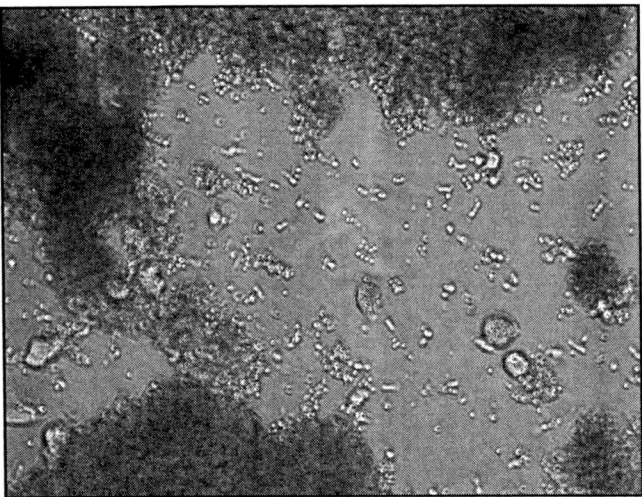

Figure 6. Colony forming unit-granulocyte-erythroid-macrophage/monocyte-megakaryocyte (CFU-GEMM) originated from expanded CD34+CD38low cells CD34+CD38low cells isolated from expansion culture (10 days) were plated in MethoCult™ H4434 semi solid medium for colony assay for 14 days. It has been observed that approximately 3% of cells retained their capability to form CFU-GEMM. The picture demonstrates the fragment of large CFU-GEMM derived from expansion culture. Macrophages as well as myeloid and erythroid cells can be clearly distinguished.

plasmid expresses VSV-G envelope. All HIV sequences required for encapsidation and reverse transcription are deleted from these constructs with the exception of a portion of gag gene. This vector system allows the generation of high titer lentivirus vectors able to transduce non-dividing cell. Clinical trials are being initiated using lentiviral-mediated HIV-1 decoy gene transfer into HSC followed by their reinfusion in HIV-positive patients.[84]

Problems Related to the Silencing of Transgenes

One of the problems encountered with clinical grade integrating vectors is silencing of transgenes, which depends on the vector type, its structure, promoter-enhancers, genome integration site, and target cell. Obviously, sustained (or inducible) expression of transgenes is crucial for therapeutic applications of the gene transfer into HSC.

Prior results demonstrated that transgene expression after transduction into murine embryonal cells or embryonal stem cells with earlier versions of oncoretroviral-based vectors was restricted to 8 to 10 days after transduction.[85-88] Both long terminal report (LTR) methylation as well as trans- and cis-acting negative regulator factors reduced gene expression.[89-93] A novel vector design has been created to override the problem. This modified retroviral vector was developed by Ostertag and colleagues.[94,95] The vector is termed murine embryonal stem cell virus (MESV) and has a disrupted binding site for the embryonic long terminal repeat binding protein (ELP) and lack a negative regulatory element (NRE) coincident with the proline tRNA primer-binding site (PBS) of MoMuLV. Furthermore, a binding site for specific protein 1 (SP1) transcription factor that is able to reactivate LTR mediated transcription was added to the enhancer region (see Fig. 2C). This vector partially alleviates the silencing problem and it has demonstrated prolonged transgene expression in HSC.[96]

Further modification of MESV vector (see Fig. 2D) has demonstrated longer expression of the neomycin phosphotransferase

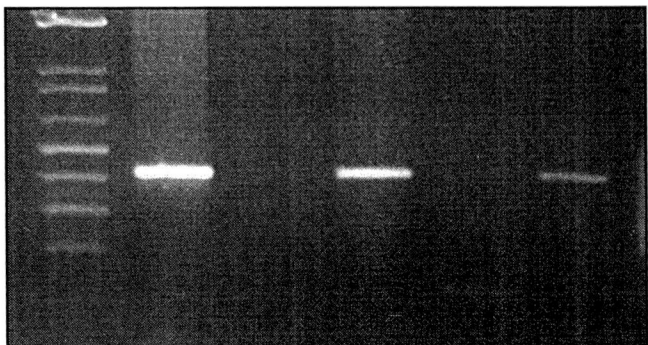

Figure 7. Integration of NeoR gene into human peripheral blood (PB) CD34+ cells and cells derived from HL-60 cell line. Mobilized human CD34+ cells were separated from PB using high speed FACS. HSC were pre-activated in X-VIVO15 medium supplemented with IL-6, IL-11, SCF, Flt-3, LIF and anti-TGF-beta1 monoclonal antibody for 24 hours. After pre- activation, cells were cultured in X-VIVO15 medium supplemented with the same set of cytokines and containing LNL vector in multiwell plates coated with Retronectin™ for 12 hours (transduction cycle). The transduction procedure has been repeated three times under same condition. Control cells were cultured in the same media without vector. At the end of transduction resulting cells were collected and NeoR gene insertion was verified by PCR. Track 1 (from left to right): DNA length ladder; Track 2: NeoR gene insertion in HL-60 cells, transduced and selected in the presence of neomycin); Track 3: Expanded PB CD34+ cells (control culture, experiment 1); Track 4: PB CD34+ cells after 3 cycles of transduction with LNL vector (experiment 1); Track 5: Expanded PB CD34+ cells (control culture, experiment 2); Track 6: PB CD34+ cells after 3 cycles of transduction with LNL vector (experiment 2).

II gene (NeoR) without transgene silencing. The vector is based on a murine embryonal stem cell virus (MESV) pR338 backbone with IRES sequence introduced upstream the NeoR gene. The vector could successfully transfer the NeoR gene into cytokine activated peripheral blood CD34+ cells and such human cell lines with immature hematopoietic phenotypes as HL-60 (see Fig. 7 and 8).

Gene Transfer into Hematopoietic Stem Cells: Pre-Clinical Studies and Clinical Trials

The prospects of clinical application of gene transfer into HSC are wide ranging. Numerous clinical trials have been initiated or developed for a variety of strategies including: transfer of genetic tags, normal genes for monogeneic hematological disorders, drug resistance genes, and decoy genes (Table 2 and 3).

HSC Marking Studies by Transgene Integration

One of the most successful clinical trials using gene marking of bone marrow stem cells was performed by Brenner and colleagues in children with acute myeloblastic leukemia and disseminated neuroblastoma.[97] The purpose of the trial was to introduce a genetic tag to trace transduced cells in vitro as well as in vivo. The NeoR gene under LTR promoter control was transferred by a LNL6 MoMuLV-based retroviral vector (Fig. 2B). The vector was successfully integrated into normal and malignant bone marrow cells during autologous bone marrow transplantation. The genetic label permitted study of the sources of disease relapses.[98]

Figure 8. Insertion of NeoR gene into HSC transduced by LNL vector prepared under good manufacturing practice (GMP) condition. Mobilized peripheral blood CD34$^+$ CD38^{low} cells were separated by FACS, pre activated by IL-6, IL-11, SCF, Flt3-L, LIF and anti TGF-beta1 McAb for 24 hours followed by 3 cycles of transduction with LNL vector in serum free X-VIVO 15 medium. At the end of transduction resulting cells were plated in MethoCult™ H4434 semi solid medium for colony assay. After 14 days of culture resulting individual colony forming unit-granulocyte-erythroid-macrophage/monocyte-megakaryocyte (CFU-GEMM) were collected and efficiency of NeoR transduction was verified by PCR: Tracks 1 and 12: DNA length ladder; Track 2: negative control (non transduced cells); Tracks 3 to 11: PCR product from single colonies. NeoR transgene insertions have been found approximately in 50% CFU-GEMM.

In these clinical studies, retrovirus-mediated ex vivo gene transfer into whole populations of non-activated BM cells were performed with efficiencies between 0 to 23.5%. The transduction rate was comparable with results of experiments performed on enriched hematopoietic progenitor cell populations activated in vitro. It is likely that the patient's age was a key factor that aided the gene transfer success. In children, the rate of spontaneous proliferation of hematopoietic stem cells and self-renewal is much higher compared to adults. Proliferation of HSC in children may not lead to as much irreversible differentiation as in adults. This proliferation may substitute for an in vitro activation step required for gene transfer into adult HSC.

Similar trials on eleven adult patients with multiple myeloma or breast cancer were performed by Dunbar and colleagues.[99] In these trials, NeoR retroviral gene transfer into bone marrow and peripheral blood-derived CD34 enriched cells was performed and followed over time for the presence of the marker gene and any adverse effects related to the gene-transfer procedure. Out of 10 evaluable patients with the marker gene detected at the time of engraftment, 3 had persistence of the marker gene for more than 18 months post transplantation. Marker gene was detected in multiple cell lineages, including both myeloid and lymphoid cells. The data suggested that better long-term marking originated from peripheral blood derived cells than from marrow. However, the levels of marking were low, with between 0.1 to 0.01% cells being positive. The marker was detected both in lymphoid and myeloid cell lineages.[99,100]

Transferring Drug Resistance Genes into HSC

Another area of interest is the introduction of drug resistance genes into HSC to allow protection from some types of chemotherapy. Upon exposure to chemotherapy, drug resistant cells should have an in vivo selective advantage. P-glycoprotein, a cellular transmembrane transporter that protects cells from cytotoxic drugs such as vinca alkaloids, anthracyclines, podophyllins, and paclitaxel, is the product of the multi-drug resistance (MDR1) gene.[101] In preclinical models, expression of MDR1 in HSC increases resistance to these cytotoxic agents.[102] Transfer of the O6-Methylguanine DNA methyltransferase (MGMT) gene under control of the phosphoglycerate kinase promoter into mouse HSC improves viability of bone marrow cells to 1,3-bis

Table 2. Summary of pre-clinical hematopoietic stem cell transduction studies

Marking or Therapeutic Gene (Reference)	Stem Cell Source and Purity	Activation Conditions Efficiency	Ex Vivo Transduction	Outcome
Glucocerebrosidase gene[115]	PB 34$^+$ cells	Centrifugation-enhanced retroviral transduction	37% of clonogenic cells and up to 50% LTC-IC	50-fold increase of GC activity in transduced cells
Glucocerebrosidase gene[116]	PB or BM CD34$^+$ cells	IL-3, IL-6, SCF with or without stroma layer	29% to 71% CFC	Correction of GC activity in LTC-IC
Alpha-galactosidase A gene Fabry disease[117]	CB CD34$^+$ cells	IL-6 and SCF in the presence of stroma layer	34% to 82%	Normalization of alpha galactosidase activity in LTC-IC
CD18 subunit to correct LAD[118,119]	PB CD34$^+$ cells	IL-3, IL-6 and SCF	31%	Restoration of CD11b and CD11c in transduced cells. Restoration of transduced cell ability to undergo respiratory burst.
FACC complementing gene at Fanconi anemia[120]	PB CD34$^+$ cells	IL-3 IL-6 and SCF	NA	Restoration of resistance to MMC by transduced CD34$^+$ cells. The initiation of clinical trial announced.

BM= bone marrow; GC= glucocerebrosidase gene; FACC= fanconi anemia (FA) C-complementing gene; IL-3= interleukin 3; IL-6= interleukin 6; LAD= leukocyte adhesion deficiency; MMC= mitomycin C; NA= not available; PB= peripheral blood; SCF= stem cell factor.

Table 3. Summary of clinical hematopoietic stem cell transduction studies

Marking or Therapeutic Gene (Reference)	Stem Cell Source and Purity	Activation Conditions	Ex Vivo Transduction Efficiency	Clinical Outcome
Neo R gene marking[97,98]	Whole bone marrow	No activation	0 to 23% BM cells	Transduced cells were detectable from 1 to 18 months and more
Neo R gene marking[99,100]	PB or BM CD34+ cells	IL-3, Il-6 and SCF, or no activation, or transduction on stroma layer	0 to 36%	More than 18 months 0.1 to 0.01% PBC
MDR1[106]	PB or BM CD34+ cells	IL-3, IL-6 and SCF	0.1-0.5%	Short-term persistence of transgene. No clinical effect.
MDR1[107]	PB or BM CD34+ cells	IL-3, IL-6 and SCF	18% to 70% CFU	Short-term persistence of transgene.
MDR1[108]	PB or BM CD34+ cells	No activation or activation by IL-3, IL-6 on stromal layer	2.8% to 5.6% of CD34+ cells	Short-term persistence of transgene. No clinical effect.
Common gamma chain in SCID patients[110]	BM CD34+	SCF, FLT-3, IL-3 and M-GDF	NA	Long-term engraftment Significant clinical effect
ADA gene in ADA SCID patients[112]	CB CD34+ cells	NA	NA	Long-term engraftment Limited clinical effect
ADA gene in ADA SCID patients[114]	BM CD34+ cells	NA	NA	Long-term engraftment Significant clinical effect
Glucocerebrosidase Gene in patients with Gaucher disease[121]	BM or PB CD34+ cells	IL-3, IL-6 and SCF With or without stroma layer	NA	Transgene persistence from 1 to 3 months postinfusion. No clinical effect.
Rev-responsive element sequence in AIDS patient[84]	BM CD34+ cells	IL-3, IL-6 and SCF with or without stroma layer	7% to 30% CFU	Short-term persistence. No clinical effect.

ADA= adenosine deaminase gene; BM= bone marrow; CB= cord blood; CFU= colony forming unit; IL-3= interleukin 3; IL-6= interleukin 6; MDR1= multi-drug resistance gene; NA= not available; Neo R= neomycin resistance gene; PB= peripheral blood; SCF= stem cell factor.

(2-chloroethyl)-1-nitrosourea (BCNU).[103] Expression of the mutant O6-benzylguanine (6-BG)-resistance MGMT gene into HSC can effectively protect transduced hematopoietic cells to 6-BG while malignant cells, expressing wild type of MGMT remain highly sensitive to 6-BG.[104] The pre-clinical data also demonstrated an unexpected complication. In a murine model, MDR-1 transduced HSC were shown to acquire a growth advantage over the normal bone marrow progenitor cells leading to myeloproliferative disorders.[105]

Three clinical trials employing HSC transduced ex vivo with MDR1 have demonstrated safety but lack convincing evidence for efficacy. MRD1 gene trandsuced HSC have shown only short-term engraftment of vector-modified cells in adult patients. After activation with SCF, IL-3 and IL-6, the ex vivo transduction efficiency was low ranging from 0.1 to 0.5%. In one study 3 of 4 patients, in another 2 of 5 patients, and in the third study 3 of 10 patients demonstrated engraftment of MDR1 modified HSC.[106-108] The percentage of circulating nucleated cells containing the MDR1 transgene has been reported to range from 0.01% to 9% for 1 to 6 cycles of post-transplantation Paclitaxel.[106,108] The lack of long term engraftment from transduced HSC might be explained by pre-activation-induced differentiation.[107]

HSC Gene Transfer for Monogeneic Diseases

One of the most important application areas of gene transfer into HSC is correction of monogeneic hematological disorders by retroviral integration of the corresponding wild type genes.

Severe Combined Immune Deficiency

Cavazzano-Calvo and colleagues describing 5 successful attempts to correct X-linked severe combined immunodeficiency (SCID-X1) by transplantation of autologous CD34+ cells transduced ex vivo with the common gamma chain (γc) gene.[109,110] The restoration of normal lymphopoiesis in patients occurred secondary to an apparent selective advantage acquired by T-cell progenitors derived from gene modified HSC expressing the wild type version of γc gene. Another key result of this clinical trial was long-term engraftment (at least 30 months) of genetically modified BM after transplantation. This study demonstrated that retroviral mediated gene transfer into whole CD34+ populations allows transduction of multipotential HSC with clinically significant expression of the therapeutic gene. The activation protocol used in the trial (SCF, FLT3-L, IL-3, and megakaryocyte growth factor) appears to allow transduction of HSC without differentiation. It should be noted that this trial was conducted in children, and it is possible that part of the success of gene transfer into HSC is dependent on the young age of the donor's HSC.

Another form of autosomal recessive SCID is a result of multiple point mutations in adenosine deaminase (ADA) gene. The defect leads to disruption of purine nucleosides metabolism followed by impaired lymphocyte development.[50,111] In 1993 Kohn and colleagues initiated the first clinical trial employing retroviral transfer of normal ADA complementary DNA into cord blood (CB) derived CD34+ HSC followed by autologous transplantation in neonates.[112] It has been demonstrated that ADA-containing peripheral blood mononuclear cells have persisted in patients from this trial up to 8 years, with T lymphocytes showing the highest prevalence of gene marking.[113] Unfortunately, the levels of transgene expression in these studies were variable and clinial effect was limited.[51] Aiuti and colleagues initiated a similar trial using modified retroviral vector transduction of bone marrow derived CD34+ HSC.[114] In this trial, the sustained engraftment of transduced HSC resulting in significant improvement of immune functions was demonstrated.

Gaucher Disease

Other clinical and experimental studies of the retrovirus mediated gene transfer into mobilized or non-mobilized CD34+ cells have been aimed at the correction of inborn defects of the glucocerebrosidase (GC) gene in patients with Gaucher disease. In vitro retroviral transduction of the CD34+ enriched cells employed centrifugation-promoted transduction protocol modified for clinical use and demonstrated a mean transduction efficiency of 37% in clonogenic cells and up to 50% in long-term culture-initiating cells (LTC-IC) at week 6. Furthermore, a 50-fold increase in the level of GC above baseline deficient levels in the patients' CD34+ enriched cells derived from peripheral blood and maintained in vitro was measured.[115] In another pre-clinical study, bone marrow and peripheral blood CD34+ cells originated from patients with Gaucher disease have been transduced with GC gene in the presence of IL-3, IL-6, and SCF. In these experiments, the transduction has been performed using a high titer retroviral supernatant, with or without irradiated allogeneic BM stromal layers. The results indicated that stromal layer support could significantly improve the efficiency of retroviral transduction into LTC-IC.[116]

Dunbar and colleagues initiated a clinical protocol to explore the safety and efficiency of retroviral transduction of G-CSF-mobilized PB or intact BM CD34+ cells with the human glucocerebrosidase cDNA under control of the LTR promoter. Three adult patients have been treated in this trial with follow-up of 6 to 15 months post-transplantation. All transduction procedures in these trials were performed in the presence of autologous or allogeneic stroma layers, with (1 patient) or without (2 patients) cytokines (IL-3, IL-6, and SCF). Gene marking was demonstrated in 2 of 3 patients. The transgene persistence in PB was documented up to 3 months post-infusion for 1 patient, while in the second patient transgene was not detected in BM cells after 1 month. The level of corrected cells (<0.02%) was too low to result in any clinical benefit, and glucocerebrosidase enzyme activity was not increased in any patient following infusion of transduced cells.[121]

Fabry Disease

Fabry disease is a human lysosomal storage disorder due to an inborn defect of the alpha-galactosidase A gene. Preliminary experiments demonstrated that cord blood or bone marrow CD34+ cells as well as long-term culture derived progenitor cells could be successfully transduced by retroviral vectors carrying the alpha-gal A transgene.[117] The efficiency of cord blood CD34+ transduction in this pre-clinical study ranged from 34% to 82%. Furthermore, in vitro experiments established that alpha-gal A transduced cells could secrete alpha-galactosidase A and complement the genetic defect of non-transduced cells derived from patients with Fabry disease.

Leukocyte Adhesion Deficiency

Leukocyte adhesion deficiency (LAD) is an autosomal recessive disease.[122] LAD is due to a defect of the integrin CD18.[123] The defect of leukocyte integrin CD18 prevents the CD11/CD18 heterodimer formation and blocks migration of leukocytes into the sites of primary infection in patients with LAD.[119] Pre-clinical studies indicated that the structural and functional defect in LAD leukocytes could be corrected by CD18 subunit gene transfer into mobilized CD34+ HSC from the peripheral blood of patients with LAD.[118]

Fanconi Anemia

Retrovirus-mediated transfer of normal Fanconi anemia (FA) C-complementing gene (FACC) into HSC has been proposed.[124,125] In 15% of patients with FA, disease arises from nonsense or missense mutations in the FACC gene. Affected patients develop progressive pancytopenia, acute leukemias and other congenital disorders. Complementation of the defective FACC gene by wild type gene has been demonstrated in pre clinical trials on cell lines and mobilized PB CD34+ cells derived from donors bearing a splice mutation. Pre clinical studies have shown that these modified hematopoietic progenitor cells could be reinfused safely into FA patients.[120]

Modification of HSC to Treat HIV-I Infection

Transfer of anti-HIV-1 genes into HSC may be useful to prevent viral replication in infected subjects. This preventive effect could allow lymphopoiesis of virus-resistant lymphocytes and monocyte differentiation after autologous transplantation of ex vivo modified HSC. Kohn and colleagues have developed a retroviral vector carrying Rev-responsive element (RRE) sequences that interfere with Rev-responsive elements by binding Rev proteins.[55] This interaction significantly inhibits HIV-1 replication in transduced cells. A clinical gene transfer trial of the decoy gene into BM CD34+ HSC has been performed in four pediatric AIDS patients. Unfortunately, the efficiency of transduction in the trial was low and the transgene persistence was transit.[84]

Safety Concerns from Retroviral Integration

Questions concerning the safety of integration of transgene into a host genome remain to be answered. The complete potential integration sites for retroviral cDNAs are unknown and thought to be mostly random. This random integration creates a number of concerns about insertional mutagenesis and integration of the transgene into either oncogenes or sequences coding the inhibitors of genes such as cyclin kinases (tumor suppressor genes) causing their disruption. These warnings are very real and serious in the context of recent reports concerning cases of acute T cell leukemia lymphoma (ATLL) that developed in two SCID children infused with common gamma chain gene transduced into autologous CD34+ cells.[126] In both cases, insertion of the transgene occurred in the first intron of rhombotin 2 (LIM-only protein 2 (LMO2)) transcription factor gene. Previous studies of LMO2 physiology and function have revealed that ATLL development is closely associated with LMO2 translocations and overexpression.[127,128] The disruption of regulatory sequences of the gene by transgene insertion followed by its uncontrolled activation is the probable reason for leukemia development in these children. Preclinical data had previously demonstrated such risks. Acute myeloid leukemia development in mice transplanted with autologous bone marrow transduced by a mutant non functional nerve growth factor receptor (NGFR) gene used as a cell surface marker.[129] Since risks do exist after stem cell transduction with integrating retroviral vectors, the development and application of vectors containing both a gene of interest and inducible suicide genes such as herpes simplex virus thymidine kinase (HSV-tk) as a safety feature are in development. The idea would be to eliminate transduced cells by activation of the suicide gene if leukemia developed. Other procedures to confirm insertion site prior to product release being discussed for human trials.

References

1. Bensinger, WI, Berenson, RJ, Andrews, RG et al. Positive selection of hematopoietic progenitors from marrow and peripheral blood for transplantation. J Clin Apheresis 1990; 5(2):74-76.
2. Hansson M, Svensson A, Engervall P. Autologous peripheral blood stem cells: Collection and processing. Med Oncol 1996; 13(2) 71-79.
3. Dick J. Gene therapy turns the corner. Natl Med 2000; 6-652.
4. Bittner RE, Schofer C, Weipoltshammer K et al. Recruitment of bone-marrow-derived cells by skeletal and cardiac muscle in adult dystrophic mdx mice. Anat Embryol 1999; 199:391-396.
5. Brazelton TR, Rossi FM, Keshet GI et al. From marrow to brain: Expression of neuronal phenotypes in adult mice. Science 2000; 290:1775-1779.
6. Eglitis MA, Mezey E. Hematopoietic cells differentiate into both microglia and macroglia in the brains of adult mice. Proc Natl Acad Sci USA 1997; 94:4080-85.
7. Mezey E, Chandross KJ, Harta G et al. Turning blood into the brain: Cells bearing neuronal antigens generated in vivo from bone marrow. Science 2000; 290:1779-1782.
8. Jackson KA, Majka SM, Wang H et al. Regeneration of ischemic cardiac muscle and vascular endothelium by adult stem cells. J Clin Invest 2001; 107:1395-1402.
9. Orlic D, Kajstura J, Chimenti S et al. Bone marrow cells regenerate infarcted myocardium. Nature 2001; 410, 701-705.
10. Krause DS, Theise ND, Collector MI et al. Multi-organ, multi-lineage engraftment by a single bone marrow-derived stem cell. Cell 2001; 105:369-377.
11. Alison MR, Poulsom R, Jeffrey R et al. Hepatocytes from non-hepatic adult stem cells. Nature 2000; 406(6793):257.
12. Lagasse E, Connors H, Al-Dhalimy M et al. Purified hematopoietic stem cells can differentiate into hepatocytes in vivo. Nat Med 2000; 6:1229-34.
13. de Wynter EA, Emmerson AJ, Testa NG. Properties of peripheral blood and cord blood stem cells. Baillieres Best Pract Res Clin Hematol 2000; 12(1-2):1-17.
14. de Wynter E, Ploemacher RE. Assays for the assessment of human hematopoietic stem cells. J Biol Regul Homeost Agents 2001; 15:23-27.
15. Civin CI, Strauss LC, Fackler MJ. Positive stem cell selection—basic science. Prog Clin Biol Res 1990; 333:387-401.
16. DiGiusto D, Chen S, Combs J. Human fetal bone marrow early progenitors for T, B, and myeloid cells are found exclusively in the population expressing high levels of CD34. Blood 1994; 84(2):421-432.
17. Ito T, Tajima F, Ogawa M. Developmental changes of CD34 expression by murine hematopoietic stem cells. Exp Hematol 2000; 28(11):1269-73.
18. Ogawa M, Tajima F, Ito T et al. CD34 expression by murine hematopoietic stem cells. Developmental changes and kinetic alterations. Ann NY Acad Sci 2001; 938:139-145.
19. Sato T, Laver JH, Ogawa M. Reversible expression of CD34 by murine hematopoietic stem cells. Blood 1999; 94(8):2548-54.
20. Bhatia M, Bonnet D, Murdoch B et al. A newly discovered class of human hematopoietic cells with SCID-repopulating activity. Natl Med 1998; 4(9):1038-1045.
21. Goodell MA, Rosenzweig M, Kim H et al. Dye efflux studies suggest that hematopoietic stem cells expressing low or undetectable levels of CD34 antigen exists in multiple species. Nat Med 1997; 3(12):1337-45.
22. Osawa M, Hanada K, Hamada H et al. Long-term lympho-hematopoietic reconstitution by a single CD34-low/negative hematopoietic stem cell. Science 1996; 273:242-45.
23. Ratajczak J, Machalinski B, Majka M et al. Evidence that human haematopoietic stem cells (HSC) do not reside within the CD34+KIT- cell population. Ann Transplant 1999; 4(1):22-30.
24. Grigg AP, Roberts AW, Raunow H et al. Optimizing dose and scheduling of filgrastim (granulocyte colony-stimulating factor) for mobilization and collection of peripheral blood progenitor cells in normal volunteers. Blood 1995; 86(12):4437-45.
25. Broxmeyer HE, Gluckman E, Auerbach A et al. Human umbilical cord blood: A clinically useful source of transplantable hematopoietic stem/progenitor cells. Int J Cell Cloning 1990; 8 suppl 1:76-89.
26. Cassel A, Cottler-Fox M, Doren S et al. Retroviral-mediated gene transfer into CD34-enriched human peripheral blood stem cells. Exp Hematol 1993; 21(4):585-591.
27. Pollok KE, van Der Loo JC, Cooper RJ et al. Differential transduction efficiency of SCID-repopulating cells derived from umbilical cord blood and granulocyte colony-stimulating factor-mobilized peripheral blood. Human Gene Ther 2001; 12(17):2095-2108.
28. Ciccutini FM, Welch K, Boyd M. Characterization of CD34+HLA-DR-CD38+ and CD34+HLA-DR-CD38- progenitor cells from human umbilical cord blood. Growth Factors 1999; 10(2):127-134.
29. de Bruyn C, Delforge A, Bron D. Comparison of coexpression of CD38, CD33 and HLA-DR antigens on CD34+ purified cells from human cord blood and bone marrow. Stem Cells 1995; 13(3):281-88.
30. Hao QL, Shah AJ, Thiemann FT et al. A functional comparison of CD34+ CD38- cells in cord blood and bone marrow. Blood 1995; 86:3745.
31. Noort WA, Wilpshaar J, Hertogh CD et al. Similar myeloid recovery despite superior overall engraftment in NOD/SCID mice after transplantation of human CD34(+) cells from umbilical cord blood as compared to adult sources. Bone Marrow Transplant 2001; 28(2):163-171.
32. Messner HA, Fauser AA, Lepine J et al. Properties of human pluripotent progenitors. Blood Cells 1980; 6(4):595-607.
33. Gartner S, Kaplan HS. Long-term culture of human bone marrow cells. Proc Natl Acad Sci USA 1980; 77:4756-59.
34. Sutherland HJ, Landsdorp PM, Henkelman DH et al. Functional characterization of individual human hematopoietic stem cells cultured at limiting dilution on supportive marrow stromal layers. Proc Natl Acad Sci USA 1990; 87:3584-88.
35. Lapidot T, Fajerman Y, Kollet O. Immune-deficient SCID and NOD-SCID mice models as functional assays for studying normal and malignant human hematopoiesis. J Mol Med 1997; 75(9):664-673.
36. Vormoor J, Lapidot T, Pflumio F et al. SCID mice as an in vivo model of human cord blood hematopoiesis. Blood Cells 1994; 20(2-3):316-20.
37. Bot FJ, Dorssers L, Wagemaker G et al. Stimulating spectrum of human recombinant multi-CSF (IL-3) on human marrow precursors: Importance of accessory cells. Blood 1988; 71(6):1609-1614.
38. Kuci S, Taylor G, Neu S et al. Phenotypic and functional characterization of mobilized peripheral blood CD34+ cells coexpressing different levels of c-Kit. Leuk Res 1998; 22(4):355-363.
39. Reisbach G, Bratke I, Kempkes B et al. Characterization of hematopoietic cell population from human cord blood expressing c-kit. Exp Hematol 1993; 21(1):74-79.
40. Mayani H, Landsdorp PM. Thy-1 expression is linked to functional properties of primitive hematopoietic progenitor cells from human umbilical cord blood. Blood 1994; 83(9):2410-17.

41. Fortunel N, Hatzfeld J, Kisselev S et al. Release from quiescence of primitive human hematopoietic stem/progenitor cells by blocking their cell-surface TGF-b type II receptor in a short-term in vitro assay. Stem Cells 2000; 18(2):102-11.
42. Srour EF, Zanjani ED, Cornetta K et al. Persistence of human multilineage, self-renewing lymphohematopoietic stem cells in chimeric sheep. Blood 1993; 82(11):3333-42.
43. LaRussa VF, Griffin JD, Kessler SW. Effects of anti-CD33 blocked ricin immunotoxin on the capacity of CD34+ human marrow cells to establish in vitro hematopoiesis in long-term marrow cultures. Exp Hematol 1992; 20(4):442-448.
44. Luence KM, Travis MA, Chen BP et al. Thrombopoietin, kit ligand, and flk2/flt3 ligand together induce increased numbers of primitive hematopoietic progenitors from human CD34$^+$Thy-1$^+$Lin$^-$ cells with preserved ability to engraft SCID-hu bone. Blood 1998; 91(4):1206-15.
45. Bhatia M, Wang JCY, Kapp U et al. Purification of primitive human hematopoietic cells capable of repopulating immune-deficient mice. Proc Natl Acad Sci USA 1997; 94:5320-5325.
46. Udomsakdi C, Eaves CJ, Sutherland HJ et al. Separation of functionally distinct subpopulations of primitive human hematopoietic cells using rhodamine-123. Exp Hematol 1991; 19(5):338-42.
47. Leemhuis T, Yoder MC, Grigsby S et al. Isolation of primitive human bone marrow hematopoietic progenitor cells using Hoechst 33342 and Rhodamine 123. Exp Hematol 1996; 24(10):1215-24.
48. Murray L, Chen B, Galy A et al. Enrichment of human hematopoietic stem cell activity in the CD34$^+$Thy-1$^+$Lin$^-$ subpopulation from mobilized peripheral blood. Blood 1995; 85(2):368-378.
49. Kollet O, Spiegel A, Peled A et al. Rapid and efficient homing of human CD34(+)CD38(-/low)CXCR4(+) stem and progenitor cells to the bone marrow and spleen of NOD/SCID and NOD/SCID/B2m(null) mice. Blood 2001; 97(10):3283-3291.
50. Agarwal RP, Crabtree GW, Parks R et al. Purine nucleoside metabolism in the erythrocytes of patients with adenosine deaminase deficiency and severe combined immunodeficiency. J Clin Invest 1976; 57(4):1025-1035.
51. Aiuti A. Advances in gene therapy for ADA-deficient SCID. Curr Opin Mol Ther 2002; 4(5):515-522.
52. Link CJ, Levy JP, Seregina T et al. Protection of murine vector producer cells from complement mediated lysis by Lovenox and sCR1. In: Mazarakis H, Swart SJ, eds. Cancer gene therapy. 135-152.
53. Young WB, Link C. Chimeric retroviral helper virus and picornavirus IRES sequence to eliminate DNA methylation for improved retroviral packaging cells. J Virol 2000; 74:5242-49.
54. Weiss RA, Tailor CS. Retrovirus receptors. Cell 1996; 82(4):531-33.
55. Bahner I, Kearns K, Hao QL et al. Transduction of human CD34$^+$ hematopoietic progenitor cells by a retroviral vector expressing an RRE decoy inhibits human immunodeficiency virus type 1 replication in myelomonocytic cells produced in long-term culture. J Virol 1996; 70(7):4352-60.
56. Orlic D, Girard LJ, Jordan CT et al. The level of mRNA encoding the amphotropic retrovirus receptor in mouse and human hematopoietic stem cells is low and correlates with the efficiency of retrovirus transduction. Proc Natl Acad Sci USA 1996; 93:11097-11102.
57. Miller von K, Kiem HP, Goehle S et al. Increased gene transfer into human hematopoietic progenitor cells by extended in vitro exposure to a pseudotyped retroviral vector. Blood 1994; 84(9):2890-97.
58. Miller DG, Miller AD. A family of retroviruses that utilize related phosphate transporters for cell entry. J Virol 1994; 68(12):8270-76.
59. Van der Loo JC, Liu BL, Goldman AI et al. Optimization of gene transfer into primitive human hematopoietic cells of granulocyte-colony stimulating factor-mobilized peripheral blood using low dose cytokines and comparison of a gibbon ape leukemia virus versus an RD114-pseudotyped retroviral vector. Hum Gene Ther 2002; 13(11):1317-30.
60. Gallardo HF, Tan C, Ory D et al. Recombinant retroviruses pseudotyped with the vesicular stomatitis virus G glycoprotein mediate both stable gene transfer and pseudotransduction in human peripheral blood lymphocytes. Blood 1997; 90(3):952-57.
61. Aiken C. Pseudotyping human immunodeficiency virus type 1 (HIV-1) by the glycoprotein of the vesicular stomatitis virus targets HIV-1 entry to an endocytic pathway and suppresses both the requirement for Nef and sensitivity to cyclosporin A. J Virol 1997; 71(8):5871-5877.
62. Rebel VI, Tanaka M, Lee JS et al. One-day ex vivo culture allows effective gene transfer into human nonobese diabetic/severe combined immune-deficient repopulating cells using high titer vesicular stomatitis virus G protein pseudotyped retrovirus. Blood 1999; 93:2217-24.
63. Beyer WR, Westphal M, Ostertag W et al. Oncoretrovirus and lentivirus vectors pseudotyped with lymphocytic choriomeningitis virus glycoprotein: Generation, concentration, and broad host range. J Virol 2002; 76(3):1488-1495.
64. Rasko JE, Battini JL, Gottschalk RJ et al. The RD114/simian type D retrovirus receptor is a neutral amino acid transporter. Proc Natl Acad Sci USA 1999; 96(5):2129-34.
65. Kelly PF, Vandergriff J, Nathwani A et al. Highly efficient gene transfer into cord blood nonobese diabetic/severe combined immunodeficiency repopulating cells by oncoretroviral vector particles pseudotyped with the feline endogenous retrovirus (RD114) envelope protein. Blood 2000; 96(4):1206-14.
66. Asada K, Uemori T, Ueno T et al. Enhancement of retroviral gene transduction on a dish coated with a cocktail of two different polypeptides: One exhibiting binding activity toward target cells, and the other toward retroviral vectors. J Biochem (Tokyo) 1998; 123(6):1041-1047.
67. Dao MA, Hashino K, Kato I. Adhesion to fibronectin maintains regenerative capacity during ex vivo culture and transduction of human hematopoietic stem and progenitor cells. Blood 1998; 92(12):4612-21.
68. Kiem HP, Andrews RG, Morris J et al. Improved gene transfer into baboon marrow repopulating cells using recombinant human fibronectin fragment CH-296 in combination with interleukin-6, stem cell factor, FLT-3 ligand, and megakaryocyte growth and development factor. Blood 1998; 92(6):1878-86.
69. Murray L, Luens K, Tushinski R et al. Optimization of retroviral gene transduction of mobilized hematopoietic progenitors by using thrombopoietin, FLT3 and Kit ligands and RetroNectin culture. Hum Gene Ther 1999; 10(11):1743-52.
70. Sanyal A, Schuening FG. Increased gene transfer into human cord blood cells by centrifugation-enhanced transduction in fibronectin fragment-coated tubes. Human Gene Therapy 1999; 10(17):2859-2868.
71. Lieber A, Kay MA, Li ZY. Nuclear import of moloney murine leukemia virus DNA mediated by adenovirus preterminal protein is not sufficient for efficient retroviral transduction in nondividing cells. J Virol 2000; 74(2):721-734.
72. Glimm H, Oh IH, Eaves CJ. Human hematopoietic stem cells stimulated to proliferate in vitro lose engraftment potential during their S/G(2)M transit and do not reenter G(0). Blood 2000; 96(13):4185-93.
73. Mayani H, Dragowska W, Landsdorp PM. Cytokine-induced selective expansion and maturation of erythroid versus myeloid progenitors from purified cord blood precursor cells. Blood 1993; 81(12):3252-58.
74. Traycoff CM, Abboud MR, Laver J et al. Rapid exit from G0/G1 phases of cell cycle in response to stem cell factor confers on umbilical cord blood CD34+ cells an enhanced ex vivo expansion potential. Exp Hematol 1994; 22(13):1264-72.
75. Piachebello W, Gammaitoni L, Bruno S et al. Negative influence of IL3 on the expansion of human cord blood in vivo long-term repopulating stem cells. J Hematother Stem Cell Res 2000; 9(6):945-956.
76. Berger F, Soliqo D, Scwarz K et al. Efficient retrovirus-mediated transduction of primitive human peripheral blood progenitor cells in stroma-free suspension culture. Gene Ther 2001; 8(9):687-696.
77. Ducos K, Hatzfeld A, Heron A. The high proliferative potential-quiescent (HPP-Q) assay allows an optimized evaluation of gene transfer efficiency into primitive hematopoietic stem/progenitor cells. Gene Ther 2000; 7(20):1790-94.

78. Yu J, Soma T, Hanazono Y et al. Abrogation of TGF-beta activity during retroviral transduction improves murine hematopoietic progenitor and repopulating cell gene transfer efficiency. Gene Ther 1998; 5(9):1265-71.
79. Barrette S, Douglas JL, Seidel NE et al. Lentivirus-based vectors transduce mouse hematopoietic stem cells with similar efficiency to Moloney murine leukemia virus based vectors. Blood 2000; 96(10):3385-3391.
80. Case SS, Price MA, Jordan CT et al. Stable transduction of quiescent CD34(+)CD38(-) human hematopoietic cells by HIV-1-based lentiviral vectors. Proc Natl Acad Sci USA 1998; 96(6):2988-93.
81. Demaison C, Parsley K, Brouns G et al. High-level transduction and gene expression in hematopoietic repopulating cells using a human imunodeficiency virus type 1-based lentiviral vector containing an internal spleen focus forming virus promoter. Hum Gene Ther 2002; 13(7):803-813.
82. Haas DL, Case SS, Crooks GM et al. Critical factors influencing stable transduction of human CD34(+) cells with HIV-1-derived lentiviral vectors. Molecular Therapy 2000; 2(1):71-80.
83. Dull T, Zufferey R, Kelly M. A third-generation lentivirus vector with a conditional packaging system. J Virol 1998; 72(11):8463-8471.
84. Kohn DB, Bauer G, Rice CR et al. A clinical trial of retroviral-mediated transfer of a rev-responsive element decoy gene into CD34(+) cells from the bone marrow of human immunodeficiency virus-1-infected children. Blood 1996; 94(1):368-71.
85. Gautsch JW. Embryonal carcinoma stem cells lack a function required for virus replication. Nature 1980; 285:110-12.
86. Palmer TD, Rosman WR, Osborne WR et al. Genetically modified skin fibroblasts persist long after transplantation but gradually inactivate introduced genes. Proc Natl Acad Sci USA 1991; 88:1330-34.
87. Scharfmann R, Axelrod JH, Verma IM. Long-term in vivo expression of retrovirus-mediated gene transfer in mouse fibroblast implants. Proc Natl Acad Sci USA 1991; 88:4626-30.
88. Seliger B, Kollek R, Stocking C et al. Viral transfer, transcription, and rescue of a selectable myeloproliferative sarcoma virus in embryonal cell lines: Expression of the mos oncogene. Mol Cell Biol 1986; 6:286-93.
89. Challita PM, Kohn DB. Lack of expression from a retroviral vector after transduction of murine hematopoietic stem cells is associated with methylation in vivo. Proc Natl Acad Sci USA 1994; 92:2567-2571.
90. Hoeben RC, Migchielsen AAJ, van der Jagt RCM et al. Inactivation of the Moloney mouse leukemia virus long terminal repeat in murine fibroblast cell lines is associated with methylation and dependent on its chromosomal position. J Virol 1991; 65:904-12.
91. Jaenisch R. Germ line integration and Mendelian transmission of the exogenous Moloney leukemia virus. Proc Natl Acad Sci USA 1976; 73:1260-64.
92. Loh TP, Sievert LL, Scott RW. Evidence for a stem cell-specific repressor of Moloney murine leukemia virus expression in embryonal carcinoma cells. Mol Cell Biol 1990; 10:4045-57.
93. Niwa O, Yokota Y, Ishida H et al. Independent mechanisms involved in the suppression of Moloney murine leukemia virus genome during differentiation of murine tetracarcinoma cells. Cell 1983; 32:1105-13.
94. Grez M, Akgun E, Hilberg F et al. Embryonic stem cell virus a recombinant murine retrovirus with expression in embryonic stem cells. Proc Natl Acad Sci USA 1990; 87:9202-06.
95. Laker C, Meyer J, Schopen A et al. Host cis-mediated extinction of a retrovirus permissive for expression in embryonal stem cells during differentiation. J Virol 1998; 72:339-48.
96. Stocking C, Grez M, Ostertag W. Regulation of retrovirus infection and expression in embryonic and hematopoietic stem cells. In: Doerfler W, Boehm P, eds. Virus Strategies. Heidelberg: VCH Verlag, 1993:433-455.
97. Brenner M, Rill DR, Moen RC et al. Gene marking and autologous bone marrow transplantation. Ann NY Acad Sci 1994; 716:204-214.
98. Brenner M. Gene transfer into human hematopoietic progenitor cells: A review of current clinical protocols. J Hematother 1994; 2(1):7-17.
99. Dunbar CE, Cottler-Fox M, O'Shaughnessy JA et al. Retrovirally marked CD34-enriched peripheral blood and bone marrow cells contribute to long-term engraftment after autologous transplantation. Blood 1995; 85(11):3048-3057.
100. Dunbar CE, Nienhuis AW, Stewart FM et al. Amendment to clinical research projects. Genetic marking with retroviral vectors to study the feasibility of stem cell gene transfer and the biology of hematopoietic reconstitution after autologous transplantation in multiple myeloma, chronic myelogenous leukemia, or metastatic breast cancer. Hum Gene Ther 1993; 4(2):205-222.
101. Deisseroth AB. Clinical trials involving multidrug resistance transcription units in retroviral vectors. Clin Cancer Res 1999; 5:1607-1609.
102. Eckert HG, Stockschlader M, Just U et al. High-dose multidrug resistance in primary human hematopoietic progenitor cells transduced with optimized retroviral vectors. Blood 1996; 88:3407-3415.
103. Maze R, Carney JP, Kelley MR et al. Increasing DNA repair methyltransferase levels via bone marrow stem cell transduction rescues mice from the toxic effects of 1,3-bis(2-chloroethyl)-1-nitrosourea, a chemotherapeutic alkylating agent. Proc Natl Acad Sci USA 1996; 93(1):206-10.
104. Ragg S, Xu-Welliver M, Bailey J et al. Direct reversal of DNA damage by mutant methyltransferase protein protects mice against dose-intensified chemotherapy and leads to in vivo selection of hematopoietic stem cells. Cancer Res 2000; 60(18):5187-95.
105. Bunting KD, Galipeau J, Topham D et al. Transduction of murine bone marrow cells with an MDR1 vector enables ex vivo stem cell expansion, but these expanded grafts cause a myeloproliferative syndrome in transplanted mice. Blood 1998; 92(7):2269-79.
106. Cowan KH, Moscow JA, Huang H et al. Paclitaxel chemotherapy after autologous stem-cell transplantation and engraftment of hematopoietic cells transduced with a retrovirus containing the multidrug resistance complementary DNA (MDR1) in metastatic breast cancer patients. Clin Cancer Res 1999; 5(7):1619-28.
107. Hesdorffer C, Ayello J, Ward M et al. Phase I trial of retroviral-mediated transfer of the human MDR1 gene as marrow chemoprotection in patients undergoing high-dose chemotherapy and autologous stem-cell transplantation. J Clin Oncol 1998; 16(1):165-72.
108. Rahman Z, Kavanagh J, Champlin R et al. Chemotherapy immediately following autologous stem cell transplantation in patients with advanced breast cancer. Clin Cancer Res 1998; 4(11):2717-21.
109. Cavvazana-Calvo M, Hacen-Bey S, de Saint Basile G et al. Gene therapy of human combined immunodeficiency (SCID)-XI disease. Science 2000; 288:669-6672.
110. Hacein-Bey-Abina S, Le Deist F, Carlier F et al. Sustained correction of x-linked severe combined immunodeficiency by ex-vivo gene therapy. N Engl J Med 2002; 346(16):1185-93.
111. Markert ML. Molecular basis of adenosine deaminase deficiency. Immunodeficiency 1994; 5(2):141-157.
112. Kohn DB, Weinberg KI, Nolta JA et al. Engraftment of gene-modified umbilical cord blood cells in neonates with adenosine deaminase deficiency. Nat Med 1996; 1(10):1017-23.
113. Schmidt M, Carbonaro DA, Speckmann C et al. Clonality analysis after retroviral-mediated gene transfer to CD34(+) cells from the cord blood of ADA-deficient SCID neonates. Nat Med 2003; 9(4):463-68.
114. Aiuti A, Slavin S, Aker M et al. Correction of ADA-SCID by stem cell gene therapy combined with nonmyeloablative conditioning. Science 2002; 296(5577):2410-2413.
115. Mannion-Henderson J, Kemp A, Mohney T et al. Efficient retroviral mediated transfer of the glucocerebrosidase gene in CD34+ enriched umbilical cord blood human hematopoietic progenitors. Exp Hematol 1996; 23(14):1628-32.
116. Xu LC, Kluepfel-Stahl S, Blanco M et al. Growth factors and stromal support generate very efficient retroviral transduction of peripheral blood CD34+ cells from Gaucher patients. Blood 1995; 86(1):141-46.

117. Takenaka T, Hendrickson CS, Tworek DM et al. Enzymatic and functional correction along with long-term enzyme secretion from transduced bone marrow hematopoietic stem/progenitor and stromal cells derived from patients with Fabry disease. Exp Hematol 1999; 27(7):1149-59.
118. Bauer TR, Schwartz BR, Liles WC et al. Retroviral-mediated gene transfer of the leukocytes integrin CD18 into peripheral blood CD34+ cells derived from a patient with leukocyte adhesion deficiency type 1. Blood 1998; 91(5):1520,1526.
119. Bauer TR, and Hickstein DD. Gene therapy for leukocyte adhesion deficiency. Curr Opin Mol Ther 2001; 2(4):383-388.
120. Liu JM, Young NS, Walsh CE et al. Retroviral mediated gene transfer of the Fanconi anemia complementation group C gene to hematopoietic progenitors of group C patients. Hum Gene Therapy 1997; 8(14):1715-30.
121. Dunbar CE, Kohn DB, Schiffmann R et al. Retroviral transfer of the glucocerebrosidase gene into CD34+ cells from patients with Gaucher disease: In vivo detection of transduced cells without myeloablation. Hum Gene Ther 1999; 9(17):2629-40.
122. Anderson DC, Springer TA Leukocyte adhesion deficiency: An inherited defect in the MAC-1, LFA-1, and p150, 95 glycoproteins. Annu Rev Med 1987; 38:175-194.
123. Root R.K. Leukocyte adhesion proteins: Their role in neutrophil function. Trans Am Clin Climatol Assoc 1991; 101:207-24.
124. Lu L, Ge Y, Li ZH et al. CD34 stem/progenitor cells purified from cryopreserved normal cord blood can be transduced with high efficiency by a retroviral vector and expanded ex vivo with stable integration and expression of Fanconi anemia complementation C gene. Cell Transplant 1995; 4(5):493-503.
125. Walsh CE, Nienhuis AW, Samulski RJ et al. Phenotypic correction of Fanconi anemia in human hematopoietic cells with a recombinant adeno-associated virus vector. J Clin Invest 1994; 94(4):1440-48.
126. Hacein-Bey-Abina S, von Kalle C, Schmidt M et al. A serious adverse event after successful gene therapy for X-linked severe combined immunodeficiency. N Engl J Med 2003; 348(3):255-56.
127. Dawenport J, Neale GA, Goorha R. Identification of genes potentially involved in LMO-2 induced leukemogenesis. Leukemia 2000; 14(11):1986-96.
128. Rabbitts TH, Axelson H, Forster A et al.Chromosomal translocations and leukemia: A role for LMO2 in T cell acute leukaemia, in transcription and in erythropoiesis. Leukemia 1997; 11(suppl)3:271-72.
129. Zhixiong Li, Dullmann J, Schiedlmeier B et al. Murine leukemia induced by retroviral gene marking. Science 2002; 296(5567):417-604.

CHAPTER 15

The Etiopathogenesis of Autoimmunity

Howard Amital and Yehuda Shoenfeld

Introduction

Diverse and complex mechanisms take part in the induction of an autoimmune condition. An appropriate genetic and hormonal setting, a concurrent infection or vaccination and many other environmental factors all converge in the pathogenesis of autoimmunity. Various processes take part in the induction of the autoimmune reaction such as antigenic spreading, antigenic mimicry, apoptosis, idiotypic network changes and many others. Some of these important factors have been pursued and targeted in order to generate modern nonhazardous therapeutic modalities that not only halt the progression of the autoimmune process but also reverse its affliction.

Autoimmune disorders affect approximately 5% of the population in the Western world and there are about 80 different autoimmune diseases.[1] However, autoimmune traits are known to exist in other clinical conditions as in malignancies, infections and in immune deficiencies. In many of these conditions, one can find an increased incidence of autoantibody production, which usually has no clear linkage to the clinical expression and outcome of the patients.[1-3] This phenomenon comes to the extreme in the case of patients with monoclonal gammopathies, a group of diseases in which an uncontrolled proliferation of plasma cells results in very high concentrations of monoclonal type immunoglobulin. In 8-10% of the cases, this immunoglobulin bears binding characteristics to various ubiquitous autoantigens as DNA, histones and others.[4-5] Only a minority of these patients develops autoimmune features that are related to the known autoantigen.[6-7] This natural concurrence clarifies that the existence of autoantibodies is insufficient to evoke an autoimmune disease. It has become well acknowledged that for an autoimmune process to develop, a proper genetic, hormonal and environmental setting is essential.

Since the introduction of the "clonal deletion" theory by Burnet,[8] the prevailing concept has turned sharply. Nowadays, we believe that autoimmunity has a role in maintaining health. Grabar[9] was the first to postulate that natural autoantibodies fulfill the role of biological scavengers binding to catabolic end products, cleansing the circulation from debris and other degraded tissue components. Concomitantly, these complexes facilitate opsonization and phagocytosis. Others outlined their role in achieving self-tolerance and balance of the idiotypic network.[10]

What Is Autoimmune?

Many individuals remain in a quiescent autoimmune state throughout their entire lives without developing an overt autoimmune disease, whereas others are destined to develop a highly aggressive disease. We believe that alteration of the immune equilibrium determines the difference between physiological and pathological. The nature of autoimmune processes have the capacity to change following an encounter of the immune system with infections, sunshine, drugs, hormones or even smoking.[11-12] We have adopted the concept of the "kaleidoscope of autoimmunity" to describe this feature of the immune system, in which alteration of the complex equilibrium between the diverse components of the immune system, leads to the induction of a remission or cure in one subject and to the acceleration of the disease in the other.[13-16]

It is difficult to define an autoimmune disorder. Even systemic lupus erythematosus (SLE), the hallmark of autoimmunity, does not fulfill the strict criteria that designate an autoimmune disease (Table 1); for instance, passive transfer of various autoantibodies (e.g., anti-DNA, anti-Sm antibodies) was not followed by disease development.[17] Although there are implications that anti-nucleic acid autoantibodies can be induced by immunization of experimental animals with nucleosomes, it is virtually impossible to stimulate an immune response to intact native DNA, and to induce a lupus induced disease by such immunization.[18] Furthermore, most passive transfer experiments employing either monoclonal or polyclonal anti-DNA antibodies failed to induce manifestations of SLE.[19]

The Mosaic of Autoimmunity

The etiology of autoimmune diseases is multifactorial.[11,12,20] Many autoimmune diseases are multigenic, with multiple susceptibility genes working in concert ending with an abnormal manifestation or disorder. In many of the different disorders, an increased incidence of B8 or DR2, DR3 or DR4 is found.[20] Some HLA alleles protect against the development of autoimmunity even when a susceptibility allele is present. For example, the HLA-DQB1*0602 allele protects against type 1 diabetes.[21]

Some of the autoimmune disorders are linked to certain functional defects of the immune system as interleukin-2 deficiency, IgA deficiency or with complete hereditary deficiencies of the

Table 1. Criteria for autoimmune disease

- A defined circulating antibody or cell-mediated immunity to autoantigens
- The ability to generate the autoantibody or self-reacting cells following the immunization with the self- antigen (with complete Freund's adjuvant)
- The ability to produce the disease in an experimental animal by passive transfer of the antibody or the self-reacting cells
- The ability to produce the disease in an experimental animal by immunization with the self- antigen (with complete Freund's adjuvant)
- Definition of a specific autoantigen

early components C1q, C2 and C4 in the classical pathway of complement system.[22,23] The high concentration of immune complexes detected in the blood of SLE has raised the suspicion that immunoglobulin Fcγ receptors polymorphism might be associated with SLE. Low affinity variants of Fcγ receptors have been shown to be associated with lupus nephritis.[23]

It is well known that the intrinsic balance within the immune system is altered following the administration of certain medications. D-penicillamine for instance is known to cause numerous adverse reactions, among them the induction of SLE, rheumatoid arthritis, pemphigus, myasthenia gravis, polymyositis, thrombocytopenia, systemic sclerosis, and Goodpasture's syndrome.[24] In a similar manner, graft versus host disease following allogeneic bone marrow transplantation bears many autoimmune features, particularly resembling conditions such as SLE, systemic sclerosis, and polymyositis.[25]

We have referred to this phenomenon as the mosaic of autoimmunity.[26] This definition implies that by reassembling the same pieces in a different order, another pattern or picture will emerge. We share this concept in a similar manner concerning diseases of the immune system; hence rearranging many of the same parameters might end in different outcomes.

Several reports have raised the attention that amelioration of one disease triggers the development of another autoimmune disorder. Levine et al[27] described two patients with immune thrombocytopenic purpura that, following a successful splenectomy, developed chronic active hepatitis probably of an autoimmune origin. From an etiological viewpoint, it may be assumed that once the immune equilibrium was altered, the clinical presentation had been rearranged. Similar events took place in patients with myasthenia gravis that, after thymectomy, developed aggressive SLE and anti-phospholipid syndrome (APS).[15,28,29] Krause et al[30] were able to block the development of diabetes in nonobese diabetes (NOD) mice (a model that spontaneously develops insulin dependent diabetes mellitus) by inducing a lupus-like–syndrome by their immunization with an anti-DNA idiotype. We have used the phrase "the kaleidoscope of the autoimmune mosaic" to describe such a switch from one abnormal immune balance to another.

Infections have also been associated with kaleidoscope phenomenon. It was pointed out for instance, that exposure to the Epstein-Barr virus (EBV) is highly associated with the development of SLE, and among 196 lupus patients all but 1 had been exposed to EBV in comparison to 22 of the 392 controls that did not have antibodies consistent with previous EBV exposure.[31] Often autoimmunity goes beyond the individual, it has been reported that family members of patients suffering from autoimmune diseases have increased titers of autoantibodies.[32,33] But are autoantigens essential for the induction of the autoimmune process? All in all, more than 120 different autoantigens were described in SLE.[34] Do all of them have a pathogenic role? Is it possible that there is no true autoantigen in a multisystemic autoimmune disease such as SLE?

There is a clear distinction between many classical, organ specific autoimmune disorders (Graves' disease, Hashimoto's disease, myasthenia gravis, pernicious anemia, insulin dependent diabetes) that have a specific autoantibody that plays a major role in the induction of the disease (such as anti-islet cell antibodies, anti thyroid stimulating hormone receptor etc.), from multi-systemic autoimmune diseases. But these distinctions do not shed light on the autoimmune kaleidoscope puzzle.

Mechanisms That Lead to Autoimmunity

Why do two patients with SLE, with similar anti-dsDNA titers present with extremely different clinical characteristics? We would like to offer an additional mechanism to resolve this enigma; the pathogenic potential of various idiotypes differs. Although two patients may have a similar autoantibody profile, one idiotype may induce hemolysis, another renal failure, and the other may be deposited in the skin or joints.[35-36]

Epitope Spread

Possible solutions to this puzzle might rise from various autoimmune models. It has been shown that immunization of animals with a given antigen can lead to spreading, of the antibody response to include not only the targeted antigen but also related epitopes. This "epitope spread" was noticed in our murine-induced lupus model. Following immunization with the anti-DNA antibody 16/6 idiotype, we recorded the production of antibodies also to Ro (SSA) and histones. On the other hand, immunization with the anti-Sm idiotype led to the production of anti-dsDNA, ssDNA, as well as anti-RNP antibodies.[35-37]

The Idiotypic Network

The basis to this theory was founded by Jerne,[38] who suggested that the immune response may be governed by unique antigenic determinants of the immunoglobulin variable regions (idiotypes). These idiotypes are regulated by other antibodies and so forth, comprising a dynamic and balanced idiotypic network with intrinsic regulatory balances. Immunization with a given antibody can elicit the spreading of an autoantibody response towards associated molecules as well as to apparently unrelated antigens. One of the mechanisms by which anti-idiotypes induce autoimmunity is via the induction of pathogenic antireceptor antibodies as in experimental myasthenia gravis, insulin dependent diabetes mellitus (IDDM), and Graves disease.[12] An anti-idiotype that combines with both the ligand-binding site at the variable region of the autoantibody can mimic the ligand by specifically combing with their receptors.

There is data pointing out that anti-idiotypes bearing internal image of viral antigens participate in the induction of several autoimmune disorders. It has been shown that myasthenia gravis has developed in five individuals a few weeks after rabies virus vaccination. Since the rabies virus binds to acetylcholine receptor, it is possible that an antiviral response can induce anti-idiotype antibodies that binds to acetylcholine receptors as well as initiating an autoimmune process. Anti-DNA antibodies and antibodies to Klebsiella have been shown to share a cross-reactive idiotype

(CRI), namely, the 16/6 idiotype. Patients with Klebsiella infections have an increased incidence of high titers of 16/6 idiotype, and both the 16/6 idiotype and anti-Klebsiella antibodies can be absorbed by Klebsiella K30 antigen.[41]

We have shown that immunization of different mice strains with either murine or human anti-DNA carrying the pathogenic 16/6 idiotype or with their antiidiotypes resulted in the production of a wide array of lupus associated antibodies in parallel to clinical findings consistent with human SLE. Similarly, immunization of naive mice with either human or mouse anticardiolipin antibodies led to the production of serological and clinical parameters suitable with antiphospholipid syndrome. Immunizing mice with idiotypes of antithyroglobulin and antiglomerular basement membrane antibodies, we were able to produce clinical characteristics of a compatible autoimmune diseases such as decreased thyroid hormone secretion and the appearance of erythrocyturia and proteinuria respectively.[36,42-44] These studies emphasize the important role that idiotypes play in the induction and pathogenesis of autoimmune diseases. However, this finding may also be used in order to divert the progression of one autoimmune condition to another.

Disruption of Immune Tolerance

Tolerance is the process that neutralizes autoreactive cells; therefore, whenever this process is disrupted, autoimmune manifestations might develop. Several mechanisms are engaged in the maintenance of this fragile equilibrium. Clonal deletion of immature B cells takes place in the bone marrow, and of autoreactive B cells within the T-cell zones of the spleen and lymph nodes.[45,46] Tolerance of B cells can be obtained via the lack of T cell assistance and by "receptor editing", a mechanism that changes the specificity of the B-cell receptor when an autoantigen is encountered.[47]

The main mechanism of T-cell tolerance is the deletion of self-reactive T cells in the thymus. Immature T cells migrate from the bone marrow to the thymus, where they bind to peptides derived from autoantigens bound to MHC molecules. Only T cell clones with an intermediate affinity for such complexes avoid apoptosis.[48] Interestingly, thymectomy or split dose-irradiation to the thymus of adult mice induced latent insulin-dependent diabetes mellitus (IDDM) in a normally resistant strain of mice, whereas administration of $CD4^+CD8^-$ thymocytes reversed this process.[49] A parallel clinical significance was reported in thymectomised mice that developed postoperative multisystemic autoimmune disorders.[28,29]

A complementary concept concerning tolerance states that a certain degree of tolerance is broken due to the dynamic response of the immune response to environmental antigens and microorganisms. Structural similarity between microbial and self-antigens (molecular mimicry) probably plays a key role in activating autoreactive T cells.[48] In addition, the release of sequestered autoantigens following tissue damage and the activation of a large fraction of the T-cell population by superantigens consequently induces the secretion of inflammatory cytokines and costimulatory molecules that enhance the autoimmune reaction.[50] Interestingly, infections are crucial for the development of protective T cell clones, depriving the immune system from combating infectious agents has also been associated with a higher frequency of autoimmune phenomena.[51]

Infections also have the capacity to induce the synthesis of class II molecules on tissue cells that do not normally generate these antigens. It has been demonstrated that not only the genetic composition of the MHC class II, but also changes in the amount of surface molecules, determine the type and magnitude of antigen presentation and of the immune response.[52] Population studies have demonstrated that patients with Graves' and Hashimoto thyroiditis, as well as with type 1 diabetes mellitus and other autoimmune disorders, are associated with specific MHC class II alleles.[52-53] The efficiency with which certain MHC class II allotypes present autoantigens may affect either failure or success to mount a productive immune response, and thereby determine the fate of the patient whether normal immune homeostasis will prevail or rather an autoimmune reaction will emerge.

Role of Cytokines

Both $CD4^+$ and $CD8^+$ T cells secrete cytokines, but the $CD4^+$ cell subtype may be subdivided according to patterns of cytokines secretion into Th1 cells that produce interleukin (IL)-2 and interferon (IFN)-γ that are critical for cell-mediated immunity and acute allograft rejection, whereas Th2 cells produce IL-4, IL-5, IL-6, IL-9 and IL-10 that promotes antibody production and humoral immunity.[54] IFN-γ and IL-12 are potent enhancers of Th1 cell differentiation, while IL-4 is a potent inducer of Th2 cells and is crucial for their maturation. IL-10 suppresses the synthesis of Th1 cytokines.[55-56]

When naive T cells first exit the thymus, they secrete both Th1 and Th2 cytokines, at that maturational phase they are termed Th0. After an antigen activates Th0 cells, they turn into memory cells and become committed to a certain T cell type.[55-56]

The discovery of this polarized immune response is of great significance in the understanding of autoimmunity and many efforts have been carried out in order to immunomodulate autoimmune disorders by affecting the dominance of one T cell type response in relation to the other. Autoimmune conditions can be classified according to their dependence on Th1 or Th2 responses; Th1 responses are primarily involved in the pathogenesis of organ-specific autoimmune disorders as experimental autoimmune encephalomyelitis, autoimmune thyroid disease and type 1 diabetes mellitus and acute allograft rejection. In contrast, allergic reactions involving IgE production and mast cell activation are typical Th2 responses. Interestingly, classical autoimmune diseases as SLE and rheumatoid arthritis (RA) share Th1 and Th2 response characteristics.[56,57]

Llorente et al[58] assessed clinical efficacy of administering an anti-IL-10 monoclonal antibody to 6 SLE patients with active and steroid-dependent disease. In all the patients, cutaneous lesions and joint symptoms improved with the beginning of anti-IL-10 administration and a decrease of the steroid dosage was made continuing all the 6 months of follow-up. Success in diverting the course of autoimmune models was achieved with recombinant IL-4 in rat experimental autoimmune uveoretinitis[59] and with anti-IFN-γ in modulating the natural history of experimental autoimmune thyroiditis and collagen-induced arthritis.[60,61] Many efforts are made in attempts to develop medications that target cytokines that are seminal to the formation of the autoimmune process. Some of these newly developed medications have reached clinical implementation as TNF-α inhibitors (Etanercept, Infliximab) and IL-1 blockers (Kineret) in the treatment of patients with long-standing refractory RA.[62-64] Recent utilization of anti-IL6 and anti-IL-15 have also been reported.[65,66] In the near future, similar agents will be part of the therapeutic arsenal of many other autoimmune conditions.

The Genetic Basis

The susceptibility to many autoimmune conditions was mapped to genes of the human and murine MHC.[11,67] Some autoimmune diseases are related to a specific genetic defect such as the autoimmune lymphoproliferative syndrome, the syndrome of autoimmune polyglandular endocrinopathy with candidiasis and ectodermal dysplasia (APECED) or to congenital deficiencies of certain complement components.[22,23,68,69] However, most autoimmune disorders are polygenic by nature with multiple susceptibility genes engaged in the generation of the autoimmune condition. Mostly, the diseases manifest with the presence of other predisposing factors. Some genes confer higher risk than others, while other are protective by nature. For instance, type 1 diabetes mellitus is associated with the presence of HLA-B8 and B-15 antigens and with the class II MHC HLA-DR3 and-DR4. Modern methods expanded these findings and today we know that the HLA-DQB1*0602 allele protects against type 1 diabetes even in individuals with coexisting HLA-DQB1*0301 or DQB1*0302 genes that act as predisposing genes.[21,70,71] Recent studies in transgenic mice demonstrated that genes that encode cytokines, cytokine ligands, molecules that promote apoptosis could stimulate or inhibit the autoimmune process.[72]

Apoptosis and Autoimmunity

Whenever normal apoptosis function is modulated, there is a substantial chance that a major deviation from the normal immune homeostasis will occur, leading to autoimmunity. It has been shown that among other factors, concentrations of self-antigens depend on the clearance rate of apoptotic cells. During apoptosis, the cell membrane forms cytoplasmic blebs, some of which are shed as apoptotic bodies. Casciola-Rosen et al[73] have shown that UV-induced apoptosis of keratinocytes induces redistribution of several autoantigens to apoptotic blebs. Mevorach et al[74] reported that immunizing mice with apoptotic cells induced autoantibodies formation. Apoptosis is one of the mechanisms that clarify the long-lasting enigma, why patients with autoimmune disorders generate antibodies that are directed against intranuclear components. The traditional cellular boundaries lose their significance since nuclear, cytosolic and membranous materials undergo structural modifications (blebs) during apoptosis, which are partially presented to the outer surface, exposed to the immune system.[75] Acceleration of apoptosis in autoimmune conditions may occur following viral infections or after sunlight exposure, these relations shed some light on environmental factors that take part in the induction of autoimmunity.[76] The association between the lupus-like syndromes of patients with deficiencies of complement components (particularly C1q and C4) may be explained by their impaired capability to clear apoptotic cells.[77] One of the most important antigens that induce apoptosis is the Fas antigen. The Fas antigen is a transmembranous receptor of the TNF/nerve growth factor receptor superfamily. In 33% of SLE patients, anti-Fas antibodies have been detected in their sera. This interesting finding implies that autoantibodies to apoptosis antigens might inhibit apoptosis, and provide a path by which autoreactive clones elude deletion.[78]

Infections, Vaccinations and Autoimmunity

The appearance of many autoimmune diseases is commonly preceded by an infectious event. A similar association might occur following the administration of a vaccine.[79,80] It is accepted that molecular mimicry (i.e., similarity between exposure antigen and self-antigen that leads to a cross reactive response) is one of the mechanisms that elucidate this linkage. One of the more familiar associations is the role which group A streptococci play in the pathogenesis of rheumatic fever, or in immunomediated acute glomerulonephritis.[81] Following the penetration of an infectious agent, besides the specific immune reaction that propagates in order to neutralize the invader, the immune system also undergoes a generalized activation that might also target self-components. This concept was thoroughly discussed in controversial reports dealing with the possible linkage between 'swine influenza vaccine' and Guillain-Barre syndrome, or between the hepatitis B vaccine and subsequent development of multiple sclerosis.[83,84]

Infections may lead to the impairment of immune tolerance through nonspecific manners. Tissue damage and necrosis may expose cryptic epitopes of diverse autoantigens to resting autoreactive T cells.[85] On the other hand, many mechanisms directly tie an infectious agent to autoimmunity. A quiescent, autoreactive T cell might be activated following an encounter with an infectious agent bearing a component that is structurally homologous to one in the host's self components. An example to this association is provided with the hepatitis B virus polymerase peptide in which six amino acids were identical to the encephalitogenic region of rabbit myelin basic protein (MBP). T cell reactivity to MBP was observed following immunization of rabbits with this peptide, and four of eleven animals showed histological signs of experimental autoimmune encephalomyelitis.[86] Microbial superantigens have the capacity to directly activate large numbers of T cells, some of them reactive with self-antigens. The Mycoplasma Arthritidis Superantigen has been associated with flares in mice with arthritis induced by type II collagen. The superantigen also triggered arthritis in mice that did not develop clinical disease following the initial immunization with type II collagen.[87]

"Epitope spreading" is a process in which a T cell response directed against a single self-peptide diversifies by priming of T cells specific for other self-peptides. This process occurs during an inflammatory reaction.[88] In the Theiler's murine encephalomyelitis virus model, epitope spreading explains the progression from CNS infection to priming to self-antigens.[89]

And, finally, direct infection of immune cells may also result in the enhancement of autoimmunity. An interesting human disease in which this mechanism is engaged is the Hepatitis C virus associated mixed cryogobulinemia. The hepatitis C virus infects both hepatocytes and B cells. Infection of B cells by hepatitis C virus results in a lymphoproliferative disease with clonal expansion of B cells. The vascular deposition of these circulating immune complexes causes a vasculitis, glomerulonephritis and alveolitis along with polyarthritis.[90]

Microchimerism

A chimera is an organism in which cells stem from two different origins. The concept of chimerism has fundamentally changed during the last decade, and what once was thought to be exceptional is acknowledged today to be common. During the normal progression of pregnancy, various cells bi-directionally migrate between maternal and fetal circulations. With the use of PCR-based techniques, and the fluorescence in-situ hybridisation (FISH), we can identify foreign DNA and islet cells of fetal origin within maternal tissues, this finding is commonly termed, microchimerism.[91] Microchimerism with fetal cells is commonly detected in the peripheral blood of healthy parous women.[92]

One cannot ignore the common denominator that ties microchimerism and autoimmunity. These two phenomena are prevalent among females during their reproductive period of life. Therefore, it has been postulated that the traffic of immune cells in female patients with systemic sclerosis might have a role in the pathogenesis of the disease. Systemic sclerosis is much more prevalent among women and reaches the ratio of 14:1 between the ages of 35-54.[93] Since the disease has many characteristics that resemble graft versus host disease, it is tempting to associate a graft versus host reaction inflicted by immune fetal cells that settled in maternal organs. Cells of fetal origin have been isolated from maternal blood and were detected in skin specimens taken from female patients.[94,95] New data indicate that microchimerism takes part also in the pathogenesis of primary biliary cirrhosis as well as in nonautoimmune liver disease and Sjögren's syndrome, and even in the very prevalent autoimmune disorder, Hashimoto's thyroiditis.[96-98] Surgical specimens that were obtained from women who underwent thyroidectomy for various thyroid disorders were examined for the presence of male fetal cells. The frequency of microchimerism in women with a male child was highest in those with Hashimoto's thyroiditis (five of six; 83%), however it was noted also in lower frequencies in other thyroidal diseases, both benign and malignant. No similar findings were detected in control specimens.[98]

Persistence of maternal cells in the offspring has also been reported in healthy subjects.[99] Hence, microchimerism may have a far-reaching role in the induction of autoimmunity. An interesting report by Tokita and colleagues[100] described how microchimerism could be exploited for therapeutic goals. A woman with poorly differentiated epithelial thymic carcinoma who was microchimeric with cells from her daughter, underwent infusion of stem-cells from the daughter that was followed with regression of the tumor that lasted a year.

References

1. Jacobson DL, Gange SJ, Rose NR et al. Epidemiology and estimated population burden of selected autoimmune diseases in the United States. Clin Immunol Immunopathol 1997; 84:223.
2. Swissa M, Amital-Teplizki H, Haim N et al. Autoantibodies in neoplasia. An unsolved enigma. Cancer 1990; 65:2554.
3. Abu-Shakra M, Buskila D, Ehrenfeld M et al. Cancer and autoimmunity: Autoimmune and rheumatic features in patients with malignancies. Ann Rheum Dis 2001; 60:433.
4. Shoenfeld Y, El-Roeiy A, Ben-Yehuda O et al. Detection of anti-histone activity in sera of patients with monoclonal gammopathies. Clin Immunol Immunopathol 1987; 42:250.
5. Buskila D, Abu Shakra M, Amital Teplizki H et al. Serum monoclonal antibodies derived from patients with multiple myeloma react with mycobacterial phosphoinositides and nuclear antigens. Clin Exp Immunol 1989; 76(3):378-83.
6. Noerager BD, Inuzuka T, Kira J et al. An IgM anti-MBP Ab in a case of Waldenstrom's macroglobulinemia with polyneuropathy expressing an idiotype reactive with an MBP epitope immunodominant in MS and EAE. J Neuroimmunol 2001; 113:163.
7. Choufani EB, Sanchorawala V, Ernst T et al. Acquired factor X deficiency in patients with amyloid light-chain amyloidosis: Incidence, bleeding manifestations, and response to high-dose chemotherapy. Blood 2001; 97:1885.
8. Burnet M. The clonal selection theory of acquired immunity. Nashville, Tennessee, USA: Vanderbilt University Press, 1959.
9. Grabar P. Autoantibodies and the physiological role of immunoglobulins. Immunol 1983; Today 4:337.
10. Shoenfeld Y, Isenberg DA. Natural autoantibodies: Their physiological role and regulatory significance. Boca Raton, FL: CRC Press, 1993.
11. Davidson A, Diamond B. Advances in immunology: Autoimmune diseases. N Engl J Med 2001; 345:340.
12. Shoenfeld Y, Schwartz RS. Immunologic studies and genetic factors in autoimmune diseases. N Engl J Med 1984; 311:1019.
13. Lorber M, Gershwin ME, Shoenfeld Y. The coexistence of systemic lupus erythematosus with other autoimmune diseases: The kaleidoscope of autoimmunity. Semin Arthritis Rheum 1994; 24:105.
14. Mevorach D, Perrot S, Buchanana NM et al. Appearance of systemic lupus erythematosus after thymectomy, four case reports and review of the literature. Lupus 1995; 4:33.
15. Alarcon-Sgovia D, Galbaith RF, Maldonado JE. Systemic lupus erythematosus following thymectomy for myasthenia gravis, report of two cases. Lancet 1963; 2:662.
16. Dalal I, Levine A, Somekh E et al. Chronic urticaria in children: Expanding the "autoimmune kaleidoscope". Pediatrics 2000; 106:1139.
17. Gilkeson GS, Grudier JP, Karounos DG et al. Induction of anti-double stranded DNA antibodies in normal mice. J Immunol 1989; 142:1482.
18. Satake F, Watanabe N, Miyasaka N et al. Induction of anti-DNA antibodies by immunization with anti-DNA antibodies: Mechanism and characterization. Lupus 2000; 9:489.
19. Madaio MP, Hodder S, Schwartz RS et al. Responsiveness of autoimmune and normal mice to nucleic acid. J Immunol 1984; 132:872.
20. Shoenfeld Y, Isenberg DA. The mosaic of autoimmunity. Immunol Today 1989; 10:123.
21. Becker KG. Comparative genetics of type 1 diabetes and autoimmune disease: Common loci, common pathways? Diabetes 1999; 48:1353.
22. Barka N, Shen GQ, Shoenfeld Y et al. Multireactive pattern of serum autoantibodies in asymptomatic individuals with immunoglobulin A (IgA) deficiency. Clin Diag Lab Immunol 1995; 2:469.
23. Johanneson B, Alarcon-Riquelme ME. An update of the genetics of systemic lupus erythematosus. Isr Med Assoc J 2001; 3:88.
24. Grove ML, Hassell AB, Hay EM et al. Adverse reactions to disease-modifying anti-rheumatic drugs in clinical practice. QJM 2001; 94:309-319.
25. Hess A, Thoburn C, Chen W et al. Autoreactive T-Cell subsets in acute and chronic syngeneic graft-versus-host disease. Transplant Proc 2001; 33:1754.
26. Shoenfeld Y, Isenberg DA. The mosaic of autoimmunity. Immunol Today 1989; 10:123.
27. Levene NA, Varon D, Shtalrid M et al. Chronic active hepatitis following splenectomy for autoimmune thrombocytopenia. Isr J Med Sci 1991; 27:199.
28. Shoenfeld Y, Lorber M, Yucel T et al. Primary antiphospholipid syndrome emerging following thymectomy for myasthenia gravis: additional evidence for the kaleidoscope of autoimmunity. Lupus 1997; 6:474-476.
29. Sherer Y, Bar-Dayan Y, Shoenfeld Y. Thymoma, thymic hyperplasia, thymectomy and autoimmune disease (review). Int J Oncol 1997; 10:939.
30. Krause I, Tomer Y, Elias D et al. Inhibition of diabetes in NOD mice by idiotypic induction of SLE. J Autoimmun 1999; 13:49.
31. James JA, Neas BR et al. Systemic lupus erythematosus in adults is associated with previous Epstein-Barr virus exposure. Arthritis Rheum 2001; 44:1122.
32. Goshen E, Livne A, Krupp M et al. Antinuclear and related autoantibodies in sera of healthy subjects with IgA deficiency. J Autoimmunity 1989; 2:51.
33. Isenberg DA, Shoenfeld Y, Walport M et al. Detection of cross-reactive anti-DNA antibody idiotype in the serum of lupus patients and their relatives. Arthritis Rheum 1985; 28:999.
34. Gorstein A, Sherer Y, Shoenfeld Y. Autoantibodies explosion in systemic lupus erythematosus. Semin Arthritis Rheum. In press.
35. Ziporen L, Shoenfeld Y, Levy Y et al. Neurological dysfunction and hyperactive behavior associated with antiphospholipid antibodies. A mouse model. J Clin Invest 1997; 100:613.
36. Mendlovic S, Brocke S, Shoenfeld Y et al. Induction of a SLE-like disease in mice by a common anti-DNA idiotype. Proc Natl Acad Sci USA 1988; 85:2260.
37. George J, Gilburd B, Shoenfeld Y. Autoantibody spread may explain multiple antibodies. The Immunologist 1999; 716:189.

38. Jerne NK. Idiotypic networks and other preconceived ideas. Immunol Rev 1984; 79:5.
39. Korn IL, Abramsky O. Myasthenia gravis following viral infection. Eur Neurol 1981; 20:435.
40. Lentz TL, Burrage TG, Smith AI et al. Is the acetylcholine receptor a rabies virus receptor? Science 1982; 215:182.
41. el Roiey A, Sela O, Isenberg DA et al. The sera of patients with Klebsiella infections contain a common anti-DNA idiotype (16/6) Id and anti- polynucleotide activity. Clin Exp Immunol 1987; 67:507.
42. Blank M, Cohen J, Toder V et al. Induction of anti-phospholipid syndrome in naive mice with mouse lupus monoclonal and human polyclonal anti-cardiolipin antibodies. Proc Natl Acad Sci USA 1991; 88(8):3069-73.
43. Tomer Y, Gilburd B, Sack J et al. Induction of thyroid autoantibodies in naive mice by idiotypic manipulation. Clin Immunol Immunopathol 1996; 78:180.
44. Shoenfeld Y, Gilburd B, Hojnik M. Induction of Goodpasture antibodies to noncollagenous domain (NC1) of type IV collagen in mice by idiotypic manipulation. Hum Antibodies Hybridomas 1996; 6:122.
45. Nemazee DA, Burki K. Clonal deletion of B lymphocytes in a transgenic mouse bearing anti-MHC class I antibody genes. Nature 1989; 337:562.
46. Rathmell JC, Townsend SE, Xu JC et al. Expansion or elimination of B cells in vivo: Dual roles for CD40- and Fas (CD95)-ligands modulated by the B cell antigen receptor. Cell 1996; 87:319.
47. Nemazee D. Receptor selection in B and T lymphocytes. Annu Rev Immunol 2000; 18:19.
48. Kamradt T, Mitchison NA. Tolerance and autoimmunity. N Engl J Med 2001; 344:655.
49. Saoudi A, Seddon B, Fowell D et al. The thymus contains a high frequency of cells that prevent autoimmune diabetes on transfer into prediabetic recipients. J Exp Med 1996; 184:2393.
50. von Herrath MG, Coon B, Lewicki H et al. In vivo treatment with a MHC class I-restricted blocking peptide can prevent virus-induced autoimmune diabetes. J Immunol 1998; 161:5087.
51. Sasazuki T, Kikuchi I, Hirayama K et al. HLA-linked immune suppression in humans. Immunol Suppl 1989; 2:21.
52. Ito T, Nakamura K, Umeda E et al. Familial predisposition of type 1 diabetes mellitus in Japan, a country with low incidence. Japan Diabetes Society Data Committee for Childhood Diabetes. J Pediatr Endocrinol Metab 14 Suppl 2001; 1:589.
53. Encinas JA, Kuchroo VK. Mapping and identification of autoimmunity genes. Curr Opin Immunol 2000; 12:691.
54. Roncarolo MG, Levings MK. The role of different subsets of T regulatory cells in controlling autoimmunity. Curr Opin Immunol 2000; 12:676.
55. Swain SL. Helper T cell differentiation. Curr Opin Immunol 1999; 11:180.
56. Barak V, Shoenfeld Y. "Cytokines in autoimmunity". In: Shoenfeld Y, ed. "The decade of autoimmunity". Amsterdam: Elsevier, 1999; 313-323.
57. Amital H, Levy Y, Blank M et al. Immunomodulation of murine experimental SLE-like disease by interferon-Y. Lupus 1998; 7:445.
58. Llorente L, Richaud Patin Y et al. Clinical and biologic effects of anti-interleukin-10 monoclonal antibody administration in systemic lupus erythematosus. Arthritis Rheum 2000; 43:1790.
59. Ramanthan S, de Kozak Y, Saoudi A et al. Recombinant IL-4 aggravates experimental autoimmune uveoretinitis in rats. J Immunol 1996; 157:2209.
60. Tang H, Mignon-Godefroy K, Meroni PL et al. The effects of a monoclonal antibody to interferon-gamma on experimental autoimmune thyroiditis (EAT): Prevention of disease and decrease of EAT-specific T cells. Eur J Immunol 1993; 23(1):275-8.
61. Boissier MC, Chiocchia G, Bessis N et al. Biphasic effect of interferon-γ in murine collagen-induced arthritis. Eur J Immunol 1995; 25:1184.
62. Weinblatt ME, Kremer JM, Bankhurst AD et al. A trial of etanercept, a recombinant tumor necrosis factor receptor:Fc fusion protein, in patients with rheumatoid arthritis receiving methotrexate. N Engl J Med 1999; 340:253.
63. Maini RN, Breedveld FC, Kalden JR et al. Therapeutic efficacy of multiple intravenous infusions of anti-tumor necrosis factor alpha monoclonal antibody combined with low-dose weekly methotrexate in rheumatoid arthritis. Arthritis Rheum 1998; 41:1552.
64. Huang CM, Tsai FJ, Wu JY et al. Interleukin-1beta and interleukin-1 receptor antagonist gene polymorphisms in rheumatoid arthritis. Scand J Rheumatol 2001; 30:225-228.
65. Choy EH, Isenberg DA, Garrood T et al. Therapeutic benefit of blocking interleukin-6 activity with an anti-interleukin-6 receptor monoclonal antibody in rheumatoid arthritis: A randomized, double-blind, placebo-controlled, dose-escalation trial. Arthritis Rheum 2002; 12: 3143.
66. Baslund B, Treda N, Danneskiold-Samsoe B et al. First use of human monoclonal antibody against IL-15 (HUMANA-IL15) in patients with active rheumatoid arthritis (RA): Results of a double-blind, placebo-controlled phase I/II trial. Fular 2003.
67. Sonderstrup G, McDevitt HO. DR, DQ, and you: MHC alleles and autoimmunity. J Clin Invest 2001; 107:871.
68. Wang CY, Davoodi-Semiromi A, Huang W et al. Characterization of mutations in patients with autoimmune polyglandular syndrome type 1 (APS1). Hum Genet 1998; 103:681.
69. Walport MJ. Complement— Second of Two Parts. N Engl J Med 2001; 344:1140.
70. Encinas JA, Kuchroo VK. Mapping and identification of autoimmunity genes. Curr Opin Immunol 2000; 12:691.
71. Whitney LW, Becker KG, Tresser NJ et al. Analysis of gene expression in mutiple sclerosis lesions using cDNA microarrays. Ann Neurol 1999; 46:425.
72. Johansson AC, Hansson AS, Nandakumar KS et al. IL-10-deficient B10.Q mice develop more severe collagen-induced arthritis, but are protected from arthritis induced with anti-type II collagen antibodies. J Immunol 2001; 167:3505.
73. Rosen A, Casciola-Rosen L. Clearing the way to mechanisms of autoimmunity. Nature Med 2001; 7:664.
74. Mevorach D, Zhou JL, Song X. Systemic exposure to irradiated apoptotic cells induces autoantibody production. J Exp Med 1998; 188: 387.
75. Rodenburg RJ, Raats JM, Pruijn GJ et al. Cell death: A trigger of autoimmunity? Bioessays 2000; 22:627.
76. Greidinger EL. Apoptosis in lupus pathogenesis. Front Biosci 2001; 6:1392.
77. Botto M. Links between complement deficiency and apoptosis. Arthritis Res 2001; 3:207.
78. Suzuki N, Ichino M, Mihara S et al. Inhibition of Fas/Fas ligand-mediated apoptotic cell death of lymphocytes in vitro by circulating anti-Fas ligand autoantibodies in patients with systemic lupus erythematosus. Arthritis Rheum 1998; 41:344.
79. Chen RT, Pless R, DeStefano F. Epidemiology of autoimmune reactions induced by vaccination. J Autoimmun 2001; 16:309.
80. Shoenfeld Y, Aron-Maor A. Vaccination and autoimmunity-'Vaccinosis': A dangerous liaison? J Autoimmun 2000; 14:1-10.
81. Cunningham MW. Pathogensis of group A streptococcal infections. Clinical Microbiology Reviews 2000; 13:470.
82. Kaplan JE, Katona P, Hurwitz ES et al. Guillain-Barre syndrome in the United States, 1979-1980 and 1980-1981. Lack of an association with influenza vaccination. JAMA 1982; 248:698.
83. Confavreux C, Suissa S, Saddier P et al. Vaccinations and the risk of relapse in multiple sclerosis. N Eng J Med 2001; 344:319.
84. Ascherio A, Zhang SM, Hernan MA et al. Hepatitis B vaccination and the risk of multiple sclerosis. N Eng J Med 2001; 344:327.
85. Horwitz MS, Bradley LM, Harbertson J et al. Diabetes induced by Coxsackie virus: Initiation by bystander damage and not molecular mimicry. Nat Med 1998; 4:781.
86. Fujinami RS, Oldstone MB. Amino acid homology between the encephalitogenic site of myelin basic protein and virus: Mechanism for autoimmunity. Science 1985; 230:1043.
87. Cole BC, Griffiths MM. Triggering and exacerbation of autoimmune arthritis by the *Mycoplasma arthritidis* superantigen MAM. Arthritis Rheum 1993; 36:994.
88. Wucherpfennig KW. Mechanisms for the induction of autoimmunity by infectious agents. J Clin Invest 2001; 108:1097.

89. Miller SD, Vanderlugt CL, Begolka WS et al. Persistent infection with Theiler's virus leads to CNS autoimmunity via epitope spreading. Nat Med 1997; 3:1133.
90. Ferri C, Zignego AL. Relation between infection and autoimmunity in mixed cryoglobulinemia. Curr Opin Rheumatol 2000; 12:53.
91. Lo YMD, Lau TK, Chan LYS et al. Quantitative analysis of the bidirectional fetomaternal transfer of nucleated cells and plasma DNA. Clin Chem 2000; 46:1301.
92. Bianchi DW, Zickwolf GK, Weil GJ et al. Male fetal progenitor cells persists in maternal blood for as long as 27 years post-partum. Proc Natl Acad Sci USA 1996; 39:191.
93. Silman AJ, Epidemiology of scleroderma. Ann Rheum Dis 1991; 50:887.
94. Nelson JL, Furst DE, Maloney S et al. Microchimerism and HLA-compatible relationships of pregnancy in scleroderma. Lancet 1998; 351:559.
95. Artlett CM, Smith JB, Jimmenez SA et al. Identification of fetal DNA and cells in skin lesions from women with systemic sclerosis. N Engl J Med 1998; 338:1186.
96. Jones DEJ. Fetal microchimerism: An aetiological factor in primary biliary cirrhosis? J Hepatol 2000; 33:834.
97. Nelson JL. Microchimerism: Expanding new horizon in human health or incidental remnant of pregnancy? Lancet 2001; 358:2011.
98. Srivatsa B, Srivatsa S, Johnson KL et al. Microchimerism of presumed fetal origin in thyroid specimens from women: A case-control study. 2001; 358:2034.
99. Maloney S, Smith AG, Furst DE et al. Microchimerism of maternal origin persists into adult life. J Clin Invest 1999; 104:41.
100. Tokita K, Terasaki P, Maruya E et al. Tumour regression following stem cell infusion from daughter to microchimeric mother. Lancet 2001; 358:20-47.

CHAPTER 16

Overview of Immune Tolerance Strategies

Charles J. Hackett and Helen Quill

Introduction

Increased understanding of immune cell activation and regulation makes plausible the development of clinical strategies to control unwanted immune responses in a specific manner. The normal safeguards collectively termed immune tolerance refer to the immune system's ability to control and prevent undesirable immune responses that lead to autoimmune diseases, asthma, or allergies, while still permitting protective immune responses to infection, vaccination, or tumor growth. The adaptive immune system, which consists of a vast repertoire of thymus-dependent T-lymphocytes and bone marrow-derived B-lymphocytes, achieves self tolerance both by eliminating dangerous cells that strongly recognize self tissues and by controlling those potentially self-reactive lymphocytes that escape deletion and are found throughout the body. Cell surface receptor molecules found on B- and T-lymphocytes determine the specificity of each cell by their ability to attach with lock-and-key accuracy to specific target molecules, termed antigens. T cell receptor molecules (TCR) bind to antigens comprised of short protein segments (peptides), derived from foreign or self proteins, that are held precisely in a pocket-like groove of self antigen presenting molecules, termed major histocompatibility complex (MHC) class I or MHC class II molecules. In this way, T cells recognize a complex of peptide-MHC molecules displayed on the surface of antigen presenting cells (APC). The B lymphocyte antigen receptor (BCR) is a membrane-bound immunoglobulin that recognizes antigen directly. Antigen binding by the TCR or BCR leads to aggregation of receptors on the cell surface, which brings together additional molecules within the cell to initiate a cascade of enzymatic reactions eventually leading to activation of genes in the nucleus. Gene activation determines how the cells respond to antigens; for example, by cell growth, secretion of factors, and activation of direct effector functions, such as the ability to destroy infected cells or secrete specific antibodies. Once activated, B cells release their immunoglobulin as antibodies that bind their antigens in blood or other tissues. Activated T cells may directly kill their target cells or help B cells to produce antibodies by direct cell-cell contact. They also produce soluble cytokines that mediate a variety of functions.[1]

A large number of potential approaches to manipulate immune tolerance are possible because of the many cellular and molecular processes that participate in immune initiation and control. Immune tolerance approaches include methods that act directly upon T or B lymphocyte activation, migration, and effector function; as well as by acting indirectly through bystander effects, such as the activation of antigen-specific regulatory cells that suppress immune effector function.

Immune tolerance is now understood not to be a static, established condition but rather an ongoing series of complex cell-cell interactions that control potentially autoreactive cells present in the circulation of every individual. Importantly, the mere presence of autoreactive cells does not indicate disease, and eliminating those cells is not necessarily beneficial. For example, a subset of autoreactive high affinity T cells is currently viewed as a regulatory cohort that actually protects against autoimmune diseases.[2] Optimal activation of antibody and T cell responses depends upon antigen processing, antigen presentation, and the molecular environment under which immune cell activation occurs. An immune tolerance strategy may be viewed as a push that tips the immune system balance in the direction of inhibition rather than activation. The balance is strongly influenced by components of innate immunity that may also serve as targets of tolerance induction protocols.

The Basics of Immune Tolerance

Central Tolerance

Developing T- and B-lymphocytes undergo rigorous selection processes based upon receptor binding strength for self-antigens. T cell selection occurs in the thymus. T cells with receptors that cannot bind at all to self-antigens die before maturing. T cells with TCRs that bind very strongly to self peptide-MHC complexes also die, while those that bind with a lesser affinity survive to populate the body. Central tolerance processes eliminate a great number of T cells in the thymus, and those that survive constitute the entire population of T cells that can recognize foreign antigens with high affinity. BCRs that bind self-antigens strongly are also eliminated. Some B cells that react with self-antigens stop growing and have the opportunity to rearrange their receptors again, permitting the B cell to reenter the development pathway if the new modification causes the receptor to lose reactivity with self.[3] The thymus and bone marrow express antigens from many tissues of the body enabling removal of the majority of potentially dangerous autoreactive cells before they mature.[1]

Peripheral Tolerance

It is not possible for every antigen from every tissue of the body to participate in the central elimination of all potentially autoreactive cells. For this reason, and because the immune system selects T cells that have a weak affinity for self, and allows mature cells to undergo BCR diversification after central tolerance occurs, cells capable of autoreactivity are believed to be present in every individual. Peripheral tolerance processes that keep autoreactive cells under control include:[4,5]

- Sequestration of antigens—for example, cells of the immune system do not regularly patrol the central nervous system, eye, and gonads unless injury or infection occurs;

- Anergy—causing lymphocytes partially triggered by antigen to become unable to respond to subsequent antigen exposure;

- Activation-induced death which limits the number of cells following a response to antigen; and,

- Regulatory T cells that inhibit immune responses discussed below.

Unlike central tolerance mechanisms that have no effect on mature lymphocytes, and are not easily developed into therapeutic approaches, peripheral tolerance processes should be much more readily exploitable as means of controlling immune responses.

Peripheral Activation

Lymphocytes leaving the thymus or bone marrow and circulating in the body are not activated until called into action by a process that involves a series of signaling steps in response to foreign antigens. For T cells, binding of the TCR to peptide-MHC alone is insufficient; a simultaneous signal from another group of cell-surface molecules, termed costimulatory molecules, must also be present.[6] These costimulatory molecules include B7-1 (CD80) and B7-2 (CD86), which are expressed on surfaces of activated APC. B cells require sufficient antigen to aggregate their BCRs, and most B cells also require signals from activated "helper" T cells in order to develop into potent antibody secreting cells.[1]

Generation of Pathological T Cells and Autoantibodies

Potentially autoreactive T cells that escape anergy or deletion mechanisms in the thymus become vigorously autoimmune when peripheral tolerance mechanisms break down. The loss of self-tolerance is incompletely understood at present. The basic requirements for initiation of an autoimmune response are: sufficient receptor affinity for self-antigen, presentation of the relevant antigen in the presence of costimulatory molecules on activated APC, and either a deficit in active regulatory cell controls or the ability to override them. Once activated, T cells may no longer require strong costimulatory signals and can respond to self-antigens presented without costimulatory molecules. For many autoimmune diseases mediated primarily by inflammatory autoreactive T cells, such as type 1 diabetes, rheumatoid arthritis, and multiple sclerosis, autoreactive cells differentiate to the Th1 (inflammatory) and cytotoxic T cell phenotypes. Antibody-mediated diseases, such as systemic lupus erythematosus and myasthenia gravis, depend upon autoreactive Th2-type T cells that help activate autoantibody producing B cells. Many factors, including genetic, environmental, and infectious diseases contribute to the quantity and antigenic specificities of potentially autoreactive cells.[7]

Innate Immunity and Tolerance

The innate immune system provides the initial responses to infection, taking both immediate defensive actions as well as activating the adaptive immune system.[8] Immediate responses include secretion of antimicrobial peptides and the upregulation of cytokines and chemokines that call additional cells into action, including specific T and B cells that can later respond to the infection. Langerhans cells of the skin, tissue dendritic cells and macrophages, and tissue-associated lymphocytes such as natural killer cells and gamma-delta receptor T cells, are components of the innate immune system that trigger these early responses to infection. Highly specialized innate receptors recognize microbial pathogen molecules. A family of such receptors, known as the Toll-like molecules, has recently been shown to function in the early detection and response to infection by recognition of bacterial cell wall components, viral double-stranded RNA, and bacterial nucleic acids. Toll-related molecules and other pathogen-specific receptors of the innate immune system are found in many species, from *Drosophila* to humans, and are inborn, rather than being selected in each individual as are the TCRs and BCRs that recognize antigens for the adaptive immune system.

In some cases, the inflammatory responses initiated by the innate immune system may contribute to the activation of autoreactive cells. For example, bacterial infection of dendritic cells leads to B7 upregulation, enabling costimulatory signaling of T cells that may interact with any antigen presented by that particular dendritic cell. Cellular necrosis, occurring as a result of infection or tissue damage, as opposed to the self-destruction process of apoptosis, can trigger immature dendritic cells to become highly activated and to function as APC for T cells.[9] Thus, self-antigens that happen to be presented in the context of B7 may activate autoimmune T cell responses. Note that several conditions must be present to initiate autoimmunity: the presentation of a sufficient amount of self antigen/MHC complexes by activated APC, availability of sufficiently high affinity autoreactive T cells in contact with the APC, and the unavailability or ineffectiveness of overriding regulatory cell activity.

Accumulating evidence suggests that, generally, the innate immune system protects against activation of autoreactive cells (reviewed in Chapter 19). For example, complement components upregulated in response to infection enhance the phagocytosis of apoptotic cells and cell debris, reducing the chance for those antigens to be presented by activated dendritic cells, which are potent immune stimulators.[11]

Immunoregulatory T Cells

Distinct types of regulatory cells prevent development of autoimmune disease when potentially autoreactive T cells are present. The ability to manipulate regulatory cells would provide enormously powerful control over immune responses. This topic is being intensively investigated and is the subject of a recent review (*Immunological Reviews* 2001; 182(1)). The prospect that methods may be devised to manipulate regulatory T cells in an antigen specific manner suggests many possibilities for new tolerogenic treatments.

T cells expressing the cell surface molecules CD4 plus CD25 (CD4$^+$CD25$^+$ T cells) have recently been defined in humans and are known to play a major regulatory role in animal models of autoimmune diseases, including control of type I diabetes, central nervous system autoimmune inflammation, and intestinal autoreactivity.[12] CD4$^+$CD25$^+$ T cells develop in the thymus like conventional T cells, and respond to as yet unknown self-antigens.

Table 1. Genetic evidence for tolerance in humans

Gene Defect(s) in:	Pathway Affected	Human Disease
Fas/FasL	Lymphocyte Apoptosis	ALPS
Perforin	CTL/NK Cell homeostatic function	FHL
CTLA-4	Negative regulation of T cell function	[implicated] thyroiditis diabetes

ALPS= Autoimmune lymphoproliferative syndrome
FHL= Familial hemophagocytic lymphohistiocytosis

When activated in vitro, they respond with a low level of cell division, but become potent suppressors of other T cells nearby. CD4+CD25+ T cells appear to act mainly by cell-to-cell contact rather than by secreted factors. Most likely, these regulatory cells respond to specific self-antigens presented by activated APC in affected tissues and act to suppress local inflammatory responses.

Regulatory T cells termed Tr1 and Th3 respond to antigens by releasing cytokines that counteract Th1 and CTL activity. Tr1 and Th3 cells release interleukins-4, -10, -13 and transforming growth factor-beta, each of which has potent anti-inflammatory effects.[13]

Another distinct regulatory population, the NK T cells, bears surface molecules commonly found on both natural killer cells and T cells.[14,15] NK T cells produce interleukin-4 (IL-4), IL-10, and interferon-gamma, probably influencing the activation of other regulatory T cell populations. NK T cells recognize the APC surface molecule CD1, which presents lipid antigens, in contrast to the peptide-presenting MHC class I or class II molecules.[16] Recently, a ligand that is recognized by NK T cells in a CD1-restricted manner, alpha-galactosylceramide, was shown to protect against the onset of type 1 diabetes in an NOD mouse model.[17,18] Analyses showed that IL-10 production by NK T cells was responsible for the protective effect.

Importantly, APC are now being recognized to play key roles in regulation of autoimmune responses. For example, immature dendritic cells may present antigen in a tolerogenic manner.[19] As with many suppressor type cells, the molecular mechanism of function is not well understood.

Potential Approaches Based on Central Tolerance

There are two approaches that target the central selection processes for T cells. The first is thymic injections of antigens,[20] an approach of quite limited potential in practical use. The second is mixed chimerism, which is more widely applicable.[21] Mixed chimerism refers to the coexistence of significant percentages of both donor and host hematopoietic cells within the body. In individuals with mixed chimerism, foreign thymic dendritic cells can tolerize new host derived T cells.[21] Thus, T cells maturing in the thymus after chimera formation will not be able to mount a strong response against the antigens present on the foreign dendritic cells that functioned in their selection. At least in some circumstances, the result is a very effective method for reducing or eliminating the potential for unwanted T cell responses. However T cells that are already present in the periphery before chimera formation need to be eliminated or controlled. Therefore, mixed chimerism without other ablative treatments is only partially effective. Potentially, a combination of central tolerance and peripheral tolerance could provide powerful control over a wide range of T cell mediated immune responses.

Genetic Evidence for Peripheral Immune Tolerance in Humans

What is the evidence that peripheral tolerance mechanisms function to control immune responses in humans? The best evidence currently comes from identification of defective genes that are associated with immune mediated diseases. Table 1 lists three genetic defects that identify Fas/Fas Ligand,[22] perforin,[23] and CTLA-4[24,25] as strong candidates for genes involved in peripheral tolerance in humans. Fas, also known as APO-1 and CD95, is a member of the tumor necrosis factor family as is Fas ligand. Strains of mice that have mutations in this system are characterized by defects in apoptosis that lead to autoimmune diseases.[26] Recently, a related human disease termed autoimmune lymphoproliferative syndrome (ALPS) was found to be characterized by massive nonmalignant lymphoadenopathy of early onset caused by the accumulation of T-cells in the periphery. Analyses show that the impaired pathway involves lymphocyte apoptosis, resulting in overpopulation of lymphocytes and consequent abnormalities.[22]

The gene that encodes perforin, a molecule that functions in the lytic mechanism of cytotoxic T-cells and natural killer cells, has recently been implicated in a homeostatic function. A disease called familial hemophagocytic lymphohistiocytosis (FHL) is associated with a genetic defect in the perforin molecule.[23] High levels of activated T cells in the blood and tissues of FHL patients lead to the overproduction of inflammation-producing factors, resulting in severe organ damage, especially of the liver, spleen, and central nervous system. Only bone marrow transplantation can successfully treat the approximately 80 children born in the United States each year with this immune disorder. Mutations in at least three different genes each can give rise to FHL. Patients with one type of FHL were lacking or had an inactive perforin molecule. As expected, the perforin-deficient FHL patients are defective in cytotoxic cell function, and presumably lack an important molecule needed to regulate cellular immune responses.

Defects in CTLA-4, a molecule that functions in the negative regulation of T-cells, has an implicated, although not fully proven role in such autoimmune diseases as thyroiditis and diabetes.[24,25] Presumably, the defective versions of CTLA-4 are unable to satisfactorily downregulate T-cells, resulting in accumulation of T-cells that would normally have been silenced by the CTLA-4 signaling pathway.

These genetic defects provide insight into the human immune system and demonstrate the profound effects that individual molecules can have on immune function related to autoimmunity.

Additional Molecules Implicated in Self-Tolerance

Table 2 lists a number of additional proteins that are strongly implicated in control of self-tolerance through in vivo studies in animal systems. By the use of gene knockout mice lacking specific molecules, or by blocking in vivo function with antibodies of other antagonists, individual molecules can be demonstrated to play major roles in self-tolerance. For example, abrogation of the key intracellular regulator SHP-1 (src homology 2 contain-

Table 2. Additional molecules implicated in self tolerance

Molecule	Consequences of in Vivo Block or Gene Knockout
CD40-L	Enhanced allograft acceptance
CD30	Enhanced experimental diabetes
ICOS	Enhanced experimental autoimmune encephalomyelitis
PD-1	Lupus-like disease
FcγRIIB	Increased susceptibility to collagen-induced arthritis
SHP-1	Widespread inflammation
CD21/CD35	Prevention of B cell anergy
CD22	Spontaneous autoantibodies
Lyn	Spontaneous autoantibodies

FcγRIIB= immunoglobulin IgG receptor; ICOS= inducible costimulator protein, a member of the CD28 family; Lyn= a src-like tyrosine kinase obtained by screening a cDNA library with the v-yes gene probe that identified a locus similar to v-yes but novel; PD-1= programmed death 1; SHP-1= Src homology 2 (SH2)-containing tyrosine phosphatase-1

ing tyrosine phosphatase-1) results in profound systemic autoreactivity,[27] while less global effects are observed by blocking inhibitory regulators that are more restricted in their cellular distribution, such as FcγIIIB,[28,29] CD30,[30,31] and CD22.[32]

Knocking out the function of the T cell costimulatory receptor CD40 ligand (CD154) improves allograft acceptance.[33-35] However, not every treatment that inhibits T cells can ameliorate autoimmune diseases. Regulatory T cells may also be affected, and certain T cell inhibitory treatments have been shown to make some autoimmune conditions much worse.[36] Therefore, although the molecules identified in animal studies provide critical insight into autoimmune disease development, their use as global targets for future therapies will depend upon how well these molecules can be targeted specifically to the autoimmune disease-causing cells.

Strategies of Manipulating Peripheral Tolerance

Table 3 summarizes major approaches that have been tested in experimental systems for controlling peripheral immune tolerance, along with the key cellular receptors involved.

Antigen Receptors

Antigen receptor targeting of pathogenic T cells generally employs antigens as highly soluble peptides or proteins, delivered without a strongly stimulatory adjuvant, for presentation by nonactivated APC. Approaches include intravenous injections, injections in incomplete Freund's adjuvant, and oral delivery of antigens.[37-39] These approaches may lead to anergy in antigen-specific T cells or the triggering of Th2 or regulatory T cells.[40,41] Although antigen targeting may appear to be the ideal approach for manipulating peripheral tolerance, there are important problems that currently limit this approach for therapy of existing disease. The major problem is practical: knowing which antigens to use. For T cells, relatively few self antigen-peptides have been described, and development of methods to develop new approaches to map such epitopes are not yet at the high

Table 3. Strategies of manipulating peripheral tolerance

Target of Strategy	Cellular Receptors
Antigen receptors	TCR; BCR
Costimulatory and other molecules	B7/CD28; CD40/CD40-L; LFA-1/ICAM; SLAM; CD4-CD8; CD52; 4-1BB/4-1BBL
Cytokine Circuits	IL-10; TGF-β
Lymphocyte Migration	L-selection; α4 integrin

BCR= B cell receptor; ICAM= intercellular adhesion molecule; LFA= leukocyte function antigen; SLAM= signaling lymphocyte activation molecule; TCR= T cell receptor; TGF= tumor growth factor; 4-1BB= an inducible T cell costimulatory receptor, also known as CD137; 4-1BBL= ligand for 4-1BB.

throughput level.[42] Knowing whether such peptides are targets of pathogenic T cells requires additional confirmation ex vivo or in situ. Further, as the autoimmune disease state progresses, an increasing number of antigenic epitopes is generated. This process, termed "epitope spreading"[43] may greatly increase the number of autoantigenic epitopes involved. The great potential for TCR diversity makes controlling existing responses by antigen targeting a formidable task. Clinical trials of antagonist analogues of self-antigens were unsuccessful for treating multiple sclerosis,[44-46] most likely because of TCR diversity. Although the peptides were antagonists for some T cells, they were agonists for others, and exacerbated autoimmunity.[47]

However, TCR targeting may have future applications in autoimmune disease prevention rather than therapy. This capability will depend on whether a few key proteins in each disease, such as insulin or GAD65 (glutamic acid decarboxylase) in type 1 diabetes, can be conclusively demonstrated to be the major instigating autoantigens involved in disease development. Potentially, antigen-specific blockade of instigating T cells, either by anergy, deletion, or induction of strong regulatory cells, may delay or protect from autoimmune damage.

Costimulatory and Other Molecules

The approach using costimulatory blockade bypasses the need to identify antigen-specific autoimmune cells. In this approach, broad inhibition of T cell costimulation will include those being activated to self-antigens, resulting in their being anergized or deleted, and thereby unable to function in autoimmunity.[48] Costimulatory blockade has been found to be highly effective against the activation of primary T cells,[49] but recent findings show that certain costimulatory molecules may also be needed to reactivate memory T cells.[50-52] Recently, von Herrath and coworkers reported that anti CD154 treatment induced regulatory cells bearing markers of both dendritic cells and NK cells that functioned in an animal model of diabetes.[53] These observations suggest that effects distinct from those believed to be the major targets of costimulatory inhibition need to be examined in greater detail in order to improve potential therapeutic approaches.

Cytokine Circuits

Treatment with anti-inflammatory cytokines provides an approach that extends beyond antigen specificity. Tissue inflamma-

Table 4. Major effects of peripheral tolerance strategies on immune system components

Target of Strategy	Effect on:	T effectors	T reg.	B Cells	APC	NK Cells	NK T Cells
Antigen receptors		Yes	Yes	Yes	Yes	?	?
Costimulatory and other molecules		Yes	Yes	Yes	?	?	Yes
Cytokine Circuits		Yes	Yes	Yes	Yes	Yes	?
Lymphocyte Migration		Yes	Yes	Yes	Yes	?	?

tion and destruction in autoimmune diseases involves macrophages and other non antigen-specific cells[54-56] which may be controlled by such counter-inflammatory cytokines as IL-10 and TGF-beta.[57] Cytokine therapy requires achieving effective local concentrations, which risks severe side effects if delivered systemically. Local cytokine expression using gene therapy approaches are potential solutions to this problem, but such approaches are in early experimental stages.[58] Novel methods, such as use of high affinity antibodies to block proinflammatory cytokines[59] have great potential, but will require considerable early stage research. Enbrel, the TNF-alpha receptor approved for rheumatoid arthritis therapy, uses receptor blocking, an approach that appears more readily manipulated than directly affecting cytokine levels.[60] Whether the ameliorating effects of cytokine circuit perturbation are strictly pharmacological or whether they may promote the development of tolerized cells or regulatory cell activation requires further investigation.

Conclusion

All strategies to manipulate peripheral tolerance affect not only the principal target, but also directly or indirectly affect other components of the immune system (Table 4). Even very specific antigen receptor targeting that affects the autoimmune disease-producing effector T cells, may hypothetically also affect T regulatory cells that share the same antigenic specificity.[61,62] T cells anergized by various treatments may directly affect the APC compartment.[63] Such "side effects" need not be peripheral to the strategy; in fact, effects beyond the primary target may contribute to their effectiveness.

Even approaches such as cytokine circuit alteration may have a variety of "indirect" effects that contribute to their effectiveness. The cytokines IL-10 and TGF-beta can directly affect T cells,[57,64] B cells,[65,66] APC,[67,68] and NK cells.[64,69] CD62L used experimentally to block T cell migration may also affect B cells and APC.[70,71] Clearly, not all of these additional effects are as important or central to effectiveness as the primary target, but their potential positive and negative contributions to the process requires careful attention as these strategies approach clinical testing. Finally, hematopoietic stem cell transplantation (HSCT) is becoming increasingly used to treat autoimmune diseases. The mechanism(s) by which autologous or allogeneic HSCT induced tolerance are only now beginning to be investigated. HSCT will quite likely alter all components of the immune system. How this re-shuffling of the immune system results in tolerance is yet to be determined.

References

1. Janeway Jr C, Travis P. Immunobiology: The immune system in health and disease. New York and London: Garland Publishing Co., 1997.
2. Shevach EM et al. Control of autoimmunity by regulatory T cells. Adv Exp Med Biol 2001; 490:21-32.
3. Nemazee D. Receptor editing in B cells. Adv Immunol 2000; 74:89-126.
4. Mackay IR. Science, medicine, and the future: Tolerance and autoimmunity. Bmj 2000; 321:93-6.
5. Kamradt T, Mitchison NA. Tolerance and autoimmunity. N Engl J Med 2001; 344:655-64.
6. Howland KC, Ausubel LJ, London CA et al. The roles of CD28 and CD40 ligand in T cell activation and tolerance. J Immunol 2000; 164:4465-70.
7. Mackay IR, Van de Water J, Gershwin ME. Autoimmunity. Thoughts for the millennium. Clin Rev Allergy Immunol 2000; 18:87-117.
8. Medzhitov R, Janeway Jr CA. How does the immune system distinguish self from nonself? Semin Immunol 2000; 12:185-188.
9. Sauter B et al. Consequences of cell death: Exposure to necrotic tumor cells, but not primary tissue cells or apoptotic cells, induces the maturation of immunostimulatory dendritic cells. J Exp Med 2000; 191:423-34.
10. Carroll M. Innate immunity in the etiopathology of autoimmunity. Nat Immunol 2001; 2:1089-90.
11. Prodeus AP et al. A critical role for complement in maintenance of self-tolerance. Immunity 1998; 9:721-31.
12. Shevach EM. Certified professionals: CD4(+)CD25(+) suppressor T cells. J Exp Med 2001; 193:F41-6.
13. Bach JF, Chatenoud L. Tolerance to islet autoantigens in type 1 diabetes. Annu Rev Immunol 2001; 19:131-61.
14. Shevach EM. Regulatory T cells in autoimmunity. Annu Rev Immunol 2000; 18:423-49.
15. Shi FD. Germ line deletion of the CD1 locus exacerbates diabetes in the NOD mouse. Proc Natl Acad Sci USA 2001; 98:6777-82.
16. Porcelli SA, Segelke BW, Sugita M et al. The CD1 family of lipid antigen-presenting molecules. Immunol Today 1998; 19:362-8.
17. Sharif S et al. Activation of natural killer T cells by alpha-galactosylceramide treatment prevents the onset and recurrence of autoimmune Type 1 diabetes. Nat Med 2001; 7:1057-62.
18. Sharif S, Arreaza GA, Zucker P et al. Regulation of autoimmune disease by natural killer T cells. J Mol Med 2002; 80:290-300.
19. Steinman RM, Nussenzweig MC. Avoiding horror autotoxicus: The importance of dendritic cells in peripheral T cell tolerance. Proc Natl Acad Sci USA 2002; 99:351-8.
20. Jones ND, Fluck NC, Mellor AL et al. The induction of transplantation tolerance by intrathymic (i.t.) delivery of alloantigen: A critical relationship between i.t. deletion, thymic export of new T cells and the timing of transplantation. Int Immunol 1998; 10:1637-46.
21. Wekerle T, Sykes M. Mixed chimerism and transplantation tolerance. Annu Rev Med 2001; 52:353-70.

22. Straus SE, Sneller M, Lenardo MJ et al. An inherited disorder of lymphocyte apoptosis: The autoimmune lymphoproliferative syndrome. Ann Intern Med 1999; 130:591-601.
23. Stepp SE et al. Perforin gene defects in familial hemophagocytic lymphohistiocytosis. Science 1999; 286:1957-9.
24. Kouki T et al. CTLA-4 gene polymorphism at position 49 in exon 1 reduces the inhibitory function of CTLA-4 and contributes to the pathogenesis of Graves' disease. J Immunol 2000; 165:6606-11.
25. Van der Auwera BJ et al. CTLA-4 gene polymorphism confers susceptibility to insulin-dependent diabetes mellitus (IDDM) independently from age and from other genetic or immune disease markers. The Belgian Diabetes Registry. Clin Exp Immunol 1997; 110:98-103.
26. Suda T, Nagata S. Why do defects in the Fas-Fas ligand system cause autoimmunity? J Allergy Clin Immunol 1997; 100:S97-101.
27. Shultz LD, Rajan TV, Greiner DL. Severe defects in immunity and hematopoiesis caused by SHP-1 protein- tyrosine-phosphatase deficiency. Trends Biotechnol 1997; 15:302-7.
28. Nakamura A et al. Fcgamma receptor IIB-deficient mice develop Goodpasture's syndrome upon immunization with type IV collagen: A novel murine model for autoimmune glomerular basement membrane disease. J Exp Med 2000; 191:899-906.
29. Yuasa T et al. Deletion of fcgamma receptor IIB renders H-2(b) mice susceptible to collagen-induced arthritis. J Exp Med 1999; 189:187-94.
30. Heath WR, Kurts C, Caminschi I et al. CD30 prevents T-cell responses to nonlymphoid tissues. Immunol Rev 1999; 169:23-9.
31. Kurts C et al. Signalling through CD30 protects against autoimmune diabetes mediated by CD8 T cells. Nature 1999; 398:341-4.
32. O'Keefe TL, Williams GT, Batista FD et al. Deficiency in CD22, a B cell-specific inhibitory receptor, is sufficient to predispose to development of high affinity autoantibodies. J Exp Med 1999; 189:1307-13.
33. Kenyon NS et al. Long-term survival and function of intrahepatic islet allografts in rhesus monkeys treated with humanized anti-CD154. Proc Natl Acad Sci USA 1999; 96:8132-7.
34. Kirk AD, Blair PJ, Tadaki DK et al. The role of CD154 in organ transplant rejection and acceptance. Philos Trans R Soc Lond B Biol Sci 2001; 356:691-702.
35. Kenyon NS et al. Long-term survival and function of intrahepatic islet allografts in baboons treated with humanized anti-CD154. Diabetes 1999; 48:1473-81.
36. Yasunami R, Bach JF. Anti-suppressor effect of cyclophosphamide on the development of spontaneous diabetes in NOD mice. Eur J Immunol 1988; 18:481-4.
37. Hutchings P, Cooke A. Protection from insulin dependent diabetes mellitus afforded by insulin antigens in incomplete Freund's adjuvant depends on route of administration. J Autoimmun 1998; 11:127-30.
38. Kennedy KJ, Smith WS, Miller SD et al. Induction of antigen-specific tolerance for the treatment of ongoing, relapsing autoimmune encephalomyelitis: A comparison between oral and peripheral tolerance. J Immunol 1997; 159:1036-44.
39. Hafler DA et al. Oral administration of myelin induces antigen-specific TGF-beta 1 secreting T cells in patients with multiple sclerosis. Ann NY Acad Sci 1997; 835:120-31.
40. Weiner HL. Induction and mechanism of action of transforming growth factor-beta- secreting Th3 regulatory cells. Immunol Rev 2001; 182:207-14.
41. Roncarolo MG, Bacchetta R, Bordignon C et al. Type 1 T regulatory cells. Immunol Rev 2001; 182:68-79.
42. Reijonen H et al. Detection of GAD65-specific T-cells by major histocompatibility complex class II tetramers in type 1 diabetic patients and at-risk subjects. Diabetes 2002; 51:1375-82.
43. McRae BL, Vanderlugt CL, Dal Canto MC et al. Functional evidence for epitope spreading in the relapsing pathology of experimental autoimmune encephalomyelitis. J Exp Med 1995; 182:75-85.
44. Martin R, Bielekova B, Gran B et al. Lessons from studies of antigen-specific T cell responses in Multiple Sclerosis. J Neural Transm 2000; Suppl 60:361-73.
45. Bielekova B et al. Encephalitogenic potential of the myelin basic protein peptide (amino acids 83-99) in multiple sclerosis: Results of a phase II clinical trial with an altered peptide ligand. Nat Med 2000; 6:1167-75.
46. Kappos L et al. Induction of a nonencephalitogenic type 2 T helper-cell autoimmune response in multiple sclerosis after administration of an altered peptide ligand in a placebo-controlled, randomized phase II trial. The altered peptide ligand in relapsing MS study group. Nat Med 2000; 6:1176-82.
47. Genain CP, Zamvil SS. Specific immunotherapy: One size does not fit all. Nat Med 2000; 6:1098-100.
48. Bluestone JA. Costimulation and its role in organ transplantation. Clin Transplant 1996; 10:104-9.
49. Viglietta V, Kent SC, Orban T et al. GAD65-reactive T cells are activated in patients with autoimmune type 1a diabetes. J Clin Invest 2002; 109:895-903.
50. Blazevic V, Trubey CM, Shearer GM. Analysis of the costimulatory requirements for generating human virus- specific in vitro T helper and effector responses. J Clin Immunol 2001; 21:293-302.
51. Sporici RA, Perrin PJ. Costimulation of memory T-cells by ICOS: A potential therapeutic target for autoimmunity? Clin Immunol 2001; 100:263-9.
52. Sporici RA et al. ICOS ligand costimulation is required for T-cell encephalitogenicity. Clin Immunol 2001; 100:277-88.
53. Homann D et al. CD40L blockade prevents autoimmune diabetes by induction of bitypic NK/DC regulatory cells. Immunity 2002; 16:403-15.
54. Kinne RW, Brauer R, Stuhlmuller B et al. Macrophages in rheumatoid arthritis. Arthritis Res 2000; 2:189-202.
55. Bar-Or A, Oliveira EM, Anderson D et al. Molecular pathogenesis of multiple sclerosis. J Neuroimmunol 1999; 100:252-9.
56. Monney L et al. Th1-specific cell surface protein Tim-3 regulates macrophage activation and severity of an autoimmune disease. Nature 2002; 415:536-41.
57. Gorelik L, Flavell RA. Transforming growth factor-beta in T-cell biology. Nature Rev Immunol 2002; 2:46-53.
58. Goudy K et al. Adeno-associated virus vector-mediated IL-10 gene delivery prevents type 1 diabetes in NOD mice. Proc Natl Acad Sci USA 2001; 98:13913-8.
59. Zagury D, Burny A, Gallo RC. Toward a new generation of vaccines: The anti-cytokine therapeutic vaccines. Proc Natl Acad Sci USA 2001; 98:8024-9.
60. LaDuca JR, Gaspari AA. Targeting tumor necrosis factor alpha. New drugs used to modulate inflammatory diseases. Dermatol Clin 2001; 19:617-35.
61. Pacholczyk R, Kraj P, Ignatowicz L. Peptide specificity of thymic selection of CD4+CD25+ T cells. J Immunol 2002; 168:613-20.
62. Jordan MS et al. Thymic selection of CD4+CD25+ regulatory T cells induced by an agonist self-peptide. Nat Immunol 2001; 2:301-6.
63. Frasca L, Scotta C, Lombardi G et al. Human anergic CD4+ cells can act as suppressor cells by affecting autologous dendritic cell conditioning and survival. J Immunol 2002; 168:1060-8.
64. Moore KW, de Waal Malefyt R, Coffman RL et al. Interleukin-10 and the interleukin-10 receptor. Annu Rev Immunol 2001; 19:683-765.
65. Malisan F et al. Interleukin-10 induces immunoglobulin G isotype switch recombination in human CD40-activated naive B lymphocytes. J Exp Med 1996; 183:937-47.
66. Lee G, Ellingsworth LR, Gillis S et al. Beta transforming growth factors are potential regulators of B lymphopoiesis. J Exp Med 1987; 166:1290-9.
67. Huang LY, Reis e Sousa C, Itoh Y et al. IL-12 induction by a TH1-inducing adjuvant in vivo: Dendritic cell subsets and regulation by IL-10. J Immunol 2001; 167:1423-30.
68. Sonoda KH, Stein-Streilein J. CD1d on antigen-transporting APC and splenic marginal zone B cells promotes NKT cell-dependent tolerance. Eur J Immunol 2002; 32:848-57.
69. Ishizaka S, Kimoto M, Kanda S et al. Augmentation of natural killer cell activity in mice by oral administration of transforming growth factor-beta. Immunology 1998; 95:460-5.
70. Erdmann I. Fucosyltransferase VII-deficient mice with defective E-, P-, and L- selectin ligands show impaired CD4+ and CD8+ T cell migration into the skin, but normal extravasation into visceral organs. J Immunol 2002; 168:2139-46.
71. Butcher EC, Picker LJ. Lymphocyte homing and homeostasis. Science 1996; 272:60-6.

CHAPTER 17

Death Receptor-Mediated Apoptosis and Lymphocyte Homeostasis

Lixin Zheng, Richard M. Siegel, Jagan R. Muppidi, Felicita Hornung and Michael J. Lenardo

Introduction

To generate specific immunity, T and B lymphocyte clones may expand massively in response to antigenic stimulation. The expansion of antigen-specific lymphocytes allows the immune system to control pathogens and possibly neoplasia. However, in order to maintain homeostasis, a clonotypic contraction of the expanded lymphocytes is often necessary during and at the conclusion of an immune response. It has been suggested that every day, 1% of mature T cells are replenished from the thymus.[1] In addition, given the rapid doubling time of activated lymphocytes, a single naïve T cell clone can expand more than 125-fold during an immune response lasting 11 days.[2,3] The limited space of the lymphocyte compartment cannot afford unlimited expansion of activated lymphocytes. Furthermore, over-activation of lymphocytes may result in detrimental effects on host tissues and autoimmune disorders.[4] Therefore, in order to maintain an immune system with a diverse antigen reactivity repertoire and self-tolerance without exceeding the limited lymphoid compartment, a negative feedback control system involving lymphocyte apoptosis, termed propriocidal regulation, has evolved.[4,5]

Apoptosis was first named by Kerr et al in 1972 to describe the morphological features of cell death with the characteristics of shrinkage and condensation of the nucleus and membrane blebbing.[6] The potential role of antigen-induced apoptosis in regulating peripheral lymphocyte populations was not generally realized until the early 1990s.[5] Nevertheless, significant advancements have been made during the last ten years in understanding the biological and molecular mechanisms of lymphocyte homeostasis. We now know that an intact mechanism of programmed death is necessary for a healthy immune system, and for the avoidance of systemic autoimmunity.[4,7]

Apoptosis of T and B cells can occur in central and peripheral lymphoid organs.[8,9] During T cell development, thymic selection controls the output of mature lymphocytes into the periphery through two major mechanisms: (1) Antigen receptor-induced apoptosis that is due to high avidity interactions between a self peptide/MHC complex and the T-cell receptor (TCR);[10,11] and (2) death by neglect of immature thymocytes that fail to encounter weak antigenic stimulation in the context of self-MHC and die off before final maturation.[12,13] Unfortunately, however, little is known about the molecular basis of the central lymphocyte apoptosis other than its clear dependence on the avidity of antigen/receptor interactions.[11,14]

In the peripheral immune system, homeostasis of mature T cells is maintained through two major mechanisms that are superficially similar to those mentioned above.[4,8] However, the molecular mechanisms of peripheral mature lymphocyte apoptosis are quite different from those in thymocytes. Cytokine withdrawal apoptosis, which occurs after the clearance of pathogens during an immune response, is dependent on a mitochondrial pathway that can be blocked by overexpression of anti-apoptotic proteins in the Bcl-2 family (see other reviews for details[4,15]). By contrast, antigen-induced cell death, which is dependent on CD95L/CD95, tumor necrosis factor (TNF)/TNF receptor (TNFR), and other death ligand/receptor interactions, acts to eradicate repeatedly stimulated T cells and prevents over-expansion of clonotypic T cells in the face of high or recurrent antigen exposure.[4,9] Interestingly, T cell growth lymphokines such as IL-2, IL-4 and IL-15 not only promote proliferation, but also predispose mature T cells to antigen-induced death, and moreover, these cytokines render T cells susceptible to death by withdrawal of those lymphokines. Mature B cells must also be activated before becoming sensitive to the death receptor-mediated apoptosis. However, unlike T cells that commit suicide, substantial evidence suggests that B cells are victims of 'murder' by CD95L-bearing T cells, because B cells do not produce CD95L themselves.[9,16]

In this review, we will discuss thymocyte and mature T cell apoptosis primarily focusing on recent developments in our understanding of mature T cell apoptosis in the healthy and diseased immune system.

Central Selections in Thymus

Since Miller first discovered that T lymphocytes develop in the thymus 40 years ago,[17] thymic selection has always been an exciting area of research in immunology. Normal thymic selection generates a mature T cell repertoire that is able to functionally distinguish "self" and "non-self" during immune responses. This "central tolerance" created by thymic selection allows the mature T cells to mount responses to stimuli that disturb the homeostatic status of the immune system (commonly by pathogen invasion), while avoiding over-reactions to self-components that would otherwise cause autoimmunity (see reviews, refs. 17, 18).

During thymocyte development, rearranged TCR chains are co-expressed with both CD4 and CD8 molecules in the same cell at the immature (the double positive, DP) stage. Starting from this stage, the fate of developing thymocytes is dependent on the avidity of clonotypic occupancy of TCR by self-peptide/MHC complexes expressed mainly in thymus epithelial cells (TEC) and dendritic cells (DC).[14,19] The DP-thymocytes with productively rearranged TCRs have the potential to develop. However, they will undergo apoptosis if they do not encounter self-peptide/MHC stimulation within thymus. These neglected thymocytes, in addition to others that lack productively rearranged TCRs and are thus unable to recognize self-peptide/MHC complexes, will be eliminated. The majority of useless thymocytes are destroyed through this process termed "death by neglect".[20] By contrast, optimal or low avidity TCR occupancy can initiate survival signals in DP and single positive (SP) thymocytes, thereby allowing these cells to overcome the default death pathway. These survival signals selectively drive the DP thymocytes to a mature phenotype, causing them to become either CD4 or CD8 SP lymphocytes, which can then enter into the periphery. Thymocyte maturation due to low-avidity self-peptide/MHC/TCR interactions is recognized as positive selection and it creates the peripheral T cell repertoire.[18,21-26]

Because of positive selection, all peripheral T cells are "educated" to recognize antigen in the context of self-MHC. To avoid generating strongly self-reactive T cells that can cause autoimmune disorders, the thymus specifically eliminates self-reactive T cells by negative selection. In this instance, clonotypic thymocyte apoptosis is triggered whenever the TCRs of developing thymocytes encounter self-peptide/MHC complexes at high avidity.[14,19] It is believed that self-tolerance of the immune system is largely dependent on thymic deletion of strong self-reactive T cells at an early stage of T cell development, a process called "central tolerance".[18]

It was generally expected that death receptor signals might be involved in TCR-mediated thymocyte apoptosis. There is some evidence indicating that strong stimulation of TCR can initiate CD95L/CD95-dependent thymocyte apoptosis in particular experimental settings in vitro[27,28] and in vivo.[29] Moreover, other data suggest that some members of the TNFR superfamily that do not belong to the death receptor subfamily may have effects on thymocyte deletion under special circumstances. For example, it was observed that in CD30 and CD40 ligand-deficient mice, negative selection of some subpopulations of thymocytes is partially impaired.[30,31] However, these data can be alternatively explained as indirect effects on the regulation of thymic APCs rather than as the delivery of death signals themselves.[31] Taking into account that negative selection of thymocytes is generally normal in gld (CD95L deficient), lpr (CD95 deficient), and TNF receptor knockout animals,[32-34] it can be concluded that these death receptors are not essential for the thymic negative selection.

Even though death-receptor signals seem to be unnecessary, the typical apoptotic appearance of the thymocytes undergoing TCR-induced death leads to the assumption that caspases (a family of cysteine aspartase proteases critical to apoptosis), a group of cysteine proteases that are involved in the execution of apoptotic cells, may play roles in this process. It has also been shown that TCR engagement can activate caspases in DP thymocytes.[35] In addition, the caspase-8/10 specific inhibitor IETD and the pan-caspase-inhibitors zVAD and DEVD can efficiently protect DP thymocytes from TCR-induced apoptosis, while the caspase-1-specific inhibitor YVAD cannot.[36] Moreover, thymocytes in mice that are deficient in caspase-9, which belongs to the caspase-8 homologous subfamily, were resistant to cell death induced by anti-CD3, dexamethasone or γ-irradiation.[37,38] None-theless, mice deficient in caspase-1 or caspase-3 are susceptible to TCR-induced apoptosis and their thymocyte development appears normal.[37,39] Despite our failure to understand how TCR stimulation can turn caspases on without requiring death-receptor signals, these data indicate that the activation of caspase-8 but not caspase-1 subfamilies can participate in the execution of the thymocyte apoptosis.

Bcl-2 (B cell leukemia/lymphoma-2 gene) family proteins, subdivided into pro- and anti-apoptotic groups that are involved in the regulation of various forms of physiological cell death through mitochondrial pathways, seem to also regulate thymocyte selection. It has been concluded that Bcl-2 proteins are upregulated during the DP to SP transition and are expressed at high levels in mature thymocytes.[40] It has also been shown that TCR stimulation induces Bcl-2 expression in DP thymocytes implying that Bcl-2 takes part in positive selection.[41] The role of Bcl-2 family in thymic selection became controversial when examining thymic selection in the Bcl-2$^{-/-}$ and Bcl-XL$^{-/-}$ animals. Bcl-2 and Bcl-XL are both dispensable for thymocyte development, although Bcl-XL seems to be crucial for extending the life span of DP thymocytes which promotes positive selection.[42-45] We would have a clearer answer on this issue if Bcl-2/Bcl-XL double knockout animals were available, unfortunately, in both cases the single knockout is perinatally lethal. Nevertheless, bcl-2 transgenic animals have provided tools to address the question from a different angle. Over-expression of anti-apoptotic Bcl-2 in transgenic mice protects thymocytes from apoptosis induced by multiple apoptotic stimuli especially by anti-CD3.[46-48] Also, there is data suggesting that TCR-stimulation can also activate the expression of pro-apoptotic Bcl-2 homology 3 (BH3)-only molecules like Bcl-2 homologue antagonist/killer (Bak), Bcl-2 inhibitor death agonist (Bid), and Bcl-2 interacting mediator (Bim) during thymocyte differentiation.[49] Recently, Strasser and colleagues have implicated the Bcl-2 family member Bim as a critical participant in negative selection (personal communication). It remains unclear how the anti-apoptotic and pro-apoptotic Bcl-2 family proteins are properly balanced by self-peptide/MHC/TCR interactions during thymic selection. The thymic microenvironments and the strength of TCR occupancy probably together convey the signals for life or death.

There are other molecules that have been suggested to have roles in thymic selection. Nur77 and Nor-1, are homologous orphan nuclear steroid receptors that have been shown to promote negative selection in a redundant fashion.[50,51] The retinoic acid-related orphan receptor (RORγ) seems to provide survival signals during thymic selection.[52] Additional molecules like thymocyte glucocorticoid receptors (GR),[53] cyclin-dependent kinase 2 (Cdk2),[54] and helix-loop-helix proteins like enhancer binding (E) proteins and inhibitor of DNA binding (Id) proteins[55,56] are also believed to play roles in thymic selection and thymocyte differentiation. How all these molecules act in concert has yet to be determined.

Overall, it is apparent that only a small fraction of developing thymocytes can struggle through the narrow bridge of TCR avidity imposed by thymic selection and become mature T cells. The majority of thymocytes undergo apoptosis either by neglect or negative selection. Many other pro- and anti-apoptotic molecules, as well as some cell cycle control molecules seem to make a joint decision for thymocytes to live or to die. Nevertheless, the molecular basis for thymocyte selection, especially for negative selection, remains unclear.

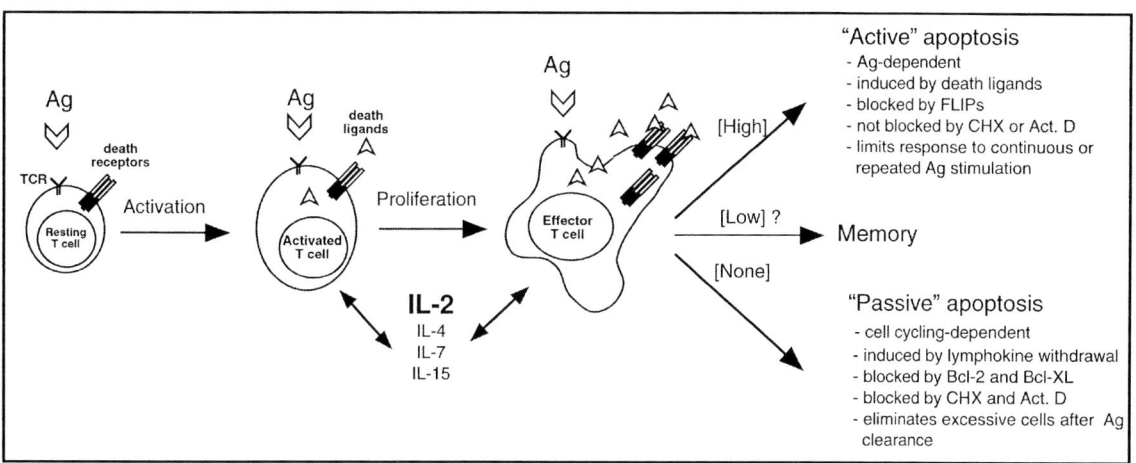

Figure 1. Propriocidal regulation of mature T lymphocytes. The T-cell response to antigen (Ag) at different phases leads to distinct consequences: Primary engagement of T cell receptor (TCR) by Ag leads to T cell activation, interleukin-2 (IL-2) and other lymphokine production and T-cell proliferation. Cycling T cells become highly susceptible to both active and passive apoptosis. Active apoptosis is triggered by Ag re-engagement of TCR and is mediated by interactions of death ligands/receptors such as CD95L/CD95 and tumor necrosis factor (TNF)/ tumor necrosis factor receptor (TNFR). Passive apoptosis is induced by deprivation of T cell growth cytokines when Ag is cleared at the end of an immune response. A small number of T cells escape both death pathways and are believed to become the "memory" T cells. Abbreviations: Act.D= actinomycin D; CHX= cyclohexamide; FLIP= Fas-associated death domain-like interleukin-1 beta-converting enzyme-inhibitory protein.

The Paradox of Cell Growth Cytokines: T Cell Proliferation Versus Apoptosis: Propriocidal Regulation

To maintain homeostasis, mature T cells undergo clonal expansions and contractions during immune responses. In response to TCR stimulation with the help of co-stimulatory signals, naïve T cells are activated and then rapidly up-regulate the expression of interleukin-2 (IL-2), IL-2 receptor, CD95 and TNFR genes.[57] These early-activated T cells are resistant to CD95-mediated apoptosis and this resistance is biologically important because it assures proliferative responses to antigen.[4,58] In contrast, the late activated and cycling T cells will undergo TCR-mediated apoptosis upon subsequent TCR re-stimulation (Fig. 1).[4] This is a paradox. From the standpoint of immunity, proliferation of the antigen-specific lymphocytes helps to eradicate invading pathogens. Therefore, it seems counterintuitive that repeated exposure to antigen should provoke T cell apoptosis. However, over-expansion of any particular clone of T cells can lead to collateral damage, as it has been shown that genetic defects leading to lymphocyte hyperactivation can result in autoimmune disorders.[59] As a consequence, negative feedback control mechanisms have evolved and are termed "propriocidal regulation". These suicide programs are achieved either by antigen-induced apoptosis, which prevents functional lymphocytes from over-expansion; or cytokine withdrawal apoptosis, which regulates the lifetime of activated effector-T cells that are potentially dangerous for self.

Antigen-induced apoptosis relies on ligand occupancy of cell surface death receptors that include CD95, TNFR1, and other members of the TNFR superfamily.[4,60-65] CD28 co-stimulatory signals are not essential.[66] There are two major characteristics of antigen-induced T-cell apoptosis: clonal specificity and the requirement of T-cell growth cytokines. The role of IL-2 in predisposing T cells towards antigen-induced death first came to light from the observation that pre-treatment of primed T cells with IL-2 led to significant apoptosis of cycling T cells following TCR re-engagement.[4,5] This unexpected result brought about a controversial new concept regarding IL-2, which had long been viewed as the major T cell-growth cytokine, has a major role in programming T-cell death. Morphological and viability studies revealed that this IL-2-driven, TCR-induced T cell death was apoptosis.[5,67] In the propriocidal model, IL-2 is considered as a key element for negative "feed-back" control. If there is repeated antigenic stimulation, the activated T cells will undergo apoptosis to avoid over-expansion. Considering these observations, T cell growth cytokines can be viewed as double-edged swords, which allow responses of mature T cells to be balanced between proliferation and apoptosis. The phenotype of the IL-2 and IL-2Rα deficient mice strongly supports this hypothesis, because severe polyclonal lymphocyte proliferation is observed in these animals. Evidence suggests that T cell expansion is uncontrolled in absence of this predisposition towards TCR-induced cell death. Thus, IL-2 is not just necessary for T cell proliferation, it is also required for controlling T cell homeostasis by apoptosis.[68-71]

One interesting unsettled question is whether IL-2 is the sole cytokine predisposing T cells to apoptosis. There are reports showing that IL-4 and IL-7, functioning as alternate T cell growth/survival cytokines, fail to induce susceptibility of purified CD4+ T cells to antigen-induced apoptosis.[72,73] However, other data argues that IL-4, IL-7, and IL-15 can all substitute to some degree for IL-2 in predisposing activated T cells to antigen-induced death.[57,74-76] In addition, it has been noticed that in polarized culture systems, IL-4-driven Th2 cells are also susceptible to antigen-induced death,[77] although these Th2 cells may not be as sensitive as the IL-2-driven Th1 cells.[78] To address this argument, we tested the roles of IL-4, IL-7, and IL-15 in activated T cells from the IL-2 receptor α-deficient mice. In this system, where IL-2/IL-2R signaling is absent, we found that IL-4, IL-7, and IL-15 all can predispose activated T cells to antigen-induced apoptosis, provided that the doses of these T-cell growth cytokines are high enough to render T-cell proliferation at a comparable level to IL-2. We also found that a CD95:Fc fusion protein, which blocks CD95-mediated apoptosis, could effectively block cell death in these circumstances.[57] Thus, multiple T-cell growth

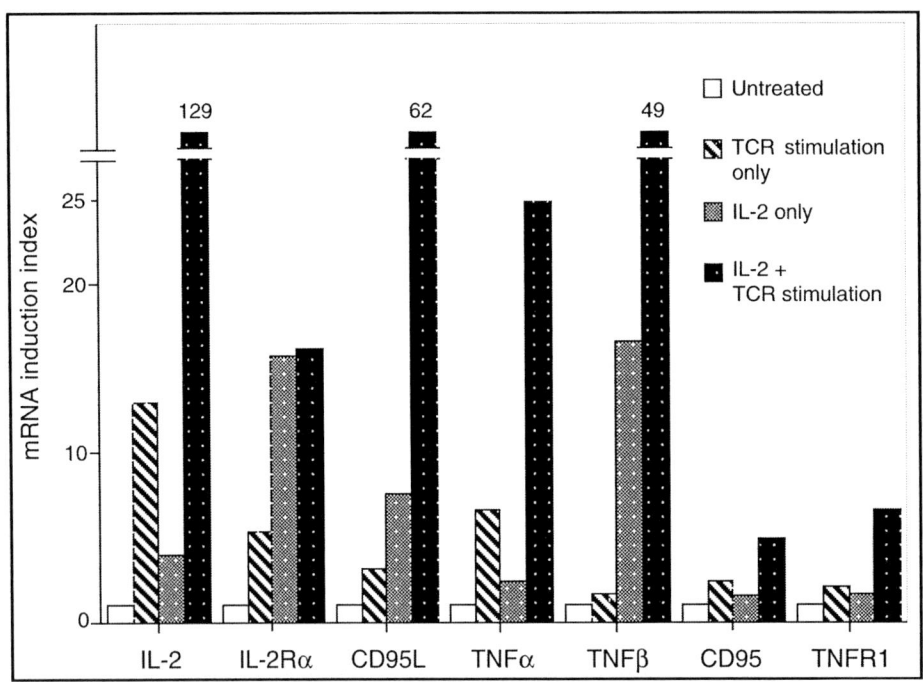

Figure 2. IL-2 up-regulates TCR-induced CD95 ligand and TNF expression. B6 mouse lymphocytes were subjected to anti-CD3 stimulation in absence or presence of IL-2. For in the absence of IL-2, T cells were stimulated with medium (unstimulated), or 5 μg/ml of anti-CD3ε for 4 hours. For in the presence of IL-2, T cells were first stimulated with anti-CD3 for 2 days followed by culturing with 50 IU/ml of IL-2 for another 2 days, and then were stimulated with medium (IL-2 only) or 5μg/ml anti-CD3 (IL-2 + TCR stimulation) for 4 hours. The collected samples were analyzed for gene expression as indicated by RT-PCR. The semi-quantitative mRNA-induction index is shown. The data has been normalized by beta-actin mRNA expression for each sample and represents three independent experiments. Abbreviations: IL-2= interleukin-1; TNF= tumor necrosis factor.

cytokines use similar signaling pathways to program T cells for proliferation as well as apoptosis. Given that continuously cycling tumor T-cell lines are also sensitive to antigen-induced death,[79] we propose that it is the ability of a cytokine to promote T cell cycling, not any intrinsic property of signaling by the particular cytokines, that determines the susceptibility of T cells to antigen-induced apoptosis.

Regulation of the Susceptibility of Lymphocytes to Death Ligand-Induced Apoptosis

It is intriguing that stimulation of the T-cell antigen receptor causes proliferation of naïve and early activated T cells but apoptosis of late activated and cycling T cells. How this process is regulated remains unclear. There is no requirement for new protein synthesis during antigen-induced cell death since protein-synthesis inhibitors cannot block this process. It is well known that the death receptors, as will be discussed later, are required to deliver death signals during TCR-induced apoptosis. So, it is apparent that the modulation of death-receptor signaling molecules needs to be investigated during TCR re-stimulation. Peter and his colleagues reported that CD95, Fas-associated death domain (FADD), and FLICE (FADD-like IL-1β-converting enzyme, or caspase 8), the three crucial components of death inducing signal complexes, are constitutively expressed through day 1 and day 6 of activated T cells. However, the activated T cells only become sensitive to CD95-mediated apoptosis after day 3 of TCR stimulation.[80,81] This suggests the existence of a regulatory mechanism that controls the conversion from resistant to ready-to-die status during T-cell activation. Such regulation may rely on either the removal of anti-apoptotic molecules or the promotion of the recruitment of pro-apoptotic components to the CD95 signaling complex.[82]

To investigate the molecular basis of the susceptibility to TCR-induced apoptosis, we examined the expression patterns of some candidate molecules involved in this death pathway (Fig. 2). We have observed that primary stimulation of naïve T cells causes less than a 6-fold induction of mRNA for CD95L, CD95, TNF, and TNFRs. In contrast, re-engagement of TCR on cycling T cells that have been treated with IL-2 caused a 25 to 60-fold superinduction of the mRNA encoding these death ligands. This induction of death ligand expression correlates well with the susceptibility of T cells to antigen-induced apoptosis.[57] There is also evidence at the gene transcriptional level showing that the CD95L gene is upregulated during IL-2 treatment, which confirms our earlier observation.[83] Other evidence also suggests that when activated T cells are driven into proliferation by IL-2, the FLICE like inhibitor protein (cFLIP) is down-modulated.[84,85] Reducing the expression of this anti-apoptotic protein may convert T cells from resistant to sensitive for antigen-induced apoptosis. So it seems that the down-modulation of apoptotic inhibitors and super-induction of death ligand expression may explain why only activated and proliferating but not naïve resting T cells are susceptible to antigen-induced apoptosis.

If the down-modulation of apoptotic inhibitors by T cell-growth cytokines and super-induction of CD95L, TNF, and other death ligands by TCR-restimulation account for all antigen-induced apoptosis, one would expect that this process should have no clonal specificity. In other words, the death-ligand expression induced by a specific TCR-engagement might be expected to kill all by-stander T cells that are surface CD95⁺. This

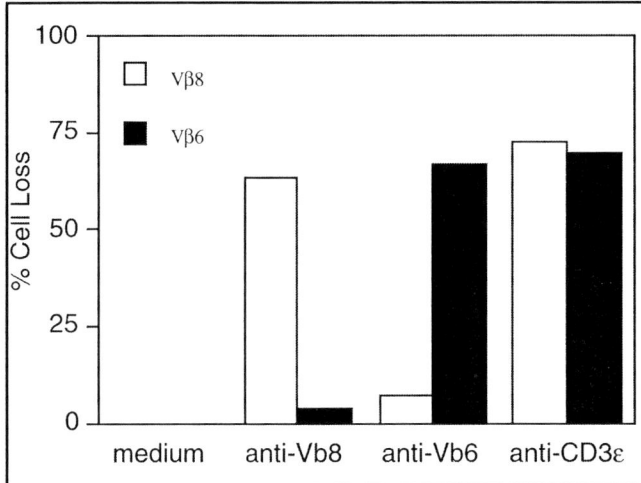

Figure 3. Clonotypic specificity of the TCR-induced T-cell death. T cells isolated for BALB/c mice were stimulated separately with anti-Vβ8- and Vβ6 antibodies for 2 days and cultured with 50 IU/ml IL-2 for 2 days. These Vβ8- and Vβ6-primed T cells were then separately labeled with different membrane dyes and the labeled cells were mixed at 1:1 ratio. The 24-hr TCR stimulation was conducted by using avidin (100ng/ml) to cross-link the biotinylated anti-Vβ8, -Vβ6, or -CD3ε Abs (200ng/ml) as indicated. The viable cells were detected by FACS analysis, and the percent cell loss was calculated. The open and solid bars represent Vβ8 and Vβ6 populations respectively. The data represents three independent experiments.

is definitely not the case. To address this issue, we performed a T-cell mixing experiment.[86] Activated Vβ6+ and Vβ8+ CD4 T cells were purified and treated with IL-2 to predispose these cells to TCR-mediated apoptosis. We found that when a mixture of Vβ6+ and Vβ8+ CD4 T cells were re-stimulated with anti-Vβ6 or anti-Vβ8 antibodies, T cell apoptosis was induced in a strict clonotypic fashion. Death only occurred in the restimulated population, not bystanders, even though both populations of T cells proliferated equally in IL-2 (Fig. 3). Further study suggests that in addition to its induction of death ligand expression, TCR stimulation may provide a death 'competency signal' that synergistically induces the susceptibility of T cells to CD95-induced apoptosis.[87] However, the molecular mechanism of this signaling pathway is still unclear.

Lymphokine Withdrawal Apoptosis

Specific contraction of expanded T cells is a general feature of immune responses when antigen clearance is accomplished. As mentioned above, there might be no further antigenic stimulation available when an immune response accomplishes its goal by clearing a pathogen. In this instance, the activated lymphocytes would have no chance to commit suicide through TCR-induced apoptosis. This hypothesis raises the question of whether TCR-induced apoptosis is the only negative feedback mechanism that prevents over-expansion of immunoreactive T cells.[80] One interesting experiment has shed light on this question. Kuroda et al put mini-osmotic pumps loaded with IL-2 into mice and challenged the animals with the superantigen SEB (staphylococcal enterotoxin B). In the control group, SEB challenge caused a significant loss of SEB-reactive Vβ8 T cells after an early expansion of the same Vβ8 population. However, in presence of osmotic pumps that release IL-2 gradually into the circulation, these SEB-activated T cells remained alive for a prolonged period of time.[88] This implies that at the late phase of an immune response, when the pathogen has been cleared, antigen-induced apoptosis will not work since there is no antigen available to restimulate the TCR. Instead, cytokine withdrawal apoptosis plays a major role in restoring the T cell homeostasis.

Lymphokine withdrawal apoptosis depends on newly synthesized proteins because it has been shown that actinomycin D and cycloheximide, inhibitors of DNA transcription and protein synthesis, respectively, can block this process.[89] The molecular mechanisms of lymphokine withdrawal apoptosis are not as clear as for antigen-induced death. Experiments with mice deficient in Fas and other death receptors have not shown any role of these molecules in lymphokine withdrawal apoptosis. However, the proto-oncogene Bcl-2 family is crucially involved in regulating lymphokine deprivation death. Bcl-2 and Bcl-XL are the major anti-apoptotic members of the family. They antagonize BAX/Bcl-Xs/BAD and other pro-apoptotic proteins of the family by stabilizing mitochondrial membrane potential and membrane integrity. This prevents cytochrome C release and activation of the caspase-9 cascade that leads to apoptosis.[90-92]

The Death Receptor Subfamily of TNF Receptors

The TNF/TNFR superfamilies are large groups of proteins that act as ligand/receptor-interacting molecules. It is well known that most of the ligand members of this family are type-II transmembrane proteins that are expressed as trimers on the cell surface or shed from the cell surface as cytokines. Many of the receptor members are type-I transmembrane proteins with 2-6 extracellular cysteine-rich domains and cytoplasmic signaling tails. TNFR family members have pleiotropic functions depending on cell type and status, as well as the availability of other signals. A subfamily of TNFR superfamily has been classified as "death receptors" based on their sharing of a homologous cytoplasmic "death domain" (DD). There are at least 8 pairs of ligands/receptors in this subfamily. The receptor members of this subfamily include CD95 (Fas), TNFR1, death receptor 3 (DR3),[93] DR4 (TRAIL receptor1, TR1),[94] DR5 (TRAIL receptor 2,TR2),[64] DR6,[65] NGFRp75 and CAR1,[95] each containing a homologous cytoplasmic death domain that is essential for apoptosis induction.[58,96-98] These DD-containing TNFR family members are thus often referred to as death receptors and their corresponding ligands are recognized as death ligands, although it is uncertain if death is the primary function of all these molecules. Some characteristics of death receptors and their associated proteins are summarized in Table 1 and Table 2. It was believed that death-ligand trimers could recruit and trimerize individual death receptors triggering apoptosis.[99] However, recent data suggests that CD95, TNFR1, and TNFR2 pre-assemble as dimers or trimers through their extracellular "pre-ligand assembling domain" (PLAD) before ligand occupancy. The pre-assembled CD95 and TNFR oligomers appear to be essential for CD95L/CD95 and TNF/TNFR interactions as well as signaling for apoptosis.[100,101] These findings support the hypothesis that pre-ligand assembly might be a general feature of the TNFR superfamily. They also shed light on strategies for developing new drugs that block the interactions between death ligands and their receptors to modulate programmed cell death as well as other TNFR-signaling-related disorders.

Table 1. Death receptors and their signaling pathways

Death Receptors	MW (kDa)	Chromosomal Location (Human)	Ligands	Interacting Molecules	Comments	References
CD95 (Fas, DR1, APO-1)	45–50	10q23	CD95L	FADD, Caspase-8	Most consistently pro-apoptotic	Trauth, 1989; Oehm, 1992; Itoh, 1991
TNFR1 (DR2, CD120a)	55–60	12q13	TNFα, LTα	TRADD, RIP (Indirect) FADD(Indirect) TRAF2 (Indirect)	Cytotoxicity can be modulated by NF-κB	Schall, 1990; Loetscher, 1990
DR3 (TRAMP, WSL, APO2, LARD)	45	12q13	TWEAK (APO-3)	TRADD, FADD (indirect), TNFR1 Caspase-10	Expression restricted to lymphocytes. Induced after DNA damage via p53-dependent mechanism	Screaton, 1997; Chinnaiyan, 1996; Kitson, 1996
TRAIL-R1 (TR1, DR4)	50	8p21	TRAIL	TRADD, FADD	May bind FADD but no requirement for FADD in DR4-induced death found in FADD knockout mice.	Schneider, 1997; Schneider, 1997; Yeh, 1998
TRAIL-R2 (TR2, DR5)	45	8p21	TRAIL	TRADD, FADD	Can heterodimerize with DR4, induced by DNA damage	Wu, 1997; Pan, 1997; Walczak, 1997
TRAIL-R3 (TR3, DcR1, TRID)	27	8p21	TRAIL	No intracellular domain	GPI-linked membrane receptor; Blocks TRAIL-induced apoptosis	Pan, 1997; Sheridan, 1997; Degli-Esposti, 1997
TRAIL-R4 (TR4, DcR2)	42	8p21	TRAIL	?	Incomplete Death Domain; Blocks TRAIL-induced apoptosis	Degli-Esposti, 1997; Marsters, 1997
DR6	72 (estimated)	6p21	?	TRADD	Induce apoptosis and activation of both NF-κB and JNK	Pan et al, 1998

Abbreviations: Caspase= cysteine proteases involved in apoptosis; DR= death receptor (i.e., DR1, DR2, etc.); FADD= Fas-associated death domain; RIP= receptor interacting protein; TNFR1= tumor necrosis factor receptor 1; TRADD= tumor necrosis factor receptor-associated death domain; TRAF2= tumor necrosis factor receptor-associated factor 2; TRAIL= tumor necrosis factor-related apoptosis-inducing ligand.

Table 2. Death receptor associated proteins

Molecules	MW (kDa)	Domain of Death Receptor	Downstream Partner	Effects of Overexpression	Phenotype of Gene Knock-out	References
FADD (Mort-1)	27	- CD95 via Death Domain - TNF-R1, DR3, TR1, and TR2 via TRADD Death domain	Caspase-8 and -10 via DED	- Death domain protects from CD95-mediated apoptosis - Death Effector Domain induces apoptosis without CD95 crosslinking	- Impaired CD95-mediated apoptosis - Block in T cell mitogenesis	Boldin, 1995; Chinnaiyan, 1995
RIP	74	- TNF-R1 and CD95 via Death Domain - TRADD	- TRAF-2 via intermediate domain - RAIDD via Death Domain	Induces cell death via death domain	Impaired NF-κB activation, hypersensitivity to TNF-induced cell death	Stanger, 1995; Ting, 1996; Kelliher, 1998; Holler, 2000
TRADD	34	- TNF-R1 via Death Domain	TRAF-2, RIP, FADD	Causes cell death probably through recruitment of FADD		Hsu, 1995
RAIDD (CRADD)	23	- RIP via Death Domain	ICH-1 via CARD domain	Induces apoptosis		Duan, 1997; Ahmad, 1997
SODD		- TNF-R1 via Death Domain		Block TNF-R1 signaling by preventing TRADD recruitment		Jiang, 1999
Sentrin (SUMO, GMP1, UBL1, PIC1)	10	- CD95 or TNF-R1 death domains	N/A	Inhibits CD95-induced apoptosis		Okura et al, 1996
UBC9 (UBC-FAP)	20	CD95 intracellular domain	Sentrin, long isoform prodomains of caspase -8 and -10	No effect on CD95 signaling	Lethal in yeast	Wright et al, 1996; Zheng et al, unpublished data
FIST/HIPK3	130	CD95 death domain	FADD	Phosphorylates FADD, activates JNK	No effect on CD95-induced death	Rochat-Steiner, 2000

Abbreviations: FADD= Fas-associated death domain; FIST/HIPK3= Fas-interating serine/threonine kinase/homeodomain-interacting protein kinase 3; ICH-1 (also known as caspase-2)= Ice and Ced-3 homolog-1; JNK= c-Jun NH2-terminal kinase; N/A= not available; RAIDD= RIP-associated ICH-1 homologous protein with a death domain; RIP= receptor interacting protein; Sentrin (SUMO)= a ubiquitin like protein; SODD= silencer of death domains; SUMO= small ubiquitin-related modifier protein; TNF= tumor necrosis factor; TRADD= tumor necrosis factor receptor- associated death domain; TRAF-2= tumor necrosis factor receptor- associated factor 2; UBC= ubiquitin conjugating; UBC-FAP= ubiquitin conjugating Fas-associated protein.

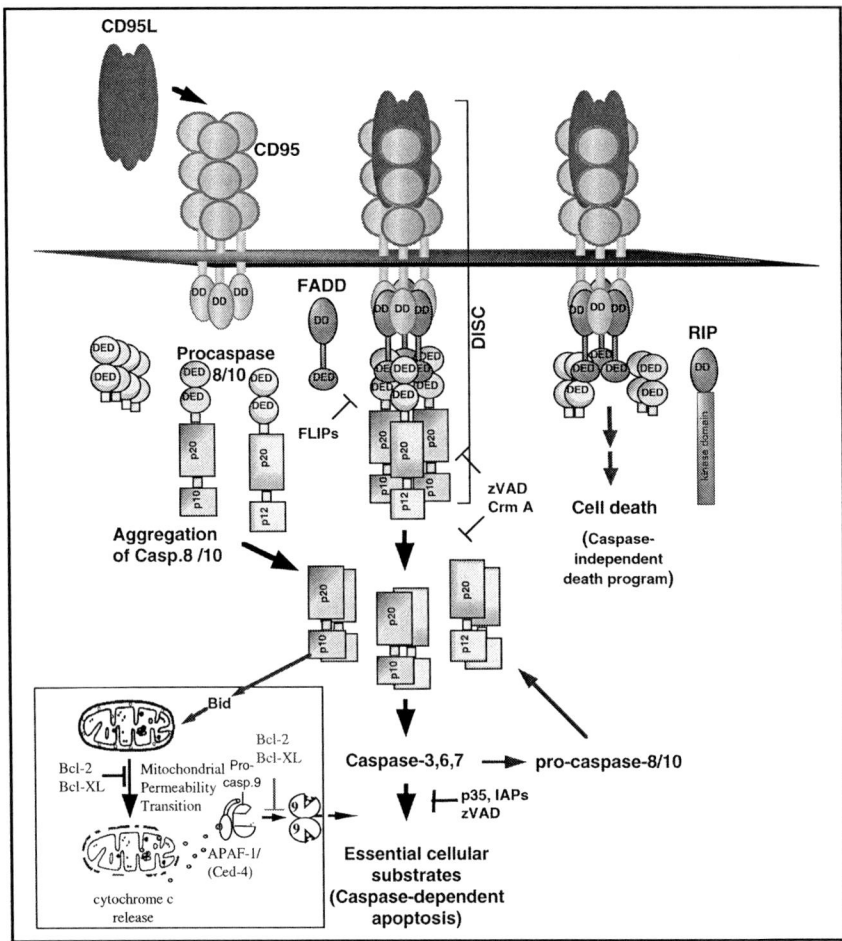

Figure 4. Essential steps of CD95-mediated cell death. The pre-ligand-association of CD95 is required for CD95 ligand binding. The engagement of pre-oligomerized CD95 by its natural ligand or cross-linking antibodies causes local aggregation of CD95, which then recruits adaptor molecule FADD and procaspase-8/10 to form the DISC. An auto-activation process is initiated by the DISC, leads to release of the active caspase-tetramer and the DED unit. The execution of committed cells by the activated caspase unit is achieved by either indirectly targeting the downstream caspase-cascade or directly striking at target molecules involved in apoptosis. Apoptosis inhibitory proteins like FLIP, IAP, CrmA and p35 interfere apoptosis signals at different steps as indicated. The pan-caspase inhibitor zVAD inhibits caspase activity at the early and final execution phase of apoptosis. A caspase-independent pathway of CD95-mediated cell death is also postulated. Abbreviations: APAF= apoptosis activating factor; Crm-A= viral caspase inhibitor; DD= death domain; DED= death effector domain; DISC= death-inducing signal complex; FADD= Fas-associated death domain; FLIP= Fas-associated death domain-like interleukin-1 beta-converting enzyme-inhibitory protein; FLIPs= FLICE-like inhibitory proteins; IAPs= inhibitor of apoptosis; p35= baculoviral caspase inhibitor; RIP= receptor interacting protein.

Death Receptor and Their Signals

Members of the TNF receptor family that contain a death domain (DD) in their cytoplasmic tail are termed death receptors. Apoptosis induced through death receptors is believed to be triggered by formation of the death inducing signaling complex (DISC)[102] (Fig. 4). The components of the DISC vary because a number of different receptors, adaptor proteins and downstream signaling molecules can participate in this signaling complex. In the case of CD95-induced apoptosis, there are at least 3 core components that come together to form the DISC;[102] they are CD95, an adapter molecule FADD/Mort-1, and caspase-8 and/or -10. CD95/FAS/APO-1 is a type-I membrane protein with a cytoplasmic tail that harbors a death domain (DD).[102] FADD/Mort-1 is a cytoplasmic adapter protein that has a C-terminal DD and an N-terminal death effector domain (DED).[103] Under appropriate conditions, cross-linking surface CD95 for example, FADD is recruited to the cytoplasmic tail of CD95 through homotypic associations between the DDs of CD95 and FADD. Caspase-8/FLICE and caspase-10/Mch-4 are two cysteine aspartate proteases (caspases). Caspase-8 and -10 are each composed of two DED domains at their N-terminus and one caspase domain with two subunits each at their C-terminus. Again, through homophilic interactions between the homologous DEDs of FADD and caspase-8/-10, these caspases are recruited into the DISC which triggers autoactivation of the enzymes[104-106] (Fig. 4). The aggregation of caspase-8 and -10 appears to be sufficient to initiate enzymatic auto-processing and activation of the enzymes. Our group and others have demonstrated that dimerization of the enzymatic domains of caspase 8 is sufficient to activate the pro-enzyme and induce apoptosis in transfected cells.[107,108]

Compared to resting T cells, activated T cells respond with much stronger up-regulation of CD95L, TNF and other death ligands as well as their receptors upon TCR re-engagement.[57] This superinduction of death ligands and death receptors may be part of the reason why only activated and cycling T cells are sensitive to TCR-mediated apoptosis. However, kinetic analysis of DISC components on the day 1- and day 6-activated T cells reveals that there is no significant difference in the level of cellular CD95, FADD, or caspase-8 expression. Moreover, cross-linking of surface CD95 can initiate DISC formation only between days 3 and 6 of the late-activated T cells but not in resting or in day-1-early activated T cells.[80] This discrepancy of DISC formation between CD95-competent (to die) and -incompetent T cells suggests the possible existence of a mechanism that controls DISC formation and caspase activation. Through this biochemical regulation, immune homeostasis is maintained by assuring appropriate immune responses before the negative feedback death of immunoresponsive T cells occurs.

DR3, DR4/TR1, and DR5/TR2 seem to utilize pathways similar to CD95 signaling. They deliver apoptotic signals through FADD/caspase 8 or 10.[93,109,110] It is not clear why there are multiple systems that appear functionally redundant. Whether this redundancy accounts for tissue specificity or has some other biological significance remains unclear.

TNF/TNFR interactions trigger lymphocyte apoptosis or necrosis through pathways that share some components with CD95 signaling.[4] There are two receptors for TNFα, referred to as TNFR1 and TNFR2. TNFR1 contains a cytoplasmic DD tail and can signal for T cell death, especially for some subpopulations of CD8 cells.[63] TNFR2 does not contain a DD itself, however, it can synergistically enhance death signals from other death receptors.[111-113] There are several molecules shown to be capable of interacting with the DD of TNFR1. The TNFR-associated death domain (TRADD) protein is believed to be an adapter for TNFR1 equivalent to FADD in CD95 signaling.[114] The receptor-interacting protein (RIP) has also been found in TNFR1-signaling complexes.[115] These two proteins each have a DD and are recruited to TNFR1 during signaling.[115] TRADD can also interact with FADD and then recruit caspase 8 and/or 10 to the signal complex, which in turn will initiate a cascade of caspase activation causing apoptosis.[104,114,116] Interestingly, recent data suggests that RIP can induce lymphocyte necrosis through CD95 and TNFR1 pathways in the absence of caspase 8 or under caspase-inhibited condition,[117] suggesting that there is more than one pathway by which a death receptor can deliver death signals (Fig. 4).

The existence of molecules that control the recruitment of death adaptors to death receptors has also been postulated. This expectation has turned out to be true for at least TNFR1 and DR3-signals. A DD-interacting protein known as SODD (silencer of death domain) has been isolated.[82] SODD associates with TNFR1 and DR3 in the absence of their ligands, thus inhibiting death signals. Upon TNF treatment, SODD dissociates from these death receptors, which allows TRADD to be recruited and to signal cell death. However, a SODD-like factor has not been identified in the CD95 system.

Death effector domains (DED) are interesting protein-interaction domains involved in apoptosis pathways. We have previously showed that overexpression of DED domains of FADD or caspase 8 causes formation of death filaments, which is well correlated with cell death.[118] The mechanism of DED-induced cell death is not clear. One explanation is that when DED is overexpressed in cells, the homophilic association of DED may cause aggregation, which activates endogenous caspase-8 and/or -10. However, this hypothesis does not hold in certain circumstances. For example, there is evidence that the expression of human DED in bacteria causes bacterial death. It is unlikely that caspase activation is occurring in this case because no caspase homologue exists in bacterial genomes. Another explanation is that many overexpressed molecules can cause generic toxicity to host cells and that over expression of DEDs is just a trivial example of this phenomenon. This seems not to be the case in mammalian cells. Our recent data suggests that in caspase 8-deficient Jurkat T cells, the prodomains of caspase-8 and -10, each containing two DEDs, can mediate CD95-triggered, FADD-dependent cell death. Moreover, this cell death process is not inhibitable by the pan-caspase inhibitors zVAD and BocD, or by specific caspase 8/10 inhibitors such as IETD, DEVD, and zAEVD. Nevertheless, the RIP is required for this DED-mediated death signaling (Zheng and Lenardo, unpublished data). The involvement of RIP in an alternative pathway of CD95-death signals has been reported recently (Fig. 4).[119] This alternative caspase-independent pathway of CD95-induced death and its physiological relevance are now under study. The biological significance of this new pathway of cell death largely depends on the confirmation of the existence of natural isoforms of caspase-8 and -10 that only contain the DEDs. Until now, these isoforms have been only observed at the mRNA level. Further work should establish if there is an alternative pathway of death receptor-mediated apoptosis, perhaps mediated by the DED-containing prodomains of caspase-8 and -10.

Deficient CD95 Signaling Leads to Lymphoproliferative Disorders in Mice and Humans

It has been found that genetic deficiencies in apoptosis pathways can lead to developmental abnormalities and disease.[120] One of the major breakthroughs in determining the physiological relevance of lymphocyte apoptosis and autoimmune disorders came from the identification of genetic mutations of CD95 (Fas) in *lpr* mice and CD95 ligand (FasL) in *gld* mice.[121,122] The *lpr* stands for "lymphoproliferative response" and the *gld* for "generalized lymphoproliferative disease". In *lpr* mice, a transposable element insertion in the 2nd intron of CD95 gene disrupts transcription of the gene while in *gld* mice, a point mutation in the CD95L gene produces a mutant protein that fails to bind CD95. In both cases, peripheral lymphocytes escape from CD95-mediated cell death. Over their lifetime, these mice develop progressive lymphoadenopathy, splenomegaly, and autoimmune disease.[123] These data indicate that death receptor-mediated apoptosis is important for shaping the immune system by limiting lymphocyte accumulation.

Sneller et al, in 1992 first described two patients with a progressive lymphoproliferative disorder associated with autoimmunity. Their clinical and immunological features resembled the lymphoproliferative/autoimmune disease seen in *lpr* and *gld* mice.[124] In 1995, CD95 mutations were identified in two series of patients. The pathogenic role of these mutations was confirmed by experiments showing that the mutant protein could dominantly interfere with wild type CD95 signaling.[125,126] Given the association between lymphoproliferation and autoimmunity as a prominent feature of this disease, it has since been designated as the Autoimmune Lymphoproliferative Syndrome (ALPS). Subsequently, other groups also identified CD95 mutations from patients with symptoms of ALPS.[127,128] Typical ALPS patients usually present with progressive lymphadenopathy, splenomegaly, and autoimmune disorders that often accompany increased levels of autoantibodies. Activated T cells from ALPS patients are

Table 3. Molecular classification of ALPS

Types	Genetic Mutations
Type Ia	Heterozygous mutations in the Fas receptor
Type Ib	Fas Ligand mutations
Type II	Caspase-10 mutations
Type III	Familial but unknown mutations (not Fas or caspase-10)

resistant to both TCR-induced apoptosis as well as CD95-induced apoptosis. Another unique feature of ALPS patients, which has been also observed in *lpr* mice, is the accumulation of polyclonal CD4⁻, CD8⁻, CD3⁺ lymphocytes in their peripheral blood.[129]

ALPS patients are currently categorized into three types. As shown in Table 3, Type-Ia and -Ib are specified as ALPS due to genetic mutations in CD95 and CD95 ligand, respectively; Type-II designates ALPS caused by mutations in caspase-8/10; while Type-III is currently reserved for those ALPS patients who have neither type-I nor –II mutations. Several interesting observations have emerged in our studies of ALPS. First, it has been noted that more than half of the mutations found in our cohort of patients (more than 100 families) are in the death domain of CD95 gene (NIH ALPS working group, unpublished data), indicating the importance of CD95 in T-cell apoptosis. Second, ALPS is inherited as an autosomal dominant disease with incomplete penetrance. Some family members of ALPS patients harboring identical CD95 mutations are apparently healthy, suggesting that environmental factors or mutations in other immunoregulatory molecules account for the penetrance of the disease. Third, an extracellular truncation mutant protein of CD95 identified in ALPS family-2 can still associate with wild-type CD95 proteins and dominantly interfere with CD95 signaling,[125] implying that pre-assembly of extracellular domains is crucial for both CD95 ligand binding and CD95 signaling. This speculation has been verified by recent experiments showing that pre-association of CD95 is required for ligand binding.[100] And last, there are many patients with typical or atypical ALPS, who do not carry CD95L, CD95, and caspase-8/10 gene mutations. The molecular etiologies of the disease in these ALPS patients need to be elucidated.

Strategies for Medical Applications Through Regulating Death Receptor Signaling

The discovery of TCR-induced apoptosis has led our group and others to determine whether harnessing this powerful immunoregulatory mechanism can achieve the therapeutic goal of ameliorating T cell mediated autoimmune diseases. Indeed, it was found that repeated doses of self-antigen could reduce autoimmunity in the mouse model of experimental autoimmune encephalomyelitis.[130] In rheumatoid arthritis, as well as inflammatory bowel disease, blocking the effects of TNF with soluble receptor compounds or anti-cytokine antibodies has significant clinical benefit.[131,132] These agents may affect both the inflammatory as well as apoptotic signals delivered through TNFR. Since TRAIL was found to be more active at inducing apoptosis in a panel of tumor lines than in normal cells, some have used exogenous TRAIL to successfully treat experimental malignancies in mice.[133] Our recent identification of pre-ligand association by members of the TNF receptor family also opens up new possibilities for therapeutic intervention by blocking receptor pre-association.[100,101] These therapeutic opportunities show how far this fast moving field has come in only the first ten years after the identification of the basic molecular events underlying immune cell homeostasis.

Undoubtedly death receptors are involved in chemotherapy induced apoptosis of hematopoietic stem cell transplant regimens. The role of death receptors in maintaining post transplant remission of autoimmune diseases is probably also important but yet to be evaluated.

References

1. Scollay RG, Butcher EC, Weissman IL. Thymus cell migration. Quantitative aspects of cellular traffic from the thymus to the periphery in mice. Eur J Immunol 1980; 10(3):210.
2. Cantrell DA, Smith KA. The interleukin-2 T-cell system: A new cell growth model. Science 1984; 224(4655):1312.
3. Gullberg M, Smith KA. Regulation of T cell autocrine growth. T4+ cells become refractory to interleukin 2. J Exp Med 1986; 163(2):270.
4. Lenardo M, Chan KM, Hornung F et al. Mature T lymphocyte apoptosis—immune regulation in a dynamic and unpredictable antigenic environment. Annu Rev Immunol 1999; 17:221.
5. Lenardo MJ. Interleukin-2 programs mouse alpha beta T lymphocytes for apoptosis. Nature 1991; 353(6347):858.
6. Kerr JF, Wyllie AH, Currie AR. Apoptosis: A basic biological phenomenon with wide-ranging implications in tissue kinetics. Br J Cancer 1972; 26(4):239.
7. Arch RH, Thompson CB. Lymphocyte survival—the struggle against death. Annu Rev Cell Dev Biol 1999; 15:113.
8. Tsubata T. Apotosis of mature B cells. Int Rev Immunol 1999; 18(4):347.
9. Ju ST, Matsui K, Ozdemirli M. Molecular and cellular mechanisms regulating T and B cell apoptosis through Fas/FasL interaction. Int Rev Immunol 1999; 18(5-6):485.
10. Pearson CI, van Ewijk W, McDevitt HO. Induction of apoptosis and T helper 2 (Th2) responses correlates with peptide affinity for the major histocompatibility complex in self- reactive T cell receptor transgenic mice. J Exp Med 1997; 185(4):583.
11. Liu CP, Crawford F, Marrack P et al. T cell positive selection by a high density, low affinity ligand. Proc Natl Acad Sci USA 1998; 95(8):4522.
12. Ashwell JD, King LB, Vacchio MS. Cross-talk between the T cell antigen receptor and the glucocorticoid receptor regulates thymocyte development. Stem Cells 1996; 14(5):490.
13. Wack A, Ladyman HM, Williams O et al. Direct visualization of thymocyte apoptosis in neglect, acute and steady-state negative selection. Int Immunol 1996; 8(10):1537.
14. Ashton-Rickardt PG, Tonegawa S. A differential-avidity model for T-cell selection. Immunol Today 1994; 15(8):362.
15. Adkins B, Nassiri M. Apoptosis of murine neonatal T cells. Int Rev Immunol 1999; 18(5-6):465.
16. Wang J, Lobito AA, Shen F et al. Inhibition of Fas-mediated apoptosis by the B cell antigen receptor through c-FLIP. Eur J Immunol 2000; 30(1):155.
17. Williams O, Brady HJ. The role of molecules that mediate apoptosis in T-cell selection. Trends Immunol 2001; 22(2):107.
18. Sprent J, Kishimoto H. The thymus and central tolerance. Philos Trans R Soc Lond B Biol Sci 2001; 356(1409):609.

19. Ashton-Rickardt PG, Bandeira A, Delaney JR et al. Evidence for a differential avidity model of T cell selection in the thymus. Cell 1994; 76(4):651.
20. von Boehmer H, Teh HS, Kisielow P. The thymus selects the useful, neglects the useless and destroys the harmful. Immunol Today 1989; 10(2):57.
21. Benoist C, Mathis D. Positive selection of the T cell repertoire: Where and when does it occur? Cell 1989; 58(6):1027.
22. Benoist C, Mathis D. T-cell development: A new marker of differentiation state. Curr Biol 1999; 9(2):R59.
23. Fink PJ, Bevan MJ. Positive selection of thymocytes. Adv Immunol 1995; 59:99.
24. Fowlkes BJ, Schweighoffer E. Positive selection of T cells. Curr Opin Immunol 1995; 7(2):188.
25. Sebzda E, Mariathasan S, Ohteki T et al. Selection of the T cell repertoire. Annu Rev Immunol 1999; 17:829.
26. Jameson SC, Hogquist KA, Bevan MJ. Positive selection of thymocytes. Annu Rev Immunol 1995; 13:93.
27. Kishimoto H, Sprent J. Negative selection in the thymus includes semimature T cells. J Exp Med 1997; 185(2):263.
28. Fisher GH, Lenardo MJ, Zuniga-Pflucker JC. Synergy between T cell Receptor and Fas (CD95/APO-1) signaling in mouse thymocyte death. Cell Immunol 1996; 169(1):99.
29. Kishimoto H, Surh CD, Sprent J. A role for Fas in negative selection of thymocytes in vivo. J Exp Med 1998; 187(9):1427.
30. Amakawa R, Hakem A, Kundig TM et al. Impaired negative selection of T cells in Hodgkin's disease antigen CD30-deficient mice. Cell 1996; 84(4):551.
31. Foy TM, Page DM, Waldschmidt TJ et al. An essential role for gp39, the ligand for CD40, in thymic selection. J Exp Med 1995; 182(5):1377.
32. Sidman CL, Marshall JD, Von Boehmer H. Transgenic T cell receptor interactions in the lymphoproliferative and autoimmune syndromes of lpr and gld mutant mice. Eur J Immunol 1992; 22(2):499.
33. Singer GG, Abbas AK. The fas antigen is involved in peripheral but not thymic deletion of T lymphocytes in T cell receptor transgenic mice. Immunity 1994; 1(5):365.
34. Page DM, Roberts EM, Peschon JJ et al. TNF receptor-deficient mice reveal striking differences between several models of thymocyte negative selection. J Immunol 1998; 160(1):120.
35. Clayton LK, Ghendler Y, Mizoguchi E et al. T-cell receptor ligation by peptide/MHC induces activation of a caspase in immature thymocytes: The molecular basis of negative selection. Embo J 1997; 16(9):2282.
36. Jiang D, Zheng L, Lenardo MJ. Caspases in T-cell receptor-induced thymocyte apoptosis. Cell Death Differ 1999; 6(5):402.
37. Kuida K, Lippke JA, Ku G et al. Altered cytokine export and apoptosis in mice deficient in interleukin- 1 beta converting enzyme. Science 1995; 267(5206):2000.
38. Hakem R, Hakem A, Duncan GS et al. Differential requirement for caspase 9 in apoptotic pathways in vivo. Cell 1998; 94(3):339.
39. Kuida K, Zheng TS, Na S et al. Decreased apoptosis in the brain and premature lethality in CPP32- deficient mice. Nature 1996; 384(6607):368.
40. Chao DT, Korsmeyer SJ. BCL-2 family: Regulators of cell death. Annu Rev Immunol 1998; 16:395.
41. Groves T, Parsons M, Miyamoto NG et al. TCR engagement of CD4+CD8+ thymocytes in vitro induces early aspects of positive selection, but not apoptosis. J Immunol 1997; 158(1):65.
42. Nakayama K, Negishi I, Kuida K et al. Disappearance of the lymphoid system in Bcl-2 homozygous mutant chimeric mice. Science 1993; 261(5128):1584.
43. Veis DJ, Sorenson CM, Shutter JR et al. Bcl-2-deficient mice demonstrate fulminant lymphoid apoptosis, polycystic kidneys, and hypopigmented hair. Cell 1993; 75(2):229.
44. Ma A, Pena JC, Chang B et al. Bclx regulates the survival of double-positive thymocytes. Proc Natl Acad Sci USA 1995; 92(11):4763.
45. Motoyama N, Wang F, Roth KA et al. Massive cell death of immature hematopoietic cells and neurons in Bcl-x- deficient mice. Science 1995; 267(5203):1506.
46. Strasser A, Harris AW, Cory S. bcl-2 transgene inhibits T cell death and perturbs thymic self- censorship. Cell 1991; 67(5):889.
47. Sentman CL, Shutter JR, Hockenbery D et al. bcl-2 inhibits multiple forms of apoptosis but not negative selection in thymocytes. Cell 1991; 67(5):879.
48. Huang DC, Strasser A. BH3-Only proteins-essential initiators of apoptotic cell death. Cell 2000; 103(6):839.
49. Bouillet P, Metcalf D, Huang DC et al. Proapoptotic Bcl-2 relative Bim required for certain apoptotic responses, leukocyte homeostasis, and to preclude autoimmunity. Science 1999; 286(5445):1735.
50. Calnan BJ, Szychowski S, Chan FK et al. A role for the orphan steroid receptor Nur77 in apoptosis accompanying antigen-induced negative selection. Immunity 1995; 3(3):273.
51. Cheng LE, Chan FK, Cado D et al. Functional redundancy of the Nur77 and Nor-1 orphan steroid receptors in T-cell apoptosis. Embo J 1997; 16(8):1865.
52. Sun Z, Unutmaz D, Zou YR et al. Requirement for RORgamma in thymocyte survival and lymphoid organ development. Science 2000; 288(5475):2369.
53. Ashwell JD, Lu FW, Vacchio MS. Glucocorticoids in T cell development and function. Annu Rev Immunol 2000; 18:309.
54. Gil-Gomez G, Berns A, Brady HJ. A link between cell cycle and cell death: Bax and Bcl-2 modulate Cdk2 activation during thymocyte apoptosis. Embo J 1998; 17(24):7209.
55. Bain G, Engel I, Robanus Maandag EC et al. E2A deficiency leads to abnormalities in alphabeta T-cell development and to rapid development of T-cell lymphomas. Mol Cell Biol 1997; 17(8):4782.
56. Bain G, Cravatt CB, Loomans C et al. Regulation of the helix-loop-helix proteins, E2A and Id3, by the Ras- ERK MAPK cascade. Nat Immunol 2001; 2(2):165.
57. Zheng L, Trageser CL, Willerford DM et al. T cell growth cytokines cause the superinduction of molecules mediating antigen-induced T lymphocyte death. J Immunol 1998; 160(2):763.
58. Scaffidi C, Kirchhoff S, Krammer PH et al. Apoptosis signaling in lymphocytes. Curr Opin Immunol 1999; 11(3):277.
59. Siegel RM, Fleisher TA. The role of Fas and related death receptors in autoimmune and other disease states. J Allergy Clin Immunol 1999; 103(5 Pt 1):729.
60. Dhein J, Walczak H, Baumler C et al. Autocrine T-cell suicide mediated by APO-1/(Fas/CD95). Nature 1995; 373(6513):438.
61. Brunner T, Mogil RJ, LaFace D et al. Cell-autonomous Fas (CD95)/Fas-ligand interaction mediates activation- induced apoptosis in T-cell hybridomas. Nature 1995; 373(6513):441.
62. Ju ST, Panka DJ, Cui H et al. Fas(CD95)/FasL interactions required for programmed cell death after T- cell activation [see comments]. Nature 1995; 373(6513):444.
63. Zheng L, Fisher G, Miller RE et al. Induction of apoptosis in mature T cells by tumour necrosis factor. Nature 1995; 377(6547):348.
64. MacFarlane M, Ahmad M, Srinivasula SM et al. Identification and molecular cloning of two novel receptors for the cytotoxic ligand TRAIL. J Biol Chem 1997; 272(41):25417.
65. Pan G, Bauer JH, Haridas V et al. Identification and functional characterization of DR6, a novel death domain-containing TNF receptor. FEBS Lett 1998; 431(3):351.
66. Boehme SA, Zheng L, Lenardo MJ. Analysis of the CD4 coreceptor and activation-induced costimulatory molecules in antigen-mediated mature T lymphocyte death. J Immunol 1995; 155(4):1703.
67. Boehme SA, Lenardo MJ. Ligand-induced apoptosis of mature T lymphocytes (propriocidal regulation) occurs at distinct stages of the cell cycle. Leukemia 1993; 7 Suppl 2:S45.
68. Sadlack B, Kuhn R, Schorle H et al. Development and proliferation of lymphocytes in mice deficient for both interleukins-2 and -4. Eur J Immunol 1994; 24(1):281.
69. Kneitz B, Herrmann T, Yonehara S et al. Normal clonal expansion but impaired Fas-mediated cell death and anergy induction in interleukin-2-deficient mice. Eur J Immunol 1995; 25(9):2572.
70. Willerford DM, Chen J, Ferry JA et al. Interleukin-2 receptor alpha chain regulates the size and content of the peripheral lymphoid compartment. Immunity 1995; 3(4):521.
71. Richardson JH, Sodroski JG, Waldmann TA et al. Phenotypic knockout of the high-affinity human interleukin 2 receptor by intracellular single-chain antibodies against the alpha subunit of the receptor. Proc Natl Acad Sci USA 1995; 92(8):3137.

72. Wang R, Rogers AM, Rush BJ et al. Induction of sensitivity to activation-induced death in primary CD4+ cells: A role for interleukin-2 in the negative regulation of responses by mature CD4+ T cells. Eur J Immunol 1996; 26(9):2263.
73. Van Parijs L, Biuckians A, Ibragimov A et al. Functional responses and apoptosis of CD25 (IL-2R alpha)-deficient T cells expressing a transgenic antigen receptor. J Immunol 1997; 158(8):3738.
74. Boehme SA, Lenardo MJ. Propriocidal apoptosis of mature T lymphocytes occurs at S phase of the cell cycle. Eur J Immunol 1993; 23(7):1552.
75. Kung JT, Beller D, Ju ST. Lymphokine regulation of activation-induced apoptosis in T cells of IL- 2 and IL-2R beta knockout mice. Cell Immunol 1998; 185(2):158.
76. Naora H, Gougeon ML. Interleukin-15 is a potent survival factor in the prevention of spontaneous but not CD95-induced apoptosis in CD4 and CD8 T lymphocytes of HIV-infected individuals. Correlation with its ability to increase BCL-2 expression. Cell Death Differ 1999; 6(10):1002.
77. Watanabe N, Arase H, Kurasawa K et al. Th1 and Th2 subsets equally undergo Fas-dependent and -independent activation-induced cell death. Eur J Immunol 1997; 27(8):1858.
78. Carter LL, Zhang X, Dubey C et al. Regulation of T cell subsets from naive to memory. J Immuno Ther 1998; 21(3):181.
79. Ashwell JD, Longo DL, Bridges SH. T-cell tumor elimination as a result of T-cell receptor-mediated activation. Science 1987; 237(4810):61.
80. Peter ME, Kischkel FC, Scheuerpflug CG et al. Resistance of cultured peripheral T cells towards activation-induced cell death involves a lack of recruitment of FLICE (MACH/caspase 8) to the CD95 death-inducing signaling complex. Eur J Immunol 1997; 27(5):1207.
81. Critchfield JM, Zuniga-Pflucker JC, Lenardo MJ. Parameters controlling the programmed death of mature mouse T lymphocytes in high-dose suppression. Cell Immunol 1995; 160(1):71.
82. Jiang Y, Woronicz JD, Liu W et al. Prevention of constitutive TNF receptor 1 signaling by silencer of death domains [published erratum appears in Science 1999 Mar 19;283(5409):1852]. Science 1999; 283(5401):543.
83. Xiao S, Matsui K, Fine A et al. FasL promoter activation by IL-2 through SP1 and NFAT but not Egr-2 and Egr-3. Eur J Immunol 1999; 29(11):3456.
84. Irmler M, Thome M, Hahne M et al. Inhibition of death receptor signals by cellular FLIP [see comments]. Nature 1997; 388(6638):190.
85. Algeciras-Schimnich A, Griffith TS, Lynch DH et al. Cell cycle-dependent regulation of FLIP levels and susceptibility to Fas-mediated apoptosis. J Immunol 1999; 162(9):5205.
86. Hornung F, Zheng L, Lenardo MJ. Maintenance of clonotype specificity in CD95/Apo-1/Fas-mediated apoptosis of mature T lymphocytes. J Immunol 1997; 159(8):3816.
87. Combadiere B, Reis e Sousa C, Trageser C et al. Differential TCR signaling regulates apoptosis and immunopathology during antigen responses in vivo. Immunity 1998; 9(3):305.
88. Kuroda K, Yagi J, Imanishi K et al. Implantation of IL-2-containing osmotic pump prolongs the survival of superantigen-reactive T cells expanded in mice injected with bacterial superantigen. J Immunol 1996; 157(4):1422.
89. Duke RC, Cohen JJ. IL-2 addiction: Withdrawal of growth factor activates a suicide program in dependent T cells. Lymphokine Res 1986; 5(4):289.
90. Reed JC. Cytochrome c: Can't live with it—can't live without it. Cell 1997; 91(5):559.
91. Vander Heiden MG, Chandel NS, Schumacker PT et al. Bcl-xL prevents cell death following growth factor withdrawal by facilitating mitochondrial ATP/ADP exchange. Mol Cell 1999; 3(2):159.
92. Vander Heiden MG, Chandel NS, Williamson EK et al. Bcl-xL regulates the membrane potential and volume homeostasis of mitochondria. Cell 1997; 91(5):627.
93. Chinnaiyan AM, O'Rourke K, Yu GL et al. Signal transduction by DR3, a death domain-containing receptor related to TNFR-1 and CD95. Science 1996; 274(5289):990.
94. Pan G, O'Rourke K, Chinnaiyan AM et al. The receptor for the cytotoxic ligand TRAIL. Science 1997; 276(5309):111.
95. Brojatsch J, Naughton J, Rolls MM et al. CAR1, a TNFR-related protein, is a cellular receptor for cytopathic avian leukosis-sarcoma viruses and mediates apoptosis. Cell 1996; 87(5):845.
96. Strasser A, O'Connor L, Dixit VM. Apoptosis signaling. Annu Rev Biochem 2000; 69:217.
97. Itoh N, Nagata S. A novel protein domain required for apoptosis. Mutational analysis of human Fas antigen. J Biol Chem 1993; 268(15):10932.
98. Tartaglia LA, Ayres TM, Wong GH et al. A novel domain within the 55 kd TNF receptor signals cell death. Cell 1993; 74(5):845.
99. Pennica D, Kohr WJ, Fendly BM et al. Characterization of a recombinant extracellular domain of the type 1 tumor necrosis factor receptor: Evidence for tumor necrosis factor- alpha induced receptor aggregation. Biochemistry 1992; 31(4):1134.
100. Siegel RM, Frederiksen JK, Zacharias DA et al. Fas preassociation required for apoptosis signaling and dominant inhibition by pathogenic mutations. Science 2000; 288(5475):2354.
101. Chan FK, Chun HJ, Zheng L et al. A domain in TNF receptors that mediates ligand-independent receptor assembly and signaling. Science 2000; 288(5475):2351.
102. Kischkel FC, Hellbardt S, Behrmann I et al. Cytotoxicity-dependent APO-1 (Fas/CD95)-associated proteins form a death-inducing signaling complex (DISC) with the receptor. Embo J 1995; 14(22):5579.
103. Chinnaiyan AM, O'Rourke K, Tewari M et al. FADD, a novel death domain-containing protein, interacts with the death domain of Fas and initiates apoptosis. Cell 1995; 81(4):505.
104. Boldin MP, Goncharov TM, Goltsev YV et al. Involvement of MACH, a novel MORT1/FADD-interacting protease, in Fas/APO-1- and TNF receptor-induced cell death. Cell 1996; 85(6):803.
105. Muzio M, Chinnaiyan AM, Kischkel FC et al. FLICE, a novel FADD-homologous ICE/CED-3-like protease, is recruited to the CD95 (Fas/APO-1) death—inducing signaling complex. Cell 1996; 85(6):817.
106. Muzio M, Stockwell BR, Stennicke HR et al. An induced proximity model for caspase-8 activation. J Biol Chem 1998; 273(5):2926.
107. Martin DA, Siegel RM, Zheng L et al. Membrane oligomerization and cleavage activates the caspase-8 (FLICE/MACHalpha1) death signal. J Biol Chem 1998; 273(8):4345.
108. Yang X, Chang HY, Baltimore D. Autoproteolytic activation of pro-caspases by oligomerization. Mol Cell 1998; 1(2):319.
109. Schneider P, Thome M, Burns K et al. TRAIL receptors 1 (DR4) and 2 (DR5) signal FADD-dependent apoptosis and activate NF-kappaB. Immunity 1997; 7(6):831.
110. Kuang AA, Diehl GE, Zhang J et al. FADD is required for DR4- and DR5-mediated apoptosis: Lack of trail- induced apoptosis in FADD-deficient mouse embryonic fibroblasts. J Biol Chem 2000; 275(33):25065.
111. Weiss T, Grell M, Siemienski K et al. TNFR80-dependent enhancement of TNFR60-induced cell death is mediated by TNFR-associated factor 2 and is specific for TNFR60. J Immunol 1998; 161(6):3136.
112. Weiss T, Grell M, Hessabi B et al. Enhancement of TNF receptor p60-mediated cytotoxicity by TNF receptor p80: requirement of the TNF receptor-associated factor-2 binding site. J Immunol 1997; 158(5):2398.
113. Chan FK, Lenardo MJ. A crucial role for p80 TNF-R2 in amplifying p60 TNF-R1 apoptosis signals in T lymphocytes. Eur J Immunol 2000; 30(2):652.
114. Hsu H, Shu HB, Pan MG et al. TRADD-TRAF2 and TRADD-FADD interactions define two distinct TNF receptor 1 signal transduction pathways. Cell 1996; 84(2):299.
115. Hsu H, Huang J, Shu HB et al. TNF-dependent recruitment of the protein kinase RIP to the TNF receptor- 1 signaling complex. Immunity 1996; 4(4):387.
116. Chaudhary PM, Eby M, Jasmin A et al. Death receptor 5, a new member of the TNFR family, and DR4 induce FADD- dependent apoptosis and activate the NF-kappaB pathway. Immunity 1997; 7(6):821.
117. Holler N, Zaru R, Micheau O et al. Fas triggers and alternative, caspase-8-independent cell death pathway using the kinase RIP as effector molecule. Nat Immunol 2000; 1(6):489.

118. Siegel RM, Martin DA, Zheng L et al. Death-effector filaments: Novel cytoplasmic structures that recruit caspases and trigger apoptosis. J Cell Biol 1998; 141(5):1243.
119. Holler N, Zaru R, Micheau O et al. Fas triggers an alternative, caspase-8-independent cell death pathway using the kinase RIP as effector molecule. Nat Immunol 2000; 1(6):489.
120. Afford S, Randhawa S. Apoptosis. Mol Pathol 2000; 53(2):55.
121. Adachi M, Watanabe-Fukunaga R, Nagata S. Aberrant transcription caused by the insertion of an early transposable element in an intron of the Fas antigen gene of lpr mice. Proc Natl Acad Sci USA 1993; 90(5):1756.
122. Takahashi T, Tanaka M, Brannan CI et al. Generalized lymphoproliferative disease in mice, caused by a point mutation in the Fas ligand. Cell 1994; 76(6):969.
123. Cohen PL, Eisenberg RA. Lpr and gld: Single gene models of systemic autoimmunity and lymphoproliferative disease. Annu Rev Immunol 1991; 9:243.
124. Sneller MC, Straus SE, Jaffe ES et al. A novel lymphoproliferative/autoimmune syndrome resembling murine lpr/gld disease. J Clin Invest 1992; 90(2):334.
125. Fisher GH, Rosenberg FJ, Straus SE et al. Dominant interfering Fas gene mutations impair apoptosis in a human autoimmune lymphoproliferative syndrome. Cell 1995; 81(6):935.
126. Rieux-Laucat F, Le Deist F, Hivroz C et al. Mutations in Fas associated with human lymphoproliferative syndrome and autoimmunity. Science 1995; 268(5215):1347.
127. Wu J, Wilson J, He J et al. Fas ligand mutation in a patient with systemic lupus erythematosus and lymphoproliferative disease. J Clin Invest 1996; 98(5):1107.
128. Drappa J, Vaishnaw AK, Sullivan KE et al. Fas gene mutations in the Canale-Smith syndrome, an inherited lymphoproliferative disorder associated with autoimmunity [see comments]. N Engl J Med 1996; 335(22):1643.
129. Sneller MC, Wang J, Dale JK et al. Clincial, immunologic, and genetic features of an autoimmune lymphoproliferative syndrome associated with abnormal lymphocyte apoptosis. Blood 1997; 89(4):1341.
130. Critchfield JM, Racke MK, Zuniga-Pflucker JC et al. T cell deletion in high antigen dose therapy of autoimmune encephalomyelitis. Science 1994; 263(5150):1139.
131. Maini RN, Taylor PC. Anti-cytokine therapy for rheumatoid arthritis. Annu Rev Med 2000; 51:207.
132. Rutgeerts P. Medical therapy of inflammatory bowel disease. Digestion 1998; 59(5):453.
133. Walczak H, Miller RE, Ariail K et al. Tumoricidal activity of tumor necrosis factor-related apoptosis- inducing ligand in vivo. Nat Med 1999; 5(2):157.

CHAPTER 18

Shifting Paradigms in Peripheral Tolerance

Jonathan D. Powell and Ronald H. Schwartz

Introduction

Through clonal selection, the immune system is able to prepare and respond to a multitude of diverse antigens. However, since the generation of diversity is a stochastic post-germline encoded event, a critical component of the immune response must subsequently be the ability to discern between harmful and innocuous antigens in a tolerance process. Early theories, (on which the foundation of modern immunology is based) described tolerance as the ability to distinguish self from nonself. In the beginning, Ehrlich and Morgenroth realized this important aspect of the system and coined the term "horror autotoxicus" to describe the consequences of tolerance being broken and the immune system unleashing its effector function on one's own body.[1] Burnet proposed that this educational skewing toward the recognition of only nonself occurred during a critical developmental period that began in utero and lasted through the perinatal period.[2] During this time, autoreactive lymphocytes would recognize self antigens and be deleted. If indeed self was defined during a critical developmental period, then a prediction of such a model was that "foreign" antigens introduced during this period would induce tolerance. Experiments by Billingham demonstrated that if cells from A mice were injected into B mice during the fetal period, the B mice would accept A-strain skin grafts but reject third party C-strain grafts.[3] Inversely, Triplett was able to show that by removing the pituitary anlage from tree frog larvae and then replacing the adult animals with their own pituitaries, the frogs rejected the autografts.[4] Since the pituitary was not present during the critical self education period, it was seen as foreign in the adult animal and rejected.

As the details of lymphocyte development became better elucidated, this concept of a critical "self/nonself" learning period became refined. Whereas Burnet and others emphasized a developmental period in terms of the organism, it is clear that the model might be applied to a critical developmental period in the life of a lymphocyte. For example, precursors to T cells home to the thymus where they are educated to self through the processes of positive and negative selection. Using T-cell receptor (TCR)-specific antibodies, Kappler and Marrack formally demonstrated the phenomenon of thymic negative selection, whereby autoreactive T lymphocytes are deleted prior to their emergence into the peripheral circulating lymphocyte pool.[5] As such, the education process to self takes place during a critical period of T cell development. Interestingly, it has been recently shown that a myriad of ostensibly peripheral self proteins are expressed in medullary thymic epithelial cells.[6-9] This "promiscuous gene expression" of self antigens, which includes insulin, somatostatin, elastase, trypsin2 (among others), appears to be general in that it is not limited to a particular set of functional genes or even exclusively secluded proteins. Since many of these genes do not have any obvious function in the thymus and are not expressed by epithelial cells elsewhere in the body, one can hypothesize that their expression in the thymus is for the purpose of inducing tolerance. Indeed, if thymic deletion could be demonstrated for most peripheral antigens, then it might be argued that mechanisms of peripheral tolerance are only minor fail-safes with the bulk of the self/nonself discrimination process taking place in the thymus.

Shifting Paradigms

Dominating concepts of tolerance then is the distinction between self and nonself. In part this is because many of the early experiments demonstrating tolerance involved alloantigens derived from different strains of mice, and in fact the readout for the presence or absence of tolerance was either a mixed lymphocyte response (MLR) or skin graft rejection. Self/nonself models of tolerance are by definition antigen driven (self antigens induce tolerance while nonself antigens lead to activation). Furthermore, the education process is determined during a critical developmental phase of the organism/lymphocyte (Table 1).

In the last decade, there have been a number of challenges to the idea that immune responses are driven by the distinction between self and nonself. In the new perspective, the outcome of an immune response is dictated by the context in which the antigen is encountered. Janeway has proposed that the immune system distinguishes between self and infectious nonself.[10] In this model, the immune system is activated by antigen in the presence of various molecules called pathogen-associated molecular patterns (PAMPS) that trigger various pattern recognition receptors (PRR) such as the Toll receptors (refer to Chapter 19).[11,12] This "adjuvant"-induced activation is the signal to the immune system to respond in a productive fashion; in its absence immune recognition leads to the induction of tolerance. Matzinger has proposed an alternative model that she has termed the "Danger Theory". Antigens presented in the presence of "danger" will lead to productive immune responses, while antigens presented in the ab-

Stem Cell Therapy for Autoimmune Disease, edited by Richard K. Burt and Alberto M. Marmont. ©2004 Landes Bioscience/Eurekah.com.

Table 1. Shifting paradigms

Old	New
• Self/Non-self • Antigen Driven Immune recognition of a self antigen leads to tolerance while foreign antigens lead to immune activation.	• Context Models • Circumstance Driven It is not the antigen that dictates the outcome of an immune response (tolerance or activation) but rather the context in which the antigen is encountered.
• Developmentally Determined Immune system is educated to self antigens during in utero/perinatal period.	• Determined by Milieu The presence or absence of "danger" signals.

sence of danger will lead to tolerance.[13] Examples of danger signals are thought to include molecules that are released during injury and necrotic cell death such as heat shock proteins. Critical to both of these models is the activation state of the antigen-presenting cells. Lipopolysaccharides (LPS), double stranded RNA, and heat shock proteins can all activate dendritic cells. Once an APC is activated, not only does it efficiently present antigen (signal 1), but it now has upregulated a variety of costimulatory molecules (signal 2). Thus, "context-based" models of immune activation are two signal hypotheses of immune activation in which the second signal is not antigen specific, but controls the nature of the immunologic decision.[14] The origin of the antigen is irrelevant; an antigen is an antigen regardless of whether it was originally derived from self or from a foreign source. What does matter is the environment in which the antigen is recognized. These new models of immune activation and tolerance can be thought of as circumstance driven rather than antigen driven, the circumstances being dictated by the milieu in which the antigen is being presented (Table 1). As such, peripheral tolerance would play a critical role in preventing horror autotoxicus. Furthermore, the role of the thymus in facilitating tolerance in these "contextual" models would become narrower. Since antigens presented by an activated dendritic cell will induce a productive immune response, it is critical that T cells are already tolerant to dendritic cell antigens. Thus, the primary role of the thymus in the "Danger" model is to eliminate T cells that are reactive with dendritic cell-derived antigens. Secondarily, the thymus would play a role in inducing tolerance to ubiquitous proteins (e.g., albumin which might be taken up and presented routinely by dendritic cells) and also perhaps facilitate tolerance to a select group of antigens that are expressed mostly in the periphery. If such is the case, however, then why (as mentioned above) do thymic epithelial cells promiscuously express such a wide variety of peripheral antigens? One intriguing possibility is that the expression of these antigens in the thymus is not to induce clonal deletion, but rather to promote the development of suppressor cells responsible for maintaining organ-specific peripheral tolerance.[7]

Activation Precedes Death and/or Anergy

What happens to a self-reactive T cell when it encounters a peripheral self antigen in the absence of costimulation? Initial studies on T cell clones had suggested that the cells do not proliferate or differentiate toward new cytokine production.[15] Clones, however, are essentially memory cell populations and the rules turn out to be different for naïve T cells. Insight into this question came largely from TCR transgenic mice and mice engineered to transgenically express, under tissue-specific promoters, the antigen that the T cells recognize.[16-22] Typically, when naïve transgenic T cells are transferred into an antigen-expressing host, the cells are activated and proliferate. In general, CD8+ T cells expand more than CD4+ T cells. In some models, the cells also differentiate to gain the ability to make IFN-γ; but in most cases the cells begin to die by 4 days. This Activation-Induced Cell Death (AICD)[23] is one mechanism employed by the immune system to promote peripheral tolerance.[24-27] For the most part, this death process is Fas-Fas Ligand (FasL)-dependent, although TNF has also been implicated (refer to Chapter 17).[23,28,29] Presumably, signal 1 leads to the upregulation of FasL which is cleaved, solubilized and then interacts with Fas on other cells to induce death. In the presence of costimulation, this process is in part prevented by the upregulation of anti-apoptotic molecules such as Bcl-2 and Bcl-x_L.[30] For completeness, it should be noted that AICD can also occur in the presence of costimulation. Such death, however, is induced by reactivation of activated (memory) T cells that have entered the S phase of the cell cycle (a process known as propriocidal death).[31] Whereas one might think of these mechanisms as primarily immunoregulatory, they play a key role in the tolerance process. Mice (lpr or gld) and humans (Autoimmune lymphoproliferative syndrome (ALPS) patients), possessing genetic inactivation of their Fas/FasL pathway, develop autoimmune disease.[32,33]

Not all T cells die following activation. In the absence of costimulation, a cohort often remains that is hyporesponsive upon rechallenge.[16-19] This state of functional inactivation is referred to as anergy. T cell clonal models in vitro have shown that anergic cells have a block in T cell receptor (TCR) signal transduction through the Ras/mitogen-activated protein (Ras-MAP) kinase pathway, a defect which in conjunction with other alterations leads to inhibition of transcription of genes involved in proliferation, such as interleukin-2 (IL-2).[34] This state is stable in the absence of antigen. In vivo, however, the mechanisms for anergy are less clear and may be multifactorial. All cytokine production seems to be down-regulated and the state is reversible in the absence of antigen.[18,21,22,35] The immune system appears to be holding these cells in check awaiting further signals, such as an increase in antigen concentration, before determining their fate. In the presence of other T cells and the persistence of antigen at the same level, the anergic cells eventually disappear.[36]

One interesting aspect of these various models of in vivo peripheral tolerance is the observation that activation seems to invariably precede the induction of the tolerant state. This fact begs the question of whether activation and proliferation are prerequisites for tolerance induction? Hayashi et al immunized ovalbumin-specific TCR transgenic mice with ovalbumin and then sorted for antigen-experienced and naïve T cells.[37] When they subsequently attempted to induce an anergic state in the T cells in vitro, they found that only the antigen-experienced T cells could be anergized. These observations are consistent with our own data using a pigeon cytochrome c-specific TCR transgenic mouse.[38] In our experiments we needed to first stimulate naïve T cells with antigen in vitro for several days, let them rest and then induce anergy; we were unable to anergize naïve T cells taken directly out of the mouse. More recently, Ragazzo et al have been able to

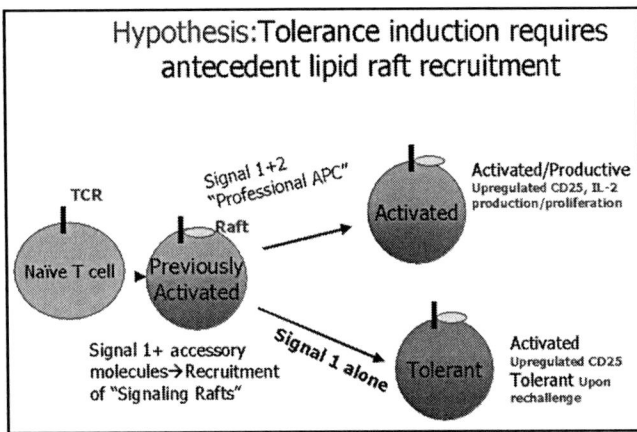

Figure 1. In our model, critical for both the activation and inactivation of T cells is the recruitment of lipid rafts to their membranes. Naïve T cells will only be able to achieve this if their TCR is activated in the context of accessory molecule engagement. Once the rafts are recruited the cell is no longer naïve and thus can either be fully activated or tolerized depending on the context in which it is stimulated.

induce anergy in naïve CD4+ T cells in the presence of lymphocyte function-associated antigen-1 (LFA-1) engagement.[39] In this model, signal 1 alone had little effect, while signal 1 + B7.2 engagement of CD28 led to activation, proliferation, differentiation and priming. In contrast, signal 1+ intracellular adhesion molecule-1 (ICAM-1) engagement of LFA-1 led to partial activation followed by anergy. Even in this case, though, a careful analysis of cell proliferation using as a marker a fluorescent dye (CFSE), revealed that most of the anergic T cells had previously divided at least once. Thus, the general consensus is that naïve T cells must divide in order to become susceptible to anergy induction. The only exception to this idea is found in the experiments of Wells et al. They sorted naïve CD4+ T cells labeled with CFSE after they had been stimulated with anti-CD3 and anti-CD28 antibodies.[40] Even in the presence of costimulation they found a large subpopulation of cells that had not divided, but which nevertheless were unresponsive. These cells failed to divide on restimulation even with the addition of IL-2, suggesting that the biochemical mechanism responsible for the block is different from what has been described for classical T cell clonal anergy.

We interpret these data to mean that naïve T cells do not have the signal transduction machinery in place to induce anergy. If they receive TCR stimulation in the presence of a strong enough costimulatory signal, then they are sufficiently activated to alter this machinery, enabling them subsequently to be susceptible to anergy induction. The biochemical changes that occur in this process are unknown. One speculative idea is shown in Figure 1. This model suggests that tolerance induction of naïve T cells first requires the antecedent recruitment of lipid rafts to the membrane. Lipid rafts have been shown to preferentially contain many critical TCR signal-transduction molecules such as the src family kinase lymphocyte specific kinase (Lck).[41] If one postulates that anergy is a negative feedback loop on TCR signaling, and that strong signaling is required to elicit this feedback program, then the premobilized rafts of an activated T cell may be a necessity for inducing the anergic state. In naïve T cells, these rafts are scattered about the cytoplasm. Once the cell becomes activated, the rafts are recruited to the membrane where they remain even after the cell has rested down and entered a memory state. From a clinical point of view, this model suggests that immunosuppressive drugs such as cyclosporin A (CSA), which prevent naïve T cells from becoming antigen-experienced T cells (and/or prevent raft recruitment to the membrane following activation of naïve T cells), will also inhibit the induction of anergy in such cells.

TCR Engagement in the Absence of Cell Cycle Progression Leads to Tolerance

As mentioned above the exact biochemical and molecular mechanisms accounting for in vivo peripheral tolerance induction have yet to be elucidated. Nonetheless, much has been learned concerning the mechanisms involved in anergy induction in vitro. Current models suggest that TCR stimulation leads to the upregulation of negative regulatory factors that mediate anergy.[34] The induction of these factors is blocked by cyclosporin as well as protein synthesis inhibitors.[42,43] Indeed, recent in vivo data suggest that cyclosporin can actually prevent transplantation tolerance.[44] In these experiments mice that were given kidney grafts in the presence of antibodies that blocked the action of B7 and CD40Ligand accepted and became tolerant to their grafts. In contrast, if CSA was added, the induction of tolerance was inhibited and the grafts were rejected.

Although anergy occurs when T cells are stimulated via the TCR in the absence of costimulation, it does not appear that costimulation alone directly inhibits anergy induction. Rather, TCR stimulation in the presence of costimulation leads to IL-2 production, which leads to T cell proliferation as well as the prevention of anergy induction.[45,46] Hence, it is the action of IL-2 that prevents anergy induction. In fact, T cell clonal anergy can be reversed if the cells are cultured in IL-2 and allowed to proliferate. Recent work has clearly linked cell cycle progression and the prevention of anergy.[47-50] Anergic cells are arrested in G1 and inhibiting cell cycle progression in the presence of signal 1 leads to anergy. In our own in vitro model, we have been able to show that the drug rapamycin, which blocks IL-2-induced cell cycle progression from G1 to S, can promote anergy even in the presence of costimulation. We have been able to exploit this property of rapamycin in an in vivo bone marrow transplantation model. If we give minimally conditioned (300 cGray Total Body Irradiation) B strain parents (BxD)F1 bone marrow under the cover of rapamycin for 4 weeks, we can induce durable long term chimerism in the absence of continued immunosuppression (Powell and Tisdale, manuscript in preparation).

As mentioned earlier, in in vivo anergy models to peripheral antigens, it has been observed that the anergic cells have all proliferated a number of times. At first glance, this observation may seem to contradict the idea that anergy induction occurs when the TCR is stimulated in the absence of cell cycle progression. But recall that a naïve cell cannot be anergized. Only after it has been activated and proliferated will it become susceptible to the negative feedback regulation accompanying a failure to progress through the cell cycle. At that point, any immunologic mediators that tend to inhibit IL-2 production and/or T cell proliferation would favor induction of tolerance (see below). Note, however, that under certain conditions, the converse is not necessarily true, i.e., that proliferation alone is sufficient to prevent anergy induction. Wells et al have isolated CD4+ T cells that have divided several times in vitro following TCR engagement with anti-CD3 and shown that they are unresponsive upon rechallenge unless they have also been signaled with anti-CD28 during the primary stimulation.[51] Such data suggest that under certain activation conditions there is also a role for anti-CD28 signaling, in addition to cell cycle progression, for the prevention of TCR-induced hyporesponsiveness.

Immunoregulation Can Lead to the Induction of Tolerance

The ultimate outcome of an encounter with antigen, that is whether it is activation inducing or tolerance inducing, is dictated by the circumstances in which the antigen is encountered. Since the process of tolerance induction is constantly occurring, molecules that might be thought of as immunoregulatory might also promote tolerance. IL-10 for example was described as a cytokine synthesis inhibitory factor (CSIF).[52] It was initially found to be produced by Th2 cells and clearly had a negative effect on cytokine production by Th1 cells. This effect of IL-10 is mediated through its ability to regulate APC function as well as a direct effect on lymphocytes.[53,54] IL-10 can inhibit the production of IL-2 and subsequent proliferation of T cells. Given this fact, it is not surprising that activation of T cells in the presence of IL-10 has been shown to lead to anergy.[55] Likewise, TGF-beta has also been shown to have a prominent role in promoting tolerance (see below). Although TGF-beta can clearly inhibit T cell apoptosis, it also directly inhibits IL-2 dependent T cell proliferation.[56,57]

Along these lines, molecules or cells that are considered to play an active effector role in one circumstance, might in other circumstances facilitate the induction of peripheral tolerance. Perhaps the best example of this is mucosal immunity and the concept of oral tolerance. The gastrointestinal tract is under constant exposure to the outside environment and a wide variety of pathogens. Consequently, gut associated lymphoid tissue (GALT) is uniquely designed for continued immune surveillance and response.[58] A key component of gut immunity is the local production of antigen-specific IgA. As a result, Peyer's patches are occupied with a large population of TGF-beta producing T cells that provide T cell help to B cells in terms of promoting IgA production. In this context, these T cells are clearly gut effector cells. However, when antigens are fed in low doses to animals and then administered systemically, the gut effector cells become mediators of oral tolerance.[59] Termed Th3 cells, these antigen-specific CD4+ T cells secrete primarily TGF-beta, and when they encounter their cognate antigen at the target organ, they mediate bystander suppression, i.e., following antigen-specific activation they suppress the activation of naïve T cells to other antigens.[60] More recently it has been shown that inhaled antigens (which are encountered by the mucosal immune system of the lungs) have the ability to induce tolerance.[61] In this case, the tolerance is mediated by another type of CD4+ suppressor cell termed a Tr1 regulatory cell.[62] These cells appear to mediate their tolerogenic effect through the elaboration of IL-10. Of note, in all cases it appears that the key player in terms of the generation of these regulatory cells is a polarizing dendritic cell.[61,63] In terms of IL-10 producing Tr1 cells, it appears that an IL-10 producing dendritic cell is involved, while for the Th3 cells it appears that a TGF-beta producing dendritic cell is involved. Along these same lines, it has been shown that the immune response in the anterior chamber of the eye can also induce peripheral tolerance.[64] In this case the tolerance induction appears to be facilitated by alpha melanoyte-stimulating hormone (α-MSH) from neurons in addition to TGF-β and IL-10 from NK T cells.[65,66]

CD4+CD25+ Suppressor Cells and CD8+ Veto Cells

Recently, there has been a resurgence of interest in the role of suppressor cells in regulating immune activation and tolerance. In particular, much work has focused on defining CD4+CD25+ suppressor cells. Nishizuka and Sakakura first noted that mice thymectomized on day 3 of life developed peripheral autoimmune disease, while mice thymectomized on day 1 or 7 did not.[67] Typically the autoimmunity is organ specific e.g., gastritis, orchitis, oophritis, and thyroiditis.[68] It was hypothesized that the prevention of autoimmunity was mediated by a thymus-derived regulatory cell that developed slightly later in ontogeny. Along these lines, the induction of autoimmunity could be prevented by reconstituting the mice with T cells.[69,70] Initially, it was determined that these regulatory T cells could be found in the CD4+ population of cells. Later, it was determined that the key regulatory cells were CD4+ CD45RB+, and subsequently further distinguished as a minor subpopulation of cells that were CD4+ CD25+.[71,72] Indeed, depletion of CD4+CD25+ cells leads to autoimmunity, while reconstitution of thymectomized mice with this population prevents autoimmunity.[73] These cells originate in the thymus and are selected on MHC class II+ cortical epithelium, suggesting that they are part of the normal T cell repertoire.[74] Perhaps, as suggested by Debinski et al, they can be selected by the promiscuous expression in the thymus of genes encoding peripheral antigens.[7] If we define tolerance as immune mechanisms designed to prevent horror autotoxicus, then it appears as if this population of cells plays an important role in maintaining peripheral tolerance.

The exact molecular mechanism by which these cells exert their inhibitory effect is unknown. What is known is that CD4+CD25+ cells need to be activated and that their effect is not mediated through antigen-presenting cells, but rather through direct T-T cell contact.[68] The suppressor cells do not need to be of the same antigen specificity as the suppressed T cells. When stimulated in vitro the CD4+CD25+ cells appear to be anergic in that they do not produce IL-2 nor proliferate. They can be made to proliferate, however, by the addition of IL-2 or anti-CD28 antibodies. CD4+CD25+ suppressor cells appear to mediate their inhibition by down-regulating the ability of the target T cell to produce IL-2, and this is accompanied by cell cycle arrest. Although there are some conflicting observations in the literature, it appears as if the function of CD4+CD25+ suppressor cells is independent of IL-4, IL-10 and TGF-beta.[75,76] In this regard they are a distinct population of cells when compared to Th3 and Tr1 regulatory cells. Interestingly, several reports have identified these cells in humans.[77-82] This raises the possibility that the manipulation of this population of T cells could be employed to control autoimmune disease.

In light of the renewed interest in suppressor cells another set of suppressors is worth noting. Early experiments by Miller and colleagues demonstrated that splenocytes from nude mice were able to inactivate T cells in an MLR if the suppressed T cells were of the same MHC as the splenocytes.[83] They termed these cells veto cells. Although initially described as cells derived from nude mice (these were most likely NK cells) it was later determined that the most potent veto cells were CD8+ cytotoxic T lymphocytes (CTLs).[84] Subsequent experiments were able to demonstrate veto cell activity vivo.[85] It was found that the primary effect of these cells was to markedly reduce CTL precursor frequency.[86] Veto cells do not utilize their TCR to induce suppression. Rather,

MHC class I and CD8 on the surface of the veto cell engage respectively the TCR and the alpha-3 domain of the MHC class I molecule on the target CTLp (the suppressed cell).[87] Although early studies suggested that the veto effect is mediated by signaling via CD8 (on the veto cell) engagement of the alpha-3 domain of MHC class I on the surface of the suppressed cell, recent experiments suggest that the veto effect is ultimately executed by the engagement of FasL on the surface of the veto cell and Fas on the surface of the target cell.[88,89] Whereas the function of CD4$^+$CD25$^+$ suppressor cells appears to be to promote organ specific peripheral tolerance, the exact role of veto cells is unclear. One hypothesis is that they prevent the activation and expansion of CTLp that are specific for activated CTLs.

Interestingly, it is believed that veto cells may be responsible for the prevention of graft rejection and tolerance induced by mismatched bone marrow transplantation utilizing large doses of CD34$^+$ stem cells. In this case the veto effect appears to be mediated by the stem cells.[90] In the case of haploidentical transplants with full conditioning, the doses of CD34$^+$ cells needed is technically feasible. However, with minimal conditioning regimens, the numbers of stem cells required to prevent graft rejection is clinically challenging. In an attempt to overcome this obstacle Reisner et al have developed a mouse model in which they infuse anti-third party CTLs as veto cells.[90] Because such cells are raised against irrelevant third party targets, they can mediate their veto effect and promote engraftment without causing graft vs. host disease.

Conclusions

Much of immunology has been based on the notion that the raison d'etre of the immune system is to distinguish between self and nonself and respond accordingly. The education to self takes place during a fixed developmental period and for the most part (at least for T cells) in a fixed place—the thymus. In current models, tolerance is an ongoing process that can take place anywhere in the periphery in addition to the negative selection process that occurs in the thymus. In such models, the antigen plays an instrumental role in terms of immune recognition. However, it is the context in which the antigen is encountered that dictates the outcome of the response following this recognition. When this ongoing peripheral process goes awry, autoimmunity can develop. Alternatively, tumors and pathogens can exploit this process to induce nonresponsiveness and thus evade the immune response.

When naïve T cells emerge from the thymus, they must encounter antigen presented by a professional APC in order to become antigen experienced. In the face of trauma, inflammation and activation of the innate immune system, the APC will have its full complement of costimulatory molecules on its surface and this will induce full activation of the naïve T cell (including recruitment of lipid rafts to the membrane). If the professional APC has not been activated and is only expressing low levels of accessory molecules, then this interaction with the T cell leads to the induction of tolerance by a number of different possible mechanisms. A common component of TCR-induced hyporesponsiveness is the engagement of the TCR in the setting of cell cycle arrest. IL-10, TGF-beta, costimulatory blockade, CTLA-4 engagement, T suppressor cells, all of which have been shown to promote tolerance, all serve to inhibit IL-2 production/T cell proliferation in response to antigen. Likewise, immunosuppressive agents such as rapamycin that block IL-2-mediated proliferation without blocking TCR-induced signaling have been shown to promote tolerance in experimental systems. However, the use of certain biologic response modifiers clinically has to be tempered by the fact that somewhere else in the immune (or another) system such molecules may have potent negative effects. For transplantation, the task ahead will be to exploit both these biologic and pharamacologic mechanisms of inducing peripheral tolerance with the goal of promoting long term tolerance in the absence of long term immunosuppression. In terms of autoimmune disease, the goal will be to reestablish tolerance. Hematopoeitic stem cell transplantation for autoimmune diseases, as discussed in this text, may allow a window to better understand the mechanisms involved in the re-establishment of tolerance.

References

1. Ehrlich P, Morgenroth J. On haemolysins: Third and fifth communications. The collected papers on Paul Ehrlich. London: Pergamon, 1957:2:205-212.
2. Burnet F, Fenner F. The production of antibodies. London: Macmillan, 1949:102-105.
3. Billingham R, Brent L, Medawar PB. Actively acquired tolerance of foreign cells. Nature 1953; 172:603-606.
4. Triplett EL. On the mechanism of immunologic self recognition. J Immunol 1962; 89:505-510.
5. Kappler JW, Roehm N, Marrack P. T cell tolerance by clonal elimination in the thymus. Cell 1987; 49(2):273-80.
6. Hanahan D. Peripheral-antigen-expressing cells in thymic medulla: Factors in self-tolerance and autoimmunity. Curr Opin Immunol 1998; 10(6):656-62.
7. Derbinski J et al. Promiscuous gene expression in medullary thymic epithelial cells mirrors the peripheral self. Nat Immunol 2001; 2(11):1032-9.
8. Klein L, Kyewski B. "Promiscuous" expression of tissue antigens in the thymus: A key to T-cell tolerance and autoimmunity? J Mol Med 2000; 78(9):483-94.
9. Klein L, Kyewski B. Self-antigen presentation by thymic stromal cells: A subtle division of labor. Curr Opin Immunol 2000; 12(2):179-86.
10. Janeway Jr CA. Approaching the asymptote? Evolution and revolution in immunology. Cold Spring Harb Symp Quant Biol 1989; 54 Pt 1:1-13.
11. Medzhitov R, Janeway Jr C. Innate immunity. N Engl J Med 2000; 343(5):338-44.
12. Medzhitov R, Janeway Jr CA. How does the immune system distinguish self from nonself? Semin Immunol 2000; 12(3):185-8. Discussion 257-344.
13. Matzinger P. Tolerance, danger, and the extended family. Annu Rev Immunol 1994; 12:991-1045.
14. Lafferty KJ et al. Immunobiology of tissue transplantation: A return to the passenger leukocyte concept. Annu Rev Immunol 1983; 1:143-73.
15. Schwartz RH. A cell culture model for T lymphocyte clonal anergy. Science 1990; 248:1349-1356.
16. Kawabe Y, Ochi A. Selective anergy of V beta 8+, CD4+ T cells in Staphylococcus enterotoxin B-primed mice. J Exp Med 1990; 172(4):1065-70.
17. Rellahan BL et al. In vivo induction of anergy in peripheral V beta 8+ T cells by staphylococcal enterotoxin B. J Exp Med 1990; 172(4):1091-100.
18. Pape KA et al. Use of adoptive transfer of T-cell-antigen-receptor-transgenic T cell for the study of T-cell activation in vivo. Immunol Rev 1997; 156:67-78.
19. Rocha B, Grandien A, Freitas AA. Anergy and exhaustion are independent mechanisms of peripheral T cell tolerance. J Exp Med 1995; 181(3):993-1003.
20. Lo D et al. Peripheral tolerance in transgenic mice: Tolerance to class II MHC and nonMHC transgene antigens. Immunol Rev 1991; 122:87-102.

21. Tanchot C et al. Adaptive tolerance of CD4+ T cells in vivo: Multiple thresholds in response to a constant level of antigen presentation. J Immunol 2001; 167(4):2030-9.
22. Adler AJ et al. In vivo CD4+ T cell tolerance induction versus priming is independent of the rate and number of cell divisions. J Immunol 2000; 164(2):649-55.
23. Russell JH et al. Mature T cells of autoimmune lpr/lpr mice have a defect in antigen-stimulated suicide. Proc Natl Acad Sci USA 1993; 90(10):4409-13.
24. Webb S, Morris C, Sprent J. Extrathymic tolerance of mature T cells: Clonal elimination as a consequence of immunity. Cell 1990; 63(6):1249-56.
25. Rocha B, von Boehmer H. Peripheral selection of the T cell repertoire. Science 1991; 251(4998):1225-8.
26. McCormack JE et al. Profound deletion of mature T cells in vivo by chronic exposure to exogenous superantigen. J Immunol 1993; 150(9):3785-92.
27. Kurts C et al. CD4+ T cell help impairs CD8+ T cell deletion induced by cross-presentation of self-antigens and favors autoimmunity. J Exp Med 1997; 186(12):2057-62.
28. Zheng L et al. Induction of apoptosis in mature T cells by tumour necrosis factor. Nature 1995; 377(6547):348-51.
29. Van Parijs L, Peterson DA, Abbas AK. The Fas/Fas ligand pathway and Bcl-2 regulate T cell responses to model self and foreign antigens. Immunity 1998; 8(2):265-74.
30. Boise LH et al. CD28 costimulation can promote T cell survival by enhancing the expression of Bcl-XL. Immunity 1995; 3(1):87-98.
31. Lenardo MJ. Interleukin-2 programs mouse alpha beta T lymphocytes for apoptosis. Nature 1991; 353(6347):858-61.
32. Watanabe-Fukunaga R et al. Lymphoproliferation disorder in mice explained by defects in Fas antigen that mediates apoptosis. Nature 1992; 356(6367):314-7.
33. Straus SE et al. An inherited disorder of lymphocyte apoptosis: The autoimmune lymphoproliferative syndrome. Ann Intern Med 1999; 130(7):591-601.
34. Powell JD et al. Molecular regulation of interleukin-2 expression by CD28 costimulation and anergy. Immunol Rev 1998; 165:287-300.
35. Rocha B, Tanchot C, Von Boehmer H. Clonal anergy blocks in vivo growth of mature T cells and can be reversed in the absence of antigen. J Exp Med 1993; 177(5):1517-21.
36. Tanchot C, Rocha B. The peripheral T cell repertoire: Independent homeostatic regulation of virgin and activated CD8+ T cell pools. Eur J Immunol 1995; 25(8):2127-36.
37. Hayashi RJ et al. Differences between responses of naive and activated T cells to anergy induction. J Immunol 1998; 160(1):33-8.
38. Lerner CG et al. Distinct requirements for C-C chemokine and IL-2 production by naive, previously activated, and anergic T cells. J Immunol 2000; 164(8):3996-4002.
39. Ragazzo JL et al. Costimulation via lymphocyte function-associated antigen 1 in the absence of CD28 ligation promotes anergy of naive CD4+ T cells. Proc Natl Acad Sci USA 2001; 98(1):241-6.
40. Wells AD et al. T cell effector function and anergy avoidance are quantitatively linked to cell division. J Immunol 2000; 165(5):2432-43.
41. Tuosto L et al. Organization of plasma membrane functional rafts upon T cell activation. Eur J Immunol 2001; 31(2):345-9.
42. Quill H, Schwartz RH. Stimulation of normal inducer T cell clones with antigen presented by purified Ia molecules in planar lipid membranes: Specific induction of a long-lived state of proliferative nonresponsiveness. J Immunol 1987; 138(11):3704-12.
43. Jenkins MK et al. Inhibition of antigen-specific proliferation of type 1 murine T cell clones after stimulation with immobilized anti-CD3 monoclonal antibody. J Immunol 1990; 144(1):16-22.
44. Larsen CP et al. Long-term acceptance of skin and cardiac allografts after blocking CD40 and CD28 pathways. Nature 1996; 381(6581):434-8.
45. Jenkins MK. The role of cell division in the induction of clonal anergy. Immunol Today 1992; 13(2):69-73.
46. Beverly B et al. Reversal of in vitro T cell clonal anergy by IL-2 stimulation. Int Immunol 1992; 4(6):661-71.
47. Powell JD, Bruniquel D, Schwartz RH. TCR engagement in the absence of cell cycle progression leads to T cell anergy independent of p27(Kip1). Eur J Immunol 2001; 31(12):3737-46.
48. Boussiotis VA et al. p27kip1 functions as an anergy factor inhibiting interleukin 2 transcription and clonal expansion of alloreactive human and mouse helper T lymphocytes. Nat Med 2000; 6(3):290-7.
49. Powell JD, Lerner CG, Schwartz RH. Inhibition of cell cycle progression by rapamycin induces T cell clonal anergy even in the presence of costimulation. J Immunol 1999; 162(5):2775-84.
50. Gilbert KM, Weigle WO. Th1 cell anergy and blockade in G1a phase of the cell cycle. J Immunol 1993; 151(3):1245-54.
51. Wells AD et al. Signaling through CD28 and CTLA-4 controls two distinct forms of T cell anergy. J Clin Invest 2001; 108(6):895-903.
52. Fiorentino DF, Bond MW, Mosmann TR. Two types of mouse T helper cell. IV. Th2 clones secrete a factor that inhibits cytokine production by Th1 clones. J Exp Med 1989; 170(6):2081-95.
53. Moore KW et al. Interleukin-10 and the interleukin-10 receptor. Annu Rev Immunol 2001; 19:683-765.
54. de Waal Malefyt R, Yssel H, de Vries JE. Direct effects of IL-10 on subsets of human CD4+ T cell clones and resting T cells. Specific inhibition of IL-2 production and proliferation. J Immunol 1993; 150(11):4754-65.
55. Groux H et al. Interleukin-10 induces a long-term antigen-specific anergic state in human CD4+ T cells. J Exp Med 1996; 184(1):19-29.
56. Kehrl JH et al. Production of transforming growth factor beta by human T lymphocytes and its potential role in the regulation of T cell growth. J Exp Med 1986; 163(5):1037-50.
57. Cerwenka A et al. Fas- and activation-induced apoptosis are reduced in human T cells preactivated in the presence of TGF-beta 1. J Immunol 1996; 156(2):459-64.
58. McGhee J, Kiyono H. The mucosal immune system. In: Paul W, ed. Fundamental Immunology. 4 ed. Philadelphia: Lippincott-Raven, 1999:909-946.
59. Weiner HL. Oral tolerance: Immune mechanisms and the generation of Th3-type TGF-beta-secreting regulatory cells. Microbes Infect 2001; 3(11):947-54.
60. Weiner HL. Induction and mechanism of action of transforming growth factor-beta-secreting Th3 regulatory cells. Immunol Rev 2001; 182:207-14.
61. Akbari O, DeKruyff RH, Umetsu DT. Pulmonary dendritic cells producing IL-10 mediate tolerance induced by respiratory exposure to antigen. Nat Immunol 2001; 2(8):725-31.
62. Roncarolo MG et al. Type 1 T regulatory cells. Immunol Rev 2001; 182:68-79.
63. Weiner HL. The mucosal milieu creates tolerogenic dendritic cells and T(R)1 and T(H)3 regulatory cells. Nat Immunol 2001; 2(8):671-2.
64. Streilein JW. Molecular basis of ACAID. Ocul Immunol Inflamm 1997; 5(3):217-8.
65. Sonoda KH et al. NK T cell-derived IL-10 is essential for the differentiation of antigen-specific T regulatory cells in systemic tolerance. J Immunol 2001; 166(1):42-50.
66. Streilein JW et al. Neural control of ocular immune privilege. Ann NY Acad Sci 2000; 917:297-306.
67. Nishizuka Y, Sakakura T. Thymus and reproduction: Sex-linked dysgenesis of the gonad after neonatal thymectomy in mice. Science 1969; 166(906):753-5.
68. Shevach EM. Suppressor T cells: Rebirth, function and homeostasis. Curr Biol 2000; 10(15):R572-5.
69. Penhale WJ et al. Thyroiditis in T cell-depleted rats: Suppression of the autoallergic response by reconstitution with normal lymphoid cells. Clin Exp Immunol 1976; 25(1):6-16.
70. Sakaguchi S, Takahashi T, Nishizuka Y. Study on cellular events in post-thymectomy autoimmune oophoritis in mice. II. Requirement of Lyt-1 cells in normal female mice for the prevention of oophoritis. J Exp Med 1982; 156(6):1577-86.
71. Fowell D et al. Subsets of CD4+ T cells and their roles in the induction and prevention of autoimmunity. Immunol Rev 1991; 123:37-64.

72. Sakaguchi S et al. Immunologic self-tolerance maintained by activated T cells expressing IL-2 receptor alpha-chains (CD25). Breakdown of a single mechanism of self-tolerance causes various autoimmune diseases. J Immunol 1995; 155(3):1151-64.
73. Salomon B et al. B7/CD28 costimulation is essential for the homeostasis of the CD4+CD25+ immunoregulatory T cells that control autoimmune diabetes. Immunity 2000; 12(4):431-40.
74. Bensinger SJ et al. Major histocompatibility complex class II-positive cortical epithelium mediates the selection of CD4(+)25(+) immunoregulatory T cells. J Exp Med 2001; 194(4):427-38.
75. Thornton AM, Shevach EM. Suppressor effector function of CD4+CD25+ immunoregulatory T cells is antigen nonspecific. J Immunol 2000; 164(1):183-90.
76. Nakamura K, Kitani A, Strober W. Cell contact-dependent immunosuppression by CD4(+)CD25(+) regulatory T cells is mediated by cell surface-bound transforming growth factor beta. J Exp Med 2001; 194(5):629-44.
77. Dieckmann D et al. Ex vivo isolation and characterization of CD4(+)CD25(+) T cells with regulatory properties from human blood. J Exp Med 2001; 193(11):1303-10.
78. Jonuleit H et al. Identification and functional characterization of human CD4(+)CD25(+) T cells with regulatory properties isolated from peripheral blood. J Exp Med 2001; 193(11):1285-94.
79. Stephens LA et al. Human CD4(+)CD25(+) thymocytes and peripheral T cells have immune suppressive activity in vitro. Eur J Immunol 2001; 31(4):1247-54.
80. Taams LS et al. Human anergic/suppressive CD4(+)CD25(+) T cells: A highly differentiated and apoptosis-prone population. Eur J Immunol 2001; 31(4):1122-31.
81. Levings MK, Sangregorio R, Roncarolo MG. Human CD25(+)CD4(+) T regulatory cells suppress naive and memory T cell proliferation and can be expanded in vitro without loss of function. J Exp Med 2001; 193(11):1295-302.
82. Baecher-Allan C et al. CD4+CD25 high regulatory cells in human peripheral blood. J Immunol 2001; 167(3):1245-53.
83. Miller RG, Derry H. A cell population in nu/nu spleen can prevent generation of cytotoxic lymphocytes by normal spleen cells against self antigens of the nu/nu spleen. J Immunol 1979; 122(4):1502-9.
84. Miller RG et al. The veto phenomenon in T-cell regulation. Ann N Y Acad Sci 1988; 532:170-6.
85. Rammensee HG, Fink PJ, Bevan MJ. Functional clonal deletion of class I-specific cytotoxic T lymphocytes by veto cells that express antigen. J Immunol 1984; 133(5):2390-6.
86. Martin DR, Miller RG. In vivo administration of histoincompatible lymphocytes leads to rapid functional deletion of cytotoxic T lymphocyte precursors. J Exp Med 1989; 170(3):679-90.
87. Sambhara SR, Miller RG. Programmed cell death of T cells signaled by the T cell receptor and the alpha 3 domain of class I MHC. Science 1991; 252(5011):1424-7.
88. Reich-Zeliger S et al. Anti-third party CD8+ CTLs as potent veto cells: Coexpression of CD8 and FasL is a prerequisite. Immunity 2000; 13(4):507-15.
89. George JF, Thomas JM. The molecular mechanisms of veto mediated regulation of alloresponsiveness. J Mol Med 1999; 77(7):519-26.
90. Reisner Y. Stem cell transplantation across major genetic barriers. Ann NY Acad Sci 2001; 938:322-6. Discussion 326-7.

CHAPTER 19

Dendritic Cells Control the Balance between Tolerance and Autoimmunity

Simon W. F. Milling and G. Gordon MacPherson

Introduction

Since their first description in 1973,[1] many important roles have been described for dendritic cells (DC) in the induction of immunity. DCs are a heterogeneous population of bone marrow derived leukocytes that efficiently link the innate and adaptive immune systems and play a crucial part in initiating, amplifying and controlling the immune response to pathogenic microorganisms. They have more recently been shown also to be important for the maintenance of tolerance to self-antigens. A more detailed understanding of the development and functions of this diverse family of cell types will doubtless provide increased opportunities for the therapeutic control of autoimmunity.

Features Common to All Dendritic Cells

DCs derive their name from the dramatic "dendritic" morphology they exhibit when fully mature. They are specialized antigen presenting cells and are highly adapted for the uptake, processing and presentation of antigens to T cells. In their immature state DCs act as sentinels, sampling antigens from their extracellular environment. After activation by inflammatory mediators or direct recognition of pathogen-associated molecules, DCs secrete pro-inflammatory cytokines and begin to mature. Mature DCs have the ability to activate naïve CD4$^+$ and CD8$^+$ T cells both in vitro and in vivo (Figs. 1 and 2). This ability makes them crucially important for the initiation of T cell responses against previously un-encountered antigens. Associated with this function is the expression of characteristic molecules that are common to DCs of all types. All mature DCs, for instance, express on their surface high levels of MHC class II molecules, which can carry pathogen-derived peptide fragments and mediate the interaction between DCs and CD4$^+$ T cells. Mature DCs also express significant amounts of the cell-surface molecules that are required for effective stimulation of naïve T lymphocytes, including CD40, CD80, and CD86. Although all DCs share some common characteristics, an array of different types have been described. For the purpose of this discussion, they may be divided into three distinct classes: the myeloid-derived interstitial DCs and Langerhans cells and the lymphoid-derived plasmacytoid DCs. There are other types of cells which share a dendritic morphology but do not have the migratory or antigen-presenting capacity of the DCs described below, and are not derived from bone marrow, including the follicular dendritic cells that facilitate antigen recognition by B cells in the follicles of germinal centres. Discussion of such cells will not form part of this review.

The Three Classes of DCs

Some of the best-described subsets of DCs have been identified in the mouse spleen. This contains three distinct populations of interstitial DCs, all of which express MHC class II and CD11c molecules at high levels and are capable of stimulating naïve lymphocytes in a mixed leukocyte culture system.[2] These three populations can be discriminated based on their expression of surface markers commonly used to differentiate between T cells, CD4 and CD8. Unlike T cells, however, the DCs express the CD8 molecule as a homodimer of two CD8α chains, while CD8$^+$ T cells express a CD8αβ heterodimer. The DCs are CD4$^-$CD8α$^-$ ("double negative" or DN), CD4$^+$CD8α$^-$, or CD4$^-$CD8α$^+$. All these subtypes in spleen have a rapid turnover, with a half-life in spleen of 1.5 to 2.9 days, all develop from bone marrow precursors, and none of the subtypes is the precursor of another. CD8α$^+$ DCs are found in the T-cell-rich areas of the spleen, lymph nodes and Peyer's patches. By contrast, CD8α$^-$ DCs are localized in the marginal zones of the spleen, the sub-capsular sinuses of the lymph nodes, and the sub-epithelial dome of the Peyer's patches. These CD8α$^-$ DCs can, however, rapidly migrate to the T cell areas after stimulus by microbial products including lipopolysaccharide[3] (LPS) or extracts from toxoplasma.[4]

Recent analysis of the three splenic CD11c^{high} DC populations using microarray technology has clarified some of the relationships between them.[5] The CD8α$^-$ and CD8α$^+$ subsets express distinctly different profiles of basal gene expression, differing by more than 200 individual genes. The microarray expression profiles of CD8α$^-$ and double negative cells, however, differed by fewer than 25 genes, suggesting a much closer phylogenetic linkage between these two subsets. The functional differences between the three subtypes of splenic DCs in the mouse are obscured by their high degree of plasticity and depend on both the type of pathogen and the cytokine microenvironment, as well as on the tissue origin of the different cells. For example, different outcomes were observed when splenic DCs were exposed to either the unicellular yeast or the filamentous forms of the fungus *Candida albicans*. In vitro, ingestion of yeasts activated DCs for IL-12 production and primed T$_H$1 cells, whereas ingestion of hyphae inhibited IL-12 and induced IL-4 production.[6]

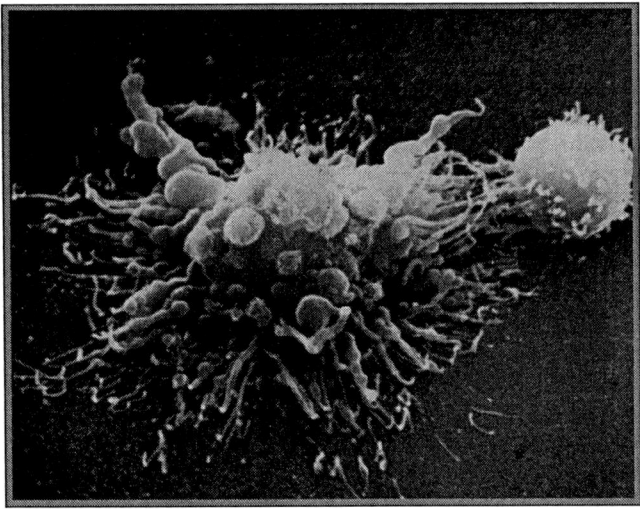

Figure 1. Scanning electron micrograph of an interaction between a dendritic cell (DC) and a T cell. Cells purified from rat pseudoefferent lymph. (GG MacPherson, unpublished data).

In mouse lymph nodes at least two more populations of CD11c+ DCs are present,[7] both of which appear to drain from the skin. These additional populations both express only low levels of CD8, but either moderate or high levels of the multilectin receptor DEC-205 (CD205), and may represent DCs derived from interstitial tissues, and mature Langerhans cells (LC) trafficking from the dermis of the skin. LCs are a well-known group of DCs which express characteristic markers such as Birkbeck granules, Langerin and E-cadherin.[8] They are typically localized in the basal and suprabasal layers of the epidermis and migrate to the T cell areas of regional lymph nodes after activation. Having migrated from the skin and matured, LCs are extremely potent antigen presenting cells.[9] LCs are present in skin-draining lymphatic vessels under steady-state conditions, and the size of the population has recently been shown to be maintained by division of LCs within the skin. Only under highly inflammatory stimuli, for example after UV irradiation, are precursors recruited from the bone marrow.[10]

As well as these interstitial and Langerhans cells, the third class of DCs, the plasmacytoid DCs, have recently begun to be systematically characterized. These cells display a plasmacytoid morphology when immature and, upon recognition of microbial antigens, they develop characteristic dendritic processes and secrete high concentrations of type 1 interferons; they had in fact been previously identified as "interferon producing cells".[11] Their precursors, plasmacytoid pre-DCs, appear to be a specialized cell lineage which circulate in the blood and enter inflamed lymph nodes at high endothelial venules, where they secrete large amounts of type I interferon.[12] Microbial recognition also triggers the plasmacytoid pre-DCs to mature into DCs able to trigger the adaptive immune response. Plasmacytoid DCs were first identified in 1958 as cells with the same morphology as immunoglobulin-producing plasma cells, but localized within the T cell areas of reactive lymph nodes.[13] Since their initial identification, they have

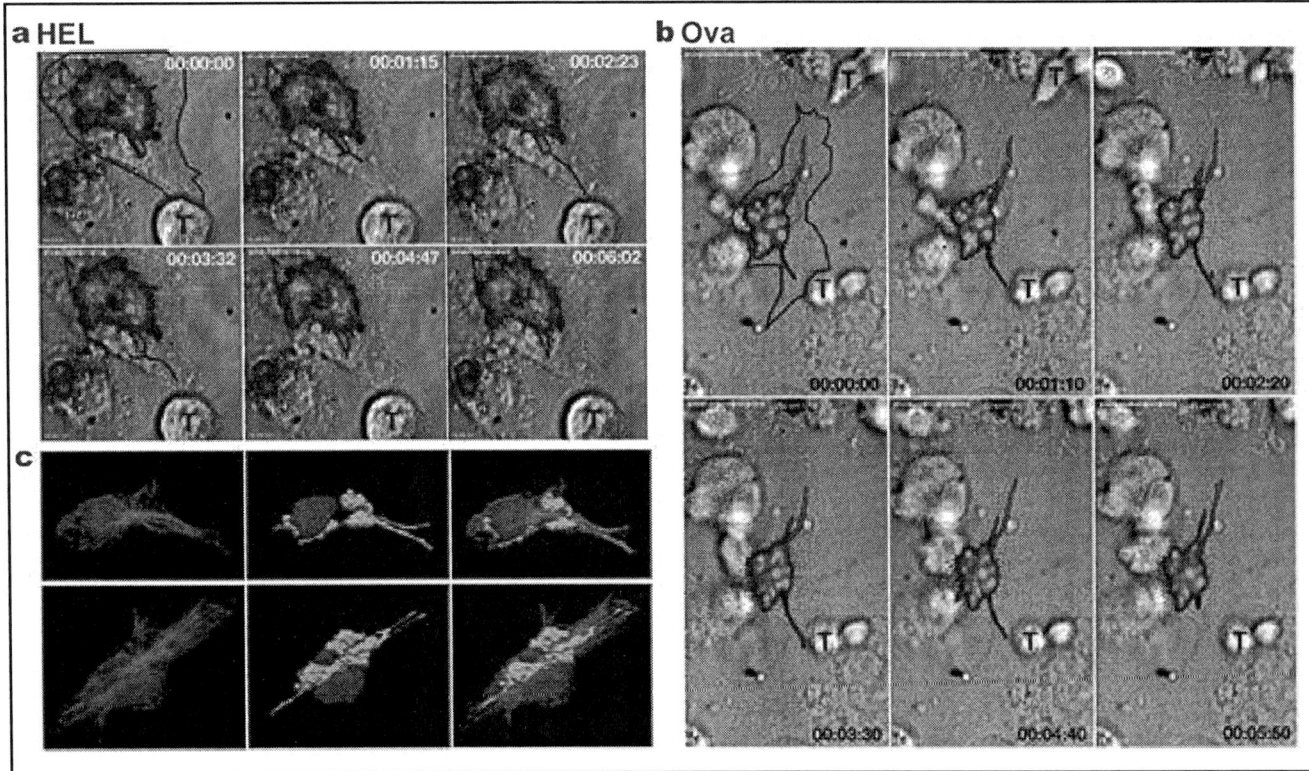

Figure 2. Microtubule-driven directional transport of endosomal compartments on T-cell contact. a) Dendritic cells presenting the hen egg lysozyme (HEL) protein and T cells specific for HEL(46–61) peptide. Time-lapse confocal microscopy in a single optical section. Scale bar, 13 μm. b) As in a, except that dendritic cells were incubated with ovalbumin and T cells specific for OVA(323–339) were used. Scale bar, 22 μm. c) T cells specific for HEL(46–61) peptide were added to HEL-loaded dendritic cells and microtubuli were visualized by staining for tubulin, nuclei are stained. Reprinted with permission from Boes et al.[70] Copyright 2002 Nature Publishing Group.

been assigned a number of labels as their phenotypic details were discovered. Feller et al,[14] reported expression of CD4 and accordingly named them "plasmacytoid T cells". Shortly afterwards, Facchetti observed that the cells lacked the CD3 component of the T cell receptor,[15] and as they express MHC class II molecules, it was inferred that they might be of the monocyte/macrophage lineage. They were thus named "plasmacytoid monocytes". Both these cell types, along with HLA-DR+, CD11c- cells from human blood which differentiate into DCs after overnight exposure to monocyte-conditioned medium[16] and CD4+ CD3- DC-like cells identified by histochemistry were shown to represent the same cell type, now called plasmacytoid pre-DC.[17]

While it has been more difficult to study the phenotype of DCs in healthy human tissues, detailed analyses have been performed after extraction of DCs and their precursors from peripheral blood, as discussed below. All three classes of DCs have, however, been identified in samples obtained after routine removal of noninflamed tonsils,[18] or from human fetal lymphoid organs.[19] Although none of these human DC subsets expresses the CD8 antigen, they can be defined on the basis of their expression of different levels of MHC class II molecules (HLA-DR), CD11c, and the IL-3 receptor, CD123. Human plasmacytoid DCs are similar to mouse plasmacytoid DCs; they express moderate levels of HLA-DR, are CD11c- and CD123+. Three subsets of human myeloid DCs have been identified.[18] All are CD11c+ and express moderate to high levels of HLA-DR. Activated myeloid DCs express high levels of HLA-DR, and the costimulatory molecules CD40 and CD86. Two subsets of less-mature myeloid DCs are also present in human tonsils and can be differentiated based on their expression of CD13. It is not yet clear how these subsets relate to the CD8α+ and CD8α- mouse lymph node DC subsets. Human tonsils also contain two other DC subsets, a novel CD11c-, CD123- population, and the follicular dendritic cells found in germinal centers.

Human and Mouse DCs Are Derived from Two Lineages

The developmental origins of the DC subsets have been the focus of much attention, particularly the question of whether subsets arise from either lymphoid or myeloid precursors. For a time it was thought that CD8α- and CD8α+ DCs were of different lineages, as CD8 was considered a defining marker for lymphoid cells.[20] Detailed analysis of the lineages of DCs that can be readily identified in human blood suggests that it is CD11c+ DCs and plasmacytoid DCs (Fig. 3), in fact, which derive differentially from myeloid and lymphoid progenitors. For instance, freshly-isolated CD11c- plasmacytoid pre-DCs not only possess a lymphoid morphology and express CD123, the IL-3 receptor, but also lack expression of the myeloid markers CD33, CD13 and the receptor for macrophage-colony stimulating factor (M-CSF).[21] After prolonged culture with cytokines known to favor the development of myeloid DCs (GM-CSF and TNFα), these cells show no evidence of the development in this direction.

Their heterogeneous nature makes it difficult to identify a single developmental pathway for the generation of myeloid DC populations. The CD11c+ lineage of DCs express the myeloid markers CD33, CD13 and M-CSF receptor and possess a characteristic hyper-lobulated nucleus.[21] These DCs also demonstrate a higher capacity for antigen uptake, by both phagocytosis and macropinocytosis. Under all culture conditions, they retain their myeloid markers and differentiate into a mixture of macrophages and DCs, facts that are all strongly suggestive of a myeloid heritage for CD11c+ DCs.

Although the developmental lineages of CD11c+ DCs and plasmacytoid DCs have been clearly defined, many aspects of their development from precursors have yet to be described. It has long been known that human peripheral blood monocytes are able to differentiate into myeloid CD11c+ DCs when cultured with IL-4 and GM-CSF.[22] It is also known that DC progenitors are present in bone marrow; a subset of CD34+ hematopoietic progenitors gives rise to all blood cells and DCs. An outline of the soluble factors required to generate human DCs from their precursors shown in Figure 4.[23] The molecular events that control the development of DCs from these precursors are largely undefined, and the relationships between DCs cultured in vitro and the subsets identified in vivo are uncertain. However, certain transcription factors have been shown to be important for the generation of myeloid DCs. The *Ikaros* transcription factor is a regulator of lymphopoiesis. Mice with a dominant-negative Ikaros gene fail to produce CD8α+ or CD8α- DCs.[24] If Ikaros or RelB are disrupted, however, CD8α+ DCs can be generated in the absence of CD8α- DCs.[25] The PU.1 gene is a myeloid regulator transcription factor and is also involved in the generation of myeloid DCs, although it is not clear whether it is required for the development of both CD8α- and CD8α+ DCs,[26] or for CD8α- DCs alone.[27] A partner gene for PU.1, ICSBP (interferon consensus sequence-binding protein) has also been shown to be critical both for the early differentiation, and for later development of CD8α+ DCs.[28]

In mice, LC require tumor growth factor beta (TGFβ) for their generation, as demonstrated by their absence from TGFβ-/- mice.[29] Human LC are also dependent on TGFβ, as they can be generated from CD34+ progenitors in vitro using a cocktail of cytokines including granulocyte macrophage colony stimulating factor (GM-CSF), stem cell factor (SCF), tumor necrosis factor alpha (TNFα), Flt3 ligand (Flt3L), and TGFβ1.[30] Although cultured and ex vivo LC share a similar phenotype, little is yet known about the molecular events or transcription factors that must be expressed to differentiate them from other DCs.

Plasmacytoid pre DCs can be generated in vitro from fetal liver, bone marrow or cord blood CD34+ progenitors in the presence of Flt3L,[31] a member of the platelet-derived growth factor (PDGF) receptor superfamily.[32] Flt3L acts very early in haematopoiesis, on haematopoietic progenitor cells of the DC lineage, causing expansion of both myeloid and lymphoid progenitor cells.[33] Unlike myeloid DCs, however, plasmacytoid pre-DCs do not require GM-CSF for their development and they carry a number of markers expressed in common lymphoid progenitors but not on myeloid cells, including the pre-T cell antigen receptor α chain,[34] and λ-like chains.[35] The lymphoid-restricted transcription factor Spi-B (spleen focus-forming provirus integration B), which impedes the development of T, B and NK cells, is also uniquely expressed in plasmacytoid DC, not monocyte-derived DC.[36] Id (inhibitor of DNA binding) proteins are negative regulators of basic helix-loop-helix (bHLH) transcription factors and inhibit differentiation. Id2 and Id3 proteins inhibit the development of plasmacytoid DCs but not myeloid DCs, or T or B cells when hyper-expressed in CD34+ CD38- progenitors from human fetal liver.[37] A greater understanding of these processes would enable better in vitro generation of specific DC subsets and facilitate improved DC-mediated therapy.

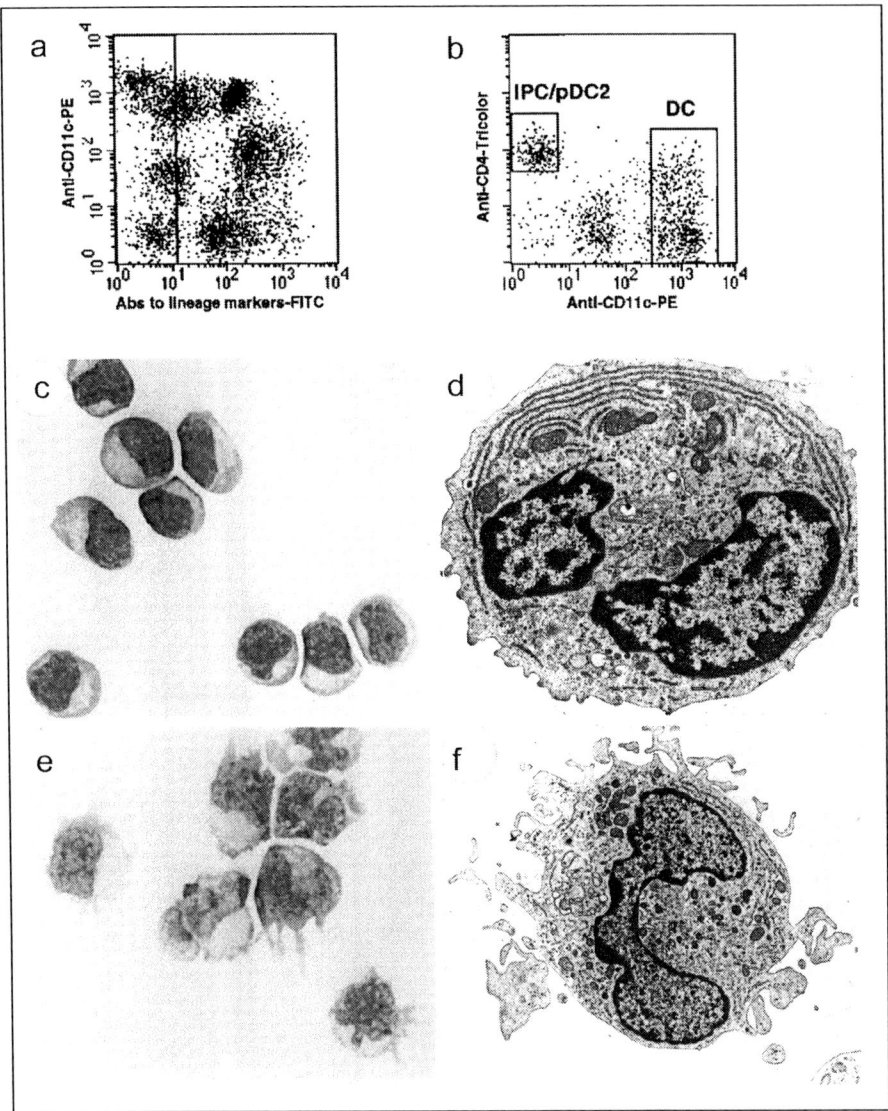

Figure 3. Tracing and isolation of immature plasmacytoid dendritic cells (IPCs)/plasmacytoid dendritic cells (pDC2s) from human peripheral blood. T cells, B cells, NK cells, and monocytes were depleted from blood mononuclear cells. The cells were stained with anti-CD4, anti-CD11c and a mixture of fluorescein isothiocyanate-labeled antibodies to CD3, CD15, CD16, CD20, CD57, CD14 and CD34. Within the lineage-negative population (a) CD4$^+$ CD11c$^-$ IPCs and CD11c$^+$ immature DCs were isolated (b) IPCs are plasmacytoid by Giemsa staining (c) and contain rough endoplasmic reticulum, and Golgi apparatus under transmission electron microscopy (d). The CD11c$^+$ blood immature DCs display dendrites (e and f). Reprinted with permission from Siegal et al.[11] Copyright 1999 American Association for the Advancement of Science.

DC Phenotypes and Functions

T cells are the middle managers of the immune system; they act on instructions from DCs to control potentially damaging situations in the most appropriate fashion. For instance, in response to intracellular microorganisms, CD4$^+$ T cells differentiate into Th1 cells, which secrete interferon-γ (IFN-γ) and support the proliferation of pathogen specific cytotoxic CD8$^+$ T cells that will kill infected cells. By contrast, the response to extracellular pathogens, such as Schistosomes, requires CD4$^+$ T cells to become Th2 cells which secrete IL-4, IL-5 and IL-10 and support IgE and eosinophil-mediated destruction of the pathogenic organism.[38]

The details of how CD4$^+$ T cells are directed to differentiate into Th1 or Th2 cells are only beginning to be elucidated, and DCs appear to play a central role. In an interaction between T cells and mature CD11c$^+$ DCs, the outcome depends on factors such as the pathogen involved, the microenvironment, and the particular DCs and T cells involved. It was initially demonstrated that CD8$^+$ DCs would elicit Th1 and CD8$^-$ DCs a Th2-type response after being pulsed with antigen and injected into the footpad of syngeneic mice.[39] In vivo studies such as this are difficult to interpret, however, as differential migration characteristics or longevity of different DC populations might alter the amount of antigen delivered to T cells and affect the results. In vitro, the effecter phenotype induced by mature myeloid DCs depends on the antigen dose, the method of DC purification, and the ratio of T cells to antigen presenting cells, but is only marginally affected by the DC subtype.[40] In response to micro-

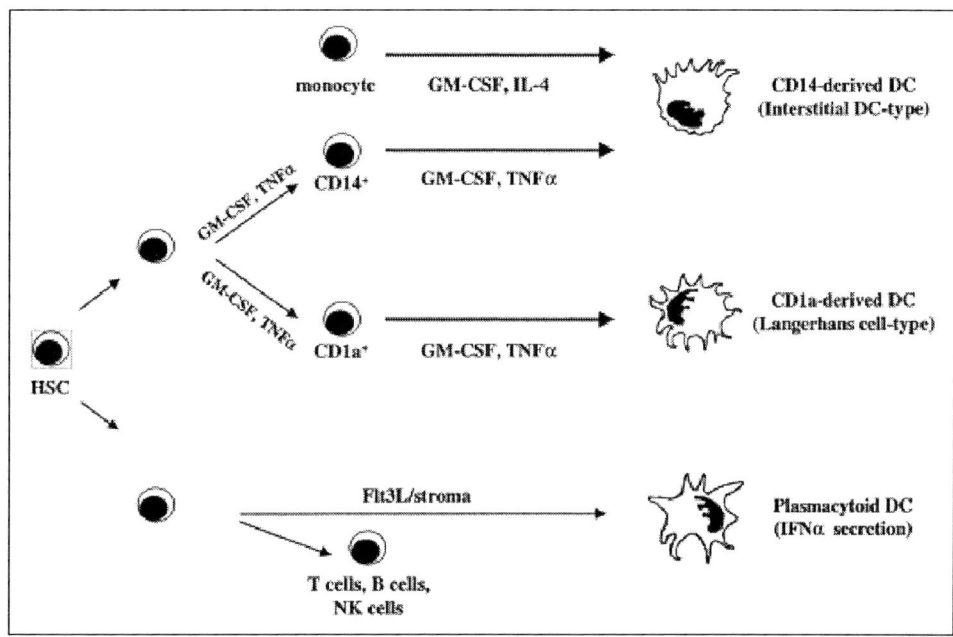

Figure 4. Different types of human dendritic cells (DC) can be generated in vitro depending on the cytokines used.[23]

bial stimuli known to induce IL-12 production, all subsets of CD11c⁺ DCs can support the development of Th1 cells, while IL-10-generating stimuli support the development of Th2 cells. Plasmacytoid DCs have also been shown to be able to support the proliferation of both Th1 and Th2 cells. In fact, higher doses of antigen tend to cause both plasmacytoid and CD11c⁺ DCs to favor the development of Th1 responses.[41]

In common with the other cells of the innate immune system, DCs express molecules that enable them to detect structures not normally found in self-tissues. These "pattern recognition receptors" include Dectin-1,[42] DC-SIGN,[43-44] and molecules of the toll-like receptor (TLR) family. TLRs 1 to 10 form a group of ten microbe-recognition receptors that are central to effective innate immunity. These transmembrane signaling receptors bind a wide variety of ligands, including proteins (flagellin), modified lipids, lipopolysaccharides (LPS), and nucleic acids (DNA and double-stranded RNA).[45] Human peripheral blood DCs can be differentiated based on the TLR molecules they express. For instance, plasmacytoid pre-DCs express high levels of TLR7 (natural ligand unknown) and TLR9 (binds CpG oligonucleotides) and low levels of TLR1 (natural ligand unknown), TLR6 (heterodimer with TLR2, and possibly others) and TLR10 (ligand unknown). On the other hand, CD11c⁺ immature DCs in blood express TLR1 (heterodimer with TLR2), TLR2 (binds peptidoglycan) and TLR3 (binds poly I:C), TLR4 (binds LPS), TLR5 (binds flagellin), TLR6 and TLR8 (ligand unknown) (see Table 1).[46] CpG, the ligand for TLR9 caused plasmacytoid DCs, which express TLR9, to stimulate a Th1 response. On the other hand, CD11c⁺ DCs stimulated the development of a Th1 response after exposure to the TLR4 ligand, LPS.[41] These data suggest that DC subsets are highly plastic in their responses to infection and that the type of the CD4⁺ T cell response depends not only on the type of DC which is initially activated, but also on the antigen dose, the state of maturation of the DC, and the ligation of pattern recognition receptors on the DC surface. Expression by DCs of overlapping groups of TLRs recognizing a wide variety of microbial structures provides important redundancy and helps ensure rapid innate recognition of diverse microbial threats.

DCs in Control of Tolerance vs. Autoimmunity

With their specialized antigen uptake and presentation abilities and their unique migratory capacity, dendritic cells are ideally positioned to play a pivotal role in both the initiation of immune responses and the maintenance of tolerance to self-antigens. Defects in the dendritic cell system can therefore not only result in the failure to mount a response to a pathogenic microorganism, but also lead to the development of autoimmunity.

T cells pass through the thymus as they develop. Thymocytes express a diverse repertoire of different T cell receptor-encoded specificities for MHC/peptide molecules. To prevent autoimmunity, it is crucial that the majority of self-reactive T cells are deleted; this occurs through a process called "negative selection" which occurs in the medulla of the thymus and appears to be mediated by dendritic cells. Thymic epithelial cells express a key transcription factor called AIRE (autoimmune regulator) that promotes ectopic expression of antigens normally restricted to peripheral tissues.[47] Dendritic cells are crucial for negative selection as they are thought to cross-present such antigens,[48] causing deletion of self-reactive thymocytes and the establishment of a safe repertoire of peripheral T cells. In fact, autoimmune CD4⁺ T cells comprise up to 5% of the peripheral repertoire in relB-/- mice, which lack CD8α⁻ and the majority of their functional CD8α⁺ DCs.[49]

It is widely believed that in the steady state, DCs contribute to the maintenance of peripheral tolerance by taking up local tissue antigens and trafficking to the local lymph nodes, even in the absence of inflammatory stimuli. A subset of DCs which contain fragments of intestinal epithelial cells are found in rat intestinal lymph and the T cell areas of the mesenteric lymph nodes under nonperturbed conditions. These cells were considered likely to play a role in inducing and maintaining peripheral self-tolerance.[50] Migrating DCs can stimulate different responses to phagocytosed material depending on whether it originated from apoptotic or necrotic cells.[51] Immature DCs are able to take up fragments from both apoptotic and necrotic tumor cells but only mature after exposure to the latter. DCs exposed to necrotic tumor cells express higher levels of CD40 and CD86 and become

Table 1. Functional TLR repertoire in human DC subsets[46]

	IPCs (Plasmacytoid DC Blood Precursors)	Blood CD11c + DCs (Immature Myeloid DCs)	Monocytes (Myeloid DC Precursors)
Expression of TLRs			
TLR1	low	+	+
TLR2	–	+	+
TLR3	–	+	–
TLR4	–	low	+
TLR5	–	low	+
TLR6	low	low	low
TLR7	+	low	low
TLR8	–	low	+
TLR9	+	–	–
TLR10	low	low	–
Cytokine production in response to ligand for TLRs			
PGN (ligand for TLR2)	–	IL-12/IL-6/TNF-α	IL-6/TNF-α
Poly I:C (ligand for TLR3)	–	IL-12/IFN-α	–
LPS (ligand for TLR4)	–	IL-12/IL-6	IL-6/TNF-α
Flagellin (ligand for TLR5)	?	?	?
Imidazoquinoline compounds (ligands for TLR7)	IFN-α	IL-12	?
CpG-ODNs (ligands for TLR9)	IFN-α	–	–

IPCs= immature plasmacytoid dendritic cells; IL-12= interleukin-12; IL-6= interleukin-6; TNF-α= tunor necrosis factor alpha; IFN-α= interferon alpha; DC= dendritic cell; TLR= toll-like receptor; PGN= petidoglycan; Poly I:C= polyinosinic polycytidylic acid; CpG-ODNs= cytosine quanine oligonucleotides; LPS= lipopolysaccarides.

potent stimulators of both CD4+ and CD8+ T cells; necrotic cells therefore provide a trigger that is essential for the initiation of immunity.

Another piece of this puzzle has recently been put in place by painstaking microscopy. MHC class II presentation of H+/K+-ATPase, an enzyme specifically expressed by gastric parietal cells, causes destructive autoimmune gastritis in BALB/c mice lacking CD4+CD25+ regulatory T cells. Under steady state conditions, however, fragments of gastric parietal cells expressing H+/K+-ATPase could be observed by immunofluorescence microscopy apparently within CD11c+ DCs in gastric mucosa.[52] The parietal (PC)-specific enzyme could also be detected within both CD8α+ and CD8α- CD11c+ cells in the gastric draining lymph nodes. These data demonstrate that DCs are not only potent initiators of immune responses, but also that they present antigens to the local lymph node under noninflammatory conditions. Steady state presentation of a tissue-specific self-antigen neither leads to full deletion of autoreactive T cells, nor induces active autoimmunity.

The interaction between DCs and CD4+ T cells dictates whether tolerance or immunity to an antigen will be generated, and this has been studied in some detail. In vitro, human naïve allogeneic T cells were either exposed to immature monocyte-derived DCs or to DCs matured by exposure to IL1β, TNFα, IL-6 and PGE$_2$.[53] Repetitive stimulation with mature DCs resulted in strong expansion of alloreactive Th1 cells. On the other hand, repetitive stimulation with immature DCs caused an up-regulation of the negative regulator cytotoxic T lymphocyte-associated molecule 4 (CTLA-4) and eventual differentiation of nonproliferating IL-10-producing T cells. In culture, these T cells inhibited antigen-driven proliferation of Th1 cells in a contact-dependent, antigen-nonspecific manner and introduced the possibility that immature DCs might be tolerogenic. In vivo, a similar phenomenon was observed; injection of immature DCs pulsed with influenza matrix peptide (MP) and keyhole limpet hemocyanin into healthy volunteers led to the production of MP-specific IL-10-producing cells and inhibited MP-specific CD8+ effecter function.[54,55] Inhibition was contact dependent and therefore independent of IL-10 secretion.

The relationship between CD11c+ DCs and T cells has also been studied using OT-1 TCR transgenic mice, specific for a chicken ovalbumin (OVA) peptide presented by MHC class II molecules of the H-2K^b haplotype.[56] When injected, OVA-loaded dying cells were phagocytosed by CD8+ DCs that drove large numbers of transgenic T cells to divide and be deleted. The animals became tolerant to challenge with OVA in complete Freunds adjuvant. In contrast, if DCs were activated at the same time as the dying cells were administered, T cells proliferated more strongly and became activated. The animals also retained responsiveness to challenge with OVA. In the steady state, the presentation of antigens by DCs therefore appears to dampen the response to self-antigens, reducing the likelihood of developing autoimmunity when dying cells are processed during an infection.

Plasmacytoid DCs have also been shown to have some tolerogenic activity.[57] If human precursor plasmacytoid DCs are matured in vitro with CD40-ligand and IL-3 they express high levels of cell-surface MHC class II, CD80 and CD86 molecules. They are also strong stimulators of CD4+ T cells, supporting production of Th2-type cytokines.[17] However, their effect on CD8+ T cells is to generate anergic, non-lytic, IL-10-producing cells. This effect is mediated by IL-10 produced during the primary stimulation. Secretion of IL-10 enables these anergic CD8+ T cells to suppress antigen-specific stimulation of CD4+ T cells in vitro. Furthermore, IL-10 production by activated anergic CD8+ T cells could also inhibit other local T cells responses, regardless of their antigen-specificity, providing yet another mechanism by which DCs may maintain peripheral tolerance.

The nature of the tolerising T cells generated after exposure to either immature CD11c+ DCs or to plasmacytoid DCs deserves further study to determine whether there are two redundant mechanisms for maintaining peripheral tolerance. There are important differences between the suppressive T cells. The regulatory subset formed after exposure to immature CD11c+ DCs suppresses through a contact-dependent mechanism, while the plasmacytoid DC (pDC)-created subset functions in a contact-independent manner through secretion of IL-10 and is antigen-specific, mediating antigen-driven bystander suppression. In fact, it appears that regulatory cells produced by immature CD11c+ DCs are similar to the previously described CD4+CD25+ (Treg) cells, and those generated after exposure to plasmacytoid DCs are most similar to the Tr1 subset (refer to Chapter 20).

A previously undescribed population of DCs may also maintain peripheral tolerance by stimulating the cytokine-dependent regulatory CD4+ T cells (Tr1 cells) that prevent autoimmunity.[58] CD45RB^{high}CD11c^{low} DCs can be generated in vitro in the presence of IL-10. These DCs have a plasmacytoid morphology and can also be identified in spleen and lymph nodes of normal mice. When OVA-pulsed CD45RB^{high}CD11c^{low} DCs are injected into BALB/c mice containing DO11.10 OVA-specific T cells an OVA-unresponsive state is generated, and when cultured with OVA-specific naïve CD4+ T cells from DO11.10 transgenic mice they induce the differentiation of IL-10-producing Tr1 cells. This DC population may therefore contribute to the prevention of autoimmunity.

DCs in Autoimmune Disease

DCs are found within the inflamed tissues of many autoimmune disorders, including the rheumatoid synovium,[59] the salivary and lacrimal glands in Sjögren's syndrome,[60] the islets of the pancreas of non-obese diabetic mice in a model of type 1 diabetes,[61] the cerebral spinal fluid in multiple sclerosis,[62] the colonic mucosa in Crohn's disease,[63] and the colonic lamina propria in mice with T cell-induced colitis.[64] During the early stages of disease, they are thought to traffic into the affected tissue in response to locally-produced chemokines and cytokines. They then mature and migrate via the lymphatics to the draining lymph nodes where they are able to present autoantigens to T cells that act to maintain the inflammatory lesion. This hypothesis is supported by experiments in which autoantigens from synovial fluid of rheumatoid arthritis patients were efficiently presented by DCs,[65] suggesting that antigen presentation by DCs plays a role in the pathogenesis of human rheumatoid arthritis.

While there is little doubt that DCs are able to contribute to the maintenance of autoimmunity, it is more difficult to define their role in the initiation of autoimmune diseases. Evidence that DCs might initiate autoimmune pathology comes from studies where patients have been injected with immunogenic DCs to treat melanoma; in a minority of patients the therapy leads to vitiligo, an autoimmune syndrome characterized by killing of melanocytes and depigmentation of the skin.[66] These symptoms only continued for the duration of the treatment, however, and did not lead to other clinical manifestations. A cautionary note has been sounded, however, after experiments in mice demonstrated that susceptible animals could be killed by the autoimmune sequelae of DC treatment.[67] DCs were purified either from mice susceptible to autoimmunity (NZB x NZW) F(1) or from nonsusceptible BALB/C animals. After purification, DCs were exposed to apoptotic thymocytes and returned to syngeneic hosts. Nonsusceptible animals developed anti-dsDNA and anti-nuclear antibodies, but these did not persist. In susceptible mice, however, antibodies persisted and the animals died of autoimmune-associated renal failure. Therapeutic vaccinations with dendritic cells may therefore have the potential to harm individuals prone to autoimmune disease.

Further evidence that DCs are able to initiate autoimmune disorders comes from transgenic mouse models. When CD40 ligand was expressed in the basal keratinocytes of the epidermis,[68] CD40 ligand expression led not only to enhanced emigration of LCs from the skin and T cell mediated dermatitis, but also led to the development of antinuclear and anti-dsDNA autoantibodies with associated renal Ig deposits, proteinuria, and lung fibrosis. In situ activation of LCs is therefore sufficient to generate systemic autoimmune disorders, which may occur through breaking immune tolerance against antigens in the skin. To demonstrate that DCs are also necessary for initiation of autoimmunity, an autoimmune myocarditis model was used. T-cell mediated inflammatory heart disease was induced by immunization with α-myosin-peptide(614-629) plus complete Freund's adjuvant.[69] Transgenic mice lacking the IL-1 receptor (IL-1R) molecule were protected against disease because IL-1R-/- DCs from the transgenic animals were inefficiently activated and produce less TNFα, IL-1, IL-6, and IL-12p70. They were therefore unable to stimulate the autoreactive CD4+ T cells that mediate myocarditis.

Conclusions

There are many subsets of dendritic cells, with different immunogenic or tolerogenic properties depending on their state of maturation. Mature myeloid DCs are required for the initiation of autoimmune diseases, while immature myeloid DCs and plasmacytoid DCs are critical for the maintenance of peripheral tolerance in the steady state. All these DCs develop from haematopoietic stem cells (HSC). Transplantation of HSC has been used to treat a number of autoimmune disorders, as described later in this book, but the immunological reasons for its success have not yet been adequately described. We propose that the therapeutic process leading to resolution of autoimmune disease can be described in 5 stages: (1) Autoimmunity may be induced by a number of genetic and pathogenic factors, but requires the presentation of self-antigens by DCs, which have undergone maturation in response to inflammatory stimuli. (2) High dose immunosupression, given at the start of treatment, starves autoimmune effector mechanisms of T cell help and ameliorates pathogenic symptoms. (3) Lack of T cell help also reduces inflammation in disease-affected tissues. (4) When the immune system is reconstituted by HSC transplantation, CD34+ progenitors re-establish a population of immature DCs in the absence of ongoing inflammation. (5) Immature DCs in the steady state are now able to maintain peripheral tolerance to "autoantigens" in the normal physiological manner. This scheme provides a series of testable hypotheses that, if explored, will increase our understanding of the causes and resolution of autoimmune diseases, and enable improved therapeutic intervention to treat these debilitating conditions.

References

1. Steinman RM, Cohn ZA. Identification of a novel cell type in peripheral lymphoid organs of mice. I. Morphology, quantitation, tissue distribution. J Exp Med 1973; 137:1142-62.
2. Kamath AT, Pooley J, O'Keeffe MA et al. The development, maturation, and turnover rate of mouse spleen dendritic cell populations. J Immunol 2000; 165:6762-70.
3. De Smedt T, Pajak B, Muraille E et al. Regulation of dendritic cell numbers and maturation by lipopolysaccharide in vivo. J Exp Med 1996; 184:1413-24.
4. Reis e Sousa C, Hieny S, Scharton-Kersten T et al. In vivo microbial stimulation induces rapid CD40 ligand-independent production of interleukin 12 by dendritic cells and their redistribution to T cell areas. J Exp Med 1997; 186:1819-29.
5. Edwards AD, Chaussabel D, Tomlinson S et al. Relationships among murine CD11c(high) dendritic cell subsets as revealed by baseline gene expression patterns. J Immunol 2003; 171:47-60.
6. d'Ostiani CF, Del Sero G, Bacci A et al. Dendritic cells discriminate between yeasts and hyphae of the fungus Candida albicans. Implications for initiation of T helper cell immunity in vitro and in vivo. J Exp Med 2000; 191:1661-74.
7. Henri S, Vremec D, Kamath A et al. The dendritic cell populations of mouse lymph nodes. J Immunol 2001; 167:741-48.
8. Ito T, Inaba M, Inaba K et al. A CD1a+/CD11c+ subset of human blood dendritic cells is a direct precursor of Langerhans cells. J Immunol 1999; 163:1409-19.
9. Schuler G, Steinman RM. Murine epidermal Langerhans cells mature into potent immunostimulatory dendritic cells in vitro. J Exp Med 1985; 161:526-46.
10. Merad M, Manz MG, Karsunky H et al. Langerhans cells renew in the skin throughout life under steady-state conditions. Nat Immunol 2002; 3:1135-41.
11. Siegal FP, Kadowaki N, Shodell M et al. The nature of the principal type 1 interferon-producing cells in human blood. Science 1999; 284:1835-37.
12. Cella M, Jarrossay D, Facchetti F et al. Plasmacytoid monocytes migrate to inflamed lymph nodes and produce large amounts of type I interferon. Nat Med 1999; 5:919-23.
13. Lennert K, Remmele W. Karyometrische Untersuchungen an lymphknottenzellen des menschen I: Mitt germinoblasten, lymphoblasten und lymphozyten. Acta Haemat 1958; 19:99.
14. Feller AC, Lennert K, Stein H et al. Immunohistology and aetiology of histiocytic necrotizing lymphadenitis. Report of three instructive cases. Histopathology 1983; 7:825-39.
15. Facchetti F, de Wolf-Peeters C, Mason DY et al. Plasmacytoid T cells. Immunohistochemical evidence for their monocyte/macrophage origin. Am J Pathol 1988; 133:15-21.
16. O'Doherty U, Peng M, Gezelter S et al. Human blood contains two subsets of dendritic cells, one immunologically mature and the other immature. Immunology 1994; 82:487-93.
17. Grouard G, Rissoan MC, Filgueira L et al. The enigmatic plasmacytoid T cells develop into dendritic cells with interleukin (IL)-3 and CD40-ligand. J Exp Med 1997; 185:1101-11.
18. Summers KL, Hock BD, McKenzie JL et al. Phenotypic characterization of five dendritic cell subsets in human tonsils. Am J Pathol 2001; 159:285-95.
19. Olweus J, BitMansour A, Warnke R et al. Dendritic cell ontogeny: A human dendritic cell lineage of myeloid origin. Proc Natl Acad Sci USA 1997; 94:12551-56.
20. Anjuere F, Martin P, Ferrero I et al. Definition of dendritic cell subpopulations present in the spleen, Peyer's patches, lymph nodes, and skin of the mouse. Blood 1999; 93:590-98.
21. Robinson SP, Patterson S, English N et al. Human peripheral blood contains two distinct lineages of dendritic cells. Eur J Immunol 1999; 29:2769-78.
22. Sallusto F, Lanzavecchia A. Efficient presentation of soluble antigen by cultured human dendritic cells is maintained by granulocyte/macrophage colony-stimulating factor plus interleukin 4 and downregulated by tumor necrosis factor alpha. J Exp Med 1994; 179:1109-18.
23. Briere F, Bendriss-Vermare N, Delale T et al. Origin and filiation of human plasmacytoid dendritic cells. Hum Immunol 2002; 63:1081-93.
24. Wu L, Nichogiannopoulou A, Shortman K et al. Cell-autonomous defects in dendritic cell populations of Ikaros mutant mice point to a developmental relationship with the lymphoid lineage. Immunity 1997; 7:483-92.
25. Wu L, D'Amico A, Winkel KD et al. RelB is essential for the development of myeloid-related CD8alpha- dendritic cells but not of lymphoid-related CD8alpha+ dendritic cells. Immunity 1998; 9:839-47.
26. Anderson KL, Perkin H, Surh CD et al. Transcription factor PU.1 is necessary for development of thymic and myeloid progenitor-derived dendritic cells. J Immunol 2000; 164:1855-61.
27. Guerriero A, Langmuir PB, Spain LM et al. PU.1 is required for myeloid-derived but not lymphoid-derived dendritic cells. Blood 2000; 95:879-85.
28. Tsujimura H, Tamura T, Gongora C et al. ICSBP/IRF-8 retrovirus transduction rescues dendritic cell development in vitro. Blood 2003; 101:961-69.
29. Borkowski TA, Letterio JJ, Farr AG et al. A role for endogenous transforming growth factor beta 1 in Langerhans cell biology: The skin of transforming growth factor beta 1 null mice is devoid of epidermal Langerhans cells. J Exp Med 1996; 184:2417-22.
30. Gatti E, Velleca MA, Biedermann BC et al. Large-scale culture and selective maturation of human Langerhans cells from granulocyte colony-stimulating factor-mobilized CD34+ progenitors. J Immunol 2000; 164:3600-7.
31. Blom B, Ho S, Antonenko S et al. Generation of interferon alpha-producing predendritic cell (Pre-DC)2 from human CD34(+) hematopoietic stem cells. J Exp Med 2000; 192:1785-96.
32. Matthews W, Jordan CT, Wiegand GW et al. A receptor tyrosine kinase specific to hematopoietic stem and progenitor cell-enriched populations. Cell 1991; 65:1143-52.
33. Maraskovsky E, Brasel K, Teepe M et al. Dramatic increase in the numbers of functionally mature dendritic cells in Flt3 ligand-treated mice: Multiple dendritic cell subpopulations identified. J Exp Med 1996; 184:1953-62.
34. Gounari F, Aifantis I, Martin C et al. Tracing lymphopoiesis with the aid of a pTalpha-controlled reporter gene. Nat Immunol 2002; 3:489-96.
35. Bendriss-Vermare N, Barthelemy C, Durand I et al. Human thymus contains IFN-alpha-producing CD11c(-), myeloid CD11c(+), and mature interdigitating dendritic cells. J Clin Invest 2001; 107:835-44.
36. Schotte R, Rissoan MC, Bendriss-Vermare N et al. The transcription factor Spi-B is expressed in plasmacytoid DC precursors and inhibits T-, B-, and NK-cell development. Blood 2003; 101:1015-23.
37. Spits H, Couwenberg F, Bakker AQ et al. Id2 and Id3 inhibit development of CD34(+) stem cells into predendritic cell (pre-DC)2 but not into pre-DC1. Evidence for a lymphoid origin of pre-DC2. J Exp Med 2000; 192:1775-84.
38. Mosmann TR, Coffman RL. TH1 and TH2 cells: Different patterns of lymphokine secretion lead to different functional properties. Annu Rev Immunol 1989; 7:145-73.
39. Maldonado-Lopez R, De Smedt T, Michel P et al. CD8alpha+ and CD8alpha- subclasses of dendritic cells direct the development of distinct T helper cells in vivo. J Exp Med 1999; 189:587-92.
40. Manickasingham SP, Edwards AD, Schulz O et al. The ability of murine dendritic cell subsets to direct T helper cell differentiation is dependent on microbial signals. Eur J Immunol 2003; 33:101-7.
41. Boonstra A, Asselin-Paturel C, Gilliet M et al. Flexibility of mouse classical and plasmacytoid-derived dendritic cells in directing T helper type 1 and 2 cell development: Dependency on antigen dose and differential toll-like receptor ligation. J Exp Med 2003; 197:101-9.

42. Ariizumi K, Shen GL, Shikano S et al. Identification of a novel, dendritic cell-associated molecule, dectin-1, by subtractive cDNA cloning. J Biol Chem 2000; 275:20157-67.
43. Tailleux L, Schwartz O, Herrmann JL et al. DC-SIGN is the major Mycobacterium tuberculosis receptor on human dendritic cells. J Exp Med 2003; 197:121-27.
44. Geijtenbeek TB, Van Vliet SJ, Koppel EA et al. Mycobacteria target DC-SIGN to suppress dendritic cell function. J Exp Med 2003; 197:7-17.
45. Akira S. Mammalian toll-like receptors. Curr Opin Immunol 2003; 15:5-11.
46. Ito T, Amakawa R, Fukuhara S. Roles of toll-like receptors in natural interferon-producing cells as sensors in immune surveillance. Hum Immunol 2002; 63:1120-25.
47. Anderson MS, Venanzi ES, Klein L et al. Projection of an immunological self-shadow within the thymus by the Aire Protein. Science 2002; 298(5597):1395-401.
48. Zhang M, Vacchio MS, Vistica BP et al. T cell tolerance to a neo-self antigen expressed by thymic epithelial cells: The soluble form is more effective than the membrane-bound form. J Immunol 2003; 170:3954-62.
49. Laufer TM, DeKoning J, Markowitz JS et al. Unopposed positive selection and autoreactivity in mice expressing class II MHC only on thymic cortex. Nature 1996; 383:81-85.
50. Huang FP, Platt N, Wykes M et al. A discrete subpopulation of dendritic cells transports apoptotic intestinal epithelial cells to T cell areas of mesenteric lymph nodes. J Exp Med 2000; 191:435-44.
51. Sauter B, Albert ML, Francisco L et al. Consequences of cell death: Exposure to necrotic tumor cells, but not primary tissue cells or apoptotic cells, induces the maturation of immunostimulatory dendritic cells. J Exp Med 2000; 191:423-34.
52. Scheinecker C, McHugh R, Shevach EM et al. Constitutive presentation of a natural tissue autoantigen exclusively by dendritic cells in the draining lymph node. J Exp Med 2002; 196:1079-90.
53. Jonuleit H, Schmitt E, Schuler G et al. Induction of interleukin 10-producing, nonproliferating CD4(+) T cells with regulatory properties by repetitive stimulation with allogeneic immature human dendritic cells. J Exp Med 2000; 192:1213-22.
54. Dhodapkar MV, Steinman RM, Krasovsky J et al. Antigen-specific inhibition of effector T cell function in humans after injection of immature dendritic cells. J Exp Med 2001; 193:233-38.
55. Dhodapkar MV, Steinman RM. Antigen-bearing immature dendritic cells induce peptide-specific CD8(+) regulatory T cells in vivo in humans. Blood 2002; 100:174-77.
56. Liu K, Iyoda T, Saternus M et al. Immune tolerance after delivery of dying cells to dendritic cells in situ. J Exp Med 2002; 196:1091-97.
57. Gilliet M, Liu YJ. Generation of human CD8 T regulatory cells by CD40 ligand-activated plasmacytoid dendritic cells. J Exp Med 2002; 195:695-704.
58. Wakkach A, Fournier N, Brun V et al. Characterization of dendritic cells that induce tolerance and T regulatory 1 cell differentiation in vivo. Immunity 2003; 18:605-17.
59. Thomas R, Davis LS, Lipsky PE. Rheumatoid synovium is enriched in mature antigen-presenting dendritic cells. J Immunol 1994; 152:2613-23.
60. Amft N, Bowman SJ. Chemokines and cell trafficking in Sjogren's syndrome. Scand J Immunol 2001; 54:62-9.
61. Green EA, Eynon EE, Flavell RA. Local expression of TNFalpha in neonatal NOD mice promotes diabetes by enhancing presentation of islet antigens. Immunity 1998; 9:733-43.
62. Pashenkov M, Huang YM, Kostulas V et al. Two subsets of dendritic cells are present in human cerebrospinal fluid. Brain 2001; 124:480-92.
63. te Velde AA, van Kooyk Y, Braat H et al. Increased expression of DC-SIGN+IL-12+IL-18+ and CD83+IL-12-IL-18- dendritic cell populations in the colonic mucosa of patients with Crohn's disease. Eur J Immunol 2003; 33:143-51.
64. Krajina T, Leithauser F, Moller P et al. Colonic lamina propria dendritic cells in mice with CD4+ T cell-induced colitis. Eur J Immunol 2003; 33:1073-83.
65. Tsark EC, Wang W, Teng YC et al. Differential MHC class II-mediated presentation of rheumatoid arthritis autoantigens by human dendritic cells and macrophages. J Immunol 2002; 169:6625-33.
66. Banchereau J, Palucka AK, Dhodapkar M et al. Immune and clinical responses in patients with metastatic melanoma to CD34(+) progenitor-derived dendritic cell vaccine. Cancer Res 2001; 61:6451-58.
67. Bondanza A, Zimmermann VS, Dell'Antonio G et al. Cutting edge: Dissociation between autoimmune response and clinical disease after vaccination with dendritic cells. J Immunol 2003; 170:24-7.
68. Mehling A, Loser K, Varga G et al. Overexpression of CD40 ligand in murine epidermis results in chronic skin inflammation and systemic autoimmunity. J Exp Med 2001; 194:615-28.
69. Eriksson U, Kurrer MO, Sonderegger I et al. Activation of dendritic cells through the interleukin 1 receptor 1 is critical for the induction of autoimmune myocarditis. J Exp Med 2003; 197:323-31.
70. Boes M, Cerny J, Massol R et al. T-cell engagement of dendritic cells rapidly rearranges MHC class II transport. Nature 2002; 418:983-88.

CHAPTER 20

CD4⁺ T Regulatory Cells and Modulation of Undesired Immune Responses

Rosa Bacchetta, Megan K. Levings and Maria-Grazia Roncarolo

Introduction

T regulatory (Tr) cells are a distinct population of T cells which modulate T helper Th1 and Th2 mediated immuno-responses and maintain immunological homeostasis. Thus far, Tr type 1 (Tr1) cells and CD4⁺CD25⁺ T cells are the best characterized subsets of CD4⁺ Tr cells. Both types of Tr cells share similar biological properties, such as poor proliferation in vitro, and suppressive capacity in vitro and in vivo. Tr1 cells act via production of interleukin (IL)-10 and transforming growth factor (TGF)-β, whereas the mechanism responsible for the suppressive function of CD4⁺CD25⁺ T cells remains to be defined. Tr1 cells have been studied in a number of settings including transplantation, allergy and chronic infectious diseases; in contrast, the role of CD4⁺CD25⁺ T cells have largely been investigated in the field of autoimmunity. Most likely, factors such as the types of antigens, the state of maturation of antigen presenting cells, and the cytokines present in the microenvironment influence the differentiation and function of the different types of Tr cells. Several studies designed to gain a better understanding of the relationship between the different subsets of Tr cells are in progress. Further characterization of Tr cells will open new therapeutic prospectives for prevention and modulation of T-cell mediated pathologies.

Immunoresponses are primarily orchestrated by CD8⁺ and CD4⁺ T cells. Within the CD4⁺ T-cell subset, two major types are distinguished based on differences in cytokine production and effector function. Th 1 cells mainly produce IL-2 and interferon (IFN)-γ and are involved in cellular mediated immunity, whereas Th2 cells secrete IL-4, IL-5, and IL-10 and mediate humoral immunity (Fig. 1). More recently, a third type of CD4⁺ T cells, Tr cells, which is specialized in controlling immunoresponses and immunological homeostasis, has been identified. Tr cells are involved not only in maintaining tolerance to non-self and self antigens, but also in regulating immunoresponses to pathogens, such as bacteria and viruses. When these regulatory circuits go awry, a number of different Th1 or Th2 mediated pathologies may arise, including autoimmunity, uncontrolled chronic infections and allergies.

The types of Tr cells which are present and functional in a defined microenvironment depend on interactions between different antigens (Ag) and antigen presenting cells (APC). Production of immunoregulatory cytokines, such as IL-10 and TGF-β, plays a central role in controlling the activation state of APCs and the effector function of many types of Tr cells.[1] T-cell mediated suppression itself may be mediated by IL-10 and/or TGF-β, but also requires direct cell-cell contact, indicating that regulation can be mediated by both cytokines and inhibitory cell-surface molecules.

Several different types of Tr cells have been described in a number of experimental models.[2] Amongst CD4⁺ Tr cells, two subsets, Tr1 and CD4⁺CD25⁺ Tr cells, have been the focus of recent intensive investigation, and much progress has been made in elucidating the mechanisms which control their differentiation and function. Here we will focus on recent advances in our understanding of the biological properties and functions of Tr1 and CD4⁺CD25⁺ Tr cells. We will also briefly discuss other types of CD4⁺ Tr cells, which have already been extensively reviewed elsewhere.[3]

Tr1 Cells

CD4⁺ Tr1 cells were originally identified based on their unique pattern of cytokine production, which was distinct from that of classical Th1 and Th2 cells.[4] Upon T-cell receptor (TCR) mediated activation, Tr1 cells produce high amounts of IL-10, considerable levels of IFN-γ, TGF-β and IL-5, but no IL-4 and very low levels of, or no, IL-2. Importantly, via production of IL-10 and TGF-β, Tr1 cells suppress primary T-cell responses.[4] It is now well established that Tr1 cells with a variety of antigen specificities arise in vivo in both mice and humans, and they can also be generated in vitro.[1] Although Tr1 cells can be specific for a number of different Ags, they all share the same profile of cytokine production and proliferate poorly in response to TCR-mediated activation.

The Importance of IL-10 and TGF-β

Both IL-10 and TGF-β have complex effects on multiple cell types in vitro and in vivo.[5-9] IL-10 has an indirect immunosuppressive effect on T-cell responses via inhibition of the antigen-presenting capacity of APCs. IL-10 down-regulates co-stimulatory molecules and inhibits secretion of cytokines and chemokines by APCs, thus influencing T-cell differentiation, proliferation and migration.[5] Moreover, IL-10 can directly suppress

Stem Cell Therapy for Autoimmune Disease, edited by Richard K. Burt and Alberto M. Marmont. ©2004 Landes Bioscience/Eurekah.com.

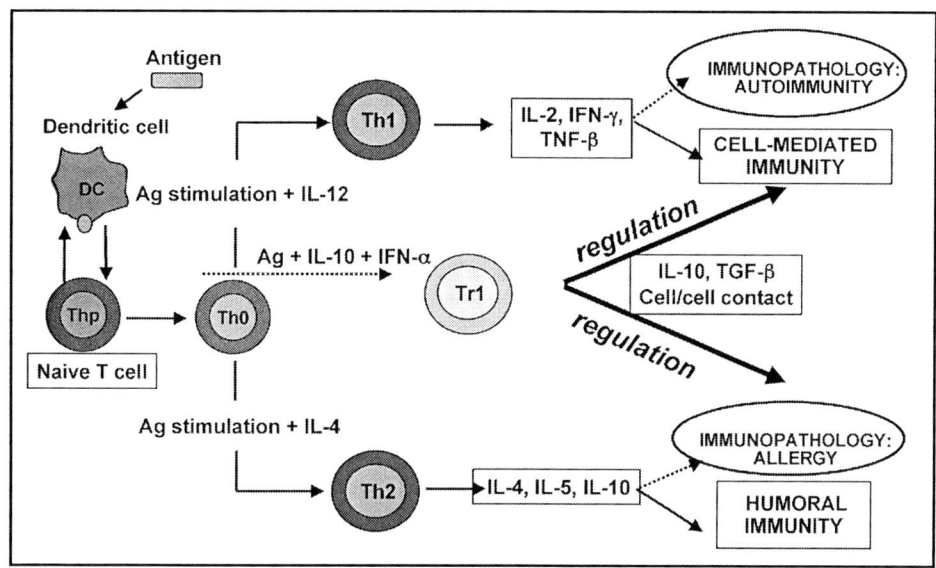

Figure 1. Cytokine-induced differentiation of CD4+ T cell subsets. When naïve CD4+ T cells encounter their antigen on dendritic cells, the cytokines produced influence their differentiation. Classically, IL-12 promotes the generation of Th1 cells, which regulate cell-mediated immunity, whereas IL-4 results in the differentiation of Th2 cells, which regulate humoral immunity. More recently, it has been shown that IL-10 promotes the differentiation of Tr1 cells, and that the presence of IFN-α can enhance this effect. Fully differentiated Tr1 cells can suppress, mainly by IL-10 and TGF-β, the activity of both Th1 and Th2 cells and thereby modulate immunopathology and maintain immune homeostasis.

T-cell proliferation by inhibiting IL-2, tumor necrosis factor-α[6] and IL-5 production.[10] Similarly, TGF-β inhibits T-cell proliferation, cytotoxicity and IFN-γ production.[8]

Additional evidence for the immunoregulatory role of IL-10 and TGF-β has come from studies of genetically deficient and/or transgenic mice. For example, IL-10-deficient mice succumb to chronic colitis due to dysregulation of Th1-mediated responses, and are more susceptible to a condition resembling rheumatoid arthritis.[11,12] Transgenic mice expressing IL-10 under control of the MHC class II promoter have a block in maturation of TCR αβ+ T cells at the pre-TCR stage, are highly susceptible to infections, and fail to mount normal Th1 or Th2 responses.[13] Mice genetically lacking TGF-β1[14] or genetically engineered to express a dominant-negative TGF-β receptor (R) II subunit specifically in T cells,[15] display a progressive infiltration of lymphocytes into multiple organs and tend to develop spontaneous autoimmune diseases. Overall, these studies indicate that T-cell homeostasis requires IL-10 and TGF-β signalling in T cells and APC.

Biological Characterization of Tr1 Cells

Following TCR-mediated activation, either by specific Ags or by anti-CD3 plus anti-CD28 mAbs, Tr1 cells have low proliferative responses[16,17] which are at least partially due to the autocrine production of IL-10.[16,18] Consequently, these cells do not expand significantly under standard T-cell culture conditions. Despite the fact that Tr1 cells do not proliferate upon TCR-mediated triggering, they express normal levels of activation markers, such as CD40L, CD69, CD25, CD28, cytotoxic T lymphocyte associated antigen (CTLA)-4 and HLA-DR.[17] Furthermore, TCR-mediated activation of mitogen activated protein (MAP)-family kinases, an intracellular signalling pathway which is generally associated with proliferation, appeared normal.[17,19] Importantly, Tr1 cells are able to proliferate in response to exogenous cytokines, in the absence of TCR activation. They prefer-

entially proliferate and expand in the presence of IL-15. In the presence of both IL-15 and IL-2, Tr1 cell clones can reach an in vitro expansion comparable to that of Th cell clones, without loosing their antigen-specificity or their characteristic pattern of cytokine production.[17] However, IL-15 does not appear to be directly involved in the in vitro differentiation of Tr1 cells from naïve CD4+ T cells, and rather acts exclusively as a growth factor.[18] This strong cytokine-induced proliferative response correlates with high constitutive expression of the IL-2/15R β and γc chains on Tr1-cell clones. In contrast, expression of the IL-2R α (CD25) chain is comparable to that of Th1 and Th2 cell clones.[17] The fact that growth and expansion of Tr1 cells can be sustained independently from TCR activation suggests that Tr1 cells could be designed to survive for prolonged periods of time in a quiescent state without being reactivated by the antigen, as previously described for memory CD8+ T cells whose survival depends on IL-15.[20,21]

Identification of T cells strictly on the basis of cytokine production is cumbersome and difficult to combine with ex-vivo purification. Thus, much effort has been focused on the identification of a Tr1-specific cell surface marker. Tr1-cell clones express a vast repertoire of chemokine receptors, including those previously associated with the Th1 phenotype, such as CXCR3 and CCR5, or the Th2 phenotype, such as CCR3, CCR4 and CCR8.[22] Interestingly, Tr1, but not Th1 or Th2, cell clones express CCR7, a receptor recently implicated in homing to lymph nodes.[23] More recently, the gene expression pattern of IL-10-producing Tr-cell clones was compared to that of Th1 and Th2 cell clones. A number of molecules, such as preproenkephalin, GITR, and perforin, were found to be upregulated in Tr1 cells. Results also showed that several of these genes were highly expressed in Tr cells which had been isolated from tolerated skin grafts, indicating the potential relevance of these molecules in vivo.[24]

Factors which Drive Tr1 Differentiation

Differentiation of Tr1 cells is strongly correlated with chronic antigenic stimulation in the periphery. However, the key factors which determine their development in vivo remain largely unknown. Among soluble factors, IL-10 is clearly essential for differentiation of Tr1 cells. Tr1 cells specific for allogeneic antigens (alloAgs) and nominal antigens (Ags), such as Tetanus Toxoid, can be obtained in vitro upon priming in the presence of IL-10.[4,17] As discussed above, IL-10 acts by down-regulating Ag-specific responses, and induces a state of Ag-specific unresponsiveness, or anergy, upon rechallenge with the same Ag.[4,17] Enforced TCR-mediated restimulation of CD4+ T cells from IL-10 anergized cultures results in the emergence of Ag-specific Tr1 cells. In the mouse, isolation of T-cell clones with a Tr1 profile of cytokine production can also be obtained from hyporesponsive cultures which were primed in the presence of exogenous IL-10.[4,25]

Although IL-10 is necessary, it is not sufficient to induce the differentiation of Tr1 cells in vitro in the absence of APCs. IFN-α, a cytokine which had previously been shown to increase IL-10 production by T cells,[26] synergises with IL-10 to promote the differentiation of naive CD4+ T cells into a T-cell population with the cytokine profile and biological activities of Tr1 cells.[18] The fact that IFN-α is a crucial cytokine for clearing viral infections, underlines the role that Tr1 cells may have in the regulation of immune responses to viruses. Interestingly, in the absence of signals from IFN-α (or IL-10),[18] the host suffers from the detrimental effects of excessive cellular immune responses elicited during acute infection.[27] Thus, it can be hypothesized that during infection and inflammation, priming in the presence of IFN-α and IL-10 may result in differentiation of IL-10-producing Tr1 cells which could down-regulate T-cell responses to pathogens.[28] Since differentiation of Tr1 requires repeated antigenic stimulation they would likely become fully functional only in the final phases of an infection and thus they would not interfere with protective immunity.[2,29]

In vivo, production of immunoregulatory cytokines and the differentiation of Tr1 cells is likely controlled by tolerogenic dendritic cells (DC).[28,30,31] In the human, at least two types of DC can influence the differentiation of CD4+ T cells: mature myeloid-derived DCs, or DC1, which induce the differentiation of Th1 cells; and lymphoid-derived DCs or DC2, which preferentially give rise to Th2 cells.[32] Interestingly, DC2 may also polarise T cells towards IL-10-producing T cells with regulatory activity.[33,34] The capacity to drive the differentiation of Tr1 cells may also depend on the maturation state of the DC.[35] Immature DC1, which typically express low levels of cytokines and costimulatory molecules, induce a population of CD4+ or CD8+ IL-10-producing T cells with suppressive capacity in vitro.[29,36,37] Treatment of DCs with immunomodulatory compounds which reduce or modulate DC maturation, such as vitamin D3 and/or dexamethasone, can also result in reduced T-cell responses and ultimately in the differentiation of Tr1 cells.[38,39]

It can be hypothesized that priming by immature DCs in the absence of inflammatory signals could be essential for the generation of Tr cells specific for self antigens or other antigens presented in a non-pathological context. Interestingly, tumors are particularly rich in immature DC, and within the lesions high concentrations of IL-10 have been observed.[40] Such an environment could affect T cells and lead to a state of anergy/excessive regulation which results in escape of tumor Ags from immune surveillance.

Tr1 Cells and Maintenance of Transplantation Tolerance

Tr1 cells induce and maintain transplantation tolerance. The first demonstration that Tr1 cells are present and functional in humans came from studies of transplantation tolerance, in which the presence of Tr1 cells was correlated with the lack of graft versus host disease (GvHD).[16] Peripheral blood from severe combined immunodeficient (SCID) patients, who had been successfully transplanted with HLA-mismatched hematopoietic stem cells, contained a high frequency of host-specific Tr1 cells of donor origin. Furthermore, high levels of endogenous IL-10 production by T cells was demonstrated ex vivo at the mRNA level.[16] More recently, analysis of kidney or liver transplanted patients with long-lived graft acceptance revealed the presence of CD4+ T cells which suppressed naive T-cell responses via production of IL-10 or TGF-β.[41]

In a mouse model of mismatched bone marrow transplantation, Blazer et al showed that ex-vivo anergized cells do not mediate GVHD[42] and co-transfer of anergized cells together with naive untreated cells could rescue a high proportion of mice from developing acute GVHD.[43] These latter data strongly support the hypothesis that IL-10-anergized cells contain the precursors of Tr1 cells which can mediate antigen-specific suppression in vivo. This ability of IL-10 to induce alloantigen-specific anergy and differentiate alloantigen-specific Tr1 cells ex vivo suggests that the adoptive transfer of IL-10-treated cells could prevent GVHD in patients undergoing allogeneic hematopoietic stem cell transplantation.[1]

Tr1 Cells and Regulation of Th1- and Th2-Mediated Pathologies

Tr1 cells have been shown to regulate immune responses in a number of different pathologies. In the mouse, Tr1 cells prevent the induction of intestinal inflammatory diseases (IBD) by suppressing the pathogenic effect of Th1 CD45RB^{high} CD4+ T cells. In this model, Tr1 cells need to be activated through the TCR in order to exert their regulatory function, since the prevention of IBD is only observed in mice which received the specific Ag recognized by the Tr1 cells.[44] However, once activated, Tr1 cells inhibit T-cell function via non-antigen-specific bystander suppression. Studies in this murine model of IBD demonstrated that Tr1 cells share many properties with CD4+CD25+CD45RB^{lo} Tr cells (see below).[45]

In addition to modulation of Th1-mediated diseases, Tr1 cells also suppress Th2-mediated pathologies. In the mouse, Ag-specific Tr1 cells confer protection against airway inflammation.[46] Recently, Zuany-Amorim et al showed that treatment of mice with a killed *Mycobacterium vaccae*-suspension induces allergen-specific regulatory T cells, which confer protection against airway inflammation via production of IL-10 and TGF-β.[47] Furthermore, induction of anergy to the major allergen of bee venom (phospholipase A$_2$) in patients who underwent specific immunotherapy, or in naturally tolerant beekeepers, correlates with a substantial increase in IL-10-producing T cells. Neutralization of endogenous IL-10 fully reconstitutes T-cell proliferative responses to phospholipase A$_2$ in vitro.[48] Tr1-cell clones specific for nickel sulfate (NiSO$_4$) can be isolated from the peripheral blood of both healthy donors and nickel-allergic patients. Via the production of IL-10, these Tr1 clones inhibit APC maturation and function, and suppress nickel-specific T-cell responses.[49]

Interestingly, Tr1 cells specific for *E. coli* proteins have been isolated from the intestinal mucosa of normal donors. These cells inhibit responses of autologous peripheral blood T cells towards the same protein via production of IL-10 and TGF-β.[50] A similar active regulation can be envisaged in response to all non harmful foreign antigens such as food antigens that we encounter daily. Thus, ex vivo and in vivo generated Tr1 cells could be used to restore the balance of the immune system to prevent and treat both Th1- and Th2-mediated diseases.

Potential Detrimental Effects of an Uncontrolled Tr1 Response

It is important to consider that in certain pathological settings, such as cancer and infectious diseases, an increase rather than a downregulation of immunoresponses is desirable and therefore a blockade or prevention of Tr1 cell responses could be beneficial. For example, the presence of IL-10-producing CD4⁺ T cells, is detrimental in *Mycobacterium tuberculosis* infection or in malaria, and correlates with increased mortality or down-regulation of antigen -specific T-cell responses.[51,52] Similarly, in *Bordetella pertussis* infections, high levels of IL-10 produced by infected cells modulate APC function, resulting in generation of Tr1 cells and prevention of protective Th1 responses.[53] Recently, it has been shown that Friend Virus infection induces an expansion of Tr cells which results in a strong immunosuppression.[54] There are a growing number of examples of IL-10-like proteins in the genomes of pathogens such as EBV, cytomegalovirus and poxvirus Orf virus.[5] Thus, it is evident that certain pathogens have evolved strategies to promote the differentiation of Tr1 cells which would then limit the anti-pathogen immune response. Similarly to Th1 or Th2 responses, the induction of a beneficial or destructive effect mediated by Tr1 cells is influenced by the pathogen. Manipulation of Tr1 cells in order to direct their function towards the most desired clinical effect will require a better understanding of the factors which control their differentiation and function in vitro and in vivo.

Naturally Occurring CD4⁺CD25⁺ T Regulatory Cells

Biological Characterization of CD4⁺CD25⁺ Tr Cells

Naturally occurring CD4⁺CD25⁺ Tr cells are defined as cells which constitutively express the IL-2Rα (CD25),[2,55,56] a marker traditionally used to identify activated T cells. CD4⁺CD25⁺ Tr cells were first identified in mice, where they comprise 5-10% of peripheral T cells. More recently, CD4⁺CD25⁺ Tr cells have been described in humans where they represent 2-5% of CD4⁺ T cells in PBMCs.[57] Experiments in thymectomized mice demonstrated that CD4⁺CD25⁺ Tr cells are generated in the thymus and that they have high affinity for self peptides.[58] CD4⁺CD25⁺ Tr cells are anergic, and do not proliferate following TCR-mediated activation. However, their ability to proliferate and expand is completely restored if these cells are activated via the TCR in the presence of IL-2 or IL-15.[59,60] CD4⁺CD25⁺ Tr cells potently suppress proliferation and production of IFN-γ by both CD4⁺ and CD8⁺ T cells.[61] Although the mechanisms which mediate these effects remain unclear, they clearly require direct cell-to-cell contact, involve inhibition of IL-2 production at the mRNA level, and promotion of cell-cycle arrest.[55] Furthermore, CD4⁺CD25⁺ Tr cells must be activated via the TCR to exert their suppressive function in vitro or in vivo, but, like Tr1 cells, once activated they can suppress in an antigen independent manner.[62]

The role of cytokines in the suppressive effects mediated by CD4⁺CD25⁺ Tr cells, and thus their relationship to Tr1 cells, is highly controversial. Some studies failed to detect production of IL-10 and/or TGF-β by human CD4⁺CD25⁺ Tr cells,[63,64] whereas others clearly detected secretion of these cytokines in response to a variety of stimuli.[59,60,65] Nevertheless, neither IL-10 nor TGF-β seemed to be directly required for their in vitro suppressive effects.[59,63,66,67] These data suggesting that IL-10 and TGF-β are dispensable are hard to reconcile with recent studies in the mouse indicating that naturally occurring CD4⁺CD25⁺ Tr cells prevent IBD via an IL-10 and/or TGF-β-dependent mechanism.[44,68] In addition, unlike their wild-type counterparts, CD25⁺CD45RB^low-CD4⁺ Tr cells from IL-10-deficient mice fail to protect from a CD45RB^high CD4⁺ T cell-induced wasting disease.[69] Furthermore, Nakamura et al have recently proposed that TGF-β produced by CD4⁺CD25⁺ Tr cells, and bound to their cell-surface, is the major mechanism by which murine CD4⁺CD25⁺ Tr cells suppress T-cell responses.[70] Further studies at the molecular level are required to better define how CD4⁺CD25⁺ Tr cells mediate their potent suppressive effects.

Naturally Occurring CD4⁺CD25⁺ Tr Cells and Regulation of Autoimmunity

The first clear demonstration that CD4⁺CD25⁺ Tr cells have regulatory functions in vivo came from experiments in murine models of autoimmunity. CD4⁺CD25⁺ Tr cells prevent the development of a variety of autoimmune diseases, such as gastritis and thyroiditis.[56] Furthermore, transfer of T cells depleted of CD4⁺CD25⁺ Tr cells into nude mice results in autoimmune diseases and skin graft rejection, which can be prevented by co-administration of CD4⁺CD25⁺ Tr cells, in a dose dependent manner. CD4⁺CD25⁺ Tr cells may also have an important role in the prevention of diabetes as NOD mice which are genetically deficient for CD28 or both B7.1 and B7.2, and therefore lack CD4⁺CD25⁺ Tr cells, succumb to diabetes earlier than control mice.[71] Together these findings indicate that CD4⁺CD25⁺ Tr cells can inhibit the activation and effector function of autoreactive T cells which failed to be deleted in the thymus. In humans, patients affected by type I diabetes have a significantly lower proportion of CD25⁺CD4⁺ T cells in their PBMC in comparison to normal donors or to patients with type II diabetes.[72] Further studies are required to demonstrate whether these CD4⁺CD25⁺ T cells have suppressive function.

CD4⁺CD25⁺ Tr Cells and Maintenance of Transplantation Tolerance

Fewer studies addressed the role of CD4⁺CD25⁺ Tr cells in inducing tolerance to alloantigens after transplantation. Recently, Blazar et al demonstrated that CD4⁺CD25⁺ Tr cells were essential for the ex vivo induction of tolerance to alloantigens by blockade of CD40/CD40L interactions.[73] Furthermore, CD4⁺CD25⁺ Tr cells play an important role in the prevention of GVHD in vivo, regardless of the strain of mice used.[74] Human CD4⁺CD25⁺ Tr cells are also able to suppress anti-allo responses in vitro, and importantly, ex-vivo expanded CD4⁺CD25⁺ Tr cells retain their capacity to suppress alloantigen-specific T cell responses in vitro[59] and in vivo.[74]

Table 1. Comparison between human CD4⁺ Tr-cell clones

Phenotype	Tr1	CD4⁺CD25⁺
Unresponsive to anti-CD3 mAbs	yes	yes
CTLA-4 high upon activation	yes	yes
IL-2Rα constitutively high	no	yes
IL-2Rβ and γc constitutively high	yes	yes
IL-10 production	yes	no
TGF-β production	yes	yes
IL-10 required for suppression	yes	no
TGF-β required for suppression	yes	?
CTLA-4 required for suppression	?	no

Studies in murine models of organ transplantation demonstrated that CD4⁺CD25⁺ Tr cells which suppress allograft rejection can be induced by using a variety of mAbs such as anti-CD3, anti-CD4, and anti-CD40L.[75,76] It remains to be shown whether these antigen-induced CD4⁺CD25⁺ Tr cells expand from the pool of naturally occurring CD4⁺CD25⁺ Tr cells, or whether, like Tr1 cells, they arise from naive CD4⁺ T cells which encountered their antigen in a tolerogenic fashion.

Relationship Between CD4⁺CD25⁺ Tr Cells and Tr1 Cells

As discussed above, the large body of controversial data regarding the role of suppressive cytokines in suppression mediated by CD4⁺CD25⁺ Tr cells, together with the observation that human CD4⁺CD25⁺ Tr cells and Tr1 cells appeared remarkably similar in many aspects (Table 1), raised the possibility that CD25⁺CD4⁺ Tr cells represented the anergic precursors of Tr1 cells. Studies on IL-10-induced antigen-specific unresponsiveness and differentiation of IL-10 producing Tr1 cells revealed that CD25⁺ T cells were dispensable for these immunomodulatory effects of IL-10.[77] Furthermore, upon analysis of human CD4⁺CD25⁺ Tr cells at the clonal level, it was revealed that unlike Tr1-cell clones,[1] suppressive CD4⁺CD25⁺ Tr-cell clones express significantly higher levels of CD25 in comparison to non-suppressive controls, do not proliferate in response to cytokines, and most importantly do not produce IL-10.[77] Together, these data strongly support the conclusion that these two types of regulatory cells are distinct, and that CD4⁺CD25⁺ Tr cells do not represent the anergic precursors of IL-10-producing Tr1 cells. These findings also support the hypothesis that a major difference between Tr1 and naturally occurring CD4⁺CD25⁺ Tr cells is that whereas CD4⁺CD25⁺ Tr cells exit the thymus as fully differentiated suppressor cells,[61] Tr1 cells can arise from any naive peripheral CD4⁺ T cell which encounters its antigen in a tolerogenic fashion.[1]

Th3 Cells

Another subset of CD4⁺ Tr cells was identified in studies of oral tolerance. Th3-cell clones, isolated from mice that were orally tolerised to myelin basic protein, suppress induction of EAE via a TGF-β-dependent mechanism.[78] CD4⁺ Tr cells which protect from EAE also arise in mice or rats, following oral administration of copolymer 1 (Cop-1).[79] Interestingly, T-cell lines generated from Cop-1-tolerant animals produce high levels of IL-10 and TGF-β, but little IL-2 or IL-4. Th3, like Tr1 cells and CD4⁺CD25⁺ Tr cells, must be activated via the TCR to be suppressive and mediate bystander suppression.[80] In patients suffering from multiple sclerosis, oral treatment with myelin basic protein (MBP) and proteolipid protein (PLP) induces a significant increase in the frequency of MBP or PLP-specific T cells that secrete TGF-β.[81] These data suggest that cells equivalent to murine TGF-β-producing Th3 cells[78] also exist in humans. The specialized environment of the oral mucosa and the selected subsets of resident DC are likely important factors for the differentiation of Th3 cells. Further studies are required to clarify whether the Th3 cells induced during oral tolerance are related to Tr1 or CD4⁺CD25⁺ Tr cells. One important difference between Th3 and Tr1 cells appears to be that Th3 cells require IL-4 for their differentiation.[80]

Concluding Remarks and Future Therapeutic Prospectives

It is now clear that responses mediated by CD4⁺ T cells represent a combination of effects mediated by Th1 or Th2 effector cells, and by T regulatory cells which maintain immune homeostasis. How these regulatory circuits evolve in vivo remains unclear, however, the type of antigen, the DCs, and the microenvironment, together with the genetic characteristics of the host, are clearly major influential factors. The ability to efficiently isolate and expand Tr cells in vitro and the knowledge of their mechanisms of action are fundamental steps for the development of new therapeutic strategies to either overcome the need for conventional immunosuppressive therapies or maintain a remission induced by hematopoietic stem cell transplants.

References

1. Roncarolo MG, Bacchetta R, Bordignon C et al. Type 1 T regulatory cells. Immunol Rev 2001; 182:68-79.
2. Roncarolo MG, Levings MK. The role of different subsets of T regulatory cells in controlling autoimmunity. Curr Opin Immunol 2000; 12:676-83.
3. Parham P. Regulatory T cells. Immunol Rev 2001; 182:1-227.
4. Groux H, O'Garra A, Bigler M et al. A CD4+ T-cell subset inhibits antigen-specific T-cell responses and prevents colitis. Nature 1997; 389(6652):737-42.
5. Moore KW, de Waal Malefyt R, Coffman RL et al. Interleukin-10 and the Interleukin-10 Receptor. Ann Rev Immunol 2001; 19:683-765.
6. de Waal Malefyt R, Yssel H, de Vries JE. Direct effects of IL-10 on subsets of human CD4+ T cell clones and resting T cells. Specific inhibition of IL-2 production and proliferation. J Immunol 1993; 150(11):4754-65.

7. Bejarano MT, de Waal Malefyt R, Abrams JS et al. Interleukin 10 inhibits allogeneic proliferative and cytotoxic T cell responses generated in primary mixed lymphocyte cultures. Int Immunol 1992; 4:1389-97.
8. Letterio JJ, Roberts AB. Regulation of immune responses by TGF-β. Ann Rev Immunol 1998; 16:137-61.
9. Bogdan C, Nathan C. Modulation of macrophage function by transforming growth factor beta, interleukin-4, and interleukin-10. Ann NY Acad Sci 1993; 685:713-39.
10. Schandene L, Alonso-Vega C, Willems F et al. B7/CD28-dependent IL-5 production by human resting T cells is inhibited by IL-10. J Immunol 1994; 152:4368-74.
11. Kuhn R, Lohler J, Rennick D et al. Interleukin-10-deficient mice develop chronic enterocolitis. Cell 1993; 75:263-74.
12. Rennick DM, Fort MM, Davidson NJ. Studies with IL-10-/- mice: An overview. J Leuk Biol 1997; 61:389-96.
13. Rouleau M, Cottrez F, Bigler M et al. IL-10 transgenic mice present a defect in T cell development reminiscent of SCID patients. J Immunol 1999; 163:1420-27.
14. Shull MM, Ormsby I, Kier AB et al. Targeted disruption of the mouse transforming growth factor-beta 1 gene results in multifocal inflammatory disease. Nature 1992; 359:693-99.
15. Gorelik L, Flavell RA. Abrogation of TGFbeta signaling in T cells leads to spontaneous T cell differentiation and autoimmune disease. Immunity 2000; 12(2):171-81.
16. Bacchetta R, Bigler M, Touraine JL et al. High levels of interleukin 10 production in vivo are associated with tolerance in SCID patients transplanted with HLA mismatched hematopoietic stem cells. J Exp Med 1994; 179(2):493-502.
17. Bacchetta R, Sartirana C, Levings MK et al. Growth and expansion of human T regulatory type 1 cells are independent from TCR activation but require exogenous cytokines. Eur J Immunol 2002; 32: In press.
18. Levings MK, Sangregorio R, Galbiati F et al. IFN-alpha and IL-10 induce the differentiation of human Type 1 T regulatory cells. J Immunol 2001; 166:5530-39.
19. Bacchetta R, Parkman R, McMahon M et al. Dysfunctional cytokine production by host-reactive T-cell clones isolated from a chimeric severe combined immunodeficiency patient transplanted with haploidentical bone marrow. Blood 1995; 85:1944-53.
20. Ku CC, Murakami M, Sakamoto A et al. Control of homeostasis of CD8+ memory T cells by opposing cytokines. Science 2000; 288:675-78.
21. Zhang X, Sun S, Hwang I et al. Potent and selective stimulation of memory-phenotype CD8+ T cells in vivo by IL-15. Immunity 1988; 8:591-99.
22. Sebastiani S, Allavena P, Albanesi C et al. Chemokine receptor expression and function in CD4+ T lymphocytes with regulatory activity. J Immunol 2001; 166:996-1002.
23. Sallusto F, Lenig D, Forster R et al. Two subsets of memory T lymphocytes with distinct homing potentials and effector functions. Nature 1999; 401(6754):708-12.
24. Zelenika D, Adams E, Humm S et al. Regulatory T cells overexpress a subset of Th2 gene transcripts. J Immunol 2002; 168(3):1069-79.
25. Groux H, Powrie F. Regulatory T cells and inflammatory bowel disease. Immunol Today 1999; 20(10):442-45.
26. McRae BL, Semnani RT, Hayes MP et al. Type I IFNs inhibit human dendritic cell IL-12 production and Th1 cell development. J Immunol 1998; 160(9):4298-4304.
27. Durbin JE, Fernandez-Sesma A, Lee CK et al. Type I IFN modulates innate and specific antiviral immunity. J Immunol 2000; 164(8):4220-28.
28. Roncarolo MG, Levings MK, Traversari C. Differentiation of T regulatory cells by immature dendritic cells. J Exp Med 2001; 193:F5-F10.
29. Jonuleit H, Schmitt E, Schuler G et al. Induction of interleukin-10-producing, nonproliferating CD4+ T cells with regulatory properties by repetitive stimulation with allogenic immature human dendritic cells. J Exp Med 2000; 192:1213-22.
30. Reid SD, Penna G, Adorini L. The control of T cell responses by dendritic cell subsets. Curr Opin Immunol 2000; 12:114-21.
31. Banchereau J, Briere F, Caux C et al. Immunobiology of dendritic cells. Ann Rev Immunol 2000; 18:767-811.
32. Rissoan MC, Soumelis V, Kadowaki N et al. Reciprocal control of T helper cell and dendritic cell differentiation. Science 1999; 283(5405):1183-86.
33. Kadowaki N, Antonenko S, Lau JY et al. Natural interferon alpha/beta-producing cells link innate and adaptive immunity. J Exp Med 2000; 192:219-26.
34. Gilliet M, Liu YJ. Generation of human CD8 T regulatory cells by CD40 ligand-activated plasmacytoid dendritic cells. J Exp Med 2002; 195(6):695-704.
35. Ito T, Amakawa R, Inaba M et al. Differential regulation of human blood dendritic cell subsets by IFNs. J Immunol 2001; 166(5):2961-9.
36. Dhodapkar MV, Steinman RM, Krasovsky J et al. Antigen-specific inhibition of effector T cell function in humans after injection of immature dendritic cells. J Exp Med 2001; 193:233-8.
37. Jonuleit H, Schmitt E, Steinbrink K et al. Dendritic cells as a tool to induce anergic and regulatory T cells. Trends Immunol 2001; 22(7):394-400.
38. Barrat FJ, Cua DJ, Boonstra A et al. In vitro generation of interleukin 10-producing regulatory CD4(+) T cells is induced by immunosuppressive drugs and inhibited by T helper type 1 (Th1)- and Th2-inducing cytokines. J Exp Med 2002; 195(5):603-16.
39. Penna G, Adorini L. 1 Alpha,25-dihydroxyvitamin D3 inhibits differentiation, maturation, activation, and survival of dendritic cells leading to impaired alloreactive T cell activation. J Immunol 2000; 164:2405-11.
40. Enk AH, Jonuleit H, Saloga J et al. Dendritic cells as mediators of tumor-induced tolerance in metastatic melanoma. Int J Cancer 1997; 73(3):309-16.
41. VanBuskirk AM, Burlingham WJ, Jankowska-Gan E et al. Human allograft acceptance is associated with immune regulation. J Clin Invest 2000; 106:145-55.
42. Zeller JC, Panoskaltsis-Mortari A, Murphy WJ et al. Induction of CD4+ T cell alloantigen-specific hyporesponsiveness by IL-10 and TGF-beta. J Immunol 1999; 163(7):3684-91.
43. Zong-ming C, O'Shaughnessy MJ, Gramaglia I et al. IL-10 and TGF-b induce alloreactive CD4+CD25- T cells to acquire regulatory cell function. Blood 2003; 101:5076-83.
44. Asseman C, Mauze S, Leach MW et al. An essential role for interleukin 10 in the function of regulatory T cells that inhibit intestinal inflammation. J Exp Med 1999; 190(7):995-1004.
45. Singh B, Read S, Asseman C et al. Control of intestinal inflammation by regulatory T cells. Immunol Rev 2001; 182:190-200.
46. Cottrez F, Hurst SD, Coffman RL et al. T regulatory cells 1 inhibit a Th2-specific response in vivo. J Immunol 2000; 165:4848-53.
47. Zuany-Amorim C, Sawicka E, Manlius C et al. Suppression of airway eosinophilia by killed Mycobacterium vaccae-induced allergen-specific regulatory T-cells. Nat Med 2002; 8(6):625-29.
48. Akdis CA, Blesken T, Akdis M et al. Role of interleukin 10 in specific immunotherapy. J Clin Invest 1998; 102:98-106.
49. Cavani A, Nasorri F, Prezzi C et al. Human CD4+ T lymphocytes with remarkable regulatory functions on dendritic cells and nickel-specific Th1 immune responses. J Invest Dermatol 2000; 114:295-302.
50. Khoo UY, Proctor IE, Macpherson AJ. CD4+ T cell down-regulation in human intestinal mucosa: Evidence for intestinal tolerance to luminal bacterial antigens. J Immunol 1997; 158(8):3626-34.
51. Boussiotis VA, Tsai EY, Yunis EJ et al. IL-10-producing T cells suppress immune responses in anergic tuberculosis patients. J Clin Invest 2000; 105(9):1317-25.
52. Plebanski M, Flanagan KL, Lee EA et al. Interleukin 10-mediated immunosuppression by a variant CD4 T cell epitope of Plasmodium falciparum. Immunity 1999; 10:651-60.
53. McGuirk P, McCann C, Mills KH. Pathogen-specific T regulatory 1 cells induced in the respiratory tract by a bacterial molecule that stimulates interleukin 10 production by dendritic cells: A novel strategy for evasion of protective T helper type 1 responses by Bordetella pertussis. J Exp Med 2002; 195(2):221-31.

54. Iwashiro M, Messer RJ, Peterson KE et al. Immunosuppression by CD4+ regulatory T cells induced by chronic retroviral infection. Proc Natl Acad Sci USA 2001; 98(16):9226-30.
55. Shevach EM. Regulatory T cells in autoimmmunity. Annu Rev Immunol 2000; 18:423-49.
56. Sakaguchi S. Regulatory T cells: Key controllers of immunologic self-tolerance. Cell 2000; 101:455-58.
57. Shevach EM. Certified professionals: CD4(+)CD25(+) suppressor T cells. J Exp Med 2001; 193(11):F41-46.
58. Jordan MS, Boesteanu A, Reed AJ et al. Thymic selection of CD4⁺CD25⁺ regulatory T cells induced by an agonist self-peptide. Nature Immunol 2001; 2:301-6.
59. Levings MK, Sangregorio R, Roncarolo MG. Human CD25+CD4+ T regulatory cells suppress naive and memory T-cell proliferation and can be expanded in vitro without loss of function. J Exp Med 2001; 193:1295-1302.
60. Dieckmann D, Plottner H, Berchtold S et al. Ex vivo isolation and characterization of CD4+CD25+ T cells with regulatory properties from human blood. J Exp Med 2001; In press.
61. Shevach EM. CD4+CD25+ suppressor T cells: More questions than answers. Nature Rev Immunol 2002; 2:389-400.
62. Thornton AM, Shevach EM. Suppressor effector function of CD4+CD25+ immunoregulatory T cells is antigen nonspecific. J Immunol 2000; 164(1):183-90.
63. Ng WF, Duggan PJ, Ponchel F et al. Human CD4(+)CD25(+) cells: A naturally occurring population of regulatory T cells. Blood 2001; 98(9):2736-44.
64. Baecher-Allan C, Brown JA, Freeman GJ et al. CD4+CD25high regulatory cells in human peripheral blood. J Immunol 2001; 167(3):1245-53.
65. Stephens LA, Mottet C, Mason D et al. Human CD4(+)CD25(+) thymocytes and peripheral T cells have immune suppressive activity in vitro. Eur J Immunol 2001; 31:1247-54.
66. Jonuleit H, Schmitt E, Stassen M et al. Identification and functional characterization of human CD4+CD25+ T cells with regulatory properties isolated from peripheral blood. J Exp Med 2001; 193: In press.
67. Taams LS, Smith J, Rustin MH et al. Human anergic/suppressive CD4(+)CD25(+) T cells: A highly differentiated and apoptosis-prone population. Eur J Immunol 2001; 31:1122-31.
68. Read S, Malmstrom V, Powrie F. Cytotoxic T lymphocyte-associated antigen 4 plays an essential role in the function of CD25(+)CD4(+) regulatory cells that control intestinal inflammation. J Exp Med 2000; 192(2):295-302.
69. Annacker O, Pimenta-Araujo R, Burlen-Defranoux O et al. CD25+ CD4+ T cells regulate the expansion of peripheral CD4 T cells through the production of IL-10. J Immunol 2001; 166:3008-18.
70. Nakamura K, Kitani A, Strober W. Cell contact-dependent immunosuppression by CD4(+)CD25(+) regulatory T cells is mediated by cell surface-bound transforming growth factor beta. J Exp Med 2001; 194(5):629-44.
71. Salomon B, Lenschow DJ, Rhee L et al. B7/CD28 costimulation is essential for the homeostasis of the CD4+CD25+ immunoregulatory T cells that control autoimmune diabetes. Immunity 2000; 12(4):431-40.
72. Kukreja A, Cost G, Marker J et al. Multiple immuno-regulatory defects in type-1 diabetes. J Clin Invest 2002; 109(1):131-40.
73. Taylor PA, Noelle RJ, Blazar BR. CD4(+)CD25(+) immune regulatory cells are required for induction of tolerance to alloantigen via costimulatory blockade. J Exp Med 2001; 193(11):1311-18.
74. Taylor PA, Lees CJ, Blazar BR. The infusion of ex vivo activated and expanded CD4(+)CD25(+) immune regulatory cells inhibits graft-versus-host disease lethality. Blood 2002; 99(10):3493-99.
75. Waldmann H, Cobbold S. Regulating the immune response to transplants: A role for CD4+ regulatory cells? Immunity 2001; 14:399-406.
76. Hara M, Kingsley CI, Niimi M et al. IL-10 is required for regulatory T cells to mediate tolerance to alloantigens in vivo. J Immunol 2001; 166(6):3789-96.
77. Levings MK, Sangregorio R, Sartirana C et al. Human CD25+CD4+ T suppressor cell clones produce transforming growth factor beta, but not interleukin 10, and are distinct from type 1 T regulatory cells. J Exp Med 2002; 196(10):1335-46.
78. Chen Y, Kuchroo VK, Inobe J et al. Regulatory T cell clones induced by oral tolerance: Suppression of autoimmune encephalomyelitis. Science 1994; 265(5176):1237-40.
79. Teitelbaum D, Arnon R, Sela M. Immunomodulation of experimental autoimmune encephalomyelitis by oral administration of copolymer 1. Proc Natl Acad Sci USA 1999; 96:3842-47.
80. Weiner HL. Induction and mechanism of action of transforming growth factor-beta- secreting Th3 regulatory cells. Immunol Rev 2001; 182:207-14.
81. Fukaura H, Kent SC, Pietrusewicz MJ et al. Induction of circulating myelin basic protein and proteolipid protein- specific transforming growth factor-beta1-secreting Th3 T cells by oral administration of myelin in multiple sclerosis patients. J Clin Invest 1996; 98(1):70-7.

CHAPTER 21

Major Histocompatibility Complex and Autoimmune Disease

Ursula Holzer and Gerald T. Nepom

Introduction

The major histocompatibility complex (MHC) encodes a large number of molecules which are key participants in the function of the immune system. Included in the human MHC, known as human leukocyte antigen (HLA), are two classes of HLA molecules which are highly polymorphic, and which provide genetic restriction for T lymphocyte responses: HLA class I and HLA class II. Several of the HLA class II molecules are highly associated with, and likely to have a major role in, susceptibility to autoimmune diseases. However, the molecular mechanisms involved are still unclear. In this chapter we address the structure of the HLA molecules and the mechanisms for presenting antigens to the immune system, as well as the correlation of specific HLA alleles with autoimmune diseases.

Inheritance of specific MHC alleles can greatly increase susceptibilty for immune-mediated diseases and autoimmune disorders. There are two classes of HLA molecules which are structurally and functionally different: HLA class I and HLA class II. Class II molecules are expressed predominantly on B-cells, dendritic cells and macrophages, whereas class I molecules are found on virtually all cells. By means of these molecules, peptides—foreign or self—can be presented to T-cells. CD4+ helper T-cells interact with the HLA class II molecule and CD8+ cytotoxic T-cells communicate with cells by interaction with class I molecules (Table 1).

Genetics and Nomenclature of the HLA System

The HLA complex is situated on the short arm of chromosome 6 (6p21.31) (Fig. 1), occupies 3.6 megabases and contains ~220 genes, 40 % of which are known to be involved in the immune response (see http://www.sanger.ac.uk/HGP/Chr6/MHC.shtml).[1] The MHC is the most gene-dense region of the human genome sequenced so far and encodes the most polymorphic human proteins, the HLA class I and II molecules.[2]

HLA class I genes encode a 2-chain structure (Fig. 2): genes for the α polypeptide chains are on chromosome 6, and for the β chain on chromosome 15, the beta$_2$-microglobulin gene. There are about 20 class I loci in the HLA region, of which three are the classical ones called class Ia genes: HLA-A, -B, -C.

The genes for the HLA class II molecule are also clustered on chromosome 6 and code for both α and β polypeptide chains. Three letters are used to designate their loci on chromosome 6: D indicates the class, the second letter (M, O, P, Q or R) the family and the last letter (A or B) the chain α or β (Fig. 3). The single genes of the HLA genes are differentiated by Arabic numbers followed by an asterisk and then the notation of the allelic variants. For example HLA-DRB1*0101 stands for the allelic variant 0101 of the gene B1 which is a class II gene of the DR family encoding the β polypeptide chain. As of January 2001, 542 different class II alleles have been named.[3]

The HLA class III region located between class I and II regions contain at least 59 genes and is the highest gene density region among the three HLA regions, including—among others—genes for complement proteins and the cytokines TNF α and β. HLA alleles are codominantly expressed; in other words, both inherited maternal and paternal HLA molecules are present in an individual. Because the class II molecules are heterodimers, heterologous molecules containing one maternal-derived chain and one paternal-derived chain are produced. Thus, combinatorial diversity can increase the number of expressed class II molecules.

Variations of the HLA Complex

It is widely accepted that selection plays an important role in shaping the variation of MHC genes present in the human population.[4] The competition of our immune system and infectious pathogens is thought to be one reason for the polymorphism of the MHC,[1] although the specific modes of selection are largely unresolved.

Among the HLA complex, the DRB1 (HLA class II) and HLA-B (HLA class I) are the most polymorphic loci with 254 and 349 current recognized alleles respectively.[4] However, the number of alleles varies greatly among populations and is fewest among isolated populations.[5]

Statistical modeling has been used to show that under neutral conditions the expected number of alleles is far lower than what is actually found at HLA loci.[6] Therefore, even a small heterozygote fitness advantage can have a large effect on polymorphism and likely accounts for the large number of alleles.

Table 1. Function and distribution of HLA molecules

	HLA Class I	HLA Class II
Distribution	Almost all cells	• Constitutive: B-cells, dendritic cells and macrophages • Inducible: most cell types
Function	Endogenous/viral antigens presented to CD8+ cells	Exogenous antigens presented to CD4+ cells; representation of endogenous antigens via exogenous pathways

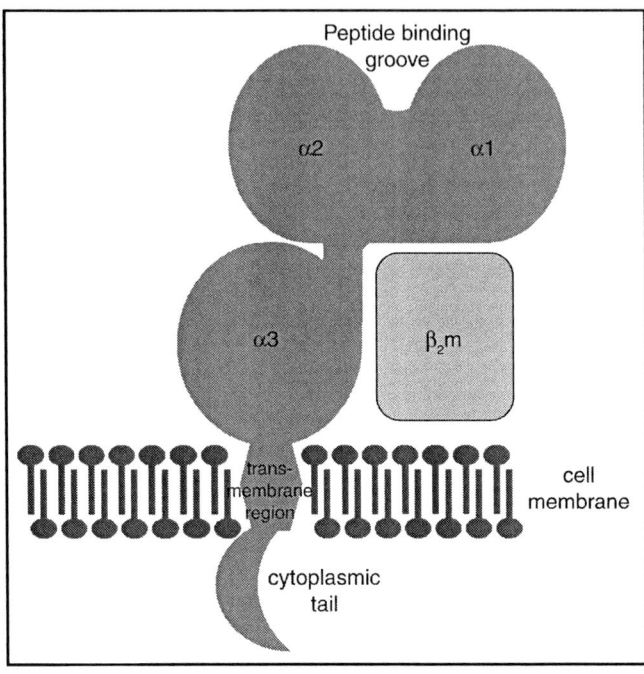

Figure 2. HLA class I molecule.

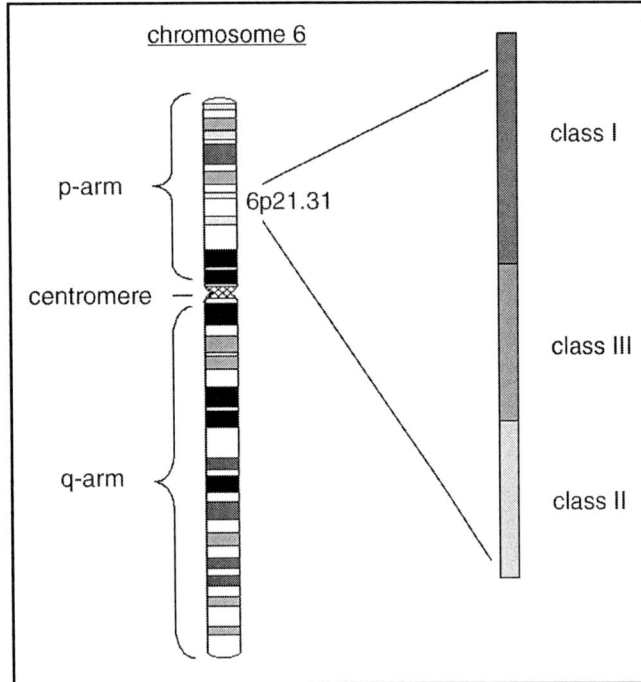

Figure 1. HLA complex on human chromosome 6.

Structure of the HLA Class I and II Molecules

Structure of the HLA Class I Molecule

The MHC class I molecule is composed of two proteins. The chromosome 6 encoded polypeptide of the α chain is about 350 amino acids long and glycosylated, giving a total molecular weight of about 45 kDa. The polypeptide folds into five separate domains: the peptide-binding domains (alpha-1 and alpha-2), an immunoglobulin-like scaffold domain (alpha-3), the transmembrane region and the cytoplasmic tail.[7]

The β2-microglobulin is a 12 kDa polypeptide that is noncovalently associated with the alpha-3 domain (Fig. 2).

Structure of the HLA Class II Molecule

MHC class II molecules are transmembrane glycoproteins consisting of two domains, an α-chain with a molecular weight of approximately 34000 and a β-chain with a molecular weight of approximately 29000. Each of the α- and β-chains has four domains: the peptide binding domains (α1 and β1), immunoglobulin like domains (α2 and β2), the transmembrane region and the cytoplasmic tail (Fig. 3).

The α-chain of HLA-DR molecules is nonpolymorphic and thus the definition of HLA-DR alleles essentially relies on the β1- domain variability. HLA-DQ molecules on the other hand contain polymorphic α and β chains, both of which influence variation in structure and function.

The MHC class II molecules are considered within the immunoglobulin superfamily (IgSF) of proteins because of their immunoglobulin homology located in the membrane-proximal domains, the region involved in the interaction with the CD4 coreceptor.

Function of the HLA Class I and II Molecules

The function of MHC molecules is to display antigens on the surface of antigen-presenting cells (APC) to T-cells. Antigen-presenting cells (APCs) degrade proteins intracellularly to generate peptides, which are then bound by MHC molecules and exposed on the surface for recognition by T cells. The T-cells utilize specific T-cell receptors (TCR) to recognize the antigen, proliferate and give signals to other cells with the aim to orchestrate a complex immune response directed to that antigen.

Function of the HLA Class I Molecule

HLA class I molecules present peptides, usually eight to nine amino acids in length, to CD8-expressing cytotoxic T-lymphocytes (CTLs) (Fig. 4). The function of the HLA class I molecules is likely primarily to alert T cells to changes inside the cell, e.g., viral infections. Therefore, the presented antigens derive mostly from newly synthesized proteins in the cytosol, e.g., viral proteins.[8] These proteins are degraded to peptides in proteolytic complexes called proteosomes. A subset of the resulting peptides are then translocated across the endoplasmatic reticulum (ER) membrane by the transporters associated with antigen processing (TAP)[9] encoded by the TAP1 and TAP2 genes, which

Major Histocompatibility Complex and Autoimmune Disease

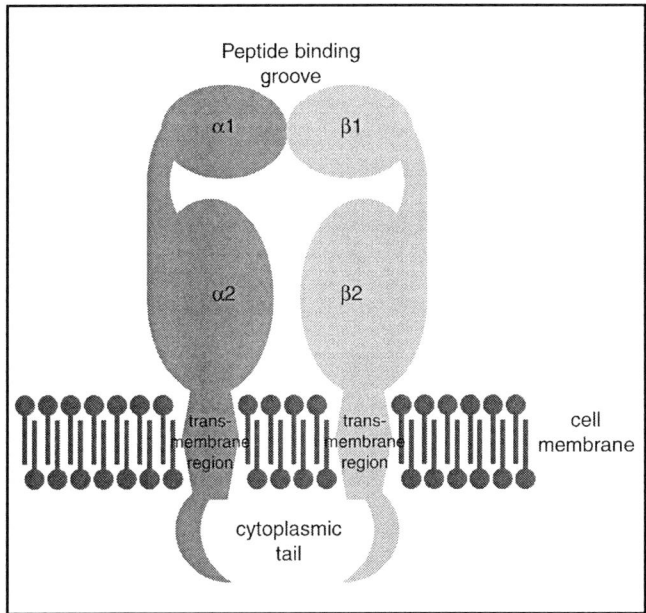

Figure 3. HLA class II molecule.

are located within the MHC class II region of chromosome 6.[2] The peptides transported by TAP are selected to match in length and sequence to the respective MHC class I molecules. Upon loading the antigen on MHC class I molecules in the ER, the whole complex is transported to the surface of the cells.[10] Figure 4 gives a summary of this process.

When considering an association of autoimmune diseases and HLA class II polymorphism, it is often considered whether other genes in the class II region may be involved in the pathogenesis of some of these diseases. TAP1 and TAP2 genes are localized in the class II region between HLA-DP and HLA-DQ,[2] and TAP gene polymorphism may influence the specificity of peptides preferentially presented by the MHC molecules and the outcome of the immune response.[11] Whereas some groups find an association of particular TAP alleles with autoimmunity e.g., with systemic lupus erythematosus (SLE) or rheumatoid arthritis,[12,13] others do not.[14,15]

Function of the HLA Class II Molecule

Antigens delivered from extracellular sources are endocytosed by the APC and degraded in endosomal compartments. These antigens will be in general presented by HLA class II molecules which are recognized by CD4 positive helper T-cells (Fig. 5).

The peptide binding site of class II αβ dimers can accommodate peptides of variable length and sequences therefore enabling APCs to present a large array of peptides using even a small set of different class II molecules.[16]

Class II αβ dimers associate with a nonpolymorphic peptide called the invariant chain (Ii)[17] which facilitates the folding and assembly of class II αβ dimers and prevents a premature loading of peptides.[18] The αβ-Ii complex meets exogenous peptides in an endosomal compartment in which the Ii chain is sequentially degraded by proteases. CLIP, the Ii fragment that remains associated with the HLA class II molecules, is ultimately dislodged and replaced by the antigenic peptide, which are likewise generated

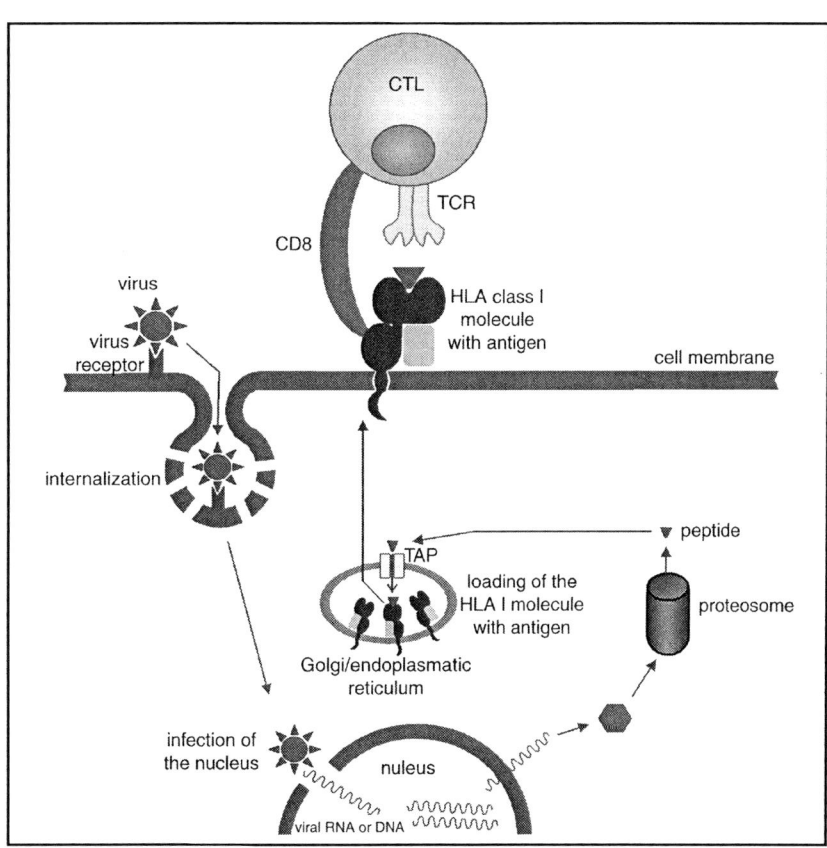

Figure 4. HLA class I antigen presentation to CD8+ T-cells.

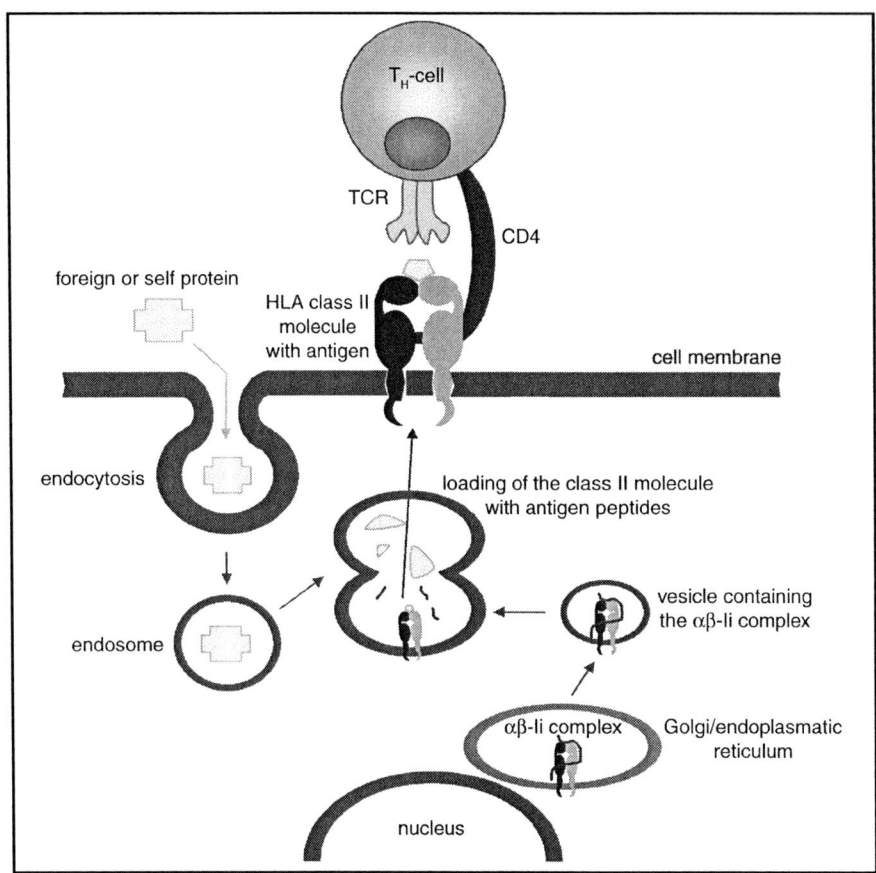

Figure 5. HLA class II antigen presentation to CD4+ T-cells.

by lysosomal proteases.[17] Therefore, in contrast to the MHC class I molecules, this favours the binding of endosomal peptides bound to class II molecules, when the antigenic peptide-binding site is no longer blocked by CLIP. Antigen presentation is therefore dependent on the collection of proteases to which αβ-Ii complexes and newly internalized antigens are exposed.[19] The peptide-laden class II molecules are then exported to the cell surface to present the foreign or self peptides to CD4-positive helper T-cells (Fig. 5).

The two main proteases involved in the degradation of the Ii chain are cathepsin L and S, which are differently distributed in tissues. Whereas cathepsin L is expressed in cortical thymic epithelial cells (cTEC), which present endogenous peptides to maturing thymocytes, cathepsin S is found in bone marrow-derived APCs like dendritic cells, macrophages or B-cells.[20] Interestingly, cathepsin S deficient mice exhibit significantly diminished susceptibility to collagen-induced arthritis, the mouse model for human rheumatoid arthritis,[21] thus suggesting that the proteolytic pathway involving antigen and the Ii chain are involved in processing and recognition of target antigens in this model of induced autoimmunity.

HLA Genes and Autoimmune Diseases

Interactions between host genes and the environment influence the susceptibility and triggering of autoimmunity by perturbing the normal regulated immune system. Within the HLA genes, certain MHC class II alleles confer the strongest susceptibility to MHC-associated disorders. Many associations between particular MHC class II alleles and autoimmune diseases have been described and they are usually different for different diseases (summarized in Table 2). Since different alleles have different abilities to present peptides from a target to CD4-positive T-cells, it is likely that in individuals with a specific HLA-type, some self-peptides are presented resulting in an autoimmune response to a specific target/organ. However, this is not the only mechanism involved. Certain class II alleles may play a role in increasing positive or decreasing negative selection of autoreactive T cells in the thymus, therefore favouring the establishment of autoimmunity and, on the other hand, some HLA class II alleles might be protective for an autoimmune disease by deleting potentially autoreactive cells. T cell recognition restricted by MHC genes thus occurs in at least two stages: in selection of T cells during development which have specificity for a presented self-antigen, as well as in subsequent activation of the responding autoreactive cells.

Individual class II alleles predispose towards autoimmunity in a disease specific way, which implies that organ-specific or disease-specific mechanisms, rather than global dysregulation, is involved. This is consistent with the notion that these alleles act via binding specific antigens. Nevertheless, regulatory elements of class II alleles can influence the levels of expression of specific disease asociated HLA class II molecules either in general or in a tissue-specific manner. This increased expression of HLA II molecules could lead in turn to enhanced antigen presentation via increased ligand density and therefore activation of autoreactive T-cells. Yet another possibility is that the association of a disease with a class II allele reflects a marker for the disease, but that other genes located within the HLA complex on chromosome 6 are actually forming the basis for the disease itself. Thus, the exact

Table 2. Selected autoimmune diseases and associated HLA alleles

Disease	Associated HLA Alleles
Organ-specific (endocrine)	
Type 1 diabetes mellitus	DRB1*0301-DQB1*0201;DRB1*04-DQB1*0302*
Graves disease	DRB1*0301-DQB1*0201
Hashimoto's thyroiditis	DRB1*11-DQB1*0301
Idiopathic Addison's disease	DRB1*0301-DQB1*0201
Organ-specific (other)	
Inflammatory bowel diseases	DRB1*0103
Crohn's disease	
Colitis ulcerosa	
Myasthenia gravis	DRB1*0301
Multiple sclerosis	DRB1*1501-DQA1*0102-DQB1*0602
Psoariasis vulgaris	C*06
Pemphigus	
Pemphigus vulgaris	DRB1*0402-DQB1*0302; DRB1*14-DQB1*0503
Pemphigus foliaceus	DRB1*0102; DRB1*0404; DRB1*1402; DRB1*1406
Narcolepsy	DQB1*0602
Coeliac disease	DRB1*0301-DQA1*0501-DQB1*0201
Dermatitis herpetiformis	DQA1*0501-DQB1*0201
Rheumatologic diseases	
Rheumatoid arthritis (RA)	DRB1*04-DQB1*0302; DRB1*01-DQB1*0501; DRB1*10-DQB1*0501
Juvenile idiopathic arthritis (JIA)	
Oligoarticular	DRB1*08-DQB1*0402; DRB1*11-DQB1*0301; DPB1*0201
Polyarticular	DRB1*01; DRB1*04
Spondyloarthopathies	B*27
Systemic lupus erythematosus (SLE)	DRB1*0301-DQB1*0201; DRB1*1501-DQB1*0602

Selected autoimmune diseases and associated HLA alleles (adapted from Erlich et al[22] with permission)

mechanism of the association of autoimmune diseases and certain MHC haplotypes may be quite diverse, and may differ for different diseases.

HLA Alleles and Type 1 Diabetes Mellitus

Type I (autoimmune) diabetes is a common disease with a very complex mode of inheritance.[23] A major part of the genetic predisposition is encoded within the HLA complex on chromosome 6. It is estimated that HLA class II gene polymorphisms account for 20% to 50% of the familial aggregation of type 1 diabetes.[24] The disease is associated with the allelic variations of several genes in the HLA region, particularly HLA-DRB1 and –DQB1. Susceptibility is dominated in white European-derived populations by the two-locus haplotypes DRB1*0301-DQB1*0201 (DR3) and DRB1*04-DQB1*0302 (DR4) reflecting the most common HLA haplotype in Caucasian diabetes type 1 patients.[25] These high risk haplotypes are present in about 90% of Caucasian type 1 diabetes patients compared to 40% of normal controls.[26] The strongest associations are with the HLA-DQ alleles DQA1*03-DQB1*0302 (DQ8) and DQA1*0501-DQB1*0201 (DQ2). In contrast, DQA1*0102-DQ0602 (DQ6) confers strong protection against diabetes type 1.[27,28] Different DRB1 alleles may modify the risk of this disease.[25] In a transgenic mouse model the DQ8 allele had a diabetogenic effect and coexpression of DR4 downregulated this effect somewhat.[29] This association of the HLA-DQ molecules with type 1 diabetes mellitus may be in part explained by an inappropriate early T-cell selection and maturation.[30]

HLA Alleles and Other Organ-Specific Endocrine Diseases

Besides Type 1 diabetes mellitus, autoimmune thyroiditis is as well an endocrine autoimmune disease which is associated with specific MHC class II alleles. Autoimmune thyroiditis can be distinguished by two clinical syndromes: Graves' disease, resulting in an overactivity of the thyroid gland (hyperthyroidism) and Hashimoto's thyroiditis, with hypothyroidism. Graves' disease is associated with the HLA haplotype DRB1*0301-DQA1*0501-DQB1*0201,[31,32] whereas for Hashimoto's thyroiditis the specific HLA relationship is less clear, with a small trend towards association with DRB1*0301 and DRB1*1101-DQB1*0301.[33-35]

For idiopathic Addison's disease, a disorder with an adrenocortical hormone deficiency due to damage to the outer layer of the adrenal gland, the association with HLA-DRB1*0301-DQA1*0501-DQB1*0201 was found in several studies.[36]

It is noteworthy that each of these autoimmune endocrine diseases associated with the HLA-DRB1*0301-DQA1*0501-DQB1*0201 (DR3) haplotype in Caucasians is characterized by tissue-specific autoantibodies (anti-islet in type 1 diabetes;

anti-thyroid in Graves' disease, anti-adrenal in Addison's disease); it is possible that the disease-predisposing element within the HLA haplotype associated with these diseases influences the disease pathogenesis through modulation of these autoantibody profiles.[37,38]

HLA Alleles and Multiple Sclerosis (MS)

The genetic association between MS susceptibility and the HLA locus has been known for almost 30 years.[39] In population studies mainly carried out in Caucasians of Northern European origin the disease is consistently associated with the HLA class II haplotype DRB1*1501-DRB5*0101-DQA1*0102-DQB1*0602.[40] Strong linkage disequilibrium between DRB1*1501 and DQB1*0602 makes it difficult to assess functional distinctions accounting for the disease risk.[41] Notably, T-cells reactive to myelin are present in MS patients, commonly restricted by the HLA-DR locus; both the HLA-DRA-DRB1*1501 and the HLA-DRA-DRB5*0101 heterodimers bind with high affinity to antigenic myelin basic protein (MBP) peptides.[42] Structural studies indicate that only two primary T cell receptor contact residues of MBP need to be conserved to stimulate antigen-specific clones,[43] thus suggesting that microbial peptides with limited sequence identity with the self-peptide may activate autoreactive cells. However, in a study carried out in Afro-Brazilian MS-patients with different MHC haplotypes, it appeared that DQB1*0602 was associated with the disease in the absence of the DRB1*1501 allele.[44]

HLA Alleles and Other Organ-Specific Diseases

For inflammatory bowel disease, such as Crohn's disease and ulcerative colitis, a genetic predisposition is strongly supposed.[45] The genes in the HLA region are candidate genes contributing to susceptibility for these diseases as they encode functionally relevant gene products in the gut epithelium and they colocalize to a linkage region on chromosome 6.[46,47] The association with HLA-DRB1*0103 and extensive ulcerative colitis has been widely replicated.[46,48,49] For Crohn's disease, an association with DR7, DRB3*0301 and DQ4 were found.[49] Nevertheless, the contribution of HLA-DQ genes is less clear-cut in these diseases and the association of HLA-DR molecules in the pathogenesis of ulcerative colitis may be threefold larger compared with Crohn's disease.[49] Non-HLA genes, such as NOD2 (a family of cytosolic proteins that regulate the host response to pathogens), are known to be contributory in this disease, as are environmental agents and bacterial gut flora, so the exact role for HLA genes in disease initiation or progression is not clear.

For myasthenia gravis, a disease with antibodies against acetylcholine receptors characterized by chronic weakness of voluntary muscles, a strong linkage disequilibrium with HLA-DRB1*03 has been reported.[50,51] Interestingly, a recent study recapitulated this correlation in a HLA-DR3 transgenic mouse model.[52]

Psoriasis vulgaris is a chronic inflammatory skin disease. An association of psoriasis vulgaris with HLA antigens was first described in 1972.[53,54] Since then, several markers at HLA-loci have been associated with the disease, with HLA-Cw6 demonstrating the strongest correlation.[55] Although the combination of HLA-Cw*0602 with HLA-B*5701 shows a high susceptibilty for this disease,[5] other studies imply that the susceptibility determinants are not encoded by HLA-A, -B, -C, -DRB1 and -DQB1 and that multiple disease alleles have arisen at the trait locus during the evolution of humans.[57]

In pemphigus, a chronic blistering disease of the skin mediated by autoantibodies, two forms can be distinguished: pemphigus vulgaris (PV) and pemphigus foliaceus (PF). The autoantibodies recognize epitopes of desmoglein 3 in PV or desmoglein 1 in PF. An association with the MHC II molecules DRB1*0402-DQB1*0302 and DRB1*14*-DQB1*0503 in PV has been shown.[58-60] The association with DRB1*14 was concluded as a secondary phenomenon due to linkage disequilibrium between the DQB1*0503 and DR14-related alleles as patients with PV carrying DQB1*0503 carried either DRB1*1401, 1405 or 1408, but all patients with DRB1*14 carried DQB1*0503.[60] This observation is particularly noteworthy, since it indicates that a singular clinical entity, such as PV, can have two completely different HLA associations. A possible explanation is that different antigenic epitopes of the desmoglien 3 protein, for example, are presented by each of the two different HLA molecules associated with the disease. In PF, a strong association was found with DRB1*0102, DRB1*0404, DRB1*1402 and DRB1*1406.[58,61,62]

Narcolepsy is a sleep disorder associated with cataplexy, with no known immunological component. However, a strong association with the HLA DQB1*0602 allele is characteristic of patients, representing multiple ethnic groups.[63,64] Individuals homozygous for DQB1*0602 have a two- to fourfold higher risk compared with that in DQB1*0602 heterozygous.[63] A study investigating other HLA alleles for additional effects in influencing the susceptibility to develop this disorder showed a relative higher risk for heterozygote combinations including DQB1*0301, DQA1*0102, DQA1*06, DRB104, DRB1*08, DRB1*11 and DRB1*12.[65]

Celiac disease is another chronic HLA-associated disease with multifactorial etiologies where genetic and environmental components are involved. The disease is an inflammatory disorder in which the mucosa of the small intestine is damaged in response to ingestion of gliaden proteins contained in wheat gluten. The primary HLA association in patients is with either the DRB1*0301-DQA1*0501-DQB1*0201 haplotype or in heterozygotes with the same HLA-DQ alleles, encoded in trans-, forming a heterodimer encoded by the DQ beta chain from the DRB1*0701-DQA1*0201-DQB1*0201 haplotype and the DQ alpha chain from the DRB1*1101-DQA1*0501-DQB1*0201 haplotype.[66] Dermatitis herpetiformis, a blistering skin disease characterized by granular IgA deposits in the papillary dermis is associated with similar HLA polymorphisms: HLA DRB1*0301 and DQA1*0501-DQB1*02.[66-68] The association with the DRB markers in both disorders is probably secondary to linkage disequilibrium with the DQ alleles mentioned above in Caucasians.

HLA Alleles and Rheumatoid Arthritis (RA)

Rheumatoid arthritis is a clinically heterogeneous disease and both genetic and environmental factors are involved in the etiology. Within the HLA class II region the HLA alleles DRB1*0401, DRB1*0404, DRB1*0408 and DRB1*0101 are associated with RA in Caucasian populations.[69] Among Eastern Asians, the most strongly associated allele is DRB1*0405, whereas DRB1*0102 predisposes to RA among Jews.[70] In some populations, there is as well an association with DRB1*1001.[71] All these DRB1 alleles encode a common amino acid sequence in the hypervariable region of the resulting DRβ1-chain known as the "shared epitope" (SE).[72] It has been proposed that the SE predisposes to RA by influencing antigenic peptide presentation as part of the peptide presenting groove of the HLA II molecule, or by selection and biased activation of specific T-cell receptors,[73] or as a peptide epitope bound and presented by HLA-DQ molecules.[74]

HLA Alleles and Other Rheumatological Diseases

Juvenile idiopathic arthritis (JIA) is an umbrella term for a group of chronic childhood arthritides in children below sixteen years of age and persisting for at least six weeks and having no known cause.[75] It includes clinically distinct forms of juvenile arthritis, which have genetic associations with different alleles of HLA class II. The most common form, oligoarticular arthritis, is defined as arthritis affecting four or less joints. The other forms are the extended oligoarthritis, polyarticular arthritis, divided in rheumatoid factor (RF) positive and negative, enthesitis related arthritis, psoriatic arthritis and "other arthritis". The polyarticular form is defined by affecting five or more joints in the first six months. For the oligoarticular form of the disease, a significant disease susceptibilty exisits for the HLA-haplotypes DRB1*08-DQB1*0401-DQB1*0402 and DRB1*11-DQA1*0501-DQB1*0301.[76,77] The class II genes HLA-DRB1*01 and DRB1*04 have been reported to increase the risk of polyarticular arthritis. DRB1*04 has a particularly high association with IgM rheumatoid factor-positive polyarticular arthritis, whereas DRB1*01 is associated with oligoarticular disease onset that converts to a polyarticular disease.[78,79] Analysis of the distribution of the DPB1 alleles showed an increase in DPB1*0201 in the oligoarticular form of the disease.[77,80,81]

The seronegative spondyloarthropathies are a cluster of inflammatory arthritides encompassing ankylosing spondylitis, psoriatic arthritis, Reiter's syndrome/reactive arthritis and the arthritis associated with inflammatory bowel disease. These disorders share similar clinical and immunogenetic features including axial arthritis and enthesopathy, a general predilection for males, the absence of rheumatoid factor and association with infections of the intestinal and genitourinary tracts. These diseases frequently show familial aggregation and are all strongly associated with HLA-B*27.[82] The strength of this assocation, however, varies not only among different ethnic groups but as well between the various diseases.[83] About 80-95% of the patients of white western European descent with ankylosing spondylitis and about 60-80% of the patients with reactive arthritis carry the MHC class I allele HLA-B*27 compared to 8% in the general population.[82]

Systemic lupus erythematosus (SLE) is a chronic, inflammatory autoimmune disorder that may affect many organ systems including the skin, joints and internal organs. Typically, antibodies against the nucleus of cells (anti-nuclear antibodies (ANAs)) are found in the serum of the patients. Studies in both animals and humans indicate that the development of this disease has a strong genetic basis — both MHC and non-MHC genes.[84] Concerning the MHC genes, the disease has been associated with the HLA haplotypes HLA-DR2 (DRB1*1501) and HLA-DR3 (DRB1*0301). HLA-DRB1*0301 is associated with SLE in Caucasians of Northern and Western European ancestry, HLA-DRB1*1501 is associated in Caucasians and Eastern Asians patients, whereas the structurally related allele DRB1*1503 is uniquely found in African Americans.[85,86] Within the DQ alleles, HLA-DQA1*0501 and DQB1*0201 are more frequent in Caucasians with SLE and DQB1*0602 is found more often in African American SLE patients compared to the respective control groups.[85]

Concluding Remarks

MHC genetic variation in the population leads to a very diverse set of immunologic elements which select, activate and amplify antigen-specific T-cells. Specific MHC molecules, alone or in combination, are associated with particular autoimmune diseases. In several cases, plausible but as yet unproven mechanisms have been proposed to provide a link between the genetics and the disease function.

References

1. Complete sequence and gene map of a human major histocompatibility complex. The MHC sequencing consortium. Nature 1999; 401(6756):921-923.
2. Forbes SA, Trowsdale J. The MHC quarterly report. Immunogenetics 1999; 50(3-4):152-159.
3. Marsh SG. Nomenclature for factors of the HLA System, update January 2001. Eur J Immunogenet 2001; 28(3):439-440.
4. Robinson J, Malik A, Parham P et al. IMGT/HLA database—a sequence database for the human major histocompatibility complex. Tissue Antigens 2000; 55(3):280-287.
5. Meyer D, Thomson G. How selection shapes variation of the human major histocompatibility complex: A review. Ann Hum Genet 2001; 65(Pt 1):1-26.
6. Potts WK, Wakeland EK. Evolution of MHC genetic diversity: A tale of incest, pestilence and sexual preference. Trends Genet 1993; 9(12):408-412.
7. Klein J, Sato A. The HLA system. First of two parts. N Engl J Med 2000; 343(10):702-709.
8. Grandea III AG, Van Kaer L. Tapasin: An ER chaperone that controls MHC class I assembly with peptide. Trends Immunol 2001; 22(4):194-199.
9. Heemels MT, Ploegh H. Generation, translocation, and presentation of MHC class I-restricted peptides. Annu Rev Biochem 1995; 64:463-491.
10. Ritz U, Seliger B. The transporter associated with antigen processing (TAP): Structural integrity, expression, function, and its clinical relevance. Mol Med 2001; 7(3):149-158.
11. Powis SJ, Deverson EV, Coadwell WJ et al. Effect of polymorphism of an MHC-linked transporter on the peptides assembled in a class I molecule. Nature 1992; 357(6375):211-215.
12. Hohler T, Weinmann A, Schneider PM et al. TAP-polymorphisms in juvenile onset psoriasis and psoriatic arthritis. Hum Immunol 1996; 51(1):49-54.
13. Martin-Villa JM, Martinez-Laso J, Moreno-Pelayo MA et al. Differential contribution of HLA-DR, DQ, and TAP2 alleles to systemic lupus erythematosus susceptibility in Spanish patients: Role of TAP2*01 alleles in Ro autoantibody production. Ann Rheum Dis 1998; 57(4):214-219.
14. Fraile A, Collado MD, Mataran L et al. TAP1 and TAP2 polymorphism in Spanish patients with ankylosing spondylitis. Exp Clin Immunogenet 2000; 17(4):199-204.
15. Vejbaesya S, Luangtrakool P, Luangtrakool K et al. Analysis of TAP and HLA-DM polymorphism in thai rheumatoid arthritis. Hum Immunol 2000; 61(3):309-313.
16. Rudensky AY, Preston-Hurlburt P, Hong SC et al. Sequence analysis of peptides bound to MHC class II molecules. Nature 1991; 353(6345):622-627.
17. Roche PA, Cresswell P. Proteolysis of the class II-associated invariant chain generates a peptide binding site in intracellular HLA-DR molecules. Proc Natl Acad Sci USA 1991; 88(8):3150-3154.
18. Bakke O, Dobberstein B. MHC class II-associated invariant chain contains a sorting signal for endosomal compartments. Cell 1990; 63(4): 707-716.
19. Villadangos JA, Ploegh HL. Proteolysis in MHC class II antigen presentation: Who's in charge? Immunity 2000; 12(3):233-239.
20. Nakagawa T, Roth W, Wong P et al. Cathepsin L: Critical role in Ii degradation and CD4 T cell selection in the thymus. Science 1998; 280(5362):450-453.

21. Nakagawa TY, Brissette WH, Lira PD et al. Impaired invariant chain degradation and antigen presentation and diminished collagen-induced arthritis in cathepsin S null mice. Immunity 1999; 10(2):207-217.
22. Erlich HA, Nepom GT, Tyan DB. Autoimmunity: Genetics and immunological mechanisms. In: Emery & Rimoin's, eds. Principles and Practice of Medical Genetics. 4th ed. Harcourt: 2001.
23. Davies JL, Kawaguchi Y, Bennett ST et al. A genome-wide search for human type 1 diabetes susceptibility genes. Nature 1994; 371(6493):130-136.
24. Noble JA, Valdes AM, Cook M et al. The role of HLA class II genes in insulin-dependent diabetes mellitus: Molecular analysis of 180 Caucasian, multiplex families. Am J Hum Genet 1996; 59(5):1134-1148.
25. Undlien DE, Friede T, Rammensee HG et al. HLA-encoded genetic predisposition in IDDM: DR4 subtypes may be associated with different degrees of protection. Diabetes 1997; 46(1):143-149.
26. Wolf E, Spencer KM, Cudworth AG. The genetic susceptibility to type 1 (insulin-dependent) diabetes: Analysis of the HLA-DR association. Diabetologia 1983; 24(4):224-230.
27. Redondo MJ, Fain PR, Eisenbarth GS. Genetics of type 1A diabetes. Recent Prog Horm Res 2001; 56:69-89.
28. Wucherpfennig KW, Eisenbarth GS. Type 1 diabetes. Nat Immunol 2001; 2(9):767-768.
29. Wen L, Chen NY, Tang J et al. The regulatory role of DR4 in a spontaneous diabetes DQ8 transgenic model. J Clin Invest 2001; 107(7):871-880.
30. Nepom GT, Kwok WW. Molecular basis for HLA-DQ associations with IDDM. Diabetes 1998; 47(8):1177-1184.
31. Hunt PJ, Marshall SE, Weetman AP et al. Histocompatibility leucocyte antigens and closely linked immunomodulatory genes in autoimmune thyroid disease. Clin Endocrinol (Oxf) 2001; 55(4):491-499.
32. Brix TH, Kyvik KO, Hegedus L. What is the evidence of genetic factors in the etiology of Graves' disease? A brief review. Thyroid 1998; 8(8):727-734.
33. Jenkins D, Penny MA, Fletcher JA et al. HLA class II gene polymorphism contributes little to Hashimoto's thyroiditis. Clin Endocrinol (Oxf) 1992; 37(2):141-145.
34. Flynn JC, Fuller BE, Giraldo AA et al. Flexibility of TCR repertoire and permissiveness of HLA-DR3 molecules in experimental autoimmune thyroiditis in nonobese diabetic mice. J Autoimmun 2001; 17(1):7-15.
35. Kong YC, David CS. New revelations in susceptibility to autoimmune thyroiditis by the use of H2 and HLA class II transgenic models. Int Rev Immunol 2000; 19(6):573-585.
36. Badenhoop K, Walfish PG, Rau H et al. Susceptibility and resistance alleles of human leukocyte antigen (HLA) DQA1 and HLA DQB1 are shared in endocrine autoimmune disease. J Clin Endocrinol Metab 1995; 80(7):2112-2117.
37. Gambelunghe G, Falorni A, Ghaderi M et al. Microsatellite polymorphism of the MHC class I chain-related (MIC-A and MIC-B) genes marks the risk for autoimmune Addison's disease. J Clin Endocrinol Metab 1999; 84(10):3701-3707.
38. Weetman AP, Zhang L, Tandon N et al. HLA associations with autoimmune Addison's disease. Tissue Antigens 1991; 38(1):31-33.
39. Opelz G, Terasaki P, Myers L et al. The association of HLA antigens A3, B7, and DW2 with 330 multiple sclerosis patients in the United States. Tissue Antigens 1977; 9(1):54-58.
40. Olerup O, Hillert J. HLA class II-associated genetic susceptibility in multiple sclerosis: A critical evaluation. Tissue Antigens 1991; 38(1):1-15.
41. Oksenberg JR, Baranzini SE, Barcellos LF et al. Multiple sclerosis: Genomic rewards. J Neuroimmunol 2001; 113(2):171-184.
42. Smith KJ, Pyrdol J, Gauthier L et al. Crystal structure of HLA-DR2 (DRA*0101, DRB1*1501) complexed with a peptide from human myelin basic protein. J Exp Med 1998; 188(8):1511-1520.
43. Hausmann S, Martin M, Gauthier L et al. Structural features of autoreactive TCR that determine the degree of degeneracy in peptide recognition. J Immunol 1999; 162(1):338-344.
44. Caballero A, Alves-Leon S, Papais-Alvarenga R et al. DQB1*0602 confers genetic susceptibility to multiple sclerosis in Afro- Brazilians. Tissue Antigens 1999; 54(5):524-526.
45. Satsangi J, Jewell DP, Rosenberg WM et al. Genetics of inflammatory bowel disease. Gut 1994; 35(5):696-700.
46. Satsangi J, Welsh KI, Bunce M et al. Contribution of genes of the major histocompatibility complex to susceptibility and disease phenotype in inflammatory bowel disease. Lancet 1996; 347(9010):1212-1217.
47. Hampe J, Shaw SH, Saiz R et al. Linkage of inflammatory bowel disease to human chromosome 6p. Am J Hum Genet 1999; 65(6):1647-1655.
48. Bouma G, Crusius JB, Garcia-Gonzalez MA et al. Genetic markers in clinically well defined patients with ulcerative colitis (UC). Clin Exp Immunol 1999; 115(2):294-300.
49. Stokkers PC, Reitsma PH, Tytgat GN et al. HLA-DR and -DQ phenotypes in inflammatory bowel disease: A meta- analysis. Gut 1999; 45(3):395-401.
50. Spurkland A, Gilhus NE, Ronningen KS et al. Myasthenia gravis patients with thymus hyperplasia and myasthenia gravis patients with thymoma display different HLA associations. Tissue Antigens 1991; 37(2):90-93.
51. Kaakinen A, Pirskanen R, Tiilikainen A. LD antigens associated with HL-A8 and myasthenia gravis. Tissue Antigens 1975; 6(4):175-182.
52. Raju R, Spack EG, David CS. Acetylcholine receptor peptide recognition in HLA DR3-transgenic mice: In vivo responses correlate with MHC-peptide binding. J Immunol 2001; 167(2):1118-1124.
53. Russell TJ, Schultes LM, Kuban DJ. Histocompatibility (HL-A) antigens associated with psoriasis. N Engl J Med 1972; 287(15):738-740.
54. White SH, Newcomer VD, Mickey MR et al. Disturbance of HL-A antigen frequency in psoriasis. N Engl J Med 1972; 287(15):740-743.
55. Elder JT, Nair RP, Guo SW et al. The genetics of psoriasis. Arch Dermatol 1994; 130(2):216-224.
56. Schmitt-Egenolf M, Windemuth C, Hennies HC et al. Comparative association analysis reveals that corneodesmosin is more closely associated with psoriasis than HLA-Cw*0602-B*5701 in German families. Tissue Antigens 2001; 57(5):440-446.
57. Jenisch S, Westphal E, Nair RP et al. Linkage disequilibrium analysis of familial psoriasis: Identification of multiple disease-associated MHC haplotypes. Tissue Antigens 1999; 53(2):135-146.
58. Loiseau P, Lecleach L, Prost C et al. HLA class II polymorphism contributes to specify desmoglein derived peptides in pemphigus vulgaris and pemphigus foliaceus. J Autoimmun 2000; 15(1):67-73.
59. Gonzalez-Escribano MF, Jimenez G, Walter K et al. Distribution of HLA class II alleles among Spanish patients with pemphigus vulgaris. Tissue Antigens 1998; 52(3):275-278.
60. Niizeki H, Inoko H, Mizuki N et al. HLA-DQA1, -DQB1 and -DRB1 genotyping in Japanese pemphigus vulgaris patients by the PCR-RFLP method. Tissue Antigens 1994; 44(4):248-251.
61. Moraes ME, Fernandez-Vina M, Lazaro A et al. An epitope in the third hypervariable region of the DRB1 gene is involved in the susceptibility to endemic pemphigus foliaceus (fogo selvagem) in three different Brazilian populations. Tissue Antigens 1997; 49(1):35-40.
62. Moraes JR, Moraes ME, Fernandez-Vina M et al. HLA antigens and risk for development of pemphigus foliaceus (fogo selvagem) in endemic areas of Brazil. Immunogenetics 1991; 33(5-6):388-391.
63. Pelin Z, Guilleminault C, Risch N et al. HLA-DQB1*0602 homozygosity increases relative risk for narcolepsy but not disease severity in two ethnic groups. US Modafinil in Narcolepsy Multicenter Study Group. Tissue Antigens 1998; 51(1):96-100.
64. Mignot E. Genetic and familial aspects of narcolepsy. Neurology 1998; 50(2Suppl1):S16-S22.
65. Mignot E, Lin L, Rogers W et al. Complex HLA-DR and -DQ interactions confer risk of narcolepsy-cataplexy in three ethnic groups. Am J Hum Genet 2001; 68(3):686-699.
66. Sollid LM, McAdam SN, Molberg O et al. Genes and environment in celiac disease. Acta Odontol Scand 2001; 59(3):183-186.
67. Hall MA, Lanchbury JS, Ciclitira PJ. HLA class II region genes and susceptibility to dermatitis herpetiformis: DPB1 and TAP2 associations are secondary to those of the DQ subregion. Eur J Immunogenet 1996; 23(4):285-296.

68. Ahmed AR, Yunis JJ, Marcus-Bagley D et al. Major histocompatibility complex susceptibility genes for dermatitis herpetiformis compared with those for gluten-sensitive enteropathy. J Exp Med 1993; 178(6):2067-2075.
69. Hiraiwa A, Yamanaka K, Kwok WW et al. Structural requirements for recognition of the HLA-Dw14 class II epitope: A key HLA determinant associated with rheumatoid arthritis. Proc Natl Acad Sci USA 1990; 87(20):8051-8055.
70. de Vries N, Ronningen KS, Tilanus MG et al. HLA-DR1 and rheumatoid arthritis in Israeli Jews: Sequencing reveals that DRB1*0102 is the predominant HLA-DR1 subtype. Tissue Antigens 1993; 41(1):26-30.
71. Sanchez B, Moreno I, Magarino R et al. HLA-DRw10 confers the highest susceptibility to rheumatoid arthritis in a Spanish population. Tissue Antigens 1990; 36(4):174-176.
72. Gregersen PK, Silver J, Winchester RJ. The shared epitope hypothesis. An approach to understanding the molecular genetics of susceptibility to rheumatoid arthritis. Arthritis Rheum 1987; 30(11):1205-1213.
73. Nepom GT. Major histocompatibility complex-directed susceptibility to rheumatoid arthritis. Adv Immunol 1998; 68:315-332.
74. van der Horst-Bruinsma I, Visser H, Hazes JM et al. HLA-DQ-associated predisposition to and dominant HLA-DR-associated protection against rheumatoid arthritis. Hum Immunol 1999; 60(2):152-158.
75. Petty RE, Southwood TR, Baum J et al. Revision of the proposed classification criteria for juvenile idiopathic arthritis: Durban, 1997. J Rheumatol 1998; 25(10):1991-1994.
76. Haas JP, Nevinny-Stickel C, Schoenwald U et al. Susceptible and protective major histocompatibility complex class II alleles in early-onset pauciarticular juvenile chronic arthritis. Hum Immunol 1994; 41(3):225-233.
77. Fernandez-Vina MA, Fink CW, Stastny P. HLA antigens in juvenile arthritis. Pauciarticular and polyarticular juvenile arthritis are immunogenetically distinct. Arthritis Rheum 1990; 33(12):1787-1794.
78. Glass DN, Giannini EH. Juvenile rheumatoid arthritis as a complex genetic trait. Arthritis Rheum 1999; 42(11):2261-2268.
79. Nepom BS, Nepom GT, Mickelson E et al. Specific HLA-DR4-associated histocompatibility molecules characterize patients with seropositive juvenile rheumatoid arthritis. J Clin Invest 1984; 74(1):287-291.
80. Begovich AB, Bugawan TL, Nepom BS et al. A specific HLA-DP beta allele is associated with pauciarticular juvenile rheumatoid arthritis but not adult rheumatoid arthritis. Proc Natl Acad Sci USA 1989; 86(23):9489-9493.
81. Odum N, Morling N, Friis J et al. Increased frequency of HLA-DPw2 in pauciarticular onset juvenile chronic arthritis. Tissue Antigens 1986; 28(4):245-250.
82. Reveille JD. HLA-B27 and the seronegative spondyloarthropathies. Am J Med Sci 1998; 316(4):239-249.
83. Gonzalez-Roces S, Alvarez MV, Gonzalez S et al. HLA-B27 polymorphism and worldwide susceptibility to ankylosing spondylitis. Tissue Antigens 1997; 49(2):116-123.
84. Vyse TJ, Kotzin BL. Genetic susceptibility to systemic lupus erythematosus. Annu Rev Immunol 1998; 16:261-292.
85. Reveille JD, Moulds JM, Ahn C et al. Systemic lupus erythematosus in three ethnic groups: I. The effects of HLA class II, C4, and CR1 alleles, socioeconomic factors, and ethnicity at disease onset. LUMINA Study Group. Lupus in minority populations, nature versus nurture. Arthritis Rheum 1998; 41(7):1161-1172.
86. Reveille JD, Barger BO, Hodge TW. HLA-DR2-DRB1 allele frequencies in DR2-positive black Americans with and without systemic lupus erythematosus. Tissue Antigens 1991; 38(4):178-180.

CHAPTER 22

Analyzing Complex Polygenic Traits: The Role of Non-HLA Genes in the Susceptibility to Autoimmune Disorders

Bernard R. Lauwerys and Edward K. Wakeland

Introduction

Genetic predisposition plays an important role in the susceptibility to autoimmune disorders. Evidence supporting the contribution of genetic factors include disease concordance in monozygotic twins, occasional familial clustering of related or even unrelated autoimmune conditions and increased incidence of several disorders in selected ethnic backgrounds.

Historically, the most striking demonstration that gene polymorphisms might be involved in the pathogenesis of autoimmune diseases comes from the analysis of patients harboring hereditary deficiencies in complement factors C1q, C2 and C4. Thus, over 90% of patients presenting with homozygous C1q deficiency (41 cases reported to date) develop a severe lupus-like disorder characterized by malar rash, glomerulonephritis and production of antinuclear antibodies.[1-3] Complete C4 deficiency is an extremely rare condition (28 cases reported) since the human genome contains about 2 to 8 copies of the C4 gene, each of them encoding a C4A or a C4B molecule according to polymorphic variations in exon 26 of the sequence. In this group of patients, the prevalence of systemic lupus erythematosus (SLE) is about 75%.[4] Finally, the most common inherited complement deficiency in humans is homozygous C2 deficiency with an estimated prevalence of 1:20,000, one third of them developing a mild form of SLE.[5]

In the majority of cases, however, the genetic predisposition to the development of autoimmune disorders cannot be attributed to a monogenic defect but is rather polygenic, involving activating or suppressor gene alleles, interacting in additive, epistatic or redundant ways. Identifying these genes and the pathways they regulate has been the target of many groups over the last years. In this chapter, we will focus on the most recent data obtained in humans and murine models of autoimmune diseases and discuss the ongoing developments in the field.

Non-HLA Genes and Autoimmunity in Human Disease

Several strategies have been developed to identify genes or genetic intervals associated with autoimmune disorders in humans. The most comprehensive approach is a genome-wide screen, performed in multiplex families (families in which more than one member is affected by the disease). This method has detected linkage of several genetic regions with the development of autoimmune diseases or some of their specific symptoms. Yet, the most commonly used approach has been candidate gene analysis which is based on the selection of specific genes because of their localization in a disease-associated interval or because of their potential physiopathological importance. Involvement of candidate genes in disease pathogenesis can be evaluated using transmission-disequilibrium tests (in multiplex families) or case-control studies (in the general population).

Genome-Wide Screens in Human Populations

Genome-wide screens analyze the allelic distribution of highly polymorphic repetitive DNA sequences (called microsatellite markers) that are found throughout the whole genome. Usually, microsatellites are amplified by the polymerase chain reaction and their alleles are differentiated on the basis of length variations as revealed by gel electrophoresis. These screens mostly target members of families with at least two affected siblings (affected sib-pair analysis). The principle of the method is that if two siblings suffer from the same disease, they probably inherited the same susceptibility alleles and, consequently, the same microsatellite markers located in their vicinity. The distribution of the microsatellites is assessed statistically and the association with disease is quantified using linkage of disequilibrium (LOD) scores. These scores indicate the probability that the linkage of a specific marker with disease has not been found by chance. Generally, a maximum LOD score greater than 2.2 is considered to be suggestive, 3.3 significant and 5.4 highly significant.

In systemic lupus erythematosus (SLE), studies performed in 4 large cohorts of multiplex families have identified not less than 50 distinct chromosomal loci that show evidence for linkage with disease or disease-related traits.[6-15] Of these, 6 regions meet criteria for significant linkage in at least one study. These regions are located on chromosomes: 1q23, 1q41-42, 2q37, 4p15, 6p21 and 16q13. In addition to these regions, there are many additional loci that show weaker but suggestive evidence for linkage in SLE. The following regions were associated with disease in at least 2 different studies: 1q31-32, 2q32, 3p21, 6q26-27, 11q23, 13q31-32, 18q21, 19q13, 20p12-13, 20q11-13 and 21q21.

While the concordant data originated from the different genome-wide screens are strongly suggestive of true genetic effects, several caveats impact the interpretation of these findings. First, the ethnic background of the patients and controls often significantly skews the results of different studies. For example, the locus at 2q37 showed very strong evidence for linkage in Swedish and Icelandic families,[8] but much weaker signals were found in Caucasian and African American populations. This could mean that the predisposing allele(s) present at this locus is only of importance in Scandinavian families. However, an alternative explanation could be that the predisposing allele is present at too high or low a frequency in other populations, thus masking its presence as a real risk factor for disease in other populations. Also, the complexity of the genetic interactions leading to a specific autoimmune phenotype might negatively impact the probability of detecting single susceptibility loci. Susceptibility genes may have only modest or even no effect on the disease phenotype when isolated from other susceptibility alleles, a phenomenon known as epistasis. Similarly, the presence of suppressor genes can modify the phenotypic expression of susceptibility genes.

Finally, both disease and genetic heterogeneity are other confounding factors that often affect the results of linkage studies. Disease heterogeneity is best illustrated in SLE in which patients may present with several combinations of symptoms (rash, arthritis, glomerulonephritis, central nervous system disease, etc) and serological findings (antibodies against dsDNA, Ro, La, etc) that usually cluster in mild versus severe forms of the disease. While both forms of SLE may represent the ends of the same disease spectrum, it is also possible that they constitute distinct disease etiologies sharing several common genetic susceptibility factors but are differentiated by specific disease-associated alleles. Dissection of the original disease into intermediate phenotypes that are used in linkage studies could overcome this problem. Similarly, healthy members of SLE multiplex families are known to be more frequently positive for the presence of antinuclear antibodies than controls. While a linkage study based on the presence or absence of disease would weaken the probability of finding genes associated with antinuclear antibody production, the use of more specific component phenotypes may enhance the power of the analysis. Finally, genetic heterogeneity indicates that multiple combinations of genes within the genome may result in identical disease phenotypes, a common feature of many genetic systems that can clearly affect the power of linkage analysis in outbred populations.

Despite these important drawbacks, genome-wide screens have been successfully used to analyze many autoimmune disorders and have led to the detection of many disease-associated chromosomal loci: type 1 diabetes, multiple sclerosis, rheumatoid arthritis, ankylosing spondylitis, Crohn's disease, asthma, etc. Interestingly, numerous positive linkages in these disorders were found to map in common clusters, a finding that could indicate that some genes in these regions might play important roles in initiating common pathogenic events leading to autoimmunity.

Candidate Genes Association Studies

The availability of large-scale genomic sequencing technologies has led to the observation that coding and non-coding regions of many genes are characterized by several polymorphisms that may affect their expression and/or function. Case-control studies and transmission-disequilibrium tests compare the distribution of these polymorphic variants or associated microsatellites between patients and healthy individuals. Transmission disequilibrium tests examine the transmission of alleles from heterozygous parents to their affected offspring. It is based on the principle that a disease-associated allele will be found more often in the diseased offspring than predicted by independent assortment. Data originating from both types of analysis are often hard to validate because of issues with sample size or confounding factors such as population admixture. Few genes have been definitely associated with autoimmune disorders and, for many of these, evidence-based description of their role in disease pathogenesis is still lacking.

In SLE, several genes located on chromosome 1 have been extensively tested for their association with disease: Fcγ receptors, T cell receptor zeta chain, interleukin-10, poly (ADP-ribose) polymerase (PARP) and tumor necrosis factor receptor 2 (TNFR2). The genes encoding the low-affinity Fcγ receptors FcγRIIa, FcγRIIb, FcγRIIc, FcγRIIIa and FcγRIIIb are tightly clustered at 1q23.[16] FcγRIIa-R/H131 polymorphism was reported in several studies to represent a significant risk factor for SLE.[17-19] Interestingly, the FcγRIIa-131R/R genotype, which is enriched in SLE patients, is also associated with a significant decrease in the clearance of immune complexes in vivo, a finding that could be relevant for the pathogenesis of the disease.[20] Other polymorphisms in FcγRIIb and FcγRIIIa genotypes have been reported as risk factors for SLE.[21,22] However, the results of these studies could be affected by linkage disequilibrium within this region.

The expression of the T cell receptor zeta chain is reduced in SLE T cells and this defect has been associated with TCR-mediated aberrant signaling in SLE (see Chapter 38).[23,24] The gene is highly polymorphic in healthy individuals as in SLE patients. Nonetheless, preliminary data indicate that the frequency of mutations in the coding or non-coding regions of the gene and alternative splicings of its cDNA are significantly more frequent in SLE patients than in controls.[25,26]

Tsao et al first reported preferential transmission of PARP alleles to SLE patients using a multiallelic transmission disequilibrium test.[27] However, other groups have not confirmed this result.[28,29] Eventually, recent fine mapping of the 1q41-42 region in humans localized two novel markers that show stronger association with disease, one just centromeric to PARP and another closer to the end of the chromosome at 1q42. Similarly, an association between the TNFR2 196 R/M polymorphism and disease was reported in Japanese SLE patients[30-32] but these data could not be reproduced by other groups,[33-35] thereby suggesting that another gene, in linkage disequilibrium with TNFR2, could explain the findings in the Japanese SLE population.

Finally, compelling evidence supports that IL-10 plays an important role in the pathogenesis of SLE. Serum levels of the cytokine are significantly higher in patients than in controls[36] and experimental data indicate that IL-10 can contribute to the B- and T-cell abnormalities that are characteristic of the disease.[37-40] A recent study reported a significant association between a single nucleotide polymorphism (SNP) at position −2763 (from the transcriptional start site) of the IL-10 promoter and disease susceptibility in a population of African-American patients. This allele is associated with increased IL-10 production. Noteworthy, randomly selected African-American individuals, when compared to Caucasians, are characterized by another SNP at position −3575 of the IL-10 promoter that is also associated with increased IL-10 production, a finding that may represent one of the risk factors for the increased prevalence of lupus in this population.[41]

Taken together, these results illustrate how genetic data from human populations are hard to interpret. Moreover, association studies do not have much impact when they do not delineate a causal relationship between a polymorphic variant and a disease pathway. Because of the rarity of autoimmune disorders, these investigations are difficult to perform using patient's material, underscoring the importance of studies on animal models of autoimmune diseases. Nonetheless, the data acquired in human studies have undoubtedly opened new perspectives in our understanding and our experimental approach of autoimmune disorders. In the future, new technological developments will allow us to improve our knowledge of differential gene expression in cell populations or tissues obtained from patients. In particular, high-density cDNAs or oligonucleotide spotted microarrays are promising tools in this context and will probably lead to the identification of novel disease-associated pathways or single genes, as is the case in animal models of autoimmune diseases.

Non-MHC Genes and Autoimmunity in Animal Models

Studies performed in animal models have dramatically improved our knowledge of the physiopathology of autoimmune disorders. Numerous models of spontaneous or induced autoimmunity are available in mice, several of which closely reproduce phenotypes characteristic of human autoimmune disorders. In the field of SLE, (NZB x NZW)$_{F1}$ (BWF1) and (NZB x SWR)$_{F1}$ (SWF1) hybrid mice spontaneously develop an autoimmune glomerulonephritis associated with anti-dsDNA antibody deposits. Both models are characterized by overt polyclonal B cell activation and T cell dependant autoantibody production and are extensively used for pathogenic or therapeutic studies. MRL/lpr mice also produce high amounts of autoantibodies, including nephritogenic anti-dsDNA antibodies, and are therefore commonly used as a murine model of SLE. The lpr trait results from an autosomal recessive mutation in Fas leading to the accumulation of non-deleted autoimmune T and B lymphocytes (lpr= lymphoproliferation). Although a similar defect has been found in a rare pediatric disorder, named Canale-Smith syndrome,[42] it has not been found in human SLE. In BXSB mice, Yaa (Y-linked autoimmune accelerator) induces severe disease in males, an uncommon feature of autoimmunity in humans.[43]

Linkage Studies in Murine Models of Autoimmunity

Although the penetrance of disease in animal models of autoimmunity is not 100%, genetic factors are the dominant element dictating disease susceptibility in inbred strains of mice. Linkage studies, performed in several murine models, have proven to be powerful tools in identifying disease associated chromosomal regions, as illustrated in SLE-prone mice. Thus, NZM/Aeg2410 is a congenic recombinant inbred strain derived from lupus-prone BWF1 animals.[44] NZM/Aeg2410 animals produce anti-dsDNA IgG antibodies beginning at about 6 months of age and develop lupus nephritis during the first 12 months of life. Both phenotypes are recessive, since (NZM/Aeg2410 x C57Bl/6(B6))$_{F1}$ hybrids do not present with glomerulonephritis and display a significant delay in autoantibody production. By contrast, 58% of the BC1 (backcrosses between (NZM/Aeg2410 x B6)$_{F1}$ and NZM/Aeg2410) develop early high titer anti-dsDNA antibody production and 42% fatal glomerulonephritis.[45] The principle of the linkage study is that, if a gene is involved in the expression of one of the recessive phenotypes, its genotype will be homozygous in the affected animals but heterozygous in the healthy mice. By contrast, alleles that are not linked to any of the phenotypes will be randomly distributed among healthy and diseased animals.

In 1994, Morel et al genotyped 77 microsatellite markers in (NZM/Aeg2410 x B6)$_{F1}$ and NZM/Aeg2410 BC1 backcrosses and identified 3 disease associated loci on chromosomes 1, 4 and 7, called "Sle1", "Sle2" and "Sle3". Interestingly, chromosomal intervals around peak association markers in the Sle1a&b, Sle1d and Sle3 regions are syntenic to human 1q22-24, 1q41-42 and 19q13.3 respectively, that have been associated with disease in several linkage studies.[45] This analysis was followed by numerous other studies that confirmed these results and reported other disease-associated chromosomal intervals, in this strain and in other lupus-prone mice[46-60] (Fig. 1). Again, variations in linkage results between the studies can be attributed to several problematic features of complex trait analysis. In particular, the mechanisms leading to autoimmunity may be different in distinct lupus-prone strains. In addition, the choice of the non-autoimmune parent will influence the results of the study.[49] For example, if both the disease-prone and the disease-resistant animals share a disease susceptibility allele, no linkage will be found.[61] Alternatively, the disease-resistant parent may carry disease suppressor or facilitating genes that will affect the expression of the disease-associated alleles derived from the lupus-prone animal. Nevertheless, genetic studies in mice have produced very strong evidence for linkage of well-defined chromosomal intervals with disease, delineating the boarders for further gene identification strategies. In particular, loci on chromosomes 1, 4, 7 and 17 have been identified in several studies, thereby suggesting that genes in these regions play important roles in mediating disease in a non-strain dependant way.

The Power of Congenic Dissection

Congenic dissection is the most powerful strategy to analyze the contribution of individual susceptibility loci to a polygenic trait. This strategy results in the production of congenic animals bearing a single disease susceptibility locus on a resistant genetic background. The component phenotypes that are mediated by each individual susceptibility locus can then be analyzed separately via the phenotypic analysis of each congenic strain. Furthermore, congenic animals are susceptible to classical gene identification strategies such as fine mapping and subsequent positional cloning.[62,63]

In 1997, the production of B6.Sle1, B6.Sle2 and B6.Sle3 congenic mice resulted in significant advances in our understanding of the pathogenesis of SLE. B6.Sle1 mice are characterized by the production of anti-H2A/H2B/DNA antibodies and spontaneously activated B and T lymphocytes, as evidenced by increased expression of B7.2/CD86 and CD69 markers respectively.[64-67] The transfer of the Sle2 interval into a B6 background results in the production of B6.Sle2 congenic mice that are associated with overproduction of polyclonal IgM and increased numbers of peritoneal and splenic B1 cells.[68] B6.Sle3 mice display phenotypic characteristics of polyclonal B and T cell activation, increased peripheral CD4/CD8 ratio, decreased T cell activation-induced cell death[69] and decreased expression of MHC II molecules on antigen-presenting cells (Lauwerys BR, Wakeland EK, unpublished observation). Interestingly, none of these strains develop full-blown SLE disease or fatal glomerulonephritis, as found in the NZM/Aeg2410 parents. However, the production of multicongenic animals bearing several disease-associated loci on a resistant B6 background showed that the combination of Sle1 with either Sle2 or Sle3 could mediate full disease expression.

Analyzing Complex Polygenic Traits

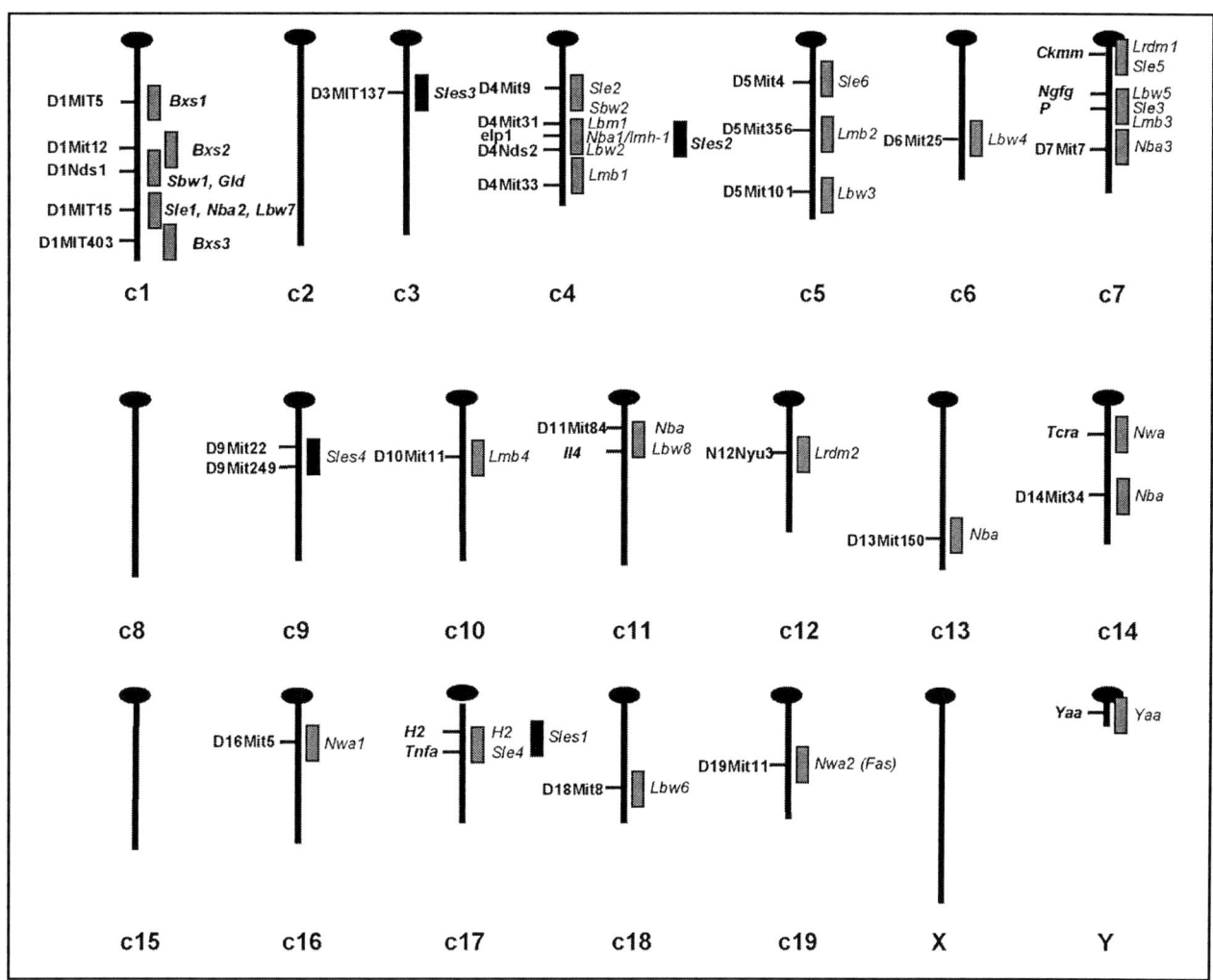

Figure 1: Susceptibility loci to SLE in the murine genome. Bxs= lupus susceptibility alleles identified in Bxs strain of mice; D#MIT#= stands for distance (D) on chromosome # microsatellite (MIT) marker #, e.g., D1MIT5= distance (D) on chromosome 1 microsatellite marker 5; Names in italic characters = SLE susceptibility loci in NZM/Aeg2410 (*Sle, Sles*), BWF1 (*Lbw, Nbw*), SWF1 (*Sbw*), BXSB (*Bxs*), NZB (*Nba*), NZW (*Nwa*) and MRL-FasLpr (*Lmb*) strains of mice. TCRα= T cell receptor alpha chain locus; TNFα= tumor necrosis factor alpha loci; Yaa= y chromosome-linked autoimmune accelerator

Thus, one-year mortality in female B6.Sle1/Sle3 bicongenics is 44%, and 18% in B6.Sle1/Sle2 mice. By contrast, the Sle2/Sle3 combination does not induce any mortality at one year, underscoring the important role played by Sle1 in the development of the disease. Noteworthy, the combination of Sle1/Sle2/Sle3 in tricongenic mice resulted in a 100% mortality in 12 months-old animals, a higher percentage than in the NZM/Aeg2410 parental strain.[70] Taken together, these experiments nicely demonstrate the power of the congenic dissection strategy. First, the susceptibility loci are identified and moved into a resistant background. Next, combination of the isolated susceptibility intervals reproduces disease in non susceptible mice, thereby demonstrating their key pathogenic role.

Surprisingly, Sle1, Sle2 and Sle3 susceptibility alleles are all derived from the NZW genome. Yet, NZW mice do not develop significant autoimmune disease nor autoantibody mediated glomerulonephritis by contrast to (NZB x NZW)$_{F1}$ hybrids, thereby suggesting that recessive genes in the NZW genome suppress the disease-facilitating effect of the susceptibility alleles. The observation that (B6.Sle1 x NZW)$_{F1}$ hybrids develop a highly penetrant (>75%) severe disease is another indication that recessive

genes displaying strong suppressor effects are present in the NZW genome. In order to identify these suppressor loci, Morel et al genotyped 122 microsatellite markers in (B6.Sle1 x NZW)$_{F1}$ x NZW BC1 backcrosses, that are all homozygous for the Sle1 susceptibility allele. If a microsatellite marker is associated with disease suppression, it will be homozygous in healthy mice and heterozygous in affected mice. By contrast, a marker that is associated with disease susceptibility would be homozygous in diseased mice and heterozygous in healthy mice. Finally, a marker that is not associated with disease will be randomly distributed among the animals. The authors identified 4 loci, named Sles1-4, which showed evidence for linkage with disease suppression. In particular, the Sles1 locus was found to completely abrogate the phenotype mediated by Sle1 in B6.Sle1/Sles1 bicongenic mice. Sles1 is located on chromosome 17, close to the MHC region. Interestingly, Sles2 is located on chromosome 4, in a region that has been reported to contain NZB-derived susceptibility alleles (Nba1 and Lbw5). The peak marker for Sles3 is on chromosome 3, close to the IL-2 gene that has been associated with susceptibility to diabetes and experimental autoimmune encephalomyelitis in other strains of mice. Finally, Sles4 is in the centromeric region of chromosome 9. Surprisingly, Sles4 displayed a suppressive effect on nephritis that was strongly gender-biased in that it was entirely male-specific.[71-73]

Fine Mapping Strategies

Fine mapping of the congenic intervals is the next step toward susceptibility gene identification. Fine mapping implies the production of recombinant subcongenic mice for phenotypic analysis and selection of the disease-associated sub-intervals. Thus, congenic lines were produced from (B6.Sle1 x B6)$_{F1}$ x B6 BC1 backcrosses bearing heterozygous recombinant chromosomal segments from the Sle1 susceptibility locus. The congenic recombinant chromosomes were bred to homozygosity and used for initial phenotypic analysis. This strategy revealed that a single susceptibility interval can be highly complex and represent a cluster of disease-associated genes. Morel et al found that Sle1 consists of four closely related loci: Sle1a, Sle1b, Sle1c and Sle1d. Sle1d plays a role in susceptibility to nephritis. Sle1a, Sle1b and Sle1c are all associated with a loss of tolerance to chromatin. However, several differences could be found between the subcongenic mice. First, the penetrance of anti-chromatin IgG antibodies is the highest in Sle1b, being even greater than that of the whole Sle1 interval. In addition, only Sle1b is associated with a significant increase in total IgM and IgG and increased number of CD69 positive T cells or B7.2 positive B cells. Next, Sle1a and Sle1b show the same response to H2A/H2B/DNA subnucleosomal particles, while Sle1c does not show a marked specificity for any chromatin component. Finally, Sle1a show characteristics that are absent in the whole Sle1 interval such as moderate increase in B220$^+$ cells, significant reduction in T cell numbers and higher proportion of CD62L^{lo}CD44^{hi} memory cells.[74] Clustering of susceptibility loci is not a specific feature of Sle1 but has been shown for other intervals in other models of autoimmune diseases. The significance of this phenomenon is unclear. It is possible that the sensitivity of linkage analysis is too low for detecting disease-associated single genes and that only clusters of functionally related genes produce phenotypes that are strong enough to be detected. Alternatively, clustering of susceptibility genes could result from functional and evolutionary pressures and illustrate a general rule in genomic organization.

Recent work by Boackle et al identified the Cr2 gene, encoding complement receptors 1 and 2, as a positional candidate for Sle1c. Comparisons of the NZM2410 and B6 alleles of Cr2 revealed several single nucleotide polymorphisms (SNP) that result in amino acid replacements in the sequence of the NZM2410 protein. One SNP introduces a novel glycosylation site in the C3d binding domain of the receptor, that reduces ligand-binding and receptor-mediated cell signaling. In addition to NZM2410 (and the parental NZW strain), two other strains share this polymorphic variant: NOD, which is a strain of diabetes-prone mice and SWR which is a parent for lupus prone SWF1 hybrids.[75] Noteworthy, targeted disruption of the Cr2 gene in B6 mice (B6 Cr2$^{-/-}$) has only limited effects on the production of autoantibodies. By contrast, mice that display a mutation of the Cr2 gene and concomitant mutation in Fas (B6/lpr Cr2$^{-/-}$) produce significantly higher levels of anti-dsDNA antibodies than B6/lpr controls, thereby indicating that the Cr2 gene might be linked to the production of autoantibodies by modifying the effect of other lupus susceptibility genes.[76]

Other candidate genes have been identified in the subcongenic intervals and the effects of SNPs on their expression and function are being thoroughly evaluated. Again, these differences do not constitute definitive evidence for involvement in disease pathogenesis. Final demonstration for establishing the pathogenic role of a candidate gene requires that its correction results in suppression of the compound phenotype present in the congenic strains. This can be achieved in transgenic approaches, using either conventional or BAC transgenes. While conventional transgenesis induces dramatic overexpression of the gene, BAC transgenes contain much larger segments of genomic DNA, also including normal regulatory sequences, thereby resulting in physiological levels of protein expression (for review, see ref. 77).

Microarray Analyses of Differential Gene Expression

Recently, high-density cDNAs or oligonucleotide spotted microarrays have been successfully introduced in the process of candidate gene identification. The availability of slides containing large arrays (>20,000 probes) and developments in the bioinformatic tools required for useful interpretation of the data, make microarrays a powerful instrument for screening differential gene expression between diseased versus control animals. Rozzo et al recently studied differential gene expression in B6 versus B6.Nba2 spleen cells. Nba2 is a NZB (New Zealand Black) autoimmunity mouse locus that has been associated with lupus traits in BWF1 lupus prone mice. High levels of IgG autoantibodies but no nephritis characterize B6.Nba2, a congenic strain of mice that harbors the Nba2 susceptibility interval in a B6 resistant background. Strikingly, (B6.Nba2 x NZW)$_{F1}$ hybrids display a high rate of autoantibody production and glomerulonephritis, while (B6 x NZW)$_{F1}$ mice do not develop any feature of autoimmunity. Using microarrays, the authors found consistent increase in Ifi202 and a reciprocal decrease in Ifi203 (Ifi are a family of interferon-activated genes) gene expression in B6.Nba2 spleen cells. Real-time PCR experiments showed that B cells and non-B, non-T cells were responsible for the increased expression of Ifi202. Interestingly, a SNP in the promoter region of the gene correlated with levels of expression: high in NZB and Balb/c mice that share the 95 C allele, low in NZW, B6 and B10 that are characterized by a 95 T allele.[61]

While heterogeneous cell populations or tissues might dilute cell specific increases in gene expression, the use of homogeneous test samples increases the power of this technique. Thus, preliminary data obtained from microarray analyses of purified B and CD4 T cells from SLE patients distinguished genes that are differentially regulated in SLE lymphocytes subsets which might not have been apparent using whole PBMC (Dao KH et al. Abstract presentation at the American College of Rheumatology meeting, Orlando 2003).

Synthetic Models of Autoimmunity

The ability to manipulate the mouse genome has allowed the creation of a variety of interesting single-gene models of autoimmunity. Several lines of mice that are transgenic or deficient for regulatory molecules in the immune system, were found to display autoimmune traits such as production of autoantibodies, polyclonal activation of B and T cells or lymphoproliferation. Thus, deficiencies in complement factors C1q and C4 have been shown to mediate the development of autoimmunity. In particular, 25% of the C1q-deficient mice have pathological evidence of glomerulonephritis with deposition of immune complexes and the presence of apoptotic bodies, thereby suggesting that C1q-deficiency mediates autoimmunity via an impaired clearance of apoptotic material.[78,79] Similarly, mice deficient in DNAse I[80] or in Serum Amyloid P[81] show modest levels of autoimmunity. Both molecules are involved in the metabolism of apoptotic debris, underscoring the potential role for impaired removal of apoptotic cells in the initiation of lupus-like autoimmunity. Surprisingly, mortality in MRL/lpr mice overexpressing a transgenic soluble complement regulator (soluble CR1-related gene/protein y) was found to be significantly lower than in wild-type MRL/lpr mice. This apparent discrepancy could be explained by reduced levels of glomerular inflammation due to reduction in complement component C3 levels. By contrast, serum levels of autoantibodies were not affected in the transgenic mice.[82]

In another group of modified animals, disruption of pathways that regulate apoptosis in the immune system results in the accumulation of autoreactive T and B lymphocytes and the production of antinuclear antibodies. This phenomenon has been extensively studied in Fas and Fas-Ligand MRL deficient mice that are classical models of SLE. Similarly, bcl-2 transgenic mice produce high levels of anti-dsDNA autoantibodies due to an increased lifespan of autoantibody producing B cells.[83] IEX-1 (Immediate Early Response gene X1) transgenic mice display reduced apoptosis rates of activated T cells, resulting in accumulation of effector/memory T cells, splenomegaly, lymphadenopathy, production of IgG2a anti-dsDNA antibodies and proteinuria.[84]

Finally, dysregulation of pathways controlling lymphocyte activation might induce various levels of autoantibody production due to polyclonal activation of the T or B cell compartment. BAFF (B cell activation factor from the TNF family), zTNF4 or BlyS, is a novel cytokine from the TNF family that induces the activation and proliferation of B cells. Mice transgenic for BlyS have elevated levels of B-1a lymphocytes, hypergammaglobulinaemia, and production of nephritogenic anti-dsDNA antibodies, presumably due to up-regulation of bcl-2 expression in B cells.[85-87] Of note, serum levels of BlyS are increased in human SLE patients and correlate with autoantibody titers.[88-89] Similar increases in the B-1 cell population and serum autoantibody titers were found in osteopontin transgenic mice.[90] In IFN-γ transgenic mice, the production of autoantibodies is T cell dependant via apoptotic keratinocytes acting as a potential source of autoantigens.[91,92] Lyn, CD22 and the thyrosine phosphatase SHP-1 are all downstream regulators of signaling through the B cell receptor. Mice deficient in these molecules display various levels of autoimmunity linked to overt polyclonal B cell activation.[93] B cell hyperactivity is also expressed by mice transgenic for an anti-CD1 TCR that activates CD1-restricted B cells or in mice expressing a CD40-ligand transgene on B cells.[94] Finally, PD-1, a cell-surface receptor belonging to the immunoglobulin superfamily and containing an immunoreceptor tyrosine-based inhibitory motif, has been shown to inhibit the proliferation of activated T cells. Mice deficient in PD-1 develop a lupus-like arthritis and glomerulonephritis with predominant IgG3 deposition.[95]

Taken together, these transgenic or knockout models nicely illustrate the principle of genetic heterogeneity. Similar phenotypes can be obtained by over-expression or targeted disruption of very different genes. It also implies that, although these experiments contribute to a better understanding of the mechanisms involved in immune regulation, they could be irrelevant for the pathogenesis of SLE or other autoimmune disorders. For instance, linkage studies did not find any evidence for association of bcl-2 polymorphisms and SLE.[96] Similarly, although serum levels of BlyS tend to be more elevated in SLE patients than in controls, no linkage between Blys alleles and SLE could be found in a recent study, thereby excluding this cytokine from the list of candidate susceptibility genes.[97] Moreover, another caveat may significantly impact the interpretation of synthetic models of autoimmunity. Several reports have clearly documented that B6 x 129 hybrid genomes are spontaneously predisposed to the development of humoral autoimmunity. Interaction of disrupted genes with susceptibility elements that are found in mice segregating a mixture of B6 and 129 genomes (a very common experimental situation) might influence the final phenotype and be wrongly attributed to the presence of the targeted gene.

Concluding Remarks

Genetic susceptibility to autoimmune disorders is a highly complex phenomenon, that is mediated by multiple interactions between disease-enhancing or suppressing genes acting in intermediate pathways leading to autoimmunity. In mice, congenic dissection of autoimmune related traits have provided important insights into the pathogenesis of autoimmune disease. It not only has allowed identification of disease-associated genes or pathways that can be targeted by specific therapies, but has also shed light on the pathophysiological steps gradually leading to autoimmune phenotypes. Thus, genes associated with susceptibility to autoimmune disorders can be categorized in hypothetical pathways that provide the mechanisms for disease development. In the first pathway, susceptibility genes mediate loss of tolerance for autoantigens. This pathway is best illustrated in Sle1 congenic mice that produce autoantibodies but do not develop glomerulonephritis. Genes involved in the amplification of the autoimmune response characterize the second pathway. These genes are present in the Sle2 or Sle3 loci. When one of these loci is combined with the Sle1 locus in bicongenic mice, severe renal damage may develop. Finally, a third pathway mediates end-organ susceptibility to damage. Evidence for such a genetic contribution to disease has been found in the Sle1d locus.

The models of congenic dissection of autoimmune traits open large avenues of research for the future. Identification of disease susceptibility genes will undoubtedly take advantage of the new developments in microarray and sequencing technologies, leading to a better understanding of disease and novel therapeutic approaches.

References

1. Walport MJ. Complement and systemic lupus erythematosus. Arthritis Res 2002; 4Suppl3:S279-93.
2. Pickering MC, Botto M, Taylor PR et al. Systemic lupus erythematosus, complement deficiency, and apoptosis. Adv Immunol 2000; 76:227-324.
3. Taylor PR, Carugati A, Fadok VA et al. A hierarchical role for classical pathway complement proteins in the clearance of apoptotic cells in vivo. J Exp Med 2000; 192:359-66.
4. Rupert KL, Moulds JM, Yang Y et al. The molecular basis of complete complement C4A and C4B deficiencies in a systemic lupus erythematosus patient with homozygous C4A and C4B mutant genes. J Immunol 2002; 169:1570-8.
5. Sullivan KE, Petri MA, Schmeckpeper BJ et al. Prevalence of a mutation causing C2 deficiency in systemic lupus erythematosus. J Rheumatol 1994; 21:1128-33.
6. Gaffney PM, Kearns GM, Shark KB et al. A genome-wide search for susceptibility genes in human systemic lupus erythematosus sib-pair families. Proc Natl Acad Sci USA 1998; 95:14875-9.
7. Moser KL, Neas BR, Salmon JE. et al. Genome scan of human systemic lupus erythematosus: Evidence for linkage on chromosome 1q in African-American pedigrees. Proc Natl Acad Sci USA 1998; 95:14869-74.
8. Lindqvist AK, Steinsson K, Johanneson B et al. A susceptibility locus for human systemic lupus erythematosus (hSLE1) on chromosome 2q. J Autoimmun 2000; 14:169-78.

9. Moser KL, Gray-McGuire C, Kelly J et al. Confirmation of genetic linkage between human systemic lupus erythematosus and chromosome 1q41. Arthritis Rheum 1999; 42:1902-7.
10. Quintero-Del-Rio AI, Kelly JA, Kilpatrick J et al. The genetics of systemic lupus erythematosus stratified by renal disease: Linkage at 10q22.3 (SLEN1), 2q34-35 (SLEN2), and 11p15.6 (SLEN3). Genes Immun 2002; 3Suppl1:S57-62.
11. Namjou B, Nath SK, Kilpatrick J et al. Genome scan stratified by the presence of anti-double-stranded DNA (dsDNA) autoantibody in pedigrees multiplex for systemic lupus erythematosus (SLE) establishes linkages at 19p13.2 (SLED1) and 18q21.1 (SLED2). Genes Immun 2002; 3 Suppl 1:S35-41.
12. Graham RR, Langefeld CD, Gaffney PM et al. Genetic linkage and transmission disequilibrium of marker haplotypes at chromosome 1q41 in human systemic lupus erythematosus. Arthritis Res 2001; 3:299-305.
13. Gray-McGuire C, Moser KL, Gaffney PM et al. Genome scan of human systemic lupus erythematosus by regression modeling: Evidence of linkage and epistasis at 4p16-15.2. Am J Hum Genet 2000; 67:1460-9.
14. Gaffney PM, Ortmann WA, Selby SA et al. Genome screening in human systemic lupus erythematosus: Results from a second Minnesota cohort and combined analyses of 187 sib-pair families. Am J Hum Genet 2000; 66:547-56.
15. Moser KL, Gray-McGuire C, Kelly J et al. Confirmation of genetic linkage between human systemic lupus erythematosus and chromosome 1q41. Arthritis Rheum 1999; 42:1902-7.
16. Su K, Wu J, Edberg JC et al. Genomic organization of classical human low-affinity Fcgamma receptor genes. Genes Immun 2002; 3 Suppl 1:S51-6.
17. Zuniga R, Ng S, Peterson MG et al. Low-binding alleles of Fcgamma receptor types IIA and IIIA are inherited independently and are associated with systemic lupus erythematosus in Hispanic patients. Arthritis Rheum 2001; 44:361-7.
18. Hatta Y, Tsuchiya N, Ohashi J et al. Association of Fc gamma receptor IIIB, but not of Fc gamma receptor IIA and IIIA polymorphisms with systemic lupus erythematosus in Japanese. Genes Immun 1999; 1:53-60.
19. Manger K, Repp R, Jansen M et al. Fcgamma receptor IIa, IIIa, and IIIb polymorphisms in German patients with systemic lupus erythematosus: Association with clinical symptoms. Ann Rheum Dis 2002; 61:786-92.
20. Dijstelbloem HM, Bijl M, Fijnheer R et al. Fcgamma receptor polymorphisms in systemic lupus erythematosus: Association with disease and in vivo clearance of immune complexes. Arthritis Rheum 2000; 43:2793-800.
21. Edberg JC, Langefeld CD, Wu et al. Genetic linkage and association of Fcgamma receptor IIIA (CD16A) on chromosome 1q23 with human systemic lupus erythematosus. Arthritis Rheum 2002; 46:2132-40.
22. Kyogoku C, Dijstelbloem HM, Tsuchiya N et al. Fcgamma receptor gene polymorphisms in Japanese patients with systemic lupus erythematosus: Contribution of FCGR2B to genetic susceptibility. Arthritis Rheum 2002; 46:1242-54.
23. Brundula V, Rivas LJ, Blasini AM et al. Diminished levels of T cell receptor zeta chains in peripheral blood T lymphocytes from patients with systemic lupus erythematosus. Arthritis Rheum 1999; 42:1908-16.
24. Liossis SN, Ding XZ, Dennis GJ et al. Altered pattern of TCR/CD3-mediated protein-tyrosyl phosphorylation in T cells from patients with systemic lupus erythematosus. Deficient expression of the T cell receptor zeta chain. J Clin Invest 1998; 101:1448-57.
25. Nambiar MP, Enyedy EJ, Warke VG et al. T cell signaling abnormalities in systemic lupus erythematosus are associated with increased mutations/polymorphisms and splice variants of T cell receptor zeta chain messenger RNA. Arthritis Rheum 2001; 44:1336-50.
26. Nambiar MP, Enyedy EJ, Warke VG et al. Polymorphisms/mutations of TCR-zeta-chain promoter and 3' untranslated region and selective expression of TCR zeta-chain with an alternatively spliced 3' untranslated region in patients with systemic lupus erythematosus. J Autoimmun 2001; 16:133-42.
27. Tsao BP, Cantor RM, Grossman JM et al. PARP alleles within the linked chromosomal region are associated with systemic lupus erythematosus. J Clin Invest 1999; 103:1135-40.
28. Criswell LA, Moser KL, Gaffney PM et al. PARP alleles and SLE: Failure to confirm association with disease susceptibility. J Clin Invest 2000; 105:1501-2.
29. Delrieu O, Michel M, Frances C et al. Poly(ADP-ribose) polymerase alleles in French Caucasians are associated neither with lupus nor with primary antiphospholipid syndrome. GRAID Research Group. Group for Research on Auto-Immune Disorders. Arthritis Rheum 1999; 42:2194-7.
30. Komata T, Tsuchiya N, Matsushita M et al. Association of tumor necrosis factor receptor 2 (TNFR2) polymorphism with susceptibility to systemic lupus erythematosus. Tissue Antigens 1999; 53:527-33.
31. Morita C, Horiuchi T, Tsukamoto H et al. Association of tumor necrosis factor receptor type II polymorphism 196R with Systemic lupus erythematosus in the Japanese: Molecular and functional analysis. Arthritis Rheum 2001; 44:2819-27.
32. Tsuchiya N, Komata T, Matsushita M et al. New single nucleotide polymorphisms in the coding region of human TNFR2: Association with systemic lupus erythematosus. Genes Immun 2000; 1:501-3.
33. Takahashi M, Hashimoto H, Akizuki M et al. Lack of association between the Met196Arg polymorphism in the TNFR2 gene and autoimmune diseases accompanied by vasculitis including SLE in Japanese. Tissue Antigens 2001; 57:66-9.
34. Lee EB, Yoo JE, Lee YJ et al. Tumor necrosis factor receptor 2 polymorphism in systemic lupus erythematosus: No association with disease. Hum Immunol 2001; 62:1148-52.
35. Tsuchiya N, Kawasaki A, Tsao BP et al. Analysis of the association of HLA-DRB1, TNFalpha promoter and TNFR2 (TNFRSF1B) polymorphisms with SLE using transmission disequilibrium test. Genes Immun 2001; 2:317-22.
36. Houssiau FA, Lefebvre C, Vanden Berghe M et al. Serum interleukin 10 titers in systemic lupus erythematosus reflect disease activity. Lupus 1995; 4:393-5.
37. Llorente L, Richaud-Patin Y, Garcia-Padilla C et al. Clinical and biologic effects of anti-interleukin-10 monoclonal antibody administration in systemic lupus erythematosus. Arthritis Rheum 2000; 43:1790-800.
38. Llorente L, Richaud-Patin Y, Couderc J et al. Dysregulation of interleukin-10 production in relatives of patients with systemic lupus erythematosus. Arthritis Rheum 1997; 40:1429-35.
39. Llorente L, Zou W, Levy Y et al. Role of interleukin 10 in the B lymphocyte hyperactivity and autoantibody production of human systemic lupus erythematosus. J Exp Med 1995; 181:839-44.
40. Lauwerys BR, Garot N, Renauld JC et al. Interleukin-10 blockade corrects impaired in vitro cellular immune responses of systemic lupus erythematosus patients. Arthritis Rheum 2000; 43:1976-81.
41. Gibson AW, Edberg JC, Wu J et al. Novel single nucleotide polymorphisms in the distal IL-10 promoter affect IL-10 production and enhance the risk of systemic lupus erythematosus. J Immunol 2001; 166:3915-22.
42. Drappa J, Vaishnaw AK, Sullivan KE et al. Fas gene mutations in the Canale-Smith syndrome, an inherited lymphoproliferative disorder associated with autoimmunity. N Engl J Med 1996; 335:1643-9.
43. Izui S, Merino R, Fossati L et al. The role of the Yaa gene in lupus syndrome. Int Rev Immunol 1994; 11:211-30.
44. Mohan C, Morel L, Yang P et al. Accumulation of splenic B1a cells with potent antigen-presenting capability in NZM2410 lupus-prone mice. Arthritis Rheum 1998; 41:1652-62.
45. Morel L, Rudofsky UH, Longmate JA et al. Polygenic control of susceptibility to murine systemic lupus erythematosus. Immunity 1994; 1:219-29.
46. Drake CG, Babcock SK, Palmer E et al. Genetic analysis of the NZB contribution to lupus-like autoimmune disease in (NZB x NZW)F1 mice. Proc Natl Acad Sci USA 1994; 91:4062-6.
47. Kono DH, Burlingame RW, Owens DG et al. Lupus susceptibility loci in New Zealand mice. Proc Natl Acad Sci USA 1994; 91:10168-72.
48. Vyse TJ, Drake CG, Rozzo SJ et al. Genetic linkage of IgG autoantibody production in relation to lupus nephritis in New Zealand hybrid mice. J Clin Invest 1996; 98:1762-72.

49. Rozzo SJ, Vyse TJ, Drake CG et al. Effect of genetic background on the contribution of New Zealand black loci to autoimmune lupus nephritis. Proc Natl Acad Sci USA 1996; 93:15164-8.
50. Vyse TJ, Rozzo SJ, Drake CG et al. Control of multiple autoantibodies linked with a lupus nephritis susceptibility locus in New Zealand black mice. J Immunol 1997; 158:5566-74.
51. Hogarth MB, Slingsby JH, Allen PJ et al. Multiple lupus susceptibility loci map to chromosome 1 in BXSB mice. J Immunol 1998; 161:2753-61.
52. Santiago ML, Mary C, Parzy D et al. Linkage of a major quantitative trait locus to Yaa gene-induced lupus-like nephritis in (NZW x C57BL/6)F1 mice. Eur J Immunol 1998; 28:4257-67.
53. Morel L, Mohan C, Yu Y et al. Multiplex inheritance of component phenotypes in a murine model of lupus. Mamm Genome 1999; 10:176-81.
54. Wither JE, Paterson AD, Vukusic B. Genetic dissection of B cell traits in New Zealand black mice. The expanded population of B cells expressing up-regulated costimulatory molecules shows linkage to Nba2. Eur J Immunol 2000; 30:356-65.
55. Haywood ME, Hogarth MB, Slingsby JH et al. Identification of intervals on chromosomes 1, 3, and 13 linked to the development of lupus in BXSB mice. Arthritis Rheum 2000; 43:349-55.
56. Rozzo SJ, Vyse TJ, Menze K et al. Enhanced susceptibility to lupus contributed from the nonautoimmune C57BL/10, but not C57BL/6, genome. J Immunol 2000; 164:5515-21.
57. Xie S, Chang S, Yang P et al. Genetic contributions of nonautoimmune SWR mice toward lupus nephritis. J Immunol 2001; 167:7141-9.
58. Rahman ZS, Tin SK, Buenaventura PN et al. A novel susceptibility locus on chromosome 2 in the (New Zealand Black x New Zealand White)F1 hybrid mouse model of systemic lupus erythematosus. J Immunol 2002; 168:3042-9.
59. Kamogawa Y, Terada M, Mizuki S et al. Arthritis in MRL/lpr mice is under the control of multiple gene loci with an allelic combination derived from the original inbred strains. Arthritis Rheum 2002; 46:1067-74.
60. Bolland S, Yim YS, Tus K et al. Genetic modifiers of systemic lupus erythematosus in FcgammaRIIB(-/-) mice. J Exp Med 2002; 195:1167-74.
61. Rozzo SJ, Allard JD, Choubey D et al. Evidence for an interferon-inducible gene, Ifi202, in the susceptibility to systemic lupus. Immunity 2001; 15:435-43.
62. Morel L, Yu Y, Blenman KR et al. Production of congenic mouse strains carrying genomic intervals containing SLE-susceptibility genes derived from the SLE-prone NZM2410 strain. Mamm Genome 1996; 7:335-9.
63. Morel L, Mohan C, Yu Y et al. Functional dissection of systemic lupus erythematosus using congenic mouse strains. J Immunol 1997; 158:6019-28.
64. Mohan C, Alas E, Morel L et al. Genetic dissection of SLE pathogenesis. Sle1 on murine chromosome 1 leads to a selective loss of tolerance to H2A/H2B/DNA subnucleosomes. J Clin Invest 1998; 101:1362-72.
65. Sobel ES, Mohan C, Morel L et al. Genetic dissection of SLE pathogenesis: Adoptive transfer of Sle1 mediates the loss of tolerance by bone marrow-derived B cells. J Immunol 1999; 162:2415-21.
66. Mohan C, Morel L, Yang P, et al. Genetic dissection of lupus pathogenesis: A recipe for nephrophilic autoantibodies. J Clin Invest 1999; 103:1685-95.
67. Sobel ES, Satoh M, Chen Y et al. The major murine systemic erythematosus susceptibility locus Sle1 results in abnormal functions of both B and T cells. J Immunol 2002; 169:2694-700.
68. Mohan C, Morel L, Yang P et al. Genetic dissection of systemic lupus erythematosus pathogenesis: Sle2 on murine chromosome 4 leads to B cell hyperactivity. J Immunol 1997; 159:454-65.
69. Mohan C, Yu Y, Morel L et al. Genetic dissection of Sle pathogenesis: Sle3 on murine chromosome 7 impacts T cell activation, differentiation, and cell death. J Immunol 1999; 162:6492-502.
70. Morel L, Croker BP, Blenman KR et al. Genetic reconstitution of systemic lupus erythematosus immunopathology with polycongenic murine strains. Proc Natl Acad Sci USA 2000; 97:6670-5.
71. Morel L, Wakeland EK. Lessons from the NZM2410 model and related strain. Int Rev Immunol 2000; 19:423-46.
72. Wakeland EK, Wandstrat AE, Liu K et al. Genetic dissection of systemic lupus erythematosus. Curr Opin Immunol 1999; 11:701-7.
73. Morel L, Tian XH, Croker BP et al. Epistatic modifiers of autoimmunity in a murine model of lupus nephritis. Immunity 1999; 11:131-9.
74. Morel L, Blenman KR, Croker BP et al. The major murine systemic lupus erythematosus susceptibility locus, Sle1, is a cluster of functionally related genes. Proc Natl Acad Sci USA 2001; 98:1787-92.
75. Boackle SA, Holers VM, Chen X et al. Cr2, a candidate gene in the murine Sle1c lupus susceptibility locus, encodes a dysfunctional protein. Immunity 2001; 15:775-85.
76. Wu X, Jiang N, Deppong C et al. A role for the Cr2 gene in modifying autoantibody production in systemic lupus erythematosus. J Immunol 2002; 169:1587-92.
77. Giraldo P, Montoliu L. Size matters: Use of YACs, BACs and PACs in transgenic animals. Transgenic Res 2001; 50:26-30.
78. Mitchell DA, Pickering MC, Warren J et al. C1q deficiency and autoimmunity: The effects of genetic background on disease expression. J Immunol 2002; 168:2538-43.
79. Botto M, Dell'Agnola C, Bygrave AE et al. Homozygous C1q deficiency causes glomerulonephritis associated with multiple apoptotic bodies. Nat Genet 1998; 19:56-9.
80. Napirei M, Karsunky H, Zevnik B et al. Features of systemic lupus erythematosus in Dnase1-deficient mice. Nat Genet 2000; 25:177-81.
81. Bickerstaff MC, Botto M, Hutchinson WL et al. Serum amyloid P component controls chromatin degradation and prevents antinuclear autoimmunity. Nat Med 1999; 5:694-7.
82. Bao L, Haas M, Boackle SA et al. Transgenic expression of a soluble complement inhibitor protects against renal disease and promotes survival in MRL/lpr mice. J Immunol 2002; 168:3601-7.
83. Kuo P, Bynoe MS, Wang C et al. Bcl-2 leads to expression of anti-DNA B cells but no nephritis: A model for a clinical subset. Eur J Immunol 1999; 29:3168-78.
84. Zhang Y, Schlossman SF, Edwards RA et al. Impaired apoptosis, extended duration of immune responses, and a lupus-like autoimmune disease in IEX-1-transgenic mice. Proc Natl Acad Sci USA 2002; 99:878-83.
85. Gross JA, Johnston J, Mudri S et al. TACI and BCMA are receptors for a TNF homologue implicated in B-cell autoimmune disease. Nature 2000; 404:995-9.
86. Khare SD, Sarosi I, Xia XZ et al. Severe B cell hyperplasia and autoimmune disease in TALL-1 transgenic mice. Proc Natl Acad Sci USA 2000; 97:3370-5.
87. Moore PA, Belvedere O, Orr A et al. BlyS: Member of the tumor necrosis factor family and B lymphocyte stimulator. Science 1999; 285:260-3.
88. Zhang J, Roschke V, Baker KP et al. Cutting edge: A role for B lymphocyte stimulator in systemic lupus erythematosus. J Immunol 2001; 166:6-10.
89. Cheema GS, Roschke V, Hilbert DM et al. Elevated serum B lymphocyte stimulator levels in patients with systemic immune-based rheumatic diseases. Arthritis Rheum 2001; 44:1313-9.
90. Iizuka J, Katagiri Y, Tada N et al. Introduction of an osteopontin gene confers the increase in B1 cell population and the production of anti-DNA autoantibodies. Lab Invest 1998; 78:1523-33.
91. Seery JP, Wang EC, Cattell V et al. A central role for alpha beta T cells in the pathogenesis of murine lupus. J Immunol 1999; 162:7241-8.
92. Seery JP, Carroll JM, Cattell V et al. Antinuclear autoantibodies and lupus nephritis in transgenic mice expressing interferon gamma in the epidermis. J Exp Med 1997; 186:1451-9.
93. Cornall RJ, Cyster JG, Hibbs ML et al. Polygenic autoimmune traits: Lyn, CD22, and SHP-1 are limiting elements of a biochemical pathway regulating BCR signaling and selection. Immunity 1998; 8:497-508.
94. Zeng D, Dick M, Cheng L et al. Subsets of transgenic T cells that recognize CD1 induce or prevent murine lupus: Role of cytokines. J Exp Med 1998; 187:525-36.

95. Nishimura H, Nose M, Hiai H et al. Development of lupus-like autoimmune diseases by disruption of the PD-1 gene encoding an ITIM motif-carrying immunoreceptor. Immunity 1999; 11:141-51.
96. Johansson C, Castillejo-Lopez C, Johanneson B et al. Association analysis with microsatellite and SNP markers does not support the involvement of BCL-2 in systemic lupus erythematosus in Mexican and Swedish patients and their families. Genes Immun 2000; 1:380-5.
97. Jiang Y, Ohtsuji M, Abe M et al. Polymorphism and chromosomal mapping of the mouse gene for B-cell activating factor belonging to the tumor necrosis factor family (Baff) and association with the autoimmune phenotype. Immunogenetics 2001; 53:810-3.

CHAPTER 23

Drug-Induced Autoimmunity

Robert L. Rubin and Anke Kretz-Rommel

Introduction

Exposure to a wide variety of synthetic compounds has been causally related to the induction of autoimmunity. The vast majority of implicated agents are deliberately ingested medications, and these are associated most often with development of lupus-like signs and/or symptoms. However, it is useful to split these phenomena into distinctive categories of xenobiotic-induced autoimmune diseases as follows (Table 1):

A. Drug exposure that is temporally related to syndromes which resemble systemic lupus erythematosus (SLE), a spontaneous, idiopathic autoimmune disease. From a clinical perspective this category encompasses both drug-induced lupus and drug-associated exacerbation of SLE or initiation of SLE flares. The latter group involves cases of pre-existing SLE which remain or recur after withdrawal of the implicated medication, while bonafide drug-induced lupus usually occurs in the setting of a previously normal immune system and disappears after discontinuation of the medication. These are mechanistically separate phenomena as discussed below.

B. Drugs or environmental agents associated with distinct autoimmune diseases, including symptoms or signs sometimes occurring in SLE. This category encompasses drug-induced hemolytic anemia, eosinophilia-myalgia syndrome associated with L-tryptophan ingestion, toxic oil syndrome associated with ingestion of aniline-adulterated cooking oil, silicosis associated with inhaled silica or asbestos dust and autoimmunity associated with vinyl chloride exposure. A so-called "adjuvant disease" associated with silicone breast implants is of dubious validity.

C. Environmental agents suspected of causing lupus-like disease based solely on studies in experimental animals. The principal examples in this category are the lupus-like syndrome associated with alfalfa sprouts (L-canavanine) in monkeys and scleroderma-like autoimmunity associated with heavy metals in rats and mice.

D. Drugs known to be immune modulating that produce autoimmune features in a minority of treated patients. Cyclosporine A has been reported to precipitate graft-vs-host disease (GVHD) after withdrawal from patients receiving autologous bone marrow transplants. Gold therapeutics might be included in this heterogeneous category. Therapeutic biologics such as interleukin-2, interferon-α,-β, and -γ, and anti-tumor necrosis factor-α have been occasionally associated with a variety of musculoskeletal manifestations and serological features of autoimmunity.

The following sections expand on these phenomena with major emphasis on drug-induced lupus and autoimmunity because there is more clinical and mechanistic information in this area. For category B items, the interested reader is referred to a recent book on autoimmune syndromes related to xenobiotic exposure;[1] some mention of drug-induce anemia is included because of the common misapplication of the underlying basis of penicillin-induced hemolytic anemia to other types of drug-induced disease. Induction of autoimmune features during exposure to therapeutic biologics is rare, and its mechanism is unknown.

Drug-Induced Lupus

Lupus-like symptoms of muscle and joint pain, fever and occasionally pleuritis and pericarditis occasionally develop as a side effect of long-term medication with 39 drugs currently in use.[2] Table 2 divides these drugs into therapeutic classes and indicates their approximate risk levels based on the number of reports. In all cases, drugs that induce lupus also induce autoantibodies in a much higher frequency. By far, the highest risk drugs are procainamide and hydralazine, with approximately 20% incidence for procainamide and 5-8% for hydralazine during one year of therapy at currently used doses. The risk for developing lupus-like disease for the remainder of the drugs is much lower, considerably less than 1% of treated patients. Quinidine can be considered moderate risk while sulfasalazine, chlorpromazine, penicillamine, methyldopa, carbamazepine, acebutalol, isoniazid, captopril, propylthiouracil and minocycline are relatively low risk. The remaining 26 drugs should be considered very low risk based on the paucity of case reports in the literature. Some drugs of very low risk may be falsely implicated or are currently of negligible risk because customary treatment doses have been decreased, but most reports on drug-induced lupus are convincing because cessation of therapy usually results in prompt resolution of symptoms and eventually autoantibodies. It should be appreciated that criteria for reaching a diagnosis of drug-induced lupus-like disease are not as rigorous as those for diagnosis of SLE. While drug-induced lupus is not a major clinical problem because it can be fully cured by discontinuing use of the medication, it is of mechanistic interest that a low molecular weight xenobiotic has capacity to initiate and sustain a well-defined systemic autoimmune disease.

The clinical and laboratory features of procainamide- and hydralazine-induced lupus are shown in Table 3 and compared with SLE. The onset of symptoms can be slow or acute, although an

Table 1. Autoimmune diseases due to ingested or injected xenobiotics

Syndrome	Main Features	Causative Agents	Exposure Duration Requirement
Drug-induced lupus	Lupus-like (Table 3)	Many medications (Table 2)	Months to years
Drug-induced lupus flares/exacerbation	SLE (Table 3)	Numerous medications	Hours to days
Autoimmune hemolytic anemia	Anemia and/or (positive Coombs)	Penicillin, stibophen, quinidine, (α-methyldopa and others)	Days to months
Eosinophilia-myalgia	Myalgia, etc.	L-tryptophan contaminants	Months
Toxic oil syndrome	Myalgia, pleuropulmonary	Oleic acid esters of aniline and/or phenylaminopropanol	Months
Dietary-induced lupus in primates	Lupus-like	L-Canavanine	Months
Murine heavy metal autoimmunity	Autoantibodies, lymphadenopathy, glomerulonephritis	Mercuric chloride, gold (I), silver nitrate	Weeks to months
Murine (human?) syngeneic graft-vs.-host disease	Fatal autoimmune syndrome	Cyclosporine A	Weeks
Cytokine toxicity	Musculoskeletal	IL-1, IL-2, interferons	Weeks

interval of 1-2 months typically passes before the diagnosis is made. Approximately 50% of patients have constitutional symptoms of fever, weight loss and fatigue. Musculoskeletal complaints are also commonly observed, with arthralgia heading the list for both drugs. Arthritis is a less common feature with lupus induced by procainamide (20%) than by hydralazine (50 to 100%), whereas serositis (pleuritis and/or pericarditis) and/or myalgia are more common presenting features of procainamide-induced lupus. By contrast, hydralazine-induced lupus is associated with a higher frequency of skin rashes. Lupus-like disease associated with quinidine or minocycline is often atypical in that the former can manifest only as polyarthalgias[4] and the latter with symmetrical polyarthritis and evidence of hepatitis and pneumonitis.[5] However, as an isolated, individual case, lupus induced by any drug cannot be distinguished by clinical features. The symptoms of drug-induced lupus usually resolve within days to weeks after discontinuing the offending drug without introduction of anti-inflammatory medications; this maneuver provides a key (although retrospective) diagnostic tool.

The frequencies of serologic abnormalities in lupus induced by procainamide and hydralazine are essentially identical (Table 3). As with idiopathic SLE, the most commonly observed abnormality is the presence of ANA, and these autoantibodies usually react with chromatin, the histone-DNA macromolecular complex in the cell nucleus. Within this structure the (H2A-H2B)-DNA subnucleosome particle shows predominate antigenicity.[6] However, unlike in patients with SLE, anti-chromatin antibodies rarely react with native (double-stranded) DNA. Other laboratory features noted in a minority of patients include a mild anemia, leukopenia, thrombocytopenia, and hypergammaglobulinemia. Although the reactivity of drug-induced antibodies is highly restricted, it is not characteristic of nor dependent on the particular drug and is similar to the anti-chromatin autoantibodies that arise spontaneously in the absence of any drugs (i.e., in SLE). This apparent paradox may be resolved by the finding that a lupus-inducing drug has capacity to non-specifically disrupt central immune tolerance to chromatin (see below), possibly related to the origin of these autoantibodies in SLE as well.

Drugs That May Exacerbate SLE

The report of Hoffman in 1945 describing a 19-year-old army recruit who developed cutaneous, hematologic, and renal disease with features of SLE after treatment with sulfadiazine[7] is often cited as the first description of drug-induced lupus. In actuality, this patient probably had a hypersensitivity-like reaction to sulfadiazine associated with exacerbation of preclinical SLE or with the onset SLE. This and subsequent similar reports helped to entrench the view that many cases of idiopathic SLE are "unmasked" during drug therapy in patients with a lupus diathesis.[8] This idea is difficult to discount or prove but now appears unlikely. Various drugs have been noted to have a temporal relationship with the exacerbation of SLE or with the onset of chronic SLE prior to diagnosis.[9] In these cases, SLE remains after withdrawal of the implicated agent. In the most recent case-controlled study of this phenomenon, 12% of SLE patients with drug allergies were considered to display disease exacerbation, predominately lupus rash.[10,11] Since SLE patients are significantly more prone to develop drug allergies, especially to antibiotics such as sulfonamides, penicillin/cephalosporin and erythromycin,[10] these agents should be avoided in patients with SLE. Lupus flares after ciprofloxacin treatment have been recently reported.[12]

Table 2. Drugs currently in use that can induce lupus-like disease

Agent*	Risk**	Agent*	Risk**
Antiarrhythmics		**Anticonvulsants**	
Procainamide (Pronestyl)	High	Carbamazepine (Tegretol)	Low
Quinidine (Quinaglute)	Moderate	Phenytoin (Dilantin)	Very Low
Disopyramide (Norpace)	Very low	Trimethadione (Tridone)	Very low
Propafenone (Rythmol)	Very low	Primidone (Mysoline)	Very low
		Ethosuximide (Zarontin)	Very low
Antihypertensives			
Hydralazine (Apresoline)	High	**Antithyroidals**	
Methyldopa (Aldomet)	Low	Propylthiouracil (Propyl-thyracil)	Low
Captopril (Capoten)	Low		
Acebutolol (Sectral)	Low	**Anti-Inflammatories**	
Enalapril (Vasotec)	Very low	D-Penicillamine (Cuprimine)	Low
Clonidine (Catapres)	Very low	Sulfasalazine (Azulfidine)	Low
Atenolol (Tenormin)	Very low	Phenylbutazone (Butazolidin)	Very low
Labetalol (Normodyne, Trandate)	Very low		
Pindolol (Visken)	Very low	**Diuretics**	
Minoxidil (Loniten)	Very low	Chlorthalidone (Hygroton)	Very low
Prazosin (Minipress)	Very low	Hydrochlorothiazide (Diuchlor H)	Very low
Antipsychotics		**Miscellaneous**	
Chlorpromazine (Thorazine)	Low	Lovastatin (Mevacor) and other statins	Very low
Perphenazine (Trilafon)	Very low	Levodopa (Dopar)	Very low
Phenelzine (Nardil)	Very low	Aminoglutethimide (Cytadren)	Very low
Chlorprothixene (Taractan)	Very low	Alpha-interferon (Wellferon)	Very low
Lithium carbonate (Eskalith)	Very low	Timolol eye drops (Timoptic)	Very low
Antibiotics			
Isoniazid (INH)	Low		
Minocycline (Minocin)	Low		
Nitrofurantoin (Macrodantin)	Very low		

*Commonly used brand names are enclosed in parentheses. **Risk refers to likelihood for lupus-like disease, not autoantibody induction, which is usually much more common.

Drugs which appear to exacerbate SLE include antibiotics, anticonvulsants, hormones, nonsteroidal anti-inflammatory drugs (NSAIDs), and dermatologic agents. Sulfonamides, tetracyclines, griseofulvin, piroxicam, and benoxaprofen are reported to be photosensitizers of varying frequency; the rash or dermatitis related to these drugs typically has a history of rapid onset and behaves as a drug hypersensitivity-type reaction that may be triggered by exposure to ultraviolet light. The majority of adverse drug reactions in previously diagnosed SLE patients are of this category.[10,11] Another, possibly related category of patients are those with acute or subacute cutaneous lupus erythematosus related to photoactive medications; these patients may have systemic disease and can fulfill criteria for a diagnosis of SLE.[13] Some of these drugs are also associated with typical drug-induced lupus and are included in Table 2. Drug-induced aseptic meningitis in SLE patients is occasionally associated with ibuprofen and other NSAIDs (e.g., sulindac, tolmetin, diclofenac). Hypersensitivity reactions that have been interpreted as initiating or aggravating factors in SLE are associated with hydralazine, sulfonamides, penicillin, para-aminosalacylic acid, hydrochlorothiazide, cimetidine, phenylbutazone, mesantoin and various NSAIDs. Unknown or suspected environmental chemicals such as hair dyes and permanent wave preparations are also occasionally implicated as aggravating agents in SLE and related diseases.

Whether or not an environmental or pharmaceutical agent might aggravate or unmask incipient SLE should be considered as a clinical problem distinct from drug-induced lupus because, by definition, symptoms of the latter resolve after discontinuation of therapy, although in severe cases full recovery may require up to one year. If drugs or environmental agents are truly causative in initiating or aggravating SLE, the mechanistic basis is probably different from that of drug-induced lupus because the steady-state blood levels of bonafide lupus-inducing drugs must generally be sustained for many months to years for development of drug-induced lupus. In contrast, for most cases believed to be exacerbated SLE, exposure is of very low level or infrequent when the suspected agent is environmental or of relatively short duration when a drug is implicated. The association between drugs and the onset of SLE resembles the lupus flares following exposure to sunlight, exercise or pregnancy.

Drug-Induced Immune Hemolytic Anemia

Long-term therapy with some drugs is associated with development of hemolytic anemia due to antibodies bound to red blood

Table 3. Prevalence of clinical and laboratory abnormalities in drug-induced lupus

Feature	Hydralazine-Induced Lupus	Procainamide-Induced Lupus	Systemic Lupus Erythematosus
Symptoms			
Arthralgia	80%	85%	80%
Arthritis	50%-100%	20%	80%
Pleuritis, pleural effusion	<5%	50%	44%
Fever, weight loss	40%-50%	45%	48%
Myalgia	<5%	35%	60%
Hepatosplenomegaly	15%	25%	5%-10%
Pericarditis	<5%	15%	20%
Rash	25%	<5%	71%
Glomerulonephritis	5%-10%	<5%	42%
Central nervous system disease	<5%	<5%	32%
Signs			
Antinuclear antibodies	>95%	>95%	97%
Lupus erythematosus cells	>50%	80%	71%
Anti-histone Abs	>95%	>95%	54%
Anti-[(H2A-H2B)-DNA] Ab	43%	96%	70%
Anti-denatured DNA Ab	50%-90%	50%	82%
Anti-native DNA Ab	<5%	<5%	28%-67%
Anticardiolipin Ab	5%-15%	5%-20%	35%
Rheumatoid factor Ab	20%	30%	25%-30%
Anemia	35%	20%	42%
Elevated erythrocyte sed. rate	60%	60%-80%	>50%
Leukopenia	5%-25%	15%	46%
+ Coombs' test	<5%	25%	25%
Elevated gammaglobulins	10%-50%	25%	32%
Hypocomplementemia	<5%	<5%	51%

For sources see reference 3.

cells (RBC) in vivo (direct Coombs test positivity). In the penicillin-type, antibody to the drug binds to RBC as a result of adsorption of the drug or its metabolite to the RBC membrane. In the methyldopa-type, the drug is not required for (and does not effect) antibody binding, and anti-RBC antibodies typically have specificity for rhesus locus or other intrinsic RBC antigens. These antibodies rarely produce frank hemolytic anemia, possibly because their isotype or low avidity does not support complement fixation. Hemolytic anemia is commonly associated with the stibophen-type of drug-induced antibodies (as is quinidine and quinine) in which immune complexes consisting of the drug or drug metabolite bind to RBC presumably via Fc or complement receptors.

The mechanism underlying the penicillin-type of anti-RBC response is frequently used as the basis for models for autoantibody elicitation in drug-induced lupus. Interestingly, the autoantibodies associated with drug-induced lupus behave more like the methyldopa-type of autoimmune response in that the likelihood for autoantibody appearance is dose-dependent (rather than a hypersensitivity-type reaction to even low dose drug exposure), but the drug is not required for antibody binding to its target antigen. In fact, many of the drugs associated with Coombs positivity of this drug-independent type (methyldopa, L-dopa, mefanamic acid, procainamide, chlorpromazine and streptomycin) are also known to cause drug-induced lupus (Table 2), although there is generally no correlation between positive Coombs test and other autoantibodies or symptoms. However, patients with methyldopa-induced hemolytic anemia have been reported to have positive lupus erythematosus cells and antinuclear antibodies (ANA). The mechanism for induction of this type of anti-RBC autoantibody is unknown. However, despite the drug-independence of anti-RBC binding, a drug-altered RBC model is commonly (inappropriately) invoked, and this model is incorrectly applied to the origin of autoantibodies associated with drug-induced lupus as well.

Xenobiotics Related to Autoimmunity in Animals

Some xenobiotics are suspected of inducing autoimmunity based solely on studies in experimental animals. Parenteral, subcutaneous or intramuscular injection of mercuric chloride, gold(I) thiomalate or silver nitrate into mice possessing the H-2^s MHC class II phenotype results in an acute, self-limited autoimmune disease characterized by ANA and immune complex glomerulonephritis.[14] The major target of mercury-induced autoantibodies is the nucleolar protein fibrillarin,[15] a specific serologic marker for a subset of patients with scleroderma.[15] While this animal model is of considerable mechanistic interest, mercury has not been definitely implicated in autoimmune disease in humans, although it has been suggested that exposure to the heavy metals such as gold, silver and mercury in dental amalgam could underly autoimmunity in genetically-susceptible individuals.[16] The mechanistic basis for mercury induction of autoantibodies is unknown, but it appears to result from disruption of peripheral tolerance machinery. Immune-mediated disease associated with gold therapy and cyclosporine are discussed below, although the vast majority of data is derived from experimental animals.

A prospective study of dietary intake of alfalfa seeds, alfalfa sprouts and L-canavanine, the suspected toxin in alfalfa, was undertaken in monkeys. Approximately half the animals developed ANA, anti-native DNA, hypocomplementemia and other symptoms and signs similar to SLE.[17] However, because this autoimmune effect required approximately one half year of ingestion of a diet containing 40-45% (dried) alfalfa, it is unlikely that legumes are responsible for disease in most SLE patients. Nevertheless, several case reports of lupus-like disease associated with the ingestion of abnormally large amounts of alfalfa seeds or tablets[18,19] add another xenobiotic (presumably L-canavanine) to the list of compounds capable of inducing or exacerbating lupus.

Autoimmunity Associated with Immune-Modulating Drugs

Cyclosporine A is used therapeutically to suppress transplantation rejection, autoimmune disease and GVHD but is paradoxically related to development of certain types of autoimmunity. Most of these observations are in experimental animals, but it has been reported that withdrawal of cyclosporine after autologous bone marrow transplantation can result in an syndrome resembling GVHD.[20,21] Treatment of neonatal mice with cyclosporine provokes organ-specific autoimmune disease,[22] and certain strains of rat and mouse subjected to lethal irradiation followed by bone marrow reconstituition and treatment with cycolsporine for three or more weeks develop GVHD after withdrawal of cyclosporine.[23]

It has been shown in experimental systems that development of autoimmunity following cyclosporine withdrawal requires a lymphopenic periphery and a thymus that retains capacity to produce functional T cells after recovery from the treatment; disease can be adoptively transferred into irradiated, syngeneic hosts with T cells from mice undergoing GVHD but can be prevented by co-infusion of normal syngeneic cells. These and other considerations argue against the importance of cyclosporine in interfering with clonal deletion of self-reactive thymocytes. The precise mechanism has not been determined, but it has been proposed that autoimmunity may be due to blockage of peripheral anergy of autoreactive T cells,[23] prevention of development of CD4+CD25+ regulatory T cells,[24] homeostatic proliferation of high affinity self-reactive T cells due to the lymphopenic periphery into which new thymic emigrants emerge,[25] and/or disruption of central T cell tolerance[26] with release of effector T cells directed against MHC class II invariant chain peptide.[27]

Gold therapeutics such as disodium aurothiomalate frequently appear on lists of drugs associated with SLE induction or exacerbation. However, although late-onset toxic reactions to gold requiring discontinuation of therapy occur in up to one-third of treated patients, the vast majority of episodes occur during the first few months of therapy and are limited to skin reactions or buccal irritation; rash occurs in about half the toxic gold reactions, and it is of a generalized or upper body distribution.[28] Approximately 1% of treated patients develop leukopenia, thrombocytopenia or proteinuria.[28] These abnormalities are generally considered immune-mediated,[29] but they fall outside the usual presenting symptoms and clinical progression of SLE[30] or drug-induced lupus, and no report of gold-induced lupus has appeared despite over six decades of treatment experience. Gold toxicity does not appear to behave as a classical delayed-type hypersensitivity reaction in that only one of 30 patients with gold reactions had rapid recurrence of skin symptoms upon re-challenge with gold therapy.[28]

Mouse Model of Drug-Induced Autoimmunity

Several lines of evidence strongly suggest that it is not the ingested compound but in vivo drug metabolites that are responsible for initiation of drug-induced lupus (reviewed in ref. 2). Procainamide-hydroxylamine (PAHA) is the principal oxidative metabolite of procainamide produced by activated neutrophils and is implicated as the initiating agent in procainamide-induced lupus. However, no autoimmune or even hyperimmune effect has been elicited by in vivo or in vitro treatment of any component of the peripheral immune system with PAHA, although infusion of activated peripheral T cells treated with procainamide or hydralazine results in a fulminate syndrome resembling GVHD.[31]

To determine whether disruption of central T cell tolerance by a lupus-inducing drug could result in autoimmunity, we injected PAHA into the thymus of young adult (C57BL/6 x DBA/2)F1 mice. Two intrathymic injections with the drug resulted in rapid appearance of IgM anti-dDNA antibodies followed by a delayed but sustained production of IgG anti-chromatin antibodies.[32] The autoantibodies in PAHA-treated mice had a specificity remarkably similar to that of patients with procainamide-induced lupus in reacting with the (H2A-H2B)-DNA complex, suggesting that the immune system had undergone similar perturbations in both species. In the peripheral immune organs such as the spleen and lymph nodes of PAHA-treated mice, chromatin-reactive T cells were detected at a time-point when anti-chromatin antibodies started to rise, suggesting that PAHA action in the thymus resulted in the export of autoreactive T cells to the periphery where they provided T-helper cell function to B cells with potential to produce autoantibodies.

Autoreactivity in Thymus Organ Culture after PAHA Exposure

Thymocytes from normal mice such as C57BL/6 do not proliferate in response to syngeneic antigen presenting cells (APC) with or without addition of chromatin, consistent with the normal state of immune tolerance to this ubiquitous antigen. In contrast, if subjected to PAHA during short-term organ culture, the recovered thymocytes showed a 6- to 20-fold proliferative response to syngeneic APC plus chromatin (Fig. 1). Acquisition of autoreactive properties while thymocytes undergo differentiation in the presence of PAHA was not unique to chromatin-reactive T cells. When thymuses from TCR-AND mice, in which the majority of T cells express a transgenic T cell receptor (TCR) specific for pigeon cytochrome c (PCC)[34] were exposed to PAHA, the emerging T cells exhibited a 4- to 25-fold increase in reactivity to PCC (Fig. 1). PCC does not normally activate transgenic T cells in C57BL/6 TCR-AND mice,[35] apparently because it is not presented by the I-A^b MHC with sufficient avidity. However, such low avidity interactions were sufficient to stimulate PCC-specific transgenic T cells if PAHA was present during their development. The capacity of PAHA to disrupt tolerance in monoclonal T cells of arbitrarily chosen specificity as well as in chromatin-reactive T cells indicates that PAHA acted in an antigen-nonspecific man-

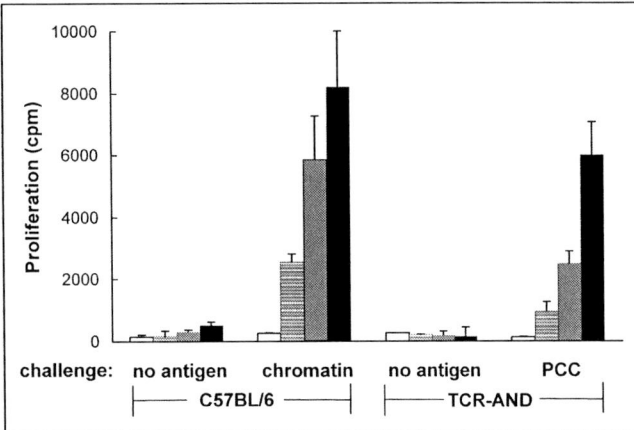

Figure 1. Appearance of autoreactive T cells after treatment of thymuses ex vivo with PAHA. Intact thymuses from neonatal mice of strains C57BL/6 or TCR-AND (C57BL/6 background) were cultured in the presence of increasing [PAHA]: 0 μM □; 5 μM ▨; 20 μM ▨; 50 μM ■. Thymocytes were harvested after 2 days and expanded on chromatin or PCC for 1 week. Proliferation of thymocytes in response to 50 mg/ml chromatin or 5 μM PCC was measured by ^{3}H-thymidine incorporation. From ref. 33, with permission.

ner. This feature allowed use of monoclonal T cell subsets and stromal cells from TCR-AND thymuses to study the underlying mechanisms leading to autoreactivity.

PAHA Affects Immature Thymocytes Undergoing Selection

When CD4$^+$CD8$^+$ double positive thymocytes are recombined with thymic epithelial cells in a reaggregate culture, they undergo differentiation and development into mature T cells over the five days culture period. On a C57BL/6 background, the emerging CD4$^+$ cells from the AND-TCR thymus are unresponsive to PCC. However, if PAHA was present during the five days of reaggregate culture of CD4$^+$CD8$^+$ cells, the recovered T cells subsequently exhibited a 40-fold increased proliferative response to PCC. In contrast, when CD4 single positive (mature) thymocytes were used in the reaggregate, PAHA presence did not result in PCC-responsive T cells, indicating that PAHA did not activate or reverse the self-tolerance status of mature T cells.[33] These results suggest that PAHA inhibits induction of self-tolerance as immature double positive thymocytes undergo selection.

No Effect of PAHA on Negative Selection of T Cells

Negative selection by clonal deletion of thymocytes with TCRs of high affinity for self-antigens is generally believed to be the principal mechanism for elimination of autoreactive T cells. To determine whether PAHA inhibits this form of immune tolerance, fetal thymuses from (TCR-AND x CBA/J) F1 mice, were exposed to PCC in the presence or absence of the drug metabolite. CBA/J mice contribute the I-E^k MHC that presents PCC with high affinity for TCR-AND T cells and initiates their death if PCC is present during thymopoiesis.[36] The presence of PAHA during thymus organ culture had no effect on deletion of PCC-reactive T cells whether measured by physically tracking the transgenic T cells by fluorescence activated cell sorter (FACS) analysis based on their Vβ3 TCR or by measuring their responsiveness to PCC, and several other systems for monitoring negative selection were unaffected by PAHA.[33] These results indicate that the appearance of chromatin- and PCC-reactive T cells after PAHA exposure of the thymus cannot be explained by the functional or physical inhibition of clonal deletion of potentially autoreactive T cells.

Effect of PAHA on Positive T Cell Selection

In the absence for a role of negative selection, the demonstration that PAHA acted during the conversion of immature CD4$^+$CD8$^+$ into mature CD4$^+$ T cells implicated the process of positive selection as the developmental step targeted by the drug. PCC peptide encompassing residues 88-104 presented by I-E^k MHC is a strong agonist for AND T cells, whereas an analogue with an amino acid substitution of alanine for lysine in position 99, one of the critical TCR contact positions, cannot activate mature TCR-AND T cells.[37,38] However, the presence of PAHA during organ culture resulted in the appearance of T cells responsive to PCC99A. Challenge with 10 μM PCC99A induced a significant increase in the proliferative response of T cells derived from PAHA-treated thymuses compared to T cells derived from untreated thymuses which did not proliferate in response to up to 100 μM PCC99A. Studies on clonal deletion are consistent with the view that the affinity for the AND-TCR of PCC99A is lower that that of PCC when presented by I-E^k.[39] Taken together, these observations suggest that the presence of PAHA during positive selection results in T cells that have a lower threshold of activation than T cells derived from an unexposed thymus, permitting them to respond to low affinity ligands such as PCC99A.

Thymus Function in Drug-Induced Lupus

A minimum requirement to relate the mouse model to the human condition is the demonstration that patients who developed drug-induced lupus have a functioning thymus. While recent studies with normal donors made likely the detection of thymus function in patients of the age typically associated with development of drug-induced lupus (reviewed in ref. 40), thymus function in a large cohort of elderly people has not been examined.

Semi-quantitative measurement of thymus function can be made by determining the content of TCR rearrangement excision circles (TRECs) in peripheral blood lymphocytes (PBLs). TRECs are by-products of the DNA rearrangement process required for the formation of the αβTCR and can be readily detected in recent thymic emigrants to the periphery using polymerase chain reaction (PCR) amplification across the new joint that creates the excision circle.[41] Eventually the TRECs disappear from peripheral T cells due to cell division or death, so TREC levels in the periphery correlate with T cell production in the thymus.

The TREC content of PBL DNA was determined in two groups of patients who developed autoimmunity secondary to treatment with procainamide, and these results were compared with their autoantibody serology. Patients who develop procainamide-induced lupus can generally be distinguished from drug-treated patients who remain asymptomatic by the presence of IgG anti-chromatin autoantibodies only in symptomatic drug-induced lupus.[42] As shown in Fig. 2, the level of these autoantibodies showed a significant correlation with the TREC content in PBLs ($p < 0.01$). The average TREC/mg PBL DNA did not differ between these patient groups (931 ± 654 vs. 976 ± 973 molecular equivalents). There was no relationship between

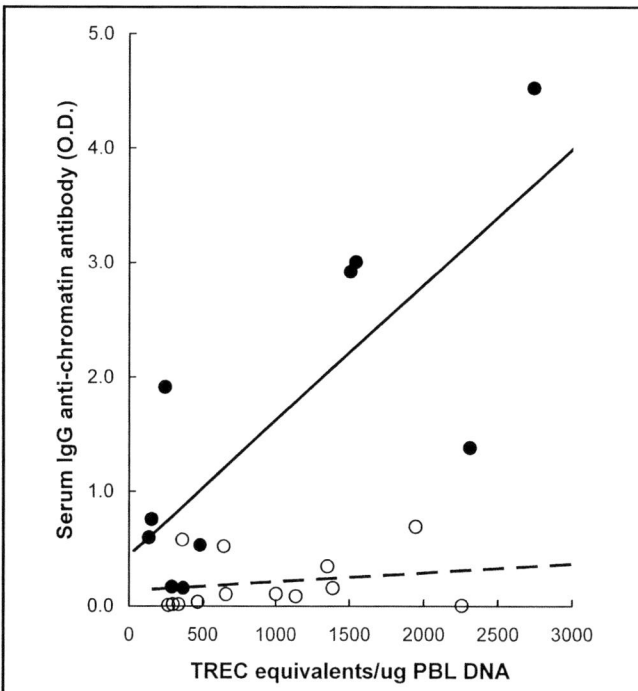

Figure 2. Relationship between the concentration of T cell receptor excision circles (TREC) in peripheral blood lymphocytes (PBL) DNA and autoantibodies in symptomatic and asymptomatic procainamide-treated patients. TREC levels in circulating PBLs, expressed as TREC equivalents, were compared to serum IgG anti-chromatin activity at the time of symptomatic drug-induced lupus (●——●) or in patients who remained asymptomatic after an average of 4.1 years of treatment with procainamide (O----O). Regression lines are the least square fit of the data; for the symptomatic patients r= 0.78, p < 0.01; for asymptomatic patients r= 0.22, p > 0.05. From ref. 43.

age of the PBL donors and TREC levels (not shown), and there was no difference in the average age of symptomatic and asymptomatic procainamide-treated patients (69 ± 9 years and 74 ± 4 years, respectively). The correlation between serum anti-chromatin antibody activity and the TREC content of circulating lymphocytes implies that thymus function is important in the appearance of disease-associated autoantibodies.

Concluding Remarks

Appearance of autoreactive T cells in the mouse model when PAHA was present in the thymus was due to the capacity of this drug metabolite to alter T cell development, interfering in the

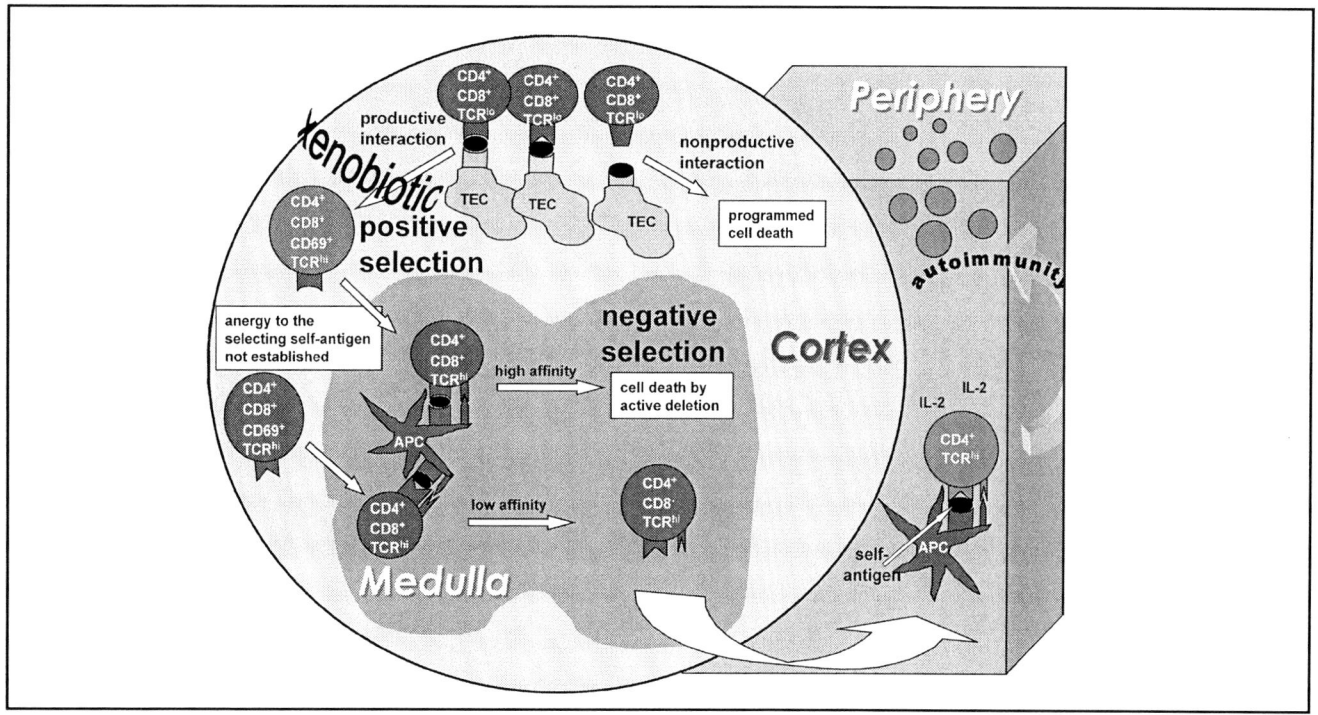

Figure 3. Proposed mechanism for procainamide-hydroxylamine (PAHA)-induced autoimmunity. At the top are depicted three possible developmental fates of CD4+CD8+ (double positive immature) T cells in the thymic cortex bearing T cell receptors (TCRs) differing in affinity for self-antigens. T cells with negligible affinity for self-peptides + the MHC die by neglect. Engagement of the TCR by self-antigens presented on thymic epithelial cells (TEC) activates signal transduction pathways, initiating a program of cellular differentiation that includes enhanced expression of the TCR and CD69. Normally, selected thymocytes acquire non-responsiveness (anergy) to the selecting self-antigens. The presence of a xenobiotic such as PAHA appears to act at this stage, preventing establishment of anergy. Because some of these cells have low affinity TCRs, they are ignored by the negative selection machinery that causes the active deletion in the thymic medulla of thymocytes with high affinity for self-antigens. After differentiating into CD4+ mature T cells and emigrating to the periphery, these non-self-tolerant T cells will encounter chromatin-derived peptides on professional APCs, similar to those that were involved in their positive selection. Because they have a reduced activation threshold as a consequence of exposure to PAHA during their development, they can respond, resulting in interleukin-2 production and clonal expansion. Upon encountering chromatin-specific B cells which concentrate the T cell antigen on their MHC, a T-B immune unit is created, initiating systemic autoimmunity.

establishment of central T cell tolerance. With normal mice, chromatin-reactive T cells were the predominate generated specificity, but we studied the cellular basis for this phenomenon using monoclonal T cells from mice transgenic for the TCR specific to the PCC peptides. PCC presented by I-E^k is a high affinity agonist for the mature AND T cell and will cause clonal deletion if present at the immature stage when thymocytes express both the CD4 and CD8 coreceptors.[37] In contrast, PCC and unknown endogenous self-peptides presented by I-A^b behave as low affinity antigens for the mature T cell and do not cause clonal deletion of immature AND thymocytes.[35] PAHA had no effect on I-E^k/PCC-mediated clonal deletion of AND CD4$^+$CD8$^+$ thymocytes as measured either by physical presence or functional properties. In contrast when AND cells were selected by endogenous self-peptides on I-A^b in either neonatal thymus organ culture or in thymus reaggregate culture using CD4$^+$CD8$^+$ thymocytes, the T cells which developed responded to PCC/I-A^b. Failure of PAHA to interfere with clonal deletion while producing T cells that respond to low affinity antigens suggests that PAHA prevents establishment of non-responsiveness during positive selection. This conclusion was consistent with the findings that thymocytes developing in the presence of PAHA responded to PCC99A presented by I-E^k, a low affinity agonist for AND T cells.[33] These data indicate that PAHA presence during positive selection results in T cells that have a lower threshold of activation than T cells derived from an untreated thymus, permitting them to respond to low affinity ligands (Fig. 3).

When non-tolerant T cells migrate to the periphery, they encounter self-antigens similar to those on which they were positively selected. Chromatin is likely to be a highly abundant self-antigen, with copious amounts released daily due to normal cell turnover. Recent evidence indicates that lymphocytes with capacity to secrete anti-chromatin antibodies are part of the constitutive B cell pool.[44] Such chromatin-specific B cells may concentrate chromatin-derived T cell epitopes onto their MHC, promoting cognate T cell help for B cell activation and autoantibody production. Although Fas-mediated activation-induced cell death eventually limits autoimmunity,[44] polyclonal expansion of both B and T chromatin-reactive lymphocytes produces a state of long-lasting systemic autoimmunity.

Despite the widely accepted view that T cells direct the autoimmune response, abnormalities in T cell development in the thymus that might underly adult onset autoimmune disease have not received much attention other than in murine lupus,[45] in part because it had been previously assumed that the thymus is non-functional in the adult. However, recent studies with normal donors demonstrated TRECs in PBLs from patients in their seventh decade[41,46] and along with other evidence supports the notion that the thymus remains functional throughout life. Individuals from 53 to 78 years old who developed drug-induced lupus during treatment with procainamide had in their circulating lymphocyte pool recent thymic emigrants marked by TRECs, and the level of these cells was significantly correlated with anti-chromatin antibody activity.[43] These observations suggest that thymus function is important in drug-induced lupus and that the intensity of the anti-chromatin antibody response depends on the amount of overall thymus function. These results link the mouse model and human drug-induced lupus, supporting the view that production of autoreactive T cells in this syndrome is initiated in the thymus by the capacity of reactive drug metabolites to disrupt central T cell tolerance.

Abbreviations

ANA= antinuclear antibody; APC= antigen presenting cell; GVHD= graft-vs.-host disease; MHC= major histocompatibility complex; NSAID= nonsteroidal anti-inflammatory drugs; PAHA= procainamide-hydroxylamine; PBL= peripheral blood lymphocytes; PCC= pigeon cytochrome c; PCR= polymerase chain reaction; RBC= red blood cell; SLE= systemic lupus erythematosus; TCR= T cell receptor; TREC= T cell receptor excision circle.

Acknowledgments

Supported by National Institutes of Health Grant ES06334.

References

1. Kaufman LD, Varga J, eds. Rheumatic Diseases and the Environment. Chapman Hall: New York, 1999.
2. Rubin RL. Etiology and mechanisms of drug-induced lupus. Curr Opinion Rheumatol 1999; 11:357-363.
3. Rubin RL. Drug-Induced Lupus. In: DJ Wallace, Hahn BH, eds. Dubois' Lupus Erythematosus. Philadelphia: Lippincott Williams & Wilkens, 2002:885-916.
4. Cohen MG, Kevat S, Prowse MV et al. Two distinct quinidine induced rheumatic syndromes. Ann Intern Med 1988; 108:369-371.
5. Christodoulou CS, Emmanuel P, Ray RA et al. Respiratory distress due to minocycline-induced pulmonary lupus. Chest 1999; 115:1471-1473.
6. Burlingame RW, Rubin RL. Autoantibody to the nucleosome subunit (H2A-H2B)-DNA is an early and ubiquitous feature of lupus-like conditions. Mol Biol Rep 1996; 23:159-166.
7. Hoffman BJ. Sensitivity to sulfadiazine resembling acute disseminated lupus erythematosus. Arch Dermatol Syphilol 1945; 51:190-192.
8. Alarcon-Segovia D. Drug-induced lupus syndrome. Mayo Clin Proc 1967; 44:664-681.
9. Wallace DJ, Dubois EL. Drugs that exacerbate and induce systemic lupus erythematosus. In: DJ Wallace, Dubois EL, eds. Dubois' Lupus Erythematosus. Philadelphia: Lea & Febiger, 1987:450-469.
10. Petri M, Allbritton J. Antibiotic allergy in systemic lupus erythematosus. J Rheumatol 1992; 19:265-269.
11. Wang C-R, Chuang C-Y, Chen C-Y. Drug allergy in Chinese patients with systemic lupus erythematosus. J Rheumatol 1993; 20:399-400.
12. Mysler E, Paget SA, Kimberly R. Ciprofloxacin reactions mimicking lupus flares. Arthritis Rheum 1994; 37:1112-1113.
13. Sontheimer RD, Provost TT. Cutaneous manifestations of lupus erythematosus. In: Wallace DJ, Hahn BH, eds. Dubois' Lupus Erythematosus. Baltimore: Williams and Wilkins, 1997:569-623.
14. Robinson CJG, Balazs T, Egorov IK. Mercuric chloride-, gold sodium thiomalate-, and D-penicillamine-induced antinuclear antibodies in mice. Toxicol Appl Pharmacol 1986; 86:159-169.
15. Hultman P, Enestrom S, Pollard KM et al. Anti-fibrillarin autoantibodies in mercury-treated mice. Clin Exp Immunol 1989; 78:470-472.
16. Hultman P, Johansson U, Turley SJ et al. Adverse immunological effects and autoimmunity induced by dental amalgam and alloy in mice. FASEB J 1996; 8:1183-1190.
17. Malinow MR, Bardana EJ, Pirofsky B et al. Systemic lupus erythematosus-like syndrome in monkeys fed alfalfa sprouts: Role of a non-protein animo acid. Science 1982; 216:415-417.
18. Malinow MR, Bardana EJ, Goodnight SH. Pancytopenia during ingestion of alfalfa seeds. Lancet 1981; 1:615-617.
19. Roberts JL, Hayashi JA. Exacerbation of SLE associated with alfalfa ingestion. N Engl J Med 1983; 308:1361.
20. Jones RJ, Vogelsang GB, Hess AD et al. Induction of graft-versus-host disease after autologous bone marrow transplantation. Lancet 1989; 1:754-757.
21. Dale BM, Atkinson K, Kotasek D et al. Cyclosporine-induced graft vs host disease in two patients receiving syngeneic bone marrow transplants. Transplant Proc 1989; 21:3816-3817.

22. Sakaguchi S, Sakaguchi N. Organ-specific autoimmune disease induced in mice by elimination of T cell subsets. V. Neonatal administration of cyclosporin A causes autoimmune disease. J Immunol 1989; 142:471-480.
23. Prud'homme GJ, Vanier LE. Cyclosporine, tolerance, and autoimmunity. Clin Immunol Immunopathol 1993; 66:185-192.
24. Sakaguchi S, Sakaguchi N, Shimizu J et al. Immunologic tolerance maintained by CD25+ CD4+ regulatory T cells: Their common role in controlling autoimmunity, tumor immunity, and transplantation tolerance. Immunol Rev 2001; 182:18-32.
25. Theofilopoulos AN, Dummer W, Kono DH. T cell homeostasis and systemic autoimmunity. J Clin Invest 2001; 108:335-340.
26. Wu DY, Goldschneider I. Cyclosporin A-induced autologous graft versus host disease: A prototypical model of autoimmunity and active (dominant) tolerance coordinately induced by rcent thymic emigrants. J Immunol 1999; 162(11):6926-33.
27. Hess AD, Bright EC, Thoburn C et al. Specificity of effector T lymphocytes in autologous graft versus host disease: role of the major histocompatibility complex class II invariant chain peptide. Blood 1997; 89(6):2203-9.
28. Lockie LM, Smith DM. Forty-seven years experience with gold therapy in 1,019 rheumatoid arthritis patients. Semin Arthritis Rheum 1985; 14:238-246.
29. Romagnoli P, Spinas GA, Sinigaglia F. Gold-specific T cells in rheumatoid arthritis patients treated with gold. J Clin Invest 1992; 89:254-258.
30. Wallace DJ. The clinical presentation of systemic lupus erythematosus. In: Wallace DJ, Hahn BH, eds. Dubois' Lupus Erythematosus. Lea & Febiger, Philadelphia, 1993. 317-321.
31. Cornacchia E, Golbus J, Maybaum J et al. Hydralazine and procainamide inhibit T cell DNA methylation and induce autoreactivity. J Immunol 1988; 140:2197-2200.
32. Kretz-Rommel A, Duncan SR, Rubin RL. Autoimmunity caused by disruption of central T cell tolerance: A murine model of drug-induced lupus. J Clin Invest 1997; 99:1888-1896.
33. Kretz-Rommel A, Rubin RL. Disruption of positive selection of thymocytes causes autoimmunity. Nat Med 2000; 6:298-305.
34. Kaye J, Hsu M-L, Sauron M-E et al. Selective development of CD4+ T cells in transgenic mice expressing a class II MHC-restricted antigen receptor. Nature 1989; 341:746-749.
35. Fink PJ, Matis LA, McElligott DL et al. Correlations between T-cell specificity and the structure of the antigen receptor. Nature 1986; 321:219-226.
36. Oehen S, Feng L, Xia Y et al. Antigen compartmentation and T helper cell tolerance induction. J Exp Med 1996; 183:2617-2626.
37. Vasquez NJ, Kane LP, Hedrick SM. Intracellular signals that mediate thymic negative selection. Immunity 1994; 1:45-56.
38. Liu C-P, Parker D, Kappler J et al. Selection of antigen-specific T cells by a single IE^k peptide combination. J Exp Med 1997; 186:1441-1450.
39. Page DM, Alexander J, Snoke K et al. Negative selection of CD4+ CD8+ thymocytes by T-cell receptor peptide antagonists. Proc Natl Acad Sci USA 1994; 91:4057-4061.
40. Haynes BF, Markert ML, Sempowski GD et al. The role of the thymus in immune reconstitution in aging, bone marrow transplantation, and HIV-1 infection. Annu Rev Immunol 2000; 18:529-560.
41. Douek DC, McFarland RD, Keiser PH et al. Changes in thymic function with age and during the treatment of HIV infection. Nature 1998; 396:690-695.
42. Burlingame RW, Rubin RL. Drug-induced anti-histone autoantibodies display two patterns of reactivity with substructures of chromatin. J Clin Invest 1991; 88:680-690.
43. Rubin RL, Salomon DR, Guerrero RS. Thymus function in drug-induced lupus. Lupus 2001; 10:795-801.
44. Kretz-Rommel A, Rubin RL. Early cellular events in systemic autoimmunity driven by chromatin-reactive T cells. Cell Immunol 2001; 208:125-136.
45. Hashimoto YDK, Montecino-Rodriquez E, Taguchi N et al. NZB mice exhibit a primary T cell defect in fetal thymic organ culture. J Immunol 2000; 164:1569-1575.
46. Jamieson BD, Douek DC, Killian S et al. Generation of functional thymocytes in the human adult. Immunity 1999; 10:569-575.

CHAPTER 24

Evidence for a Role of Infections in the Activation of Autoreactive T Cells and the Pathogenesis of Autoimmunity

J. Ludovic Croxford and Stephen D. Miller

Introduction

The immune system provides protection against infection by microorganisms. The first line of defense against infectious pathogens is the innate immune response by neutrophils, NK cells, and macrophages, which differentiates between foreign organisms (nonself) and self-tissue, by recognizing "pattern recognition markers" using a limited number of receptors from germ line genes, resulting in the nonspecific elimination of invading microorganisms. Unlike innate immunity, the adaptive immune system of T and B cells are able to rearrange a limited number of germ line VJD (variable, joining and diversity) genes to produce a highly diversified repertoire of somatically mutated T cell (immunoglobulin-like) receptors and B cell immunoglobulin receptors. Following the phagocytosis of microorganisms during the innate immune response, specific peptides from the infectious pathogen are processed and presented to antigen-specific T cells by antigen-presenting cells (APC), thus activating the adaptive immune system (T and B cells). This results in the neutralization of the infectious pathogen, death of infected host cells, and the induction of antigen-specific memory T cells. In rare instances, T cells that recognize self-peptides may become activated, and induce autoimmune disease.

Normally, T cells which have a high-affinity for self-tissue, undergo clonal deletion during their development in the thymus. However, autoreactive T cells have been observed in normal, healthy humans suggesting that self-reactive T cells can escape thymic deletion.[1,2] Other protective mechanisms, including peripheral tolerance and/or peripheral deletion of autoreactive T cells, serve to inhibit the activation of autoreactive T cells to specialized antigens in peripheral organs which are not expressed in the thymus. In addition, ignorance of the self-antigen, by a physical barrier, e.g., the blood brain barrier, which separates the CNS, an "immune privileged site", from the periphery, can limit the activation of an autoreactive T cell by its cognate self-antigen. Autoimmune disease occurs when autoreactive T cells escape these tolerance mechanisms and become activated by mechanisms yet to be fully elucidated. Recent studies show that T cell receptor recognition of peptide-MHC complexes is degenerate, and experimental models of autoimmunity have shown that activated, but not resting, autoreactive T cells can transfer clinical disease.[3] Therefore, the activation of autoreactive T cells appears to be a critical step in the initiation of autoimmune disease.

Genetic factors are also important in determining the susceptibility of an individual, although it is unclear whether genetic susceptibility is linked to specific autoimmune diseases or to autoimmunity in general.[4] Susceptibility to autoimmune disease may be a polygenic trait involving multiple disease susceptibility loci and may indicate complex interactions amongst gene products.[4] Elucidation of the mechanism(s) of activation of autoreactive T cells therefore is key to a greater understanding of autoimmune disease pathogenesis. Many studies suggest that infectious pathogens can induce activation of autoreactive T cells and may play a role in the etiology of autoimmune diseases. This review article will discuss the mechanisms by which pathogens can activate autoreactive T cells and induce autoimmune disease, and review the evidence for a viral pathogenesis in both human autoimmune disease and animal models.

Mechanisms of Autoreactive T Cell Activation by Infectious Pathogens

Differences in geographical locations of the incidence of multiple sclerosis (MS), suggests that environmental factors play a role in the development of autoimmune disease. Epidemiological evidence suggests there is a higher incidence of MS in Caucasians living in the northern latitudes, which increases in temperate zones with moderate or cold climates. In addition, migration studies have determined that migrants moving from high- to low-risk areas retain their original risk factor, if they move after the age of fifteen years.[5] The age of acquisition of "inducing factors" may also play a role in determining how infection may induce autoimmunity. It is rare for young children to be diagnosed with MS, which may suggest a latency period following viral infection, where clinical disease is silent. Further support for a role of infectious agents in MS is described in a number of studies, which suggest that the Faroe Islands, Iceland and the Shetney-Orkland Islands have experienced epidemics of MS.[6-8] These data suggest a correlation between the development of an autoimmune disease and infection, with a region-specific pathogen at a young age in a genetically susceptible population.

There are a number of putative mechanisms by which infectious pathogens can activate autoreactive T cells, either directly or indirectly (Table 1 and Fig. 1) and have recently been reviewed.[9,10] The direct mechanisms, such as molecular mimicry and superantigens, describe the direct contact between the infectious agent and/or its peptides and the autoreactive lymphocyte. In contrast, the indirect mechanisms, such as bystander activation and epitope spreading, involve chronic inflammation, which arises subsequent to the infection, as the activating signal for autoimmune T cells.

Molecular Mimicry

The major putative mechanism for the activation of autoimmune T cells by infectious pathogens is molecular mimicry (Fig. 1A), and has recently been reviewed.[9] By this mechanism, it is proposed that pathogens may encode "mimic" peptides, which share homology with a host's self-antigens. Following infection with the pathogen, these mimic sequences may be presented by the host APC in an inflammatory context, which can subsequently induce the activation of autoreactive T cells specific for the mimic/self peptide. Following activation of self-reactive T cells, an immune response may be directed towards self-tissue and lead to the progression of autoimmune tissue destruction. It is becoming clear that degeneracy in the T cell receptor (TCR) allows the recognition of peptides with varying sequences to be recognized by the same T cell clone. There appears to be degeneracy in both MHC class II HLA-DR2 and TCR peptide recognition, in relation to the self-myelin peptide myelin basic protein epitope 85-99 (MBP_{85-99}).[11] A few residues within the peptide sequence were shown to be critical in determining whether the TCR bound MBP_{85-99}.[12] A number of sequences from human pathogens, which shared critical residues with MBP_{85-99} were shown to cross-react with MBP_{85-99}-specific T cell clones derived from MS patients.[13] This provides evidence to suggest that, theoretically at least, infection with a pathogen, which shares sequence and structural similarity to self-peptides, can cross-activate self-reactive T cell clones and induce autoimmunity.

Superantigens

Superantigens are immunostimulatory molecules, which can stimulate T cells displaying particular TCR Vβ subsets, independent of antigen recognition. They bind outside of the TCR cleft and therefore are capable of activating a wide range of T cell clones.[14] Although studies have not confirmed the ability of superantigens to induce autoimmunity, they can play a role in the exacerbation and induction of relapse, once clinical disease has been induced.[15] Superantigens have been shown to play an important role in animal models of MS and arthritis, and in human Crohn's disease.[15-17]

Dual T Cell Receptor Expression

Usually T cells express one specific TCR composed of a variable α (Vα) and β (Vβ) chain. In some instances, however, T cells may express two alleles from either the Vα or Vβ chain resulting in a T cell with two specificities.[18,19] Therefore, it is possible that a T cell with dual specificity for both a viral antigen and self-tissue may become activated following viral infection and induce autoimmunity. However, there is little supportive evidence for the phenomena of dual TCR in vivo, during experimental animal models of disease or during human autoimmune disease.

Table 1. Hypothetical mechanisms of autoreactive T cell activation by infectious pathogens

	Mechanism of Autoreactive T Cell Activation
DIRECT	Molecular mimicry
	Superantigens
	Dual T cell receptor expression
INDIRECT	Epitope spreading
	Bystander activation
	Cryptic antigen expression
	Reactivation of memory T cells
	Immune dysfunction
	Enhanced antigen processing/presentation

Epitope Spreading

Epitope spreading (Fig. 1B) describes a mechanism by which autoimmunity is initiated via the de novo activation of autoreactive T cells, subsequent to self-tissue damage following infection.[9] Damage may occur directly, via virus-mediated tissue destruction or indirectly, as a result of virus-specific T cell immune responses. Following tissue destruction, self-peptides may be processed by activated host APC in the infected organ and presented to autoreactive CD4+ T cells, which can initiate the autoimmune response. In addition, multiple self-tissue specificities may arise with time leading to a further progression of autoimmune disease. Mice infected with Theiler's murine encephalomyelitis virus (TMEV) exhibit a chronic progressive clinical disease with similarities to MS.[20] During the initial asymptomatic phase of disease, the initial immune response is to the TMEV capsid. However, 50 days post-infection, self-myelin proteolipid protein epitope 139-151 ($PLP_{139-151}$)-specific CD4+ Th1 responses arise, concurrent with clinical symptoms.[21,22] In late stage disease, T cell responses to other PLP epitopes (intra-molecular spread) and other myelin antigens such as MBP (inter-molecular spread), arise.[22] Therefore, the initial response to the infectious pathogen can "spread" from pathogen-specific epitopes to self-epitopes following chronic tissue damage, and can lead to organ-specific autoimmune disease.

Bystander Activation/Immune Dysfunction

Similar to epitope spreading, bystander activation describes an indirect or nonspecific mechanism of autoreactive T cell activation, caused by the inflammatory environment present during infection (Fig. 1C). Immunostimulatory cytokines and other inflammatory mediators can directly activate T cells in the absence of TCR signaling and provide an alternate mechanism of autoreactive T cell activation.[23,24] In addition, the inflammatory environment serves to activate local APC, which can process self-antigens released following tissue damage from the immune responses to the infection. These self-antigens can be presented by APC with a full complement of costimulatory molecules present and serve to activate autoreactive T cells. Evidence for bystander activation has been demonstrated in IFN-γ transgenic mice.[25,26] Over expression of IFN-γ in the pancreas or central nervous system (CNS) can lead to the spontaneous development of diabetes or CNS demyelination, respectively. Activated autoreactive T cells specific for their self-antigens were observed

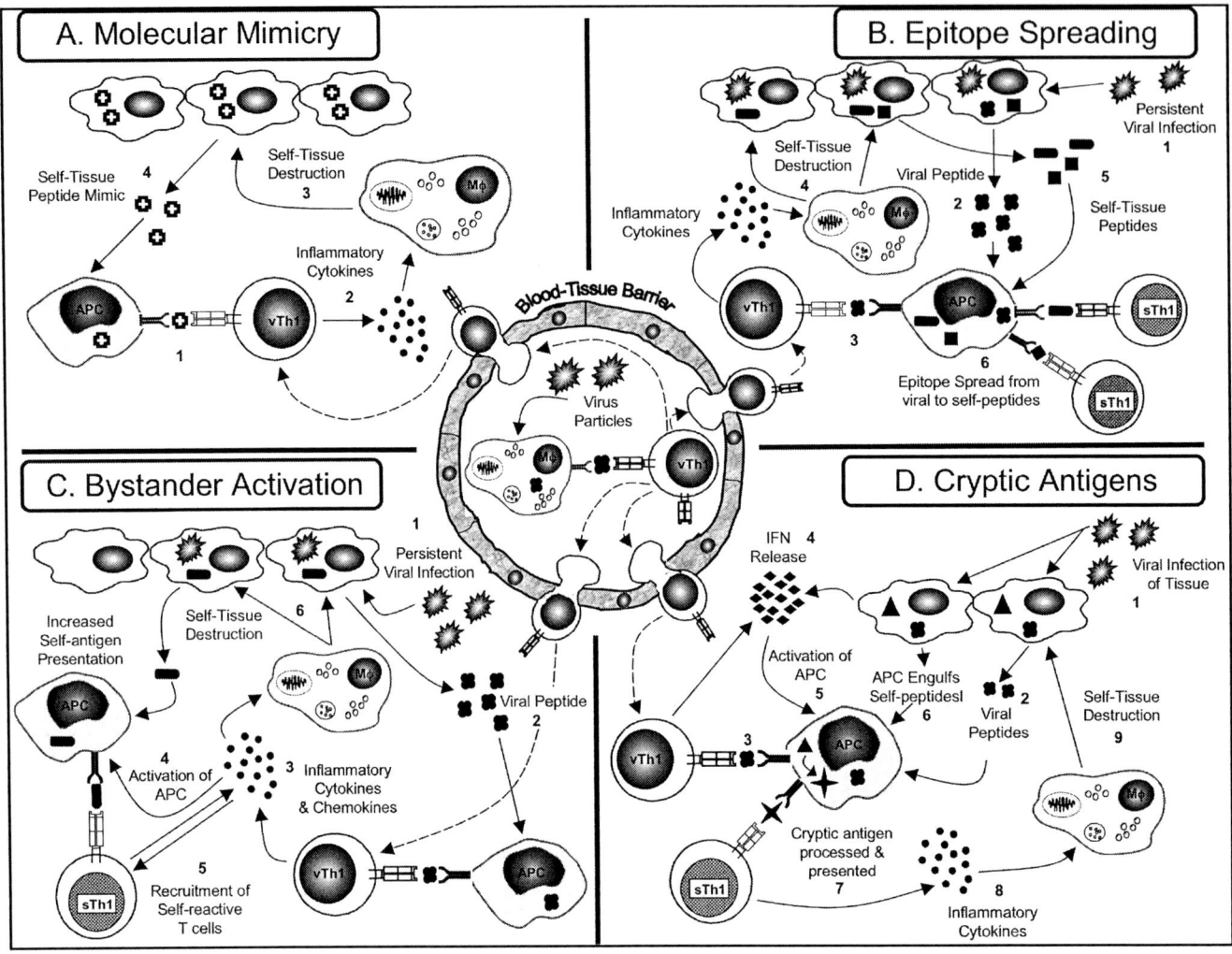

Figure 1. Potential mechanisms of autoimmune disease induction. Following a viral infection, activated virus-specific Th1 (vTh1) cells migrate through a blood-tissue barrier to the infected organ. Molecular mimicry (Panel A) describes the activation of cross-reactive Th1 cells that recognize both the viral epitope and the self epitope.[1] Activation of the cross-reactive T cells results in the release of cytokines and chemokines[2] that recruit and activate resident and peripheral monocyte/macrophage cells which can mediate self-tissue damage.[3] The subsequent release of self-tissue antigens and their uptake by APCs perpetuates the autoimmune disease.[4] In epitope spreading (Panel B), persistent viral infection[1] results in the activation of virus-specific Th1 cells,[2,3] which mediate self-tissue damage[4] resulting in the release of self-peptides,[5] which are engulfed by APCs and presented to self-reactive Th1 cells (sTh1).[6] Continual damage and release of self peptides results in the spread of the self-reactive immune response to multiple self epitopes.[1] The bystander activation model (Panel C) describes the nonspecific activation of self-reactive Th1 cells. Activation of virus-specific Th1 cells[1,2] and the up-regulation of immune functions throughout the tissue[3,4] results in the increased infiltration of T cells to the site of infection and the activation of self-reactive Th1 cells by TCR-dependent and -independent mechanisms.[5] Self-reactive T cells activated in this manner can then mediate self-tissue damage and further perpetuate the autoimmune response.[6] The cryptic antigen model (Panel D) describes the initiation of the autoimmune response by differential processing of self-peptides. Following viral infection[1] IFNs are secreted both by activated virus-specific Th1 cells[2,3] and virus-infected cell.[4] This up-regulates the immune functions of APCs[5] and can lead to APC engulfing self-peptides[6] (triangle). Cytokine activation of APC can induce increased protease production and different processing of captured self-epitope (triangle) resulting in "cryptic" epitopes (star). The presentation of these "cryptic" epitopes can activate self-reactive Th1 cells[7] and lead to self-tissue destruction.[8,9] Published with permission from reference 10.

in both models.[25,26] In contrast to molecular mimicry where the disease-inducing infectious agent is likely to be specific, as it must contain similar sequences to self-antigen and also be recognized by the MHC-TCR complex of the infected host, activation of autoreactive T cells by bystander activation requires only a persistent infection in the target organ. In addition, individuals with an immune dysfunction may also have a higher predisposition to autoimmunity, following infection. Dysfunction of an immune response in the clearance of an infection, failure of activated T cells to undergo apoptosis, or the over expression of pro-inflammatory mediators in response to infection may all induce autoreactive T cell responses. Therefore, multiple different infectious agents may induce a similar organ-specific autoimmune disease in individuals with different genetic backgrounds. This underscores the difficulty that faces researchers investigating infectious pathogens in the induction of autoimmunity.

Table 2. Human autoimmune diseases with a putative infectious etiology

Disease	Target Organ	Associated Infectious Pathogen
Myocarditis	Heart	AV, Coxsackie virus B3, hepatitis C virus, enterovirus
Chagas disease	Heart	*Trypanosoma cruzi*
Multiple sclerosis	CNS	Coronavirus, HSV, EBV, HHV-6, measles virus, mumps virus, HTLV-1
Lyme arthritis	Large joints	*Borrelia burgdorferi*
Type I diabetes	Islet cells	Coxsackie virus B4, CMV, mumps virus, reovirus, rubella virus
Stromal keratitis	Cornea	HSV-1

Abbreviations: AV= adenovirus; HSV= herpes simplex virus; EBV= Epstein-Barr virus; HHV-6= human herpes virus 6; HTLV-1= human T-lymphocyte virus 1; CMV= cytomegalovirus.

Cryptic Antigens/Enhanced Antigen Processing

The term cryptic antigen describes antigenic determinants of proteins, which in contrast to dominant determinants, remain invisible to the immune system under normal circumstances (Fig. 1D). The lack of presentation of cryptic self-antigens by APC in the thymus or periphery, may allow autoreactive T cells to escape clonal deletion and peripheral tolerance induction. Cryptic antigens may also normally be sequestered in organs where naïve autoreactive T cells cannot penetrate, thereby providing immune ignorance of the self-epitope. However, following viral infection of an organ, the release of pro-inflammatory mediators may lead to the upregulation of cryptic antigen processing and presentation by activated APC, the activation of autoreactive T cells and their trafficking to the site of inflammation and cryptic self-antigen expression. In addition, inflammation can lead to the increased expression of proteases in APC and lead to the processing of cryptic antigens that are normally inactive.

Reactivation of Memory T Cells

An initial infection outside of the target organ may prime autoreactive T cells via previously described mechanisms, and lead to a population of self-reactive memory T cells. This population of autoreactive T cells may survive the initial infection and persist following pathogen clearance. Memory T cells may become reactivated either directly by reinfection with the initial pathogen or indirectly, by potent nonspecific mechanisms. Evidence suggests that memory T cells can be reactivated in the absence of signaling through the TCR by type I interferon (IFN) molecules secreted in response to infection.[27] In addition, the activation of memory T cells is less costimulatory dependent than for naïve T cells.[28] Although evidence for the role of infectious pathogens in the induction of autoimmune disease is scarce, both exacerbation and relapse of autoimmune disease is often preceded by viral infection. This suggests that reactivation of autoreactive memory T cells by the release of self-peptides following virus-induced damage in the target organ, may play an important role in autoimmunity.

Human Autoimmune Diseases with a Putative Infectious Etiology

Although many human autoimmune diseases are purported to have an infectious etiology, there remains a paucity of direct evidence in support of this theory. The main obstacle to prove that an infectious pathogen can induce human autoimmune disease is that many individuals will have experienced numerous viral and bacterial infections in their lifetime. Therefore, isolation and identification of a pathogen responsible for induction of autoimmune disease is problematic. In addition, it is entirely possible that only a transient infection is required to initiate autoimmunity, and therefore, the responsible organism may have been cleared by the immune response prior to testing for its presence. Therefore, some of the best evidence for the infectious etiology of autoimmune disease comes from animal models of infectious pathogen-induced autoimmune disease, as described in detail below. However, there is growing evidence for a viral etiology in a number of human autoimmune diseases (Table 2) and this section will discuss the evidence that exists for the presence of infectious pathogens in autoimmune disease and the putative mechanisms of disease induction where possible.

Multiple Sclerosis

MS is the most common human demyelinating disease of the CNS in humans with a frequency of between 5-300 per 100,000 in developed countries. It is widely believed that MS is an immune-mediated disease of the CNS with blood brain barrier breakdown, edema and inflammation. The pathogenesis of MS is thought to involve demyelination of neurons and the subsequent loss of neuronal function in MS patients. Although the etiology of MS is still unknown, epidemiological evidence suggests an infectious pathogen is involved in the development of disease. Although numerous infectious agents have been isolated from MS patients (Table 2), none of these have definitively been proven to be the causative agent. The most recent viral agent associated with MS is human herpes virus 6 (HHV-6). Increased HHV-6 IgG titers have been detected in MS patients,[29] and IgM titers (indicative of a recent infection) in both sera and cerebrospinal fluid (CSF).[30,31] Moreover, HHV-6 DNA has been detected in the CSF and plaques from MS patients (11-17%) but not in controls.[32-34] Immunohistochemical analysis detected HHV-6 in oligodendrocytes in MS lesions and not in control brains. Herpes virus can become reactivated by stress and/or infection, which is consistent with evidence suggesting these factors are involved in relapse episodes of MS.[35] MS is thought to be an autoimmune disease, as adoptive transfer of myelin-specific CD4+ Th1 cells can induce CNS autoimmunity in animal models of disease similar to MS.[3] In humans, autoreactive myelin-specific memory T cells recognizing epitopes of PLP and MBP, have been observed at a higher frequency in MS patients than in normal healthy individuals.[1,2]

One way in which HHV-6 may activate cross-reactive T cells and induce autoimmune disease is by molecular mimicry. To test this hypothesis, a number of T cell lines from MS patients and healthy controls were established, and their ability to respond to either HHV-6 viral peptides or MBP myelin peptides analyzed. Many T cell lines responded to both viral and myelin peptides, indicating that T cells can have dual specificity for both environmental pathogens and myelin antigens.[36] Although there was no statistical difference between the numbers of cross-reactive T cell lines from either MS patients or controls, there was a trend for higher numbers of MBP cross-reactive T cell lines from MS patients (40%) than controls (21%).[36] If HHV-6 does play a role in the pathogenesis of MS, this study suggests that molecular mimicry might be one of the mechanisms involved. However, this study did not address the activation state of cross-reactive cells from either MS patients or controls. Therefore, although most healthy individuals harbor cross-reactive T cells, they are usually in a quiescent state. However, it has previously been shown that myelin-specific T cells from MS patients have an activated phenotype,[37] therefore we cannot rule out the MBP cross-reactive T cells in MS patients were previously activated by HHV-6 by molecular mimicry. Functional studies, such as IFN-γ secretion from T cells isolated from MS patients and controls may help further elucidate the role of cross-reactive T cells in MS.

Although EBV is rarely found in MS lesions, serological studies have found higher titers of EBV-specific antibodies in MS patients compared to controls.[38] Human T-lymphocyte virus 1 (HTLV-1) is a retrovirus associated with HTLV-1 associated myelopathy/tropical spastic paraparesis (HAM/TSP), a human demyelinating disease of the CNS.[39] However, many HTLV-1 seropositive individuals are asymptomatic, suggesting that genetic predisposition and immunological dysfunction may also play a role in HTLV-1-induced disease. Although HTLV-1 has been detected in MS patients, this has not been confirmed in other studies.[40,41]

The mechanisms by which a virus infection can induce demyelination have recently been reviewed[9,10] and include: a direct cytopathic effect of the virus on infected cells such as oligodendrocytes, lysis of infected cells by antibody-mediated mechanisms and/or cytotoxic T cells, or bystander demyelination due to toxic levels of proinflammatory mediators such as TNF-α or reactive oxygen species. It is possible that the induction of MS may involve a nonspecific infection of the CNS and, therefore, the genetic susceptibility of each individual to a variety of infectious pathogens holds the key as to whether the infectious agent is capable of inducing autoimmune disease. Furthermore, studies need to address whether the presence of certain infectious agents correlate to the different temporal characteristics associated with relapsing-remitting or primary-progressive MS. Most of the evidence supporting viral infection in the pathogenesis of MS has been elucidated from animal models and will be discussed later.

Heart Disease

Numerous viruses including the enterovirus, Coxsackie virus B3 (CVB3), adenovirus and hepatitis C virus have been detected in the hearts of patients with myocarditis and cardiomyopathy (Table 2),[42-45] which are the leading causes of heart disease in young people. CVB3 can be detected in 40-50% of patients with dilated cardiomyopathy and it has been linked by epidemiological studies to both cardiomyopathy and juvenile diabetes. Although patients develop autoantibody responses to cardiac myosin antigens and other cardiac antigens, it is unknown to what extent they play a role in the pathogenesis of disease.[46]

Chagas disease is a form of heart disease linked to inflammatory cardiomyopathy, and occurs in individuals chronically infected with *Trypanosoma cruzi*. The continuation of disease in the supposed absence of parasitic infiltration has led to the proposal that Chagas disease has an autoimmune component.[47] A number of studies have described potential cross-reactive sequences between *T. cruzi* and numerous cardiac antigens such as cardiac myosin, mammalian nerve, ribonucleoprotein, microtubule-associated protein of the cytoskeleton and the β1-adrenergic receptor, suggesting that *T. cruzi* infection may induce an immune response towards self-tissue via molecular mimicry.[48-51] Furthermore, T cells isolated from the hearts of Chagas patients were reactive to both cardiac myosin heavy chain and a *T. cruzi* antigen, as measured by proliferation and Th1 cytokine production.[47] However, in most cases the evidence for a role of molecular mimicry in Chagas disease is circumstantial. An alternative theory suggests that it is the persistent infection by the parasite in the target organ, which induces chronic inflammation and leads to tissue damage.[52]

Lyme Arthritis

Lyme arthritis is a multi-system disease resulting from infection by the spirochete, *Borrelia burgdorferi*, following a tick bite. Early symptoms occur within days and include secondary skin lesions, mild hepatitis and cardiac disease. The main symptoms appear months after the initial infection, and include large joint inflammation, swelling and pain, similar to rheumatoid arthritis. Most infected people can be treated with a course of antibiotics, whereby the spirochete is eliminated, as determined by PCR.[53] However, around 10% of patients are resistant to treatment, and even though the spirochete is eliminated, chronic joint inflammation persists. A current hypothesis suggests that this may be due to autoimmunity. The majority of antibiotic-resistant individuals with Lyme arthritis express the HLA-DR1*0401, 0404, 0101, 0102 alleles.[54] A number of spirochete T cell epitope specificities arise in the early phase of infection.[55] In late stage disease, CD4+ Th1 cells and serum IgG specific for the *Borrelia burgdorferi* outer surface protein A (OspA) are detected in the synovial fluid of Lyme arthritis patients concomitant with prolonged attacks of arthritis.[55,56] Interestingly, T cell responses specific for $OspA_{154-173}$ were rare in antibiotic-treatable cases of Lyme arthritis compared to resistant Lyme arthritis patients, suggesting that this epitope may be involved in the pathogenesis of disease.[55] A possible mechanism of autoimmune pathogenesis by this epitope, is molecular mimicry between the foreign OspA epitope and an unknown self-tissue epitope. Computer searches determined that $OspA_{165-173}$ shared sequence homology with a human protein epitope, leukocyte function-associated antigen $1\alpha_{L326-345}$ (LFA-$1\alpha_{L326-345}$).[57] Interestingly, T cells from antibiotic-resistant Lyme arthritis patients displayed Th1-type responses to both $OspA_{165-173}$ and LFA-$1\alpha_{L326-345}$ confirming cross-reactivity between the spirochete and self-epitopes.[57] Although LFA-1α is not joint-specific, it is expressed by lymphocytes, which may be present in the joints following spirochete infection. Current thinking suggests that the large number of IFN-γ secreting OspA-specific T cells activate APCs in the joints, which may phagocytize apoptotic T cells and present LFA-1α. Following eradication of the spirochete infection, APC presenting LFA-1α may cross-activate OspA-specific T cells, which then presumably mediate joint damage. Although the present evidence is suggestive for a role of molecular mimicry in the pathogenesis of Lyme arthritis, further elucidation of the pathogenic role of LFA-1α specific T cells is required.

Type I Diabetes

Type I diabetes is an immune-mediated disease wherein the insulin-producing β cells of the pancreas are destroyed. Although the genetic phenotype of an individual undoubtedly plays a role in the susceptibility to type I diabetes, only 10% of genetically susceptible individuals progress to disease. In addition, concordance rates of type I diabetes between monozygotic twins are between 25-60%.[58-60] Together, this implies that other factors, such as environmental agents, may also be necessary to trigger disease pathogenesis. The development of type I diabetes has long been associated with a viral etiology, based on epidemiological and clinical evidence. A number of viruses have been linked to type I diabetes, by serological, epidemiologic or immunologic studies, including mumps virus,[61] Coxsackie virus (CV), cytomegalovirus (CMV), reovirus and rubella virus (Table 2). As with epidemiological studies in MS, type I diabetes has significant incidence in certain geographical regions, and migration studies suggest that high-risk individuals retain their risk level upon migration to low-risk areas.[62,63] Another study observed that patients who had experienced congenital rubella infections were more likely to develop type I diabetes than those who had not.[64] A number of viruses have been isolated from type I diabetes patients. The most striking evidence for the viral induction of type I diabetes, is the presence of Coxsackie virus B4 (CVB4) in the pancreas of acute-onset patients, which upon isolation and subsequent infection of susceptible mice, could induce diabetes.[65] In addition, high levels of IgM, indicative of a recent infection, specific for Coxsackie B virus could be detected in children recently diagnosed with type I diabetes.[66,67] The presence of autoantibodies and T cells which recognize islet cell antigens suggest that there is an autoimmune component to type I diabetes.[68] A number of autoantigens have been identified in type I diabetes, including insulin, glutamic acid decarboxylase-65 (GAD-65), and protein tyrosine phosphatases IA-2 and IA-2β. Interestingly, protein 2C from CVB4 has been shown to share sequence homology with glutamate decarboxylase (GAD)-65, found on islet cells, and T cells isolated from type I diabetes patients show cross reactivity to GAD-65 (self) and protein 2C (P2-C) (nonself).[69] This suggests that CVB4 may induce type I diabetes by molecular mimicry. However, another study using GAD-65-specific T cell lines isolated from recent onset type I diabetes patients observed no cross reactivity to homologous proinsulin or P2-C peptides.[70] Furthermore, T cells isolated from diabetic patients or healthy controls proliferated equally in response to rechallenge, by pairs of peptides which shared sequence homology, i.e., GAD-65$_{250-273}$ and P2-C$_{28-50}$ or GAD-65$_{506-518}$ and proinsulin$_{24-36}$ peptides.[71] However, similar responses were seen in rechallenge with the nonhomologous peptide or to tetanus toxoid suggesting that the peripheral autoreactive T cells are hyper-reactive.[71] Therefore, the presence of autoreactive T cells in type I diabetes may not be specific to disease pathogenesis. However, further study is required to fully elucidate the role of the autoreactive T cell in type I diabetes. An alternative scenario to the molecular mimicry hypothesis is bystander destruction, where infection of the β cells by CVB4 leads to damage and presentation of self-antigen to autoreactive T cells, in an inflammatory context, and has been studied extensively in animal models of diabetes, as discussed later. Furthermore, in vitro studies have shown that mumps virus, CMV, CVB4 and reovirus type 3 can infect human β cells.[72-74] CMV infection has also been linked to type I diabetes, as persistent CMV infection has been detected in 15% of newly diagnosed patients.

Herpetic Stromal Keratitis

Infection of the cornea with herpes simplex virus 1 (HSV-1) induces herpetic stromal keratitis (HSK), a sight-threatening disease, which involves edema, tissue destruction by CD4$^+$ T cells and subsequent scarring of the cornea (Table 2). The pathogenesis of HSK by HSV-1 needs further elucidation and most evidence for autoimmune involvement has been determined in animal models of HSK, where HSV-1 isolated from patient's cornea can induce HSK in ocularly infected mice.[75] Interestingly, the peak of opacity in the cornea occurs when the HSV-1 titer is low and viral transcripts are not detectable, suggesting that autoimmune responses may play a role in perpetuating disease. Molecular mimicry is one postulated mechanism by which HSV-1 may induce autoimmune corneal tissue destruction. Studies in mouse models suggested that the immunopathogenic T cell response in HSK is directed towards an HSV-1 protein, UL6, that shares homology with a antigen expressed in the murine cornea, the C$_H$3 region of immunoglobulin γ2a^b (IgG2a^b) isotype.[76] Additionally, pretolerance with UL6 prevented HSK induction. Identification of the human homologue to the murine corneal autoantigen may elucidate further the mechanism of autoimmune pathogenesis induced by HSV-1.

An alternative hypothesis suggests that bystander activation of autoreactive T cells occurs following the tissue destruction induced by HSV-1 infection of the cornea, and the resulting release of self-epitopes. Evidence to support this was demonstrated in a study of 12 HSK patients, where T cell lines isolated from the cornea showed no reactivity to HSV-1 UL6 peptide, harboring the cross-react sequence, or other human corneal antigens (human soluble cornea protein extract), although 9 out of 12 T cell lines had HSV-1-specificity.[77] Upon antigenic stimulation, the HSV-1-specific T cell lines secreted IFN-γ and IL-2, both of which are important mediators in inflammation and in the pathology seen in murine models of HSV-1-induced HSK.[78,79] This would suggest that the HSV-1-induced keratitis is not an autoimmune disease.

Mouse Models of Autoimmunity Induced by Infectious Pathogens

The putative mechanism(s) of pathogenesis of autoimmune disease are unknown. Infectious agents have long been associated with the onset of autoimmunity, but evidence for the role of pathogen-induced autoimmunity requires further elucidation. The use of animal models has allowed detailed research into the numerous putative mechanisms of infectious pathogen-induced autoimmune disease (Table 3). Evidence from animal models can provide significant data, which can be correlated to autoimmune disease seen in humans, and can suggest novel parameters to be studied. This section will review the animal models available for the study of infectious pathogen-induced autoimmune disease, and the proposed mechanisms involved in the induction of disease.

Multiple Sclerosis Models

Theiler's murine encephalomyelitis virus (TMEV) is a single stranded RNA virus belonging to the cardiac group of the *Picornaviridae* family and is an enteric pathogen of mice.[20] Following intracerebral infection with the BeAn strain of TMEV, susceptible strains of mice develop a chronic-progressive inflammatory disease of the CNS, TMEV-induced demyelinating disease (TMEV-IDD), similar to primary-progressive MS. Disease is associated with viral persistence, chronic inflammation and

Table 3. Animal models of human autoimmune disease with an infectious pathogenesis mediated by autoreactive T cells

Human Autoimmune Disease	Infectious Animal Model	Proposed Mechanism of Pathogen-Induced Autoimmunity
Multiple Sclerosis	1. Theiler's murine encephalomyelitis virus (TMEV)	Bystander activation/epitope spreading
	2. Recombinant TMEV	Molecular mimicry
	3. Semliki forest virus	Bystander activation, molecular mimicry
Diabetes	1. Coxsackie virus B4	Bystander activation
	2. Transgenic mouse	Bystander activation, molecular mimicry
	3. Lymphocyte choriomeningitis virus	
Myocarditis	1. Mouse cytomegalovirus	Bystander activation
	2. *Chlamydia*	Molecular mimicry
	3. Coxsackie virus B3	Bystander activation, molecular mimicry
Chagas Disease	*Trypanosoma cruzi*	Bystander activation, molecular mimicry
Herpes Stromal Keratitis	Herpes simplex virus-1	Bystander activation, molecular mimicry
Lyme Arthritis	*Borrelia burgdorferi*	Molecular mimicry

demyelination of spinal cord neurons in the white matter.[21] Early disease is asymptomatic and T cell responses are directed towards TMEV capsid epitopes. Clinical disease develops between 35-50 days post-infection and presents as abnormal hind limb gait and eventually spastic hind-limb paralysis.[21,80] Concomitant with the onset of clinical disease is the presence of autoreactive T cells which recognize the immunodominant self-peptide, proteolipid protein residues 139-151 ($PLP_{139-151}$).[22] Following the destruction of self-tissue in the CNS, the release and presentation of myelin antigens by APC in the CNS and peripheral lymphoid tissue leads to the priming of autoreactive T cells. This mechanism of induction of autoimmunity from a viral to myelin immune response is termed epitope spreading. Further epitope spreading occurs to other PLP epitopes, PLP_{56-70} and $PLP_{178-191}$, and to other myelin antigens, myelin oligodendrocyte glycoprotein, MOG_{92-106}.[22,81,82] Peripheral tolerance to viral antigens in the early but not late phase of disease can inhibit the onset of TMEV-IDD, suggesting that the viral-specific T cell response plays an important role in the initiation of disease.[83,84] Conversely, tolerance to myelin antigens in early disease has no effect upon the onset, incidence or severity of TMEV-IDD, whereas myelin tolerance during late-stage disease can significantly reduce the severity of TMEV-IDD.[85] Importantly, computer searches have not shown sequence homology between TMEV and PLP, and furthermore, the late temporal appearance of myelin-specific responses would suggest that molecular mimicry is not a potential mechanism of TMEV-IDD.[22]

Although the mechanism of TMEV-IDD pathology appears to be bystander activation and epitope spread, molecular mimicry is still a favored hypothesis for the initiation of MS. Computer searches have identified a number of peptides from infectious pathogens which share sequence homology to $PLP_{139-151}$.[86] To test the molecular mimicry hypothesis, a recombinant TMEV, engineered to express a $PLP_{139-151}$ mimic sequence from *Haemophilus influenzae* ($HI_{574-586}$) was produced.[87-89] Following intracerebral infection with the $HI_{574-586}$-expressing virus, susceptible mice developed an early mild clinical disease, which was concomitant with the cross-reactive activation of $PLP_{139-151}$-specific T cells, as measured by DTH responses, T cell proliferation and IFN-γ secretion.[87,88] Peripheral tolerance of experimental autoimmune encephalomyelitis in SJL/J mice induced by immunization with $PLP_{139-151}$ in CFA, with the $HI_{574-586}$ peptide significantly inhibited the onset and severity of disease, providing supporting evidence for a role of $HI_{574-586}$-induced cross-reactive $PLP_{139-151}$ T cells in the pathogenesis of disease[87] (unpublished observations). In addition, tolerance of mice to $PLP_{139-151}$ inhibited the onset of $HI_{574-586}$-expressing virus-induced disease and the accompanying T cell cross-activation.[87] Interestingly, immunization of mice with the $HI_{574-586}$ peptide in CFA did not induce clinical disease, although cross-reactive $PLP_{139-151}$ T cells responses were observed[87] (unpublished observations). In this instance, although $PLP_{139-151}$ T cells were shown to proliferate in response to $HI_{574-586}$, Th1 differentiation was reduced as measured by IFN-γ secretion[87] (unpublished observations). Therefore, it appears that activation of the host innate immune response by TMEV is integral to disease induction by molecular mimicry with the $HI_{574-586}$ mimic peptide.

Semliki forest virus (SFV) is a single stranded RNA alpha virus of the *Togaviridae* family, which induces an encephalitis, followed by a demyelinating disease of the central nervous system (CNS) in susceptible strains of mice.[90] SFV is neurotropic and systemic injection leads to the virus crossing the blood brain barrier. Following systemic injection of the A774 SFV strain, it enters the CNS and remains in small foci around the sites of entry.[91] Following SFV infection of BALB/c mice, the subsequent immune response clears the virus, and maximal demyelination can be observed 14 days post-infection. In contrast, lesions and active demyelination SFV infection of SJL/J mice can be observed up to one year post-infection.[92,93] Depletion of CD8+ T cells in SFV-infected BALB/c mice completely protects mice from demyelination, suggesting that CD8+ T cells induce demyelination by killing SFV-infected oligodendrocytes.[94] In addition, MBP and MOG-specific T cell responses have been observed in SFV-infected SJL/J mice, suggesting a role for autoreactive myelin T cells in the pathogenesis of SFV-induced demyelination. In addition, there is sequence homology between MOG_{18-32} and a SFV $E2_{115-129}$ peptide suggesting molecular mimicry may be involved in the pathogenesis of SFV-induced demyelination of susceptible strains of mice.[90]

Diabetes Models

Type I diabetes in humans is primarily associated with infection of pancreatic β cells by CVB4. A viral protein, P2-C, shares amino acid sequence similarity with an islet antigen, glutamate decarboxylase (GAD)-65, and molecular mimicry has been proposed as a potential mechanism of CVB4-induced diabetes.[69,95] Nonobese diabetic (NOD) mice develop spontaneous diabetes via the destruction of pancreatic β cells, and the induction of T cell responses to GAD-65 appear to be important in the pathogenesis of disease and immunity to β cells. The immune response may then become targeted towards other autoantigens such as insulin by epitope spreading.[96] Moreover, immunization of NOD mice with the P2-C CV protein, or CV peptide containing the mimic sequence, could induce cross-reactive T cells that recognized GAD-65 or GAD peptides.[97] Contrasting studies observed opposite results when infecting NOD mice with CVB4. In one case, there was no enhancement of disease or cross-reactive T cell responses following CVB4 infection,[98] whereas infection of NOD mice with a pancreatic CV isolate induced the early onset of diabetes.[99]

Another scenario includes the exposure of cryptic antigens following damage to the β cells following viral infection, or the bystander activation of autoreactive T cells by the inflammatory response to the virus. Following treatment with streptozotocin, an islet damaging agent, BDC2.5 mice, which harbor a transgene (Tg) coding for a diabetogenic T cell receptor (TCR), developed diabetes similar to that seen following CVB4 infection.[100] In contrast, BDC2.5 TCR Tg mice treated with poly I:C, a nonspecific T cell activator, did not develop diabetes, suggesting that it is the presentation of a sequestered, or cryptic, epitope contained in the islet cells following infection which leads to the stimulation of islet-specific T cells and the initiation of disease.[100]

A recent study determined that Tg mice that lack type I interferons due to a transgene coding for the cytokine-signaling suppressor (SOCS-1), were highly susceptible to CVB4 infection. β cell death by apoptosis was induced by the innate immune system, and not by T or B cells, suggesting that the adaptive immune response may not be required for the induction of diabetes.[101]

Heart Disease Models

Animal models of myocarditis with similar disease progression can be induced by two viruses from different families. Both CVB3, a single stranded positive RNA virus from the *Picornavirus* family, and murine cytomegalovirus (MCMV), a double-stranded linear DNA virus of the beta herpesvirus family, induce acute myocarditis (day 7 to 14 p.i.) in both susceptible and "resistant" strains of mice, although only susceptible strains of mice develop chronic disease (day 28 to 100 p.i.).[43] T cells have been shown to play an important role in the pathogenesis of both disease models. Athymic mice and T cell depleted mice exhibit significantly reduced severity of disease following infection, and autoimmune T cells can transfer disease to naïve recipient mice.[46,102-105] Autoantibodies to cardiac myosin heavy chain develop following CBV3 infection, and suggest that this is the target of the autoimmune component of disease.[106,107] Furthermore, immunization of susceptible mice with cardiac myosin in complete Freund's adjuvant induced an immune mediated heart disease with inflammatory lesions similar to CBV3-induced disease.[107] However, other studies have not observed T and B cells cross-reactive to cardiac myosin and CVB3 following infection.[108] Furthermore, sequence analysis of murine cardiac myosin failed to find any mimic sequences within the CVB3 capsid.[108] These observations suggest that bystander activation, and not molecular mimicry, may be the putative mechanism of CVB3-induced myocarditis. In contrast, sequence analysis of MCMV determined that it shared 6 out of 9 amino acids with a protein 48 of murine cardiac myosin heavy chain.[98]

The A/J strain of mouse is susceptible to *T. cruzi* infection-induced myocarditis, and interestingly to both the CVB3 and MCMV-induced models of myocarditis. In addition, the C57Bl/6 strain is resistant to all three pathogen-induced heart diseases. This suggests a common mechanism may be involved in all three models. Two main hypotheses have been used to explain the pathology observed in Chagas disease. The parasite persistence model suggests that disease severity, progression and perpetuation directly correlates to the presence of parasite DNA in muscle tissue.[109] The immunosuppression of mice infected with *T. cruzi* leads to exacerbation of disease severity and infection,[110,111] whereas alternatively, enhancing the anti-parasite response leads to a decreased severity of disease.[112] This evidence supports the parasite presence hypothesis.

In contrast, some investigators suggest that pathology of disease occurs in the absence of parasitic presence, and therefore the inflammatory responses seen are autoimmune in nature. The presence of T cells reactive to autoantigens supports this hypothesis.[52] In support of this, *T. cruzi* infected humans and mice develop cardiac myosin responses.[47,113,114] Cardiac myosin delayed type hypersensitivity (DTH) responses are observed 7 days post *T. cruzi* infection and suggest that molecular mimicry might play a role in the cross-react activation of autoreactive T cells.[115,116] The B13 peptide of *T. cruzi* has been shown to have sequence homology with cardiac myosin and is a candidate autoantigen for Chagas disease.[117] However, there is no experimental evidence to suggest that the B13 epitope is cross-reactive and capable of inducing myosin autoimmunity.

Other evidence suggests that bystander activation of cardiac myosin reactive T cells may be induced due to the inflammatory response directed towards the *T. cruzi*-infected heart muscle cells. Autoimmunity has been shown to occur following cardiac surgery, cardiac transplant rejection and viral infection of the heart.[107,118,119]

Herpes Stromal Keratitis Models

BALB/c strains of mice are susceptible to herpes stromal keratitis (HSK) following infection with Herpes simplex virus 1 (HSV-1) isolated from infected human corneal tissue.[75] Following ocular infection of susceptible strains of mice with HSV-1, inflammatory cells initially infiltrate the underlying stroma and mice develop HSK.[120] One to two weeks following HSV-1 infection, the cornea becomes opaque, and after three weeks, lesions develop in the cornea and are associated with massive inflammatory cell infiltration of the stroma. In some instances, mice can make a spontaneous recovery.[121] Immunopathogenesis of HSK by infection with HSV-1 is similar to the equivalent disease identified in humans. The importance of CD4+ Th1 cells in the pathogenesis of HSK was determined by infecting athymic mice with HSV-1. The mice were protected from disease, however, the transfer of virus-specific T cells to HSV-1 infected athymic mice could restore disease susceptibility.[122,123] As discussed previously, evidence suggests that molecular mimicry between an HSV-1 peptide, UL6, and corneal antigen, IgG2a^b may be involved in the pathogenesis of stromal HSK.[76] In contrast, other studies have been unable to isolate T cell clones that show cross-reactivity to both UL6 and IgG2a^b peptides, and could not detect any T cell clones from HSV-1 infected mice that recognized either UL6 or IgG2a^b.[124] In addition, mice infected with vaccinia virus engineered to express UL6 did not induce HSK, although UL6-specific reactivity was observed.[124] Furthermore, HSV-1 corneal infection can induce HSK in OT2 x RAG1$^{-/-}$ mice (OVA transgenic mice), in the absence of HSV CD4+ T cell responses, suggesting that corneal damage induced by HSV-1 infection may be sufficient to induce HSK. Therefore, autoimmune responses that occur during HSK may do so by bystander activation, and may play a minor role in the pathogenesis of HSK.[124]

Lyme Arthritis Models

Unfortunately, there is a lack of good rodent models for the study of Lyme arthritis. Although HLA-DRB1*0401 Tg mice immunized with OspA, a protein found in *Borrelia burgdorferi*, could elicit T cell responses to both OspA and human LFA-1, these mice did not develop arthritis.[57] This is probably due to sequence differences between human and murine LFA-1α.[125] Furthermore, TCR degeneracy of OspA$_{164-175}$ reactive T cells is extensive, and therefore the presence of autoreactive T cells in Lyme arthritis joints may just reflect the degeneracy of the TCR in OspA$_{164-175}$ reactive T cells.[126] HLA-DRB1*0401 Tg mice engineered to express human LFA-1α may help further elucidate the role of LFA-1α autoreactive T cells in *Borrelia burgdorferi*-induced Lyme arthritis.

Summary

Although the wealth of evidence for the role of pathogen-induced autoimmunity in human disease discussed here is attractive, further elucidation of the mechanisms involved are required. Specifically, further evidence is needed for the induction of autoimmune responses that arise: (1) directly by the infectious pathogen, rather than by bystander activation; and (2) that the ensuing autoimmune response plays a role in the pathology of disease. Animal models can help test the number of hypotheses available, but ultimately, evidence in the human condition will be required. The case for autoimmune responses playing a role in the pathology of pathogen-induced disease is stronger in human disease where the initial pathogen has been cleared and yet clinical symptoms are ongoing, in the presence of a sustained autoimmune response. In other situations, autoreactive T cell responses in the presence of an infectious disease may merely reflect the bystander activation and the degeneracy of TCR/epitope recognition. However, the fact that TCR may be more degenerate than previously thought also lends credence to the possibility that molecular mimicry could be a feasible mechanism of pathogen-induced autoimmunity. However, care must be taken that although, undoubtedly, numerous epitopes from infectious pathogens will be found to share sequence homology with self-peptides by computer sequence analysis, cross-reactivity in in vivo model systems must be analyzed. Furthermore, in vitro T cell assays or immunizing mice with cross-reactive peptides is not sufficient, as this bypasses the checkpoints inherent in an in vivo system whereby the host APC must be able to process and present the cross-reactive epitope from the pathogen. Finally, the cross-reactive T cells induced by mimic peptides must be shown to induce pathology in the absence of the infectious pathogen, thereby demonstrating their role in the pathology of disease.

It is likely that in some instances, infection early in life may induce a cross-reactive population of autoreactive memory T cells that may become activated later in life either by specific or nonspecific mechanisms, and lead to autoimmunity. In these cases, the inducing pathogen may be dormant or long since cleared by the immune system. The genetic susceptibility of individuals may also compound the situation, so certain viral infections may be persistent and lead to autoimmunity in one individual but not another. This underscores the difficulty searching for the role of pathogens in the induction of autoimmune disease.

References

1. Ota K, Matsui M, Milford EL et al. T-cell recognition of an immunodominant myelin basic protein epitope in multiple sclerosis. Nature 1990; 346:183-187.
2. Pette M, Fujita K, Wilkinson D et al. Myelin autoreactivity in multiple sclerosis: Recognition of myelin basic protein in the context of HLA-DR2 products by T lymphocytes of multiple-sclerosis patients and healthy donors. Proc Natl Acad Sci USA 1990; 87:7968-7972.
3. Zamvil SS, Steinman L. The T lymphocyte in experimental allergic encephalomyelitis. Ann Rev Immunol 1990; 8:579-621.
4. Croxford JL, O'Neill JK, Baker D. Polygenic control of experimental allergic encephalomyelitis in Biozzi ABH and BALB/c mice. J Neuroimmunol 1997; 74:205-211.
5. Kurtzke JF. Epidemiologic evidence for multiple sclerosis as an infection. Clin Microbiol Rev 1993; 6:382-427.
6. Kurtzke JF. Multiple sclerosis in time and space-geographic clues to cause. J Neurovirol 2000; 6 Suppl 2:S134-S140.
7. Kurtzke JF, Gudmundsson KR, Bergmann S. Multiple sclerosis in Iceland. I. Evidence of a postwar epidemic. Neurology 1993; 32:143-150.
8. Kurtzke JF, Beebe GW, Norman JE. Epidemiology of multiple sclerosis in U.S. veterans. I. Race, sex, and geographic distribution. Neurology 1979; 29:1228-1235.
9. Croxford JL, Olson JK, Miller SD. Epitope spreading and molecular mimicry as triggers of autoimmunity in the theiler's virus induced demyelinating disease model of multiple sclerosis. Autoimmun Rev 2002; 1:251-260.
10. Olson JK, Croxford JL, Miller SD. Virus-induced autoimmunity: Potential role of viruses in initiation, perpetuation, and progression of T cell-mediated autoimmune diseases. Viral Immunol 2001; 14:227-250.
11. Wucherpfennig KW, Sette A, Southwood S et al. Structural requirements for binding of an immunodominant myelin basic protein peptide to DR2 isotypes and for its recognition by human T cell clones. J Exp Med 1994; 179:279-290.
12. Wucherpfennig KW, Hafler DA, Strominger JL. Structure of human T-cell receptors specific for an immunodominant myelin basic protein peptide: Positioning of T-cell receptors on HLA- DR2/peptide complexes. Proc Natl Acad Sci USA 1995; 92:8896-8900.
13. Wucherpfennig KW, Strominger JL. Molecular mimicry in T cell-mediated autoimmunity: Viral peptides activate human T cell clones specific for myelin basic protein. Cell 1995; 80:695-705.
14. Kotzin BL, Leung DY, Kappler J et al. Superantigens and their potential role in human disease. Adv Immunol 1993; 54:99-166.
15. Brocke S, Gaur A, Piercy C et al. Induction of relapsing paralysis in experimental autoimmune encephalomyelitis by bacterial superantigen. Nature 1993; 365:642-644.

16. Cole BC, Griffiths MM. Triggering and exacerbation of autoimmune arthritis by the Mycoplasma arthritidis superantigen MAM. Arthritis Rheum 1993; 36:994-1002.
17. Dalwadi H, Wei B, Kronenberg M et al. The Crohn's disease-associated bacterial protein I2 is a novel enteric t cell superantigen. Immunity 2001; 15:149-158.
18. Padovan E, Casorati G, Dellabona P et al. Expression of two T cell receptor alpha chains: Dual receptor T cells. Science 1993; 262:422-424.
19. Padovan E, Giachino C, Cella M et al. Normal T lymphocytes can express two different T cell receptor beta chains: Implications for the mechanism of allelic exclusion. J Exp Med 1995; 181:1587-1591.
20. Theiler M. Spontaneous encephalomyelitis of mice, a new virus disease. J Exp Med 1937; 65:705-719.
21. Lipton HL. Theiler's virus infection in mice: An unusual biphasic disease process leading to demyelination. Infect Immun 1975; 11:1147-1155.
22. Miller SD, Vanderlugt CL, Begolka WS et al. Persistent infection with Theiler's virus leads to CNS autoimmunity via epitope spreading. Nature Med 1997; 3:1133-1136.
23. Tough DF, Borrow P, Sprent J. Induction of bystander T cell proliferation by viruses and type I interferon in vivo. Science 1996; 272:1947-1950.
24. Unutmaz D, Pileri P, Abrignani S. Antigen-independent activation of naive and memory resting T cells by a cytokine combination. J Exp Med 1994; 180:1159-1164.
25. Horwitz MS, Bradley LM, Harbetson J et al. Diabetes induced by Coxsackie virus: Initiation by bystander damage and not molecular mimicry. Nat Med 1998; 4:781-786.
26. Sarvetnick N, Liggitt D, Pitts SL et al. Insulin-dependent diabetes mellitus induced in transgenic mice by ectopic expression of class II MHC and interferon-gamma. Cell 1988; 52:773-782.
27. Nahill SR, Welsh RM. High frequency of cross-reactive cytotoxic T lymphocytes elicited during the virus-induced polyclonal cytotoxic T lymphocyte response. J Exp Med 1993; 177:317-327.
28. Croft M, Bradley LM, Swain SL. Naive versus memory CD4 T cell response to antigen: Memory cells are less dependent on accessory cell costimulation and can respond to many antigen-presenting cell types including resting B cells. J Immunol 1994; 152:2675-2685.
29. Sola P, Merelli E, Marasca R et al. Human herpesvirus 6 and multiple sclerosis: Survey of anti-HHV-6 antibodies by immunofluorescence analysis and of viral sequences by polymerase chain reaction. J Neurol Neurosurg & Psych 1993; 56:917-919.
30. Soldan SS, Berti R, Salem N et al. Association of human herpes virus 6 (HHV-6) with multiple sclerosis: Increased IgM response to HHV-6 early antigen and detection of serum HHV-6 DNA. Nature Med 1997; 1394-1397.
31. Ongradi J, Rajda C, Marodi CL et al. A pilot study on the antibodies to HHV-6 variants and HHV-7 in CSF of MS patients. J Neurovirol 1999; 5:529-532.
32. Liektke W, Faustmann PM, Eis-Hubinger AM. Human herpesvirus 6 polymerase chain reaction findings in human immunodeficiency virus associated neurological disease and multiple sclerosis. J Neurovirol 1995; 1:253-258.
33. Wilborn F, Schmidt CA, Brinkmann V et al. A potential role for human herpesvirus type 6 in nervous system disease. J Neuroimmunol 1994; 49:213-214.
34. Challoner PB, Smith KT, Parker JD et al. Plaque-associated expression of human herpesvirus 6 in multiple sclerosis. Proc Natl Acad Sci USA 1995; 92:7440-7444.
35. Steinman L. Multiple sclerosis: A coordinated immunological attack against myelin in the central nervous system. [Review]. Cell 1996; 85:299-302.
36. Cirone M, Cuomo L, Zompetta C et al. Human herpesvirus 6 and multiple sclerosis: A study of T cell cross-reactivity to viral and myelin basic protein antigens. J Med Virol 2002; 68:268-272.
37. Scholz C, Patton KT, Anderson DE et al. Expansion of autoreactive T cells in multiple sclerosis is independent of exogenous B7 costimulation. J Immunol 1998; 160:1532-1538.
38. Sanders VJ, Felisan S, Waddell A et al. Detection of herpesviridae in postmortem multiple sclerosis brain tissue and controls by polymerase chain reaction. J Neurovirol 1996; 2:249-258.
39. Osame M, Usuku K, Izumo S et al. HTLV-I associated myelopathy, a new clinical entity. Lancet 1986; 1:1031-1032.
40. Koprowski H, DeFreitas EC, Harper ME et al. Multiple sclerosis and human T-cell lymphotropic retroviruses. Nature 1985; 318:154-160.
41. Richardson JH, Newell AL, Newman PK et al. HTLV-I and neurological disease in South India. Lancet 1989; 1:1079.
42. Bowles NE, Richardson PJ, Olsen EG et al. Detection of Coxsackie-B-virus-specific RNA sequences in myocardial biopsy samples from patients with myocarditis and dilated cardiomyopathy. Lancet 1986; 1:1120-1123.
43. Fairweather D, Kaya Z, Shellam GR et al. From infection to autoimmunity. J Autoimmun 2001; 16:175-186.
44. Matsumori A, Matoba Y, Sasayama S. Dilated cardiomyopathy associated with hepatitis C virus infection. Circulation 1995; 92:2519-2525.
45. Pauschinger M, Bowles NE, Fuentes-Garcia FJ et al. Detection of adenoviral genome in the myocardium of adult patients with idiopathic left ventricular dysfunction. Circulation 1999; 99:1348-1354.
46. Woodruff JF. Viral myocarditis. A review. Am J Pathol 1980; 101:425-484.
47. Cunha-Neto E, Coelho V, Guilherme L et al. Autoimmunity in Chagas' disease. Identification of cardiac myosin-B13 Trypanosoma cruzi protein crossreactive T cell clones in heart lesions of a chronic Chagas' cardiomyopathy patient. J Clin Invest 1996; 98:1709-1712.
48. Cossio PM, Diez C, Szarfman A et al. Chagasic cardiopathy. Demonstration of a serum gamma globulin factor which reacts with endocardium and vascular structures. Circulation 1974; 49:13-21.
49. Bonfa E, Viana VS, Barreto AC et al. Autoantibodies in Chagas' disease. An antibody cross-reactive with human and Trypanosoma cruzi ribosomal proteins. J Immunol 1993; 150:3917-3923.
50. Kerner N, Liegeard P, Levin MJ et al. Trypanosoma cruzi: Antibodies to a MAP-like protein in chronic Chagas' disease cross-react with mammalian cytoskeleton. Exp Parasitol 1991; 73:451-459.
51. Ferrari I, Levin MJ, Wallukat G et al. Molecular mimicry between the immunodominant ribosomal protein P0 of Trypanosoma cruzi and a functional epitope on the human beta 1-adrenergic receptor. J Exp Med 1995; 182:59-65.
52. Tarleton RL. Parasite persistence in the aetiology of Chagas disease. Int J Parasitol 2001; 31:550-554.
53. Nocton JJ, Dressler F, Rutledge BJ et al. Detection of Borrelia-burgdorferi Dna by polymerase chain-reaction in synovial-fluid from patients with lyme arthritis. N Eng J Med 1994; 330:229-234.
54. Steere AC, Baxter-Lowe LA. Association of chronic, treatment-resistant Lyme arthritis with rheumatoid arthritis (RA) alleles. Arthritis Rheum 1998; 41:S81.
55. Chen J, Field JA, Glickstein L et al. Association of antibiotic treatment-resistant lyme arthritis with T cell responses to dominant epitopes of outer surface protein A of Borrelia burgdorferi. Arthritis Rheum 1999; 42:1813-1822.
56. Akin E, Mchugh GL, Flavell RA et al. The immunoglobulin (IgG) antibody response to OspA and OspB correlates with severe and prolonged lyme arthritis and the IgG response to P35 correlates with mild and brief arthritis. Infect Immun 1999; 67:173-181.
57. Gross DM, Forsthuber T, Tary-Lehmann M et al. Identification of LFA-1 as a candidate autoantigen in treatment-resistant lyme arthritis. Science 1998; 281:703-706.
58. Christie MR, Tun RY, Lo SS et al. Antibodies to GAD and tryptic fragments of islet 64K antigen as distinct markers for development of IDDM. Studies with identical twins. Diabetes 1992; 41:782-787.
59. Kaprio J, Tuomilehto J, Koskenvuo M et al. Concordance for type 1 (insulin-dependent) and type 2 (noninsulin-dependent) diabetes mellitus in a population-based cohort of twins in Finland. Diabetologia 1992; 35:1060-1067.
60. Kyvik KO, Green A, Beck-Nielsen H. Concordance rates of insulin dependent diabetes mellitus: A population based study of young Danish twins. BMJ 1995; 311:913-917.
61. Gamble DR. Relation of antecedent illness to development of diabetes in children. Br Med J 1980; 281:99-101.

62. Green A. Nutrition and environmental factors in insulin-dependent diabetes mellitus: A genetic-epidemiological perspective. Proc Nutr Soc 1997; 56:225-231.
63. Serrano-Rios M, Goday A, Martinez LT. Migrant populations and the incidence of type 1 diabetes mellitus: An overview of the literature with a focus on the Spanish-heritage countries in Latin America. Diabetes Metab Res Rev 1999; 15:113-132.
64. McIntosh ED, Menser MA. A fifty-year follow-up of congenital rubella. Lancet 1992; 340:414-415.
65. Yoon J, Onodera T, Notkins AL. Virus-induced diabetes mellitus: VIII. Passage of encephalomyocarditis virus and severity of diabetes in susceptible and resistant strains of mice. J Gen Virol 1977; 37:225-232.
66. Hyoty H, Hiltunen M, Knip M et al. A prospective study of the role of coxsackie B and other enterovirus infections in the pathogenesis of IDDM. Childhood Diabetes in Finland (DiMe) Study Group. Diabetes 1995; 44:652-657.
67. Frisk G, Fohlman J, Kobbah M et al. High frequency of Coxsackie-B-virus-specific IgM in children developing type I diabetes during a period of high diabetes morbidity. J Med Virol 1985; 17:219-227.
68. Atkinson MA, Bowman MA, Campbell L et al. Cellular immunity to a determinant common to glutamate decarboxylase and coxsackie virus in insulin-dependent diabetes. J Clin Invest 1994; 94:2125-2129.
69. Kaufman DL, Erlander MG, ClareSalzler M et al. Autoimmunity to two forms of glutamate decarboxylase in insulin- dependent diabetes mellitus. J Clin Invest 1992; 89:283-292.
70. Schloot NC, Willemen SJ, Duinkerken G et al. Molecular mimicry in type 1 diabetes mellitus revisited: T-cell clones to GAD65 peptides with sequence homology to Coxsackie or proinsulin peptides do not crossreact with homologous counterpart. Hum Immunol 2001; 62:299-309.
71. Sarugeri E, Dozio N, Meschi F et al. T cell responses to type 1 diabetes related peptides sharing homologous regions. J Mol Med 2001; 79:213-220.
72. Yoon JW, Onodera T, Jenson AB et al. Virus-induced diabetes mellitus XI. Replication of coxsackie B3 virus in human pancreatic beta cell cultures. Diabetes 1978; 27:778-781.
73. Yoon JW, Onodera T, Notkins AL. Virus-induced diabetes mellitus XV. Beta cell damage and insulin- dependent hyperglycemia in mice infected with coxsackie virus B4. J Exp Med 1978; 148:1068-1080.
74. Yoon JW, Selvaggio S, Onodera T et al. Infection of cultured human pancreatic B cells with reovirus type 3. Diabetologia 1981; 20:462-467.
75. Panoutsakopoulou V, Sanchirico ME, Huster KM et al. Analysis of the relationship between viral infection and autoimmune disease. Immunity 2001; 15:137-147.
76. Zhao Z-S, Granucci F, Yeh L et al. Molecular mimicry by herpes simplex virus-type 1: Autoimmune disease after viral infection. Science 1998; 279:1344-1347.
77. Verjans GM, Remeijer L, Mooy CM et al. Herpes simplex virus-specific T cells infiltrate the cornea of patients with herpetic stromal keratitis: No evidence for autoreactive T cells. Invest Ophthalmol Vis Sci 2000; 41:2607-2612.
78. Tang Q, Chen W, Hendricks RL. Proinflammatory functions of IL-2 in herpes simplex virus corneal infection. J Immunol 1997; 158:1275-1283.
79. Hendricks RL, Tumpey TM, Finnegan A. IFN-gamma and IL-2 are protective in the skin but pathogenic in the corneas of HSV-1-infected mice. J Immunol 1992; 149:3023-3028.
80. Dal Canto MC, Lipton HL. Primary demyelination in Theiler's virus infection. An ultrastructural study. Lab Invest 1975; 33:626-637.
81. Katz-Levy Y, Neville KL, Padilla J et al. Temporal development of autoreactive Th1 responses and endogenous antigen presentation of self myelin epitopes by CNS-resident APCs in Theiler's virus-infected mice. J Immunol 2000; 165:5304-5314.
82. Vanderlugt CL, Begolka WS, Neville KL et al. The functional significance of epitope spreading and its regulation by costimulatory molecules. Immunol Rev 1998; 164:63-72.
83. Karpus WJ, Pope JG, Peterson JD et al. Inhibition of Theiler's virus-mediated demyelination by peripheral immune tolerance induction. J Immunol 1995; 155:947-957.
84. Miller SD, Gerety SJ, Kennedy MK et al. Class II-restricted T cell responses in Theiler's murine encephalomyelitis virus (TMEV)-induced demyelinating disease. III. Failure of neuroantigen-specific immune tolerance to affect the clinical course of demyelination. J Neuroimmunol 1990; 26:9-23.
85. Neville KL, Padilla J, Miller SD. Myelin-specific tolerance attenuates the progression of a virus-induced demyelinating disease: Implications for the treatment of MS. J Neuroimmunol 2002; 123:18-29.
86. Carrizosa AM, Nicholson LB, Farzan M et al. Expansion by self antigen is necessary for the induction of experimental autoimmune encephalomyelitis by T cells primed with a cross-reactive environmental antigen. J Immunol 1998; 161:3307-3314.
87. Croxford JL, Olson JK, Miller SD. The role of molecular mimicry in the initiation of central nervous system autoimmune disease. FASEB J 2002; 16:A1216.
88. Olson JK, Croxford JL, Calenoff M et al. A virus-induced molecular mimicry model of multiple sclerosis. J Clin Invest 2001; 108:311-318.
89. Miller SD, Olson JK, Croxford JL. Multiple pathways to induction of virus-induced autoimmune demyelination: Lessons from Theiler's virus infection. J Autoimmun 2001; 16:219-227.
90. Mokhtarian F, Zhang Z, Shi Y et al. Molecular mimicry between a viral peptide and a myelin oligodendrocyte glycoprotein peptide induces autoimmune demyelinating disease in mice. J Neuroimmunol 1999; 95:43-54.
91. Fazakerley JK, Khalili-Shirazi A, Webb HE. Semliki Forest virus (A7[74]) infection of adult mice induces an immune-mediated demyelinating encephalomyelitis. Ann NY Acad Sci 1988; 540:672-673.
92. Donnelly SM, Sheahan BJ, Atkins GJ. Long-term effects of Semliki Forest virus infection in the mouse central nervous system. Neuropathol Appl Neurobiol 1997; 23:235-241.
93. Smyth JM, Sheahan BJ, Atkins GJ. Multiplication of virulent and demyelinating Semliki Forest virus in the mouse central nervous system: Consequences in BALB/c and SJL mice. J Gen Virol 1990; 71:2575-2583.
94. Subak-Sharpe I, Dyson H, Fazakerley J. In vivo depletion of CD8+ T cells prevents lesions of demyelination in Semliki Forest virus infection. J Virol 1993; 67:7629-7633.
95. Vreugdenhil GR, Geluk A, Ottenhoff TH et al. Molecular mimicry in diabetes mellitus: The homologous domain in coxsackie B virus protein 2C and islet autoantigen GAD65 is highly conserved in the coxsackie B-like enteroviruses and binds to the diabetes associated HLA-DR3 molecule. Diabetologia 1998; 41:40-46.
96. Tisch R, Yang XD, Singer SM et al. Immune response to glutamic acid decarboxylase correlates with insulitis in nonobese diabetic mice. Nature 1993; 366:72-75.
97. Tian J, Lehmann PV, Kaufman DL. T cell cross-reactivity between coxsackie virus and glutamate decarboxylase is associated with a murine diabetes susceptibility allele. J Exp Med 1994; 180:1979-1984.
98. Lawson CM. Evidence for mimicry by viral antigens in animal models of autoimmune disease including myocarditis. Cell Mol Life Sci 2000; 57:552-560.
99. Serreze DV, Ottendorfer EW, Ellis TM et al. Acceleration of type 1 diabetes by a coxsackie virus infection requires a preexisting critical mass of autoreactive T-cells in pancreatic islets. Diabetes 2000; 49:708-711.
100. Horwitz MS, Ilic A, Fine C et al. Presented antigen from damaged pancreatic beta cells activates autoreactive T cells in virus-mediated autoimmune diabetes. J Clin Invest 2002; 109:79-87.
101. Flodstrom M, Maday A, Balakrishna D et al. Target cell defense prevents the development of diabetes after viral infection. Nat Immunol 2002; 3:373-382.
102. Lawson CM, O'Donoghue H, Reed WD. The role of T cells in mouse cytomegalovirus myocarditis. Immunol 1989; 67:132-134.
103. Hashimoto I, Komatsu T. Myocardial changes after infection with coxsackie virus-B3 in nude-mice. Brit J Exp Pathol 1978; 59:13-20.

104. Guthrie M, Lodge PA, Huber SA. Cardiac injury in myocarditis induced by coxsackie-virus group-B, type-3 in Balb/C mice is mediated by Lyt-2+ cytolytic lymphocytes. Cell Immunol 1984; 88:558-567.
105. Huber SA. Coxsackie virus-induced myocarditis is dependent on distinct immunopathogenic responses in different strains of mice. Lab Invest 1997; 76:691-701.
106. Alvarez FL, Neu N, Rose NR et al. Heart-specific autoantibodies induced by Coxsackie virus B3: Identification of heart autoantigens. Clin Immunol Immunopathol 1987; 43:129-139.
107. Neu N, Craig SW, Rose NR et al. Coxsackie virus induced myocarditis in mice: Cardiac myosin autoantibodies do not cross-react with the virus. Clin Exp Immunol 1987; 69:566-574.
108. Rose NR, Hill SL. The pathogenesis of postinfectious myocarditis. Clin Immunol Immunopathol 1996; 80:S92-99.
109. Zhang L, Tarleton RL. Parasite persistence correlates with disease severity and localization in chronic Chagas' disease. J Inf Dis 1999; 180:480-486.
110. Andrade ZA, Andrade SG, Sadigursky M. Enhancement of chronic trypanosoma-cruzi myocarditis in dogs treated with low-doses of cyclophosphamide. Am J Pathol 1987; 127:467-473.
111. Tarleton RL, Grusby MJ, Postan M et al. Trypanosoma cruzi infection in MHC-deficient mice: Further evidence for the role of both class I- and class II-restricted T cells in immune resistance and disease. Int Immunol 1996; 8:13-22.
112. Tarleton RL, Grusby MJ, Zhang L. Increased susceptibility of Stat4-deficient and enhanced resistance in Stat6-deficient mice to infection with Trypanosoma cruzi. J Immunol 2000; 165:1520-1525.
113. Rizzo LV, Cunha-Neto E, Teixeira AR. Autoimmunity in Chagas' disease: Immunomodulation of autoimmune and *T. cruzi*-specific immune responses. Mem Inst Oswaldo Cruz. 1988; 83Suppl1:360-362.
114. Tibbetts RS, McCormick TS, Rowland EC et al. Cardiac antigen-specific autoantibody production is associated with cardiomyopathy in Trypanosoma cruzi-infected mice. J Immunol 1994; 152:1493-1499.
115. Leon JS, Engman DM. Autoimmunity in Chagas heart disease. Int J Parasitol 2001; 31:555-561.
116. Leon JS, Godsel LM, Wang K et al. Cardiac myosin autoimmunity in acute Chagas' heart disease. Infect Immun 2001; 69:5643-5649.
117. Cunha-Neto E, Duranti M, Gruber A et al. Autoimmunity in Chagas disease cardiopathy: Biological relevance of a cardiac myosin-specific epitope crossreactive to an immunodominant Trypanosoma cruzi antigen. Proc Natl Acad Sci USA 1995; 92:3541-3545.
118. Descheerder IK, Debuyzere ML, Delanghe JR et al. Anti-myosin humoral immune-response following cardiac injury. Autoimmunity 1989; 4:51-58.
119. Fedoseyeva EV, Zhang F, Orr PL et al. De novo autoimmunity to cardiac myosin after heart transplantation and its contribution to the rejection process. J Immunol 1999; 162:6836-6842.
120. Thomas J, Gangappa S, Kanangat S et al. On the essential involvement of neutrophils in the immunopathologic disease: Herpetic stromal keratitis. J Immunol 1997; 158:1383-1391.
121. Babu JS, Kanangat S, Rouse BT. T cell cytokine mRNA expression during the course of the immunopathologic ocular disease herpetic stromal keratitis. J Immunol 1995; 154:4822-4829.
122. Metcalf JF, Hamilton DS, Reichert RW. Herpetic keratitis in athymic (nude) mice. Infect Immun 1979; 26:1164-1171.
123. Russell RG, Nasisse MP, Larsen HS et al. Role of T-lymphocytes in the pathogenesis of herpetic stromal keratitis. Invest Ophthalmol Vis Sci 1984; 25:938-944.
124. Deshpande SP, Lee S, Zheng M et al. Herpes simplex virus-induced keratitis: Evaluation of the role of molecular mimicry in lesion pathogenesis. J Virol 2001; 75:3077-3088.
125. Gross DM, Huber BT. Cellular and molecular aspects of Lyme arthritis. Cell Mol Life Sci 2000; 57:1562-1569.
126. Maier B, Molinger M, Cope AP et al. Multiple cross-reactive self-ligands for Borrelia burgdorferi-specific HLA-DR4-restricted T cells. Eur J Immunol 2000; 30:448-457.

CHAPTER 25

Molecular Analysis of Immunity

Daniel Douek

Introduction

The molecular analysis of immunity means just that: the analysis, at the molecular level, of the ability to mount an immune response. It is an analysis of function rather than status. As such, it is critical to understand that although we are able to identify many aspects of the immune system in intricate detail (especially in mice), these aspects may not translate into 'immunity'. The concept of immunity entails whether an individual can respond to a herpes virus infection, whether immunization against a pathogen will confer protection, whether a broad T cell repertoire returns after a stem cell transplant, and so on. When considering the molecular analysis of immunity in humans, especially in the settings of lymphopenia and immune reconstitution, it is the adaptive arm of the immune system that draws the most attention from immunologists and clinicians. The T and B lymphocytes that make up this arm use their diverse and unique antigen receptors to recognize and respond to intracellular and extracellular pathogens respectively. Not only can adaptive immunity respond to insults, but it can also amplify itself through clonal expansion to generate increased numbers of differentiated memory cells. Lymphocytes use two different but related mechanisms to recognize extracellular and intracellular pathogens. B cells use the immunoglobulin molecules, either on their surface or as a secreted antibody for their activation and defense against pathogens in the extracellular space, respectively.[1] T cells express receptors on their surface that recognize peptide fragments derived from pathogens that are presented on the surface of major histocompatibility (MHC) molecules.[2] The central role of adaptive immunity in fighting infection is nowhere better illustrated than in the morbidity and mortality associated with the immunodeficiency of AIDS or after chemotherapy.[3,4] When it comes to cells such as T and B cells, the analysis of immunity can be summarized in three questions:

(1) How many cells are there in the individual?

(2) What do they look like? and

(3) How well do they function?

Thus, the molecular analysis of immunity entails the enumeration of the cells involved, the assessment of their phenotypes, and the analysis of their functions in physiologically relevant assays. Particular emphasis will be laid on current and novel technologies available to assess immune function in humans, and their application to clinical settings.

T Cell Immunity

The majority of T cells express their antigen receptor as a surface molecule comprising an α and a β chain.[2,5-7] They recognize antigen as a peptide expressed on the surface of one of the two classes of MHC molecules, both of which are highly polymorphic. Class I MHC molecules are present on almost all cell types; they bind the CD8 molecule on the surface of T cells that express it and generally present intracellular peptide antigen that has been produced in an infected cell, for example viral proteins. Class II MHC molecules are present only on the surface of specialized antigen presenting cells (APC); they bind the CD4 molecule on the surface of T cells that express it and generally present extracellular peptide antigen derived from proteins which have been taken up and processed in lysosomes. CD8+ and CD4+ T cells generally have separate functions, the former acting to kill infected cells, and the later secreting cytokines and other soluble molecules with various effector functions to modulate immune responses. Thus, CD8+ T cells are often referred to as cytotoxic T cells (CTL) and CD4+ T cells as T helper cells (Th). A minority of T cells, often located in specialized sites such as the intestines, express a structurally related surface receptor consisting of a γ and a δ chain.[5] It does not require antigens to be presented by classical polymorphic MHC molecules, but can also recognize non-peptide ligands on the surface of non-polymorphic MHC molecules and CD1.[8] The T cell receptor chains each consist of a variable (V) region and a constant (C) region.[2,7] The great diversity of T cell receptor molecules is generated through the recombination of a great number of gene segments. These segments rearrange during T cell development to form complete V domain exons. The V domains of TCRβ chains comprise variable (V), joining (J) and diversity (D) gene segments. There are up to 52 TCRBV genes and two sets of D, J and C genes. The V domains of TCRα chains comprise only variable (V), joining (J) gene segments.[9-11] The area of the V domain which has the greatest variability is the region at the junctions of the V D and J loci, the so-called complementarity determining region 3 (CDR3).[7,12] Nucleotides are randomly inserted and deleted from these junctions to increase the diversity already generated by the random recombination of the various V, D and J loci. Figure 1 shows how the many V, D and J genes of the *TCRB* locus rearrange to form a coding TCRβ chain. The principle is exactly the same for the V and J genes of the *TCRA* locus.

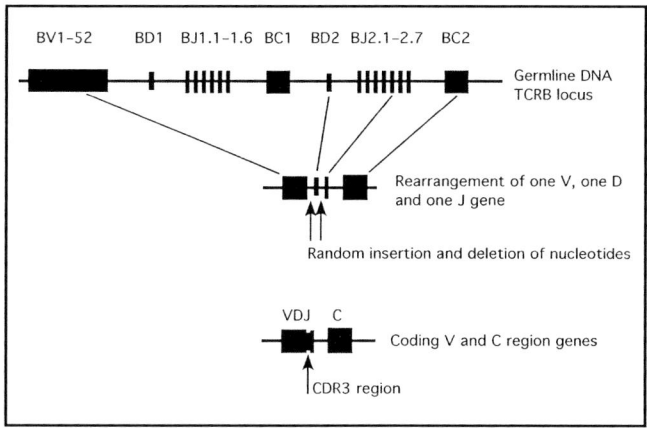

Figure 1. TCRB gene rearrangement.

Development of T Cells and Thymocyte Surface Phenotypes

T cells develop from the same common lymphoid progenitor in the bone marrow that gives rise to B cells. However, those cells destined to become T cells circulate to the thymus where they begin a process of maturation and TCR gene rearrangement which will ultimately result in the export of mature T cells to the periphery.[13-19] The thymus consists of two major regions defined by different types of epithelial stroma — the outer cortex and the inner medulla. Immature thymocytes reside in the cortex in which the large part of differentiation occurs, whereas the more mature thymocytes enter the medulla before exiting the thymus. Maturation of thymocytes can be defined by the sequential rearrangement of the genes that encode the TCR α and β chains, along with the expression and removal of various cell surface molecules.[20-24] At the $CD4^+$ $CD8^+$ double positive stage of maturation, the majority of thymocytes undergo death by apoptosis, some from neglect due to lack of expression of a TCR or insufficient stimulation for survival through their TCR, and others through high affinity stimulation of potentially autoreactive TCR.[25-31] The remainder of the thymocytes, the minority which experience a low affinity interaction between their TCR and self-MHC/peptide sufficient for survival, mature to become $CD4^+$ or $CD8^+$ single positive (SP) thymocytes.[25,32-36] SP thymocytes then exit the thymus as mature naïve T cells, and may subsequently be activated by their cognate antigen to become antigen-experienced memory/effector T cells. Figure 2 shows the phenotypes of thymocytes and the extent of their TCR gene rearrangement at each stage of their development. Although complex, this schema is by no means exhaustive.

Measurement of Thymic Output

Since naïve T cells are the emigrant product of the thymus, one could estimate its output by the analysis of phenotypically naïve T cells in peripheral blood. However, there are concerns about estimating thymic function based on naïve T-cell phenotype alone, as T cells expressing a naïve phenotype are not necessarily accurate surrogate markers of thymic function. Following thymic emigration, $CD45RA^+$ naïve T-cells can have a long quiescent lifespan, may proliferate in an antigen-independent manner, or may rapidly convert to $CD45RO^+$ memory/effector phenotype T-cells.[37-40] Furthermore, naïve T-cell markers may be acquired by memory T-cells, especially among $CD8^+$ T-cells, and this compounds the difficulty in accurately defining naïve T-cells.[41,42] This ambiguity limits the utility of naïve T cell phenotype alone as a measurement of human thymic function.

The phenomenon underlying the development of a genotypic biomarker for human recent thymic emigrants is that during thymocyte development and TCR gene rearrangement, the DNA segments between the recombined genes are excised to form circular episomal DNA fragments.[43,44] These products, termed TCR rearrangement excision circles (TREC), are stable, are not duplicated during mitosis, and are therefore diluted out with each cellular division. In TCR-αβ T cells, rearrangement of both *TCRA* and *TCRB* genes produce TREC. Because of the enormous diversity of *TCRAVJ* and *TCRBVDJ* recombination events, and thus of the TREC produced, no single TREC can be used as a marker to assess overall thymic function. However, a common requirement for productive rearrangement of the *TCRA* locus is deletion of the *TCRD* locus which it encompasses. The rearrangement event that occurs during this process is identical in both alleles in approximately 70% of TCR-αβ T cells. The δ-rec locus recombines to the ψ-Jα locus, forming a single TREC. This unique TREC generated is common to the majority of TCR-αβ T cells.[45,46] The new sequence generated by the inverse ligation during the formation of the TREC can be detected in peripheral blood T cells by quantitative PCR and used to quantify thymic output.[47] An increase in peripheral blood TREC indicates the thymus is producing new T cells. The quantification of TREC has been used to show that thymic output persists in adults and that the thymus can contribute to immune reconstitution after chemotherapy, hematopoietic stem cell transplantation and in the treatment of HIV infection.[47-49] Figure 3 shows the δ-rec locus to ψ-Jα locus recombination event responsible for generating the TREC.

An immunophenotypic measure of human thymic function also exists that uses the cell surface expression of a specific integrin (CD103) within specific thymic and extrathymic T cell subsets, namely $CD3^+$ $CD8^+$ thymocytes and $CD8^+$ naïve T cells.[50] However, CD103 expression is essentially non-existent within the $CD4^+$ T cell subpopulation and therefore quantification of CD103 expressing naïve CD4 T cells is not feasible. Since the thymus does not regulate $CD4^+$ and $CD8^+$ T cell output independently, the restriction of expression of CD103 to naïve $CD8^+$ T cells does not reduce its validity as a phenotypic marker of overall thymus function. The analysis of CD103 expression must be performed by four color flow cytometry followed by careful multi-parameter cluster analysis. Furthermore, due to the dim and labile expression of CD103 on the surface of the naïve cells, fresh blood specimens are necessary for the immunophenotypic analysis.

Analysis of T Cell Function and Phenotypes

Antigen-inexperienced naïve T cells circulate through the lymphoid tissues and organs until antigen is presented to them by professional APCs such as B cells, macrophages and dendritic cells.[40,51] The recognition of antigen in association with costimulatory molecules, such as B7 expressed on the APC surface,[52-55] results in the activation of naïve T cells, their rapid proliferation, and elaboration of 'effector' T cell function and phenotype.[56,57] After the antigenic load has been controlled by the immune response, effector T cells die off and a low frequency of antigen-specific resting 'memory' T cells remains. These cells are able to be subsequently stimulated by any cell expressing their cognate antigen in association with the appropriate class of MHC molecule.[56] Although the majority of $CD8^+$ T cells perform cytotoxic functions, CD4 helper T cells may divide into Th1 and

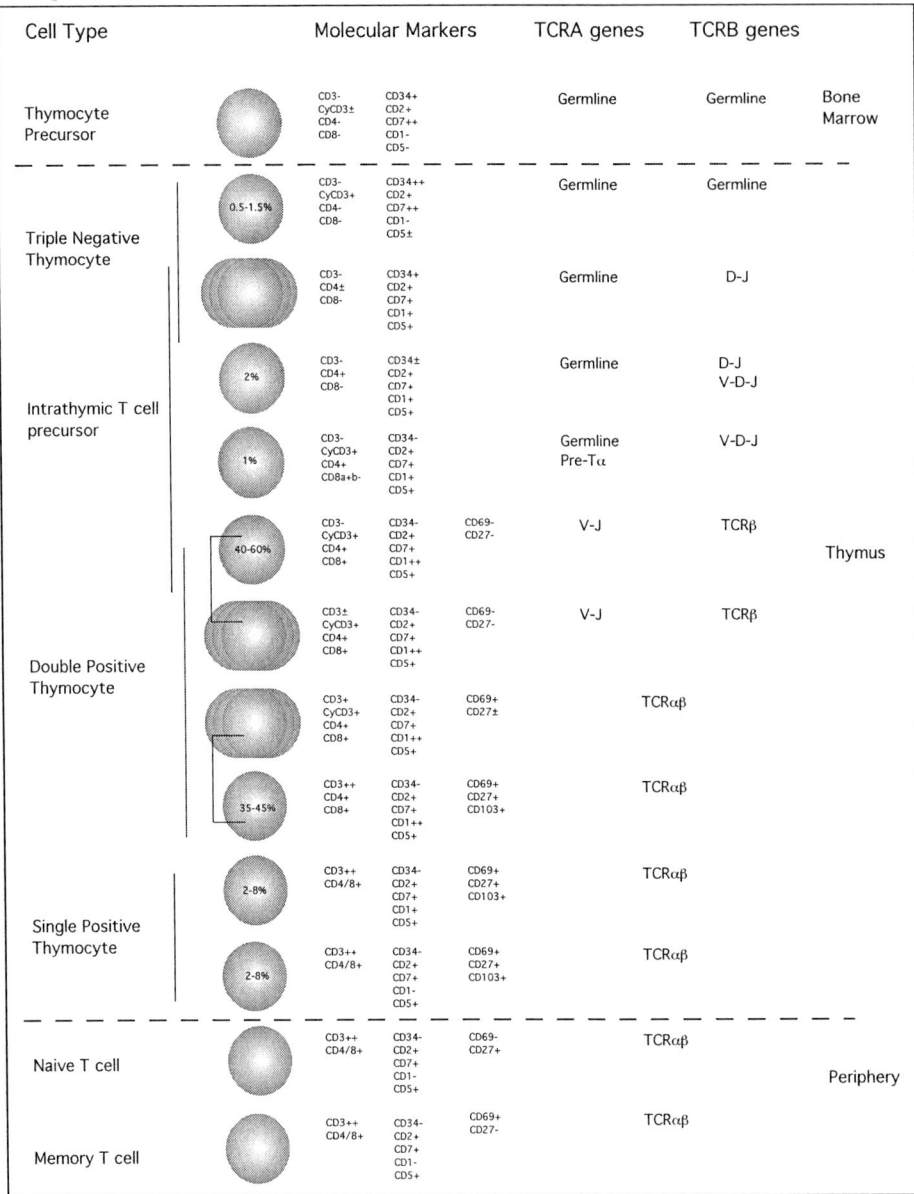

Figure 2. Thymocyte development and phenotypes.

Th2 types.[58] These can be differentiated by expression of different surface makers and their function. Whereas Th1 cells produce cytokines and other factors which aid in the activation of macrophages and CTL,[58-60] Th2 cells produce cytokines involved in the activation of B cells and the generation of humoral immunity.[61-64] T cell phenotype is typically measured by flow cytometry using antibodies specific for the various cells surface markers.[65] A summary of the phenotypic and functional attributes of naïve, effector and memory T cells is given in Table 1. The terms 'bright' and 'dim' refer to their appearance by flow cytometry. As one can see, this table differs from the rigid definitions given in most textbooks; this is to emphasize that expression of these phenotypic and functional markers is rarely all-or-none, but rather fluid and interchangeable.

Critical to the analysis of T cell immunity is the measurement of the frequency and function of antigen-specific T cells. There are a number of ways to perform this, the most recent of which have revolutionized our understanding of the molecular events that define antigen-specific T cells.

The Limiting Dilution Assay

The frequency of antigen specific memory T cells has been measured by the limiting dilution assay (LDA), which uses the Poisson distribution to determine the probability that a limiting dilution of T cells contains one antigen specific cell.[66] The read out after stimulation of the cultures with antigen has generally been proliferation (for Th responses) or cytotoxicity (for CTL responses).[67,68] Although used extensively in the past, this assay has been superceded by simpler assays. Furthermore, it has become clear that the LDA tends to underestimate antigen-specific precursor T cell frequencies by at least 100-fold.[68-72] For these reasons, we will not detail this outdated method of analysis.

Tetramers

The past use of antibodies specific for unique TCRs has allowed the analysis of antigen-specific T cell frequency. But such antibodies are rare and their use in outbred species such as humans is unfeasible. The ability to quantify CD8⁺ T cells with

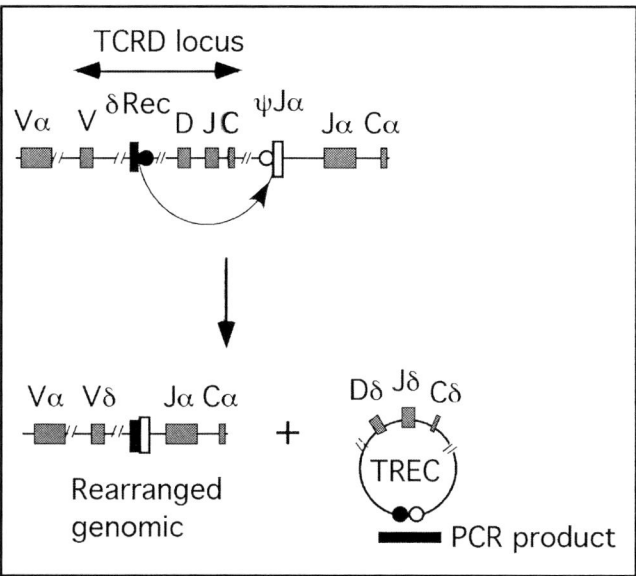

Figure 3. TCR rearrangement to generate TREC.

TCRs that recognize specific peptide-MHC complexes was recently revolutionized by the use of tetramers.[73] Tetramers comprise 4 peptide antigen-MHC complexes and a fluorochrome-conjugated streptavidin molecule. They have sufficient specificity and avidity for the TCR that they can be used to stain specifically the surface of T cells for flow cytometric analysis.[71,73-76] The quantity of CD8[+] T cells with MHC-peptide specificities has been shown to be far greater than was previously estimated by LDA, and illustrates the large expansion of circulating CD8[+] T cells specific for single viral epitopes during acute and chronic infections.[74,76] Thus, in theory, the frequency of any antigen-specific T cell can be determined by using a tetramer consisting of the MHC-peptide combination that its TCR recognizes. In practice, however, this is not the case. While this assay has greatly increased our ability to quantify the CD8[+] T cell immune response to virus infections, it is limited by the requirement for prior knowledge of peptide-MHC complexes to which the T-cells respond. Thus, an overly narrow view of an immune response may arise through tetramer analysis, since it is restricted only to available tetramers.[77,78] Furthermore, the application of tetramer technology to the analysis of CD4[+] T cells has been curbed, since the antigenic targets of CD4[+] T cell responses to HIV have not been defined as completely at the peptide level.[79] The other aspect to consider is that this assay measures the phenotype of the reactivity of a cell, but does not measure its functional capacity. However, the great advantage of tetramer analysis is that it is relatively simple and may be used in conjunction with any of the phenotypic markers outlined in the table above. Figure 4 shows the flow cytometric profile of staining of CD8[+] T cells in an infected individual with a tetramer specific for TCRs which recognize HLA-A2 combined with an HIV-derived peptide.

The ELISPOT Assay

As shown in the table of T cell phenotype and function, effector T cells secrete cytokines when stimulated by cognate antigen.[80,81] Thus, the frequency of antigen-specific T cells in an individual may be assessed by counting the number of cytokine-secreting T cells in an in vitro culture after antigenic stimulation. One method of achieving this is the enzyme-linked immunospot (ELISPOT) assay, a modification of the ELISA assay used for measuring antibody titers.[71,72] This assay is elegant in its simplicity and high accuracy especially at low precursor antigen-specific T cell frequencies. It can be used for either CD4 or CD8 T cell responses. Cultures of T cells are stimulated with the test antigen in wells of plastic plates coated with antibodies specific for the cytokine to be assayed (usually IL-2 or IFNγ). If a T cell is activated by the antigen, it will secrete cytokine which will be bound by the antibody. After an incubation period, the T cells are washed off and the remaining plate-bound cytokine may be probed and visualized with a secondary antibody. Usually, an enzymatic reaction will deposit a colored precipitate revealing the exact site of the antigen-specific T cells. Their frequency can thus be determined by counting the spots and expressing this value as a fraction of the number of T cells added to the culture. The advantage of the ELISPOT is that it uses few cells, is very sensitive, and can analyze Th1 and Th2 responses separately by using different cytokine capture antibodies. Furthermore, if the T cells are depleted of CD8[+] T cells beforehand, then CD4[+] and CD8[+] antigen-specific T cell frequencies can be enumerated separately.

The Intracellular Cytokine Staining (ICS) Assay

The ICS assay offers the most information in terms of the analysis of the frequency and function of antigen-specific T cells. This assay similarly uses the property of stimulation-induced cytokine secretion to quantify antigen-specific T cells.[82-84] However, in this case, the cytokines are prevented from being secreted and detected within the T cells. The stimulated cultures are thus analyzed by flow cytometry. Cytokine secretion is blocked with metabolic poisons such as brefeldin A and monensin, while T cells are being stimulated with antigen. Thus, cytokines accumulate within the cell for the period of incubation, after which the cells are fixed and permeabilized. This allows fluorochrome-labeled antibodies specific for the cytokine to be detected to enter the cells and stain them if cytokine is present. Typically the cells are also stained with antibodies specific for the cell surface markers CD3, CD69 (an activation marker) and CD4 or CD8, and then analyzed by four-color flow cytometry. Clearly this assay offers an enormous range of advantages over the ELISPOT assay in that the responding T cells can be accurately phenotyped by using different antibodies specific for different cell surface markers. In fact, many studies have combined ICS with staining for markers specific for the naïve, effector and memory T cell subsets shown in Table 1 to delineate the function of the different stages to T cell activation.[85-87] Th1 and Th2 responses can be differentiated by their different cytokine production profiles. Other markers of function can be assessed, such as the expression of perforin, granzymes and fas ligand by antigen-specific CD8[+] T cells. Finally, even apoptosis and proliferative capacity in antigen-specific T cells can be monitored. Apoptotic cells may be defined by staining with annexin V or antibodies to activated caspase 3. Proliferation may be very elegantly demonstrated by using the fluorescent membrane dye carboxyfluorescein succinidimyl ester (CFSE), which is incorporated into the cell membranes of living T cells.[88] If stimulated T cells divide, the CFSE is evenly distributed between the progeny, and thus the cellular fluorescence decreases 2-fold with each division. Thus, with a single assay it is possible to examine simultaneously and at the molecular level many properties of antigen specific T cells—phenotype, cytokine production, cytolytic potential, apoptosis and proliferation.[89-96] Figure 5 shows flow cytometric plots of intracellular cytokine (IFNγ) staining of activated CD69[+]CD4[+] T cells specific for a variety of antigens.

Table 1. Intracellular cytokine staining (ICS) assay combined with staining for specific markers

Phenotypic and Functional Markers	Naïve CD4/8	Memory CD4	Effector CD4	Memory CD8	Effector CD8
CD45RA	+	Few +	Few +	Few +	Few +
CD45RO	–	Most +	Most +	Most +	Most +
CD62L (L-selectin)	+	Few +	–	Few +	–
CD27	+	Most +	–	Most +	–
CD28 (Leu 8)	+	Most +	Few +	Most +	Few +
CD11a (LFA-1)	Dim	Bright	Bright	Bright	Bright
HLA-DR (MHC-II)	–	Few +	Most +	Few +	Most +
CD69	–	Few +	+	Few +	+
CD134 (Ox-40)	–	Few +	+	Few +	+
CD152 (CTLA-4)	–	Few +	+	Few +	+
CD57 (Leu 7)	–	Few +	Few +	Few +	Few +
CD154 (CD40 ligand)	–	–	+	–	–
CD95 (fas)	–	Most +	+	Most +	+
CD25 (IL-2αR)	–	Few +	+	Few +	+
IL-12β2R	–	Th1	Th1	Few+	+
IL-18αR	–	Th1	Th1	Few+	–
GATA3	–	Th2	Th2	–	–
BCL-2	+	+	All dim	+	All dim
ICOS	–	Few +	Bias to Th2	Few +	Few +
CCR7	+	Few +	–	Few +	–
CCR5	–	Few +	Few +	Few +	Few +
CXCR4	+	+	+	+	+
IFN-γ	–	Few +	Th1	–	+
TNF-α	–	Few +	Th1	Few +	+
TNF-β	–	Few +	Th1	Few +	+
IL-2	+	+	+	Few +	Few +
GM-CSF	–	Few +	+	Few+	+
IL-3	–	Few Th2	+	–	–
IL-4	–	Few Th2	Th2	–	–
IL-5	–	Few Th2	Th2	–	–
IL-10	–	Few Th2	Th2	–	–
TGF-β	–	Few Th2	Th2	–	–
Perforin	–	Few Th1	Few Th1	Few +	+
Granzymes	–	Few Th1	Few Th1	Few +	+
CD95L (Fas Ligand)	–	Few +	Few Th1	Few +	+

T Cell Receptor Spectratyping

As discussed above, the TCR repertoire is extraordinarily diverse as a consequence of the recombination events between the multitude of V, D and J genes to form coding TCR V region domains.[97] This diversity is concentrated at the region of greatest genetic randomness, the CDR3 region. Because the V domains must encode in-frame, the CDR3 regions of all TCRs are necessarily either the same length or three nucleotides different.[98] CDR3 regions can vary in length by as much as 27 nucleotides (9 amino acids).[98] A useful, if a little crude, estimate of the diversity of the TCR repertoire in an individual can be made by measuring the CDR3 lengths of the TCR V regions encoding all the TCRβ chains. Although there are many different Vβ gene segments, they can be easily grouped into 24 Vβ families. This allows for a specific oligonucleotide primer to be designed for each Vβ family, and these can be used in a PCR in combination with a primer specific for the Cβ region (common to all TCRs) to amplify all the CDR3 regions within each family. When run on a gel, the different length bands can be resolved and their distribution, or 'spectrum', can be analyzed for the intensity of each band; hence the term 'TCR spectratyping'.[98-103] Since the generation of the different CDR3 regions is random, the distribution of CDR3 lengths in a normal individual will be Gaussian. If the spectratype is skewed (non-Gaussian), this may indicate the expansion of a particular T cell clone—as may occur in a T cell response, or the lack of T cell clones—as may occur during immune reconstitution after bone marrow transplantation.[104,105] Figure 6 shows two polyacrylamide gels of TCR spectratypes from two individuals one year after bone marrow transplantation. The gel on the left is clearly very skewed with many oligoclonal expansions (dark bands) and missing clones. The gel on the right is near normal with each TCR Vβ family having a Gaussian distribution of CDR3 lengths. The TCR Vβ family analyzed is shown on the gel.

mRNA Analysis and DNA Microarrays

All the cell surface and intracellular phenotypic markers, cytokines and other molecules involved in T cell function discussed above may be also analyzed at the level of their mRNA.

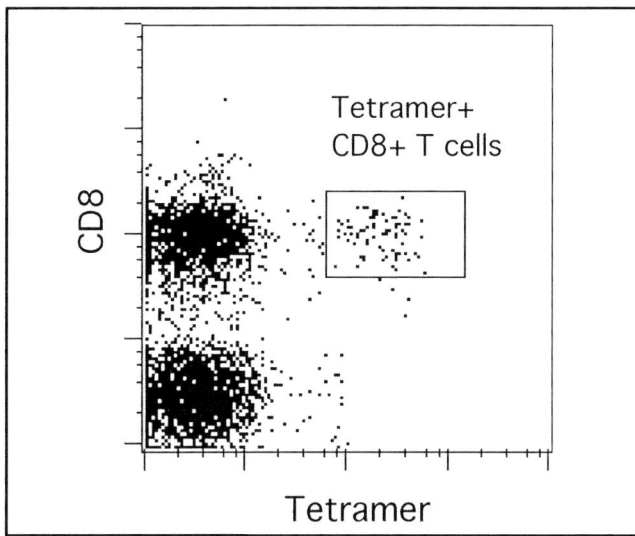

Figure 4. Tetramer staining of T cells.

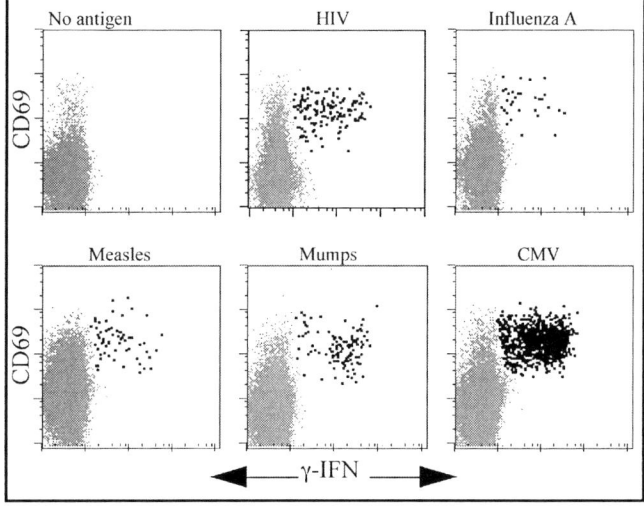

Figure 5. Intracellular cytokine staining of antigen-specific T cells.

This is often useful when antibodies to the molecule are not available, or when the kinetics of gene activation are to be analyzed. The mRNA from a particular cell population is extracted and subjected to reverse transcription-PCR for the gene of interest. This is increasingly being performed in a truly quantitative manner using real-time quantitative assays such as the "Taqman™" method.[106,107] The advantage of this technique is that it is highly sensitive, quantitative and specific, but the sequence of the gene of interest must already be known for the design of the PCR primers. Analysis of the expression of the potentially hundreds or thousands of molecules that play various roles in T cell function requires a different level of molecular analysis. DNA microarrays consist of thousands of DNA sequences, all derived from known genes, bound to glass or silicon surfaces (hence their alternative name of 'DNA chip'). Each sequence has a known, fixed position on the microarray. Fluorochrome-labeled mRNA or cDNA derived from a T cell population of interest is then hybridized to the microarray. Thus, the pattern of binding to the microarray reveals the pattern of gene expression within that cell population.[108-112] Although not fully quantitative, this method is enormously useful in terms of its analytical power. It can viewed as a first line screen for subsequent quantitative RT-PCR and protein analysis. Figure 7 shows a microarray probed with cDNA derived from T cells. The bright points correspond to expressed genes.

B Cells and Humoral Immunity

B cells have one major function in vivo—the production of antibodies. All antibodies consist of paired heavy and light polypeptide chains, and exist in five different classes—IgM, IgD, IgG, IgA and IgE. IgG is by far the most abundant immunoglobulin, and is also the most important when it comes to protection against pathogens such as viruses and bacteria. An IgG molecule comprises four polypeptide chains, two light chains and two heavy chains, arranged so that they form a flexible Y-shape.[1] The variable (V) regions of the antibody that may contribute to the antigen-binding site are formed from the amino-terminal ends, whereas the constant (C) regions determine the isotype.[113-115] The great diversity of antibody molecules is generated through the recombination of a great number of gene segments. These segments rearrange during B cell development to form complete V domain exons in much the same way as TCR rearrangement.[116-118] The analysis of their function is relatively straightforward, and limited to the analysis of serum antibody levels and

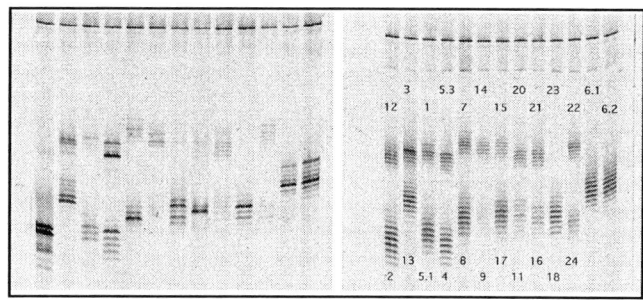

Figure 6. TCR spectratypes.

Figure 7. DNA Microarray.

diversity of antigenic specificity. The repertoire of antibodies is enormous, at least 10^{11} individual specificities in humans, achieves its remarkable diversity by a number of means.[119] Most importantly, V regions are encoded through the somatic recombination of a multitude of different gene segments (V, D and J) which come together to increase the potential diversity of each complete encoded V region.[116-118] The V domains of heavy chains comprise variable (V), joining (J) and diversity (D) gene segments, whereas the V domains of light chains comprise only variable (V), joining (J) gene segments. There are two light chain loci, λ and κ.[120-122] Random insertions and deletions occur where these gene segments join,[123] and further random mutations occur after an immunoglobulin has been expressed through somatic hypermutation seen after B cell stimulation.[124,125] The most variable part of the V domains which contact the antigen are made up from three hypervariable regions (HV1, 2 and 3), also known as complementarity determining regions (CDRs), which are present in each heavy and light chain. Thus, the incredible diversity in antibody recognition is in part generated by the combinatorial diversity of the CDRs. The analysis of such rearrangements is relatively straightforward and consists of PCR and RT-PCR amplification of genomic and expressed Ig sequences, using primers that bind to consensus sequences within variable and constant regions of the heavy and light chains.[126-128] When combined with DNA sequencing, ongoing Ig gene rearrangement may be monitored. Furthermore, as with CDR3 spectratyping for TCR genes, CDR3 length analysis may be performed in this way for Ig genes to give a measure of the breadth of the B cell antibody repertoire. Figure 8 shows a scheme of light chain rearrangement (upper panel) and heavy chain rearrangement (lower panel).

Development of B Cells and Surface Phenotypes

B cell development from hematopoietic stem cell precursors occurs in the bone marrow, and is defined by separate stages which may be distinguished by the differential and sequential expression of immunoglobulin chains and specific cell surface proteins.[129,130] The molecular analysis of immunoglobulin gene expression and rearrangement by PCR, coupled with flow cytometric analysis of cell surface protein expression allows for the analysis of B cell development and the reconstitution of B cell numbers.[126-128] The analysis of B cell antibody production will follow. The earliest, pro-B cell, stage of development is characterized by rearrangement of the immunoglobulin heavy chain locus.[131,132] If VDJ recombination is successful, intact μ heavy chain is expressed intracellularly in association with a surrogate light chain at the large pre-B cell stage.[133] After several rounds of division, light chain rearrangement begins at the small pre-B cell stage, and an immature B cell arises once a complete IgM molecule is expressed at the cell surface. It is important to understand that analysis of Ig gene rearrangement does not simply define the stages of B cell development, it is also a gauge of the dynamic process of B cell development itself, as the expression of differentially rearranged Ig chains regulates the progression from one stage to the next.[134] Figure 9 shows the phenotypes of developing B cells and the extent of their Ig gene rearrangement at each stage of their development.

B Cell Function and Antibodies

Immunoglobulins comprise 5 classes of isotypes, defined by their heavy chain constant regions, each encoded by a separate C region gene. Initially, a productively rearranged V region is expressed in association with μ and δ C_H regions. After a process

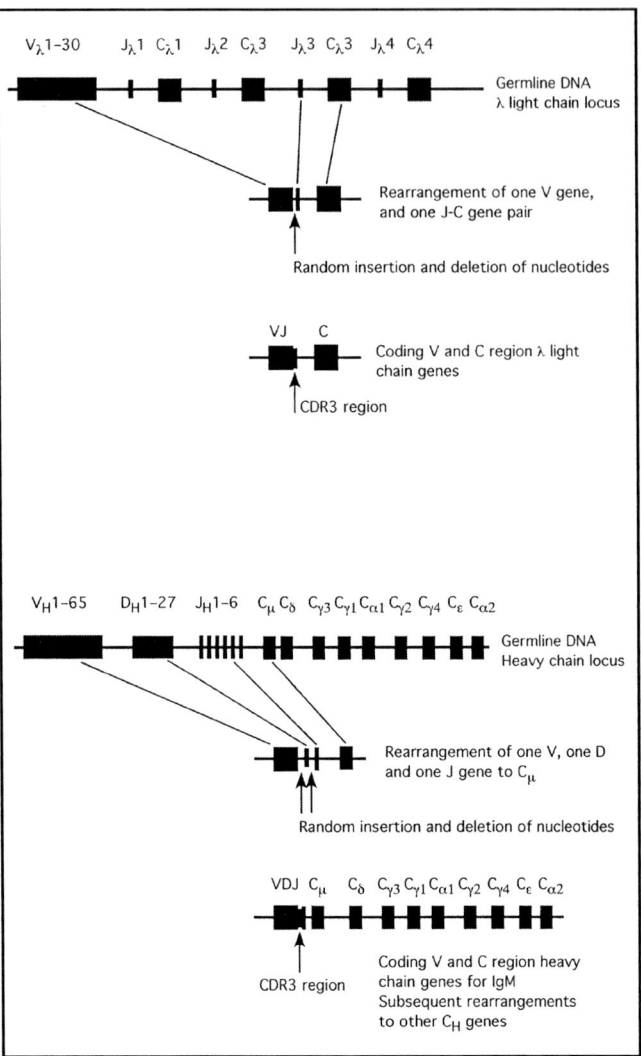

Figure 8. B cell receptor (BCR) gene rearrangement.

known as class switching, the same V region may become associated to any of the other C region genes (δ, α or ε) to form an antibody of a different isotype.[135] Class switching occurs after B cells have been activated by antigen in the presence of help from T cells.[136,137] Thus, on initial exposure to an antigen, a B cell will produce and secrete IgM, and on subsequent stimulation, class switching will occur resulting in the production of IgG antibodies with the same specificity. The various isotypes confer different functions, including complement activation (IgM), opsonization of bacteria and neutralization of toxins and virus particles (IgG and IgA) and local reactions (IgE).[138] Table 2 shows a summary of these functions.

Thus, the analysis of humoral immunity involves the characterization of circulating serum or plasma antibodies, usually IgG, specific for particular antigens.

The most important aspects of an antibody response, in terms of functional immunity, are the amount and specificity of particular antibodies, as well as their isotypes and affinities. The amount, or 'titer', of antibody will give an estimate of the degree of immunity to a particular antigen. For example, after immunization with recombinant hepatitis B surface antigen, specific IgG levels may be monitored until an increase in titer, sufficiently substantial to confer sterilizing immunity, is achieved.[139-141] The

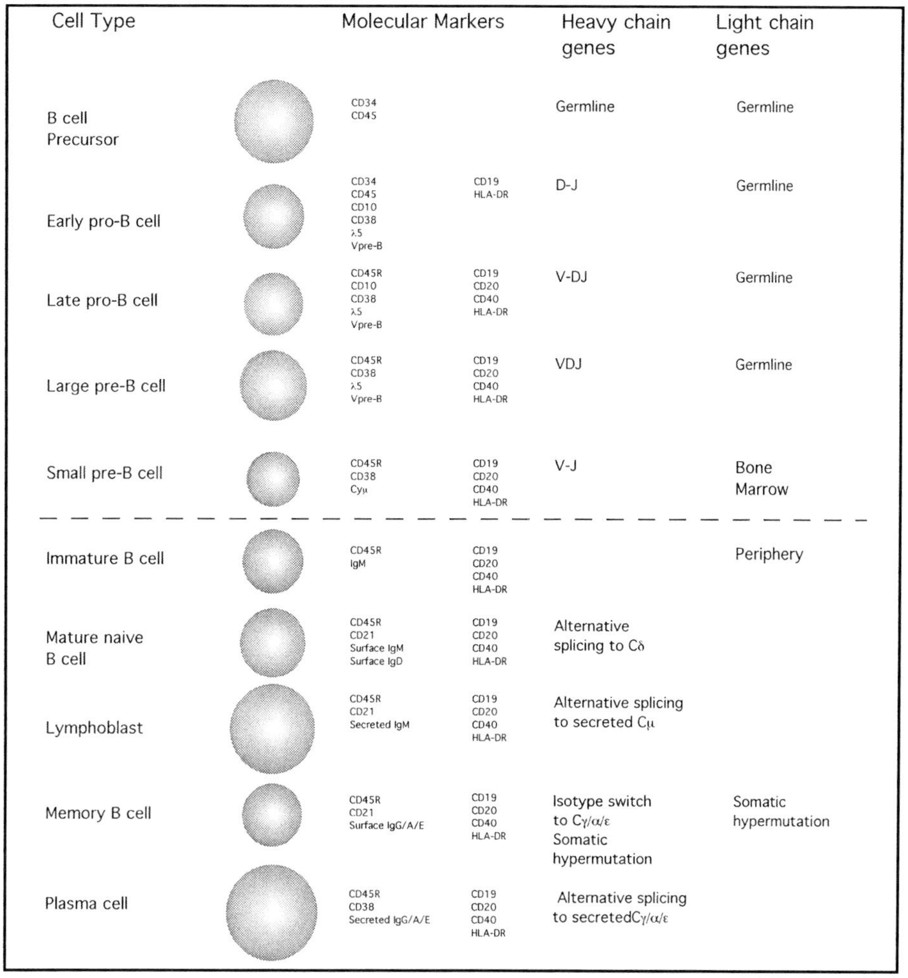

Cell Type		Molecular Markers		Heavy chain genes	Light chain genes
B cell Precursor		CD34 CD45		Germline	Germline
Early pro-B cell		CD34 CD45 CD10 CD38 λ5 Vpre-B	CD19 HLA-DR	D-J	Germline
Late pro-B cell		CD45R CD10 CD38 λ5 Vpre-B	CD19 CD20 CD40 HLA-DR	V-DJ	Germline
Large pre-B cell		CD45R CD38 λ5 Vpre-B	CD19 CD20 CD40 HLA-DR	VDJ	Germline
Small pre-B cell		CD45R CD38 Cμ	CD19 CD20 CD40 HLA-DR	V-J	Bone Marrow
Immature B cell		CD45R IgM	CD19 CD20 CD40 HLA-DR		Periphery
Mature naive B cell		CD45R CD21 Surface IgM Surface IgD	CD19 CD20 CD40 HLA-DR	Alternative splicing to Cδ	
Lymphoblast		CD45R CD21 Secreted IgM	CD19 CD20 CD40 HLA-DR	Alternative splicing to secreted Cμ	
Memory B cell		CD45R CD21 Surface IgG/A/E	CD19 CD20 CD40 HLA-DR	Isotype switch to Cγ/α/ε Somatic hypermutation	Somatic hypermutation
Plasma cell		CD45R CD38 Secreted IgG/A/E	CD19 CD20 CD40 HLA-DR	Alternative splicing to secreted Cγ/α/ε	

Figure 9. B cell development.

specificity of an antibody confers the ability to distinguish one antigen from another, and the isotype determines the half-life and general distribution among the tissues of an antibody.[138]

Molecular analysis of the titer, specificity and isotype of an antibody present within serum may all be performed using the enzyme-linked immunosorbent assay (ELISA).[142-145] The principle of ELISA is relatively simple. The antigen against which specific antibodies are being measured is bound to wells in plastic plates. Serum at different dilutions is then added to the wells and incubated until binding reaches equilibrium (usually about 30 mins). The presence of antigen-bound specific antibody is detected using a secondary layer of monoclonal antibodies specific for one of Ig isotypes. This secondary layer antibody is conjugated to an enzyme such as horseradish peroxidase or alkaline phosphatase, which then reacts with an added substrate to produce a colored soluble product, the intensity of which is measured by light absorbance and which is proportional to the amount of antigen-specific antibody present in serum. The serial dilution of serum at which binding falls to 50% of the maximum is referred to as the 'titer' of that antibody in serum. Thus, the result

Table 2. Functions conferred by antibody isotypes

Isotype	Complement Activation	Opsonization	Neutralization	Mast Cell Sensitization	NK Cell Sensitization
IgM	+++	–	+	–	–
IgD	–	–	–	–	–
IgG1	++	+++	++	+	++
IgG2	+	+/–	++	–	–
IgG3	+++	++	++	+	++
IgG4	–	+	++	–	–
IgA	+	+	++	–	–
IgE	–	–	–	+++	–

of the ELISA measures the amount and isotype of antibodies specific for a particular antigen in serum. This type of molecular analysis forms the basis of the assessment of immune reconstitution of B cell function in situations of lymphopenia such as HIV infection[146,147] or after chemotherapy.[148,149] Furthermore, it forms a crucial part of the assessment of antigen-specific immunity after immunization.[139-141]

mRNA Analysis and DNA Microarrays

In exactly the same way that was discussed above for T cells, powerful molecular analysis of B cell function can be performed with quantitative RT-PCR (150-154) and DNA microarrays.[155] The principles and practice are the same.

References

1. Edelman GM. Antibody structure and molecular immunology. Scand J Immunol 1991; 34:1.
2. Garcia KC, Teyton L, Wilson IA. Structural basis of T cell recognition. Annu Rev Immunol 1999; 17:369.
3. Levy JA. HIV pathogenesis and long-term survival. AIDS 1993; 7:1401.
4. Vescio R, Schiller G, Stewart AK et al. Multicenter phase III trial to evaluate CD34(+) selected versus unselected autologous peripheral blood progenitor cell transplantation in multiple myeloma. Blood 1999; 93:1858.
5. Winoto A, Baltimore D. Separate lineages of T cells expressing the alpha beta and gamma delta receptors. Nature 1989; 338:430.
6. Garcia CK, Degano M, Stanfield RL et al. An $\alpha\beta$ T cell receptor structure at 2.5 Å and its orientation in the TCR-MHC complex. Science 1996; 274:209.
7. Garcia KC, Degano M, Stanfield RL et al. An alphabeta T cell receptor structure at 2.5 A and its orientation in the TCR-MHC complex. Science 1996; 274:209.
8. Shawar SM, Vyas JM, Rodgers JR et al. Antigen presentation by major histocompatibility complex class I-B molecules. Ann Rev Immunol 1994; 12:839.
9. Petrie HT, Livak F, Schatz DG et al. Multiple rearrangements in T cell receptor alpha chain genes maximize the production of useful thymocytes. J Exp Med 1993; 178:615.
10. Dudley EC, Petrie HT, Shah LM et al. T cell receptor beta chain gene rearrangement and selection during thymocyte development in adult mice. Immunity 1994; 1:83.
11. Rowen L, Koop BF, Hood L. The complete 685-kilobase DNA sequence of the human beta T cell receptor locus. Science 1996; 272:1755.
12. Davis MM, Boniface JJ, Reich Z et al. Ligand recognition by alpha beta T cell receptors. Annu Rev Immunol 1998; 16:523.
13. Miller J. The thymus and the development of the immunological responsiveness. Science 1964; 144:1544.
14. Janeway C Jr, Travers P. The Thymus and the development of T lymphocytes. In: Immuno Biology. Robertson M, ed. New York: Garland Publishing, Inc., 1994: 6:1.
15. Carlyle JR, Zuniga-Pflucker JC. Requirement for the thymus in alphabeta T lymphocyte lineage commitment. Immunity 1998; 9:187.
16. Bodey B. Development of lymphopoiesis as a function of the thymic microenvironment. Use of CD8+ cytotoxic T lymphocytes for cellular immunotherapy of human cancer. In Vivo 1994; 8:915.
17. Dunon D, Courtois D, Vainio O et al. Ontogeny of the immune system: Gamma/delta and alpha/beta T cells migrate from thymus to the periphery in alternating waves. J Exp Med 1997; 186:977.
18. Haynes BF, Denning SM, Singer KH et al. Ontogeny of T-cell precursors: A model for the initial stages of human T-cell development. Immunol Today 1989; 10:87.
19. Spits H. Early stages in human and mouse T-cell development. Curr Opin Immunol 1994; 6:212.
20. Petrie HT, Hugo P, Scollay R et al. Lineage relationships and developmental kinetics of immature thymocytes: CD3, CD4, and CD8 acquisition in vivo and in vitro. J Exp Med 1990; 172:1583.
21. Petrie H, Livak F, Burtrum D et al. T cell receptor gene recombination patterns and mechanisms: cell death, rescue, and T cell production. J Exp Med 1995; 182:121.
22. Anderson G, Moore NC, Owen JJ et al. Cellular interactions in thymocyte development. Annu Rev Immunol 1996; 14:73.
23. Junta CM, Passos J GA. Emergence of TCR alpha/beta V(D)J recombination and transcription during ontogeny of inbred mouse strains. Mol Cell Biochem 1998; 187:67.
24. Wang F, Huang CY, Kanagawa O. Rapid deletion of rearranged T cell antigen receptor (TCR) Valpha-Jalpha segment by secondary rearrangement in the thymus: Role of continuous rearrangement of TCR alpha chain gene and positive selection in the T cell repertoire formation. Proc Natl Acad Sci USA 1998; 95:11834.
25. Ashton-Rickardt PG, Tonegawa S. A differential-avidity model for T-cell selection. Immunol Today 1994; 15:362.
26. Bonomo A, Matzinger P. Thymus epithelium induces tissue-specific tolerance. J Exp Med 1993; 177:1153.
27. Douek DC, Corley KT, Zal T et al. Negative selection by endogenous antigen and superantigen occurs at multiple thymic sites. Int Immunol 1996; 8:1413.
28. Hengartner H, Odermatt B, Schneider R et al. Deletion of self-reactive T cells before entry into the thymus medulla. Nature 1998; 336:388.
29. Falk I, Nerz G, Haidl I et al. Immature thymocytes that fail to express TCRbeta and/or TCRgamma delta proteins die by apoptotic cell death in the CD44(-)CD25(-) (DN4) subset. Eur J Immunol 2001; 31:3308.
30. Capone M, Romagnoli P, Beermann F et al. Dissociation of thymic positive and negative selection in transgenic mice expressing major histocompatibility complex class I molecules exclusively on thymic cortical epithelial cells. Blood 2001; 97:1336.
31. Trobridge PA, Forbush KA, Levin SD. Positive and negative selection of thymocytes depends on Lck interaction with the CD4 and CD8 coreceptors. J Immunol 2001; 166:809.
32. Alam SM, Travers PJ, Wung JL et al. T-cell-receptor affinity and thymocyte positive selection. Nature 1996; 381:616.
33. Anderson G, Owen JJT, Moore NC et al. Thymic epithelial cells provide unique signals for positive selection of CD4+CD8+ thymocytes in vitro. J Exp Med 1994; 179:2027.
34. Hogquist KA, Gavin MA, Bevan MJ. Positive selection of CD8+ T cells induced by major histocompatibility complex binding peptides in fetal thymic organ culture. J Exp Med 1993; 177:1469.
35. Hogquist KA, Jameson SC, Heath WR et al. T cell receptor antagonist peptides induce positive selection. Cell 1994; 76:17.
36. Zuniga Pflucker JC, Longo DL, Kruisbeek AM. Positive selection of CD4-CD8+ T cells in the thymus of normal mice. Nature 1989; 338:76.
37. Tough DF, Sprent J. Turnover of naive- and memory-phenotype T cells. J Exp Med 1994;179:1127.
38. Mclean A, Michie C. In vivo estimates of division and death rates of human T lymphocytes. Proc Natl Acad Sci USA 1995; 92:3707.
39. Soares MV, Borthwick NJ, Maini MK et al. IL-7-dependent extrathymic expansion of CD45RA+ T cells enables preservation of a naive repertoire. J Immunol 1998; 161:5909.
40. Picker L, Treer J, Ferguson-Darnell B et al. Control of lymphocyte recirculation in man. J Immunol 1993; 150:1105.
41. Young JL, Ramage JM, Gaston JS et al. In vitro responses of human CD45R0brightRA- and CD45R0-RAbright T cell subsets and their relationship to memory and naive T cells. Eur J Immunol 1997; 27:2383.
42. Hamann D, Baars PA, Rep MH. Phenotypic and functional separation of memory and effector human CD8+ T cells. J Exp Med 1997; 186:1407.
43. Livak F, Schatz D. T-cell receptor α locus V(D)J recombination by-products are abundant in thymocytes and mature T cells. Mol Cell Biol 1996; 16:609.
44. Kong F-K, Chen C-L, Cooper M. Thymic function can be accurately monitored by the level of recent T cell emigrants in the circulation. Immunity 1998; 8:97.
45. Verschuren M, Wolvers-Tettero I, Breit T et al. Preferential rearrangements of the T cell receptor-δ–deleting elements in human T cells. J Immunol 1997; 158:1208.

46. Verschuren M, Wolvers-Tettero I, Breit T et al. T-cell receptor Vδ–Jα rearrangements in T-cell receptor-δ gene deletion. Immunology 1998; 93:208.
47. Douek DC, McFarland RD, Keiser PH et al. Changes in thymic function with age and during the treatment of HIV infection. Nature 1998; 396:690.
48. Douek D, Vescio R, Betts M et al. Assessment of thymic output in adults after haematopoietic stem cell transplant and prediction of T cell reconstitution. Lancet 2000; 355:1875.
49. Weinberg K, Blazar BR, Wagner JE et al. Factors affecting thymic function after allogeneic hematopoietic stem cell transplantation. Blood 2001; 97:1458.
50. McFarland RD, Douek DC, Koup RA et al. Identification of a human recent thymic emigrant phenotype. Proc Natl Acad Sci USA 2000; 97:4215.
51. Picker LJ, Siegelman MH. Lymphoid tissues and organs. In: Fundamental Immunology. WE Paul, ed. New York: Raven Press Ltd., 1993: 152.
52. Gonzalo JA, Delaney T, Corcoran J et al. Cutting edge: The related molecules CD28 and inducible costimulator deliver both unique and complementary signals required for optimal T cell activation. J Immunol 2001; 166:1.
53. Grakoui A, Bromley SK, Sumen C et al. The immunological synapse: A molecular machine controlling T cell activation. Science 1999; 285:221.
54. Gunzer M, Schafer A, Borgmann S et al. Antigen presentation in extracellular matrix: Interactions of T cells with dendritic cells are dynamic, short lived, and sequential. Immunity 2000; 13:323.
55. Banchereau J, Steinman RM. Dendritic cells and the control of immunity. Nature 1998; 392:245.
56. Gudmundsdottir H, Wells AD, Turka LA. Dynamics and requirements of T cell clonal expansion in vivo at the single-cell level: Effector function is linked to proliferative capacity. J Immunol 1999; 162:5212.
57. Harty JT, Tvinnereim AR, White DW. CD8+ T cell effector mechanisms in resistance to infection. Annu Rev Immunol 2000; 18:275.
58. O'Garra A, Arai N. The molecular basis of T helper 1 and T helper 2 cell differentiation. Trends Cell Biol 2000; 10:542.
59. Paulnock, DM. Macrophage activation by T cells. Curr Opin Immunol 1992; 4:344.
60. Andreasen SO, Christensen JE, Marker O. Role of CD40 ligand and CD28 in induction and maintenance of antiviral CD8+ effector T cell responses. J Immunol 2000; 164:3689.
61. Parker DC. T cell-dependent B cell activation. Annu Rev Immunol 1993; 11:331.
62. Croft M, Swain SL. B cell response to T helper cell subsets. II. Both the stage of T cell differentiation and the cytokines secreted determine the extent and nature of helper activity. J Immunol 1991; 147:3679.
63. Croft M, Swain SL. B cell response to fresh and effector T helper cells. Role of cognate T- B interaction and the cytokines IL-2, IL-4, and IL-6. J Immunol 1991; 146:4055.
64. Jaiswal AI, Croft M. CD40 ligand induction on T cell subsets by peptide-presenting B cells: Implications for development of the primary T and B cell response. J Immunol 1997; 159:2282.
65. De Rosa SC, Herzenberg LA, Roederer M. 11-color, 13-parameter flow cytometry: Identification of human naive T cells by phenotype, function, and T-cell receptor diversity. Nat Med 2001; 7:245.
66. Lefkovits I, Waldmann H. Limiting dilution analysis of cells in the immune system. Cambridge: Cambridge University Press, 1979:204.
67. Meyaard L, Otto SA, Hooibrink B et al. Quantitative analysis of CD4+ T cell function in the course of human immunodeficiency virus infection. Gradual decline of both naive and memory alloreactive T cells. J Clin Invest 1994; 94:1947.
68. Koup RA, Pikora CA, Luzuriaga K et al. Limiting dilution analysis of cytotoxic T lymphocytes to human immunodeficiency virus gag antigens in infected persons: In vitro quantitation of effector cell populations with p17 and p24 specificities. J Exp Med 1991; 174:1593.
69. Gotch FM, Nixon DF, Alp N et al. High frequency of memory and effector gag specific cytotoxic T lymphocytes in HIV seropositive individuals. Int Immunol 1990; 2:707.
70. Moss PA, Rowland-Jones SL, Frodsham PM et al. Persistent high frequency of human immunodeficiency virus-specific cytotoxic T cells in peripheral blood of infected donors. Proc Natl Acad Sci USA 1995; 92:5773.
71. Murali-Krishna K, Altman JD, Suresh M et al. Counting antigen-specific CD8 T cells: A reevaluation of bystander activation during viral infection. Immunity 1998; 8:177.
72. Goulder PJ, Tang Y, Brander C et al. Functionally inert HIV-specific cytotoxic T lymphocytes do not play a major role in chronically infected adults and children. J Exp Med 2000; 192:1819.
73. Altman JD, Moss PA, Golder R et al. Phenotypic analysis of antigen-specific T lymphocytes. Science 1996; 274:94.
74. Callan MF, Tan L, Annels N. Direct visualization of antigen-specific CD8+ T cells during the primary immune response to Epstein-Barr virus In vivo. J Exp Med 1998; 187:1395.
75. Kuroda MJ, Schmitz JE, Barouch DH et al. Analysis of Gag-specific cytotoxic T lymphocytes in simian immunodeficiency virus-infected rhesus monkeys by cell staining with a tetrameric major histocompatibility complex class I-peptide complex. J Exp Med 1998; 187:1373.
76. Ogg GS, Jin X, Bonhoeffer S et al. Quantitation of HIV-1-specific cytotoxic T lymphocytes and plasma load of viral RNA. Science 1998; 279:2103.
77. Korber BTM, Brander C, Walker BD et al. 1995. HIV molecular immunology database. Los Alamos National Laboratory.
78. Betts MR, Casazza JP, Patterson BA et al. Putative immunodominant human immunodeficiency virus-specific CD8(+) T-cell responses cannot be predicted by major histocompatibility complex class I haplotype. J Virol 2000; 74:9144.
79. Novak EJ, Liu AW, Nepom GT et al. MHC class II tetramers identify peptide-specific human CD4(+) T cells proliferating in response to influenza A antigen. J Clin Invest 1999; 104:R63.
80. Romagnani S. Th1 and Th2 in human diseases. Clin Immunol Immunopathol 1996; 80:225.
81. Constant SL, Bottomly K. Induction of Th1 and Th2 CD4+ T cell responses: The alternative approaches. Annu Rev Immunol 1997; 15:297.
82. Pitcher CJ, Quittner C, Peterson DM et al. HIV-1-specific CD4+ T cells are detectable in most individuals with active HIV-1 infection, but decline with prolonged viral suppression. Nat Med 1999; 5:518.
83. Kern F, Surel IP, Brock C et al. T-cell epitope mapping by flow cytometry. Nat Med 1998; 4:975.
84. Betts MR, Ambrozak DA, Douek DC et al. Analysis of total human immunodeficiency virus (HIV)-specific CD4 and CD8 T-cell responses: Relationship to viral load in untreated HIV infection. J Virol 2001; 75:11983.
85. Picker LJ, Singh MK, Zdraveski Z et al. Direct demonstration of cytokine synthesis heterogeneity among human memory/effector T cells by flow cytometry. Blood 1995; 86:1408.
86. Waldrop SL, Pitcher CJ, Peterson DM et al. Determination of antigen-specific memory/effector CD4+ T cell frequencies by flow cytometry: Evidence for a novel, antigen-specific homeostatic mechanism in HIV-associated immunodeficiency. J Clin Invest 1997; 99:1739.
87. Waldrop SL, Davis KA, Maino VC et al. Normal human CD4+ memory T cells display broad heterogeneity in their activation threshold for cytokine synthesis. J Immunol 1998; 161:5284.
88. Kaech SM, Ahmed R. Memory CD8+ T cell differentiation: Initial antigen encounter triggers a developmental program in naive cells. Nat Immunol 2001; 2:415.
89. Krupnick AS, Kreisel D, Szeto WY et al. Multiparameter flow cytometric approach for simultaneous evaluation of T lymphocyte-endothelial cell interactions. Cytometry 2001; 46:271.
90. Gudmundsdottir H, Turka LA. A closer look at homeostatic proliferation of CD4+ T cells: Costimulatory requirements and role in memory formation. J Immunol 2001; 167:3699.

91. Honda M, Mengesha E, Albano S et al. Telomere shortening and decreased replicative potential, contrasted by continued proliferation of telomerase-positive CD8+CD28(lo) T cells in patients with systemic lupus erythematosus. Clin Immunol 2001; 99:211.
92. Sheehy ME, McDermott AB, Furlan SN et al. A novel technique for the fluorometric assessment of T lymphocyte antigen specific lysis. J Immunol Methods 2001; 249:99.
93. Suchin EJ, Langmuir PB, Palmer E et al. Quantifying the frequency of alloreactive T cells in vivo: New answers to an old question. J Immunol 2001; 166:973.
94. Lyons AB, Hasbold J, Hodgkin PD. Flow cytometric analysis of cell division history using dilution of carboxyfluorescein diacetate succinimidyl ester, a stably integrated fluorescent probe. Methods Cell Biol 2001;63:375.
95. Hasbold J, Gett AV, Rush JS et al. Quantitative analysis of lymphocyte differentiation and proliferation in vitro using carboxyfluorescein diacetate succinimidyl ester. Immunol Cell Biol 1999; 77:516.
96. Wells AD, Gudmundsdottir H, Turka LA. Following the fate of individual T cells throughout activation and clonal expansion. Signals from T cell receptor and CD28 differentially regulate the induction and duration of a proliferative response. J Clin Invest 1997; 100:3173.
97. Arstila TP, Casrouge A, Baron V et al. A direct estimate of the human alphabeta T cell receptor diversity. Science 1999; 286:958.
98. Pannetier C, Cochet M, Darche S et al. The sizes of the CDR3 hypervariable regions of the murine T-cell receptor beta chains vary as a function of the recombined germ-line segments. Proc Natl Acad Sci USA 1993; 90:4319.
99. Pannetier C, Even J, Kourilsky P. T-cell repertoire diversity and clonal expansions in normal and clinical samples. Immunol Today 1995; 16:176.
100. Puisieux I, Bain C, Merrouche Y et al. Restriction of the T-cell repertoire in tumor-infiltrating lymphocytes from nine patients with renal-cell carcinoma. Relevance of the CDR3 length analysis for the identification of in situ clonal T-cell expansions. Int J Cancer 1996; 66:201.
101. Gorski J, Yassai M, Zhu X et al. Circulating T cell repertoire complexity in normal individuals and bone marrow recipients analyzed by CDR3 size spectratyping. Correlation with immune status. J Immunol 1994; 152:5109.
102. Maslanka K, Piatek T, Gorski J et al. Molecular analysis of T cell repertoires. Spectratypes generated by multiplex polymerase chain reaction and evaluated by radioactivity or fluorescence. Hum Immunol 1995; 44:28.
103. Sourdive DJ, Murali-Krishna K, Altman JD et al. Conserved T cell receptor repertoire in primary and memory CD8 T cell responses to an acute viral infection. J Exp Med 1998; 188:71.
104. Hirokawa M, Horiuchi T, Kitabayashi A et al. Delayed recovery of CDR3 complexity of the T-cell receptor-beta chain in recipients of allogeneic bone marrow transplants who had virus-associated interstitial pneumonia: Monitor of T-cell function by CDR3 spectratyping. J Allergy Clin Immunol 2000; 106:S32.
105. Verfuerth S, Peggs K, Vyas P et al. Longitudinal monitoring of immune reconstitution by CDR3 size spectratyping after T-cell-depleted allogeneic bone marrow transplant and the effect of donor lymphocyte infusions on T-cell repertoire. Blood 2000; 95:3990.
106. Pongers-Willemse MJ, Verhagen OJ, Tibbe GJ et al. Real-time quantitative PCR for the detection of minimal residual disease in acute lymphoblastic leukemia using junctional region specific TaqMan probes. Leukemia 1998; 12:2006.
107. Kammula US, Lee KH, Riker AI et al. Functional analysis of antigen-specific T lymphocytes by serial measurement of gene expression in peripheral blood mononuclear cells and tumor specimens. J Immunol 1991; 163:6867.
108. Schena M, Shalon D, Heller R et al. 1996. Parallel human genome analysis: Microarray-based expression monitoring of 1000 genes. Proc Natl Acad Sci USA 93:10614.
109. Ollila J, Vihinen M. Stimulation of B and T cells activates expression of transcription and differentiation factors. Biochem Biophys Res Commun 1998; 249:475.
110. Geiss GK, Bumgarner RE, An MC et al. Large-scale monitoring of host cell gene expression during HIV-1 infection using cDNA microarrays. Virology 2000; 266:8.
111. Hamalainen H, Zhou H, Chou W et al. Distinct gene expression profiles of human type 1 and type 2 T helper cells. 2001; Genome Biol 2.
112. Galon J, Franchimont D, Hiroi N et al. Gene profiling reveals unknown enhancing and suppressive actions of glucocorticoids on immune cells. FASEB J 2002; 16:61.
113. Yamaguchi Y, Kim H, Kato K et al. Proteolytic fragmentation with high specificity of mouse immunoglobulin G. Mapping of proteolytic cleavage sites in the hinge region. J Immunol Methods 1995; 181:259.
114. Gerstein M, Lesk AM, Chothia C. Structural mechanisms for domain movements in proteins. Biochemistry 1994; 33:6739.
115. Han W, Mou J, Sheng J. Cryo atomic force microscopy: A new approach for biological imaging at high resolution. Biochemistry 1995; 34:8215.
116. Hozumi N, Tonegawa S. Evidence for somatic rearrangement of immunoglobulin genes coding for variable and constant regions. Proc Natl Acad Sci USA 1976; 73:3628.
117. Matthyssens G, Hozumi N, Tonegawa S. Somatic generation of antibody diversity. Ann Immunol (Paris) 1976; 127:439.
118. Early P, Huang H, Davis M. An immunoglobulin heavy chain variable region gene is generated from three segments of DNA: VH, D and JH. Cell 1980; 19:981.
119. Weigert M, Perry R, Kelley D et al. The joining of V and J gene segments creates antibody diversity. Nature 1980; 283:497.
120. Davis MM, Calame K, Early PW et al. An immunoglobulin heavy-chain gene is formed by at least two recombinational events. Nature 1980; 283:733.
121. Joho R, Weissman IL, Early P et al. Organization of kappa light chain genes in germ-line and somatic tissue. Proc Natl Acad Sci USA 1980; 77:1106.
122. Matsuda F, Honjo T. Organization of the human immunoglobulin heavy-chain locus. Adv Immunol 1996; 62:1.
123. Gauss GH, Lieber MR. Mechanistic constraints on diversity in human V(D)J recombination. Mol Cell Biol 16:258.
124. Neuberger MS, Ehrenstein MR, Klix N et al. Monitoring and interpreting the intrinsic features of somatic hypermutation. Immunol Rev 1998; 162:107.
125. Milstein C, Neuberger MS, Staden R. Both DNA strands of antibody genes are hypermutation targets. Proc Natl Acad Sci USA 1998; 95:8791.
126. Kuppers R, Hansmann ML, Diehl V et al. Molecular single-cell analysis of Hodgkin and Reed-Sternberg cells. Mol Med Today 1995; 1:26.
127. Ehlich A, Kuppers R. Analysis of immunoglobulin gene rearrangements in single B cells. Curr Opin Immunol 1995; 7:281.
128. ten Boekel E, Melchers F, Rolink A. The status of Ig loci rearrangements in single cells from different stages of B cell development. Int Immunol 1995; 7:1013.
129. Rolink A, Melchers F. B lymphopoiesis in the mouse. Adv Immunol 1993; 53:123.
130. Hardy RR, Hayakawa K. B-lineage differentiation stages resolved by multiparameter flow cytometry. Ann NY Acad Sci 1995; 764:19.
131. Loffert D, Schaal S, Ehlich A et al. Early B-cell development in the mouse: Insights from mutations introduced by gene targeting. Immunol Rev 1994; 137:135.
132. Hardy RR, Carmack CE, Shinton SA et al. Resolution and characterization of pro-B and pre-pro-B cell stages in normal mouse bone marrow. J Exp Med 1991; 173:1213.
133. Loffert D, Ehlich A, Muller W et al. Surrogate light chain expression is required to establish immunoglobulin heavy chain allelic exclusion during early B cell development. Immunity 1996; 4:133.
134. Melchers F, ten Boekel E, Yamagami T et al. The roles of preB and B cell receptors in the stepwise allelic exclusion of mouse IgH and L chain gene loci. Semin Immunol 1999; 11:307.
135. Stavnezer J. Immunoglobulin class switching. Curr Opin Immunol 1999; 8:199.

136. Noelle RJ, Roy M, Shepherd DM et al. A 39-kDa protein on activated helper T cells binds CD40 and transduces the signal for cognate activation of B cells. Proc Natl Acad Sci USA 1992; 89:6550.
137. Bartlett WC, McCann J, Shepherd DM et al. Cognate interactions between helper T cells and B cells. IV. Requirements for the expression of effector phase activity by helper T cells. J Immunol 1990; 145:3956.
138. Ward ES, Ghetie V. The effector functions of immunoglobulins: Implications for therapy. Ther Immunol 1995; 2:77.
139. Mele A, Tancredi F, Romano L et al. Effectiveness of hepatitis B vaccination in babies born to hepatitis B surface antigen-positive mothers in Italy. J Infect Dis 2001; 184:905.
140. Styczynski J, Wysocki M, Koltan S et al. A nine-year experience of immunoprophylaxis against hepatitis B virus infection in children with cancer: Results from a single institution in Poland. J Hosp Infect 2001; 48:298.
141. Heijtink RA, van Bergen P, van Roosmalen MH et al. Anti-HBs after hepatitis B immunization with plasma-derived and recombinant DNA-derived vaccines: Binding to mutant HBsAg. Vaccine 2001; 19:3671.
142. Plested JS, Gidney MA, Coull PA et al. Enzyme linked immunosorbent assay (ELISA) for the detection of serum antibodies to the inner core lipopolysaccharide of Neisseria meningitidis group B. J Immunol Methods 2000; 237:73.
143. Pospisilova S, Brazda V, Amrichova J et al. Precise characterisation of monoclonal antibodies to the C-terminal region of p53 protein using the PEPSCAN ELISA technique and a new non-radioactive gel shift assay. J Immunol Methods 2000; 237:51.
144. Devito C, Levi M, Broliden K et al. Mapping of B-cell epitopes in rabbits immunised with various gag antigens for the production of HIV-1 gag capture ELISA reagents. J Immunol Methods 2000; 238:69.
145. Wedege E, Bolstad K, Wetzler LM et al. IgG antibody levels to meningococcal porins in patient sera: Comparison of immunoblotting and ELISA measurements. J Immunol Methods 2000; 244:9.
146. Barassi C, De Santis C, Pastori C et al. Early production of HIV-1 neutralising antibodies in patients following highly active antiretroviral treatment (HAART) during primary HIV infection. J Biol Regul Homeost Agents 2000; 14:68.
147. Price P, Mathiot N, Krueger R et al. Immune dysfunction and immune restoration disease in HIV patients given highly active antiretroviral therapy. J Clin Virol 2001; 22:279.
148. Storek J, Dawson MA, Storer B et al. Immune reconstitution after allogeneic marrow transplantation compared with blood stem cell transplantation. Blood 2001; 97:3380.
149. Kamani N, Kattamis A, Carroll A et al. Immune reconstitution after autologous purged bone marrow transplantation in children. J Pediatr Hematol Oncol 2000; 22:13.
150. Giulietti A, Overbergh L, Valckx D et al. An overview of real-time quantitative PCR: Applications to quantify cytokine gene expression. Methods 2001; 25:386.
151. Xiang SD, Benson EM, Dunn IS. Tracking membrane and secretory immunoglobulin alpha heavy chain mRNA variation during B-cell differentiation by real-time quantitative polymerase chain reaction. Immunol Cell Biol 2001; 79:472.
152. Rasmussen T, Poulsen TS, Honore L et al. Quantitation of minimal residual disease in multiple myeloma using an allele-specific real-time PCR assay. Exp Hematol 2000; 28:1039.
153. Schwartz GN, Kammula U, Warren MK et al. Thrombopoietin and chemokine mRNA expression in patient post-chemotherapy and in vitro cytokine-treated marrow stromal cell layers. Stem Cells 2000; 18:331.
154. Verhagen OJ, Willemse MJ, Breunis WB et al. Application of germline IGH probes in real-time quantitative PCR for the detection of minimal residual disease in acute lymphoblastic leukemia. Leukemia 2000; 14:1426.
155. Walker J, Flower D, Rigley K. Microarrays in hematology. Curr Opin Hematol 2002; 9:23.

CHAPTER 26

Immune Reconstitution after Hematopoietic Stem Cell Transplantation

Andreas Thiel, Tobias Alexander, Christian A. Schmidt, Falk Hiepe, Renate Arnold, Andreas Radbruch, Larissa Verda and Richard K. Burt

Introduction

Qualitative and quantitative changes occur within the immune system during aging. Normal age-related T cell alterations include a decline in CD4⁺ cells, loss of naïve (antigenic virgin) cells, increase in memory (antigen experienced) cells, decline in T cell proliferative responses, and narrowing of the T cell receptor repertoire. T cell age-associated alterations are secondary to increased extrathymic T cell reconstitution following post puberty thymic involution combined with peripheral antigen driven T cell maturation and expansion. Age-related B cell changes are predominately qualitative and include loss of high affinity antibodies, increase in low affinity and autoantibodies, impaired isotype switching, and hindered antibody responses to vaccination. B cell related changes appear secondary to age-related impairment of T cell function. Immune reconstitution after hematopoietic stem cell transplantation (HSCT), whether from autologous or allogeneic stem cells, initially arises by means of extrathymic reconstitution. It is, therefore, at onset dominated by changes similar to those of an aged immune system. After HSCT, it may take 1 to 2 or more years for thymic reconstitution of a T and B cell phenotype and function normal for the recipient's age. The younger the recipient, the more rapid thymic reconstitution occurs. The health of the graft as well as patient survival appears to correlate with the rapidity of shift from extrathymic to thymic T cell reconstitution following HSCT.

Innate versus Adaptive Immunity

Two types of immune system, innate and adaptive, co-exist in humans. The evolutionarily more primitive innate immune system consists of cells using germ line genes to express receptors that recognize specific bacterial, viral, or otherwise foreign antigens. These cells include granulocytes, macrophages, and natural killer (NK) cells, as well as proteins such as C3-like complement. There are approximately 33,000 genes in the human genome.[1] Therefore, while specific for nonself determinants, the innate immune system has limited diversity. Receptors for cells of adaptive immunity (T and B lymphocytes) arises by somatic recombination of germ line variable (V), joining (J) and diversity (D) genes leading to a highly diverse number (10^{14} to 10^{19}) of possible T and B cell receptors from a limited number of V(D)J genes. This provides adaptive immunity with diversity but also allows for generation of self-reactive repertoires.

Phylogeny of Adaptive Immunity

The most primitive living vertebrates such as modern jawless fish (agnathans like hagfish and lampreys) generally lack T cell receptors (TCR), immunoglobulin (IgG) receptors and major histocompatibility (MHC) molecules. Some hagfish and lampreys have been reported to have lymphocytes and plasma cells despite absence of an identifiable thymus, spleen, or lymph nodes. These arise within lymphoid clusters in the pronephros (kidney) and intestinal lamina propria that may be phylogenetic precursors of a thymus and spleen, respectively.[2-3]

Cartilaginous fish (chondrichthyans) such as sharks are the first vertebrates to have adaptive immunity. They have demonstrable lymphocytes with TCR, MHC class I and II molecules, and immunoglobulins.[4] While lymph nodes are absent, the primary lymphoid organ or thymus appears for the first time in evolution with chondrichthyans. The thymus arises with occurrence of jawed fish as a part of the pharyngeal pouch. The major site of T cell production is the thymus. In comparison, the major site of hematopoietic and B cell production is the bone marrow. Early primitive vertebrates may have no bone marrow. In these animals, hematopoietic cell and B cell production occurs in various organs (liver, kidney, gonads, meninges) in which there is a stromal environment similar to marrow. Compared to the thymus and bone marrow, secondary lymphoid structures such as the spleen and lymph nodes are later developments of evolution.[3]

Generation of T Cells

Thymus

T cell differentiation occurs within the thymus and is characterized by ordered expression of various CD surface molecules and V, D, and J gene rearrangements. Progenitor cells originating in the bone marrow migrate to the thymus. These early pre-T cells are CD3⁻CD4⁻CD8⁻ triple negative (TN) cells. TN cells differentiate into CD3⁻CD4⁺CD8⁻ intrathymic T progenitor (ITTP) cells. ITTP subsequently differentiate into CD3⁺CD4⁺CD8⁺ double positive (DP) T cells. DP thymocytes undergo apoptosis if their TCR fails to recognize an antigen. Negative selection or apoptosis also occurs if the antigen binding avidity is too strong. Positive selection or survival appears to occur if antigen binding is of moderate avidity. Along with positive and negative selec-

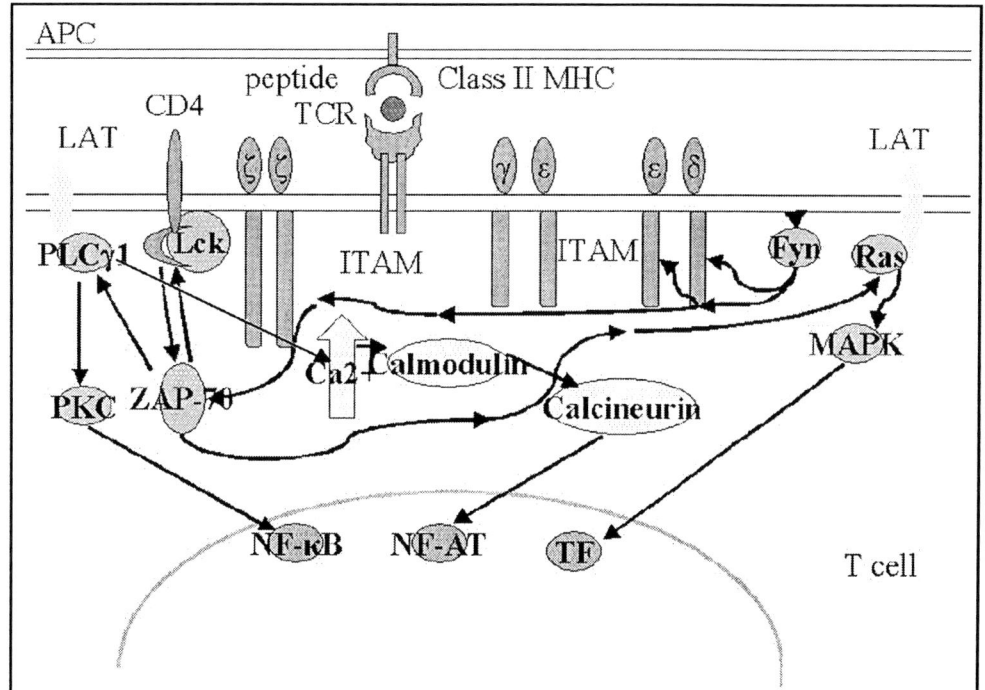

Figure 1. Scheme of signaling pathways during T-cell activation. Modified from ref. 188. TCR= T-cell receptor; APC= antigen presenting cell; ITAM= immunoreceptor tyrosine-based activation motif; ζ (zeta), γ (gamma), ε (epsilon), δ (delta)= invariant polypeptide chains of CD3 complex; MAPK= mitogen activated protein kinase; PKC= protein kinase C; PLCγ1= phospholipase C; ZAP-70= zeta associated protein of 70 kDa; Ras= from raus sarcoma virus oncogene, here: small G protein Ras; LAT= linker for activation of T-cells; Fyn= Fgr yes novel protein; Lck= lymphocyte specific tyrosine kinase; NF-κB= nuclear factor κB; NF-AT= nuclear factor of activated T cells; TF= transcription factor. (The TCR-mediated signal results in the activation of various signal transduction pathways with consequent activation of transcription factors as NF-κB, NF-AT, and TF. Phosphorylation of tyrosines in the ITAMs by Fyn and Lck is an early consequence of crosslinking the TCR and results in the rapid phosphorylation and activation of ZAP-70. Other substrates for Fyn and Lck are ITAMs themselves, thus provide the recruitment of ZAP-70 to the TCR-CD3 signaling complex. In one pathway, ZAP-70 activates PLCγ1 that lead finally to the Ca^{2+} release, activation of calcineurin and translocation of NF-AT to the nucleus. In another pathway, activated ZAP-70 phosphorylates LAT resulting in Ras activation and the initiation of MAPK pathway distal events as well as via the activation of PKC causing the release of NF-κB.)

tion, DP thymocytes differentiate into single positive (SP) cells, either $CD3^+CD4^+CD8^-$ (CD4 SP) or $CD3^+CD4^-CD8^+$ (CD8 SP), and exit the thymus as recent thymic emigrants into the blood and lymphoid tissues. Via their TCRs, $CD4^+$ SP and $CD8^+$ SP T cells recognize peptide bound to MHC class II and I molecules, respectively.[5-7]

The TCR is usually composed of an alpha (α) beta (β) heterodimer (TCRαβ). In a minority of peripheral blood T lymphocytes, the TCR receptor is composed of a gamma (γ) delta (δ) heterodimer (TCRγδ). Either the TCRγδ or TCRαβ heterodimer is coupled to the CD3 molecule, which is composed of epsilon (ε), gamma (γ), delta (δ), and zeta (ζ) chains (CD3γδεζ). The TCR recognizes antigen by binding to peptide bound within the cleft of an MHC molecules on the surface of cells. Following TCR engagement, signal transduction begins with phosphorylation of CD3 chains at their immunoreceptor tyrosine activation motifs (ITAM). Each CD3 γ, δ, and ε chain contains one ITAM, while the ζ chain contains three ITAM (Fig. 1).[8-10]

Using the membrane protein CD45 to differentiate naïve (CD45RA) from memory (CD45RO) T cells, thymic dependent T cell production appears to diminish markedly after puberty presumably due to thymic atrophy.[11-12] If the thymus involutes, new adult T cells would then be derived exclusively from peripheral expansion of existing memory cells. However, cells with a CD45RO phenotype may revert to CD45RA and visa versa.[13] Despite the existence of CD45 phenotype switch between RA and RO isotypes, $CD4^+CD45RA^+$ remains a common surrogate marker for naïve $CD4^+$ cells.[13-15] Memory $CD4^+$ T cells may be denoted by a $CD4^+CD29^+$ rather than $CD4^+CD45RO^+$ phenotype.[16] Since cord blood is enriched for $CD8^+CD11a^{low}$ T cells, this phenotype may be used as a marker for naïve $CD8^+$ T cells.[17-18] Alternatively, newer DNA assays of T cell receptor excision circles are being utilized as an assay for recent thymic emigrants (RTE).

T Cell Receptor Excision Circles (TREC)

There is no cell surface phenotypic marker uniquely specific for recent thymic emigrants. A DNA assay may be used to help separate recent thymic emigrants from the rest of the T cell pool. T cell receptor excision circles (TREC) are extra chromosomal excision DNA segments generated within the thymus during rearrangement of V, D and J genes to generate TCRδ and TCRβ chains or V and J genes to generate TCRγ and TCRα chains. TRECs cannot replicate and are diluted during peripheral T cell expansion. Peripheral blood TREC level is, therefore, dependent on both thymic output and longevity of naïve T cells that in turn is dependent on T cell division and death.[19-20] The order and segments of TCR genes rearranged is tightly regulated. TCRδ chain rearrangement occurs first followed by TCRγ, then TCRβ

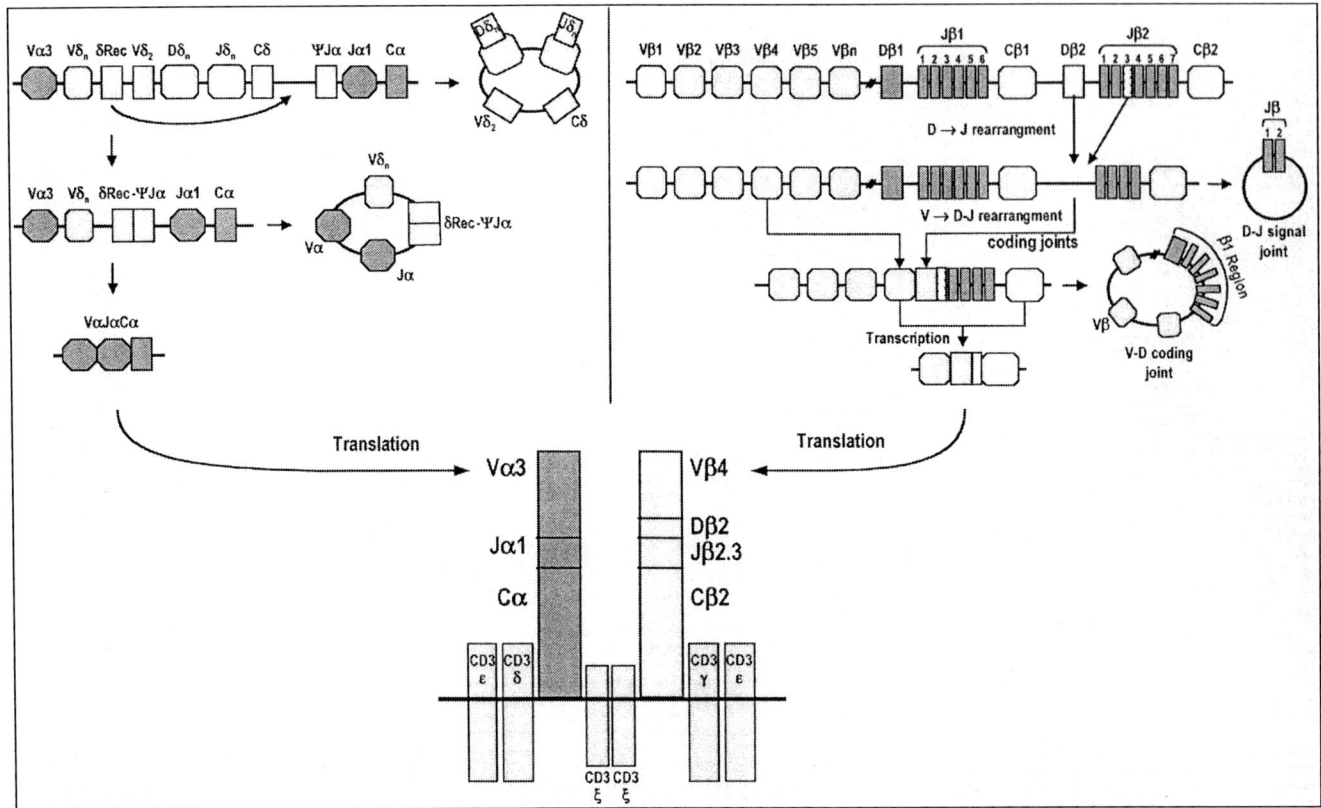

Figure 2. Thymic T-cell receptor rearrangement. Modified from ref.15. V=variable; D= diversity; J= joining; C= constant.

and finally TCRα. The TCRβ chain is rearranged during the stage of DN thymocytes (CD3+CD4-CD8-).[21] TCRα chain rearrangement occurs in double positive thymocytes.[22]

TCR gene rearrangement to generate the αβ TCR creates two episomal circular DNA fragments, a signal joint TREC (sjTREC) and a coding joint TREC (cjTREC) (Fig. 2). The δ locus is located within the α locus and is excised in the process of α gene recombination. The first recombination event to remove Dδ and Jδ segments leads to the sjTREC. The second recombination to unite Vα and Jα genes leads to the cjTREC (Fig. 2). For Vβ gene rearrangement, there is no intervening δ locus to excise. Similar to the α chain, Dβ and Jβ recombination results in a sjTREC, while Vβ recombined with the prior DJ rearrangement results in a cjTREC (Fig. 2). If gene rearrangement occurs within both alleles, then a maximum of two sjTREC and two cjTREC may be present for each α and β chain rearrangement for each αβ TCR positive T cell. Besides the type of signal or coding TREC joint analyzed, PCR methodology and measurements units vary. TREC concentrations have been reported as TREC per million peripheral blood mononuclear cells, TREC per CD45RA+ T cells, TREC per microgram of T cell DNA, and TREC per 100,000 CD4+ T cells. As will be discussed later, using TREC as a surrogate for recent thymic emigrants, thymic T cell maturation appears to continue although at diminished rates with aging.[15,23-25]

Extrathymic T Cells

Limited T cell differentiation may occur in some extra thymic sites. For example, intraepithelial lymphocytes (IEL) may be TCRγδ CD8α/α positive (the normal thymic CD8 molecule is an α/β heterodimer). Many TCRαβ IEL also present the unique CD8α/α heterodimer. IEL are located between gut mucosa epithelium and form a pool of lymphocytes equivalent in size to the peripheral lymph nodes and spleen.[26] TCRγδ IEL T cells appear to develop within the intestinal tract.[27] Epithelial cells (enterocytes) within the gut produce IL-7, which appears necessary and sufficient for γδ T cell development.[28] TCRαβ CD8α/α cells have also been considered to arise in extrathymic sites.[29-33] However, more recent data suggests that TCRαβ CD8α/α cells may arise by down regulation of the CD8β chain during antigen driven terminal differentiation of thymic-derived polyclonal TCRαβ CD8α/β cells resulting in oligoclonal memory effector TCRαβ CD8α/α cells.[34]

IEL are present in athymic nude mice.[35] Thymectomized, irradiated, and marrow reconstituted mice also have IEL suggesting an extra thymic development.[36] In nude mice, T cells with a restricted TCR repertoire may be demonstrated in the peripheral blood. However, in these murine experiments, residual rudimentary thymic tissue contributing to IEL development cannot be completely excluded.[37] A patient with DiGeorge syndrome, a triad of cardiac malformations, hypocalcaemia, and congenital athymia, has been reported to have a small number of peripheral blood CD3+CD4+ T cells also with a restricted TCR repertoire.[38]

The necessity of a thymus for development of T cells from hematopoietic stem cells (HSC) has been questioned by the finding that CD3+ T cells may be generated ex vivo from CD34+ HSC in the absence of a thymus or thymic derived factors.[39] Culturing highly purified CD34+ human HSC with flt-3 ligand, stem cell factor (SCF) and interleukin-2 (IL-2) may cause differentiation into CD3+CD4+ T cells that proliferate to mitogens and have a polyclonal TCR repertoire.[40] It has also been demonstrated in mice that under non-physiologic cytokine stimulation, lymph nodes may generate functional TCRαβ T cells. Oncostatin M is

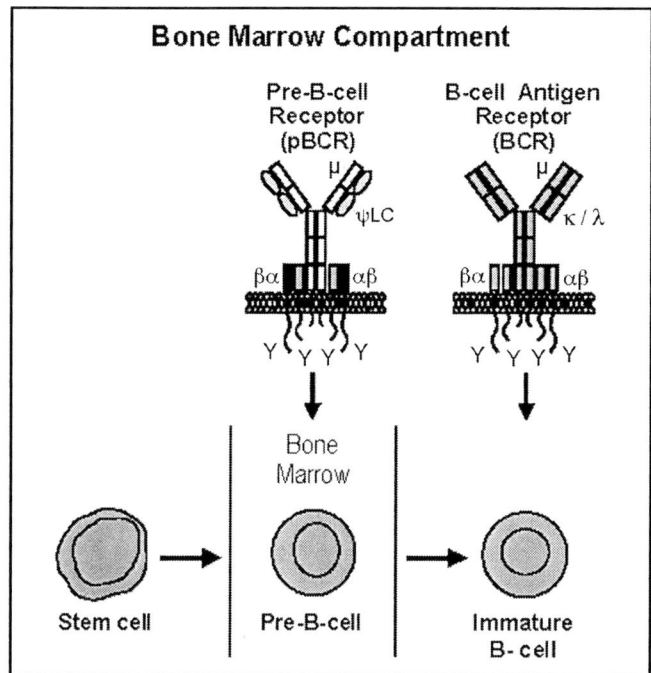

Figure 3. B-cell maturation (part 1). Modified from ref. 67. μ= heavy chain; αβ= signal-transducing Igα/Igβ heterodimer also known as a CD79α/β; κ= kappa light chain; λ= lambda light chain; ΨLC= pseudo light chain; Y= immunoreceptor tyrosine activation motif (ITAM).

an interleukin-6 like cytokine normally produced by hematopoietic cells. When transgenic mice express oncostatin M under a lymphocyte specific kinase (lck) promoter (p56^{lck}), extrathymic T cell development occurs in mesenteric lymph nodes.[41-44] Administration of oncostatin M protein in non-transgenic mice also produces a similar result.[42-43] Whether any human T cells may be generated in vivo but outside of a thymus under normal or cytokine (oncostatin M or IL-7)[45] stimulated conditions and what phenotype, number and functional characteristics these cells will have remains controversial. However, currently, a fully functional and polyclonal TCR repertoire requires a thymus and thymic function declines with age.

Regulatory T Cells

T cells may suppress or down regulate the immune system. An identifiable phenotype for suppressor T cells was often vague and for this reason the terminology "suppressor T cell" fell out of favor only to be resurrected under the terminology "T regulatory cell".[33] As more easily identifiable phenotypic characteristics were discovered, several types of regulatory T cells such as NKT cells and CD4$^+$CD25$^+$ T cells have been described.

NKT cells express both NK markers (CD56, and CD16) and T cell markers (CD3 and an αβ TCR). NKT cells have an unusually restricted TCR being invariably Vα14 positive in mice (Vα24 in humans).[46-47] The human Vα24 and murine Vα14 share 90% sequence homology in the CDR3 TCR antigen binding region.[48] In mice, as well as in humans, these cells may be either single positive (CD4$^+$) or double negative (CD4$^-$CD8$^-$).[49-50] Vα14 NKT produce high titers of IL-4 resulting in a suppressor or regulatory function. NKT cells are present in gut Peyer's patches and in the liver, and athymic nude mice have NKT cells suggesting extrathymic development of NKT cells.[51-53] In fact, NKT cells may be an evolutionary bridge between extrathymic and thymic lymphopoeisis.

CD4$^+$CD25$^+$ T cells contain the high affinity interleukin-2 receptor alpha (IL-2Rα) also known as CD25, a T cell activation marker.[54-55] From experiments with thymectomized mice, CD4$^+$CD25$^+$ T cells are thymic-derived.[56] Once activated via their TCR, CD4$^+$CD25$^+$ cells exert an antigen independent inhibition of IL-2 production and promote cell cycle arrest.[57] CD4$^+$CD25$^+$ T cells have been reported to inhibit by means of IL-10 and/or TGF-β.[58-59] However, this remains controversial and while the mechanism CD4$^+$CD25$^+$ T cell suppression is unclear, the effect requires cell-cell contact.[60-61] CD4$^+$CD25$^+$ T cells have been shown to be important in preventing several animal autoimmune disorders.[62-64]

Generation of B Cells

Immunoglobulins, similar to TCR genes, are generated by somatic recombination of V(D)J genes.[65] Similar to development of the thymus, immunogobulins are found only in vertebrates and first arise in the evolutionarily more primitive jawless vertebrates (hagfish and lamprey).[3] Immunoglobulin class switching (e.g., IgM to IgG) appears later in amphibians, and somatic hypermutation occurs even later with development of germinal centers that first arise in warm-blooded vertebrates (birds and mammals). For all vertebrates, TCRab repertoire diversity occurs in the thymus. In contrast, B cell receptor (BCR) diversity occurs in different lymphoid tissues for different vertebrate species. For birds, immunoglobulin diversification occurs in the bursa of Fabricius, a cloacal gut associated lymphoid tissue.[3] For humans, B cell development begins in the bone marrow and is completed in the germinal center of lymph nodes (Figs. 3-4). Within the bone marrow, stem cell differentiation into B lymphocyte lineage has two major developmental checkpoints, a pre-B cell that expresses the pre-B cell receptor (pBCR) and an immature B cell that expresses the mature BCR. At both B cell selection checkpoints, signaling through either the pBCR or BCR causes either apoptosis or proliferation.[66]

The pBCR and BCR consist of membrane immunoglobulin (mIg) and a heterodimer of Igα/Igβ (CD79 α/β) subunits (Fig. 3). The pBCR mIg is a heavy chain combined with a pseudo light chain (ψLC). The BCR mIg is composed of a heavy chain combined with either a lambda (λ) or kappa (κ) light chain. The pBCR and BCR are functionally divided into the ligand-binding mIg and the signal-transducing IgαIgβ heterodimer (Fig. 3). Similar to the T cell γδεζ chains of the CD3 molecule, the cytoplasmic domains of CD79α and CD79β, contain immune receptor tyrosine-based activation motifs (ITAMs) which, following tyrosine kinase phosphorylation, transduces the stimulus produced by crosslinking of mIg molecules into an intracellular signal (Fig. 3).[67-70]

B cells undergo further affinity maturation within lymph node germinal centers by a process of somatic hypermutation (SHM), gene conversion, and class switching recombination (CSR) (Fig. 4). SHM is the term for insertion of point mutations in the vicinity of the variable region exon (Fig. 4) and results in generation of antigen specific high affinity antibodies. Gene conversion is the transfer of a pseudovariable (ψV)gene sequence into the variable region exon (Fig. 4). Both SHM and gene conversion alters the antigen binding site of the immunoglobulin.[71-72] CSR involves switching the constant region heavy change (e.g., IgM to IgG$_1$) that alters the effector function of the antibody (Fig. 4). The mechanisms involved in DNA SHM, gene conversion, and

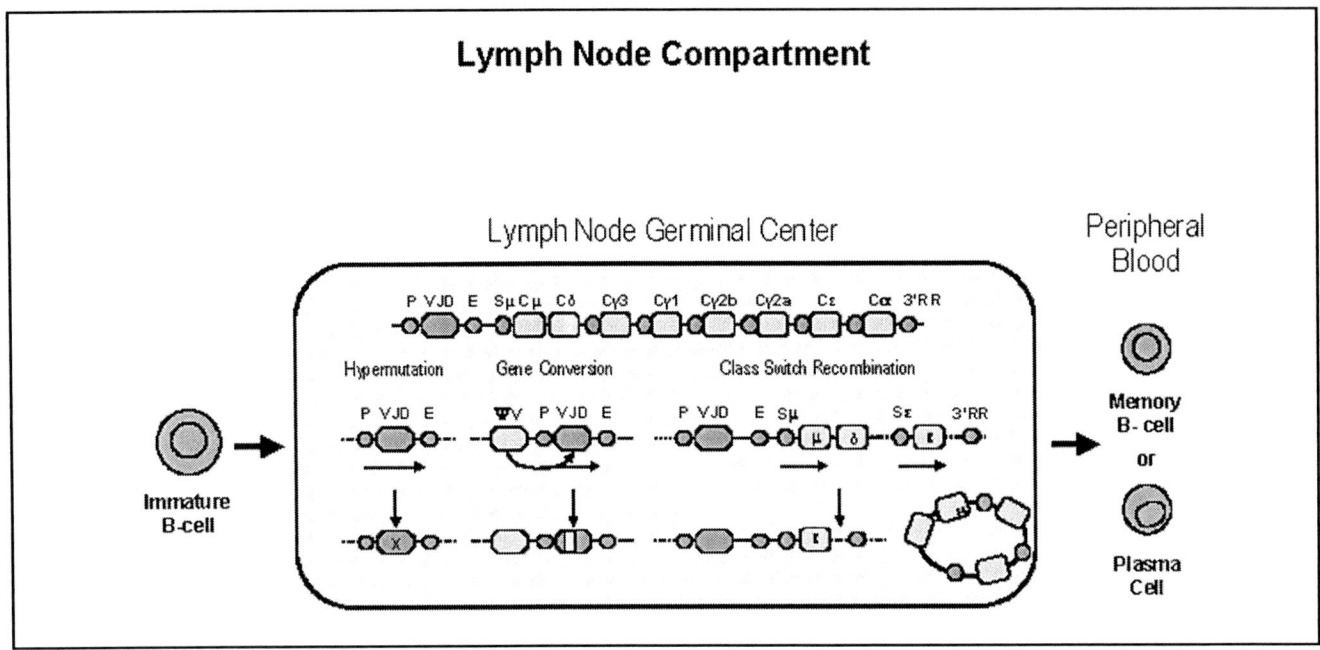

Figure 4. B-cell maturation (part 2). Modified from ref. 74. ΨV= pseudogene; VJD= variable; and C= constant region exons; P= promoter; E= enhancer; S= switch region; 3'RR= 3' regulatory regions.

CSR although incompletely understood probably involve common mechanisms of DNA recognition, targeting, cleavage, and repair.[73] The enzyme activation-induced cytidine deaminase (AID) is involved in all three reactions by helping to create the DNA cut or cleavage.[65,74-75]

SHM, gene conversion, and CSR take place in germinal center B cells in response to BCR antigen stimulation. This results in high affinity antibodies with specialized effector functions. The reactions are dependent on T cell and germinal center follicular (epithelial) cell help.[71] T cells communicate with B cells via a number of membrane bound receptor ligand complexes such as B cell CD40 and T cell CD40 ligand (CD40L). Interaction of CD40 with CD40L promotes B cell proliferation and CSR.[76] If the CD40-CD40L interaction is blocked, B cells cannot proliferate and produce immunoglobulin.[77] Interestingly, most lymphomas are B cells arising in the lymph node germinal center and involve oncogene translocations possibly due to errors in targeting of the SHM and CSR enzymes.[78]

Immune System Aging

Immune changes influenced by age-related diseases such as cancer, dementia, or autoimmune disorders should be differentiated from the immune system of healthy aging individuals. In fact, lymphocyte alterations have been associated with infections, cancer and autoimmune disorders. Data on lymphocyte subsets may be presented in the literature as percentages that may remain normal, while absolute numbers have declined or are altered. Therefore, if percentages are given, it is important to also evaluate absolute numbers.

NK cells are elevated at birth and then normalize within weeks postpartum. Absolute numbers of NK cells remain stable until old age when their number increases (Fig. 5). In the first year of life, total T and B cell number as well as $CD4^+$ $CD45RA^+$ naïve T cells increase to provide a large pool of cells ready to recognize and respond to antigens.[12,79-81] The first year of life is, therefore, marked by rapid development of a thymic-derived immune system. With onset of puberty, the size and architecture of the thymus involutes and thymic lymphopoiesis declines. Age-related reduction of thymic T cell development appears to be the key event associated with subsequent age-related change that has been correctly or incorrectly termed "immunosenescence".[82-85]

Absolute numbers of T cells and $CD4^+$ and $CD8^+$ subsets decline with age (Fig. 5).[80,86] Numbers of phenotypically naïve $CD4^+CD45RA^+$ T cells decline and memory $CD4^+CD29^+$ T cells increase especially after 40 years of age (Fig. 5).[87-88] Due to an approximately equal decrease in $CD4^+$ than $CD8^+$ cells, the CD4/CD8 is unchanged or increases with aging (Fig. 5).[16,83,89] Both sjTREC and cjTREC in $CD4^+$ and $CD8^+$ cells demonstrate an age-related decline (Fig. 6). Age-related loss of $CD4^+CD45RA^+$ cell number and TREC concentration are consistent with an age-related decline in recent thymic emigrants. The diversity of T cell antigen recognition as measured by the heterogeneity within TCR Vβ families appears intact until advanced adulthood when clonal expansion of $CD8^+$ memory cells occurs with aging.[90-91] Activated peripheral T cells presenting HLA-DR also increase with aging.[92] In old age, declining thymic output combined with antigen driven activation and extrathymic expansion may cause emergence of $CD8^+$ clonal expansion resulting in TCR Vβ restriction. TCR skewing may be viewed as the equivalent of a B cell benign monoclonal gammopathy, an occurrence that is also age-related.[93-94]

Both ex vivo and in vivo T cell function declines with age.[85,86,95-97] Ex vivo proliferation to mitogens such as phytohemagglutinin (PHA) or to TCR engagement by anti-CD3 antibody, T cell IL-2 production, and T cell IL-2 high affinity receptors decline with age (Fig. 6).[98-99] In vivo delayed type hypersensitivity (DTH) responses to injected antigens also diminish with age.[100] Decline of T cell function is probably related to loss of naïve cells (naïve T cells preferentially produce IL-2), increase in antigen experienced memory T cells (memory cells preferentially produce interferon-γ, IL-4 and/or IL-6), and possi-

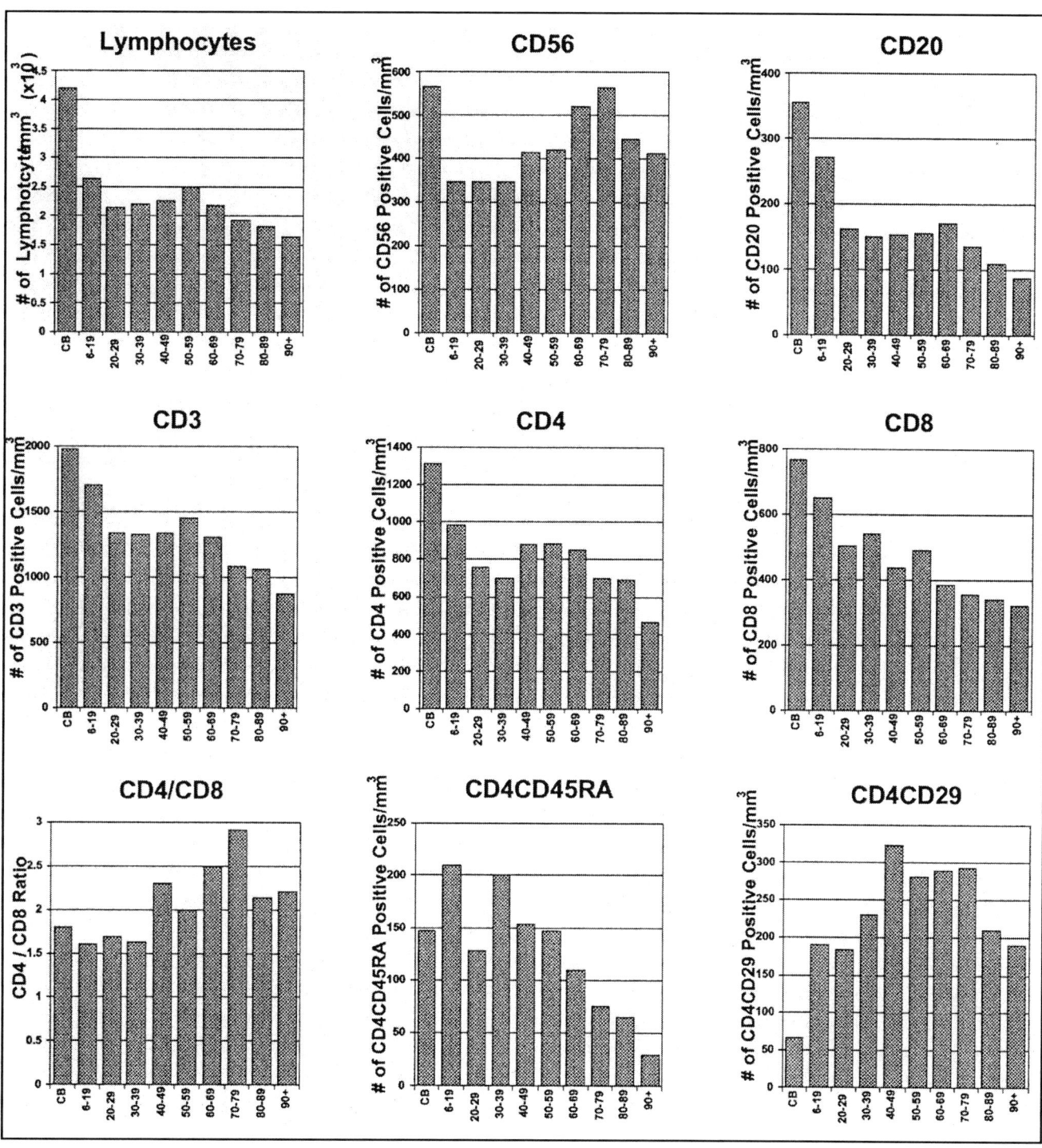

Figure 5. Age-related changes in lymphocyte subpopulations. Modified from ref.16. CD3= pan T-cell marker; CD56= natural killer (NK) cell marker; CD20= B-cell marker; CD4= T helper (Th) cell marker; CD8= T cytotoxic (Tc) marker; CD4CD45RA= naïve CD4 T cell marker; CD4/CD8 ratio; CD4CD29= memory CD4 T cell marker; CB= cord blood.

bly alterations in signal transduction pathways between T cells from younger versus older individuals.[101] While IL-2 production declines with age (Fig. 6),[102] consistent results have not always been reported for other cytokines. In general, IL-6, INF-γ, IL-1 and TNF-α production has been reported to increase with age consistent with a shift towards differentiated memory effector cells (Fig. 6).[99,103]

Numerous membrane protein kinases are crucial for TCR signaling. After T cell activation, tyrosine kinases such as lck (lymphocyte specific tyrosine kinase), fyn (a novel kinase with homology to viral oncogenes V-Fgr and V-yes, and is thus termed fyn for Fgr yes novel protein), and Zap-70 (zeta chain associated protein-70) phosphorylate the immunoreceptor tyrosine based activation motifs (ITAM) on the γ, δ, ε, ζ CD3 chains initiating signal transduction (Fig. 1).[8-9,104] Aged T cells have been reported to have altered distribution of kinases such as pp56^{lck} and protein kinase C (PKC) compared to T cells from young individuals.[105] Whether these changes are due to differences in protein kinases

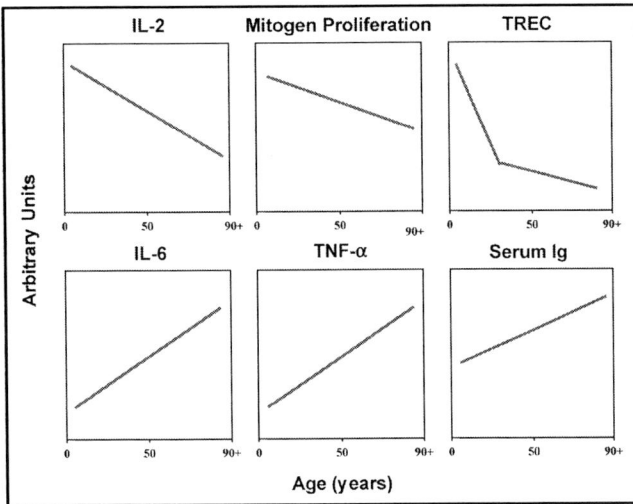

Figure 6. Age-related changes. Modified from ref. 92. IL-2= interleukin-2; TREC= T cell receptor excision circle; IL-6= interleukin-6; TNFα= tumor necrosis factor α; Ig= immunoglobulin.

between memory cells and naïve T cells, or are related to T cell longevity independent of phenotype, remains unclear.[9,106-108]

As one ages, B cell number and in vivo antibody, especially high affinity antibody, response to immunization decreases,[86-87] but autoantibodies and serum IgG and IgA increase.[109] Marrow cellularity and marrow B cell production decreases with age. It is possible that marrow B cell production is in part a feedback down regulation to prevent age-related accumulation of long-lived B cells. However, B cell number is not maintained but rather diminishes, suggesting gradual marrow exhaustion in B cell production.[110-111] In vivo transplantation studies in *scid* mice,[112-114] and ex vivo mixing of aged and young T and B cells, demonstrated that defects in B cell function from aged individuals could usually be corrected with T cells from young individuals.[85,115-117] Decreased high affinity antibody production appears related to inability of aged T cells to provide signal for SHM in lymph node germinal centers.[93] Increased memory T cells that produce IL-4 and IL-6 could explain the increase in IgG and IgA despite lower numbers of B cells. Chronic antigenic stimulation of terminal B cells results in age-related immunoglobulin repertoire skewing that may contribute to oligoclonality.[118-119]

T cell changes due to diminished thymic function play a dominant role in age-related T and B cell changes. However, many areas merit further investigation, including differential gene expression, signal transduction and co-stimulatory molecule alterations and changes in suppressor or regulatory T cells and anti-idiotypic antibody networks between the lymphocytes of young and old individuals.

Immune Reconstitution After HSCT

Immune reconstitution in adults occurs after retroviral therapy for HIV infection, and following high dose chemotherapy or chemo-radiotherapy with or without autologous or allogeneic HSCT. Although the trends are similar, this chapter will not discuss immune reconstitution in patients with HIV infection. There are multiple factors that could theoretically influence immune reconstitution after HSCT including type of graft (cord blood, bone marrow, peripheral blood, T cell depleted, unmanipulated), disease, conditioning regimen, graft versus host disease, recipient or donor age, cytokines, post transplant immune suppressive or immune modulating medications, and infections. Of these conditions, the factors that influence development of healthy immune reconstitution are factors that affect thymic function. For autologous HSCT, the dominant factor is patient age. For allogeneic HSCT, the predominant factors are recipient age, GVHD, and number of T cells infused in the graft.

Autologous HSCT

Post transplant recovery is the inverse of aging with initial expansion of memory T cells and delayed recovery from naïve T cells. Immune reconstitution after autologous HSCT is not affected by graft versus host disease or immune suppressive medications and will be reviewed first. NK cells which regenerate independent of a thymus recover normal number and function within one month of autologous bone marrow transplantation (BMT) or peripheral blood stem cell transplants (PBSCT).[120-126] After autologous BMT, CD3+ T cell number normalize within 3 months.[127] The number of CD4+ cells is decreased for 12 or more months.[128,129] CD4+CD45RA+ naïve T cells may take 1-2 years to reach normal numbers. By 3 months, CD8+ cell numbers usually return to normal, leading to an inverted CD4/CD8 ratio for 12 or more months.[128,129] The CD8+ cells tend to be CD28– consistent with expansion of memory T cells.[124,127,130-132]

Autologous PBSC grafts contain 1-log greater T cells than BM grafts and may be expected to hasten lymphopoiesis.[133] However, CD4+ T cell recovery and T cell proliferative responses to mitogens are similar between autologous PBSCT and BMT as well as between unmanipulated PBSCT and CD34+ selected (i.e., T cell depleted) PBSCT.[124,131,136] There is, however, a statistically significant correlation between CD4+ cell count reconstitution and age at time of transplant. The younger the patient, the more rapid the recovery of the CD4+ T cell count.[124,130,134-136]

B cell number is normal by 3 months and IgM returns to normal in 6 months.[124,128,137-140] T cell proliferative responses to mitogens may be reduced for 12 or more months,[139,141-142] indicating lack of effective T cell help in germinal center class switch rearrangements that results in subnormal IgG and IgA for 12 and 24 months, respectively.[131,138-139] The patient's disease and type of conditioning regimen, particularly TBI versus non-TBI, has not been reported to influence immune reconstitution. Radiation could cause more damage to thymic epithelium than chemotherapy, but little is available in the literature for comparison.

Allogeneic HSCT

Allogeneic grafts may be unmanipulated, T cell depleted, or from cord blood, HLA matched sibling, a one, two, or three (haploidentical) antigen mismatched related donor or an unrelated donor. The post transplant course is often complicated by immunosuppressive medications and graft versus host disease (GVHD). TREC assays have confirmed T cell regeneration after allogeneic HSCT. Both age and extensive chronic GHVD are associated with thymic involution and are strongly correlated with lower TREC recovery.[143-147] TREC recovery is more rapid and higher for younger patients (age < 19 years old) (Fig. 7) and correlates with CD4+CD45RA+ T cell recovery.[143] One manifestation of extensive chronic GVHD is lymphoid and thymic atrophy and low TREC values are associated with extensive chronic GVHD (Fig. 7).[144,148-149] TREC values, as well as naïve CD4+CD45RA+ T cell recovery, may also be delayed early after HSCT with TCD allografts compared to unmanipulated allografts, although differences diminish by 9 months after HSCT.[144,150] When corrected for GVHD and age, TREC recov-

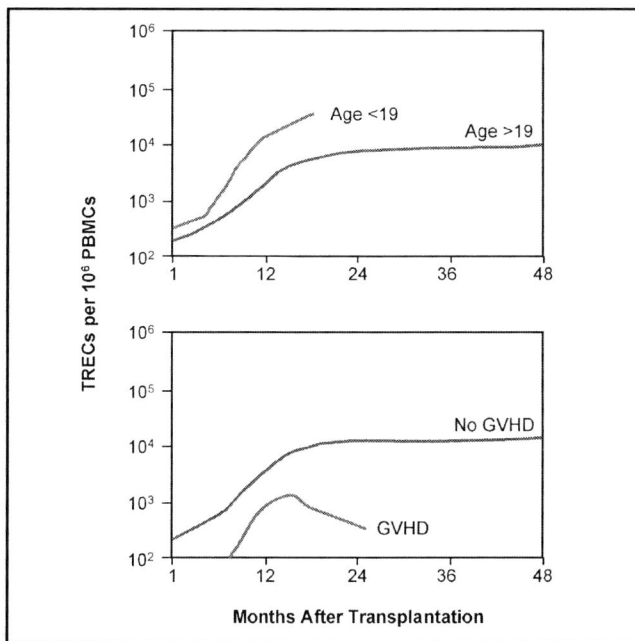

Figure 7. Influence of age and GVHD on TREC reconstitution after allogeneic stem cell transplantation. Adapted from ref. 144. TRECs= T cell receptor excision circles; PBMCs= peripheral blood mononuclear cells; GVHD= graft-versus-host disease.

ery is not affected by whether the donor is HLA identical or HLA mismatched.[123,143,148] Low TREC values correlate with both a decrease in TCR diversity and increase in opportunistic infections.[144] Absent, monoclonal, or oligoclonal profiles for the majority of Vβ TCR repertoires persists up to 6 months after allogeneic BMT, when normalization of TCR complexity starts.[151]

As mentioned above for autologous transplants, immune reconstitution occurs at a similar pace for both BMT and PBSCT. In contrast for allogeneic HSCT, immune reconstitution for adults is faster with PBSC compared to bone marrow (Fig. 8).[152,153] Peripheral blood provides a 10 times greater inoculum of lymphocytes with the graft compared to a marrow graft.[152] It may be that during the first year after an allogeneic HSCT, most immune reconstitution arises from the T cells infused with the graft. De novo generation of new T cells from the stem cell compartment is hindered by thymic atrophy from clinical or subclinical GVHD and chronic immune suppressive therapy (i.e., cyclosporin).[154-157] Total CD4$^+$ number and naïve CD4$^+$-CD45RA$^+$ T cells recover more rapidly after PSCT compared to BMT (Fig. 8). Similarly, when compared to marrow, total CD8$^+$ and naïve CD8$^+$CD11a^{low} cells recover more rapidly with PBSC (Fig. 8).[152,153] Independent of graft (marrow or blood), total CD4$^+$ and CD4$^+$CD45RA$^+$ numbers remain depressed at 12 months after transplantation (Fig. 8).[120,121,152,153,158,159] Similar to an autologous HSCT, CD8$^+$ cell numbers, most of which are infused memory cells, reach normal levels after day 100 (Fig. 8).[121] Thus, the thymic pathway may appear more important for CD4$^+$ T cell than CD8$^+$ T cell regeneration. However, most of the CD8$^+$ T cells are expanded graft derived memory cells since naïve CD8$^+$CD11a^{low} cells that depend on thymic reconstitution remain low for more than 12 months.[153] Monocytes and NK cells arise extrathymically and are within normal range by 30 days independent of marrow or blood stem cell transplantation (Fig. 8).[120,122,123,145,152-153] Total B cell numbers as well as naïve B cells (IgD$^+$) normalize before 1 year with no difference between mar-

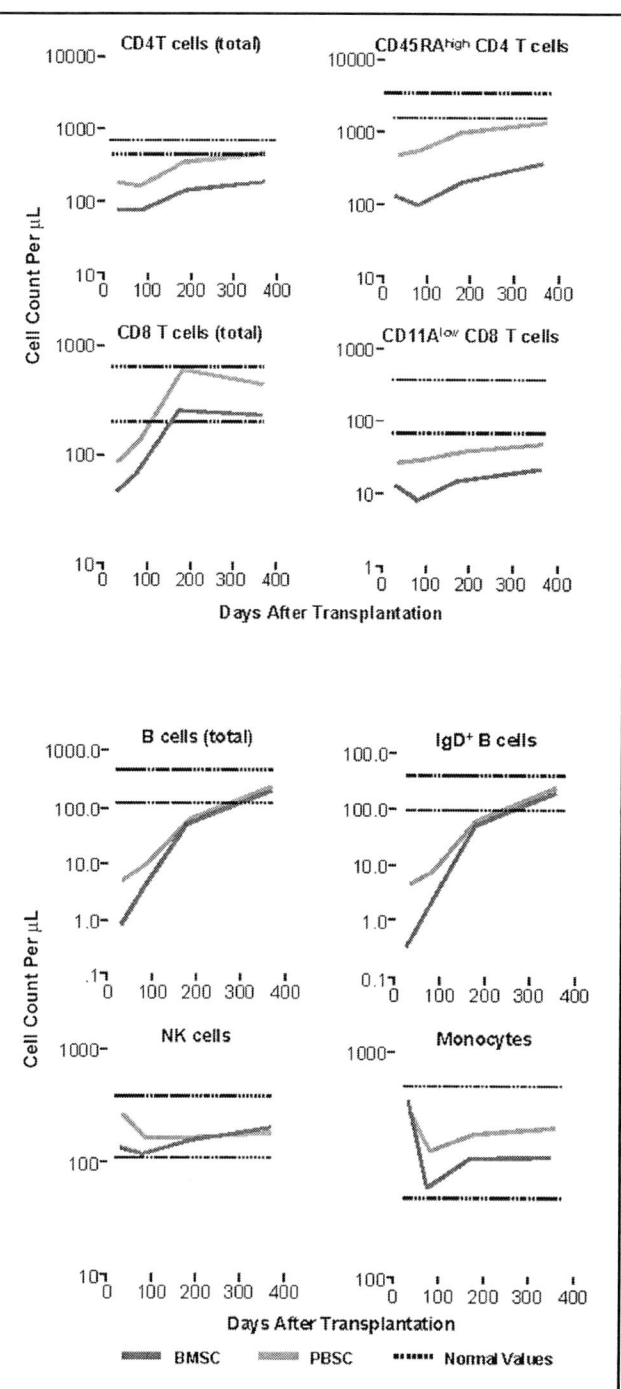

Figure 8. Schematic diagrams of immune reconstitution after allogeneic bone marrow or allogeneic peripheral blood stem cell transplantation. Modified from ref. 153. CD4 T cells= T helper (Th) cells; CD45^{high}CD4 T cells= naïve T helper cells; IgD$^+$ B cells= naïve B cells; CD8 T cells= T cytotoxic (Tc) cells; CD11^{low}CD8 cells= naïve T cytotoxic (Tc) cells; NK cells= natural killer cells.

row and peripheral blood stem cell B cell number after day 100 (Fig. 8).[122,153,160]

Lymphocyte dose in the graft has also been reported to be important for immune reconstitution when T cell depleted grafts are compared to unmanipulated donor grafts.[122,150,161] T-cell spectratypes analysis showed delayed T-cell repertoire normalization after T-cell depleted allogeneic BMT compared to unmanipulated BMT.[162-163] For pediatric and adult patients receiving unrelated donor transplants, T cell depleted marrow grafts manifest slower recovery of CD3[+], CD4[+], and CD4[+]CD45RA[+] T cells compared to unmanipulated grafts from an unrelated donor.[123,164] Immune reconstitution has been reported to be similarly slower in unrelated or mismatched related HSCT that received higher doses of lymphocyte depleting anti-thymocyte globulin compared to those receiving lower doses.[123,165] It has been observed in one study that immune recovery seemed to be more effective in highly CD34[+] purified haploidentical allografts compared with CD34[+] selected matched unrelated grafts.[166] However, while age, chronic GVHD, and the number of lymphocytes in the allograft affect immune reconstitution, in general, HLA-match/mismatch related or unrelated grafts appear to have similar recovery of CD3[+], CD4[+], and CD8[+] T cells.[167-168]

Due to limited number of cells, cord blood is generally restricted to children or young adults and is generally thought to have a lower risk of GVHD, both of which may be favorable for immune reconstitution. In general, immune reconstitution after unrelated cord blood stem cell transplantation has been reported to be similar to other allogeneic stem cell sources.[149,158] Natural killer, CD19[+] and CD4[+] cell numbers are normal by 2 months, 6 months, and 12 months, respectively. Mitogen responses are normal by 6 to 9 months. The exception is that CD8[+] T cells recovery may be more delayed than other stem cell sources with normal counts delayed until 9 months. Consequently, the CD4/CD8 ratio may be normal as early as 6 months after cord blood transplantation. Delayed CD8[+] cells recovery may be secondary to fewer CD8[+] antigen stimulated memory T cells in cord blood.[149,160,169]

In general, the disease for which a stem cell recipient undergoes HSCT does not affect immune reconstitution. The exception is severe combined immune deficiency (SCID). SCID is marked by an absence of T cells. B cells may (B+ SCID) or may not (B- SCID) be present. Patients with B+ SCID have faster T and B cell reconstitution than B- SCID following HSCT. B+ SCID may engraft without a conditioning regimen. NK cells are absent in B+ SCID but are present in B- SCID. The presence of host NK cells increases the risk of graft rejection unless an immune suppressive conditioning regimen is used. Therefore, patients with B- SCID often receive an immune suppressive conditioning regimen prior to stem cell infusion. Conditioning regimens may damage the thymus and impair T cell regeneration in B- SCID.[146,148,170,171] Evaluating immune responses for the first 3 months after HSCT, a study compared mini-conditioning regimens, also known as non-myeloablative transplant (NST) regimens, to intense myeloablative conditioning regimens. The myeloablative regimen markedly diminished T cell mitogen responses. In contrast, early after a NST, the mitogen response was maintained.[172] The type (e.g., total body irradiation) or intensity of the conditioning regimens effect upon immune reconstitution has generally not been investigated or reported, perhaps because the intensity of the different regimens have been basically similar. Minor differences in conditioning regimen-related thymic injury may be over shadowed by more important factors such as recipient age and presence or absence of GVHD.[162,172-176]

While CD19[+] or CD20[+] B cell number generally normalizes within the first few months after HSCT, B cell function appears to be largely dependent on recovery of T cell helper function. B cell immunoglobulin production is diminished in patients with chronic GVHD compared to those without GVHD.[121,137,138,177] Antigen specific titers after immunization are diminished or absent if GVHD is present. For this reason, after an allogeneic HSCT, re-immunization to childhood vaccines is delayed until the patient is off immune suppression without GVHD.[130,178]

Cytokines (predominately G-CSF) are used to mobilize peripheral blood stem cells and to shorten the duration of post conditioning neutropenia. G-CSF promotes IL-4 and IL-10 Th2 cytokine production.[179] In addition, G-CSF preferentially mobilizes lymphoid dendritic (DC2) cells.[180,181] G-CSF mobilized PBSC contain higher doses of DC2 compared to myeloid dendritic (DC1) cells. DC1 cells produce IL-12 and promote Th1 T cell differentiation. DC2 cells promote Th2 T cell differentiation.[182,183] It has been suggested that adoptive transfer of DC2 cells in the PBSC graft diminishes the risk of acute GVHD.[184,185] However, it is possible that DC2 cells may also increase the risk of chronic GVHD that may offset any potential immune reconstitution benefit. The role of cytokines and/or adoptive cell transfer in post transplant immune reconstitution has yet to be fully appreciated. For example, interleukin-7 (IL-7) and oncostatin M are thymic cytokines that may be beneficial in post transplant immune reconstitution.[41-42] In murine models, post transplant IL-7 accelerated memory CD4[+] and CD8[+] T cell regeneration without worsening GVHD. However, IL-7 had little effect on naïve CD4[+] or CD8[+] T cells.[186,187] As mentioned earlier, Oncostatin M appears to induce extrathymic T cell lymphogenesis in animal models and may be beneficial for naïve CD4[+] and CD8[+] recovery following HSCT.

Autoimmune Diseases

In the past 10 years, increasing evidence has been provided that HSCT can improve or even cure refractory autoimmune diseases and induce long-lasting complete clinical remission.[189-203] By far, most HSCT for autoimmune diseases have been with CD34[+] selected autologous grafts, although selected and unselected allogeneic HSCT have recently been performed (refer to Chapter 57) for autoimmune diseases. It has been suggested that long-term remissions depend on both complete in vivo immune ablation and efficient ex vivo stem cell purification.[190,192,200,201] Following immune ablation, the immune reconstitution would then occur in a unique "immune vacuum" starting almost completely from "zero" or "rebirth". Others have suggested that the immune system is dynamic and normally fluctuates between tolerance and immunity, and that intense immune suppression without immune ablation may be sufficient to "reset" the immune system resulting in durable remissions (refer to Chapter 31).[204] The first approach results in more intense conditioning regimen with greater risk of lethal infections and regimen related organ dysfunction. The latter approach results in more mild and better tolerated conditioning regimens. Which philosophy is correct is unknown and may differ depending on the type of autoimmune disease. In clinical practice, the conditioning regimen, whether myeloablative or immune suppressive but non-myeloablative, may induce an incomplete elimination of circulating immune cells from the periphery. In particular, it has been emphasized that certain subsets of antigen-experienced, i.e., memory/effector lymphocytes, are more refractory to the conditioning agents, while antigen-unexperienced, naïve lymphocytes are highly susceptible.

Overall, data about the immune reconstitution in autoimmune disease patients receiving HSCT can only be compared amongst a few studies published so far. In particular, small clinical trials initiated in Chicago, Leiden and Berlin have been designed based on quite similar combinations of autologous HSCT, a non-

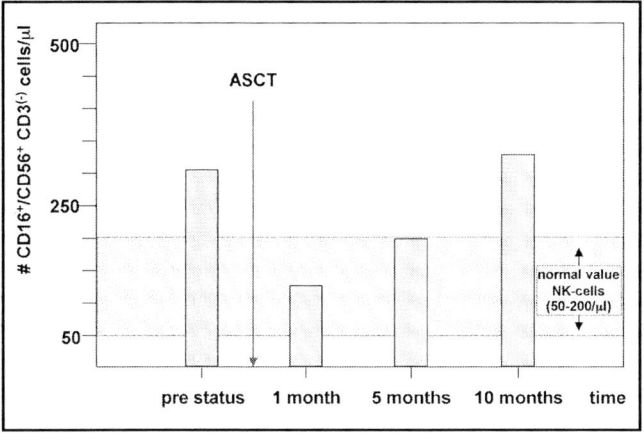

Figure 9. NK-cell reconstitution in a SLE-patient after autologous HSCT/absolute counts during therapy (HSCT was performed according to the Berlin-Study-Protocol (conditioning: cyclophosphamide + ATG, transplantation of > 2×10^6 highly purified, autologous CD34$^+$ cells/kg).

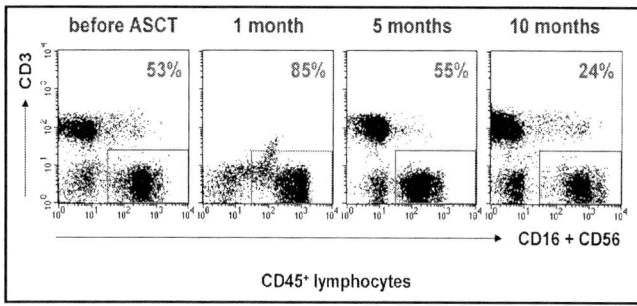

Figure 10. NK-cell reconstitution in a SLE-patient after HSCT/ cytometric evaluation of CD16/CD56$^+$ NK-cell frequencies during therapy.

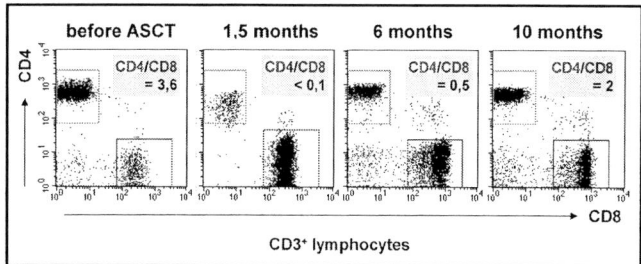

Figure 11. Alteration of CD4/CD8 ratio in a SLE-patient after HSCT.

myeloablative cyclophosphamide based regimen, and ex vivo depletion of autoreactive lymphocytes by CD34$^+$ selection.[192,200,201] The results of these studies are the basis of our current knowledge on reconstitution of the immune system after autologous HSCT for autoimmune disease. The results are similar to immune reconstitution after HSCT for malignancies. One important difference is that autoimmune diseases are often associated with pre transplant abnormalities such as Th1/Th2 or other cytokine deviations, TCR skewing, and/or increased expression of lymphocyte activation markers. In the limited studies performed to date, these abnormalities appear to normalize or improve after HSCT along with clinical resolution of disease activity. HSCT also offers an opportunity to follow precursor frequency of autoreactive clones and disease manifestations. If the autoantigen is unknown, this approach would allow determination of initiating and spread epitopes for human autoimmune diseases. To date, no data has been investigated on change in autoreactive precursor frequencies before and after HSCT. Both the Berlin and Chicago groups have limited current publications addressing the immune changes after HSCT for systemic lupus erythematosus (SLE). These results will be discussed below.

NK cells which arise outside the thymus have no known etiologic role in autoimmune diseases such as SLE and, similar to HSCT for malignancies, rapidly recover after CD34$^+$ selected autologous HSCT for SLE. Figure 9 shows a typical example of NK-cell regeneration in the course of autologous HSCT in a patient with refractory SLE. Frequency analysis of CD56$^+$/CD16$^+$ NK-cells among CD45$^+$ lymphocytes often suggests a dramatic expansion of NK-cell percentage within one month after HSCT as shown in Figure 10. Although NK cells can represent more than 80% of CD45$^+$ lymphocytes at 1 month after HSCT (Fig. 9), the absolute counts of the reappearing CD56$^+$/CD16$^+$ NK cells are normal one month after HSCT (Fig. 10). This emphasizes the importance of reporting not only percentage but also actual number of cells during immune reconstitution. Our results that NK-cells are among the first immune cells to recover after HSCT for SLE are in line with other types of non-autoimmune transplants.[176,205-208]

Similar to malignancies, CD8$^+$ T cells reappear very fast after transplantation.[136,176,201,205-214] Since, like malignancies, restoration of the peripheral CD4$^+$ T cell pool occurs more slowly, this results in an initial inversion of the normal CD4/CD8 ratio. Figure 11 shows a typical example from a SLE patient undergoing HSCT. CD8$^+$ T cells detectable early after HSCT consist exclusively of memory/effector type subset. Figure 12 shows an example of a typical phenotype of CD8$^+$ T cells from healthy donors and SLE patients 3, 6 and 36 months after HSCT, respectively. Whereas comparable rates of naïve, memory and effector CD8$^+$ T cell subsets are detected in the healthy donor and the SLE patient 3 years after HSCT, only very low frequencies of peripheral CD45RA$^+$ CD27^{++} naïve CD8$^+$ T cells can be identified at early occasions after HSCT. Since the autograft was extensively T cell depleted, most likely the first "wave" of memory and effector CD8$^+$ T cells after HSCT reflects memory effector cells that survived the conditioning regimen. Like an aged immune system, a high percentage among these cells expresses HLA-DR characterizing these cells as recently in vivo activated.[176]

CD4$^+$ T cells are thought to be key regulators of autoimmune reactions. They induce autoantibody production, are present in high frequencies among cellular infiltrates in organ-specific autoimmunity (e.g., rheumatoid arthritis, multiple sclerosis, systemic scleroderma), and adoptively transfer disease in animal models such as experimental autoimmune encephalomyelitis, an animal model of multiple sclerosis. CD4$^+$ T cells are one of the main sources of proinflammatory effector cytokines (e.g., IFNγ, TNFα) that are involved in sustaining chronic inflammation. Therefore, it is one of the key issues in HSCT-based treatments for autoimmune diseases to eliminate autoreactive CD4$^+$ T cells both in vivo by immunoablation and ex vivo by efficient purging strategies. For the patients, this leads to a dilemma, since reconstitution of the peripheral CD4$^+$ T cell pool, the central immune cells also in protective immunity, is relatively slow as compared to other lymphocyte subsets. After HSCT, CD8$^+$ memory/effector subsets of T cells can recover fast within the first 6 months.

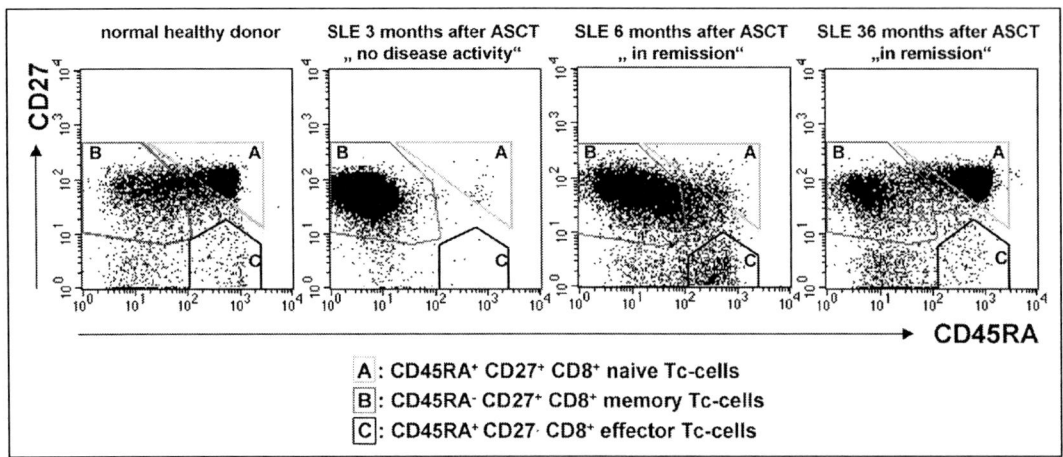

Figure 12. Cytometric analysis of different CD8 subsets in healthy donors and during HSCT treatment.

Interestingly, memory/effector CD4+ Th-cells reappear much slower as compared to memory/effector CD8+ Tc-cells. This could be because early after HSCT, MHC-II expressing antigen-presenting cells are more limited than MHC-I expressing cells. Long-term broad immune responses, however, depend on a polyclonal naïve T-cell repertoire. Reconstitution of the naïve T cell pool after HSCT is thymus dependent and usually takes 1 year before absolute counts of naïve CD45RA+ CD4+ T cells reach lower levels of normal values.[135,200-202,209,215] For malignancies, re-establishment and maintenance of polyclonal TCR repertoires is important for immune surveillance and patient survival. Recently, this has also been proposed for patients receiving autologous stem cell grafts for autoimmune diseases.[202-203] Figure 13 shows the reconstitution of CD45RO-expressing memory/effector T-cells and CD45RA-expressing naïve T cells in a SLE patient during the course of HSCT. Reconstitution with naïve CD45RA-expressing T cells is slow within the first year after HSCT. However, frequencies of naïve T cells continue to increase and stabilize thereafter, comprising usually more than 60% of the peripheral T cell pool 3 years after HSCT.

As previously discussed, there is no phenotypic marker unique for recent thymic emigrants. The group in Berlin has proposed that the subset of human peripheral CD45RA+ CD4+ T cells co-expressing CD31 contains recent thymic emigrants. The counterpart population lacking CD31 expression may represent T cells which have extensively proliferated during ageing.[47] The majority of the CD45RA+ CD4+ T cells recurring after HSCT in SLE patients co-expressed CD31 and thus resembled naïve CD4+ T cells (Fig. 14). As shown in Figure 14b, even 22 months after HSCT, high numbers of TREC were detected in purified CD31+ thymic naïve CD4+ T-cells. In the future, most likely a combination of phenotypic quantification of naïve CD4+ T cells and molecular analysis of TREC levels will be necessary for detailed monitoring of thymic activity after HSCT patients. Especially regarding HSCT for autoimmune diseases, this might be of prognostic value, since a successful "thymic take" followed by a stable "rebuilding" of a polyclonal diverse naïve CD4+ T cell repertoire would stabilize a new tolerant immune system and therefore could favor long term remissions. Although speculative, thymic dependent regulatory T cells (CD4+CD25+ T cells) may also have to be replenished in the course of reactivation of thymic activity and long-term disease free remission of autoimmune disease.

Residual memory B-cells cannot expand efficiently in the T cell lymphopenic environment after HSCT. The peripheral B-cell pool and protective antibody titers have to be re-established after HSCT which depends on return of CD4+ naïve T cells.[136,176,200-202,205-210,212,214] Disturbed peripheral B-cell homeostasis is a typical feature of pre transplant lupus. SLE patients are characterized by elevated frequencies of CD27++ plasmablasts and by a reversed ratio of memory and naïve B-cells.[217] These abnormalities normalize during the course of reconstitution after HSCT for SLE. At first, only extremely low frequencies of CD19+ B-cells can be detected for up to 6 months. Thereafter, IgD+ CD27- naïve B-cells continuously increase in absolute numbers in the responding SLE patients. Interestingly, frequencies of CD27+ memory B-cells among the total CD19+ B-cell pool have remained below 10% in the responding patients for more than 3 years after HSCT. Figure 15 illustrates a typical B cell reconstitution profile of a SLE patient during HSCT.

SLE manifests a Th2 cytokine skewing and restricted CDR3 T cell repertoire. The Chicago group has reported normalization of these abnormalities following HSCT.[202] Multiple sclerosis patients have an elevated production of interleukin-12 and TNF-α by resting peripheral blood monocytes. The Chicago group has also found that non-stimulated monocyte cytokine production

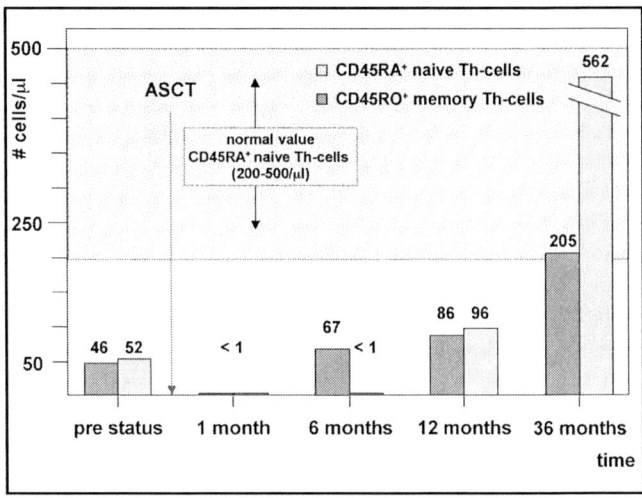

Figure 13. Reconstitution with naïve and memory Th-cell subsets in a SLE-patient after autologous HSCT/absolute counts during therapy.

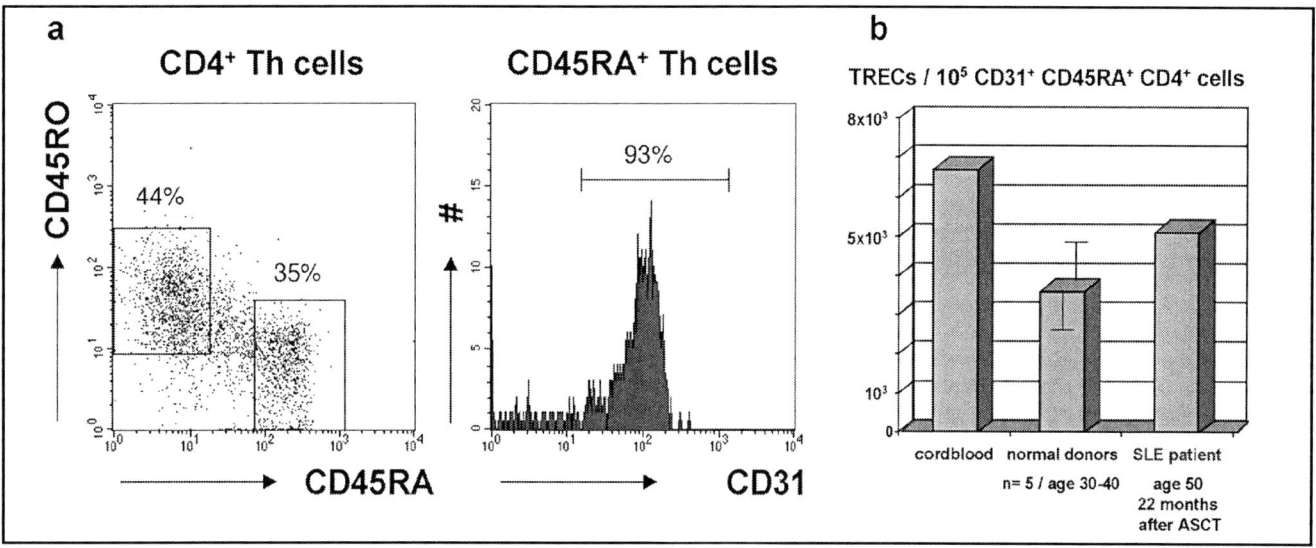

Figure 14. a) Naïve Th-cells recurring in the course of autologous HSCT express CD31. b) Detection of T-cell receptor excision circles (TREC) among CD31+ thymic naive Th-cells recurring 22 months after HSCT.

normalizes after HSCT (unpublished data). Multiple sclerosis is characterized by immune mediated demyelination from T cells clones specific to myelin protein epitopes. In collaboration with the Neuroimmunology Branch of the National Institutes of Health, the Chicago group is attempting to determine myelin precursor clone frequency and phenotype before and after HSCT and correlation, if any, with recurrence of demyelinating events. Finally, The National Institutes of Health is funding several upcoming HSCT studies for autoimmune diseases and is in the process of establishing an immune reconstitution working group.

Summary

Lymphopoiesis changes with age from thymic production to dependence on extra-thymic mechanisms. After HSCT, the reverse occurs with early extra-thymic lymphopoiesis followed by thymic regeneration of naïve T cells. The ultimate health of the graft is determined by the ability of the thymus to regenerate naïve T cells. After autologous HSCT, immune reconstitution is predominately affected by recipient age. Following an allogeneic HSCT, immune reconstitution correlates with recipient age, GVHD, and number of infused donor T cells, all of which are related to thymic reconstitution. In general, immune reconstitution is not affected by donor age, type of graft, or disease. Conditioning regimen intensity may result in thymic injury and affect immune reconstitution, although this needs further investigation. Patient survival is associated with a healthy graft, which correlates with return of thymic reconstitution of T cells. Poor thymic reconstitution as demonstrated by low TRECs, correlates with diminished T cell receptor diversity, opportunistic infections, and lower patient survival. After autologous HSCT for autoimmune disorders, disease-specific immune abnormalities normalize, although as yet unproven, regeneration of naïve CD4+ recent thymic emigrants may correlate with long-term remission of autoimmune disorder. Future trials may be designed to improve immune reconstitution by using post transplant cytokines such as IL-7 or Oncostatin M. The importance of thymic regeneration of T regulatory cells such as CD4+CD25+ T cells following HSCT is unclear. Control of GVHD or autoimmune disease by CD4+CD25+ T cells may offer new opportunities for post transplant adoptive immunotherapy and/or preservation and cytokine enhancement of thymic function.

References

1. Venter JC, Adams MD, Myers EW et al. The sequence of the human genome. Science 2001; 291:1304-51.
2. Tomanaga S, Hirokane H, Awaya K. The primitive spleen of the hagfish. Zool Mag 1973; 82:215-17.
3. Zapata A, Amemiya CT. Phylogeny of lower vertebrates and their immunological structures. Curr Topics Microbiol & Immunol 2000; 248:67-107.
4. Du Pasquier L, Flajnik M. Origin and evolution of the vertebrate immune system. In: Paul WJ, ed. Fundamental Immunology, 4th ed. Philadelphia: Lippincott-Raven Publishers, 1999:605-650.
5. Sprent J, Webb SR. Intrathymic and extrathymic clonal deletion of T cells. Curr Opin Immunol 1995; 7:196-205.
6. Von Boehmer H. Positive selection of lymphocytes. Cell 1994; 76:219-28.
7. Nossal GJV. Negative selection of lymphocytes. Cell 1994; 76:229-40.

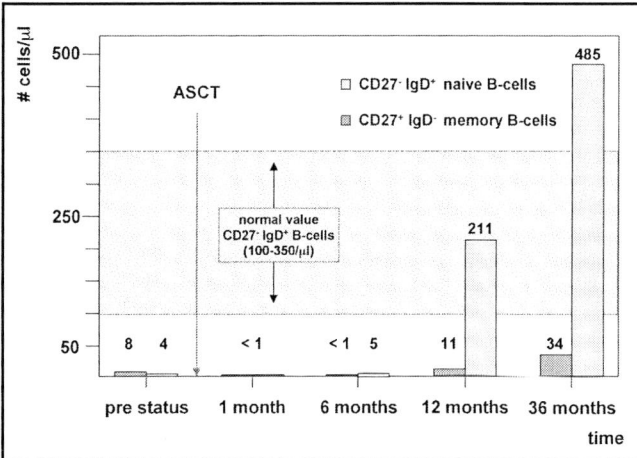

Figure 15. Reconstitution with naïve and memory B-cell subsets in a SLE-patient after autologous HSCT/absolute counts during therapy.

8. Nel AE. T-cell activation through the antigen receptor. Part 1: Signaling components, signaling pathways, and signal integration at the T-cell antigen receptor synapse. J Allergy Clin Immunol 2002; 109:758-70.
9. Nel AE, Slaughter N. T-cell activation through the antigen receptor. Part 2: Role of signaling cascades in T-cell differentiation, anergy, immune senescence, and development of Immunotherapy. J Allergy Clin Immunol 2002; 109:901-15.
10. Sebzda E, Mariathasan S et al. Selection of the T cell repertoire. Annu Rev Immunol 1999; 17:829-874.
11. Kay MM, Mendoza J, Diven J et al. Age-related changes in the immune system of mice of eight medium and long-lived strains and hybrids. Mech Ageing Devel 1979; 11:295-346.
12. de Vries E, de Groot R, de Bruin-Versteeg et al. Analyzing the developing lymphocyte system of neonates and infants. Eur J Pediatr 1999; 158:611-617.
13. Bell EB, Sparshott SM. Interconversion of CD45R subsets of CD4 T cells in vivo. Nature 1990; 348:163-166.
14. Brod SA, Rudd CE, Purvee M et al. Lymphokine regulation of CD45R expression on human T cell clones. J Exp Med 1989; 170:2147-2152.
15. Steffens CM, Al-Harthi L, Shott S et al. Evaluation of thymopoiesis using T cell receptor excision circles (TRECs): Differential correlation between adult and pediatric TRECs and naïve phenotypes. Clin Immunol 2000; 97:95-101.
16. Utsuyama M, Hirokawa K, Kurashima C et al. Differential age-change in the numbers of CD4+CD45RA+ and CD4+CD29+ T cell subsets in human peripheral blood. Mech Ageing Devel 1992; 63:57-68.
17. Reason DC, Ebisawa M, Saito H et al. Human cord blood lymphocytes do not simultaneously express CD4 and CD8 cell surface markers. Biol Neonate 1990; 58:87-90.
18. Griffiths-Chu S, Patterson JAK, Berger CL et al. Characterization of immature T cell subpopulations in neonatal blood. Blood 1984; 64:296-300.
19. McFarland RD, Douek DC, Koup RA et al. Identification of a human recent thymic emigrant phenotype. Proc Natl Acad Sci USA 2000; 97:4215-20.
20. Sempowski GD, Thomasch JR, Gooding ME et al. Effect of thymectomy on human peripheral blood T cell pools in myasthenia gravis. J Immunol 2001; 166:2808-17.
21. Reimann J. Double-negative (CD4-CD8-), TCRαβ-expressing, peripheral T cells. Scand J Immunol 1991; 34:679-88.
22. Hazenberg MD, Verschuren MCM, Hamann D et al. T cell receptor excision circles as markers for recent thymic emigrants: Basic aspects, technical approach, and guidelines for interpretation. J Mol Med 2001; 79:631-640.
23. Douek DC, McFarland RD, Keiser PH et al. Changes in thymic function with age and during the treatment of HIV infection. Nature 1998; 396:690-95.
24. Ye P, Kirschner DE. Reevaluation of T cell receptor excision circles as a measure of human recent thymic emigrants. J Immunol 2002; 169:4968-79.
25. Hazenberg MD, Borghans JAM, de Boer RJ et al. Thymic output: A bad TREC record. Nature Immunol 2003; 4:97-99.
26. Howie D, Spencer J, DeLord D et al. Extrathymic T cell differentiation in the human intestine early in life. J Immunol 1998; 161:5862-72.
27. Guy-Grand D, Azogui O, Celli S et al. Extrathymic T cell lymphopoiesis: Ontogeny and contribution to gut intraepithelial lymphocytes in athymic and euthymic mice. J Exp Med 2003; 197:333-41.
28. Lary K, Lefrancois L, Lingenheld EG et al. Enterocyte expression of Il-7 induces development of γδ T cell and Peyer's patches. J Exp Med 2000; 191:1569-80.
29. Peschon JJ, Morrissey PJ, Grabstein KH et al. Early lymphocyte expansion is severely impaired in interleukin 7 receptor deficient mice. J Exp Med 1994; 180:1955-60.
30. Rocha B, Guy-Grand D, Vassalli P. Extrathymic T cell differentiation. Curr Opin Immunol 1995; 7:235-42.
31. Mowat AMI, Viney JL. The anatomical basis of intestinal immunity. Immunol Rev 1997: 156:145-66.
32. Guy-Grand D, Cerf-Bensussan N, Malissen B et al. Two gut intraepithelial CD8+ lymphocyte populations with different T cell receptors: A role for the gut epithelium in T cell differentiation. J Exp Med 1991; 173:471-81.
33. Poussier P, Julius M. Thymus independent T cell development and selection in the intestinal epithelium. Annu Rev Immunol 1994; 12:521-53.
34. Konno A, Okada K, Mizuno K et al. CD8αα memory effector T cells descend directly from clonally expanded CD8α+β^{low} TCRαβ T cells in vivo. Blood 2002; 100:4090-97.
35. Kennedy JD, Pierce CW, Lake JP. Extrathymic T cell maturation. J Immunol 1992; 148:1620-29.
36. Antica M, Scollay R. Development of T lymphocytes at extrathymic sites. J Immunol 1999; 163:206-11.
37. Maleckar JR, Sherman LA. The composition of the T cell receptor repertoire in nude mice. J Immunol 1987; 138:3873-76.
38. Collard HR, Boeck A, Mc Maughlin TM et al. Possible extrathymic development of nonfunctional T cells in a patient with complete DiGeorge syndrome. Clin Immunol 1999; 91:156-62.
39. Wang X, Dao MA, Kuo I et al. Phenotypic comparison of extrathymic human bone-marrow-derived T cells with thymic-selected T cells recovered from different tissues. Clin Immunol 2001; 100:339-48.
40. Pawelec G, Muller R, Rehbein A et al. Extrathymic T cell differentiation in vitro from human CD34+ stem cells. J Leukocyte Biol 1998; 64:733-39.
41. Boileau C, Houde M, Dulude G et al. Regulation of extrathymic T cell development and turnover by Oncostatin M. J Immunol 2000; 164:5713-20.
42. Clegg CH, Rulffes JT, Wallace PM et al. Regulation of an extrathymic T-cell development pathway by Oncostatin M. Nature 1996; 384:261-63.
43. Clegg CH, Haugen HS, Rulffes JT et al. Oncostatin M transforms lymphoid tissue function in transgenic mice by stimulating lymph node T-cell development and thymus autoantibody production. Exp Hematol 1999; 27:712-25.
44. Terra R, Labrecque N, Perreault C. Thymic and extrathymic T cell development pathways follow different rules. J Immunol 2002; 169:684-92.
45. Tsark EC, Dao MA, Wang X et al. IL-7 enhances the responsiveness of human T cells that develop in the bone marrow of athymic mice. J Immunol 2001; 166:170-181.
46. Wilson SB, Kent SC, Patton KT et al. Extreme Th1 bias of invariant Vα25JαQ T cells in type 1 diabetes. Nature 1998; 391:177-81.
47. Bendelac A, Killeen N, Littman D et al. A subset of CD4+ thymocytes selected by MHC class I molecules. Science 1994; 263:1774-78.
48. Godfrey DI, Hammond KJL, Poulton LD et al. NKT cells: Facts, functions and fallacies. Immunol Today 2000; 21:573-83.
49. Lee PT, Benlagha K, Teyton L et al. Distinct functional lineages of human Vα24 natural killer T cells. J Exp Med; 2002; 637-41.
50. Bendelac A, Rivera MN, Park S-H et al. Mouse CD1-specific NK1 T cells: Development, specificity, and function. Annu Rev Immunol 1997; 15:535-62.
51. Abo T, Kawamura T, Watanabe H. Physiological responses of extrathymic T cells in the liver. Immunol Rev 2000; 174:135-49.
52. Makino Y, Yamagata N, Sasho T et al. Extrathymic development of V[alpha]14-positive cells. J Exp Med 1993; 177:1399-1408.
53. Taniguchi M, Makino Y, Cui J et al. V[alpha]14+ NK T cells: A novel lymphid cell lineage with regulatory function. J Allergy Clin Immunol 1996; 98:263-69.
54. Roncarolo MG, Levings MK. The role of different subsets of T regulatory cells in controlling autoimmunity. Curr Opin Immun 2000; 12:676-83.
55. Shevach EM. Regulatory T cells in autoimmunity. Annu Rev Immunol 2000; 18:423-49.
56. Jordan MS, Boesteanu A, Reed AJ et al. Thymic selection of CD4+CD25+ regulatory cells induced by an agonist self-peptide. Nat Immunol 2001; 2:301-306.
57. Thornton AM, Shevach EM. Suppressor effector function of CD4+CD25+ immunoregulatory T cells is antigen nonspecific. J Immunol 2000; 164:183-90.

58. Levings MK, Sangregorio R, Roncarolo MG. Human CD25+CD4+ T regulatory cells suppress naïve and memory T-cell proliferation and can be expanded in vitro without loss of function. J Exp Med 2001; 193:1295-1302.
59. Dieckmann D, Plottner H, Berchtold S et al. Ex vivo isolation and characterization of CD4+CD25+ T cells with regulatory properties from human blood. J Exp Med 2001; 193:1303-10.
60. Ng WF, Duggan PJ, Ponchel F et al. Human CD4(+)CD25(+) cells: A naturally occurring population of regulatory T cells. Blood 2001; 98:2736-44.
61. Baecher-Allan C, Brown JA, Freeman GJ et al. CD4+CD25high regulatory cells in human peripheral blood. J Immunol 2001; 167:1245-53.
62. Shevach EM. Certified professionals: CD4(+)CD25(+) suppressor T cells. J Exp Med 2001; 193:F41-46.
63. Sakaguchi S. Regulatory T cells: Key controllers of immunologic self-tolerance. Cell 2000; 101:455-58.
64. Salomon B, Lenschow DJ, Rhee L et al. B7/CD28 costimulation is essential for the homeostasis of the CD4+CD25+ immunoregulatory T cells that control autoimmune diabetes. Immunity 2000; 12:431-40.
65. Longo NS, Lipsky PE. Somatic hypermutation in human B cell subsets. Springer Semin Immunopathol 2001; 23:367-85.
66. Hasler P, Zouali M. B cell receptor signaling and autoimmunity. FASEB J 2001; 14:2085-98.
67. Rajewsky K. Clonal selection and learning in the antibody system. Nature 1996; 381:751-58.
68. Reth M. Antigen receptors on B lymphocytes. Annu Rev Innumol 1992; 10:97-121.
69. Kurosaki T, Johnson SA, Pao L et al. Role of Syk autophosphorylation site and SH2 domains in B cell receptor signaling. J Exp Med 1995; 182:1815-23.
70. Gold MR, Matsuuchi L, Kelly RB et al. Tyrosine phosphorylation of components of the B-cell antigen receptors following receptor crosslinking. Proc Natl Acad Sci USA 1991; 88:3436-40.
71. Honjo T, Kinoshita K, Muramatsu M. Molecular mechanism of class switch recombination: Linkage with somatic hypermutation. Annu Rev Immunol 2002; 20:165-96.
72. Diaz M, Casali P. Somatic immunoglobulin hypermutation. Curr Opin Immunol 2002; 14(2):235-40.
73. Flajnik MF. Comparative analyses of immunoglobulin genes: Surprises and portents. Nature Rev Immunol 2002; 2:688-98.
74. Papavasiliou FN, Schtaz DG. Somatic hypermutation of immunoglobulin genes: Merging mechanisms for genetic diversity. Cell 2002; 109:S35-S44.
75. Jacobs H, Bross L. Towards an understanding of somatic hypermutation. Curr Opin Immunol 2001; 13:208-18.
76. Arpin C, Dechanet J, Van Kooten C et al. Generation of memory B cells and plasma cells in vitro. Science 1995; 268:720-22.
77. Clark EA, Ledbetter JA. How B and T cells talk to each other. Nature 1994; 367:425-28.
78. Kuppers R, Klein U, Hansmann ML et al. Cellular origin of human B-cell lymphomas. N Engl J Med 1999; 341:1520-29.
79. de Vries E, de Bruin-Versteeg S, Comans-Bitter WM et al. Longitudinal survey of lymphocyte subpopulations in the first year of life. Pediatr Research 2000; 47:528-37.
80. Comans-Bitter MW, de Groot R, van den Beemd R et al. Immunophenotyping of blood lymphocytes in childhood: Reference values for lymphocyte subpopulations. J Pediatr 1997; 130:388-93.
81. Erkeller-Yuksel FM, Deneys V, Hannet I et al. Age-related changes in human blood lymphocyte subpopulations. J Pediatr 1991; 120:216-22.
82. Ben-Yehuda A, Weksler ME. Immune senescence: Mechanisms and clinical implications. Cancer Invest 1992; 10:525-31.
83. Cossarizza A, Ortolani C, Monti D et al. Cytometric analysis of immunosenescence. Cytometry 1997; 27:297-313.
84. Tarazona R, Solana R, Ouyang Q et al. Basic biology and clinical impact of immunosenescence. Exp Gerontol 2002; 37:183-89.
85. Thoman ML, Weigle WO. The cellular and subcellular bases of immunosenescence. Adv Immunol 1989; 46:221-61.
86. Ferguson FG, Wikby A, Maxson P et al. Immune parameters in a longitudinal study of a very old population of Swedish people: A comparison between survivors and nonsurvivors. J Gerontol 1995; 50A:B378-B382.
87. Ginaldi L, De Martinis M, D'Ostilio A et al. Changes in the expression of surface receptors on lymphocyte subsets in the elderly: Quantitative flow cytometric analysis. Am J Hematol 2001; 67:63-72.
88. Thoman ML. Early steps in T cell development are affected by aging. Cell Immunol 1997; 178:117-123.
89. Mackall CL, Gress RE. Thymic aging and T-cell regeneration. Immunol Rev 1997; 160:91-102.
90. LeMaoult J, Messaoudi I, Manavalan JS et al. Age-related dysregulation in CD8 T cell homeostasis: Kinetics of a diversity loss. J Immunol 2000; 165:2367-73.
91. Pannetier C, Even J, Kourilsky P. T-cell repertoire diversity and clonal expansions in normal and clinical samples. Immunol Today 1995; 16:176-81.
92. Franceschi C, Monti D, Sansoni P et al. The immunology of exceptional individuals: The lesson of centenarians. Immunol Today 1995; 16:12-16.
93. Song H, Price PW, Cerny J. Age-related changes in antibody repertoire: Contribution from T cells. Immunol Rev 1997; 160:55-62.
94. Liu YJ, de Bouteiller O, Fugier-Vivier I. Mechanisms of selection and differentiation in germinal centers. Curr Opin Immunol 1997; 9:256-62.
95. Mackall CL, Punt JA, Morgan P et al. Thymic function in young/old chimeras: Substantial thymic T cell regenerative capacity despite irreversible age-associated thymic involution. Eur J Immunol 1998; 28:1886-93.
96. Douziech N, Seres I, Larbi A et al. Modulation of human lymphocyte proliferative response with aging. Exp Gerontol 2002; 37:369-87.
97. Huang Y-P, Pechere J-C, Michel M et al. In vivo T cell activation, in vitro defective IL-2 secretion, and response to Influenza vaccination in elderly women. J Immunol 1992; 148:715-22.
98. Guidi L, Bartoloni C, Frasca D et al. Impairment of lymphocyte activities in depressed aged subjects. Mech Ageing Devel 1991; 60:13-24.
99. McNerlan SE, Rea IM, Alexander HD. A whole blood method for measurement of intracellular TNF-α, IFN-γ and Il-2 expression in stimulated CD3+ lymphocytes: Differences between young and elderly subjects. Exp Gerontol 2002; 37:227-34.
100. Wayne SJ, Rhyne RL, Garry PJ et al. Cell-mediated immunity as a predictor of morbidity and mortality in subjects over 60. J Gerontol 1990; 45:M45-48.
101. Dennett NS, Barcia RN, McLeod JD. Age associated decline in CD25 and CD28 expression correlate with an increased susceptibility to CD95 mediated apoptosis in T cells. Exp Gerontol 2002; 37:271-83.
102. Pahlavani MA, Harris MD, Richardson A. The age-related decline in the induction of the IL-2 transcription is correlated to changes in the transcription factor NFAT. Cell Immunol 1995; 165:84-91.
103. Colonna-Romano G, Potestio M, Aquino A et al. Gamma/delta T lymphocytes are affected in the elderly. Exp Gerontol 2001; 37:205-211.
104. Kawakami T, Kawakami Y, Aaronson SA et al. Acquisition of transforming properties by FYN, a normal SRC-related human gene. Proc Natl Sci USA 1988; 85:3870-74.
105. Guidi L, Antico L, Bartoloni C et al. Changes in the amount and level of phosphorylation of p56(lck) in PBL from aging humans. Mech Ageing Devel 1998; 102:177-86.
106. Pawelec G, Hirokawa K, Fulop T. Altered T cell signalling in ageing. Mech Ageing Devel 2001; 122:1613-37.
107. Miller RA. The aging immune system: Primer and prospectus. Science 1996; 273:70-74.
108. Tamir A, Eisenbraun MD, Garcia GG et al. Age-dependent alterations in the assembly of signal transduction complexes at the site of T cell/APC interaction. J Immunol 2000; 165:1243-53.
109. Fernandez-Gutierrez B, Jover JA, De Miguel S et al. Early lymphocyte activation in elderly humans: Impaired T and T-dependent B cell responses. Exp Gerontol 1999; 34:217-29.
110. Klinman NR, Kline GH. The B-cell biology of aging. Immunol Rev 1997; 160:103-14.

111. LeMaoult J, Szabo, P, Weksler ME. Effect of age on humoral immunity, selection of the B-cell repertoire and B-cell development. Immunol Rev 1997; 160:115-26.
112. Waldschmidt TJ, Panoskaltsis-Mortaro A, McElmurry RT et al. Abnormal T-cell dependent B-cell responses in SCID mice receiving allogeneic bone marrow in utero. Severe combined immune deficiency. Blood 2002; 100:4557-64.
113. Hinkley KS, Chiasson RJ, Prior TK et al. Age-dependent increase of peritoneal B-1b B cells in SCID mice. Immunology 2002; 105:196-203.
114. Weksler ME, Russo C, Siskind GW. Peripheral T cells select the B cell repertoire in old mice. Immunol Rev 1989; 110:173-85.
115. Gottesman SR, Walford RL, Thorbecke GJ. Proliferative and cytotoxic immune function in aging mice. II. Decreased generation of specific suppressor cells in alloreactive cultures. J Immunol 1984; 133:1782-87.
116. Gottesman SR, Walford RL, Thorbecke GJ. Proliferative and cytotoxic immune function in aging mice. III. Exogenous interleukin-2 rich supernatant only partially restores alloreactivity in vitro. Mech Ageing Devel 1985; 31:103-13.
117. Thoman ML, Weigle WO. Deficiency in suppressor T cell activity in aged animals. Reconstitution of this activity by interleukin 2. J Exp Med 1983; 157:2184-89.
118. Wang X, Stollar BD. Immunoglobulin VH gene expression in human aging. Clin Immunol 1999; 93:132-42.
119. Banerjee M, Mehr R, Belelovsky A et al. Age- and tissue-specific differences in human germinal center B cell selection revealed by analysis of IgV_H gene hypermutation and lineage trees. Eur J Immunol 2002; 32:1947-57.
120. Shenoy S, Mohanakumar T, Todd G et al. Immune reconstitution following allogeneic peripheral blood stem cell transplants. Bone Marrow Transplant 1999; 23:335-46.
121. Fujimaki K, Maruta A, Yoshida M et al. Immune reconstitution assessed during five years after allogeneic bone marrow transplantation. Bone Marrow Transplant 2001; 27:1275-81.
122. Novitzky N, Davison GM, Hale G et al. Immune reconstitution at 6 months following T-cell depleted hematopoietic stem cell transplantation is predictive for treatment outcome. Transplantation 2002; 74:1551-59.
123. Small TN, Papadopoulos EB, Boulad F et al. Comparison of immune reconstitution after unrelated and related T-cell depleted bone marrow transplantation: Effect of patient age and donor leukocyte infusions. Blood 1999; 93:467-80.
124. Hoepfner S, Haut PR, O'Gorman M et al. Rapid immune reconstitution following autologous hematopoietic stem cell transplantation in children: A single institution experience. Bone Marrow Transplant 2003; 31:285-90.
125. Lamb Jr LS, Gee AP, Henslee-Downey PJ et al. Phenotypic and functional reconstitution of peripheral blood lymphocytes following T-cell depleted bone marrow transplantation from partially mismatched related donors. Bone Marrow Transplant 1998; 21:461-71.
126. Moretta A, Maccario R, Fagioli F et al. Analysis of immune reconstitution in children undergoing cord blood transplantation. Exp Hemat 2001; 29:371-79.
127. Sugita K, Soiffer RJ, Murray C et al. Morimoto C. The phenotype and reconstitution of immunoregulatory T cell subsets after T-cell-depleted allogeneic and autologous bone marrow transplantation. Transplantation 1994; 57:1465-73.
128. Anderson KC, Ritz J, Takvorian T et al. Hematologic engraftment and immune reconstitution posttransplantation with anti-B1 purged autologous bone marrow. Blood 1987; 69:597-604.
129. Schlenke P, Sheikhzadeh M, Weber K et al. Immune reconstitution and production of intracellular cytokines in T lymphocyte populations following autologous peripheral blood stem cell transplantation. Bone Marrow Transplant 2001; 28:251-57.
130. Guillaume T, Rubinstein DB, Symann M. Immune reconstitution and immunotherapy after autologous hematopoietic stem cell transplantation. Blood 1998; 92:1471-90.
131. Mackall CL, Stein D, Fleisher TA et al. Prolonged CD4 depletion after sequential autologous peripheral blood progenitor cell infusions in children and young adults. Blood 2000; 96:754-62.
132. Leblond V, Othman TB, Blanc C et al. Expansion of CD4+CD7- T cells, a memory subset with preferential interleukin-4 production, after bone marrow transplantation. Transplantation 1997; 64:1453-59.
133. Weaver CH, Longin K, Buckner CD et al. Lymphocyte content in peripheral blood mononuclear cells collected after the administration of recombinant human granulocyte colony-stimulating factor. Bone Marrow Transplant 1994; 13:411-15.
134. Rutella S, Pierelli L, Bonanno G et al. Immune reconstitution after autologous peripheral blood progenitor cell transplantation: Effect of interleukin-15 on T-cell survival and effector functions. Exp Hemat 2001; 29:1503-16.
135. Mackall CL, Fleisher TA, Brown MR et al. Age, thymopoiesis, and CD4+ T-lymphocyte regeneration after intensive chemotherapy. N Engl J Med 1995; 332:143-49.
136. Bomberger C, Singh-Jairam M, Rodey G et al. Lymphoid reconstitution after autologous PBSC transplantation with FACS-sorted CD34+ hematopoietic progenitors. Blood 1998; 91:2588-2600.
137. Storek J, Ferrara S, Ku N et al. B cell reconstitution after human bone marrow transplantation: Recapitulation of ontogeny? Bone Marrow Tranpslant 1993; 12:387-98.
138. Small TN, Keever CA, Weiner-Fedus S et al. B-cell differentiation following autologous, conventional, or T-depleted bone marrow transplantation: A recapitulation of normal B-cell ontogeny. Blood 1990; 76:1647-56.
139. Akpek G, Lenz G, Lee SM et al. Immunologic recovery after autologous blood stem cell transplantation in patients with AL-amyloidosis. Bone Marrow Transplant 2001; 28:1105-1109.
140. Avigan D, Wu Z, Joyce R et al. Immune reconstitution following high-dose chemotherapy with stem cell rescue in patients with advanced breast cancer. Bone Marrow Transplant 2000; 26:169-76.
141. Wulffraat NM, Kuis W. Autologous stem cell transplantation: A possible treatment for refractory juvenile chronic arthritis? Rheumatology 1999; 38:764-66.
142. Heitger A, Winklehner P, Obexer P et al. Defective T-helper cell function after T-cell-depleting therapy affecting naïve and memory populations. Blood 2002; 99:4053-4062.
143. Storek J, Ansamma J, Dawson MA et al. Factors influencing T-lymphopoiesis after allogeneic hematopoietic cell transplantation. Transplantation 2002; 73:1154-58.
144. Lewin SR, Heller G, Zhang L et al. Direct evidence for new T-cell generation by patients after either T-cell-depleted or unmodified allogeneic hematopoietic stem cell transplantation. Blood 2002; 100:2235-41.
145. Savage WJ, Bleesing JJH, Douek D et al. Lymphocyte reconstitution following non-myeloablative hematopoietic stem cell transplantation follows two patterns depending on age and donor/recipient chimerism. Bone Marrow Transplant 2001; 28:463-71.
146. Myers LA, Patel DD, Puck JM et al. Hematopoietic stem cell transplantation for severe combined immunodeficiency in the neonatal period leads to superior thymic output and improved survival. Blood 2002; 99:872-78.
147. Hochberg EP, Chillemi AC, Wu CJ et al. Quantitation of T-cell neogenesis in vivo after allogeneic bone marrow transplantation in adults. Blood 2001; 98:1116-21.
148. Haddad E, Landais P, Friedrich W et al. Long-term immune reconstitution and outcome after HLA-nonidentical T-cell-depleted bone marrow transplantation for severe combined immunodeficiency: A European retrospective study of 116 patients. Blood 1998; 91:3646-53.
149. Weinberg K, Blazar BR, Wagner JE et al. Factors affecting thymic function after allogeneic hematopoietic stem cell transplantation. Blood 2001; 97:1458-66.
150. Lowdell MW, Craston R, Ray N et al. The effect of T cell depletion with Campath-1M on immune reconstitution after chemotherapy and allogeneic bone marrow transplant as a treatment for leukaemia. Bone Marrow Transplant 1998; 21:679-86.
151. Wu CJ, Chillemi A, Alyea EP et al. Reconstitution of T-cell receptor repertoire diversity following T-cell depleted allogeneic bone marrow transplantation is related to hematopoietic chimerism. Blood 2000; 95:352-59.

152. Ottinger HD, Beelen DW, Scheulen B et al. Improved immune reconstitution after allotransplantation of peripheral blood stem cells instead of bone marrow. Blood 1996; 88:2775-79.
153. Storek J, Dawson MA, Storer B et al. Immune reconstitution after allogeneic marrow transplantation compared with blood stem cell transplantation. Blood 2001; 97:3380-89.
154. Beschorner WE, Di Gennaro KA, Hess AD et al. Cyclosporine and the thymus: Influence of irradiation and age on thymus immunopathology and recovery. Cell Immunol 1987; 110:350-64.
155. Dulude G, Roy D-C, Perreault C. The effect of graft-versus-host disease on T cell production and homeostasis. J Exp Med 1999; 189:1329-41.
156. Rondelli D, Re F, Bandini G et al. Different immune reconstitution in multiple myeloma, chronic myeloid leukemia and acute myeloid leukemia patients after allogeneic transplantation of peripheral blood stem cells. Bone Marrow Transplant 2000; 26:1325-31.
157. Lum LG, Seigneuret MC, Storb RF et al. In vitro regulation of immunoglobulin synthesis after marrow transplantation. I. T-cell and B-cell deficiencies in patients with and without chronic graft-versus-host disease. Blood 1981; 58:431-39.
158. Talvensaari K, Clave E, Douay C et al. A broad T-cell repertoire diversity and an efficient thymic function indicate a favorable long-term immune reconstitution after cord blood stem cell transplantation. Blood 2002; 99:1458-64.
159. Honda K, Takada H, Nagatoshi Y et al. Thymus-dependent expansion of T lymphocytes in children after allogeneic bone marrow transplantation. Bone Marrow Transplant 2000; 25:647-52.
160. Abu-Ghosh A, Goldman S, Slone V et al. Immunological reconstitution and correlation of circulating serum inflammatory mediators/cytokines with the incidence of acute graft-versus-host disease during the first 100 days following unrelated umbilical cord blood transplantation. Bone Marrow Transplant 1999; 24:535-44.
161. Roux E, Durmont-Girard F, Starobinski M et al. Recovery of immune reactivity after T-cell-depleted bone marrow transplantation depends on thymic activity. Blood 2000; 96:2299-2303.
162. Verfuerth S, Peggs K, Vyas P et al. Longitudinal monitoring of immune reconstitution by CDR3 size spectratyping after T-cell-depleted allogeneic bone marrow transplant and the effect of donor lymphocyte infusions on T-cell repertoire. Blood 2000; 95:3990-95.
163. Roux E, Helg C, Dumont-Girard F et al. Analysis of T-cell repopulation after allogeneic bone marrow transplantation: Significant differences between recipients of T-cell depleted and unmanipulated grafts. Blood 1996; 87:3984-92.
164. Godthelp BC, van Tol MJD, Vossen JM et al. T-cell immune reconstitution in pediatric leukemia patients after allogeneic bone marrow transplantation with T-cell-depleted or unmanipulated grafts: Evaluation of overall and antigen-specific T-cell repertoires. Blood 1999; 94:4358-69.
165. Duval M, Pedron B, Rohrlich P et al. Immune reconstitution after haematopoietic transplantation with two different doses of pre-graft antithymocyte globulin. Bone Marrow Transplant 2002; 30:421-26.
166. Handgretinger R, Lang P, Schumm M et al. Immunological aspects of haploidentical stem cell transplantation in children. Ann NY Acad Sci 2001; 938:340-57.
167. Beelen DW, Ottinger HD, Elmaagacli A et al. Transplantation of filgrastim-mobilized peripheral blood stem cells from HLA-identical sibling or alternative family donors in patients with hematologic malignancies: A prospective comparison on clinical outcome, immune reconstitution, and hematopoietic chimerism. Blood 1997; 90:4725-35.
168. Paloczi K. Immune reconstitution: An important component of a successful allogeneic transplantation. Immunol Letters 2000; 74:177-81.
169. Thomson BG, Robertson KA, Gowan D et al. Analysis of engraftment, graft-versus-host disease, and immune recovery following unrelated donor cord blood transplantation. Blood 2000; 96:2703-11.
170. Dror Y, Gallagher R, Wara DW et al. Immune reconstitution in severe combined immunodeficiency disease after lectin-treated, T-cell depleted haplocompatible bone marrow transplantation. Blood 1993; 81:2021-30.
171. Brugnoni D, Airo P, Pennacchio M et al. Immune reconstitution after bone marrow transplantation for combined immunodeficiencies: Down-regulation of Bcl-2 and high expression of CD95/Fas account for increased susceptibility to spontaneous and activation-induced lymphocyte cell death. Bone Marrow Transplant 1999; 23:451-57.
172. Morecki S, Gelfand Y, Nagler A et al. Immune reconstitution following allogeneic stem cell transplantation in recipients conditioned by low intensity vs myeloablative regimen. Bone Marrow Transplant 2001; 28:243-49.
173. Roberts MM, To LB, Gillis D et al. Immune reconstitution following peripheral blood stem cell transplantation, autologous bone marrow transplantation and allogeneic bone marrow tranplantation. Bone Marrow Transplant 1993; 12:469-75.
174. Lum LG, Seigneuret MC, Storb RF et al. In vitro regulation of immunoglobulin synthesis after marrow transplantation. I. T-cell and B-cell deficiencies in patients with and without chronic graft-versus-host disease. Blood 1981; 58:431-39.
175. Parra C, Roldan E, Rodriguez C et al. Immunologic reconstitution of peripheral blood lymphocytes in patients treated by bone marrow transplantation. Med Clin Brac 1996; 106:169-73.
176. Steingrimsdottir H, Gruber A, Bjorkholm M et al. Immune reconstitution after autologous hematopoietic stem cells transplantation in relation to underlying disease, type of high-dose therapy and infectious complications. Haematologica 2000; 85:832-38.
177. Paulin T, Ringden O, Nilsson B. Immunological recovery after bone marrow transplantation: Role of age, graft-versus-host disease, prednisolone treatment and infections. Bone Marrow Transplant 1987; 1:317-28.
178. Li Volti SL, Di Gregorio F, Romeo MA et al. Immune status and the immune response to hepatitis B virus vaccine in thalassemic patients after allogeneic bone marrow transplantation. Bone Marrow Transplant 1997; 19:157-60.
179. Singh RK, Ino K, Varney ML et al. Immunoregulatory cytokines in bone marrow and peripheral blood stem cell products. Bone Marrow Transplant 1999; 23:53-62.
180. Klangsinsirikul P, Russell NH. Peripheral blood stem cell harvests from G-CSF-stimulated donors contain a skewed Th2 CD4 phenotype and a predominance of type 2 dendritic cells. Exp Hematol 2002; 30:495-501.
181. Volpi I, Perruccio K, Tosti et al. Postgrafting administration of granulocytes colony-stimulating factor impairs functional immune recovery in recipients of human leukocyte antigen haplotype-mismatched hematopoietic transplants. Blood 2001; 97:2514-20.
182. Arpinati M, Green CL, Heimfeld S et al. Granulocyte-colony stimulating factor mobilizes T helper 2-inducing dendritic cells. Blood 2000; 95:2484-90.
183. Rissoan M-C, Soumelis V, Kadowaki N et al. Reciprocal control of T helper cell and dendritic cell differentiation. Science 1999; 283:1183-86.
184. Pan L, Delmonte J Jr, Jalonen CK et al. Pretreatment of donor mice with granulocytes colony-stimulating factor polarizes donor T lymphocytes toward type-2 cytokine production and reduces severity of experimental graft-versus-host disease. Blood 1995; 86:4422-29.
185. Rondelli D, Raspadori D, Anasetti C et al. Alloantigen presenting capacity, T cell alloreactivity and NK function of G-CSF-mobilized peripheral blood cells. Bone Marrow Transplant 1998; 22:631-37.
186. Alpdogan O, Schmaltz C, Muriglan SJ et al. Administration of interleukin-7 after allogeneic bone marrow transplantation improves immune reconstitution without aggravating graft-versus-host disease. Blood 2001; 98:2256-65.
187. Schluns KS, Kieper WC, Jameson SC et al. Interleukin-7 mediates the homeostasis of naïve and memory T cells in vivo. Nat Immunol 2000; 1:426-32.
188. Clements JL, Boerth NJ, Lee JR et al. Integration of T cell receptor-dependent signaling pathways by adapter proteins. Annu Rev Immunol 1999; 17:89-108.
189. Snowden JA, Patton WN, O'Donnell JL et al. Prolonged remission of longstanding systemic lupus erythematosus after autologous bone marrow transplant for non-Hodgkin's lymphoma. Bone Marrow Transplant 1997; 19:1247-50.

190. Burt RK, Traynor A, Ramsey-Goldman R. Hematopoietic stem-cell transplantation for systemic lupus erythematosus. N Engl J Med 1997; 337:1777-78.
191. Tyndall A, Black C, Finke J et al. Treatment of systemic sclerosis with autologous haemopoietic stem cell transplantation. Lancet 1997; 349:254.
192. Marmont AM, van Lint MT, Gualandi F et al. Autologous marrow stem cell transplantation for severe systemic lupus erythematosus of long duration. Lupus 1997; 6:545-48.
193. Schachna L, Ryan PF, Schwarer AP. Malignancy-associated remission of systemic lupus erythematosus maintained by autologous peripheral blood stem cell transplantation. Arthritis Rheum 1998; 41:2271-72.
194. Burt RK, Traynor AE, Pope R et al. Treatment of autoimmune disease by intense immunosuppressive conditioning and autologous hematopoietic stem cell transplantation. Blood 1998; 92:3505-14.
195. Burt RK, Traynor AE, Cohen B et al. T cell-depleted autologous hematopoietic stem cell transplantation for multiple sclerosis: Report on the first three patients. Bone Marrow Transplant 1998; 21:537-41.
196. Joske DJ, Ma DT, Langlands DR et al. Autologous bone-marrow transplantation for rheumatoid arthritis. Lancet 1997; 350:337-38.
197. Burt RK, Georganas C, Schroeder J et al. Autologous hematopoietic stem cell transplantation in refractory rheumatoid arthritis: Sustained response in two of four patients. Arthritis Rheum 1999; 42:2281-85.
198. Durez P, Toungouz M, Schandene L et al. Remission and immune reconstitution after T-cell-depleted stem-cell transplantation for rheumatoid arthritis. Lancet 1998; 352:881.
199. McColl G, Kohsaka H, Szer J et al. High-dose chemotherapy and syngeneic hemopoietic stem-cell transplantation for severe, seronegative rheumatoid arthritis. Ann Intern Med 1999; 131:507-509.
200. Wulffraat N, van Royen A, Bierings M et al. Autologous haemopoietic stem-cell transplantation in four patients with refractory juvenile chronic arthritis. Lancet 1999; 353:550-53.
201. Rosen O, Thiel A, Massenkeil G et al. Autologous stem-cell transplantation in refractory autoimmune diseases after in vivo immunoablation and ex vivo depletion of mononuclear cells. Arthritis Res 2000; 2:327-36.
202. Traynor AE, Schroeder J, Rosa RM et al. Treatment of severe systemic lupus erythematosus with high-dose chemotherapy and haemopoietic stem-cell transplantation: A phase I study. Lancet 2000; 356:701-707.
203. Burt RK, Slavin S, Burns WH et al. Induction of tolerance in autoimmune diseases by hematopoietic stem cell transplantation: Getting closer to a cure? Blood 2002; 99:768-84.
204. Burt RK, Traynor A, Oyama Y et al. High-dose immune suppression and autologous hematopoietic stem cell transplantation in refractory Crohn disease. Blood 2003; 101:2064-66.
205. Charbonnier A, Sainty D, Faucher C et al. Immune reconstitution after blood cell transplantation. Hematol Cell Ther 1997; 39:261-64.
206. Koehne G, Zeller W, Stockschlaeder M et al. Phenotype of lymphocyte subsets after autologous peripheral blood stem cell transplantation. Bone Marrow Transplant 1997; 19:149-56.
207. Laurenti L, Sica S, Sora F et al. Long-term immune recovery after CD34+ immunoselected and unselected peripheral blood progenitor cell transplantation: A case-control study. Haematologica 1999; 84:1100-1103.
208. Kalwak K, Gorczynska E, Toporski J et al. Immune reconstitution after haematopoietic cell transplantation in children: Immunophenotype analysis with regard to factors affecting the speed of recovery. Br J Haematol 2002; 118:74-89.
209. Nachbaur D, Kropshofer G, Heitger A et al. Phenotypic and functional lymphocyte recovery after CD34+-enriched versus non-T cell-depleted autologous peripheral blood stem cell transplantation. J Hematother Stem Cell Res 2000; 9:727-36.
210. Eyrich M, Croner T, Leiler C et al. Distinct contributions of CD4(+) and CD8(+) naïve and memory T-cell subsets to overall T-cell-receptor repertoire complexity following transplantation of T-cell-depleted CD34-selected hematopoietic progenitor cells from unrelated donors. Blood 2002; 100:1915-18.
211. Mackall CL, Hakim FT, Gress RE. Restoration of T-cell homeostasis after T-cell depletion. Semin Immunol 1997; 9:339-46.
212. Martinez C, Urbano-Ispizua A, Rozman C et al. Immune reconstitution following allogeneic peripheral blood progenitor cell transplantation: Comparison of recipients of positive CD34+ selected grafts with recipients of unmanipulated grafts. Exp Hematol 1999; 27:561-68.
213. Toubert A, Charron D. Immune reconstitution after hematopoietic stem cell transplantation: Gaining experience without losing naivety. Hum Immunol 2001; 62:500-503.
214. Rutella S, Rumi C, Laurenti L et al. Immune reconstitution after transplantation of autologous peripheral CD34+ cells: Analysis of predictive factors and comparison with unselected progenitor transplants. Br J Haematol 2000; 108:105-15.
215. Dumont-Girard F, Roux E, van Lier RA et al. Reconstitution of the T-cell compartment after bone marrow transplantation: Restoration of the repertoire by thymic emigrants. Blood 1998; 92:4464-71.
216. Kimmig S, Przybylski GK, Schmidt CA et al. Two subsets of naive T helper cells with distinct T cell receptor excision circle content in human adult peripheral blood. J Exp Med 2002; 195:789-94.
217. Odendahl M, Jacobi A, Hansen A et al. Disturbed peripheral B lymphocyte homeostasis in systemic lupus erythematosus. J Immunol 2000; 165:5970-79.

CHAPTER 27

Historical Perspective and Rationale of HSCT for Autoimmune Diseases

Alberto M. Marmont

Introduction

Autoimmune diseases have been defined as a fascinating but still poorly understood group of diseases,[1] which pose "some of the most baffling scientific questions and daunting clinical challenges in internal medicine."[2] This intricated and still imperfectly elucidated background must be considered before reviewing the history and, even more so, the rationale of hematopoietic stem cell (HSC) therapy for severe autoimmune diseases (SADs). Their etiology is clearly multifactorial, as reflected in the concept of an integrated fabric of components known as the "mosaic of autoimmunity."[3] There are, however, some distinctions which bear directly on the rationale of HSC transplantation for SADs.

Firstly, a distinction must be made between autoimmune conditions produced by lymphoproliferative diseases which may be overt but also occult (autoimmune MGUS) and include paraproteinemic (IgM) polyneuropathy and chronic cold agglutinin disease (CAD), where autoantibodies display a monoclonal pattern (monoclonal autoimmunity),[4] and where successful HSCT may be clearly curative, and the more common organ-specific and systemic autoimmune diseases, which generally follow the pattern of antigen-driven immune reactions.[5,6] Another distinction must be made between autoimmune diseases (ADs) with a postulated impaired immune system[7] and diseases which are believed to be antigen-driven, this last being a subcellular particle or a multimolecular complex involved in important cellular functions.[5,8] A bias has also been suggested in the development of the immunoglobulin (Ig) repertoire, which might play a role in the tendency to autoimmunity.[8,9] Perhaps the single autoimmune human disease arising from a one gene mutation is APECED (autoimmune polyendocrinopathy candidiasis ectodermal dystrophy), where the gene encodes an autoimmune regulator gene (AIRE).[10]

In general, however, both immune stimulation by infections[11] and molecular mimicry[12] mechanisms play an important role. When considering the interaction among genetic and nongenetic factors, it must be kept in mind that the concordance rate for identical twins in most ADs is between 15% and 30%[13,14] with a genetic risk, in the case of SLE, of 20 for siblings and 250 for monozygotic twins.[15] The availability of multiple murine models of SLE has been pivotal for our understanding of many genetic and environmental factors.[15,16] Thus far, at least 45 named loci have been reported to be linked with one or more lupus traits in murine lupus,[17] while the presence of susceptibility genes within several chromosomal regions has been confirmed also in humans.[18] Recent studies of genetic reconstitution with polycongenic murine strains have characterized three susceptibility genes,[19] of which sle 1 mediates the loss of tolerance to chromatin, sle 2 lowers the activation threshold of B cells, and sle 3 mediates a dysregulation of CD4+ T cells, which in human patients have been shown to display features of a secondary antigen driven immune response.[20] The loci identified in a number of experimental animals and in humans overlap, although a true sharing of autoimmune genes is still unproven. For most primary ADs, multiple genetic, environmental, and hormonal factors conspire to instigate etiopathogenesis.[21]

While therapeutic strategies in gene therapy for ADs are still in their infancy,[22,23] these notions are fundamental for the understanding of the history and rationale of hematopoietic stem cell transplantation (HSCT) in this area. The history of HSCT for autoimmune disease is complicated, and a simplified overview is presented in Table 1 and 2.

Experimental Animal Studies

Animal models of autoimmunity are important for our understanding and treatment of human ADs,[24] and it is typical that HSCT therapy started with mouse experiments. These models are shown on Table 3, taken from a recent review of Ikehara.[25] Other exhaustive reviews on animal experiments have been published[26-30] and are presented by van Bekkum and Ikehara in Chapters 29 and 30, respectively.

The first animal experiments demonstrated that the transfer of spleen cells or whole bone marrow cells to anti-lymphocyte globulin treated irradiated mice was capable of reproducing the donors' murine lupus.[31,32] This adoptive transfer with marrow grafts was subsequently confirmed for many other ADs in animals, including the antiphospholipid syndrome, insulin-dependent diabetes mellitus (IDDM), experimental autoimmune encephalomyelitis (EAE) and many others (reviewed in refs. 25, 28, 29).

Transgenic and/or knock-out mice have also been utilized, and a number of clinical ADs have been transferred into severe combined immune deficient (SCID) and SCID/non-obese diabetic (NOD) mice by means of xenogeneic lymphocytes from human patients. They include autoimmune thyroiditis (Graves and Hashimoto), primary biliary cirrhosis, pemphigus vulgaris and rheumatoid arthritis (reviewed in refs. 25-30).

Table 1. Time line of 1st events in development of hematopoietic stem cell transplantation for autoimmune diseases

Date	Author / Reference	Comment
1977	Baldwin JL. Arthritis & Rheumatism 1977; 20(5):1043-8.	1st case report of remission of autoimmune disease in patients undergoing HSCT for another reason.
1979	Morton JI, Siegel BV. Transplantation 1979; 27(2):133-4.	1st report of reversal of autoimmune syndrome in animals (NZB mice) by allogeneic bone marrow transplantation.
1983	Pestronk A. Annals of Neurology 1983; 14(2):235-41.	1st report of animals cured of myasthenia gravis in EAMG model by syngeneic bone marrow transplant.
1985	Ikehara S, Good RA. PNAS 1985 Apr; 82(8):2483-7.	1st report that allogeneic bone marrow transplant. cures SLE like disease in murine MRL/lpr/lpr model.
1985	Ikehara S, Good RA. PNAS 1985; 82(22):7743-7.	1st report that allogeneic bone marrow transplant prevents diabetes in NOD mouse model.
1989	van Bekkum DW. PNAS 1989; 86(24):10090-4.	1st report that syngeneic bone marrow transplantation cures arthritis in animal model of adjuvant-induced arthritis.
1992	Karussis DM, Slavin S. Journal of Neuroimmunology 1992; 39(3):201-10.	1st report that syngeneic bone marrow transplantation cures MS like disease in experimental autoimmune encephalomyelitis animal model.
1993	Marmont AM. Lupus 1993; 2(3):151-6.	1st editorial suggesting that HSCT should be used as treatment for SLE.
1995	Burt RK. Bone Marrow Transplantation 1995; 16(1):1-6.	1st editorial suggesting that HSCT should be used as treatment for MS.
1995	Committee Chair—Alan Tyndall	European Group for Blood and Marrow Transplantation (EBMT) and European League Against Rheumatism (EULAR) establish a working committee on stem cell transplant for autoimmune diseases for Europe and Asia.
1995	Committee Chair—Richard Burt	International Bone Marrow Transplantation Registry (IBMTR) and Autologous Blood and Marrow Transplant Registry (ABMTR) establish a working committee on stem cell transplant for autoimmune diseases for the Americas.
1997	Fassas A. Bone Marrow Transplantation 1997; 20(8):631-8.	1st report of HSCT for MS in the world.
1997	Marmont AM. Lupus 1997; 6(6):545-8.	1st report of autologous HSCT for SLE in the world.
1997	Burt RK. New England Journal of Medicine 1997; 337(24):1777-8.	1st report of autologous HSCT for SLE in America.
1997	Joske DJ. Lancet 1997; 350(9074):337.	1st report of autologous HSCT for RA in the world.
1997	Tyndall A. Lancet 1997; 349(9047):254.	1st report of autologous HSCT for scleroderma.
1998	Burt RK. Bone Marrow Transplantation 1998; 21(6):537-41.	1st report of autologous HSCT for multiple sclerosis in America.
1999	Wulffraat N, Lancet 1999; 353(9152):550-3.	1st report of autologous HSCT for juvenile chronic arthritis in the world.
1999	Burt RK. Arthritis & Rheumatism 1999; 42(11):2281-5.	1st report of autologous HSCT for RA in America.
1999	Contract principal investigators—Burt RK, Burns WH, Sullivan K	National Institutes of Health (NIH) awards three contracts to develop phase III trials of HSCT for autoimmune diseases.
2000	Slavin S. Exp Hematol 2000; 28(7):853-7.	1st clinical evidence of allogeneic graft versus autoimmunity effect (GVA).
2001	Snowden J, principal investigator	EBMT/EULAR opens phase III trial of autologous HSCT for RA.
2001	van Laar J, principal investigator	EBMT/EULAR opens phase III trial of autologous HSCT for scleroderma.
2002	Burt RK. Blood 2003; 101(5):2064-6.	1st report of autologous HSCT for Crohn's disease in the world.

EBMT= European Bone Marrow Transplant Registry; EULAR= European League Against Rheumatism; HSCT= Hematopoietic stem cell transplantation; MRL/lpr= Murphy Roth laboratory lymphoproliferative; MS= multiple sclerosis; NZB= New Zealand Black; NOD= non-obese diabetic; RA= rheumatoid arthritis; SLE= systemic lupus erythematosus

Table 2. Main features of the history of hematopoietic stem cell transplant for clinical autoimmune diseases

1) First transfer of autoimmunity (murine SLE) following marrow/spleen allo-BMT (1960 on)
2) Identification of cellular elements carrying autoimmunity (1960s on)
3) Cure of animal autoimmune disease by HSCT (1970s on)
4) Cure of patients with coincidental autoimmune diseases following HSCT for another reason (1970s—1990s)
5) Positive results following clinical studies of autologous HSCT (1990s on)
6) Complete remissions following intense immunosuppression without HSC (1990s on)
7) Development of data registries and inception of phase III studies (1990s)
8) Utilization of donor stromal cells, osteoblasts and portal or intraosseous accesses of HSC for animal autoimmune diseases (2000 on)
9) Suggestion of a Graft-versus-Autoimmunity effect (2000 on)
10) Application of non-myeloablative allogeneic HSCT to autoimmune diseases (2001 on)

The identification of the cellular elements carrying the autoimmune information came later, since unmanipulated bone marrow was utilized initially. It was shown subsequently that T-cell depleted marrow cells from B/W F1 mice transferred their autoimmune potential.[33] This was confirmed in the murine subtype elicited by the monoclonal anti 16/6 idiotypic antibody.[34] B lymphoid precursors from B/W F1 marrow cultures reproduced the disease in SCID mice.[35]

In more recent investigations, Ikehara and his group showed that HSC deriving from autoimmune animals are more "resilient" than those coming from healthy ones, in as much as they can proliferate in major histocompatibility complex (MHC) incompatible microenvironments.[36] They have also proven the importance of stromatic and osteoblastic cells as engraftment facilitators, and indicated the interest of utilizing portal and intraosseous accesses to treat ADs.[25] On the basis of these investigations, Ikehara has proposed the concept of ADs as polyclonal stem cell disorders.[26] He has also indicated that there are two types of animal models of ADs: in one type, the disease is antigen-induced and transferred by mature lymphocytes, such as in EAE and adjuvant arthritis (AA); in the other, genetic factors predominate, the disease is transferred by the hematopoietic stem cell, and it develops spontaneously such as listed in Table 3. The first type has been considered to be curable following autologous SCT, but the second not.

A wide spectrum of animal experimental ADs were cured following allogeneic HSCT, including autoimmune thrombocytopenic purpura, crescentic glomerulonephritis and others. In MRL/lpr mice, which relapsed after conventional bone marrow transplantation (BMT), the integration with donor stromal cells was found to be curative.[25] A scarcely cited but interesting finding was the demonstration that human cord blood HSC were capable of suppressing autoantibody production in lupus mice,[37] perhaps the first demonstration of a curative effect by xenogeneic SC.

One of the most important findings in allogeneic BMT for leukemic and other malignancies, both in animal experiments and in human disease, is the well-known Graft-vs-Leukemia (GVL) effect. As stated recently, this GVL effect "is very real, and thousands of patients are alive because of it."[38] Evidence is now accumulating, both in experimental and clinical settings, that a similar Graft-vs-Autoimmunity (GVA) effect might also exist, consisting in the substitution of normal T, B and lymphoid progenitor cells in the place of an autoimmune lymphoid system.[39] This GVA effect is supported by experiments showing that mixed chimerism obtained utilizing a sublethal irradiation conditioning regimen followed by allogeneic BMT can prevent the onset of diabetes and even reverse pre-existing autoimmune insulitis in

Table 3. Some animal models for spontaneous autoimmune diseases

Mouse Strain	Sex	Autoimmune Disease
NZB	F	Hemolytic anaemia, SLE
(NZB x NZW)F1:B/WF1	F	SLE, Sjögren syndrome
MRL/lpr(MRL/MP-lpr/lpr)	F	SLE, rheumatoid arthritis
MRL/+ (MRL/MP-+/+)	F	SLE, rheumatoid arthritis, pancreatitis, sialoadenitis
BXSB	M	SLE, autoimmune hepatitis
(NZW x BXSB)F1:W/BF1	M	SLE, ITP, autoimmune hepatitis, anti-phospholipid antibody syndrome, myocardial infarction
NZB/KN	M	Rheumatoid arthritis
NOD (Non-obese diabetic)	F	Insulin-dependent (type I) diabetes mellitus
KK-Ay	F = M	Non insulin-dependent (type II) diabetes mellitus
FGS	F = M	FGS (focal segmental glomerular sclerosis)

BXSB= Black-6 mouse crossed to an SB mouse; FGS= focal glomerular sclerosis; KK= non-obese hereditary diabetic mouse (KK strain); MRL/lpr= Murphy-Roth laboratory lymphoproliferative; NOD= non-obese diabetic; NZB= New Zealand Black; NZW= New Zealand White; SLE= systemic lupus erythematosus

non-obese diabetic (NOD) mice, whereas the same radiation protocol without allogeneic SCT is insufficient.[40] A similar effect has been shown using sublethal conditioning and an anti-CD154 monoclonal antibody.[41] These findings are being currently reproduced in the clinic (see later).

The apparently paradoxical idea of curing ADs utilizing their own SC following irradiation protocols originates from the elegant provocative animal experiments pioneered by several investigators including D.W. van Bekkum and his group.[22,42] First they demonstrated that autologous BMT was capable of curing adjuvant arthritis in rats,[43] and then that closely superimposable results could also be obtained in experimental allergic (autoimmune) encephalomyelitis (EAE),[44] in which relapses could be prevented only following allogeneic transplants. These results ignited the outgoing interest and the widespread trials with autologous procedures in clinical autoimmunity.

Clinical Results

New approaches to treatment of severe autoimmune diseases (SADs) based on immunosuppressive dose escalation without or with either autologous or allogeneic stem cell transplantation are shown in Figure 1.

The transfer of an autoimmune disorder from a bone marrow donor to a recipient has been considered one of the best criteria for establishing adoptive autoimmunity following allogeneic BMT.[45] The transfer of type I diabetes is still the most demonstrative case.[46] This group of patients must be distinguished from those developing autoimmune events following allogeneic BMT without any evidence that the donors were harboring unrecognized autoimmune lymphoid clones. In this latter group, which is by far more common, and where autoimmune cytopenias, thyroid diseases and myasthenia gravis are the most frequent cases, multifactorial causes predominate. The concept of a veritable "immunological chaos" was proposed in 1998.[47]

In other cases, a pathogenic clonal B expansion of donor origin could be demonstrated. A most demonstrative case is the one of a Japanese patient who underwent allogeneic-BMT for AML and experienced the transfer of autoimmune thyroiditis and the resolution of palmoplantar pustular psoriasis.[48] Donors with autoimmune rheumatic diseases, however, did not transmit their conditions to the recipients.[49,50]

There are serendipitous cases, in which allogeneic BMT was performed because of a supervening severe hematologic disease, and in which a complete remission of a pre-existing AD was an additional (and probably unexpected) benefit from the procedure. The non-hematologic AD with the highest number of allogeneic BMT is rheumatoid arthritis (RA). In almost all patients, the reason for transplantation was iatrogenic: gold-induced severe aplastic anemia (SAA). They are analysed in detail elsewhere.[51,52] All patients enjoyed complete remissions of both diseases, but three died of transplant-related mortality (TRM).[53] Of the remaining 5 patients, 3 are in complete remission, one developed a positive rheumatoid factor, and one relapsed of her RA even though the patient's immune system was 98.5% of donor origin.[54] Relapse was also observed in a patient with psoriatic arthropathy.[55] In another case of chronic myelogenous leukemia (CML) coincident with severe psoriatic arthritis, there was an initial complete remission following a non-myeloablative allogeneic HSCT. A Philadelphia chromosome-positive relapse of CML occurred coincident with relapse of psoriatic polyarthritis. Discontinuation of immunosuppression was then followed by all male DNA (the patient was female), molecular remission (bcr-abl negativity), grade II GVHD and clearing of the psoriatic disease.[56]

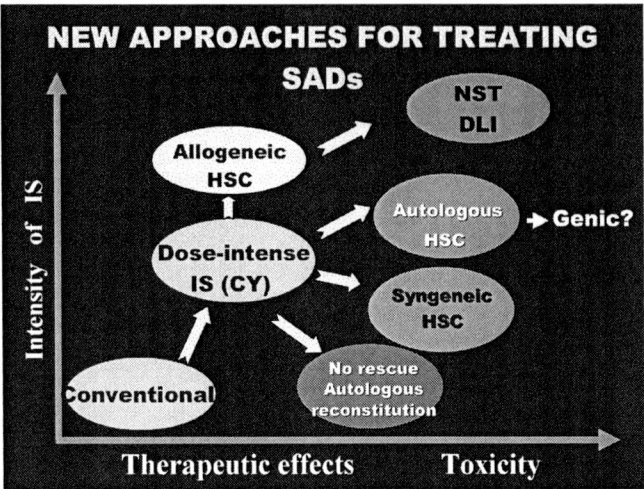

Figure 1. A hypothetical representation of some new approaches for the treatment of severe autoimmune diseases (SADs). The intensity of immunosuppression (IS) is proportional to its efficacy but also to its potential toxicity. The allogenic approach is clearly founded on non-myeloablative procedures (NST), but will probably require donor lymphocyte infusions (DLI) to achieve full chimerism. The autologous approach may be conducive in future to genic therapy. For other aspects see text. From Marmont AM. Immunoablation followed or not by hematopoietic stem cells as an intense therapy for severe autoimmune diseases. New perspectives, new problems. Haematologica 2001; 86: 337-345 (with permission).

Six patients with leukemia and Crohn's disease underwent allogeneic BMT in Seattle.[57] One patient died of TRM; 4 of the surviving 5 patients had no signs or symptoms of Crohn's disease for several years after transplant. One patient with mixed chimerism had a relapse both of Crohn's and of chronic myeloid leukemia (CML) 1.5 years post transplantation.

The most significant data in syngeneic transplants reside in the area of aplastic anemia (AA), which will be briefly discussed later. In a study of 40 patients who received BMT from their genetically identical twins between 1964 and 1992, it was found that most of them recovered marrow function, and that pretransplantation conditioning increased response rate but did not seem to improve survival.[58] Leaving aside AA, there are two cases in the literature in which complete remission of a SAD followed a syngeneic transplant from a healthy identical twin. The first is a patient with severe seronegative rheumatoid arthritis who received cyclophosphamide (CY) and anti-thymocyte globulin (ATG), minimal conditioning and a peripheral blood stem cell (PBSC) graft with no transplant-related toxicity. Post-transplant T cell receptor V-beta cell usage of circulating lymphocytes demonstrated evidence of full donor T cells, which corresponded to complete remission of the disease[59] at 36 months follow-up.[60] Another syngeneic PBSC transplant has been successfully performed for a 20 year old patient with chronic refractory (splenectomised) autoimmune thrombocytopenic purpura (AITP).[61]

Aplastic anemia is widely considered an autoimmune disease, and is covered by an enormous literature. An excellent review has been published most recently.[62] AA and the related autoimmune myelopathies[63] have been and are treated successfully by allogeneic HSCT, but because of the subject's vastity they will not be specifically covered in this volume, although they will be surveyed in a single chapter.

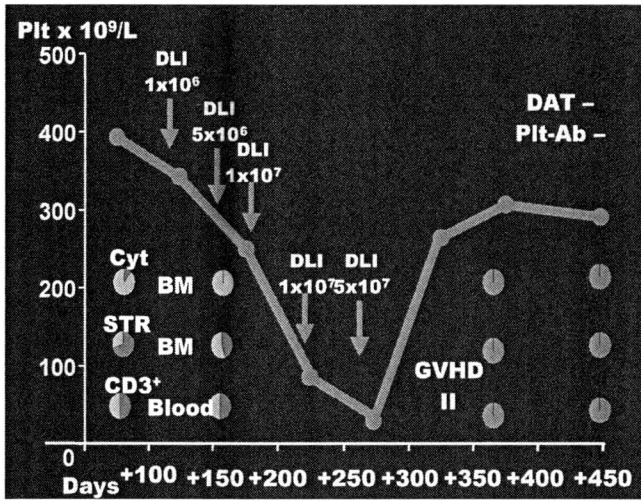

Figure 2. The effect of graded incremental donor lymphocyte infusions on engraftment and final clinical results in a patient with intractable Evans syndrome treated with allogeneic BMT from his HLA-identical sister. From Marmont AM et al. Bone Marrow Transplant 2003; 31:399-402 (with permission).

Patients with other SADs having received allogeneic HSCT are few, obviously in consideration of the unacceptably high TRM in non-malignant diseases. Perhaps the most significant results up to now have been obtained in SADs of the blood, with the combination of haemolytic anemia and thrombocytopenic purpura (Evans syndrome) in the foreground. Four cases have been treated with allogeneic transplants, one with cord blood.[64] In two of them, complete remissions were obtained only after the onset of GVHD.[65,66] Non-myeloablative procedures relying on a postulated GVA effect corresponding to the progressive elimination of the patients' immune system, harboring the autoimmune clones, by donor lymphocytes are being considered for very refractory patients. One of the most demonstrative cases is the one of a patient with Evans syndrome who relapsed following a non-myeloablative BMT (NST), but finally responded to a series of incremental donor lymphocyte infusions (DLI)[66] (see Fig. 2). A response to allo-BMT after relapse of autoimmune haemolytic anemia following autologous SCT has been reported.[67] As already stated, the graft-versus malignancy effect is fundamental to cure of some malignancies,[68,69] but also a GVA effect, first detected experimentally,[40,41] may be operative in the clinic.[70,71]

The idea of utilizing autologous HSC following intense immunosuppression (immunoablation) was proposed for refractory-relapsing SLE in 1993.[72] Two main reasons are at the origin of the wide acceptance of this procedure worldwide: the encouraging experimental results from Rotterdam[27,43,44] and from Jerusalem,[73,74] and the greater safety of the autologous procedure. A consensus report written on behalf of the European League against Rheumatism (EULAR) and the European Group for Blood and Marrow Transplantation (EBMT) was published in 1997.[75] A retrospective study of the same group found a 2-year post-transplant mortality of 8.6%.[76] Contributing factors to this perhaps higher than expected mortality may have been a learning curve for utilizing HSCT in new diseases such as SADs, and unique organ dysfunctions, such as heart and lung failure in systemic sclerosis.[77]

Following single case reports (the first HSCT for SLE were published in 1997)[78,79] the number of SADs registered by the European Working Party of the EBMT coordinated by Tyndall are rapidly approaching 500. The most extensive single center experience with HSCT for autoimmune diseases and SLE in particular have been performed in Chicago.[80] The Chicago SLE studies are still ongoing (see later), but the first group of patients were still in complete clinical and immunologic remission 2 years after transplantation. Other long-lasting remissions in SLE have been reported from Berlin,[81] while in other cases there were gradual immunologic and clinical relapses,[81] easily controlled by previously inefficient routine therapy. Contrary to the results following allogeneic HSCT for coincidental diseases, two cases have been reported in which lymphoma was cured following autologous HSCT, but AD was not.[82,83]

Many other SADs have been treated accordingly. Perhaps the most encouraging results have been obtained in multiple sclerosis (MS),[84,85] where a long-lasting abrogation of active, gadolinium enhancing lesions as evidenced by magnetic resonance imaging (MRI) could be achieved.[86] Other important multicentric retrospective clinical studies have been published on systemic sclerosis[87] (SSc), rheumatoid arthritis[88-90] (also of the juvenile subtypes),[91,92] recurrent polychondritis[81] and Behçet's disease.[93] Comprehensive reviews have been published recently.[29,60,94]

This volume is dedicated to HSC transplant therapy, but it is appropriate to mention the interesting high-dose chemotherapy approach with no HSC rescue which is an ongoing program at Johns Hopkins Oncology Center at Baltimore, and which will be discussed later. This immunosuppressive protocol, consisting of 200 mg/kg of cyclophosphamide over 4 days, originated in 1996 for severe aplastic anemia[95] (SAA), and, notwithstanding other differing results,[96] has been recently confirmed by the authors.[97] The program was extended to include SADs[98] and in particular SLE,[99] with good remissions and no mortality. A theoretical advantage consists in the lack of reinfusion of autoreactive lymphocytes. However, stem and progenitor cell expansion techniques may render the post-transplant neutropenic phase shorter,[100,101] thus conferring a certain degree of advantage to the transplant procedures.

Advantages and Problems

The majority of ADs are controlled, more or less satisfactorily, by conventional manipulation of the immune system. There is, however, a hard core of refractory/relapsing, treatment-resistant ADs for which the definition of "malignant autoimmunity"[102] has been proposed quite appropriately. The "holy grail" of therapy should be a targeted treatment that would specifically destroy the pathogenetic clones responsible for autoimmunity.[103] However, even this would not necessarily prevent recurrence ignited by resensitization to autoantigens. The real goal remains the achievement of tolerance.[104]

When considering the autologous procedures, a distinction must be made between the underlying philosophy and its clinical results. The fundamental question is, whether their effect is mainly immunosuppressive, or whether the conditioning regimens might provide "a window of time free of memory T cell influence, during which the maturation of new lymphocyte progenitors may occur without recruitment to anti-self reactivity."[80] Studies on immune reconstitution are numerous,[105,106] and those on thymic output following autologous HSCT in adults utilizing TREC (T cell receptor excision circles) as a measure of recent thymic emigrants have shown that the increase in TREC post-transplant indicates a substantial contribution of adult thymus.[107] In order to elucidate whether, if relapses occur, disease is reinitiated by lymphocytes surviving the conditioning regimens, or from the HSC compartment, marker studies have been performed.[108] However, a great wealth of observations, including the Johns Hopkins experience,[97,98] indicate the important role of powerful conditioning regimens. The observation that patients are more responsive to conventional treatments to which they were originally refractory has been thought to suggest an immunomodulation process[60] ("resetting"), but debulking of autoimmune clones also may be conducive to similar effects.

From a practical point of view, the dramatic effect of this comparatively new procedure has been amply demonstrated in an extensive spectrum of SADs.[29,60,94,109] Two reasons are in favor of modifying the conditioning regimens: the need to switch from the old myeloablative regimens to more specifically immunosuppressive ones, and the demonstration that high intensity conditioning has been shown to be associated with higher TRM.[110] This is particularly true in patients with advanced multisystem disease, so that selection of better-risk candidates could improve outcome. Also, dose-intensive mobilization regimens should be avoided: deaths have occurred following mobilisation,[111] and some diseases such as SLE may be poor mobilizers also independently from former cytotoxic therapy.[112] Finally, the question of late malignant complications such as myelodysplastic syndromes and leukaemia (MDS/AML) must be kept in mind,[113-116] although the severity of disease in HSCT candidates might eventually force the issue.

The paucity of reports of allogeneic HSCT for SADs has already been discussed. However, the idea (or the hope) of obtaining a genuine eradication of a faulty immune system harboring autoimmune lymphoid clones is attractive, and, as already remarked, it may have a significant edge on the autologous procedures. It remains to be seen whether the GVA effect will prove to be comparable to the Graft-versus-Malignancy effects, and in which ADs it could be clinically useful. A long-term complete remission was obtained in a female patient with classical SLE who underwent allogeneic BMT from an HLA-identical brother having developed marrow aplasia. Fifteen years later, the patient's disease is still in a complete remission (CR) and requires no treatment,[117] making this case the one with the longest follow ups (FU) in the literature. A weak ANA test of the speckled subtype is still present, thus proving that the persistence of minimal residual autoimmune disease, probably a residue of oligoclonal B cells, is not inevitably associated with relapse in the presence of a healthy immune system.

The case for allogeneic non-myeloablative stem cell transplants has been recently stressed as a possible platform for future developments in transplant immunology to avoid graft versus host disease (GVHD) and enhance the allograft GVA effect.[118,119] However, the separation between GVHD and GVA, which is being considered as feasible in a number of situations,[120] will probably remain hard to achieve. Connective tissue diseases, and most particularly scleroderma (SSc), cannot afford to be exposed to any kind of GVHD. Finally, the unresolved question whether resensitization of a "new" immune system may occur owing to the antigenic rechallenge of endogenous epitopes/peptides can only be answered by a more extensive clinical experience, which should be conduced following phase III trials, and will hopefully bring about reliable results supplying sufficient data for evidence-based programs and algorithms.

A new trend area in the treatment of many ADs resides in the expanding area of biologic response modifiers, and more specifically, monoclonal antibodies (MoAbs). Two anti-tumor necrosis factor (TNF) agents, etanercept and infliximab, are being utilized with good results in RA,[121] Crohn's disease and others. The chimeric murine/human MoAb Rituximab has been shown to be active in autoimmune thrombocytopenic purpura, and the CAMPATH-1 MoAb has controlled severe vasculitis and refractory Wegener's disease,[124,125] besides various autoimmune disorders of the blood.[125,126] Their beneficial effect in the various ADs depends on their antigen-directed specificity, and from the pathogenic role sustained by antigen-bearing cells. It is clear that an anti-CD20 MoAb will exert a specific therapeutic effect not only in the B-cell malignancies, but also in the hematologic, antibody B cell dependent ADs. It is accordingly possible that determined MoAbs, such as Rituximab, may be advantageously utilized pre- or post-HSCT, as it has been conclusively demonstrated in B non-Hodgkin's lymphomas.[127]

A new important issue that is expanding impressively is the field of regenerative medicine. After the resolution of an autoreactive immune system against self antigens has been achieved, the next task consists of providing new cells in the place of the irreversibly damaged ones. It has been remarked recently that "cell replacement is taking a central part in medicine."[128] Both "hype and hope" conspire to the issue's complexity. A novel dynamic model of hematopoietic stem cell organization has been proposed recently.[129] The fascinating breakthroughs in stem cell plasticity[130] are an area under intensive investigation (refer to Chapter 11), and it will certainly become a key issue of regenerative medicine also within the spectrum of autoimmune diseases.

References

1. Davidson A, Diamond B. Autoimmune diseases. N Engl J Med 2001; 345:340-350.
2. Diamond B, Bluestone J, Wofsy D. The immune tolerance network and rheumatic disease. Immune tolerance comes to the clinic. Arthritis Rheum 2001; 44:1730-1735.
3. Shoenfeld Y, Isenberg. The mosaic of autoimmunity. Amsterdam, Elsevier 1989.
4. Marmont AM, Merlini G. Monoclonal autoimmunity in haematology. Haematologica 1991; 76:449-459.
5. Boitard C. B cells and autoantibody production in autoimmune diseases. Austin: Springer-Landes, 1996.
6. Tan EM. Autoantibodies and autoimmunity: A three decade perspective. In: Chiorazzi N, Lahita RG, Pavelka K et al, eds. B-lymphocytes and autoimmunity. Ann NY Acad Sci 1997; 815:1-14.
7. Feltkamp TEW. The mystery of autoimmune diseases. In: Shoenfeld Y, ed. The decade of autoimmunity. Amsterdam: Elsevier 1999; 1-5.
8. Hiepe F, Dorner T, Burmester G-D. Editorial overview, Antinuclear antibody and extractable nuclear antigen-related diseases. Int Arch Allergy Immunol 2000; 123:6-9.
9. Coutinho A. An outsider's view on SLE research. Lupus 1999; 8:171-173.
10. The Finnish-German APECED Consortium. An autoimmune disease, APECED, caused by mutations in a novel gene featuring two PHD-type zinc-finger domains. Nat Genet 1997; 17:393-398.
11. Shoenfeld Y, Cervera R. Innovations in autoimmunity in the last decade. In: Shoenfeld Y, ed. The decade of autoimmunity. Amsterdam: Elsevier 1999:7-18.
12. Albert LJ, Inman RD. Molecular mimicry and autoimmunity. N Engl J Med 1999; 341:2068-2074.
13. Theofilopoulos AN, ed. Genes and genetics of autoimmunity. Curr Dir Autoimmunity. Basel: Karger, 1999:1:1-296.
14. Encinas Ja, Kuchroo VK. Mapping and autoimmunity genes. Autoimmunity 2001; 91-697.
15. Gregersen PK. Genetics. In: Tsokos GC, ed. Principles of molecular rheumatology. Totowa NJ: Humana Press, 2000:3-14.
16. Hahn BH. Lessons in lupus: The mighty mouse. Lupus 2001; 10:589-593.

17. Kono DH, Theophilopoulus AN. The genetics of murine systemic lupus erythematosus. In: Wallace DJ, Hahn BH, eds. Dubois lupus erythematosus. Philadelphia: Lippincott William & Wilkins, 2002:121-143.
18. Tsao BP. The genetics of human lupus. In: Wallace DJ, Hahn BH, eds. Dubois' lupus erythematosus. Philadelphia: Lippincott William & Wilkins, 2002:97-120.
19. Morel L, Croker BP, Blenman KR et al. Genetic reconstitution of systemic lupus erythematosus immunopathology with polycongenic murine strains. PNAS 2000; 97:6870-6875.
20. Kolowos W, Gaip US, Vole RE et al. CD4 positive peripheral T cells from patients with systemic lupus erythematosus (SLE) are clonally expanded. Lupus 2001; 10:321-331.
21. Liossis S-N G, Tsokos GC. Systemic lupus erythematosus. In: Tsokos GC, ed. Principles of molecular rheumatology. Totowa NJ: Humana Press 2000:311-323.
22. Fathman CG, ed. Biologic and gene therapy for autoimmune disease. Curr Dir in Autoimmunity. Basel: Karger 2000:2:1-23.
23. Kimberly RP. Gene therapy. In: Tsokos CG, ed. Principles of molecular rheumatology. Totowa NJ: Humana Press 2000:515-542.
24. Sakaguchi S. Animal models of autoimmunity and their relevance to human diseases. Autoimmunity 2001; 684-690.
25. Ikehara S. Treatment of autoimmune diseases by hematopoietic stem cell transplantation. Exp Hematol 2001; 29:661-669.
26. Ikehara S. Autoimmune diseases as stem cell disorders. Normal stem cell transplant for their treatment. Int J Mol Med 1998; 1:5-15.
27. Marmont AM. Stem cells transplantation for severe autoimmune diseases: Progress and problems. Haematologica 1998; 83:733-743.
28. Van Bekkum DW. Stem cell transplantation in experimental models of autoimmune disease. J Clin Immunol 2000; 20:11-17.
29. Marmont AM. New horizons in the treatment of autoimmune diseases: Immunoablation and stem cell transplantation. Annu Rev Med 2000; 51:115-134.
30. Good RA, Verjee T. Historical and current perspectives on bone marrow transplantation for prevention and treatment of immunodeficiencies and autoimmunities. Biol Blood Marrow Transplant 2001; 7:123-135.
31. Denman AM, Russell AS, Denman EJ. Adoptive transfer of the diseases of New Zealand Black mice to normal mouse strains. Clin Exper Immunol 1969; 5:269.
32. Morton H, Siegel BV. Transplantations of autoimmune potential. I. Development of antinuclear antibodies in H-2" compatible recipients of bone marrow from New Zealand Black mice. Proc Natl Acad Sci USA 1974; 71:2162-2166.
33. Akizuki M, Reeves JP, Steinberg AD. Expression of autoimmunity by NZB/NZW marrow. Clin Immunol Immunopathol 1978; 10:247.
34. Levite M, Zinger H, Mozes I et al. Systemic lupus erythematosus-related antibody production in mice is determined by marrow-derived cells. Bone Marrow Transplant 1993; 12:179-183.
35. Reininger I, Radkiewiecs T, Kosco F. Development of autoimmune disease in SCID mice populated with long-term "in vitro" proliferating (NZBxNZW) F1 pre-B cells. J Exp Med 1992; 176:1343-1353.
36. Kawamura M, Hisha H, Li Y et al. Distinct qualitative differences between normal and abnormal hemopoietic stem cells in vivo and in vitro. Stem Cells 1997; 15:56-62.
37. Ende N, Czarnesky J, Raveche E. Effect of human cord blood transfer on survival and disease activity in MRL-lpr/lpr mice. Clin Immunol Immunopathol 1995; 75:190-195.
38. Sondel PM. The graft-versus-leukemia effect. In: Barrett J, Jiang Y-Z, eds. Allogeneic immunotherapy for malignant disease. New York: Dekker 2000:1-12.
39. Marmont AM. Eminence-based medicine. Lupus: Tinkering with haematopoietic stem cells. Lupus 2001; 10:769-774.
40. Li H, Kaufman CL, Boggs SS et al. Mixed allogeneic chimerism induced by a sublethal approach prevents autoimmune diabetes and reverses insulitis in non-obese diabetic (NOD) mice. J Immunol 1996; 156:380-387.
41. Seung E, Iwakoshi N, Woda BA et al. Allogeneic hematopoietic chimerism in mice treated with sublethal myeloablation and anti-CD154 antibody: Absence of graft-versus host disease, induction of skin allograft tolerance, and prevention of recurrent autoimmunity in islet-allografted NOD/Lt mice. Blood 2000; 95:2175-2182.
42. van Bekkum DW. Review: BMT in experimental autoimmune diseases. Bone Marrow Transplant 1993; 11:183-187.
43. Knaan-Shanzer S, Houben P, Kinwel-Bohrè EBM et al. Remission induction of adjuvant arthritis in rats by total body irradiation and autologous bone marrow transplantation. Bone Marrow Transplant 1992; 8:333-338.
44. van Gelder M, van Bekkum DW. Effective treatment of relapsing autoimmune encephalomyelitis with pseudoautologous bone marrow transplantation. Bone Marrow Transplant 1996; 18:1029-1034.
45. Marmont AM. Defining criteria for autoimmune diseases. Immunol Today 1994; 15:388.
46. Lampeter EF, Homberg M, Quabeck et al. Transfer of insulin-dependent diabetes between HLA-identical sibling by bone marrow transplantation. Lancet 1993; 341:1243-1244.
47. Sherer Y, Shoenfeld Y. Autoimmune diseases and autoimmunity post-bone marrow transplantation. Bone Marrow Transplant 1998; 22:873-881.
48. Kishimoto Y, Yamamoto Y, Ito T et al. Transfer of autoimmune thyroiditis and resolution of palmoplantar psoriasis following allogeneic bone marrow transplantation. Bone Marrow Transplant 1997; 19:1041-1043.
49. Snowden JA, Atkinson K, Kearney P et al. Allogeneic bone marrow transplantation from a donor with severe active rheumatoid arthritis not resulting in adoptive transfer of disease to recipient. Bone Marrow Transplant 1997; 20:71-73.
50. Sturfeldt G, Lenhoff S, Sullefors B et al. Transplantation with allogeneic bone marrow from a donor with systemic lupus erythematosus: Successful outcome in the recipient and induction of a SLE flare in the donor. Ann Rheum Dis 1996; 55:638-641.
51. Marmont AM. Stem cell transpolantation for severe autoimmune diseases, with special reference to rheumatic diseases. J Rheumatol 1997; 24(48):13-18.
52. Nelson JL, Torres L, Louie FM et al. Pre-existing autoimmune disease in patients with long-term survival after allogeneic bone marrow transplantation. J Rheumatol 1997; 24(48):23-29.
53. Baldwin JL, Storb R, Thomas ED et al. Bone marrow transplantation in patients with gold-induced marrow aplasia. Arthritis Rheum 1977; 20:1043-1048.
54. Mckendry RJR, Huebsch L, Le Clair B. Progression of rheumatoid arthritis following bone marrow transplantation: A case report with a 13-years follow-up. Arthritis Rheum 1996; 96:1246-1253.
55. Snowden JA, Kearney P, Kearney A et al. Long-term outcome of autoimmune disease following allogeneic bone marrow transplantation. Arthritis Rheum 1998; 41:453-459.
56. Slavin S, Nagler A, Varadi G et al. Graft vs autoimmunity following allogeneic non-myeloablative blood stem cell transplantation in a patient with chronic myelogenous leukaemia and severe systemic psoriasis and psoriatic polyarthritis. Exp Hematol 2000; 28:853-857.
57. Lopez Cubero SO, Sullivan KM, McDonald GB. Course of Crohn's disease after allogeneic bone marrow transplantation: Report of 6 cases. Gastroenterology 1998; 114:433-440.
58. Hinterberger W, Rowlings PA, Hinterberger-Fisher M et al. Results of transplanting bone marrow from genetically identical twins into patients with aplastic anemia. Ann Intern Med 1997; 26:116-122.
59. McColl G, Koshaka H, Wicks I. High-dose chemotherapy and syngeneic hemopoietic stem cell transplantation for severe seronegative rheumatoid arthritis. Ann Intern Med 1999; 131:550-553.
60. Moore J, Brooks P. Stem cell transplantation for autoimmune diseases. In: Harrison WB, Dijkmans BAC, eds. Combination treatment in autoimmune diseases. Berlin-Heidelberg-New York: Springer 2002:193-213.
61. Zaydan MA, Turner C, Miller AM. Resolution of chronic idiopathic thrombocytopenia purpura following syngeneic peripheral blood progenitor transplant. Bone Marrow Transplant 2002; 29:87-89.
62. Young NS. Acquired aplastic anemia. Ann Intern Med 2002; 136:534-546.
63. Marmont AM. The autoimmune myelopathies. Acta Haematol 1983; 69:73-97.
64. Raetz E, Beatty PG, Rose J. Treatment of severe Evans' syndrome with an allogeneic cord blood transplant. Bone Marrow Transplant 1999; 20:427-429.

65. Oyama Y, Papadopoulos EB, Miranda M et al. Allogeneic stem cell transplantation for Evans syndrome. Bone Marrow Transplant 2001; 28:1903-1905.
66. Marmont A, Gualandi F, van Lint MT et al. Refractory Evans syndrome treated with allogeneic SCT followed by DLI. Demonstration of a Graft-versus-Autoimmunity effect. Bone Marrow Transplant 2003; 31:399-402.
67. Di Stefano P, Lecca M, Giorgiani G et al. Resolution of immune haemolytic anaemia with allogeneic bone marrow transplantation after unsuccessful autograft. Br J Hematol 1999; 100:1063-1064.
68. Barrett AJ, Childs R. Annotation. Non-myeloablative stem cell transplant. Br J Haematol 2000; 111:6-17.
69. Appelbaum FR. Hematopoietic stem cell transplantation as immunotherapy. Nat Med 2001; 411:385-389.
70. Hinterberger W, Hinterberger-Fischer M, Marmont AM. Clinically demonstrable anti-autoimmunity mediated by allogeneic immune cells favourably affects outcome after stem cell transplantation in human autoimmune diseases. Bone Marrow Transplant 2002; 30:753-759.
71. Burt RK, Arnold R, Emmons R et al. Stem cell therapy for autoimmune disease: Overview of concepts from the Snowbird 2002 tolerance and tissue regeneration meeting. Bone Marrow Transplant 2003; 32(1):S3-5.
72. Marmont AM. Perspective: Immunoablation with stem cell rescue: A possible cure for systemic lupus erythematosus. Lupus 1993; 2:151-156.
73. Slavin S. Treatment of life-threatening autoimmune diseases with myeloablative doses of immunosuppressive agents: Experimental background and rationale for ABMT. Bone Marrow Transplant 1993; 12:201-210.
74. Slavin S, Nagler A, Naparstek E et al. Non-myeloablative stem cell transplantation and cell therapy as an alternative to conventional bone marrow transplantation with lethal cytoreduction for the treatment of malignant and non-malignant disease. Blood 1998; 91:756-763.
75. Tyndall A, Gratwohl A. Blood and marrow stem transplant in auto-immune disease: A consensus report written behalf of the European League against Rheumatism (EULAR) and the European Group for Blood and Marrow Transplantation (EBMT). Bone Marrow Transplant 1997; 19:643-645.
76. Tyndall A, Fassas A, Passweg J et al. Autologous hematopoietic stem cell transplantation fro autoimmune disease-feasibility and transplant-related mortality. Bone Marrow Transplant 1999; 24:729-734.
77. McSweeney PA, Furst DE, West SG. High-dose immunosuppressive therapy for rheumatoid arthritis: Some answers, more questions. Arthritis Rheum 1999; 42:2269-2274.
78. Marmont AM, van Lint MT, Gualandi F et al. Autologous marrow stem cell transplantation for systemic lupus erythematosus of long duration. Lupus 1997; 2:151-156.
79. Burt RK, Traynor AE, Ramsey-Goldman R. Hematopoietic stem cell transplantation for systemic lupus erythematosus (Letter). N Engl J Med 1997; 357:1777-1778.
80. Traynor AE, Schroeder J, Rosa RM et al. Treatment of severe systemic lupus erythematosus with high dose chemotherapy and haemopoietic stem-cell transplantation: A phase I study. Lancet 2000; 356:701-707.
81. Rosen O, Thiel A, Massenkeil G et al. Autologous stem-cell transplantation in refractory autoimmune diseases after in vivo immunoablation and ex vivo depletion of mononuclear cells. Arthritis Res 2000; 2:237-336.
82. Rosler W, Wanger B, Repp R et al. Autologous PBPCT in a patient with lymphoma and Sjogren's syndrome: Complete remission of lymphoma without control of the autoimmune disease. Bone Marrow Transplant 1998; 22:211-213.
83. Ferraccioli G, Damato R, De Vita S et al. Haematopoietic stem cell transplantation (HSCT) in a patient with Sjogren's syndrome and lung malt lymphoma cured lymphoma not the autoimmune disease. Ann Rheum Dis 2001; 60:174-176.
84. Fassas A, Passweg JR, Anagnostopoulos A et al. Hemopoietic stem cell transplantation for multiple sclerosis: A retrospective multicenter study. J Neurology 2002; 249:1088-97.
85. Kozak T, Hardova E, Pit'ha J et al. High-dose immunosuppressive therapy with PRBL support in the treatment of poor risk systemic sclerosis. Bone Marrow Transplant 2000; 25:525-531.
86. Mancardi GL, Saccardi R, Filippi M et al. Autologous hematopoietic stem cell transplantation suppresses Gd-enhanced MRI activity in MS. Neurology 2001; 57:62-67.
87. Binks M, Passweg JR, Furst D et al. Phase I/II trial of autologous stem cell transplantation in systemic sclerosis: Procedure-related mortality and impact on skin disease. Ann Rheum Dis 2001; 60:577-584.
88. Snowden JA, Biggs JC, Milliken ST et al. A phase I-II escalation study of intensified cyclophosphamide and autologous bone marrow rescue in severe, active rheumatoid arthritis. Arthritis Rheum 1999; 42:2286-2292.
89. Burt RK, Georganas C, Schroeder J et al. Autologous hematopoietic stem cell transplantation in refractory rheumatoid arthritis. Arthritis Rheum 1999; 42:2281-2285.
90. Verburg RJ, Kruize AA, van den Hoogen FHJ et al. High-dose chemotherapy and autologous hematopoietic stem cell transplantation in patients with rheumatoid arthritis. Arthritis Rheum 2001; 44:754-760.
91. Barron S, Wallace C, Woolfrey CEA et al. Autologous stem cell transplantation for pediatric rheumatoid diseases. J Rheumatol 2001; 28:2337-2358.
92. Wulfraat NM, Sanders EAM, Kamphcris SSM et al. Prolonged remission without treatment after autologous stem cell transplantation for refractory childhood systemic lupus erythematosus. Arthritis Rheum 2000; 44:728-734.
93. Hensel M, Breitbart A, Ho AD. Autologous hematopoietic stem-cell transplantation for Behçet's disease with pulmonary involvment. N Engl J Med 2001; 344:69.
94. Gratwohl A, Passweg J, Gerber I et al. Stem cell transplantation for autoimmune diseases. Best Practice Research Clin Hematol 2001; 14:755-776.
95. Brodsky RA, Sensenbrenner LL, Jones RJ. Complete remission in severe aplastic anemia after high-dose cyclophosphamide without bone marrow transplantation. Blood 1996; 87:491-494.
96. Tisdale JF, Dunn DE, Geller N et al. High-dose cyclophosphamide in severe aplastic anaemia: A randomised trial. Lancet 2000; 356:1554-1559.
97. Brodsky RA, Sensenbrenner LL, Douglas-Smith B et al. Durable treatment-free remission after high-dose cyclophophamide therapy in previously untreated severe aplastic anemia. Ann Intern Med 2001; 135:477-483.
98. Brodsky RA, Petri M, Smith BD et al. Immunoablative high dose cyclophosphamide without stem cell rescue for refractory, severe autoimmune disease. Ann Intern Med 1998; 129:1031-1035.
99. Petri M, Jones RJ, Brodsky RA. High dose cyclophosphamide without stem cell transplantation in systemic lupus erythematosus. Arthritis Rheum 2003; 48:166-173.
100. Reiffers J, Caillot C, Dazey B et al. Abrogation of post-myeloablative chemotherapy neutropenia by ex vivo expanded autologous CD34 positive cells. Lancet 1999; 354:1002-1003.
101. McNiece I, Briddell R. Ex vivo expansion of hematopoietic precursor cells and mature cells. Exp Hematol 2001; 29:3-11.
102. Lafferty KL, Gaza LS. Costimulation and the regulation of autoimmunity. In Rose NR, Mackay IR, eds. The autoimmune diseases. San Diego: Academic Press 1999:849-872.
103. Mackay IR, Rose NR. Autoimmunity yesterday, today and tomorrow. In Rose NR, Mackay IR, eds. The autoimmune diseases. San Diego: Academic Press 1999:849-872.
104. Burt RK, Slavin S, Burns WH et al. Induction of tolerance in autoimmune diseases by haematopoietic stem cell transplantation: Getting closer to a cure? Blood 2002; 99(3):768-84.
105. Guillaume T, Rubinstein DB, Symann M. Immune reconstitution and immunotherapy after autologous hematopoietic stem cell transplantation. Blood 1998; 92:1471-1490.
106. Mackall CL. T-cell immunodeficiency following cytotoxic antineoplastic therapy: A review. Stem Cells 2000; 18:10-18.
107. Douek DC, Vescio RA, Betts MR et al. Assessment of thymic output in adults after haematopoietic stem cell transplantation and prediction of T-cell reconstitution. Lancet 2000; 355:1875-1881.
108. Burt RK, Brenner M, Burns W et al. Gene-marked autologous hematopoietic stem cell transplantation of autoimmune disease. J Clin Immunol 2000; 9:481-483.

109. Marmont AM. Immunoablation followed or not by hematopoietic stem cells as an intense therapy for severe autoimmune diseases. New perspectives, new problems. Haematologica 2001; 86:337-345.
110. Kashyap A, Passweg J, Tyndall A. Autologous hematopoietic stem cell transplant conditioning regimens for the treatment of severe autoimmune diseases. In: Dicke KA, Keating A, eds. Autologous blood and marrow transplantation. Charlottesville: Carden-Jennings, 2001:219-225.
111. Martino R, Sureda A, Brunet S. Peripheral stem cell mobilization in a refractory autoimmune Evans syndrome: A cautionary case report. Bone Marrow Transplant 1997; 20:521.
112. Burt RK, Fassas A, Snowden JA et al. Collection of hematopoietic stem cells from patients with autoimmune diseases. Bone Marrow Transplant 2001:28:1-12.
113. Armitage J. Myelodysplasia and acute leukemia after autologous bone marrow transplantation. J Oncol 2000; 18:945-946.
114. Cibere J, Sibley J, Haga M. Systemic lupus erythematosus and the risk of malignancy. Lupus 2001; 10:394-400.
115. Curtis RE. Solid cancers after bone marrow transplantation. N Engl J Med 1997; 336(13):897-904.
116. Ramsey-Goldman R, Clarke AE. Double trouble: Are lupus and malignancy associated? Lupus 2001; 10:388-391.
117. Gur-Lavi M. Long-term remission with allogeneic bone marrow transplantation in systemic lupus erythematosus. Arthritis Rheum 1999; 42:1777.
118. Battiwalla M, Barrett J. Allogeneic transplantation using non-myeloablative transplant regimens. Best Practice Research Clin Hematol 2001; 14:701-722.
119. Peggs CS, Mackinnon S. Exploiting graft-versus-tumor responses using donor leukocyte infusions. Best Practice Research Clin Hematol 2001; 14:723-739.
120. Dickinson A, Cant A. Haemopoietic stem-cell transplantation: Improving immune reconstitution, avoiding graft-versus-host disease. Lancet 2002; 360:98-99.
121. Moreland LW, Schiff MH, Baumgartner SW et al. Etanercept therapy in rheumatoid arthritis. A randomised, controlled trial. Ann Intern Med 1999; 130:478-486.
122. Stasi R, Pagano A, Stipa E et al. Rituximab chimeric anti-CD20 monoclonal antibody treatment for adults with chronic idiopathic thrombocytopenic purpura. Blood 2001; 98:952-7.
123. Lockwood CM, Thiru S, Isaacs JD et al. Long-term remission of intractable systemic vasculitis with monoclonal antibody therapy. Lancet 1993; 341:1620-2.
124. Lockwood CM. Refractory Wegener's granulomatosis: A model for shorter immunotherapy of autoimmune disease. J R Coll Physicians Lond 1998; 32: 473-8.
125. Lim SH, Hale G, Marcus RE et al. CAMPATH-1 monoclonal antibody therapy in severe refractory autoimmune thrombocythopenic purpura. Br J Haematol 1993; 84:532-4.
126. Willis F, Marsh JCW, Bevan DH et al. The effect of treatment with Campath-1 in patients with autoimmune cytopenias. Br J Haematol 2001; 114:891-8.
127. Coiffer B, Lepage E, Briere J et al. CHOP chemotherapy plus Rituximab compared with CHOP alone in elderly patients with diffuse large –B-cell lymphoma. N Engl J Med 2002; 346:235-242.
128. McKay R. Stem cells: Hype and hope. Nature 2000; 406:361-364.
129. Roeder I, Loeffler M. A novel dynamic model of hematopoietic stem cell organization based on the concept of within-tissue plasticity. Exp Hematol 2002; 30:853-861.
130. Lemischka I. A few thoughts about the plasticity of stem cells. Exp Hematol 2002; 30:848-852.

CHAPTER 28

High-Dose Immune Suppression without Hematopoietic Stem Cells for Autoimmune Diseases

Robert A. Brodsky

Introduction

Hematopoietic stem cell transplantation holds great promise for treating autoimmunity. The source of hematopoietic stem cells can be from a normal donor (allogeneic) or from the patient (autologous). Due to the high morbidity and mortality of traditional allogeneic stem cell transplantation, most centers are pursuing the use of autologous stem cell transplantation for the treatment of autoimmune diseases. There are three major components to autologous stem cell transplantation: (1) stem cell collection (mobilization); (2) conditioning with high-doses of chemotherapy; and (3) stem cell infusion. However, for treating autoimmunity with autologous stem cell transplantation, the therapeutic efficacy (high-dose immunosuppression) is derived from the conditioning regimen; hematopoietic stem cells are used as a rescue procedure. Most centers utilize mobilized peripheral blood progenitors as a source of stem cells. Cyclophosphamide (2 to 4 g/m^2) combined with granulocyte-colony stimulating factor (G-CSF) is the most frequently employed regimen to mobilize hematopoietic stem cells, but other regimens including G-CSF and corticosteroids have been used successfully. A concern with autologous stem cell transplantation for autoimmunity is that the mobilized product contains several logs of effector cells (lymphocytes) which may, theoretically, re-establish the disease. Therefore, most groups purge the autograft of contaminating lymphocytes. Due to its potent immunosuppressive activity, high-dose cyclophosphamide forms the basis of most conditioning regimens used for autoimmunity. High-dose cyclophosphamide is non-myeloablative; thus, high-dose cyclophosphamide without stem cell rescue has been employed to treat autoimmune disease.

Aplastic Anemia as an Autoimmune Disease

Aplastic anemia is a potentially fatal bone marrow failure disorder that manifests as pancytopenia in conjuction with a hypocellular bone marrow. The disease is classified as moderate, severe and very severe. Severe aplastic anemia (SAA) is defined as bone marrow cellularity of less than 25% and markedly decreased values of at least two of three hematopoietic lineages (neutrophil count <0.5 x 10^9/L, platelet count <20 x 10^9/L and absolute reticulocyte count of <60,000). Very severe aplastic anemia satisfies the above criteria except the neutrophil count is <0.2 x 10^9, while moderate aplastic anemia is characterized by a hypocellular bone marrow but with cytopenias that do not meet the criteria for severe disease. The 2 year mortality rate with supportive care alone of patients with severe or very severe aplastic is roughly 80%,[1] necessitating prompt therapeutic intervention. In contrast, moderate aplastic anemia is seldom life-threatening and in many instances requires no therapy.

Acquired aplastic anemia was originally thought to result from a quantitative deficiency of hematopoietic stem cells initiated by a direct toxic effect on stem cells. However, attempts to treat aplastic anemia patients by simple transfusion of hematopoietic stem cells from an identical twin failed to reconstitute hematopoieisis in most patients. This suggested that the pathophysiology of aplastic anemia was more complicated.[2] In the early 1970s, Mathé and colleagues were among the first to postulate an immune basis for aplastic anemia.[3] They performed marrow transplantation using partially mismatched donors in patients with aplastic anemia after administering anti-lymphocyte globulin as an immunosuppressive conditioning regimen. Although the patients failed to engraft, the investigators witnessed autologous recovery of hematopoiesis, suggesting that functional hematopoietic stem cells exist in patients with aplastic anemia and that the immune system in these patients was somehow suppressing the growth and differentiation of hematopoietic stem cells.

The cardinal event triggering aplastic anemia is seldom identified; hence, most cases of acquired aplastic anemia are classified as idiopathic, with less than 20% of cases attributed to specific drugs or viruses. Regardless of the initiating event, the pathophysiology of most cases of aplastic anemia can be attributed to an organ-specific, T cell-mediated attack. The immune destruction of hematopoietic stem cells in aplastic anemia appears to result from the direct effects of cytotoxic T cells mediated by inhibitory Th1 cytokines and the Fas-dependent cell death pathway. In aplastic anemia, the increase in cytotoxic T cells is more conspicuous in the bone marrow than in the peripheral blood.[4-6] These cytotoxic T cells overproduce interferon-γ and tumor necrosis factor (TNF),[7,8] two cytokines known to inhibit hematopoiesis. In turn, TNF and interferon-γ appear to induce the expression of Fas on CD34$^+$ cells from aplastic anemia patients leading to Fas-mediated killing.[9] While these data do not directly authenticate the autoimmune pathophysiology of aplastic anemia, they are suggestive, especially when coupled with the high response rate of the disease to immunosuppressive therapy and the association of the disease with certain HLA alleles such as HLA-DR2.[10,11]

More direct data implicating aplastic anemia as an autoimmune disease comes from studies examining T cell diversity using complementarity-determining region (CDR3) spectratyping.[12-14] Mature T cells interact through the T cell receptor, a heterodimer comprised of α and β chains. Normally, the diversity of the T cell repertoire is the consequence of somatic recombination events and random nucleotide additions that occur during the joining of the variable (V), diverse (D) and joining (J) segments from the β-chain, and of V and J segments from the α-chain. The hypervariable regions of V-D-J segments termed, CDR3, are responsible for T cell receptor antigen specificity. During normal differentiation, T cells encounter myriad different antigens which results in a polyclonal profile of T cell receptors. When T cells encounter an incriminating foreign antigen or an autoantigen, the preferential expansion of antigen-specific T cells leads to skewing of the T cell repertoire and this can be detected by CDR3 spectratyping. In aplastic anemia, CDR3 spectratyping demonstrates limited heterogeneity of the T cell receptor β-chain (BV), suggesting that there is oligoclonal or even clonal expansion of T cells in response to a specific antigen.[12,13,15] Isolated T cell clones from aplastic anemia patients secrete Th1 cytokines and are capable of lysing autologous CD34 cells.[14,16] Hence, there is now compelling data to classify acquired aplastic anemia as an autoimmune disease.

High-Dose Cyclophosphamide: Pharmacology and Stem Cell Biology

The unique pharmacology of cyclophosphamide accounts for its potent immunosuppressive properties and its ability to spare hematopoietic stem cells. Cyclophosphamide is a pro-drug that is converted in the liver to 4-hydroxycyclophosphamide and its tautomer aldophosphamide (Fig. 1). Aldophosphamide diffuses freely into the cell and is converted to the active compound phosphoramide mustard, or inactivated by aldehyde dehydrogenase to form carboxyphosphamide.[17-19] Lymphoid cells, including natural killer cells, B and T lymphocytes, have low levels of aldehyde dehydrogenase and are rapidly killed by high doses of cyclophosphamide. Interestingly, primitive hematopoietic stem cells possess high levels of aldehyde dehydrogenase rendering them highly resistant to cyclophosphamide.[17-19] Although the CD34 pool is markedly reduced in severe aplastic anemia, the high response rate in aplastic anemia to immunosuppressive therapy attests to the fact that healthy hematopoietic stem cells remain in most cases of aplastic anemia.

High-Dose Cyclophosphamide for Severe Aplastic Anemia

In 1972, the first successful bone marrow transplant in a human being was performed in a patient with severe aplastic anemia.[20] The conditioning regimen in this case was high-dose cyclophosphamide (50 mg/kg/day x 4 days) and the stem cell source was an HLA-identical sibling. Thus, the first successful bone marrow transplant was for an autoimmune disease and used a non-myeloablative conditioning regimen.

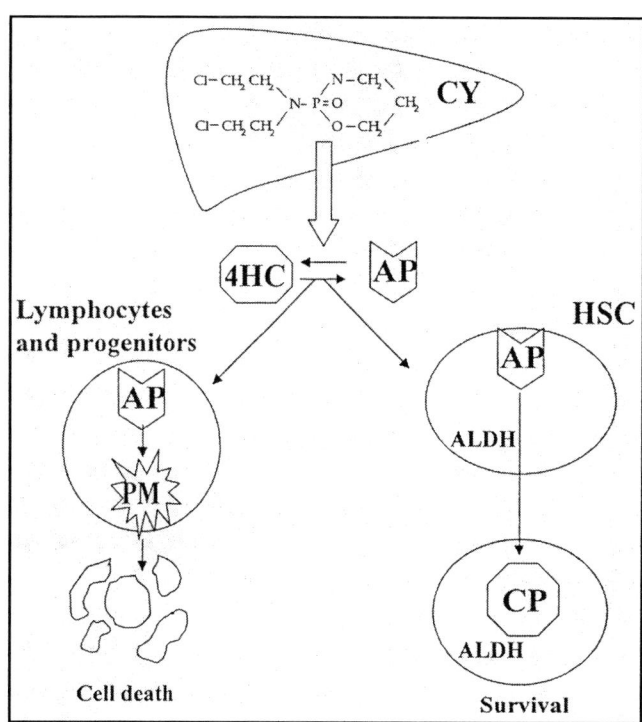

Figure 1. Pharmacology of high-dose cyclophosphamide. Cyclophosphamide (CY) is converted in the liver to 4 hydrodroxycyclophosphamide (4HC) and its tautomer aldophosphamide (AP) which diffuses into cells. Hematopoietic stem cells (HSC) have high-levels of aldehyde dehydrogenase (ALDH) which converts AP into the inert compound carboxyphosphamide (CP). Lymphocytes and progenitor cells have low levels of ALDH. In these cells, AP is converted into the active compound phosphoramide mustard (PM) leading to cell death.

By the mid-1970s, conditioning with high-dose cyclophosphamide followed by allogeneic bone marrow transplantation became the treatment of choice for young patients with severe aplastic anemia. Shortly thereafter, reports of autologous hematopoietic reconstitution following allogeneic bone marrow transplantation for SAA appeared, suggesting that high-dose cyclophosphamide alone could restore normal hematopoiesis in aplastic anemia.[21-23] In 1976, a case report in the *New England Journal of Medicine*, described a patient with aplastic anemia who was successfully treated with high-dose cyclophosphamide without bone marrow transplantation.[24] Despite the apparent success of high-dose cyclophosphamide in treating SAA, the approach received little attention for another two decades due to the improving results with allogeneic BMT and the emergence of immunosuppressive therapy with antithymocyte globulin and cyclosporine for patients not eligible for BMT. However, the high rate of relapse and secondary clonal disease (myelodysplastic syndromes and paroxysmal nocturnal hemoglobinuria) associated with immunosuppressive therapy and the high risk of severe graft-versus-host-disease associated with unrelated and mismatched bone marrow transplantation meant that durable treatment-free remission from aplastic anemia was achieved by a small percentage of patients with severe aplastic anemia, principally young patients (less than 40 years-old) who had an HLA-matched sibling donor.

In 1996, Brodsky et al, reported that high-dose cyclophosphamide without bone marrow transplantation produced durable complete remissions in seven of ten patients with SAA.[25] This trial was initiated in the late 1970s during a time period when antithymocyte globulin was temporarily unavailable. With a median follow-up exceeding 10 years, none of the patients in the trial relapsed, required additional therapy, or acquired a secondary clonal disease demonstrating that high-dose cyclophosphamide has the potential to cure patients with severe aplastic anemia. The trials small size and the fact that the patients were not treated consecutively evoked concern that the patients were highly selected and that these data would not apply to most patients with severe aplastic anemia. A group of Chinese investigators next reported on the use of high-dose cyclophosphamide in two adult patients with SAA.[26] The first patient was previously untreated and achieved a complete remission. The second patient, who was refractory to cyclosporine plus androgens and had an acute pulmonary infection prior to cyclophosphamide, died of an intracerebral hemorrhage. In Mexico, Jaime-Perez et al, treated five children with high-dose cyclophosphamide therapy who were previously unresponsive to androgen therapy (four patients) or cyclosporine therapy (1 patient).[27] Normal hematopoiesis was restored in three of the five children with follow-up of 23, 24 and 40 months, respectively. Two patients died; one on day 8 of mucormycosis and the other at three months from a cerebral hemorrhage. There were no relapses or late clonal diseases.

The initial enthusiasm for high-dose cyclophosphamide to treat severe aplastic anemia was tempered after The National Institutes of Health (NIH) reported a high incidence of fungal infections when they combined 6 months of oral cyclosporine with high-dose cyclophosphamide. These investigators initiated a randomized study of high-dose cyclophosphamide / cyclosporine versus ATG / cyclosporine with an intent to enroll 91 patients to each arm.[28] The study was terminated early due to greater toxicity in the cyclophosphamide/cyclosporine arm after enrolling 31 total patients over three years; none of the stopping rules, primary or secondary endpoints were reached. Only 13 patients in the cyclophosphamide/cyclosporine arm and 12 patients in the ATG/CSA arm were evaluable at 6 months. Mortality at 6 months in the cyclophosphamide plus cyclosporine arm (3/13) was not significantly different from the mortality in the ATG arm (1/12), (p=0.3 %). There were 5 relapses and one case of MDS in the ATG/cyclosporine arm, and one relapse in the cyclophosphamide/cyclosporine arm. The study showed slower hematopoietic recovery and a greater need for blood products and antibiotics in the cyclophosphamide/cyclosporine arm, but it was not sufficiently powered to address the important endpoints of the study: response rate, response duration, overall survival, and evolution to secondary clonal disease. The concomitant use of cyclosporine with cyclophosphamide may have increased toxicity.[29] Moreover, because cyclosporine blocks the induction of tolerance, a potential mechanism of action for cyclophosphamide, it should not be used with high-dose cyclophosphamide when treating aplastic anemia.[30]

More recently, investigators at Johns Hopkins, Hahnemann/MCP and The University of Maryland reported the results of the largest trial to date studying the efficacy of high-dose cyclophosphamide for the treatment of SAA.[31] A total of 19 patients with a median age of 47 (range, 18 to 68) years were enrolled. The survival probability was 84% at 24 and 48 months, with a

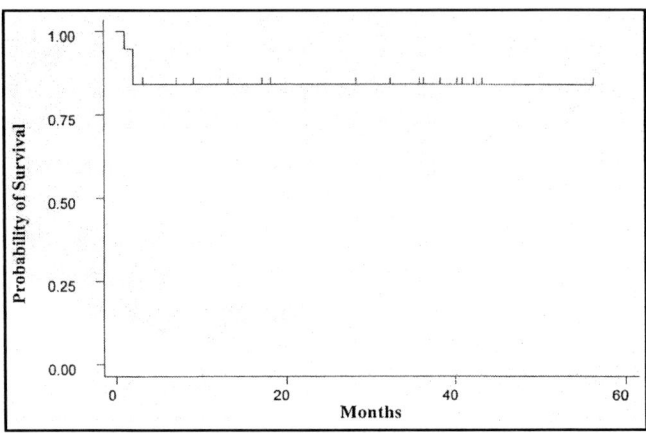

Figure 2. Kaplan-Meier probability of survival in patients with severe aplastic anemia (SAA) given high-dose cyclophosphamide as first-line therapy. Reprinted with permission from: Ann Intern Med 2001; 135:477-483.

median follow-up in surviving patients of 34 months (Fig. 2). The probability of transfusion independence was 73% at 24 months and the probability of complete response was 65% at 50 months. Three patients died; two from presumed fungal infections on days 26 and 57 after cyclophosphamide, and one from an intracerebral hemorrhage on day 67 after cyclophosphamide. All three of these patients were greater than 50 years of age and had active infections at the time of treatment. No other patients developed invasive fungal infections. All responding patients remain in a treatment-free remission with Karnofsky scores of 100. Hematopoietic reconstitution in these patients was gradual, taking a median of 49 days to achieve 0.5×10^9 neutrophils/L and median 36 months to achieve a complete remission. The median time to transfusion independence of red cells and platelets was 11 months. Of the thirteen patients with greater than one year follow-up, twelve are in treatment-free remissions (nine complete responses and three partial responses). No patient has relapsed, required additional immunosuppressive therapy, developed paroxysmal nocturnal hemoglobinuria, myelodysplasia, or other malignancies. Thus, while the early toxicity of high-dose cyclophosphamide may exceed ATG/CSA, the quality and duration of response in uncontrolled trials appears to be superior. Further study of high-dose cyclophosphamide in treating severe aplastic anemia is necessary before accepting this approach as standard therapy.

High-Dose Cyclophosphamide for Other Autoimmune Diseases

The ability of high-dose cyclophosphamide to induce durable complete remissions in one autoimmune disease, aplastic anemia, has broad implications for the management of other autoimmune and even alloimmune conditions. Increasingly, centers around the world are using high-dose cyclophosphamide followed by peripheral blood stem cell transplantation to treat a variety of refractory severe autoimmune diseases.[32-34] A concern with this approach is that autoreactive effector cells that are re-infused with the autologous graft will lead to relapse; thus, many investigators are studying ways to purge the autograft of lymphocytes.[35] Peripheral blood stem cells are clearly necessary as a rescue procedure with myeloablative conditioning regimens,

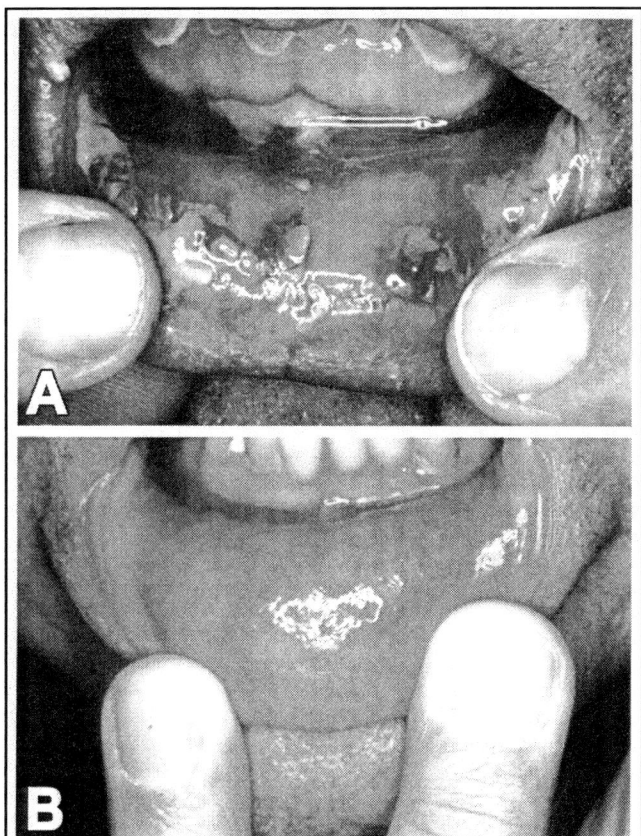

Figure 3. Inferior labial mucosa of a 70 year-old male with paraneoplastic pemphigus before (A) and 2 months after (B) high-dose cyclophosphamide. Reprinted with permission from Harcourt Health Sciences Company, J Am Acad Dermatol 1999; 40:750-754.

such as, cyclophosphamide/total body irradiation or busulfan/cyclophosphamide. While duration of cytopenia is shortened by infusion of hemtopoietic stem cells (HSC), conditioning regimens that are non-myeloablative such as high-dose cyclophosphamide alone or high-dose cyclophosphamide combined with antithymocyte globulin do not require infusion of HSC to prevent long term aplasia.

Investigators at Johns Hopkins and MCP/Hahnemann treated eight patients with a variety of severe refractory autoimmune disorders (2 systemic lupus erythematosus, 2 Felty syndrome, 1 immune thrombocytopenia, 2 autoimmune hemolytic anemia, and 1 chronic inflammatory demyelinating polyneuropathy) with high-dose cyclophosphamide alone.[36] Seven patients showed marked clinical improvement: five achieved a complete remission and 2 achieved a partial remission. Hematopoietic reconstition was rapid with median times to a neutrophil count of 500 per µL and platelet transfusion independence of 17 and 16 days, respectively. High-dose cyclophosphamide without stem cell transplantation has also been reported to induce durable complete remissions in patients with paraneoplastic pemphigus (Fig. 3).[37] and pemphigus vulgaris.[38] The patient with paraneoplastic pemphigus did not require blood products and recovered to an neutrophil count of greater than 500 per µL by day 15. The patient with pemphigus vulgaris began to recover neutrophils by day 9; he received 2 platelet transfusions, but did not require red cell transfusions. In both patients, the pathogenic autoantibodies specific for the disease became undetectable after high-dose cyclophosphamide. Durable remission following high-dose cyclophosphamide has also been reported in refractory autoimmune hemolytic anemia.[39] Moyo et al, treated 9 patients with high-dose cyclophosphamide for refractory autoimmune hemolytic anemia; 7 had an IgG warm autoantibody, one had an IgM cold agglutinin and one had both warm and cold agglutinin disease. The median hemoglobin at the time of treatment was 6.7 (range; 5-10) g/dl and 8 of the 9 patients were dependent on erythrocyte transfusions. The median times to a neutrophil count of 500 per µL and to platelet transfusion independence after high-dose cyclophosphamide was 16 and 15 days, respectively. All patients responded and became transfusion independent; 6 patients achieved a complete remission (normal untransfused hemoglobin for age and sex) and 3 patients achieved a partial remission (hemoglobin >10.0 g/dl without support of transfusions). There were no relapses at a median follow-up of 15 (range; 4-29) months and 7 of the 9 patients were able to discontinue steroids.

High-dose cyclophosphamide also has the potential to eradicate alloimmunization, a major problem in patients who require chronic blood transfusions and in patients being considered for organ transplantation.[40] Five patients with severe aplastic anemia who were refractory to platelet transfusions due to HLA-specific antibodies were studied before and after treatment with high-dose cyclophosphamide. Complete remission of aplastic anemia was achieved in four of these five patients. All four responders demonstrated a marked reduction in anti-HLA antibody titer after high-dose cyclophosphamide; in three of these patients the antibody was completely eradicated.

Conclusion

High-dose cyclophosphamide is a potent, non-myeloablative, immunosuppressive regimen that leads to durable treatment-free remissions in the majority of patients with untreated SAA. Durable complete remissions after high-dose cyclophosphamide have also been observed in a variety of other autoimmune and even alloimmune conditions. The duration of aplasia following high-dose cyclophosphamide in autoimmune diseases other than SAA is brief, usually 10 to 14 days. This approach holds great promise for the treatment of SAA and a wide spectrum of other severe autoimmune disorders. Nevertheless, more experience and longer follow-up are necessary in individual autoimmune disease such as systemic lupus erythematosus, pemphigus, and autoimmune hemolytic anemia etc., in order to determine where in the natural history of these disorders to employ high-dose cyclophosphamide. A phase III randomized controlled trial of high-dose cyclophosphamide versus monthly pulse dose cyclophosphamide for moderate to severe lupus is currently underway.

References
1. Camitta BM, Thomas ED, Nathan DG et al. A prospective study of androgens and bone marrow transplantation for treatment of severe aplastic anemia. Blood 1979; 53:504-514.
2. Hinterberger W, Rowlings PA, Hinterberger-Fischer M et al. Results of transplanting bone marrow from genetically identical twins into patients with aplastic anemia. Ann Intern Med 1997; 126(2):116-122.
3. Mathe G, Amiel JL, Schwarzenberg L et al. Bone marrow graft in man after conditioning by antilymphocytic serum. Br Med J 1970; 2:131-136.

4. Zoumbos NC, Gascón P, Djeu JY et al. Circulating activated suppressor T lymphocytes in aplastic anemia. N Engl J Med 1985; 312:257-265.
5. Maciejewski JP, Hibbs JR, Anderson S et al. Bone marrow and peripheral blood lymphocyte phenotype in patients with bone marrow failure. Exp Hematol 1994; 22:1102-1110.
6. Melenhorst JJ, van Krieken JHJM, Dreef E et al. T cells selectively infiltrate bone marrow areas with residual haemopoiesis of patients with acquired aplastic anaemia. Br J Haematol 1997; 99:517-519.
7. Nakao S, Yamaguchi M, Shiobara S et al. Interferon-gamma gene expression in unstimulated bone marrow mononuclear cells predicts a good response to cyclosporine therapy in aplastic anemia. Blood 1992; 79(10):2532-2535.
8. Nistico A, Young NS. Gamma-Interferon gene expression in the bone marrow of patients with aplastic anemia. Ann Intern Med 1994; 120(6):463-469.
9. Maciejewski JP, Selleri C, Sato T et al. Increased expression of Fas antigen on bone marrow CD34+ cells of patients with aplastic anaemia. Br J Haematol 1995; 91(1):245-252.
10. Chapuis B, Von Fliedner VE, Jeannet M et al. Increased frequency of DR2 in patients with aplastic anaemia and increased DR sharing in their parents. Br J Haematol 1986; 63(1):51-57.
11. Nakao S, Takamatsu H, Chuhjo T et al. Identification of a specific HLA class II haplotype strongly associated with susceptibility to cyclosporine-dependent aplastic anemia. Blood 1994; 84(12):4257-4261.
12. Manz CY, Dietrich PY, Schnuriger V et al. T-cell receptor beta chain variability in bone marrow and peripheral blood in severe acquired aplastic anemia. Blood Cells Mol Dis 1997; 23(1):110-122.
13. Zeng W, Nakao S, Takamatsu H et al. Characterization of T-cell repertoire of the bone marrow in immune-mediated aplastic anemia: Evidence for the involvement of antigen-driven T-cell response in cyclosporine-dependent aplastic anemia. Blood 1999; 93(9):3008-3016.
14. Zeng W, Maciejewski JP, Chen G et al. Limited heterogeneity of T cell receptor BV usage in aplastic anemia. J Clin Invest 2001; 108(5):765-773.
15. Melenhorst JJ, Fibbe WE, Struyk L et al. Analysis of T-cell clonality in bone marrow of patients with acquired aplastic anaemia. Br J Haematol 1997; 96(1):85-91.
16. Nakao S, Takami A, Takamatsu H et al. Isolation of a T-cell clone showing HLA-DRB1*0405-restricted cytotoxicity for hematopoietic cells in a patient with aplastic anemia. Blood 1997; 89(10):3691-3699.
17. Hilton J. Role of aldehyde dehydrogenase in cyclophosphamide-resistant L1210 leukemia. Cancer Res 1984; 44:5156-5160.
18. Jones RJ, Barber JP, Vala MS et al. Assessment of aldehyde dehydrogenase in viable cells. Blood 1995; 85:2742-2746.
19. Kastan MB, Schlaffer E, Russo JE et al. Direct demonstration of elevated aldehyde dehydrogenase in human hematopoietic progenitor cells. Blood 1990; 75:1947-1950.
20. Thomas ED, Storb R, Fefer A et al. Aplastic anaemia treated by marrow transplantation. Lancet 1972; 1(7745):284-9.
21. Sensenbrenner LL, Steele AA, Santos GW. Recovery of hematologic competence without engraftment following attempted bone marrow transplantation for aplastic anemia: Report of a case with diffusion chamber studies. Exp Hematol 1977; 77(1):51-58.
22. Speck B, Cornu P, Jeannet M et al. Autologous marrow recovery following allogeneic marrow transplantation in a patient with severe aplastic anemia. Exp Hematol 1976; 4:131-137.
23. Thomas ED, Storb R, Giblett ER et al. Recovery from aplastic anemia following attempted marrow transplantation. Exp Hematol 1976; 4:97-102.
24. Baran DT, Griner PF, Klemperer MR. Recovery from aplastic anemia after treatment with cyclophosphamide. N Engl J Med 1976; 295:1522-1523.
25. Brodsky RA, Sensenbrenner LL, Jones RJ. Complete remission in acquired severe aplastic anemia following high-dose cyclophosphamide. Blood 1996; 87:491-494.
26. Li Z, Yin S, Xie S et al. Treatment of severe aplastic anemia using high-dose cyclophosphamide alone in China. Haematologica 2000; 85:E06.
27. Jaime-Perez JC, Gonzalez-Llano O, Gomez-Almaguer D. High-dose cyclophosphamide in the treatment of severe aplastic anemia in children. Am J Hematol 2001; 66(1):71.
28. Tisdale JF, Dunn DE, Geller N et al. High-dose cyclophosphamide in severe aplastic anaemia: A randomised trial. Lancet 2000; 356:1554-1559.
29. Deeg HJ, Shulman HM, Schmidt E et al. Marrow graft rejection and veno-occlusive disease of the liver in patients with aplastic anemia conditioned with cyclophosphamide and cyclosporine. Transplantation 1986; 42(5):497-501.
30. Nomoto K, Eto M, Yanaga K et al. Interference with cyclophosphamide-induced skin allograft tolerance by cyclosporin A. J Immunol 1992; 149(8):2668-2674.
31. Brodsky RA, Sensenbrenner LL, Smith BD et al. Durable treatment-free remission following high-dose cyclophosphamide for previously untreated severe aplastic anemia. Ann Intern Med 2001; 135:477-483.
32. Passweg J, Fassas A, Furst D et al. Autologous stem cell transplantation for autoimmune disease. A preliminary report. Bone Marrow Trans 1998; 21((Suppl):51):S51-S51.
33. Verburg RJ, Kruize AA, van den Hoogen FH et al. High-dose chemotherapy and autologous hematopoietic stem cell transplantation in patients with rheumatoid arthritis: Results of an open study to assess feasibility, safety, and efficacy. Arthritis Rheum 2001; 44(4):754-760.
34. Traynor AE, Schroeder J, Rosa RM et al. Treatment of severe systemic lupus erythematosus with high-dose chemotherapy and haemopoietic stem-cell transplantation: A phase I study. Lancet 2000; 356(9231):701-707.
35. Burt RK, Traynor AE, Pope R et al. Treatment of autoimmune disease by intense immunosuppressive conditioning and autologous hematopoietic stem cell transplantation. Blood 1998; 92(10):3505-3514.
36. Brodsky RA, Petri M, Smith BD et al. Immunoablative high-dose cyclophosphamide without stem cell rescue for refractory severe autoimmune disease. Ann Intern Med 1998; 129(12):1031-1035.
37. Nousari HC, Brodsky RA, Jones RJ et al. Immunoablative high-dose cyclophosphamide without stem cell rescue in paraneoplastic pemphigus: Report of a case and review of this new therapy for severe autoimmune disease. J Am Acad Dermatol 1999; 40:750-754.
38. Hayag MV, Cohen JA, Kerdel FA. Immunoablative high-dose cyclophosphamide without stem cell rescue in a patient with pemphigus vulgaris. J Am Acad Dermatol 2000; 43(6):1065-1069.
39. Moyo VM, Smith BD, Brodsky I et al. High-dose cyclophosphamide for refractory autoimmune hemolytic anemia. Blood 2002; 100:704-706.
40. Brodsky RA, Fuller AK, Ratner LE et al. Elimination of alloantibodies by immunoablative high-dose cyclophosphamide. Transplantation 2001; 71(3):482-484.

CHAPTER 29

Autologous Stem Cell Transplantation in Animal Models of Autoimmune Diseases

D.W. van Bekkum

Introduction

Autologous bone marrow transplantation was introduced for the treatment of severe intractable autoimmune disease (AD) following the demonstration that impressive responses could be obtained with this modality in two different animal models: adjuvant arthritis (AA) and experimental autoimmune encephalomyelitis (EAE). The exploration of autologous stem cell grafts for treating these diseases was by no means rational, considering the prevailing view that experimental AD are stem cell associated diseases and as a consequence are expected to be curable only with allogeneic bone marrow grafts from resistant donors.

The association of the spontaneous as well as the induced types of AD in laboratory rodents with hematopoietic stem cells (HSC) was suggested more than 30 years ago by Morton and Siegel.[1] They described the development of antinuclear antibodies in normal mice following the transplantation of bone marrow from NZB mice. NZB is an inbred strain of mice that spontaneously develop a syndrome resembling systemic lupus erythematosus (SLE). Adoptive transfer of the potential to develop AD with bone marrow from susceptible animals as well as its prevention with bone marrow grafts from resistant donors was subsequently demonstrated for many other AD in experimental animals. These publications have been reviewed elsewhere.[2] The transfer of AD and allergic disorders with donor bone marrow to patients treated for leukemia or aplastic anemia has also been reported sporadically. These transfers include: thrombocytopenic purpura, thyroiditis, diabetes type I, celiac disease, and myasthenia gravis.[3]

After several groups of investigators had shown that allogeneic bone marrow transplantation (BMT) was highly effective in treating full-blown AD in animals, striking similarities were recognized between these experimental results and anecdotal clinical observations of the regression of an autoimmune affection associated with an allogeneic bone marrow transplant for treating leukemia. As a consequence, the records of long term survivors of allogeneic BMT were searched for patients with coexisting AD at the time of transplantation. So far, about 2 dozen patients were identified who suffered from rheumatoid arthritis (RA), psoriasis, Crohn's disease, ulcerative colitis, systemic lupus erythematosus (SLE) or insulin dependent diabetes type 1 (IDD1), all of whom experienced a complete remission of their autoimmune disorder.[4] Yet, allogeneic BMT with the intention to treat AD has not thus far been used in the clinic because its transplantation associated risks are considered to be unacceptably high for patients whose disease is not immediately life-threatening. The only exception being refractory idiopathic aplastic anemia where allogeneic BMT is a long-established treatment modality, primarily intended to replace the lost bone marrow.

It was only after the discovery that experimental AD could also be cured with autologous BMT, that this modality was introduced into the clinic as a treatment option for refractory AD. Thus far, the results of clinical studies, which comprise more than 500 patients, have largely confirmed the results obtained with laboratory animals and thus established a high predictive value of the disease models that were employed.

The experiments with autologous grafts were initiated after the unexpected finding that arthritic rats responded just as well to treatment with syngeneic bone marrow from healthy donors as to allogeneic marrow. The experiments with syngeneic bone marrow were part of a study on therapy with allogeneic BMT. A group of rats receiving syngeneic bone marrow was included because of the (mistaken) assumption that they would serve as negative controls. Even more surprisingly, reimmunization of these cured rats, either at 24 hours or at 28 days following the treatment, did not induce any relapses.[5] These challenging observations made it imperative to investigate what would happen when bone marrow from syngeneic rats that were actually suffering from overt AD was used. These donors were called pseudoautologous at that time. As to cellular composition and biological properties, pseudoautologous and autologous bone marrow are obviously identical. The collection of real autologous bone marrow necessitates a surgical intervention that causes additional suffering and mortality of these very sick animals. This can be avoided by the use of pseudoautologous marrow which proceeds as follows. For each experiment, about 100 rats are immunized and when the disease is fully developed, each animal is scored using a grading scale for the clinical symptoms. The animals are distributed over the various experimental groups and one group of rats are sacrificed to serve as bone marrow donors, thus assuring that the average score of all groups is similar. Animals without symptoms (10-20%) are excluded. For the sake of completeness, an experiment with real autologous marrow was also performed, with—as expected—identical results.[6] In this review, the term autologous will be used to indicate both autologous and pseudoautologous marrow. Also, bone marrow transplantation will be used for all forms of stem cell transplantation as most of the experimental and many of the clinical data referred to were obtained with bone marrow grafts.

Relevance of Animal Models of AD for the Clinic

The predictive value of experiments with animal disease models is generally felt to be greater if the causes and nature of the model disease have more resemblance to that of the human disease. Many of the human autoimmune diseases are T cell initiated or T cell mediated. In the majority, the presence of auto-antibodies in the serum is considered to be an associated- or epiphenomenon. The mechanisms of activation of T lymphocytes against self-antigens are largely unknown but certainly diverse. It may be induced by the release of an excessive amount of tissue specific antigens to which the organism has not become tolerant during development, for instance as in sympathetic ophthalmia.

Tissue inflammation and/or excessive cell death as induced by infectious agents and drugs are widely implicated as causes of normal tissue antigen release. Although the target antigens have been identified for many of the human AD, the inducing agents are unknown for most, despite extensive epidemiological research. This suggests that exposure to such antigens is ubiquitous and that affected individuals must have an unusually high responsiveness – a predisposition, that is genetically determined. However, though numerous reports describe the linkage of certain MHC genes of both the HLA and the D/DR regions with specific AD, the linkages are far from absolute.[7] Also, concordance in monozygotic twins is limited: in MS 20-30%, in Crohn's disease 44% and in RA 11%,[8] indicating that both genetic and environmental factors are involved.

The animal models of AD are of two distinct categories: the spontaneous (or hereditary) and the induced forms. In the first category, the disease develops spontaneously in a large proportion or in all of the individuals of a so-called autoimmune strain of mice or rats. Well known examples are the lupus-like syndromes in several inbred mouse strains and diabetes in NOD mice and BB rats. The induced AD require immunization with certain antigens and develop only in selected inbred strains -the susceptible ones or responders—and not in others—the resistant or nonresponder strains. The best described models are adjuvant arthritis (AA) in rats and experimental autoimmune encephalitis (EAE) which can be induced in many species. The latter is considered to be the most appropriate animal model for multiple sclerosis (MS). It was shown by cross breeding that susceptibility and resistance to inducible AD are genetically determined.

The question is what type of disease model may translate best in designing new therapeutic modalities for AD. At first sight, the induced diseases appear to be the favored candidates as they share a dual etiology—genetic and environmental—with their human counterparts. However, some of the spontaneous AD of animals may also be strongly influenced by environmental factors, such as microbes, nutrients and hormones.

In view of these data, that are reviewed elsewhere,[9] the differences between the spontaneous and the inducible AD may well be less fundamental than is generally assumed. In both disease categories, there is a genetic disposition that allows activation of anti-self immunity; in the case of the spontaneous diseases, activation is by relatively weak ubiquitous antigens; in the case of the inducible AD by strong specific antigens. Both mechanisms leave ample room for a role of infectious agents acting as the initiating stimulus, either by providing antigens that resemble tissue targets (so-called mimicry), or by acting as regulators of the immune reactivity, similar to adjuvants, hormones and dietary factors. All these determinants, genetic as well as environmental, have also been implicated in the pathogenesis of human AD. The ethiology of human AD is undisputedly multifactorial. In almost all autoimmune conditions there is a familial tendency. Many AD are induced by drugs: more than 70 different drugs have been reported to induce SLE. Furthermore, many xenobiotics e.g food supplements, heavy metals and environmental toxins, have been linked to the development of SLE-like illnesses. Although in many patients with AD the causative agent remains unknown, the low concordance in identical twins seems to argue in favor of the inducible disease models as being the more appropriate tools for translational research.

Results of Autologous and Syngeneic BMT in Experimental AD

In contrast to the preclinical experiments with autologous grafts, those with syngeneic transplants are of little practical value as identical twin donors are rarely available. However, any difference between the results of treatment with autologous and syngeneic cells is of great interest as it may shed light on the impact of activated T cells and autoimmune memory cells in the graft. As can be seen from Table 1, few published data are available on the treatment of spontaneous AD with syngeneic or autologous BMT. An early publication by Morton and Siegel[10] contains an experiment with syngeneic BMT in lethally irradiated 6 months old (NZBxDBA/2) F1 mice. At that time, 45% of the recipients were positive for anti-nuclear antibodies (ANA). This proportion did not decline after the treatment, in contrast to the near complete disappearance of ANA after allogeneic BMT. This study did not include an assessment of the clinical condition of the animals. The second publication[11] concerns the failure to cure HLA-B27 transgenic rats that suffer from AD dominated by arthritis and colitis, with syngeneic BMT as contrasted to allogeneic bone marrow. It is likely that Good and Ikehara, who extensively studied treatment with allogeneic BMT of spontaneous immune disorders, have also at some time investigated syngeneic grafts because they wrote "Our preclinical studies do not support autologous or syngeneic BMT for treatment of mice which may already have developed systemic autoimmune disease".[12] However, the data supporting this statement have not been published.

Obviously, the scarcity of investigations on the use of autologous BMT for treating spontaneous AD is based on the paradigm that holds AD as a disease of stem cells. However, there are some strong indications that some of the spontaneous SLE-like syndromes in mice might respond favorably to high dose immunomyeloablation and rescue with autologous bone marrow. One of these is from a paper by Karussis et al[20] who treated MLR/lpr mice aged 9-10 weeks with high dose cyclophosphamide or with 9 Gy total body irradiation (TBI) followed by syngeneic BMT. These mice remained disease free for at least 36 weeks, while untreated controls began to die at week 16. Unfortunately, it is not stated whether the recipients were already sick at the time of transplantation, nor was the age and disease status of the bone marrow donors reported.

There are also several reports—as reviewed by Loor et al[21]— of complete and lasting remissions in mice with spontaneous lupus-like diseases after treatment with sublethal TBI. Furthermore, long-term treatment with relatively high dose cyclophosphamide (100mg/kg weekly for 16 weeks) caused reversal of adenopathy, prevented the development of arthritis and glomerulonephritis and prolonged survival of sick MRL/lpr mice.[22] In the latter study, the follow-up was only 6 weeks after the last administration of the drug, which is not long enough. Notably, both sublethal TBI and high dose cyclophophamide cause massive destruction of the lymphatic and hematopoietic cells,

Table 1. Treatment of fully developed spontaneous and induced autoimmune disease with syngeneic or autologous BMT

Autoimmune Disease (Strain)	Bone Marrow Origin (Conditioning)	Effect	Authors
Spontaneous Onset Autoimmune Disease			
(NZBxDBA/2)F1 *lupus like (mice)*	syngeneic (TBI)	none	Morton and Siegel 1979[10]
HLA-B27 *arthritis-colitis (rats)*	syngeneic (TBI)	none	Breban et al 1993[11]
Induced Autoimmune Disease			
EAMG *(Lewis rats)*	autologous (CY plus TBI)	reduction of anti-acetylcholine receptor antibody titer, elimination of anamnestic response	Pestronk et al 1983[13]
AA *(BUF rats)*	syngeneic (TBI)	complete remission	van Bekkum et al 1989[5]
	autologous (TBI)	complete remission	Knaan-Shanzer et al 1991[6]
			van Bekkum et al 2000[14]
CIA *(DBA/1 mice)*	syngeneic (TBI)	no remission prevention of progression	Kamiya et al 1993[15]
EAE *(BUF rats)*	syngeneic (TBI)	complete remission	van Gelder et al 1993[16]
	autologous (TBI)	complete remission	van Gelder et al 1996[17]
EAE *(SJL/J mice)*	syngeneic (CY)	complete remission	Karussis et al 1992[18]
EAE *(Biozzi mice)*	syngeneic	complete remission	van Gelder et al 1995[19]
EAE *(Lewis rats)*	syngeneic (TBI)	elimination of disease causing T cell repertiore from brain	Burt et al 1995[45]
EAE *(SJL/J mice)*	syngeneic (TBI)	complete remission in early disease, no effect in chronic disease	Burt et al 1998[23]
TMEV *(SJL/J mice)*	syngeneic (TBI)	viral hyperinfection of CNS and disease exacerbation	Burt et al 1999[24]

Abbreviations: AA= adjuvant arthritis; CIA= collagen induced arthritis; EAE= experimental autoimmune encephalomyelitis; EAMG= experimental autoimmune myasthenia gravis; TBI= total body irradiation; CY= cyclophosphamide; TMEV= Theiler's murine encephalomyelitis virus-induced demyelinating disease.
Induced autoimmune disease include disease induced by adoptive transfer of lymphocytes.

which is followed by endogenous repopulation of the bone marrow and the immune system from primitive precursors. That regeneration closely resembles the repopulation following lethal TBI and autologous BMT. Hence, there is good reason to resume investigations into the responses of spontaneous AD to treatment with autologous BMT.

In contrast to the unexplored spontaneous AD, several of the induced AD have been subject to extensive studies concerning treatment with autologous and syngeneic BMT. Excellent therapeutic effects on fully developed disease were obtained in all models except for collagen induced arthritis (CIA) in mice. It should be noted that failure of these mice to enter complete remission was also observed after allogeneic BMT,[15] which suggests that either the conditioning was inadequate, or that the swelling of the joints at the time of treatment was already due to irreversible exostosis.

In addition to the results of treatment of spontaneous and induced AID as collected in Table 2, mention should be made of the experiments reported by Burt et al[23] who treated mice suffering from adoptively transferred EAE with syngeneic bone marrow. Adoptively transferred EAE is a relapsing form of EAE that can be induced in SJL/J mice by transfer of lymph node cells from sensitized syngeneic donors. Prior to the transfer, these cells are stimulated in vitro with the disease-initiating peptide called proteolipid protein 139-151(PLP). The mice were treated at the peak of the acute phase of the disease at 14 days after transfer or at day 74 during the chronic phase, with a lethal dose of TBI or with TBI and cyclophosphamide followed by rescue with syngeneic BM. Treatment in the early phase caused a significant clinical and histological improvement, but there was no effect of treatment in the chronically ill animals. This is reminiscent of the lack of responses of chronically ill Biozzi mice to treatment with allogeneic BM[19] and supports the notion that BMT cannot cure lesions that represent scar tissue. Burt et al, also reported in a CNS demyelinating disease due to ongoing viral infection, Theiler's murine encephalomyelitis virus (TMEV)-induced demyelinating disease, that syngeneic BMT caused disease exacerbation due to central nerous system hyperinfection with TMEV.[24] This suggests that if an occult or ongoing infection is causing an AD, autologous BMT would be contraindicated. Current results of HSCT for MS have not indicated the persistence of a neuropathic virus like TMEV that could exacerbate disease post-HSCT.

Table 2. Incidence of relapses after treatment of experimental AD with BMT following high dose conditioning

Disease	Remission	Spontaneous Relapse	Induced Relapse	Authors
AA rats[1]				
syngeneic	100%	0%	0%	van Bekkum et al[5]
autologous	100%	0%	6%	Knaan-Shanzer et al[6]
allogeneic	100%	0%	Not done	van Bekkum et al[5]
EAE mice[2]				
syngeneic	100%	7%	25%	Karussis et al[18]
EAE rats[3]				
syngeneic	100%	29%	44%	van Gelder et al[16]
autologous	100%	30%	72%	van Gelder et al[17]
allogeneic	100%	5%	11%	van Gelder et al[25,26]
EAMG rats[4]				
autologous	100%	Does not apply	11%[5]	Pestronk et al[13]

Abbreviations: see Table 1 Conditioning: [1]TBI 9 Gy, [2]CY, [3]TBI 10 Gy, [4]CY+TBI 6 Gy, [5]In this disease model the criterium for remission is decrease of the anti-acetylcholine receptor titer. The equivalent of relapse is a secondary antibody response following reimmunization.

Translational Research into Optimal Treatment Protocols

General

Most of the recommendations for the clinical protocols of autologous BMT were derived from the results obtained with two models of induced AD namely AA and EAE both in Buffalo (BUF) rats. The two models have a lot in common as regards the responses to treatment, but there are also notable differences. In trying to translate the results of these experiments into the clinic, one can choose to adhere strictly to each specific model e.g., use the results of the AA model only for RA and SLE treatment strategies and the results obtained with EAE only for MS. This approach is of course subject to the restrictions imposed by the imperfections of each model. A more general way to apply the results is to select from each model the conditions that seem to be most favorable and translate these into all clinical protocols until shown otherwise.

Conditioning with high dose TBI (9-10 Gy) followed by autologous BMT causes remissions of both diseases in all treated animals. In AA, 70% are complete responders and 30% partial responders. Spontaneous relapses or exacerbations are extremely rare in AA and relapses are hardly ever inducible (6%). In this disease model, the outcome seems to be dominated by the intensity of the conditioning. Even the addition of large numbers of autologous spleen cells, or lymphocytes from the lymph nodes or from the peripheral blood, did not adversely influence the responses, nor did it evoke relapses.[14] So far, this can not be explained, but it seems to be in line with the failure to passively transfer AA in BUF rats with lymphoid cells from diseased animals.

Rats with EAE respond to autologous BMT with a rapid regression of the neurological symptoms, but one or more spontaneous relapses occur in 30% of the animals.[17] When syngeneic BM instead of autologous BM was used for rescue, the spontaneous relapse rate was the same, suggesting that these relapses are initiated by autoreactive cells that have survived the conditioning. This is in accordance with the finding that T cell depletion of autologous or syngeneic BM grafts—which reduced the T cells to 0.1%—did not diminish the spontaneous relapse rate. Rat bone marrow contains 2-3% T lymphocytes. In these experiments, the number of T lymphocytes (5×10^5) that were reinfused with the unmanipulated bone marrow was in the same range as the estimated number of residual T lymphocytes (10^6), which explains the futility of T cell depletion in this situation. This does not imply that the number of T lymphocytes returned with the stem cells is irrelevant. On the contrary, the addition of autologous spleen cells containing 3×10^7 T lymphocytes to the BM graft raises the proportion of T cells to over 50% and causes the spontaneous relapse rate to rise to 93%. For comparison: unmanipulated human bone marrow grafts may contain 20-30% T cells and PBSC up to 50%.

These findings in EAE are in sharp contrast with those in arthritic animals referred to above, where addition of autologous lymphocytes had no influence whatsoever on the outcome of the treatment. This discrepancy underlines the dilemma of translational preclinical research: Is T cell depletion to be recommended for clinical transplants or not? Or only for the treatment of patients with MS? As will be explained later, the decision in this case may be made on pragmatic grounds.

Interestingly, the spontaneous relapse incidence in EAE was only 5% after allogeneic BMT, as compared to 30% after autologous and syngeneic BMT.[25,26] The difference is ascribed to elimination of residual autoimmune lymphocytes of the recipient by an immunological reaction of the grafted lymphocytes. The term graft-versus-autoimmunity reaction was proposed by A.M. Marmont in reference to the well known graft-versus-leukemia effect of allogeneic BM grafts. To enlarge upon this analogy, one might suggest that the treatment of severe AD should also be aimed at eradication of as many autoreactive cells as possible, those in the patient by conditioning, and those in the autologous graft by ex vivo purging prior to reinfusion.

The Conditioning Regimen

In both models, the best results have been obtained with the strongest lympho-myeloablative regimens e.g., the highest tolerated dose of 9-10 Gy of TBI.[14,17] Irradiation of the affected tissues only (the brain and spinal cord in the case of EAE, or the legs and tail in the case of AA) or shielding of those parts while irradiating the rest of the body, resulted at best in a limited and temporary remission.[5,16] Fractionated irradiation was studied in the AA model[14] and was shown to be as effective as single dose TBI, provided the total dose was properly adjusted upwards. In both AA and EAE, cyclophosphamide (CY) alone or busulfan (BU) alone at highest tolerated doses were less effective than high dose TBI; the combination of CY and BU was equally effective. The combination of a lower dose of TBI (4 Gy or 7 Gy) with a lower dose of CY (2 x 60 mg/kg) was also as effective as the highest dose of TBI.

In rats with experimental allergic myasthenia gravis (EAMG), conditioning with a high dose of CY alone (200 mg/kg, which requires hematopoietic rescue) induced a normalization of the high anti-acetylcholine receptor antibody titers, but had no effect on memory cells as evidenced by complete failure to prevent the anamnestic response following reimmunization. Elimination of the latter required a moderate dose of TBI (6 Gy) to be added to the conditioning.[13] Interestingly, this dose of irradiation by itself did not influence the antibody levels at all. Notable features of high dose CY as the sole conditioning agent in AA were not only the lower incidence of complete responders, but also the higher incidence (36%) of spontaneous exacerbations. In contrast, among 155 arthritic rats treated with high dose TBI or the combination regimens, only 1 relapse occurred.

The conditioning regimen of adult patients with refractory rheumatoid arthritis has consisted so far of high dose CY (200 mg/kg), either as the sole agent or combined with ATG. The complete response rate was roughly 50%, but around two thirds relapsed usually within one year.[27] Apparently, this regimen is not sufficiently lympho-ablative. It also appears to be incompletely myeloablative, as it was associated with rapid hematological recovery when used without stem cell rescue for treatment of refractory AD.[28] In this study comprising 25 patients, the complete remission rate was also 50%, which suggests that the short-term responses at least are determined by the intensity of the conditioning and not by the reinfusion of autoreactive cells with the autologous stem cell grafts. The considerations outlined above demand for more effective conditioning than is possible with CY alone.

The selection of conditioning agents should ideally be guided by their specificity for the target cells, but unfortunately these are yet poorly characterized. The most likely candidate target cells in overt AD are activated T lymphocytes and memory T lymphocytes. The phenotypical properties of these subpopulations are still incompletely defined, let alone their sensitivities to cytotoxic agents. Yet, there are some indications of differences in target specificity between radiation and CY. The study on EAMG cited above[13] showed CY to be ineffective in eliminating immunological memory. In this case, the cells involved were most likely B memory cells, but T memory cells (against Mycobacterium tuberculosis) were also reported to be CY-resistant.[29]

In addition, there is compelling clinical evidence that CY as a single agent is inadequate to ablate memory T cells involved in transplant rejection. In nonsensitized patients with aplastic anemia, the allograft failure rate is low after conditioning with high dose CY alone. However, patients sensitized by multiple blood transfusions tend to reject allogeneic bone marrow grafts, and more intensive conditioning with a combination of TBI and CY is required to obtain takes. The superiority of high dose TBI over CY alone was also convincingly demonstrated in the AA and EAE models. Accordingly, the most obvious addition to the conditioning regimen is TBI either as a single dose or in fractions. Fractionated TBI was investigated only in AA and found to be equally effective as single dose TBI.[14] The outcome of these experiments can be safely extrapolated to other AD because extensive radiobiological data show that fractionation does not produce different effects from single dose TBI, provided the total dose is adjusted for the so-called fractionation effect. This adjustment increases the total dose as the number of fractions go up. Initially, application of TBI in MS patients was not envisaged because irradiation induced an acute exacerbation of the neurological symptoms in rats with EAE. This reaction recedes after 24 hours, but was fatal in a small proportion of cases. It occurred even after a dose as low as 1.5 Gy. Fortunately, such adverse effects have not been encountered so far in MS patients after treatment with high dose TBI and CY for conditioning.[30] One may thus conclude that this particular side effect of irradiation is species specific.

In spite of the well documented superiority of irradiation, many teams engaged in treating refractory AD patients with autologous stem cell transplants avoid the use of TBI in fear of its acute toxicity and because of its several delayed side effects, especially the development of excess tumors. As regards the acute morbidity associated with TBI, its use in conditioning for allogeneic BMT for several decades has resulted in schedules that are well tolerated, as to the increased risk of tumor development, this is not an exclusive property of radiation. Cyclophosphamide, busulfan and melphalan are alkylating agents that cause similar damage to DNA as radiation, and there is no uncertainty that these agents are carcinogenic in experimental animals as well as in humans.[31-33] The exceptional position of radiation is that its adverse delayed effects are far better documented and analyzed than that of any other conditioning agent. Risk estimates of radiation have been universally accepted and provide the foundations of national laws regulating the permissible exposure of the population. It is calculated that a dose of 4 Gy TBI as used in combination with CY for the conditioning of children with juvenile chronic arthritis (JCA) carries a lifetime risk of 20% excess tumors. Such risks have to be viewed in the light of the risks involved in the continued immunosuppressive treatment given to refractory JCA patients. One survey of kidney transplant patients receiving immunosuppression with cyclosporin A reported an incidence of 25% malignancies within 6 years after transplantation[34] A similar risk calculation cannot be made for CY for lack of follow-up data of human patients. One life span study in rats conditioned with either high dose CY or high dose TBI followed by autologous BMT suggests that CY carries at least the same risk for excess tumor development as TBI.[32]

Those considerations clearly emphasize the need for more specific and less toxic conditioning. Anti-lymphocyte antibodies (ATG (anti-T cell globulin) and ALG (anti-lymphocyte globulin)) are currently used as part of the conditioning regimen in clinical protocols of autologous BMT for MS[35] and JCA.[36] ALS (anti-lymphocyte serum) was previously shown to protect against allogeneic GvHD in mice and monkeys even when administered before the bone marrow.[37,38] These observations suggest that for conditioning of AD patients, it is best to inject the last dose of ATG 24 hours or shorter before the stem cell reinfusion. It may then react with residual lymphocytes in the recipient as well as with lymphocytes that are introduced with the graft.[38] Unfortunately, the merits of ATG and its optimal application in conditioning for AD could not be investigated properly in animal models so far. The available polyclonal anti-lymphocyte antibodies against rat T lymphocytes cross-reacted with hematopoietic stem cells, which precluded reliable dose finding and made extrapolations to the clinical antibodies very difficult. In view of the experience with current ATG preparations, there is no doubt that highly specific anti-human T cell monoclonal antibodies, especially if specificity against autoreactive cells could be achieved, would be a great asset for current conditioning regimens.

One promising new drug is fludarabine, which may be more specifically immunosuppressive. It was recently used successfully (120 mg/m^2 over 4 days) in place of TBI in combination with CY and ATG for conditioning of 6 transfusion dependent patients with severe aplastic anemia for grafting of allogeneic bone marrow.[39,40] There were no take failures and all patients achieved full donor chimerism. In Cynomolgus monkeys, fludarabine (250 mg/m^2 over 5 days) induced T and B cell lymphopenia and prolonged the survival of allogeneic skin grafts both in naive and presensitized monkeys.[41] The drug schedule employed caused transient neutropenia as the only side-effect. Treatment of patients with refractory severe rheumatoid arthritis with pulsed fludarabine induced a reduction of both naive and memory CD4$^+$T cells.[42] High dose fludarabine (300 mg/m^2 in 2 courses of 5 days) was used in combination with ALG followed by autologous stem cells in a pilot study for treating patients with various severe AD.[43] This regimen was not toxic and the immediate responses resembled those after treatment with CY plus ATG, but follow-up was not long enough for other conclusions. There is definitely an urgent need for sorting out in the appropriate animal models what advantages this drug has to offer and how it can be applied best. Also, it is important to collect data on its possible delayed side effects in particular tumorgenicity.

Clearly, high priority should also be given to a search for agents with an effective therapeutic window between lymphotoxicity and myelotoxicity, and with sufficient penetration into affected tissues in the various AD such as spinal cord, brain and joints. Treatment with such agents should leave more of the stem cell population intact which is an advantage, but whether tolerance would also develop under these conditions remains to be seen. The same argument applies to the discussion about the need for nonmyeloablative regimens. With the current agents, myeloablation is largely a side effect of the immuno-ablation that is the primary objective. As long as decreasing the toxicity to the hemopoiesis is associated with a decreased kill of lymphocytes, the price for less morbidity of the conditioning is likely to be less responses and more relapses.

The Composition of the Stem Cell Graft

As mentioned earlier, rodent bone marrow contains 10 times less T lymphocytes than human bone marrow, and the T cell content of unmanipulated peripheral blood stem cells may be twice as much of that of human bone marrow harvests. For translational exercises, the most important demonstration is in EAE that excess of T lymphocytes in the autologous graft induces an excess of relapses. One can also use the finding that in the relapse prone EAE model, T cell depletion of the rat bone marrow graft did not decrease the incidence of spontaneous relapses. This would imply that a 2 log T cell depletion of blood stem cells would be sufficient in clinical practice. However, it is not known if the proportions of autoreactive cells in the human AD and the animal models are in the same range, nor even if these proportions differ per disease entity.

Fortunately there is another, more pragmatic approach to the question how many T cells should be allowed to be reinfused. It is based on the doctrine that reinfusing amounts of T lymphocytes that add substantially to the residual population, should be avoided. A bone marrow graft for an adult patient may contain as many as 4×10^9 T cells and a mobilized peripheral blood cell graft up to 2×10^{10} T lymphocytes, as compared to an estimated 3×10^8 residual T lymphocytes. The latter estimate is based on conditioning regimes equivalent to 9 Gy TBI (single dose) which causes roughly a 3 log kill of lymphocytes. The total T cell population in an adult is taken as 3×10^{11}. In view of the uncertainties of these estimates, and in analogy with the policy of maximal purging of tumor cells in autologous bone marrow grafts used for the treatment of leukemia, it was recommended to T cell deplete the autograft as completely as current techniques allow. This strategy was generally adopted following the relapse of all of the first 5 patients who received unmanipulated autologous bone marrow or mobilized peripheral blood cells.[44] The recommended maximum number of reinfused T cells was less than 10^5 per kg,[45] requiring 3 and 4 log depletion for bone marrow cells and mobilized peripheral blood cells respectively. In practice, 10^6 T cells/kg seems more realistic as a maximum; this number being well below the estimates of residual T cells ($3-5 \times 10^6$/kg). Such calculations provide the best available approach at present, with the restriction that it is unknown which subpopulations of T lymphocytes are involved in the development of relapses, and what their radio- and chemosensitivity is.

Rescue with highly purified CD34$^+$ stem cells is likely to cause a extended period of severe immunosuppression with increased risks of infections and lymphoproliferative malignancies. Several cases of a fatal "activated macrophage syndrome" have occurred recently in children following treatment for juvenile chronic arthritis with highly purified autologous stem cells. It should be noted that the CD34$^+$ cell positive selection techniques that are currently employed also remove B cells and NK cells and macrophages, which may be unnecessary and possibly harmful. That has been the rationale for some clinical teams to change to purging methods that specifically deplete T lymphocytes only. Finally, there is the option of using allogeneic stem cells, which in the EAE rat minimizes relapses and also showed a presumed graft versus autoimmune effect. Considering the higher risks of transplantation associated mortality of allogeneic BMT, its exploration should be postponed until it becomes clear from the ongoing studies with autologous stem cells, which patients might benefit from allogeneic grafts. The most threatening side effect of allogeneic BMT is graft versus host disease, which represents an anti-self immune reaction par excellence. One risks therefore to replace one severe disease with another. The use of allogeneic stem cells should specifically be discouraged in the treatment of SLE, systemic sclerosis and related syndromes because certain lesions induced by the graft versus host reactions—especially the ones associated with chronic GvHD—will be very hard to differentiate from lesions due to a relapse of the original AD.

Mechanisms of Curative Action of Autologous Stem Cell Transplantation

Successful treatment of AD is generally defined as induction of a complete or partial remission and the absence of recurrences. In MS, stabilization of progressive disease without further exacerbations is also considered as a favorable outcome. The improved chances of inducing a remission with the intensive conditioning employed in the treatment with autologous stem cell transplants came not as a surprise in view of the positive experience of the past decade with moderate doses of cytoreductive drugs such as cyclophosphamide and methotrexate in connective tissue AD. Apparently, in the refractive cases, more intensive lymphotoxicity is needed and this can be provided by higher doses of these drugs and/or irradiation.

More difficult to understand is that the remissions persist in many cases notwithstanding the reinfusion of autologous stem cells. The incidence of spontaneous and induced relapses as observed in the animal models is listed in Table 2. These rates were obtained with maximal tolerated conditioning regimens. Very few relapses were seen under those conditions in the AA model, but in the EAE model, both the spontaneous and the induced relapse rates are substantial except when allogeneic bone marrow is used. The equal incidence of spontaneous relapses following autologous and syngeneic BMT indicates that these relapses have to be ascribed to residual autoreactive lymphocytes in the host and not to cells reinfused with the autologous bone marrow. As mentioned in the previous section, the autologous graft can contribute to the relapse rate of EAE when its T cell content is raised by adding spleen cells. The incidence of induced relapses is roughly double that of the spontaneous ones. The rate is higher after autologous than after syngeneic BMT, which means that the autologous graft contributes cells that are involved in the anamnestic response. However, it is not known how to translate induced relapses in these models to the clincal situation, as the pathogenesis of the relapses in patients remains unsolved.

Following less intense conditioning, the relapse rate, both spontaneous and induced, is higher in both AA and EAE. The general conclusion is therefore that the main cause of relapse following autologous stem cell transplantation is inadequate conditioning, but that the graft may contribute if it contains more than a critical number of autoreactive lymphocytes. When the two requirements for optimal treatment i.e., intensive conditioning and T cell depleted grafts are met, the incidence of relapses can be kept low in the animal models as well as in many of the patients with AD so far treated. This outcome remains puzzling, because if autoimmunity were predominantly a disease of stem cells, one would expect recurrences to be the unavoidable consequence of rescue with an autologous graft. The most favored explanation at present is that the reconstitution of the immune system from a few stem cells is actually a recapitulation of ontogenesis which entails the acquisition of self-tolerance. Evidence for such mechanism was provided by Burt et al.[46] They found persistence of T lymphocytes that react with fragment 68-82 of myelin basic protein (MBP) in the spinal cord of Lewis rats in spontaneous clinical remission from acute EAE, but not in rats that had remitted as a result of treatment with high dose TBI and syngeneic bone marrow. Karussis et al have also claimed evidence for the development of tolerance in EAE mice after treatment with a 30% lethal dose of CY and rescue with syngeneic T cell depleted bone marrow.[47] The spontaneous relapse rate was low in the treated group (1 relapse in 15 mice) as compared to 21 relapses in 15 non treated controls. Two out of 8 treated mice suffered a relapse after rechallenge as compared to 9/9 controls. The induction of EAE and rechallenge were with mouse spinal cord homogenate in complete Freund's adjuvant. They measured the proliferation response of lymphocytes from the lymph nodes to guinea pig myelin basic protein (GMBP) and tuberculin-purified protein derivative (PPD). The responses to GMBP were weak or negative both before and after rechallenge in the treated animals as well as in the controls. The stimulation index with PPD increased from 8 before to 46 after rechallenge in the controls and from 2.3 to 3.8 in the treated group. It is doubtful if this can be regarded as evidence in support of self- tolerance, as the role of PPD in the induction of the encephalomyelitis is that of an adjuvant only.

The causes of the different responses of AA and EAE to autologous BMT remain the subject of speculation. In both models, a high percentage of complete remissions are obtained with the maximal tolerated conditioning by TBI. In AA, relapses are virtually absent, even if autologous lymphocytes are added to the bone marrow graft. In contrast, there as 30% spontaneous relapses in EAE, and the addition of autologous lymphocytes to the transplant increases this rate to over 90%. One attractive hypothesis to explain these different reactions is that self-tolerance is not broken easily if the genes determining susceptibility to AD are weakly expressed. This may be the case in AA where the disease can not even be reinduced after complete remission has been obtained. In EAE on the other hand, although the majority of the treated rats remains free of spontaneous relapses, reinduction results in a relapse in more than two thirds of the animals. At the extreme high end of the susceptibility range is the HLA-B27 transgenic rat that cannot even be brought into remission with syngeneic stem cells.[11] These animals bear up to 150 copies of the B27 gene, which assumedly make them respond to a large variety of environmental antigenic stimuli with autoimmune reactions, thereby precluding the development of self-tolerance.

As clinical experience with autologous stem cell grafts in AD patients has been generally very favorable in that remissions are obtained in the majority of patients, the main focus of interest is now on relapses. So far, the incidence of recurrences seems to vary a great deal. As yet, it cannot be concluded if each separate AD has its own specific relapse rate because the conditioning regimens have been notably suboptimal for some diseases e.g., refractory adult RA. Moreover, there are differences in conditioning regimens between different centers. Finally, especially for diseases like MS a longer follow -up is needed to allow a proper evaluation of this issue.

References

1. Morton JI, Siegel BV. Transplantation of autoimmune potential I: Development of antinuclear antibodies in H-2 histocompatible recipients of bone marrow from New Zealand Black mice. Proc Nat Acad Sci 1974; 71(6):2162-65.
2. van Bekkum DW. New opportunities for the treatment of severe autoimmune disease: Bone marrow transplantation. Clin Immunol Immunopathol 1998; 89(1):1-10.
3. Marmont AM. Immune ablation followed by allogeneic or autologous bone marrow transplantation: A new treatment for severe autoimmune disease? Stem Cells 1994; 12:125-135.
4. Nelson JL. Pre-existing autoimmune disease in patients with long term survival after allogeneic bone marrow transplantation. J Rheumatol 1997; Suppl 48:23-29.
5. van Bekkum DW. Regression of adjuvant-induced arthritis in rats following bone marrow transplantation. Proc Nat Acad Sci USA 1989; 86:10090-10094.
6. Knaan-Shanzer S. Remission induction of adjuvant arthritis in rats by total body irradiation and autologous bone marow transplantation. Bone Marrow Transplant 1991; 8:333-338.
7. Charon D. Molecular basis for human leukocyte antigen class II associations. Adv Immunol 1990; 48:107-159.
8. Silman AJ. Twin concordance rates for rheumatoid arthritis results from a nationwide study. Br J Rheumatol 1993; 32:903-907.
9. van Bekkum DW. Experimental basis of hematopoietic stem cell therapy for treatment of autoimmune diseases. J Leukocyte Biol 2002. in press.
10. Morton IL, Siegel BV. Transplantation of autoimmune potential IV. Reversal of the NZB autoimmune syndrome by bone marrow transplantation. Transplantation 1979; 2:133-134.
11 Breban M. Transfer of inflammatory disease of HLA-B27 transgenic rats by bone marrow engraftment. J Exp Med 1993; 178:1607-1616.

12. Good RA, Ikehara S. Preclinical investigations that subserve efforts to employ bone marrow transplantation for rheumatoid or autoimmune diseases. J Rheumatol 1997; 24(Suppl 48):5-12.
13. Pestronk A. Combined short-term immunotherapy for experimental autoimmune myasthenia gravis. Ann Neurol 1983; 14:235-241.
14. van Bekkum DW. Conditioning regimens for the treatment of experimental arthritis with autologous bone marrow transplantation. Bone Marrow Transplant 2000; 25:357-64.
15. Kamiya M. Effective treatment of mice with type II collagen induced arthritis with lethal irradiation and bone marrow transplantation. J Rheumatol 1993; 20:225-230.
16. van Gelder M. Treatment of experimental allergic encephalomyelitis in rats with total body irradiation and syngeneic bone marrow transplantation. Bone Marrow Tranplant 1993; 11:233-241.
17. van Gelder M. Effective treatment of relapsing experimental autoimmune encephalomyelitis with pseudoautologous bone marrow transplantation. Bone Marrow Transplant 1996; 18:1029-34.
18. Karussis DM. Chronic-relapsing experimental autoimmune encephalomyelitis (CR-EAE): Treatment and induction of tolerance, with high dose cyclophosphamide followed by syngeneic bone marrow transplantation. J Neuroimmunol 1992; 39:201-210.
19. Van Gelder M. Bone marrow transplantation for treatment of experimental autoimmune encephalomyelitis in rats. Prospects for therapy of severe multiple sclerosis. Thesis. Leiden: 1995.
20. Karussis DM. Immunomodulation of autoimmunity in MRL/lpr mice with syngeneic bone marrow transplantation (sBMT). Clin Exp Immunol 1995; 100:111-17.
21. Loor F. Radiation therapy of spontaneous autoimmunity. A review of mouse models. Int J Radiat Biol 1988; 53:119-36.
22. Smith. Cyclophosphamide-induced changes in the MRL/lpr/lpr mouse: Effects upon cellular composition, immune function, and disease. Clin Immunol Immunopathol 1984; 30:51-61.
23. Burt RK. Effect of disease stage on clinical outcome after syngeneic bone marrow transplantation for relapsing experimental autoimmune encephalomyelitis. Blood 1998; 91:2609-16.
24. Burt RK, Padilla J, Pal Canto MC, Miller SC. Viral hyperinfection of the central nervous system and high mortality after hematopoietic stem cell transplantation for treatment of Theiler's murine encephalomyelitis virus-induced demyelinating disease. Blood 1999, 94(8):2915-22.
25. van Gelder M. Treatment of relapsing experimental autoimmune encephalomyelitis in rats with allogeneic bone marrow transplantation from a resistant strain. Bone Marrow Transplant 1995; 16:343-351.
26. van Gelder M, van Bekkum DW. Treatment of relapsing experimental autoimmune encephalomyelitis with largely MHC-matched allogeneic bone marrow transplantation. Transplantation 1996; 62(6):810-81.
27. Tyndall A. Autologous hematopoietic stem cell transplantation for severe autoimmune disease with special reference to rheumatoid arthritis. J Rheumatol 2001; 28 Suppl 64:5-7.
28. Brodsky RA. Immunoablative treatment of autoimmune diseases without stem cell transplantation support. Symposium on autoimmune disease, immunoablation and stem cells: Towards YK2. 28th Annual meeting ISEH. Monte Carlo July 10 1999.
29. Orme IM. Characteristics and specificity of acquired immunological memory to Mycobacterium tuberculosis infection. J Immunol 1988; 140:3589-93.
30. Burt RK. Treatmernt of autoimmune disease by intense immunosuppressive conditioning and autologous hematopoietic stem cell transplantation. Blood 1998; 92(10):3505-14.
31. van Bekkum DW. Effectiveness and risks of total body irradiation for conditioning in the treatment of autoimmune disease with autologous bone marrow transplantation. Rheumatology 1999; 38:757-61.
32. Zurcher C. Late effects of cyclophosphamide and total body irradiation as a conditioning regimen for bone marrow transplantation in rats. Int J Radiat Biol 1987; 51:1059-68
33. Radis CD. Effects of cyclophosphamide on the development of malignancy and on long term survival of patients with rheumatoid arthritis. Arthritis Rheum 1995; 38:1120-27.
34. Dantal J. Effect of long-term immunosuppression in kidney-graft recipients on cancer incidence: Randomized comparison of two cyclosporin regimens. Lancet 1998; 351:623-28.
35. Fassas A. Peripheral blood stem-cell transplantation for treatment of progressive multiple sclerosis: First results of a pilot study. Bone Marrow Transplant 1997; 20:631-638.
36. Wulffraat NM. Autologous hemopoietic stem-cell transplantation for children with refractory autoimmune disease. Curr Rheumatol Rep 2000; 2(4):316-23.
37. van Bekkum DW. Mitigation of acute secondary disease by treatment of the recipient with antilymphocyte serum before grafting of hemopoietic cells. Exp Hematol 1970; 20:3-4.
38. van Bekkum. The effect of pretreatment of allogeneic bone marrow recipients with antilymphocyte serum on the acute GVH reaction in monkeys. Transplantation 1972; 13:400-07.
39. Chan KW. A fludarabine-based conditioning regimen for severe aplastic anemia. Bone Marrow Transplant 2001; 27:125-29.
40. Nishio M. Successful nonmyeloablative stem cell transplantation for a heavily transfused woman with severe aplastic anemia complicated by heart failure. Bone Marrow Transplant 2001; 28:783-85.
41. Goodman ER. Fludarabine phosphate, a DNA synthesis inhibitor with potent immunosuppressive activity and minimal clinical toxicity. Am Surg 1996; 62(6):435-42.
42. Davis JC Jr, Fessler BJ, Tassiulas IO et al. High dose versus low dose fludarabine in the treatment of patients with severe refractory rheumatoid arthritis. J Rheumatol 1998; 25(9):1694-704.
43. Rabusin M. Immunoablation followed by autologous hematopoietic stem cell infusion for the treatment of severe autoimmune disease. Haematologica 2000; 85(11 Suppl):81-85.
44. Euler HH. Early recurrence or persistence of autoimmune disease after unmanipulated autologous stem cell transplantation. Blood 1996; 88(9):3621-3625.
45. Tyndall A, Gratwohl A. Consensus statement on blood and stem cell transplantation in autoimmune disease. Br J Rheumatol 1997; 36:3 90-92.
46. Burt RK. Syngeneic bone marrow transplantation eliminates VB8.2 T lymphocytes from the spinal cord of Lewis rats with experimental allergic encephalomyelitis. J Neurosci Res 1995; 41:526-31.
47. Karussis DM. Prevention of experimental autoimmune encephalomyelitis and induction of tolerance with acute immunosuppression followed by syngeneic bone marrow transplantation. J Immunol 1992; 148(6):1693-98.

CHAPTER 30

Allogeneic Hemopoietic Stem Cell Transplantation in Animal Models of Autoimmune Disease

Susumu Ikehara

Introduction

Using animal models for autoimmune diseases, we show that allogeneic bone marrow transplantation (allo BMT) can be used to treat autoimmune diseases and, in addition, provide evidence that autoimmune diseases are stem cell disorders. To apply allo BMT to humans, we have very recently established a new method for allo BMT using cynomolgus monkeys. In this method, bone marrow cells (BMCs) are harvested from the long bones using a "Perfusion Method" (PM) and the whole BMCs are then injected directly into the intra-bone marrow (IBM).

We have previously found using animal models for autoimmune diseases, that allo BMT can be used to treat autoimmune diseases.[1-6] These findings have recently been confirmed even in humans.[7-15] However, in humans, the success rate of BMT across major histocompatibility complex (MHC) barriers is lowered by: (i) graft-versus-host disease (GvHD); (ii) graft rejection; and (iii) incomplete T-cell recovery. Therefore, autologous BMT (auto BMT) or peripheral blood stem cell transplantation (auto PBSCT) are the currently preferred treatments for human autoimmune diseases.[16-19] There have, however, been reports on recurrence or persistence of autoimmune diseases after auto BMT or auto PBSCT.[20] Therefore, it is important to establish a safe new method for allo BMT. In this chapter, we review previous data on the effects of BMT on autoimmune diseases, and show a new method for allo BMT, PM harvest and IBM injection, that could become a powerful strategy for the treatment of intractable diseases, including autoimmune diseases.

Historical Review of Articles on BMT in Spontaneous Animal Models of Autoimmune Diseases

In 1969, Denman et al[21] demonstrated that the transfer of spleen cells or whole bone marrow cells (BMCs) from NZB (H-2^d) mice to antilymphocyte globulin-treated BALB/c (H-2^d) mice leads to the development of autoimmune diseases in the recipients.

In 1974, Morton and Siegel[22] succeeded in transferring autoimmune diseases to irradiated BALB/c or DBA/2 (H-2^d) mice by transplanting whole BMCs from NZB mice. Based on this result, they have proposed that the hemopoietic stem cell (HSC) population of NZB mice has a primary role in the etiology of autoimmune diseases.[23] However, as they used whole BMCs (not T-cell-depleted BMCs), the possibility remains that autoreactive T cells present in the bone marrow of NZB mice are responsible for the transfer of these diseases.

In 1978, Akizuki et al[24] showed that T-cell-depleted BMCs of (NZB × NZW)F1 mice have the capacity to express their autoimmune potential in lethally-irradiated normal recipients (BALB/c or DBA/2). However, the possibility remains that abnormal B cells and antigen-presenting cells (APCs) present in the bone marrow of (NZB × NZW)F1 mice are responsible for transfer of autoimmune diseases.

In 1980, based on experiments using male and female BXSB mice, Eisenberg et al[25] showed that the rate of progress of autoimmune diseases in BXSB mice is determined by the donor BMCs, not by the environment in which these cells develop. The correction of the hematopoietic and immunologic abnormalities expressed in NZB (H-2^d) mice was obtained by BMT from DBA/2 mice.[26]

In 1985, we demonstrated that allo (not syngeneic) BMT across MHC barriers can be used to treat established autoimmune diseases in autoimmune-prone mice.[1] To our knowledge, this was the first report indicating that glomerular damage induced by lupus nephritis is reversible. Similar studies have been performed in many other experimental autoimmune disease models, such as NOD mice,[27-29] NZB/KN mice,[30] and BB rats.[31]

Historical Review of Articles on BMT in Antigen-Induced Animal Models of Autoimmune Diseases

In 1989, van Bekkum et al[32] reported that syngeneic BMT could be used to treat adjuvant-induced arthritis in rats. In 1991, Knaan-Shanzer et al[33] demonstrated that even auto BMT was effective in adjuvant-induced arthritis in rats. In 1992, Karussis et al[34] showed that syngeneic BMT could be used to prevent experimental allergic encephalitis (EAE) in mice. In 1993, van Gelder et al[35] showed that syngeneic BMT could be used to treat EAE in rats. For a complete review on syngeneic and autologous BMT in animal models (refer to Chapter 29).

Differences Between Spontaneous Autoimmune Diseases and Antigen-Induced Autoimmune Diseases

There are two types of animal models for autoimmune diseases: in one type, the disease is antigen-induced, whereas it develops spontaneously in the other. The animal models for rheumatoid arthritis and EAE are well known as belonging to the former type. However, it seems likely that the abnormality in this model of autoimmune disease resides in immunocompetent cells, particularly T cells and B cells (but not HSCs). Therefore, auto BMT, auto PBSCT, or mixed allo BMT can be used to treat antigen-induced autoimmune diseases (as described above), since autoreactive clones will be eliminated by irradiation when the BMT is carried out. In contrast, auto BMT, auto PBSCT, or mixed allo BMT cannot be used to treat spontaneous autoimmune diseases, since spontaneous autoimmune diseases are "stem cell disorders."[36] Whether human autoimmune diseases are more like antigen-induced or spontaneous onset animal diseases remains unclear.

Prevention and Treatment of Diabetes Mellitus by Allo BMT and Organ Transplantation

It is thought that autoimmune mechanisms are involved in the etiopathogenesis of Type I diabetes mellitus: humoral and cellular autoimmune responses specific for insulin-producing β cells in both humans and animal models have been well documented.[37-39] An animal model for Type I diabetes, the non-obese diabetic (NOD) mouse, was established by Makino et al.[40] More than 90% of both male and female NOD mice develop insulitis by the age of 200 days. This is followed by overt diabetes that has been shown to be due to the destruction of β cells in the pancreatic islets in 80% of the females and 20% of the males by the age of 210 days. Non-treated NOD mice die within 1 month of the development of glycosuria.

First, we attempted to prevent insulitis and overt diabetes by allo BMT. NOD mice (>4 months) were lethally irradiated and then reconstituted with T cell-depleted BALB/c BMCs. The mice were sacrificed more than 3 months after BMT. No lymphocyte infiltration was observed in the islets of the BMT-treated NOD mice. Immunohistochemical studies revealed the presence of intact β cells as well as α and δ cells. The BMT-treated NOD mice showed a normal pattern in glucose tolerance tests (GTTs). Diabetic nephropathy was also corrected by BMT.[2] Thus, BMT can prevent insulitis and overt diabetes. However, we could not treat overt diabetes in NOD mice by BMT, because mice with overt diabetes have no β cells.

We therefore performed a combined transplantation of fetal or newborn pancreas plus allogeneic BMCs, since we know that organ allografts are accepted if the organ is transplanted from the same donor as the BMCs at the same time.[41] NOD mice that had already developed overt diabetes were lethally irradiated and then reconstituted with BALB/c BMCs. The pancreatic tissues from fetal or newborn BALB/c mice were then engrafted under the renal capsules of the NOD diabetic mice. Three months after the transplantation, the mice exhibited a normal GTT pattern, and insulin levels in the sera were also normalized. Immunohistochemical studies revealed the presence of β cells in the islets engrafted under the renal capsules of the NOD mice (Fig. 1). It should be noted that neither insulitis nor rejection occurred. Thus, we succeeded in treating diabetes by the combined transplantation of the pancreas and BMCs.[42]

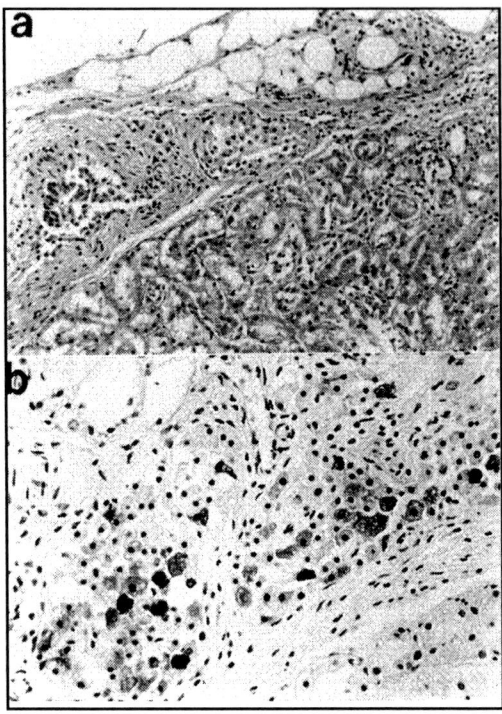

Figure 1. Histology of engrafted pancreas. Clusters of islet cells are observed under the renal capsule by hematoxylin-eosin staining (a). These cells are shown to contain insulin by means of immunohistological staining (b).

We next examined the effects of allo BMT on Type II diabetes mellitus in KK-Ay (H-2^b) mice, which are considered to be the animal model for Type II diabetes mellitus.[43] Several abnormalities have been found in this mouse strain, including impaired glucose tolerance, hyperglycemia, insulin resistance of peripheral tissue, hyperinsulinemia, and glomerular changes. KK-Ay mice reconstituted with KK-Ay BMCs [KK-Ay→KK-Ay] show glycosuria, hyperinsulinemia, and hyperlipidemia. However, KK-Ay mice (H-2^b) that were lethally irradiated (9.0 Gy) and then reconstituted with T cell-depleted BMCs from normal BALB/c mice (H-2^d) [BALB/c→KK-Ay] showed negative urine sugar with decreases in serum insulin and lipid levels 4 months after BMT (Fig. 2). Morphological recovery of islets and glomeruli was also noted after allo BMT. These findings suggest that BMT can be used to treat not only a certain type of Type II diabetes mellitus, but also its complications such as hyperlipidemia and diabetic nephropathy.[6]

Since it is technically quite difficult to transplant the pancreas in humans, we next attempted to transplant pancreatic islets (PIs) in conjunction with allo BMT in rats. We have previously found that the administration of allogeneic cells via the portal vein (PV) induces donor-specific tolerance across MHC barriers,[44] and that donor hemopoietic stem cells (HSCs), which are trapped in the liver after PV injection, induce anergy to host CD8$^+$ T cells due to the absence of costimulatory signals.[45] In addition, we have found that the injections of BMCs via the PV plus IV are effective in inducing persistent tolerance not only in chimeric-resistant MRL/lpr mice[46] but also in skin allografts of mice.[47]

We therefore attempted to examine the effect of PV injection of BMCs on tolerance induction using the PI transplantation system. To detect the diabetic condition, nonfasting blood glucose levels were monitored every other day in F344 rats in which

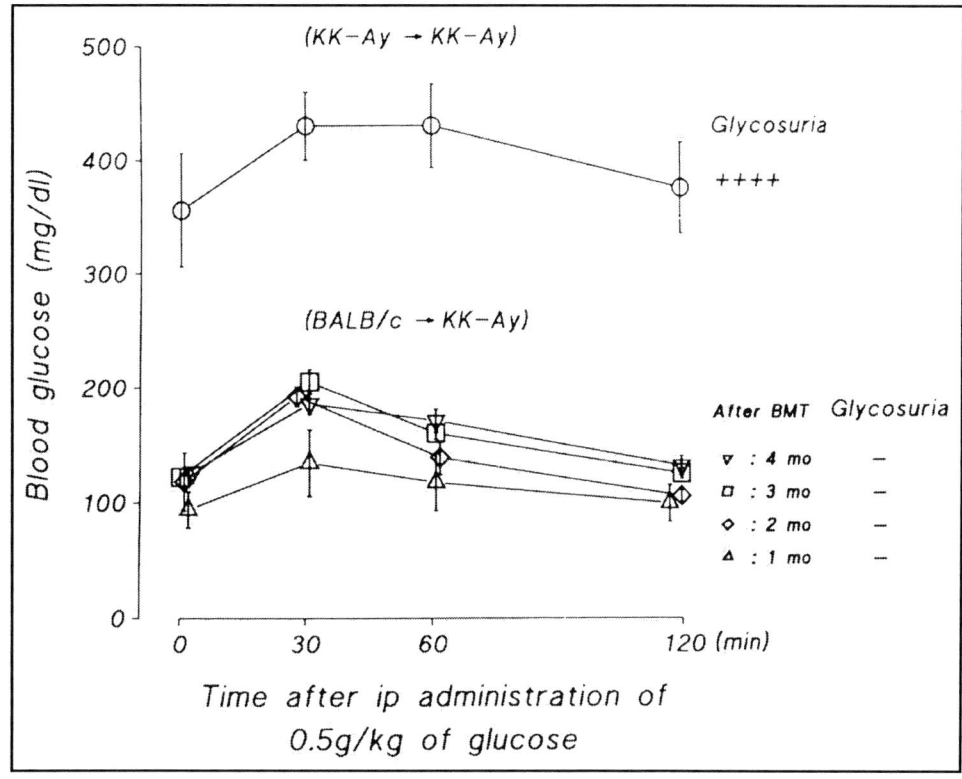

Figure 2. Glucose tolerance tests (GTTs) in (KK-Ay→KK-Ay) and (BALB/c→KK-Ay) mice; syngeneic group (O) and allogeneic group; 1 mo (Δ), 2 mo (◊), 3 mo (□), and 4 mo (∇) after BMT.

diabetes mellitus had been induced by streptozotocin (STZ). Plasma glucose levels in all recipient rats returned to normal within 24 hours after the PI transplantation. We monitored the graft survival by measuring nonfasting blood glucose levels. In the "8Gy+PV+IV", "8.5Gy+IV+IV", and "9Gy+IV+IV" groups, all grafts were rejected within 30 days. The "9Gy+PV" group showed a 70% graft survival rate 320 days after transplantation, whereas the "9Gy+PV+IV" group showed a 100% graft survived rate more than one year after the PI transplantation. All rats (10/10) in the "9Gy+PV+IV" group showed normoglycemia for more than one year. The graft survival rate in the "8.5Gy+PV+IV" group was 44% 150 days after transplantation (Fig. 3). In the "8.5Gy + PV" group, only one in seven rats accepted the PIs until day 185. These results suggest that simultaneous PV injection of BMCs induces tolerance to allo PIs more easily than the IV injection of BMCs, and that the additional IV injection of BMCs maintains the tolerance induced by the PV injection of BMCs.[48] We thus succeeded in preventing and treating both Type I and Type II diabetes mellitus by allo BMT alone or in conjunction with pancreas or islet transplantation.

Prevention and Treatment of Both Organ-Specific and Systemic Autoimmune Diseases by Allo BMT

We have found that (NZW x BXSB)F1 (W/BF1) mice, which develop lupus nephritis with myocardial infarction, show thrombocytopenia with age, and that this can be attributed to the presence of both platelet-associated and circulating anti-platelet Abs. In addition, we have found that myocardial infarction in W/BF1 mice is associated with the presence of anti-cardiolipin antibodies (Abs); this mouse seems to be an animal model for anti-phospholipid syndrome.[49]

The transplantation of BMCs from normal to W/BF1 mice was found to exert preventative and curative effects on lupus nephritis, thrombocytopenia and anti-phospholipid syndrome; the platelet counts were normalized, and circulating anti-platelet Ab levels as well as anti-cardiolipin Ab levels were reduced.[49]

Evidence for Autoimmune Diseases as Stem Cell Disorders

We first attempted to transfer Type I diabetes mellitus to normal mice by transplanting NOD BMCs to C3H/HeN mice. Female C3H/HeN (H-2^k) mice were lethally irradiated (9.5Gy) at the age of 8 weeks and then reconstituted with the T cell-depleted BMCs from young (<8 weeks) female NOD (K^d, 1-Ag^7, D^b) mice. Two of four [NOD→C3H/HeN] chimeric mice developed both insulitis and overt diabetes more than 40 weeks after the BMT; β cells were selectively destroyed by the infiltration of T cells. These mice exhibited elevated glucose levels and abnormal glucose tolerance curves.

The next step was to investigate whether both systemic (SLE) and organ-specific (ITP) autoimmune diseases in W/BF1 mice could be transferred to normal mice by BMT. We used W/BF1 (H-2^z/H-2^b) mice as donors and C3H/HeN (H-2^k) or C57BL/6J (H-2^b) mice as recipients. C3H/HeN or C57BL/6J mice were lethally irradiated (9.5Gy) and then reconstituted with T cell-depleted BMCs from young (<8 weeks) male W/BF1 mice. The [W/BF1→C57BL/6J] mice showed thrombocytopenia by 5 months after BMT.[50]

To confirm that the defective HSCs were indeed the elements responsible for the development of the autoimmune diseases, we transferred the cells in a HSC-enriched fraction (wheat germ agglutinin [WGA]-binding [WGA$^+$] cells) of W/BF1 BMCs to C3H/HeN mice, since both Visser et al,[51] and we[52] have found

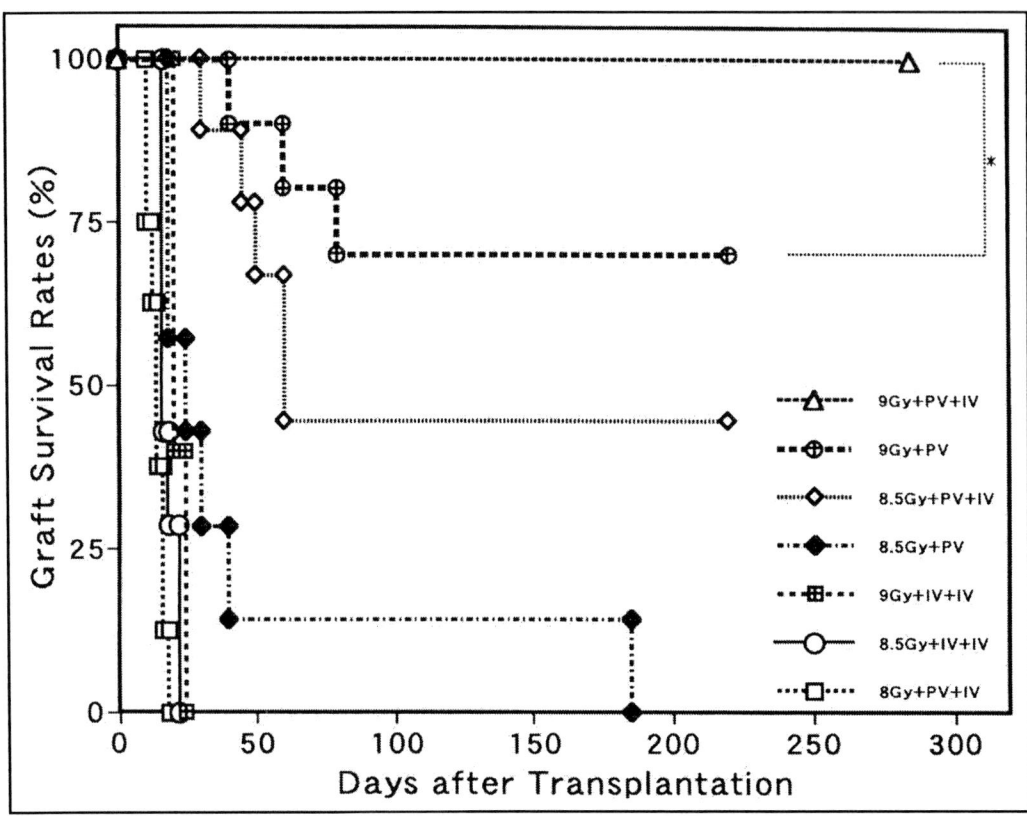

Figure 3. Graft survival of transplanted pancreatic islets (PIs). After irradiation, the PIs of BN rats were transplanted into F344 rats with bone marrow cells (BMCs) of BN rats via the portal vein (PV) or intravenous (IV). Additional IV injections of BMCs from BN rats were performed in some groups. (*$p<0.01$).

that spleen colony-forming units are enriched in WGA$^+$ cells. We therefore attempted to transfer autoimmune diseases to normal mice by the transplantation of partially-purified HSCs (WGA$^+$ cells) from W/BF1 mice. Although the C3H mice that had received 1×10^5 WGA$^+$ cells (without bone grafts) from W/BF1 mice died due to graft failure, those that received WGA$^+$ HSCs plus bone grafts from the W/BF1 mice began to show proteinuria (++) and thrombocytopenia three months after the transplant (Fig. 4). H-2 typing revealed that the hematolymphoid cells were donor-derived. All the mice died of renal failure due to lupus nephritis by 300 days after the transplants. The survival rates were similar to those in C3H mice that had received T cell-depleted (TCD) BMCs from W/BF1 mice. Anti-platelet Abs were detected in the sera of these mice. We thus succeeded in inducing autoimmune diseases (SLE and ITP) in normal mice by transplanting partially purified HSCs with bone grafts from W/BF1 mice.[53]

Treatment of Autoimmune Diseases by Allo BMT in Chimeric-Resistant MRL/lpr Mice

It is well known that MRL/lpr mice develop SLE and rheumatoid arthritis (RA). We have recently found that the MRL/lpr mouse is a suitable model for establishing a safe new strategy for allo BMT, since the MRL/lpr mouse itself is radio-sensitive (<8.5 Gy), while the abnormal hemopoietic stem cells of the MRL/lpr mouse are radio-resistant (>8.5 Gy). Conventional BMT (8.5 Gy irradiation plus allo BMT) has a transient effect on the autoimmune disease, which recurs 3 months after the BMT.[54] However, we have found that BMT plus bone grafts (to recruit donor stromal cells) completely prevents the recurrence of autoimmune diseases in MRL/lpr mice.[55] Donor-derived stromal cells seem to play a crucial role in successful allo BMTs,[55,56] since there is an MHC restriction between HSCs and stromal cells.[57] We have, however, found that the combination of BMT plus bone grafts has no effect on the treatment of autoimmune diseases in MRL/lpr mice,[58] since MRL/lpr mice become more radiosensitive after the onset of lupus nephritis due to uremic enterocolitis. To reduce the cytotoxic effect of radiation on the intestine, we carried out fractionated irradiation and attempted to devise a new strategy.

Recently, we have found that most donor HSCs are trapped and retained in the liver when they are injected either portal venously (PV) or even intravenously (IV), and that the HSCs induce anergy to host CD8$^+$ T cells.[45] In addition, we have found a strategy (the PV plus the supplemental IV injections of donor whole BMCs) that induces persistent tolerance in the skin allograft system.[47] On the basis of these findings, we have very recently established a new strategy for allo BMT: fractionated irradiation (5.5Gy x 2) and the PV plus IV injections of whole BMCs[46] (the data are shown in the next section). However, this method has two demerits for human patients: (i) laparoscope-guided injection of BMCs via the PV is necessary; and (ii) an additional IV injection is essential for obtaining a 100% success rate.[46] We have analyzed the mechanism underlying the tolerance induced by the PV injection of BMCs and noted the importance of donor-derived stromal cells trapped in the liver, which facilitate the proliferation and differentiation of donor HSCs. Based on these findings, we attempted to inject whole BMCs (including stromal cells) directly into the bone marrow (intra-bone marrow [IBM] injection).

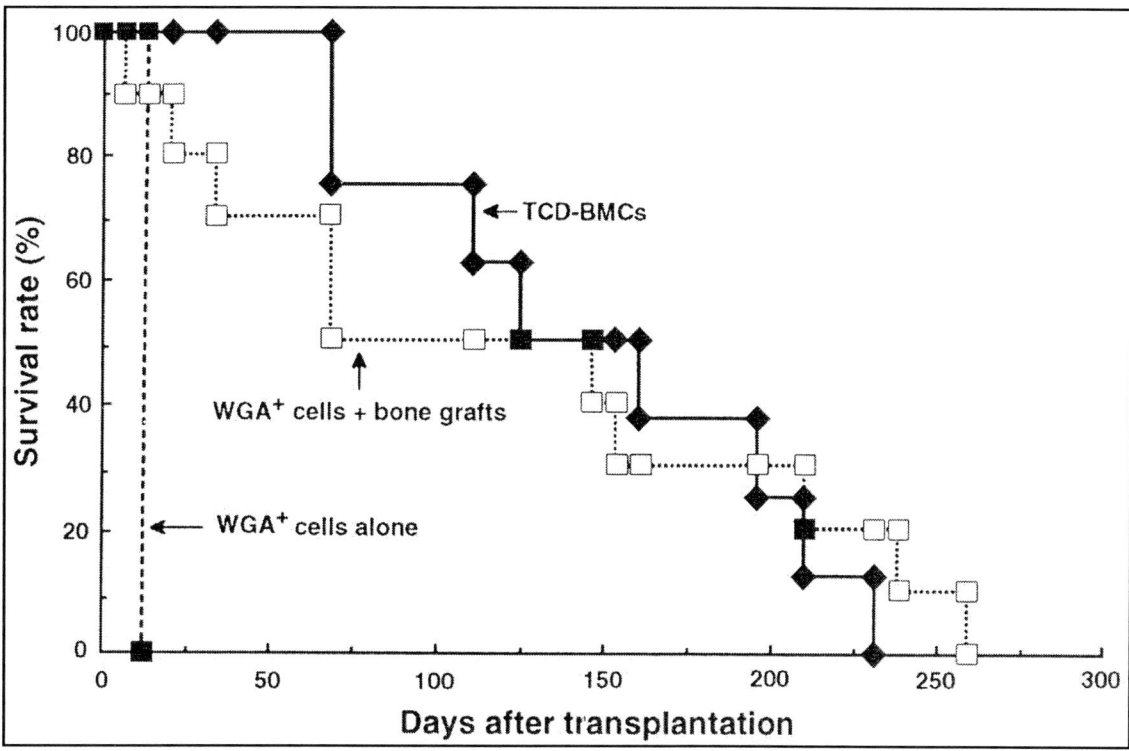

Figure 4. Survival rates in (W/BF1→C3H) chimeric mice. C3H/HeN mice were irradiated (9.5 Gy) and then reconstituted with either 1-2 × 10^7 T cell depleted (TCD)-bone marrow cells (BMCs) (♦—♦) or 1-2 × 10^5 WGA⁺ (wheat germ agglutin binding, hematopoietic stem cell enriched fraction) cells plus 40 Gy-irradiated bone grafts (□ — □) or WGA⁺ cells alone from male W/BF1 mice (< 2 mo) (■ — ■).

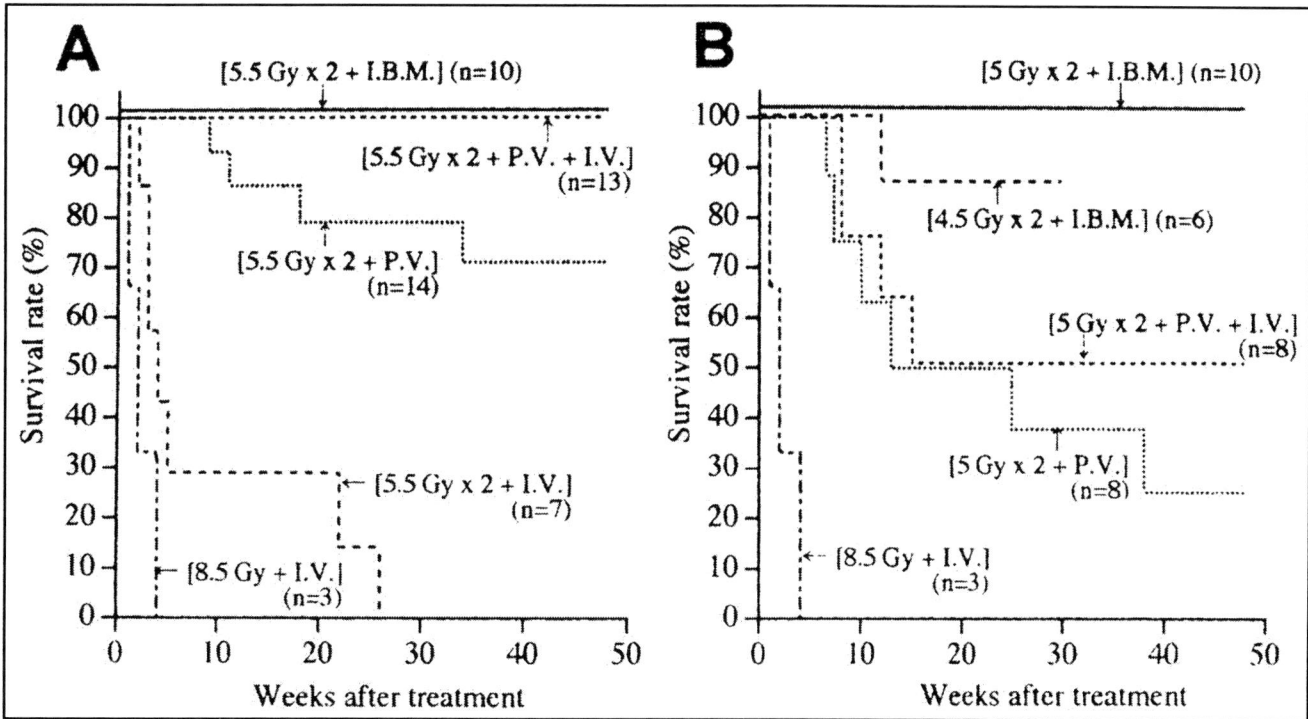

Figure 5. Survival rates in MRL/lpr mice treated by various methods. A) MRL/lpr mice irradiated with 5.5 Gy × 2 were injected with bone marrow cells (BMCs) via the peripheral vein (intravenously, IV), via the portal vein (PV), or into the bone marrow cavity (intra-bone marrow, IBM). As a negative control, the mice irradiated with 8.5 Gy received BMCs via the IV. Numbers in parentheses represent the numbers of mice used in each group. Statistical analyses were carried out by a log-rank test: $P < .001$, 5.5Gy × 2 + IBM versus 5.5 Gy × 2 + PV. B) MRL/lpr mice irradiated with 5 Gy × 2 were injected with BMCs IV, via the PV, or via the IBM. Furthermore, the mice that had been irradiated with 4.5 Gy × 2 were injected with BMCs via the IBM.

Treatment of Autoimmune Diseases in MRL/lpr Mice by PV-BMT or IBM-BMT

The MRL/lpr mice with lymphadenopathy and a high level of proteinuria (after the onset of autoimmune diseases) were treated with BMT via the IBM, PV, or IV As shown in Fig. 5A, all the mice treated with [8.5 Gy + IV] died within 4 weeks due to the side effects of radiation, as we previously reported.[58] The ineffectiveness of BMT via the IV route was again observed when the recipients received fractionated irradiation to reduce the side effects of radiation [5.5 Gy x 2 + IV]: all the mice died within 30 weeks (Fig. 5A). In contrast, more than 70% of MRL/lpr mice treated with fractionated irradiation and BMT via the PV [5.5 Gy x 2 + PV] survived more than one year after the treatment. These findings suggest that the PV injection of BMCs is more effective in prolonging survival than the IV injection. The supplemental injection via the IV [5.5 Gy x 2 + PV + IV] completely cured the autoimmune diseases in the MRL/lpr mice; 100% of the mice survived one year after the treatment, indicating that the supplemental IV injection is helpful for successful engraftment, as we previously reported.[46]

We next examined whether the radiation dose could be reduced. MRL/lpr mice treated with either [5 Gy x 2 + PV] or [5 Gy x 2 + PV + IV] showed survival rates of 30% and 50%, respectively (Fig. 5B). These findings indicate that the 5 Gy x 2 irradiation alone (without using CY) is insufficient to prevent graft rejection.[58] Therefore, we attempted to establish a new method for BMT that prevents the graft rejection even under the reduced radiation dose. BMCs were injected directly into the bone marrow (IBM). Surprisingly, all the recipients that received "IBM-BMT" survived 48 weeks after the treatment without showing any signs of graft rejection or recurrence of autoimmune diseases when treated with [5 Gy x 2 + IBM] (Fig. 5B) as well as [5.5Gy x 2 + IBM] (Fig. 5A). Furthermore, more than 85% of the MRL/lpr mice survived 30 weeks after the treatment even when treated with [4.5 Gy x 2 + IBM] (Fig. 5B). When ≥ 5 x 10^6 cells were injected directly into the femur cavity, the same result was obtained as when 3 x 10^7 cells were injected.[59]

Prospects for "IBM-BMT" as Treatment of Various Intractable Diseases in Humans

IBM-BMT seems to be the best strategy for allogeneic BMT for the following reasons: (i) no GvHD develops even if the T cells are not depleted from the bone marrow; (ii) no graft failure occurs even if the dose of radiation as the conditioning for BMT is reduced to 5Gy x 2; (iii) hemopoietic recovery is rapid; and (iv) the restoration of T cell functions is complete even in donor-recipient combinations across the MHC barriers, since donor-derived stromal cells migrate into the thymus where they are engaged in positive selection.[60] This "IBM-BMT" is therefore applicable to humans, since intraosseous (i.o.) infusion (IBM-injection) is an established method for administering fluids, drugs, and blood to critically ill patients, and particularly infants.[61] Indeed, Hagglund et al have recently compared the effectiveness of i.o. infusion with that of i.v. infusion in human allogeneic BMT;[62] they have concluded that allogeneic BMT can be safely performed by i.o. infusion, but the incidences of acute and chronic GvHD, transplantation-related mortality, and survival rates are similar. However, they aspirated the donor BMCs from the iliac bones and infused these BMCs into the iliac bones of the recipients.

Using cynomolgus monkeys, we have just established a new method ("Perfusion Method [PM]") for collecting BMCs from the long bones (humerus, femur, etc.) without contaminating the peripheral blood (Fig. 6).[63] This method has three main advantages: (i) no GvHD develops even in cynomolgus monkeys, since the percentage of T cells in the BMCs thus collected is less than 6%; (ii) a large number of BMCs can be collected quickly and safely; and (iii) the BMCs thus collected contain stromal cells, including mesenchymal stem cells. Indeed, we have very recently found using TsK mice[64] that IBM-BMT can be used to treat not only scleroderma but also emphysema in TsK mice.[65] We, therefore, believe that this method (IBM-BMT in conjunction with the "perfusion method") will become a powerful new strategy for not only allo BMT, but also organ transplantation in conjunction with BMT. Furthermore, this method would become a valuable strategy in regeneration therapy for injured organs and tissues (myocardial infarction, cerebral infarction, Alzheimer's disease, etc.), since IBM-BMT can efficiently reconstitute the recipient with both donor-derived hemopoietic stem cells and mesenchymal stem cells.

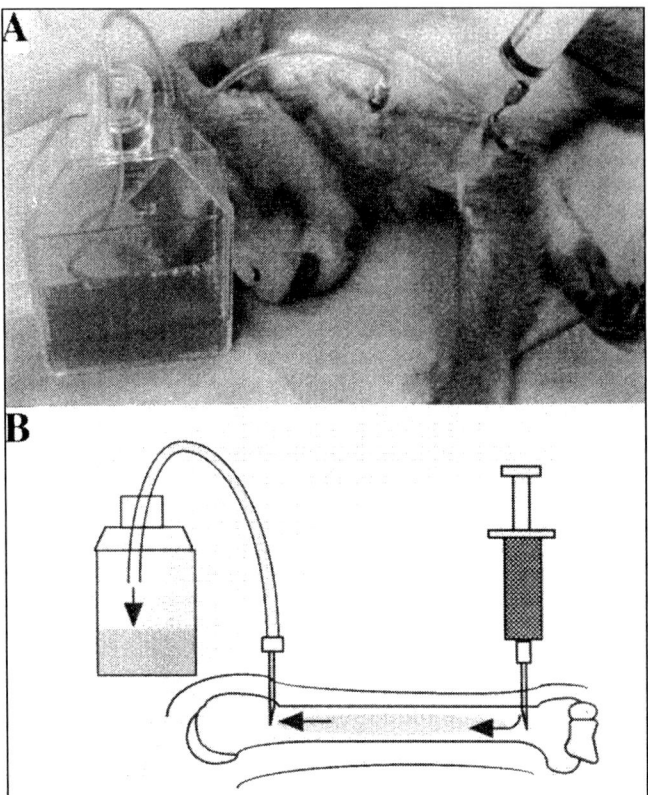

Figure 6. Perfusion method. Two needles were inserted into a humerus. One needle was connected to an extension tube and the end of the tube was inserted into a culture flask to collect the bone marrow (BM) fluid. The other needle was connected to a syringe containing 30 ml of phosphate buffered saline (PBS). The solution was pushed gently from the syringe into the BM cavity. The medium containing BM fluid was collected into the flask.

Acknowledgments

Supported by a grant from "Haiteku Research Center" of the Ministry of Education, grant-in-aid for scientific research (B) 11470062, grants-in-aid for scientific research on priority areas (A) 10181225 and (A) 11162221; a grant from "Millennium" of Ministry of Education, Culture, Sports, Science and Technology; and a grant from the "Science Frontier" program of the Ministry of Education, Culture, Sports, Science and Technology; a grant from the "21st Century COE Program" of the Ministry of Education, Culture, Sports, Science and Technology; Health and Labour Sciences research grants (Research on Human Genome, Tissue Engineering, Food Biotechnology); and also a grant from the Department of Transplantation for Regeneration Therapy (sponsored by Otsuka Pharmaceutical Co., Ltd.); a grant from Molecular Medical Science Institute, Otsuka Pharmaceutical Co.; and a grant from Japan Immunoresearch Laboratories Co., Ltd., (JIMRO).

We thank Mr. Hilary Eastwick-Field and Ms. K. Ando for their help in the preparation of the manuscript.

References

1. Ikehara S, Good RA, Nakamura T et al. Rationale for bone marrow transplantation in the treatment of autoimmune disease. Proc Natl Acad Sci USA 1985; 82:2483-2487.
2. Ikehara S, Ohtsuki H, Good RA et al. Prevention of type I diabetes in nonobese diabetic mice by allogeneic bone marrow transplantation. Proc Natl Acad Sci USA 1985; 82:7743-7747.
3. Oyaizu N, Yasumizu R, Inaba-Miyama M et al. (NZW x BXSB)F1 mouse, a new model of idiopathic thrombocytopenic purpura. J Exp Med 1988; 167:2017-2022.
4. Nakagawa T, Nagata N, Hosaka H et al. Prevention of autoimmune inflammatory polyarthritis in male New Zealand black/KN mice by transplantation of bone marrow cells plus bone (stromal cells). Arthritis Rheum 1993; 36:263-268.
5. Nishimura M, Toki J, Sugiura K et al. Focal segmental glomerular sclerosis, a type of intractable chronic glomerulonephritis, is a stem cell disorder. J Exp Med 1994; 179:1053-1058.
6. Soe Than, Ishida H, Inaba M et al. Bone marrow transplantation as a strategy for treatment of non-insulin-dependent diabetes mellitus in KK-Ay mice. J Exp Med 1992; 176:1233-1238.
7. Baldwin JL, Storb R, Thomas ED et al. Bone marrow transplantation in patients with gold-induced marrow aplasia. Arthritis Rheum 1977; 20:1043-1048.
8. Jacobs P, Vincent MD, Martell RW. Prolonged remission of severe refractory rheumatoid arthritis following allogeneic bone marrow transplantation for drug-induced aplastic anemia. Bone Marrow Transplant 1986; 1:237-239.
9. Lowenthal RM, Cohen ML, Atkinson K et al. Apparent cure of rheumatoid arthritis by bone marrow transplantation. J Rheumatol 1993; 20:137-140.
10. Eedy DJ, Burrows D, Bridges JM et al. Clearance of severe psoriasis after allogeneic bone marrow transplantation. Br Med J 1990; 300:908.
11. Yin JA, Jowitt SN. Resolution of immune-mediated diseases following allogeneic bone marrow transplantation for leukaemia. Bone Marrow Transplant 1992; 9:31-33.
12. Marmont AM. Immune ablation followed by allogeneic or autologous bone marrow transplantation. A new treatment for severe autoimmune diseases? Stem Cells 1994; 12:125-135.
13. Sullivan KM, Furst DE, eds. Role of Hematopoietic Stem Cell Transplantation for Autoimmune Diseases. Seattle: Fred Hutchinson Cancer Research Center, J Rheumatol 1997; 24:1-102.
14. Nelson JL, Torrez R, Louie FM et al. Pre-existing autoimmune diseases in patients with long-term survival after allogeneic bone marrow transplantation. J Rheumatol 1997; 24:23-29.
15. Snowden JA, Kearney P, Kearney A et al. Long-term outcome of autoimmune disease following allogeneic bone marrow transplantation. Arthritis Rheum 1998; 41:453-459.
16. Marmont AM. Stem cell transplantation for severe autoimmune diseases: Progress and problems. Haematologica 1998; 83:733-743.
17. Burt RK. BMT for severe autoimmune diseases: An idea whose time has come. Oncology 1997; 11:1001-1024.
18. Tyndall A, Black C, Finke J et al. Treatment of systemic sclerosis with autologous haematopoietic stem cell transplantation. Lancet 1997; 349:254.
19. Traynor AE, Schroeder J, Rosa RM et al. Treatment of severe systemic lupus erythematosus with high-dose chemotherapy and haemopoietic stem-cell transplantation: A phase I study. Lancet 2000; 356:701-707.
20. Euler HH, Marmont AM, Bacigalupo A et al. Early recurrence or persistence of autoimmune diseases after unmanipulated autologous stem cell transplantation. Blood 1996; 88:3621-3625.
21. Denman Am, Russell SA, Denman EJ. Adoptive transfer of the diseases of New Zealand Black mice to normal mouse strains. Clin Exp Immunol 1969; 5:567-595.
22. Morton JI, Siegel BV. Transplantation of autoimmune potential. I. Development of antinuclear antibodies in H-2 histocompatible recipients of bone marrow from New Zealand Black mice. Proc Natl Acad Sci USA 1974; 71:2162-2165.
23. Morton JI, Siegel BV. Transplantation of autoimmune potential IV. Reversal of the NZB autoimmune syndrome by bone marrow transplantation. Transplant 1979; 2:133-134.
24. Akizuki M, Reeves JP, Steinberg AD. Expression of autoimmunity by NZB/NZW marrow. Clin Immunol Immunopathol 1878; 10:247-250.
25. Eisenberg RA, Izui S, McConahey PJ et al. Male determined accelerated autoimmune disease in BXSB mice: Transfer by bone marrow and spleen cells. J Immunol 1980; 125:1032-1036.
26. Jyonouchi H, Kincade PW, Good RA et al. Reciprocal transfer of abnormalities in clonable B lymphocytes and myeloid progenitors between NZB and DBA/2 mice. J Immunol 1981; 127:1232-1238.
27. Serreze DV, Leiter EH, Worthen SM et al. NOD marrow stem cells adoptively transfer diabetes to resistant (NOD x NOD)F1 mice. Diabetes 1988, 37:252-255.
28. Wicker LS, Miller BJ, Chai A et al. Expression of genetically determined diabetes and insulitis in the nonobese diabetic (NOD) mouse at the level of bone marrow-derived cells: Transfer of diabetes and insulitis to nondiabetic (NOD x B10)F1 mice with bone marrow cells from NOD mice. J Exp Med 1988; 167:1801-1810.
29. LaFace DM, Peck AB. Reciprocal allogeneic bone marrow transplantation between NOD mice and diabetes-nonsusceptible mice associated with transfer and prevention of autoimmune diabetes. Diabetes 1989; 38:894-901.
30. Nakagawa T, Nagata N, Hosaka N et al. Prevention of autoimmune inflammatory polyarthritis in male New Zealand Black/KN mice by transplantation of bone marrow cells plus bone (stromal cells). Arthritis Rheum 1993; 36:263-268.
31. Naji A, Silvers WK, Kimura H et al. Influence of islet and bone marrow transplantation on the diabetes and immunodeficiency of BB rats. Metabolism 1983; 32:62-68.
32. Van Bekkum DW, Bohre EOM, Houben PFJ et al. Regression of adjuvant-induced arthritis in rats following bone marrow transplantation. Proc Natl Acad Sci USA 1989; 86:10090-10094.
33. Knaan-Shanzer S, Houben P, Kinwel-Bohre EOM et al. Remission induction of adjuvant arthritis in rats by total body irradiation and autologous bone marrow transplantation. Bone Marrow Transplant 1991; 8:333-338.
34. Karussis DM, Slavin S, Lethman D et al. Prevention of experimental autoimmune encephalomyelitis and induction of tolerance with acute immunosuppression followed by syngeneic bone marrow transplantation. J Immunol 1992; 148:1693-1698.
35. Van Gelder M, Kinwel-Bohre EPM, Van Bekkum DW. Treatment of experimental allergic encephalomyelitis with bone marrow transplantation. Exp Hematol 1993; 21:1155.
36. Ikehara S, Kawamura M, Takao F et al. Organ-specific and systemic autoimmune diseases originate from defects in hematopoietic stem cells. Proc Natl Acad Sci USA 1990; 87:8341-8344.
37. Lernmark A. Molecular biology of type 1 (insulin-dependent) diabetes mellitus. Diabetologia 1985; 28:195-203.
38. Wong FS, Janeway CA Jr. Insulin-dependent diabetes mellitus and its animal models. Curr Opin Immunol 1999; 11:643-647.

39. Gottlieb PA, Eisenbarth GS. Mouse and man: Multiple genes and multiple autoantigens in the aetiology of type I DM and related autoimmune disorders. J Autoimmun 1996; 9:277-281.
40. Makino S, Kunimoto K, Muraoka Y et al. Exp Anim 1980; 29:1-13.
41. Nakamura T, Good RA, Inoue S et al. Successful liver allografts in mice by combination with allogeneic bone marrow transplantation. Proc Natl Acad Sci USA 1986; 83:4529-4532.
42. Yasumizu R, Sugiura K, Iwai H et al. Treatment of type 1 diabetes mellitus in non-obese diabetic mice by transplantation of allogeneic bone marrow and pancreatic tissue. Proc Natl Acad Sci USA 1987; 84:6555-6557.
43. Diani AR, Sawada GA, Hannah BA et al. Analysis of pancreatic islet cells and hormone content in the spontaneously diabetic KKAy mouse by morphometry, immunocytochemistry and radio-immunoassay. Virchows Arch Pathol Ant Histol 1987; 412:53-61.
44. Zhang Y, Yasumizu R, Sugiura K et al. Fate of allogeneic or syngeneic cells in intravenous or portal vein injection: Possible explanation for the mechanism of tolerance induction by portal vein injection. Eur J Immunol 1994; 24:1558-1565.
45. Sugiura K, Kato K, Hashimoto F et al. Induction of donor-specific T cell anergy by portal venous injection of allogeneic cells. Immunobiol 1997; 197:460-477.
46. Kushida T, Inaba M, Takeuchi K et al. Treatment of intractable autoimmune diseases in MRL/lpr mice using a new strategy for allogeneic bone marrow transplantation. Blood 2000; 95:1862-1868.
47. Jin T, Toki J, Inaba M et al. Novel strategy for organ allografts using sublethal (7Gy) irradiation followed by injection of donor bone marrow cells via portal vein. Transplant 2001; 71:1725-1731.
48. Ikebukuro K, Adachi Y, Yamada Y et al. Treatment of streptozotocin-induced diabetes mellitus by transplantation of islet cells plus bone marrow cells via portal vein in rats. Transplant In press.
49. Adachi Y, Inaba M, Amoh Y et al. Effect of bone marrow transplantation on anti-phospholipid antibody syndrome in murine lupus mice. Immunobiol 1995; 192:218-230.
50. Ikehara S, Kawamura M, Takao F et al. Organ-specific and systemic autoimmune diseases originate from defects in hematopoietic stem cells. Proc Natl Acad Sci USA 1990; 87:8341-8344.
51. Visser JWM, Bauman JGJ, Mulder AH et al. Isolation of murine pluripotent hemopoietic stem cells. J Exp Med 1984; 159:1576-1590.
52. Miyama-Inaba M, Ogata H, Toki J et al. Isolation of murine pluripotent hemopoietic stem cells in the Go phase. Biochem Biophys Res Commun 1987; 75:1809-1812.
53. Kawamura M, Hisha H, Li Y et al. Distinct qualitative differences between normal and abnormal hemopoietic stem cells in vivo and in vitro. Stem Cells 1997; 15:56-62.
54. Ikehara S, Yasumizu R, Inaba M et al. Long-term observations of autoimmune-prone mice treated for autoimmune disease by allogeneic bone marrow transplantation. Proc Natl Acad Sci USA 1989; 86:3306-3310.
55. Ishida T, M. Inaba M, Hisha H et al. Requirement of donor-derived stromal cells in the bone marrow for successful allogeneic bone marrow transplantation. J Immunol. 1994; 152:3119-31271.
56. Hisha H, Nishino T, Kawamura M et al. Successful bone marrow transplantation by bone grafts in chimeric-resistant combination. Exp Hematol 1994; 23:347-352.
57. Hashimoto F, Sugiura K, Inoue K et al. Major histocompatibility complex restriction between hematopoietic stem cells and stromal cells in vivo. Blood 1997; 89:49-54.
58. Takeuchi K, Inaba M, Miyashima S et al. A new strategy for treatment of autoimmune diseases in chimeric resistant MRL/lpr mice. Blood 1998; 91:4616-4623.
59. Kushida T, Inaba M, Hisha H et al. Intra-bone marrow injection of allogeneic bone marrow cells: A powerful new strategy treatment of intractable autoimmune diseases in MRL/lpr mice. Blood 2001; 97:3292-3299.
60. Li Y, Hisha H, Inaba M et al. Evidence for migration of donor bone marrow stromal cells into recipient thymus after bone marrow transplantation plus bone grafts; a role of stromal cells in positive selection. Exp Hematol 2000; 28: 50-960.
61. Spivey WH. Intraosseous infusions. J Pediatr 1987; 111:639-643.
62. Hagglund H, Ringden O, Agren B et al. Intraosseous compared to intractable infusion of allogeneic bone marrow. Bone Marrow Transplant 1998; 21:331-335.
63. Kushida T, Inaba M, Ikebukuro K et al. A new method for bone marrow cell harvesting. Stem Cells 2000; 18:453-456.
64. Green MC, Sweet HO, Bunker LE. Tight-skin, a new mutation of the mouse causing excessive growth of connective tissue and skeleton. Am J Pathol 1976; 82:493-512.
65. Oyaizu H, Adachi Y, Taketani S et al. Treatment of emphysema in TsK mice by a new bone marrow transplantation method. Submitted for publication.

CHAPTER 31

Mobilization and Conditioning Regimens in Stem Cell Transplant for Autoimmune Diseases

Ewa Carrier and Richard K. Burt

Introduction

Hematopoietic stem cell transplantation (HSCT) of autoimmune diseases has been performed as phase I/II studies in Europe, USA, Asia, and Australia. Phase III studies for systemic lupus erythematosus (SLE), multiple sclerosis (MS) and systemic sclerosis (SSc) are being designed and implemented. This chapter will review and discuss the rationale and use of mobilization and conditioning regimens for both autologous and allogeneic HSCT.

Philosophy of Autologous HSCT for Autoimmune Diseases: Immune Ablation Versus Immune Balance

There are two slightly different philosophical approaches in developing autologous HSCT for autoimmune disease. One notion is to ablate all auto-reactive immune cells with the conditioning regimen. Advocates of this position use the terminology: high dose immune suppressive therapy (HDIT).[1] It is a notion borrowed from the field of malignancies in which every single viable malignant cell is considered pathologic. Autologous HSCT conditioning regimens are designed to maximally reduce or hopefully eliminate all malignant clones in cancer and, in analogy, all activated auto-reactive lymphocytes in autoimmune diseases. In autoimmune disorders, there is animal data from adjuvant arthritis (AA), a model of rheumatoid arthritis, to support the concept of HDIT.[2,3] In AA, the more intense the conditioning regimen, the lower the relapse rate. Relapse is significantly higher in rats receiving a nonmyeloablative regimen such as high dose cyclophosphamide alone compared with those treated with a myleoablative regimen of either the combination of busulfan and cyclophosphamide or total body irradiation (TBI) as a single agent. The peculiarities and intricacies of a single animal model have limitations for broader application to human diseases or even other animal model systems. Outside of recognition that some animal models require an allogeneic HSCT for cure (refer to Chapters 29 and 30), animal models with the exception of AA, in general, have not explored the relapse rate in relationship to conditioning regimen intensity.

Theoretically, high dose chemotherapy or chemo-radiotherapy can eliminate every lymphocyte, but in practice this may not be feasible. As learned from most malignancies, conditioning regimen chemo-radiotherapy can reduce by several logs but not completely eliminate, tumour cells. In many cases, cure is possible only by an allogeneic HSCT with adoptive transfer of the donor's lymphocytes that mediate an immuno-therapeutic anti-tumour effect termed graft versus leukaemia (GVL).[4,5] Similarly, complete immune ablation, the philosophical goal of HDIT and autologous HSC support, would probably require the graft versus autoimmune (GVA) effect of an allogeneic HSCT to be realized.[6,7]

A second or alternative philosophical concept is one of the immune "reset" or immune "balance".[8] In this notion, auto-reactive cells, unlike a malignant cell, are "normal". During development, T cells that bind self with high avidity undergo apoptosis.[9-11] However, T cells that fail to recognize a self-epitope also undergo apoptosis.[9-11] Therefore, circulating T cells in a healthy person normally posses a T cell receptor repertoire selected to self. In this view, immune cells, unlike malignant cells, are not intrinsically bad but rather in a dynamic equilibrium that maintains steady state by constantly fluctuating between tolerance and immunity. A dynamic state that is best demonstrated by the intermittent clinical course of some autoimmune diseases such as relapsing-remitting multiple sclerosis that flares and remits spontaneously even without treatment. This indicates that physiologic mechanisms may exist to down-regulate activated auto-reactive cells. These mechanisms and their regulation are poorly understood, but include activation induced cell death (AICD),[12] regulatory cells such as $CD25^+CD4^+$ cells,[13-15] costimulatory signals,[16,17] peripheral antigen avidity (i.e., antigen persistence, concentration, and affinity),[18,19] anti-idiotypic immunoglobulins, anti-idiotypic T cell receptor repertoires,[20,21] and cytokine shifts.[22,23] An analogy would be the dynamic homeostasis of blood coagulation. Blood is in a dynamic state maintained by a balance between lysis and clotting, which appears static unless perturbed by a clinical event such as a thrombotic stroke or bleeding. Using the notion of immune balance, the conditioning regimen is not intended to destroy every immune cell, but rather be sufficient enough to restore immune "balance".

In practice, the philosophy of HDIT leads to a maximal immune suppressive regimen and lymphocyte depletion of the graft. These regimens have been accompanied by infection-related, as well as regimen-related, mortality.[24,25] In comparison, the notion of immune balance leads to less intense regimens that are more easily tolerated and have less infection-related risk. Whether one concept or the other is correct remains unclear. There is currently no data to support more intense regimens over less toxic regimens in terms of disease remission or relapse rate. The appropriate regimen intensity may vary by disease. For example, cyclophosphamide +/- anti-thymocyte globulin (ATG) appears inadequate for complete responses or sustained untreated partial responses in rheumatoid arthritis.[26,27] Yet, the same regimen appears highly effective in systemic lupus erythematosus.[28,29] Whatever the most appropriate concept for a given disease, it is probably prudent to determine outcome with less intense regimens before testing more intense and potentially more toxic conditioning regimens.

Autoimmune Diseases Likely to Respond to Autologous HSCT

For malignancies, autologous HSCT is generally appropriate if the disease is chemotherapy responsive. The concept is to dose escalate disease responsive drugs to myeloablation, followed by rescue with autologous hematopoietic stem cells (HSC). Cancers nonresponsive to chemotherapy, such as pancreatic cancer or squamous cell lung cancer, would not be appropriate for consideration of autologous HSCT. Similarly, autoimmune diseases that are responsive to immunosuppressive therapy appear to be responsive to dose escalation and autologous HSCT. Examples of immune responsive diseases are: systemic lupus erythematosus (SLE), Crohn's disease (CD), pemphigus vulgaris (PV), relapsing-polychondritis, relapsing-remitting multiple sclerosis (RRMS), rheumatoid arthritis (RA), and juvenile idiopathic arthritis (JRA). Response does not seem to be dependent on cytokine profile, since SLE and Crohn's disease that are Th2 and Th1 skewed, respectively, seem to respond equally well.[28-30]

Traditional immune nonresponsive diseases such as primary progressive MS and late secondary progressive MS show little or no improvement following autologous HSCT.[31-34] Marginally immune responsive diseases, such as scleroderma, have generally demonstrated improved skin flexibility and quality of life.[1,35] Therefore, autoimmune diseases that traditionally respond to standard doses of immune suppression appear to be the best candidates for dose escalation of immune suppression with autologous HSC support.

Mobilization Regimens

Mobilization methods are quite heterogeneous and methods used to collect HSC are center specific. Unprimed, unpurified bone marrow has generally been preferred in children and in China, while mobilized peripheral blood mononuclear cells (PBMC) were used in adults in Europe and North America. Methods of PBSC mobilization included: G-CSF (5,10 and 16 μg/kg), GM-CSF (5 μg/kg), cyclophosphamide alone (2 or 4 g/m^2), G-CSF and corticosteroids, or cyclophosphamide with either G-CSF (5 or 10 μg/kg) or GM-CSF (5μg/kg).[36]

Increasing the G-CSF dose increases stem cell recovery, and the addition of cyclophosphamide tends to further increase the progenitor yield.[36] In a retrospective multi-center analysis, when evaluated by disease, stem cell yield appeared highest for scleroderma and lowest for SLE and idiopathic thrombocytopenic purpura (ITP), independent of the mobilization regimen used.[36] This observation would need to be confirmed in a prospective study, using identical apheresis instruments and procedures, before drawing definitive conclusions.

Although only a small number have been performed, bone marrow harvests have been without complications and have not been reported to affect disease activity. In some diseases, such as RA and MS, the use of G-CSF for PBSC mobilization caused disease flare or exacerbation.[36] The complications arising from G-CSF mobilization appear to be disease specific. For example, in MS, clinically significant flares and, in some cases, irreversible neurological deterioration occurred in patients who received G-CSF for mobilization.[36,37] In scleroderma, G-CSF has been associated with cutaneous telangiectasia but not disease flares per se. Flares of swollen and/or tender joint count have occurred in patients with RA, but symptoms resolved without treatment or were easily controlled with corticosteroids.[36]

G-CSF-related disease flares appear to be prevented by the combination of cyclophosphamide and G-CSF. In some diseases such as MS, SLE, and RA, the combination of cyclophosphamide and G-CSF temporarily ameliorated symptoms.[36] The time of initiating G-CSF after cyclophosphamide is often not reported. Some centers begin G-CSF the day after cyclophosphamide, other centers waited 72 to 96 hours in order to ensure cyclophosphamide-induced lymphotoxic effects before initiating G-CSF. Whether the timing of G-CSF initiation after cyclophosphamide influences disease activity is unknown. The absolute neutrophil count (ANC) nadir is usually reached 8-9 days after cyclophosphamide with an ANC <1000/μl for 1-2 days. Apheresis is initiated on ANC rebound, usually 10 days after cyclophosphamide. Some patients with MS have been mobilized by combining daily corticosteroids (1.0 mg/kg) and G-CSF without exacerbation of disease, although this approach has not been reported to ameliorate disease activity. Apheresis is initiated on either day 4 or 5 of daily G-CSF. If G-CSF is used for mobilization, it should be combined with either cyclophsophamide and/or corticosteroids in order to avoid disease flares.

With the exception of GM-CSF, which was combined with cyclophosphamide in a small number of MS patients, no other cytokines have been evaluated for PBSC mobilization in autoimmune disorders.[36] In order to investigate the consequences of growth factors in autoimmune diseases, varies cytokines have been used in the animal autoimmune disease models. In experimental autoimmune encephalomyelitis (EAE), a model of multiple sclerosis, G-CSF, stem cell factor (SCF), and flt-3 ligand all caused EAE to flare, with flt-3 ligand causing lethal exacerbations (Burt, unpublished data). Flt-3 ligand-related EAE flares were accompanied by increased numbers of circulating monocytes and pro-inflammatory cytokines. The only cytokine tested, which did not cause a disease flare, was thrombopoietin (TPO) (Burt, unpublished data). If these results in EAE are applicable to MS or other autoimmune diseases, this implies that TPO should be the preferred cytokine for PBSC mobilization from patients with autoimmune diseases.

The majority of patients mobilized with chemotherapy developed transient neutropenia. One out of 187 patients reported by Burt et al died of an infection related to mobilization.[36] This patient with SLE had been on high dose corticosteroids for several years, and one week after mobilization developed seizures due to cerebral mucormycosis. Prophylactic anti-microbial agents, including anti-fungal agents, should be considered during any neutropenic interval in patients with a history of chronic high dose corticosteroid exposure.

Table 1. Toxicity of mobilization*

Disease	Mobilization Regimen	No. of Patients	Deaths	No. of Patients/Other Toxicities
RA	G5 or G10	16	0	8-Myalgia
	Cy2 + G10	11	0	2-Fever neutropenia/1-Abdominal pain/1-myalgia
	Cy4 + G10	10	0	5-Fever neutropenia/1 bacteremia
MS	G10	12	0	0
	G16	8	0	0
	Cy2 + G10	3	0	0
	Cy4 + G5	20	0	1-Seizures
	Cy4 + G7	4	0	0
	Cy4 + G10	24	0	0
	Cy4 + GM	7	0	1-Hypotension
SLE	G5	1	0	0
	Cy2 + G10	12	1	1-Death (cerebral mucormycosis)/ 1-CMV pneumonitis/2-Intubation and dialysis for pulmonary edema/1-pericardial effusion.
	Cy4 + G5	1	0	0
SSc	G16	10	0	1-Join pain / 2-Telangectasia
	Cy2 + G10	4	0	0
	Cy4 + G10	12	2	2-Died
ITP	G15	1	0	0
	G10	8	0	1-Chest pain requiring ICU admission
Behçets	G5	1	0	0
	Cy4 + G5	2	0	0
Polymyositis	Cy2 + G10	1	0	0
Polychondritis	Cy2 + G10	1	0	0

Cy2= Cyclophosphamide at 2 g/m^2; Cy4= Cyclophosphamide at 4 g/m^2; G5= G-CSF 5 μg/kg/day; G10= G-CSF 10 μg/kg/day; G16= G-CSF 16 μg/kg/day; ITP= idiopathic thrombocytopenic purpura; MS= multiple sclerosis; RA= rheumatoid arthritis; SSc= Scleroderma; SLE= systemic lupus erythematosus. *Adapted from Burt et al.[36]

Noninfectious mobilization regimen related mortality has also been reported in systemic sclerosis patients receiving 4 g/m^2 cyclophosphamide. The deaths occurred in patients with scleroderma associated with pulmonary artery hypertension. The cause of death was alveolar hemorrhage and/or myocardial infarction. Pulmonary artery hypertension should be considered a contraindication to either HSCT or cyclophosphamide stem cell mobilization. Patients with autoimmune cytopenias, such as ITP-related thrombocytopenia, have a compensatory marrow hyperplasia. In one patient, cyclophosphamide mobilization caused lethal bleeding by suppressing marrow production of megakaryocytes, while antibody immune-mediated platelet destruction was still present.[38] While no patient receiving only G-CSF died or became bacteremic, some patients mobilized with G-CSF experienced chest pain of unclear etiology, transient elevation of transaminases, headaches, joint pain, myalgias, and as mentioned above, disease exacerbation.[36] Therefore, the symptoms and organ dysfunction due to the preexisting disease process must be taken into consideration in selecting candidates and designing the mobilization regimens. Mobilization regimens used in autoimmune transplants for autoimmune diseases are shown in Table 1.

In a retrospective analysis, patients with MS who received more than 2 years of interferon-γ had lower stem cell recovery than patients who were exposed to interferon for less than 2 years.[36] For patients with RA, the use of gold and methotrexate prior to the mobilization affected the stem cell yield. This suggests that prior medications may affect the outcome of mobilization, although further prospective studies would be needed for confirmation. While no data exists to support this practice, some centers discontinue medications that may diminish progenitor cell collection such as γ-interferon, gold, or methotrexate, several weeks prior to mobilization. In patients who fail to mobilize sufficient stem cell numbers, increasing a dose of G-CSF from 10 μg/kg to 15 μg/kg or repeating mobilization with cyclophosphamide and G-CSF in those who fail G-CSF, has achieved successful progenitor cells mobilization. Additionally, in poor mobilizers, a bone marrow harvest may be performed.

Stem Cell Purification

Although it is unclear whether or not CD34$^+$ selection of an autologous graft is beneficial, most autologous grafts have undergone a selection process to remove lymphocytes. Positive enrichment for CD34+ cells has been performed using CEPRATE (CellPro, Bothell, WA, USA), Isolex (Baxter, Chicago, USA), or CliniMACS (Miltenyi, Bergish Gladbach, Germany). Negative selection has been performed with T-cell antibodies by E-rosetting or Nexell Isolex CD4/CD8 depletion. Most patients have not had unmanipulated back up cells cryopreserved. Because the US Food and Drug Administration (FDA) views the processed cells as a new drug, in the USA, the use of any of these devices requires that the investigator obtain an Investigational New Drug (IND) exemption from the FDA. In Europe, the therapeutic agent is the conditioning regimen, and the infused autologous cells are viewed as a supportive care regimen, similar to a packed red blood cell transfusion, and require no special government permission.

The only study comparing a CD34+ enriched graft versus an unmanipulated graft was in a small number of patients with RA whose conditioning regimen was cyclophosphamide. The results indicated no benefit in terms of relapse from CD34+ selection. If more lymphocytes survive the in vivo exposure to a nonmyeloablative cyclophosphamide that are purged ex vivo by CD34+ selection, then T cell depletion of the graft would not be anticipated to diminish relapse. Alternatively, T cell depletion of the graft, depending on the intensity of the conditioning regimen, could increase the risk of serious infections. Therefore, the benefit of CD34+ selection can only be determined by future trials that randomize patients to selected and unselected grafts.

Autologous HSCT Conditioning Regimens

Conditioning regimens used in patients with malignancy or aplastic anemia and a coincidental autoimmune disease, who have undergone HSCT, are listed in Table 2. These regimens were designed for the malignancies, but demonstrated the ability of patients with autoimmune disorders to tolerate high dose chemotherapy or radiotherapy. Subsequently, at least 80 centers are now involved worldwide in HSCT for severe autoimmune diseases and many of them used center-specific conditioning, resulting in quite heterogeneous data.

In malignancies, selection of a conditioning regimen agent is based on dose escalation of a drug effective at a standard doses. Dose escalation of a noneffective drug would not be included in the conditioning regimen, since it would result in increased toxicity without efficacy. Ideally, selection of autoimmune conditioning agents should be based on dose escalation of agents effective at standard doses. SLE is highly responsive to cyclophosphamide and a dose escalation of cyclophosphamide, as a conditioning agent, has achieved impressive responses.[28,29] Systemic sclerosis, a disease unresponsive to virtually all therapies, has been shown to be somewhat responsive to cyclophosphamide. Skin scores and quality of life have improved following transplant doses of cyclophosphamide.[1,35] In comparison, radiation has never been demonstrated to benefit systemic sclerosis, and in fact can cause disease exacerbation.[38-40] A study using total lymphoid irradiation (TLI), despite profound immune suppression, and avoiding radiation to the lung, exacerbated scleroderma pulmonary disease and caused a gastrointestinal scleroderma-related death.[38] In breast cancer patients who have coincidental scleroderma, localized breast radiation is considered a relative contraindication.[39] Using total body irradiation (TBI) in the transplant regimen for scleroderma, two patients died from pulmonary complications, and despite subsequent lung shielding, pulmonary function tests declined acutely in other patients.[1] Radiation may act synergistically with scleroderma to damage blood vessels. Designing conditioning regimens utilizing disease-exacerbating agents seems like a singularly inconsistent idea. Due to the unique aspects of each disease, conditioning regimens (Table 3) will be discussed in relation to each specific autoimmune disease.

SLE

The principle treatment modality used in refractory SLE is cyclophosphamide; therefore, dose escalation of this compound is a logical approach in severe diseases. The conditioning regimens used for HSCT in the USA is predominately cyclophosphamide/ATG.[28,29] In Europe, a variety of cyclophosphamide-based SLE conditioning regimens have been used: a Thiotepa/cyclophosphamide combination,[41] cyclophosphamide alone, cyclophosphamide/ATG, cyclophosphamide/TBI,[42] or BEAM.[43] Two patients died, one of mucormycosis following mobilization, and one as a result of graft failure after HSCT.[44] In a case of graft failure, infusion of allogeneic stem cells from an HLA-matched sibling was unsuccessful in restoring hematopoiesis. Subsequent analysis of stored serum demonstrated a presence of a soluble factor suppressing hematopoiesis.

Patients with SLE may have a long history of high dose immune suppression prior to the referral for HSCT. Prophylactic anti-fungal, anti-viral, and anti-bacterial therapy should be considered during any neutropenia interval. HSCT studies indicate that high-dose cyclophosphamide may be used safely in patients with end stage renal failure (ESRF). While the pharmacokinetics of cyclophosphamide and its metabolites are not understood in ESRF, if dialysis is performed the morning after each cyclophosphamide infusion, treatment appears to be well tolerated. Patients in renal failure may produce residual urine that may allow stagnation of cyclophosphamide metabolites in the bladder, resulting in hemorrhagic cystitis. In order to minimize this, a Foley catheter with bladder irrigation and intravenous Mesna may be used during and for 24 hours after cyclophosphamide. SLE nephritis with renal insufficiency may be unusually sensitive to diuresis resistant volume overload, requiring aggressive volume-status monitoring and correction by dialysis. SLE patients may also be prone to thrombus formation from anti-phospholipid antibodies. Prophylactic anti-coagulation with subcutaneous Lovenox or fragmin has generally been well tolerated without bleeding during the HSCT.

Multiple Sclerosis

The HSCT conditioning regimens used for MS in Europe tend to be BEAM-based[32,33] and in the USA cyclophosphamide/TBI based.[25,45] The largest report on HSCT for MS comes from Thessaloniki, Greece, by Fassas et al.[33] There is currently no evidence of superiority of one regimen over the other. However, several MS patients have suffered infectious-related mortality from multi-agent conditioning regimens combined with CD34+ selection of the graft.

The City of Hope used an intensive conditioning regimen of busulfan (16 mg/kg), cyclophosphamide (120 mg/kg), and anti-thymocyte globulin (30 mg/kg) along with CD34+ selection to deplete lymphocytes from the graft.[24] Two out of 5 treated patients died from infections. One patient died 22 days after transplant from influenza and the second died 19 months after HSC transplant from streptococcus pneumonia sepsis. The Fred Hutchinson Cancer Center Consortium treated 26 patients with progressive MS. The conditioning regimen was total body irradiation (TBI) (800 cGy given 200 cGy BID with lung shields to 650 cGy), cyclophosphamide (120 mg/kg divided 60 mg/kg/day)

Table 2. HSCT regimens in patients with coincidental autoimmune disease

Autoimmune Disease	HSCT Indication	Type of Transplant/Conditioning	Reference
RA	AA	Allogeneic/ Cy, TBI, Prednisone	Baldwin JL. Arthritis Rheum 1977; 20:1043-1048
RA	AA	Allogeneic/ Cy, TBI, Prednisone	Snowden JA. Arthritis Rheum 1998; 41:453-459
RA	AA	Allogeneic/ Cy, TBI, Prednisone	McKendry RJ. Arthritis Rheum 1996; 39:1246-1253
RA	NHL	Autologous/ BEAM	Cooley HM. Arthritis Rheum 1997; 40:1712-1715
RA	NHL	Autologous/ Bu, Cy, VP-16	Euler HH. Blood 1996; 88:3621-3625
Psoriasis	CML	Allogeneic	Snowden JA. Arthritis Rheum 1998; 41:453-459
Psoriasis	CML	Allogeneic	Yin JA. Bone Marrow Transplant 1992; 9:31-33
Psoriasis	NHL, APL, Plasma Cell Leukemia	Autologous	Cooley HM, Arthritis Rheum 1997; 40:1712-1715
SLE	AA	Allogeneic	Gur-Lavi M. Arthritis Rheum 1999; 42:1777
SLE	Red Cell Aplasia	Allogeneic/ Cy, ATG	Roychowdhury DF. Bone Marrow Transplant 1995; 16:471-472
SLE	NHL	Autologous/ Cy, BCNU, VP-16	Snowden JA. Bone Marrow Transplant 1997; 19:1247-1250
SLE	NHL	Autologous/ BEAM	Euler HH. Blood 1996; 88:3621-3625
SLE	CML	Autologous/ BCNU, Ara-C, m-AMSA, VP-16	Meloni G. Blood 1997; 89:4659
SLE	HD	Autologous	Schachna L. Arthritis Rheum 1998; 41:2271-2272
MS	CML	Allogeneic/ Cy, TBI	McAllister LD. Bone Marrow Transplant 1997; 19:395-397
Ulcerative Colitis	AML	Autologous	Yin JA. Bone Marrow Transplant 1992; 9:31-33
Ankylosing Spondylitis	NHL	Allogeneic/ BEAM	Jantunen E. Rheumatology (Oxford) 2000; 39:563-564
Sjögren's disease	NHL	Allogeneic/ Ara-C, BCNU, Cy	Ferraccioli G. Ann Rheum Dis 2001; 60:174-176
Sjögren's disease	NHL	Autologous/ BCNU, VP-16, Cy, Mel	Rosler W. Bone Marrow Transplant 1998; 22:211-21
MG	Ovarian Cancer	Autologous/ VP-16, Thiotepa, Cisplatin	Euler HH. Blood 1996; 88:3621-3625
Crohn's disease	Leukemia	Allogeneic	Lopez-Cubero SO. Gastroenterology 1998;114:433-440
Autoimmune Hepatitis	ALL	Allogeneic, Cy, TBI	Vento S. Lancet 1996; 348:544-545

Legend: AA= Aplastic Anemia; AML= Acute Myelogenous Leukemia; ATG= antithymocyte glubulin; CML= Chronic Myelogenous Leukemia; Cy= cyclophosphamide; HSCT= Hematopoietic Stem Cell Transplantation; Mel= melphalan; MG= Myasthenia Gravis; MS= multiple sclerosis; RA= rheumatoid arthritis; SLE= systemic lupus erythematosus; TBI= total body irradiation; VP-16= etoposide.

Table 3. Conditioning regimens for autoimmune disease

Disease Reference	Number of Patients	Conditioning Regimen	Reference
RA	4	Cy, ATG, 1 patient TBI (400 cGy)	Burt RK. Arthritis Rheum 1999; 42:2281-2285
RA	1	Cy 200 mg/kg	Joske DJ, Lancet 1997; 350:337-338
RA	8	Cohort I – Cytoxan 100 mg/kg Cohort II – Cytoxan 200 mg/kg	Snowden JA. Arthritis Rheum 1999; 42:2286-2292
RA	1	Busulfan, Cy	Durez P. Lancet 1998; 352:881
RA	1	Cy, ATG	McColl G. Ann Intern Med 1999; 131:507-509
SLE	1	Thiotepa, Cy	Marmont AM. Lupus 1997; 6:545-548
SLE	7	Cy, ATG	Traynor AE. Lancet 2000; 356:701-707
SLE	1	BEAM	Fouillard L. Blood Lupus 1999; 8:320-323
SLE	1	Cy, ATG	Musso M. Lupus 1998; 7:492-494
MS	15	BEAM	Fassas A, Bone Marrow Transplant 1997; 20:631-638
MS	6	Cy, TBI	Burt RK. Blood 1998; 92:3505-3514
MS	8	BEAM	Kozak T. Cas Lek Cesk 2000; 139:329-333
MS	85	BEAM	Fassas A. J Neurol 2002; 249 (8): 1088-97
SSc	1	Cy, CAMPATH	Martini A. Arthritis Rheum 1999; 42:807-811
SSc	1	Cy	Tyndall A. Lancet 1997; 349:254
SSc	27	Cy/TBI	McSweeney PA. Blood 2002; 100(5); 1602-10
JRA	12 1	12 Cases: Cy, ATG, TBI 1 Case: Cy, ATG	Wulffraat NM. Curr Rheumatol Rep 2000; 2:316-323.
Polymyositis	1	Busulfan, Cy, ATG	Baron F. Br J Haematol 2000; 110:339-342
Polychondritis	1	Cy, ATG	Rosen O. Arthritis Res 2000; 2:327-336
Wegener's Granulomatosis	1	Cy	Jantunen EJ. Scand J Rheumatol 1999; 28:67-74
PAN	1	CAMPATH-1H, Fludarabine, Cy	Rheumatology (Oxford) 2001; 40(11):1299-307
Still's Disease	1	Cy, ATG	Lanza F. Bone Marrow Transplant 2000; 25:1307-1310
ITP	2	Cy	Lim SH. Lancet 1997; 349:475
Pemphigus Vulgaris	1	Cy	Hayag MV. J Am Acad Dermatol 2000; 43:1065-1069

ATG= anti-thymocyte globulin; BEAM= carmustine, etoposide, cytosine arabinoside, melphalan; Cy= cyclophosphamide; MS= multiple sclerosis; PAN= polyarthritis nodosa; RA= rheumatoid arthritis; SLE= systemic lupus erythematosus; SSc= scleroderma; TBI= total body irradiation

and ATG (either 90 mg/kg equine or 15 mg/kg rabbit) given for 6 days (days -5, -3, -1, +1, +3, and +5). The graft was lymphocyte depleted by CD34+ positive selection. The only patient in whom rabbit ATG had been given instead of equine ATG, died from Epstein-Barr virus (EBV)-associated post-transplant lymphoproliferative disorder (PTLD).[25] The patient who developed PTLD had received 6 days of rabbit ATG at 2.5 mg/kg/day (total dose of 15 mg/kg). The Thessaloniki group in Greece has reported that BEAM (carmustine, etoposide, cytarabine, melphalan), when combined with ATG and lymphocyte depletion of the graft, was complicated by mortality from an opportunistic infection (aspergillosis).[33] As phase I studies, these results suggest caution in combining lymphocyte-depleted grafts with aggressive immune suppressive conditioning regimens. This may be accomplished by decreasing the dose intensity of conditioning agents, eliminating one of the conditioning agents, or infusing an unmanipulated graft not depleted of lymphocytes.

Fever may cause exacerbation of neurologic deficits in patients with multiple sclerosis, a phenomenon termed pseudo-exacerbation. Therefore, infections and drugs that cause fever should be avoided. An engraftment syndrome of fever, rash, and fatigue with exacerbation of symptoms has been described in MS and is treated by a short course of peri-engraftment oral corticosteroids.[46] Finally, when HSCT regimens were first designed, MS was recognized to be an immune-mediated demyelinating disease. Since then, MS has been recognized to be both an immune-mediated demyelineating disease and an axonal degenerative process. This raised questions about TBI or radiation inducing axonal injury and/or TBI-related inhibition of CNS repair by neural stem cells or oligodendrocyte progenitors. Future conditioning regimens should be designed towards immune suppression while minimizing risks of further axonal injury. Future studies will probably focus on patients with less disability and relapsing disease (refer to Chapter 37). Such patients have a five-year MS-related mortality of virtually zero. It will, therefore, be important to utilize safer conditioning regimens.

Systemic Sclerosis

Multiple conditioning regimens were used in Europe for HSCT of scleroderma including: cyclophosphamide, cyclophosphamide/ATG, cyclophosphamide/Fludarabine/Thiotepa,[47] or cyclophosphamide/CAMPATH-1H.[48] In the United States, McSweeney et al transplanted 27 patients with SSc, two of which died of pulmonary complications secondary to TBI.[1] This prompted the adjustment of the irradiation dose and lung shielding, and since these changes were made, no mortality was observed although pulmonary function tests continued to decline despite shielding.[1] TBI may, therefore, accelerate scleroderma vasculopathy. Lethal organ toxicity such as myocardial infarction or pulmonary hemorrhage occurred in patients with pulmonary artery hypertension (PAH) during either cyclophosphamide mobilization or conditioning.[36] These data indicate that future studies should exclude patients with PAH and have raised concerns about TBI-based regimens. Corticosteroids used during conditioning may precipitate scleroderma related renal crises. Prophylactic use of angiotensin converting enzyme inhibitors (ACEI) may prevent renal crises. Conversely, ACEI should be avoided during apheresis because they block bradykinin degradation. During apheresis, use of ACEI may result in hypotension.

Rheumatoid Arthritis

HSCT conditioning regimens for RA include: cyclophosphamide 100 mg/kg,[49] cyclophosphamide 200 mg/kg,[49,50] Busulfan 16 mg/kg and cyclophosphamide 200 mg/kg,[51] or cyclophosphamide 200 mg/kg plus ATG 90 mg/kg.[26,52] By far, the most common regimen was either cyclophosphamide or cyclophosphamide and ATG. No deaths have occurred in 70 so far reported HSCT cases of RA. RA is associated with rheumatoid interstitial lung disease that may be of concern if pulmonary toxic agents such as Busulfan (without dose adjustment) or TBI (without lung shielding) are employed. TBI-based regimens should probably be avoided in RA due to a history of myelodysplasia from a nontransplant RA study utilizing nonablative total nodal irradiation. Cautious dosing of Busulfex at lower doses (not to exceed 9.6 mg/kg divided 0.8 mg every 6 hours) should avoid pulmonary and liver toxicity in a combined cyclophosphamide/Busulfan regimen.

Juvenile Idiopathic Arthritis (JIA)

Most transplants for JIA have been performed in Europe. The conditioning regimens have generally been either ATG/Cyclophosphamide/Total Body Irradiation[53] or ATG and cyclophosphamide.[53,54] Two patients, both of whom received TBI, died of a Macrophage Activation Syndrome (MAS) (also known as Infection Associated Hemophagocytic Syndrome, IAHS), which is a complication unique to HSCT and arises from an immune dysregulation following too aggressive immune suppressive regimens.[54] Outcome, in terms of swollen joints and Child Health Assessment Questionnaire, show no advantage to addition of TBI, which has been omitted in presently designed transplants for JIA (refer to Chapter 44 on JIA).[54]

Other Diseases

Patients with diseases such as polymyositis,[55] polychondritis,[56] Wegener's granulomatosis,[57] Behçet's syndrome,[58] Still's disease,[57] immune thrombocytopenic purpura (ITP),[59] and pemphigus vulgaris[60] who received HSCT have also been reported. Cytoxan/ATG, busulfan/cytoxan/ATG, cytoxan/ALG, cytoxan alone, melphalan alone, and cytoxan/thiotepa were used for conditioning. No mortality was associated with these reports.

Allogeneic HSCT

Allogeneic HSCT from an HLA matched sibling offers the possibility to completely replace the immune system with that of a healthy sibling, transferring numerous nonHLA autoimmune disease-resistant genes to the donor. The donor's lymphocytes would also eliminate residual host hematopoietic and immune cells. In malignancies, donor lymphocyte induced recipient hematopoietic aplasia is termed graft versus leukemia (GVL), while the same phenomena in autoimmune diseases is termed graft versus autoimmunity (GVA).[3-6] For autoimmune diseases, this approach would have to be modified to diminish the risk of donor lymphocyte mediated graft versus host disease (GVHD). Diseases and patient selection should initially focus on patients at a high risk of disease-related mortality. Allogeneic HSCT may also allow replacement of epithelial cells, endothelial cells, and fibroblasts from a disease-resistant donor's hematopoietic stem cell compartment.[61]

A nonmyeloablative stem cell transplant (NST) regimen has been used for patients with high risk RA. NST is used to minimize regimen related toxicity due to the older age of RA patients. Again, due to the older age, in order to minimize the risk of GVHD, the donor graft is lymphocyte depleted. The goal of this approach is to create stable mixed chimerism. Rheumatoid factor negative HLA matched sibling must be available as a donor. The conditioning regimen is fludarabine (125 mg/m^2), cyclophosphamide (150 mg/kg), and CAMPATH-1H (20 mg). The donor graft is CD34$^+$ selected (Isolex, Baxter) with a goal of >10 million CD34$^+$ cells/kg. The first patient treated on this study is a mixed chimera, 100 days post NST, without infections or GVHD.[62]

On the assumption that the high disease-related mortality of scleroderma could justify the risk of some treatment-related GVHD, an unmanipulated graft to achieve complete donor chimerism is being used for patients with scleroderma who have a healthy HLA matched sibling. The conditioning regimen is cyclophosphamide (200 mg/kg) and CAMPATH-1H (100 mg).

CAMPATH-1H was chosen because it has been reported to significantly reduce the risk of GVHD in both matched sibling and unrelated HSCT for malignancies.[63] Since the severity of GVHD is generally thought to increase with age, GVHD associated complications are further minimized by limiting eligibility to patients less than 45 years old. Corticosteroids may precipitate renal crisis in patients with scleroderma. Therefore, cyclosporin and mycophenolate mofetil are used for GVHD prophylaxis. The pretransplant creatinine should be normal to avoid further exacerbating renal injury from cyclosporin. Patients with severe scleroderma-related pulmonary involvement develop pulmonary artery hypertension (PAH) and are at high risk for cardiovascular complications. Similar to autologous HSCT, PAH should be a contraindication to NST. The first patient on this study has just been enrolled (Burt et al, unpublished data).

Conclusions and Future Directions

There is no consensus among the clinical community as to what is the best mobilization and conditioning regimen for HSCT in autoimmune disease. Data on toxicity has been collected and analyzed and demonstrates that these toxicities may be disease and organ-specific. There currently is no available data on efficacy of specific conditioning regimens in terms of inducing and/or increasing duration of disease remission. G-CSF alone used for mobilization may exacerbate symptoms in some diseases during mobilization. Current data suggest that G-CSF combined with either cyclophosphamide or corticosteroids appears to prevent mobilization related flares. Newer protocols, presently in development, will address the issue of allogeneic NST in treatment of these diseases. It must be remembered that the data presented herein were gathered from phase I studies. Safety studies should be viewed as providing information on how to minimize toxicity when designing future efficacy trials.

References

1. McSweeney PA, Nash RA, Sullivan KM et al. High-dose immunosuppressive therapy for severe systemic sclerosis: Initial outcomes. Blood 2002; 100(5):1602-10.
2. Van Bekkum DW. Conditioning regimens for the treatment of experimental arthritis with autologous bone marrow transplantation. Bone Marrow Transplant 2000; 25:357-364.
3. Van Bekkum DW. Experimental basis for the treatment of autoimmune diseases with autologous hematopoietic stem cell transplantation. Bone Marrow Transplantation 2003; suppl. (In press).
4. Sondel PM. The graft-versus-leukemia effect. In: Barrett J, Jiang Y-Z, eds. Allogeneic immunotherapy for malignant disease. New York: Delker, 2000:1-12.
5. Appelbaum FR. Hematopoietic stem cell transplantation as immunotherapy. Nat Med 2001; 411:385-389.
6. Hinterberger W, Hinterberger-Fischer M, Marmont AM. Clinically demonstrable anti-autoimmunity mediated by allogeneic immune cells favourably affects outcomes after stem cell transplantation in human autoimmune diseases. Bone Marrow Transplant 2003; In press.
7. Slavin S, Nagler A, Varadi G et al. Graft vs autoimmunity following allogeneic nonmyeloablative blood stem cell transplantation in a patient with chronic myelogenous leukaemia and severe psoriasis and psoriatic polyarthritis. Exp Hematol 2000; 28:853-857.
8. Burt RK, Slavin S, Burns WH et al. Induction of tolerance in autoimmune diseases by hematopoietic stem cell transplantation: Getting closer to a cure? Blood 2002; 99(3):768-84.
9. Bevan MJ, Hogquist KA, Jameson SC. Selecting the T cell receptor repertoire. Science 1994; 264(5160):796-7.
10. Ashton-Rickardt PG, Tonegawa S. A Differential-avidity model for T-cell selection. Immunol Today 1994; 15(8):362-6.
11. Sebzda E, Wallace VA, Mayer J et al. Positive and negative thymocyte selection induced by different concentrations of a single peptide. Science 1994; 263(5153):1615-8.
12. Pender MP. Activation-induced apoptosis of autoreactive and alloreactive T lymphocytes in the target organ as a major mechanism of tolerance. Immunol Cell Biol 1999; 77(3):216-23.
13. Shevach EM. Suppressor T cells: Rebirth, function and homeostasis. Curr Biol 2000; 10(15):R572-5.
14. Shevach EM. Regulatory T cells in autoimmmunity. Ann Rev Immunol 2000; 18:423-49.
15. Roncarolo MG, Levings MK. The role of different subsets of T regulatory cells in controlling autoimmunity. Curr Opin Immunol 2000; 12(6):676-83.
16. Harding FA, McArthur JG, Gross JA et al. CD28-mediated signalling costimulates murine T cells and prevents induction of anergy in T-cell clones. Nature 1992; 356(6370):607-9.
17. Shahinian A, Pfeffer K, Lee KP et al. Differential T cell costimulatory requirements in CD28-deficient mice. Science 1993; 261(5121):609-12.
18. Garza KM, Agersborg SS, Baker E et al. Persistence of physiological self-antigen is required for the regulation of self tolerance. J Immunol 2000; 164(8):3982-9.
19. Critchfield JM, Racke MK, Zuniga-Pflucker JC et al. T cell deletion in high antigen dose therapy of autoimmune encephalomyelitis. Science 1994; 263(5150):1139-43.
20. Shoenfeld Y, Amital H, Ferrone S et al. Anti-idiotypes and their application under autoimmune, neoplastic, and infectious conditions. Int Arch Allergy Immunol Nov 1994; 105(3):211-23.
21. Poskitt DC, Jean-Francois MJ, Turnbull S et al. The nature of immunoglobulin idiotypes and idiotype-anti-idiotype interactions in immunological networks. Immunol Cell Biol 1991; 69(Pt2):61-70.
22. Gallagher R. Tagging T cells: TH1 or TH2? Science 1997; 275(5306):1615.
23. Forsthuber T, Yip HC, Lehmann PV. Induction of TH1 and TH2 immunity in neonatal mice. Science 1996; 271(5256):1728-30.
24. Openshaw H, Lund BT, Kashyap A et al. Peripheral blood stem cell transplantation in multiple sclerosis with busulfan and cyclophosphamide conditioning: Report of toxicity and immunological monitoring. Biol Blood Marrow Transplant 2000; 6(5A):563-75.
25. Nash RA, Dansey R, Storek J et al. Epstein Barr virus (BBV)-associated post-transplant lymphoproliferative disorder (PTLD) after high dose immunosuppressive therapy (HDIT) and autologous CD34-selected stem cell transplantation (SCT) for severe autoimmune diseases. (abst 1747) Blood 2000; 96(supplA):406a.
26. Burt RK, Georganas C, Schroeder J et al. Autologous hematopoietic stem cell transplantation in refractory rheumatoid arthritis: Sustained response in two of four patients. Arthritis Rheum 1999; 42:2281-2285.
27. Moore JJ, Snowden J, Pavletic S et al. Hematopoietic stem cell transplantation (HSCT) for severe rheumatoid arthritis. Bone Marrow Transplant 2003; suppl(In press).
28. Traynor AE, Schroeder J, Rosa RM et al. Treatment of severe systemic lupus erythematosus with high-dose chemotherapy and haemopoietic stem-cell transplantation: A phase I study. Lancet 2000; 356:701-707.
29. Traynor AE, Barr WG, Rosa RM et al. Hematopoietic stem cell transplantation for severe and refractory lupus. Analysis after five years and fifteen patients. Arthritis Rheum 2002; 46(11):2917-23.
30. Burt RK, Traynor AE, Oyama Y et al. High-dose immune suppression and autologous hematopoietic stem cell transplantation in refractory Crohn's Disease. Blood 2003; in press.
31. Fassas A, Passweg JR, Anagnostopoulos A et al. Hematopoietic stem cell transplantation for multiple sclerosis. A retrospective multicenter study. J Neurol Aug 2002; 249(8):1088-97.
32. Kozak T, Havrdova E, Pit'ha J et al. High-dose immunosuppressive therapy with PBPC support in the treatment of poor risk multiple sclerosis. Bone Marrow Transplant 2000; 25:525-531.
33. Fassas A, Anagnostopoulos A, Kazis A et al. Autologous stem cell transplantation in progressive multiple sclerosis—an interim analysis of efficacy. J Clin Immunol 2000; 20(1):24-30.
34. Burt RK, Kozak T. Hematopoietic stem cell transplantation for multiple sclerosis: Finding equipoise. Bone Marrow Transplant 2003; suppl(In press).

35. Binks M, Passweg JR, Furst D et al. Phase I/II trial of autologous stem cell transplantation in systemic sclerosis: Procedure related mortality and impact on skin disease. Ann Rheum Dis 2001; 60:577-584.
36. Burt RK, Fassas A, Snowden J et al. Collection of hematopoietic stem cells from patients with autoimmune diseases. Bone Marrow Transplant 2001; 28:1-12.
37. Openshaw H, Stuve O, Antel JP et al. Multiple sclerosis flares associated with recombinant granulocyte colony-stimulating factor. Neurology 2000; 54:2147-2150.
38. O'Dell JR, Steigerwald JC, Kennaugh RC et al. Lack of clinical benefit after treatment of systemic sclerosis with total lymphoid irradiation. J Rheumatol Aug 1989; 16(8):1050-4.
39. Breast radiotherapy after breast-conserving surgery. The Steering Committee on Clinical Practice Guidelines for the Care and Treatment of Breast Cancer. Can Assoc Radiation Oncol CMAJ 1998; 158 Suppl 3:S35-42.
40. Abu-Shakra M, Lee P. Exaggerated fibrosis in patients with systemic sclerosis (scleroderma) following radiation therapy. J Rheumatol 1993; 20(9):1601-3.
41. Marmont AM, van Lint MT, Gualandi F et al. Autologous marrow stem cell transplantation for severe systemic lupus erythematosus of long duration. Lupus 1997; 6:545-548.
42. Musso M, Porretto F, Crescimanno A et al. Autologous peripheral blood stem and progenitor (CD34+) cell transplantation for systemic lupus erythematosus complicated by Evans syndrome. Lupus 1998; 7:492.
43. Fouillard L, Gorin NC, Laporte JP et al. Control of severe systemic lupus erythematosus after high-dose immunosuppressive therapy and transplantation of CD34+ purified autologous stem cells from peripheral blood. Lupus 1999; 8:320-323.
44. Shaughnessy PJ, Ririe DW, Ornstein DL et al. Graft failure in a patient with systemic lupus erythematosus (SLE) treated with high-dose immunosuppression and autologous stem cell rescue. Bone Marrow Transplant 2001; 27:221-224.
45. Burt RK, Traynor AE, Pope R et al. Treatment of autoimmune disease by intense immunosuppressive conditioning and autologous hematopoietic stem cell transplantation. Blood 1998; 92:3505-3514.
46. Oyama Y, Cohen B, Traynor A et al. Engraftment syndrome: A common cause for rash and fever following autologous hematopoietic stem cell transplantation for multiple sclerosis. Bone Marrow Transplant 2002; 29(1):81-5.
47. Tyndall A, Black C, Finke J et al. Treatment of systemic sclerosis with autologous haemopoietic stem cell transplantation. Lancet 1997; 349:254.
48. Martini A, Maccario R, Ravelli A et al. Marked and sustained improvement two years after autologous stem cell transplantation in a girl with systemic sclerosis. Arthritis Rheum 1999; 42:807-811.
49. Snowden JA, Biggs JC, Milliken ST et al. A phase I/II dose escalation study of intensified cyclophosphamide and autologous blood stem cell rescue in severe, active rheumatoid arthritis. Arthritis Rheum 1999; 42:2286-2292.
50. Joske DJ, Ma DT, Langlands DR et al. Autologous bone-marrow transplantation for rheumatoid arthritis. Lancet 1997; 350:337-338.
51. Durez P, Toungouz M, Schandene L et al. Remission and immune reconstitution after T-cell-depleted stem-cell transplantation for rheumatoid arthritis. Lancet 1998; 352:881.
52. McColl G, Kohsaka H, Szer J et al. High-dose chemotherapy and syngeneic hemopoietic stem-cell transplantation for severe, seronegative rheumatoid arthritis. Ann Intern Med 1999; 131:507-509.
53. Wulffraat NM, Sanders LA, Kuis W. Autologous hemopoietic stem-cell transplantation for children with refractory autoimmune disease. Curr Rheumatol Rep 2000; 2:316-323.
54. Wulffraat NW, Brinkman D, Ferster A et al. Long term follow up of autologous stem cell transplantation for refractory juvenile idiopathic arthritis. Bone Marrow Transplantation 2003; suppl.(in press).
55. Baron F, Ribbens C, Kaye O et al. Effective treatment of Jo-1-associated polymyositis with T-cell-depleted autologous peripheral blood stem cell transplantation. Br J Haematol 2000; 110:339-342.
56. Rosen O, Thiel A, Massenkeil G et al. Autologous stem-cell transplantation in refractory autoimmune diseases after in vivo immunoablation and ex vivo depletion of mononuclear cells. Arthritis Res 2000; 2:327-336.
57. Jantunen EJ, Myllykangas-Luosujarvi RA. Stem cell transplantation— A treatment for severe rheumatic diseases?—A review. Scand J Rheumatol 1999; 28:69-74.
58. Hensel M, Breitbart A, Ho AD. Autologous hematopoietic stem-cell transplantation for Behcet's disease with pulmonary involvement. N Engl J Med 2001; 344:69.
59. Lim SH, Kell J, al-Sabah A et al. Peripheral blood stem-cell transplantation for refractory autoimmune thrombocytopenic purpura. Lancet 1997; 349:475.
60. Hayag MV, Cohen JA, Kerdel FA. Immunoablative high-dose cyclophosphamide without stem cell rescue in a patient with pemphigus vulgaris. J Am Acad Dermatol 2000; 43:1065-1069.
61. Kocher AA, Schuster MD, Szabolcs MJ et al. Neovascularization of ischemic myocardium by human bone-marrow-derived angioblasts prevents cardiomyocyte apoptosis, reduces remodeling and improves cardiac function. Nat Med 2001; 7(4):430-6.
62. Burt RK, Oyama Y, Barr W et al. Allogeneic nonmyeloablative stem cell transplantation (NST) for refractory rheumatoid arthritis. Bone Marrow Transplant 2003; suppl (abst)(in press)
63. Kottaridis PD, Milligan DW, Chopra R et al. In vivo CAMPATH-1H prevents GvHD following nonmyeloablative stem-cell transplantation. Cytotherapy 2001; 3(3):197-201.

CHAPTER 32

Infection in the Hematopoeitic Stem Cell Transplant Recipient with Autoimmune Disease

Valentina Stosor and Teresa R. Zembower

Introduction

Infection is a well-recognized complication of hematopoietic stem cell transplantation (HSCT). Based on decades of cumulative experience from HSCT for hematologic malignancy, there are relatively defined periods of risk for infection in the transplanted host.[1-5] Vulnerability to infection is dynamic and is dependent upon the baseline immune status and underlying diseases, specific procedures and treatments related to the transplant regimen, and post-transplantation factors that influence immune reconstitution such as the development of graft versus host disease, the presence of opportunistic infection (OI), or the need for ongoing immunomodulating therapies.

Differences in the spectrum and timing of infections in persons undergoing stem cell transplantation for severe autoimmune diseases (SADS), as opposed to hematologic malignancy, are likely to emerge because of the chronic nature of the underlying autoimmune diseases, the end organ and functional damage caused by these diseases, the inherent immune defects associated with autoimmune diseases, and the acquired immunodeficiencies that result from disease modifying therapies. Infection risk is compounded by the specific practices related to the stem cell transplant procedure itself. While the general risk associated with autologous HSCT is less than with allogeneic transplantation, the immunoablation techniques of graft manipulation and anti-lymphocyte antibodies administration, often employed in HSCT for autoimmune disease, increase the host susceptibility to infection.[3,6-8] Indeed, there are multiple reports of OI occurring during and after the immediate transplant period in recipients with autoimmune diseases.

Determination of Infection Risk in the Transplant Recipient

Each transplant recipient is unique with respect to risk for infection throughout the transplantation process. The determination of such risk begins with a systematic review of the cumulative medical problems and interventions that have occurred in the recipient (Table 1). The underlying disease processes and the baseline immune status of the recipient are important factors that influence the severity, timing, and type of infections seen throughout the transplant process. The components and intensity of the conditioning regimen are important determinants of transplant-related infection risk, particularly if combined with manipulation of the stem cell graft.[9]

Underlying Diseases

Even without a transplant, infections are frequently encountered in patients with autoimmune diseases. The underlying diseases and the immunosuppressive medications directed against these diseases synergistically predispose to infection. One goal of the HSCT, by correcting underlying immune abnormalities and discontinuing chronic immune suppressive medications, is to ultimately decrease infection-related morbidity and mortality. Clinicians evaluating patients with severe autoimmune diseases (SADS) prior to HSCT must familiarize themselves with all aspects of their patient's disease and condition to properly assess the risk of infection.[10] Some autoimmune diseases will be discussed as examples.

Multiple Sclerosis

Loss or alteration of sensations, motor deficits, gait disturbances, and ultimately restriction to bed can lead to unappreciated skin trauma and pressure decubiti. Urinary dysfunction manifesting as urgency, frequency, urge incontinence, or retention, with self catheterization or chronic indwelling catheters, can lead to urosepsis. Although the life expectancy in patients with multiple sclerosis (MS) is only slightly decreased compared to that of the general population, infectious complications are responsible for nearly 50 percent of the disease-related mortality. The infectious complications include pneumonia, aspiration, urosepsis, and decubitus ulcers with resulting wound infections and osteomyelitis.[11,12]

In addition to contributing to mortality in MS, infections have been associated with exacerbations of disease activity. One study demonstrated that 35 percent of patients admitted to the hospital with an MS exacerbation had a significant bacterial infection compared to 11 percent of MS patients admitted without exacerbation. When presumptive viral and bacterial infections diagnosed before admission were included, almost 50 percent of patients with disease exacerbation had a concurrent infectious process.[13] This phenomenon is observed in MS patients who develop infections while undergoing HSCT.[14,15]

Table 1. Assessment of immunodeficiency in autologous HSCT recipients for SADS*

Phase of HSCT	Time Period	Range of Host Immune Defects	Precipitating Factors
Pretransplantation	Variable	· Functional asplenia · Altered complement activation · Cell-mediated-immunity · Neutropenia/neutrophil dysfunction · Abnormal anatomic barriers	· Underlying disease (SLE, RA, JCA) · Cytotoxic chemotherapy · Previous immunosuppressive therapies
Pre-engraftment	Days 0 to 30	· Neutropenia · Phagocyte function · Breakdown of anatomic barriers · Cell mediated immunity	· Cytotoxic chemotherapy · Total body irradiation · Anti-thymocyte globulin · Autograft manipulation
Early Post-engraftment	Days 30 to 100	· Cell mediated immunity · Humoral immunity · Phagocyte function	· Immaturity of new immune system · Anti-thymocyte globulin · Autograft manipulation
Late Post-engraftment	After Day 100	· Cell mediated immunity · Humoral immunity	· Delayed recovery of immune system · Relapse of autoimmune disease · Need for immunosuppressive therapies · Opportunistic infection (CMV)

Systemic Sclerosis

Systemic sclerosis (SSc) is a generalized disorder that affects the connective tissue of the skin and internal organs such as the gastrointestinal tract, lungs, heart, and kidneys. It is characterized by alterations of the microvasculature, disturbances of the immune system and by massive deposition of collagen.[16]

Skin loses elasticity and is easily broken and infected at points of contact. Esophageal dysfunction can lead to aspiration with pneumonia, and reduced gut motility often leads to bacterial overgrowth syndromes. Calcinosis can lead to skin ulcerations that become secondarily infected, occasionally resulting in soft tissue infections and gangrene.[16,17] One case of pyomyositis in a patient with progressive SSc has been reported. Interstitial lung disease and pulmonary fibrosis can mimic pneumonia and complicate the diagnosis of infection in these patients.[16,17]

Rheumatoid Arthritis

Rheumatoid arthritis (RA) is a systemic inflammatory disease of unclear etiology. It is characterized by chronic synovitis leading to devastating joint dysfunction by cartilage and subchondral bone destruction.[18] Several immunologic abnormalities have been described in RA patients, including altered complement system activation, abnormal chemotaxis and phagocytosis of polymorphonuclear leukocytes (PMNs), reduced gamma interferon production, and blunted NK cell activity.[10] Infections play a major role in both the morbidity and mortality of patients with RA. One prospective study from the Mayo Clinic, during which cohorts of both RA and nonRA patients were followed for a mean of 12.7 and 15.0 years, respectively, demonstrated an increased risk of infection for RA patients, with bone, joint, skin, soft tissue, and respiratory tract infections predominating.[19] Strong predictors of infection include increasing age, presence of extra-articular manifestations of disease, leukopenia, the use of corticosteroids and other comorbid conditions such as chronic lung disease, alcoholism, organic brain disease, and diabetes mellitus.[20]

Staphylococcus aureus is the major cause of septic arthritis.[21] Other reports have described joint infections caused by encapsulated organisms such as *Streptococcus pneumoniae* and *Haemophilus influenzae* or more atypical pathogens such as *Pasteurella multocida*, *Candida* species and *Mycobacteria* species.[22] Skin and soft tissue abscesses, intra-abdominal abscesses, and empyemas are caused primarily by *S. aureus*, enteric gram-negative bacilli, and *S. pneumoniae* respectively.[23]

Patients with RA frequently undergo total joint arthroplasty (TJA). Although the overall rate of infection after TJA is about one to two percent in most series, the infection risk to RA patients is two to three times that of the osteoarthritic undergoing this procedure.[24] In the first year following TJA, infections stem from nosocomial surgical complications, and *S. aureus* is the most commonly isolated pathogen. After three years, however, the likelihood increases that an infected joint has been seeded hematogenously from a distant site. These late infections exhibit increased numbers of *Enterobacteriaceae*, *Pseudomonas* species, and mixed gram-positive and gram-negative bacterial pathogens.[10]

Juvenile Chronic Arthritis

Juvenile chronic arthritis (JCA) is a group of systemic inflammatory disorders affecting children below the age of 16. The mortality rate in JCA is estimated at less than 0.3% in North America, and two-thirds of deaths occur in children with systemic onset JCA. The most important causes of death are macrophage activation syndrome, infections that are related to underlying immunosuppressive therapy, and cardiac complications.[25-28] Krugman and colleagues describe the case of a 53-month-old girl with JCA treated with methotrexate and cyclosporine A, whose course was complicated by Epstein-Barr virus-associated Hodgkin's lymphoma and *Legionella* pneumonia.[29]

Macrophage activation syndrome (MAS) is a potentially life-threatening complication of JCA that is seen almost exclusively in the systemic-onset type. MAS may be precipitated by viral infections, changes in medications, and/or autologous stem cell transplantation. This syndrome, which often mimics overwhelming infection, is characterized by the sudden onset of sustained fever, generalized lymphadenopathy, hepatosplenomegaly, and coagulopathy. Encephalopathy, respiratory distress, and renal failure can occur. Associated laboratory abnormalities include a falling ESR, pancytopenia, elevated transaminases, low fibrinogen, elevated d-dimer, prolonged prothrombin and partial thromboplastin times, increased fasting triglyceride levels, and markedly increased serum ferritin levels. Demonstrating active phagocytosis of red blood cells and platelets by histiocytes in the bone marrow, lymph nodes, liver, or spleen supports the diagnosis. Prompt treatment with high dose corticosteroids may be successful, but cyclosporine or etoposide may be required. Successful outcome has also been reported with etanercept therapy. Antimicrobial therapy may be helpful in treating an inciting infectious agent, which should be rigorously sought.[25]

Systemic Lupus Erythematosus

Systemic lupus erythematosus (SLE) is a complex disease characterized by immune dysregulation in the form of excessive autoantibody production, immune complex formation, and immunologically-mediated tissue injury.[30]

Infections are the leading cause of mortality in SLE patients. This is particularly true for patients with active SLE who are receiving treatment with immunosuppressive medications. SLE is associated with widespread immunologic dysfunction of the complement, humoral, and cell-mediated immune systems, and such defects render these patients susceptible to a wide range of infectious pathogens.[10,31]

A broad spectrum of bacterial infections have been reported in SLE patients. Low complement levels, dysfunction of the monocyte/macrophage system, and decreased opsonic capacity for both *Escherichia coli* and *S. aureus* make these two pathogens the leading causes of bacterial infections. Apart from these, *Pseudomonas aeruginosa* is the leading cause of major infections among hospitalized SLE patients.[31]

Abnormalities of the reticuloendothelial system result in enhanced susceptibility to infections with intracellular and encapsulated bacteria. For example, infection caused by *Salmonella* species, *Listeria monocytogenes*, *S. pneumoniae*, *Neisseria meningitides*, and *N. gonorrhea* are well described in patients with SLE.[32-35] Both higher carrier and relapse rates with *Salmonella* are observed in this patient population.[32] SLE patients, regardless of vaccination status, have presented with progressive fatal pneumococcal septicemia.[34,35] *Nocardia asteroides* pneumonia and central nervous system (CNS) infections have been described in SLE patients; however, almost half were diagnosed postmortem due to the fact that the clinical and laboratory findings often mimic primary lupus pathology.[36] For example, bacterial peritonitis mimics lupus serositis, and CNS infections may be confused with lupus cerebritis.

Impaired cell-mediated immunity and especially macrophage system defects render SLE patients susceptible to *Mycobacterium tuberculosis*. Tuberculosis in SLE patients commonly presents in the miliary form with high mortality, especially in those receiving immunosuppressive therapy.[31,37]

Due to the defects in T lymphocytes, interferon production, and NK cell number and function, herpes zoster infection is the leading viral opportunistic infection in SLE patients.[38] In contrast, cytomegalovirus (CMV) infections are not common in this population. One reason may be that CMV infection requires a more advanced level of immunosuppression similar to that found in AIDS patients with very low $CD4^+$ lymphocytes. Nevertheless, CMV pneumonia, vasculitis, and disseminated disease have all occurred in patients with SLE.[31,39] Retinitis caused by human herpesvirus-7 has been described in a 14-year old SLE patient.[40]

Candida albicans is the most common opportunistic pathogen diagnosed at the time of death in lupus patients, although it may not be the primary cause of death. *Candida* infections can manifest clinically as oral, vaginal, invasive, or disseminated disease. It is frequently found in combination with other pathogens.[31] *Crytpococcus neoformans* infections, primarily meningitis, have been described. Like nocardiosis, cryptococcal meningitis may be difficult to diagnosis as its symptoms can mimic CNS lupus.[31] Several cases of pulmonary and CNS aspergillosis, primarily due to *Aspergillus fumigatus*, have been reported, most of which have been fatal. Often the diagnosis is made postmortem, again because the symptoms may mimic lupus.[41,42]

Patients with SLE, because of underlying immunologic defects coupled with the immunosuppressive effects of corticosteroid and cytotoxic therapies, are at risk for the development of OIs such as *Toxoplasma gondii*, a parasite, and *Pneumocystis carinii*, formerly classified as a parasite and more recently as a fungal pathogen.[43,44]

Idiopathic Thrombocytopenic Purpura

Idiopathic thrombocytopenic purpura (ITP) is an immunoregulatory disorder in which antibodies damage platelets leading to their removal by the reticuloendothelial system.[45,46] In children, a history of preceding viral illness or live virus immunization has frequently been described.[47] Because of the significant association with HIV, eliciting risk factors and testing for HIV in any patient presenting with ITP is warranted.[48]

Mortality in patients afflicted with ITP is primarily due to hemorrhage.[49] Infectious complications are uncommon and are chiefly related to treatment with glucocorticoids, IVIG, and splenectomy. Overwhelming pneumococcal sepsis is occasionally reported in patients with ITP related to splenectomy, but when preoperative pneumococcal vaccine is given, pneumococcal infections are rare.[50,51]

Polymyositis/Dermatomyositis

Polymyositis and dermatomyositis are idiopathic inflammatory myopathies characterized by proximal limb and neck weakness, sometimes associated with muscle pain.[52] Malignancy, cardiac and pulmonary dysfunction, and infections are the most common causes of death.[53,54] The most common infectious complication is aspiration pneumonia due to respiratory muscle weakness. A case series of four patients with fulminant PCP is described in patients receiving corticosteroids. Three of the patients died in the first month of steroid therapy of overwhelming PCP.[55] Herpes zoster is reported to occur with high frequency in these patients. Interestingly, it occurs more commonly in the inactive stages of disease and is not associated with steroid therapy.[56] Two cases of nosocomial pneumonia with *Stenotrophomonas maltophilia* and one case of disseminated *N. brasiliensis* infection have been described.[57,58]

Disease Modifying Therapies

Immune modulating medications are commonly employed in the treatment of autoimmune diseases and increase the risk of infections.

Glucocorticoids

The mechanism of action by which corticosteroids achieve their profound short-term effects on inflammation is not completely understood. They may interrupt the inflammatory and immune cascade at several levels including (1) impairment of antigen opsonization (2) interference with inflammatory cell adhesion and migration through vascular endothelium (3) interruption of cell-cell communication by altering release or antagonizing cytokines (4) impairing leukotriene and prostaglandin synthesis and (5) inhibiting neutrophil superoxide production.[59] These effects, as well as corticosteroid induced diabetes, predispose to a wide range of infections including those caused by typical and atypical bacteria, viruses, fungi, and parasites.[60]

Cytotoxic Agents

There are a number of antimetabolite and alkylating agents that are used for the treatment of autoimmune disorders. These include the antimetabolites, azathioprine and methotrexate, and the alkylating agents, cyclophosphamide and chlorambucil.[61-64] The primary side effects common to these agents include myelosuppression and gastrointestinal mucositis, both of which can lead to infection, particularly, of bacterial and fungal etiologies. In the case of azathioprine, patients should be screened for thiopurine-S-methyltransferase (TPMT) deficiency (Caucasian prevalence 0.3-11 percent) as patients with low TPMT activity are at high risk for severe and potentially fatal hematopoietic toxicity if treated with conventional doses of this agent.[64] Cyclophosphamide is also associated with hemorrhagic cystitis that can mimic urinary tract infections caused by opportunistic viral pathogens.

Cyclosporine

Cyclosporine A, a calcineurin inhibitor, predisposes to infection by blocking calcium-dependent signal transduction in T-lymphocytes, thus inhibiting secretion of IL-2 and other cytokines.[53,65] The overall effect is a depression in cell-mediated immunity that leads to viral, fungal, and atypical bacterial infections.

Biological Agents

The biological agents are a diverse therapeutic category with increasing indications for the treatment of autoimmune disorders. Reports of infection occurring in treated patients are not surprising, given the immunomodulatory effects of these agents.

The interferons are produced naturally by various cells including fibroblasts and macrophages.[66] Binding of interferon to its receptors initiates a complex cascade of intracellular events. The specific interferon-induced proteins and mechanisms by which interferons exert their effects are not fully known. Side effects include injection site discomfort, flu-like illness, abdominal pain, autoimmune hepatitis, depression, and hematologic abnormalities.[67,68] Neutropenia is the most common hematologic toxicity. Necrosis and soft tissue infection or abscess can occur at the site of injection.[69]

Etanercept and infliximab both inhibit tumor necrosis factor α (TNF-α). Etanercept is a form of soluble TNF receptor (a decoy receptor for TNF-α), and infliximab is a monoclonal antibody that binds to TNF-α. Infliximab-bound TNF-α is thus prevented from binding with the TNF receptor. The side effects of these medications include infusion reactions, headache, flushing, nausea, vomiting, diarrhea, rash, chest pain, development of antinuclear antibodies and anti-DNA antibodies, clinical symptoms of SLE, cytopenias, demyelination, tuberculosis, bacterial and fungal infections, and PCP.[70-73]

Plasmapheresis

The process of plasmapheresis nonselectively removes immunoglobulins from serum; such effects are measurable for at least three weeks after treatment, or longer, if combined with cytotoxic therapies. Severe infections, such as those caused by bacteria and herpesvirus, are associated with plasmapheresis in patients with SLE. Cytomegalovirus (CMV) infections must be sought in febrile patients undergoing chronic plasmapheresis.[74]

Radiation

As a rule, patients with autoimmune diseases have not done well with radiation as a disease treatment modality. Psoriasis and RA treated with either total nodal or total body irradiation was abandoned due to a high incidence of myelodysplasia and leukemia.[75-77] In systemic sclerosis, radiation exacerbates disease.[78,79] With the exception of myasthenia gravis for which refractory cases may respond to total nodal irradiation, radiation is not considered a standard therapy for autoimmune diseases.

Transplant-Related Procedures

Because of the complex nature of HSCT, clinicians must consider each component of this procedure to determine infection risk and develop a differential diagnosis of infection for each patient.

Disruption of Normal Anatomic Barriers

The placement of central venous catheters and other devices such as indwelling bladder catheters causes disruption of normal anatomic barriers to infection.[2,3] Such instrumentation results in heightened susceptibility to bacterial and fungal infection. This is particularly true in the acute setting of chemotherapy-induced myelosuppression, although such infections can occur even after engraftment if indwelling lines remain in place.

Cytotoxic Chemotherapy

Typically, one or more cytoxic chemotherapeutic agents are administered prior to reinfusion of hematopoeitic stem cells. These agents cause myelotoxicity and periods of absolute neutropenia that significantly contribute to infection risk in the acute transplant setting. In addition, these agents have the propensity to cause gastrointestinal mucositis, thus, even further predisposing to early infection with gut-associated bacteria.[3] The clinician involved in the care of the stem cell transplant recipient must become familiar with the toxicity profiles of the regimens being used (refer to Chapter 31).

Anti-Lymphocyte Therapies

Anti-lymphocyte therapies are a common component of immunoablative conditioning regimens for stem cell transplants for autoimmune diseases.[80] These agents add to the overall infection risk, and patients undergoing such therapies need to be placed on appropriate anti-infective prophylaxis regimens.

Antithymocyte globulin (ATG) preparations act as potent immunosuppressants by reducing the number of circulating T-lymphocytes and altering T-lymphocyte activation, cytotoxic T-lymphocyte activity, and ultimately, the cell-mediated and humoral immune responses. Lymphocyte counts are reduced up to 90% after ATG administration.[81] Rabbit-derived ATG (Thymoglobulin®) is considerably more potent than equine-derived ATG (Atgam®), with 1.0 mg of rabbit ATG equivalent to 10-15 mg/kg of equine ATG.[9,81] Although ATG-exposed patients are susceptible to a wide spectrum of opportunistic pathogens, infections caused by viruses, namely CMV and the other herpesviruses, are important causes of morbidity after treatment with ATG.[81,82]

Alemtuzumab, also known as CAMPATH-1H which is named after where it was developed, Cambridge University's department of pathology, is a humanized monoclonal anti-CD52 antibody increasingly used in conditioning regimens.[80] Although its exact mechanism of action is unknown, treatment with alemtuzumab results in profound B- and T-lymphopenia.[83] For this reason, when combined with aggressive immune suppressive conditioning agents, alemtuzumab-treated patients are at risk for the development of OIs such as those caused by herpesviruses, *Candida* species, invasive molds, *P. carinii*, and bacterial pathogens.[83-86]

Ex Vivo Manipulation of Stem Cell Grafts

The practice of ex vivo T-lymphocyte depletion of stem cell autografts either by positive CD34$^+$-cell enrichment or negative selection (i.e., depletion) of CD4$^+$/CD8$^+$ lymphocytes is prevalent in transplant protocols for autoimmune disease. Recipients of manipulated autografts have delayed lymphocyte engraftment as compared to those who receive unmanipulated grafts, and this is clinically manifested by both T cell and B cell lymphopenia and inversion of the CD4/CD8 ratio.[87] When CD34$^+$ selected autografts are used in combination with aggressive immune conditioning regimens containing cytotoxic chemotherapy, irradiation and/or ATG for autoimmune disease, immune reconstitution studies have demonstrated CD4$^+$ lymphopenia and inverse CD4/CD8 ratios for at least one year to more than 18 months post-transplantation.[14,88] It is not surprising, therefore, that serious viral infections have been reported in the first 100 days following transplantation in patients receiving the combination of multi-agent conditioning drugs and CD34$^+$ selected autografts.[7,8] While CMV disease is a relatively uncommon complication of autologous stem cell transplantation, Holmberg and colleagues reported the development of CMV disease during the first 100 days after transplant in 22.6 percent of patients undergoing CD34$^+$ selected stem cell transplantation for a variety of conditions including autoimmune disease versus 4.2 percent of a retrospective cohort of patients that underwent unselected autologous transplantation for malignancy.[7] Because of such risks, recipients of manipulated autografts should undergo surveillance for CMV reactivation and receive prophylaxis directed against *Candida* and *P. carinii*.[3]

Epidemiology of Infection in the Stem Cell Transplant Recipient

Traditionally, HSCT-related infections are categorized according to the relative risk periods that correspond to the evolution of immune deficiencies that render the host susceptible to infection before, during, and after the transplant (Fig. 1).[1-5] While it is important to consider infectious etiologies within this context, it is of utmost importance to consider a broad differential diagnosis in the stem cell transplant recipient, as atypical pathogens and presentations of disease can occur in these hosts throughout the transplantation process. This may be particularly true in the setting of stem cell transplantation for autoimmune disease given the heightened preexisting susceptibility and exposure to infection in these patient populations and the use of aggressive immunoablative conditioning regimens in the current transplant protocols. Compared to malignancies, further studies are needed to highlight important differences that are emerging in the epidemiology of infection after stem cell transplantation for autoimmune disease.

Pre-Transplantation Period

Chemotherapy for stem cell mobilization results in absolute neutropenia during which patients often experience episodes of fever and, occasionally, documented infection. Infections during this time period are primarily of bacterial etiology, particularly coagulase-negative staphylococci, enteric gram-negative bacilli, and *Clostridium difficile*.[89]

In a multi-center review of mobilization procedures for HSCT for SADS,[90] thirteen of 116 patients developed fever or infection during the mobilization neutropenic period after administration of cyclophosphamide, 2-4 mg/kg. The extent of anti-infective prophylaxis administered to the patients in this series is not described. Smaller, single center series have reported a higher incidence of neutropenic fever (47-73 percent).[91,92] Mobilization-related infections in autoimmune disease patients are predominately caused by bacteria,[88,90,91,93] although fungal and viral infections have also been reported (Table 2).[90,91,94,95] These somewhat unexpected opportunistic infections have included CMV pneumonitis and fatal rhinocerebral mucormycosis, highlighting the probable contribution of preexisting subclinical infection or previous immunosuppressive therapy to infection risk during this period. Infection-related mortality associated with mobilization is low, but one death has been reported after cyclophosphamide mobilization in a patient with SLE.[90] Such events underscore the need for aggressive infection surveillance and prophylaxis during any period of neutropenia in patients with a history of long-term immunosuppression.

Pre-Engraftment Period (Days 0 to +30)

Infections that occur during the period beginning with stem cell infusion to the time of engraftment most often relate to chemotherapy-induced granulocytopenia and breakdown of physical and mucosal barriers.[1-3] Unless prophylactic or preemptive anti-infective therapy is administered, most patients develop fever during this time period.[91-93,96-105]

Bacterial infections predominate during the preengraftment phase, and gram positive aerobic bacteria such as staphylococci and gut-associated streptococci are more commonly isolated than other bacteria.[106,107] Enteric gram-negative bacilli are also isolated. The predominance of gram-positive bacterial infections stems from the widespread use of indwelling venous catheters and prophylactic antimicrobial agents with activity against gram-negative bacteria.[3,14,15,91,92,95-97,101,106-112] In addition to blood stream and central venous catheter infections, respiratory, urinary tract, and soft tissue infections occur during this period and, in some cases, may be related to physical impairments caused by the underlying autoimmune disease.[14,91,111]

Before the introduction of fluconazole prophylaxis, early superficial and invasive candidal infections were prevalent among transplant recipients with cancer, occurring in 33.3 percent and

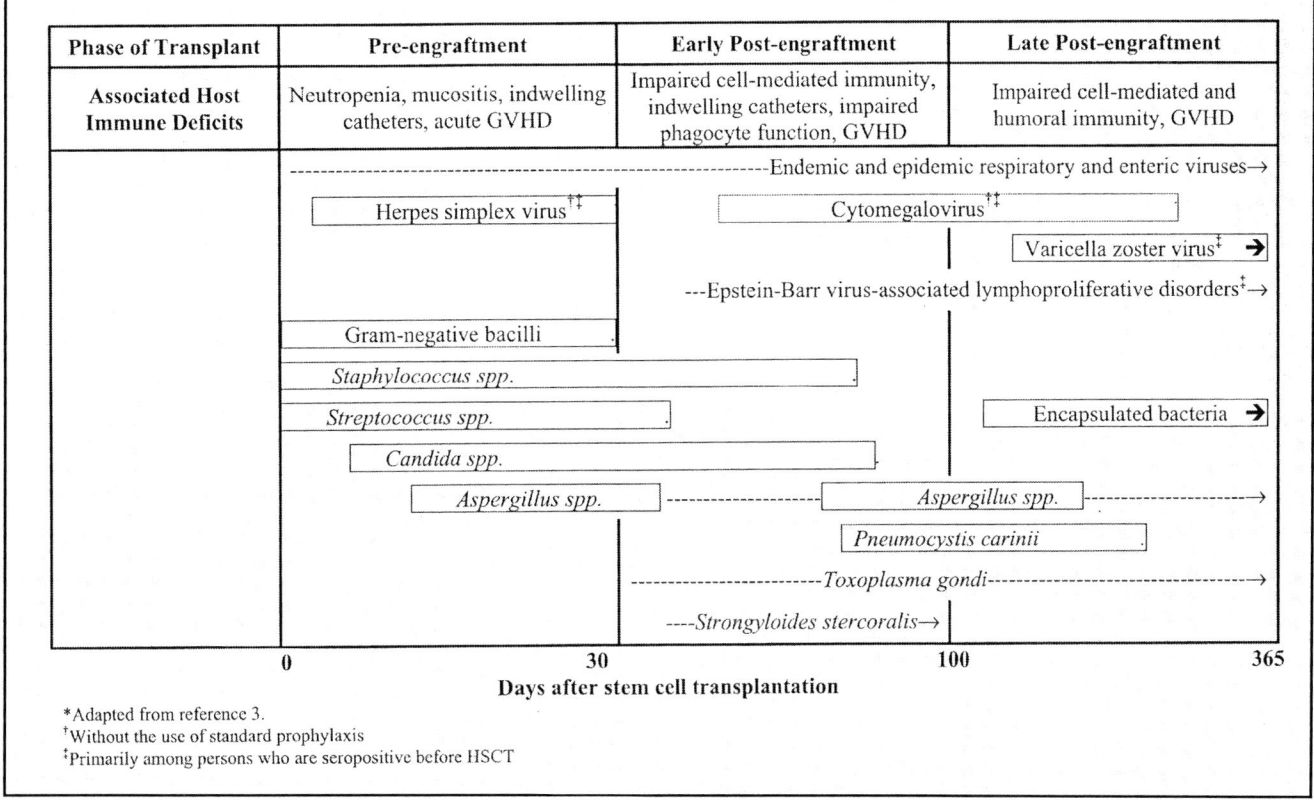

Figure 1. Timing of infections after marrow and stem cell transplantation. Adapted from CDC Guidelines for preventing opportunistic infections among hematopoietic stem cell transplant recipients. MMWR Morb Wkly Rep 2001; 49:101-125.

40 percent of recipients, respectively.[4,113] While antifungal prophylaxis has significantly reduced the frequency of *Candida* infections, the epidemiology has been impacted, such that, azole-resistant, nonalbicans species of *Candida* are diagnosed increasingly.[114,115] To date, there is only one episode of early candidemia reported after HSCT for MS.[14,91]

Transplant recipients, especially those who have experienced prolonged periods of neutropenia, are at risk for early invasive mold infections.[1] *Aspergillus* species account for more than half of such infections.[116] Invasive aspergillosis is one of the most dreaded complications of stem cell transplantation because the reported mortality rate approaches 80-100% in some series.[117-119] No pre-engraftment invasive mold infections have yet been described in transplant recipients for autoimmune disease.

Historically, herpes simplex virus (HSV) is the predominant viral pathogen encountered during the preengraftment phase of HSCT, but this virus is now diagnosed only infrequently because of antiviral prophylaxis.[2-3] Other common viral pathogens that affect the respiratory or gastrointestinal tracts, such as influenza, parainfluenza, adenovirus, respiratory syncytial virus, the enteroviruses, and rotavirus, can afflict HSCT recipients during the appropriate endemic or epidemic settings throughout all phases of the transplant process.[3,120-124]

Multiple reports of early viral infection have emerged in recipients of HSCT for autoimmune disease, although the true incidence of such infections is not known.[15,111,125-130] Of note, herpesviruses, especially CMV, are the most commonly reported early viral pathogens; however, respiratory and enteric viral pathogens are also isolated from these patients.

Early protozoal infections are uncommon after bone marrow and stem cell transplantation. It is noteworthy, then, that a fatal case of disseminated toxoplasmosis occurred in a patient prior to engraftment after undergoing T-lymphocyte depleted stem cell transplantation for JCA.[131]

Early Post-Engraftment Period (Day +30 to +100)

After engraftment until Day +100 post-transplant, the occurrence of infection is related to ongoing immunologic dysfunction such as defective neutrophil function, immunoglobulin deficiency, and absolute T-cell lymphopenia and dysfunction.[87,132-136] In the setting of HSCT for autoimmune disease, the persistent effects of immunoablation such as lymphopenia and lymphocyte dysfunction contribute to infection risk.

Herpes viruses are important opportunistic pathogens during the early post-engraftment phase.[1-4,14,91,94,95,111,126-128,130,137-139] Two pathogens, in particular, warrant special attention because of the potential for significant post-transplant morbidity and mortality and their occurrence during this period after HSCT for SADS.

Serious CMV infection and disease most commonly occur after allogeneic transplantation, although it is described, albeit much less frequently, after receipt of an autograft.[140,141] An important exception is in the case of CD34+-selected or T-lymphocyte depleted autologous stem cell transplantation which, for patients with malignancies, has been associated with the development of CMV infections.[7,8] There are multiple reports of CMV infection and disease occurring in HSCT recipients for autoimmune disease.[14,91,102,111,127,128,138] The exact incidence of CMV infection and disease is not reported for recipients of autografts for SADS.

Epstein Barr virus (EBV)-associated post transplant lymphoproliferative disease (PTLD) is an infrequent event after

Table 2. Infections reported in recipients of autologous HSCT for SADS

Phase of HSCT	Timing	Reported Infections*	References
Mobilization	Variable	· Neutropenic fever	· 90-92,101
		· Bacterial infections Bacteremia Lower respiratory tract infection Skin and soft tissue urinary tract infection	· 88,90,91,93
		· Fungal infections Candidemia Cerebral mucormycosis	· 90,91,94,95
		· Viral infections CMV pneumonitis	· 90,94,95
Pre-engraftment	Days 0 to +30	· Neutropenic fever	· 91,92,96-105,172,173
		· Bacterial infections Bacteremia (gram-positive > gram-negative) Central venous catheter infections Lower respiratory tract infections (LRTI) Skin and soft tissue infections Urinary tract infection Intestinal infections	· 14,91,92,94-97,101,108-112
		· Viral infections CMV infection and disease Other herpesvirus infections (HSV, VZV) Other viral infections (influenza, RSV, enterovirus)	· 15,111,125-130
		· Fungal infections Candidemia	· 14,91
		· Protozoal infections Disseminated toxoplasmosis	· 131
Early Post-engraftment	Days +30 to 100	· Viral infections Dermatomal VZV CMV infection Other herpesvirus infections (HSV, EBV-PTLD) Other viral infections (adenovirus)	· 14,91,94,95, 111,126-128,130, 137-139
		· Bacterial infections Upper and lower RTIs Skin and soft tissue infections Bacteremia	· 14,91,92,130
		· Fungal infections Invasive aspergillosis Disseminated fusariosis PCP	· 14,15,91,94,95,99
Late Post-engraftment	Beyond Day +100	· Viral infections Dermatomal VZV Other herpesvirus infections (CMV, EBV)	· 14,88,92,109,137,146
		· Bacterial infections	· 15,91,129
		· Fungal infections PCP	· 139,147

*in relative order of frequency reported

HSCT and usually occurs in the setting of nonHLA identical or T-cell depleted allografts and is only rarely described after autologous transplantation.[142] However, two cases of aggressive and fatal EBV-related PTLD are associated with the use of aggressive multi-drug agent conditioning regimens along with CD34+-selection of the autograft.[111,126,127]

Pneumonia caused by *P. carinii* is an important syndrome during the post-engraftment phases after allogeneic transplanta-

Table 3. Recommended evaluation for autologous HSCT candidates

- **Screening Tests Recommended for all HSCT Candidates**
    - Physical examination
    - Dental evaluation
    - Chest radiograph
    - Tuberculin skin test
    - Serologic testing for syphilis (RPR or VDRL)
    - Serum anti-CMV Ig G
    - Serum anti-HSV Ig G
    - Serum anti-*Toxoplasma* Ig G
    - Serum anti-HIV-1,2 antibodies
    - Serum anti-HTLV-1,2 antibodies
    - Serum hepatitis B surface antigen
    - Serum hepatitis B total core antibody
    - Serum anti-hepatitis C virus antibody
- **Consider Additional Screening Tests**
    - Serum anti-human herpes virus-6 IgG (for MS patients)
    - Serum Anti-EBV IgG (assess risk of PTLD if given ATG, manipulated grafts)
    - Urinalysis and urine culture
- **Screening Tests Recommended for HSCT Candidates (with suggestive travel or residence history or symptoms and signs):**
    - Serologic testing for *Stronglyloides stercoralis* (or 3 screening stools)
    - Serum anti-*Trypanosoma cruzi* Ig G
    - Serologic screening for endemic mycoses (*Histoplasma, Coccidioides, Blastomyces*)
- **Consider these screening studies for infection control of unit (based on local nosocomial infection patterns)**
    - Anterior nares, axillary, and inguinal cultures for MRSA
    - Rectal swab or fecal culture for VRE

tion. Recipients of autologous transplants who have undergone intensive conditioning regimens, or received a manipulated graft, are also at risk for the development of PCP.[1-3] One such case, reported in an autograft recipient with SLE, occurred on post-transplant day 60.[94,95]

Invasive mold infections occur in HSCT and BMT recipients in the early post-engraftment period,[1-3] although typically, the risk for autograft recipients diminishes significantly after engraftment.[143] Even so, fatal invasive aspergillosis occurred post-transplant day +65 in the recipient of an autologous HSCT for MS.[14,91] A case of disseminated fusariosis was also described after CD34⁺-selected stem cell transplantation for SLE that was complicated by graft failure.[99]

Late Post-Engraftment Period (Beyond Day +100)

After Day +100, infections in recipients of stem cell transplants result from residual immune defects from the stem cell transplant, whether or not ongoing immunomodulatory therapies are required, and for allogeneic HSCT, with the occurrence and degree of GVHD.[5,144] Due to GVHD-related risks, allogeneic HSCT are rarely performed for autoimmune diseases and then only under controlled trials designed to minimize or prevent GVHD. In general, recipients of autologous grafts have more rapid immune reconstitution,[135] so infection risk is considerably lower among autograft versus allograft recipients. However, after T-cell depleted autologous HSCT, CD4 lymphopenia and dysfunction into this late period may persist and, ultimately, contribute to infection risk after transplant for autoimmune disease.

Dermatomal infection caused by reactivation of varicella zoster virus (VZV), occurs in up to 22 percent of cancer transplant recipients and is among the most frequently reported infection of the late post-transplantation period.[5,144,145] To date, dermatomal VZV is the most commonly reported infection after post-transplant day +100 in autograft recipients with SADS (Table 2).[88,92,109,137]

While late infections, other than dermatomal VZV, occur after HSCT for autoimmune disease, such reports are limited to bacterial infection, an episode of EBV reactivation, and one case of PCP.[129,146,147] Death caused by pneumococcal sepsis occurred in an MS patient 19 months after HSCT, making the relationship to HSCT uncertain.[129] More long-term outcome data are clearly needed to better assess the risk of late infection in autoimmune disease patients who undergo HSCT.

Prevention of Infection in the HSCT Recipient

A collaborative effort between the Centers for Disease Control and Prevention, the Infectious Diseases Society of America, and the American Society for Blood and Marrow Transplantation resulted in publication of national guidelines for preventing opportunistic infections among HSCT recipients in October 2000.[3] These should be viewed as guidelines, since there may be unique features or infection risks for autoimmune diseases yet to be appreciated.

The Pre-Transplantation Period

Patient and family education begins in this period and includes general instructions for preventing exposure to potential pathogens. Strict attention to hand hygiene and avoidance of sick contacts are essential throughout the entire transplant process. Proper food handling instructions are provided for both HSCT candidates and family members involved in food preparation. Recipients must avoid sharing eating utensils to minimize exposure to viruses.[148] Safe sexual practices, including condom use, are important for recipients if not in a long-term monogamous relationship or, possibly, for couples discordant for diseases such as CMV and HSV. Pet safety instructions include avoiding contact with stray or ill animals, and, especially reptiles, ducklings, chicks, or exotic pets such as nonhuman primates. Ideally, HSCT recipients must avoid contact with animal excreta and refrain from cleaning litter boxes and fish tanks, or, alternatively, wear disposable gloves during these activities. Finally, to prevent exposure to fungal pathogens, HSCT recipients must avoid areas with high dust exposures such as construction or renovation areas and avoid occupations or activities involving soil.[3]

A complete history and physical examination is performed during the pre-transplantation screening evaluation (Table 3). Thorough social, sexual, occupation, vaccination, travel and residence, and pet histories are included. Certain workplace settings, such as healthcare facilities, prisons, and homeless shelters, increase the risk for exposure to infectious agents, especially tuberculosis. Prior to transplantation, patients must undergo thorough dental evaluation and relevant treatments.

The pre-transplantation laboratory evaluation includes a complete blood count (CBC) with differential, blood chemistries, a tuberculin skin test (PPD), and a chest radiograph (CXR). Urinalysis (UA) and urine culture are especially important for patients with MS. Serologic evaluation for toxoplasmosis, CMV, HSV, and VZV are routinely obtained.[3] Additional screening for viruses includes serologic testing for HIV, HTLV-1 and -2, hepatitis B virus (HBV), and hepatitis C virus (HCV). For research interest, serologies for human herpes virus 6 (HHV-6) may be considered for MS patients, as there is some evidence correlating HHV-6 reactivation with exacerbation of disease activity.[149-153] Patients with the appropriate endemic exposures should have serologies obtained for *Strongyloides stercoralis*, *Trypanosoma cruzi*, and the endemic mycoses *Histoplasma capsulatum*, *Blastomyces dermatitidis*, and *Coccidioides immitis*.[3]

The Transplantation Period

Aggressive infection prevention efforts are required during the period encompassing the time between the beginning of conditioning and engraftment. All HSCT candidates are admitted to private rooms with directed airflow. Upon admission, nasal, axillary, and inguinal swabs are often obtained to screen for methicillin-resistant *S. aureus* (MRSA) and rectal swabs or stool cultures to screen for vancomycin-resistant enterococci (VRE). Detection of the carrier state allows institution of appropriate isolation precautions to prevent nosocomial transmission and alerts clinicians to potential pathogens should infection occur during the transplantation process.[3,154-156] Stringent adherence to hand, oral, and perineal hygiene are essential.[3] Dried or fresh plants and flowers are prohibited in patient rooms.[157] To reduce the risk of CMV infection in CMV-seronegative recipients, only CMV-negative or leukocyte-reduced blood products are administered.[158]

The Post-Transplantation Period

HSCT recipients require instruction on methods to minimize exposure to pathogens during the activities of daily life after discharge from the hospital until immunologic recovery occurs.[3] Instructions for dietary restrictions, food and water preparation, household pet handling and care, and safe sexual, recreational, and travel precautions need review and reinforcement. Recipients are encouraged to seek professional medical advice before participating in new activities that may have unforeseen risks. Patients are also educated about signs and symptoms that should prompt medical consultation such as fever, upper respiratory, gastrointestinal, or urinary symptoms, or rash.

For HSCT recipients with cancer, there are published guidelines for revaccination after immune reconstitution.[3] A concern among physicians caring for autoimmune disease patients is the potential risk of disease relapse or exacerbation caused by vaccination-evoked immune stimulation. While studies examining this issue have failed to demonstrate that vaccination increases such risk for patients with MS, RA, and SLE,[159,160] until guidelines are developed for patients with autoimmune diseases, the individual transplant center must weigh the risk and benefits of vaccination for its patient populations.

Anti-Infective Prophylaxis Strategies

Evidence-based guidelines for the use of anti-infective agents for the prevention of infection in HSCT recipients are published.[3] When developing institutional protocols, each transplant center must consider the characteristics of the patient population, the intensiveness of the conditioning regimen, graft manipulation, and local epidemiology and antimicrobial susceptibility patterns to determine the type and duration of prophylaxis required. While the basic goal of prophylaxis is to prevent infection, the inevitable risk of such strategies is the development and spread of antimicrobial resistance, and this, too, should influence the ultimate design of prophylaxis protocols.

Table 4 outlines the published guidelines for anti-infective prophylaxis for autologous HSCT. The authors have chosen to highlight autologous transplantation guidelines since the vast majority of autoimmune disease patients have received autografts to date. These guidelines were developed for patients with cancer and may need to be modified as data accumulates for HSCT for autoimmune diseases. For example, patients with refractory and active SLE have a two and five year mortality due predominately to infection of 20 and 35%, respectively. Therefore, for SLE patients it may be prudent to use prophylatic antibiotics during periods of transplant related neutropenia.

Antibacterial Prophylaxis

Most transplant centers use single or combination antibacterial agents to prevent infections during the neutropenic period. In patients with cancer, this stategy prevents bacteremia, although not overall infection-related mortality, in neutropenic hosts.[161-163] If prophylaxis is administered, nosocomial epidemiology and antimicrobial susceptibility patterns should guide the selection of the agent(s). Prophylaxis regimens also influence the treatment choices for empiric therapy of fever and infections during the transplant, as patients may develop infections caused by organisms that have inherent or acquired resistance to prophylactic agents.[161,163-165]

Antiviral Prophylaxis

Prophylaxis to prevent infections with HSV is recommended for seropositive recipients from the beginning of conditioning and continued until the time of engraftment.[3,166] The efficacy of extending antiviral prophylaxis in the setting of delayed lymphocyte recovery has not been studied. However, it is common to extend oral acyclovir prophylaxis for 6 to 12 months after HSCT for SADS.

Because of the risk of CMV infection after CD34$^+$-selected autologous HSCT, a preemptive CMV treatment strategy is recommended[3] that involves weekly screening of CMV-seropositive recipients for antigenemia by the CMV pp65 antigen assay, or alternatively, the quantitative CMV DNA PCR assay,[167] from the time of engraftment until the first 100 days after transplantation. Surveillance for CMV infection may be continued for a longer period in the setting of significant ongoing immunosuppressive therapy or in the event of the development of CMV infection. Preemptive treatment with intravenous ganciclovir is administered for any level of antigenemia. Oral ganciclovir and the new antiviral, valganciclovir, are promising but not currently approved for the treatment or prevention of CMV infection in HSCT recipients. It should be emphasized that the true incidence and timing of CMV infections in recipients of ATG and manipulated autografts for autoimmune diseases are currently unknown, and recommendations for prevention and surveillance for these infections may require modification once more complete data emerges.

Antifungal Prophylaxis

Recipients of autografts for SADS generally receive manipulated autografts and/or intensive conditioning regimens such that prophylactic antifungal therapy with fluconazole is warranted from the time of conditioning until engraftment,[3] or longer. Fluconazole prophylaxis reduces the incidence of both superficial fungal in-

Table 4. Anti-infective prophylaxis strategies for autologous HSCT recipients*

Pathogen	Recommended Agent or Strategy	Duration	Alternative Agents
Bacteria			
General	Routine use of anti-bacterials for asymptomatic, afebrile neutropenic patients not recommended. If anti-bacterial prophylaxis desired, choice of agent based on local antimicrobial susceptibility data. Vancomycin is <u>not</u> recommended for routine prophylaxis. Routine use of IVIG not recommended.		
H. influenzae type B	Rifampin for acute exposures	4 days	
MRSA	Topical mupirocin ointment	5 days	Bacitracin TMP-sulfa Rifampin
M. tuberculosis	INH and pyridoxine for exposures or positive skin test without active disease	9 months	
Viruses			
CMV[a,b]	Pre-emptive treatment strategy as follows for first 100 days after transplant: Weekly screening by the CMV pp65 antigenemia assay or quantitative PCR; Treatment with ganciclovir for any level of antigenemia.		Foscarnet IVIG <u>not</u> routinely recommended
HSV[1]	Acyclovir	30 days post-transplant	Valacyclovir
VZV	Long-term anti-viral agents not recommended. Passive immunization for acute varicella exposures in VZV-seronegative HSCT recipients.		
Influenza A	Rimantidine for nosocomial and community outbreaks.	14 days after influenza vaccination or until the end of the outbreak if the HSCT recipient is unvaccinated.	Amantidine Oseltamivir
Fungi			
Yeasts[b]	Fluconazole	30 days post-transplant	
Aspergillus spp.	Prophylaxis not recommended.		
PCP[b]	TMP-sulfa	6 months post-transplant[c]	Dapsone Aerosolized pentamidine
Protozoa			
Toxoplasma[a]	Adults: Prophylaxis not recommended. Adolescents and pediatrics: TMP-sulfa	6 months post-transplant	Clindamycin, pyrimethamine and leucovorin
Strongyloides[d]	Ivermectin	Prior to HSCT	Albendazole Thiabendazole

*Adapted from CDC. Guidelines for preventing opportunistic infections among hematopoietic stem cell transplant recipients. MMWR Morb Mortal Rep 2001; 49:101-125. [a] For seropositive recipients; [b] For autologous HSCT recipients in the setting of intense conditioning regimens or graft manipulation; [c] Extend duration of prophylaxis in setting of ongoing corticosteroid therapy; [d] For HSCT recipients with suggestive travel or residence history, unexplained eosinophilia or evidence of Strongyloides infection by screening test.

fections and candidemia in HSCT recipients.[168,169] The major pitfall of widespread fluconazole use has been a shift in the epidemiology of candidal infections to fluconazole-resistant species such as *C. glabrata* and *C. krusei*, or even fluconazole-resistant *C. albicans*.[114,169-171] While specific guidelines for antifungal prophylaxis against invasive molds are not available, some centers prefer prophylactic antifungal agents with activity directed against *Aspergillus* sp. throughout the periods of risk, especially in higher risk patients groups, such as those with SLE. Agents such the lipid-based formulations of amphotericin, caspofungin, itraconazole and voriconazole are approved for the treatment of invasive aspergillosis, but data regarding their efficacies as prophylactic agents are not yet available.

Anti-Pneumocystis carinii *Prophylaxis*

Recipients require antimicrobial therapy to prevent pneumocystis carinii pneumonia (PCP) for the first six months after transplant, or longer, if there is a need for ongoing corticosteroid treatment. The most effective prophylactic agent is trimethoprim-sulfamethoxazole (TMP/SMX) initiated after engraftment.[3] Many transplant centers refrain from TMP-SMX use because of concerns for myelosuppression or TMP-SMX related SLE flares. In this case, aerosolized pentamidine is a less effective alternative. TMP-SMX has the added benefit of providing both anti-bacterial and anti-*Toxoplasma* coverage.

Summary

Infection remains a major cause of transplant-related morbidity and mortality, and available data suggest that recipients of manipulated autografts for autoimmune diseases are at significant risk for infection before, during, and after stem cell transplantation. The impact of infectious diseases on HSCT recipients can be minimized by aggressive screening efforts prior to transplantation, and appropriate prophylaxis and prevention strategies after transplantation. When infection does occur, attempts at etiologic diagnosis require an aggressive but rational approach. It is, therefore, essential that clinicians involved in the care of HSCT recipients have a thorough understanding of the risks for infection in this patient population, as well as the accepted strategies for the prevention and treatment of infections during and after transplantation.

The combination of patient and disease characteristics and immunoablative techniques involved in the HSCT process for autoimmune diseases results in a unique scenario with respect to infection risk that is not comparable to transplantation for other disease states. Controlled studies designed specifically to examine this issue are needed to better define the epidemiology of infections in this patient population. Such information, when available, will direct the development of preventive and treatment measures that are more specific and appropriate for patients with autoimmune disorders undergoing stem cell transplantation.

References

1. Sable A, Donowitz G. Infections in bone marrow transplant recipients. Clin Infect Dis 1994; 18:273-284.
2. Bowden R, Meyers JD. Infection complicating bone marrow transplantation. In: Rubin RH, ed. Clinical approach to infection in the compromised host. 3rd ed. New York: Plenum Publishing Corporation, 1984:601-628.
3. CDC. Guidelines for preventing opportunistic infections among hematopoeitic stem cell transplant recipients. MMWR Morb Mortal Wkly Rep. 2001; 49:101-125.
4. Winston DJ, Gale RP, Meyer DV et al. Infectious complications of human bone marrow transplantation. Medicine (Baltimore) 1979; 58:1-31.
5. Atkinson K, Storb R, Prentice RL et al. Analysis of late infections in 89 long-term survivors of bone marrow transplantation. Blood 1979; 53:720-731.
6. Anderson KC, Soiffer R, DeLage R et al. T-cell-depleted autologous bone marrow transplantation therapy: Analysis of immune deficiency and late complications. Blood 1990; 76:235-244.
7. Holmberg LA, Boeckh M, Hooper H et al. Increased incidence of cytomegalovirus disease after autologous CD34-selected peripheral blood stem cell transplantation. Blood 1999; 94:4029-4035.
8. Miyamoto T, Gondo H, Miyoshi Y et al. Early viral complications following CD34-selected autologous peripheral blood stem cell transplantation for nonHodgkin's lymphoma. Br J Haematol 1998; 100:348-350.
9. Burt RK. Hematopoietic stem cell transplantation for multiple sclerosis: Finding equipoise. Stem Cell Therapy for Tolerance and Tissue Regeneration. Snowbird, Utah:2002.
10. Payan D. Evaluation and management of patients with collagen vascular disease. In: Rubin RH, Young LS, eds. Clinical approach to infection in the compromised host. 3rd ed. New York: Plenum Publishing Corporation, 1984:581-600.
11. Hogancamp WE, Noseworthy JH. Demyelinating disorders of the central nervous system. In: Goetz CG, Pappert EJ, eds. Textbook of Clinical Neurology. 1st ed. Philadelphia: W.B. Saunders Company, 1999:971-983.
12. O'Connor P. Key issues in the diagnosis and treatment of multiple sclerosis. An overview. Neurology 2002; 59:S1-33.
13. Rapp NS, Gilroy J, Lerner AM. Role of bacterial infection in exacerbation of multiple sclerosis. Am J Phys Med Rehabil 1995; 74:415-418.
14. Fassas A, Anagnostopoulos A, Kazis A et al. Autologous stem cell transplantation in progressive multiple sclerosis—an interim analysis of efficacy. J Clin Immunol 2000; 20:24-30.
15. Fassas A, Passweg JR, Anagnostopoulos A et al. Hematopoietic cell transplantation for multiple sclerosis. A retrospective multicenter study. J Neurol 2002; 249:1088-1097.
16. Haustein UF. Systemic sclerosis-scleroderma. Dermatol Online J 2002; 8:3.
17. Cossio M, Menon Y, Wilson W et al. Life-threatening complications of systemic sclerosis. Crit Care Clin 2002; 18:819-839.
18. Blaschke S, Schwarz G, Moneke D et al. Epstein-Barr virus infection in peripheral blood mononuclear cells, synovial fluid cells, and synovial membranes of patients with rheumatoid arthritis. J Rheumatol 2000; 27:866-873.
19. Doran MF, Crowson CS, Pond GR et al. Frequency of infection in patients with rheumatoid arthritis compared with controls: A population-based study. Arthritis Rheum 2002; 46:2287-2293.
20. Doran MF, Crowson CS, Pond GR et al. Predictors of infection in rheumatoid arthritis. Arthritis Rheum 2002; 46:2294-2300.
21. Mitchell WS, Brooks PM, Stevenson RD et al. Septic arthritis in patients with rheumatoid disease: A still underdiagnosed complication. J Rheumatol 1976; 3:124-133.
22. Myers AR. Septic arthritis caused by bacteria. In: Kelley WN, ed. Textbook of Rheumatology. 2nd ed. Philadelphia, PA: Saunders, 1985:1870-1907.
23. Huskisson EC, Hart FD. Severe, unusual, and recurrent infections in rheumatoid arthritis. Ann Rheum Dis 1972; 31:118-121.
24. Poss R, Thornhill TS, Ewald FC et al. Factors influencing the incidence and outcome of infection following total joint arthroplasty. Clin Orthop 1984; 117-126.
25. Schneider R, Passo MH. Juvenile rheumatoid arthritis. Rheum Dis Clin North Am 2002; 28:503-530.
26. Ansell BM, Wood PH. Prognosis in juvenile chronic polyarthritis. Clin Rheum Dis 1979; 2:397-412.
27. Stoeber E. Prognosis in juvenile chronic arthritis. Follow-up of 433 chronic rheumatic children. Eur J Pediatr 1981; 135:225-228.
28. Hull RG. Outcome in juvenile arthritis. Br J Rheumatol 1988; 27Suppl 1:66-71.

29. Krugmann J, Sailer-Hock M, Muller T et al. Epstein-Barr virus-associated Hodgkin's lymphoma and legionella pneumophila infection complicating treatment of juvenile rheumatoid arthritis with methotrexate and cyclosporine A. Hum Pathol 2000; 31:253-255.
30. Al-Mayouf SM, Al-Jumaah S, Bahabri S et al. Infections associated with juvenile systemic lupus erythematosus. Clin Exp Rheumatol 2001; 19:748-750.
31. Iliopoulos AG, Tsokos GC. Immunopathogenesis and spectrum of infections in systemic lupus erythematosus. Semin Arthritis Rheum 1996; 25:318-336.
32. Pablos JL, Aragon A, Gomez-Reino JJ. Salmonellosis and systemic lupus erythematosus. Report of ten cases. Br J Rheumatol 1994; 33:129-132.
33. Kraus A, Cabral AR, Sifuentes-Osornio J et al. Listeriosis in patients with connective tissue diseases. J Rheumatol 1994; 21:635-638.
34. Hill MD, Karsh J. Invasive soft tissue infections with Streptococcus pneumoniae in patients with systemic lupus erythematosus: Case report and review of the literature. Arthritis Rheum 1997; 40:1716-1719.
35. Scerpella EG. Functional asplenia and pneumococcal sepsis in patients with systemic lupus erythematosus. Clin Infect Dis 1995; 20:194-195.
36. Mok CC, Yuen KY, Lau CS. Nocardiosis in systemic lupus erythematosus. Semin Arthritis Rheum 1997; 26:675-683.
37. Hernandez-Cruz B, Sifuentes-Osornio J, Ponce-de-Leon Rosales S et al. Mycobacterium tuberculosis infection in patients with systemic rheumatic diseases. A case-series. Clin Exp Rheumatol 1999; 17:289-296.
38. Manzi S, Kuller LH, Kutzer J et al. Herpes zoster in systemic lupus erythematosus. J Rheumatol 1995; 22:1254-1258.
39. Rider JR, Ollier WE, Lock RJ et al. Human cytomegalovirus infection and systemic lupus erythematosus. Clin Exp Rheumatol 1997; 15:405-409.
40. Taccetti G, Repetto T, Marianelli L et al. Human herpesvirus-7 infection in a patient with systemic lupus erythematosus. Clin Exp Rheumatol 1999; 17:126.
41. Gonzalez-Crespo MR, Gomez-Reino JJ. Invasive aspergillosis in systemic lupus erythematosus. Semin Arthritis Rheum 1995; 24:304-314.
42. Katz A, Ehrenfeld M, Livneh A et al. Aspergillosis in systemic lupus erythematosus. Semin Arthritis Rheum 1996; 26:635-640.
43. Zamir D, Amar M, Groisman G et al. Toxoplasma infection in systemic lupus erythematosus mimicking lupus cerebritis. Mayo Clin Proc 1999; 74:575-578.
44. Porges AJ, Beattie SL, Ritchlin C et al. Patients with systemic lupus erythematosus at risk for Pneumocystis carinii pneumonia. J Rheumatol 1992; 19:1191-1194.
45. Wong GC, Lee LH. A study of idiopathic thrombocytopenic purpura (ITP) patients over a ten-year period. Ann Acad Med Singapore 1998; 27:789-793.
46. Bussel JB. Autoimmune thrombocytopenic purpura. Hematol Oncol Clin North Am 1990; 4:179-191.
47. Rand ML, Wright JF. Virus-associated idiopathic thrombocytopenic purpura. Transfus Sci 1998; 19:253-259.
48. Scaradavou A. HIV-related thrombocytopenia. Blood Rev 2002; 16:73-76.
49. Silverman MA. Idiopathic thrombocytopenic purpura: eMedicine. 2001. http://emedicine.com/EMERG/topic282.htm.
50. Naouri A, Feghali B, Chabal J et al. Results of splenectomy for idiopathic thrombocytopenic purpura. Review of 72 cases. Acta Haematol 1993; 89:200-203.
51. Bell WR Jr. Long-term outcome of splenectomy for idiopathic thrombocytopenic purpura. Semin Hematol 2000; 37:22-25.
52. Plotz PH, Leff RL, Miller FW. Inflammatory and metabolic myopathies. In: Schumacher HR, Klippel JH, Koopman WJ, eds. Primer on the rheumatic diseases. 10th ed. Atlanta, GA: Arthritis Foundation, 1993:127-131.
53. Koler RA, Montemarano A. Dermatomyositis. Am Fam Physician 2001; 64:1565-1572.
54. Rosenkranz H. Polymyositis: eMedicine. 2002. http://www.emediine.com/EMERG/topic474.htm.
55. Bachelez H, Schremmer B, Cadranel J et al. Fulminant Pneumocystis carinii pneumonia in 4 patients with dermatomyositis. Arch Intern Med 1997; 157:1501-1503.
56. Nagaoka S, Tani K, Ishigatsubo Y et al. Herpes zoster in patients with polymyositis and dermatomyositis. Kansenshogaku Zasshi 1990; 64:1394-1399.
57. Amano K, Maruyama H, Takeuchi T. Nosocomial pneumonia likely caused by Stenotrophomonas maltophilia in two patients with polymyositis. Intern Med 1999; 38:910-916.
58. Klein-Gitelman MS, Szer IS. Disseminated Nocardia brasiliensis infection: An unusual complication of immunosuppressive treatment for childhood dermatomyositis. J Rheumatol 1991; 18:1243-1246.
59. Paulus HE, Bulpitt KJ. Nonsteroidal anti-inflammatory agents and corticosteroids. In: Schumacher HR, Klippel JH, Koopman WJ, eds. Primer on the Rheumatic Diseases. 10th ed. Atlanta, GA: Arthritis Foundation, 1993:298-303.
60. Fekety R. Infections associated with corticosteroids and immunosuppressive therapy. In: Gorbach SL, Bartlett JG, Blacklow NR, eds. Infectious Diseases. Philadelphia: W. B. Saunders Company, 1992:1050-1057.
61. Brooks P. Slow-acting antirheumatic drugs and cytotoxic agents. In: Schumacher HR, Klippel JH, Koopman WJ, eds. Primer on the Rheumatic Diseases. 10th ed. Atlanta, GA: Arthritis Foundation, 1993:303-306.
62. Cytoxan. The Physicians' Desk Reference. Montvale, NJ: Medical Economics Company Inc, 1999.
63. Imuran. The Physicians' Desk Reference. Montvale, NJ: Medical Econocmics Company Inc, 2002:2893.
64. Yates CR, Krynetski EY, Loennechen T et al. Molecular diagnosis of thiopurine S-methyltransferase deficiency: Genetic basis for azathioprine and mercaptopurine intolerance. Ann Intern Med 1997; 126:608-614.
65. Neoral. The Physicians' Desk Reference. Montvale, NJ: Medical Economics Inc 2002:2380-2387.
66. Drugs aproved by the FDA: Rebif (interferon beta-1a). Center Watch. 2002. http://www.centerwach.com/patient.drugs/dru767.html.
67. Vial T, Descotes J. Clinical toxicity of the interferons. Drug Saf 1994; 10:115-150.
68. Raanani P, Ben-Bassat I. Immune-mediated complications during interferon therapy in hematological patients. Acta Haematol 2002; 107:133-144.
69. Sheremata WA, Taylor JR, Elgart GW. Severe necrotizing cutaneous lesions complicating treatment with interferon beta-1b. N Engl J Med 1995; 332:1584.
70. Arend WP. Cytokines in rheumatoid arthritis. Bulletin on the Rheumatic Diseases. Arthritis Foundation 2002; 51.
71. Keane J, Gershon S, Wise RP et al. Tuberculosis associated with infliximab, a tumor necrosis factor alpha-neutralizing agent. N Engl J Med 2001; 345:1098-1104.
72. Infliximab. In: Reents S, ed. Clinical Pharmacology 2000. Tampa, FL: Gold Standard Multimedia 2000. http://cp.gsm.com.
73. Etanercept. In: Reents S, ed. Clinical Pharmacology 2000. Tampa, FL: Gold Standard Multimedia 2000. http://cp.gsm.com.
74. Aringer M, Smolen JS, Graninger WB. Severe infections in plasmapheresis-treated systemic lupus erythematosus. Arthritis Rheum 1998; 41:414-420.
75. Weiss HA, Darby SC, Fearn T et al. Leukemia mortality after X-ray treatment for ankylosing spondylitis. Radiat Res 1995; 142:1-11.
76. Darby SC, Doll R, Gill SK et al. Long term mortality after a single treatment course with X-rays in patients treated for ankylosing spondylitis. Br J Cancer 1987; 55:179-190.
77. Urowitz MB, Rider WD. Myeloproliferative disorders in patients with rheumatoid arthritis treated with total body irradiation. Am J Med 1985; 78:60-64.
78. O'Dell JR, Steigerwald JC, Kennaugh RC et al. Lack of clinical benefit after treatment of systemic sclerosis with total lymphoid irradiation. J Rheumatol 1989; 16:1050-1054.
79. Varga J, Haustein UF, Creech RH et al. Exaggerated radiation-induced fibrosis in patients with systemic sclerosis. Jama 1991; 265:3292-3295.

80. Burt RK, Slavin S, Burns WH et al. Induction of tolerance in autoimmune diseases by hematopoietic stem cell transplantation: Getting closer to a cure? Blood 2002; 99:768-784.
81. Antithymocyte Globulin. In: Reents S, ed. Clinical Pharmacology 2000. Tampa, FL: Gold Standard Multimedia, 2000. http://cp.gsm.com.
82. Rubin RH, Cosimi AB, Hirsch MS et al. Effects of antithymocyte globulin on cytomegalovirus infection in renal transplant recipients. Transplantation 1981; 31:143-145.
83. Alemtuzumab. In: Reents S, ed. Clinical Pharmacology 2000. Tampa, FL: Gold Standard Multimedia, 2000. http://cp.gsm.com.
84. Keating MJ, Flinn I, Jain V et al. Therapeutic role of alemtuzumab (Campath-1H) in patients who have failed fludarabine: Results of a large international study. Blood 2002; 99:3554-3561.
85. Rai KR, Freter CE, Mercier RJ et al. Alemtuzumab in previously treated chronic lymphocytic leukemia patients who also had received fludarabine. J Clin Oncol 2002; 20:3891-3897.
86. Perez-Simon JA, Kottaridis PD, Martino R et al. Nonmyeloablative transplantation with or without alemtuzumab: Comparison between 2 prospective studies in patients with lymphoproliferative disorders. Blood 2002; 100:3121-3127.
87. Bomberger C, Singh-Jairam M, Rodey G et al. Lymphoid reconstitution after autologous PBSC transplantation with FACS-sorted CD34+ hematopoietic progenitors. Blood 1998; 91:2588-2600.
88. Burt RK, Traynor AE, Pope R et al. Treatment of autoimmune disease by intense immunosuppressive conditioning and autologous hematopoietic stem cell transplantation. Blood 1998; 92:3505-3514.
89. Toor AA, van Burik JA, Weisdorf DJ. Infections during mobilizing chemotherapy and following autologous stem cell transplantation. Bone Marrow Transplant 2001; 28:1129-1134.
90. Burt RK, Fassas A, Snowden J et al. Collection of hematopoietic stem cells from patients with autoimmune diseases. Bone Marrow Transplant 2001; 28:1-12.
91. Fassas A, Anagnostopoulos A, Kazis A et al. Peripheral blood stem cell transplantation in the treatment of progressive multiple sclerosis: First results of a pilot study. Bone Marrow Transplant 1997; 20:631-638.
92. Kozak T, Havrdova E, Pit'ha J et al. High-dose immunosuppressive therapy with PBSC support in the treatment of poor risk multiple sclerosis. Bone Marrow Transplant 2000; 25:525-531.
93. Binks M, Passweg JR, Furst D et al. Phase I/II trial of autologous stem cell transplantation in systemic sclerosis: Procedure related mortality and impact on skin disease. Ann Rheum Dis 2001; 60:577-584.
94. Traynor AE, Schroeder J, Rosa RM et al. Treatment of severe systemic lupus erythematosus with high-dose chemotherapy and haemopoietic stem-cell transplantation: A phase I study. Lancet 2000; 356:701-707.
95. Traynor A, Burt RK. Haematopoietic stem cell transplantation for active systemic lupus erythematosus. Rheumatology (Oxford) 1999; 38:767-772.
96. Snowden JA, Biggs JC, Milliken ST et al. A phase I/II dose escalation study of intensified cyclophosphamide and autologous blood stem cell rescue in severe, active rheumatoid arthritis. Arthritis Rheum 1999; 42:2286-2292.
97. Verburg RJ, Kruize AA, van den Hoogen FH et al. High-dose chemotherapy and autologous hematopoietic stem cell transplantation in patients with rheumatoid arthritis: Results of an open study to assess feasibility, safety, and efficacy. Arthritis Rheum 2001; 44:754-760.
98. McColl G, Kohsaka H, Szer J et al. High-dose chemotherapy and syngeneic hemopoietic stem-cell transplantation for severe, seronegative rheumatoid arthritis. Ann Intern Med 1999; 131:507-509.
99. Shaughnessy PJ, Ririe DW, Ornstein DL et al. Graft failure in a patient with systemic lupus erythematosus (SLE) treated with high-dose immunosuppression and autologous stem cell rescue. Bone Marrow Transplant 2001; 27:221-224.
100. Mohren M, Amberger C, Daikeler T et al. Autologous stem cell transplantation in a patient with psoriatic arthropathy. Ann Rheum Dis 2001; 60:713. (Abstract P15)
101. Bingham S, Veale D, Fearon U et al. High-dose cyclophosphamide with stem cell rescue for severe rheumatoid arthritis: Short-term efficacy correlates with reduction of macroscopic and histologic synovitis. Arthritis Rheum 2002; 46:837-839.
102. Carreras E, Saiz Z, Graus F et al. Autologous CD34+ selected haemopoietic stem cell transplantation for multiple sclerosis: A single center experience. Ann Rheum Dis 2001; 60:713. (Abstract P19)
103. Ledziowski PT, Mensah PA, Pitkowska-Jackubas B. High dose chemotherapy with autologous blood stem cell transplantation for treatment of multiple sclerosis: A case presentation. Ann Rheum Dis 2001; 60:715. (Abstract P28).
104. Lisukov I, Sizikova S, Kulagin A et al. High dose melphalan with stem cell support for the treatment of refractory immune thrombocytopenic purpura: A case report. Ann Rheum Dis 2001; 60:710. (Abstract P8).
105. Lim SH, Kell J, al-Sabah A et al. Peripheral blood stem-cell transplantation for refractory autoimmune thrombocytopenic purpura. Lancet 1997; 349:475.
106. Collin BA, Leather HL, Wingard JR et al. Evolution, incidence, and susceptibility of bacterial bloodstream isolates from 519 bone marrow transplant patients. Clin Infect Dis 2001; 33:947-953.
107. Ninin E, Milpied N, Moreau P et al. Longitudinal study of bacterial, viral, and fungal infections in adult recipients of bone marrow transplants. Clin Infect Dis 2001; 33:41-47.
108. Burt RK, Georganas C, Schroeder J et al. Autologous hematopoietic stem cell transplantation in refractory rheumatoid arthritis: Sustained response in two of four patients. Arthritis Rheum 1999; 42:2281-2285.
109. Brunner M, Greinix HT, Redlich K et al. Autologous blood stem cell transplantation in refractory systemic lupus erythematosus with severe pulmonary impairment: A case report. Arthritis Rheum 2002; 46:1580-1584.
110. Espigado I, Garcia A, Rodriguez J et al. Immunoablation and T-cell depleted autologous peripheral blood stem cell rescue as effective treatment for long term progressive rheumatoid arthritis refractory to multiple conventional treatment. Ann Rheum Dis 2001; 60:708. (Abstract P2).
111. McSweeney PA, Nash RA, Sullivan KM et al. High-dose immunosuppressive therapy for severe systemic sclerosis: Initial outcomes. Blood 2002; 100:1602-1610.
112. Musso M, Porretto F, Crescimanno A et al. Autologous peripheral blood stem and progenitor (CD34+) cell transplantation for systemic lupus erythematosus complicated by Evans syndrome. Lupus 1998; 7:492-494.
113. Goodrich JM, Reed EC, Mori M et al. Clinical features and analysis of risk factors for invasive candidal infection after marrow transplantation. J Infect Dis 1991; 164:731-740.
114. Bodey GP, Mardani M, Hanna HA et al. The epidemiology of Candida glabrata and Candida albicans fungemia in immunocompromised patients with cancer. Am J Med 2002; 112:380-385.
115. Marr KA, Seidel K, White TC et al. Candidemia in allogeneic blood and marrow transplant recipients: Evolution of risk factors after the adoption of prophylactic fluconazole. J Infect Dis 2000; 181:309-316.
116. Baddley JW, Stroud TP, Salzman D et al. Invasive mold infections in allogeneic bone marrow transplant recipients. Clin Infect Dis 2001; 32:1319-1324.
117. Ribaud P, Chastang C, Latge JP et al. Survival and prognostic factors of invasive aspergillosis after allogeneic bone marrow transplantation. Clin Infect Dis 1999; 28:322-330.
118. Denning DW, Stevens DA. Antifungal and surgical treatment of invasive aspergillosis: Review of 2,121 published cases. Rev Infect Dis 1990; 12:1147-1201.
119. Jantunen E, Ruutu P, Niskanen L et al. Incidence and risk factors for invasive fungal infections in allogeneic BMT recipients. Bone Marrow Transplant 1997; 19:801-808.
120. Ljungman P. Respiratory virus infections in bone marrow transplant recipients: The European perspective. Am J Med 1997; 102:44-47.
121. Ljungman P. Respiratory virus infections in stem cell transplant patients: The European experience. Biol Blood Marrow Transplant 2001; 7 Suppl:5S-7S.

122. Ljungman P, Ward KN, Crooks BN et al. Respiratory virus infections after stem cell transplantation: A prospective study from the Infectious Diseases Working Party of the European Group for Blood and Marrow Transplantation. Bone Marrow Transplant 2001; 28:479-484.
123. Nichols WG, Gooley T, Boeckh M. Community-acquired respiratory syncytial virus and parainfluenza virus infections after hematopoietic stem cell transplantation: The Fred Hutchinson Cancer Research Center experience. Biol Blood Marrow Transplant 2001; 7 Suppl:11S-15S.
124. Whimbey E, Elting LS, Couch RB et al. Influenza A virus infections among hospitalized adult bone marrow transplant recipients. Bone Marrow Transplant 1994; 13:437-440.
125. Martini A, Maccario R, Ravelli A et al. Marked and sustained improvement two years after autologous stem cell transplantation in a girl with systemic sclerosis. Arthritis Rheum 1999; 42:807-811.
126. Nash RA, Dansey R, Storek J et al. Epstein-Barr virus (EBV)-associated post-transplant lymphoproliferative disorder (PTLD) after high-dose immunosuppressive therapy and autologous CD34-selected stem cell transplantation for severe autoimmune diseases. Blood 2000; 96(Suppl):406a (Abstract1747).
127. Nash RA, Kraft GH, Bowen JD et al. Treatment of severe multiple sclerosis with high-dose immunosuppressive therapy and autologous stem cell transplantation. Blood 2000; 96(Suppl):842a-843a (Abstract 3640).
128. Saccardi R, Mancardi GL, Bacigalupo A et al. Autologous peripheral blood progenitor cell transplantation in severe progressive multiple sclerosis. (Abstract P21). Ann Rheum Dis 2001; 60:713.
129. Openshaw H, Lund BT, Kashyap A et al. Peripheral blood stem cell transplantation in multiple sclerosis with busulfan and cyclophosphamide conditioning: Report of toxicity and immunological monitoring. Biol Blood Marrow Transplant 2000; 6:563-575.
130. Wulffraat NM, Sanders EA, Kamphuis SS et al. Prolonged remission without treatment after autologous stem cell transplantation for refractory childhood systemic lupus erythematosus. Arthritis Rheum 2001; 44:728-731.
131. Quartier P, Prieur AM, Fischer A. Haemopoietic stem-cell transplantation for juvenile chronic arthritis (letter). Lancet 1999; 353:1885-1886.
132. Clark RA, Johnson FL, Klebanoff SJ et al. Defective neutrophil chemotaxis in bone marrow transplant patients. J Clin Invest 1976; 58:22-31.
133. Olsen GA, Gockerman JP, Bast Jr RC et al. Altered immunologic reconstitution after standard-dose chemotherapy or high-dose chemotherapy with autologous bone marrow support. Transplantation 1988; 46:57-60.
134. Storek J, Dawson MA, Storer B et al. Immune reconstitution after allogeneic marrow transplantation compared with blood stem cell transplantation. Blood 2001; 97:3380-3389.
135. Guillaume T, Rubinstein DB, Symann M. Immune reconstitution and immunotherapy after autologous hematopoietic stem cell transplantation. Blood 1998; 92:1471-1490.
136. Forman SJ, Nocker P, Gallagher M et al. Pattern of T cell reconstitution following allogeneic bone marrow transplantation for acute hematological malignancy. Transplantation 1982; 34:96-98.
137. Wulffraat N, van Royen A, Bierings M et al. Autologous haemopoietic stem-cell transplantation in four patients with refractory juvenile chronic arthritis. Lancet 1999; 353:550-553.
138. Baron F, Ribbens C, Kaye O et al. Effective treatment of Jo-1-associated polymyositis with T-cell-depleted autologous peripheral blood stem cell transplantation. Br J Haematol 2000; 110:339-342.
139. Fouillard L, Gorin NC, Laporte JP et al. Control of severe systemic lupus erythematosus after high-dose immunosuppressive therapy and transplantation of CD34+ purified autologous stem cells from peripheral blood. Lupus 1999; 8:320-323.
140. Wingard JR, Chen DY, Burns WH et al. Cytomegalovirus infection after autologous bone marrow transplantation with comparison to infection after allogeneic bone marrow transplantation. Blood 1988; 71:1432-1437.
141. Reusser P, Fisher LD, Buckner CD et al. Cytomegalovirus infection after autologous bone marrow transplantation: Occurrence of cytomegalovirus disease and effect on engraftment. Blood 1990; 75:1888-1894.
142. Jenkins D, DiFrancesco L, Chaudhry A et al. Successful treatment of post-transplant lymphoproliferative disorder in autologous blood stem cell transplant recipients. Bone Marrow Transplant 2002; 30:321-326.
143. Meyers JD. Fungal infections in bone marrow transplant patients. Semin Oncol 1990; 17(Suppl 6):10-13.
144. Ochs L, Shu XO, Miller J et al. Late infections after allogeneic bone marrow transplantations: Comparison of incidence in related and unrelated donor transplant recipients. Blood 1995; 86:3979-3986.
145. Schuchter LM, Wingard JR, Piantadosi S et al. Herpes zoster infection after autologous bone marrow transplantation. Blood 1989; 74:1424-1427.
146. Daikeler T, Mohren M, Amberger C et al. High dose immunosuppression with autologous stem cell support of a patient with refractory Wegener's granulomatosis. Ann Rheum Dis 2001; 60:712. (Abstract P716).
147. Durez P, Toungouz M, Schandene L et al. Remission and immune reconstitution after T-cell-depleted stem-cell transplantation for rheumatoid arthritis. Lancet 1998; 352:881.
148. Dykewicz CA. Guidelines for preventing opportunistic infections among hematopoietic stem cell transplant recipients: Focus on community respiratory virus infections. Biol Blood Marrow Transplant 2001; 7 Suppl:19S-22S.
149. Chapenko S, Millers A, Nora Z et al. Correlation between HHV-6 reactivation and multiple sclerosis disease activity. J Med Virol 2003; 69:111-117.
150. Alvarez-Lafuente R, Martin-Estefania C, de Las Heras V et al. Active human herpesvirus 6 infection in patients with multiple sclerosis. Arch Neurol 2002; 59:929-933.
151. Gutierrez J, Vergara MJ, Guerrero M et al. Multiple sclerosis and human herpesvirus 6. Infection 2002; 30:145-149.
152. Knox KK, Brewer JH, Henry JM et al. Human herpesvirus 6 and multiple sclerosis: Systemic active infections in patients with early disease. Clin Infect Dis 2000; 31:894-903.
153. Soldan SS, Berti R, Salem N et al. Association of human herpes virus 6 (HHV-6) with multiple sclerosis: Increased IgM response to HHV-6 early antigen and detection of serum HHV-6 DNA. Nat Med 1997; 3:1394-1397.
154. Stosor V, Peterson LR, Postelnick M et al. Enterococcus faecium bacteremia: Does vancomycin resistance make a difference? Arch Intern Med 1998; 158:522-527.
155. Herwaldt LA. Control of methicillin-resistant Staphylococcus aureus in the hospital setting. Am J Med 1999; 106:11S-18S. Discussion 48S-52S.
156. Sample ML, Gravel D, Oxley C et al. An outbreak of vancomycin-resistant enterococci in a hematology-oncology unit: Control by patient cohorting and terminal cleaning of the environment. Infect Control Hosp Epidemiol 2002; 23:468-470.
157. Kusne S, Krystofiak S. Infection control issues after bone marrow transplantation. Curr Opin Infect Dis 2001; 14:427-431.
158. Chang H, Hawes J, Hall GA et al. Prospective audit of cytomegalovirus-negative blood product utilization in haematology/oncology patients. Transfus Med 1999; 9:195-198.
159. Confavreux C, Suissa S, Saddier P et al. Vaccinations and the risk of relapse in multiple sclerosis. Vaccines in Multiple Sclerosis Study Group. N Engl J Med 2001; 344:319-326.
160. Elkayam O, Paran D, Caspi D et al. Immunogenicity and safety of pneumococcal vaccination in patients with rheumatoid arthritis or systemic lupus erythematosus. Clin Infect Dis 2002; 34:147-153.
161. Cruciani M, Rampazzo R, Malena M et al. Prophylaxis with fluoroquinolones for bacterial infections in neutropenic patients: A meta-analysis. Clin Infect Dis 1996; 23:795-805.
162. Cruciani M. Antibacterial prophylaxis. Int J Antimicrob Agents 2000; 16:123-125.
163. Murphy M, Brown AE, Sepkowitz KA et al. Fluoroquinolone prophylaxis for the prevention of bacterial infections in patients with cancer—is it justified? Clin Infect Dis 1997; 25:346-348.

164. Cometta A, Calandra T, Bille J et al. *Escherichia coli* resistant to fluoroquinolones in patients with cancer and neutropenia. N Engl J Med 1994; 330:1240-1241.
165. Kirkpatrick BD, Harrington SM, Smith D et al. An outbreak of vancomycin-dependent Enterococcus faecium in a bone marrow transplant unit. Clin Infect Dis 1999; 29:1268-1273.
166. Management of herpes virus infections following transplantation. A report from the British Society for Antimicrobial Chemotherapy Working Party on antiviral therapy. J Antimicrob Chemother 2000; 45:729-748.
167. Solano C, Munoz I, Gutierrez A et al. Qualitative plasma PCR assay (AMPLICOR CMV test) versus pp65 antigenemia assay for monitoring cytomegalovirus viremia and guiding preemptive ganciclovir therapy in allogeneic stem cell transplantation. J Clin Microbiol 2001; 39:3938-3941.
168. Goodman JL, Winston DJ, Greenfield RA et al. A controlled trial of fluconazole to prevent fungal infections in patients undergoing bone marrow transplantation. N Engl J Med 1992; 326:845-851.
169. MacMillan ML, Goodman JL, DeFor TE et al. Fluconazole to prevent yeast infections in bone marrow transplantation patients: A randomized trial of high versus reduced dose, and determination of the value of maintenance therapy. Am J Med 2002; 112:369-379.
170. Marr KA, White TC, van Burik JA et al. Development of fluconazole resistance in Candida albicans causing disseminated infection in a patient undergoing marrow transplantation. Clin Infect Dis 1997; 25:908-910.
171. Perea S, Patterson TF. Antifungal resistance in pathogenic fungi. Clin Infect Dis 2002; 35:1073-1080.
172. Joske DJ, Ma DT, Langlands DR et al. Autologous bone-marrow transplantation for rheumatoid arthritis. Lancet 1997; 350:337-338.
173. Espigado I, Rodriguez J, Carmona M et al. Immunoablation with T-cell depleted autologous peripheral blood stem cell rescue produces complete remission in thrombocytopenic purpura recidivant and refractory to standard treatment. (Abstract P1). Ann Rheum Dis 2001; 60:708.

CHAPTER 33

Immunological Aspects of Multiple Sclerosis with Emphasis on the Potential Use of Autologous Hemopoietic Stem Cell Transplantation

Paolo A. Muraro, Henry F. McFarland and Roland Martin

Introduction

Multiple sclerosis (MS) is one of the most intensively studied human autoimmune diseases, and belongs to the few for which immunomodulatory therapies have been approved during the last decade. Nevertheless, these treatments that include interferon-β, glatiramer acetate and mitoxantrone are only moderately effective in reducing disease exacerbations and brain inflammation. Further, although we have gained a better understanding of the disease pathogenesis of MS, we are far from achieving therapeutic specific immune intervention through the elimination of autoreactive T cells or by other means for restoring immune tolerance, and we have no MS-specific therapy to offer to patients with more advanced disease and long-standing disability.

Autologous hemopoietic stem cell transplantation (HSCT) is one of the many strategies that are being explored in autoimmune diseases such as MS. HSCT combines a step of purging the existing immune repertoire with subsequent reconstitution by the infusion of autologous hemopoietic stem cells that had been harvested prior to the procedure. HSCT offers the prospect of resetting the immunological clock and restoring immune tolerance. HSCT for autoimmune diseases is, however, still in its infancy, and a number of questions need to be addressed before its widespread use can be advocated. In this short review, we will summarize important clinical, genetic and immunological aspects of MS and discuss how these specifically relate to HSCT as a potentially curative treatment for MS

Clinical Presentation of Multiple Sclerosis— Variation in Phenotype and Course

MS is a chronic inflammatory demyelinating disorder of the central nervous system (CNS) that predominantly affects young adults.[1,2] Approximately two million individuals worldwide are affected, 250,000 to 350,000 only in the United States according to estimates reported by the US National Multiple Sclerosis Society.[3] Females are affected twice as frequently compared to males. Impairment of motor function with paralysis and spasticity, incoordination (ataxia), loss of vision, sensitive disturbances, paroxysmal pain, sphincter dysfunction and cognitive impairment are the most common symptoms. The clinical course is very heterogeneous, the most typical being a relapsing/remitting form with exacerbations followed by complete or partial recovery of neurological function (relapsing-remitting MS). More advanced stages of the disease are characterized by accumulation of disability, either from incomplete recovery from relapses or by continuous progression (secondary progressive MS). Less commonly, the disease presents from the beginning with a progressive course, usually without relapses (primary progressive MS) but occasionally with superimposed relapses (progressive-relapsing).[4]

The diagnosis of MS remains primarily clinical, based on neurological history and physical findings suggestive of dissemination in space and time of CNS lesions.[5] New diagnostic criteria,[6] however, have formally integrated abnormalities in radiological and laboratory investigations as paraclinical evidence to substantiate dissemination in space and time, attributing a particular importance to magnetic resonance imaging (MRI) findings.[7]

Epidemiology and Genetics of the Disease

Population, family and twin studies have firmly established a genetic component in MS.[8] The disease is more prevalent in certain ethnic groups, particularly in Caucasians in Northern Europe and America, and much less prevalent in other populations such as African Americans, or completely absent in others, such as in native Americans of pure ancestry.[9] The prevalence of MS ranges between 10 and 130 per 100,000 in north American and north European populations.[10] First degree relatives of MS patients carry a 15 to 25-fold higher risk to develop MS than that of the general population or of adopted siblings.[11] Twenty-five to thirty percent of monozygotic twins are concordant for the disease, while the concordance rate in dizygotic twins is similar to that observed in first degree relatives.[12] Hence, while there is no question that genetic factors contribute to disease susceptibility, the search for candidate genes has been difficult for MS, similar to other autoimmune diseases.

Major histocompatibility complex (MHC; or HLA= human leukocyte antigen in humans) class II genes, in particular HLA-DRB1*1501 and—DRB5*0101 (both are expressed in the HLA-DR2 haplotype), and HLA-DQA1*0102 and—DQB1*0602, have reproducibly shown the strongest association with MS.[13,14] It is believed that the surface receptors encoded by these genes contribute via their role as antigen-presenting molecules for CD4+ T lymphocytes. The exact mechanisms by which these genes confer susceptibility to disease, however, remain to be elucidated. A number of other candidate genes has been examined including those encoding for cytokines, chemokines, chemokine receptors, T cell receptors, and complement components.[15] Often, positive disease associations that had been reported in one study were not confirmed by others. This disagreement is largely due to disease heterogeneity and the fact that frequently small patient groups have been examined, or the patient cohorts were composed of individuals with different course, MRI presentation, and brain pathology. Recently, whole genome screens in North America and Europe,[16-19] confirmed that the HLA region on chromosome 6 harbors an important susceptibility locus, but that other genes probably contribute to disease as well. It is currently assumed that the genetic susceptibility to MS, as well as to other autoimmune diseases, is based on a quantitative trait to which multiple genes contribute weakly and in an additive fashion.[20] Depending on which and how many of the susceptibility loci are carried by an individual in the heterogeneous and outbred human population, the risk of this particular individual to develop disease will be lower or higher. Furthermore, the individual's genetic trait will likely contribute to disease heterogeneity, i.e., whether a patient primarily shows prominent inflammation due to the production of excess proinflammatory cytokines, or presents with more vulnerable cells of the target tissue such as oligodendrocytes or neurons, to mention only two possibilities. Unfortunately, the identification of genes associated with different disease phenotypes will require not only defeating the complexities of genetic analysis of susceptibility to MS, but also overcoming the difficulty in measuring disease course and severity.[15] Nevertheless, improved genetic techniques such as high density microsatellite mapping of susceptibility alleles and single nucleotide polymorphisms, as well as expression profiling by microarrays and proteomics techniques, will probably advance our understanding of the genotype-phenotype interactions in the near future.

Experimental studies in congenic mouse strains, in which the number of susceptibility alleles for one autoimmune disease was systematically increased, indicated that more susceptibility alleles translate into stochastically higher susceptibility for disease independent of the environment.[21] While this observation might be interpreted as evidence that environmental factors are not relevant, epidemiological studies that examined the migration of ethnic groups from geographic areas with low to areas with higher prevalence indicated that environmental influences are important as well.[22] Furthermore, MS exacerbations occur more often in the context of viral infections,[23] and a number of common pathogens, particularly human herpes virus type 6[24] and Epstein Barr virus,[25] as well as Chlamydia pneumoniae,[26-28] are currently being examined as potential triggers of MS.

It is likely that both genetic as well as environmental factors contribute to disease susceptibility and to the resulting clinical and immunopathological phenotype of a patient, i.e., whether disease is mild or severe and whether it is primarily driven by immune dysfunction or vulnerability of the target tissue, or both.

The Stages of Lesion Development

MS pathology affects selectively the CNS. Focal demyelinating lesions can be found predominantly in the deep white matter of the brain, of the brainstem, of the cerebellum and of the spinal cord. Lesions involving the cortical gray matter are frequently seen at postmortem histopathological examination, but are more difficult to visualize in vivo by conventional imaging. The pathological hallmark of MS is the demyelinated plaque, characterized by varying grades of myelin loss and gliosis. Prominent intra-parenchymal and peri-vascular inflammatory cell infiltrates, predominantly lymphocytes and macrophages, are commonly found in active lesions in the white matter.[29] Axonal damage has been recognized as an additional important component of MS pathology.[30] Interestingly, cortical lesions show less inflammation and predominant apoptotic loss of neurons compared to white matter plaques.[31]

Different histological patterns have recently been associated to different possible stages[32] and pathways[33] of immune-mediated tissue destruction. The inter-individual heterogeneity of MS lesions suggests that different mechanisms may act in different patients, accounting for the variability of clinical courses, of immunological findings, and of responses to treatments. A major breakthrough for monitoring in vivo the evolution of lesions in patients with MS has been the development of MRI techniques that allow not only the detection of subclinical disease activity and the identification of active lesions,[34] but also quantitative assessment of lesion burden, of brain atrophy and of structural changes in the brain that can be used as secondary outcome measures in clinical trials.[35,36]

Antigen-Specific Cellular Immune Responses in Multiple Sclerosis

The cause of MS is unknown and the pathogenic process is incompletely understood. Several features of the disease, which include the presence of immune cells in lesions, cerebrospinal fluid abnormalities with intrathecal production of oligoclonal IgG, the response to immune-modifying treatments and the increased disease susceptibility of individuals with certain HLA class II haplotypes, point to an immune-mediated pathogenesis. Current knowledge supports a T cell-dependent, T and/or B cell-mediated autoimmune pathogenesis targeting myelin components or myelin-producing cells.[37] This hypothesis is supported by observations in experimental animal models. Immunization of susceptible animal strains with myelin antigens or transfer of myelin antigen-reactive T cells induces experimental autoimmune encephalomyelitis (EAE), an inflammatory disorder of the CNS which resembles MS. EAE studies have demonstrated that inflammatory demyelinating CNS disorders can be mediated by CD4+ myelin-specific T cells. In this context, disease-inducing T cells are clearly biased toward a proinflammatory T helper type 1 (Th1) phenotype. Similar to EAE, in MS myelin-reactive Th1, CD4+ T cells may become activated in secondary lymphoid organs, migrate through the blood-brain barrier, recognize myelin antigens and trigger inflammation. The point in time and the event triggering autoreactive T cell activation are unknown. One hypothesis is that such activation may take place in genetically susceptible individuals as a result of crossreactivity (molecular mimicry) toward environmental antigens, which normally stimulate protective immune responses.[38]

MRI findings in the early stages of MS lesion development suggest an important role of blood-brain barrier (BBB) dysfunction in the initiation of an inflammatory response in the CNS.[34] BBB dysfunction alone, however, is not sufficient to start an inflammatory response toward myelin constituents. In fact, resident CNS cells in normal conditions lack MHC molecules, thus preventing the initiation of antigen-specific immune responses. This state of immunological ignorance can be altered by the induction of MHC expression,[39] which can occur during inflammation. Following the activation of autoreactive T cells in the CNS, secretion of proinflammatory cytokines and chemokines may recruit additional T cells and B cells into the lesion. Different mechanisms may be involved in the effector stage of demyelination, such as cytotoxicity and the action of cytokines and tumor necrosis factors. Activated macrophages can contribute to myelin destruction through several mediators including cytokines, reactive oxygen intermediates, proteases and lipases.

Myelin proteins that could be involved as targets of the inflammatory immune response in MS mainly include myelin basic protein (MBP), proteolipid protein (PLP), and myelin oligodendroglia glycoprotein (MOG), although other minor myelin components may be involved.[37] MBP is considered the major candidate autoantigen in MS.[40] The evidence supporting this claim includes the following well established observations: (1) Degradation products of MBP are present in macrophages surrounding MS lesions; (2) MBP content in plaques is reduced; (3) The protein is relatively abundant (~30%) in CNS myelin; (4) The immunodominant T cell epitope clusters of MBP [N-terminus (1-11, 1-19), central (87-106, 83-99, 87-99, 84-102), central (111-129, 115-125), C-terminus (143-168, 131-159)] are predominantly recognized in the context of MS-associated HLA-DR molecules.[41-43] Further supporting a possible role of T cell responses to MBP in MS, MBP epitopes corresponding to human immunodominant regions are encephalitogenic in EAE models.[37] Epitopes that can induce EAE vary among different animal species and strains, depending on the immunogenetic background. Consistent with this observation in EAE, we have previously shown in humans that a different HLA background of a population may result in the recognition of different immunodominant epitopes.[43,44]

Further supporting a possible role of T cell responses to MBP in MS, a transgenic mouse model, in which a human MBP (84-102)-specific TCR and HLA-DR2 molecules were expressed as chimeric transgenes, demonstrated that these mice spontaneously developed EAE.[45] In addition, complexes of MBP (84-102) bound to HLA-DR2 could be visualized in brain tissue of MS patients.[46] The most direct evidence probably stems from a therapeutic trial with an altered peptide ligand (APL) of MBP(83-99).[47] We observed MS exacerbations in 2 of 8 patients, and these exacerbations could clearly be linked to the expansion of MBP (83-99)-specific cells unwillingly induced by the administration of the altered peptide ligand.[47,48] Observations in untreated MS patients also showed that oligoclonal expansions of T cell receptor (TCR) Vβ gene families and increased cellular responses to MBP were correlated with the presence of subclinical disease activity detected by MRI.[49]

Taken together, these findings link the expansion of potentially autoreactive, myelin-specific T cell populations in the peripheral circulation with ongoing inflammation in the CNS, representing an important clue for deciphering the pathogenesis of the disease. However, the highly dynamic nature of anti-myelin T cell reactivity demonstrated in several studies,[50-52] as well as the difficulty in safely obtaining tolerization via APL-based treatment,[47] raise questions about the applicability of antigen-specific immunotherapy in MS.

Antigen-Specific Humoral Immune Responses in Multiple Sclerosis

Recent observations showing the presence of myelin-reactive autoantibodies in brain lesions from MS patients and nonhuman primate models renewed the interest on T cell-B cell interactions in the pathogenesis of MS.[53,54] In addition, oligoclonal B cell accumulations have been detected in the cerebrospinal fluid (CSF) of patients with MS.[55,56] Myelin-specific autoantibodies may provoke tissue damage by antibody-dependent cell-mediated cytotoxicity and by the activation of complement.[57] Affinity-purified autoantibodies to MBP from MS lesions were found to be specific for the same MBP(82-99) peptide that is immunodominant for HLA-DR2-restricted T cell recognition.[58,59] Interestingly, the key amino acid residues important for antibody binding were also key contacts for T cell recognition of the immunodominant MBP epitope.[60]

These data suggested that identity of immunodominant T and B cell epitopes of MBP may play a role in the autoimmune pathogenesis of MS. However, the poor or limited results of treatments aiming to eliminate or compete with autoantibodies, such as plasma exchange and high-dose intravenous immunoglobulin,[61] indicate that autoantibodies may only in part—or only in a subset of patients—contribute to mediating inflammatory demyelination in MS.

Engraftment of Autologous Hematopoietic Stem Cells as a Therapy in Multiple Sclerosis

Neither approved treatments,[61] nor experimental immunotherapies,[62] currently offer curative options for patients with MS, particularly for patients in a secondary progressive stage of disease. The results of Phase I studies have suggested that the ablation of the immune system and reconstitution with hemopoietic precursors can arrest the progression of disease in a majority of patients with secondary progressive MS and high risk features.[63-65] Recent results of European studies have demonstrated a striking suppression of MRI brain lesion activity after autologous BMT as well as stabilization of clinical disease.[66,67] These encouraging results prompted the design of new randomized controlled trials of HSCT vs. standard therapy (mitoxantrone) which are planned to start soon in Europe and in the USA.

Which changes in the immune system induced by HSCT are responsible for favorably affecting the course of MS remains to a large extent unknown. Three nonmutually exclusive hypothesis can be formulated: (1) Abrogation of a pathogenic immune response; (2) Reconstitution of an immune system with a new or restored immune tolerance; (3) Potential for CNS repair provided by differentiation of hematopoietic stem cells into glial and neural precursors that may contribute to restoring nervous function in damaged tissues. A careful, long-term follow-up of patients may offer clues as to whether HSCT provides only a profound, yet temporary immune suppression or actually restores tolerance on a permanent basis. As exemplified in Figure 1, in the case of a transient immune suppression, after a latency of undetermined duration, autoimmune pathogenesis may resume resulting in reactivation of the disease process (Fig. 1A). In contrast, a stable reconstitution of a tolerant immune repertoire would result in a permanent resolution of the pathogenic process, with arrest of disease progression and potential for functional (rehabilitation) and anatomic (spontaneous and treatment-induced remyelination) recovery (Fig. 1B). Only basic immunological investigations, however, might provide definitive proof for any of these hypotheses.

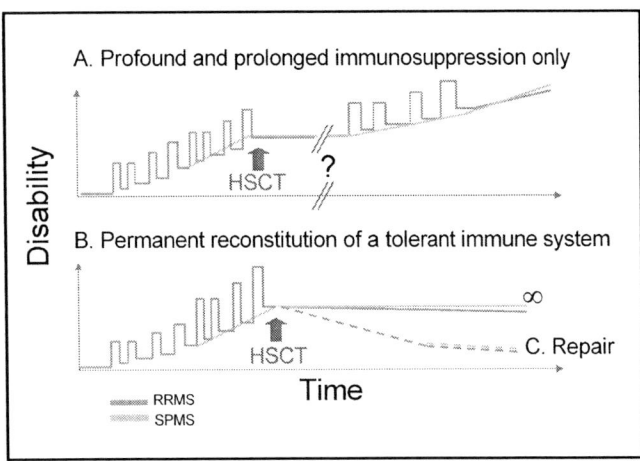

Figure 1. HSCT for MS: different immunological mechanisms may result in different clinical outcomes. Long-term follow-up of patients treated with autologous HSCT for MS may show (A) only a temporary efficacy resulting from a profound immune suppression, with disease activity resuming after a period of undetermined duration; or (B) a permanent remission, suggestive of a curative treatment possibly resulting from reconstitution of a newly tolerant immune system. In the latter case, functional improvement of neurological deficits could be obtained through spontaneous recovery and/or through therapeutic interventions promoting neural repair, which may become available in the future (C).

Few data are available on the changes of immune functions after HSCT in MS. Fassas et al treated 15 patients with progressive MS (8 primary progressive, 7 secondary progressive) with autologous HSCT.[68] A profound fall of CD4+ cell counts and increase in CD8+ cells was observed for all patients. Surprisingly, CD4+ cells were almost exclusively CD45 RA- (a feature of memory cells) early after transplant. Openshaw et al treated with HSCT and conducted immunological studies in 5 patients with secondary progressive MS.[69] Lymphocyte proliferation assays to myelin antigens were performed in 3 patients using a panel of 14 MBP peptides and a set of 5 immunodominant myelin protein peptides (MBP 84-106; MBP 142-163; PLP 104-117; PLP 142-153; MOG 42-53). Some responses to myelin peptides were suppressed after HSCT. CSF obtained from two patients at one year after transplant showed persistence of oligoclonal bands. In another report, the same oligoclonal band pattern of one MS patient's CSF was shown to persist post-transplant.[67] These data altogether provide preliminary evidence of immune suppression during the lymphopenic (particularly CD4 depleted) post-transplant phase. There are currently no data, however, offering insight on how immune tolerance may be reconstituted in MS patients after transplant. An analysis of myelin antigen-specific T cell repertoires and of their thymic vs. peripheral origin before and after HSCT may provide important clues on whether the beneficial effects of HSCT in patients with MS only depend on a profound and prolonged immune suppression or derive from the stable reconstitution of a tolerant immune system.

A potential for promotion of CNS repair by HSCT can be at present only hypothesized on the basis of elegant experiments in animal models. The potential migration of hemopoietic stem cells in the CNS has been shown by work by Wu et al, who used green fluorescent protein (GFP)-transfected bone marrow graft to investigate the distribution of hematogenous cells after bone marrow transplantation in the twitcher mice.[70] The twitcher mouse, bearing a mutation in the galactosylceramidase gene (twi-/twi-) is a murine model of a human genetic demyelinating disease, globoid cell leukodystrophy (Krabbe's disease). The affected mice usually die before reaching age 45 days, showing apoptotic death of oligodendrocytes and demyelination associated with extensive glial activation. Following bone marrow transplant, many GFP positive cells were detected in the white matter of the spinal cord, brainstem and brain, demonstrating the migration of progenitor cells to CNS sites. Furthermore, work by Mezey and colleagues showed that bone marrow cells from adult mice can migrate into the brain and originate cells that express neural-specific antigens, strongly supporting the potential for transdifferentiation of hemopoietic stem cells into neural lineages.[71] Whether, and to which extent, these processes may take place and play a significant role in humans with MS are exciting questions that still need to be addressed.

Special Considerations Regarding the Transfer of Autologous Hematopoietic Stem Cells

Although myelin-specific cells can mediate EAE under the appropriate conditions, extensive data in humans, including our own studies, indicate that myelin-reactive cells are part of the normal T cell repertoire also in healthy individuals.[43,72-74] In addition, we have recently shown that the majority of MBP-specific T cells originate from the naïve subset even in patients with MS.[75] These data are compatible with the observation that MBP-specific T cells obtained from either naïve or immunized animals are equally able to induce EAE upon adoptive transfer.[76] An implication is that autoimmune attacks in MS may not depend on the reactivation of memory cells, as autoreactive T cells could be constantly recruited from the naïve pool. This notion raises important issues with respect to the use of HSCT as a treatment for MS. Even naïve, nonactivated lymphocytes contaminating the hematopoietic graft may carry the risk of resuming an autoimmune process in a susceptible host. Indeed, HSCT with reinfusion on unmanipulated grafts has been associated with relapses of autoimmune disease.[77] In this context, allogeneic transplant would be in principle preferable to an autologous one as it would not carry the risk of reinfusing cells causing autoimmunity. However, the higher treatment-related mortality of allogeneic HSCT is considered unacceptable for autoimmune diseases,[78] and efforts have been focused to identify the best approach to minimize the risk of reintroduction of disease-mediating cells via the autologous graft in HSCT. Current strategies employ either in vitro CD34+ cell selection or in vivo purging with antithymocyte globulin for lymphocyte depletion.

Immunoablative conditioning without hemopoietic stem cell support is a possible alternative that does not carry the risk of reinfusing pathogenic T cells. Brodsky et al reported on patients with severe AID treated with immunosuppression by high-dose cyclophosphamide.[79] Hematologic reconstitution was found to be similar to patients treated with autologous HSC. It has been appropriately pointed out, however, that individual sensitivity of bone marrow to high dose cyclophosphamide may vary and that the risk of protracted cytopenia, requiring administration of leukocyte growth factors, may increase the risk of disease exacerbations.[80] An additional possible disadvantage of this approach may be an irreversible damage to the hemopoietic stem cell compartment by the cytotoxic conditioning regimen, resulting in decrease of potential for neural repair, as shown in animal models.[81]

A crucial point to be addressed in future studies is to define which patients should be treated with HSCT and when they should be treated. There is increasing evidence that in later stages of the disease, particularly in the secondary progressive phase, progression continues to a large extent independent of inflammation. A large clinical study reported that while the time required to reach a moderate level of disability (expanded disability status scale, EDSS 4.0) varies greatly, patients with higher disability (from EDSS 4.0 to EDSS 7.0) progressed at a similar rate.[82] These data suggest that irreversible pathogenic processes leading to inexorable clinical progression probably start when a certain threshold of neurological damage is reached. The exact pathophysiology of this process is poorly understood, however, axonal damage and loss is thought to be a likely component.[30] This hypothesis is indirectly supported by data from clinical trials with immunomodulatory drugs for the treatment of secondary progressive multiple sclerosis. While a European clinical trial with interferon-β1b in secondary progressive MS showed reduced progression in the active treatment group,[83,84] North American studies on trials with IFN-β1a[85,86] and IFN-β1b[87] failed to show a significant effect on progression. A possible explanation of this discrepancy is that the European trial included a greater proportion of patients with active inflammatory disease than the North American trials. Taken together, these observations suggest that immune intervention can be most effective during the early, predominantly inflammatory phases of disease and be less useful in more advanced, progressive stages.

Previous phase I and II clinical trials of HSCT in MS included patients who in most cases had established severe disability. For phase III trials, while aggressiveness of disease course remains a key criterion in order to outweigh treatment risks, early treatment before the onset of irreversible progression may be essential for the evaluation of efficacy. Meeting this condition will require inclusion of patients with lower disability scores.

Future Directions—Conclusions

In our opinion, systematic studies in three main areas are required to determine the role of HSCT as therapy for MS: (1) controlled clinical trials to assess the efficacy of the treatment; (2) long-term follow up of treated patients including appropriate paraclinical evaluations, most importantly MRI; (3) defining the immunological mechanisms of HSCT.

Indeed, integrating careful follow-up of patients treated with HSCT with immunological studies may offer an extraordinary opportunity to better understand the role of autoreactive T cells in the pathogenesis of the disease. The identification of different pre and post-transplant myelin-specific T cell repertoires or phenotypes in patients who received effective HSCT may help to identify disease-mediating cell populations. In contrast, if the same myelin-reactive T cell repertoire (including its functional phenotype) is reconstituted in patients for whom HSCT resulted in successful clinicopathological remission, its importance in the disease process would become dubious. This scenario may also suggest that an original triggering event which may have activated pathogenic cells with a different, unknown specificity is no longer ongoing, and that immunological memory had played a role in the persistence of the disease process. In this case, the study of candidate pathogenic T cells could be focused on the memory lymphocyte pool and the search for the initial triggering event oriented to an earlier point in time. If, on the other hand, disease progression or relapses resume after transplant in some patients, investigating the origin of autoreactive T cells in the naïve and memory repertoires may answer the critical question, whether persistence or reoccurrence of disease was due to spontaneous reconstitution or to incomplete immunoablation of pathogenic T cells. In the former case, future research should be aimed to clarify the reasons of disease susceptibility, which could possibly be genetically determined. In the event that disease-related myelin-reactive memory T cells are found in MS patients after transplant, one could envision that either the immuno-ablative regimen may have been inadequate, or that pathogenic cells may have been reintroduced via the autologous hemopoietic graft. Greater efforts should then be aimed to identify technical improvements of the transplant procedure that allow a complete eradication of pathogenic T cells. However, it should be kept in mind that even after successful HSCT, the previous extensive tissue damage in the CNS and the persistence of a susceptible genetic background clearly pose a risk of resumption of the disease process during long-term follow up. Although in most studies none of the patients with MS treated with HSCT required immunosuppressive treatments after transplant, available immunomodulatory drugs could be considered as a tool to manage late risk in patients showing MRI signs of resumed disease activity.

Once the efficacy has been confirmed in a controlled setting, one could envision that HSCT may represent a curative option to be considered early during the disease course for patients with a strong inflammatory component. Later during the disease process, i.e., if disability has accumulated, HSCT may offer the most powerful means to achieve complete immunopathological disease remission. This may be a precondition to attempt and evaluate stem cell therapy for neural repair, an approach that may become available in the near future.[88]

References

1. McFarlin DE, McFarland HF. Multiple Sclerosis (First of Two Parts). N Engl J Med 1982; 307:1183-1188.
2. Whitaker JN, Mitchell GW. Clinical features of multiple sclerosis. In: Raine CS, Mcfarland HF, Tourtellotte WW, eds. Multiple sclerosis: Clinical and pathogenetic basis. London: Chapman & Hall, 1997:3-19.
3. NMSS. Research Fact Sheet. New York: National Multiple Sclerosis Society, 2002.
4. Lublin FD, Reingold SC. Defining the clinical course of multiple sclerosis: Results of an international survey. National Multiple Sclerosis Society (USA) Advisory Committee on Clinical Trials of New Agents in Multiple Sclerosis. Neurology 1996; 46:907-911.
5. Poser CM, Paty DW, Scheinberg LC et al. New diagnostic criteria for multiple sclerosis: Guidelines for research protocols. Ann Neurol 1983; 13:227-231.
6. McDonald WI, Compston A, Edan G et al. Recommended diagnostic criteria for multiple sclerosis: Guidelines from the International Panel on the diagnosis of multiple sclerosis. Ann Neurol 2001; 50:121-127.
7. McFarland HF. The emerging role of MRI in multiple sclerosis and the new diagnostic criteria. Mult Scler 2002; 8:71-72.
8. Ebers GC, Sadovnick AD. The role of genetic factors in multiple sclerosis susceptibility. J Neuroimmunol 1994; 54:1-17.
9. Oger J, Lai H. Demyelination and ethnicity: Experience at the University of British Columbia Multiple Sclerosis Clinic with special reference to HTLV-I-associated myelopathy in British Columbian natives. Ann Neurol 1994; 36 Suppl:S22-24.
10. Kurtzke JF. Epidemiologic contributions to multiple sclerosis: An overview. Neurology 1980; 30:61-79.
11. Sadovnick AD, Dircks A, Ebers GC. Genetic counselling in multiple sclerosis: Risks to sibs and children of affected individuals. Clin Genet 1999; 56:118-122.

12. Ebers GC, Bulman DE, Sadovnick AD et al. A population-based study of multiple sclerosis in twins. N Engl J Med 1986; 315:1638-1642.
13. Hillert J. Human leukocyte antigen studies in multiple sclerosis. Ann Neurol 1994; 36 Suppl:S15-17.
14. Sawcer S, Maranian M, Setakis E et al. A whole genome screen for linkage disequilibrium in multiple sclerosis confirms disease associations with regions previously linked to susceptibility. Brain 2002; 125:1337-1347.
15. Kantarci OH, de Andrade M, Weinshenker BG. Identifying disease modifying genes in multiple sclerosis. J Neuroimmunol 2002; 123:144-159.
16. Ebers GC, Kukay K, Bulman DE et al. A full genome search in multiple sclerosis. Nat Genet 1996; 13:472-476.
17. Sawcer S, Jones HB, Feakes R et al. A genome screen in multiple sclerosis reveals susceptibility loci on chromosome 6p21 and 17q22. Nat Genet 1996; 13:464-468.
18. Haines JL, Ter-Minassian M, Bazyk A et al. A complete genomic screen for multiple sclerosis underscores a role for the major histocompatability complex. The Multiple Sclerosis Genetics Group. Nat Genet 1996; 13:469-471.
19. Kuokkanen S, Gschwend M, Rioux JD et al. Genomewide scan of multiple sclerosis in Finnish multiplex families. Am J Hum Genet 1997; 61:1379-1387.
20. Chataway J, Feakes R, Coraddu F et al. The genetics of multiple sclerosis: Principles, background and updated results of the United Kingdom systematic genome screen. Brain 1998; 121 (Pt 10):1869-1887.
21. Wakeland EK, Liu K, Graham RR et al. Delineating the genetic basis of systemic lupus erythematosus. Immunity 2001; 15:397-408.
22. Ebers GC, Sadovnick AD. The geographic distribution of multiple sclerosis: A review. Neuroepidemiology 1993; 12:1-5.
23. Sibley WA, Bamford CR, Clark K. Clinical viral infections and multiple sclerosis. Lancet 1985; 1:1313-1315.
24. Soldan SS, Berti R, Salem N et al. Association of human herpes virus 6 (HHV-6) with multiple sclerosis: Increased IgM response to HHV-6 early antigen and detection of serum HHV-6 DNA. Nat Med 1997; 3:1394-1397.
25. Wandinger K, Jabs W, Siekhaus A. Association between clinical disease activity and Epstein-Barr virus reactivation in MS. Neurology 2000; 55:178-184.
26. Yao SY, Stratton CW, Mitchell WM et al. CSF oligoclonal bands in MS include antibodies against Chlamydophila antigens. Neurology 2001; 56:1168-1176.
27. Sriram S, Stratton CW, Yao S et al. Chlamydia pneumoniae infection of the central nervous system in multiple sclerosis. Ann Neurol 1999; 46:6-14.
28. Derfuss T, Gurkov R, Then Bergh F et al. Intrathecal antibody production against Chlamydia pneumoniae in multiple sclerosis is part of a polyspecific immune response. Brain 2001; 124:1325-1335.
29. Raine CS. The Dale E. McFarlin Memorial Lecture: The immunology of the multiple sclerosis lesion. Ann Neurol 1994; 36 Suppl:S61-72.
30. Trapp BD, Peterson J, Ransohoff RM et al. Axonal transection in the lesions of multiple sclerosis. [see comments.]. N Engl J Med 1998; 338:278-285.
31. Peterson JW, Bo L, Mork S et al. Transected neurites, apoptotic neurons, and reduced inflammation in cortical multiple sclerosis lesions. Ann Neurol 2001; 50:389-400.
32. Gay FW, Drye TJ, Dick GW et al. The application of multifactorial cluster analysis in the staging of plaques in early multiple sclerosis. Identification and characterization of the primary demyelinating lesion. Brain 1997; 120 (Pt 8):1461-1483.
33. Lucchinetti C, Bruck W, Parisi J et al. Heterogeneity of multiple sclerosis lesions: Implications for the pathogenesis of demyelination. Ann Neurol 2000; 47:707-717.
34. McFarland HF, Frank JA, Albert PS et al. Using gadolinium-enhanced magnetic resonance imaging lesions to monitor disease activity in multiple sclerosis. Ann Neurol 1992; 32:758-766.
35. Miller DH, Albert PS, Barkhof F et al. Guidelines for the use of magnetic resonance techniques in monitoring the treatment of multiple sclerosis. US National MS Society Task Force. Ann Neurol 1996; 39:6-16.
36. McFarland HF, Barkhof F, Antel J et al. The role of MRI as a surrogate outcome measure in multiple sclerosis. Mult Scler 2002; 8:40-51.
37. Martin R, McFarland HF. Immunology of multiple sclerosis and experimental allergic encephalomyelitis. In: Raine CS, McFarland HF, Tourtellotte WW, eds. Multiple Sclerosis: Clinical and pathogenetic basis. London: Chapman & Hall, 1997:221-242.
38. Wucherpfennig KW, Strominger JL. Molecular mimicry in T cell-mediated autoimmunity: Viral peptides activate human T cell clones specific for myelin basic protein. Cell 1995; 80:695-705.
39. Neumann H, Schmidt H, Wekerle H. Interferon gamma gene expression in sensory neurons: Evidence for autocrine gene regulation. J Exp Med 1997; 186:2023.
40. Hohlfeld R, Wekerle H. Immunological update on multiple sclerosis. Curr Opin Neurol 2001; 14:299-304.
41. Martin R, Howell MD, Jaraquemada D et al. A myelin basic protein peptide is recognized by cytotoxic T cells in the context of four HLA-DR types associated with multiple sclerosis. J Exp Med 1991; 173:19-24.
42. Meinl E, Weber F, Drexler K et al. Myelin basic protein-specific T lymphocyte repertoire in multiple sclerosis. Complexity of the response and dominance of nested epitopes due to recruitment of multiple T cell clones. J Clin Invest 1993; 92:2633-2643.
43. Muraro PA, Vergelli M, Kalbus M et al. Immunodominance of a low-affinity MHC binding myelin basic protein epitope (residues 111-129) in HLA-DR4 (B1*0401) subjects is associated with a restricted T cell receptor repertoire. J Clin Invest 1997; 100:339-349.
44. Vergelli M, Kalbus M, Rojo S et al. T cell response to myelin basic protein in the context of the multiple sclerosis-associated HLA-DR15 haplotype: Peptide binding, immunodominance and effector functions of T cells. J Neuroimmunol 1997; 77:195-203.
45. Madsen LS, Andersson EC, Jansson L et al. A humanized model for multiple sclerosis using HLA-DR2 and a human T-cell receptor. Nat Genet 1999; 23:343-347.
46. Krogsgaard M, Wucherpfennig KW, Cannella B et al. Visualization of myelin basic protein (MBP) T cell epitopes in multiple sclerosis lesions using a monoclonal antibody specific for the human histocompatibility leukocyte antigen (HLA)-DR2-MBP 85-99 complex. J Exp Med 2000; 191:1395-1412.
47. Bielekova B, Goodwin B, Richert N et al. Encephalitogenic potential of myelin basic protein peptide (83-99) in multiple sclerosis - Results of a phase II clinical trial with an altered peptide ligand. Nat Med 2000; 6:1167-1175.
48. Muraro PA, Wandinger K-P, Bielekova B et al. Molecular tracking of antigen-specific T cell clones in neurological immune-mediated disorders. Brain 2002; In press.
49. Muraro PA, Bonanni L, Mazzanti B et al. Short-term dynamics of circulating T cell receptor V beta repertoire in relapsing-remitting MS. J Neuroimmunol 2002; 127:149-159.
50. Goebels N, Hofstetter H, Schmidt S et al. Repertoire dynamics of autoreactive T cells in multiple sclerosis patients and healthy subjects - Epitope spreading versus clonal persistence. Brain 2000; 123:508-518.
51. Vergelli M, Mazzanti B, Traggiai E et al. Short-term evolution of autoreactive T cell repertoire in multiple sclerosis. J Neurosci Res 2001; 66:517-524.
52. Hellings N, Gelin G, Medaer R et al. Longitudinal study of antimyelin T-cell reactivity in relapsing-remitting multiple sclerosis: Association with clinical and MRI activity. J Neuroimmunol 2002; 126:143-160.
53. Raine CS, Cannella B, Hauser SL et al. Demyelination in primate autoimmune encephalomyelitis and acute multiple sclerosis lesions: A case for antigen-specific antibody mediation. Ann Neurol 1999; 46:144-160.
54. Raine CS, Cannella B, Hauser SL et al. Demyelination in primate autoimmune encephalomyelitis and acute multiple sclerosis lesions: A case for antigen-specific antibody mediation. Ann Neurol 1999; 46:144-160.
55. Qin Y, Duquette P, Zhang Y et al. Clonal expansion and somatic hypermutation of V(H) genes of B cells from cerebrospinal fluid in multiple sclerosis. J Clin Invest 1998; 102:1045-1050.

56. Colombo M, Dono M, Gazzola P et al. Accumulation of clonally related B lymphocytes in the cerebrospinal fluid of multiple sclerosis patients. J Immunol 2000; 164:2782-2789.
57. Noseworthy JH, Lucchinetti C, Rodriguez M et al. Multiple sclerosis. N Engl J Med 2000; 343:938-952.
58. Warren KG, Catz I. Synthetic peptide specificity of anti-myelin basic protein from multiple sclerosis cerebrospinal fluid. J Neuroimmunol 1992; 39:81-89.
59. Warren KG, Catz I, Steinman L. Fine specificity of the antibody response to myelin basic protein in the central nervous system in multiple sclerosis: The minimal B-cell epitope and a model of its features. Proc Natl Acad Sci USA 1995; 92:11061-11065.
60. Wucherpfennig KW, Catz I, Hausmann S et al. Recognition of the immunodominant myelin basic protein peptide by autoantibodies and HLA-DR2-restricted T cell clones from multiple sclerosis patients. Identity of key contact residues in the B-cell and T-cell epitopes. J Clin Invest 1997; 100:1114-1122.
61. Goodin DS, Frohman EM, Garmany Jr GP et al. Disease modifying therapies in multiple sclerosis: Report of the Therapeutics and Technology Assessment Subcommittee of the American Academy of Neurology and the MS Council for Clinical Practice Guidelines. Neurology 2002; 58:169-178.
62. Martin R, Calabresi PA, McFarland HF. Experimental immunotherapies of multiple sclerosis. In: Zhang J, Hafler D, Hohlfeld R et al, eds. Immunotherapies in Neuroimmunological Diseases. Blackwell Science, 1998:29-73.
63. Burt RK, Traynor AE, Pope R et al. Treatment of autoimmune disease by intense immunosuppressive conditioning and autologous hematopoietic stem cell transplantation. Blood 1998; 92:3505-3514.
64. Fassas A, Anagnostopoulos A, Kazis A et al. Autologous stem cell transplantation in progressive multiple sclerosis—an interim analysis of efficacy. J Clin Immunol 2000; 20:24-30.
65. Kozak T, Havrdova E, Pit'ha J et al. High-dose immunosuppressive therapy with PBPC support in the treatment of poor risk multiple sclerosis. Bone Marrow Transplant 2000; 25:525-531.
66. Mancardi GL, Saccardi R, Filippi M et al. Autologous hematopoietic stem cell transplantation suppresses Gd-enhanced MRI activity in MS. Neurology 2001; 57:62-68.
67. Saiz A, Carreras E, Berenguer J et al. MRI and CSF oligoclonal bands after autologous hematopoietic stem cell transplantation in MS. Neurology 2001; 56:1084-1089.
68. Fassas A, Anagnostopoulos A, Kazis A et al. Peripheral blood stem cell transplantation in the treatment of progressive multiple sclerosis: First results of a pilot study. Bone Marrow Transplant 1997; 20:631-638.
69. Openshaw H, Lund BT, Kashyap A et al. Peripheral blood stem cell transplantation in multiple sclerosis with busulfan and cyclophosphamide conditioning: Report of toxicity and immunological monitoring. Biol Blood Marrow Transplant 2000; 6:563-575.
70. Wu YP, McMahon E, Kraine MR et al. Distribution and characterization of GFP(+) donor hematogenous cells in Twitcher mice after bone marrow transplantation. Am J Pathol 2000; 156:1849-1854.
71. Mezey E, Chandross KJ, Harta G et al. Turning blood into brain: Cells bearing neuronal antigens generated in vivo from bone marrow. Science 2000; 290:1779-1782.
72. Martin R, Jaraquemada D, Flerlage M et al. Fine specificity and HLA restriction of myelin basic protein-specific cytotoxic T cell lines from multiple sclerosis patients and healthy individuals. J Immunol 1990; 145:540-548.
73. Ota K, Matsui M, Milford EL et al. T-cell recognition of an immunodominant myelin basic protein epitope in multiple sclerosis. Nature 1990; 346:183-187.
74. Pette M, Fujita K, Kitze B et al. Myelin basic protein-specific T lymphocyte lines from MS patients and healthy individuals. Neurology 1990; 40:1770-1776.
75. Muraro PA, Pette M, Bielekova B et al. Human autoreactive CD4+ T cells from naïve CD45RA+ and memory CD45RO+ subsets differ with respect to epitope specificity and functional antigen avidity. J Immunol 2000; 164:5474-5481.
76. Uccelli A, Giunti D, Mancardi G et al. Characterization of the response to myelin basic protein in a non human primate model for multiple sclerosis. Eur J Immunol 2001; 31:474-479.
77. Euler H, Marmont A, Bacigalupo A et al. Early recurrence or persistence of autoimmune diseases after unmanipulated autologous stem cell transplantation. Blood 1996; 88:3621-3625.
78. Marmont AM. New horizons in the treatment of autoimmune diseases: Immunoablation and stem cell transplantation. Annu Rev Med 2000; 51:115-134.
79. Brodsky RA, Petri M, Smith BD et al. Immunoablative high-dose cyclophosphamide without stem-cell rescue for refractory, severe autoimmune disease. Ann Intern Med 1998; 129:1031-1035.
80. Kozak T, Havrdova E, Pitha J. High dose immunosuppression with hemopoietic stem cell support in the treatment of multiple sclerosis. Isr Med Assoc J 2000; 2:610-614.
81. Smith PM, Franklin RJ. The effect of immunosuppressive protocols on spontaneous CNS remyelination following toxin-induced demyelination. J Neuroimmunol 2001; 119:261-268.
82. Confavreux C, Vukusic S, Moreau T et al. Relapses and progression of disability in multiple sclerosis. N Engl J Med 2000; 343:1430-1438.
83. Kappos L, Polman C, Pozzilli C et al. Final analysis of the European multicenter trial on IFNbeta-1b in secondary-progressive MS. Neurology 2001; 57:1969-1975.
84. Placebo-controlled multicentre randomised trial of interferon beta-1b in treatment of secondary progressive multiple sclerosis. European Study Group on interferon beta-1b in secondary progressive MS. Lancet 1998; 352:1491-1497.
85. Li DK, Zhao GJ, Paty DW. Randomized controlled trial of interferon-beta-1a in secondary progressive MS: MRI results. Neurology 2001; 56:1505-1513.
86. SPECTRIMS Study Group. Randomized controlled trial of interferon- beta-1a in secondary progressive MS: Clinical results. Neurology 2001; 56:1496-1504.
87. Goodkin DE, Group NASS. The North American Study of interferon beta-1b in secondary progressive multiple sclerosis. 52nd Annual Meeting of the American Academy of Neurology. San Diego, CA: 2000.
88. Kim JH, Auerbach JM, Rodriguez-Gomez JA et al. Dopamine neurons derived from embryonic stem cells function in an animal model of Parkinson's disease. Nature 2002; 418:50-56.

CHAPTER 34

Axonal Injury and Disease Progression in Multiple Sclerosis

Carl Bjartmar and Bruce D. Trapp

Introduction

Multiple sclerosis (MS) is a chronic inflammatory neurodegenerative disease of the central nervous system (CNS). Approximately one million people are affected worldwide and women outnumber men 2:1. MS is a major cause of nontraumatic neurological disability among young adults. The etiology is unknown, although the disease commonly is regarded an autoimmune process triggered in susceptible individuals by an early environmental exposure. A majority of MS patients exhibit a relapsing-remitting course (RR-MS) for 10 to 15 years, which later transform into a secondary-progressive course (SP-MS) characterized by irreversible neurological deficits.[1]

The histopathology of MS is complex and involves focal inflammation, demyelination, loss of oligodendrocytes, some early remyelination, reactive astrogliosis, and axonal pathology.[2,3] Of these hallmarks, myelin loss has attracted most interest and MS research has historically focused on mechanisms associated with inflammatory demyelination and remyelination. Axonal pathology in MS has been noted in the literature for more than a century (for review see ref. 4). Charcot,[5] for example, described MS lesions in terms of demyelination and astrogliosis, but he also discussed axonal degeneration. Present data on axonal injury in MS has been provided through various approaches including magnetic resonance imaging (MRI),[6] magnetic resonance spectroscopy (MRS),[7] and morphological analysis of sections from MS brains.[8,9] In addition, axonal degeneration has been described in animal models of MS[10,11] and in dysmyelinating mice.[12,13] Together, these data demonstrate that axonal injury constitutes a significant component of the pathogenesis in MS, and suggest that cumulative axonal loss is the main determinant of the progressive neurological disability seen in these patients.[7,14]

Axonal injury is the major determinant of irreversible neurological disability in patients with multiple sclerosis (MS). Loss of axons begins at disease onset and correlates with the degree of inflammation within lesions, indicating that inflammatory demyelination influences axon pathology at the relapsing-remitting stage of MS. Axonal loss from disease onset can remain clinically silent for many years, and irreversible neurological disability develops when a threshold of axonal loss is reached and compensatory CNS resources are exhausted. Lack of myelin-derived trophic support due to long-term demyelination may cause axonal degeneration in chronic inactive lesions during secondary-progressive MS. In addition, it is possible that reduced trophic support from damaged targets or degeneration of efferent fibers may trigger preprogrammed degeneration of some neurons. The concept of MS as a neurodegenerative disorder has important clinical implications regarding therapeutic approaches, monitoring of patients, and the development of neuroprotective treatment strategies.

Axonal Damage at Early Stages of Disease

Ferguson and colleagues demonstrated axonal amyloid precursor protein (APP) in active MS lesions and at the border of chronic active lesions.[8] Accumulation of APP is considered a marker for axonal dysfunction or injury since it is detected immunohistochemically only in axons with impaired axonal transport.[15] Many APP-immunoreactive structures resembled axonal ovoids, characteristic of newly transected axons. The results suggested axonal dysfunction within inflammatory MS lesions and indicated that many of these axons were transected. These observations were confirmed and extended by a morphological investigation on lesions from MS brains with various degrees of inflammation and disease duration.[9] Axonal ovoids were identified through confocal microscopy as terminal ends of transected axons immunostained for nonphosphorylated neurofilaments. Over 11,000 transected axons were found per mm^3 in active lesions and over 3,000 per mm^3 at the edge of chronic active lesions. The core of chronic active lesions contained on average 875 transected axons per mm^3. In contrast, less than one axonal ovoid per mm^3 was detected in control white matter. Kornek and colleagues reported a similar correlation between activity of MS lesions and density of APP-positive axons.[16] The occurrence of terminal ends in active lesions, detected in patients with short disease duration, support axonal transection from onset of MS.

The pathophysiology of axonal damage during early stages of MS is unknown, and we can only speculate on possible mechanisms. Since the extent of axonal damage in active MS lesions correlates with inflammatory activity within the lesion, axonal injury could be associated with the inflammation per se. Inflammatory substances such as proteolytic enzymes, cytokines, oxidative products and free radicals produced by activated immune and glial cells are potential mediators of such damage.[17] Oxidative damage to mitochondrial DNA and impaired activity of mitochondrial enzyme complexes in MS lesions indicate that inflammation can affect energy metabolism, ATP synthesis, and viability of affected cells.[18] Treatment with the AMPA/kaniate

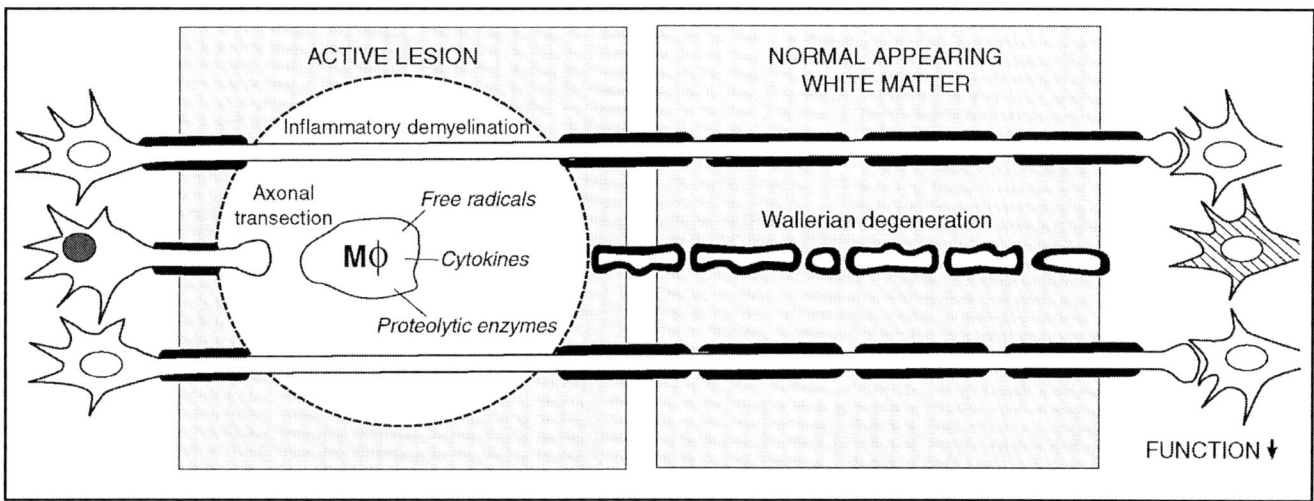

Figure 1. Axonal transection caused by inflammatory demyelination in an active MS lesion. Substances produced by activated immune and glial cells may mediate tissue damage, including irreversible axonal injury. Axons degenerate rapidly distal to the site of transection, whereas CNS myelin can persist for a long time and form empty tubes or degenerating ovoids. Distal to the lesion, the white matter may appear normal on conventional MRI images or histological examination. Deafferentation of target neurons may cause functional loss.

glutamate receptor antagonist NBQX resulted in increased oligodendrocyte survival and reduced axonal damage in experimental autoimmune encephalomyelitis (EAE), an animal model of MS, suggesting that excitotoxicity mediated by glutamate is involved in tissue damage in acute lesions.[19] Recently, data indicating that cytotoxic CD8+ T cells can mediate axonal transection in active MS lesions were provided in MS tissue,[20] in EAE mice[21] and in vitro.[22] Also, inflammatory edema may cause increased extracellular pressure that results in axonal damage, particularly in anatomical locations of the CNS where space for tissue expansion is limited such as the spinal cord.[23] In support of this hypothesis, the spinal cord cross-sectional area of relapsing-remitting EAE mice increased by 9% at first attack, but returned to normal at end-stage disease.[11] Finally, genes involved in axonal responses to inflammation and demyelination could influence the extent of axonal injury in individual patients.[1]

Considerable Axonal Loss in Long Term MS

The presence of axonal transection from early stages of disease, in the absence of obvious progressive disability between relapses, raises questions regarding the magnitude of cumulative axonal loss during long term MS. In order to quantify total axonal loss in MS lesions, an axonal sampling protocol that accounts for both tissue atrophy and reduced axonal density was developed using spinal cord cross sections.[24] Total axonal loss was quantified in 10 chronic inactive lesions from 5 quadriplegic MS patients (EDSS ≥ 7.5) with long disease duration. These lesions contained on average 68% (45-84%) axonal loss compared to controls. Axonal loss, therefore, constitutes a significant pathology of many chronic MS lesions (Fig. 2). Given the severe permanent neurological disability of the examined patients, the data also support axonal degeneration as the main cause of irreversible disability (Fig. 1) in nonambulatory MS patients. Average axonal density (number of axons per unit area) in these lesions was decreased by 58%. A similar reduction in axonal density, 61%, was reported in spinal cord lesions from patients with secondary progressive multiple sclerosis (SP-MS).[25]

Extensive axonal loss and progression of disability, even in the absence of overt inflammatory activity, suggests that mechanisms other than inflammatory demyelination contribute to axonal degeneration during SP-MS. A number of genes coding for myelin related proteins contribute to long-term viability of axons.[26,27] For example, late onset axonal pathology such as atrophy or swelling, cytoskeleton alterations, organelle accumulation and degeneration was observed in mice lacking myelin associated glycoprotein (MAG)[12] and proteolipid protein (PLP).[13] In PLP-null mice, the axonal pathology was accompanied with progressive clinical disability including impaired gait, tremor and spasticity. Hence, chronically demyelinated axons may undergo degeneration due to lack of trophic support from myelin or myelin forming cells. Recently, it was proposed that abnormal expression of sodium channel subtypes—maladaptive responses to demyelination—may render axons vulnerable to degeneration, conforming with the interesting possibility that MS may involve an acquired channelopathy.[28]

Axonal Loss in Nonlesion White Matter

Transected CNS axons undergo rapid degeneration distal to the site of transection. In contrast to axons, CNS myelin can persist for a long time after proximal fiber transection. Histologically, such remaining myelin sheaths will form empty tubes, or later degenerating ovoids (Fig. 1). The white matter, however, may appear normal on conventional MRI images or routine histological examination by luxol fast blue. A number of postmortem studies have addressed the extent of axonal loss in MS normal appearing white matter (NAWM). Lovas and colleagues reported reduced axonal densities by as much as 57% in spinal cord NAWM from patients with SP-MS.[25] Ganter and colleagues found reductions in axonal density by 19-42% in the lateral corticospinal tract of MS patients with lower limb weakness.[29] A study that accounted for both changes in tissue volume and axonal density examined total axonal loss in MS patients with disease duration between 5 and 34 years and various degree of functional impairment.[30] The average axonal loss in normal appearing corpus callosum of these patients was 53%. Since the reduction in axonal density in the same material was only 34%, the data also demonstrate that measures of both tissue volume and axonal density are necessary to determine total axonal loss.

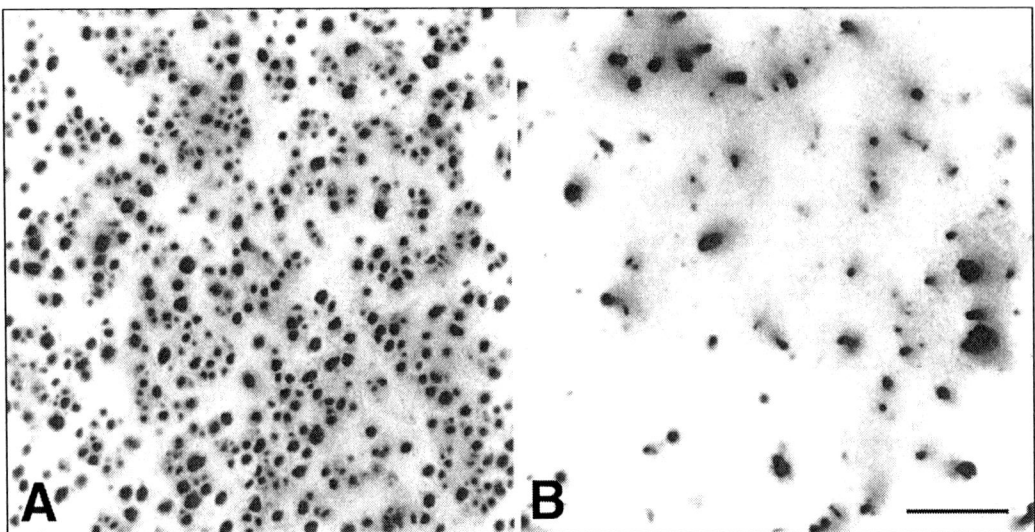

Figure 2. Axonal loss in a chronic inactive spinal cord lesion from a paralyzed patient with secondary-progressive MS and long disease duration. Staining for neurofilaments demonstrates axonal density in control (A) and in a demyelinated area in the gracile fasciculus of MS cervical spinal cord (B). This MS lesion exhibits considerable loss of axons. Scale bar= 25 μm. (From Bjartmar et al, 1999)

Together, these studies suggest that NAWM, as seen on MRI scans or after immunohistochemistry against myelin, might contain considerable axonal loss, especially in chronic patients with long disease duration.

Morphological evidence for axonal degeneration in NAWM distal to an acute lesion was reported in an MS case with a 9 month history of relapsing remitting multiple sclerosis (RR-MS).[31] Postmortem analysis of the ventral spinal cord column, containing descending tracts distal to a terminal brain stem lesion, demonstrated a 22% tract specific axonal loss in spite of grossly normal immunostaining for myelin. Confocal microscopy revealed empty myelin sheaths, myelin ovoids, and signs of myelin degradation by activated microglia, findings characteristic for fiber degeneration caused by proximal transection. There was no sign of primary demyelination and adjacent axons were morphologically intact. Other descending and ascending fiber tracts exhibited normal axon numbers. These data suggest that irreversible axonal loss with retention of slowly degenerating myelin occurs early in disease and is one histopathological correlate to NAWM abnormalities observed in MS patients by some MR techniques.[31,32]

Tissue Atrophy and Neurodegeneration

MRI reports have indicated correlations between clinical disability and atrophy of cerebellum,[33] spinal cord,[34] and the brain[35] in MS. Loss of axons is a plausible contributor to atrophy, although demyelination and reduced axon diameter may also decrease tissue volume. Such correlations have considerable clinical interest since measurements of CNS atrophy may be used as a surrogate marker for disease progression in MS patients. It is generally accepted that total brain lesion volume, as measured on T2 weighted MRI scans, has poor correlation to clinical disability.[6] Motor performance has a relatively high impact on measurements of clinical disability in MS, such as EDSS. The spinal cord is therefore considered a suitable model to study the correlation between tissue atrophy, as revealed by MRI, and clinical progression.[34,36,37] In SP-MS patients, cervical spinal cord atrophy averages 25-30%.[24,34]

The periventricular white matter is frequently affected by MS lesions, which might explain the progressive enlargement of lateral ventricles observed in many MS patients. In a group of RR-MS patients with mild to moderate disability followed over 2 years, brain atrophy increased yearly.[38,39] The course of brain atrophy appears to be influenced by general inflammatory disease activity, as indicated by the occurrence of gadolinium-enhanced lesions in these brains. A new sensitive measure of whole-brain atrophy was applied to this population of relapsing patients.[38] The brain parenchymal fraction (BPF), which constitutes the ratio of brain parenchyma to the total volume within the brain surface contour, was highly reproducible thus allowing precise comparison of individual brain volumes. The BPF declined at a highly significant rate and was significantly reduced compared with age- and sex-matched control individuals during each of 2 years follow-up of these patients.

Cumulative Axonal Loss Determines Permanent Disability in EAE and MS

Episodes of reversible symptoms during RR-MS are primarily associated with acute inflammatory lesions in articulate parts of the CNS. It is believed that four mechanisms contribute to clinical remission: resolution of the inflammation, redistribution of axolemmal sodium channels, remyelination, and adaptive cortical changes.[26,40,41] A recent combined functional MRI and MRS study of RR-MS patients without overt permanent functional disability demonstrated a fivefold increase in sensorimotor cortex activation after simple hand movements when compared with control individuals.[41] These results suggest that compensatory cortical changes, possibly involving reorganization of functional pathways, may contribute to maintained motor function after axonal damage during early MS. The extent of axonal loss in severely disabled patients with long disease duration, and the reduction of levels of the neuronal/axonal marker N-acetyl aspartate (NAA) in MS brains over time,[42] support the hypothesis that the transition from RR-MS to SP-MS occurs when a neurodegenerative threshold is reached.[7,14,26,40] The time-point when a patient develops SP-MS vary between individuals, and probably reflect a number of factors such as location of lesions, disease activity, medication and various aspects of genetic susceptibility.

Figure 3. Permanent neurological disability correlates with axonal loss in spinal cords from mice with chronic experimental autoimmune encephalomyelitis (EAE). (A-D): Axons in the dorsolateral spinal cord from control (A) and EAE mice with clinical scores of increasing disability from 0 (B), 2 (C) and 4 (D) are visualized with neurofilament staining. (E) Density of axons in cervical (white bars) and lumbar (black bars) spinal cord is expressed as percent axonal loss relative to controls. Loss of axons increased significantly for each higher clinical score (Spearman's rank correlation test; cervical cord: r= 0.75; p= 0.0001, and lumbar cord: r= 0.63; p= 0.004). Scale bars (A-D)= 10 μm. (From Wujek et al, 2002)

Most neurodegenerative diseases typically have an initial clinically silent stage of neuronal loss. In amyotrophic lateral sclerosis (ALS), and in Parkinson's disease, it has been estimated that patients may lose 50 to 80% of target neurons before they present with neurological symptoms.[43,44] These figures raise questions regarding how many CNS axons that can be destroyed in MS patients before permanent functional disability occurs. With the exception for in vivo measurements of N-acetylaspartate (NAA),[7] tract specific correlations between axonal loss and disability are difficult to obtain in MS patients. In addition, such quantifications may not be straightforward since axonal loss threshold levels probably vary between different CNS regions. However, experimental support for the hypothesis that the extent of axonal loss determines disability in MS patients was recently described in chronic relapsing-remitting EAE mice.[11] At initial attack, clinical score correlated with spinal cord inflammation but not with axonal loss, indicating that reversible disability in relapsing EAE results from mechanisms other than axonal transection. In contrast, clinical score correlated significantly with axon loss, but not with inflammation, at a chronic end-stage of disease in both cervical (r= 0.75; p= 0.0001) and lumbar spinal cords (r= 0.63; p= 0.004). The data support a causal relationship between axonal loss and permanent neurological disability in these mice (Fig. 3). In end-stage EAE mice lacking clinical impairment, axonal loss compared to controls was 30 and 15% in cervical and lumbar spinal cords respectively. In end-stage mice with permanent paralysis, cervical and lumbar axonal loss was 60 and 45% respectively. These numbers approach the 68% axonal loss in spinal cord lesions of paralyzed MS patients.[24] Axonal loss in mice with moderate clinical impairment (poor righting reflex) was intermediate between that found in nonsymptomatic and paralyzed mice. Therefore, the data also support the hypothesis that a threshold of axonal loss must be surpassed before permanent neurological disability occurs.

The observation that most MS patients progress from moderate to severe disability, even in the absence of overt inflammatory disease activity,[45] support the concept that different mechanisms cause axonal loss at different stages of the disease. Axonal transection caused by inflammatory demyelination is the main cause of initial moderate disability in most MS patients. In addition, chronically demyelinated axons may degenerate due to lack of myelin-derived trophic support (see above; refs. 27,46). Neurodegeneration caused by these two mechanisms will affect remaining neurons due to withdrawal of pre- and/or postsynaptic trophic signals. It is possible that depletion of trans-synaptic support triggers degenerative pathways in some neurons. An epidemiologic study by Confavreux and colleagues including 1844 MS patients showed that although the time from disease onset to EDSS score 4 varied between 1 and 33 years, the time course from EDSS score 4 to EDSS score 7 was similar among the patients.[47] These data support the hypothesis that many MS patients eventually enter a final common pathway of progressive neuronal degeneration once a clinical threshold is reached. Hence, the extent of inflammatory demyelination and tissue damage may influence the time course to EDSS score 4. In contrast, preprogrammed neurodegeneration could cause functional decline at subsequent stages of disease since the progression to EDSS score 7 continued in the absence of inflammatory lesions.[47] Such preprogrammed neurodegeneration dissociate acute inflammatory damage observed early in MS from progressive tissue degeneration during chronic stages of the disease, and the identification of such mechanisms will provide new therapeutic opportunities.

MS Lesions in Gray Matter

It is possible that mechanisms other than loss of white matter axons contribute to neurological decline in MS patients. Although MS traditionally is regarded a white matter disease, demyelinated lesions can occur also in gray matter.[48-51] Many axons originating from and terminating on cortical neurons are myelinated. Compared to white matter lesions, however, cortical MS lesions are less obvious macroscopically, histologically, and on conventional T2-weighted MRI scans. Although involvement of cerebral cortex in MS has attracted interest recently,[52] the significance of cortical lesions may have been underestimated previously.[50,53] Kidd and colleagues reported that the use of gadolinium-enhancement increased the detection of cortical lesions on MRI scans by 140%.[50] Of all active brain lesions examined in this study, 26% arose within or adjacent to the cerebral cortex.

A recent portmortem study on MS brains using immunohistochemistry and confocal microscopy examined inflammation and neuronal pathology in gray matter lesions.[51] Cortical lesions contained 13 times fewer lymphocytes and 6 times fewer microglia/macrophages than white matter lesions, suggesting reduced inflammation in gray matter lesions. Also, cortical lesions contained extensive neuronal injury including neuritic swellings, and dendritic or axonal transection. The density of transected neurites (axons and dendrites) was 4,119 per mm^3 in active cortical lesions, 1,107 per mm^3 in chronic active cortical lesions and 25 per mm^3 in chronic inactive cortical lesions. In contrast, myelinated MS cortex and control gray matter contained 8 and 1 transected neurite per mm^3 respectively. In both active and chronic active cortical lesions, activated microglia were closely associated with neurites and neuronal perikarya. In addition, the number of apoptotic neurons was significantly increased in cortical lesions compared to myelinated cortex.[51]

It is possible that cortical MS lesions have several functional consequences. Neuronal damage in motor and sensory cortex may, for example, contribute to ambulatory decline. Various aspects of cognitive deficits occur frequently in MS, affecting 40-70% of all patients.[54,55] Most commonly involved are functions related to learning, memory and information processing.[54] Positron emission tomography (PET) studies demonstrated that decreased cerebral metabolism correlates with MRI lesion load and cognitive dysfunction in MS.[56] Cognitive impairment in MS patients has generally been attributed to subcortical white matter lesions. However, considering the extent and nature of damage to cortical neurites in many MS brains, damage to neurons in cortical lesions may provide an additional biological substrate for this functional impairment.

Conclusion

The data reviewed in this chapter suggest that axonal injury begins at onset of MS, and that cumulative axonal loss provides the pathological substrate for permanent disability in affected patients. Since different mechanisms may contribute to axonal damage during different stages of disease, it is crucial to clarify the pathogenesis of neurodegeneration in MS. Since axonal injury may be mediated by molecules produced by inflammatory cells, anti-inflammatory and immunomodulatory treatment should be neuroprotective. However, lesions can outnumber clinical relapses by as much as 10:1.[57] Continuous inflammation at early stages of MS may therefore cause considerable tissue damage in the absence of obvious clinical manifestations. A number of drugs with documented effect during RR-MS are now available, for example interferon beta and glatiramer acetate.[1,58] Disease modifying therapy should therefore be used early and continuously in order to prevent and delay accumulating axonal degeneration, and thereby prevent and delay development of permanent functional disability.[45]

The development of specific neuroprotective drugs aimed for MS patients constitute an important goal for MS research that requires appropriate animals models. The relapsing EAE model discussed above resembles MS in many aspects and may be suitable for testing the efficacy of such drugs.[11,59] Promoting remyelination at an early stage of MS will restore conduction and, more important, may be neuroprotective since chronic demyelination can cause axonal degeneration. Possible avenues to obtain remyelination include enhancing endogenous CNS cells to repopulate and remyelinate lesions, or transplanting stem or progenitor cells into lesions. Bone marrow cells that can give rise to neuronal cells may provide an effective source of donor cells for transplantation that, in addition, will avoid controversial political issues associated with stem cells. Finally, the silent nature of early axonal loss and the lack of reliable surrogate markers of disease progression during RR-MS are major obstacles for testing future therapeutics. Surrogate markers of axonal loss are therefore needed to monitor patients with MS.

Acknowledgements

This work was supported by NIH grants NS35058, NS38667 and by a pilot study grant (B.D.T.) and a postdoctoral fellowship (C.B.) from the National Multiple Sclerosis Society.

References

1. Noseworthy JH, Lucchinetti C, Rodriguez M et al. Multiple sclerosis. N Engl J Med 2000; 343:938-952.
2. Lassmann H. Neuropathology in multiple sclerosis: New concepts. Multiple Sclerosis 1998; 4:93-98.
3. Prineas J. Pathology of multiple sclerosis. In: Cook S, ed. Handbook of multiple sclerosis. 3rd ed. Marcel Dekker, 2001:289-324.
4. Kornek B, Lassmann H. Axonal pathology in multiple sclerosis. A historical note. Brain Pathol 1999; 9:651-656.
5. Charcot M. Histologie de le sclerose en plaques. Gazette Hopitaux 1868; 141:554-558.
6. Stevenson VL, Miller DH. Magnetic resonance imaging in the monitoring of disease progression in multiple sclerosis. Multiple Sclerosis 1999; 5:268-272.
7. Matthews PM, De Stefano N, Narayanan S et al. Putting magnetic resonance spectroscopy studies in context: Axonal damage and disability in multiple sclerosis. Semin Neurol 1998; 18:327-336.
8. Ferguson B, Matyszak MK, Esiri MM et al. Axonal damage in acute multiple sclerosis lesions. Brain 1997; 120:393-399.
9. Trapp BD, Peterson J, Ransohoff RM et al. Axonal transection in the lesions of multiple sclerosis. N Engl J Med 1998; 338:278-285.
10. Njenga MK, Murray PD, McGavern D et al. Absence of spontaneous central nervous system remyelination in class II-deficient mice infected with Theiler's virus. J Neuropathol Exp Neurol 1999; 58:78-91.
11. Wujek JR, Bjartmar C, Richer E et al. Axon loss in the spinal cord determines permanent neurological disability in an animal model of multiple sclerosis. J Neuropathol Exp Neurol 2002; 61:23-32.
12. Yin X, Crawford TO, Griffin JW et al. Myelin-associated glycoprotein is a myelin signal that modulates the caliber of myelinated axons. J Neurosci 1998; 18:1953-1962.
13. Griffiths I, Klugmann M, Anderson T et al. Axonal swellings and degeneration in mice lacking the major proteolipid of myelin. Science 1998; 280:1610-1613.
14. Bjartmar C, Trapp BD. Axonal and neuronal degeneration in multiple sclerosis: Mechanisms and functional consequences. Curr Opin Neurol 2001; 14:271-278.
15. Koo EH, Sisodia SS, Archer DR et al. Precursor of amyloid protein in Alzheimer disease undergoes fast anterograde axonal transport. Proc Nat Acad Sci USA 1990; 87:1561-1565.
16. Kornek B, Storch MK, Weissert R et al. Multiple sclerosis and chronic autoimmune encephalomyelitis. A comparative quantitative study of axonal injury in active, inactive, and remyelinated lesions. Am J Pathol 2000; 157:267-276.
17. Hohlfeld R. Biotechnological agents for the immunotherapy of multiple sclerosis. Principles, problems and perspectives. Brain 1997; 120:865-916.
18. Lu F, Selak M, O'Connor J et al. Oxidative damage to mitochondrial DNA and activity of mitochondrial enzymes in chronic active lesions of multiple sclerosis. J Neurol Sci 2000; 177:95-103.
19. Pitt D, Werner P, Raine CS. Glutamate excitotoxicity in a model of multiple sclerosis. Nat Med 2000; 6:67-70.

20. Babbe H, Roers A, Waisman A et al. Clonal expansions of CD8+ T cells dominate the T cell infiltrate in active multiple sclerosis lesions as shown by micromanipulation and single cell polymerase chain reaction. J Exp Med 2000; 192:393-404.
21. Huseby ES, Liggitt D, Brabb T et al. A pathogenic role for myelin-specific CD8+ T cells in a model for multiple sclerosis. J Exp Med 2001; 194:669-676.
22. Medana I, Martinic MA, Wekerle H et al. Transection of major histocompatibility complex class I-induced neurites by cytotoxic T lymphocytes. Am J Pathol 2001; 159:809-815.
23. Shi R, Blight AR. Compression injury of mammalian spinal cord in vitro and the dynamics of action potential conduction failure. J Neurophysiol 1996; 76:1572-1580.
24. Bjartmar C, Kidd G, Mörk S et al. Neurological disability correlates with spinal cord axonal loss and reduced N-acetyl aspartate in chronic multiple sclerosis patients. Ann Neurol 2000; 48:893-901.
25. Lovas G, Szilagyi N, Majtenyi K et al. Axonal changes in chronic demyelinated cervical spinal cord plaques. Brain 2000; 123:308-317.
26. Bjartmar C, Yin X, Trapp BD. Axonal pathology in myelin disorders. J Neurocytol 1999; 28:383-395.
27. Waxman SG. Acquired channelopathies in nerve injury and MS. Neurology 2001; 56:1621-1627.
28. Ganter P, Prince C, Esiri MM. Spinal cord axonal loss in multiple sclerosis: A post-mortem study. Neuropathol Appl Neurobiol 1999; 25:459-467.
29. Evangelou N, Esiri MM, Smith S et al. Quantitative pathological evidence for axonal loss in normal appearing white matter in multiple sclerosis. Ann Neurol 2000; 47:391-395.
30. Bjartmar C, Kinkel RP, Kidd G et al. Axonal loss in normal-appearing white matter in a patient with acute MS. Neurology 2001; 57:1248-1252.
31. Simon JH, Kinkel RP, Jacobs L et al. A Wallerian degeneration pattern in patients at risk for MS. Neurology 2000; 54:1155-1160.
32. Davie CA, Barker GJ, Webb S et al. Persistent functional deficit in multiple sclerosis and autosomal dominant cerebellar ataxia is associated with axon loss. Brain 1995; 118:1583-1592.
33. Losseff NA, Wang L, Lai HM et al. Progressive cerebral atrophy in multiple sclerosis. A serial MRI study. Brain 1996; 119:2009-2019.
34. Losseff NA, Webb SL, O'Riordan JI et al. Spinal cord atrophy and disability in multiple sclerosis. A new reproducible and sensitive MRI method with potential to monitor disease progression. Brain 1996; 119:701-708.
35. Filippi M, Colombo B, Rovaris M et al. A longitudinal magnetic resonance imaging study of the cervical cord in multiple sclerosis. J Neuroimaging 1997; 7:78-80.
36. Kidd D, Thorpe JW, Thompson AJ et al. Spinal cord MRI using multi-array coils and fast spin echo. II. Findings in multiple sclerosis. Neurology 1993; 43:2632-2637.
37. Rudick RA, Fisher E, Lee JC et al. Use of the brain parenchymal fraction to measure whole brain atrophy in relapsing-remitting MS. Multiple Sclerosis Collaborative Research Group. Neurology 1999; 53:1698-1704.
38. Simon JH, Jacobs LD, Campion MK et al. A longitudinal study of brain atrophy in relapsing multiple sclerosis. Neurology 1999; 53:139-148.
39. Waxman SG. Demyelinating diseases - new pathological insights, new therapeutic targets. N Engl J Med 1998; 338:223-225.
40. Trapp BD, Ransohoff RM, Fisher E et al. Neurodegeneration in multiple sclerosis: Relationship to neurological disability. Neuroscientist 1999; 5:48-57.
41. Reddy H, Narayanan S, Arnoutelis R et al. Evidence for adaptive functional changes in the cerebral cortex with axonal injury from multiple sclerosis. Brain 2000; 123:2314-2320.
42. Gonen O, Catalaa I, Babb JS et al. Total brain N-acetylaspartate. A new measure of disease load in MS. Neurology 2000; 54:15-19.
43. Lloyd KG. CNS compensation to dopamine neuron loss in Parkinson's disease. Adv Exp Med Biol 1977; 90:255-266.
44. Bradley WG. Recent views on amyotrophic lateral sclerosis with emphasis on electrophysiological studies. Muscle Nerve 1987; 10:490-502.
45. Rudick RA, Goodman A, Herndon RM et al. Selecting relapsing remitting multiple sclerosis patients for treatment: The case for early treatment. J Neuroimmunol 1999; 98:22-28.
46. Scherer S. Axonal pathology in demyelinating diseases. Ann Neurol 1999; 45:6-7.
47. Confavreux C, Vukusic S, Moreau T et al. Relapses and progression of disability in multiple sclerosis. N Engl J Med 2000; 343:1430-1438.
48. Brownell B, Hughes JT. The distribution of plaques in the cerebrum in multiple sclerosis. J Neurol Neurosurg Psychiatry 1962; 25:315-320.
49. Lumsden CE. The neuropathology of multiple sclerosis. In: Vinken PJ, Bruyn GW, eds. Handbook of clinical neurology. Vol 9. Amsterdam: North-Holland, 1970:217-309.
50. Kidd D, Barkhof F, McConnell R et al. Cortical lesions in multiple sclerosis. Brain 1999; 122:17-26.
51. Peterson JW, Bö L, Mörk S et al. Transected neurites, apoptotic neurons, and reduced inflammation in cortical multiple sclerosis lesions. Ann Neurol 2001; 50:389-400.
52. Filippi M. Multiple sclerosis: A white matter disease with associated gray matter damage. J Neurol Sci 2001; 185:3-4.
53. Sharma R, Narayana PA, Wolinsky JS. Grey matter abnormalities in multiple sclerosis: Proton magnetic resonance spectroscopic imaging. Mult Scler 2001; 7:221-226.
54. Rao SM, Leo GJ, Bernardin L et al. Cognitive dysfunction in multiple sclerosis. I. Frequency, patterns, and prediction. Neurology 1991; 41:685-691.
55. Beatty WW, Paul RH, Wilbanks SL et al. Identifying multiple sclerosis patients with mild or global cognitive impairment using the Screening Examination for Cognitive Impairment (SEFCI). Neurology 1995; 45:718-723.
56. Blinkenberg M, Rune K, Jensen CV et al. Cortical cerebral metabolism correlates with MRI lesion load and cognitive dysfunction in MS. Neurology 2000; 54:558-564.
57. McFarland HF, Frank JA, Albert PS et al. Using gadolinium-enhanced magnetic resonance imaging lesions to monitor disease activity in multiple sclerosis. Ann Neurol 1992; 32:758-766.
58. Rudick RA, Cohen JA, Weinstock-Guttman B et al. Management of multiple sclerosis. N Engl J Med 1997; 337:1604-1611.
59. Yu M, Nishiyama A, Trapp BD et al. Interferon-beta inhibits progression of relapsing-remitting experimental autoimmune encephalomyelitis. J Neuroimmunol 1996; 64:91-100.

CHAPTER 35

Monitoring Disease Activity in Multiple Sclerosis

Lorri Lobeck

Introduction

The search for tools to monitor disease activity in multiple sclerosis (MS) has expanded in recent years. Availability of new therapies has prompted reassessment of clinical examination scales and development of new magnetic resonance imaging (MRI) techniques to improve trial efficiency such that effective therapies are promptly recognized. This chapter will review the use of standardized examination scales, MRI, cerebrospinal fluid (CSF) and evoked potentials (EP) for monitoring disease activity in MS.

Measurement of disease activity in an individual with MS is often difficult. This is, in part, due to the unpredictable nature of the disease and variability between individuals. Further, there are no easily obtained markers of disease activity. In recent years, large clinical treatment trials have prompted reassessment of older tools and development of new tools for more accurate assessment of disease activity and progression. These include standardized examination scales, magnetic resonance imaging and electrophysiologic markers. This chapter will review the use of these parameters to monitor disease activity.

Clinical Outcome Measures

The use of defined measures of disease relapses or disability progression became increasingly important as large scale studies were done to assess the efficacy of new therapies. Instruments used for clinical outcome measures must be assessed for reliability, validity and sensitivity. One must also consider their clinical usefulness including time to administer the test, ease of use, patient tolerability and training requirements. Interrater and intrarater reliability is especially important in MS clinical trials, as change in neurologic examination becomes the primary endpoint of these studies. It is especially important that the instrument not falsely contribute to measures of disease progression or improvement. The instrument must detect small but significant changes.[1]

In 1985, the National MS Society published the Minimum Record of Disability (MRD).[2] This is the most widely used MS outcome measure. The MRD followed the World Health Organization (WHO) classifications for 3 types of dysfunction. Dysfunction due to MS was defined as:

1. Impairment as evidenced by clinical signs and symptoms of neurologic disease;
2. Disability indicative of the personal limitations imposed on the activities of daily living by the neurologic impairment; and
3. Handicap defining the social and environmental effects of the disability or impairment on the individual.[3]

The MRD includes:

1. Expanded Disability Status Scale (EDSS) measuring impairment
2. Incapacity Status Scale measuring disability
3. Environmental Status Scale measuring handicap

Other scales not described here are listed in Table 1. The most widely used impairment outcome measure is the Expanded Disability Status Scale (EDSS). Kurtzke originally developed the Disability Status Scale (DSS) in 1955.[4] This scale rated impairment due to MS on a 1 to 10 point scale. Subsequently, the scale was expanded to include half point steps.[5] (Table 2) The scale is actually two scales. A single item scale assesses individual functional systems including visual, brainstem, pyramidal, cerebellar, sensory, cerebral, and bowel and bladder functions from 0 to 5 or 6. (Table 3) From these subsets a single EDSS score is obtained. On the lower end of the scale (0 to 3.5), the EDSS score is based on scores from the subsets. In the range of 4.0 to 6.5, scoring is based on walking distance and/or the need for an assistive device such as a cane or walker. At 7.0 and above, patients are essentially nonambulatory and scoring is based on upper extremity and bladder function. Administration of the EDSS takes 10 to 20 minutes and is performed by a neurologist or health care professional trained to administer the exam. An advantage of the EDSS includes evaluation of all major areas of the nervous system as they apply to MS. However, it is more heavily weighted toward effects on ambulation and provides limited assessment of upper extremity function, cognitive function and fatigue. EDSS scores are bimodal, clustering at 3 to 4 and 6 to 7. It is not a linear scale, thus a change from 1 to 2 is not equal in disability as a change from 6 to 7.[6] Widespread use of the scale provides familiarity among examiners. Unfortunately, interrater reliability is variable, in part, because of the definition of significant change. Correlation coefficients reach 0.96 if 1.5 EDSS points are allowed as an acceptable difference between raters.[7] Other studies have reported coefficients as low as 0.32.[8] Interrater variability may equal one standard deviation from the true EDSS, thus a two-step change (one point on the EDSS and 2 points on the FS) has been recommended as providing definite evidence of change.[9] The EDSS is further limited in that it does not detect change over a short period of time.[10] A large sample size and a 2 to 3 year study is

Table 1. Clinical measures of MS

IMPAIRMENT MEASURES
    Kurtzke's Expanded Disability Status Scale
    Scripps Neurological Rating Scale
    Ambulation Index
    Self-administered Kurtzke
    Quantitative exam of neurologic function
        (Neuro-performance Testing)
    Troiano Functional Exam
    MS Impairment Scale

DISABILITY MEASURES
    Barthel Index
    Incapacity Status Scale
    Functional Independence Measure

HANDICAP MEASURES
    Environmental Status Scale
    London Handicap Scale

QUALITY OF LIFE MEASURES
    MS Quality of Life Inventory
    MS Quality of Life - 54
    Short Form (SF-36)
    Short Form (SF-12)
    Sickness Impact Profile
    Medical Rehabilitation Follow Along

QUANTITATIVE FUNCTIONAL MEASURES
    MS Functional Composite
        9-Hole Peg Test
        Timed 25 foot walk
        Paced Auditory Serial Addition Test
    Quantitative Evaluation of Neurologic Function
    Quantitative Motor Testing
    Box and Block Test

Table 2. Expanded disability status scale in multiple sclerosis[5]

A mental function grade of 1 does not enter into FS scores for EDSS steps

- 0.0 = Normal neurological exam (all grade 0 in FS).
- 1.0 = No disability, minimal signs in one FS (i.e., grade 1).
- 1.5 = No disability, minimal signs in more than one FS (more than 1 grade 1).
- 2.0 = Minimal disability in one FS (one FS grade 2, others 0 or 1).
- 2.5 = Minimal disability in two FS (two grade 2, others 0 or 1).
- 3.0 = Moderate disability in one FS (one FS grade 3, others 0 or 1) or mild disability in three or four FS (three/four FS grade 2, others 0 or 1) though fully ambulatory.
- 3.5 = Fully ambulatory but with moderate disability in one FS (one grade 3) and one or two FS grade 2; or two FS grade 3; or five FS grade 2 (others 0 or 1).
- 4.0 = Ambulatory without aid or rest for ≥ 500 M.
- 4.5 = Ambulatory without aide for rest for ≥ 300 M.
- 5.0 = Ambulatory without aide or rest for ≥ 200 M.
- 5.5 = Ambulatory without aid or rest for ≥ 100 M.
- 6.0 = Unilateral assistance (cane or crutch) required to walk at least 100 M with or without resting.
- 6.5 = Constant bilateral assistance (canes or crutches) required to walk at least 20 M without resting.
- 7.0 = Unable to walk 20 M even with aid, essentially restricted to wheelchair; wheels self and transfers alone.
- 7.5 = Unable to take more than a few steps; restricted to wheelchair; transfers alone and wheels self.
- 8.0 = Essentially restricted to bed or chair or perambulated in wheelchair, but out of bed most of day; retains many self care functions; generally has effective use of arms.
- 8.5 = Essentially restricted to bed much of the day; has some effective use of arm(s); retains some self-care functions.
- 9.0 = Helpless bed patient; can communicate and eat.
- 9.5 = Totally helpless bed patient; unable to communicate effectively or eat/swallow.
- 10.0 = Death due to MS.

necessary to confirm change.[11] A change of 1.0 point for 3 to 6 months has been used in recent large trials as indicative of sustained progression in disease change.[12-14] The EDSS is significantly correlated with the Scripps Neurologic Rating Scale.[10] Change in the EDSS and Ambulation Index is significantly correlated.[15]

The Scripps Neurological Rating Scale (SNRS) or the Neurological Rating Scale (NRS)[16] (Table 4) is similar to the EDSS scale in that it is based on the neurologic examination. Categories include mentation and mood, key cranial nerves, lower cranial nerves, motor, reflexes (including Babinski), sensory, cerebellar, gait, bowel bladder and sexual dysfunction. This 100-point scale has variable point values for each system. For example, the visual cranial nerves have a maximum score of 21; reflexes have a maximum of 8 points. This scale puts more emphasis, than the EDSS, on the upper extremity function. In general, the changes in clinical disease are reflected more gradually on the SNRS as the 100 point system is much broader than 20 points of the EDSS.[10] Unfortunately, the grading options of mild, moderate and severe impairment are not defined, thus left to the examiner's opinion resulting in inherent variability. Cognitive deficits are minimally described. Interrater reliability is high.[16] Correlation between the EDSS and SNRS is high (r= 0.78).[10]

The Ambulation Index (AI) was developed to quantify changes in gait on an ordinal scale.[17] (Table 5) Ambulatory patients are timed as they walk 25 feet quickly, but safely. Use of assistive devices are recorded. A 10-point scale (0-9) grades mobility from asymptomatic to restricted to wheelchair. Because restrictions of gait are more specifically defined, use with the EDSS increases sensitivity for change in the 4.0 to 6.0 range of the EDSS.[18] The test does not require a Neurologist to perform and is used frequently in MS clinical trials to assess lower extremity function. The test has high interrater reliability (r= 0.98).[19]

In 1994, the National MS Society convened an international meeting of experts in the field to develop recommendations for

Table 3. Kurtzke's functional system

A. PYRAMIDAL FUNCTIONS
- 0 = normal
- 1 = abnormal signs without disability
- 2 = minimal disability
- 3 = mild or moderate paraparesis or hemiparesis, severe monoparesis
- 4 = marked paraparesis or hemiparesis, moderate quadriparesis, or monoplegia
- 5 = paraplegia, hemiplegia, or marked quadriparesis
- 6 = quadriplegia
- V = unknown

B. CEREBELLAR FUNCTIONS
- 0 = normal
- 1 = abnormal signs without disability
- 2 = mild ataxia
- 3 = moderate truncal or limb ataxia
- 4 = severe ataxia, all limbs
- 5 = unable to perform coordinated movements due to ataxia
- V = unknown
- X = used throughout after each number when weakness (grade or more on pyramidal) interferes with testing

C. BRAINSTEM FUNCTIONS
- 0 = normal
- 1 = signs only
- 2 = moderate nystagmus or other mild disability
- 3 = severe nystagmus, marked extraocular weakness or moderate disability of other cranial nerves
- 4 = marked dysarthria or other marked disability
- 5 = inability to swallow or speak
- V = unknown

D. SENSORY FUNCTIONS
- 0 = normal
- 1 = vibration or figure-writing decrease only in 1 or 2 limbs
- 2 = mild decrease in touch or pain or position sense, and/or moderate decrease in vibration in 1 or 2 limbs; or vibratory (c/s figure writing) decrease alone in 3 or 4 limbs
- 3 = moderate decrease in touch or pain or position sense, and/or essentially lost vibration in 1 or 2 limbs; or mild decrease in touch or pain and/or moderate decrease in all proprioceptive tests in 3 or 4 limbs
- 4 = marked decrease in touch or pain or loss of proprioception, alone or combined, in 1 or 2 limbs; or moderate decrease in touch or pain and/or severe proprioceptive decrease in more than 2 limbs
- 5 = loss (essentially) of sensation in 1 or 2 limbs; or moderate decrease in touch or pain and/or loss of proprioception for most of the body below the head
- 6 = sensation essentially lost below the head
- V = unknown

E. BOWEL AND BLADDER FUNCTIONS
- 0 = normal
- 1 = mild urinary hesitancy, urgency or retention
- 2 = moderate hesitancy, urgency, retention of bowel or bladder, or rare urinary incontinence
- 3 = frequent urinary incontinence
- 4 = in need of almost constant catheterization
- 5 = loss of bladder function
- 6 = loss of bowel and bladder function
- V = unknown

F. VISUAL (OR OPTIC) FUNCTIONS
- 0 = normal
- 1 = scotoma with visual acuity (corrected better than 20/30)
- 2 = worse eye with scotoma with maximal visual acuity (corrected) of 20/30 to 20/59
- 3 = worse eye with large scotoma, or moderate decrease in fields, but with maximal visual acuity (corrected) of 20/60 to 20/99
- 4 = worse eye with marked decrease of fields and maximal visual acuity (corrected) of 20/100 to 20/200; grade 3 plus maximal acuity of better eye of 20/60 or less
- 5 = worse eye with maximal visual acuity (corrected) less than 20/200; grade 4 plus maximal acuity of better eye of 20/60 or less
- 6 = grade 5 plus maximal visual acuity of better eye of 20/60 or less
- V = unknown
- X = added to grades 0-6 for presence of temporal pallor

G. CEREBRAL (OR MENTAL) FUNCTIONS
- 0 = normal
- 1 = mood alteration only (does not affect DSS score)
- 2 = mild decrease in mentation
- 3 = moderate decrease in mentation
- 4 = marked decrease in mentation (chronic brain syndrome moderate)
- 5 = dementia or chronic brain syndrome - severe or incompetent
- V = unknown

---

the appropriate assessment of MS. This group recommended that clinical outcome measure for the treatment of MS should:

1. Reflect the extent of disease process.

2. Be multidimensional to include components which assess gait and lower extremity function, arm function, visual function, and neuropsychological outcome. This should not be redundant.

3. Have high reliability, practicality, acceptability to patients and cost-effectiveness. Interval, rather than categorical or ordinal, data are preferred.

4. Change over time as the severity of the disease worsens or improves.

5. Be capable of demonstrating treatment effects.

6. Be predictive of clinically meaningful change.[20]

The National MS Society Clinical Outcomes Assessment Task Force reviewed multiple measures to determine which were reliable, correlated well with disease duration, changed over time and had concurrent and predictive validity based on the EDSS.[21] Development of the MS Functional Composite followed. (Table 6)

Three measures, which assess disability and impairment, were recommended. This included the timed 25-foot walk, the

Table 4. Scripps Neurologic Rating Scale (SNRS)[16]

System Examined	Maximum Points	Degree of Impairment			
		Normal	Mild	Moderate	Severe
Mentation and Mood	10	10	7	4	0
Cranial Nerves:					
Visual Acuity	21	5	3	1	0
Fields, Discs, Pupils		6	4	2	0
Eye Movements		5	3	1	0
Nystagmus		5	3	1	0
Lower Cranial Nerves	5	5	3	1	0
Motor:					
RU	20	5	3	1	0
LU		5	3	1	0
RL		5	3	1	0
LL		5	3	1	0
DTRS:					
UE	8	4	3	1	0
LE		4	3	1	0
Babinski: R; L	4	4	—	—	0
Sensory:					
RU	12	3	2	1	0
LU		3	2	1	0
RL		3	2	1	0
LL		3	2	1	0
Cerebellar:					
UE	10	5	3	1	0
LE		5	3	1	0
Gait; Trunk and Balance	10	10	7	4	0
Special Category:					
Bladder/Bowels/ Sexual Dysfunction	0	0	-3	-7	-10
TOTALS: 100					

nine-hole peg test and Paced Auditory Serial Addition Test—3-second version (PASAT-3). Individual standardized Z-scores combine to a single composite score. The composite score has the advantage of being a continuous scale, rather than an ordinal scale.

The timed 25-foot walk is performed like the AI, but scored using the actual amount of time to walk the distance rather than the ordinal scale. The test can be administered by a trained person and takes only a few minutes to perform. An increase of more than 20% in the time of the walk indicates significant change.[22] Minimally affected patients require between 3 and 5 seconds to walk 25 feet. In patients who complained of walking difficulties, but had no change on routine neurologic examination, gait was slowed. Assessing 3 trials and recording the 2 fastest had a variability of 10%.[22] Spasticity, fatigue and body temperature may impact on the time of the walk and should be taken into consideration when assessing change.

The 9-hole peg test requires a patient to place nine wooden 9 mm diameter, 32 mm long, dowels, into 9 holes spaced 15 mm apart on a 3 X 3 array.[23] This test measures upper extremity disability and has been used in several clinical trials. It is an excellent assessment of fine manual dexterity with high interrater reliability and moderate test-retest reliability.[24] Without impairment, the test can be completed in 18 to 20 seconds.[25] Variations of the test include timing of placement and removal of the pegs or counting number of pegs placed per 50 seconds.[26] A 20% decline in the score has been suggested as significant change in the upper extremity function.[27] Its use is limited in those with severe upper extremity weakness or tremor.

The PASAT-3 is a cognitive screening tool to estimate informational processing speed and attention.[28] The numbers 1 to 9 are presented in random order, every 3 seconds, to a subject by audiotape. The subject is told to add pairs of numbers. With each new number presented, the first number is omitted and the new number added to the second number. For example, if the numbers 3, 7, 9 and 4 are presented, the patient adds 3 + 7 = 10, then 7 + 9 = 16 and then 9 + 4, etc. The test requires 15 to 20 minutes to perform, including practice trials. Scoring is based on number correct, number of incorrect, and no responses. Comparison of the odd numbered items to the even (split-half reliability) is necessary to assess reliability because there is a significant practice effect.[29] Reliability of the PASAT is high (r= 0.96). The PASAT is useful in detecting mild cognitive impairment.[30] The test can be frustrating for normal controls and thus often distressing for those with mild impairment.

The Z-score is the common measure used to combine the results of tests.[20] This standardized number represents how close

Table 5. Ambulation index[17]

INSTRUCTIONS: Time the patient's walk of 25 feet, after telling the patient to walk as quickly as he/she safely can.

0   Asymptomatic; fully active.

1   Walks normally but reports fatigue that interferes with athletic or other demanding activities.

2   Abnormal gait or episodic imbalance; gait disorder is noticeable to family and friends. Able to walk 25 feet in 10 seconds or less.

3   Walks independently; able to walk 25 feet in between 11-20 seconds.

4   Requires unilateral support (cane, single crutch) to walk; uses support more than 80% of the time. Walks 25 feet in 20 seconds or less.

5   Requires unilateral support but walks 25 feet in greater than 20 seconds or requires bilateral support (canes, crutches, walker) and walks 25 feet in 20 seconds or less.

6   Requires bilateral support and walks 25 feet in greater than 20 seconds. (May use wheelchair on occasion*).

7   Walking limited to several steps with bilateral support; unable to walk 25 feet. (May use wheelchair most of the time).

8   Restricted to wheelchair; able to transfer independently.

9   Restricted to wheelchair; unable to transfer independently.

10.   Bedridden.

Table 6. MS functional composite[21]

Performance Test	Scoring
Nine hole peg test	Two trials of the left and two trials of the right hand are averaged separately. The reciprocals of the left and right hands' results are averaged.
Timed 25-foot walk	Two trials are averaged
PASAT-3	Number of correct answers

the test result is to the mean of a reference population to which the result is compared. It is a relative measure. A Z-score of 3 means the result is 3 standard deviations from the mean. A Z-score is calculated for each test, then combined in the following formula for the MSFC score:

MFSC score =
$$\frac{Z \text{ arm, average} - Z \text{ leg, average} + Z \text{ cognitive, average}}{3.0}$$

Additional information regarding scoring is provided in the scoring manual.[21] The MSFC provides a continuous measure of impairment and disability. This allows for the more precise measure and smaller sample size. The increased sensitivity of this continuous scale becomes a means to decrease trial time and size, thus containing costs. The MFSC is currently being incorporated into large clinical trials. Comparison to older methods will clarify its use for future trials.

Neurophychological Function

Cognitive impairment is common in patients with MS. A community based MS sample showed 43% of patients were impaired on formal neuropsychological testing.[31] A clinic based sample revealed similar rates.[32] Cognitive functions most likely to be impaired included learning and recall of new information, deficits in information processing speed and flexibility, visuospatial deficits and executive functions such as reasoning, problem solving and planning sequencing. Auditory attention span and verbal abilities tend not to be involved.[31] As a result, these individuals are less likely to be employed or engage in social activities. They are more likely to require assistance with the activities of daily living including household management.[33] Unfortunately, the usual clinical outcome measures do not assess these important aspects of disability.[6] Several problems exist in the evaluation of cognitive dysfunction as a measure of progression in MS. First, the natural history of MS cognitive impairment is not thoroughly understood. A 3 year longitudinal study showed significant deterioration in 25% of individuals with MS. The patients showed a broad range of decline in neuropsychological dysfunction.[33] The 8 year data is currently being analyzed (Julie Bobholz, personal communication). A recent study reported 56% of patients were cognitively impaired after 10 years. Physical disability, progressive disease and increasing age predicted cognitive decline.[34] Others have shown the physical disability and cognitive dysfunction are weakly correlated.[31,32] There is significant heterogeneity among patients. Studies have identified subgroups of patients ranging from normal to isolated cognitive deficits to dementia.[35]

The relationship of MRI and neuropsychological dysfunction has shown cognitive impairment is correlated with the extent of T2 weighted brain lesions.[36] Decline in neuropsychological measures correlates with increase in disease burden.[37] Other measures, including brain atrophy and magnetization transfer ratio (MTR), have shown a similar pattern.[38,39]

Since the early 1990s, measures of cognitive function have been incorporated into clinical trials. The interferon beta 1a (Avonex) trial identified subgroups of relapsing-remitting MS patients based on their cognitive dysfunction. Assessing individuals every 6 months and with a comprehensive evaluation at 2 years, a treatment effect was identified for memory and information processing variable, executive functions and visuospatial abilities.[40] Beta interferon 1b also showed a treatment effect on delayed visual reproduction.[41] The glatiramer acetate trial had neither beneficial or detrimental effects.[42]

For clinical trials, it has been important to include measures that are sensitive to change and, specifically, the changes related to MS. Tests used in trials are variable and thus comparison is difficult. Because the patient population is heterogenous in severity and type of cognitive dysfunction, patient selection and selection of measures used to assess for progression is crucial. Performance will fluctuate over time due to measurement error or disease activity. Many neuropsychological measures are subject to practice effect. This must be incorporated into planning the clinical trial. Finally, tools that detect impairment may not be sensitive to change in the condition. Because of the wide patient variability, the use of one test for cognitive assessment is limited. More extensive neuropsychological testing provides data about diverse areas of cognitive dysfunction, but is time consuming and may blur a significant cognitive change in one area. For large studies, a limited but directed cognitive battery might be most appropriate.[43]

Evoked Potentials

Evoked potentials (EP) are electrical potentials that are induced by sensory stimuli. Scalp electrodes are used to record the potentials. When demyelination occurs in the central or peripheral nervous system, the potentials are delayed, decreased in size or absent. Evoked potentials have been used for the diagnosis of MS, particularly with clinically silent lesions. For example, "tingling" in the arms or legs without findings on the examination may be associated with abnormal somatosensory evoked potentials. Similarly, "blurred vision" without a central scotoma or afferent pupillary defect may be indicative of optic neuritis, confirmed by visual evoked potentials. The most recent criteria for diagnosis of MS suggest visual evoked potentials are the only useful evoked potential for the confirmation of disease symptoms.[44]

The use of evoked potentials for monitoring disease activity has been controversial. Many studies supporting their use were performed before the widespread availability of MRI.[45,46] Multimodality (visual, brainstem and median somatosensory evoked potentials) were used in a clinical trial of azathioprine. A statistically significant difference in EP stability was evident with one treatment, not evident with placebo or the other treatment arm. A similar pattern was seen in the clinical examination, but did not reach statistical significance.[45] Correlation between clinical and evoked potential abnormalities has been poor in other studies.[47] Visual evoked potentials, generally easy to perform and well tolerated, may be the most sensitive evoked potential for therapeutic trial.[45] Further consideration of EPs for clinical trial may be dependent on the specific question of the clinical trial. For example, visual evoked potentials (VEPs) may be most useful in assessing therapy aimed at treating or stabilizing visual loss. Most large trial do not incorporate evoked potentials because MRI has been considered superior to quantify extent of disease, including "silent disease," which traditionally was detected by evoked potentials.

Cerebrospinal Fluid

Cerebrospinal fluid abnormalities are useful in the diagnosis of MS, but like evoked potentials are of limited value for monitoring disease activity. A high percentage of patients with MS will ultimately have elevated immunoglobulins and oligoclonal bands in the cerebrospinal fluid. While oligoclonal bands may be absent early in the disease, once present they remain. There is no correlation with severity, disease activity or duration.[48] Immunoglobulin levels may increase over time, but there is no correlation of intrathecal IgG production and disease activity.[6] Intravenous methylprednisolone decreased IgG synthesis in the CSF.[49] CSF myelin basic protein levels fluctuate, including reports on increased levels during clinical activity.[50] Myelin basic protein levels correlate with MRI disease activity.[51] Myelin basic protein-like material in the CSF may predict response to glucocorticoids. However, myelin basic protein is not disease specific. Further, assays for myelin basic protein and oligoclonal bands are quite variable between laboratories limiting their use for disease monitoring.

Accessing fluid is cumbersome and not without complications. Cytokines, chemokines and adhesion molecules are being investigated as potential markers to follow disease activity. At this time, there are no standardized, reliable markers available for large clinical trials, but many warrant further investigation.

Magnetic Resonance Image

Magnetic resonance (MR) imaging is the optimal technique for providing paraclinical data for the diagnosis of MS. In recent years, MR imaging has played a major role in assessing new treatment, including phase III trials of immunomodulating agents. It is recognized that the changes on MR imaging are more frequent and fluid than changes detected clinically. It remains a challenge for researchers, as it is not yet a perfect tool to monitor the disease. Newer techniques have led to a greater understanding of the pathology of the disease, yet at this time are only beginning to be incorporated into clinical trials as a measure of disease activity.

MR images result when energy is released from protons that have been aligned in the axis of a strong magnet. Water, lipids and other molecules have hydrogen atoms, which spin or precess around the main axis or magnetization vector. Radio frequency pulses cause protons to rotate and spin away from the main axis, until the pulse is removed and protons return to their original rotation resulting in signal decay. Signals decay exponentially with specific time constants or relaxation times known as T1 (longitudinal) and T2 (transverse) relaxation time. These relaxation times reflect the environment of the protons. For example, CSF has long T1 and long T2 relaxation times compared to fat. Radio frequency pulses and magnetic field gradient changes are given in different times and lengths generating pulse sequences such as spin echo, gradient recalled and inversion recovery sequences. Spin echo sequences produce images that are "weighted" toward T1 and T2 by changing the repetition time (TR) and echo time (TE).[53] MS lesions are characterized by increased T1 and T2 relaxation times. The lesions of MS appear hyperintense (bright) on T2 weighted images and hypointense (dark relative to surrounding white matter), or isointense on T1 weighted images (Fig. 1 and 2—T1 and T2 images, respectively). The typical ovoid lesions are perpendicular to the ventricle or align on the inner surface of the corpus callosum. Periventricular and deep white matter areas are more often involved than peripheral areas.[54] Gray matter lesions, documented pathologically, are present but difficult to detect with routine MR sequences. These lesions are detected more frequently with T2 weighted inversion recovery sequences such as fluid attenuated inversion recovery (FLAIR) sequences.[55] (Fig. 3) FLAIR images increase detection of white matter lesions by canceling bright signal of the cerebrospinal fluid. However, the technique is less sensitive to lesions in the posterior fossa and is prone to artifact from vascular structures.

Gadolinium (Gd), a paramagnetic contrast agent, is used to assess the blood-brain barrier. If this barrier is disrupted, presumably by inflammation, bright signal is evident in brain parenchyma and spinal cord on T1 weighted images. (Fig. 4) This is an early event in the development of an MS lesion.[56] T2 lesions not seen on Gd-enhanced images are felt to represent chronic lesions. T2 lesions may extend beyond the area of enhancement seen on T1 Gd-enhancing scans. This is likely due to edema, but occasionally, the rim of T2 hypointensity may be related to macrophage infiltration.[57]

Lesions are also detected at all levels of the spinal cord, most common in the dorsal or lateral surface of the cervical cord.[58]

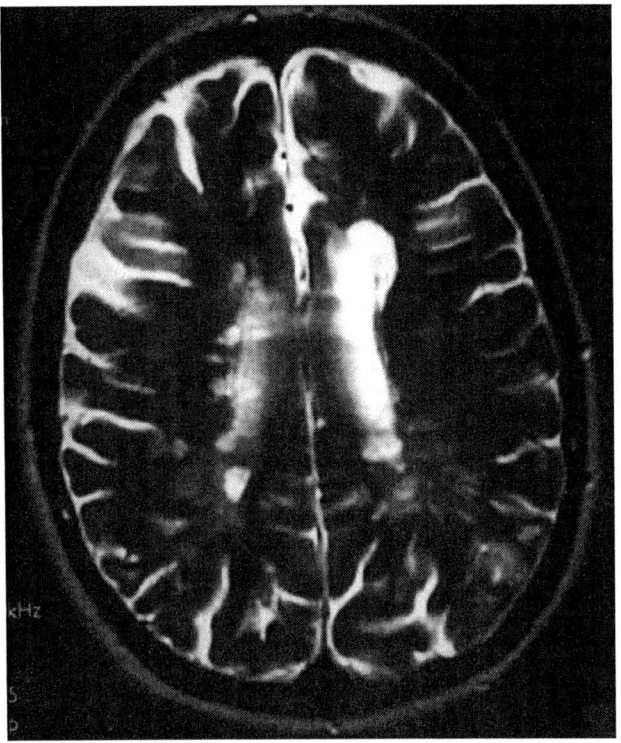

Figure 1. MS lesions are hyperintense (white) on T2 weighted MRI.

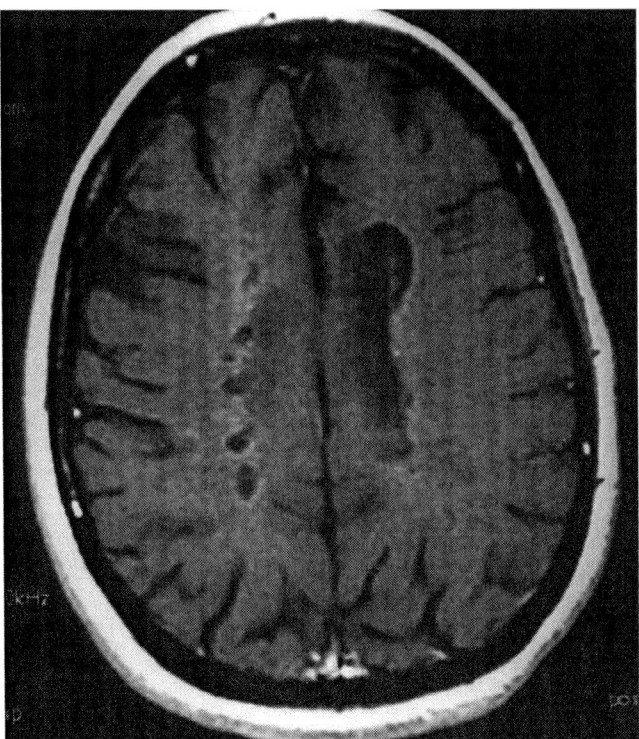

Figure 2. MS lesions are hypotense (dark) on T1 weighted MRI.

Like the brain, Gd-enhancement occurs during the acute phase of lesion development and in some cases, spinal cord swelling may occur. Over time, cord atrophy may develop at the site of the lesion or diffusely. The most recent MRI criteria for the diagnosis of MS are described in Table 7.

MRI Finding and Implications for Assessing Disease Activity

Understanding the evolution of MR imaging abnormalities is necessary for incorporation of this tool into therapeutic trials and for monitoring disease activity clinically. The greater frequency of MR lesions, as compared to clinical events, makes it a useful tool for disease monitoring. While many lesions are silent, those in clinically evident areas parallel the clinical course. Guidelines have been suggested for the use of MRI in clinical trials.[59,60] Phase I trials designed to assess the safety of a therapy have not usually included MR imaging, although toxicity could be considered if enhancing lesion number were to increase greater than expected. Phase II trials designed for an initial assessment of efficacy and further safety data include MR imaging data, often by monthly scans for up to 12 months. Outcome measures include enhancing lesions, or new T2 hyperintense lesions.[59,61] Because the natural history of MS is one of decreasing clinical relapse frequency and MR Gd-enhancement activity, the use of crossover design may be problematic. Adequate sample size is necessary for treatment versus placebo arms in parallel study design. The definitive phase III clinical trials require clinical outcomes (e.g., relapse rate or disability progression) to determine efficacy. However, MRI is included as a secondary outcome measure. Measures include new or enlarging T2 lesions, Gd-enhancing lesion number and size. Scanning frequency varies from every 6 to 12 months with a smaller group studied on a more frequent basis. Further details of each of these methods are outlined below.

Gd Enhanced T1 Images

Gd-enhancing lesions present on T1 images represent disruption of the blood brain barrier, as Gd is normally unable to pass the tight endothelial junction of the blood-brain barrier. These abnormalities have been confirmed with biopsy or autopsy speci-

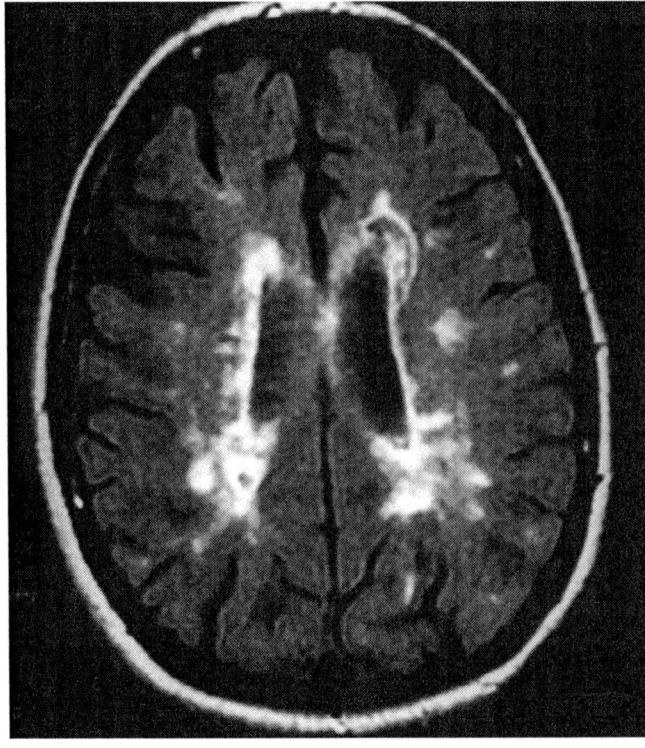

Figure 3. FLAIR image-fluid attenuated inversion recovery.

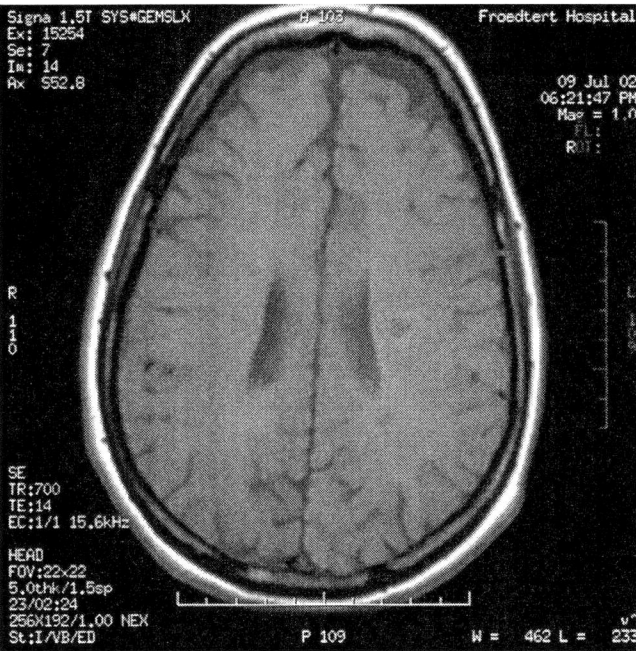

Figure 4A. T1 weighted MRI pre gadolinium.

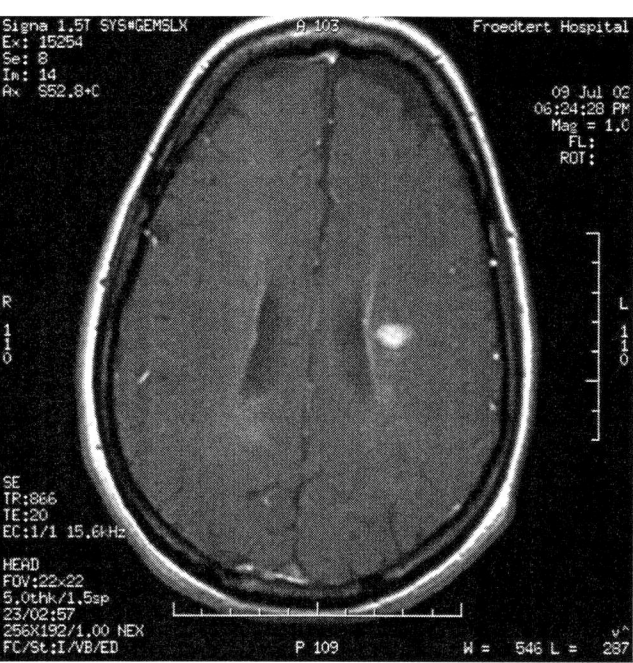

Figure 4B. T1 weighted MRI post gadolinium.

mens as areas of inflammation.[57] This enhancement may be present for 2 to 16 weeks.[62] Lesion size varies from 2-10 mm and appears homogeneous or progresses to ring-enhancing lesions.[63] In roughly 4 to 8 weeks, lesion size reaches a maximum and then begins to decline. Gd-enhancement becomes less evident and disappears.[56] While this would suggest closure of the blood brain barrier, biopsy of chronic MS plaques would suggest otherwise.[57] Gd-enhancing lesions are 5-10 times more frequent than clinical events,[61,64] thus provide insight to subclinical disease activity. Gd-enhancing lesions are predictive of new and increasing T-2 lesion number and volume.[65] A meta-analysis has shown that Gd-enhancing lesion development correlates with relapse rate over a few months.[66] Patterns of enhancement appear stable within individual patients over at least 1-2 years.[67] The correlation between Gd-enhancing lesions and clinical relapse is better than the correlation with subsequent clinical disability.[66] Some, but not all, studies have suggested enhancing lesions become less frequent in secondary progressive disease.[68] This change in MR enhancement correlates with decreased frequency of relapses in progressive disease. Whether this is due to change in immunologic pattern, or is due to technical features is unknown. Gd-enhancing lesions also correlate with cerebral atrophy, in some studies, especially in mild to moderate relapsing-remitting MS.[69,70] Others found that enhancing lesions are not predictive of cerebral atrophy.[71]

Typically, Gd is dosed 0.1 mmol/kg. Triple dose Gd can increase lesion detection by 50 to 100%.[72] This may reflect a heterogenous nature of the enhancing plaque. The pattern of enhancement is also heterogenous with more extensive tissue damage occurring in ring enhancing lesions.[70] Unfortunately, triple doses of Gd are significantly more costly, thus they are used less frequently.

Corticosteroids induce a marked decline in Gd-enhancement as they effectively close the blood brain barrier within several minutes to hours of infusion.[73] This is a short-lived response with enhancement recurring in days to weeks, unless therapy is continued. Because of this, corticosteroid use must be carefully monitored and incorporated into analysis of MRI images during clinical trials.

In summary, Gd-enhancing T1 weighted imaging is most useful for assessing inflammatory aspects of MS, correlates with clinical relapse activity, but at this time does not provide adequate information predicting future outcome.

T-2 Weighted Imaging

T2 lesions correlate with the plaques detected at autopsy in individuals with MS.[57] However, T2 lesions are nonspecific and represent inflammation, edema, demyelination, gliosis and axonal loss. There is a weak correlation of T2 lesion volume with clinical disability as determined by the EDSS.[74,75] This may be explained by the nonspecific nature of the T2 lesion. It is likely that axonal loss and/or permanent demyelination contribute to accumulation of disability, while the inflammation and edema may cause transient symptoms. Further, the EDSS reflects lesions located in the brainstem and spinal cord, which are more likely to cause clinically recognized neurologic disability. As discussed earlier, there is a stronger correlation between T2 lesion load and cognitive dysfunction than with the EDSS.[36] This may reflect the lesion burden present in frontal lobes and deep white matter

---

Table 7. MR imaging criteria for multiple sclerosis[44]

Three of the four following features:
1. At least 1 Gd-enhancing lesion or 9 T2 hyperintense lesions, if no Gd-enhancing lesion present
2. At least 1 juxtacortical lesion
3. At least 3 periventricular lesions
4. At least 1 infratentorial lesion, one spinal cord lesion may be substituted for one brain lesion.

From McDonald[44] et al as adapted from Barkhof et al[101] and Tintoré et al;[102] Gd= gadolinium

not generally detected on the neurologic examination. Gd enhancing and T2 abnormalities will fluctuate over time, but in general there is a gradual accumulation of T2 white matter lesions.[75] T2 weighted lesions may appear confluent over time. Progression of the EDSS over time has a mild correlation with increasing disease burden on T2 weighted MR imaging.[76] Variability between patients is significant. Lesion volumes may actually decrease as large lesions shrink. Simultaneous new lesion development occurs, indicative of ongoing disease activity.[77,78] The average T2 lesion load of individuals with secondary progressive MS is higher than seen with relapsing-remitting disease.[74] While recent studies have incorporated T2 lesion volume as a marker of disease activity, the nonspecific nature of these lesions and poor correlation with disability have prompted the search for a better MRI secondary outcome measure.

T1 Weighted Imaging

During early lesion development, contrast enhancement is present. Lesions on T1-weighted images are isointense, occasionally hypointense, secondary to edema.[57] Over time, these areas may become hypointense and are considered "black holes" or T1 holes.[79] These areas are felt to represent axonal destruction as evident from autopsy specimens.[80] "Black holes" occur more frequently in severely affected secondary progressive MS patients than in relapsing-remitting disease.[79] Further, the areas of hypointensity appear to have a better correlation of baseline disability and disease progression than T2 lesion volume or Gd-enhancing lesions.[79] T1 hypointensity does not correlate with the extent of demyelination or reactive gliosis.[80] Thus, quantifying T1 hypotense lesion load has a better potential to become a surrogate marker for MS disability than T2 lesion load. In addition, T1-hole lesion load at baseline predicts future T1 holes. This may be a useful tool to predict future disability.[80]

Primary Progressive MS

The unique clinical and MRI features of primary progressive disease (PPMS) require special consideration when monitoring disease activity. Clinically, this disease is progressive from onset without relapse-type events. There is controversy as to whether PPMS is a subtype of MS, or whether it is a distinct illness. Clinically, the disease is more common in men and occurs later is life. MR imaging shows fewer and smaller lesions, which may correlate with less inflammation seen by histopathology.[81] These patients often have progressive myelopathy, but do not show an increase in the number or extent of spinal cord lesions as compared to secondary progressive MS.[82] Brain MR imaging may be entirely normal in this group of patients. Thus, incorporation of this patient population into a clinical trial must include consideration of the MRI characteristics with expectation that changes may be less robust as compared to relapsing-remitting or secondary progressive MS.

Use of MRI in Recent Trials

Therapeutic trials with use of beta-interferon (beta-IFN)[14,75,76,83] and glatiramer acetate[85] have further advanced the understanding of MR imaging (Table 8) in monitoring disease activity and treatment effect. Therapy with 8 million international units (MIU) of Beta-IFN 1b (Betaseron)[83] slowed the increase in T2 weighted disease burden and reduced the development of lesions as compared to placebo. Beta-IFN 1a (Avonex)[76] had a less robust effect on T2 burden, but showed a reduction of new or enlarging lesions as compared to placebo. Correlation between relapse rate and Gd-enhancing lesion number was poor. Beta-IFN 1a (Rebif),[14] in two doses 22 µg and 44 µg showed a reduction in T2 lesion burden in both dosage groups with an increase in the placebo treated group. The number of active lesions decreased significantly in the low and high dose groups respectively, as compared to placebo. Beta-IFN appears to have a much greater effect on the blood brain barrier, as compared to glatiramer acetate, yet decline in relapse rates is similar. The effect on MR imaging appears delayed with glatiramer acetate. This data suggests beta-IFN has a pronounced effect on the blood-brain barrier, whereas glatiramer acetate's mode of action is likely different and delayed effect on Gd-enhancement.[84] This raises concern that Gd-enhancement alone is a poor indicator of the clinical effects of glatiramer acetate, if not other drugs. Clinical trials with agents, such as monoclonal antibodies, have shown a significant effect on enhancing lesions, without change in relapses or disability.[85] At this time, the evidence supports that measurement of Gd-enhancement and T2 lesion volume may not be sufficient to monitor disease evolution in MS. However, this does not make MRI an invalid tool. Changes on MRI reflect disease activity, but at this time cannot be considered primary outcome measure.

Newer Techniques

There are a number of newer techniques that are not currently standardized for treatment trials in MS, but hold potential for more sensitive and specific measure of disease activity. Magnetization transfer imaging is useful in distinguishing the type and extent of tissue injury. This technique measures the exchange of magnetization between bound and free protons in a given tissue.[86] The magnetization transfer ratio is calculated by comparing reduction in signal of images obtained with a saturation pulse as compared to those without a pulse. A mild decrease in the magnetization transfer ratio (MTR) is consistent with edema and inflammation, while greater reductions are associated with demyelination and axonal injury. Normal appearing white matter on T2 weighted images, may not be "normal" based on abnormal MTR.[87] Areas of future lesion development may have a decrease in MTR before the lesion appears.[88] T1 black holes have very low MTR values.[89] MTR is decreased in secondary progressive MS[90] and correlated with disability. Histograms of the MTR values of the entire brain may be useful as a global measure of the white matter disease.[91] Fluctuations in the histogram in beta-interferon treated patient suggests inflammatory processes persist, which was not evident on traditional imaging techniques.[92] Magnetization transfer imaging is not currently standardized, as there is considerable variability in image acquisition, however, because of what appears to be ability to distinguish type of injury, it holds significant potential as a tool to measure decreased activity.

Magnetic resonance spectroscopy measures biochemical changes that occur as a result of tissue metabolites. Four major resonance peaks, N-acetylaspartate (NAA), choline, creatine and lactate are measured. NAA is present in neurons and when this is decreased, axonal or neuronal injury is present.[93] Increased choline is associated with myelin and lipid breakdown products in a lesion.[94] Creatine is usually stable and can be used to compare the other peaks. Lactate, often not evident, is present with an acute inflammatory lesion.[95] NAA in acute lesions can be reversed, but when persistent, correlates with clinical disability. T1 black holes have decreased NAA, all which continues to support the hypothesis of irreversible axonal injury reflected as black holes and resultant clinical disability.[96] Decreasing NAA in relapsing

Table 8. Evaluating the effect of therapy for MS by MRI

Therapy and Dosage	Effect on T-2 Lesion Burden	New T-2 or Enhancing Lesions	Clinical Relapse Rate Versus Placebo
Beta-IFN 1b[75, 103] (Betaseron)		75% reduction in new T2 lesions versus placebo	34% decrease
8 MIU qod	9.3% decrease		
placebo	15.0% increase		
Beta-IFN 1a (Avonex)[76] 30 mg weekly	No significant difference at 1 or 2 years		18% decrease
Beta-IFN 1a (Rebif)[14]			
22 μg × 3/week	1.2% reduction	67% decrease	27% decrease
44 μg × 3/week	3.8% reduction	78% decrease versus placebo	33% decrease
placebo	10.9% increase		
Glatiramer acetate[12, 184] (Copaxone) 20 mg/day	40% reduction in mean percentage change	29% reduction in mean total number of Gd-enhanced lesions at 9 months	29% decrease

N/A= not available; MIU= million international units; qad= every other day

remitting MS patients correlates better than T2 disease burden with clinical disability.[97] NAA is reduced in normal appearing white matter of MS patients, more so in secondary progressive disease than relapsing remitting disease.[97] Magnetic resonance spectroscopy may become the tool to monitor axonal destruction, thus clinical disability, in future trials.

Brain Atrophy

Early computerized tomography scans provided the first evidence that brain atrophy occurs in MS. Brain and spinal cord atrophy is thought to represent irreversible tissue injury related to demyelination, axonal injury, both or other yet to be identified features. Measurement of atrophy is difficult because of wide variability between individuals. Changes are small, often gradual and external factors such as dehydration, corticosteroid use, and alcohol may contribute to brain atrophy. Edema and inflammation may increase brain volume. Relative and absolute measures have been developed to measure brain atrophy. Absolute measures include ventricle width or volume to measure cerebrospinal fluid space. Brain parenchymal fraction (BPF) measures the ratio of the whole brain parenchyma to the total volume of the brain as measured by the outer surface contour of the brain.[71] Using this technique, atrophy was evident in early relapsing-remitting MS. A correlation between EDSS and BPF was present at baseline, but progression of atrophy and change in EDSS were not correlated. T1 hole volume and T2 lesion load also correlated with baseline BPF. The correlation between atrophy and MR measures such as enhancing lesions was poor in that study.[71] Two studies have shown a positive correlation between ventricle size and number and frequency of Gd-enhanced lesions.[69,70] While these findings are conflicting, it may be that inflammatory disease as measured by T1 weighted Gd-enhanced images reflects a different pathologic process than that which is measured by cerebral atrophy. Another technique to measure atrophy, the brain boundary shift integral, uses image coregistration to measure the difference in brain volume between two MRI scans at different times. MS patients showed more than twice as much atrophy and five times greater ventricular enlargement as compared to normal controls. Change in brain or ventricular volume did not correlate with change in disability. This may be related to a small change clinically, or the inability to measure physical or cognitive change by usual clinical evaluation. Large scale use of these techniques to measure brain atrophy is not yet possible as they require special segmentation routine, vary with different pulse sequences and require normal controls for comparison. Because of its potential to reflect axonal injury, use of this technique may be more useful to assess for disease progression than current available techniques.

Spinal Cord Atrophy

Initial studies of cord area at C5, T1, T7 and T11 by manual techniques showed a correlation with the EDSS.[82] Technically, the procedure was difficult. More recently, a semi-automated method to measure spinal cord diameter at C-2 has been shown to correlate with the EDSS and disease duration.[99] This patient group had progressive decline in cord area at C-2, but this did not correlate with progression in the EDSS.[100] Cord diameter was smaller in secondary progressive multiple sclerosis than relapsing-remitting multiple sclerosis, but progressed more quickly in relapsing-remitting multiple sclerosis.[100]

Summary

Currently, multiple modalities are available to monitor disease activity in MS. Unfortunately, there is not one marker or tool that becomes a surrogate marker for clinical trials. Continued search for the optimal clinical, neurophysiologic or MRI parameter is necessary. Because MS is a heterogeneous process, continued use of multiple tools will be necessary. Specifically, clinical evaluations will expand on cognitive and autonomic parameters. MRI techniques will focus on individual aspects of the disease, such as edema, inflammation, demyelination or axonal loss. Incorporation of multiple MRI techniques into clinical trials may be necessary to provide evidence of effectiveness and define the mechanism of therapeutic benefit. Body fluid markers may allow for subcategorizing disease types and therefore define appropriate treatment choices. Fortunately, the expansion of these tools has been rapid in recent years and leads to greater confidence that ultimately the cause, and therefore appropriate treatment of this illness, will be identified.

References

1. Hobart JC, Freeman JA, Lamping DK. The evaluation of outcome measurement instruments. MS Management 1995; 2:6-12.
2. International Federation of Multiple Sclerosis Societies. M.R.D. Minimal Record of Disability of multiple sclerosis. New York: National Multiple Sclerosis Society, 1985.
3. World Health Organization (WHO). International classification of impairments, disabilities and handicaps. Geneva: WHO, 1980.
4. Kurtzke JF. A new scale for evaluating disability in multiple sclerosis. Neurology 1955; 5:580-3.
5. Kurtzke, JF. Rating neurologic impairment in multiple sclerosis: An Expanded Disability Status Scale (EDSS). Neurology 1983; 33:1444-1452.
6. Whitaker JN, McFarland HF, Rudge P et al. Outcomes assessment in multiple sclerosis clinical trials: A critical analysis. Multiple Sclerosis 1995; 1:37-47.
7. Goodkin DE, Cookfair D, Wende K et al. Inter- and intra rater scoring agreement using grades 1.0 to 3.5 of the Kurtzke expanded disability status scale (EDSS). Neurology 1992; 42:859-863.
8. Francis DA, Bain P, Swan AV et al. An assessment of disability rating scales used in multiple sclerosis. Arch Neurol 1991; 48:299-301.
9. Noseworthy JH, Vandervoort MK, Wong CJ et al. Inter-rater variability with the Expanded Disability Status Scale (EDSS) and Functional Systems (FS) in multiple sclerosis clinical trials. Neurology 1990; 40:971-975.
10. Koziol JA, Frutos A, Sipe JC et al. A comparison of two neurologic scoring instruments for multiple sclerosis. J Neurol 1996; 243:209-213.
11. Rudick R, Antel J, Confavreux C et al. Clinical outcomes assessment in multiple sclerosis. Ann Neurol 1996; 40:469-479.
12. Johnson KP, Brooks BR, Cohen CC et al. Copolymer/reduces relapse rate and improves disability in relapsing-remitting multiple sclerosis: Results of a phase III multicenter, double-blind, placebo-controlled trial. Neurology 1995; 45:1268-1276.
13. Jacobs LD, Cookfair DL, Rudick RA et al. Intramuscular interferon Beta-1a for disease progression in relapsing multiple sclerosis. Ann Neurol 1996; 39:285-294.
14. PRISMS. Prevention of relapses and disability by Interferon B-1a subcutaneously in multiple sclerosis study group. Randomized double-blind, placebo-controlled study of Interferon B-1a in relapsing-remitting multiple sclerosis. Lancet 1998; 352:1498-1504.
15. Rudick RA, Medendorp SV, Namey M et al. Multiple sclerosis progression in natural history study: Predictive value of cerebrospinal fluid free kappa light chains. Multiple Sclerosis 1995; 1:150-155
16. Sipe JC, Knobler RL, Braheny SL et al. A neurological rating scale (NRS) for use in multiple sclerosis. Neurology 1984; 34:1368-1372.
17. Hauser SL, Dawson MD, Lehrich JR et al. Intensive immunosuppression in progressive multiple sclerosis. N Engl J Med 1983; 308:173-180.
18. Willoughby E. Impairment in multiple sclerosis. MS Management 1995; 2:13-16.
19. Ravnborg M, Gronbech-Jensen M, Jonsson A. The MS Impairment Scale: A pragmatic approach to assessment of impairment in patients with multiple sclerosis. Multiple Sclerosis 1997; 3:31-42.
20. Rudick R, Antel J, Confavreux C et al. Recommendations from the National Multiple Sclerosis Society Clinical Outcomes Assessment Task Force. Ann Neurol 1997; 42:379-382.
21. Fischer JS, Jak AJ, Kniker JE et al. Administration and scoring manual for the Multiple Sclerosis Functional Composite Measure (MSFC). New York: Demos Medical Publishing, 1999.
22. Kaufman M, Moyer D, Norton J. The significant change for the timed 25-foot walk in multiple sclerosis functional composite. Multiple Sclerosis 2000; 6: 286-290.
23. Mathiowetz V, Weber K, Kashman N et al. Adult norms for Nine Hole Peg Test of finger dexterity. Occ Ther J Res 1985; 5:24-38.
24. Mathiowetz V, Volland G, Kashman N et al. Adult norms for box and block test of manual dexterity. Am J Occ Ther 1985; 39:386-391.
25. Wade DT. Measurement in neurologic rehabilitation. New York: Oxford Medical Publications, 1992.
26. Wade DT. Measuring arm impairment and disability after stroke. Intl Disability Studies 1989; 11:89-92.
27. Goodkin DE, Hertsgaard D, Senuary J. Upper extremity function in multiple sclerosis. Improving assessment sensitivity with box-and-block and nine-hole peg tests. Arch Phys Med Rehabilitation 1988; 69:850-854.
28. Gronwall DM. Paced auditory serial addition task: A measure of recovery from concussion. Percept Motor Skills 1977; 44:367-373.
29. Egan V. PASAT: Observed correlations with IQ. Personality and Individual Differences 1988; 9:179-180.
30. Litvan I, Grafman J, Vendrell P et al. Slowed information processing in multiple sclerosis. Arch Neurol 1988; 45:281-285.
31. Rao SM, Leo GJ, Bernardin L et al. Cognitive dysfunction in multiple sclerosis. I Frequency, patterns and prediction. Neurology 1991; 41:685-691.
32. Heaton RK, Nelson LM, Thompson DS et al. Neuropsychological findings in relapsing-remitting and chronic-progressive multiple sclerosis. J Consult Clin Psychol 1985; 53:103-110.
33. Rao SM, Leo GJ, Ellington L et al. Cognitive dysfunction in multiple sclerosis. II. Impact on employment and social functioning. Neurology 1991; 41:692-696.
34. Amato MP, Ponziani G, Siracusa G et al. Cognitive dysfunction in early-onset multiple sclerosis. A reappraisal after ten years. Arch Neurol 2001; 58:1602-1606.
35. Fischer JS, Jacobs LD, Cookfair DL et al. Heterogeneity of cognitive dysfunction in multiple sclerosis. Clin Neuropsychol 1998; 12:286.
36. Rao SM, Leo GJ, Haugton VM et al. Correlation of magnetic resonance imaging with neuropsychological testing in multiple sclerosis. Neurology 1989; 39:161-166.
37. Hohol M, Guttmann CRG, Orav J et al. Serial neuropsychological assessment and magnetic resonance imaging analysis in multiple sclerosis. Arch Neurol 1997; 54:1018-1025.
38. van Buchem MA, Grossman RI, Armstrong C et al. Correlation of volumetric magnetization transfer imaging with clinical data on multiple sclerosis. Neurology 1998; 50:1609-1617.
39. Rovaris M, Fillipi M, Falautano M et al. Relation between MR abnormalities and patterns of cognitive impairment in multiple sclerosis. Neurology 1998; 50:1601-1608.
40. Fischer JS, Priore RL, Jacobs LD et al. Neuropsychological effects of interferon beta 1a in relapsing multiple sclerosis. Ann Neurology 2000; 48:885-892.
41. Pliskin NH, Hamer DP, Goldstein DS et al. Improved delayed visual reproduction test performance in multiple sclerosis patients receiving interferon, 1b. Neurology 1996; 47:1463-1468.
42. Weinstein A, Schwid SR, Schiffer RB et al. Neuropsychological status in multiple sclerosis after treatment with glatiramer acetate (Copaxone). Arch Neurology 1999; 56:319-324.
43. Fischer J. Assessment of neuropsychological function in Multiple Sclerosis Therapeutics. Rudick R, Goodkin D, eds. London: Martin Dunitz Ltd, 1999:31-47.
44. McDonald WI, Compston A, Edan G et al. Recommended diagnostic criteria for multiple sclerosis: Guidelines from the International Panel on the Diagnosis of Multiple Sclerosis. Ann Neurol 2001; 50:121-127.
45. Nuwer MR, Packwood JW, Myers LW et al. Evoked potentials predict the clinical changes in multiple sclerosis drug study. Neurology 1987; 37:1754-1761.
46. Mertin J, Rudge P, Kremer M et al. Double-blind controlled trial of immunosuppression in the treatment of multiple sclerosis: Final report. Lancet 1982; 2:351-354.
47. Smith T, Zeeberg I, Sjo O. Evoked potentials in multiple sclerosis before and after high-dose methylprednisolone infusion. Eur Neurol 1986; 25:67-73.

48. Thompson AJ, Hutchinson M. Martin IA et al. Suspected and clinically definite multiple sclerosis: The relationship between CSF immunoglobulins and clinical course. J Neurol Neurosurg Psychiatry 1985; 48:989-994.
49. Frequin STFM, Lamers KJB, Barkhof F et al. Follow-up study of MS patients treated with high-dose intravenous methylprednisolone. Acta Neurol Scand 1994; 90:105-110.
50. Lamers KKJB, diReres HPM, Jongen PJH. Myelin basic protein in CSF as indicator of disease activity in multiple sclerosis. Multiple Sclerosis 1998; 4:124-126.
51. Barkhoff F, Frequin STFM, Hommes OR et al. A correlative trial of gadolinium-DTPA MRI, EDSS and CSF MBP in relapsing multiple sclerosis patients treated with high dose intravenous methylprednisolone. Neurology 1992; 42:63-67.
52. Whitaker JN, Layton BA, Herman PK et al. Correlation of myelin basic protein-like material in cerebrospinal fluid of multiple sclerosis patients with their response to glucocorticoid treatment. Ann Neurology 1993; 33:10-17.
53. Stone LA, Richert N, McFarland HF. Neuroimaging and the use of magnetic resonance in multiple sclerosis. In: Cook S, ed. Handbook of Multiple Sclerosis. New York: Marcell Decker; 2001:403-431.
54. Brownell B, Hughes JT. The distribution of the plaques in the cerebrum in multiple sclerosis. J Neurol Neurosurg Psychiatry 1962; 25:315-320.
55. Hajnal TV, Bryant DJ, Kasuboski L et al. Use of fluid attenuated inversion recovery (FLAIR) pulse sequences on MRI of the brain. JCAT 1992; 16:841-844.
56. Kermode AG, Tofts PS, Thompson AJ et al. Heterogenicity of blood-brain barrier changes in multiple sclerosis: An MRI study with gadolinium-DTPA enhancement. Neurology 1990; 40(2):229-235.
57. Bruck W, Bitsch A, Kolenda H et al. Inflammatory central nervous system demyelination: Correlation of magnetic resonance imaging findings with lesion pathology. Ann Neurol 1997; 42:783-793.
58. Oppenheimer DR. The cervical cord in multiple sclerosis. Neuropath Appl Neurobiol 1978; 4:151-162.
59. Miller DH, Albert PS, Barkhof F et al. Guidelines for the use of magnetic resonance techniques in monitoring the treatment of multiple sclerosis. Ann Neurol 1996; 39:6-16.
60. Filippi M, Horsfield MA, Ader HJ et al. Guidelines for using quantitative measures of brain magnetic resonance imaging abnormalities in monitoring the treatment of multiple sclerosis. Ann Neurol 1998; 43:499-506.
61. McFarland HF, Frank JA, Albert PS et al. Using gadolinium-enhanced magnetic resonance imaging lesions to monitor disease activity in multiple sclerosis. Ann Neurol 1992; 32(6):758-766.
62. Lai HM, Hodgson T, Guwne-Cain M et al. A preliminary study into the sensitivity of disease activity detection by serial weekly magnetic resonance imaging in multiple sclerosis. J Neurol Neurosurg Psychiatry 1996; 60:334-341.
63. Guttman CRG, Ahn SS, Hsu L et al. The evolution of multiple sclerosis lesions on serial MR. AJNR 1995; 16:1481-1491.
64. Miller DH, Barkhof F, Nauta JJ. Gadolinium enhancement increases the sensitivity of MRI in detecting disease activity in multiple sclerosis. Brain 1993; 116:1077-1094.
65. Molyneux PD, Filipi M, Burhof F et al. Correlations between monthly enhanced MRI lesion rate and changes in T2 lesion volume in multiple sclerosis. Ann Neurol 1998; 43:332-339.
66. Kappos L, Moeri D, Radue EW et al. Predictive value of gadolinium-enhanced magnetic resonance imaging for relapse rate and changes in disability or impairment in multiple sclerosis: A meta-analysis. Gadolinium MRI Meta-Analysis Group. Lancet 1999; 353:964-969.
67. Harris JO, Frank JA, Patronas N et al. Serial gadolinium-enhanced magnetic resonance imaging scans in patients with early, relapsing-remitting multiple sclerosis: Implications for clinical trials and natural history. Ann Neurol 1991; 29:548-555.
68. Goodkin DE, Rudick RA, VanderBrug-Medendorps et al. Low dose oral methotrexate in chronic progressive multiple sclerosis: Analyses of serial MRIs. Neurology 1996; 47:1153-1157.
69. Simon JH, Jacobs LD, Campion M et al. brain atrophy in relapsing multiple sclerosis. The Multiple Sclerosis Collaborative Research Group (MS CRG). Neurology 1999; 53:139-148.
70. Leist TP, Gobbini Ml, Frank JA et al. Enhancing magnetic resonance imaging lesions and cerebral atrophy in patients with relapsing multiple sclerosis. Arch Neurol 2000; 57:57-60.
71. Rudick RA, Fisher E, Lee JC et al. Use of the brain parenchymal fraction to measure whole brain atrophy in relapsing-remitting MS. Neurology 1999; 53:1698.
72. Filippi M, Yousry T, Campi A et al. Comparison of triple dose versus standard dose gadolinium-DTPA for detection of MRI enhancing lesions in-patients with MS. Neurology 1996; 46:379-384
73. Barkhof F. Hommes OR, Scheltens P et al. Quantitative MRI changes in gadolinium-DPTA enhancement after high-dose intravenous methylprednisolone in multiple sclerosis. Neurology 1991; 41(8):1219-1222.
74. Filippi M, Paty DW, Kappos L et al. Correlations between changes in disability and T2-weighted brain MRI activity in multiple sclerosis: A follow-up study. Neurology 1995; 45:255-260.
75. Paty DW, Li DK. Interferon beta-1b is effective in relapsing-remitting multiple sclerosis. II. MRI analysis results of a multicenter, randomized, double blind, placebo-controlled trial. UBC MS/MRI Study Group and the IFNB Multiple Sclerosis Study Group. Neurology 1993; 43:662-667.
76. Simon JH, Jacobs LD, Campion M et al. Magnetic resonance studies of intramuscular interferon β-1a for relapsing multiple sclerosis. Ann Neurol 1998; 43:79-87.
77. Wiebe S, Lee DH, Karlik SJ et al. Serial cranial and spinal cord magnetic resonance imaging in multiple sclerosis. Ann Neurol 1992; 32:643-650.
78. Miller DH, Rudge P, Johnson G et al. Serial gadolinium enhanced magnetic resonance imaging in multiple sclerosis. Brain 1988; 111:927-939.
79. Truyen L, van Waesberghe JHTM, van Walderveen MAA et al. Accumulation of hypointense lesions ("black holes") on T1 spin echo MRI correlates with disease progression in multiple sclerosis. Neurology 1996; 47:1469-1476.
80. van Walderveen MAA, Barkhot F, Hommes OR et al. Correlating MRI and clinical disease activity in multiple sclerosis. Relevance of hypointense lesions in short-TR/short-TE (T1-weighted) spin echo images. Neurology 1995; 45-1684-1690.
81. Thompson AJ, Polman CH, Miller DH et al. Primary progressive multiple sclerosis. Brain 1997; 120:1085-1096
82. Kidd D, Thorpe JW, Thompson AJ et al. Spinal cord MRI using multi-array coils and fast spin echo. II. Findings in multiple sclerosis. Neurology 1993; 43:2632-2637.
83. The IFNB Multiple Sclerosis Study Group and the University of British Columbia. MS/MRI Analysis Group. Interferon beta-1b in the treatment of multiple sclerosis: Final outcome of the randomized controlled trial. Neurology 1995; 45:1277-1285.
84. Comi G, Fillippi M, Wolinsky JS and European/Canadian Glatiramer Acetate Study Group. The European/Canadian multicenter, double-blind, randomized, placebo controlled study of the effects of glatiramer acetate on magnetic resonance imaging-measured disease activity and burden in patients with relapsing multiple sclerosis. Ann Neurol 2001; 49:290-297.
85. Coles A, Puolili A, Molyneux P et al. Monoclonal antibody treatment exposes three mechanisms underlying the clinical course of multiple sclerosis. Ann Neurol 1999; 46:296-304.
86. Dousset V, Grossman RI, Ramer KN et al. Experimental allergic encephalomyelitis and multiple sclerosis: Lesion characterization with magnetization transfer imaging. Radiology 1992; 182:483-491.
87. Fillippi M, Campi A, Dousset V et al. A magnetization transfer imaging study of normal-appearing white matter in multiple sclerosis. Neurology 1995; 45:478-482.
88. Fillippi M, Rocca MA, Martino G et al. Magnetization transfer changes in the normal appearing white matter precede the appearance of enhancing lesions in patient with multiple sclerosis. Ann Neurol 1998; 43:809-814.

89. Hiehle JF Jr, Grossman RI, Rumer KN et al. Magnetization transfer effects in MR-detected multiple sclerosis lesions. Comparison with gadolinium-enhanced spin-echo images and nonenhanced T1-weighted images. AJNR 16:69-77.
90. Gass A, Barker GJ, Kidd D et al. Correlation of magnetization transfer ratio with clinical disability multiple sclerosis. Ann Neurol 1994; 36:62-67.
91. van Buchem MA, Udupa JK, McGowan JC et al. Global volumetric estimation of disease burden in multiple sclerosis based on magnetization transfer imaging. AJNR Am J Neuroradiol 1997; 18:1287-1290.
92. Richert ND, Ostuni JL, Bash CN et al. Serial whole brain magnetization transfer imaging in patients with relapsing-remitting multiple sclerosis at baseline and during treatment with interferon beta 1b. AJNR Am J Neuroradiol 1998; 19:1705-1713.
93. McDonald WI, Rachelle Fishman-Matthew Moore Lecture. The pathological and clinical dynamics of multiple sclerosis. J Neuro Pathol Exp Neurol 1994; 53:338-343.
94. Arnold DL, Wolinsky J, Matthews PM et al. The use of magnetic resonance spectroscopy in the evaluation of the natural history of multiple sclerosis. J Neurol Neurosurg Psychiatry 1998; 64(suppl1):594-5101.
95. Clanet M, Berry I. Magnetic resonance imaging in multiple sclerosis. Curr Opin Neurol 1998; 11:299-303.
96. Matthews PM, DeStefano N, Narayanan S et al. Putting magnetic resonance spectroscopy studies in context: Axonal damage and disability in multiple sclerosis. Semin Neurol 1998; 18:327-336.
97. DeStefano N, Matthews PM, Fu L et al. Axonal damage correlates with disability in patients with relapsing-remitting multiple sclerosis: Results of a longitudinal magnetic resonance spectroscopy study. Brain 1998; 121:1469-1477.
98. Fox NC, Jenkins R, Leary SM et al. Progressive cerebral atrophy in MS. A serial study using registered, volumetric MRI. Neurology 2000; 54:807-812.
99. Stevenson VL, Leary SM, Lossef NA et al. Spinal cord atrophy and disability in MS. A longitudinal study. Neurology 1998; 51:234-238.
100. Liu C, Edwards S, Gong Q et al. Three-dimensional MRI estimates of brain and spinal cord atrophy in multiple sclerosis. J Neurol Neurosurg Psychiatry 1999; 66:323-330.
101. Barkhof F, Filippi M, Miller DH et al. Comparison of MR imaging criteria at first presentation to predict conversion to clinically definite multiple sclerosis. Brain 1997; 120:2059-2069.
102. Tintore' M, Rovira A, Martinez M et al. Isolated demyelinating syndromes: comparison of different MR imaging criteria to predict conversion to clinically definite multiple sclerosis. Am J Neuroradiol 2000; 21:702-706.
103. The IFNB Multiple Sclerosis Study Group. Interferon beta-1b is effective in relapsing-remitting multiple sclerosis. I. Clinical results of a multicenter, randomized, double-blind, placebo-controlled trial. Neurology 1993; 43:655-661.

CHAPTER 36

Intense Immunosuppression Followed by Autologous Stem Cell Transplantation in Severe Multiple Sclerosis Cases: MRI and Clinical Data

G.L. Mancardi, R. Saccardi, A. Murialdo, F. Pagliai, F. Gualandi, A. Marmont, M. Inglese, P. Bruzzi, M.P. Sormani, M.G. Marrosu, G. Meucci, L. Massacesi, A. Bertolotto, A. Lugaresi, E. Merelli, M. Filippi and the Italian Gitmo-Neuro Intergroup on ASCT for Multiple Sclerosis

Introduction

Multiple sclerosis (MS) is an autoimmune disease characterized by attacks against components of the myelin sheath of the central nervous system (CNS) mediated by immuno-competent cells. In the early stages, distinct relapses are followed by remission and are often trailed by a chronic phase of disease with a progressive clinical course. Inflammation and demyelination are the most relevant physiopathological events of this disorder. Later in the disease course, axons, already damaged and devoid of the enveloping myelin sheath, degenerate through unknown mechanisms causing progressive worsening of the neurological condition.

Currently available therapies, namely interferons and glatiramer acetate, are only of partial and modest efficacy, especially in more severe cases and in the progressive phase of the disease. The utility of conventional immunosuppressive drugs (such as azathioprine or methotrexate) is not convincing. Only mitoxantrone has a demonstrated effect on magnetic resonance imaging (MRI) activity[1] and the capacity to slow the progression of disability in progressive MS which follows a relapsing-remitting clinical course (secondary progressive MS)[2] and is the only immunosuppressive drug approved by the FDA for the treatment of MS. Mitoxantrone has, however, several drawbacks, including the fact that it is of limited efficacy and many patients continue to worsen in spite of treatment. Therefore, the search for more effective immunosuppressive therapies is necessary and must be firmly pursued.

Intense immunosuppression followed by autologous stem cell transplantation (ASCT) is a treatment of particular interest in autoimmune disorders, especially in MS.[3-7] After the pioneering work of Fassas et al[8] and Burt et al,[9] we established an Italian group of neurological and hematological teams (Gitmo-Neuro intergroup on ASCT for MS), whose aim was to evaluate whether intense immunosuppression followed by ASCT has the capacity to abrogate the inflammatory phase of the disease, detectable on MRI, and if the duration of this effect is sustained for a period of time which justifies its toxicity and potential risk of death. Preliminary data on the first 10 cases have been elsewhere reported.[9] Presently we describe the clinical and MRI findings of an additional 6 cases. The median follow up period is now long enough (24 months) to allow for some general considerations regarding the application of this method in MS cases unresponsive to conventional therapies.

Three additional severe MS cases, which did not fulfill inclusion criteria for the frequent MRI study, were treated with the identical therapeutic schedule during the same period. The MRI and clinical data of these cases are described separately in the present report.

Results from the Italian Intergroup Study

Frequent MRI Cohort

This phase II study started in July 1998. Seven neurological and hematological centers of Italy combined efforts, with the aim to investigate the effect of intense immunosuppression followed by ASCT on magnetic resonance imaging (MRI) and clinical course in severe MS cases, unresponsive to conventional therapies.

The inclusion criteria were clinically and laboratory definite secondary progressive MS with or without relapses, Expanded Disability Status Scale (EDSS) score between 5 and 6.5, documented rapid progression over the previous year unresponsive to conventional therapies (worsening of 1 point between EDSS 5 and 6, or of 0.5 between 6 and 6.5), age between 18 and 55, absence of cognitive disturbances and the presence of at least one enhancing lesion on brain MRI, using a triple dose (TD) of gadolinium-diethylenetriamine penta-acetic acid (Gd), (0.3 mmol/kg of Gadodiamide, Omniscan, Nycomed Amersham). Ethical Committees approved the study. Patients' cognitive functioning was measured utilizing the Mini Mental Status Examination and a battery of standard neuropsychological tests. All subjects scored within the standard normal ranges on these measures at the time the informed consent was signed.

Sixteen cases have undergone ASCT with a median follow up of 24 months (range 4-42 months). Ten subjects were female and 6 male, with a median age of 34.5 years (range 26-52), median disease duration of 12 years (range 6-19) and a secondary progressive course for a median of 3 years (range 1-8). Relapses were superimposed in 7 of 16 cases. Ten subjects had been previously treated with interferon beta 1b, and 6 subjects with immunosuppressive therapies only (azathioprine, cyclophosphamide or mitoxantrone) or steroids, all without clinical response. At screening, the median EDSS was 6.5 (range 5.5-6.5) and the Scripps Neurological Rating Scale (SNRS) was 59 (range 41-66). At baseline, just before stem cell mobilization, the neurological con-

---

Stem Cell Therapy for Autoimmune Disease, edited by Richard K. Burt and Alberto M. Marmont. ©2004 Landes Bioscience/Eurekah.com.

Table 1. Patient's characteristics

Patient/Gender	Age	Clinical Course	Disease Duration (Years)	Duration of Progressive Phase (Years)	Follow-up after ASCT (Months)
1/F	37	SP	12	7	42
2/F	52	SP with relapses	6	2	30
3/F	32	SP with relapses	10	2	30
4/F	33	SP	16	6	24
5/M	35	SP	12	4	24
6/F	37	SP with relapses	17	8	24
7/M	32	SP	SP	3	24
8/M	38	SP with relapses	14	5	24
9/F	50	SP	19	6	24
10/M	26	SP	11	1	15
11/M	29	SP	12	3	9
12/F	36	SP	11	5	15
13/M	28	SP with relapses	8	2	6
14/F	34	SP	12	1	12
15/F	37	SP	15	4	6
16/M	34	SP	12	6	6

Legends: F= female; M= male; SP= secondary progressive

dition had deteriorated in 6 cases (median EDSS was 6.5, range 5.5-8). Patient characteristics are reported in Table 1. The effect of ASCT was evaluated with serial monthly TD Gd-enhanced MRI for a pretreatment period of three months, and compared with serial monthly TD Gd-enhanced MRI imaging for the following 6 months. Subsequently, MR scans were obtained every three months until month 24 and then every 6 months. TD was utilized instead of a standard dose of Gd, since TD-enhanced MRI is much more sensitive in detecting disease activity.[10] The MRI protocol was standardized in all Centers and was performed according to established criteria.[9] All images were hard copied using a laser imager and stored on magnetic tapes. Images were sent to the Neuroimaging Research Unit, Scientific Institute and University, S. Raffaele Hospital, Milan, where they were examined and quantified by an experienced observer, blind to the period in which they were obtained, who counted the number of Gd-enhancing lesions and the number of new T2-weighted lesions. The images were then sequentially ordered, obtaining the number of Gd-enhancing and of T2-positive lesions before and after treatment.

Neurological examinations (EDSS, Scripps) were scheduled at screening, baseline, before and after the mobilization regimen, after the conditioning regimen, every month for the following 6 months and then every three months until 24 months. Subsequently, patients were examined every 6 months. Peripheral blood progenitor cells (PBPCs) were mobilized with cyclophosphamide (CY) (4 gr/m^2) followed by Filgrastim 5 μg/kg/day until the completion of the cell harvests. The immunoablative regimen was carried out within 30-40 days after mobilization. According to the protocol previously utilized by Fassas et al,[8] the regimen consisted of BCNU 300 mg/m^2 at day –7; Cytosine-Arabinoside 200mg/m^2 and Etoposide 200 mg/m^2 from day –6 to day –3 and Melphalan 140 mg/m^2 at day –2. Rabbit ATG (Thymoglobulin, Sangstat), 5 mg/kg/day was administered at +1 and +2 as in vivo T-depletion. Intravenous cyclosporin A (1 mg/kg) was given during the conditioning regimen to prevent disease exacerbation due to cytokine release.

Mobilization was successful in all cases and generally well tolerated and a median number of 9.06x10^6/kg of CD34$^+$ cells were collected (range 3.51-26.02x10^6/kg). However, adverse effects were observed in some cases: one patient experienced subclavian phlebitis, one transient inappropriate secretion of ADH and a third hemorrhagic cystitis. This last patient (n 13) was treated in the previous year with CY 1 gr. IV every month for 6 months, without significant clinical results. The hemorrhagic cystitis was severe and lasted almost 2 months. Nadir of polymorphonuclear cells (PMN) occurred 8 days after mobilization (range 7-11 days) and of platelets (Plt) on day 10 (range 3-13). Median days with PMN<0.5x10^9/L and Plt<50x10^9/L were 4 (range 2-4) and 0 (range 0-4), respectively.

Adverse effects were common following the transplantation. Fever occurred in the neutropenic post-stem cell (SC) infusion

Table 2. MRI TD Gd-enhancing lesions

	Months Before ASCT					Months After ASCT															
Pt	Sc	-3	-2	-1	30 days after CY	BEAM	1	2	3	4	5	6	9	12	15	18	21	24	30	36	42
1	2	6	8	29	10		0	0	0	0	0	0	0	0	0	0	0	0	0	0	0
2	7	8	14	28	12		0	0	0	0	0	0	0	0	0	0	0	0	0		
3	1	3	11	38	0		0	0	0	0	0	0	0	0	0	0	0	0			
4	nd	nd	4	2	1		0	0	0	0	0	0	nd	0	0	0	nd	0			
5	5	7	7	18	13		1	1	1	0	0	0	0	0	0	0	0	0			
6	7	6	17	23	7		0	0	0	0	0	0	0	0	0	0	0	0			
7	7	6	7	2	1		0	0	0	0	0	0	0	0	0	0	0	0			
8	nd	12	2	3	0		0	0	0	0	0	0	0	0	0						
9	6	5	2	2	0		0	0	0	0	0	0	0	0	0	0					
10	nd	4	10	22	6		2	0	0	0	0	0	0	0	0						
11	nd	8	10	10	15		0	0	0	0	0	0	0								
12	nd	17	8	3	0		0	0	0	0	0										
13	nd	36	26	15	0		0	0	0	0											
14	nd	nd	66	27	nd		0	nd	0	0	0	0	0								
15	nd	1	0	0	0		0	0	0	0											
16	nd	12	9	9	3		0	0	0	0	0	0									

nd= not done; Pt= patient; Sc= screening visit; TD Gd= triple dose gadolinium

period in 15 out of 16 patients, successfully treated with IV antibiotic therapy. In 4 cases a urinary tract infection, in 3 cases a mucous enteritis and in 5 cases a reactivation of CMV occurred within 15-35 days post transplantation. In 3 patients, CMV reactivation was symptomatic, with fever, arthralgia and diarrhea, requiring treatment with ganciclovir. Herpes Zoster appeared in 3 cases and was successfully treated with acyclovir. A median number of 6.85×10^6/kg of CD34$^+$ cells were infused (range $3.01\text{-}16.46 \times 10^6$/kg). Following transplant, median days with PMN<0.5×10^9/L and Plt<50×10^9/L were 7 (range 6-12) and 9 (range 6-14) respectively. Infusion of peripheral blood progenitors and rabbit ATG was well tolerated with minor adverse effects. In one case, however, ATG infusion was followed by high fever, cutaneous rash, and diffuse myalgia. The patient was treated with steroids, with complete recovery. In general, due to fever and concomitant infections with fever, neurological conditions deteriorated in the first 1 or 2 months and severe asthenia was common. This worsening was not due to a flare of disease or to the appearance of new symptoms, but rather to fever and infections, and patients completely recovered with the improvement of their general conditions.

No adverse effects or laboratory abnormalities were recorded with the monthly prolonged use of TD Gd. During the 3 month pretreatment period, 598 Gd-enhancing lesions were detected in our series of patients. The mean number of Gd-enhancing lesions/month/patient in the pretreatment period was 12 (range 1-66). The number of Gd-positive lesions decreased dramatically, immediately after mobilization with CY, with complete disappearance of enhancing areas in 6 out of 16 cases 30 days after therapy with CY. In the month following the conditioning therapy, the number of Gd-enhancing lesions dropped to zero in 14 out of 16 cases. In two cases (see Table 2), new enhancing lesions were detected at month +1, or in the first 3 months after BEAM therapy, followed by complete suppression of MRI activity, which remained inactive for the following months. The number of Gd-enhancing lesions per month before, and after mobilization and following ASCT is detailed in Table 2.

The number of new T2-weighted positive lesions paralleled data previously described for Gd-enhanced MRI. In 14 out of 16 cases, no new T2 lesions were observed after ASCT. In case 5 and 10, the same cases in which MRI showed Gd-enhancing activity after conditioning regimen, one new T2 lesion was detected in each of the first 3 months or in the month after therapy. For comparison, 247 new T2-weighted lesions were detected in the three month pretreatment period. The difference between pre and post-treatment period was highly significant, both for Gd-enhancing lesions and for new T2-weighted positive lesions, with a $p < 0.000001$ using the Mantel Haenszel test.

Clinically, all patients slightly improved or remained stable. In two patients (case 1 and 4), after initial improvement which lasted for 30 and 9 months respectively, the clinical disturbances, mainly characterized by spastic paraparesis, resumed worsening and the patients currently have an EDSS of 6.5 and 7 as compared to 6.5 prior to therapy (Table 3). In the other cases, in addition to an MRI effect, a clinical effect was also evident and they improved or at least remained stable. Clinical data, before and after transplantation, are reported in Table 3.

Cases Not Included in the Frequent MRI Study Cohort

In the period 1998-2001 three additional severe MS cases were treated by the teams involved in the study and not included in the frequent MRI study because they did not fulfill inclusion criteria.

Case 1 is a 50 year old male, with a 19 years history of MS, and a secondary progressive clinical course for 15 years. He was

Table 3. Clinical Course: EDSS in the months before and after treatment

Pt	-12	-3	-2	-1	30 days after CY	BEAM	1	2	3	4	5	6	9	12	15	18	21	24	30	36	42
1	6	6.5	6.5	6.5	6.5		6.5	6.5	6	6	6	6	6	6	6	6	6	6	6	6,5	6,5
2	6	6.5	6.5	6.5	6		6	6	6	6	6	6	6	6	6	6	6	6			
3	6	6.5	8	8	7.5		7.5	7.5	7.5	7.5	7	7	7	7	7	7	7	7			
4	5	6.5	6.5	6.5	6.5		6	6	6	6	6	6	6	6.5	6,5	6,5	6,5	7			
5	4	5.5	5.5	5.5	5.5		5	5	4	4	4	4	4	4	4	4	4	4			
6	5	6.5	6.5	6.5	6.5		6.5	6	6	6	6	6	6	6	6	6	6				
7	5.5	6	6.5	6.5	6.5		6.5	6.5	6.5	6.5	6.5	6.5	6.5	6.5	6,5	6,5	6,5	6,5			
8	6	6.5	6.5	7	7		7	7	7.5	6.5	6.5	6.5	6.5	6,5	6,5	6,5					
9	5.5	6	6	6	6		6	6	6	6	6	6	6	6	6	6					
10	4.5	6	6.5	6.5	6.5		6.5	6.5	6.5	6.5	6,5	6,5	6,5	6,5	6,5						
11	4	5	5,5	5,5	5,5		5,5	5,5	5,5	5,5	5,5	5,5	5,5								
12	6	6,5	6,5	6,5	6,5		6	6	6	6	6										
13	6	6,5	6,5	7	7		6,5	6,5	6,5	6,5											
14	6	6,5	6,5	7	7		7	6,5	6,5	6,5		6,5	6,5	6,5							
15	6	6,5	6,5	6,5	6,5		6,5	6,5	6,5	6,5											
16	6	6,5	7	7,5	7,5		7,5	7,5	7	7	7	7									

previously treated with steroids, then with CY, without significant results. At screening, EDSS was 7, Ambulation index (AI) was 8 and SNRS was 45. No relapses occurred in the last 2 years, although there was a worsening on EDSS of 0.5 (from 6.5 to 7) in the previous year. MRI with TD of Gd (repeated after 1 month) did not show any enhancing lesion and, for this reason as well as for the high EDSS score, the patient was not included in the frequent MRI study. Following the patient's request and after careful evaluation of the neurological situation and possible therapeutic alternatives, he was treated with intense immunosuppression followed by ASCT, applying the same immunoablative regimen previously reported.

The mobilizing and conditioning regimen were tolerated well. In the postconditioning period, an acute enteritis with fever occurred and was successfully treated with antibiotics. The neurological condition worsened in the initial months after ASCT, mainly due to fever. Subsequently, the patient improved, with an EDSS of 7.5 four months after ASCT, compared to 7 at the premobilization examination. Currently, 16 months post transplantation, his EDSS remains at 7.5. MRI, which was negative for Gd-enhancing areas before treatment, was also negative after 1 year (with triple dose of Gd).

Case 2 is a 16 year old male who experienced a right hemiparesis of sudden onset when he was 15 years old. MRI showed a large T2 positive lesion in the left pons and midbrain and a number of hyperintense lesions in the hemispheric white matter. CSF examination showed oligoclonal bands in the CSF and not in the serum. At that time he was treated with steroids, followed by a severe relapse with cerebellar signs 2 months later. The patient then started therapy with interferon beta 1a but within the next 5 months he had 3 relapses with right hemianopsia, left hemiparesis and diplopia, and beta interferon was interrupted. He was treated with plasma-exchange without results and subsequently with CY (1 gram every month for 2 months). At the end of the second cycle, the patient relapsed again. MRI showed dramatic enlargement of the previous lesions and the appearance of new lesions after the administration of a single dose of Gd. At that time, partial epileptic seizures with secondary generalization occurred, and the patient was treated successfully with phenobarbital. When he reached an EDSS of 7.5, one year after the onset of symptoms, and following informed consent of the parents, the patient was treated with the same previously described regimen. He was not included in the frequent MRI study due to the age requirement.

Fever occurred in the post-conditioning period. The patient was confined to bed and a thrombosis of the deep cava vein occurred, with marked edema of the right leg. The patient was treated with anticoagulant therapy, first with heparin, then with dicumarol, with a gradual improvement of the edema and the appearance of efficacious collateral circulation. The neurological condition dramatically improved in the months following ASCT. Currently, 6 months after therapy, the patient can walk with a cane for 100 meters and is independent in activities of daily living. MRI showed a reduction in the size of previous lesions and no contrast enhancement was detected (single dose of Gd).

Case 3 is a 40 year old female who reported numbness in the lower limbs followed by visual loss in the right eye occurring three years earlier. Visual evoked potentials (VEPs) were delayed on both sides and MRI of the brain was normal. After a few months, she experienced loss of leg strength. Again MRI was negative, but cerebro-spinal fluid (CSF) examination showed oligoclonal bands in the CSF and serum, with additional bands in the CSF. The patient improved and remained in almost normal condition for 2 years, at which time she experienced a severe tetraparesis. MRI of the brain and spinal cord showed multiple T2 lesions areas, some of which enhanced after a single dose of Gd. After 1 month, paraparesis worsened and visual disturbances recurred. A diagnosis of neuromyelitis optica was established. The patient was treated with CY (1.5 grams IV) and then with plasma exchange, without any significant improvement. At that time, EDSS was 6.5. After 4 months, her neurological condition rapidly worsened, reaching an almost complete paraplegia. When the patient's EDSS was 8 it was decided to treat her with intense immunosuppression and ASCT, similarly to the other previously reported cases. The patient signed an informed consent. This patient was not included in the frequent MRI study because EDSS was higher than the upper limit established in the protocol.

A severe enteritis with subsequent paralytic ileus occurred within days following the conditioning regimen. The patient remained in a critical situation and needed an intensive support therapy for at least 5 days, then, with the normalization of the white blood count, gradually improved. Neurological status ameliorated as well. Currently, 6 months post ASCT, the patient can walk without aid. Her EDSS passed from 8 before therapy to 6 after treatment. MRI, performed 4 months post ASCT with single dose Gd, was negative for enhancing areas.

Conclusions from the Italian Intergroup Study

We have demonstrated in patients with severe MS, that the sequence of CY-BEAM-ASCT-ATG has the capacity to completely suppress inflammatory activity and that this effect was sustained for a median time of 15 months.[10] These 16 patients currently have a median follow-up of 2 years (range 4- 42 months). In all cases, lesion activity, determined with monthly MRI with TD of Gd for the first 6 months and then every three months, completely disappeared, within one month following the conditioning therapy in 14 out of 16 cases and in the subsequent months in the other 2 cases. This result is remarkable, considering that it was obtained in severe MS cases unresponsive to conventional therapies, with a significant degree of MRI activity in the months prior to ASCT (mean of 12 Gd-enhancing lesions/month, range 0-66) and that a TD of Gd was utilized, which detects 70-80% more lesions than the standard dose.[11] The areas of Gd-enhancement correspond to areas of breakdown of the blood-brain-barrier and inflammation. Therefore, our data indicate that intense immunosuppression, followed by ASCT, abrogates blood brain barrier damage and the inflammatory phase of the disease for a considerable length of time. The first three cases have now been followed for 3 years or more and MRI continues to be negative for the presence of enhancing areas.

Clinically, the majority of patients remained stable or ameliorated. All patients had an aggressive form of MS with deterioration of at least one point on EDSS in the previous year. The treatment apparently had the capacity to halt the deterioration of the neurological condition. However, in two cases, after initial improvement, which lasted 30 and 9 months, the disease resumed progression (Table 3). In these cases, however, MRI did not show any Gd enhancing area during the long period of follow-up. In general, patients who responded better were those with very rapid clinical worsening in the previous year and those with a very high number of Gd enhancing areas in the months prior to therapy.

Adverse events occurred in some cases following mobilization and were more common after transplant. The most significant complication after mobilization was severe hemorrhagic cystitis in case number 13. The hemorrhagic cystitis resolved two months after the mobilization procedure and consequently, in this case, the conditioning therapy was carried out, without complications, almost 3 months after mobilization. Post-ASCT toxicity was characterized by fever during the neutropenic period in almost all cases. ATG was well tolerated, although following the first infusion, fever, cutaneous rash and dyspnea occurred in one patient. The second infusion of ATG at day +2 was successfully carried out after increasing the steroid premedication. In the first month after transplantation, the neurological condition often deteriorated, due to asthenia, worsening of the general condition, or fever. A relatively high incidence of documented infections, including CMV reactivations, was shown, as compared to other high dose chemotherapy trials using unmanipulated HST as a rescue, possibly due to a profound immunosuppression following the procedure. Therefore, patients with a medical history of bacterial/fungal colonization and/or with severe disability have a high risk of serious transplant related complications and their inclusion in such an aggressive treatment should be discouraged.

An additional 3 patients were treated according to the same protocol as the other patients, although not included in the frequent MRI study because they did not fulfill inclusion criteria. The first of these cases, not included in the study for the absence of Gd-enhancing areas at MRI, apparently did not obtain any advantage from the treatment and showed worsened neurological status within a few months following therapy, without any further improvement. This patient's EDSS is now 7.5 as compared to 7 prior to treatment. In the other two cases, improvement after therapy was dramatic. Both cases were characterized by a brief clinical history (1 year and 3 years from onset, respectively), severe clinical course with frequent relapses without improvement after each relapse (relapsing-progressive cases), unresponsiveness to traditional therapy and Gd-enhancing lesions on MRI. Currently, at 6 months post-conditioning, both patients can walk for 100 and 300 meters, respectively, with unilateral aid, while their pretreatment EDSS were 7.5 and 8, respectively. While it is not possible to formally exclude a spontaneous improvement in these cases, this eventuality is, in our opinion, very unlikely. Indeed, case 2 experienced relapses almost every 2 or 3 months from onset, without recovery or improvement after each relapse, and with enlargement and appearance of new lesions on every MRI scan. After treatment, no relapse occurred and for the first time the patient's neurological condition improved, while MRI showed a reduction in size of preexisting lesions without the appearance of new T2 positive or Gd-enhancing lesions. Case 3 had a very severe form of neuromyelitis optica, with at least 3 relapses without recovery or improvement in the year prior to ASCT, and the appearance of new Gd-enhancing lesions on every MRI. After therapy, she regained the ability to walk with aid and MRI did not show any T2-positive or Gd-enhancing area.

The robust data obtained in the frequent MRI study and further information gained with the single cases allow some more general consideration regarding this promising therapy. Intense immunosuppression followed by ASCT (the sequence CY-BEAM-ASCT-ATG) has the capacity to completely abrogate inflammation in severe, active MS, and the effect is not transitory and persists. We do not know the length of the period of suppression of inflammation, but in our series of patients undergoing frequent MRI, the median follow up is now 2 years. The duration of the effect is, of course, a crucial point because toxicity and the risk of mortality can be justified only if the suppression of MRI activity and the effect on clinical course persists for a considerable period of time. There are, however, cases of rapid and dramatic deterioration of the neurological condition in a relatively short period of time, despite the use of conventional therapies. In these cases, intense immunosuppression followed by ASCT has the capacity to halt the progression of the disease.

At the moment, there is no other treatment with such a profound impact on inflammation in severe cases of MS. A few of our patients who responded to the sequence CY-BEAM-ASCT-ATG, failed the treatment with CY at a lower dosage (1gram IV every month for 6 month, for example, in case 13).

The dosage and type of immunosuppressive drug utilized is therefore crucial in significantly impacting the inflammatory phase of the disease. There are other immunosuppressive drugs utilized in MS and among them, certainly, the most interesting and promising is mitoxantrone. Mitoxantrone has the capacity to decrease at least 90% of MRI activity, has an effect on clinical course, and has a very good safety profile, although side effects are often present and a total cumulative dosage of 100-120 mg per m^2 cannot be exceeded due to its cardiotoxicity. Presently, it is not clear if intense immunosuppression followed by ASCT is more efficacious than standard therapy with mitoxantrone and if the possible greater efficacy of this treatment justifies its side effects and the risk of mortality. Studies using a control arm are in preparation to answer this question.

ASCT is effective on inflammation while it is probably useless if degeneration is the most important pathophysiological mechanism of disease progression.[12] In our series of cases, patients who responded better were those with very active MRI or rapid clinical worsening. For example, case 1 not included in the frequent MRI Cohort, who was treated with ASCT in spite of an "inactive" MRI, slightly deteriorated in the 16 months after transplantation. Therefore, it appears that ASCT may only be effective in severe MS cases where the inflammatory phase is the most important mechanism of disease activity. Relapsing remitting cases without improvement after relapse (relapsing-progressive) or secondary progressive cases with rapid clinical worsening and Gd-enhancing MRI, should be the types of patients eligible for ASCT. Patients with primary progressive MS or with slow deterioration, without signs of MRI activity, should not be treated with ASCT. Considering that axonal damage is due not only to inflammation but also to progressive degeneration of axons devoid of the myelin sheath, it is reasonable to consider that, in any case, degeneration of axons will not be influenced by intense immunosuppression. We are now evaluating whether brain atrophy continues to progress in our series of patients despite the complete suppression of inflammation, similar to that which occurs with beta interferons. If intense immunosuppression has a future in the management of severe MS, it should be utilized early in the course of the disease, as soon as the failure of conventional therapies become evident.

Finally, the sequence of CY-BEAM-ASCT-ATG is probably more than an intense immunosuppressive therapy with stem cell rescue. In fact, treatment with CY at high dosage and BEAM induces an almost complete immunoablation, and is followed by the infusion of hematopoietic stem cells, which reconstitute the hematopoietic and, later, the immune system. Beyond its immunosuppressive potential, ASCT could also have some beneficial effect for the substitution of a defective immune system with stem cells which could reset the autoaggressive reaction at a lower level. Moreover, it has been demonstrated, in a murine model, that after bone marrow transplantation, hematopoietic stem cells enter the CNS and differentiate into microglia, especially in sites of neuronal damage.[13] Intense immunosuppression followed by ASCT could, therefore, be a method utilized both for the eradication of an abnormal autoimmune response and for the modification of a harmful milieu at the site of tissue damage.

Acknowledgements

Supported in part by the Italian Multiple Sclerosis Foundation.

The Italian GITMO-NEURO Intergroup that has actively contributed to this study also includes the following persons: G. La Nasa, E. Cocco, and V. Derchi (Cagliari); A. Bosi, A. Repice, and A. Konze, (Firenze); A. Uccelli, O. Figari, E. Capello, and L. Dogliotti, (Genova); F. Papineschi, S. Mosti, and A. Abbruzzese (Pisa); P. Di Bartolomeo, D. Farina, C. Iarlori, and A. Tartaro (Chieti); G. Saglio, S. Duca and M. Capobianco (Torino); F. Casoni, F. Cavalleri and A. Donelli (Modena).

References

1. Edan G, Miller D, Clanet M et al. Therapeutic effect of mitoxantrone combined with methylprednisolone in multiple sclerosis: A randomised multicentre study of active disease using MRI and clinical criteria. J Neurol Neurosurg Psychiatry 1997; 62:112-118.
2. Hartung HP, Gonsette R and the MIMS Study Group. Mitoxantrone in progressive multiple sclerosis: A placebo controlled, randomised, observer blind phase III trial: Clinical results and three year follow up. Neurology 1999; 52(suppl 1): A290. Abstract.
3. Marmont AM. New horizons in the treatment of autoimmune diseases: Immunoablation and stem cell transplantation. Annu Rev Med 2000; 51:115-134.
4. Burt RK, Rowlings PH, Traynor A. Hematopoietic stem cell transplantation for severe autoimmune disease: Know thyself. In: Ball E, Lister J, Law P, eds. Hematopoietic Stem Cell Therapy. New York: Churchill Livingstone, 2000:203-215.
5. Kozak T, Havrdova E, Pit'ha J et al. High-dose immunosuppressive therapy with PBPC support in the treatment of poor risk multiple sclerosis. Bone Marrow Transplant 2000; 25:525-531.
6. Fassas A, Anagnostopoulos A, Kazis A et al. Autologous stem cell transplantation in progressive multiple sclerosis. An interim analysis of efficacy. J Clin Immunol 2000; 20:24-30.
7. Saiz A, Carreras E, Berenguer J et al. MRI and CSF oligoclonal bands after autologous hematopoietic stem cell trasplantation in MS. Neurology 2001; 56:1084-9.
8. Fassas A, Anagnostopoulos A, Kazis A et al. Peripheral blood stem cell transplantation in the treatment of progressive multiple sclerosis: First results of a pilot study. Bone Marrow Transplant 1997; 20:631-638.
9. Burt RK, Traynor AE, Cohen B et al. T cell depleted autologous hematopoetic stem cell transplantation for multiple sclerosis: Report on the first three patients. Bone Marrow Transplant 1998; 21:537-41.
10. Mancardi GL, Saccardi R, Filippi M et al. Autologous hematopoietic stem cell transplantation suppresses Gadolinium-enhanced MRI activity in multiple sclerosis. Neurology 2001; 57:62-69.
11. Filippi M, Rovaris M, Capra R et al. A multi-centre longitudinal study comparing the sensitivity of monthly MRI after standard and triple dose gadolinium-DTPA for monitoring disease activity in multiple sclerosis. Implications for phase II clinical trials. Brain 1998; 121:2011-2020.
12. Trapp BD, Peterson J, Ransohoff RM et al. Axonal transection in the lesions of multiple sclerosis. N Engl J Med 1998; 29:338:323-5.
13. Priller J, Flugel A, Wehner T et al. Targeting gene-modified hematopoietic cells to the central nervous system: Use of green fluorescent protein uncovers microglial engraftment. Nat Med 2001; 7:1356-61.

CHAPTER 37

Hematopoietic Stem Cell Transplantation for Multiple Sclerosis: Finding Equipoise

Athanasios Fassas and Richard K. Burt

Introduction

In view of the lack of curative treatments, high-dose myelo/immunosuppression with autologous stem cell transplantation (HSCT) has been utilized as therapy of multiple sclerosis (MS). HSCT has immunosuppressive and immunomodulatory effects which may tip the immunological balance towards disease quiescense. Does support for this hypothesis exist from current animal and human trials and if so, what information do they tell us in design of a future trial?

Animal Models of Multiple Sclerosis

Two animal models have been employed to determine the effect and mechanism of HSCT on immune-mediated demyelinating diseases. One model, experimental autoimmune encephalomyelitis (EAE), is an autoimmune demyelinating disease that, depending on species and circumstances, has a variable course but is most often relapsing remitting.[1] The second, Theiler's murine encephalitis virus (TMEV) demyelinating disease, is a viral initiated demyelinating disease.[2] Experimental transplantation using these model systems has emphasized a number of crucial points.

Experimental Autoimmune Encephalomyelitis

EAE is induced by immunization with myelin proteins and an adjuvant such as Freund's adjuvant. Once immunized, lymph nodes draining injection sites may be removed and reprimed ex vivo with myelin peptide. These ex vivo activated lymphocytes, when injected into a normal mouse, adoptively transfer disease. Adoptive transfer of disease by syngeneic lymphocytes activated to self epitopes is proof for an autoimmune pathogenesis. The myelin epitope used to induce disease is species specific. Lewis rats develop EAE following immunization with myelin basic protein (MBP) fragment sequence 68 to 82 (MBP 68-82).[3] Lewis rat EAE is a monophasic acute disease with complete recovery and is caused by V beta 8.2 T cell receptor positive CD4+ T cells. Swiss Jackson Laboratory/Jackson (SJL/J) mice develop a relapsing remitting form of MS following immunization with proteolipid protein peptide (PLP) fragment sequence 139-151 (PLP 139-151). Unlike MBP which is present in both the peripheral and central nervous system, PLP is located only within the central nervous system and is the dominant peptide in myelin. Using different myelin protein epitopes, EAE can be induced in a variety of animals including primates. These animal models suggest that auto-reactive myelin repertoires of T cells exist in normal animals, but are unable to initiate disease unless primed in vivo by immunization or reprimed ex vivo and adoptively transferred into a naïve animal. Why such myelin-reactive repertoires exist is unclear. Nevertheless, autoreactive T cell repertoires are part of normal lymphocyte development.

While T cells from normal mice have receptors for myelin peptides, lymphocytes from normal mice do not proliferate to PLP 139-151. Within 10 days of immunization and with onset of disease, lymphocytes proliferate to PLP 139-151. Proliferative responses to other myelin peptides such as PLP 178-191 as well as MBP occur with disease progression. Intra- and inter-determinant or epitope spreading suggests that peptide or T cell receptor specific therapies may be ineffective and/or impractical.

HSCT of EAE with a variety of conditioning regimens, cyclophosphamide alone, total body irradiation (TBI) alone or cyclophophamide and TBI ameliorates neurologic manifestations.[4] If performed early after disease onset, a long term remission ensues (Fig. 1). The mechanism of remission is not simply return of a naïve PLP unresponsive phenotype, since post transplant splenocytes proliferate to PLP epitopes similar to untreated animals with EAE (Fig. 2).[4] This implies that aggressive immune ablative regimens designed to destroy every pretransplant lymphocyte (which increase the risk of serious infections) may not be necessary. It also suggests that self reactive cells are not necessarily associated with clinical disease. While clinical remission does not seem to correlate with proliferative responses to myelin epitopes, it does correlate with delayed type hypersensitivity (DTH) responses (Fig. 3).[4] The mechanism of HSCT induced remission in EAE remains unknown, but future evaluation should include regulatory or suppressor cells, cytokine shifts, ability of disease causing

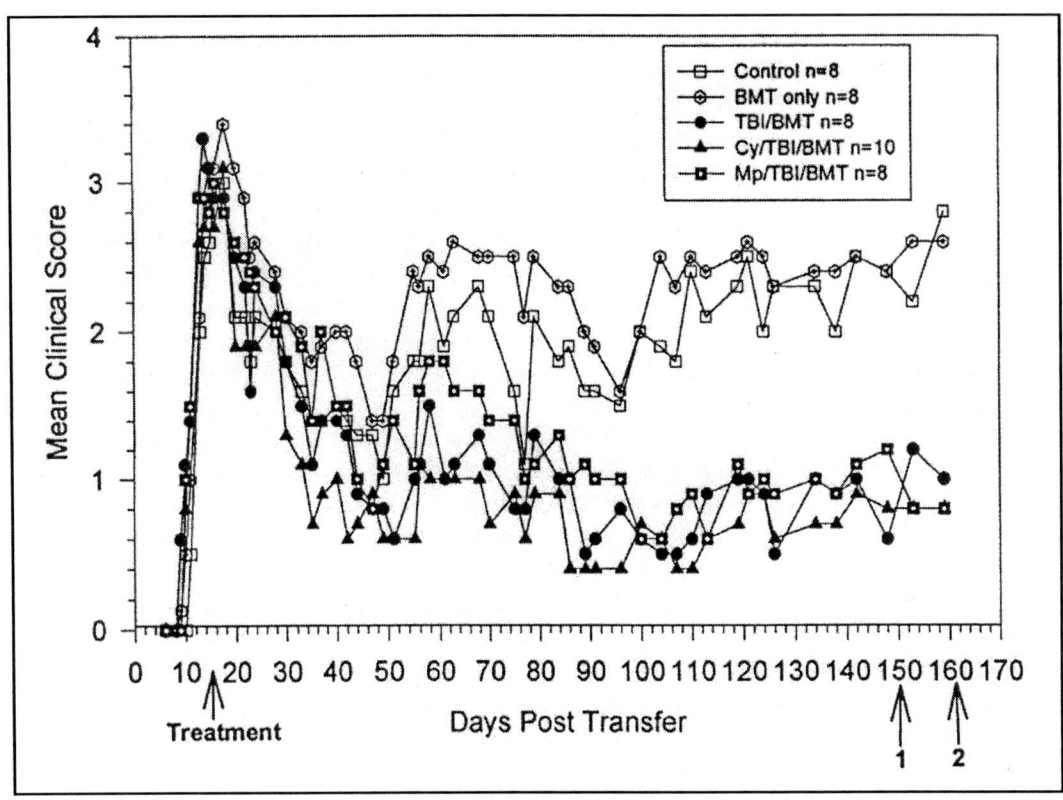

Figure 1. Bone marrow transplantation from naïve (unimmunized) syngeneic mice early in course of EAE. Control= mice with EAE; BMT= infusion of marrow cells from normal syngeneic mouse without a conditioning regimen; TBI= total body irradiation (1100cGy); Cy= cyclophosphamide 120 mg/kg; MP= methylprednisolone 1 gram prior to TBI; Treatment= time of conditioning and transplant; 1 and 2= time of lymphocyte assay shown in Figures 2 and 3. Mean clinical score 0= normal; 1= tail weakness; 2= ataxia; 3= hind limb weakness; 4= hind limb paralysis. (Reprinted from Burt RK et al. Blood 1998; 97(7):2609-2616.)

cells to home or localize to the target sites, and function of cellular subsets other than T cells such as macrophages.

In EAE, different high dose immunosuppressive (cyclophosphamide) versus myeloablative (TBI or TBI and cyclophosphamide) regimens may have different clinical toxicities. While Burt et al found TBI to be safe without acute toxicities, van Bekkum et al reported that TBI could cause acute neurologic exacerbations in EAE. The effect of conditioning regimens, and in par-

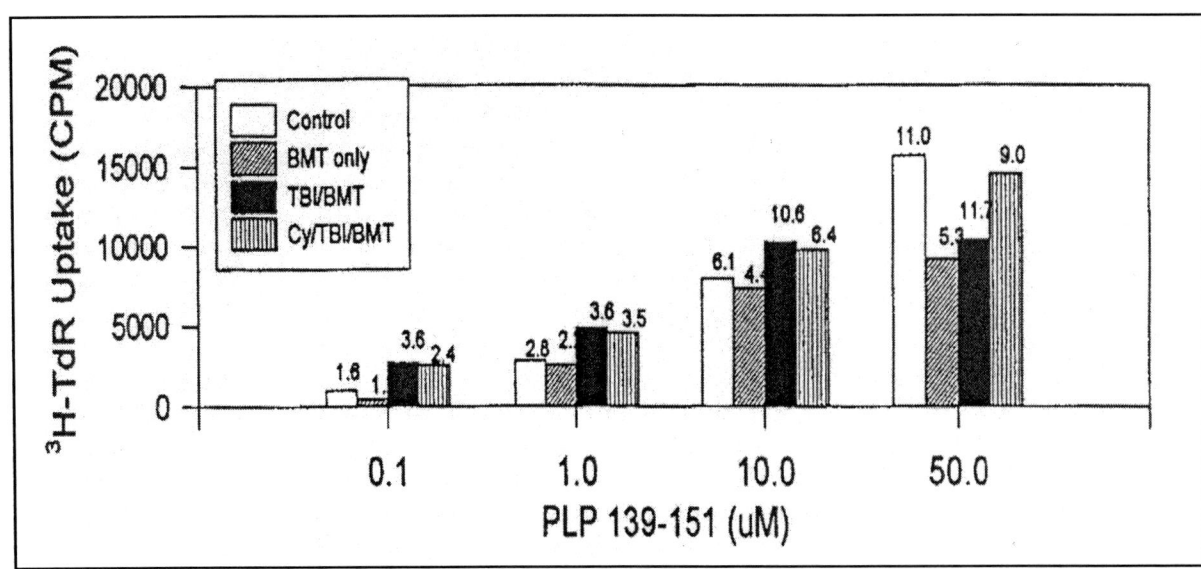

Figure 2. Proliferative responses to disease initiating PLP 139-151. Control= mice with EAE; BMT= infusion of marrow cells from normal syngeneic mouse without a conditioning regimen; TBI= total body irradiaion (1100cGy); Cy= Cyclophosphamide 120 mg/kg. Normal naïve mice have no proliferative response to PLP 139-151.

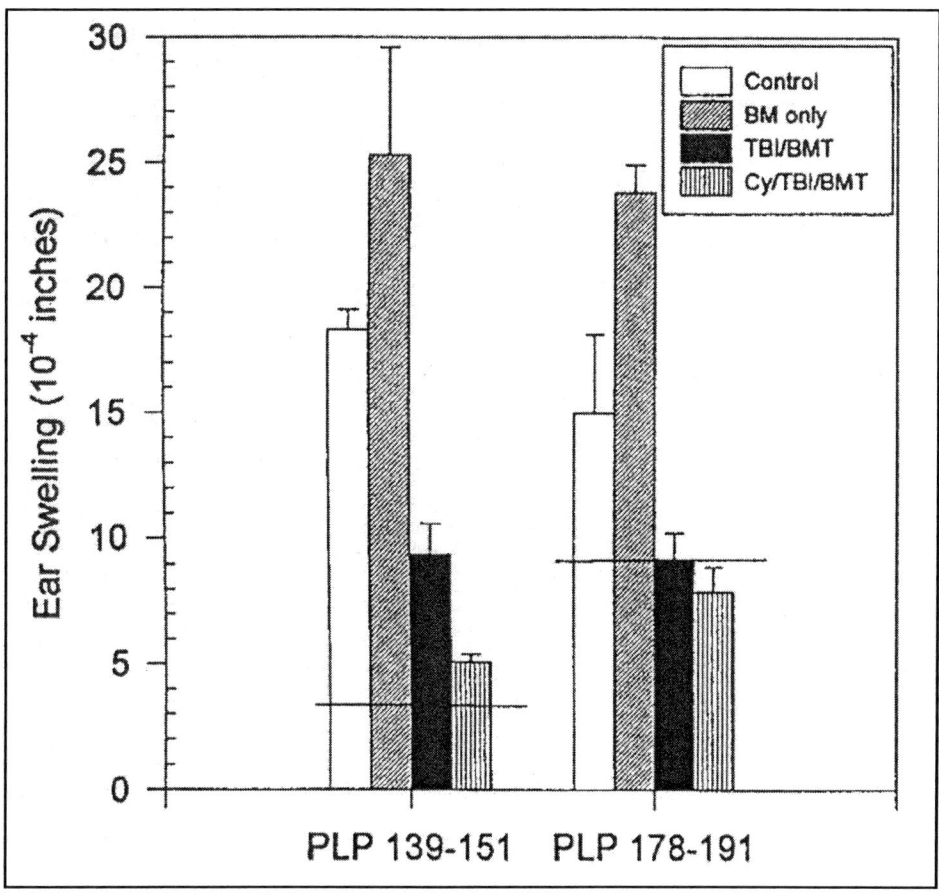

Figure 3. Delayed type hypersensitivity (DTH) to proteolipid protein (PLP) epitopes. Control= mice with EAE; BMT= infusion of marrow cells from normal syngeneic mouse without any conditioning regimen; TBI= total body irradiaion (1100 cGy); Cy= Cyclophosphamide 120 mg/kg. Horizontal lines are DTH response of normal unimmunized mice. (Reprinted from Burt RK et al. Blood 1998; 97(7):2609-2616).

ticular TBI, on long term axonal integrity and chronic progressive disability, an important consideration for patients with multiple sclerosis, has not to date been investigated in animal models.

While HSCT is an effective therapy for early onset EAE, it is of no benefit in late EAE (Fig. 4).[4] This may be important in selection of patients with MS for HSCT. The lessons from EAE would suggest effectiveness in early inflammatory disease, but not late progressive disease. HSCT is a form of immune suppression, and similar to other immune suppressive or immune modulating interventions, may be of little or no effectiveness in progressive disease.

Theiler's Murine Encephalomyelitis Virus (TMEV)

TMEV is a small RNA virus (picornavirus) that establishes persistent infection within the CNS and causes a progressive neurologic decline in susceptible mice similar to primary progressive MS. In the wild mice, infection occurs via ingestion, while in the laboratory, infection is usually accomplished by intracerebral innoculation. Disease resistant strains clear virus within 2 weeks of infection, while disease susceptible strains of mice have a persistent CNS infection. Both virus- and myelin-specific T cell responses occur in TMEV demyelinating disease. The appearance of viral specific responses before myelin-specific responses, and the lack of cross reactivity between viral and myelin epitopes, indicates that demyelination arises by epitope spreading and bystander activation, rather than molecular mimicry between viral and myelin epitopes.[5]

Syngeneic HSCT of TMEV-infected disease susceptible mice results in high neurologic mortality.[6] Allogeneic HSCT of TMEV-infected disease susceptible hosts by disease resistant but uninfected donors caused an equally high neurologic-related mortality associated with an increase in CNS viral titers (Figs. 5 and 6).[6] HSCT using marrow from disease resistant, but previously infected, donors ameliorated neurologic related mortality (Fig. 6) presumably by adoptive transfer of viral specific cytotoxic T cells along with the marrow graft.[6] Therefore, a functional immune system appears essential to prevent lethal neuropathic effects from persistent viral mediated CNS demyelinating disease. This raises concerns about accelerating neurologic disability using autologous HSCT if MS or some subset of MS patients have a viral etiology and persistent infection.

Based on the results in animals, clinical HSCT trials were suggested in 1995[7] and have been conducted ever since, although on small scales, all over the world, mainly in Europe and the U.S.A. and as phase I-II studies of autologous HSCT in MS. The results suggest that EAE is a better model for MS than TMEV demyelinating disease. However, primary progressive MS patients have generally been excluded from HSCT protocols, and it remains possible that TMEV induced demyelinating disease is a reliable model for primary progressive MS. If correct, patients with primary progressive MS would be anticipated to do significantly worse following HSCT. In the few patients with primary progressive MS undergoing autologous HSCT, as will be discussed below, neurologic progression appears to be worse.

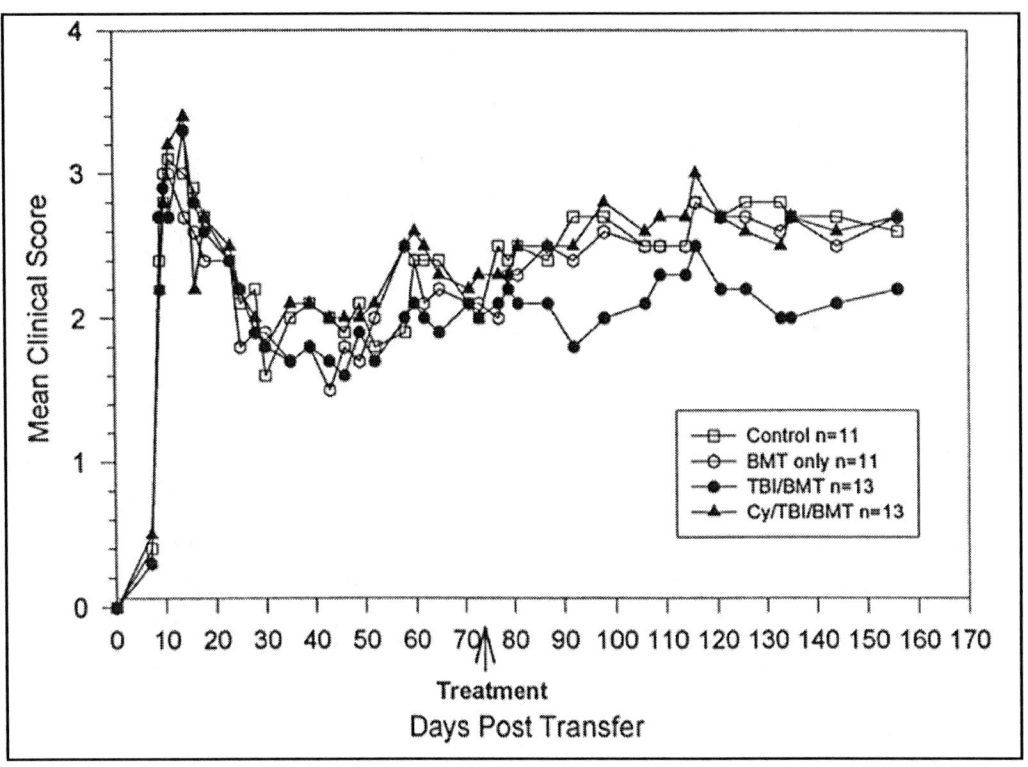

Figure 4. Bone marrow transplantation from naïve (unimmunized) syngeneic mice late in the course of EAE. Control= mice with EAE; BMT= infusion of marrow cells from normal syngeneic mouse without a conditioning regimen; TBI= total body irradiation (1100 cGy); Cy= Cyclophosphamide 120 mg/kg. Treatment= time of conditioning and transplant. Mean clinical score 0= normal; 1= tail weakness; 2= ataxia; 3= hind limb weakness; 4= hind limb paralysis. (Reprinted from Burt RK et al. Blood 1998; 97(7):2609-2616.)

Clinical Hematopoietic Stem Cell Transplant Trials

European Results

In Europe, about 30 centers have treated small numbers of patients and reported their results to the Autoimmune Disease Working Party Registries. Four centers (Thessaloniki, Praque, Barcelona, Italy) have reported individually (Table 1). These studies will be briefly reviewed below.

The Thessaloniki Study

A total of 35 progressing patients, of a median age of 40 years (range, 19-54), who had failed a variety of therapies, were treated for secondary progressive MS (SPMS) (20 patients), primary progressive MS (PPMS) (14 patients), or relapsing/remitting MS (RRMS) (1 patient).[8,9] Median interval from diagnosis to transplant was 7 years (range, 1-16) and the median score on the extended disability status scale (EDSS) was 6 (assistance needed to walk 100 meters) with a range of 4.5 (ability to walk without aid for 300 meters) to 8 (restriction to bed or chair). The patients were selected because they had declined on the EDSS by at least 1 step over the preceding year. Twelve patients (34%) had evidence of disease activity on MRI, i.e., gadolinium-enhancing, new, or enlarging, lesions on serial scans.

Two conditioning regimens were employed: (a) Twenty-five patients received BEAM (BCNU 300 mg/m^2—etoposide 800 mg/m^2-cytaribine 800 mg/m^2—melphalan 140 mg/m^2). (b) Ten patients were treated with busulfan at 16 mg/kg over 4 days. The patients were rescued with peripheral blood stem cells, mobilized with cyclophosphamide (CY) at 4 g/m^2 and G- or GM-CSF at 5 or 10 μg/kg/day. No steroids were given during mobilization. In 10 cases, a degree of ex vivo T cell-depletion was performed by CD34$^+$ cell-selection of the grafts. All 35 patients also received rabbit antithymocyte globulin (ATG, Merieux/Sangstat) at total doses ranging from 5 mg/kg to 20 mg/kg, mainly soon after the infusion of stem cells (in vivo T depletion).

During mobilization, there was no mortality. However, a limited number of cytopenic infections occurred, and, an incident of epilepsy, ascribed to G-CSF, occurred in one patient, and another worsened significantly while on G-CSF, with appearance of new lesions in the MRI scan. Early post-transplant toxicity included reactions to ATG, mild mucositis, cytopenic infections, one case of mild liver veno-occlusive disease, one case of mild thrombotic thrombocytopenic purpura, and one fatal case of pulmonary and cerebral aspergillosis. There were no delayed hematological recoveries or graft failures. In all, the rate of grade 3 plus 4 medical toxicity was 89% and the therapy-related mortality (TRM) was 3% (1/35). A degree of mild neurological deterioration developed in 13 patients (37%), owing to fever, infection, or possibly to G-CSF given post transplant for hematological recovery. Later toxicities included viral reactivations occurring in the first hundred days and leading to disease in two patients (one CMV pneumonitis, one zoster), and a case of autoimmune thyroiditis diagnosed eleven months post HSCT. CD4$^+$ cytopenia was profound and prolonged for up to two years after HSCT. Median follow-up time at last assessment was 35 months (range, 3-67).

In terms of efficacy, the probability of confirmed-progression free survival (cPFS) was 90% (± 7%) for SPMS, but lower 65% (± 14%) for PPMS, at 5 years. Progression (of disability) was defined as an increase of 0.5 EDSS points if baseline (enrollment)

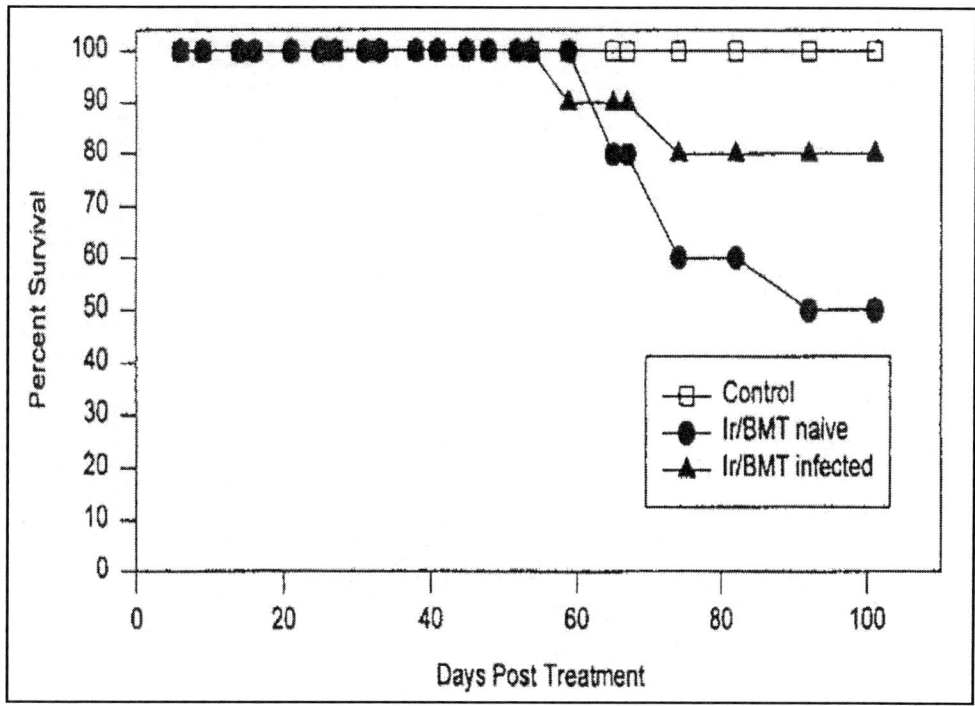

Figure 5. Hematopoietic stem cell transplantation of TMEV-induced demyelinating disease. Control= mice with TMEV-induced demyelinating disease. Ir/BMT naïve= mice with TMEV-induced demyelinating disease treated by HSCT using 1100 cGy irradiation and infusion of bone marrow cells from an unimmunized disease resistant strain. Ir/BMT infected= mice with TMEV-induced demyelinating disease treated by HSCT using 1100 cGy irradiation and infusion of bone marrow cells from a previously TMEV infected disease resistant strain. (Reprinted from Burt RK et al. Blood 1999; 94(8):2915-2922.)

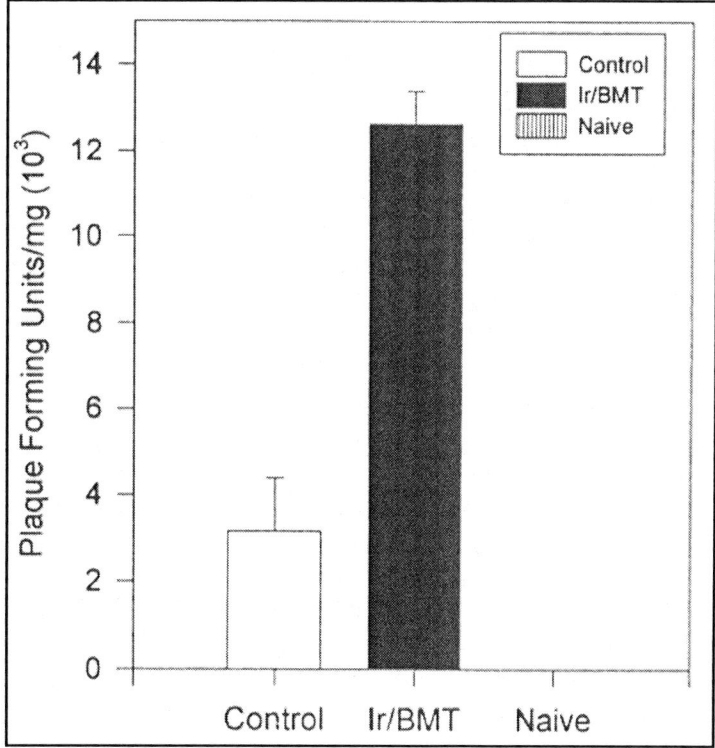

Figure 6. CNS viral titer after hematopoietic stem cell transplantation of TMEV-induced demyelinating disease. Control= mice with TMEV-induced demyelinating disease. Ir/BMT naïve= mice with TMEV-induced demyelinating disease treated by HSCT using 1100 cGy irradiation and infusion of bone marrow cells from an unimmunized disease resistant strain. Ir/BMT infected= mice with TMEV-induced demyelinating disease treated by HSCT using 1100 cGy irradiation and infusion of bone marrow cells from a previously TMEV infected disease resistant strain. (Reprinted from Burt RK et al. Blood 1999; 94(8):2915-2922.)

Table 1. Autologous hematopoietic stem cell transplantation in patients with multiple sclerosis

Group	Regimen	EDSS Range	# of Patients (# of Deaths)	Cause and Date of Death After Transplant	Comment
European					
Fassas (Greece) (8,9)	BEAM + ATG or Bu + ATG	4.5 - 8.0	35 (1)	1 from aspergillosis (day 65)	90% PFS for SPMS 65% PFS for PPMS
Kozak (Prague) (10, 11)	BEAM + ATG	6.0 - 7.5	15 (0)	1 died of disease progression 3 years after HSCT	11 patients stable or improved, 3 progressed by 0.5 EDSS steps
Saiz (Spain) (12)	BCNU + Cy + ATG	5.0 - 6.5	5 (0)	No mortality	Generally clinically stable, fewer relapses, reduced MRI activity
Mancardi (Italian) (14,15)	BEAM + ATG	5.0 - 6.5	18 (0)	No mortality	Marked reduction in MRI activity
EBMT retrospective study (17)	BEAM or Bu/Cy based or TBI based	4.5 - 8.5	85 (5)	8% mortality due to infection in 3 and progressive disease in 2	78% three year PFS if PPMS excluded
North American					
Burt (Chicago, Milwaukee) (18,19)	Cy/TBI	3.0 - 8.5	28 (0)	2 died of progressive disease at 12 and 18 months respectively	50% progression in patients with pre-transplant EDSS >6.0
Nash (Seattle consortium) (21,22)	Cy/TBI + ATG		26 (1)	1 from Epstein Barr virus post-transplant lymphoproliferative disease (PTLD) (day 53)	N/A
Openshaw (City of Hope) (23)	BU/Cy + ATG		5 (2)	1 from influenza (day 22) and 1 from streptoccocus (19 months)	Study stopped
Atkins (Canadian)	Bu / Cy + ATG	3.0 - 6.0	8 (1)	Hepatic failure	Total of 24 patients and 8 controls

Actual patient numbers may be more than reported in reference and are based on updated communication with author; ATG= anti-thymocyte globulin; BEAM= carmustine, etoposide, cytarabine, melphalan; BU/CY= busulfan and cyclophosphamide; CY/TBI= cyclophosphamide and total body irradiation.

score was above 5, or as an increase of 1 point if baseline score was below 5.5. In 15 patients (43%), the disability scores were firmly improved by at least 1 step on the EDSS during the first six months post transplant. Twelve to 45 months (median, 25) after transplant, the scores started to increase (worsen) again in four of the 15 improved patients, but in only one did the score exceed baseline.

The analysis of MRI scans performed before and regularly after transplantation yielded encouraging results. Considering only gadolinium-enhancing lesions, their number dropped dramatically: 40 scans yielded 87 lesions in eleven patients at enrollment; after HSCT, in a total of 197 scans, only 25 gadolinium positive lesions were found in five patients.

The Prague Study

Kozak and Havrdova enrolled only SPMS patients, aged 18 to 55, with EDSS scores of 6 to 7.5, refractory to standard therapies, and with evidence of clinical deterioration of at least 1.5 EDSS points in the year preceding enrollment.[10,11] Nineteen patients were mobilized with cyclophosphamide at $4g/m^2$ plus G-CSF at 7-10 µg/kg/day. One patient had a significant improvement after cyclophosphamide mobilization, clinically and in MRI activity. Eventually, 15 patients were transplanted using BEAM for conditioning. Ex vivo T cell-depletion by CD34+ cell-selection plus anti CD2/3 depletion was performed in 9 cases; the remaining 6 patients received ATG (Fresenius), at 8mg/kg post transplant, for in vivo T cell-depletion. Toxicity included one case of severe respiratory infection, and one case of late zoster. There was no transplant-related mortality. Mild, but transient deterioration of mobility developed early post transplant. Median follow-up at last assessment was 20 months (range, 4-38). Eleven patients have been stable or improved, three worsened by 0.5 EDSS point, and one patient who had not responded died of disease progression three years after transplantation.

The Barcelona Study

The Spanish experience now includes 22 MS patients from various centers.[12,13] Saiz and Graus published on the MRI activity and cerebrospinal fluid (CSF) oligoclonal band (OB) changes in a series of 15 autotranspants performed by Carreras et al in the Hospital Clinic, Barcelona. Three patients with SPMS and two with RRMS, all resistant to therapy, aged 24 to 44 years (median, 30), were treated with BCNU 300 mg/m^2, cyclophosphamide 150 mg/kg, ATG at high-dose (Merieux/Sangstat; 60 mg/kg), and were rescued with CD34$^+$ cell-selected blood stem cells, mobilized with cyclophosphamide 3.0 g/m^2 and G-CSF 5 µg/kg/day. The patients had EDSS scores 5 to 6.5 (median, 6.5) at enrollment, and a mean disease duration of nine years (range, 6-14); they had multiple relapses despite therapy, and a median increase in disability of 0.5 EDSS step in the year preceding HSCT. One patient worsened slightly, but transiently, on G-CSF mobilization. Another patient had a reaction to ATG administration and developed severe paraparesis which improved only partially in the following months. At a median follow-up time of 18 months (range, 12-24), this patient was reported stable, and the other four were reported improved or stable clinically. Very few and brief, sensory relapses occurred after transplant, which did not require treatment. Regular MRI studies after HSCT showed no enhanced T1 lesions, no new or enlarging T2 lesions, and compared to baseline, a decrease in T2 lesional load by 4 to 26.6% (median change, -11.8%). However, atrophy of the corpus callosum increased during the first months after HSCT. At one year, the change from baseline was -7.8 to -20.5% (median, -12.4%), but no further reduction in the corpus callosum area was noted in the second year. CSF assays did not show any decrease in OB, which could still be detected one year after HSCT.

The Italian Study

This is a cooperative study of the Italian Group for Bone Marrow Transplantation (GITMO) and a group of neurology teams, with the main objective to investigate MRI changes after HSCT.[14,15] Eighteen patients have been treated with the same protocol as in the Thessaloniki study, i.e., with BEAM, ATG (Sangstat; 10 mg/kg), and blood stem cells mobilized with Cyclophosphamide at 4 g/m^2 and G-CSF at 5 µg/kg/day. No ex vivo graft manipulation was performed. Only patients with SPMS were included with presence of at least one gadolinium-enhanced area on MRI, using triple dose of gadolinium which increases the sensitivity in detecting disease activity.[16] All patients had long-standing (6-19 years), unresponsive disease with documented rapid progression over the previous year, and with EDSS scores between 5 and 6.5. MRI scans were performed monthly for three months before transplantation, and serially every month for the first six months post transplant, and then once every three months.

An analysis of the first 10 patients performed at a median follow-up of 15 months (range, 4-30)[14] showed that the enhancing activity began to decrease after cyclophosphamide mobilization and the number of lesions was zero in all patients after transplantation from 341 detected before mobilization. The number of T2-weighted positive lesions was also reduced: no new T2 lesions were observed in 9 of 10 patients after transplant, whereas 62 new lesions were detected during the three months of the pretreatment period. In a recent update of this study, brain atrophy was still present and ongoing at one year, but it appeared to slow down after the 6th month. Oligoclonal bands were still detected in CSF at 2 years after HSCT. Clinically, the patients were stable or slightly improved.[15] No major adverse events were recorded except for a case of hemorrhagic cystitis after cyclophosphamide. After transplant, CMV reactivation occurred in 5 cases, one symptomatic, and zoster developed in two cases. Infections or fever were common in the early period, as was transient worsening of the previous neurological condition.

The EBMT Retrospective Study

There are about 120 MS patients treated with HSCT in Europe, Asia, and Australia and reported to the EBMT registry. The first 85 have recently been published.[17] Most of the patients just described above were also included in this comprehensive analysis using the same criteria of response for all patients. Conditioning regimens like BEAM (mainly), busulfan-based, cyclophsophamide-based, and also TBI were employed. The majority received blood stem cells for rescue, and ATG and/or ex vivo purged grafts for lymphocyte depletion. Five patients died as a result of the therapy (6%), owing mainly to infection, and two more patients died of progressive disease. The probability of progression free survival at 3 years was 72%, and was higher (78%) for all disease types except PPMS. Moreover, MRI activity was found markedly reduced after HSCT.

North American Studies

Chicago/Milwaukee

The first HSCT for multiple sclerosis in North America was performed in 1996 at Northwestern University Medical Center (Chicago, IL) and Medical College of Wisconsin (Milwaukee) (Table 1).[18-20] In order to ensure good CNS penetration across the blood brain barrier, a TBI based regimen was chosen for the initial HSCT trial consisting of cyclophosphamide 120 mg/kg and TBI 1200 cGy with 50% lung and 30% right liver shielding with CD34$^+$ selection (either CEPRATE, CellPro or Isolex, Baxter) of stem cells. Mobilization of HSC was initially performed with G-CSF, but due to disease flare on the 6th patient, was switched to cyclophosphamide 2.0 g/m^2 and G-CSF starting 72 hours later. No subsequent patient had a flare of disease during mobilization.

The primary endpoint of the Northwestern/Milwaukee phase I trial was safety. The primary outcome results in 28 patients indicated that such a regimen may be used with sufficient safety, since there was no mortality or unexpected toxicity. Secondary outcomes included durability of disease remission as assessed by change in Kurtzke extended disability status scale (EDSS). With the longest follow-up approaching 6 years, only one patient has experienced a flare of MS (although unconfirmed) since the HSCT. Disease progression (EDSS increased 1.0 points) was confined to patients whose pretransplant EDSS was greater than 6.0 (Fig, 7). However, most patients with pretransplant EDSS scores <6.0 were enrolled later in the study after gaining experience with more disabled patients and have a shorter post transplant follow-up duration.

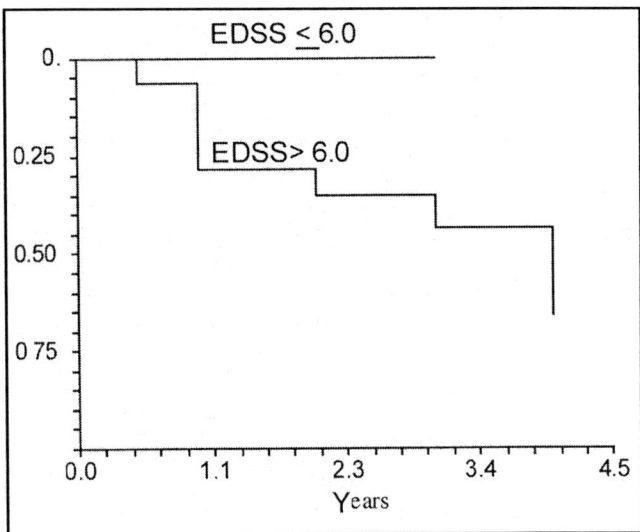

Figure 7. Time to treatment failure on the Northwestern (Chicago)/Milwaukee HSCT study. EDSS progression by 1.0 or more points separated by pre-HSCT EDSS ≤ 6.0 (10 patients) or EDSS > 6.0 (16 patients).

Longer follow-up on most of the 28 patients enrolled on this study will be required to confirm these preliminary observations. HSCT appears to have slowed clinical and MRI evidence of relapse and demyelination, respectively. However, it appears that insidious progression of disability continues, especially in patients with higher pretransplant EDSS scores. The phase I studies suggest that HSCT, like other immune-based therapeutic approaches to the treatment of MS, and similar to EAE, is likely to offer the most benefit to individuals in earlier, more active inflammatory stages of disease.

Seattle Consortium

The Fred Hutchinson Cancer Center multi-center trial treated 26 patients with progressive MS, 17 being treated at the Fred Hutchinson Cancer Center.[21] The conditioning regimen was total body irradiation (TBI) (800 cGy given 200 cGy BID with lung shields to 650 cGy), cyclophosphamide (120 mg/kg divided 60 mg/kg/day) and ATG (either 90 mg/kg equine or 15 mg/kg rabbit) given for 6 days (days -5, -3, -1, +1, +3, and +5). G-CSF mobilized peripheral blood stem cells (PBSC) were lymphocyte depleted by CD34⁺ positive selection (Isolex, Baxter). The only patient in whom rabbit ATG had been given instead of equine ATG died from Epstein-Barr virus (EBV) associated post-transplant lymphoproliferative disorder (PTLD).[22] The patient who developed PTLD had received 6 days of rabbit ATG at 2.5 mg/kg/day (total dose of 15 mg/kg). PTLD is a complication not unique to rabbit ATG, but rather secondary to the extent of immune suppression. While not known at the time, the rabbit ATG dose (15 mg/kg) is roughly equivalent to an equine ATG dose of 150-225 mg/kg.

City of Hope

The City of Hope (Duarte, California) used an intensive conditioning regimen of busulfan (16 mg/kg), cyclophosphamide (120 mg/kg), and anti-thymocyte globulin (30 mg/kg) along with CD34⁺ selection (Isolex, Baxter) to deplete lymphocytes from the graft.[23] Two out of 5 treated patients died from infections.

One patient died 22 days after transplant from influenza and the second died 19 months after HSC transplant from streptococcus sepsis. This would indicate caution in combining CD34 selection with very intense immunosuppresive regimens especially in a disease with low mortality such as multiple sclerosis.

Canadian Study

A Canadian multi-center trial is being conducted by Harry Atkins in Ottawa, Toronto, and Montreal. The regimen is busulfan 16 mg/kg, cyclophosphamide 200 mg/kg, and rabbit ATG 6.0 mg/kg with CD34⁺ selection using CLINIMACS (Miltenyi). Eligibilty are an EDSS of 3.0 to 6.0, a functional score of 3 within 5 years of onset or >5 relapses within first 2 years, and deterioration of the EDSS by 1.0 point in 18 months prior to enrollment if less than 5.5 or, if >5.5, deterioration by 0.5 points or more than 2 significant relapses in last year or 3 relapses in the last 2 years. A significant relapse is defined as occurring despite interferon or copaxone therapy and needing corticosteroid treatment. The study will enroll 24 patients and 8 controls. One patient died from regimen-related liver failure. This would indicate caution in using intense regimens like "big" busulfan/cytoxon known to cause organ damage especially in MS patients with low risk of disease related mortality.

Toxicities from Phase I Studies

In the EBMT analysis, the risk of death of any cause at three years was 10% (± 7%), and was higher in older patients, in patients with high EDSS scores and, probably, with ex vivo lymphocyte purging. This is a significant risk for a nonmalignant disease, especially if a comparative trial is proposed, considering the absence of mortality of the currently available therapies for MS. Nevertheless, taken together, it seems that, with proper selection of patients and conditioning regimens, the TRM can be and should be brought down significantly lower.

HSCT has an infection-associated mortality risk that is related to the intensity of the conditioning regimen. Three or more intense conditioning agents such as busulfan, cyclophosphamide, and ATG or cyclophosphamide, TBI, and ATG, or BEAM and ATG when combined with CD34⁺ selection may cause opportunistic infectious deaths. For example, with a conditioning regimen of total body irradiation (TBI) (800 cGy), cyclophosphamide (120 mg/kg), rabbit ATG (15 mg/kg) and CD34⁺ positive selection, a patient died from Epstein-Barr virus (EBV) associated post-transplant lymphoproliferative disorder (PTLD).[22] A late death from pneumoccocus bacteremia has been reported after busulfan (16 mg/kg), cyclophosphamide (120 mg/kg), and anti-thymocyte globulin (30 mg/kg) combined with CD34⁺ selection.[23] This may have been secondary to prolonged immune deficiency-related hypogammaglobulinemia. Aspergillosus, which is also an opportunistic infection related to immune suppression, occurred following the combination of BEAM/ATG and CD34⁺ selection.[9] Therefore, less intense regimens that omit at least one of these agents, such as CD34⁺ selection or ATG or TBI and/or continues the same drugs at lower doses, should be considered for phase II/III trials.

There is also a risk for disease worsening by the procedure. Early virus reactivation[24] and infections are common. It is known that MS patients deteriorate with fever and infection and that G-CSF and ATG may increase neurotoxicity which at times can be serious.[25-27] While G-CSF and ATG toxicity may be ameliorated with concurrent corticosteroids, reactivation of cytomegalovirus or herpes zoster may be associated with fatigue and decline in neurologic performance status not appreciated in HSCT of non MS patients. Extensive immunosuppression is associated with post transplant infections and may not yield better results than less immunosuppressive regimens.

Drugs that commonly cause fever should be avoided or premedicated with antipyretics and/or anti-inflammatory drugs prior to infusion. A cytokine release engraftment syndrome of fever, rash, and nonfocal neurologic deterioration or pseudo-exacerbation of MS that may be accompanied by eosinophilia has also been described in patients receiving TBI based regimens with CD34+ selected grafts.[28] Skin biopsy is histologically identical to graft versus host disease. The engraftment syndrome may be avoided or minimized by a short course of corticosteroids from day 5 to 22 post HSC infusion or by not using TBI.

When phase I studies were initiated, MS was recognized as an immune-mediated demyelinating disease. Since then, further evidence indicates that MS is also a neuronal degenerative process.[29-32] HSCT conditioning agents should be designed to minimize neuronal injury which could accelerate the slow insidious degeneration seen in patients with progressive disease. To date there is no evidence that radiation or chemotherapy (in the doses given) worsens axonal degeneration. However, phase I MRI studies demonstrate continued widening of ventricular dimensions and progressive disability post HSCT despite marked improvement in demyelinating activity. Finally, HSCT trials for MS should be designed to minimize neuronal and oligodendrocyte progenitor cells injury that could result from agents like TBI.

Recently, it has been reported in a rat model that cranial irradiation adversely affects neural stem cells and mechanisms of brain repair through apoptosis, alteration in cell cycle progression, or destruction of a favorable neural mileu by invasion of macrophages and microglia.[33] The radiation dose studied was 10 cGy, a dose similar to the Northwestern and Seattle clinical trials.

TBI at higher doses is known to be neurotoxic in normal individuals. Whether chronically demyelinated and already injured neurons are more susceptible to radiation injury is unknown. Theoretically, axonal degeneration may be accelerated by conditioning regimen-related neuronal injury. This is particularly important since permanent disability in MS correlates with axonal degeneration not demyelination.

Efficacy of HSCT from Phase I/II Studies

All the studies reviewed were unblinded neurological assessments in which the clinical benefits can easily be overestimated. It is well-known that MS is a most difficult disease in which to show the efficacy of a therapy, owing to subjective influences and optimistic expectations. Any of the current clinical results, therefore, should be received with caution and should be treated as hypothesis driven, not efficacy trials that provide information on patient and regimen selection to minimize morbiity of future efficacy trials.

All centers appear to have similar results: HSCT is feasible in treatment of MS, it has a remarkable, sustained, impact on MRI evidence of inflammation, but there currently is little evidence for clinical benefit. CSF oligoclonal bands, persist after HSCT, irrespective of the conditioning regimen used, e.g., high-dose cyclophosphamide plus BCNU,[12] BEAM,[15] or BuCY, suggesting that lymphocytes are not totally eliminated in the CNS. It is still not known whether TBI would be more effective in this respect, but complete elimination of lymphocytes in the CNS is of unclear significance and probably unnecessary as lymphocytes are activated in the peripheral lymphoid organs. Mobilization doses of cyclophosphamide can, too, reduce the number of MRI lesions, repress inflammation in the CNS and, consequently, improve the patients clinically by more than 1 EDSS step (6% in the EBMT study). It is unknown whether the dose used for mobilization, which is considerably less than transplant doses of cyclophosphamide, can offer sustained benefit. It does suggest that transplant doses of cyclophosphamide when used without radiation or other chemotherapy drugs may be an effective conditioning regimen in suppressing further inflammation.

The nearly total extinction of gadolinium-enhancing areas is an objective criterion of excellent response to the high-dose therapy. No standard therapy has given such a result. However, it remains to be established whether HSCT has also an impact on the course of the disease, since progression of disability does not depend on inflammation and demyelination only, but also on axon degeneration that may continue despite suppression of inflammation. Brain atrophy, which reflects degeneration, seems to continue for at least the first year after transplantation. Longer follow-up will show whether the process continues. However, this argues for intiating aggressive immunosuppressive therapy earlier in the course of MS while still a relapsing-remitting disease.

Future Trials

North America

For a therapy designed to inhibit immune-mediated demyelination, candidates should be selected early in disease course to minimize the impact of already established irreversible and progressive axonal degeneration. Selection of patients with minimal (EDSS < 3.0) or moderate (EDSS 3.0 to 5.5) disability would mandate a less intense conditioning regimen in order to minimize morbidity and mortality. Since MS is an inflammatory neuro-degenerative disease, the conditioning regimen should also be designed to minimize further neuronal injury.

Patients for the new North American studies will be selected earlier in disease and the protocol would be aimed at suppressing relapses in patients at risk for progressive disability. Unfortunately, there are no good clinical or MRI markers predictive for worsening disability in patients with relapsing-remitting disease. Weinshanker has reported that the number of relapses within the first 2 years correlates with late disability.[34-36] Confavreux et al reported that for patients with an EDSS of 4.0 or greater (attained after a longer disease duration), the number of relapses does not correlate with progression of irreversible disability.[37] An EDSS of 4.0 or more may already be too late for therapy aimed at inflammatory demyelinating events to prevent progression of subsequent disability. However, some level of sustained disability would be required to justify the procedure until more evidence on safety and efficacy have accumulated. Efficacy of earlier intervention in MS is supported by the CHAMPS study, in which over a three year interval, treatment with interferon following the first clinical event significantly lowered the probability of developing clinically definite MS.[38] Unlike the phase I/II studies previously performed, instead of selecting patients for rapidly increasing disability which is related to axonal injury, patients will be selected for active relapses which are related to inflammatory demyelination despite conventional therapy.

Since patients in this study would be earlier in the disease course, a safer conditioning regimen than used in the phase I trials should be chosen. Cyclophosphamide at 200 mg/kg with or without ATG has been used safely with no mortality in over

100 autologous HSCT for a variety of autoimmune diseases including systemic lupus erythematosus, Crohn's disease, and rheumatoid arthritis. Another 100 patients with autoimmune disease have been treated with 200 mg/kg at Johns Hopkins without infusion of stem cells and no mortality (verbal communication, Robert Brodsky). Therefore, the anticipated mortality of this regimen should be less than 0.5 to 1.0%.

As mentioned, MS is probably both an axonal degenerative and demyelinating disease. While no data exists that TBI is associated with axonal injury at the doses used in the American phase I regimens, radiation's effect is via generation of intracellular free radicals causing damage to DNA and protein. TBI may, theoretically, accelerate axonal degeneration despite preventing or slowing demyelination. TBI may also injure neural stem cells or inhibit oligodendrocyte remyelination impairing mechanisms of central nervous system repair. The sensitivity of neural stem cells to cyclophosphamide is unknown, but hematopoietic stem cells are resistant to cyclophosphamide despite sensitivity to even low doses of radiation. TBI has been associated with a high incidence of late myelodysplasia and leukemia,[39-41] as well as solid tumors,[42] hypothyroidism, and cataracts which would be unlikely from pulse cyclophosphamide. TBI will cause infertility in 100% of patients. In contrast, cyclophosphamide induced infertility is age related. Females under age 26 regain normal ovarian function, while 1/3 of females over age 26 regain ovarian function.[43] Phase I HSCT studies using triple agent immune suppressive regimens and CD34+ selection have been associated with lethal opportunistic infections including PTLD, influenza and pneumoccocus. No lethal infections have been reported for cyclophosphamide ATG with or without stem cells in any of approximately 200 patients treated for autoimmune disorders. Unlike TBI containing regimens, a regimen of cyclophosphamide +/- either ATG or CAMPATH is not myeloablative and even if autologous stem cells did not engraft, hematopoiesis would recover spontaneously.[44]

Intravenous cyclophosphamide used at doses of 500 to 1000 mg/m^2 monthly over 6 months (total 3.0 to 6.0 grams) has been reported to markedly diminish relapses and MRI activity for at least 18 months in aggressive RRMS.[45,46] Transplantation allows the safe administration of 2.0 grams/m^2 for mobilization and 14 grams at conditioning (total 18 grams) in one treatment with a more pronounced immune suppressive effect.

Phase I HSCT trials, as well as animal data, suggest that to be effective, HSCT should be performed earlier in disease onset while still an immune mediated inflammatory process. However, such patients will have an MS-related 10 year mortality of zero percent. Therefore, regimens should be designed to achieve a treatment related mortality of less than 0.5 to 1.0 % while also minimizing any late toxicities. Since the development of the initial phase I studies, MS has been recognized to be both an immune-mediated and neuro-degenerative disease. MS is a unique disease in which we must find equipoise by balancing the benefits of immune suppression against the risks of infection and neuronal toxicity. Based on these concepts, future American trials are being developed under the auspice of the National Institutes of Health. Patients will be selected for 2 or more relapses that required treatment with intravenous solumedrol despite at least 6 months of interferon and moderate disability (an EDSS of 3.0 to 5.5).

European

The Autoimmune Disease Working Party of the EBMT decided to demonstrate the efficacy of HSCT in a prospective, randomized trial which is named ASTIMS, i.e., Autologous Stem Cell Transplantation International Multiple Sclerosis Trial. The protocol is being finalized and will compare the sequence CY-BEAM-ATG without CD34 selection in patients with SPMS and poor-risk RRMS with the "best standard" treatment of MS, which is currently regarded to be mitoxantrone.[47,48] In order to decrease infections seen in the earlier studies, CD34 selection will be omitted. It is estimated that about 200 patients are needed to enroll in the study which will take five years to be completed.

References

1. Steinman L, Schwartz G, Waldor M et al. In: EAE: A Good Model for MS. San Diego: Academic Press, 1984:393-397.
2. Lipton HL. Theiler's virus infection in mice: An unusual biphasic disease process leading to demyelination. Infect Immun 1975; 11:1147.
3. Burt RK, Ruvollo P, Hess A et al. Syngeneic bone marrow transplantation in lewis rats with experimental autoimmune encephalitis. J Neurosci Res 1995; 41:526-531.
4. Burt RK, Padilla J, Begolka WS et al. Effect of disease stage on clinical outcome after syngeneic bone marrow transplantation for relapsing experimental autoimmune encephalomyelitis. Blood 1998; 91:2609-2616.
5. Miller SD, Vanderlugt CL, Begolka WS et al. Persistent infection with Theiler's virus leads to CNS autoimmunity via epitope spreading. Nat Med 1997; 3(10):1133-6.
6. Burt RK, Padilla J, Dal Canto MC et al. Viral hyperinfection of the central nervous system and high mortality following hematopoietic stem cell transplantation for treatment of Theiler's murine encephaleomyelitis virus (TMEV)–induced demyelineating disease. Blood 1999; 94(8), 2915-2922.
7. Burt RK. Bone marrow transplantation in multiple sclerosis. Bone Marrow Transplant 1995; 16:1-6.
8. Fassas A, Anagnostopoulos A, Kazis A et al. Peripheral blood stem cell transplantation in the treatment of progressive multiple sclerosis: First results of a pilot study. Bone Marrow Transplant 1997; 20:631-638.
9. Fassas A, Anagnostopoulos A, Kazis A et al. Autologous stem cell transplantation in progressive multiple sclerosis – an interim analysis of efficacy. J Clin Immunol 2000; 20:24-30.
10. Kozak T, Havrdova E, Pitha J et al. High-dose immunosuppressive therapy with PBPC support in the treatment of poor risk multiple sclerosis. Bone Marrow Transplant 2000; 25:525-531.
11. Kozak T, Havrdova E, Pitha J et al. Immunoablative therapy with PBPC support with in vitro or in vivo T-cell depletion in patients with poor risk multiple sclerosis. Bone Marrow Transplant 2002; 29(Suppl2):S15 (abstract O129).
12. Saiz A, Carreras E, Berenguer J et al. MRI and CSF oligoclonal bands after autologous hematopoietic stem cell transplantation in MS. Neurology 2001; 56:1084-1089.
13. Espigado I, Rovira M, Julia A et al. Phase I/II trials of autologous peripheral blood stem cell transplantation (ASCT) in autoimmune diseases (AID) resistant to conventional therapy: Preliminary results of the Spanish experience. Bone Marrow Transplant 2002; 29(Suppl2):S15(abstract O130).
14. Mancardi GL, Saccardi R, Filippi M et al. Italian GITMO-NEURO Intergroup on Autologous Hematopoietic Stem Cell Transplantation for Multiple Sclerosis. Autologous hematopoietic stem cell transplantation suppresses Gd-enhanced MRI activity in MS. Neurology 2001; 57:62-68.
15. Saccardi R, Mancardi G, Bacigalupo A et al. Autologous hematopoietic stem cell transplantation in secondary progressive MS: Clinical, MRI and laboratory findings. Bone Marrow Transplant 2002; 29(Suppl2): S13(abstract O125).
16. Filippi M, Rovaris M, Capra R et al. A multi-centre longitudinal study comparing the sensitivity of monthly MRI after standard and triple dose gadolinium-DTPA for monitoring disease activity in multiple sclerosis. Implications for phase II clinical trials. Brain 1998; 121:2011-2020.

17. Fassas A, Passweg JR, Anagnostopoulos A et al. For the Autoimmune Disease Working Party of the EBMT (European Group of Blood and Marrow Transplantation). Hematopoietic stem cell transplantation for multiple sclerosis: A retrospective multicenter study. J Neurol 2002; 249:1088-1097.
18. Burt RK, Traynor AE, Cohen B et al. T cell-depleted autologous hematopoietic stem cell transplantation for multiple sclerosis: Report on the first three patients. Bone Marrow Transplant 1998; 21(6):537-41.
19. Burt RK, Traynor AE, Pope R et al. Treatment of autoimmune disease by intense immunosuppressive conditioning and autologous hematopoietic stem cell transplantation. Blood 1998; 92(10):3505-14.
20. Burt RK, Burns W, Miller SD. Bone marrow transplantation for multiple sclerosis, returning to Pandora's box, Immunol Today 1997; 18(12):559-562.
21. Nash RA, Kraft GH, Bowen JD et al. Treatment of severe multiple sclerosis (MS) with high dose immunosuppressive therapy (HDIT) and autologous stem cell transplantation (SCT). (abst 3640) Blood 2000; (suppl):842a.
22. Nash RA, Dansey R, Storek J et al. Epstein Barr virus (EBV)-associated post-transplant lymphoproliferative disorder (PTLD) after high dose immunosuppressive therapy (HDIT) and autologous CD34-selected stem cell transplantation (SCT) for severe autoimmune diseases. (abst 1747) Blood 2000; 96(suppl A):406a.
23. Openshaw H, Lund BT, Kashyap A et al. Peripheral blood stem cell transplantation in multiple sclerosis with busulfan and cyclophosphamide conditioning: Report of toxicity and immunological monitoring. Biol Blood & Marrow Transplant 2000; 6(5A):563-75.
24. Moreau T, Coles A, Wing M et al. Transient increase in symptoms associated with cytokine release in patients with multiple sclerosis. Brain 1996; 119:225-237.
25. Openshaw H, Stuve O, Antel JP et al. Multiple sclerosis flares associated with recombinant granulocyte colony-stimulating factor. Neurology 2000; 54:2147-2150.
26. Burt RK, Fassas A, Snowden JA et al. Collection of hematopoietic stem cells from patients with autoimmune diseases. Bone Marrow Transplant 2001; 28:1-12.
27. te Boekhorst PAW, Schipperus MR, Samijn JPA et al. Total body irradiation (TBI) and high-dose chemotherapy combined with T-cell depleted marrow grafts in multiple sclerosis (MS). Bone Marrow Transplant 2002; 29(Suppl2):S104(abstract P471).
28. Oyama Y, Cohen B, Traynor A et al. Engraftment syndrome: A common cause for rash and fever following autologous hematopoietic stem cell transplantation for multiple sclerosis. Bone Marrow Transplant 2002; 29(1):81-5.
29. Trapp BD, Peterson J, Ransohoff RM, Rudick R, Mork S, Bo L. Axonal transection in the lesions of multiple sclerosis. N Engl J Med 1998; 338: 278-85.
30. De Stefano N, Matthews PM, Fu L et al. Axonal damage correlates with disability in patients with relapsing-remitting multiple sclerosis. Results of a longitudinal magnetic resonance spectroscopy study. Brain 1998; 121 (Pt 8):1469-77.
31. Fu L, Matthews PM, De Stefano N et al. Imaging axonal damage of normal-appearing white matter in multiple sclerosis. Brain 1998; 121(1):103-13.
32. Ferguson B, Matyszak MK, Esiri MM et al. Axonal damage in acute multiple sclerosis lesions. Brain 1997; 120(3):393-9.
33. Monje ML, Mizumatsu S, Fike JR et al. Irradiation induces neural precursor-cell dysfunction. Nat Med 2002; 8(9):955-62.
34. Weinshenker BG. The natural history of multiple sclerosis: Update 1998. Semin Neurol 1998; 18(3):301-7.
35. Weinshenker BG. The natural history of multiple sclerosis. Neurol Clinics 1995; 13(1):119-46.
36. Weinshenker BG. Natural history of multiple sclerosis. Ann Neurology 1994; 36 Suppl:S6-11
37. Confavreux C, Vukusic S, Moreau T et al. Relapses and progression of disability in multiple sclerosis. N Engl J Med 2000; 343(20):1430-8.
38. Jacobs LD, Beck RW, Simon JH et al. Intramuscular interferon beta-1a therapy initiated during a first demyelinating event in multiple sclerosis. CHAMPS Study Group. N Engl J Med 2000; 343(13):898-904.
39. Stone RM, Neuberg D, Soiffer R et al. Myelodysplastic syndrome as a late complication following autologous bone marrow transplantation for non-Hodgkin's lymphoma. J Clin Oncol 1994; 12(12):2535-42.
40. Krishnan A, Bhatia S, Slovak ML et al. Predictors of therapy-related leukemia and myelodysplasia following autologous transplantation for lymphoma: An assessment of risk factors. Blood 2000; 95(5):1588-93.
41. Milligan DW, Ruiz De Elvira MC, Kolb HJ et al. Secondary leukaemia and myelodysplasia after autografting for lymphoma: Results from the EBMT. EBMT Lymphoma and Late Effects Working Parties. European Group for Blood and Marrow Transplantation. [Journal Article. Multicenter Study] Brit J Haematol 1999; 106(4):1020-6.
42. Curtis RE, Rowlings PA, Deeg HJ et al. Solid cancers after bone marrow transplantation. New Engl J Med 1997; 336(13):897-904.
43. Sanders JE, Hawley J, Levy W et al. Pregnancies following high-dose cyclophosphamide with or without high-dose busulfan or total-body irradiation and bone marrow transplantation. Blood 1996; 87(7):3045-52.
44. Brodsky RA, Petri M, Smith BD et al. Immunablative high dose cyclophosphamide without stem cell rescue for refractory severe autoimmune disease. Ann Intern Med 1998; 129:1031-1035.
45. Khan OA, Zvartau-Hind M, Caon C et al. Effect of monthly intravenous cyclophosphamide in rapidly deteriorating multiple sclerosis patients resistant to conventional therapy. Mult Scler 2001; 7(3):185-8.
46. Patti F, Cataldi ML, Nicoletti F et al. Combination of cyclophosphamide and interferon-beta halts progression in patients with rapidly transitional multiple sclerosis. J Neurol, Neurosurg Psychiatry 2001; 71(3):404-7.
47. Edan G, Miller D, Clanet M et al. Therapeutic effect of mitoxantrone combined with methylprednisolone in multiple sclerosis: A randomised multicentre study of active disease using MRI and clinical criteria. J Neurol Neurosurg Psychiatry 1997; 62:112-118.
48. Hartung HP, Gonsette R. The MIMS-Study Group. Mitoxantrone in progressive multiple slcerosis: A placebo controlled, double-blind, multicentre trial. Lancet 2002; 360:2018-2025.

CHAPTER 38

Molecular and Cellular Pathogenesis of Systemic Lupus Erythematosus

George C. Tsokos, Yuang-Taung Juang, Christos G. Tsokos and Madhusoodana P. Nambiar

Introduction

Systemic lupus erythematosus (SLE) is a prototypic autoimmune disease that afflicts women during their child-bearing years. The heterogeneous clinical disease is characterized by abnormal production of autoantibodies and immune complexes in association with a diverse array of clinical manifestations. Investigation into the etiopathogenesis has been directed to understanding the susceptibility genes, intrinsic biochemical abnormalities, mechanisms of autoimmunity, as well as the basis for clinical heterogeneity. Emerging immune cell signaling abnormalities in SLE may be more clearly understood by studying samples before and after hematopoietic stem cell transplantation, comparing abnormalities in the same individual with and without active disease.

Systemic lupus erythematosus (SLE) is an idiopathic autoimmune disease characterized by disorders of cellular and humoral immune response leading to autoantibody production. SLE predominantly afflicts women (9:1 compared to men) in their child-bearing years. People of African, Afro-Caribbean, Asian, Native American and East Asian descent are more likely to develop SLE and may have more severe course of disease and prognosis than Caucasians. The autoimmune response leads to abnormal production of a wide spectrum of autoantibodies forming immune complexes that deposit on tissues propagating a chronic inflammatory process that destroys organ parenchyma and results in end-stage organ failure. The precise pathogenic mechanisms leading to abnormal T cell functions in SLE remain incompletely understood. Although tissue bound antibody and immune complexes are the principal effectors of this inflammatory process, currently it is held that fundamental abnormalities of the immune system exist at the cellular level that authorize the improper production of autoantibodies. The current treatment of SLE centers on refined regimens of corticosteroids and cytotoxic drugs.

Despite considerable research, the etiology of SLE remains elusive. Genetic, hormonal, environmental and immunomodulatory factors contribute to the pathogenesis of SLE. A central role for T cells is suggested in the pathogenesis of the disease and immune aberrations of SLE T cells are considered to be the primary event in the pathologic process. Identifying the underlying genetic and biochemical mechanisms will contribute to our understanding of the etiopathogenesis of SLE. Here we discuss the newest developments in the molecular and cellular pathogenesis of human lupus with emphasis on abnormal immune cell signaling.

Factors Involved in Human SLE

Genome-wide searches in families with multiple affected members revealed the involvement of a large number of genes contributing to the expression of the disease and are distributed through out the genome, although most are clustered in the 1q and 6p chromosomes (1-3;3) (Table 1). The latter include genes for HLA and complement factors. Genes determine susceptibility and no particular gene is necessary or sufficient for disease expression. However, almost all of the patients with deficiency of the early classical complement pathway, such as C1 (C1q, C1r, or C1s subunits), C2 and C4 genes that are integral to the immune response, develop SLE. In African Americans, chromosome region 1q was found to have the strongest linkage. Candidate genes in this interval include the receptors for binding the Fc portion of immunoglobulin (FcR) such as FcγRIIA, FcγRIIIA, FcγRIIIB and the ζ (zeta) chain of the T cell receptor complex.[4] Environmental factors include ultraviolet light, a large list of medications, and infections including viruses and other pathogens. Predominant female preponderance of SLE suggests that hormonal factors play a critical role in SLE pathogenesis. Hormonal factors (estrogens) are very important and are involved in the regulation of the transcription of genes central to the expression of SLE. In T cells of females with SLE, estradiol augments expression and activity of calcineurin, suggesting that there is a gender-specific estrogen receptor dysfunction that might modulate immune response and contribute to the autoimmune diathesis.[5] No single factor can cause the disease; multiple factors should act simultaneously or sequentially to cause the disease. Relative contribution of each factor varies among patients.

Immune Cell Aberrations

Immune system abnormalities recorded in SLE are multiple and diverse.[6,7] Central among them is the presence of overactive B cells responsible for the production of autoantibodies. B cells operate under the control of T cells, which provide cognate help by engaging costimulatory pairs of molecules on the surface of interacting immune cells. Accumulating evidence suggests that a

Table 1. Possible susceptibility loci for systemic lupus erythematosus

Complement pathway
    C2, C4, C1q
    Complement receptor 1 and 2
    Mannose binding protein
Human major histocompatibility complex
    DR3, DR2, DQ2, DQ6 containing haplotypes
Lymphokines
    Tissue necrosis factor (α gene promoter)
    Tumor necrosis factor receptor
    Interleukin-6, Interleukin-10
Fc receptors
    FcγRIIA, FcγRIIIA
Apoptsois factors
    Poly (ADP-ribose) polymerase
    Bcl2
    Fas ligand
Cell signaling
    Protein kinase A-I and Protein kinase A-II

fundamental disorder of both CD4⁺ T helper (Th) and CD8⁺ T cell cytotoxic/suppressor functions exist in SLE.[7] The physiologic ratio of circulating CD4⁺ Th1 to Th2 is skewed toward Th2 cells in SLE, resulting in diminished generation of interleukin-2 and interferon-γ by Th1 cells, and augmented production of interleukin-6 and interleukin-10.[8] Activated CD4⁺ Th cells bearing increased cell surface CD40 ligand and CD11a/CD18 (lymphocyte function associated antigen-1) overproduce interleukin-6 and interleukin-10. In turn, CD4⁺ Th2 cells interact with B cells driving overproduction of immunoglobulins by B cell clones, directed against components of cell nucleus in human SLE. Simultaneously, in response to antigens such as tetanus toxoid and viruses, T cells display decreased cellular cytotoxic responses and decreased production of interleukin-2 that may contribute to the increased rate of infections in SLE patients.

T Cell Signaling Abnormalities

Immunocompetent cells respond to external antigens following their engagement to the antigen receptor through a series of biochemical processes involving protein tyrosine phosphorylation, calcium mobilization and activation of transcription factors, known as cell signaling[9,10] (Fig. 1).

Following engagement of the antigen receptor, T and B lymphocytes from SLE patients respond rapidly by hyperphosphorylating a number of cytosolic signaling intermediates and increasing their concentration of free calcium (Table 2).[9,10] This abnormal response occurs in both principal T cell subsets and in cells culture propagated in vitro. In SLE T cells, the more sustained increased calcium response has a faster kinetics and is observed under deficient T cell receptor zeta (ζ) chain expression.[11-13] The deficient T cell receptor ζ chain was also associated with decreased activation-induced cell death in SLE T cells.[14] The T cell receptor ζ chain is considered to be the limiting factor in T cell receptor assembly, transport and surface expression and is crucial to receptor signaling function.[15,16] A vast majority of the SLE patients also display decreased expression of T cell receptor ζ chain messenger RNA. The T cell receptor ζ chain transcript is generated as the spliced product of 8 exons that are separated by distances of 0.7 kb to more than 8 kb.[17] T cell receptor ζ chain gene is located in chromosome 1q23.[18,19] Genes encoded in chromosome 1q have been suggested to contribute to genetic predisposition and susceptibility to SLE by genome-wide scan of mulitplex SLE families.[2,3,20-22]

Although the precise molecular mechanisms underlying ζ chain deficiency is still being examined, current evidence supports the possibility of a transcriptional defect. In SLE, T cells that expressed low levels of T cell receptor ζ chain transcripts, cloning and sequencing revealed more frequent heterogeneous polymorphisms/ mutations and alternative splicing of T cell receptor ζ chain.[12,23,24] Most of these mutations are localized to the three immunoreceptor tyrosine activation motifs (ITAM) or guanosine triphosphate (GTP) binding domain and could functionally affect the ζ chain providing a molecular basis to the known T cell signaling abnormalities in SLE T cells. Absence of the mutations/polymorphisms in the genomic DNA suggests that these are the consequence of irregular RNA editing. SLE patients also showed significant increase in the splice variation of the ζ chain. The splicing abnormality included two insertion splice variants of 145 bases and 93 bases between exons I and II, and also several deletion splice variants of T cell recceptor ζ chain resulting from the deletion of individual exons II, VI, VII, or a combined deletion of exons V and VI; VI and VII; II, III and IV; and V, VI and VII in SLE T cells.

SLE T cells also showed significant increase in the ζ chain promoter polymorphism.[25] Reverse transcription-polymerase reaction analysis of the T cell receptor ζ chain 3'untranslated region showed an alternatively spliced 344 bp product with both splicing donor and acceptor sites, resulting from deletion of nucleotides from 672 to 1233 of T cell receptor ζ chain mRNA. Unlike the normal T cell receptor ζ chain, the expression of T cell receptor ζ chain with the alternatively spliced 344 bp 3'untranslated region was higher in SLE T cells compared to non-SLE controls. Preliminary studies show that the stability of the T cell receptor ζ chain with alternatively spliced 3'untranslated region is unstable leading to its downregulation in SLE T cells (Tsokos et al, unpublished data). Since many of the principal antinuclear antibodies in SLE are directed against spliceosome components, it is not clear whether these autoantibodies play any role in the defective messenger RNA synthesis, splicing and processing in SLE T cells.

Abnormalities of T Cell Receptor ζ Chain Transcription Factor, ELF-1 Expression

Analysis of the T cell receptor ζ chain transcription factor, Elf-1, indicates that the p98 form of the protein was either decreased or not functional in a majority of the lupus patients, whereas the p80 form was increased or comparable. Elf-1 is a member of the Ets (E twenty-six specific) transcription factor family. Elf-1 has a calculated molecular weight of 68 kDa while we found that it existed in T cells as both 80 and 98 kDa forms. We found the molecular basis for this conversion involving dual post-translational processes, i.e., glycosylation and phosphorylation.[26] Shift assays and ultraviolet cross-linking studies indicated that the phosphorylated 98 kDa Elf-1 is the functional form that binds to the T cell receptor ζ chain promoter. Nuclear proteins from approximately 25% of SLE T cells displayed decreased production of the 98 but not the 80 kDa form, which correlated with their defective T cell receptor ζ chain promoter binding. Another 30% of SLE T cells nuclear proteins are defective in T cell receptor ζ chain promoter binding, despite normal expression of both 98 and 80 forms of Elf-1. The defective formation of

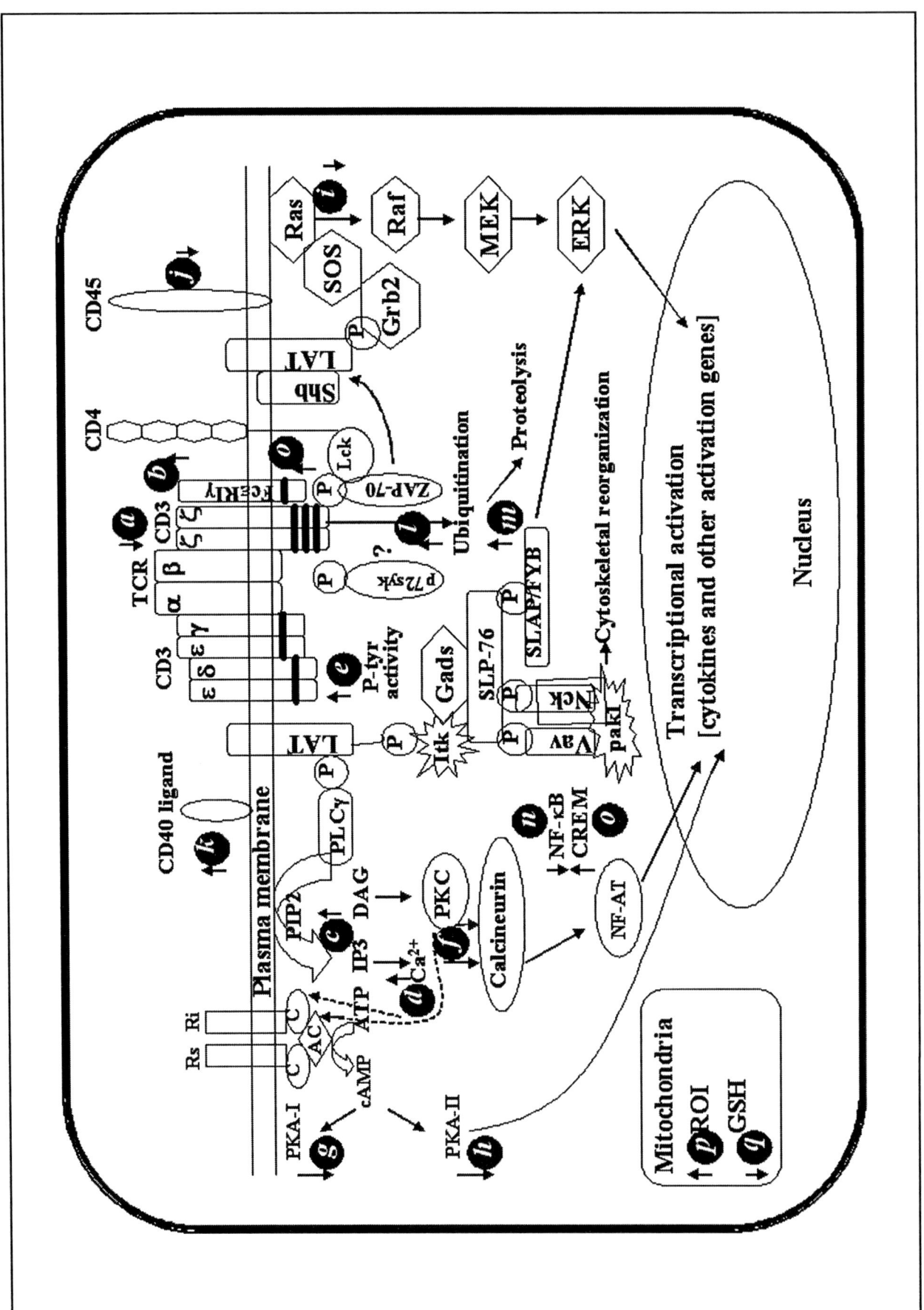

Figure 1. Legend on opposite page.

Table 2. Molecular defects of the T cell signaling pathways in systemic lupus erythematosus

Molecular Defect	Functional Effect	Reference
CD45 tyrosyl phosphatase	Deficient	28
T cell receptor ζ chain	Deficient, increased lipid raft association	11-13,27
Fc epsilon receptor type I gamma-chain	Increased	29
Lymphocyte-specific tyrosine protein kinase	Deficient	54
Inositol triphospate	Increased	11
Intracellular calcium response	Increased and more sustained	11
Protein kinase A-I and protein kinase A-II isozymes	Deficiency	50,64
Protein kinase C	Deficiency	53
Nuclear factor kappa B, p65 Rel A subunit	Deficiency	60
T cell receptor ζ chain transcription factor, Elf-1	Deficiency	26a
Cyclic AMP response element modulator	Upregulation and increased nuclear binding to −180 site of IL-2 promoter	63
Ras-mitogen activated protein kinase	Deficiency	57

functional 98 kDa Elf-1 thus underlies the defective T cell receptor ζ chain expression in SLE patients.[26a]

The T cell receptor ζ chain exists in multiple forms and membrane fractions with distinct function in Ag-mediated signaling process. Studies on the complete spectrum of expression of various molecular forms of the T cell receptor ζ chain has shown that the phosphorylated 21 and 23 kDa forms of the T cell receptor ζ chain are significantly decreased in SLE T cells compared to normal T cells. In contrast, major ubiquitinated forms of the T cell receptor ζ chain were increased in SLE T cells suggesting that T cell receptor ζ chain undergoes an enhanced ubiquitin-mediated degradation in SLE T cells. The level of T cell receptor ζ chain was also significantly decreased in the detergent-insoluble membranes in SLE T cells. Similarly, the expression of T cell receptor η chain, an alternatively spliced form of the ζ chain, was diminished in SLE T cells. Recently, we identified upregulation of a novel 14 kDa form of the ζ chain, a potential alternatively spliced or degraded species of the T cell receptor ζ chain in SLE T cells.[27]

Mechanisms of Increased T Cell Receptor/CD3-Mediated Intracellular Calcium Response in SLE

How does the T cell receptor engagement induce hyperphosphorylation of cytosolic proteins and increased intracellular calcium response in the milieu of deficient T cell receptor ζ chain has been a topic of intense research for the last couple of years. In addition to the possible gain-of-function mutations of the T cell receptor ζ chain, our investigation has proposed two mechanisms involving increased expression of Fc epsilon receptor type I gamma chain and increased membrane lipid-raft association of the residual T cell receptor ζ chain, that could explain

Figure 1. Current model of signal transduction in T lymphocytes and the molecular defects identified in human systemic lupus erythematosus. Upon T cell receptor activation by antigen, the tyrosine residues within the immunotyrosine activation motifs become phosphorylated by lymphocyte specific protein kinase (Lck) and Fyn, leading to the association and activation of zeta chain associated protein-70 (ZAP-70). Once activated, Fyn, lymphocyte-specific tyrosine kinase, spleen tyrosine kinase and zeta chain associated protein-70 cooperate in the tyrosine phosphorylation, activation and juxtaposition of downstream signal transducers that contribute to the initiation of mitogen activated protein (MAP) kinase cascades, phosphatidyl inositol 3 (IP3) -kinase activation, intracellular calcium flux and activation of transcription factors. Eventually these transcription factors translocate to the nucleus and modulate the expression of genes that regulate lymphocyte activation, cellular proliferation, anergy or apoptosis, secretion of soluble mediators and effector functions. Signaling abnormalities currently identified in SLE T cells are indicated in black circles and the direction of the arrow indicates up or downregulation. a) T cell receptor z chain deficiency; b) increased Fc epsilon receptor type I gamma (FceRIg) chain that replace the deficient T cell receptor z chain; c) moderate increase in inositol triphostate (IP3); d) increased and sustained intracellular calcium (Ca^{++}) levels; e) T cell receptor/CD3 mediated increased phosphorylation of cellular protein substrates with faster kinetics; f) reduced protein kinase C (PKC) phosphorylation; g) deficient protein kinase A I (PKA-I) activity; h) Impaired protein kinase A II (PKA-II) activity and increased nuclear translocation of free RIIb subunit; i) reduced Ras-mitogen activated protein (Ras MAP) kinase signaling; j) deficient CD45 phosphatase; k) increased expression of CD40 ligand; l) increased ubiquitination; m) increased proteolytic acitivity; n) defective nuclear factor kappa B (NF-kB) ; o) upregulation of cyclic adenosine monophosphate response element modulator (CREM); p) increased reactive Oxygen species; and q) decreased level of glutathione. Abbreviations: Gads= Grb2 related adaptor downstream of Shc; Grb2= growth factor receptor binding protein 2; SLP-76= SH2 containing leukocyte phosphoprotein of 76 kDa; SOS= Son of Sevenless; Syk= spleen tyrosine kinase; ERK= extracellular signal regulated kinase; Itk= interleukin-2-inducible T-cell kinase; MEK= mitogen activated protein kinase/extracellular signal regulated kinase; Vav= a guanine nucleotide exchange factor for Ras; Fyn= a src-like protein tyrosine kinase; LAT= linker for activation in T cells; NF-AT= Nuclear factor of activated T cells; Lck= lymphocyte-specific tyrosine protein kinase; Shb= src homology 2 protein; SLAP= src like adaptor protein; FYB= Fyn binding protein; GSH= glutathione; ROI= reactive Oxygen species.

the increased T cell receptor /CD3-mediated intracellular calcium response in SLE T cells. Also, the protein tyrosine phosphatase activity of CD45 on peripheral blood lymphocytes is reduced in SLE.[28]

The hypothesis that other members of the ζ chain family, such as Fc epsilon receptor type I gamma (FcεRIγ) chain, may substitute for the deficient T cell receptor ζ chain, has been investigated by Enyedy et al.[29] Immunoprecipitation/immunoblotting and confocal microscopy experiments demonstrated that a large proportion of SLE T cells express very high levels of FcεRIγ chain that is functionally associated with the T cell receptor and takes part in antigen receptor mediated signal transduction. Expression of FcεRIγ, in lieu of T cell receptor ζ chain, has been reported in mouse large granular lymphocytes.[30] T lymphocytes from tumor-bearing mice expressed T cell receptor that completely lacked T cell receptor ζ chain and was replaced by FcεRIγ chain.[31] Also, T cell receptor ζ deficient mice have been shown to express FcεRIγ chain as part of the T cell receptor -γδ complex.[32-34] Unlike T cell receptor ζ chain, which mediates signaling through zeta chain-associated protein-70 (ZAP-70), Fc epsilon receptor type I gamma chain mediates signaling by associating with 100-fold more potent phosphorylated protein spleen tyrosine kinase (SyK).[35,36] Presently, it is unknown whether SyK is upregulated in SLE T cells and FcεRIγ chain signal transduction occurs by associating with SyK or other downstream signaling molecules. High-level expression of tyrosine kinase activity and alternative antigen receptor mediated signaling has been described in T cells of patients with ZAP-70 deficiency.[37] Overexpression of the FcεRIγ gamma chain in normal T cells also leads to increased T cell receptor/CD3 mediated intracellular calcium response suggesting that the single immunotyrosine activation motif in the FcεRIγ chain, compared to the three in T cell receptor ζ chain, does not hinder the level of intracellular calcium response. The high level expression of FcεRIγ chain could replace the defective T cell receptor ζ chain and contribute to the aberrant antigen receptor-initiated signaling in SLE T cells.

Increased Membrane Lipid-Raft Association of the Residual T Cell Receptor ζ Chain

The T cell receptor ζ chain associated with the detergent-insoluble fraction is distributed between cytoskeleton as well as lipid-rich membrane microdomains, composed primarily of sphingolipids and cholesterol, and an enriched subset of proteins that float laterally as 'rafts' within the plasma membrane.[38] Lipid rafts are preformed functional modules that serve as platforms for signal transduction and membrane trafficking. Recent data indicate that lipid rafts are crucial for effecting T cell receptor signal transduction.[39,40] T cell receptor engagement leads to translocation and concentration of tyrosine phosphorylated T cell receptor ζ chain and downstream signal transduction molecules within lipid rafts.[41] Conversely, perturbation of the structural integrity of lipid rafts inhibits T cell receptor-induced protein tyrosine phosphorylation and calcium flux.[40,42]

Dissociation of the lipid-rafts by cholesterol depletion using methyl-β-cyclodextrin, showed increased percentage of the residual membrane bound T cell receptor ζ chain in the lipid-rafts in resting SLE T cell membranes.[27] Fluorescence microscopy indicated that the residual T cell receptor ζ chain is more clustered on the cell membranes of SLE T cells compared to normal. Faster kinetics of T cell receptor/CD3 mediated intracellular calcium response in SLE T cells also support the clustering or cross-talk between signaling pathways in SLE. Upon T cell receptor/CD3 activation, ζ chain clusters became more prominent in SLE T cells and they superimpose with linker for activation of T cells suggesting that they are colocalized to lipid rafts. Increased lipid-raft association and surface clustering of the residual T cell receptor ζ chain may explain the increased T cell receptor/CD3-mediated intracellular response in SLE T cells. Based on these data, we have proposed a model that suggests that in SLE T cells, although there is a deficiency, the residual T cell receptor ζ chain is more associated with membrane lipid rafts resulting in more preformed T cell receptor clustering. Increased membrane T cell receptor clustering, as well as replacement of the deficient T cell receptor ζ chain by FcεRIγ chain, could lead to increased intracellular calcium response under T cell receptor ζ chain deficiency and decreased tolerance in SLE T cells. In support of this view, recently, it has been reported that mice with N-acetylglucosaminyltransferase deficiency show decreased glycosylation of T cell membrane proteins that prevent galectin binding and thereby disrupting the galectin-glycoprotein lattice, leading to increased clustering of T cell receptor.[43] Increased T cell receptor clustering in these autoimmune mice had a very similar phenotype to human SLE, with lowered T cell activation thresholds and increased T cell receptor signaling.[44] Similarly, it has also been suggested that genetic remodeling of protein glycosylation by mutation of α mannosidase II induce autoimmune disease.[45] As the association between the residual ζ chain with the lipid rafts, T cell receptor clustering and T cell effector functions is further explored, it is likely that important new insights will emerge that will explain ζ chain abnormalities and autoreactivity of T cells.

Deficient Protein Kinase A-I and RAS-Mitogen Activated Protein Kinase Activity

Studies on adenyl cyclase (AC)/cyclic adenosine monophosphate (cAMP)/protein kinase A (PKA) system, a key metabolic pathway integral to cellular homeostasis, identified impaired cyclic adenosine monophosphate-dependent protein phosphorylation in SLE T cells.[46,47] Abnormal cyclic adenosine monophosphate-dependent signaling pathway is associated with deficient CD8 suppressor T cell function and altered cytoskeletal regulation of CD3, CD4 and CD8 receptor mobility within the plane of the plasma membrane.[48,49] Recently, it has been demonstrated that abnormal cyclic adenosine monophosphate-dependent signaling pathway reflects a profound reduction in protein kinase A-I (PKA-I) isozyme.[50] Deficient PKA-I activity is the consequence of reduced type I regulatory subunit. Deficient T cell PKA-I activity reflects reduction of both holoenzymes RIβ$_2$C$_2$ >RIα$_2$C$_2$.[51] Deficient PKA-I activity possibly contributes to altered T cell effector function by altering the protein phosphorylation that regulates cellular pathways that promote cell growth and differentiation. Similar to T cell receptor ζ chain, increased mutation/polymorphism of PKA-I regulatory subunit α has been reported in a SLE patient.[52] The kinase activity of protein kinase C (PKC)[53] and lymphocyte-specific protein tyrosine kinase (LcK) is also impaired in SLE T cells.[54] The activity of other kinases, such as protein kinase R (PKR) that is involved in the phosphorylation of translation initiation factors, is increased in SLE T cells.[55]

Interrelationship between the calcium and adenyl cyclase/cyclic adenosine monophosphate/protein kinase A pathway raise the possibility that defective protein kinase A (PKA) activity in part contribute to the impaired calcium homeostasis. T cell receptor/CD3-mediated increase in inositol 1,4,5-trisphosphate

(IP_3) and intracellular calcium is downregulated by inactivation of phospholipase C-γ1 (PLC-γ1) through PKA-dependent phosphorylation.[56] Deficient PKA catalyzed phosphorylation may retain the activity of phospholipase C-γ1 and contribute to the increased and sustained intracellular calcium response in SLE T cells.

Environmental factors, including drugs and ultraviolet light induce SLE-like disorders. Because these agents induce hypomethylation of DNA, and modify the expression of the affected genes, a relationship between DNA hypomethylation and SLE has been sought. Upregulation of lymphocyte function associated antigen 1 (LFA-1/CD11a) generate autoreactive state by exaggerated help for the production of autoantibodies. The activity of the DNA methylation enzyme DNA (cytosine-5)-methyltransferase (Dmnt) appears to be regulated in part by the Ras-Mitogen Activated Protein (Ras-MAP) kinase pathway and it has been suggested that the reduced (cytosine-5)-methyltransferase-I activity in SLE T cells is a function of deficient Ras-MAP kinase signaling.[57]

Abnormalities in Transcription Factor Expression

Early signaling abnormalities are followed by altered activation of transcription factors and abnormal gene transcription. It is notable that while certain genes are transcribed at low rates (the T cell receptor ζ chain and interleukin-2 genes), others are transcribed at increased rates (the genes for the γ chain of the Fc receptor for IgE and the CD40 ligand).[58] In addition to the T cell receptor ζ chain, transcription factor Elf-1, described above, defects have been identified in the expression/activation of other transcription enhancers and repressors including nuclear factor κ B (NF-κB) p65-Rel A subunit and cyclic adenosine monophosphate response element modulator.

The transcription factor, NF-κB plays a profound role in immune and proinflammatory responses and interleukin-2 production.[59] The possibility that reduced interleukin-2 production by SLE T cells may be a product of altered NF-κB activity has been analyzed by electrophoretic mobility shift assays.[60] NF-κB activity in the nuclear extracts is significantly decreased in SLE T cells. In the group of SLE patients with decreased NF-κB activity, the transcriptionally active, heterodimeric p65/p50 complex was not formed in the cytosol. The deficiency of nuclear factor κB heterodimeric complex could be responsible for the downregulation of interleukin-2 and may have wide extensive pathophysiological significance in the expression of the disease.

SLE T cells stimulated in vitro in response to antigens or mitogens, proliferate significantly less than T cells from normal donors.[61] Activated SLE T cells also secrete low interleukin-2 in vitro that may reflect increased T cell anergy. Another mechanism for the low interleukin-2 production by SLE T cells is inhibition of interleukin-2 enhancer/promoter transcriptional activation. Cyclic adenosine monophosphate response element modulator (CREM) and inducible cyclic adenosine monophosphate early repressor (ICER) are two transcriptional repressors that bind to cyclic adenosine monophosphate response elements (CRE) and downregulate genes containing this binding site. Recently, it has been demonstrated that CREM binds to the -180 region of the interleukin-2 enhancer/promoter and contributes to T cell anergy.[62] This raised the question that CREM upregulation may contribute to the downregulation of interleukin-2 secretion and contribute to T cell anergy. Nuclear extracts from resting SLE T cells showed significantly increased binding of CREM/phosphorylated CRE-binding protein to the -180 site of the interleukin-2 enhancer/promoter.[63] Some patients revealed both phosphorylated cyclic adenosine monophosphate response element modulator (p-CREM) and phosphorylated cyclic adenosine monophosphate response element-binding protein (p-CREB), although p-CREM was the main factor. We have found that activated normal T cells increase p-CREB attachment by 10-fold in contrast to negligible p-CREM binding to the -180 site. By contrast, activated SLE T cells bound both p-CREM and p-CREB in 2:1 ratio. Thus, preferential phosphorylation and predominant occupancy of the -180 site of the interleukin-2 enhancer/promoter by p-CREM during activation may hamper optimum transcriptional activation by p-CREB binding resulting in very low interleukin-2 production in SLE-T cells.

Abnormal Nuclear Translocation of RIIβ Subunit

The RIIβ subunit of PKA-II is not classified currently as a nuclear factor, although its translocation to nucleus and binding to p-CREB supports the role of a transcription factor. Upon activation of the PKA-II holoenzyme, RIIβ/β$_2$C$_2$ by reversible cyclic adenosine monophosphate binding to the RII subunits release RIIβ that reversibly translocates to nucleus. Although it has been suggested that nuclear RIIβ binds p-CREB, at present its function in the nucleus still remains uncertain. RIIβ subunit is expressed constitutively in the nucleus suggesting an ongoing activation of RIIβ$_2$C$_2$ during homeostasis. The mechanism of deficient PKA-II activity in SLE T cells differs from PKA-I, and involves increased nuclear translocation of RIIβ to the nuclear fraction, diminishing the capacity to form RIIβ$_2$C$_2$ holoenzyme.[64]

Summary and Future Perspectives

SLE is a challenging disease with varied manifestations resulting from widespread immune complex deposition. Our understanding of the role of genetics and environmental agents in the pathogenesis of SLE has improved over the past ten years.[63] In addition, the past ten years have seen tremendous improvements in identifying molecular defects in immune cells to unravel the mechanism of autoimmunity. Emphasis has been placed in identifying the genetic basis of these molecular defects and restoration of their functions. Abnormalities of transcription factors contribute to immune response by linking signaling pathways and gene expression. Identification of several signaling defects suggests successive components of the signaling cascade may be intrinsically defective in SLE T cells. In regards to the cellular and molecular pathogenesis of SLE, it is important to seek all molecular defects at the cellular level and establish the genetic basis for this complex autoimmune disease in a timely fashion.[63]

References

1. Tsao BP, Cantor RM, Kalunian KC et al. Evidence for linkage of a candidate chromosome 1 region to human systemic lupus erythematosus. J Clin Invest 1997; 99:725-731.
2. Gaffney PM, Kearns GM, Shark KB et al. A genome-wide search for susceptibility genes in human systemic lupus erythematosus sib-pair families. Proc Natl Acad Sci USA 1998; 95:14875-14879.
3. Gray-McGuire C, Moser KL, Gaffney PM et al. Genome scan of human systemic lupus erythematosus by regression modeling: Evidence of linkage and epistasis at 4p16-15.2 [In Process Citation]. Am J Hum Genet 2000; 67:1460-1469.
4. Harley JB, Moser KL, Gaffney PM et al. The genetics of human systemic lupus erythematosus. Curr Opin Immunol 1998; 10:690-696.
5. Rider V, Foster RT, Evans M et al. Gender differences in autoimmune diseases: Estrogen increases calcineurin expression in systemic lupus erythematosus. Clin Immunol Immunopathol 1998; 89:171-180.

6. Tsokos GC, Balow JE. Cellular immune responses in systemic lupus erythematosus. Prog Allergy 1984; 35:93-161.
7. Tsokos GC. Overview of cellular immune function in systemic lupus erythematosus. In: Lahita RG, ed. Systemic Lupus Erythematosus. New York: Churchill Livingstone Inc, 1992:15-50.
8. Hagiwara E, Gourley MF, Lee S et al. Disease severity in patients with systemic lupus erythematosus correlates with an increased ratio of interleukin-10:Interferon-gamma-secreting cells in the peripheral blood. Arhtritis Rheum 1996; 39:379-385.
9. Vassilopoulos D, Kovacs B, Tsokos GC. TCR/CD3 complex-mediated signal transduction pathway in T cells and cell lines from patients with systemic lupus erythematosus. J Immunol 1995; 155:2269-2281.
10. Liossis SN, Kovacs B, Dennis G et al. B cells from patients with systemic lupus erythematosus display abnormal antigen receptor-mediated early signal transduction events. J Clin Invest 1996; 98:2549-2557.
11. Liossis SN, Ding DZ, Dennis GJ et al. Altered pattern of TCR/CD3-mediated protein-tyrosyl phosphorylation in T cells from patients with systemic lupus erythematosus. Deficient expression of the T-cell receptor zeta chain. J Clin Invest 1998; 101:1448-1457.
12. Takeuchi T, Tsuzaka K, Pang M et al. TCR zeta chain lacking exon 7 in two patients with systemic lupus erythematosus. Int Immunol 1998; 10:911-921.
13. Brundula V, Rivas LJ, Blasini AM et al. Diminished levels of T cell receptor zeta chains in peripheral blood T lymphocytes from patients with systemic lupus erythematosus. Arthritis Rheum 1999; 42:1908-1916.
14. Kovacs B, Vassilopoulos D, Vogelgesang SA et al. Defective CD3-mediated cell death in activated T cells from patients with systemic lupus erythematosus: Role of decreased intercellular TNF-alpha. Clin Immunol Immunopathol 1996; 81:293-302.
15. Weiss A, Littman DR. Signal transduction by lymphocyte antigen receptors. Cell 1994; 76:263-274.
16. Wange RL, Samelson LE. Complex complexes: Signaling at the TCR. Immunity 1996; 5:197-205.
17. Jensen JP, Hou D, Ramsburg M et al. Organization of the human T cell receptor zeta/eta gene and its genetic linkage to the Fc gamma RII-Fc gamma RIII gene cluster. J Immunol 1992; 148:2563-2571.
18. Weissman AM, Baniyash M, Hou D et al. Molecular cloning of the zeta chain of the T cell antigen receptor. Science 1988; 239:1018-1021.
19. Stacey M, Barlow A, Hulten M. Human T-cell receptor zeta chain gene Map position 1q23.1. Chromosome Res 1997; 5:279.
20. Moser KL, Neas BR, Salmon JE et al. Genome scan of human systemic lupus erythematosus: Evidence for linkage on chromosome 1q in African-American pedigrees. Proc Natl Acad Sci USA 1998; 95:14869-14874.
21. Gaffney PM, Ortmann WA, Selby SA et al. Genome screening in human systemic lupus erythematosus: Results from a second Minnesota cohort and combined analyses of 187 sib-pair families. Am J Hum Genet 2000; 66:547-556.
22. Shai R, Quismorio Jr FP, Li L et al. Genome-wide screen for systemic lupus erythematosus susceptibility genes in multiplex families. Hum Mol Genet 1999; 8:639-644.
23. Nambiar MP, Enyedy EJ, Warke VG et al. T cell signaling abnormalities in systemic lupus erythematosus are associated with increased mutations/polymorphisms and splice variants of T cell receptor zeta chain messenger RNA. Arthritis Rheum 2001; 44:1336-1350.
24. Tsuzaka K, Takeuchi T, Onoda N et al. Mutations in T cell receptor zeta chain mRNA of peripheral T cells from systemic lupus erythematosus patients. J Autoimmun 1998; 11:381-385.
25. Nambiar MP, Enyedy EJ, Warke VG et al. Polymorphisms/mutations of TCR-zeta-chain promoter and 3' untranslated region and selective expression of TCR zeta-chain with an alternatively spliced 3' untranslated region in patients with systemic lupus erythematosus. J Autoimmun 2001; 16:133-142.
26. Juang YT, Solomou EE, Rellahan B et al. Phosphorylation and O-linked glycosylation of Elf-1 leads to its translocation to the nucleus and binding to the promoter of the TCR zeta-chain. J Immunol 2002; 168:2865-2871.
26a. Juang YT, Tenbrock K, Nambiar NP et al. Defective production of the 98 kDa form of Elf-1 is responsible for the decreased expression of TCR zeta chain in patients with systemic lupus erythematous. J Immunol 2002; 169:6048-6055.
27. Nambiar MP, Enyedy EJ, Fisher CU et al. Abnormal expression of various molecular forms and distribution of T cell receptor zeta chain in patients with systemic lupus erythematosus. Arthritis Rheum 2002; 46:163-174.
28. Takeuchi T, Pang M, Amano K et al. Reduced protein tyrosine phosphatase (PTPase) activity of CD45 on peripheral blood lymphocytes in patients with systemic lupus erythematosus. Clin Exp Immunol 1997; 109:20-26.
29. Enyedy EJ, Nambiar MP, Liossis SN et al. Fc epsilon receptor type I gamma chain replaces the deficient T cell receptor zeta chain in T cells of patients with systemic lupus erythematosus. Arthritis Rheum 2001; 44:1114-1121.
30. Russell JH, Rush BJ, Abrams SI et al. Sensitivity of T cells to anti-CD3-stimulated suicide is independent of functional phenotype. Eur J Immunol 1992; 22:1655-1658.
31. Alava MA, DeBell KE, Conti A et al. Increased intracellular cyclic AMP inhibits inositol phospholipid hydrolysis induced by perturbation of the T cell receptor/CD3 complex but not by G-protein stimulation. Association with protein kinase A-mediated phosphorylation of phospholipase C-gamma 1. Biochem J 1992; 284:189-199.
32. Trump BF, Berezesky IK. The role of cytosolic Ca2+ in cell injury, necrosis and apoptosis. Curr Opin Cell Biol 1992; 4:227-232.
33. Klingmuller U, Bergelson S, Hsiao JG et al. Multiple tyrosine residues in the cytosolic domain of the erythropoietin receptor promote activation of STAT5. Proc Natl Acad Sci USA 1996; 93:8324-8328.
34. Weissman AM, Bonifacino JS, Klausner RD et al. T cell antigen receptor: Structure, assembly and function. Year Immunol 1989; 4:74-93.
35. Bossu P, Singer GG, Andres P et al. Mature CD4+ T lymphocytes from MRL/lpr mice are resistant to receptor-mediated tolerance and apoptosis. J Immunol 1993; 151:7233-7239.
36. DeBell KE, Conti A, Alava MA et al. Microfilament assembly modulates phospholipase C-mediated signal transduction by the TCR/CD3 in murine T helper lymphocytes. J Immunol 1992; 149:2271-2280.
37. Noraz N, Schwarz K, Steinberg M et al. Alternative antigen receptor (TCR) signaling in T cells derived from ZAP-70-deficient patients expressing high levels of Syk. J Biol Chem 2000; 275:15832-15838.
38. Simons K, Ikonen E. Functional rafts in cell membranes. Nature 1997; 387:569-572.
39. Ilangumaran S, He HT, Hoessli DC. Microdomains in lymphocyte signalling: Beyond GPI-anchored proteins. Immunol Today 2000; 21:2-7.
40. Montixi C, Langlet C, Bernard AM et al. Engagement of T cell receptor triggers its recruitment to low-density detergent-insoluble membrane domains. EMBO J 1998; 17:5334-5348.
41. Kosugi A, Saitoh S, Noda S et al. Translocation of tyrosine-phosphorylated TCRzeta chain to glycolipid- enriched membrane domains upon T cell activation. Int Immunol 1999; 11:1395-1401.
42. Xavier RM, Nakamura M, Tsunematsu T. Isolation and characterization of a human nonspecific suppressor factor from ascitic fluid of systemic lupus erythematosus. Evidence for a human counterpart of the monoclonal nonspecific suppressor factor and relationship to the T cell receptor alpha- chain. J Immunol 1994; 152:2624-2632.
43. Demetriou M, Granovsky M, Quaggin S et al. Negative regulation of T-cell activation and autoimmunity by Mgat5 N- glycosylation. Nature 2001; 409:733-739.
44. Demetriou M, Granovsky M, Quaggin S et al. Negative regulation of T-cell activation and autoimmunity by Mgat5 N- glycosylation. Nature 2001; 409:733-739.
45. Chui D, Sellakumar G, Green R et al. Genetic remodeling of protein glycosylation in vivo induces autoimmune disease. Proc Natl Acad Sci USA 2001; 98:1142-1147.

46. Mandler R, Birch RE, Polmar SH et al. Abnormal adenosine-induced immunosuppression and cAMP metabolism in T lymphocytes of patients with systemic lupus erythematosus. Proc Natl Acad Sci USA 1982; 79:7542-7546.
47. Hasler P, Schultz LA, Kammer GM. Defective cAMP-dependent phosphorylation of intact T lymphocytes in active systemic lupus erythematosus. Proc Natl Acad Sci USA 1990; 87:1978-1982.
48. Kammer GM. Impaired T cell capping and receptor regeneration in active systemic lupus erythematosus. Evidence for a disorder intrinsic to the T lymphocyte. J Clin Invest 1983; 72:1686-1697.
49. Kammer GM, Mitchell E. Impaired mobility of human T lymphocyte surface molecules during inactive systemic lupus erythematosus. Relationship to a defective cAMP pathway. Arthritis Rheum 1988; 31:88-98.
50. Kammer GM, Khan IU, Kammer JA et al. Deficient type I protein kinase A isozyme activity in systemic lupus erythematosus T lymphocytes: II. Abnormal isozyme kinetics. J Immunol 1996; 157:2690-2698.
51. Laxminarayana D, Khan IU, Mishra N et al. Diminished levels of protein kinase A RI alpha and RI beta transcripts and proteins in systemic lupus erythematosus T lymphocytes. J Immunol 1999; 162:5639-5648.
52. Laxminarayana D, Kammer GM. mRNA mutations of type I protein kinase A regulatory subunit alpha in T lymphocytes of a subject with systemic lupus erythematosus. Int Immunol 2000; 12:1521-1529.
53. Tada Y, Nagasawa K, Yamauchi Y et al. A defect in the protein kinase C system in T cells from patients with systemic lupus erythematosus. Clin Immunol Immunopathol 1991; 60:220-231.
54. Matache C, Stefanescu M, Onu A et al. P56lck activity and expression in peripheral blood lymphocytes from patients with systemic lupus erythematosus. Autoimmunity 1999; 29:111-120.
55. Grolleau A, Kaplan MJ, Hanash SM et al. Impaired translational response and increased protein kinase PKR expression in T cells from lupus patients. J Clin Invest 2000; 106:1561-1568.
56. Park DJ, Min HK, Rhee SG. Inhibition of CD3-linked phospholipase C by phorbol ester and by cAMP is associated with decreased phosphotyrosine and increased phosphoserine contents of PLC-gamma 1. J Biol Chem 1992; 267:1496-1501.
57. Deng C, Kaplan MJ, Yang J et al. Decreased Ras-mitogen-activated protein kinase signaling may cause DNA hypomethylation in T lymphocytes from lupus patients. Arthritis Rheum 2001; 44:397-407.
58. Tsokos GC, Kammer GM. Molecular aberrations in human systemic lupus erythematosus. Mol Med Today 2000; 6:418-424.
59. Ghosh S, May MJ, Kopp EB. NF-kappa B and Rel proteins: Evolutionarily conserved mediators of immune responses. Annu Rev Immunol 1998; 16:225-260.
60. Wong HK, Kammer GM, Dennis G et al. Abnormal NF-kappaB activity in T lymphocytes from patients with systemic lupus erythematosus is associated with decreased p65-relA protein expression [In Process Citation]. J Immunol 1999; 163:1682-1689.
61. Suciu-Foca N, Buda JA, theim T et al. Impaired responsiveness of lymphocytes in patients with systemic lupus erythematosus. Clin Exp Immunol 1974; 18:295.
62. Powell JD, Lerner CG, Ewoldt GR et al. The -180 site of the IL-2 promoter is the target of CREB/CREM binding in T cell anergy. J Immunol 1999; 163:6631-6639.
63. Solomou EE, Juang YT, Gourley MF et al. Molecular basis of deficient IL-2 production in T cells from patients with systemic lupus erythematosus. J Immunol 2001; 166:4216-4222.
64. Mishra N, Khan IU, Tsokos GC et al. Association of deficient type II protein kinase A activity with aberrant nuclear translocation of the RII beta subunit in systemic lupus erythematosus T lymphocytes. J Immunol 2000; 165:2830-2840.

CHAPTER 39

Definition, Classification, Activity and Damage Indices in Systemic Lupus Erythematosus

Jennifer M. Grossman and Kenneth C. Kalunian

Systemic lupus erythematosus (SLE) is a multisystem disease that is caused by antibody production and complement fixing immune complex deposition that results in tissue damage. As potentially many different antibodies can be produced in SLE patients, the different organ specific targets of these antibodies can cause a wide spectrum of clinical presentations, which are characterized by remissions and exacerbations. The pathogenic immune responses probably result from environmental triggers acting in the setting of certain susceptibility genes. Ultraviolet light and certain drugs are the only known environmental triggers identified to date.

SLE Classification Criteria

In 1971, the American Rheumatism Association (ARA) published preliminary criteria for the classification of SLE. These criteria were developed for clinical trials and population studies rather than for diagnostic purposes.[1] The criteria were based on information from 52 rheumatologists in clinics and hospitals in the United States and Canada; each physician provided 74 items of information on five of their own patients in each of the following categories in which they had classified these patients using their own clinical judgment: unequivocal SLE, probable SLE, classic RA, and medical patients with nonrheumatic diseases.

Based on computer analysis of the data, 14 manifestations were selected. The ARA committee proposed that a person can be said to have SLE if any four or more of the following manifestations are present, either serially or simultaneously, during any period of observation:

1. Facial erythema (i.e., butterfly rash): Diffuse erythema, flat or raised, over the malar eminence(s) and/or bridge of the nose; may be unilateral.
2. Discoid lupus: Erythematous raised patches with adherent keratotic scaling and follicular plugging; atrophic scarring may occur in older lesions; may be present anywhere on the body.
3. Raynaud's phenomenon: Requires a two phase color reaction, by patient's history or physician's observation.
4. Alopecia: Rapid loss of a large amount of scalp hair, by patient's history or physician's observation.
5. Photosensitivity: Unusual skin reaction from exposure to sunlight, by patient's history or physician's observation.
6. Oral or nasopharyngeal ulceration.
7. Arthritis without deformity: One or more peripheral joints involved with any of the following in the absence of deformity: (a) pain on motion, (b) tenderness, (c) effusion or periarticular soft-tissue swelling. (Peripheral joints include feet, ankles, knees, hips, shoulders, elbows, wrists, metacarpophalangeal, proximal interphalangeal, and terminal interphalangeal and temporomandibular joints.)
8. LE cells: Two or more classical LE cells seen on one or more occasions, or one cell seen on two or more occasions, using an accepted, published method.
9. Chronic false positive serologic test for syphilis (STS): Known to be present for at least 6 months and confirmed by Treponema pallidum immobilization (TPI) or Reiter's tests.
10. Profuse proteinuria: Greater than 3.5 g/d.
11. Urinary cellular casts: May be red cell, hemoglobin, granular, tubular, or mixed.
12. One or both of the following: (a) pleuritis, good history of pleuritic pain; or rub heard by a physician; or radiographic evidence of both pleural thickening and fluid; and/or (b) pericarditis, documented by electrocardiogram (ECG) or rub.
13. One or both of the following: (a) psychosis, and/or (b) convulsions, by patient's history or physician's observation in the absence of uremia and offending drugs.
14. One or more of the following: (a) hemolytic anemia; (b) leukopenia, white blood cell count of less than 4000/mL on two or more occasions; and/or (c) thrombocytopenia, platelet count less than 100,000/mL.

These criteria were chosen because of their high sensitivity and specificity when the gold standard was physicians' clinical judgment of the diagnosis; the committee noted a 90% sensitivity and 99% specificity against rheumatoid arthritis and a 98% specificity against a miscellany of nonrheumatic diseases.[1] In a retrospective pilot study of 500 male veterans with scleroderma, only 10 patients satisfied the SLE criteria at the time of diagnosis.[1]

These criteria were subsequently tested in other centers; in these various studies, sensitivities varied between 57.2 to 98.0%.[2-6]

The studies with the lowest sensitivities involved patients who were seen either initially or at only one particular point in time;[4,7] these investigators noted that a higher proportion of their patients eventually demonstrated four or more criteria over time. Lom-Orta et al[8] studied 31 patients who were thought to have SLE but who did not fulfill the ARA criteria; 21 of them fulfilled the criteria within a few years.

Numerous suggestions were made for improvement of the classification criteria, including the inclusion of antinuclear antibody (ANA) and other autoantibodies,[9-11] and the use of a weighted scoring system in which certain criteria are given more weight than others based on clinical importance.[12] An ARA subcommittee was created to evaluate these considerations; their study led to the publication of revised criteria in 1982.[13] Thirty potential criteria were studied, including numerous serologic tests and histologic descriptions of skin and kidney, as well as each of the original 1971 criteria. These variables were compared in SLE patients and matched controls. Eighteen investigators representing major clinics contributed patient report forms; these forms indicated the presence or absence of each variable at the time of examination or at any time in the past. Abnormalities that could be attributed to comorbid conditions or concurrent medications were not reported.[14] Each investigator was instructed to report prospective data on 10 consecutive patients and the next age-, race-, and sex-matched patient with a nontraumatic, nondegenerative, connective tissue disease seen at that clinic. This generated data from 177 patients with SLE and 162 control patients from 18 institutions. Cluster and other multivariate analysis techniques were used in studying the variables; numerous potential criteria sets were analyzed.

The resulting revised criteria consist of 11 items, compared with 14 in the preliminary criteria; 5 of the criteria are composites of one or more abnormalities. As in the preliminary data, patients must fulfill 4 or more criteria; no single criterion is absolutely essential. These criteria are the same as the 1997 revised criteria shown in Table 1, with the exception of criterion 10, which according to the revised 1982 criteria read: immunologic disorder (a) positive LE-cell preparation; OR (b) anti-double-stranded DNA; OR (c) anti-Sm; (d) BFP (false–positive serologic test for syphilis positive for at least 6 months with negative TPI or FTA).

Skin and kidney biopsies were not used in the final criteria set because of the infrequency with which they are obtained. Raynaud's phenomenon and alopecia were eliminated because their combined sensitivity/specificity scores were low. Renal criteria were consolidated. In the preliminary criteria set, cellular casts and proteinuria were separate criteria; in the revised set, there is only a single renal criterion, which is satisfied if a patient has cellular casts and/or proteinuria. In addition, the revised criteria reduced the amount of proteinuria that is needed for fulfillment, from greater than 3.5 g/d in the preliminary set, to more than 0.5 g/d (or >3 + if quantitation is not performed) in the revised set.

ANA, anti-DNA, and anti-Sm antibodies were included in the revised set, and the importance of false-positive serology for syphilis and LE-cell preparations were downgraded. ANA was felt to be the most important addition to the criteria set, because this serological marker was present at some point during the course of disease in 176 of the 177 patients. Despite the nonspecificity (the marker was present in 51% of the controls studied), the subcommittee felt its almost universal positivity made it a necessary criterion.

Using the patient database on which they were based, the revised criteria were 96% sensitive and specific, compared with 78% and 87%, respectively, for the 1971 criteria.[13] The subcommittee further tested the revised criteria against an ARA database of 590 patients with SLE, scleroderma, or dermatomyositis/polymyositis. Using the revised criteria against this database population, sensitivity in SLE patients was 83%, and specificity against the combined scleroderma and dermatomyositis/polymyositis patients was 89%. Using the preliminary criteria, sensitivity for SLE was only 78% and specificity only 87%.[14]

In a subsequent comparison of the relative sensitivities of the 1971 and 1982 criteria, Levin et al[15] studied 156 SLE patients at the University of Connecticut. Eighty-eight percent met the preliminary criteria, whereas 83% met the revised criteria when arthritis was strictly defined (i.e., nonerosive arthritis). Ninety-one percent met the revised criteria when arthritis was more liberally defined (i.e., nondeforming arthritis). These differences were not statistically significant. Their analysis also noted that of the three serologic tests added in the revised criteria (i.e., ANA, anti-Sm, and anti-DNA antibodies), ANA accounted for the increased sensitivity of the revised criteria. Levin et al noted that both the preliminary and the revised criteria were inappropriate for diagnostic purposes, in that over 50% of their patients fulfilled neither set of criteria when tested at the time of diagnosis. These patients subsequently fulfilled both sets of criteria at the same rate (77.5% fulfilled preliminary criteria and 78.5% revised criteria 5 years after diagnosis, and 84.5% and 83.0% for preliminary and revised criteria, respectively, at 7 years).

Passas et al[16] compared specificity of the preliminary and revised criteria in 207 University of Connecticut patients with non-SLE rheumatic diseases that are important in the differential diagnosis of SLE. The specificity was 98% for the preliminary criteria and 99% for the revised criteria. The preliminary and revised criteria also were tested on 285 Japanese SLE patients and 272 control patients with non-SLE connective tissue diseases.[17] The preliminary criteria had a sensitivity of 78% and a specificity of 98%, compared with a sensitivity of 89% and specificity of 96% for the revised criteria. Davis and Stein[18] applied the preliminary and revised criteria to 18 Zimbabwean patients with SLE reported up to 1989; they noted a sensitivity of 83% for the preliminary and 94% for the revised criteria. When serologic criteria were excluded, the sensitivity of the revised criteria was only 78%. They concluded that in many areas of Zimbabwe, where serologic tests are not readily available, the preliminary criteria may be more valuable than the revised criteria in the classification of patients with SLE, because the preliminary criteria rely more on clinical rather than serologic variables.

Because the presence of antiphospholipid antibodies and the antiphospholipid syndrome was increasingly recognized in SLE patients, the Diagnositic and Therapeutic Criteria Committee of the ACR reviewed the 1992 revised criteria for SLE.[19] They recommended that the immunologic criteria be modified with the removal of the LE cell preparation and the addition of IgG or IgM anticardiolipin antibodies or a lupus anticoagulant.

None of the methods for classifying patients with SLE were intended for diagnostic purposes. The findings of Levin et al[15] underscore the problems that are associated with use of classification criteria for diagnostic purposes. Over 50% of their SLE patients did not fulfill the criteria at one particular point in time, and while all eventually did, it required 9 to 20 years in some cases. In addition, the sensitivity of these classification criteria for milder cases of SLE is not known. In a study of Swedish patients

Table 1. The 1997 Revised Criteria for the Classification of systemic lupus erythematosus (SLE)

Criterion	Definition
1. Malar rash	Fixed malar erythema, flat or raised
2. Discoid rash	Erythematous– raised patches with keratotic scaling and follicular plugging; atrophic scarring may occur in older lesions
3. Photosensitivity	Skin rash as an unusual reaction to sunlight, by patient history or physician observation
4. Oral ulcers	Oral or nasopharyngeal ulcers, usually painless, observed by physician
5. Arthritis	Nonerosive arthritis involving two or more peripheral joints, characterized by tenderness, swelling, or effusion
6. Serositis	a. Pleuritis (convincing history of pleuritic pain or rub heard by physician or evidence of pleural effusion) OR b. Pericarditis (documented by ECG, rub, or evidence of pericardial effusion)
7. Renal disorder	a. Persistent proteinuria (>0.5 g/d or >3+) OR b. Cellular casts of any type
8. Neurologic disorder	a. Seizures (in the absence of other causes) OR b. Psychosis (in the absence of other causes)
9. Hematologic disorder	a. Hemolytic anemia OR b. Leukopenia (<4,000/mL on two or more occasions) OR c. Lymphopenia (<1,500/mL on two or more occasions) OR d. Thrombocytopenia (<100,000/mL in the absence of offending drugs)
10. Immunologic disorder	a. Anti-double-stranded DNA OR b. Anti-Sm OR c. Positive finding of antiphospholipid antibodies based on (1) an abnormal serum level of IgG or IgM anticardiolipin antibodies, (2) a positive test result for lupus anticoagulant using a standard method, or (3) a false positive serologic test for syphilis known to be positive for at least 6 months and confirmed by *Treponema pallidum* immobilization or fluorescent treponemal antibody absorption test
11. Antinuclear antibody	An abnormal titer of antinuclear antibody (ANA) by immunofluroscence or an equivalent assay at any time and in the absence of drugs known to be associated with "drug-induced lupus syndrome"

For identifying patients in clinical studies, a person shall be said to have SLE if any four or more of the 11 criteria are present, either serially or simultaneously, during any interval of observation. Ig= immunoglobulin. From Hochberg MG. Updating the American College of Rheumatology revised criteria for the classification of systemic lupus erythematosus (letter). Arthritis Rheum 1997; 40:1725.

with SLE, Jonsson et al[20] noted that the number of criteria in the 1982 revised set fulfilled by their patients was similar to or higher than that in other reported series despite overall mild disease. However, no strict measure of disease activity was applied, and no comparisons of disease activity in other populations were made.

SLE Activity Indices

Defining the degree of disease activity is essential in quantitating changes in patients, standardizing differences between patients, and evaluating clinical responses to therapy, especially in therapeutic trials. Although over 60 systems to assess disease activity in SLE exist, agreement on a definition for SLE activity has not been reached.[21] Most studies of the usefulness of the various available indices have focused on:

1. the British Isles Lupus Assessment Group scale (BILAG),

2. the Systemic Lupus Erythematosus Disease Activity Index (SLEDAI),

3. the Systemic Lupus Activity Measure (SLAM), and

4. the University of California, San Francisco/Johns Hopkins University Lupus Activity Index (LAI).

The BILAG system (see Table 2), which was developed by clinical investigators from four centers in the United Kingdom and one in the Republic of Ireland, rates SLE activity in eight organ systems.[22] Scoring in each organ system is based on the principle of intention to treat using the following ratings: (A) disease that requires urgent, disease-modifying therapy; (B) disease that demands close attention and, perhaps, modification of minor therapy (e.g., addition of such medications as low--dose corticosteroids or hydroxychloroquine) with maintenance, but

Table 2. The BILAG Index for assessment of disease activity in SLE

General
1. Pyrexia (documented)
2. Weight loss-unintentional >5% in one month
3. Lymphadenopathy/splenomegaly
4. Fatigue/malaise/lethargy
5. Anorexia/nausea/vomiting

Mucocutaneous
6. Maculopapular eruption—severe, active (discoid/bullous)
7. Maculopapular eruption—mild
8. Active discoid lesions—generalized, extensive
9. Active discoid lesions—local, including lupus profundus
10. Alopecia—severe, active
11. Alopecia—mild
12. Panniculitis, severe
13. Angio-oedema
14. Extensive mucosal ulceration
15. Small mucosal ulcers
16. Malar erythema
17. Subcutaneous nodules
18. Perniotic skin lesions
19. Peri-ungual erythema
20. Swollen fingers
21. Sclerodactyly
22. Calcinosis
23. Telangiectasia

Neurological
24. Impaired level of consciousness
25. Psychosis or delirium or confusional state
26. Seizures
27. Stroke or stroke syndrome
28. Aseptic meningitis
29. Mononeuritis multiplex
30. Ascending or transverse myelitis
31. Peripheral or cranial neuropathy
32. Disc swelling/cytoid bodies
33. Chorea
34. Cerebellar ataxia
35. Headaches—severe unremitting
36. Organic depressive illness
37. Organic brain syndrome including pseudotumor cerebri
38. Episodic migrainous headaches

Musculoskeletal
39. Definite myositis
40. Severe polyarthritis—with loss of function
41. Arthritis (definite synovitis)
42. Tendonitis
43. Mild chronic myositis
44. Arthralgia
45. Myalgia
46. Tendon contractures and fixed deformity
47. Aseptic necrosis

Cardiovascular and Respiratory
48. Pleuropericardial pain
49. Dyspnea
50. Cardiac failure
51. Friction rub
52. Effusion (pericardial or pleural)
53. Mild or intermittent chest pain
54. Progressive CXR changes—lung fields
55. Progressive CXR chages—heart size
56. ECG evidence of pericarditis or myocarditis
57. Cardiac arrhythmias including tachycardia >100 in absence of fever
58. Pulmonary function fall by >20%
59. Cyto-histological evidence of inflammatory lung disease

Vasculitis
60. Major cutaneous vasculitis including ulcers
61. Major abdominal crisis due to vasculitis
62. Recurrent thromboembolism (excluding stroke)
63. Raynaud's
64. Livedo reticularis
65. Superficial phlebitis
66. Minor cutaneous vasculitis (nailfold, digital, purpura, urticaria)
67. Thromboembolism (excluding stroke), first episode

Renal
(Answer with value or Y/N)
68. Systolic blood pressure mmHg
69. Diastolic blood pressure mmHg
70. Accelerated hypertension
71. Dipstick urine (0, 1+, 2+, 3+, 4+)
72. 24hour urine protein (g)
73. Newly documented proteinuria of >1 g/24h
74. Nephrotic syndrome
75. Creatinine (plasma/serum)
76. Creatinine clearance/GFR
77. Active urinary sediment
78. Histological evidence of active nephritis (within 3 months)
If abnormal value(above), was this due to lupus?

Hematology
(Answer with value or Y/N)
79. Hemoglobin (g/dL)
80. Total white cell count
81. Neutrophils
82. Lymphocytes
83. Platelets
84. Evidence of active hemolysis
85. Coomb's test positive
86. Evidence of circulating anticoagulant
If abnormal value(above), was this due to lupus?

It is implicit in this scoring system that all features scored are thought to be due to active lupus. If a new feature has developed in the last month (or since the last assessment if less than one month ago), it should be scored as new, even if it has subsequently improved or resolved. Each manifestation is scored using the following guidelines: 1= improving; 2= same; 3= worse; 4= new.

From Symmons DPM, Coopock JS, Bacon PA et al. Development and assessment of a computerized index of clinical disease activity in systemic lupus erythematosus. Q J Med 1988; 68:927-937.

not institution, of new major modalities (including medications such as high-dose corticosteroids or cytotoxic agents); (C) static or inactive disease requiring no or only sy-mptomatic therapy (including pain medications and nonsteroidal anti--inflammatory agents); and (D) absence of symptoms or laboratory abnormalities. Ratings are made based on the patient's clinical condition within the last month before evaluation. For statistical comparison with other numerically based indices, the following weights have been given to the four categories: A=9, B= 4, C= 1, and D= 0 (23). Possible scores with this system vary from a minimum of 0 to a maximum of 72.

The SLEDAI was developed at the University of Toronto. Several rheumatologists with extensive experience in the management of lupus patients rated the importance of 37 variables in defining SLE activity.[24] Using the highest-ranking 24 variables, 39 fictitious patients were created, and 14 rheumatologists ordered these patients in terms of disease activity. The implied weights of each variable in contributing to the judgment of activity in the group of fictitious patients were derived from multiple regression analysis. Real patients were then used to compare the instrument with the physician's global assessment of activity, and significant correlations were seen. The index was then modified to reflect persistent, active disease in those descriptors that had previously only considered new or recurrent occurrences; the modified index has been validated against the original SLEDAI as a measure of global disease activity and as a predictor of mortality.[25] The modified index, the SLEDAI-2K, is a one-page form with 24 items with definitions of the items provided on the form (see Table 3). Items that are present are noted, and scoring is calculated by summing the predetermined weights for the items that are present. Items that are life threatening have higher weights. Possible scores using this instrument vary from 0 to 105. Manifestations must be present in the 10 days preceding evaluation.

Two modifications of SLEDAI have been proposed: MEX–SLEDAI, and SELENA-SLEDAI. MEX-SLEDAI[26] was developed for use in Third World countries where immunologic and complement assays are costly and/or unavailable. The instrument uses most aspects of SLEDAI with some modifications, but it does not include anti-DNA antibodies or complement descriptors. MEX-SLEDAI also does not include the following SLEDAI clinical descriptors: visual disturbance, lupus headache, and pyuria. MEX-SLEDAI descriptors that are not part of SLEDAI include creatinine increase of greater than 5 mg/dL, hemolysis, peritonitis, fatigue, and lymphopenia. Whereas the proteinuria descriptor in the SLEDAI is defined as greater than 0.5 g per 24 hours of new onset or a recent increase of more than 0.5 g per 24 hours, the proteinuria descriptor of MEX-SLEDAI is new onset of greater than 0.5 g/L on random specimen. MEX-SLEDAI requires that significant proteinuria be new to denote activity, whereas active renal involvement in SLEDAI can be interpreted as significant new proteinuria or a significant increase in existing proteinuria. In a prospective study of 39 patients representing a spectrum of disease activity, five physicians scored disease activity using SLEDAI and MEX-SLEDAI on three consecutive patient visits.[26] Both instruments demonstrated validity and responsiveness.

SELENA-SLEDAI was adapted from the SLEDAI for use in a multicenter safety study of estrogens in women with SLE; it has been validated through its prospective use in the ongoing study.[27] SELENA-SLEDAI differs from SLEDAI in the definitions of some descriptors for clarification and attribution. The definition of the seizure descriptor has been expanded to exclude seizures resulting from past, irreversible central nervous system damage. Scleritis and episcleritis have been added to the definition of the visual disturbance descriptor, and vertigo has been added to the cranial nerve disorder descriptor. The cerebrovascular accident descriptor excludes hypertensive causes in SELENA-SLEDAI. The rash, alopecia, and mucosal ulcer descriptors have been modified to include ongoing presence of the descriptors, rather than the SLEDAI definitions of new or recurrent activity. The alopecia and mucosal ulcer descriptors also have been modified in SELENA-SLEDAI to attribute the manifestations to active lupus; attribution is meant to be a clinical decision. The pleurisy and pericarditis descriptors have been modified as well; rather than defining pleurisy as pleuritic chest pain with pleural rub or effusion or pleural thickening, SELENA-SLEDAI defines this descriptor as classic and severe pleuritic chest pain, pleural rub, effusion, or new pleural thickening with attribution to lupus. Pericarditis is similarly redefined; instead of pericardial pain with rub, effusion, ECG, or echocardiographic confirmation, SELENA-SLEDAI defines this descriptor as classic and severe pericardial pain, rub, effusion, or ECG confirmation. In addition, the proteinuria descriptor in SELENA-SLEDAI simplifies the SLEDAI descriptor. The SELENA-SLEDAI definition of proteinuria is the new onset or recent increase of more than 0.5 g per 24 hour, whereas the SLEDAI definition can be interpreted to include all patients with proteinuria of greater than 0.5 g per 24 hours.

The SLAM, which was developed at Brigham and Women's Hospital,[28] lists 33 clinical and laboratory manifestations of SLE, and each manifestation is assessed as either active or inactive. Graded estimates of activity are based on severity of increasing disability, organ destruction, need to follow the patient more closely, or need to consider major treatment change. Possible scores with this instrument vary from 0 to 86. Manifestations must be present in the month before evaluation. This index includes subjective symptoms such as fatigue, myalgias, arthralgias and abdominal pain felt to be attributable to SLE and therefore may detect smaller changes in disease activity.

The LAI[29] is a five-part scale. Part one is the physician's global disease activity assessment on a 0- to 3-point visual analog scale (VAS). Part two is an assessment of four symptoms (i.e., fatigue, rash, arthritis, serositis), each on a 0- to 3-point VAS. Part three scores the activity of four organ systems (i.e., neurologic, renal, pulmonary, hematologic), each on a 0- to 3-point VAS, and part four involves medication, that is, prednisone (1 point for 0-15 mg/d, 2 points for 16-39 mg/d, 3 points for 40 mg/d) and cytotoxic agents (3 points for use of cyclophosphamide, chlorambucil, azathioprine, or methotrexate). Part five scores for three laboratory parameters:

1. proteinuria (0 points for negative or trace, 1 point for 1+, 2 points for 2-3+, and 3 points for 4+ on urine dipstick);

2. anti-DNA antibodies (0-3 points assigned according to range used in the local laboratory); and

3. C3, C4, or CH50 (0-3 points assigned according to range used in the local laboratory).

The LAI summary score is the arithmetic mean of the part one score, the mean of the four values in part two, the maximum of the four values in part three, the mean of the two values in part four, and the mean of the three laboratory values. Possible LAI scores range from 0 to 3. Scores reflect the manifestations, laboratory abnormalities, and medications during the 2-week period before scoring.

Table 3. The Systemic Lupus Erythematosus Disease Activity Index-2K (SLEDAI-2K)

Descriptor is scored if present at the time of the visit or in the preceding 10 days.

Weight	Descriptor	Definition
8	Seizure	Recent onset, exclude metabolic, infectious or drug causes.
8	Psychosis	Altered ability to function in normal activity due to severe disturbance in the perception of reality. Include hallucinations, incoherence, marked loose associations, impoverished thought content, marked illogical thinking, bizarre, disorganized, or catatonic behavior. Exclude uremia and drug causes.
8	Organic brain syndrome	Altered mental function with impaired orientation, memory, or other intellectual function, with rapid onset and fluctuating clinical features, inability to sustain attention to environment, plus at least 2 of the following: perceptual disturbance, incoherent speech, insomnia or daytime drowsiness, or increased or decreased psychomotor activity. Exclude metabolic, infectious, or drug causes.
8	Visual disturbance	Retinal changes of SLE. Include cytoid bodies, retinal hemorrhages, serous exudates or hemorrhages in the choroid, or optic neuritis. Exclude hypertension, infection, or drug causes.
8	Cranial nerve disorder	New onset of sensory or motor neuropathy involving cranial nerves.
8	Lupus headache	Severe, persistent headache, may be migrainous, but must be nonresponsive to narcotic analgesia.
8	CVA	New onset of cerebrovascular accident(s). Exclude arteriosclerosis.
8	Vasculitis	Ulceration, gangrene, tender finger nodules, periungual infarction, splinter hemorrhages, or biopsy or angiogram proof of vasculitis.
4	Arthritis	2 or more joints with pain and signs of inflammation (i.e., tenderness, swelling or effusion).
4	Myositis	Proximal muscle aching/weakness, associated with elevated creatine phosphokinase/aldolase or electromyogram changes or biopsy showing myositis.
4	Urinary casts	Heme-granular or red blood cell casts.
4	Hematuria	>5 red blood cells/high power field. Exclude stone, infection, or other cause.
4	Proteinuria	>0.5 gram/24 hours.
4	Pyuria	>5 white blood cells/high power field. Exclude infection.
2	Rash	Inflammatory type rash.
2	Alopecia	Abnormal, patchy or diffuse loss of hair.
2	Mucosal ulcers	Oral or nasal ulcerations.
2	Pleurisy	Pleuritic chest pain with pleural rub or effusion, or pleural thickening.
2	Pericarditis	Pericardial pain with at least 1 of the following: rub, effusion, or electrocardiogram or echocardiogram confirmation.
2	Low complement	Decrease in CH50, C3, or C4 below the lower limit of normal for testing laboratory.
2	Increased DNA binding	Increased DNA binding by Farr assay above normal range for testing laboratory.
1	Fever	>38 degrees C. Exclude infectious cause.
1	Thrombocytopenia	<100,000 platelets, exclude drug causes.
1	Leukopenia	<3,000 white blood cells, exclude drug causes.

From Gladman DD, Ibanez D, Urowitz MB. Systemic lupus erythematosus disease activity index 2000. Arthritis Rheum 2002; 29:288-91.

In a study that did not include the LAI, Liang et al[21] looked at six indices for their reliability and validity in assessing disease activity in patients with SLE at one center. Twenty-five patients, who were selected to represent a spectrum of disease activity, were independently evaluated by two physicians on two occasions approximately 1 month apart. Validity of the six instruments was demonstrated by significant correlations of scores among the different indices (r= 0.81-0.97). BILAG, SLEDAI, and SLAM demonstrated the best inter-visit and inter-rater reliability.

An international group sponsored by the North Atlantic Treaty Organization (NATO) studied the operational validity, reliability, and sensitivity to change of several indices when used by physicians at different centers.[23] This group chose to limit their studies to BILAG, SLEDAI, and SLAM because of the prior work of Liang et al,[21] who did not study LAI. The NATO group[23] initially studied the validity of the three indices when used by physicians from eight different centers to assess the same patients; the indices were tested using data on patients from chart review.

Indices were compared to the clinician's judgment of disease activity using a VAS. The three indices correlated significantly with each other and with the VAS (P ≤ .05 for all correlation coefficients). However, the activity scores for the same patients using different indices varied widely among two of the eight physician scorers. This suggested that the indices are complex and require familiarity for effective use, and that the considerable intra- and interobserver variation may make the indices difficult to correlate in multicenter comparison studies of patient groups or treatment protocols in which different investigators use different indices to assess activity. The group suggested that uniform use of one or two of these indices might improve our ability to compare the results of studies from different centers.

The NATO group next studied the same three indices for their reproducibility and validity in the assessment of patients in an actual clinical setting.[30] Seven patients, representing a spectrum of disease activity and disease manifestations, were each examined by four of seven physicians from seven different centers; physicians from the center where the study patients received their care were excluded. Each observer completed the three indices and a VAS of disease activity on each of the examined patients. All the indices significantly correlated with each other, and there was no significant interobserver variation. All three indices detected differences among patients. This study suggests that physicians from different countries and health-care systems can evaluate patients reproducibly, regardless of the instruments used and the disease activity of the patients and without significant interobserver variation. Differences in the findings of the two NATO studies may be attributable to the methodologies that were used. The problems with inter- and intraobserver variations seen in the first study may have resulted from difficulties associated with chart abstraction of clinical information, and these problems may not exist when the indices are used in an actual clinical setting.

Petri et al[31] characterized the validity and reliability of the SLEDAI and LAI using patients from the Johns Hopkins lupus cohort. Validity was assessed by comparing the indices with the physician's global assessment of disease activity; the correlation of M-LAI (LAI modified so as not to contain part one, which assesses the physician's global assessment) and SLEDAI with the physician's global assessment was 0.64 and 0.55, respectively. Reliability was tested in six patients who were seen twice, 1 week apart, by nine physicians; the inter-rater reliability and test/retest reliability was greater for LAI than for SLEDAI. This study demonstrates that the indices can be readily assimilated into routine clinical practice.

To establish the reliability of SLEDAI among less experienced clinicians who were not familiar with it as a tool for patient assessment, Hawker et al[32] studied the reliability of SLEDAI by having three second-year rheumatology fellows apply the instrument on nine outpatients with SLE; each fellow independently interviewed and examined the patients. SLEDAI distinguished between patients (P= 0.0009), and physician variability was not statistically significant (P= 0.27). Inter- and intraobserver agreement was 78.7 and 98.0%, respectively.

Disease Indices and Sensitivity to Change

The NATO group studied the comparative ability of BILAG, SLEDAI and SLAM to assess change in disease activity over time.[33] Clinical and laboratory features of eight patients with SLE who were seen on three consecutive visits were abstracted and sent to eight physicians at different centers in three separate packages. Order of the patient-visit summaries were randomized, and the three indices were rated in one of six specific sequences. The three indices were significantly correlated (P ≤0.01 for all comparisons); the sequence presented, patient order, and order of index scoring did not significantly contribute to the variation of any of the indices. All three indices detected differences among patients (P ≤0.01); differences between visits were detectable with SLEDAI (P = 0.04), but not with SLAM or BILAG.

Petri et al[34] have demonstrated that both LAI and SLEDAI are sensitive to change. As part of an ongoing, prospective study, the physician's global assessment of disease activity, LAI, and SLEDAI have been completed at least quarterly for 185 patients with SLE followed by rheumatologists at Johns Hopkins Hospital. Using a definition of disease flare as a change of greater than 1.0 in the physician's global assessment of disease activity (measured on a 0 to 3 scale) from the previous visit or one within the prior 93 days, mean SLEDAI scores increased by 3.0, and mean LAI scores (modified to omit the physician's global assessment) increased by 0.26 at times of flare. These increases in the SLEDAI and LAI scores were found to be significant changes. These findings suggest that both SLEDAI and LAI can detect changes in activity with time, and that they may be useful in following changes of disease activity in the clinic setting.

Fortin and colleagues[35] evaluated 96 patients monthly for 5 months, completing at each visit a SLAM-R, a SLEDAI, a physician's global assessment and a physician's transition score that coded a patient as stable, improved or worse. Using multiple statistically methods, they found that both SLEDAI and SLAM-R were sensitive to change, but that the SLAM-R was consistently more sensitive. Ward and colleagues[36] further evaluated the disease activity indices for sensitivity to change. 23 patients were evaluated prospectively at 2-week intervals for up to 40 weeks. SLEDAI, SLAM, BILAG, LAI, and ECLAM (European Consensus Lupus Activity Measure), as well as a physician's and patient's global assessment, were determined at each visit. Compared to the physician's global assessment, all indices were sensitive to change (r= .52 to .75). LAI and ECLAM were the most sensitive to change, however, the LAI score incorporates the PGA and thus may be artificially elevated. The SLEDAI was the least sensitive to change. When compared to the patient's global assessment, only SLAM correlated with changes in disease activity and this correlation was weak. Patients in this study had mildly to moderately active lupus, and patients with more severe disease may have correlated better with the disease activity indices. These results are important when considering the design of clinical trials.

The above studies demonstrate that there are numerous indices that are valid, sensitive measures of disease activity in lupus. However, further research is needed to determine the most responsive instrument to maximize the ability to evaluate outcome in clinical trials. Ongoing studies are underway to define response for the purposes of determining whether an intervention is effective. Once a consensus is reached on this definition, these activity indices will be tested for their ability to assess response.

Disease Flare

Petri et al[34] have defined disease flare as an increase in the LAI part one physician's global assessment of 1.0 or greater. In following 185 patients with SLE at least quarterly over several years, 98 (53%) had at least one flare; the total number of flares was 146. The incidence of flare was 0.65 per each patient-year of follow-up;

the median time from the first study visit to flare was 12 months. Flares were frequently characterized by constitutional symptoms, musculoskeletal involvement, cutaneous involvement, and hypocomplementemia. At the time of flare, mean SLEDAI scores increased by 3.0 and mean LAI scores by 0.26. Overall, 44.8% of flares prompted a change in treatment. Patients who experienced flares fulfilled more of the SLE criteria at entry and had been followed for a longer duration after entry into the cohort compared with those who did not have flares. No specific clinical or laboratory variables present at entry were found to predict time to first flare. These data demonstrate that quantification of flare is possible, that flares are frequent in patients with disease of long duration, and that most flares involve minor organ systems.

An alternative means of identifying important changes in lupus activity is the physician's decision to initiate or increase treatment. This is the basis of the BILAG.

TerBorg et al[37] defined criteria for both major and minor exacerbations based on a Dutch activity index. A patient is considered to have a major exacerbation if one or more criteria for major flare are fulfilled; these include severe manifestations of disease with specific definitions or severe changes in laboratory parameters without improvement after prednisolone therapy at a maximum of 30 mg/d for at least 1 week. Criteria for minor exacerbation include an increase of the activity index by at least two points, with a minimum activity index of three points, with the clinical necessity of beginning a regimen of prednisolone at a dosage of at least 10 mg/d, of increasing the prednisolone dosage by at least 5 mg/d, or of starting an antimalarial or immunosuppressive drug.

Criteria for mild/moderate and severe flare have been developed for use in the SELENA study; these criteria will be prospectively validated as part of that study. Criteria for mild/moderate flare include a change in SLEDAI of 3 points or more, new or worse discoid rash, photosensitivity, lupus profundus, cutaneous vasculitis or bullous lesions, nasopharyngeal ulcers, pleuritis, pericarditis, arthritis, or fever attributable to lupus. Other criteria of mild/moderate flare include a need to increase prednisone, but not to doses greater than 0.5 mg/kg/d, because of lupus activity and an increase in the physician's global assessment of 1.0 to 2.5 (on a 0- to 3-point VAS). Criteria for a severe flare include a change in SLEDAI to greater than 12 points, new or worse central nervous system lupus activity, vasculitis, nephritis, myositis, thrombocytopenia of less than 60,000 per mL, or hemolytic anemia with hemoglobin levels of less than 7 or a decrease in hemoglobin of greater than 3 requiring a doubling of prednisone dosage or the need for prednisone doses of greater than 0.5 mg/kg/d, or hospitalization. Other criteria for severe flare include the institution of cyclophosphamide, azathioprine, or methotrexate, prednisone >0.5 mg/kg/d or hospitalization because of lupus activity, or an increase in the physician's global assessment to a level greater than 2.5 (on a 0- to 3-point VAS).

Abrahamowicz et al[38] evaluated the relationship between SLAM and SLEDAI scores of 30 paper patients based on actual cases and 38 lupus experts' decision to start therapy. They found that the disease activity as measured by SLAM-R or SLEDAI predicted the ultimate institution of steroid or alternative treatments. Using modeling, the authors determined that at least 70% of the physicians would initiate treatment at a score of 10 for both instruments. These indices do have limitations. SLAM-R scores include subjective symptoms that may be difficult to determine attribution to lupus, while SLEDAI does not include many manifestations of SLE. Furthermore, different patients with similar scores had markedly different percentages of physicians who would start therapy. Thus, these instruments can be useful in clinical practice and in clinical trials, however, they are not meant to replace physician judgment.

Gladman and colleagues[39] compared the SLEDAI score of 230 patients with 5 visits determined at the time of the visit to the score on a 5 point scale (no activity, mild activity but no change in treatment, mild activity but improvement, persistent activity and flare with a preset definition) determined by review of the medical record by a non-treating clinician. Based on this analysis, they propose that a flare be defined as an increase in SLEDAI >3, improvement as a decrease in SLEDAI >3, persistent active disease as a change in SLEDAI ± 3, and remission as a SLEDAI score of 0.

Measuring response to therapy is another important aspect of monitoring SLE. This aspect of clinical activity is not captured in the current disease activity indices. Validation of a new instrument RIFLE (Responder Index for Lupus Erythematosus) is underway. This instrument, which was developed by members of SLICC (Systemic Lupus International Collaborating Clinics), characterizes numerous manifestations of lupus activity and rates the activity of these manifestations at different points in time as improved or worsened with degrees of change. With this instrument, a patient's response to a therapeutic intervention can be characterized as a complete response, partial response, or nonresponse.

Damage Index

To compare patient groups and measure outcome in treatment protocols, the SLICC/ACR Damage Index for SLE (see Table 4) was developed by the Systemic Lupus International Collaborating Clinics and accepted by the American College of Rheumatology as a valid measure of damage in SLE patients.[40] This instrument measures accumulated organ damage occurring since the onset of SLE. Damage can result from either the disease process, its sequelae, or treatment, because attribution often is difficult in patients with SLE. The index is assessed irrespective of current disease activity, amount, or duration of any therapy and/or disability of the patient.

In developing the index, a list of items considered to reflect damage in SLE was generated through a group process. This group, representing international clinicians who were considered to be experts in lupus, reached a consensus as to which items should be included in an index. Each clinician submitted clinical information on four patients at two points in time (average interval between visits, 5 years). Two patients from each center had active disease and two patients inactive disease; some patients had increased damage with time and others stable damage. Nineteen clinicians completed the index on 42 case scenarios. Analysis of variance revealed that the index could identify changes in damage seen in patients with both active and inactive disease.

The index includes descriptors in 12 organ systems. For the purposes of the SLICC/ACR index, damage is considered only if present for at least 6 months.

The instrument has been demonstrated to have construct validity using clinical data on patients abstracted from chart review.[40] The SLICC/ARC Damage Index also was noted to have reliability and validity when used by 10 physicians from five countries in the assessment of 10 actual patients with SLE representing a spectrum of damage and activity; each of these patients was assessed by 6 of the 10 physicians.[41] The SLICC/ACR Damage Index detected differences among patients ($P < 0.001$); there was no detectable observer difference ($P = 0.933$) and no order effect

Table 4. The Systemic Lupus International Collaborating Clinics/American College of Rheumatology (SLICC/ACR) Damage Index for SLE

Item	Score
Ocular (either eye, by clinical assessment)	
Any cataract ever	1
Retinal change or optic atrophy	1
Neuropsychiatric	
Cognitive impairment (e.g., memory deficit, difficulty with calculation, poor concentration, difficulty in spoken or written language, impaired performance level or major psychosis)	1
Seizure requiring therapy for 6 months	1
Cerebrovascular accident ever (score 2 if >1)	1 or 2
Cranial or peripheral neuropathy (excluding optic)	1
Transverse myelitis	1
Renal	
Estimated or measured glomerular filtration rate <50%	1
Proteinuria 3.5 gm/24 hours or greater *OR*	1
End-stage renal disease (regardless of dialysis or transplantation)	3
Pulmonary	
Pulmonary hypertension (right ventricular prominence or loud P2)	1
Pulmonary fibrosis (physical and radiograph)	1
Shrinking lung (radiograph)	1
Pleural fibrosis (radiograph)	1
Pulmonary infarction (radiograph)	1
Cardiovascular	
Angina *or* coronary artery bypass	1
Myocardial infarction ever (score 2 if >1)	1 or 2
Cardiomyopathy (ventricular dysfunction)	1
Valvular disease (diastolic murmur, or systolic murmur >3/6)	1
Pericarditis for 6 months, *or* pericardiectomy	1
Peripheral vascular	
Claudication for 6 months	1
Minor tissue loss (pulp space)	
Significant tissue loss ever (e.g., loss of digit or limb) (score 2 if >1 site)	1 or 2
Venous thrombosis with swelling, ulceration, *or* venous stasis	1
Gastrointestinal	
Infarction or resection of bowel below duodenum, spleen, liver or gall bladder ever, for any cause (score 2 if >1 site)	1 or 2
Mesenteric insufficiency	1
Chronic peritonitis	1
Stricture *or* upper gastrointestinal tract surgery ever	1
Musculoskeletal	
Muscle atrophy or weakness	1
Deforming or erosive arthritis (including reducible deformities, excluding avascular necrosis)	1
Osteoporosis with fracture or vertebral collapse (excluding avascular necrosis)	1
Avascular necrosis (score 2 if >1)	1 or 2
Osteomyelitis	1
Skin	
Scarring chronic alopecia	1
Extensive scarring or panniculum other than scalp and Pulp space	1
Skin ulceration (excluding thrombosis) for >6 months	1
Premature gonadal failure	1
Diabetes (regardless of treatment)	1
Malignancy (exclude dysplasia (score 2 if >1 site)	1 or 2

Damage (nonreversible change, not related to active inflammation) occurring since onset of lupus, ascertained by clinical assessment and present for at least 6 months unless otherwise stated. Repeat episodes must occur at least 6 months apart to score 2. The same lesion cannot be scored twice.

From Gladman DD, Ginzler E, Goldsmith C et al. The development and initial validation of the Systemic Lupus International Collaborating Clinics/American College of Rheumatology Damage Index for systemic lupus erythematosus. Arthritis Rheum 1996; 39:363-369.

(P= 0.261). There was concordance in the SLICC/ACR Damage Index among observers despite a wide spectrum of disease activity detected by the SLEDAI. The authors concluded that physicians from different centers are able to reproducibly assess patients with SLE using the SLEDAI to assess disease activity and the SLICC/ACR Damage Index to assess accumulated damage.

In a study of 200 lupus patients from 5 centers, 61% were noted to have damage within 7 years of disease onset with a mean of 3.8 years.[42] Furthermore, in a study of 1297 patients from 8 centers, the SLICC/ACR Damage Index has also been shown to increase over time.[43] 99 patients died, and these patients had significantly higher DI early in their illness compared to those patients who had survived (1.56 versus .99, p= .0003). This index may be useful in clinical practice as a further means of identifying patients with a poor prognosis.

Health Status

In addition to disease activity and damage, quality of life is another important component in the assessment of SLE patients. Assessment of health status has been shown to be an important independent outcome measure in lupus.[44] This was confirmed in a second study where SLEDAI, SLICC/ACR DI and the SF-20 did not correlate well with each other.[45] Fortin and colleagues[46] found that in a cross-sectional analysis, the SLAM-R score correlated with most subscales of the SF-36, but SLEDAI did not. However, longitudinal changes in both disease activity scales did correlate with changes in the SF-36. Sutcliffe and colleagues[47] found that higher disease activity is associated with worse physical and emotional function, pain and general health. The study also found that patients who were more satisfied with health care, and had greater social support, had a better general view of their health, suggesting a potential non-pharmacological intervention to improve health care of lupus patients. The Medical Outcomes Study Short Form 36 (SF-36) is an instrument that has been validated and is one of the most commonly used in lupus.[48]

In 1998, OMERACT (Outcome Measures in Rheumatology) IV was held and included a module on SLE.[49] The investigators considered 21 domains and concluded that randomized controlled trials and longitudinal observational series should include a minimum of 4 domains: a measure of disease activity, a measure of health-related quality of life, a measure of damage, and toxicity/adverse events. It was thought that by the institution of these core domains, the quality of clinical trials and the efficacy of evaluating new therapies would improve.

References

1. Cohen AS, Reynolds WE, Franklin EC et al. Preliminary criteria for the classification of systemic lupus erythematosus. Bull Rheum Dis 1971; 21:643-648.
2. Cohen AS, Canoso JJ. Criteria for the classification of systemic lupus erythematosus status 1972 (editorial). Arthritis Rheum 1972; 15:540-543.
3. Davis P, Atkins B, Josse RG et al. Criteria for classification of SLE. Br Med J 1973; 3:90--91.
4. Fries JF, Siegel RC. Testing the preliminary criteria for classification of SLE. Ann Rheum Dis 1973; 32:171-177.
5. Gibson TP, Dibona GF. Use of the American Rheumatism Association's preliminary criteria for the classification of systemic lupus erythematosus. Ann Intern Med 1972; 77:754-756.
6. Lin CY. Improvement in steroid and immunosuppressive drug resistant lupus nephritis by intravenous prostaglandin E1 therapy. Nephron 1990; 55:258-264.
7. Lie JT. Medical complications of cocaine and other illicit drug abuse simulating rheumatic disease. J Rheumatol 1990; 17:736-737.
8. Lom-Orta H, Alarcon-Segovia D, Diaz-Jouanen E. Systemic lupus erythematosus. Differences between patients who do and who do not fulfill classification criteria at the time of diagnosis. J Rheumatol 1980; 7:831-837.
9. Canoso JJ, Cohen AS. A review of the use, evaluations, and criticisms of the preliminary criteria for the classification of systemic lupus erythematosus. Arthritis Rheum 1979; 22:917-921.
10. Liang M, Rogers M, Swafford J et al. The psychological impact systemic lupus erythematosus and rheumatoid arthritis. Arthritis Rheum 1984; 27:13-19.
11. Weinstein A, Bordwell B, Stone B et al. Antibodies to native DNA and serum complement (C3) levels. Application to diagnosis and classification of systemic lupus erythematosus. Am J Med 1983; 74:206-216.
12. Tan PLJ, Borman GB, Wigley RD. Testing clinical criteria for systemic lupus erythematosus in other connective tissue disorders. Rheumatol Int 1981; 1:147-149.
13. Tan EM, Cohen AS, Fries JF et al. Special article: The 1982 revised criteria for the classification of systemic lupus erythematosus. Arthritis Rheum 1982; 25:1271-1277.
14. Fries JF. Methodology of validation of criteria for systemic lupus erythematosus. Scand J Rheumatol 1987; 65(Suppl):25-30.
15. Levin DL, Roenigk HH, Caro WA et al. Histologic, immunofluorescent, and antinuclear antibody findings in PUVA-treated patients. J Am Acad Dermatol 1982;6:328-333.
16. Passas CM, Wond RL, Peterson M et al. A comparison of the specificity of the 1971 and 1982 American Rheumatism Association criteria for the classification of systemic lupus erythematosus. Arthritis Rheum 1985; 28:620-623.
17. Yokohari R, Tsunematsu T. Application to Japanese patients, of the 1982 American Rheumatism Association revised criteria for the classification of systemic lupus erythematosus. Arthritis Rheum 1985; 28:693-698.
18. Davis P, Stein M. Evaluation of criteria for the classification of SLE in Zimbabwean patients (letter). Br J Rheumatol 1989; 28:546-556.
19. Hochberg, MC. Updating the American College of Rheumatology Revised Criteria for the Classification of Systemic Lupus Erythematosus (letter). Arthritis Rheum 1997; 40:1725.
20. Jonsson H, Nived O, Sturfelt G. Outcome in systemic lupus erythematosus: A prospective study of patients from a defined population. Medicine 1989; 68:141-150.
21. Liang MH, Socher SA, Larson MG et al. Reliability and validity of six systems for the clinical assessment of disease activity in systemic lupus erythematosus. Arthritis Rheum 1989; 32:1107-1118.
22. Symmons DPM, Coopock JS, Bacon PA et al. Development and assessment of a computerized index of clinical disease activity in systemic lupus erythematosus. Q J Med 1988; 68:927-937.
23. Kalunian KC, Gladman DD, Bacon PA et al. Development and assessment of a computerized index of clinical disease activity in systemic lupus erythematosus. Unpublished manuscript.
24. Bombardier C, Gladman DD, Urowitz MB et al. Derivation of the SLEDAI. A disease activity index for lupus patients. The Committee on Prognosis Studies in SLE. Arthritis Rheum 1992; 35:630-640.
25. Gladman DD, Ibanez D, Urowitz MB. Systemic lupus erythematosus disease activity index 2000. J Rheumatol 2002; 29:288-291.
26. Guzman J, Cardiel MH, Arce-Salinas A et al. Measurement of disease activity in systemic lupus erythematosus. Prospective validation of 3 clinical indices. J Rheumatol 1992; 19:1551-1558.
27. Petri M, Buyon J, Skovron ML et al. Reliability of SELENA SLEDAI and flares as a clinical trial outcome measure. Arthritis Rheum 1998; 41:S218.
28. Liang MH, Socher SA, Roberts WN et al. Measurement of systemic lupus erythematosus activity in clinical research. Arthritis Rheum 1988; 31:817-825.
29. Petri M, Bochemstedt L, Colman J et al. Serial assessment of glomerular filtration rate in lupus nephropathy. Kidney Int 1988; 34:832-839.
30. Gladman DD, Goldsmith CH, Urowitz MB et al. Cross--cultural validation and reliability of three disease activity indices in systemic lupus erythematosus. J Rheumatol 1992; 19:608-611.
31. Petri M, Hellman D, Hochberg M. Validity and reliability of lupus activity measures in the routine clinic setting. J Rheumatol 1992;19:53-59.

32. Hawker G, Gabriel S, Bombardier C et al. A reliability study of SLEDAI: A disease activity index for systemic lupus erythematosus. J Rheumatol 1993; 20:657-660.
33. Gladman DD, Goldsmith CH, Urowitz MB et al. Sensitivity to change of 3 systemic lupus erythematosus disease activity indices: International validation. J Rheumatol 1994; 21:1468-1471.
34. Petri M, Genovese M, Engle E et al. Definition, incidence and clinical description of flare in systemic lupus erythematosus: A prospective cohort study. Arthritis Rheum 1991; 34:937-944.
35. Fortin PR, Abrahamowicz M, Clarke AE et al. Do lupus disease activity measures detect clinically important change? J Rheumatol 2000; 27:1421-1428.
36. Ward MM, Marx AS, Barry NN. Comparison of the validity and sensitivity to change of 5 activity indices in systemic lupus erythematosus. J Rheumatol 2000; 27:664-670.
37. TerBorg EJ, Horst G, Huymmel EJ et al. Measurement of increases in anti-double--stranded DNA antibody levels as a predictor of disease exacerbation in systemic lupus erythematosus. Arthritis Rheum 1990; 33:634-643.
38. Abrahamowicz M, Fortin P, du Berger R et al. The relationship between disease activity and expert physician's decision to start major treatment in active systemic lupus erythematosus: A decision aid for development of entry criteria for clinical trials. J Rheumatol 1998; 25:277-284.
39. Gladman DD, Urowitz ME, Kagal A et al. Accurately describing changes in disease activity in systemic lupus erythematosus. J Rheumatol 2000; 27:377-379.
40. Gladman DD, Ginzler E, Goldsmith C et al. The development and initial validation of the Systemic Lupus International Collaborating Clinics/American College of Rheumatology Damage Index for systemic lupus erythematosus. Arthritis Rheum 1996; 39:363-369.
41. Gladman D, Urowitz, Goldsmith C et al. The reliability of the Systemic Lupus International Collaborating Clinics/American College of Rheumatology damage index in patients with systemic lupus erythematosus. Arthritis Rheum 1997; 40:809-813.
42. Rivest C, Lew RA, Welsing PMJ et al. Association between clinical factors, socioeconomic status, and organ damage in recent onset systemic lupus erythematosus. J Rheumatol 2000; 27:680-4.
43. Gladman DD, Goldsmith CH, Urowitz MB et al. The Systemic Lupus International Collaborating Clinics/American College of Rheumatology (SLICC/ACR) damage index for systemic lupus erythematosus international comparison. J Rheumatol 2000; 27:373-376.
44. Gladman DD, Urowitz MB, Ong A et al. Lack of correlation among the 3 outcomes describing SLE: Disease activity, damage and quality of life. Clin Exp Rheumatol 1996; 14:305-308.
45. Hanly JG. Disease activity, cumulative damage and quality of life in systematic (sic) lupus erythematosus: Results of a cross-sectional study. Lupus 1997; 6:243-247.
46. Fortin PR, Abrahamowicz M, Neville C et al. Impact of disease activity and cumulative damage on the health of lupus patients. Lupus 1998; 7:101-107.
47. Sutcliffe N, Clarke AE, Levinton C et al. Associates of health status in patients with systemic lupus erythematosus. J Rheumatol 1999; 26:2352-2356.
48. Stoll T, Gordon C, Seifert B et al. Consistency and validity of patient administered assessment of quality of life by the MOS SF-36; its association with disease activity and damage in patients with systemic lupus erythematosus. J Rheumatol 1997; 24:1608-1614.
49. Smolen J, Strand V, Cardiel M et al. Randomized clinical trials and longitudinal observational studies in systemic lupus erythematosus: consensus on a preliminary core set of outcome domains. J Rheumatol 1999; 26:504-507.

CHAPTER 40

Lupus Nephritis

Annie Y. Suh and Robert M. Rosa

Introduction

Systemic lupus erythematosus (SLE) is a clinically heterogeneous autoimmune disease that can affect multiple organs. Studies of new therapies for SLE often involve, or may even be confined, to individuals with renal involvement because an unambiguous endpoint may be defined, i.e., time to dialysis. Although only 25-50% of patients with lupus demonstrate clinical and laboratory evidence of nephropathy early in the course of disease, 60-75% subsequently develop overt renal abnormalities. The clinical presentation of lupus nephritis is variable, ranging from minimal proteinuria and hematuria, to nephrotic syndrome and depressed renal function in severe cases. The classification of lupus nephritis is based upon light microscopy (WHO), immunoflorescence, and electron microscopy findings on renal biopsy.

Lupus nephritis contributes significantly to the morbidity and mortality of patients with SLE. Almost fifty years ago, the five year survival of patients with the most severe forms of lupus nephritis was less than 20%. In the ensuing years, survival has improved dramatically with advances in therapy. Nonetheless, when aggressive immunosuppressive therapy is used in lupus nephritis, there is considerable morbidity and mortality related to the treatment itself. The treatment of lupus nephritis, therefore, needs to balance carefully the therapeutic benefits with the toxicity of therapy.

Steroids are the cornerstone of therapy for lupus nephritis, either alone or in combination with cytotoxic agents. While mild lupus nephritis often responds to steroids alone, the addition of cytotoxic agents confers greater benefit in patients with severe nephritis. In particular, both cyclophosphamide and azathioprine have demonstrated efficacy in retarding the decline of renal function in lupus. More recently, cyclosporine and mycophenolate mofetil have been used in refractory disease, with promising preliminary results. Finally, stem cell transplantation may hold promise as a beneficial treatment for patients with lupus nephritis.

Diagnosis

SLE, an autoimmune disease characterized by antibodies to components of the cell nucleus, is associated with diverse systemic clinical manifestations. Although it can occur at any age, 65% of patients with SLE have disease onset between 16 and 55 years of age.[1] In the general population, the prevalence of SLE is approximately 1 in 2000, but there is a higher frequency in Latino and African-American populations. Women predominate, with a ratio of at least 10:1 in adults.[2] In addition, there appears to be a hereditary component — 5-12% of relatives of patients with SLE have the disease[3] and there is a concordance rate of 14-57% of SLE in monozygotic twins.[4,5]

Because no one factor is diagnostic of SLE, the American Rheumatology Association developed a set of criteria for the diagnosis of SLE in 1971. These criteria, last revised in 1982, include clinical manifestations, laboratory features and serologic abnormalities. Traditionally, 4 out of 11 criteria are required to meet the diagnosis (refer to Chapter 39). Currently, however, the diagnosis is less strictly dependent on the number of criteria met.[6,7]

A major feature of SLE is the production of antibodies directed against components of the cell nucleus. Among these, the anti-nuclear antibody (ANA) is most characteristic and is positive in significant titer in virtually all patients with SLE.[8] These antibodies bind DNA, RNA, nucleic proteins and protein-nucleic acid complexes. Two such antibodies appear to be virtually unique to lupus: anti-double stranded DNA (anti-dsDNA), which is associated with nephritis, and anti-Smith.[9,10] Hypocomplementemia is found at presentation in more than 75% of untreated patients with lupus and is commonly associated with active renal disease.[11]

Pathogenesis

In lupus nephritis, nuclear antigens (especially DNA) form immune complexes with antinuclear antibodies (predominantly IgG), either in the circulation or in situ.[12,13] Once deposited in the mesangial or subendothelial part of the glomerular basement membrane (GBM), the immune complexes activate the complement system, which in turn generates chemotactic factors. These factors result in leukocyte and monocyte infiltration, releasing mediators such as cytokines that cause and sustain glomerular inflammation. With continued immune complex deposition chronic inflammation ensues, which can lead to fibrosis, scarring, and decreased renal function.[14]

Renal Involvement in SLE

While renal involvement is common, only 25-50% of patients with lupus have an abnormal urinalysis or impaired renal function early in the course of disease. Subsequently, however, 60-75% may develop renal abnormalities.[1] The clinical manifestations of renal lupus are listed in Table 1. The most common clinical renal

Table 1. Clinical features of patients presenting with lupus nephritis

Feature	% of Those With Nephritis
Proteinuria	100
Nephrotic syndrome	45-65
Granular casts	30
Red cell casts	10
Microscopic hematuria	80
Macroscopic hematuria	1 to 2
Reduced renal function	46 to 80
Rapidly declining renal function	30
Acute renal failure	1 to 2
Hypertension	15 to 50
Hyperkalemia	15
Tubular abnormalities	60 to 80

Cameron JS. Lupus Nephritis. J Am Soc Neph 1999; 10:413.

manifestation is proteinuria. Microscopic hematuria is also frequently seen, but usually in combination with proteinuria. Although 30-50% of all patients with SLE ultimately develop an elevated plasma creatinine concentration, acute renal failure is an unusual initial presentation.[1] The total incidence of renal involvement among patients with SLE probably exceeds 90%. When patients without clinical evidence of renal disease undergo a biopsy, the pathology is often that of mesangial glomerulonephritis; rarely, proliferative glomerulonephritis may be seen.[15,16] There are a number of different types of renal disease in SLE. Immune-complex mediated glomerular disease is most common and will be discussed below.

Renal Biopsy

There is controversy regarding which patients with lupus should undergo a renal biopsy. Patients with mild renal involvement (minimal proteinuria without hematuria) will likely show mild histologic changes on biopsy, while biopsy in patients with active serology, active urine sediment (RBCs, WBCs, casts), and an elevated serum creatinine will likely show diffuse proliferative lupus nephritis. Nonetheless, in a given patient, it is difficult to predict with any certainty the type of nephritis based on the clinical presentation alone. Renal biopsy is the only tool that allows correct pathologic classification of lupus nephritis and thereby provides critical information necessary to determine the appropriate treatment.

Less commonly, some patients present with renal disease without any systemic symptoms or signs of SLE. In such cases, renal biopsy may reveal pathologic changes consistent with lupus nephritis. For instance, SLE with membranous nephritis can be distinguished from idiopathic membranous nephropathy by a biopsy showing a typical pattern of immune deposition (mesangial and subendothelial as well as subepithelial deposits). Also, deposition of all isotypes of immunoglobulin is characteristic of SLE, as opposed to idiopathic membranous nephritis where only IgG is commonly present. These patients may exhibit symptoms and serologic markers for SLE only years after the diagnosis of renal lupus.[17]

Repeat biopsies may also be useful in the setting of worsening proteinuria or deteriorating renal function. Information may be obtained regarding transformation to another type of lupus nephritis, disease advancement requiring more aggressive treatment, or evidence of predominant renal scarring instead of active inflammation. For these reasons, repeat biopsies are often warranted in patients with lupus nephritis.[17]

World Health Organization (WHO) Classification of Lupus Nephritis

Table 2 summarizes the WHO classifications for lupus nephritis based on appearance on light microscopy. Each class has distinct histologic, clinical and prognostic characteristics, though overlap of different classes exists and 15-50% of patients can evolve from one form to another.[18-20]

When patients with lupus nephritis undergo a biopsy, more than half will show class III or IV nephritis. With respect to immunohistology, multiple isotypes of immunoglobulin are seen, but IgG is most commonly predominant. A few patients, however, will demonstrate predominantly IgA or IgM. Early components of complement such as C4 and C1q are usually present along with C3. Approximately 25% of patients exhibit all three isotypes of immunoglobulin together with C3, C4, and C1q, referred to as a "full house" pattern.

Class I (Normal)

The glomeruli appear normal on light microscopy (LM), but deposits can be seen by electron microscopy (EM).

Class II (Mesangial Proliferation)

Mesangial proliferation, seen in 10-20% of cases, represents a mild form of glomerular involvement. Immune deposits are seen on EM. Granular immunoglobulins and complement are seen on immunofluorescence. Further classification into A and B are made based on the degree of hypercellularity. Clinically, up to 25% have a normal urinalysis. The remainder have microscopic hematuria and/or a mild degree of proteinuria.[11] Hypertension or renal insufficiency is rare. Because the renal prognosis is excellent, no specific therapy for the renal disease is usually recommended. It is possible, however, for class II to transform to a more advanced class of nephritis.

Class III (Focal Proliferative Lupus Nephritis, FPLN)

Class III (focal proliferative lupus nephritis) and class IV (diffuse proliferative lupus nephritis) exhibit renal lesions that are qualitatively similar, but distinguished by the amount of proliferation seen in the glomerular capillaries.

Focal proliferative lupus nephritis, which is present in 10-20% of cases, represents a more advanced form than class II. By definition, less than 50% of glomeruli are affected on light microscopy. There are segmental areas of hypercellular lesions with proliferation of glomerular cells and occasional necrosis. On EM, there are immune deposits in the subendothelial space and in the mesangium. Almost all patients have hematuria and proteinuria with some exhibiting nephrotic syndrome, hypertension and decreased renal function.

In focal proliferative lupus nephritis, the renal prognosis is variable. It appears that progressive renal dysfunction is less common if fewer than 25% of the glomeruli are involved with only segmental proliferation in these glomeruli.[20] Indeed, most of these patients receive treatment for extrarenal disease without specific additional treatment for their renal disease.[20] For patients with more severe involvement, (40-50% glomeruli affected with areas

Table 2. Pathologic classification for lupus nephritis

Class	Clinical Renal Manifestation	Renal Pathology
I. Normal	Usually asymptomatic	Normal appearance on light microscopy Mesangial immune deposits
II. Mesangial Proliferation	Low grade hematuria and/or proteinuria, normal renal function	Mesangial expansion and proliferation but mostly patent capillaries Mesangial immune deposits
III. FPLN	Nephritic urine sediment Variable but usually non-nephrotic proteinuria	Predominantly segmental proliferation, occasional necrosis, crescents in <50% of glomeruli Predominantly mesangial and subendothelial immune deposits
IV. DPLN	Nephritic and nephrotic syndromes Hypertension Variable renal insufficiency	Predominantly global proliferation, necrosis, crescents in >50% glomeruli Variable sclerosis, atrophy, fibrosis Predominantly mesangial and subendothelial immune deposits
V. Membranous Lupus Nephritis	Nephrotic syndrome	Uniform capillary loop thickening Subepithelial, subendothelial, and mesangial immune deposits
VI. Sclerosing Nephropathy	Inactive urinary sediment: broad, waxy casts	Glomerular obsolescence, tubular atrophy, and interstitial fibrosis Few if any immune deposits

Adapted from Balow J. Renal manifestations of systemic lupus erythematosus and other rheumatic disorders. Primer on Kidney Diseases. National Kidney Foundation 1998:210. FPLN= focal proliferative lupus nephritis; DPLN= diffuse proliferative lupus nephritis.

of necrosis or crescent formation, nephrotic range proteinuria, hypertension), however, the long-term prognosis may be similar to that of diffuse proliferative lupus nephritis.[21] As with class II, class III nephritis can transform into class IV or V.

Class IV (Diffuse Proliferative Lupus Nephritis, DPLN)

The histologic changes found on renal biopsy in diffuse proliferative lupus, the most severe form of lupus nephritis, are similar to focal proliferative lupus nephritis but more global, involving by definition more than 50% of the glomeruli on LM.[22] Typically, there is also thickening of the glomerular capillary wall caused by marked deposition of immunoglobulin and complement. Large electron dense subendothelial deposits are prominent on EM. Almost all patients have hematuria and proteinuria, and nephrotic syndrome, hypertension and renal insufficiency are frequently seen. Serum complement levels are usually depressed and anti-DNA levels are elevated.[23] Immunosuppressive therapy is generally required to prevent or retard the progression of active DPLN to end stage renal disease (ESRD).

Class V (Membranous Lupus Nephritis)

Membranous lupus nephritis, which affects 10-20% of patients, has clinical and histologic features that are similar to idiopathic membranous nephritis.[11] On LM, diffuse basement membrane thickening, on immunofluorescence granular deposits of immunoglobulin and complements in the capillary walls, and on EM multiple subepithelial, subendothelial, and mesangial deposits are observed.[24]

These patients may present with no other manifestations of SLE (clinical or serologic), but there are distinguishing characteristics on biopsy suggestive of lupus rather than idiopathic membranous nephritis. These include concurrent subendothelial or prominent mesangial deposits (such as those seen in the proliferative forms of lupus), immune deposits along the tubular basement membranes and in the small blood vessels, and tubuloreticular structures in the endothelial cells on EM.[24]

A predominant clinical feature is nephrotic syndrome, which is seen in two-thirds of patients. Microscopic hematuria is present in over 50% of the patients. Serum creatinine concentration is usually normal or only slightly elevated.[11]

Tubulointerstitial Disease

While not part of the WHO classification, tubulointerstitial disease with interstitial infiltrates and tubular injury with or without immune deposits along the tubular basement membrane is a common finding in lupus nephritis. In about 50% of patients (more in those with class IV), immune complexes are present in the tubular basement membrane and T lymphocytes (predominantly CD8 cells) and monocytes are seen in the interstitium.[11] The severity of tubulointerstitial involvement appears to be predictive of prognosis and correlates with hypertension, renal dysfunction and a progressive clinical course.[25,26] Tubular basement membrane deposits alone, however, only correlate with severity of disease but not with prognosis.[25] In such instances, proteinuria

Table 3. Five-year actuarial survival for lupus, lupus nephritis, and WHO class IV nephritis over the past 40 years

Period	% 5-Year Survival		
	All Lupus	Lupus Nephritis	Class IV Nephritis
1953-1969	49%	44%	17%
1970-1079	82%	67%	55%
1980-1989	86%	82%	80%
1990-1995	92%	82%	82%

Cameron JS. Lupus Nephritis. J Am Soc Neph 1999; 10:413.

Table 4. Attributed causes of death in lupus with nephritis

Main Cause of Death	Percent
Renal failure	28
Infections	31
Active lupus, other organs	13
CVA	4
Pulmonary embolus	3
Pulmonary hypertension	0
Malignancy	0
Unrelated causes	6
Other and unknown	16

Cameron JS. Lupus Nephritis. J Am Soc Neph 1999; 10:413.

is usually minimal. Urinalysis often reveals only a few red cells and/or white cells.[27] There may, however, be evidence of tubular dysfunction such as metabolic acidosis due to distal renal tubular acidosis and hyperkalemia due to impaired distal potassium secretion.[28,29]

Vascular Disease

Lupus can affect the renal vasculature in the form of immune complex deposition, hyaline and noninflammatory necrotizing lesions, thrombotic microangiopathy, and true vasculitis with lymphocyte and monocyte infiltration of the vessel wall. Glomerular and vascular thrombi can also be seen in association with antiphospholipid antibodies such as lupus anticoagulant and anticardiolipin antibody.[30] Necrotizing vasculopathy has a poor renal prognosis.[30] Immunosuppressive agents do not appear efficacious in cases with thrombi though, unlike in other forms of lupus nephritis, there is some evidence that plasmapheresis may be beneficial as a treatment for vascular disease associated with a thrombotic thrombocytopenic purpura (TTP)-like syndrome.[30]

Drug-Induced Lupus

A lupus-like syndrome has been reported following the administration of hydralazine, procainamide and isoniazid. Renal involvement is uncommon in such cases, but proliferative glomerulonephritis and nephrotic syndrome has been reported.[31,32]

Renal Prognosis

Thirty years ago, almost half of the patients with class IV lupus nephritis died within 5 years.[11] Since that time there has been a marked improvement in outcome for patients with class IV nephritis (Table 3). Although survival has improved in all patients with lupus with and without renal involvement, when death occurs it has been attributed to several different causes (Table 4). Often it is related to complications from treatment, of which infection is most common.[11]

Clinical factors associated with a poor outcome in some but not all series have included age at onset greater than 55, childhood onset, black race, raised serum creatinine, hypertension, a greater number of ARA criteria present at onset, and certain laboratory findings (i.e., anemia, thrombocytopenia, hypocomplementemia, and high anti-dsDNA titer at onset).[11] In addition, flares of active lupus and a lack of response to therapy are associated with an increased risk of progressive renal disease.[33] In one study, the likelihood of a persistent doubling of creatinine increased in patients with nephritic flares (relative risk 6.8).

As noted above, conventional wisdom has long held that renal pathology is a reliable prognostic indicator in lupus nephritis. Older studies found a good correlation between the types of lupus nephritis and short-term renal outcome.[17] A group at the National Institutes of Health (NIH), moreover, reported that the activity index (measuring potentially reversible lesions such as cell proliferation and interstitial inflammation) and chronicity index (measuring irreversible lesions such as glomerular sclerosis and interstitial fibrosis) could predict the risk of worsening renal failure.[35] Subsequent studies, however, have suggested that these classifications may correlate less well with outcome in severe nephritis when aggressive treatment is administered.[36,37]

Treatment

The decision to treat lupus nephritis and the therapeutic regimen to be used depend on the clinical and pathologic features. Clinical features that are more common and often warrant treatment include nephritic urine sediment, worsening nephrotic syndrome, and deteriorating renal function. As noted, a renal biopsy can be of critical help in making the appropriate therapeutic decision.

For WHO class II, there is little evidence that treatment changes the subsequent evolution. Therefore, unless there is progression to a more severe form of nephritis, most clinicians will direct treatment at the extrarenal manifestations of disease.[38]

Optimal treatment for class III is uncertain. Progression to advanced renal failure within 5 years is infrequent with mild focal involvement (<25% glomeruli affected primarily with segmental areas of proliferation).[20] It has, however, been observed that when patients are not specifically treated for renal disease but given steroids for extrarenal manifestations of SLE, the creatinine appears to stabilize for at least 5 years.[16] With more widespread and severe focal disease (40-50% glomeruli affected with necrosis or crescent formation, significant subendothelial immune deposits, nephrotic range proteinuria), there is a worse long-term outcome that is probably similar to that of class IV (diffuse proliferative lupus nephritis).[22] Since the incidence of death or progression to renal failure in this form is closer to 15-25%, many clinicians will treat these patients aggressively.[2,16] Similar to those with severe class III, patients with untreated class IV lupus nephritis are at a significant risk for development of progressive renal failure. Furthermore, treatment has been shown to improve outcome. Thus, finding severe class III or IV lupus nephritis on renal biopsy warrants aggressive therapy.[21,39]

For WHO class V, renal prognosis is variable,[18,19] and the risk of ESRD is fairly low (25% in 15 years). Indications for treatment are controversial as there is no compelling evidence that outcome improves after treatment. Nonetheless, many patients are treated with steroids for nephrotic syndrome or for extrarenal manifestations. In this setting, partial or complete remission of proteinuria has been observed, with serum creatinine remaining normal or near normal for 5 or more years. Therefore, an argument can be made that worsening renal function, severe nephrotic syndrome, or other characteristics associated with a poor outcome (black race, hypertension, increased creatinine at presentation, and evidence of some proliferative lesions on biopsy) may warrant a course of treatment.[40]

If a decision is made to treat membranous lupus nephritis, it should be appreciated that optimal therapy is uncertain. In a retrospective study, Ponticelli demonstrated that a regimen of alternating steroids and chlorambucil (similar to the regimen for idiopathic membranous nephropathy) was superior to steroids alone.[41] Recently, a prospective, randomized NIH trial comparing three regimens(steroid alone; steroid plus cyclophosphamide; and steroid plus cyclosporine) has been reported in abstract form. There appeared to be a higher remission rate (46% vs. 13%) and less persistent nephrotic syndrome (19% vs. 60%) in the combination groups.[42] Although the initial study did not find a difference between the cyclophosphamide and the cyclosporine group, the follow-up suggested a trend towards more relapses in the cyclosporine as compared with the cyclophosphamide group.[42]

Treatment Regimens

Steroids

Corticosteroids are the most frequently prescribed drugs for the treatment of lupus, but are not without significant concerns. Long-term use is associated with numerous complications including osteoporosis, avascular necrosis of bone, cataracts and atherosclerosis. Multiple studies have shown, moreover, that monotherapy is not as effective in more severe renal disease as is a combination with cytotoxic agents.[43-45] Mesangial proliferation and some mild forms of proliferative lupus, however, may respond to steroids alone. To minimize the complications of long-term use, prednisone is tapered to an alternate-day regimen and the lowest possible maintenance dose is established as soon as there is evidence of remission. In patients with severe active disease who develop acute renal failure, initial steroid therapy may be high dose intravenous pulse methylprednisolone (0.5 to 1g x 3 days), which can produce a rapid improvement.[46]

Cyclophosphamide

As noted above, cyclophosphamide has become one of the most important therapies for lupus nephritis. Data from a meta-analysis of multiple controlled trials demonstrated its superiority when combined with steroids over prednisone alone in slowing the rate of decline of the glomerular filtration rate and preventing ESRD by 40% in all classes of lupus.[47] High-risk patients may derive even greater benefit. A recent NIH study showed that the probability of avoiding renal failure at 10-12 years in these high-risk patients was 90% with cyclophosphamide, 60% with azathioprine (not statistically significant different from cyclophosphamide), and 20% with prednisone.[43]

The optimal duration of therapy with cyclophosphamide is not known. Another NIH trial compared three regimens in patients with proliferative lupus nephritis with a mean creatinine of 1.6-2.0 mg/dl. The regimens were: monthly pulse intravenous cyclophosphamide for six months; a similar regimen followed by maintenance intravenous cyclophosphamide every 3 months for another 24 months; and monthly pulses of intravenous methylprednisolone for six months, which was then tapered. They found that there was a 48% chance of doubling of creatinine in 3 years in the steroid only group compared with 25% in the cyclophosphamide group. Also, the likelihood of relapse was significantly higher in the short-course as opposed to the long-course cyclophosphamide regimen.[48]

In response to the criticism that the course of steroid therapy was too short, the NIH did a follow-up study to compare cyclophosphamide and a longer-course steroid regimen. They treated 65 patients with diffuse proliferative membranous nephritis with three different regimens: monthly pulse methylprednisolone for 12 months, monthly pulse cyclophosphamide (given for six months then quarterly), and combination therapy of cyclophosphamide and steroids. Renal remission occurred in 85% patients in combination therapy, 29% in monthly methylprednisolone (p= 0.03), and 65% in cyclophosphamide therapy (not significantly different from combination therapy).[49] In addition, there was a significantly increased rate of relapse in the steroid only group compared with the combination group (36% vs. 0%). This study showed that cyclophosphamide-based regimen was superior to prednisone alone in both inducing remission and preventing relapse.

More recently, the NIH group published results of an extended follow-up (median, 11 years) of a randomized controlled trial. Eighty-two patients with proliferative lupus nephritis were treated with monthly intravenous boluses of methylprednisolone, cyclophosphamide, or both. Compared with the methylprednisolone group, both the cyclophosphamide and the combination groups had a significantly lower likelihood of treatment failure (defined as need for supplemental immunosuppressive therapy or doubling of serum creatinine, or death). The combination therapy, furthermore, was associated with a lower risk of doubling of serum creatinine than the cyclophosphamide group without adding additional toxicity.[50]

Despite its efficacy in treating lupus nephritis, cyclophosphamide therapy is not without significant side effects. There is a greater incidence of amenorrhea and premature ovarian failure with cyclophosphamide therapy. In addition, there is a risk of hemorrhagic cystitis and transitional cell cancer of the uroepithelium.[44,49] Combination therapy, moreover, carries an increased risk of major infection and herpes zoster. There is evidence, however, from trials that monthly intravenous cyclophosphamide may be less toxic to the bone marrow, gonads, and bladder than daily oral therapy.[48,49]

Because of the potential complications of longer cytotoxic therapy, the Lupus Nephritis Collaborative Study examined a shorter course of treatment. Patients were treated with 60-80mg of prednisone per day for 4 wks followed by tapering to 20-25mg over 32 weeks. Oral cyclophosphamide was given for 8 weeks. Those patients with initial Cr <1.2 mg/dl showed remission or stable disease. However, among those patients with a higher creatinine, 40% of them at 3-4 years follow-up had a creatinine greater than 3 mg/dl.[51] This finding is consistent with those seen by the NIH studies.[48,52]

Azathioprine

Azathioprine is another cytotoxic agent that has been shown in clinical trials of lupus nephritis to be efficacious in inducing remission when used in conjunction with steroids.[43,47] Because cyclophosphamide has shown superior results in clinical trials in inducing remission,[38,43,53] however, azathioprine is now more commonly used as a maintenance agent after induction with cyclophosphamide.[54]

In maintenance regimens, both cyclophosphamide and azathioprine have a steroid sparing effect, and there is evidence that azathioprine does not increase the risk of infection beyond that of steroids.[11] Doses of 2-2.5 mg/kg/d have been shown to be safe in the long-term.[11] Azathioprine can, however, cause macrocytosis, and induce leukopenia. Furthermore, the risk of developing a malignancy with azathioprine is small and tends to be a late occurrence and unlike cyclophosphamide, pregnancy during maintenance azathioprine is considered safe.[11] At present, the recommendations for the treatment of class IV nephritis are steroids in combination with cyclophosphamide (usually intravenous pulse monthly for 6 months) to induce remission, followed by quarterly intravenous cyclophosphamide or oral azathioprine.

Cyclosporine

Cyclosporine has a powerful effect on the clonal expansion of helper T cells through inhibition of calcineurin, which impairs the synthesis of interleukin-2 and other lymphokines by activated T lymphocytes. It is thought that by blocking this lymphokine production, autoreactive cytotoxic T cell activation is inhibited with subsequent decreased production of autoreactive antibodies by B cells. This pharmacological effect could potentially diminish immune complex formation and deposition, thereby ultimately decreasing cell mediated immune injury of the kidney.[55]

Limited evidence has suggested that 5 mg/kg/day of cyclosporine results in a good response in some patients with lupus nephritis. Two uncontrolled studies[55,56] of 13 and 26 adults with steroid-resistant or steroid-dependent nephritis showed a significant reduction in proteinuria and a stable creatinine with cyclosporine therapy. Enthusiasm for this drug is tempered, however, by the fact that serologic markers did not change and that relapse followed taper or withdrawal of cyclosporine. Cyclosporine, moreover, is nephrotoxic, and some have argued that the decrease in proteinuria seen with cyclosporine may reflect changes in local glomerular hemodynamics rather than modulation of the activity of lupus nephritis.

Mycophenolate Mofetil (MMF)

Mycophenolate mofetil has been extensively used for immunosuppression in renal transplantation. Mycophenolic acid, the metabolite of MMF, inhibits de novo purine synthesis by inhibiting inosine monophosphate dehydrogenase. This action results in decreased lymphocyte production and a subsequent decrease in T and B cell proliferation and antibody production.[57]

Recently, a study from Hong Kong randomly assigned 42 patients with diffuse proliferative lupus nephritis to receive oral MMF at an initial dose of 1gm twice a day and prednisolone, or oral cyclophosphamide (initial dose 2.5 mg/kg/d) and prednisolone. At six months, the dose of MMF was decreased by 50% in the first group and cyclophosphamide was changed to azathioprine in the second group. At year one, complete remission occurred in 81 and 76 percent, respectively. Partial remissions occurred in 14% of both groups. Infection rates were not significantly different.[58]

This study suggested that MMF was equally efficacious (and with fewer potential side effects) than cyclophosphamide followed by azathioprine. It was, however, criticized for its limited patient population (Asians only in the study), small sample size, short-term follow up without addressing the issue of maintenance therapy, and lack of comparison of MMF to intravenous cyclophosphamide. Nonetheless, it suggested a potentially effective new therapeutic regimen for lupus nephritis that might avoid some of the toxic effects of cyclophosphamide.[58] MMF has been well tolerated both in studies of lupus nephritis and transplantation. Its most common side effects are gastrointestinal upset and diarrhea, which usually respond to dose reduction. Unlike cyclophosphamide, it does not have the potential to cause cystitis or cancer, and the risk of leukopenia is much less.[57]

Intravenous Immunoglobulin (IVIg)

IVIg has been used to treat a number of autoimmune conditions. IVIg is thought to solubilize immune complexes deposited in the kidney or to block Fc receptors. Studies involving a small number of children and adults (2-9 patients) have shown encouraging results, including a decrease in proteinuria and an improvement in renal function. Randomized, controlled trials of its efficacy, however, are lacking.[11,59]

Plasmapheresis

Plasmapheresis may potentially remove autoreactive antibodies, antigens, and immune complexes from circulation. To examine the therapeutic efficacy of this modality, the Lupus Nephritis Collaborative Study Group randomized 86 patients with Class III, IV, or V nephritis to 8 weeks of cyclophosphamide and steroids with or without plasmapheresis.[60] After a 3-year follow-up, there was no significant difference in the two groups in terms of ESRD, death, renal remission, serum creatinine, or proteinuria. Another trial using synchronized treatment (to counter the "rebound synthesis" of auto antibodies) also did not show a beneficial effect on clinical outcomes.[61] At this time, therefore, there is no role for plasmapheresis in the treatment of lupus nephritis except in selected patients with a TTP-like syndrome who may benefit from it.

Stem Cell Transplantation

Since the pathology of SLE is likely mediated by abnormally reactive T cells, extensive immunosuppression followed by stem cell transplantation may result in prolonged or even permanent remission. Traynor and Burt et al administered a regimen of high dose methylprednisolone, cyclophosphamide, and anti-thymocyte globulin followed by autologous stem cell infusion to 7 patients with persistent active lupus refractory to aggressive therapy. After a median follow-up period of 25 months, all patients remained in clinical remission with normal serologic markers. Prior to transplantation, CD69 expression had been increased and the T cell profile skewed, suggesting T cell hyper-responsiveness. After transplantation, both CD69 expression and the T cell profile normalized.[62] The mechanism of remission induced by autologous stem cell transplantation may be transient severe immune suppression or a fundamental alteration of the immune system after transplantation. Longer follow-up will be necessary to evaluate the durability and toxicity of this novel form of treatment of lupus nephritis (see Chapter 41).

Monitoring Response

With effective immunosuppressive therapy, there is typically a decrease and then a cessation of inflammation that is manifested by a decreased activity of the urine sediment with fewer red and white cells and casts, and a decreased or stabilized creatinine. A stable serum creatinine over a period of time, however, does not necessarily imply a lack of disease. Proteinuria should be reduced as well. In the future, it may be possible to monitor other factors such as the urinary excretion of cytokines and chemokines. In this context, a recent report found an increase in the urinary excretion of monocyte chemoattractant protein-1 in patients with active lupus nephritis.[63]

Relapse

Relapse of lupus nephritis is usually manifested by an active urine sediment (red cell and/or white cell casts especially), an increase in proteinuria and/or an increase in the serum creatinine. The role of serologic markers in predicting renal relapse is less clear. An increase in anti-dsDNA titer and a fall in complement has, however, been associated with a subsequent relapse within 8-10 weeks.[64]

Conclusion

Lupus nephritis is a disease with a wide range of clinical and histologic presentations. For this reason, a combination of clinical findings, laboratory evidence, and biopsy results are essential for accurate diagnosis and prognosis as well as optimal treatment. Significant therapeutic advances have been achieved over the last several years, and it is likely that this trend will continue with the introduction of HSCT. Stem cell transplantation may hold substantial promise in this regard, though more experience with this modality is required.

References

1. Rothfield N. Clinical features of systemic lupus erythematosus. In: Kelley WN, Harris ED, Ruddy S et al, eds. Textbook of Rheumatology. Philadelphia: WB Saunders, 1981.
2. Lahita RG. The role of sex hormones in systemic lupus erythematosus. Current Opin Rheumatol 1999; 11:352.
3. Arnett FC, Reveille JD, Wilson RW et al. Systemic lupus erythematosus: Current state of the genetic hypothesis. Semin Arthritis Rheum 1984; 14:24.
4. Block SR, Winfield JB, Lockshin MD et al. Twin studies in systemic lupus erythematosus. A review of the literature and presentation of 12 additional sets. Am J Med 1975; 59:533.
5. Deapen D, Escalante A, Weinrib L et al. A revised estimate of twin concordance in SLE: A servery of 138 pairs. Arthritis Rheum 1992; 35:311.
6. Hochberg MC. Updating the American College of Rheumatology Revised Criteria for the classification of systemic lupus erythematosus. Arthritis Rheum 1997; 40:1725.
7. Tan EM, Cohen AS, Fries JF et al. The 1982 revised criteria for the classification of systemic lupus erythematosus. Arthritis Rheum 1982; 25:1271.
8. Von Muhlen CA, Tan EM. Autoantibodies in the diagnosis of systemic rheumatic diseases. Semin Arthritis Rheum 1995; 24:323.
9. Smeenk R, Brinkman K, Van der Brink H et al. Antibodies to DNA in patients with systemic lupus erythematosus. Their role in the diagnosis, the follow up and the pathogenesis of the disease. Clin Rheumatol 1990; 9:100.
10. Menves EF, Schur PH. Antibodies to Am and RNP: Prognosticators of disease involvement. Arthritis Rheum 1983; 26:848.
11. Cameron JS. Lupus Nephritis. J Am Soc Neph 1999; 10:413.
12. Tojo T, Frous GJ. Lupus Nephritis: Varying complement-fixing properties of immunoglobulin G antibody to antigen of cell nuclei. Science 1968; 161:904.
13. Termaat RM, Assman KJ, Oijkman HG et al. Anti-DNA antibodies can bind to the glomeruli via two distinct mechanisms. Kidney Int 1992; 42:1363.
14. Elkon K. Autoantibodies in systemic lupus erythematosus. Curr Opi Rheumatol 1955; 7:384.
15. Leekey DJ, Katz AI, Azaran AH et al. Silent diffuse lupus nephritis: Long-term follow up. Am J Kidney Dis 1983; 2(1 suppl1):188.
16. Huong DL, Papo T, Beaufils H et al. Renal involvement in systemic lupus erythematosus. Medicine (Baltimore) 1999; 78:148.
17. Ponticelli C, Moroni G. Renal biopsy in lupus nephritis—what for, when, and how often? Nephol Dial Transplant 1998; 13:2452.
18. Appel GB, Silva FG, Pirani CL et al. Renal involvement in systemic lupus erythematosus: A study of 56 patients emphasizing histologic classification. Medicine (Baltimore) 1978; 57:371.
19. Lee HS, Mujais SK, Kasinath BS et al. Course of renal pathology in patients with systemic lupus erythematosus. Am J Med 1984; 77:612.
20. Schwartz MM, Kawala KS, Corwin H et al. The prognosis of segmental glomerulonephritis in systemic lupus erythematosus. Kidney Int 1987; 32:274.
21. Appel GB, Cohen DJ, Pirani CL et al. Long-term follow-up of lupus nephritis: A study based on the WHO classification. Am J Med 1987; 83:877.
22. Schwartz MM, Lan SL, Bonsib SM et al. Clinical outcome of three discrete histologic patterns of injury in severe lupus glomerulonephritis. Am J Kidney Dis 1989; 13:273.
23. Lloyd W, Schur PH. Immune complexes, complement and anti-DNA in exacerbation of systemic lupus erythematosus. Medicine 1981; 60:208.
24. Jeanette JC, Iskandar SS, Dalldorf FG. Pathologic differentiation between lupus and nonlupus membranous glomerulopathy. Kidney Int 1983; 24:377.
25. Park NH, D'Agati V, Appel GB et al. Tubulointerstitial disease in lupus nephritis: Relationship to immune deposits, interstitial inflammation, glomerular changes, renal function, and prognosis. Nephron 1986; 44:309.
26. Alexopoulos E, Seron D, Hartley RB et al. Lupus nephritis: Correlation of interstitial cells with glomerular function. Kidney Int 1990; 37:100.
27. Singh AJ, Ucci A, Madias NE. Predominant tubulointerstitial nephritis. Am J Kidney Dis 1996; 27:273.
28. Kozeny GA, Bau W, Bansal VK et al. Occurrence of renal tubular dysfunction in lupus nephritis. Arch Intern Med 1987; 147:891.
29. DeFonzo RA, Cooke CR, Goldberg M et al. Impaired renal tubular potassium secretion in systemic lupus erythematosus. Ann Intern Med 1977; 86:268.
30. Appel GB, Pirani CL, D'Agait V. Renal vascular complications of systemic lupus erythematosus. J Am Soc Nephrol 1994; 4:1499.
31. Fitzler MJ. Drugs recently associated with lupus syndromes. Lupus 1994; 3:455.
32. Hess E. Drug-induced lupus. N Engl J Med 1988; 318:1460.
33. Korbet SM, Lewis EJ, Schwartz MM et al. Factors predictive of outcome in severe lupus nephritis. Lupus Nephritis Collaborative Study Group. Am J Kidney Dis 2000; 35:904.
34. Appel GB. Cyclophosphamide therapy of severe lupus nephritis. Am J Kidney Dis 1997; 30:872.
35. Austin HA, Muenz LR, Joyce KM et al. Prognostic factors in lupus nephritis. Contributions of renal histological data. Am J Med 1983; 75:382.
36. Ponticelli C, Zucchelli P, Moroni G et al. Long-term prognosis of diffuse lupus nephritis. Clin Nephrol 1987; 28:263.
37. Schwartz MM, Bernstein J, Hill GS et al. Predictive value of renal pathology in diffuse proliferative lupus glomerulonephritis. Kidney Int 1989; 361:891.
38. Lee HS, Mujais SK, Kasinath BS et al. Course of renal pathology in patients with systemic lupus erythematosus. Am J Med 1984; 77:612.
39. Schwartz MM, Lan SP, Berstein et al. The risk of pathology indices in the management of severe lupus glomerulonephritis. Kidney Int 1992; 42:743.
40. Sloan RP, Schwartz MM, Korbet SM et al. Long-term outcome in systemic lupus erythematosus membranous glomerulonephritis. J Am Soc Nephrol 1996; 7:299.

41. Moroni G, Maccario M, Bangi G et al. Treatment of membranous lupus nephritis. Am J Kidney Dis 1998; 31:681.
42. Austin HA, Vaughan EM, Coumpas DT et al. Lupus membranous nephropathy: Controlled trial of prednisone, pulse cyclophosphamide, and cyclosporine A (abstract). J Am Soc Nephrol 2000; 11:81A.
43. Steinberg AD. The treatment of lupus nephritis. Kidney Int 1986; 30:769.
44. Austin HA, Klippel JH, Barlow JE et al. Therapy of lupus nephritis. Controlled trial of prednisone and cytotoxic drugs. N Engl J Med 1986; 314:614.
45. Felson DT, Anderson J. Evidence for the superiority of immunosuppressive drugs and prednisone over prednisolone alone in lupus nephritis. Results of a pooled analysis. N Engl J Med 1984; 311:1528.
46. Kimberly RP, Lockshin MD, Sherman RL et al. High-dose intravenous methylprednisolone pulse therapy in systemic lupus erythematosus. Am J Med 1981; 70:817.
47. Bansal VK, Beto JA. Treatment of lupus nephritis: A meta-analysis of clinical trials. Am J Kidney Dis 1997; 29:193.
48. Boumpas DT, Austin HA, Vaughn EM et al. Controlled trial of pulse methylprednisolone versus two regimens of pulse cyclophosphamide in severe lupus nephritis. Lancet 1992; 340:741.
49. Gourley MF, Austin III HA, Scott D et al. Methylprednisolone and cyclophosphamide, alone or in combination, in patients with lupus nephritis. A randomized, controlled trial. Ann Intern Med 1996; 125:549.
50. Illei GG, Austin HA, Craine M et al. Combination therapy with pulse cyclophosphamide plus pulse methylprednisolone improves long-term renal outcome without adding toxicity in patients with lupus nephritis. Ann Intern Med 2001; 135:248.
51. Levey AS, Lan SP, Corwin HL et al. Progression and remission in renal disease in the Lupus Nephritis Collaborative Study. Results of treatment with prednisone and short-term oral cyclophosphamide. Ann Intern Med 1992; 116:114.
52. Austin HA, Fessler BJ, Boumpas DT et al. Prognostic indicators supporting use of short courses of pulse immunosuppression for severe lupus nephritis (abstract). J Am Soc Nephrol 1995; 6:411.
53. Steinberg AD, Steinberg SC. Long-term preservation of renal function in patients with lupus nephritis receiving treatment that includes cyclophosphamide versus those treated with prednisone only. Arthritis Rheum 1991; 34:945.
54. Contreras G, Roth D, Berho M et al. Immunosuppressive treatment for proliferative lupus nephritis: Preliminary report of a prospective, randomized clinical trial with Mycophenolate mofetil (abstract). J Am Soc Neph 1999; 10:99A.
55. Feutren G, Querin S, Noel LH et al. Effects of cyclosporine in severe systemic lupus erythematosus. J Pediatr 111:1063.
56. Favre H, Miescher PA, Huang YP et al. Cyclosporine in the treatment of lupus nephritis. Am J Nephrol 1989; 9(Suppl.1):57.
57. Zimmerman R, Radhakrishnan J, Valeri A. Advances in the treatment of lupus nephritis. Ann Rev Med 2001; 52:63.
58. Chan TM, Fu LK, Coiln SO et al. Efficacy of Mycophenolate mofetil in patients with diffuse proliferative lupus nephritis. N Eng J Med 2000; 343:1156.
59. Boletis JN, Ioannidis JP, Boki KA et al. IVIg compared with cyclophosphamide for proliferative lupus nephritis. Lancet 1999; 354:569.
60. Lewis EJ, Hunsicker LG, Lan SP et al. A controlled trial of plasmapheresis therapy in severe lupus nephritis. The Lupus Nephritis Collaborative Study Group. N Engl J Med 1992; 326:1373.
61. Wallace DJ, Goldfinger D, Pepkowitz SH et al. Randomized controlled trial of pulse/synchronized cyclophosphamide/apheresis for proliferative lupus nephritis. J Clin Apheresis 1998; 13:163.
62. Traynor AE, Schroeder J, Rosa R et al. Treatment of severe systemic lupus erythematosus with high-dose chemotherapy and haemopoietic stem-cell transplantation: A phase I study. Lancet 2000; 356:701.
63. Wada T, Yoloyama H, Su SB et al. Monitoring urinary levels of monocyte chemotactic and activating factor reflects disease activity of lupus nephritis. Kidney Int 1996; 49:761.
64. ter Borg EJ, Horst G, Hummel EJ et al. Predictive values of rise in anti-double stranded DNA antibody levels for disease exacerbation in systemic lupus erythematosus: A long term prospective study. Arthritis Rheum 1990; 33:634.

CHAPTER 41

Hematopoietic Stem Cell Transplantation for Systemic Lupus Erythematosus

Ann E. Traynor, Richard K. Burt and Alberto Marmont

Introduction

Systemic lupus erythematosus (SLE) is a systemic, clinically and genetically heterogenous disease characterized by diffuse tissue damage mediated in great part by autoimmune pathogenic reactions. Notwithstanding impressive therapeutic progress and significant amelioration of quantity and quality of life, there are still intractable cases which demand more incisive interventions. Hematopoietic stem cell transplantation (HSCT) was first proposed in 1993,[1] and the first two patients underwent autologous transplants (ASCT) in 1997.[2,3] Since then, the procedure has been performed worldwide, including the largest ongoing single center study at Northwestern University, Chicago[4-6] and a multicentric registry study by the European Group for Blood and Marrow Transplantation (EBMT) and the European League against Rheumatism (EULAR);[7] summarizing individual center experience in Europe. A National Institutes of Health (NIH) funded phase III clinical trial of HSCT for refractory SLE is anticipated to begin in 2003/2004. Encouraging results are raising new questions about the role of HSCT in SLE, as well as in other severe autoimmune diseases.[8]

Etiology

An interplay of genetic and environmental factors (including hormones such as estrogens and prolactin) permits the (auto) immune cascade which gives rise to SLE[9] (see also Chapter 38). The relative genetic risk of developing clinical SLE has been estimated as 20 for siblings of individuals with lupus and 250 for their monozygotic twins.[10] The concordance rate for identical twins is approximately 24%, compared to about 2% for dizygotic twins.[11] The low penetrance of each contributing gene is the main reason why diseases such as SLE are so genetically complex.[12]

The availability of multiple highly inbred murine models has been pivotal to our understanding of genetic and environmental factors.[13-15] Thus far, at least 45 named loci are linked to one or more lupus traits in murine lupus,[16] while the presence of susceptibility genes in several chromosomal regions has been confirmed in humans.[17] Recent studies of genetic reconstitution in polycongenic murine strains have characterized three susceptibility genes.[18] Sle 1 mediates the loss of tolerance to chromatin; sle 2 lowers the activation threshold of B cells, and sle 3 mediates a dysregulation of CD 4+ T cells, which in humans have been shown to mediate a secondary antigen-driven immune response[19] and to stimulate the B-cell dependent production of antinuclear antibodies.[20] The whole process has been imaginatively equated by Alarcón-Segovia to the effect of three troikas, Russian carriages in which teams of three horses pull independently but concertedly.[21]

Prognosis

Over the last several decades, prolongation of survival has been the most important progress made in the treatment of SLE. Survival has improved from 50% at 2 years in 1939, to 5- and 10-year survival of 70% and 50% in the 1950s, to 90% and 80% 5- and 10-year survivals in the 1980s, and to almost 70% survival at 20 years in the 1990s.[22,23] Within the first 5 years, the main cause of death is active disease (neurologic, renal, systemic) and/or infection. Thereafter, causes of death tend to be, in addition to infections, cardiovascular events (strokes and/or myocardial infarction), often associated with uncontrolled hypertension, and complications of diabetes and hyperlipidemia, induced by chronic corticotherapy. This bimodal pattern of mortality was demonstrated as early as 1976.[24] In addition to delayed treatment related deaths, a spectrum of disability is often exchanged for the short term life-saving benefit of corticotherapy, including avascular necrosis, tendon rupture, vertebral and long bone fractures, obesity and cushingoid habitus, reducing the independence and capacity of the individual for routine exercise. Therefore, despite pulse cyclophosphamide and advances in supportive care such as new anti-hypertensive medications, patients with active SLE involving visceral organs have a 2 and 5 year mortality of approximately 20% and 35%, respectively (Fig. 1).

Poor prognostic factors reported in the literature for systemic lupus include the presence of nephritis, pneumonitis, alveolar hemorrhage, cerebritis, carditis, or hypoalbuminemia. Also of prognostic value are: (1) accumulated damage as assessed by the Systemic Lupus International Collaborating Clinics (SLICC) score; (2) disease activity, as assessed by activity scores (refer to Chapter 39); (3) low hemoglobin; (4) antiphospholipid syndrome; (5) socioeconomic status; and (6) chronic corticosteroid therapy.[22-29] Serology titers of anti-nuclear antibodies and HLA phenotype do not correlate with mortality.[30] In summary, high risk lupus is an active multi-organ disease requiring chronic and/or high dose corticosteroids. Since SLE is predominantly a disease of young women, improvements in disease–related morbidity and mortality are sought.

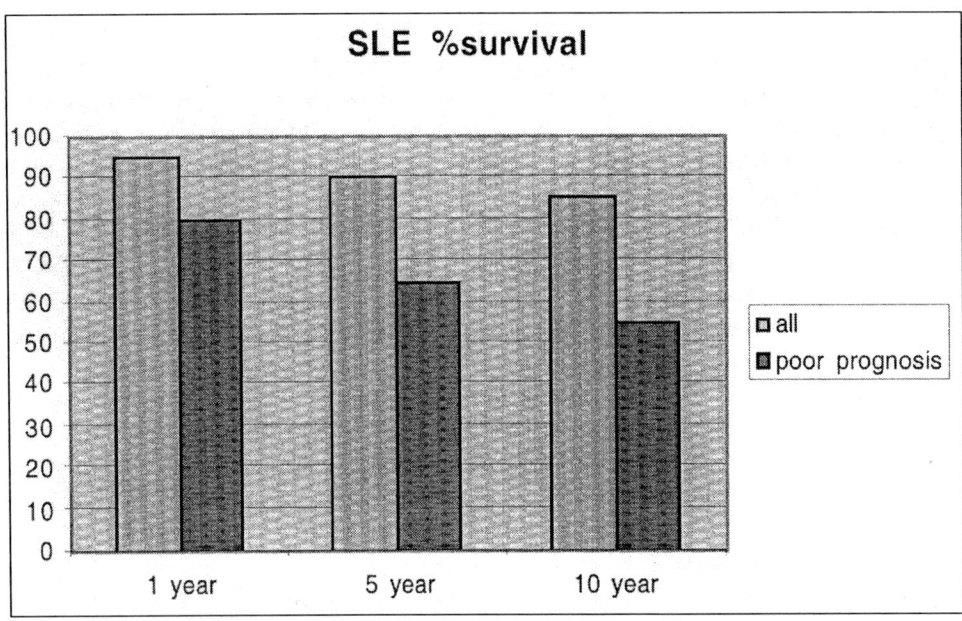

Figure 1. Current survival for SLE for all patients (light) and high risk patients (dark) at 1, 5 and 10 years. Figure provided courtesy of Dr. Bevra Hahn (UCLA) as summary of all articles summarizing mortality in SLE.

Therapy for SLE is well known, assiduously updated,[31-33] founded on various immunosuppressive modalities, and generally successful. However, there still are refractory/relapsing patients with high corticosteroid requirements and unsatisfactory response to current immunosuppressive therapy ("intractable").[34] For such patients, on the basis of the encouraging experimental evidence and anecdotal case reports of remissions following aggressive treatment for concurrent diseases, immunoablative conditioning followed by stem/progenitor transplantation (allogeneic, syngeneic, autologous) was proposed in 1993.[1]

Experimental Allogeneic Transplantation Studies

The literature treating murine and other lupus-like experimental animal models is extensive[35,36] and is discussed in Chapter 30. The principal mouse models of lupus are shown on Table 1. As already remarked, the availability of these models has been pivotal to our understanding of genetic and environmental factors.

The first experiments demonstrated that the transfer of spleen cells or whole bone marrow cells from New Zealand Black (NZB;H-2^d) mice to anti-lymphocyte globulin treated BALB-L,H-2^d (ref. 37) SCID or irradiated mice was capable of reproducing the donors' murine lupus.[38] This adoptive transfer of clinical disease with cellular grafting was subsequently confirmed for many other autoimmune diseases in animals, including the antiphospholipid syndrome, insulin-dependent diabetes mellitus (IDDM), experimental autoimmune encephalomyelitis and many others (refer to Chapters 29 and 30).

The identity of the cellular elements carrying the autoimmune susceptibility is important for the rationale of HSC transplantation. In the original experiments, unmanipulated marrow was utilized,[37,38] but it was shown subsequently that murine lupus of (NZB x NZW) F$_1$ mice could be transplanted (adoptive autoimmunity) utilizing T-cell depleted marrow.[39] B lymphoid precursors from (NZB/NZW) F1 marrow cultures reproduced the disease in SCID mice.[40] Human lymphocytes were also capable of transferring SLE in SCID mice.[41] In other extensive investigations, Ikehara and his group showed that HSC deriving from autoimmune animals were more "resilient" than those coming from healthy ones,[42] and have demonstrated the importance of stromatic and osteoblastic cells as engraftment facilitators,[43] leading finally to their postulate that experimental autoimmune diseases are polyclonal stem cell diseases.[44] Because of this overwhelming genetic predisposition among these highly inbred murine models, allogeneic HSCT is curative in most subtypes of murine lupus, when compatible, but nonidentical and nondiseased donors are used (Table 1). Human cord blood HSCT was capable of suppressing autoantibody production in lupus mice,[45] probably the first demonstration of a curative effect of a xenogeneic HSCT in an experimental autoimmune diseasess. However the greater penetrance of genetic predisposition in inbred animals cannot automatically be generalized to humans.

Clinical Experience: Autologous Transplantation

Mobilization of Stem Cells

In the great majority of cases, mobilized peripheral blood stem cells (rather than bone marrow) were utilized. It is common knowledge that HSC may be harvested from the bone marrow or mobilized into the peripheral blood using either a hematopoietic colony stimulating factor such as G-CSF, a chemotherapeutic drug such as cyclophosphamide, or both. Since G-CSF is a pro-inflammatory cytokine which by itself may exacerbate disease, the stem cells are generally mobilized into the peripheral blood using cyclophosphamide in dosages varying from 2 to 4 g/m^2 and G-CSF usually beginning 72 hours after cyclophosphamide, with harvest by apheresis started upon white blood cell rebound, usually 10 days following cyclophosphamide. It should be emphasized that SLE patients, due to disease as well as chronic high dose corticosteroid therapy, are prone to infections and prophylactic antibiotics and/or inpatient mobilization may be prudent. These cells may then be positive selected, using an antibody to CD34, a progenitor cell antigen, resulting in a 4 log depletion of lymphocytes. The product, either unmanipulated or CD34

Table 1. Mouse models of lupus

Mouse Strain	Onset/Manifestation	Predominant Sex with Disease	Genetics	Result of Hematopoietic Stem Cell Transplantation
NZB/NZW (B/W)	Spontaneous anti-ds DNA antibody and GN contribute	Female	Polygenic at least 8 loci on multiple chromosomes	Allogeneic transplant from normal strain cures
NZB/SWR (SNF)	Spontaneous anti-ds DNA antibody and GN	Female	Polygenic	N/A
MRL-lpr	Spontaneous/lymph-adenopathy, splenomegaly anti-ds DNA antibody, and GN	Female	Single defect of Fas gene prevents apoptosis of autoreactive lymphocytes in polygenic susceptible strain	Syngeneic T cell depleted transplant prolongs survival
BXSB	Spontaneous anti-ds DNA antibody and GN	Male	In polygenic susceptible background a single Y chromosome gene (Yaa)	Allogeneic transplant from normal strain cures
NZW/BXSB	Spontaneous anti-ds DNA antiphospholipid antibody and early death from coronary artery disease	Male	Polygenic	N/A

ds= double stranded; GN= glomerulonephritis; N/A= not applicable

selected, is cryopreserved, thawed, and reinfused after the conditioning regimen has been given. It has been suggested that some lupus patients may be poor mobilizers, independently from having undergone previous cytotoxic treatments such as cyclophosphamide.[46]

Conditioning Regimens

Autologous conditioning regimens for patients with malignancies are based on dose escalation of chemotherapeutic drugs that demonstrate effectiveness at standard dosing. Borrowing this concept from the field of oncology, Northwestern University dose escalated cyclophosphamide, the most effective anti-lupus chemotherapeutic agent, to 200 mg/kg. Also borrowing from experience with a HSCT conditioning regimen for aplastic anemia, itself an autoimmune disease, either rabbit or equine anti-thymocyte globulin (ATG) is added to the cyclophosphamide. The conditioning regimen used at Northwestern University is cyclophosphamide (50 mg/kg/day for 4 days) and equine ATG (30 mg/kg/day for 3 days). European centers have employed numerous regimens, including the BEAM (BCNU, etoposide, cytosine arabinoside and melphalan) and TBI based regimens. The choice of the conditioning regimen has generally been empiric, reflecting institutional preferences. However, the utilization of TBI in a disease such as lupus, where radiation has never been shown to have a beneficial therapeutic effect and could impart further injury to disease involved organs such as the lung and kidney, must be approached with caution. The variability in European regimens is reported in the multicentric European Bone Marrow Transplant/European League against Rheumatism (EBMT/Eular) registry.[7]

Results

The clinical experience for severe, intractable SLE can be divided in four groups: HSCT, in which SLE was a coincidental disease, case reports of HSCT for SLE, single center clinical studies, and registry reports.

Case Reports of HSCT for SLE as a Coincidental Disease

The first includes case reports of coincidental diseases. They include chronic myelogenous leukemia (CML) and SLE,[47] and malignant lymphoma and SLE. The first patient eventually died of her leukemia without evidence of active lupus. In a patient with nonHodgkins lymphoma (NHL), the lymphoma had not relapsed at the time of publication, but autoimmune thrombocytopenic purpura appeared in conjunction with an anticentromere antibody.[48] In other cases, there was again the coincidence of malignant lymphomas and SLE, with remission of both diseases following autologous HSCT.[49,50]

Case Reports of HSCT for SLE

This group includes case reports of patients with intractable SLE.[2,51-57] In all cases but one, where the conditioning regimen was BEAM, cyclophosphamide was the principal immunosuppressant, with TBI or ATG as potentiators. CD34+ selection was performed in 2 cases and lymphocyte depletion (T and B) in one. In a case of treatment-resistant central nervous system (CNS) lupus, autologous HSCT with peripheral T and B depletion was followed by prompt amelioration which persisted during 15 months of follow-up.[55] Dramatic remissions of massive proteinuria

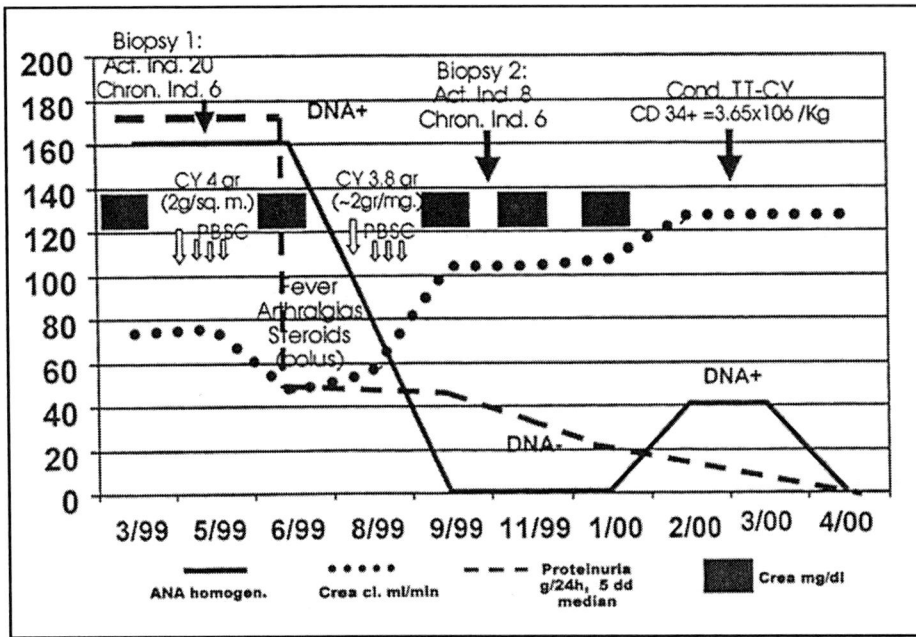

Figure 2. SLE with class IV diffuse proliferative glomerulonephritis and severe nephrotic syndrome treated with autologous HSCT. This 25-year old male patient started in 1999 with heavy proteinuria and strongly positive ANA and anti-ds DNA antibodies. Two biopsies showed class IV diffuse proliferative glomerulonephritis. Plasma exchange, high-dose corticosteriods and oral cyclophosphamide (3 g) failed to change the situation. A first mobilization with G-CSF and cyclophosphamide 4 g/m^2 gave a yield of 1.59 x 10^6/kg CD34$^+$ progenitor cells. There was some clinical amelioration followed by a flare, proteinuria and anti-ds ANA antibodies started to recur. A second mobilization with Cy 2 g/mg yielded 3.65 x 10^6/kg CD34$^+$ cells. Conditioning was performed with thiotepa 10 mg/kg, followed by CD34$^+$ cell reinfusion. The therapeutic result was dramatic. Reapperance of mild to moderate (~2 g/24hr) protenuria 6 months later is being successfully treated with low-dose corticosteroids and mycophenolate mofetil (MMP).

caused by lupus nephritic syndrome have also been reported[57] (see Fig. 2). Another significant aspect was the disappearance of antiphospholipid antibodies in a secondary antiphospholipid syndrome (APS).[53] Up to now, this type of autoimmunity had been abrogated only following allogeneic HSCT,[59] and both reports conjunctly point out that there are more decisive treatments for APS other than the conventional, long-term anticoagulation protocols. Still another significant result was the complete normalization of pulmonary function in a patient with 3 biopsy-proven flares of SLE pneumonitis that had required mechanical ventilation.[58] A synopsis of these cases is shown on Table 2. All these cases have been included in the multicentric EBMT/EULAR retrospective study which will be discussed subsequently.

In the United States, two autologous stem cell transplants have been performed for severe systemic lupus at the University of California at San Diego with success (Ewa Carrier, personal communication). The preparative regimen consisted of high dose cyclophosphamide only. One patient has only minor manifestations of disease two years post transplant. The other has no disease manifestations. Brazil has performed one successful HSCT for SLE, now followed for two years; a second patient was treated successfully and there was one transplant related death.[60] This experience is shown on Table 3. In China, the first SLE patient was autotransplanted at Nanjing University Medical College in 1998.[61] Another six patients were similarly treated at the same Institution, utilizing cyclophosphamide 200 mg/kg, ATG and selected peripheral CD34$^+$ cells. The Zhengzhou group reported its experience in 18 patients,[62] but a detailed analysis has not yet been performed.

Single Center Studies

The third and most significant group is the most important single center clinical study, which is still ongoing and is being performed at Northwestern University, Chicago. Summarizing a series of consecutive publications,[63,64] a total of 35 patients have been transplanted with no mortality. Fifteen patients are beyond 1 year following autologous HSCT. These figures are progressively increasing. No patient died as a consequence of transplantation. Since candidates were highly immune-suppressed prior to study entry, aggressive antimicrobial prophylaxis, including lipid amphotericin formulations were undertaken during periods of neutropenia regardless of fever. Treatment-related complications were initially higher in patients with nephritis, predominately caused by electrolyte disturbances, fluid shifts, and volume overload leading to pulmonary edema and intubation. Subsequently, for patients with nephritis, early initiation of dialysis or ultrafiltration to maintain dry weight prevented pulmonary edema and intubation. Patients without nephritis who were oxygen-dependent, either because of pulmonary interstitial fibrosis or pulmonary hemorrhage, had few transplant-related complications. Following HSCT, patients gradually improved and were slowly weaned off corticosteroids. By 12-18 months after HSCT, patients are corticosteroids free, often for the first time since disease onset that occurred years or a decade or more earlier. This phase I/II trial, which began 6 years ago, has provided the data and impetus for a randomized phase III trial of HSCT vs. pulse cyclophosphamide in patients with less severe disease at the time of entry.

Table 2. European reports of HSCT for SLE

Site/Institution	First Author	Number SLE/HSC	Number of Remissions	Preparative Regimen	Organs Specified	Follow Up at Last Publication
Hôpital S. Antoine	Fouillard (54)	1	1	BEAM	Glomerulo-nephritis	12 months
University of Goteborg	Trysberg (55)	1	1	Cy/ATG	Cerebritis and myelitis	18 months
University of Vienna	Brunner (58)	1	1	Cy/ATG	Recurrent respiratory failure	21 months
University of Utrecht	Wulffraat (52)	2	2	Cy/ATG and LDTBI	Pneumonitis, GN and retinal vasculitis/GN	12 months 15 months
University of Palermo	Musso (53)	3	3 taking ≤10 mg prednisone	Cy 100 mg/kg/ATG 90	Evan's/SLE and SLE	4 months 18 months 32 months
Theagerion Hospital of Thessaloniki	Fassas (51)	1	0	BEAM	Pancytopenia	Transient
University Hospital Charité Berlin	Rosen (56)	3	3	CY/ATG	SLE	10-21 months

Table 3. North and South American reports of HSCT for SLE

Institution Performing Autologous SCT for SLE	Preparative Regimen Used	Number of Patients Registered	Number of Patients Transplanted	100 day Mortality/ and Cause	Patients in Sustained Remission >1yr
MD Anderson	CY 120 mg/kg and low dose TBI	1	1	0	0
Northwestern University[2]	CY 200 mg/kg and ATG 90 mg/kg	37	35	0	15/18 followed from 1-6 years post HSCT
San Antonio	CY 200, ATG 90 mg/kg	1	1	1	0
Brazil	CY 200 mg/kg ATG 90 mg/kg	2	2	1 pulmonary edema	1 followed 24 months
Tulane	TBI 550, CY 40 mg/kg x 3 ATG 80 mg/kg	2	2	1 pulmonary death	1 followed over 30 months
UCSD	CY 200 mg/kg	2	2	0	2 followed 12, 24 months

Registry Data

A retrospective registry survey by the European Blood and Marrow Transplant (EBMT) and the European League against Rheumatism (EULAR) has collected data from 53 patients with SLE treated by autologous HSCT in 23 European centers.[7] Mobilized peripheral stem cells were utilized, 42% of them with CD34+ selection, and conditioning regimens employed cyclophosphamide in 84%, anti-thymocyte globulin in 76% and lymphoid irradiation in 22%. The mean duration of follow-up after autologous HSCT was 26 (0-78) months. Remission of disease activity (SLEDAI <3) was seen in 33/50 (66%) evaluable patients by 6 months, of which 10/31 (32%) subsequently relapsed after 6 (3-40) months. There were 12 deaths (0-48) months, of which 7 (12%) were related to the procedure. This registry data demonstrated the efficacy of autologous HSCT for remission induction of refractory SLE, although mortality was excessively high. The difference in the procedure's safety between the single center Northwestern University mortality of 0 out of 35 patients and the European multi-center registry data with a mortality of 12 out of 53 emphasizes the importance of experience and a focused team approach in selection of patients, conditioning regimens, and supportive care.

Allogeneic Transplantation

A single, little-known but most informative case report concerns a 30 year old female patient who underwent allogeneic BMT in 1984 (time of last follow-up was 1999), and who was still in complete remission (for a discussion of this concept see later) at the time of this writing.[65] This patient had severe SLE since the age of 15, went on to present neuropsychiatric symptoms and seizures, and finally developed pancytopenia. She underwent allogeneic BMT from an HLA-identical brother after conditioning with cyclophosphamide, 50 mg/day for 4 consecutive days. There was a total resolution of all clinical symptoms, and 15 years later the patient is still in complete remission and requires no treatment. There is still a weak (1:80) ANA positivity of the speckled subtype, probably the expression of a residual lymphoplasmacytic host microclone.

High-Dose (HD) Immunosuppression Alone

HD immunosuppression (200 mg/kg divided over 4 days), without subsequent utilization of stem cells of any kind, is the well-known therapeutic protocol for autoimmune diseases utilized at John Hopkins University (see Chapter 28). In their last contribution, 14 lupus patients were treated in this way, and 5 of them achieved a complete response, which was found to be durable at a median follow-up of 32 months.[66-68] Eradication (or near eradication) of autoreactive lymphocytes and lesser costs are considered main advantages of this procedure, which is discussed elsewhere. Another little known clinical event is the sustained remission which followed the inadvertent administration of a single high dose of cyclophosphamide (5 g) in an SLE patient.[69]

Discussion

Mortality and Potential Oncogenicity

The 12% transplant related mortality (TRM) found in the EBMT/EULAR study contrasts with the absence of fatalities reported by Northwestern University (Chicago);[63,64] and Johns Hopkins (Baltimore);[66] with their cyclophosphamide +/- ATG protocols.[66-68] This is a well known difference between retrospective multicentric studies and single centers of excellence studies. Accurate selection of patients (not too terminally ill, not severe enough to enter the protocol) is essential. A learning curve and a center effect are also distinct possibilities.

Even if the problem of excessive TRM will almost certainly solved, as it has been in centers of excellence, the problem of late oncogenicity must also be taken into consideration, especially in younger patients with nonmalignant diseases. The risk of developing solid cancers was 3-4 times higher in patients treated with combined modality therapy during allogeneic marrow transplantation than in controls.[70] In another study, a higher risk of acute myeloid leukemia (AML) was found following autologous HSCT when the conditioning regimen included TBI.[71] However, in a large group of predominantly Caucasian SLE patients, the rates of overall malignancies were increased significantly above the expected rates in the general population, but the use of cytotoxic agents was not associated with increased malignancy.[72] Studies to evaluate if SLE patients are indeed in "double trouble", that is whether lupus and malignancy are really associated, have been proposed.[73] In conclusion, however, patients with severe intractable SLE, exactly as it happens with other otherwise incurable diseases, should consider how to overcome their present danger rather than conjecture about distant, relatively improbable complications.

Efficacy of Treatment

There is no doubt that high-dose immunosuppression, followed or not by stem cell transplantation (autologous, syngeneic, allogeneic) is an intense procedure capable of exerting a powerful therapeutic effect on SLE, even the cyclophosphamide-refractory kind, as well as in the majority of severe autoimmune diseases (see Chapter 27). A still debatable point concerns the definition of complete remission (CR). Two different points of view may be considered.

A long-term complete remission (CR) should probably be considered as an equivalent of "cure". It is a well established axiom that the absence of minimal residual disease after allogeneic HSCT is the best demonstration of a cure. In analogy to oncohematology, CR in lupus has been defined as a long-term and treatment-free complete remission with restoration of normal blood counts and a normal immune system.[74] The already reported patient who underwent allogeneic HSCT and is (still) in CR 15 years after allo-HSCT, with no treatment and only a residual ANA (1:40) trace, would seem to fit in this definition. Since probably no patient would meet these stringent requirements, in the EBMT/EULAR study CR, which was utilized for calculating the disease-free survival (DFS), was defined as a SLEDAI index <3. However, for a disease such as SLE, it is more practical for a patient to utilize the concept of sustained remission from clinical disease activity and no residual, irreversible damage, without therapy. There is a growing consensus that a CR should (or might) be defined as no clinical evidence of active disease and on no immune suppressive medications. Small (<10 mg) doses of prednisone are accepted in this context, but it is uncertain whether they are useful because of prolonged hypocorticism (in which case ACTH would be more indicated) or to control minimal residual autoimmune disease.

Organ dysfunction, such as renal or pulmonary insufficiency, is considered a contraindication to HSCT among many patients with malignancies. For patients with SLE, in contrast, organ com-

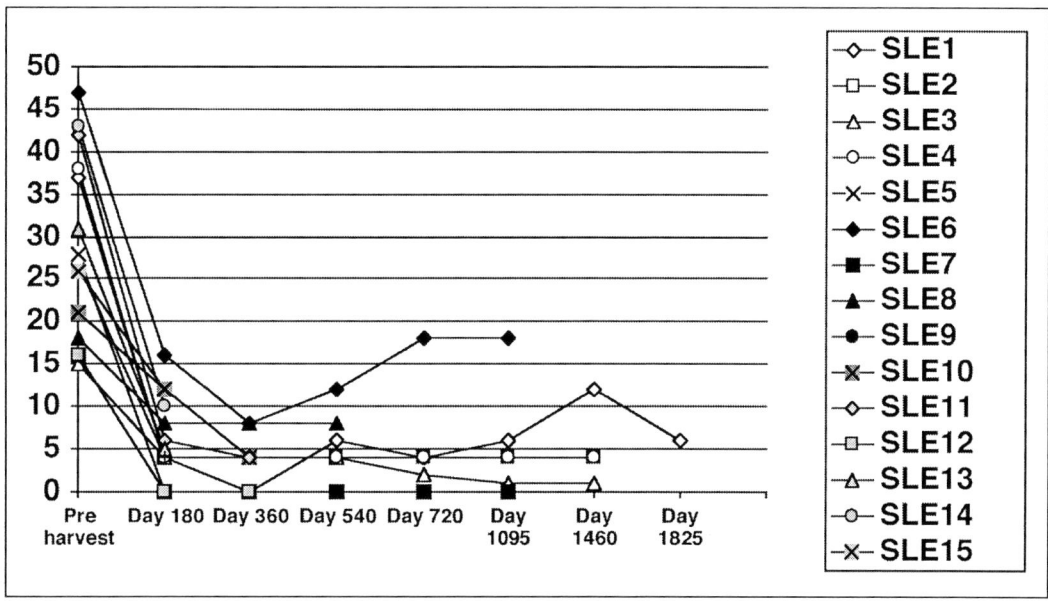

Figure 3. Systemic lupus disease activity index (SLEDAI) before and after HSCT in 15 patients. (Reprinted with permission from: Traynor A, Burt RK. Arthritis Rheum 2002; 46(11)2917-2923.)

promise is due to lupus. Therefore, impaired visceral organ function is not necessarily a contraindication and may even be the major indication for HSCT. While this makes the transplant procedure more complicated, eventually organ improvement, particularly of the lung, kidney and central nervous system, typically occurs. Practical issues to consider include the fact that patients with nephritis are unusually susceptible to electrolyte disturbances and to volume overload that may lead to pulmonary edema and respiratory failure, in spite of aggressive diuresis and electrolyte replacement. Early and expectant dialysis or continuous veno-venous hemofiltration (CVVH) is crucial to prevent or correct these problems. For this reason, it is imperative that at least one nephrologist and at least one intensivist be a part of the transplant protocol and team and become acquainted with the patients before high dose therapy is begun. Pharmacokinetics of drugs such as cyclophosphamide metabolites are incompletely understood in patients with acute and chronic renal failure. For this reason, in patients with either renal insufficiency or renal failure, dialysis is performed the morning after each cyclophosphamide infusion. Many patients have an indication for on going anticoagulation at the time of accrual. They may have catastrophic anti-phospholipid syndrome, or a history of recurrent deep venous thrombosis, pulmonary embolus or valvular thrombi. While initially this seemed a potential obstacle to safe autologous HSCT, it has not proven to be a problem. The approach followed at Northwestern University has been to stop coumadin and start therapeutic fragmin the week prior to mobilization. Fragmin has been continued at full therapeutic dose until the platelet count falls below $60 \times 10^9/l$. Until it reaches $60 \times 10^9/l$ again, the patient is maintained on prophylactic dose fragmin. In the event that fragmin is not tolerated, lovenox would be a second choice drug that has worked well in this setting.

The best efficacy data currently comes from Northwestern's single center trial and is reviewed in Figures 3 to 6. Disease activity was monitored by the systemic lupus disease activity index (SLEDAI) (refer to Chapter 39). A remission is considered a SLEDAI of less than either 3 or 5, depending on the author. Severely active disease is considered a SLEDAI >20 to 25. As shown in Figure 3 most patients had very active disease (SLEDAI >20) despite intravenous cyclophosphamide. Following HSCT, most patients have gradually improved and maintained a SLEDAI <5, off immune suppression, with the exception of prednisone doses <10 mg/day. Creatinine clearance has either improved or remained unchanged following HSCT (Fig. 4). While recent renal biopsies to determine chronicity of disease were not obtained, as a generalization, independent of the creatinine level, recent deterioration in creatinine clearance within 3 months prior to HSCT is generally reversible, while more chronic renal insufficiency may not improve. Proteinuria universally improves (Fig. 5), however, improvement may not be appreciated for 3 to 6 months. Pulmonary function monitored by forced vital capacity (Fig. 6) or diffusion capacity (DLCO) gradually normalizes over 6 to 18 months following HSCT. Some patients have become oxygen free for the first time in several years following HSCT. Finally, neurocognitive function including reasoning and intellect (i.e., IQ) have improved following HSCT (paper in preparation).

Mechanism of Action

The fundamental but still unresolved question is, whether intense immunosuppression followed by autologous HSCT is indeed capable of eradicating autoimmunity and thus inducing tolerance,[8] or if the immune system remains substantially unaltered, even if drastically depleted, and the so-called transplant is little more than a hematopoietic rescue. The first goal appears to have been achieved experimentally,[75,76] but in clinical settings it has not been conclusively demonstrated. It has been proposed, however, that the conditioning, including the utilization of ATG, might provide a window of time free of memory T cell influences, during which the maturation of new lymphocyte progenitors may occur without recruitment to anti-self reactivity.[63] Studies with gene-marked autologous stem cells haven't answered this question definitively.[77] To the dynamics of the T cell subpopulations, which are analyzed elsewhere,[76] new studies on regulatory T cells are clearly indicated.[79] The immunologic perspective changes totally when coming to allogeneic transplantation, as will be discussed later.

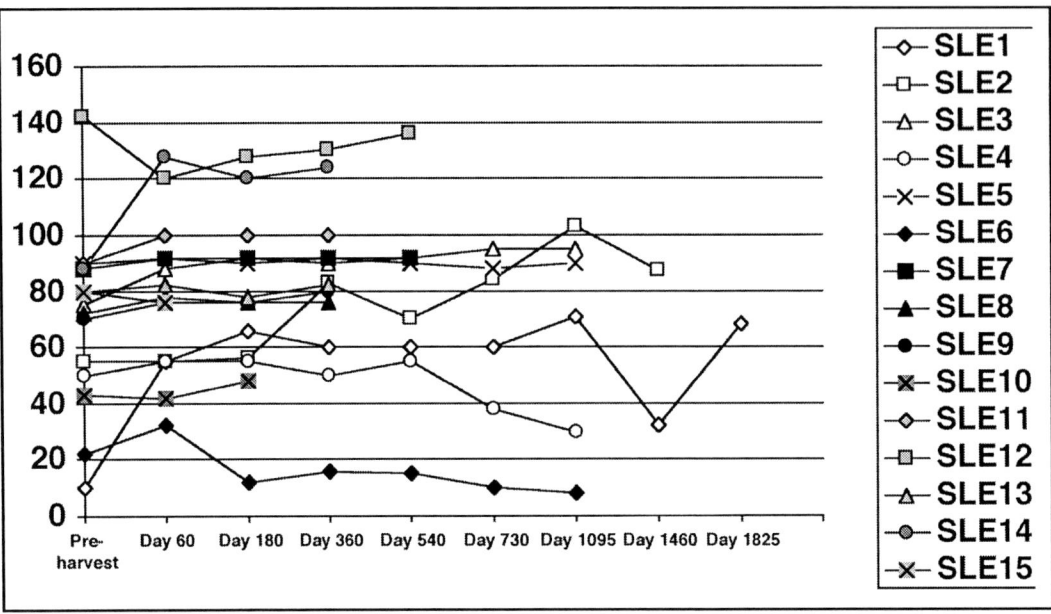

Figure 4. Creatinine clearance before and after HSCT in 15 patients. (Reprinted with permission from: Traynor A, Burt RK. Arthritis Rheum 2002; 46(11)2917-2923.)

Future Developments

Autologous Transplants

There certainly will be further refinements verging on conditioning regimens, more specifically lympholytic, such as the combination of cyclophosphamide and Rituximab to achieve B-cell depletion. Fludarabine may also be considered, but it would be ill advised to use fludarabine due to known CNS toxicity of fludarabine in patients with renal insufficiency. A phase III NIH funded trial is being developed to compare the efficacy of HSCT to a continuation of the currently accepted standard of care, intravenous monthly cyclophosphamide in persistent, active SLE. The same mobilization and conditioning regimen will be used in the phase III trial as in the phase I/II Northwestern trial discussed above. Stem cells will mobilized with cyclophosphamide (2.0 g/m^2) followed 48-72 hours later with daily G-CSF (10 mcg/kg/day). The conditioning regimen will be cyclophosphamide (200 mg/kg) and equine ATG (90 mg/kg). In selecting candidates for this Phase III trial, it was imperative that rheumatologists feel that the standard approach appeared to be failing before they randomize their patient. Therefore, patients with nephritis as their eligibility criterion will need to have received 6 cycles of cyclophosphamide and still have active disease. For other organ systems, the standard of treatment is less well established.

The current standard of care in the USA for patients with SLE nephritis or organ-threatening disease is intravenous pulse cyclo-

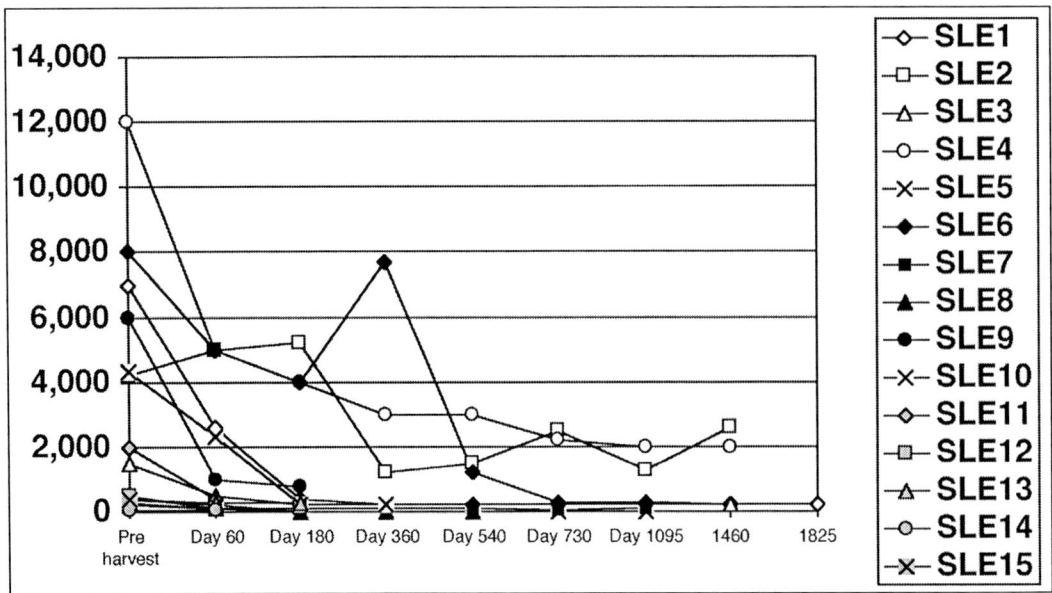

Figure 5. Proteinuria (in grams) before and after HSCT in 15 patients more than 1 year post HSCT. (Reprinted with permission from: Traynor A, Burt RK. Arthritis Rheum 2002; 46(11)2917-2923.)

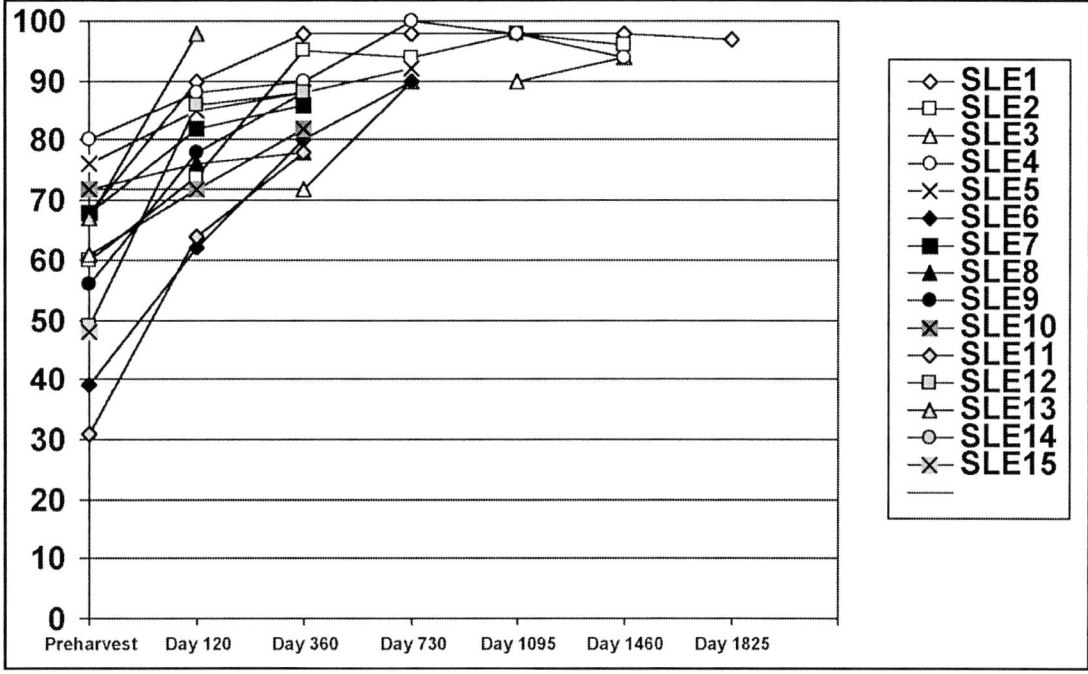

Figure 6. Percent forced vital capacity (FVC) before and after HSCT. (Repritned with permission from: Traynor A, Burt RK. Arthritis Rheum 2002; 46(11)2917-2923.)

phosphamide (500-1000mg/m^2) per month for six months, and then every three months, which is currently being planned for the control arm. Newer immune suppressive agents have more recently been used to treat SLE including cyclosporine and mycophenolate mofetil. The efficacy of the latter has only been shown in small studies. In a study from Hong Kong, twelve-month survival rates and response rates were similar in patients with lupus nephritis (approximately 80%) treated with either daily oral cyclophosphamide for 6 months (followed by daily oral azathioprine for 6 months) compared to oral mycophenolate mofetil daily for 12 months.[61] However, there are no comparative trials demonstrating superiority or equivalency of cyclosporine, mycophenolate mofetil, or oral cyclophosphamide to IV monthly cyclophosphamide, the standard of care in America.

If the question to answer is "What is the best salvage therapy for cyclophosphamide-refractory SLE?", then comparison between HSCT versus mycophenolate mofetil or cyclosporine or even transplant doses of cyclophosphamide without stem cell infusion might be appropriate. However, the question for the current NIH study is: " Is there a therapy better than the current standard of care, i.e., pulse cyclophosphamide continuation for disease which is still requiring therapy?" Currently, patients with active disease despite pulse cyclophosphamide are often continued on monthly IV cyclophosphamide therapy. If patients with lupus were offered HSCT at onset of disease, many patients who would otherwise have been successfully treated with pulse cyclophosphamide would be unnecessarily exposed to the more dangerous and aggressive procedure of HSCT. On the other hand, it would be difficult to enroll patients into a trial of continued "failed therapy" versus HSCT. Therefore, candidates must have active SLE despite exposure to some pulse cyclophosphamide therapy, but yet be able to remit with continued pulse cyclophosphamide. If the HSCT arm proves superior to the current standard of care, i.e., monthly pulse cyclophosphamide, the next study would probably compare 200 mg/kg cyclophosphamide without stem cells to 200 mg/kg cyclophosphamide and ATG with infusion of cyclophosphamide mobilized and selected CD34$^+$ stem cells.

Another future area of investigational research as applied to autologous HSCT is an effort to utilize autologous HSCT in association with gene therapy. This strategy consists of transplantation of autologous stem cells engineered with identified self-antigens, with the aim of promoting the generation within the thymus of dendritic cells specifically "armed" with culprit self-antigen in order to tolerize developing T cells.[80] This has been achieved in experimental gastric autoimmunity[81] and autoimmune diabetes,[82] but this approach is applicable only to autoimmune diseases in which culprit antigens are identified, also independently from the new concerns with the safety of gene therapy trials (refer to Chapter 14).

Allogeneic Transplants

If allogeneic SCT will be explored in refractory SLE, it is likely that nonmyeloablative procedures[83] will be utilized preferentially.[84] This will be not only because of their sparing effects on conditioning-regimen related mortality, but also because they provide a platform for a Graft-versus-Autoimmunity effect which has been clearly demonstrated in experimental autoimmune diseases,[85] but is probably also present in humans.[86-89]

References

1. Marmont AM. Perspective Immunoablation with stem-cell rescue: A possible cure for systemic lupus erythematosus? Lupus 1993; 2:151-6.
2. Marmont AM, van Lint MT, Gualandi F et al. Autologous marrow stem cell transplantation for severe systemic lupus erythematosus of long duration. Lupus 1997; 6(6):545-8.
3. Burt RK, Traynor A, Ramsey-Goldman R. Hematopoietic stem-cell transplantation for systemic lupus erythematosus. N Engl J Med 1997; 337(24):1777-8.
4. Traynor A, Burt RK. Haematopoietic stem cell transplantation for active systemic lupus erythematosus. Rheumatology (Oxford) 1999; 38(8):767-72.

5. Traynor AE, Schroeder J, Rosa RM et al. Treatment of severe systemic lupus erythematosus with high-dose chemotherapy and haemopoietic stem-cell transplantation: A phase I study. Lancet 2000; 356(9231): 701-7.
6. Traynor AE, Barr WG, Rosa RM et al. Hematopoietic stem cell transplantation for severe and refractory lupus. Analysis after five years and fifteen patients. Arthritis Rheum 2002; 46(11)2917-23.
7. Jayne D, Passweg J, Marmont A et al. (for the European Group for Blood and Marrow Transplantation and European League against Rheumatism): Autologous stem cell transplantation for systemic lupus erythematosus. Lupus, In press.
8. Burt RK, Slavin S, Burns WH et al. Induction of tolerance in autoimmune disease by hematopoietic stem cell transplantation: Getting closer to a cure? Blood 2002; 99:870-87.
9. Liossis NC, Tsokos GC. Systemic lupus erythematosus. In: Tsokos GC, ed. Principles of molecular rheumatology. Totowa NJ: Humana Press 2000:311-23.
10. Gregersen PK. Genetics. In: Tsokos GC, ed. Principles of molecular rheumatology. Totowa NJ: Humana Press, 2000:3-14.
11. Deapen D, Escalante A, Weinrib L et al. A revised estimate of twin concordance in systemic lupus erythematosus. Arthritis Rheum 1992; 35:311-18.
12. Kotzin BL, West SG. Systemic lupus erythematosus. In: Rich RR, Fleisher TA, Shearer WT et al, eds. Clinical immunology.
13. Hahn BH. Eminence based medicine. Lessons in lupus: The mighty mouse. Lupus 2001; 10:589-93.
14. Kono DH, Theophilopoulos AN. The genetics of murine systemic lupus erythematosus. In: Wallace DJ, Hahn BH, eds. Dubois' lupus erythematosus. Philadelphia: Lippincott Williams & Wilkins, 2002:97-120.
15. Theophilopoulos AN, ed. Genes and genetics of autoimmunity. Curr Dir Autoimmunity 1999; Basel: Karger, 1:1-296.
16. Encinas JA, Kuchroo VK. Mapping and autoimmunity genes. Autoimmunity 2001; 12:91-7.
17. Tsao BP. The genetics of human lupus. In: Wallace DJ, Hahn BH, eds. Dubois lupus erythematosus. Philadelphia: Lippincott Williams & Wilkins, 2002:97-120.
18. Morel L, Croker BP, Blenman KR et al. Genetic reconstitution of systemic lupus erythematosus immunopathology with polycongenic murine strains. PNAS 2000; 97:6870-75.
19. Kolowos W, Gaip US, Vole RE et al. CD4 positive peripheral T cells from patients with systemic lupus erythematosus (SLE) are clonally expanded. Lupus 2001; 10:321-31.
20. Herrmann M, Winkler T, Gaipl U et al. Etiopathogenesis of systemic lupus erythematosus. Int Arch Allergy Immunol 2000; 123:28-35.
21. Alarcón-Segovia D, Alarcón-Riquelme ME. Etiopathogenesis of systemic lupus erythematosus: A tale of three troikas. In: Lahita RG, ed. Systemic Lupus Erythematosus. San Diego: Academic Press, 1999: 56-65.
22. Fraenkel L, MacKenzie T, Joseph L et al. Response to treatment as a predictor of long term outcome in patients with lupus nephritis. J Rheumatol 1994; 21(11):2052-7.
23. Esdaile JM. Prognosis in systemic lupus erythematosus. Springer Semin Immunopathol. 1994;16(2-3):337-55.
24. Esdaile JM, Joseph L, MacKenzie T et al. The pathogenesis and prognosis of lupus nephritis: Information from repeat renal biopsy. Semin Arthritis Rheum 1993; 23(2):135-48.
25. Ginzler EM, Schorn K. Outcome and prognosis in SLE. Rheum Dis Clin North Am 1988; 14:67-78.
26. Bakir AA, Levy PS, Dunea G. The prognosis of lupus nephritis in African-Americans: A retrospective analysis. Am J Kidney Dis 1994; 24:159.
27. Ward MM, Pyun E, Studenski S. Long-term survival in SLE: Patient characteristics associated with poorer outcomes. Arthritis Rheum 1995; 38:274.
28. Jacobsen S, Petersen J, Ullman S et al. A muticentre study of 513 Danish patients with SLE. II. Disease mortality and clinical factors of prognostic value. Clin Rheumatol 1988; 17:478.
29. Urowitz MB, Gladman DD, Abu-Shakra M et al. Mortality studies in SLE. Results from a single center. III. Improved survival over 24 years. J Rheumatol 1997; 24:1061.
30. Alarcòn GS, McGwin G Jr, Bastian HM et al. Systemic lupus erythematosus in three ethnic groups. Predictors of early mortality in the LUMINA cohort. LUMINA Study Group. Arthritis Rheum 2001; 45(2):191-202.
31. Strand V. New therapies for systemic lupus erythematosus. Rheum Dis Clin North Am 2000; 26:389-405.
32. Schur PH, Kammer GA. Treatment of systemic lupus erythematosus. In: Tsokos GC, ed. Modern therapeutics in rheumatic diseases. Totowa NJ: Humana Press, 2002:259-95.
33. Moroni G, Della Casa Alberighi O, Ponticelli C. Combination treatment in autoimmune diseases: Systemic lupus erythematosus. In: Harrison WB, Dijkmans BAC, eds. Combination treatment in autoimmune diseases. Berlin-Heidelberg-New York: Springer, 2002:75-89.
34. Cash JM, Wilder RL, eds. Treatment-resistant rheumatic disease. Rheum Dis Clin N Am 1995; 21:1-170.
35. Ehrenstein MR, Horsfall A, Isenberg D. Immunopathology of lupus. In: Schur PH, ed. The clinical management of systemic lupus erythematosus. Philadelphia-New York: Lippincott-Raven 1996:17-34.
36. Hahn BH. Animal models of systemic lupus erythematosus. In: Wallace DJ, Hahn BH, eds. Dubois' lupus erythematosus. Lippincott Williams & Wilkins 2002:339-88.
37. Denman AM, Russel AS, Denman EJ. Adoptive transfer of the disease of New Zealand black mice to normal mouse strains. Clin Exp Immunol 1969; 5:567-70.
38. Morton JC, Siegel BV. Transplantation of autoimmune potential. Development of antinuclear antibodies in H-2 histocompatible recipients of bone marrow from New Zealand Black mice. PNAS 1974; 71:2162-66.
39. Levite M, Zinger H, Mozes E et al. Systemic lupus erythematosus-related autoantibody production in mice is determined by bone marrow derived cells. Bone Marrow Transplant 1993; 12:79-83.
40. Reininger L, Radaszkiewics T, Kosco M et al. Development of autoimmune disease in SCID mice populated with long-term "in vitro" proliferating(NZBxNZW)F1 pre-B cells. J Exper Med 1992; 176:1343-53.
41. Duchosal MA, McConahey PJ, Robinson CA et al. Transfer of human systemic lupus erythematosus in severe combined immunodeficient (SCID) mice. J Exper Med 1990; 172:985-88.
42. Kawamura M, Hisha H, Li Y et al. Distinct qualitative differences between normal and abnormal hemopoietic stem cells in vivo and in vitro. Stem Cells 1997; 15:56-62.
43. Hisha H, Nishino T, Kawamura M et al. Successful bone marrow transplantation by bone grafts in chimeric-resistant combination. Exp Hematol 1994; 23:347-52.
44. Ikehara S. Autoimmune diseases as stem cell disorders: Normal stem cell transplant for their treatment (review). Int J Mol Med 1998; 1:5-16.
45. Ende N, Czarnesky J, Raveche R. Effect of human cord blood transfer on survival and disease activity in MRL/lpr/lpr mice. Clin Immunol Immunopathol 1995; 75:190-5.
46. Burt RK, Fassas A, Snowden J et al. Collection of hematopoietic stem cells from patients with autoimmune diseases. Bone Marrow Transplant 2001; 28:1-12.
47. Meloni G, Capria S, Vignetti M et al. Blast crisis of chronic myelogenous leukaemia in long-lasting systemic lupus erythematosus: Regression of both diseases after autologous bone marrow transplantation. Blood 1997; 89:4659.
48. Euler HH, Marmont AM, Bacigalupo A et al. Early recurrence or persistence of autoimmune diseases after unmanipulated autologous stem cell transplantation. Blood 1996; 88:3621-25.
49. Snowden JA, Patton WN, O'Donnell JAL et al. Prolonged remission of long-standing lupus erythematosus after autologous bone marrow transplant for non-Hodgkin lymphoma. Bone Marrow Transplant 1997; 19:1247-50.
50. Schacha L, Ryan PF, Schwarer AP. Malignancy-associated remission of systemic lupus maintained by autologous peripheral blood stem cell transplantation. Arthritis Rheum 1998; 41:2271-72.
51. Fassas A, Anagnastopoulos A, Giannaki C et al. Autologous peripheral blood stem cell therapy for autoimmune pancytopenia due to systemic lupus erythematosus. Bone Marrow Transplant 1998; 21:53 (abstract).

52. Wulffraat NM, Sanders EA, Kamphuis SS et al. Prolonged remission without treatment after autologous stem cell transplantation for refractory childhood systemic lupus erythematosus. Arthritis Rheum 2001; 44(3):728-31.
53. Musso M, Porretto F, Crescimanno A et al. Autologous peripheral blood stem and progenitor (CD34+) cell transplantation for systemic lupus erythematosus complicated by Evans syndrome. Lupus 1998; 7:492-94.
54. Fouillard L, Gorin NC, Laporte JPH et al. Control of severe systemic lupus erythematosus after high-dose immunosuppressive therapy and transplantation of CD34+ purified autologous stem cells from peripheral blood. Lupus 1999; 8:320-23.
55. Trysberg E, Lindgreen I, Tarkowski A. Autologous stem cell transplantation in a case of treatment-resistant central nervous system lupus. Ann Rheum Dis 2000; 59:236-38.
56. Rosen O, Thiel A, Massenkeil G et al. Autologous stem-cell transplantation in refractory autoimmune diseases after in vivo immunoablation and ex vivo depletion of mononuclear cells. Arthritis Res 2000; 2:327-36.
57. Marmont AM. Eminence based medicine. Lupus tinkering with haematopoietic stem cells. Lupus 2001; 10:769-74.
58. Brunner M, Greinix HT, Redlich K et al. Autologous blood stem cell transplantation in refractory systemic lupus erythematosus with severe pulmonary involvement. Arthritis Rheum 2002; 46:1580-84.
59. Olalla JI, Ortin M, Hermida G et al. Disappearance of lupus anticoagulant after allogeneic bone marrow transplantation. Bone Marrow Transplant 1999; 23:83-85.
60. Voltarelli JC. Transplante de células tronco hematopoéticas para doenças auto-imunes no Brasil (Haematopoietic stem cell transplantation for autoimmune diseases in Brazil). Rev bras hematol hemoter 2002; 24: 9-13.
61. Ouyang J, Sun LY, Yang YG et al. First results of a pilot study in treatment of SLE by ABMSCT. Chinese J Intern Med 2001; 40:229-31.
62. Zhao J, Fu Y, Peng X. Autologous hematopoietic stem cell transplantation in the treatment of systemic lupus erythematosus. Bone Marrow Transplant 2002; 29(2):S15.
63. Traynor AE, Schroeder J, Rosa RM et al. Treatment of severe lupus with high dose chemotherapy and haematopoietic stem cell support. Lancet 2000; 356:701-7.
64. Traynor AE, Barr WG, Rosa RM et al. Hematopoietic stem cell transplantation for severe and refractory lupus. Arthritis Rheum 2002; 46:2917-23.
65. Gur-Lavi M. Long-term remission with allogeneic bone marrow transplantation in systemic lupus erythematosus. Arthritis Rheum 1999; 42:1777.
66. Brodsky RA, Petri M, Smith BD et al. Immunoablative high-dose cyclophosphamide without stem-cell rescue for refractory, severe autoimmune disease. Ann Intern Med 1998; 129(12):1031-5.
67. Gladstone DE, Prestrud AA, Pradhan A et al. High-dose cyclophosphamide for severe systemic lupus erythematosus. Lupus 2002; 11:405-10.
68. Petri M, Jones RJ, Brodsky RA. High dose cyclophosphamide without stem cell transplantation in systemic lupus erythematosus. Arthritis Rheum 2003; 48:166-73.
69. Mittal G, Balarishna C, Mangat G et al. "Sustained remission" in a case of SLE following megadose cyclophosphamide. Lupus 1998; 8:77-80.
70. Curtis RE, Rowlings PA, Deeg HJ et al. Solid cancers after bone marrow transplantation. N Engl J Med 1997; 336:897-904.
71. Dorrington DL, Vase JM, Anderson R et al. Incidence and characterization of secondary myelodysplastic syndrome and acute myelogenous leukemia following high-dose radiochemotherapy and autologous stem cell transplantation for lymphoid malignancies. J Clin Oncol 1994; 12:2527-34.
72. Cibere J, Sibley J, Haga M. Systemic lupus erythematosus and the risk of malignancy. Lupus 2001; 10:394-400.
73. Ramsey-Goldman R, Clarke AE. Double trouble: Are lupus and malignancy associated? Lupus 2001; 10:388-91.
74. Euler HH. A preliminary definition of the term "cure" as applied to systemic lupus erythematosus. Arthritis Rheum 1999; 42:1552.
75. van Bekkum DW. Stem cell transplantation in experimental models of autoimmune diseases. J Clin Immunol 2000; 20:11-17.
76. Ikehara S. Bone marrow transplantation for autoimmune diseases. Acta Haematol 1998; 99:116-32.
77. Marmont AM. New horizons in the treatment of autoimmune diseases: Immunoablation and stem cell transplantation. Ann Rev Med 2000; 51:115-34.
78. Burt RK, Brenner M, Burns W et al. Gene-marked autologous hematopoietic stem cell transplantation of autoimmune disease. J Clin Immunol 2000; 20:1-9.
79. von Herrath MG, Harrison LC. Antigen-induced regulatory T cells in autoimmunity. Nature Immunol 2003; 10:223-32.
80. Alderuccio F, Murphy K, Toh B-H. Stem cells engineered to express self-antigen to treat autoimmunity. Trends Immunol 2003; 24:176-79.
81. Alderuccio F et al. Animal models of human disease: Experimental autoimmune gastritis –a model for autoimmune gastritis and pernicious anemia. Clin Immunol 2002; 102:48-58.
82. Steptoe RJ, Ritchie JM, Harrison LC. Transfer of hematopoietic stem cells encoding autoantigen prevents autoimmune diabetes. J Clin Invest 2003; 111:1357-63.
83. Giralt S, Slavin S, eds. Non-myeloablative stem cell transplantation (NST). Abington: Darwin Scientific Publications, 2000.
84. Burt RK, Traynor AE, Craig R et al. The promise of hematopoietic stem cell transplantation for autoimmune diseases. Bone Marrow Transplant 2003; 31:521-24.
85. Li H, Kaufman CL, Boggs S et al. Mixed allogeneic chimerism induced by sublethal approach prevents autoimmune diabetes and reverses insulitis in nonobese diabetic (NOD) mice. J Immunol 1996; 156:380-87.
86. Slavin S, Nagler A, Varadi G et al. Graft vs autoimmunity following allogeneic nonmyeloablative blood stem cell transplantation in a patient with chronic myeloid leukaemia and severe systemic psoriasis and psoriatic polyarthritis. Exp Hematol 2000; 28:853-57.
87. Oyama Y, Papadopoulos EB, Miranda M et al. Allogeneic stem cell transplantation for Evans syndrome. Bone Marrow Transplant 2001; 28:903-5.
88. Hinterberger M, Hinterbereger-Fischer M, Marmont AM. Clinically demonstrable anti-autoimmunity mediated by allogeneic immune cells favourably affects outcome after stem cell transplantation in human autoimmune diseases. Bone Marrow Transplant 2002; 30:753-59.
89. Marmont AM, Gualandi F, van Lint MT et al. Refractory Evans syndrome treated with allogeneic SCT followed by DLI. Demonstration of a graft-versus-autoimmunity effect. Bone Marrow Transplant 2003; 31:525-30.

CHAPTER 42

Treatment of Rheumatoid Arthritis

Stuart Weisman and Arthur Kavanaugh

Introduction

During the past decade, the approach to the treatment of patients with rheumatoid arthritis (RA) has undergone remarkable change. Non steroidal anti-inflammatory drugs (NSAIDs), corticosteroids, and conventional disease modifying anti-rheumatic drugs (DMARDs), including methotrexate, sulfasalazine, hydroxychloroquine, and others continue to be used successfully in some patients. However, the introduction of highly effective treatments, including leflunomide and the biological response modifiers, etanercept, infliximab, and anakinra, have dramatically altered therapy for patients with rheumatoid arthritis. Some recent studies evaluating combination of therapies have also shown substantial efficacy. These advances will be discussed in this chapter. Hematopoietic stem cell transplantation will be discussed in Chapter 43.

Therapeutic Approach

The traditional approach to treatment of rheumatoid arthritis involved conservative management, initially with gradual progression to more aggressive therapy, if patients worsened or failed to improve over time. NSAIDs were the first line of therapy. If they were ineffective, or the disease showed progression over the course of years, DMARDs might be added in sequential fashion. The traditional DMARDs, including gold salts, D-penicillamine, hydroxychloroquine, sulfasalazine, and methotrexate, were typically used as monotherapy. Because of their slow onset of action, efficacy could not be assessed for many months. If one DMARD was judged ineffective, not tolerated, or caused serious side effects, it would be discontinued and the next DMARD tried. With this approach, patients often had unsatisfactory outcomes with the development of further joint destruction and decreased functional status. Because of these poor outcomes, it was realized a different approach to therapy was needed. Over the past decade, earlier, more aggressive treatment with DMARDs, and the use of combination therapy have become the standard of care for those patients with the potential for progressive, erosive disease. Several trials have shown the importance of this early intervention.[1,2] The indicators for potentially progressive disease include the presence of rheumatoid factor in the serum, evidence of erosions on radiographs, extensive polyarticular disease, and impaired functional status. Earlier, more aggressive intervention is particularly important in those patients with severe progressive disease.

Outcome Measures

The criteria for a diagnosis of rheumatoid arthritis are listed in Table 1. To assess treatments for rheumatoid arthritis, several standardized and validated clinical outcome measurements are used in addition to radiographic assessments. The commonly used indices in clinical trials are the American College of Rheumatology (ACR) response criteria[3] (Table 2) and the Disease Activity Score (DAS).[4] These are composite indices of the important variables in patients with rheumatoid arthritis including: activities of daily living health assessment questionnaires (ADL-HAQ) (Table 3), tender and swollen joint counts (Table 4), patient's assessment of pain, patient's global assessment of disease activity, acute phase reactant values, physician's global assessments of disease activity, and patient's assessment of physical function and disability. Radiographic assessments are part of the evaluation because structural damage is important in rheumatoid arthritis and cannot always be predicted by clinical variables. Standard radiographic assessment is with plain X-rays films. Different scoring systems have been developed including the Sharp[5] and Larsen[6] scoring methods and their subsequent modifications. These instruments are used to assess the severity of erosions and joint space narrowing. More recent clinical trials incorporate radiographic evaluations as part of the assessment of treatments for rheumatoid arthritis.

Conventional Therapy

Prior to the introduction of the most recent therapies, numerous medications were available for the treatment of rheumatoid arthritis; however, the efficacy and tolerability of these medications were variable. The options included NSAIDs, corticosteroids, and DMARDs. NSAIDs are useful for symptomatic treatment; however, they do not have disease-modifying properties. They continue to be widely used since they are effective for relief of pain and stiffness. The overall role of corticosteroids in the treatment of rheumatoid arthritis remains controversial. Corticosteroids are used to promptly relieve the symptoms of inflammation and decrease disease activity. Corticosteroids may also protect against progressive radiological damage in the short term.[7] Significant debate persists regarding the ability of this group of medications to prevent long-term damage. Corticosteroids have also never been shown to provide long-term improvement in func-

Table 1. The American Rheumatism Association 1987 revised criteria. For classification of rheumatoid arthritis*

Criterion	Definition
1. Morning stiffness	Morning stiffness in and around the joints, lasting at least 1 hours before maximal improvement.
2. Arthritis of 3 or more joint areas	At least 3 joint areas simultaneously have had soft tissue swelling or fluid (not bony overgrowth alone) observed by a physician. The 14 possible areas are right or left PIP, MCP, wrist, elbow, knee, ankle, and MTP joints.
3. Arthritis of hand joints	At least 1 area swollen (as defined above) in wrist, MCP, or PIP joint.
4. Symmetric arthritis	Simultaneous involvement of the same joint areas (as defined in 2) on both sides of the body (bilateral involvement of PIPs, MCPs, or MTPs is acceptable without absolute symmetry).
5. Rheumatoid nodules	Subcutaneous nodules, over bony prominences, or extensor surfaces, or juxtaarticular regions observed by a physician.
6. Serum rheumatoid factor	Demonstration of abnormal amounts of serum rheumatoid factor by any method for which the result has been positive in <5% of normal control subjects.
7. Radiographic changes	Radiographic changes typical of rheumatoid arthritis on posterior hand and wrist radiographs, which must include erosions or unequivocal bony decalcification localized in or most marked adjacent to the involved joints (osteoarthritis changes alone do not qualify).

*For classification purposes, a patient shall be said to have rheumatoid arthritis if he/she has satisfied at least 4 of these 7 criteria. Criteria 1 through 4 must have been present for at least 6 weeks. Patients with 2 clinical diagnoses are not excluded. Designation as classic, definite, or probable rheumatoid arthritis is not to be made. MCP= metacarpal pharyngeal; MTP= metatarsal pharyngeal; PIP= proximal interpharyngeal

tional outcome. Because of the side effects and toxicity of corticosteroids, they are often used only until other medications become effective.

The traditional DMARDs have poorly understood mechanisms of action in rheumatoid arthritis; yet, the medications are effective for a subset of patients. In some patients, clinical measures improved and the rate of radiological progression was decreased. It still remains difficult predicting which patients will respond to each of these various medications. In a meta-analysis of trials of the conventional DMARDs, the efficacy and toxicity of methotrexate, injectable gold, D-penicillamine, sulfasalazine, auranofin, and antimalarial drugs were analyzed.[8] The efficacy of methotrexate, injectable gold, D-penicillamine, and sulsalazine were comparable. The efficacy of antimalarials and auranofin (oral gold) were significantly less than the other treatments. In the meta-analysis of these therapies in regard to toxicity, 30% of patients dropped out of the trials.[8] Of the patients who did not complete the trials, greater than 50% of patients discontinued because of drug toxicity. Injectable gold had the highest toxicity rates, while antimalarial drugs and auranofin had the lowest rates of toxicity. Unfortunately, longterm adherence to any of these medications is low. In clinical practice, the median duration of DMARD monotherapy was less than two years for agents other than methotrexate.[9,10] Discontinuation of the medication may be due to limited efficacy, the slow onset of action, intolerable side effects, or significant toxicity. The majority of the older agents have significant side effect profiles or toxicity and many require extensive laboratory monitoring.[11] Despite these medications' shortcomings, numerous studies and clinical experience have proven that they can be effective and valuable treatment options for rheumatoid arthritis patients.

Methotrexate

Since the early 1980s, methotrexate has been the most widely used therapy for rheumatoid arthritis. It has proven to be effective and tolerable to patients in high doses. Methotrexate is an antifolate and inhibits dihydrofolate reductase (DHFR), thereby decreasing the intracellular supply of reduced folates that are needed for nucleotide synthesis. It also has effects on other folate dependent enzymes in the purine biosynthesis pathway. The exact mechanism of action in rheumatoid arthritis remains uncertain.

Many studies indicated effectiveness of weekly methotrexate in patients not responding to conventional DMARDs or in comparison to placebo.[12-16] Most patients do not obtain remissions with methotrexate, but studies have shown consistent clinical responses of >50% relative to baseline, accompanied by improvement in functional status.[17-19] Methotrexate has also been shown to slow the rate of joint destruction as measured by radiography.[20] Several long-term observational studies have demonstrated continued efficacy with good tolerability and low discontinuation rates.[16,17,21,22] In the long-term studies, it has been suggested that between 46-64% of patients continued to use the drug for five years and 34-38% continued to use it for more than 10 years.[16,17,20]

Because of its efficacy and tolerability, methotrexate has become the most widely prescribed therapy for rheumatoid arthritis. Effective doses range between 7.5 and 25 mg weekly given orally or by subcutaneous or intramuscular injections. If higher oral doses are ineffective or not tolerated, subcutaneous or intramuscular administration can be used to increase bioavailability. Adjustments to dosage need to be made in patients with decreased renal function.[23] Methotrexate is contraindicated in the setting of severe renal insufficiency. The side effect and toxicity profile of

Table 2. American College of Rheumatology (ACR) preliminary definition of improvement in rheumatoid arthritis

Components of the American College of Rheumatology of Improvement in Rheumatoid Arthritis
Tender joint count
Swollen joint count
Patient's assessment of pain
Patient's global assessment of disease activity
Physician's global assessment of disease activity
Patient's assessment of physical function
Acute phase reactant value

ACR 20 response is defined as 20% reduction in tender joint count and swollen joint count in addition to a 20% improvement in at least three of the remaining activity measures. An ACR 50 or ACR 70 would require 50% and 70% improvements, respectively.

Disease Activity Measure	Method of Assessment
1. Tender joint count	ACR tender joint count, an assessment of 28 or more joints. The joint count should be done by scoring several different aspects of tenderness, as assessed by pressure and joint manipulation on physical examination. The information on various types of tenderness should then be collapsed into a single tender-versus-nontender dichotomy.
2. Swollen joint count	ACR swollen joint count, an assessment of 28 or more joints. Joints are classified as either swollen or not swollen.
3. Patient's assessment of pain	A horizontal visual analog scale (usually 10 cm) or Likert scale assessment of the patient's level of pain.
4. Patient's global assessment of disease activity	The patient's overall assessment of how the arthritis is doing. One acceptable method for determining this is the question from the AIMS instrument: "Considering all the ways your arthritis affects you, mark "**X**" on the scale for how well you are doing". An anchored horizontal, visual analog scale (usually 10 cm) should be provided. A Likert scale response is also acceptable.
5. Physician's global assessment of disease activity	A horizontal visual analog scale (usually 10 cm) or Likert scale measure of the physician's assessment of the patient's current disease activity.
6. Patient's assessment of physical function	Any patient self-assessment instrument which has been validated, has reliability, has been proven in RA trials to be sensitive to change, and which measures physical function in RA patients is acceptable. Instruments which have been demonstrated to be sensitive in RA trials include the AIMS, the HAQ, the Quality (or Index) of Well Being, the MHIQ, and the MACTAR.
7. Acute-phase reactant level value	A Westergren erythrocyte sedimentation rate or a C-reactive protein level.

AIMS= Arthritis Impact Measurement Scales; ACR= American College of Rheumatology; CRP= C-reactive protein; ESR= erythrocyte sedimentation rate; HAQ= Health Assessment Questionnaire; MHIQ= McMaster Health Index Questionnaire; MACTAR= McMaster Toronto Arthritis Patient Preference Disability Questionnaire; RA= Rheumatoid Arthritis.

methotrexate is well known. Rarely, methotrexate can cause pneumonitis, severe cytopenias, and hepatic dysfunction. Careful monitoring of blood counts, creatinine, hepatic aminotransferases, albumin, and pulmonary symptoms can minimize these potential toxic effects. Methotrexate is also teratogenic; therefore, it is contraindicated during pregnancy. An increased incidence of infections has also been noted with the use of methotrexate.[24,25] The addition of folic acid (usually 1 mg/day) or folinic acid reduces minor side effects such as alopecia, stomatitis, gastrointestinal intolerance, and hemopoetic toxic effects without significantly lowering efficacy.[26,27]

Methotrexate has become the benchmark agent for treatment of rheumatoid arthritis. The efficacy of methotrexate is well established and it's safety profile, short and long term, is well known. Safety and efficacy of new therapies are measured against those of methotrexate and moreover, most combination therapy trials include the use of methotrexate.

Leflunomide

In 1999, the US Food and Drug Administration approved leflunomide for the treatment of rheumatoid arthritis. Leflunomide is an immunomodulatory agent that acts by inhibiting de-novo pyrimidine synthesis by selective inhibition of dihydro-orotate dehydrogenase. Cells such as activated T-lymphocytes that require pyrimidine synthesis via the de novo pathway are especially sensitive to the effect of leflunomide. The exact mechanism of action in rheumatoid arthritis patients is not clear. Leflunomide is taken orally and converted to the active metabolite A771726. It has an extensive enterohepatic circulation and a very long half-life (between 15-18 days). Typically,

Table 3. Activities of Daily Living Health Assessment Questionnaire. The Health Assessment Questionnaire (HAQ) and a modified HAQ (MHAQ)

Are You Able To:	Degree of Difficulty Without Any	With Some	With Much	Unable To Do
Dressing and grooming				
1. Dress yourself, including tying shoelaces, and doing buttons?				
2. Shampoo your hair?				
Arising				
3. Stand up straight from an armless straight chair?				
4. Get in and out of bed?				
Eating				
5. Cut your meat?				
6. Lift a full cup or glass to your mouth?				
7. Open a new milk carton?				
Walking				
8. Walk outdoors on flat ground?				
9. Climb up 5 steps?				
Hygiene				
10. Wash and dry your entire body?				
11. Take a tub bath?				
12. Get on and off the toilet?				
Reaching				
13. Reach and get down a 5-pound object from just above your head?				
14. Bend down to pick up clothing from the floor?				
Gripping				
15. Open car door?				
16. Open jars which have been previously opened?				
17. Turn faucets on and off?				
Other activities				
18. Run errands and shop?				
19. Get in and out of a car?				
20. Do chores such as vacuuming or yardwork?				

patients receive a loading dose of 100 mg/day x 3 days because of the length of time to achieve steady state. Subsequently, patients receive an oral dose of 10-20 mg/day.

Several double blind randomized controlled trials have demonstrated the efficacy of leflunomide. In a large US study of 482 patients comparing leflunomide to methotrexate and placebo treatments for one year, ACR 20 response rates for patients receiving leflunomide treatment (52%) and methotrexate treatment (46%) were significantly higher than those patients receiving placebo (26%) (Fig. 1).[28] ACR 50 response (at least 50% improvement) rates were 34% for leflunomide, 23% for methotrexate, and 8% for placebo. No significant differences were noted between patients receiving methotrexate and leflunomide. X-ray analysis demonstrated less progression of joint damage in patients as a result of treatment with methotrexate or leflunomide compared to placebo. In a large European study, 999 RA patients with active disease received either leflunomide or methotrexate (10-15 mg/wk) for one year to evaluate the efficacy and safety of leflunomide compared to methotrexate.[29] The results of this study also demonstrated similar efficacy of these two agents. Both groups had significant improvement in most clinical parameters including tender and swollen joint counts and patient and physician global assessments. When the ACR criteria were applied, an ACR 20 response was noted in 62% and 54% of the leflunomide and methotrexate treated patients, respectively, at 12 weeks.[28] Both medications decreased the rate of progression of joint damage when assessed by x-rays. Many of these patients from these two studies were followed for an additional year of treatment. Evaluation following the second year of treatment showed maintenance of the efficacy and safety of both leflunomide and methotrexate.[30] There was sustained retardation of radiologic progression in both the leflunomide and methotrexate treated groups. Another European study compared the efficacy and safety of leflunomide to sulfasalazine and placebo. In this study of twenty-four weeks duration, leflunomide and sulfasalazine were significantly superior to placebo in regards to tender and swollen joint counts and physician and patient assessments.[31] Using the ACR criteria, ACR 20 responses were noted in 55%, 29% and 56% in the leflunomide, placebo, and sulfasalazine treated groups, respectively (Fig. 2). Radiographic disease progression was significantly slower with leflunomide and sulfasalazine than with placebo. In summary, these major studies have shown that the efficacy of leflunomide is equivalent to methotrexate and sulfasalazine.

Table 4. Joint count (check to indicate present)

Right Pain/Tender	Right Swelling	Joint	Left Pain/Tender	Left Swelling
		Shoulder		
		Elbow		
		Wrist		
		MCP 1		
		MCP 2		
		MCP 3		
		MCP 4		
		MCP 5		
		IP (thumb)		
		PIP 2		
		PIP 3		
		PIP 4		
		PIP 5		
		Knee		
		Ankle		
		Tarus		
		MTP 1		
		MTP 2		
		MTP 3		
		MTP 4		
		MTP 5		

Total number of painful/tender joints (maximum=44) _____
Total number of swollen joints (maximum=42) _____
MCP= metacarpal pharyngeal; MTP= metatarsal pharyngeal; PIP= proximal interpharngeal

Leflunomide has also been evaluated as an add-on therapy in patients with active disease despite the use of methotrexate. In a study of 266 patients, the combination of leflunomide and methotrexate conferred a significant therapeutic advantage in patients who had active arthritis while on methotrexate alone.[32] The ACR 20 response rate at 24 weeks was 46.2% in the leflunomide plus methotrexate group vs. 19.5% for the placebo plus methotrexate group. The combination was well tolerated. Two abstracts have reported on the efficacy and safety of the combination of leflunomide and infliximab in a small number of patients.[33,34]

In regards to safety, leflunomide has been found to have a similar incidence of side effects and toxicity as methotrexate and sulfasalazine in the clinical trials.[27,29,30,35] The most common adverse events include diarrhea, nausea, allergic reactions (rash and pruritus), reversible alopecia, hypertension, and elevated transaminases. Hepatic aminotransferases require monitoring with leflunomide therapy. Dosage reduction or cessation of medication is recommended for repeated elevations. In the US study, increased levels of aminotransferases were noted in 15% of patients receiving leflunomide and 12% of patients receiving methotrexate.[27] In the combination study with methotrexate, increases in hepatic aminotransferases were noted in 63%; however, only 10% discontinued treatment due to transaminase elevations.[34] Long-term studies are needed to obtain further information on the risks of hepatotoxicity and other potential toxicities. Similar to methotrexate, leflunomide is contraindicated in pregnancy. It has shown teratogenic effects in animal studies.

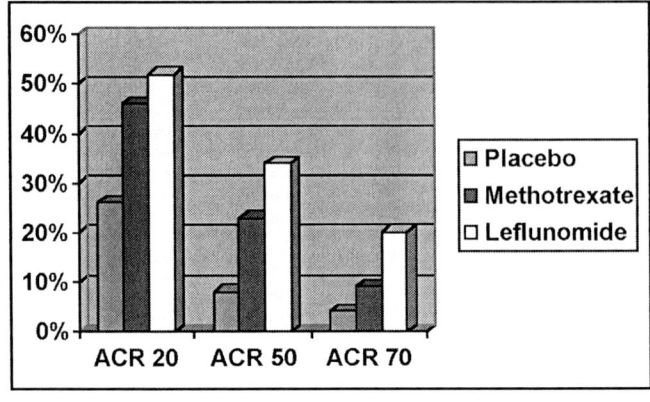

Figure 1. Rheumatoid arthritis patients with ACR 20, 50, and 70 responses in the leflunomide compared with placebo and methotrxate (Strand 1999).

Biologic Response Modifiers

In the late 1990s, two biologic response modifiers, etanercept and infliximab, were approved for the treatment of rheumatoid arthritis. These agents function by neutralizing the effect of tumor necrosis factor-alpha (TNF-α), a major proinflammatory cytokine in rheumatoid arthritis. Both drugs bind to TNF-α; therefore, limiting it's bioavailability. TNF-α is overproduced in the joints of rheumatoid arthritis patients.[36] TNF-α stimulates the production of other proinflammatory cytokines, increases cell migration by increasing the production of cellular adhesion molecules, and increases the rate of tissue remodeling by matrix-degrading proteases.[37,38] Studies with both of these agents clearly demonstrate its remarkable efficacy. Other anti-TNF therapies are currently being developed or evaluated in clinical trials. In 2001, the US Food and Drug Administration approved another biologic response modifier. This agent, anakinra, is a receptor antagonist for IL-1, another major proinflammatory cytokine in rheumatoid arthritis. The biologic response modifiers represent a major advance in therapy. These "directed" therapies have been designed based on the understanding of the pathogenesis of rheumatoid arthritis.

Etanercept

Etanercept is a bioengineered construct of two p75 TNF receptors attached to the Fc portion of an IgG1 molecule. It binds

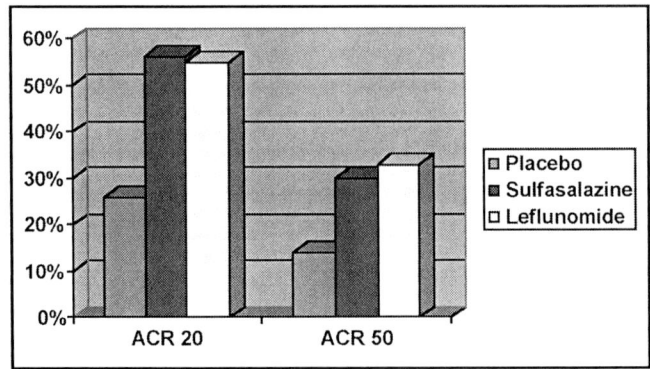

Figure 2. ACR 20, 50, and 70 responses in rheumatoid arthritis patients receiving leflunomide, sulfasalazine or placebo (Smolen 1999).

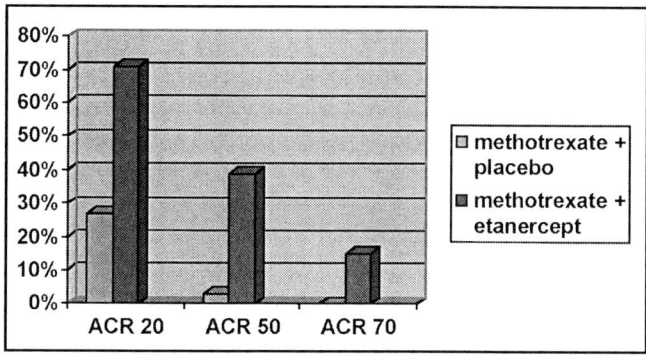

Figure 3. Patients with ACR 20, 50, 70 response in the etanercept trial in patients with rheumatoid arthritis receiving methotrexate (Weinblatt 1999).

to both TNF-α and TNF-β (lymphotoxin-α). Etanercept is administered by subcutaneous injections at a dose of 25 mg twice a week. It has a median half-life of 115 hours.

The efficacy and safety of etanercept have been evaluated in studies in comparison with placebo, and also in studies of patients with persistent, active disease while taking methotrexate. In a 3 month placebo controlled trial, etanercept showed superior clinical efficacy as compared to placebo in active, refractory RA patients.[39] At 3 months, ACR 20 response rates were 75% vs. 14% in the etanercept and placebo groups. In a subsequent 24 week study of 89 patients with active rheumatoid arthritis on methotrexate, etanercept or placebo was added. At 24 weeks, 74% of patients receiving etanercept and methotrexate and 27% of patients receiving placebo and methotrexate met the ACR 20 criteria[40](Fig. 3). ACR 50 (39% vs. 3%) and ACR 70 (15% vs. 0%) response rates were also very impressive. In a study of 632 patients with early rheumatoid arthritis, etanercept showed superior efficacy in terms of clinical measures and slowing of radiographic progression at 6 months compared to methotrexate.[41] However, after six months, the efficacy of etanercept was marginally superior to methotrexate. The rate of radiographic progression was similar in the second six months. In long-term follow-up of patients from clinical trials that continued on etanercept (median duration of therapy 25 months), the clinical benefit of etanercept was sustained.[42]

Because of the potential of TNF-α in controlling infections and in immune surveillance for malignancies, the safety of the TNF inhibitors is an area of importance. Results from clinical trials with etanercept have not shown an increased risk for serious infections or malignancies. However, based on some individual reports of serious infections, the FDA in 1999 changed the package insert to include a directive warning against its use in patients with diabetes, active infection, or a history of chronic infections. Post marketing reports of life threatening infections in patients receiving etanercept and infliximab have appeared. There have also been reports of demyelinating disorders, cytopenias, and development of antibodies to DNA in some patients treated. The early clinical trials showed a slight increased risk for upper respiratory infections in etanercept versus placebo treated patients.[38] Trials involving methotrexate have demonstrated a decreased number of infections in etanercept treated patients compared to methotrexate treated patients.[39,40] The medication is generally well tolerated. The most common side effect is injection site reactions (>40%) that occur frequently on initiation but decrease with repeated injections. Although short-term evaluation of etanercept has shown a good safety profile, long-term safety remains an area where further data are needed.

Infliximab

Infliximab is a chimeric antibody that specifically targets TNF-α. It does not bind TNF-β. The variable region of the antibody is murine in origin, while the remainder of the immunoglobulin is a human IgG1. It is administered intravenously. The initial studies that compared infliximab to placebo demonstrated substantial clinical responses with infliximab treatment.[43] Further studies with multiple infusions of infliximab at varying doses with and without methotrexate confirmed the clinical efficacy of infliximab. In studies of patients on methotrexate monotherapy, addition of infliximab at 3 mg/kg or 10 mg/kg every four or eight weeks for a total of 30 weeks led to ACR 20 response rates of 50-59% compared to 20% in placebo treated patients[44] (Table 5). The ACR 50 response rates were 27-31% and 4%, respectively. This level of response was maintained when the study was extended to 54 weeks.[45] In the study of 54-week duration, repeated doses of infliximab with methotrexate also halted the progression of joint destruction as noted with radiography. Another study of 40-week duration confirmed the findings of sustained clinical benefits in patients receiving infliximab and methotrexate.[46] In these early studies, patients responded favorably to infliximab irrespective of whether a patient was taking methotrexate; however, patients who received the combination had higher response rates and increased duration of response to therapy. From the early studies, development of antibodies against infliximab (human antichimeric antibodies) was detected in the patients. Subsequent studies with methotrexate showed a decreased prevalence of these antibodies if patients continued on methotrexate while receiving infliximab.[43,44]

The US FDA approved infliximab for use in combination with methotrexate in patients with active disease despite the use of methotrexate. The typical starting dose is 3 mg/kg at weeks 0, 2,

Table 5. Patients with ACR 20, 50, and 70 responses at 54 weeks in the infliximab and methotrexate trial for the treatment of rheumatoid arthritis (Lipsky 2000)

ACR Response	MTX + Placebo	MTX + Infliximab 3 mg/kg Every 8 Weeks	MTX + Infliximab 3 mg/kg Every 4 Weeks	MTX + Infliximab 10 mg/kg Every 8 Weeks	MTX + Infliximab 10 mg/kg Every4 Weeks
ACR 20	17	42	48	59	59
ACR 50	8	21	34	39	38
ACR 70	2	10	17	25	19

and 6, followed by maintenance dosing every eight weeks thereafter. More recently, the FDA approved an increase in the dose of infliximab of up to 10 mg/kg in order to achieve the desired clinical response. In the clinical trials, adverse events with infliximab were infrequent. The most frequent problem was infusion reactions characterized by fever, chills, urticaria, chest pain, dyspnea, or hypotension. This did not lead to discontinuation of therapy for the majority of these patients. Infusion reactions are treated by slowing the infusion and by pretreating with acetaminophen or antihistamines. In the larger studies, there was a greater frequency of upper respiratory tract infections compared to patients receiving methotrexate alone.[43] Although the clinical trials did not report increased numbers of serious infections, subsequent reports show a number of serious infections, including tuberculosis and opportunistic fungal infections, in patients treated with infliximab. The risk of serious infections appears to be a concern with the use of any of the TNF inhibitors. Because of the reports of serious infections, the FDA has revised its recommendations for screening for tuberculosis prior to treatment with infliximab, and does not recommend its use in patients with diabetes, active infection, or with a history of chronic infections. In regards to malignancies, several malignancies were reported in the clinical trials; however, the incidence was not greater than expected.[43,44] During the clinical studies of both etanercept and infliximab, the development of anti DNA antibodies was detected in many patients; yet the clinical significance of this finding is unknown. Patients with these antibodies did not develop SLE or lupus like illnesses. Similar to etanercept, the long-term safety of infliximab remains unknown.

Anakinra

In rheumatoid arthritis, IL-1β is present in large quantities in the synovial fluid and synovial tissue.[47] Interleukin-1β can stimulate metalloproteinase expression, adhesion molecule expression, secretion of other cytokines, and prostaglandin production. Interleukin 1 receptor antagonist is a naturally occurring competitive inhibitor for IL-1α and IL-1β. IL-1Ra binds to the IL-1R1 (interleukin-1 receptor) without transducing a signal, thus blocking the ability of interleukin 1α or 1β to bind to the receptor. In November 2001, anakinra was approved for the treatment of rheumatoid arthritis. Anakinra is a recombinant form of the naturally occurring interleukin-1 receptor antagonist (IL1-Ra) produced using *E. coli*.

The initial placebo controlled clinical trial of 472 patients with severe rheumatoid arthritis demonstrated modest clinical efficacy of IL1-Ra at the highest dose and significant reduction in radiological progression at all doses.[48] The ACR 20 response was 43% in the 150mg dose group compared to 27% in the placebo group. All clinical parameters were significantly improved in the high dose anakinra group. From this study, 309 patients enrolled for an additional 24 weeks in an extension study with the patients that had received placebo randomized into one of the three treatment doses.[49] Similar efficacy was noted in that study along with maintenance of response in those continuing at the same dosage. Reduced rate of radiological progression was maintained in the extension study.[48] Another study evaluated 419 patients with active rheumatoid arthritis receiving methotrexate. In this study, patients were randomized into six groups with different doses of anakinra. ACR 20 response rates were 46% with the 1 mg/kg dose, 38% with the 2mg/kg dose, and 19% with placebo medication[48] (Fig. 4).

Anakinra is administered by subcutaneous injection at a dose of 100 mg/day. The most common adverse event noted in the

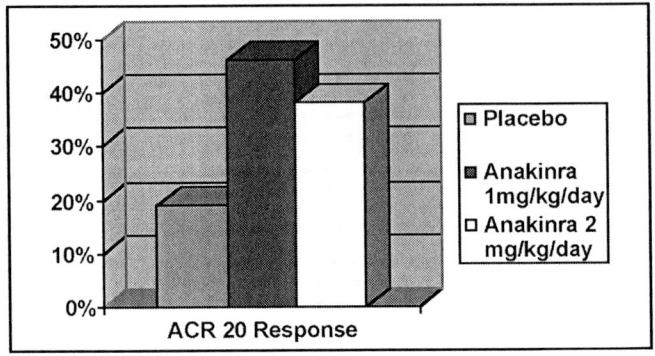

Figure 4. ACR 20 response to IL1-RA compared to placebo (Bresnihan 2000).

clinical trials was an injection site reaction (up to 80%), although this frequently subsided within weeks of continued daily injections. This infrequently led to patient withdrawals from the study. Anakinra was well tolerated in the clinical trials. No significant differences were noted in the frequency of infections. The frequency of neoplasms was no greater than expected for the population under study. Several patients developed asymptomatic reversible neutropenia while receiving anakinra. Early studies suggest that anakinra is safe; however, long-term safety data need to be accumulated. Anakinra appears to have modest clinical effects and as well as effects on slowing the progression of joint damage. The safety and efficacy of anakinra in combination with anti TNF therapy is presently under study.

Combination Therapy

Over the last two decades, greater than twenty studies have assessed the efficacy and safety of combination therapy in rheumatoid arthritis. In early trials with conventional DMARDs, combination therapy did not demonstrate advantages over monotherapy. However, several studies in the mid 1990s demonstrated clear clinical benefit with no increased toxicity with the use of combination therapy. In a six month study of 148 patients with active rheumatoid arthritis on treatment with methotrexate, the addition of cyclosporin led to ACR 20 response rates of 48% vs. 16% in the placebo treated patients.[50] The main side effect was increased serum creatinine in the cyclosporin treated group. In an extension study of 24 weeks, patients who had initially received the combination of cyclosporin and methotrexate continued to obtain clinical benefit.[51] Patients who had received placebo in the initial trial and subsequently cyclosporin in the extension study, also showed significant clinical improvement. Another study evaluated the use of methotrexate alone, sulfasalazine and hydroxychloroquine, or a combination of all three medications in patients with active rheumatoid arthritis. In this study, 50 percent improvement was noted at nine months in 77% of patients treated with the three drugs, 33%, with methotrexate alone, and 40%, with sulfasalazine and hydroxychloroquine.[52] These therapies were tolerated with minimal toxicity. Another study showed the efficacy of combined step-down prednisolone, methotrexate, and sulfasalazine compared to sulfasalazine alone in early rheumatoid arthritis. Clinical response in the combination therapy was almost double that of monotherapy at 28 weeks (prior to discontinuation of prednisolone or methotrexate).[53] The group that received combination therapy showed less radiographic progression than the group that received sulfasalazine. Decreased radiographic progression was still observed at 80 weeks. All of these studies demonstrated increased efficacy with no worsening toxicity when patients were treated with combination therapy.

The most recently approved medications, which include leflunomide, etanercept, infliximab, and anakinra, have all been evaluated with concurrent methotrexate. In clinical trials, combination therapy, especially with the newer medications, continues to display superior clinical efficacy and greater prevention of joint destruction compared to monotherapy. In clinical practice today, the majority of rheumatologists prescribe combination therapy for treatment of rheumatoid arthritis.

Conclusion

Advances in the treatment of rheumatoid arthritis have been remarkable over the past decade. These advances have shaped a markedly different treatment approach to patients diagnosed with rheumatoid arthritis. The present approach is to start aggressive disease modifying therapies early in the disease course, especially in those patients most likely to develop progressive, disabling disease. Often the initial choice of medication is methotrexate or sulfasalazine in addition to a NSAID. Corticosteroids may also be used initially until the DMARD shows efficacy. If patients have a partial response or no response within a period of a several months, combination therapy with other DMARDs is often started. The choice of combination therapy may include the older agents or the newer medications such as leflunomide, etanercept, infliximab, or anakinra. Patients are very likely to begin therapy with the biologic response modifiers if conventional DMARDs have not been effective. With the myriad of very effective treatments that are currently available, patients with rheumatoid arthritis will likely improve or maintain their quality of life. With the extensive ongoing research and development of other therapeutics including hematopoietic stem cell transplantation (see Chapter 43), the future holds even more promise for people with rheumatoid arthritis.

References

1. Van der Heide A, Jacobs JW, Bijlsma JW et al. The effectiveness of early treatment with "second-line" antirheumatic drugs: A randomized, controlled trial. Ann Intern Med 1996; 124:699-707.
2. Egsmose C, Lund B, Borg G et al. Patients with rheumatoid arthritis benefit from early 2nd line therapy: 5 year followup of a prospective double blind placebo controlled study. J Rheumatol 1995; 22:2208-13.
3. Felson DT, Anderson JJ, Boers M et al. American College of Rheumatology: Preliminary definition of improvement in rheumatoid arthritis. Arthritis Rheum 1995; 38:727-35.
4. Van der Heijde DM, van't Hof M, van Riel PL et al. Development of a disease activity score based on judgement in clinical practice by rheumatologists. J Rheumatol 1993; 20:579-81.
5. Sharp JT. Radiologic assessment as an outcome measure in rheumatoid arthritis. Arthritis Rheum 1989; 32:221-29.
6. Larsen A. A radiological method for grading the severity of rheumatoid arthritis. Scand J Rheumatol 1975; 4:225-33.
7. Kirwan JR. Arthritis and Rheumatism Council Low-dose Glucocorticoid Study Group. The effect of glucocorticosteroids on joint destruction in rheumatoid arthritis. N Engl J Med 1995; 333:142-46.
8. Felson DT, Anderson JJ, Meenan RF. The comparative efficacy and toxicity of second-line drugs in rheumatoid arthritis. Arthritis Rheum 1990; 33:1449-61.
9. Morand EF, McCloud PI, Littlejohn GO. Life table analysis of 879 treatment episodes with slow acting antirheumatic drugs in community rheumatology practice. J Rheumatol 1992; 19:704-08.
10. Pincus T, Marcum SB, Callahan LF. Longterm drug therapy for rheumatoid arthritis in seven rheumatology private practices: II, second line drugs and prednisone. J Rheumatol 1992; 19:1885-94.
11. American College of Rheumatology Ad Hoc Committee on Clinical Guidelines. Guidelines for monitoring drug therapy in rheumatoid arthritis. Arthritis Rheum 1996; 39:723-31.
12. Wilkens RF et al. Low-dose pulse methotrexate in rheumatoid arthritis. J Rheumatol 1980; 7:501-05.
13. Wilke WS, Calabrese LH, Scherbel AL. Methotrexate in the treatment of rheumatoid arthritis. Pilot study. Cleve Clin Q 1980; 47:305-09.
14. Hoffmeister RT. Methotrexate therapy in rheumatoid arthritis: 15 years experience. Am J Med 1983; 75:69-73.
15. Weinblatt ME, Coblyn JS, Fox DA et al. Efficacy of low-dose methotrexate in rheumatoid arthritis. N Engl J Med 1985; 312:818-22.
16. Williams HJ, Willkens RF, Samuelson CO et al. Comparison of low-dose oral pulse methotrexate and placebo in the treatment of rheumatoid arthritis – A controlled clinical trial. Arthritis Rheum 1985; 28:721-29.
17. Weinblatt ME, Maier AL, Fraser PA et al. Longterm prospective study of methotrexate in rheumatoid arthritis: Conclusion after 132 months of therapy. J Rheumatol 1998; 25:238-42.
18. Kremer JM. Safety, efficacy, and mortality in a long-term cohort of patients with rheumatoid arthritis taking methotrexate: Followup after a mean of 13.3 years. Arthritis Rheum 1997; 40:984-85.
19. Tugwell P, Wells G, Strand V et al. Clinical improvement as reflected in measures of function and health-related quality of life following treatment with leflunomide compared with methotrexate in patients with rheumatoid arthritis; sensitivity and relative efficiency to detect a treatment effect in a twelve-month, placebo-controlled trial. Arthritis Rheum 2000; 43:506-14.
20. Weinblatt ME, Polisson R, Blotner SD et al. The effects of drug therapy on radiographic progression of rheumatoid arthritis: Results of a 36-week randomized trial comparing methotrexate and auranofin. Arthritis Rheum 1993; 36:613-19.
21. Sany J, Anaya JM, Lussiez V et al. Treatment of rheumatoid arthritis with methotrexate: A prospective open longterm study of 191 cases. J Rheumatol 1991; 18:1323-27.
22. Rau R, Schleusser B, Herborn G et al. Long-term treatment of destructive rheumatoid arthritis with methotrexate. J Rheumatol 1997; 24:1881-89.
23. Rheumatoid Arthritis Clinical Trial Archive Group. The effect of age and renal function on the efficacy and toxicity of methotrexate in rheumatoid arthritis. J Rheumatol 1990; 22:218-23.
24. van der Veen MJ, van der Heijde A, Kruize AA et al. Infection rate and use of antibiotics in patients with rheumatoid arthritis treated with methotrexate. Ann Rheum Dis 1994; 53:224-28.
25. Boerbooms AM, Kerstens PJ, van Loenhout JW et al. Infections during low-dose methotrexate treatment in rheumatoid arthritis. Semin Arthritis Rheum 1995; 24:411-21.
26. Morgan SL, Daggott JE, Vaugh WH et al. The effect of folic acid supplementation on the toxicity of low-dose methotrexate in patients with rheumatoid arthritis. Arthritis Rheum 1990; 33:9-18.
27. Weinblatt ME, Maier AL, Coblyn JS. Low dose leucovorin does not interfere with the efficacy of methotrexate in rheumatoid arthritis: An 8 week randomized placebo controlled trial. Arthritis Rheum 1993; 20:950-52.
28. Strand V, Cohen S, Schiff M et al. Treatment of active rheumatoid arthritis with leflunomide compared with placebo and methotrexate. Arch Intern Med 1999; 159:2542-50.
29. Emery P, Breedveld FC, Lemmel et al. A comparison of the efficacy and safety of leflunomide and methotrexate for the treatment of rheumatoid arthritis. Rheumatol 2000; 39:655-65.
30. Cohen S, Cannon GW, Schiff M et al. Two-year, blinded, randomized, controlled trial of treatment of active rheumatoid arthritis with leflunomide compared with methotrexate. Arthritis Rheum 2001; 44:1984-92.
31. Smolen JS, Kalden JR, Scott DL et al. Efficacy and safety of leflunomide compared with placebo and sulphasalazine in active rheumatoid arthritis: A double-blind, randomized multicentre trial. Lancet 1999; 353:259-66.

32. Kremer JM, Caldwell JR, Cannon GW et al. The combination of leflunomide and methotrexate in patients with active rheumatoid arthritis who are failing on methotrexate treatment alone: A double-blind placebo controlled study. Arthritis Rheum 2000; 44:S224.
33. Patel S, Bergen W, Kraemer A et al. Efficacy and safety of remicade™ (infliximab) plus arava™ (leflunomide) in rheumatoid arthritis (RA). Arthritis Rheum 2001; 44:S189.
34. Hansen KE, Cush J, Singhal A et al. The safety and efficacy of leflunomide in combination with infliximab in rheumatoid arthritis. Arthritis Rheum 2001; 44:S188.
35. Weinblatt ME, Kremer JM, Coblyn JS et al. Pharmacokinetics, safety, and efficacy of combination treatment with methotrexate and leflunomide in patients with active rheumatoid arthritis. Arthritis Rheum 1999; 42:1322-28.
36. Saxne T, Palladino Jr MA, Heinegard D et al. Detection of tumor necrosis factor α but not tumor necrosis factor β in rheumatoid arthritis synovial fluid and serum. Arthritis Rheum 1988; 31:1041-45.
37. Paleolog EM, Hunt M, Elliott MJ et al. Deactivation of vascular endothelium by monoclonal anti-tumor necrosis factor α antibody in rheumatoid arthritis. Arthritis Rheum 1996; 39:1082-91.
38. Tak PP, Taylor PC, Breedveld FC et al. Decrease in cellularity and expression of adhesion molecules by anti-tumor necrosis factor α monoclonal antibody treatment in patients with rheumatoid arthritis. Arthritis Rheum 1996; 39:1077-81.
39. Moreland LW, Baumgartner SW, Schiff MH et al. Treatment of rheumatoid arthritis with a recombinant human tumor necrosis factor receptor (p75)-Fc fusion protein. N Engl J Med 1997; 337:141-47.
40. Weinblatt ME, Kremer JM, Bankhurst AD et al. A trial of etanercept, a recombinant tumor necrosis factor receptor: Fc fusion protein, in patients with rheumatoid arthritis receiving methotrexate. N Engl J Med 1999; 340:253-59.
41. Bathon JM, Martin RW, Fleischmann RM et al. A comparison of etanercept and methotrexate in patients with early rheumatoid arthritis. N Engl J Med 2000; 343:1586-93.
42. Moreland LW, Cohen SB, Baumgartner SW et al. Longterm safety and efficacy of etanercept in patients with rheumatoid arthritis. J Rheumatol 2001; 28:1238-44.
43. Elliott MJ, Maini RN, Feldman M et al. Randomised double-blind comparison of chimeric monoclonal antibody to tumor necrosis factor alpha (cA2) versus placebo in rheumatoid arthritis. Lancet 1994; 344:1105-10.
44. Maini R, St Clair EW, Breedveld F et al. Infliximab (chimeric anti-tumour necrosis factor alpha monoclonal antibody) versus placebo in rheumatoid arthritis patients receiving concomitant methotrexate: A randomized phase III trial. Lancet 1999; 354:1932-39.
45. Lipsky PE, van der Heijde DM, St Clair EW et al. Infliximab and methotrexate in the treatment of rheumatoid arthritis. N Engl J Med 2000; 343:1594-1602.
46. Kavanaugh A, St Clair EW, McCune WJ et al. Chimeric anti-tumor necrosis factor-alpha monoclonal antibody treatment of patients with rheumatoid arthritis receiving methotrexate therapy. J Rheumatol 2000; 27:841-50.
47. Kahle P, Saal JG, Schaudt K et al. Determination of cytokines in synovial fluids: Correlation with diagnosis and histomorphological characteristics of synovial tissue. Ann Rheum Dis 1992; 51:731-34.
48. Bresnihan B, Alvaro-Garcia JM, Cobby M et al. Treatment of rheumatoid arthritis with recombinant human interleukin-1 receptor antagonist. Arthritis Rheum 1998; 41:2196-2204.
49. Bresnihan B. The safety and efficacy of interleukin-1 receptor antagonist in the treatment of rheumatoid arthritis. Semin Arthritis Rheum 2001; 30(5 suppl 2):17-20.
50. Tugwell P, Pincus T, Yocum D et al. Combination therapy with cyclosporine and methotrexate in severe rheumatoid arthritis. N Engl J Med 1995; 333:137-41.
51. Stein CM, Pincus T, Yocum D et al. Combination treatment of severe rheumatoid arthritis with cyclosporine and methotrexate for forty-eight weeks. Arthritis Rheum 1997; 40:1843-51.
52. O'Dell JR, Haire CE, Erikson N et al. Treatment of rheumatoid arthritis with methotrexate alone, sulfasalazine and hydroxychloroquine, or a combination of all three medications. N Engl J Med 1996; 334:1287-91.
53. Boers M, Verhoeven AC, Markusse HM et al. Randomised comparison of combined step-down prednisolone, methotrexate and sulphasalazine with sulphasalazine alone in early rheumatoid arthritis. Lancet 1997; 350:309-18.

CHAPTER 43

Haemopoietic Stem Cell Transplantation for Rheumatoid Arthritis—World Experience and Future Trials

John A. Snowden, John J. Moore, Sarah J. Bingham, Steve Z. Pavletic and Richard K. Burt

Introduction

Rheumatoid arthritis (RA) affects approximately 1% of the population. It is rarely life threatening in the short term, but, in the long term, RA and the side effects of therapy shorten life. RA causes significant disability. Of those who have the disease for more than 12 months, a majority will continue to have exacerbations for the rest of their lives, and about half of affected individuals are unable to work 10 years after contracting the disease. The associated economic costs are considerable, both to the individual and to the community.[1-3]

Optimal management of RA involves an interaction of medical, surgical, psychological and physical approaches.[4] In previous years, medical therapy consisted of a stepwise approach in which first-line non-steroidal anti-inflammatory drugs were followed by corticosteroids, and disease modifying anti-rheumatic agents (DMARDs) were introduced at a relatively late stage. In recent years, it has become increasingly apparent that this approach is unsatisfactory, with many patients developing radiological evidence of erosion soon after contracting the disease.[5] This has led to a change in management with a more aggressive approach and early use of multiple anti-rheumatic drugs including methotrexate soon after diagnosis.[6] The most significant advance in recent years has been the development of blockers of TNF-α.[7,8] These provide disease control in the majority of resistant patients with low toxicity. Undoubtedly, such targeted therapy is a major achievement, although there remain a significant proportion (25%) of treatment failures.[9]

On the background of increasing safety of haemopoietic stem cell transplantation (HSCT) with the advent of peripheral blood stem cells, promising data from animal studies and 'coincidental' human reports, pilot studies of autologous HSCT were commenced in patients with RA at risk of significant long-term morbidity and mortality. Severe resistant rheumatoid arthritis (RA) can be considered as a good candidate disease for this experimental approach. Even in the 'anti-TNF' era, there are many patients with resistant disease. Efficacy may be easily and non-invasively assessed, and it is possible to select patients with good vital organ function who would be expected to tolerate high dose therapy well.

Since 1996, over 70 patients have received autologous HSCT for severe rheumatoid arthritis. Stem cell mobilisation using either G-CSF alone or in combination with cyclophosphamide has proved effective in collecting sufficient stem cells for transplant and has resulted in only a small percentage of flares. Studies in Australia, USA and Europe have established cyclophosphamide 200 mg/kg alone or in combination as a safe and effective conditioning regimen. Overall, ACR 50 responses have been achieved in around half of these therapy-resistant patients. The major problem has been sustaining responses in the long term. No significant benefit has been observed with CD34 selection of grafts with this dose of conditioning. A common observation has been renewed sensitivity of RA to Disease Modifying Anti-rheumatic Drugs (DMARD) therapy, although this needs to be confirmed in a formal clinical trial setting. In addition, two further patients have been treated more intensively with the myeloablative busalfan/cyclophosphamide (BuCy) regimen and highly purified grafts with more profound responses but greater toxicity. The world experience has recently been combined into a single European Bone Marrow Transplant (EBMT) and Autologous Bone Marrow Transplant Registry (ABMTR) analysis.

In addition to the patients treated with autologous HSCT, one patient with severe RA has been treated with cyclophosphamide 200 mg/kg and anti-thymocyte globulin followed by syngeneic HSCT and remains in complete remission with over 5 years long term follow up.

Clinical trials are necessary to clarify the utility of autologous HSCT in RA. In Europe, the ASTIRA trial aims to investigate the efficacy of high dose cyclophosphamide with autologous HSCT followed by maintenance DMARD treatment. In North America, Northwestern University and their collaborators are performing further phase I/II studies using myeloablative therapy and selected autologous rescue and reduced intensity allogeneic transplantation.

In summary, RA is a significant source of morbidity and, in the long term, mortality. Although many treatments are effective in suppressing disease activity, they usually require chronic administration and are often toxic, and do little to cure the disease

Table 1. Treatment of animal arthritis with BMT

Animal	Type	BMT	Conditioning	Response	Reference
NZB/KN mouse	Spontanteous	Allogeneic	TBI	Prevention	11
DBA/1J mouse	Induced with collagen	Allogeneic	TBI	Regression	12
Buffalo rat	Induced with Mycobacterium tuberculosis	Allogeneic/Syngeneic	TBI	Regression	13
Buffalo rat	Induced with Mycobacterium tuberculosis	Autologous/Pseudoautologous	TBI	Regression	14

or to prevent irreversible end organ damage. There is a significant minority of patients with disease who respond poorly and/or only temporarily to anti-rheumatics and progress inexorably to severe disability and increased mortality. Some have likened mortality for subsets of patients with RA to that of triple vessel coronary artery disease or Hodgkin's disease.[10] It is in this resistant group of patients that HSCT is most appropriate.

Background Data

Animal Studies

Animal models of autoimmune disease can be 'spontaneous' (or 'hereditary') when the disease develops with high frequency in inbred strains of rodents and 'induced' where autoimmunity requires immunisation of a genetically susceptible strain.

Allogeneic, syngeneic and autologous bone marrow transplantation (BMT) have been shown to modify and reverse a variety of animal models of RA (Table 1). The term 'pseudoautologous' refers to transplantation from a syngeneic animal brought to an identical stage of autoimmune disease as the recipient and is technically easier and more humane than true autologous grafting.

In Japan, Ikehara's group treated NZB/KN mice, a 'spontaneous' model of inflammatory polyarthritis, with myeloablative therapy followed by infusion of allogeneic bone marrow resulting in remission of disease. The addition of bone marrow stroma was found to be useful in promoting donor engraftment and preventing relapse.[11] Kamiya et al demonstrated the curative effect of allogeneic BMT in a collagen induced arthritis model.[12]

In Rotterdam, van Bekkum's group used Buffalo rats, which develop polyarthritis one month after immunization with Mycobacterium tuberculosis and Freund's adjuvant. Irradiation and allogeneic transplantation from a non-susceptible strain resulted in regression of the arthritis. Syngeneic transplantation was equally effective. The treatment was most effective when performed shortly after the induction of arthritis.[13] A similar degree of success was achieved with true and pseudo-autologous BMT, and also with the highest tolerated doses of irradiation (8 Gy) without haematological rescue,[14] but not with lower doses of TBI nor with equivalent local radiotherapy given only to the limbs. Addition of spleen cells to the graft did not influence the outcome. They hypothesized that remissions were achieved by ablation of the autoreactive immune and haemopoietic systems and regeneration of new non-autoreactive lymphocytes.

Accepting obvious differences between animal arthritis and human RA, these animal models illustrated the potential for ablative therapy and BMT. Encouraging results are seen with syngeneic and autologous BMT, which are less toxic options for clinical practice. These findings provided a basis for, and, in many respects, have been remarkably predictive of, later studies in humans.

Human Data

Additional evidence that HSCT might offer promise for the treatment of RA has been provided by a handful of cases in which patients with RA have received HSCT or intermediate doses of chemotherapy for the conventional indications of malignancy or severe aplastic anaemia (SAA).

Patients with coincidental RA undergoing HSCT for other resasons are listed in Table 2. In 1975, a 30 year old woman with a 2 year history of seronegative RA was treated with cyclophosphamide 200mg/kg and allogeneic transplantation for oxyphenbutazone related severe aplastic anemia (SAA). The post-transplant course was complicated by graft-versus-host disease (GVHD) and hepatitis C infection, but the patient has remained free of RA for over 20 years.[15]

Three further patients with RA who received allogeneic transplantation for therapy induced SAA have been reported to be in complete remission of their arthritis for up to 13 years. All patients had a degree of GVHD. One patient had a recurrence of disease at 2 years, and recommenced disease modifying anti-rheumatic drugs for 2 years, and then the disease subsequently ran an attenuated course and re-entered remission for a further 11 years.[16]

The case by McKendry et al relapsed after a two year remission and continued to have progressive disease thereafter. It is noteworthy that this case did not develop GVHD, suggesting that, like HSCT for malignancy, GVHD might have a role in eradication of disease in the host.[17] This concept of "graft-versus-autoimmune" (GVA) effect was applied recently when a 'mini-allograft' was used to treat chronic myeloid leukaemia in a patient with coexistent psoriatic arthritis. In this case, recurrence of recipient cells resulted in reappearance of both the leukaemia and the arthritis after an initial remission. Reduction in immunosuppression to induce GVHD resulted in disappearance of both leukaemia and arthritis and recovery of donor haemopoiesis.[18]

In line with the work of Ikehara[19] and Ochi,[20] supportive of the hypothesis that RA is a haemopoietic stem cell disorder, the ability of allogeneic transplantation to cure RA may lie in its ability to replace an aberrant 'rheumatoid' haemopoiesis. However, there may be other factors as adoptive transfer of RA by bone marrow donation has not been observed.[21]

Table 2. Long-term outcome of patients with RA treated with sibling allogeneic BMT for severe aplastic anaemia

Conditioning	Post BMT therapy	GVHD present	Remission	Reference
Cy 200 mg/kg	Methotrexate, ATG	Acute and chronic	>20 years	15
Cy 200 mg/kg	Cyclosporin A	Chronic	2 years, DMARD treatment then remission >11 years	16
Cy 200 mg/kg	Cyclosporin A, steroids	Acute and chronic	>10 years	16
Cy 200 mg/kg	Cyclosporin A, steroids	Acute and chronic	>13 years	16
Cy 200 mg/kg, TBI 4 Gy, Prednisolone 400 mg	Methotrexate, steroids	Nil	2 years then progressive disease	17

There is clearly potential for allogeneic HSCT to cure RA, but the morbidity and mortality of the procedure would need to be reduced substantially before it could be realistically considered as a widely used treatment option, even in the most aggressive RA.

In contrast to the cases of 'coincidental' allogeneic transplantation, the few reported cases of RA receiving autologous HSCT for malignancy (Table 3)[22-24] were not so strongly associated with long-term remissions, although significant amelioration of disease was observed. Similar responses were observed in two patients, who entered 8 and 13 month remissions following m-AMSA, daunorubicin and cytarabine treatment for acute myeloid leukaemia.[25]

Clinical Studies

Although data from 'coincidental' cases suggested that autologous HSCT offers a lesser chance of cure, substantial remissions are possible. Since 1996, increasing numbers of patients with RA have undergone stem cell mobilisation and HSCT, either as sporadic cases or in pilot studies. To date, all procedures have been autologous, except one where syngeneic and another where allogeneic stem cells were used.

Mobilisation Data

Studies of stem cell mobilisation were considered necessary in RA, as animal models and anecdotal clinical data suggested that colony-stimulating factors might cause flare. In addition, there was the possibility that RA and its treatment might affect effective stem cell mobilisation. In 1997, the Leeds group reported a pilot study of granulocyte colony stimulating factor (G-CSF, filgrastim) at 5 µg/kg/day for stem cell mobilisation in five patients.[26] Efficacy, measured using peripheral blood CD34 count, was considered adequate. Disease activity remained stable, although the pre-administration of intramuscular or intra-articular methylprednisolone (median 80 mg, range 40-120 mg) may have inhibited any pro-inflammatory effect of filgrastim.

In Australia, a phase I placebo controlled study investigated the safety and efficacy of G-CSF in patients with severe active RA for the purpose of stem cell collection.[27] In a minority of patients, G-CSF administration was associated with an early or late transient flare of RA, which settled spontaneously or was responsive to an increase in prednisolone. Progenitor cell yields were satisfactory in all patients based on both CD34 counts and CFU-GM assays. In all patients receiving G-CSF at 10 µg/kg/day, the target threshold of 2×10^6/kg CD34 cells was achieved with one leukapheresis. Comparison of PBSC harvests from RA patients with those from healthy donors showed less efficient mobilisation of CD34 cells but normal in vitro progenitor cell function and a relative increase in monocytes.[28]

In Paris, four patients received mobilisation with cyclophosphamide 4 g/m^2, followed by G-CSF 5 µg/kg/day.[29] As expected, CD34 cell yields were higher than with G-CSF alone and were sufficient for CD34 selection to be performed in three of the patients. The incorporation of cyclophosphamide also resulted in improvement in parameters of disease activity with one patient achieving ACR 70, two patients achieving ACR 50 and one patient ACR 20. Improvements were noted for both arthritis and extra-articular manifestations. However, after initial improvement, relapse of arthritis occurred in all patients, reaching a peak at 4-6 months. Persistent disease activity was seen in three patients, although this never reached baseline levels even 2 years after the

Table 3. Outcome of RA post autologous peripheral blood stem cell transplantation for non-Hodgkin's lymphoma

Conditioning	Outcome	Reference
BuCy	Relapse 5 weeks	22
BEAM	Relapse 20 months	23
BEAM	Remission >19 months	24

procedure. In one patient, the disease gradually remitted without additional treatment. Similar observations were made by a group in Pavia, Italy,[30] with clinical responses and improved CD34 yields using cyclophosphamide 4 g/m^2 and G-CSF 10 μg/kg/day.

A number of other reports have included data on stem cell mobilisation using chemotherapy. Burt et al reported yields sufficient for CD34 selection in four patients using cyclophosphamide 2 g/m^2 and G-CSF.[31] In Leeds, Bingham et al were able to perform double selection on harvests in six patients mobilised with cyclophosphamide 2 g/m^2 and G-CSF 263 μg daily.[32] Durez et al successfully mobilised and performed double selection using cyclophosphamide 1.5 g/m^2, etoposide 300 mg/m^2 and G-CSF 5 μg/kg/day.[33,34]

Reports of flare seem to be rare when cyclophosphamide is used in mobilisation and in some cases it seems to have resulted in sustained improvement of disease. However, the case of Joske et al, which was mobilised with cyclophosphamide 4 g/m^2 and G-CSF flared on neutrophil recovery.[35]

Burt et al reviewed peripheral blood stem cell mobilisation data from 174 individuals from 24 transplant centres.[36] A total of 37 cases of RA were analysed. Greater CD34 yields were achieved with the combination of cyclophosphamide and G-CSF, with the greatest yield associated with cyclophosphamide at 4g/m.^{2} Three out of 16 patients (19%) were considered to have had an exacerbation of RA, and all were cases mobilised with G-CSF alone. In contrast, in 12/21 cases (57%) mobilised with cyclophosphamide and G-CSF, there was improvement. All patients produced sufficient stem cells for transplant. An attempt was made to identify factors which might influence stem cell mobilisation. Although no statistically significant conclusion could be drawn, results suggested a trend in recent administration of methotrexate or gold and reduced CD34 yield. Reassuringly, there seemed to be no relationship between CD34 yield and corticosteroids.

Transplant Data

Australian Experience

Joske et al reported the first autologous HSCT performed specifically for RA.[35] A wheelchair bound patient who had received >10 second line therapies underwent stem cell collection with cyclophosphamide 4 g/m^2 and G-CSF. He then received 200 mg/kg of cyclophosphamide as conditioning followed by infusion of an unmanipulated peripheral blood stem cells (PBSC) graft. Transplant related morbidity was minimal and the patient attained an ACR 70 remission for 25 months. Reintroduction of methotrexate 10mg weekly (which had previously been unsuccessful) has maintained his disease under substantial control for a further 12 months at last follow up (D. Joske, personal communication).

In Sydney, two cohorts of four patients who fulfilled criteria for severe active resistant RA were recruited into a dose escalation study of cyclophosphamide at 100 mg/kg (cohort 1) or 200 mg/kg (cohort 2) followed by unmanipulated peripheral blood stem cell rescue.[37] Disease modifying drugs were discontinued before treatment, but corticosteroids were maintained and later tapered where possible. The procedure was tolerated well by all patients. Patients in cohort 1 all had an initial response to therapy, but by 3-4 months disease activity returned to baseline or worse.

Cohort 2 had more profound and sustained responses and has now been followed for over two years (Fig. 1). Patient 2.1 (refer to Fig. 1), a 25 year old female with 4 years of seronegative RA, fulfilled the ACR criteria for complete remission. At 9 months post treatment, she became pregnant. Her RA became more ac-

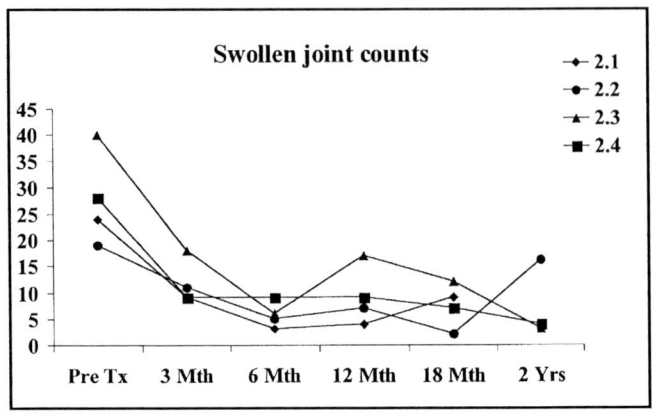

Figure 1. Swollen joint counts of 4 patients receiving 200 mg/kg cyclophosphamide and unmanipulated PBSCT, St Vincent's Hospital, Sydney

tive post-partum at 19 months, but was successfully controlled with intra-articular steroids and methotrexate. Patient 2.4 (refer to Fig. 1), a 41 year old female with a short duration of RA with poor prognostic features, had the second best response. At 18 months, she had reductions in swollen joint count from 28 to 7, tender joint count from 40 to 3, morning stiffness for most of the day to none, pain score from 65 to 12 mm and RF from 273 to <20 i.u./ml. At two years, she flared but the disease came under rapid disease control with leflunamide. In patient 2.2 (see Fig. 1), a 37 year old female, reduction in prednisolone dosage at three months post-therapy was followed by generalised flare of disease between four and five months post therapy which was treated with increased corticosteroids and a single intramuscular injection of gold. At 17 months, her swollen joint count was reduced from 19 to 2, tender joint count from 25 to 6. Patient 2.3 (see Fig. 1), a 51 year old female who had the most advanced and erosive disease in this cohort, had a good response initially, but by 3 months had experienced localised flare in one knee necessitating intra-articular corticosteroid injection and reintroduction of methotrexate therapy. She subsequently underwent left then right knee replacements (at 8 and 11 months) and at 2 years her disease parameters remained significantly improved, although she remained on methotrexate and prednisolone.

The study continued as an Australian multicentre trial in which patients were randomised to either unmanipulated or CD34$^+$ selected grafts.[38] The trial recruited 33 patients who had failed therapy with methotrexate and at least one other disease modifying agent. Patients received high dose immunosuppression with cyclophosphamide 200 mg/kg followed by randomised rescue with unmanipulated or CD34 selected cells. Thirty-one patients proceeded to transplant. There were no deaths and no major unexpected toxicities. On an intention to treat basis (including 2 patients who failed to mobilise stem cells), ACR 20, 50, and 70 responses were achieved in 70%, 45% and 39% of patients. There were no significant differences between the unmanipulated and the CD34 selected groups in terms of response and time to relapse and re-introduction of disease modifying therapy. Although not part of the trial, a minority of patients were observed closely after re-introduction of disease modifying therapy. Responses, some as profound as ACR 70, were observed in two thirds of the patients followed up, supporting the hypothesis that the transplant procedure provides disease debulking such that subsequent control is possible with low dose agents.

Table 4. Treatment of relapses in 6 patients after autologous stem cell transplantation for severe RA (adapted from reference 41)

Centre	Patient Details	Initial Response	First Salvage*	Response	Second Salvage	Response	Third Salvage	Response	Last Followup
North-western	F 46	ACR 70	22 months MTX, SSZ, HCQ	NR	36 months Anti-TNF, MTX	ACR 30	N/A	N/A	42 months ACR 30 continues Anti-TNF, MTX
North-western	F 42	<ACR 20	3 months MTX, HCQ	NR	9 months CSA, MTX, HCQ	NR	36 months Anti-TNF, MTX, HCQ	ACR 40	42 months ACR 40 continues Anti-TNF, MTX
North-western	M 48	ACR 70	18 months Leflunomide	NR (leflunomide intolerance)	30 months Anti-TNF	ACR 50	N/A		32 months ACR 50 continues Anti-TNF
North-western	F 49	ACR 80	6 months MTX	NR	12 months iv CTX 1g, MTX	NR	NR	NR	33 months Disabled Leflunomide
Omaha	F 25	ACR 80	6 months MTX, SSZ, HCQ	ACR 80	18 months Synthroid, MTX, SSZ, HCQ	ACR 80 Response to thyroxine	NR	NR	36 months Leflunomide added, added at 29 months without benefit
Omaha	M 45	ACR 80	6 months MTX, SSZ, HCQ	ACR 60	14 months Leflunomide, MTX, SSZ, HCQ	NR	NR	NR	24 months Disabled Steroid pulses

Abbreviations: ACR= American College of Rheumatology; Anti-TNF= anti-tumor necrosis factor; CSA= cyclosporine; CTX= cyclophosphamide (cytoxan); F= female; HCQ= hydroxychloroquine; M= male; MTX= Methotrexate; NA= not applicable; NR= no response; SSZ= sulfasalazine

In contrast to the autologous setting, McColl et al[39] reported the first syngeneic HSCT for autoimmune disease. A 46 year old man with a 7 year history of severe seronegative RA received conditioning with cyclophosphamide and anti-thymocyte globulin (ATG) and an unmanipulated PBSC graft with minimal transplant related toxicity. Post transplant T cell receptor V-Beta gene usage of circulating lymphocytes demonstrated evidence of full donor T cells which corresponded to complete remission of the disease, at over 5 years follow up (J. Szer, personal communication).

North American Experience (Table 4)

In Chicago, four patients with RA have been treated with cyclophosphamide 200 mg/kg, ATG 90 mg/kg, methylprednisolone 3 g followed by a CD34+ cell enriched (2.5-2.7 log T-cell depleted) autograft.[31,40] In one patient, low dose TBI (4 Gy) was also used. The regimen was well tolerated in all and produced sustained responses in two of the four patients. A 46 year old woman with seven years of resistant seropositive RA maintained ACR 70 criteria throughout the first 20 months of follow up. Swollen joint count fell from 27 to 2, tender joint count from 41 to 4 and she had no morning stiffness. Corticosteroids were discontinued at 8 months, but she continued on hydroxychloroquine. At 22 months, relapse was treated with hydroxychloroquine, sulfasalazine and methotrexate. At 3 years post transplant, her disease was poorly controlled and she was treated with etanercept. By 42 months, the number of tender and swollen joints had fallen to 15 and 4 and she continued on etanercept and methotrexate.

A 48 year old male with 4 years of RA fulfilled ACR 70 at 1 and 3 months, but flared at 5 months requiring an increase in oral prednisolone and intra-articular corticosteroids. His disease subsequently settled and at 9 months he discontinued anti-rheumatic drugs with ACR 70. At 18 months, he required treatment with leflunomide for increasing disease activity, but was intolerant. He subsequently refused treatment except for intra-articular steroids until 30 months post-transplant when he commenced etanercept, which resulted in ACR 50.

Two other patients had limited responses. A 42 year old woman with 7 years of resistant RA failed to achieve ACR 20. She continued on prednisolone and hydroxychloroquine and at 3 months methotrexate 20 mg weekly was added. At 9 months, she was commenced on cyclosporin A. At 30 months, her tender joint count was 30 and swollen joint count was 19, she was enrolled on a randomised study of TNF inhibitor against placebo, but at 36 months was taken off the study and started etanercept. By 42 months post-transplant, the numbers of tender and swollen joints were 13 and 5 respectively on etanercept and methotrexate.

A 49 year old female with 6 years of resistant RA, who received TBI 4 Gy in addition to the above conditioning, showed a significant improvement in the first three months with ACR 70, but at six months she relapsed with similar joint counts to baseline. From 6-33 months, she received a number of failed attempts at treatment with oral methotrexate, intravenous cyclophosphamide, leflunomide and infliximab. At 2 years post-transplant, tender and swollen joint counts were 34 and 30 respectively.

In Omaha, Nebraska, two patients with severe RA were mobilised with cyclophosphamide 2 g/m² and G-CSF, which in itself resulted in a ACR 50 response. They were then treated with cyclophosphamide 200 mg/kg and horse ATG on days –3 to +3. Prednisone was given at a dose of 0.5 mg/kg/day in a single dose until day +14, and tapered to 10 mg daily on day 28 and subsequent taper at the discretion of the attending rheumatologist. G-CSF was given from day +1 until neutrophil recovery. Both patients achieved an ACR 80, but both relapsed at 6 months. They were then treated with combination DMARD therapy (sulphasalazine, hydroxychloroquine and oral methotrexate) that had been previously ineffective, and at 12 months the ACR responses were 80 and 60%. Thereafter, one patient relapsed at 14 months and progressed despite introduction of further leflunomide and etanercept, and at 24 months post-transplant he had returned to full disability and was dependent on frequent pulses of high dose prednisone. The other patient relapsed at 18 months in association with hypothyroidism and regained an ACR 80 response at two years following commencement of thyroxine. At 25 months, she was commenced on etanercept for increased disease activity, but with no improvement. Similarly, leflunomide at 29 months resulted in no benefit. At 36 months, she failed to meet ACR 20 criteria.[40,41]

European Experience

In Belgium, a 22 year old patient with a 7 year history of refractory systemic and erosive seronegative RA was treated with busulfan 16 mg/kg and cyclophosphamide 120 mg/kg, followed by rescue with a highly purified autograft (98.4% CD34⁺ cells and no detectable T-cells).[33] At three months, she was in complete remission off steroids. At four months, she was successfully treated for pneumocystis pneumonitis. She remains in remission at 56 months (P Durez, personal communication). The remission has been maintained despite complete reconstitution of the T cell repertoire to pre-transplant levels. One other patient treated with this protocol also achieved complete remission by three months, but died of multiresistant staphylococcus and carcinoma of the lung at 5 months.[34]

The Dutch collaborative group has reported the progress of 14 patients with active, progressively erosive RA who had failed at least 4 DMARDs.[42] Mobilisation was with cyclophosphamide 4 g/m² and G-CSF. Twelve patients went on to receive cyclo-

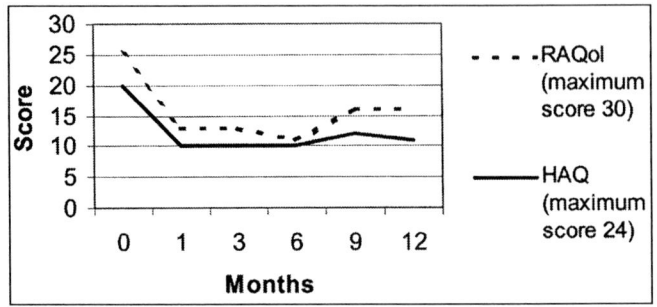

Figure 2. Sustained reduction in mean HAQ and RAQol in seven patients undergoing autologous HSCT for RA in Leeds UK.

phosphamide 200 mg/kg before haemopoietic rescue with CD34 enriched harvests. With a follow up from 7-21 months, there were no deaths and marked improvements in disease activity were observed in 8 of 12 patients at >50% of assessments. The disease activity scores were significantly improved from baseline at 3, 6, 9 and 12 months. Notably, two of the responders were patients who had failed TNF antagonists. In 7/12 patients, disease modifying agents were reinstituted because of active disease, and there was amelioration of disease activity in 3/7, including 2 patients who had been non-responders to the transplant procedure.

In Leeds, seven patients with RA resistant to at least 4 DMARDs were mobilised with cyclophosphamide 2 g/m² and G-CSF[32]. In six patients, the graft underwent CD34 selection and CD4 and CD8 negative purging in order to achieve at least a 5-log reduction in T-lymphocytes. In the seventh patient, insufficient cells were harvested to allow graft manipulation. All patients were conditioned with cyclophosphamide 200 mg/kg. Minimal transplant-related toxicity was observed. Patients have been followed-up for 36 months. All patients responded well initially with 4 of the 7 patients achieving at least a 50% reduction in ACR without disease modifying agents and this was maintained for up to 9 months (Fig. 2). All seven patients relapsed and were commenced on cyclosporin A. Three patients subsequently achieved ACR 70, one patient, ACR 50, one patient showed improvement but did not satisfy ACR 20 and two patients did not improve (Fig. 3). One patient continued to have minimal disease activity (but did not quite satisfy ACR remission criteria on account of a tender joint due to secondary osteoarthritis) at 36 months whilst only taking cyclosporin A 100 mg daily and prednisolone 5 mg daily. All patients had an improved quality of life and a reduction in disability score. Only one serious infective episode has occurred during follow-up: an infected toe wound at 18 months that required intravenous antibiotics. No other serious long-term complications have occurred.

The European Group for Blood and Marrow Transplantation (EBMT) and Autologous Bone Marrow Transplant Registry (ABMTR) Analysis

The EBMT and ABMTR registries have analysed the combined world experience of RA treated with autologous HSCT.[43] The analysis includes both published and unpublished cases. Seventy-six patients were registered from 15 centres; 73 patients had received autologous HSCT. Patients were 74% female and had a median age of 42 years. Eighty-six percent were rheumatoid factor positive. Patients had been previously treated with an average of 5 (range 2-9) DMARDs. High dose immunosuppressive therapy was cyclophosphamide alone in 85% patients, mostly

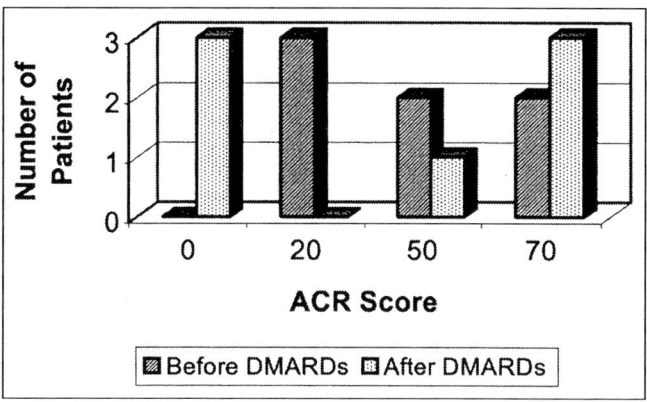

Figure 3. ACR Response Score before and after the reintroduction of DMARDs (cyclosporin A) at relapse in patients treated in Leeds, UK.

200 mg/kg. Seven patients received cyclophosphamide 200 mg/kg with ATG, two patients received busulfan 16 mg/kg and cyclophosphamide 120 mg/kg, one patient received cyclophosphamide 200 mg/kg with total body irradiation and ATG, and one patient received fludarabine with ATG. All but one patient received PBSC rescue. Some form of graft manipulation, mainly CD34 selection, was performed in 62% patients. At a median follow up of 16 months (range, 3-55 months), 67% of patients had achieved at least ACR 50 at some point following transplant and there was a significant reduction in HAQ disability score (p≤ 0.005). Most patients were re-started on DMARDs within six months for persistent or recurrent disease activity, which provided disease control in about half the cases. Response was significantly related to seronegative RA (p= 0.02), but not to disease duration, number of previous DMARDs, HLA-DR4, mobilisation chemotherapy or graft manipulation. The analysis included the patient described above who died from infection and incidental lung cancer at 5 months post-transplant, but there were no other deaths following transplant. The authors concluded that autologous HSCT is a relatively safe form of salvage treatment in severe, resistant RA resulting in profound responses in most patients, albeit with the reintroduction of DMARDs in most cases to control recurrent or persistent disease activity.

Mechanism of Action

At the EBMT Meeting in 1998, Alberto Marmont described his 'high grade' and 'low grade' hopes for HSCT in autoimmune disease. His 'high grade' hopes were for a re-education of the autoreactive immune system, with the elimination of the autoreactive components, producing tolerance. The 'low grade' hopes were that this approach would at least produce clinical efficacy by delivering intensive immunosuppression, but without re-education or tolerisation.

Based on the clinical results discussed, it seems reasonable to state that the high grade hope of tolerance induction has not been achieved in RA, at least with the doses of conditioning used, which have been almost entirely non-myeloablative. Instead, the effect has been variously described as 'turning back the immunological clock', or 'debulking of inflammation'. Indeed, early observations suggest that changes induced by the autologous transplant are associated with renewed ability to control disease activity.[32,37-42]

The mechanism of action probably has two essential aspects. Firstly, the high dose cytotoxic therapy has a direct effect on the rheumatoid synovium, a vascular proliferating and invasive tissue, similar to a tumour. The study of Bingham et al has shown that clinical improvement is associated with reduction of macroscopic and microscopic synovitis following cyclophosphamide 200 mg/kg and selected autograft.[44] Interestingly, low level inflammation may persist despite this treatment, illustrated by persistence of a single inflamed synovial villus, perhaps analogous to 'minimal residual disease' in the cancer setting. Eventually, there was clinical relapse and this was associated with a re-invasion of the synovium by CD4 cells and florid macroscopic synovitis.

Secondly, the process of high dose cytotoxic therapy and autologous rescue results in profound and sustained changes in the peripheral blood milieu. Study of such changes in peripheral blood in association with remission and relapse might provide insights into the pathophysiology of RA.

Analysis of the immune reconstitution data confirm that similar patterns of immune recovery occur in RA patients post HSCT as in subjects undergoing the procedure for malignancy, including significant reductions in CD4 cells and relatively rapid regeneration of CD8 and NK cells.[45,46] However, naïve CD4+ CD45+ cells seem to be reduced more profoundly and for longer in RA than in the malignant setting where naïve T cell regeneration usually returns to baseline at 12-18 months.[32,38,42,47] Thymic function is known to be impaired in RA and might explain delayed reconstitution of naïve CD4 cells.[48]

T-cell dynamics following treatment with cyclophosphamide 200mg/kg and selected autologous stem cell transplantation were studied by Ponchel et al.[49] Thymic activity measured by T-receptor excision circle (TREC) analysis was evident after therapy in most patients. Total CD4 counts were still low two years after therapy. Naïve CD4 cells accumulated slowly but decreased at times of flare, associated with a decrease in TREC-content and an increase in CRP levels. Total CD8 counts regained baseline level by one month via large expansion of memory cells. There was a correlation between inflammation, circulating levels of IL-6 and the proliferation and differentiation of naïve CD4 cells. Persistent systemic inflammation may additionally contribute to the poor reconstitution following HSCT for RA. Interestingly, in the long-term remitter who received the myeloablative BuCy regimen, T-cell numbers returned rapidly to normal, perhaps providing additional support that absence of inflammation might encourage normal immune reconstitution.

Post-transplant peripheral blood immune recovery at the time of relapse or flare has been observed with the hope of identifying re-emergence of potentially pathogenetic cell populations. To date, there has been little consistency between studies. Despite an influx of CD4 cells into the rheumatoid joint at relapse, there seems to be little correlation with peripheral blood CD4 count. Indeed, the Australian randomised trial also noted a greater reduction in CD4 cells in the CD34 selected arm with no apparent clinical advantage.[38] The use of non-myeloablative chemotherapy in the majority of cases has been associated with predominant regeneration of CD4+CD45RO+ cells from the peripheral pool, which may include a proportion of pathogenetic memory cells. The failure to destroy potentially autoreactive memory cells might explain recrudescence of RA following non-myeloablative treatment.

In the Australian study, there was a trend between relapse and recovery of B cells and rheumatoid factor production. In this study, non-responders had higher pre-transplant levels of rheumatoid factor. Although the Dutch study[42] showed no correlation with peripheral blood cell numbers, non-responders tended to have higher levels of total IgG pre-transplant, possibly a reflection of disease related B-cell activity. An association of relapse and B-cell

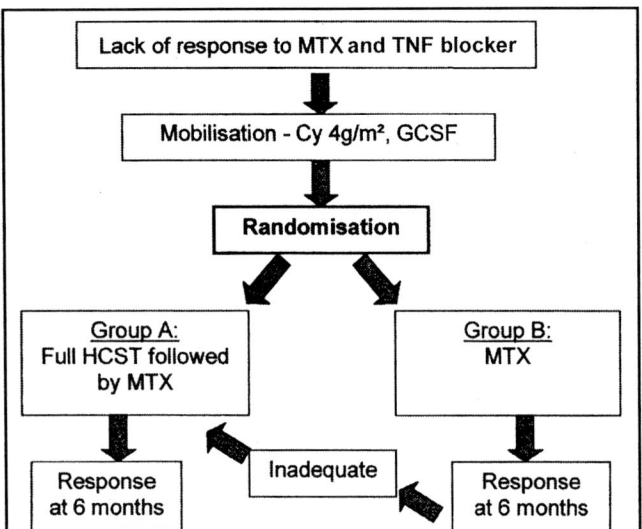

Figure 4. Outline of ASTIRA trial.

recovery has previously been observed following treatment of RA with CAMPATH (anti-CD52), which depletes B and T cells.[50] Recently, use of the specific B-cell depleting antibody, rituximab (anti-CD20), has shown promise in RA,[51] and is the focus of ongoing studies in Australia, where it has been incorporated in post-transplant salvage treatment.

Clinical Trials

There are several avenues for clinical trials including further intensification of the cytotoxic regimen and post transplant maintenance.[52,53] Ultimately, randomised phase III trials against the best alternative treatment are necessary to evaluate the utility of this approach.

European Clinical Trial: Autologous Stem Cell Transplantation International Rheumatoid Arthritis (ASTIRA) Trial

In contrast to using myeloablative regimens with their attendant risks, it may be reasonable to accept that, in the interests of safety, that dose intensification of cytotoxic therapy should not increase significantly and that other strategies should be considered. High dose cyclophosphamide 200 mg/kg, which has been shown to be relatively safe and produce profound responses, at least in the short term, might be best regarded as a form of 're-mission induction' or 'debulking' therapy, but, as suggested by the study of Bingham et al,[44] leaving 'minimal residual disease'. Using this regimen, there seems to be no major clinical benefit to be gained from CD34 purification and lymphocyte depletion for RA.[38] However, there is now a body of data, albeit uncontrolled, to support increased disease control by the re-introduction of disease modifying agents. In most instances, disease modifying agents have been re-introduced at a time of relapse. By logical extension, given that disease activity is not completely eradicated in the majority of patients, it would seem reasonable to introduce such treatment immediately post-transplant as maintenance treatment.[52] This approach is analogous to that of low grade lymphoproliferative disorders, such as myeloma and follicular lymphoma, which are incurable but are routinely treated with autologous transplant with the aim of inducing profound remis-

Table 5. American phase II autologous myeloablative HSCT trial regimen

Day	-8	-7	-6	-5	-4	-3	-2	-1	0
Busulfex (0.8 mg/kg q 6 hours)	X	X	X						
Cyclophosphamide (45 mg/kg/day)				X	X	X	X		
Autologous CD34+ enriched stem cells infused									X

sions. Disease progression may be delayed further with the use of maintenance agents such as interferon or thalidomide.

The EBMT/EULAR Working Party have produced the ASTIRA trial (Fig. 4). Based in Leeds, UK, patients must fulfil entry criteria for severe, resistant RA, which include failure of TNF blockers and methotrexate. Mobilisation is with cyclophosphamide 4 g/m² and G-CSF in all patients, and grafts will not be manipulated. Patients are randomised to receive either immediate maintenance treatment with methotrexate or receive cyclophosphamide 200 mg/kg and unmanipulated autologous stem cell rescue followed by maintenance treatment with methotrexate.

American Clinical Trials

Phase II Myeloablative

While the European's have initiated a phase III trial using non-myeloablative cytoxan (200 mg/kg) pioneered in the phase I studies, the Americans (Burt et al, Northwestern University, Chicago) have developed a myeloablative phase II protocol. This is based on suggestions of a dose response effect to high dose chemotherapy with the observation that a myeloablative regimen (busulfan and cyclophosphamide) seems to indicate more durable remissions.

The philosophy is that for such an intense and expensive treatment to be considered for rheumatoid arthritis, sustained complete remissions, or 70% improvement as defined by the American College of Rheumatology (ACR 70), should be achieved without use of post transplant immune suppression maintenance. This approach views RA as a disease that is potentially curable by re-induction of self tolerance through autologous HSCT.

The regimen includes cyclophosphamide (135 mg/kg) and intravenous busulfan (2.4 mg/kg with a CD34+ selected autograft (Table 5). While this regimen is more intense than cyclophosphamide alone, the busulfan was decreased from 16 to 12 doses and administered in the more predictable intravenous formulation ('Busulfex') in order to diminish the risk of regimen related toxicities. Busulfan was selected due to its cytotoxicity to monocytes amd macrophages. This may be important since the inflammatory cytokine pattern within RA synovium is predominantly derived from tissue macrophages (not T cells). Patients are selected for failure of infliximab and either methotrexate or leflunomide, failure being defined as more than 6 swollen and 20 involved joints. If a high percentage of patients continue to relapse following a myeloablative regimen, an autologous transplant will probably be unlikely to cure, and the regimen-related toxicity of more intense myeloablative conditioning regimens would probably be unacceptable.

Phase I Allogeneic Hematopoietic Stem Cell Transplantation for RA

As discussed, the rationale for allogeneic HSCT is to change the genetic susceptibility to disease by replacing an aberrant lymphohaemopoietic system. The most significant toxicity of allogeneic HSCT is graft versus host disease (GVHD). As a consequence, compared with autologous HSCT, allogeneic transplants are complicated by significantly higher morbidity and mortality, predominantly due to GVHD.

Lymphocyte depletion or CD34 enrichment of an allogeneic graft will markedly decrease and, depending on the degree of lymphocyte depletion, even prevent the risk of GVHD. A lymphocyte depleted graft may be associated with an increased risk of graft failure. Engraftment is affected by intensity of the conditioning regimen, and the number of donor stem cells infused, as well as the number of donor T cells facilitating engraftment. Therefore, the risk of graft failure may be partially offset by maximizing donor stem cell numbers. Even if full donor engraftment does not occur, mixed chimerism (both donor and host hematopoiesis) may be beneficial in ameliorating disease. In order to minimize the morbidity of allogeneic stem cell transplantion, non-myeloablative, yet strongly immunoablative, conditioning could be combined with CD34 enrichment of donor stem cells.

Based on the above considerations, the Chicago group have opened a phase I allogeneic HSCT trial using a regimen of CAMPATH-1H, cyclophosphamide and fludarabine that would be intensely immune suppressive but not myeloablative (Table 6). In case the allograft would be rejected, autologous hematopoietic reconstitution is anticipated within two weeks with at least transient or perhaps even durable improvement or resolution of RA. If engraftment occurs with low level (microchimerism) or mixed chimerism, indicative of host versus graft tolerance, and the patient's RA returns, donor lymphocyte infusions (DLI) could be given at escalating doses until full donor chimerism is accomplished or RA resolves. In order to minimize the risk of GVHD, DLI would have to be carefully titrated with gradual dose escalation over extended periods.

The view at Northwestern University is that candidates for allogeneic HSCT should have predicted 5 year mortality of 30-70% and have failed both TNF inhibitors and either methotrexate or leflunomide. These patients may be identified by the number of involved joints or functional status as assessed by a questionnaire on activities of daily living (Health Assessment Questionnaire, HAQ). Failure is defined as patients with active high risk disease such as 12 swollen joints, more than 20 involved joints, and being unable to answer more than 70% of the HAQ "with no difficulty". In summary, to justify the risk benefit of allogeneic HSCT, candidates should be selected for refractory disease (despite TNF inhibitor and methotrexate) that are at high risk for RA related mortality (determined by number of involved joints and HAQ).

Conclusion

A number of individual reports, case series and studies in Australia, Europe and North America have confirmed relative safety of autologous HSCT in severe RA. Treatment is often followed by profound responses, but eventually disease relapse occurs. With cyclophosphamide 200 mg/kg, relapse seems to occur irrespective of whether the graft is purged of T-cells or not. A common observation has been apparent renewed sensitivity of RA to DMARDs. Such reports suggest that autologous HSCT may be associated with a 'debulking of inflammation', although this needs to be investigated prospectively. Given that relapses may occur relatively quickly after treatment, it seems logical to consider the introduction of maintenance treatment early in the post transplant course. Alternatively, the intensity of the conditioning could be increased, which may result in a more thorough eradication of the autoimmune diathesis, but may be associated with greater toxic risks.

Table 6. American phase I allogeneic nonmyeloablative HSCT trial regimen

Day	-6	-5	-4	-3	-2	-1	0
Fludarabine (25 mg/m^2/d)	X	X	X	X	X		
Cyclophosphamide (50 mg/kg)				X	X	X	
CAMPATH-1H 4 mg/d	X	X	X	X	X		
Allogeneic CD34$^+$ enriched stem cells infused							X

Most of the studies to date have been performed in the 'pre-TNF blocker era' of rheumatology. The introduction of these drugs has represented a major therapeutic advance in the treatment of resistant RA with relatively low toxicity. Clearly, given its potential morbidity and mortality, autologous HSCT can only be considered in the small group of patients who have failed both conventional treatments and TNF antagonists.

Even in such patients, decision making is difficult. Verburg et al have attempted to resolve these issues systematically using a Markov decision analysis model.[55] Such a model can be used to trade the potential benefits and risks of HSCT, including transplant related mortality, against the risks of resistant RA treated with conventional treatment. These results suggest that HSCT may be acceptable to patients with severe RA, provided the responses are substantial and prolonged (Table 7).

The curative potential of allogeneic SCT has been supported by animal studies and anecdotal reports, with evidence for a graft-versus-autoimmune effect, but the attendant risks have meant that its use has been limited to date. Only a small number of allogeneic transplants have been performed in autoimmune disease, and, to date only one in RA with short term followup.[55] Low intensity allografting techniques have reduced mortality and morbidity and it seems inevitable that this approach will be applied to severe RA. In the first instance at least, procedures are likely to be limited to the minority of younger patients with HLA matched siblings.

Overall, investigators should tread carefully, erring on the side of caution as opposed to exposing patients to risks in excess of the risks of their disease. Ultimately, destroying the aberrant immune system and rebuilding or replacing it may not only result in clinical benefit but may also provide useful insights into the aetiology and pathogenesis of RA.

Acknowledgements

Dr. Snowden and Dr. Bingham acknowledge the support of the Arthritis Research Campaign U.K. Dr. Moore acknowledges the support of the National Health and Medical Research Council of Australia. Dr. Pavletic acknowledges the support by the University of Nebraska Clinical Research Center. We also thank the

Table 7. Summary of Markov decision analysis of autologous haemopoietic stem cell transplantation in RA[54]

If TRM 1%	14.6% need to have ACR >20 at 2.5 years
	5.4% need to have ACR >50 at 2.5 years
	1.4% need to have ACR 70 at 2.5 years
If TRM 10%	43% need to have ACR >20 at 2.5 years
	27.3% need to have ACR >50 at 2.5 years
	15.3% need to have ACR 70 at 2.5 years
If TRM rises to 31%	100% need to have ACR 70 at 5.5 years

Based on the sample population interviewed, in order for the quality adjusted life years (QALY) associated with HSCT to exceed those associated with conventional treatment over a 5.5 year period, the transplant related mortality (TRM) has to be <3.3%. Minimally desired effectiveness i.e., point at which QALY's for autologous HSCT exceed those for conventional treatment.

editorial staff of the *Journal of Rheumatology* for allowing us to derive and reproduce Figure 1 and Table 4 from earlier publications.

References

1. Reilly PA, Cosh JA, Maddison PJ et al. Mortality and survival in rheumatoid arthritis: A 25 year prospective study of 100 patients. Ann Rheum Dis 1990; 49:363-369.
2. Markenson JA. World wide trends in the socioeconomic impact and long term prognosis of rheumatoid arthritis. Semin Arthritis Rheum 1991; 212(1):4-12.
3. Wong JB, Ramey DR, Singh G. Long-term morbidity, mortality, and economics of rheumatoid arthritis. Arthritis Rheum 2001; 44:2746-2749.
4. American College of Rheumatology Subcommittee on Rheumatoid Arthritis Guidelines. Guidelines for the Management of Rheumatoid Arthritis. 2002 Update. Arthritis Rheum 2002; 46:328-346.
5. Wilske KR, Healey LA. The need for aggressive therapy of rheumatoid arthritis. Rheum Dis Clin North Amer 1993; 19:153-161.
6. Emery P, Salmon M. Early rheumatoid arthritis: Time to aim for remission. Ann Rheum Dis 1995; 54:944-947.
7. Weinblatt ME, Kremer JM, Bankhurst AD et al. A trial of etanercept, a recombinant tumor necrosis factor receptor:Fc fusion protein, in patients with rheumatoid arthritis receiving methotrexate. N Engl J Med 1999; 340:253.
8. Maini R, St. Clair EW, Breedveld F et al. Infliximab (chimeric anti-tumour necrosis factor alpha monoclonal antibody) versus placebo in rheumatoid arthritis patients receiving concomitant methotrexate: A randomised phase III trial. Lancet 1999; 354: 1932-1939.
9. Emery P, Buch M. Treating rheumatoid arthritis with tumour necrosis factor a blockade. Brit Med J 2002; 324:312-313.
10. Pincus T, Callahan LF. What is the natural history of rheumatoid arthritis? Rheum Dis Clin North Amer 1993; 19:123.
11. Nakagawa T, Nagata N, Hosaka N et al. Prevention of autoimmune inflammatory polyarthritis in male New Zealand black/KN mice by transplantation of bone marrow plus bone (stromal cells). Arthritis Rheum 1993; 36:263-268.
12. Kamiya M, Sohen S, Yamane T et al. Effective treatment of mice with type II collagen induced arthritis with lethal irradiation and bone marrow transplantation. J Rheumatol 1993; 20:225-230.
13. van Bekkum DW, Bohre EP, Houben PF et al. Regression of adjuvant-induced arthritis in rats following bone marrow transplantation. Proc Natl Acad Sci 1989: 86:10090.
14. Knaan-Shanzer S, Houben P, Kinwel-Bohre EP et al. Remission induction of adjuvant arthritis in rats by total body irradiation and autologous bone marrow transplantation. Bone Marrow Transplant 1991; 8(5):333.
15. Nelson JL, Torrez R, Louie FM et al. Pre-existing autoimmune disease in patients with longterm survival after allogeneic bone marrow transplantation. J Rheumatol 1997; 24 suppl 48:23.
16. Snowden JA, Kearney P, Kearney A et al. Long term outcome of autoimmune disease following allogeneic transplantation. Arthritis Rheum 1998; 41:453-459.
17. McKendry RJR, Huebsch L, Le Clair B. Progression of rheumatoid arthritis following bone marrow transplantation: A case report with a 13 year follow up. Arthritis Rheum 1996; 39:1246-1253
18. Slavin S, Nagler A, Varadi G et al. Graft vs autoimmunity following allogeneic non-myeloablative blood stem cell transplantation in a patient with chronic myelogenous leukemia and severe systemic psoriasis and psoriatic polyarthritis. Exp Hematol 2000; 28(7):853.
19. Ikehara S, Kawamura M, Takao F et al. Organ-specific and systemic autoimmune diseases originate from defects in hematopoietic stem cells. Proc Natl Acad Sci 1990; 87:8341.
20. Hirohata S, Yanagida T, Nagai T et al. Induction of fibroblast-like cells from CD34(+) progenitor cells of the bone marrow in rheumatoid arthritis. J Leukoc Biol 2001; 70:413-21.
21. Snowden JA, Atkinson K, Kearney P et al. Allogeneic bone marrow transplantation from a donor with severe active rheumatoid arthritis without adoptive transfer of disease to recipient. Bone Marrow Transplant 1997; 20:71-73.
22. Euler HH, Marmont AM, Bacigalupo A et al. Early recurrence or persistence of autoimmune diseases after unmanipulated autologous stem cell transplantation. Blood 1996; 88:3621.
23. Cooley HM, Snowden JA, Grigg AP et al. Outcome of rheumatoid arthritis and psoriasis following autologous stem cell transplantation for hematologic malignancy. Arthritis Rheum 1997; 40:1712.
24. Jondeau K, Job-Deslandre C, Bouscary D et al. Remission of nonerosive polyarthritis associated with Sjogren's syndrome after autologous hematopoietic stem cell transplantation for lymphoma. J Rheumatol 1997; 24:2466.
25. Roubenoff R, Jones RJ, Karp JE et al. Remission of RA with the successful treatment of acute myelogenous leukaemia with cytosine arabinoside, daunorubicin and m-AMSA. Arthritis Rheum 1987; 30:1187-90.
26. McGonagle D, Rawstron A, Richards S et al. A phase I study to address the safety and efficacy of granulocyte colony stimulating factor for the mobilisation of hematopoietic progenitor cells in active rheumatoid arthritis. Arthritis Rheum 1997; 40:1838-1842.
27. Snowden JA, Biggs JC, Milliken ST et al. A randomised, blinded, placebo-controlled, dose escalation study of the tolerability and efficacy of filgrastim for haemopoietic stem cell mobilisation in patients with severe active rheumatoid arthritis. Bone Marrow Transplant 1998; 22:1035-1041.
28. Snowden JA, Nink V, Cooley M et al. Composition and function of peripheral blood stem and progenitor cell harvests from patients with severe active rheumatoid arthritis. Br J Haematol 1998; 103:601-609.
29. Breban M, Dougadous M, Picard F et al. Intensified-dose cyclophosphamide and granulocyte colony-stimulating factor administration for hematopoietic stem cell mobilization in refractory rheumatoid arthritis. Arthritis Rheum 1999; 42:2275-2280.
30. Ponchio L, Pedrazzoli P, Da Prada GA et al. High efficiency of hematopoietic progenitor cell mobilisation with cyclophosphamide (CTX) plus G-CSF (but not with G-CSF alone) in patients with severe rheumatoid arthritis (SARA). Bone Marrow Transplant 2000; suppl 1:S112.
31. Burt RK, Georganas C, Schroeder J et al. Autologous hematopoietic stem cell transplantation in refractory rheumatoid arthritis. Sustained response in two of four patients. Arthritis Rheum 1999; 42:2281-2285
32. Bingham SJ, Snowden J, McGonagle D et al. Autologous stem cell transplantation for rheumatoid arthritis – report of six patients. J Rheumatol 2001; 28 suppl;21-24.
33. Durez P, Toungouz M, Schandene L et al. Remission and immune reconstitution after T-cell-depleted stem-cell transplantation for rheumatoid arthritis. Lancet 1998; 352:881.
34. Durez P, Ferster A, Toungouz M et al. Autologous T cell depleted CD34+ peripheral blood stem cell transplantation in four patients with refractory rheumatoid arthritis and juvenile chronic arthritis. Arthritis Rheum 1999; 42 suppl:S77.

35. Joske DJL, Ma DT, Langlands DR et al. Autologous bone marrow transplantation for rheumatoid arthritis. Lancet 1997; 350:337-8.
36. Burt R, Fassas A, Snowden J et al. Special report. Collection of hematopoietic stem cells from patients with autoimmune diseases. Bone Marrow Transplant 2001; 28:1-12.
37. Snowden JA, Biggs JC, Milliken ST et al. A phase I/II dose escalation study of intensified cyclophosphamide and autologous blood stem cell rescue in severe, active rheumatoid arthritis. Arthritis Rheum 1999; 42:2286-2292.
38. Moore JJ, Brooks P, Milliken S et al. A pilot randomised trial comparing CD34 selected versus unmanipulated haemopoietic stem cell transplantation for severe resistant rheumatoid arthritis. Arthritis Rheum 2002; 46:2301-2309.
39. McColl G, Kohsaka H, Szer J et al. High-dose chemotherapy and syngeneic hemopoietic stem-cell transplantation for severe, seronegative rheumatoid arthritis. Ann Intern Med 1999; 131:507-9.
40. Pavletic SZ, O'Dell JR, Pirruccello SJ et al. Intensive immunoablation and autologous blood stem cell transplantation in patients with refractory rheumatoid arthritis:The University of Nebraska Experience. J Rheumatol 2001; 28 Suppl 64:13-20.
41. Pavletic SZ, Klassen LW, Pope R et al. Treatment of relapse after autologous blood stem cell transplantation for severe rheumatoid arthritis. J Rheumatol 2001; 28 Suppl 64:28-31.
42. Verburg RJ, Kruize AA, van den Hoogen FH et al. High-dose chemotherapy and autologous hematopoietic stem cell transplantation in patients with rheumatoid arthritis: Results of an open study to assess feasibility, safety, and efficacy. Arthritis Rheum 2001; 44:754-760.
43. Snowden JA, Passweg J, Moore JJ et al. Autologous haemopoeitic stem cell transplantation in severe rheumatoid arthritis: a report from the EBMT and ABMTR. J Rheumatology 2004; In press.
44. Bingham S, Veale D, Fearon U et al. High dose cyclophosphamide with stem cell rescue for severe RA: Short term efficacy correlates with reduction in macroscopic and histological synovitis. Arthritis Rheum 2002;46:837-39
45. Roberts MM, To LB, Gillis D et al. Immune reconstitution following peripheral blood stem cell transplantation, autologous bone marrow transplantation and allogeneic bone marrow transplantation. Bone Marrow Tranplant 1993; 12:469.
46. Guillaume T, Rubinstein DB, Symann M. Immune reconstitution and immunotherapy after autologous hematopoietic stem cell transplantation. Blood 1998; 92:1471-90.
47. Hakim FT, Cepeda R, Kaimei S et al. Constraints on CD4 recovery post chemotherapy in adults: Thymic insufficiency and apoptotic decline of expanded peripheral CD4 cells. Blood 1997; 90:3789-98.
48. Koetz K, Bryl E, Spickschen K et al. T cell homeostasis in patients with rheumatoid arthritis. Proc Natl Acad Sci 2000; 97:9203-8.
49. Ponchel F, Morgan A, Quinn M et al. Thymic activity and T-cell differentiation in rheumatoid arthritis. Arthritis Rheum 2001 44; 9suppl:S301.
50. Isaacs JD, Manna VK, Rapson N et al. CAMPATH-1H in rheumatoid arthritis – an intravenous dose-ranging study. Br J Rheumatol 1996; 35:231-40.
51. Edwards JC, Cambridge G. Sustained improvement in rheumatoid arthritis following a protocol designed to deplete B lymphocytes. Rheumatology 2001; 40:205-11
52. Snowden JA. High dose therapy and autologous haemopoietic stem cell transplantation for rheumatoid arthritis-the feasibility of phase III trials. J Rheumatol 2001; 28 suppl 64:21-24.
53. Burt RK, Barr W, Oyama Y et al. Future strategies in hematopoietic stem cell transplantation for rheumatoid arthritis. J Rheumatol 2001; 28 Suppl 64:42-8.
54. Verburg RJ, Sont JK, Thea PM et al. High dose chemotherapy followed by autologous peripheral blood stem cell transplantation or conventional pharmacological treatment for refractory rheumatoid arthritis? A Markov decision analysis. J Rheumatol 2001;28:719-27.
55. Burt R, Omaya T, Barr W et al. Allogeneic non-myeloablative stem cell transplantation (NST) for refractory rheumatoid arthritis. Presented July 2003; EBMT, Istanbul.

Autologous Stem Cell Transplantation for Refractory Juvenile Idiopathic Arthritis (JIA)

Nico Wulffraat

Introduction

Although the overall prognosis for most children with chronic arthritis is good, a small proportion of children with systemic onset or polyarticular Juvenile Idiopathic Arthritis (JIA) are refractory to combinations of non-steroidal anti-inflammatory drugs (NSAIDS) and immunosuppressive drugs such as methotrexate (MTX), cyclosporin (CsA), and prednisone.[1-4] Such children often have severe joint destruction, growth retardation and adverse drug effects from long-term treatment with second-line anti-rheumatic drugs. In the evaluation of new treatments, one needs to balance a possible significant improvement of the quality of life with risk of severe side effects. Recently, the introduction of anti-Tumor Necrosis Factor receptor (TNF-r) treatment has had a major impact on outcome of children with polyarticular JIA who were unresponsive to methotrexate, with a persistent response of up to 80%.[5] Use of anti-TNF-r treatment in children with systemic disease has been discussed extensively at several Pediatric Rheumatology meetings and the general impression (reflecting experience in some 40 patients with systemic JIA, treated for more than 4 months) is that, in active systemic disease this treatment is less effective with a clear response in a minority of patients only.

Autologous hematopoietic stem cell transplantation (HSCT) has been described recently as a possible treatment for patients with severe autoimmune disease such as Systemic Scleroderma, Rheumatoid Arthritis (RA), Multiple Sclerosis (MS) and Systemic Lupus Erythematosus (SLE) refractory to conventional treatment.[6-9] The first 4 children with severe JIA treated with HSCT were published earlier.[10] We report here an extension of this study, which at present includes 18 children with JIA, treated in the Netherlands, and 13 children from other European pediatric centers, with a follow up of 5 to 51 months.

Definition of Chronic Childhood Arthritis

Childhood chronic arthritides may be referred to as JIA or juvenile rheumatoid arthritis (JRA) or juvenile chronic arthritis (JCA) due to three overlapping but different terminologies developed by three different committees. The classification system developed by the American College of Rheumatology (ACR) was termed JRA,[11] by the European League Against Rheumatism (EULAR) was termed JCA,[12] and by the International League of Associations for Rheumatology (ILAR) was termed JIA[13,14] (Table 1). For this review, the terminology JIA will be used. JIA has an incidence of 2-20 per 100,000 people and a prevalence of 16-150 per 100,000. There are approximately 32,000 children in the United States with JIA.[15] The peak age of onset is 1-3 years old. By definition, JIA must begin before 16 years of age but may persist into adulthood. For all patients mortality is approximately 1%.[16] However, mortality may be as high as 10-15% in high risk groups with systemic onset and severe functional status limitations.[17] Death is generally a result of infections from immune suppression, amyloid renal or heart failure, or myocardial infarction from corticosteroid related atherosclerosis.[16]

Common clinical manifestations are morning stiffness, joint pain or aches, and anorexia, weight loss and growth failure or deformity especially with systemic onset disease. Joints may show swelling, tenderness, erythema and loss of function. Although any joint may be affected, including the small joints of the hands, large joints (e.g., knees) are most commonly affected. Cervical or thoracic apophyseal joint arthritis may cause torticollis or scoliosis, respectively. Cicoarytenoid arthritis may cause acute airway obstruction. Less common manifestations include a chronic uveitis, which is more common in oligoarthritis, and manifests as ocular pain, reddness, photophobia, headache, and change or loss of vision. Rheumatoid nodules and vasculitis are more common in RF positive polyarthritis. Pericarditis, pericardial effusions, myocarditis, interstitial pulmonary fibrosis, hepatomegaly, splenomegaly, lymphadenopathy, and amyloidosis are more frequent in systemic onset disease (Table 2).

Patients Selection for HSCT

Eligibility criteria for HSCT are generally deemed to be failure of polyarticular or systemic JIA to respond to high dose methotrexate (MTX) (1mg/kg/wk, intramuscular), failure to respond or unacceptable toxicity to at least 2 disease modifying anti-rheumatic drugs (DMARDs), cyclosporine (CsA) (2.5 mg/kg/day) and anti-tumor necrosis factor alpha (anti-TNFα) therapy, and steroid dependency. Exclusion criteria are cardio-respiratory insufficiency, chronic active infection such as Epstein-Barr virus (EBV), cytomegalovirus (CMV), toxoplasmosis, spiking fever despite steroids, end stage disease or poor compliance.

Table 1. Classification of childhood arthritis

Classification Committee	ACR	EULAR	ILAR
Name of chronic childhood arthritis	Juvenile rheumatoid arthritis (JRA)	Juvenile chronic arthritis (JCA)	Juvenile idiopathic arthritis (JIA)

Subtypes	Systemic
	Pauciarticular
	Polyarticular
		Systemic
		Polyarticular JCA
		Juvenile rheumatoid arthritis
		Pauciarticular
		Juvenile ankylosing spondylitis
		Juvenile psoriatic arthritis
			Systemic
			Polyarticular RF negative
			Polyarticular RF positive
			Oligoarticular
			   Persistent
			   Extended
			Psoriatic arthritis
			Enthesitis-related arthritis

Age of onset of arthritis	≤16 years old	≤16 years old	≤16 years old
Duration of arthritis	≥6 weeks	≥3 months	≥6 weeks
Includes JAS, JPsA	no	yes	yes
Exclusion of other causes	yes	yes	yes

ACR= American College of Rheumatology; EULAR= European League Against Rheumatism; ILAR= International League of Associations for Rheumatology; JAS= juvenile ankylosing spondylitis; JPsA= juvenile psoriatic arthritis; RF= rheumatoid factor.

Outcome Measures

A growing number of functional outcome scales have become available for paediatric rheumatology research. Instruments to measure health status and function specifically developed for juvenile arthritis include the Childhood Arthritis Impact Measurement Scales (CHAIMS),[18] the Childhood Health Assessment Questionnaire (CHAQ),[19] the Juvenile Arthritis Functional Assessment Report (JAFAR),[20] the Childhood Arthritis Health Profile (CAHP),[21] the Juvenile Arthritis Quality of Life Questionnaire (JAQQ),[22] and the Juvenile Arthritis Self Report Index (JASI).[23] Of these, the instrument tested most widely is the CHAQ. For evaluation after HSCT, we used the core set of outcome variables for clinical trials in childhood arthritis as proposed by Giannini and the Pediatric Rheumatology International Trials Organisation group (PRINTO)[24-27] (Table 3). PRINTO rheumatological outcome measures include a physician's global assessment of disease activity, the child health assessment questionnaire, CHAQ (a parent/patient assessment of overall well-being, the functional ability and disease severity) (Table 4), the number of joints with active arthritis (Fuchs Swelling Index, FSI), the number of joints with limited range of motion, (EPM-ROM) (Table 5) and the erythrocyte sedimentation rate (ESR). Giannini proposed a 30% improvement from baseline of 3 out of 6 variables, with no more than one of the remaining variables worsening more than 30%.[24] The evolution of the disease was followed at 3 month intervals.

Hematopoietic Stem Cell Transplantation

In the Netherlands, we conducted a pilot study in 1997 and 1998, that was reported in the Lancet in 1999. Since then, 31

Table 2. Juvenile chronic arthritis signs and symptoms by subtype

	Polyarticular	Oligoarticular	Systemic Onset
Fever	++ (30%)	-	+++ (100%)
Rheumatoid nodules	+ (10%)	-	+ (5%)
Hepatospenomegaly	+ (10%)	-	+++ (85%)
Lymphadenopathy	+ (5%)	-	+++ (70%)
Chronic uveitis	+ (5%)	++ (20%)	+ (1%)
Pericarditis	+ (5%)	-	+ (35%)
Abdominal pain	+ (1%)	-	+ (10%)
Anemia	+	-	++
Leukocytosis	+	-	+++
Thrombocytosis	+	-	++
Anti-nuclear antibodies	+	++	-
Elevated heaptic enzymes	+	-	++

Table 3. Preliminary definition of improvement in juvenile arthritis

Core set of outcome variables:
1) physician global assessment of disease activity
2) parent/patient assessment of overall well being
3) functional ability
4) number of joints with active arthritis
5) number of joints with limited range of motion
6) erythrocyte sedimentation rate.

To establish a definition of improvement using this core set, at least 30% improvement from baseline in 3 of any 6 variables in the core set, with no more than 1 of the remaining variables worsening by >30%.

From—Giannini EH, Ruperto N, Ravelli A et al. Preliminary definition of improvement in juvenile arthritis. Arthritis Rheum 1997; 40(7):1202-9.

patients with a follow-up of 5 to 60 months (median 33 months) were transplanted in 8 different European pediatric centers. Although forty-one cases were identified in the registry of the European Blood and Marrow Transplantation Group (EBMT), of these, 31 were reported in detail that enabled evaluation of follow up and these 31 from multiple European centers are included in this manuscript.

Bone Marrow Harvest and T Cell Depletion

Unprimed bone marrow was harvested in 23 cases and peripheral stem cells in 8. In all patients, the graft was depleted of T lymphocytes with either 2 cycles of purging with CD2 and CD3 antibodies, or using positive stem cell selection by means of CD34 selection devices. These techniques yielded a final suspension with a CD34 positive stem cell count of at least 0.5×10^6 cells per Kg recipient weight with 1 to 5×10^5 CD3 cells per Kg recipient weight, which was stored in liquid nitrogen.[28] Initially, we used bone marrow, not mobilized peripheral blood stem cells, because mobilisation of peripheral blood progenitor cells with G-CSF has been associated with reactivation of rheumatoid arthritis (RA) in adults and development of leukocytoclastic vasculitis, but peripheral stem cell mobilisation using G-CSF performed in 6 JIA patients in other centers was uneventful. At present, it is not clear whether T cell depletion of the marrow is crucial to the process of the transplant, or that an intense but non-myeloablative regimen without stem cell support would be just as effective.

Conditioning for Autologous HSCT

The conditioning regimen included 4 days of Anti-Thymocyte Globulin (ATG, IMTIX, France) in a dosage of 5 mg per Kg recipient weight from day -9 to -6, Cyclophosphamide in a dose of 50 mg/kg/day from day -5 to -2; and low dose Total Body Irradiation (TBI, 4 gray, single fraction) on day -1. Ten children did not receive TBI as a part of their conditioning. On day 0, the frozen stem cell suspension was thawed and infused. Anti TNF-r therapy, MTX and CsA were stopped before autologous HSCT, prednisone was tapered over 2 months.

Patient Characteristics

The transplanted patients included 25 children with systemic JIA (sJIA) and 6 with polyarticular JIA (polyJIA), all with progressive disease activity for more than 5 years despite the use of NSAIDS, prednisone (both maintenance dose and pulses), cyclophosphamide pulses (750 mg/m^2), MTX up to 1 mg/kg/wk and CsA (2.5 mg/kg/day). The clinical characteristics in all children were a polyarticular course with erosions and osteoporosis, stunted growth and, in those with systemic onset JIA (sJIA), periods of spiking fever and exanthema. Most of them suffered from steroid related side effects. The mean time interval between diagnosis and transplant was 6 years (range 13–137 months). The clinical follow-up of these children ranges from 8 to 60 months (median 31 months).

Rheumatologic Outcome After HSCT

Rheumatological follow-up for up to 48 months showed a marked decrease in arthritis severity. Seventeen patients showed a drug free follow-up of 4 to 60 months with a marked decrease in the scores of the CHAQ, the physician's global assessment and joint swelling (Figs. 1 to 3). The measurement of the limitation of motion (EPM-ROM) largely reflects permanent erosive destruction to the joints and, as expected, is not subject to change after HSCT (Fig. 4). ESR, CRP and hemoglobin returned to near normal values within 6 weeks. In 2 of these patients, the ESR increased again after 3 months, with mild and transient synovitis of the hip and knee, following a varicella zoster virus (VZV) infection and tonsillitis. Most patients in remission show an impressive growth catch up (1 to 4 Standard Deviations) and gained profound general well being. Seven patients showed only a partial response, while 4 were resistant to this treatment and showed only a transient response (with ACR responses ranging from 0 to 30%) within a range of 4 to 12 months after HSCT. These children were again treated with low dose prednisone, MTX and in one case, Enbrel and pulsed cyclophosphamide (750mg/m^2/month) without success.

A relapse was noted in 7 children 18 months after HSCT. This relapse is so far mild with oligoarthritis and sporadic fever, that could be controlled easily with a 3-month course of low dose prednisone and NSAID. These patients showed nevertheless a 30% improvement of their disease, using the Giannini criteria for improvement of disease. Four children were resistant to HSCT and showed a persistent recurrence of the disease that was as severe as before.

During the 36 months of follow-up, the first patient showed a catch-up growth of 22cm (and a corresponding increase in shoe size), in contrast to the minimal gain of only 2 cm in the 3 preceding years. The second patient also showed a rapid drug free remission of the disease that persisted at 30 months follow-up. She is on a physical therapy program to improve muscle strength after years of immobilization and prednisone induced obesity. Since HSCT she has grown 18 cm in 30 months. For each age, a mean length and standard deviations have been described. A given length can thus be expressed as a Standard Deviation Score (SDS) of height for age. Prior to the onset of their disease, the children in this study had length between -0.2 and +2 SD of the mean length for their age. During the course of their disease, these children lost 3 to 5 SDS. After HSCT some, mostly younger children, show a catch up growth of 1-2 SDS, but the older children in our study, with the longest disease duration did not show catch up growth, but their SDS did not decrease any further.

In our study, HSCT induced a remission of disease in all children with severe and drug resistent JCA. Prolonged prednisone free growth catch up and general well being is a major therapeutic gain in such children. Since this approach was introduced only 4 years ago, the current experience includes only case reports of

Table 4. Childhood Health Assessment Questionnaire (CHAQ)

	Without Difficulty	With Some Difficulty	With Much Difficulty	Unable To Do	Not Applicable
Dressing and grooming					
Dress, including tying shoelaces and doing buttons					
Shampoo hair					
Remove socks					
Cut fingernails					
Arising					
Stand up from low chair or floor					
Get in or out of bed or stand up in crib					
Eating					
Cut own meat					
Lift cup or glass to mouth					
Open a cereal box					
Walking					
Walk outdoors on flat ground					
Climb up 5 steps					
Hygiene					
Wash and dry entire body					
Get in and out of tub					
Get on and off the toilet or potty chair					
Brush teeth					
Comb/brush hair					
Reach					
Reach and get a heavy object such as a game from above head					
Bend down and pick up clothing or paper from the floor					
Pull on a sweater over his/her head					
Turn neck to look back over shoulder					
Grip					
Write or scribble with pen or pencil					
Open car door					
Open previously opened jars					
Turn faucets on and off					
Turn a doorknob and push open a door					
Activities					
Run errands and shop					
Get in and out of car or school bus					
Ride bike or tricycle					
Do household chores e.g., make bed, vaccuum, wash dishes					
Run and play					

From— Singh G, Athreya BH, Fries JF et al. Measurement of health status in children with juvenile rheumatoid arthritis. Arthritis Rheum 1994; 37:1761-9.

selected patients. At the 28th annual EBMT meeting, early 2002, we summarized the European experience, including a total of 32 cases, performed by 9 centers. Overall, in about 50%, complete remission was observed even after prolonged withdrawal of anti-rheumatic drugs. This was also the case in the 10 children that did not receive total body irradiation (TBI) as part of their conditioning. Given the obvious concerns over the use of TBI, and since TBI did not induce higher response rates, it will probably be eliminated from future conditioning regimens. Erosive joint destruction that existed prior to HSCT cannot be cured by this procedure and remains a concern. The actual follow-up is however, too short to conclude that these children are completely cured from their disease.

Table 5. The Escola Paulista de Medicina range of motion (EPM-ROM) scale

3 [#]	2 [#]	1 [#]	0 [#]				0 [#]	1 [#]	2 [#]	3 [#]
+70	110-80	30-70	0-30	Right elbow	Extension	Left elbow	0-30	30-70		+70
-80		130-110	150-130		Flexion		150-130	130-110	110-80	-80
-30	55-30	70-55	90-70	Right wrist	Flexion	Left wrist	90-70	70-55	55-30	-30
-30	55-30	70-55	80-70		Extension		80-70	70-55	55-30	-30
-20		35-20	45-35	Right thumb	Abduction	Left thumb	45-35	35-20		-20
-30	50-30	70-50	90-70		Flexion IP		90-70	70-50	50-30	-30
-30	50-30	70-50	90-70	Right fingers (average)	Flexion MCP	Left fingers (average)	90-70	70-50	50-30	-30
-30	90-30	120-30	130-120	Right hip	Flexion	Left hip	130-120	120-90	90-30	-30
+30	10-25	5-10	0	Right knee	Extension	Left knee	0	5-10	10-25	+30
-10	25-10	35-25	45-35	Right ankle	Extension	Left ankle	45-35	35-25	25-10	-10

ROM= range of motion of joint. The score ranges from 0 (normal range of motion) to 3 (severe limitation). [#]The cut of degrees of motion for each joint to obtain that score (between 0 and 3).

From—Oliveira LM, Araujo PM, Atra E et al. EPM-ROM Scale: An evaluative instrument to be used in rheumatoid arthritis trials. Clin Exp Rheumatol 1990; 8(5):491-4.

Toxicity

All children developed chills, fever and malaise during infusion of ATG. Neutrophil recovery (>0.5 x 10^9/l) occurred at day +12 to +30, and the platelet count reached 20 x 10^9/l after 16 to 35 days post HSCT. Five to 9 months after HSCT, the numbers of circulating T cells were normal, with normal in vitro mitogenic responses at 6 to 18 months after HSCT. Due to the prolonged (CD4) lymphopenia lasting from 6 to 9 months, infectious complications were seen frequently. Epstein Barr virus (EBV) reactivation, one case of an atypical mycobacterial infection, a case of *Legionella pneumoniae*, 2 septicemia's and 12 cases of varicella Zoster virus (VZV) infection were seen. All infections were treated successfully. As reported earlier, there were 2 cases of transplantation related mortality due to a Macrophage Activation Syndrome (MAS) 17 days and 4 months after HSCT and a case of disseminated toxoplasmosis 12 days after autologous HSCT. This led us to study in more detail a possible role of infections and activated macrophages in the pathogenesis of JIA (see below).

Infectious Complications

During the aplastic period, blood cultures were positive for *S. epidermidis* in 2 children. They responded favorably to i.v. antibiotics. Seven patients developed a limited VZV eruption, 3 to 18 months after HSCT, which was treated by acyclovir. In addition, 1 case of localized atypical Mycobacterial infection and 1 case of *Legionella pneumoniae* were seen, that resolved completely. Two patients died of a Macrophage Activation Syndrome (also known as Infection associated hemophagocytic syndrome). The first case was induced by an EBV 4 months after HSCT. At the time of the EBV infection, her JIA was in remission. The other fatal MAS case (patient 11) occurred 18 days post transplant, while he was still in complete aplasia.[29] The occurrence of MAS in sJIA after HSCT may be caused by the T-cell depletion resulting in inadequate control of macrophage activation. A third fatal case resembling MAS occurred shortly after HSCT and was reported in Paris to be caused by a disseminated toxoplasmosis infection.[30] The occurrence of this MAS could be induced by a severe macrophage dysfunction or an imbalance between macrophages and regulatory T cells, as is postulated in Familial Haemophagocytic Lymphohistiocytosis (FHL).[31-32a] Accumulation of activated macrophages and CD8$^+$DR$^+$ lymphocytes characterize this disease.

Macrophage Activation Syndrome

MAS is a unique complication of HSCT for JIA and has not been reported following HSCT for other autoimmune diseases. MAS which occurred in 2 of our 18 children, is a very serious problem, and illustrates that this procedure in severely affected children currently carries a mortality risk between 5 and 12%. As mentioned, a third MAS-like case shortly after HSCT for JCA was reported by the Paris group.[30] The occurrence of MAS was induced by an infection (EBV, adenovirus, toxoplasmosis). It must be stressed here that MAS and Infection Associated Haemophagocytic Syndrome (IAHS) are to be regarded as synonyms. In the rheumatological literature this condition is well known as Macrophage activation syndrome, whereas (pediatric) hematologists/immunologists usually refer to this condition as (virus induced) hemophagocytic lymphohistiocytosis (HLH). It is important to avoid confusion and to use the same terminology. We feel it is important for the pediatric rheumatology community to use the same terminology as pediatric hematologists to avoid confusion, and we thus refer to it as MAS/HLH.[32a] There is evidence for a relation between MAS in sJIA independent of HSCT. By far the most cases of MAS occurring in JIA are seen in sJIA.[31] MAS and sJIA share symptoms such as the spiking fever.

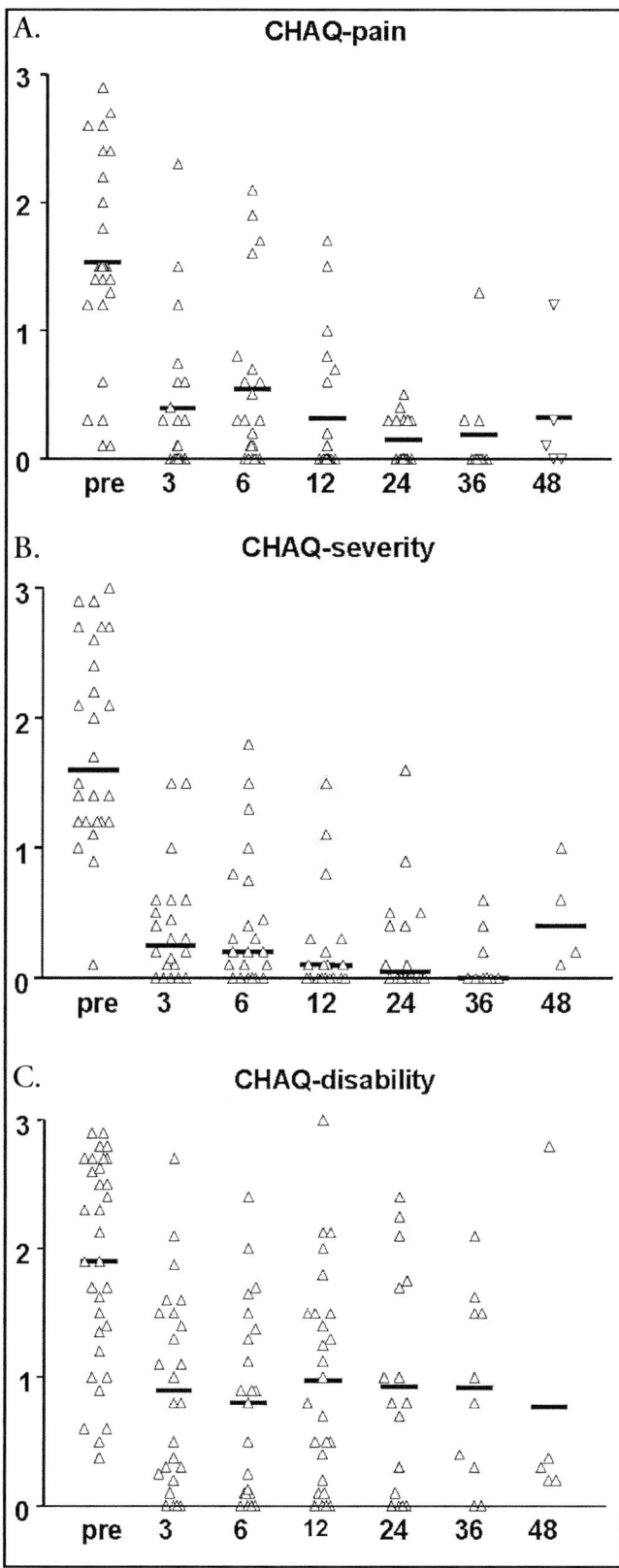

Figure 1. The mean parent/patient assessment of overall well-being (the child health assessment questionnaire (CHAQ) before and after HSCT. The CHAQ contains 3 domains: Pain (upper panel), Disability (middle panel) and Severity (lower panel). The score ranges between 0-3. Patients with a follow-up of at least 12 months are given only.

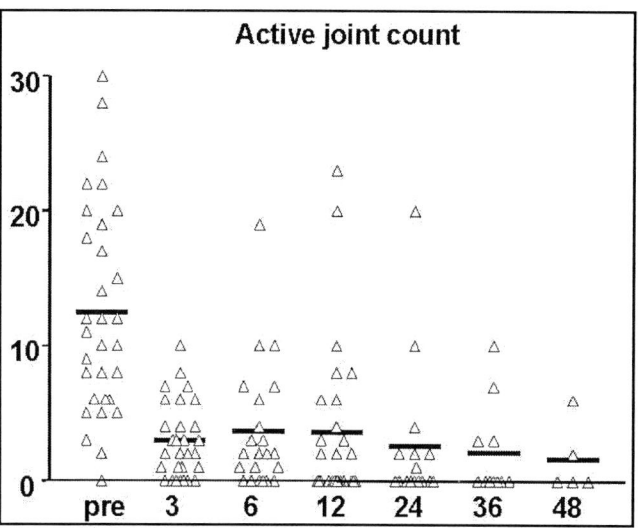

Figure 2. Number of joints with active arthritis before and after HSCT. The Fuchs Swelling Index (FSI) refers to 18 joints, with a cumulative score of 0-3 for each joint (scores ranging 0-54). Patients with a follow-up of at least 12 months are given only. The children that did not receive TBI as part of their conditioning show the same response to treatment.

The occurrence of MAS in sJIA after HSCT may be caused by a profound T-cell depletion (leaving no regulatory T-cells) in the graft that can result in activation of macrophages, leading to hemophagocytosis. It is advised that patients with active disease (fever), that cannot be controlled by steroids, should be excluded from HSCT. The immune suppression after HSCT must be tapered more slowly. In case of unexplained fever >39°Celsius for 48 hours, MAS must be considered and treatment with methylprednisolone 20 mg/kg/day (in 4 divided dosages) and CsA 2 mg/kg/day must be started immediately.

MAS can be considered as a reactive haemophagocytic lymphohistiocytic disorder. Other diseases or syndromes that are characterized by haemophagocytosis are Familial Haemophagocytic Lymphohistiocytosis (FHL), Griscelli Syndrome, Purtillo's syndrome and the Virus Associated Haemophagocytic Syndrome (VAHS).[33,33a] FHL is a rare, autosomal recessive disease with a rapid fatal outcome, which occurs in previously healthy infants or young children. The disease presents itself with fever, hepato-splenomegaly, pancytopenia, coagulation disorders, neurological abnormalities, and high serum

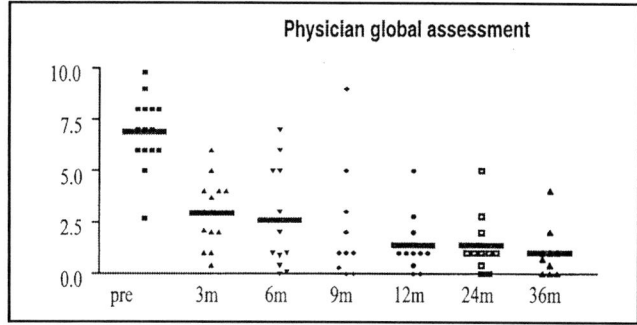

Figure 3. The mean score of the physician's global assessment of disease activity (represented as a visual analogue scale from 0-10 cm).

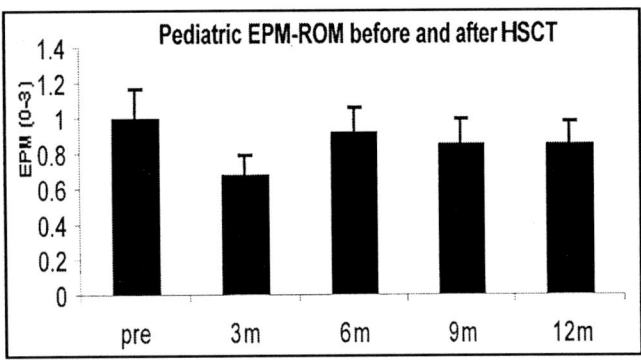

Figure 4. Limitation of joint movement (EPM-ROM) before and after HSCT.

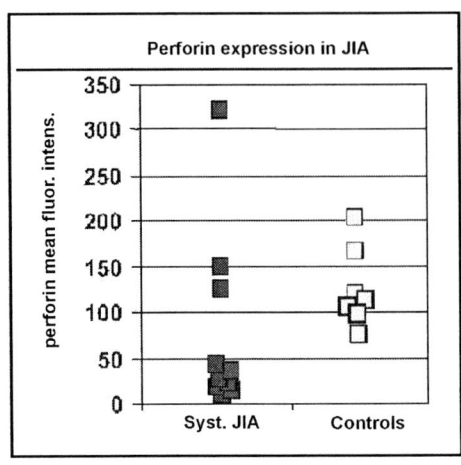

Figure 5. Mean fluorescence Index of perforin expression in CD8+ CD28-CD45RA+ cells from patients with systemic onset JIA (n= 18) as compared to CD8 positive cells from healthy controls (n= 7).

levels of ferritin, γ-interferon and Tumor Necrosis Factor-α (TNF-α). Accumulation of activated macrophages and lymphocytes characterize this disease. The lymphocytes are mainly positive for CD8 and DR. The hallmark of the diagnosis is the haemophagocytosis in the bone marrow, spleen, liver, lymph nodes or central nervous system. The function of the T-cells and NK-cells have often been reported to be defective. Antigens present on a target cell (such as an infected T cell) are presented to cytotoxic T cells that will lyse the target cell by a perforin dependent process. A defect in the negative control of T lymphocyte activation is hypothesised.[34] When these activated cytotoxic T cells are not down regulated, they will continue to activate macrophages, thus causing the characteristic clinical and laboratory features.

Sequencing of the coding region of the perforin gene, which is located on chromosome 10, showed various mutations in the perforin gene in patients with FHL.[35,36] Lymphocytes of these patients have a defective cytotoxic activity and perforin is not or hardly demonstrable on these lymphocytes. The absence of perforin in CD8+ T-cells and NK-cells explains the defective cytotoxic function of these cells. Because of the clinical similarity between FHL and systemic JIA, we investigated the role of perforin in systemic JIA. Perforin and granzyme expression levels in cytotoxic effector cells and in natural killer cells were determined by 3 or 4-color immunofluorescence. For FACS analysis of cytotoxic effector cells, cells were cell surface stained with PerCP conjugated CD8, CD28, and CD45RA. For natural killer cells, surface staining was performed with CD16 and CD56 antibodies. After fixation (formalhedyde) and permeabilisation, cells were stained with FITC conjugated monoclonal anti-human perforin antibody or granzyme A antibody. In CD8 positive cells obtained from 15 out of 18 patients with systemic onset JIA under conventional therapy, the expression of perforin was severely impaired (Fig. 5), while granzyme expression was normal.[37,38] Perforin is preferentially expressed in cytotoxic effector cells (CD8+CD28-CD45RA- and CD8+CD28-CD45RA+) and in NK cells, so in subsequent experiments perforin expression was determined in these phenotypically distinct subsets. Both CD8+CD45RA- and CD8+CD45RA+ cells of sJIA patients express significant lower (2 sided T test, p<0.05) levels of perforin as pJIA patients (MFI in sJIA: 34.6; in pJIA 87.6; in healthy control donors 120.7). Also NK cells from sJIA patients expressed significantly less (2 sided T test, p<0.01) intracellular perforin than healthy controls or patients with other forms of JIA.[38]

Although mean perforin expression levels were decreased in sJIA, considerable variation did exist between individual patients. We found no clear correlation between the level of perforin expression and disease duration or medication use, since 3 newly diagnosed patients that were not yet treated with corticosteroids or methotrexate also showed decreased perforin expression. The number of patients studied, however, is too low to conclude on an association of disease activity and perforin expression. In 4 patients with sJIA who were treated with HSCT, perforin expression was analyzed before and 12 months after transplantation.[25] In all 4 patients, a clear increase in perforin expression was found in both subsets of the cytotoxic effector cells (Figs. 6 and 7). If indeed low perforin expression does induce hemophagocytosis in FHL, our finding may explain why sJIA can be complicated by macrophage activation syndrome and hemophagocytosis. Interestingly, we were able to monitor perforin expression in a patient before HSCT, during remission shortly after HSCT and during relapse. Here the decreased expression found before HSCT normalized within 3 months after HSCT, but decreased again at the same time as the clinical symptom recurred. A decrease of perforin expression was not found to precede the clinical relapse, but this may be due to the increased blood sampling interval during remission. MAS can be considered as a reactive haemophagocytic disorder that resembles in some aspects the Familial Haemophagocytic Lymphohistiocytosis (FHL), a disease caused by a mutation in the perforin gene. We found that CD8+ lymphocytes and NK cells of patients with systemic JIA have a decreased expression of perforin that could be up regulated after HSCT when the disease is in remission.

Myeloid Related Protein

Additional immunological monitoring was performed by testing for serum myeloid related protein (MRP) levels before and at regular intervals after HSCT. MRP is a protein of the S100 family related with neutrophil and monocyte activation. Myeloid Related Protein 8 (MRP8) and MRP14 are two S100 proteins specific to myeloid cells and expressed in neutrophils and macrophages. MRP8 an MRP14 play a distinct role in neutrophil and monocyte activation.[39] They are specifically released during the interaction of monocytes with inflammatory activated endothelium, probably at sites of local inflammation.[40] Elevated MRP appear specific for sJIA when compared to normals and other autoimmune diseases (Fig. 8). In 12 JIA patients (9 with systemic JIA and 3 with polyarticular JIA), serial determination of MRP serum concentrations before and after HSCT was per-

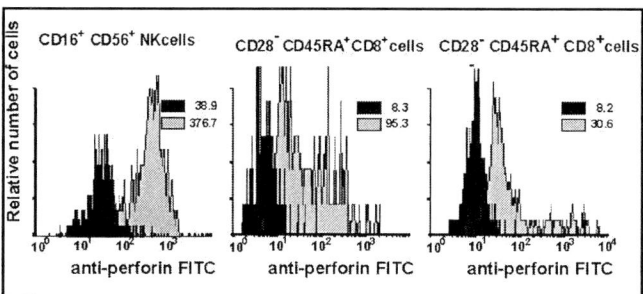

Figure 6. FACS analysis of perforin expression in cytotoxic effector cells (CD8⁺CD28⁻CD45RA⁻ and CD8⁺CD28⁻CD45RA⁺) and in NK cells of a patient with systemic JIA before (black histogram) and 1 year after HSCT (gray histogram). Cytotoxic effector cells obtained before HSCT express significantly lower cells of perforin, that return to normal values after HSCT. Highest expression is normally found in NK cells and in CD8⁺CD28⁻CD45RA⁺ subsets.

formed. MRP8/MRP14 serum concentrations were determined by sandwich ELISA system as described previously.[39] For calibration, different amounts (0.25-250 ng/ml) of the native complex of human MRP8 and MRP14 were used, which were isolated from human granulocytes as described previously. The assay has a sensitivity of <0.5 ng/ml and a linear range between 1 and 30 ng/ml. MRP8 and MRP14 form noncovalently associated complexes in the presence of extracellular calcium concentrations, which are detected by the sandwich ELISA system. Therefore, the ELISA is calibrated with the native MRP8/MRP14 complex and the data are expressed as ng/ml MRP8/MRP14.

After HSCT, MRP8/MRP14 serum concentrations in JIA showed a positive correlation with the CHAQ (r= 0.80) and ESR (r= 0.45), but not with the total leukocyte count (r= 0.26). Mean MRP8/MRP14 serum concentrations dropped dramatically in the first three months post HSCT (p= 0.0039) together with a marked improvement of the clinical parameters of disease activity such as CHAQ (p= 0.0039). Mean MRP8/MRP14 concentration dropped from 33029 ng/ml (248600-660 ng/ml) to 5820 ng/ml (48200-100 ng/ml) in the first 3 months post HSCT (p= 0.0039) (Fig. 9). In 6 out of 12 patients tested, MRP8/MRP14 serum concentrations reached normal values within the first 3 months. Parallel to the decrease in MRP8/MRP14 serum concentration, there is a decrease in parameters of clinical disease activity. During transient relapse, there is an increase in MRP8/MRP14 (patients 1 and 3 in Fig. 9). This is further illustrated by serial determinations of MRP8/MRP14 in a patient with a persistent relapse (a less than 30% improvement) after 7 months (patient 12 in Fig. 9; the same patient as shown in Fig. 6). He showed an initial decrease in MRP8/MRP14 concentration from 25000 ng/ml before HSCT to 3730 ng/ml 5 months after HSCT (corresponding with a 50% improvement). Then his MRP8/MRP14 concentration increased concurrent with a clinical relapse to 23000 ng/ml at 8 to 11 months follow up, comparable to his pre transplant MRP8/MRP14 values.

It appears from these preliminary experiments that MRP8/MRP14 serum concentration may be used as a marker for disease activity in patients who received an HSCT for refractory JIA. The occurrence of the complication of Macrophage Activation Syndrome during lymphopenia after HSCT may indicate a role of macrophage activation or defective lymphocyte-macrophage interaction in the pathogenesis of JIA. We were able to test MRP8/MRP14 values in 3 patients with MAS shortly after HSCT. The

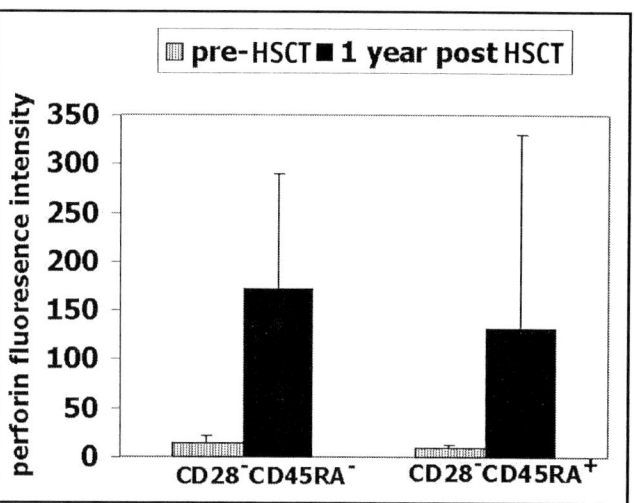

Figure 7. Autologous Stem Cell Transplantation restores perforin expression in sJIA. Shown here are the mean values of perforin expression (MFI) of 4 patients in CD8⁺CD28⁻CD45RA⁻ and CD8⁺CD28⁻CD45RA⁺ lymphocytes before HSCT (striped) and 1 year after HSCT (black) when clinical symptoms of systemic JIA were absent.

occurrence of MAS in these patients was, however, not preceded by significant changes in MRP8/MRP14 concentration.[41] MRP has, however, proved to be useful as a marker for disease activity, and as an early indicator for relapse of Juvenile Idiopathic Arthritis. After HSCT, MRP serum concentrations in JIA showed a

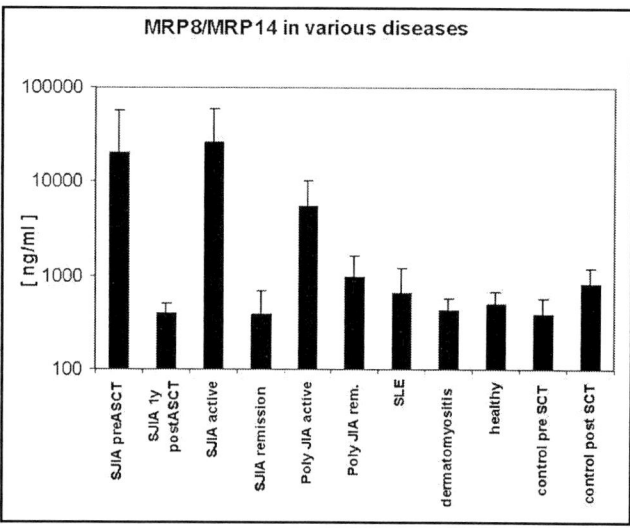

Figure 8. Myeloid related protein (MRP)8/MRP14 serum concentrations in different auto-immune diseases. Serum concentrations of MRP8 and MRP14 were determined by ELISA. Data are presented as mean ± SEM. There is a significant decrease in MRP8/MRP14 concentration after transplantation in the systemic JIA group. The control group of patients with active systemic JIA has comparable values of the SJIA patients before HSCT. The MRP values of SJIA in remission with conventional therapy are the same as the 1 year post HSCT samples. MRP8/MRP14 concentrations in SLE (n= 9) and dermatomyositis (3 active, 4 in remission) do not differ significantly from healthy controls (n= 30). Please notice the logarithmic scale of the y-axis.

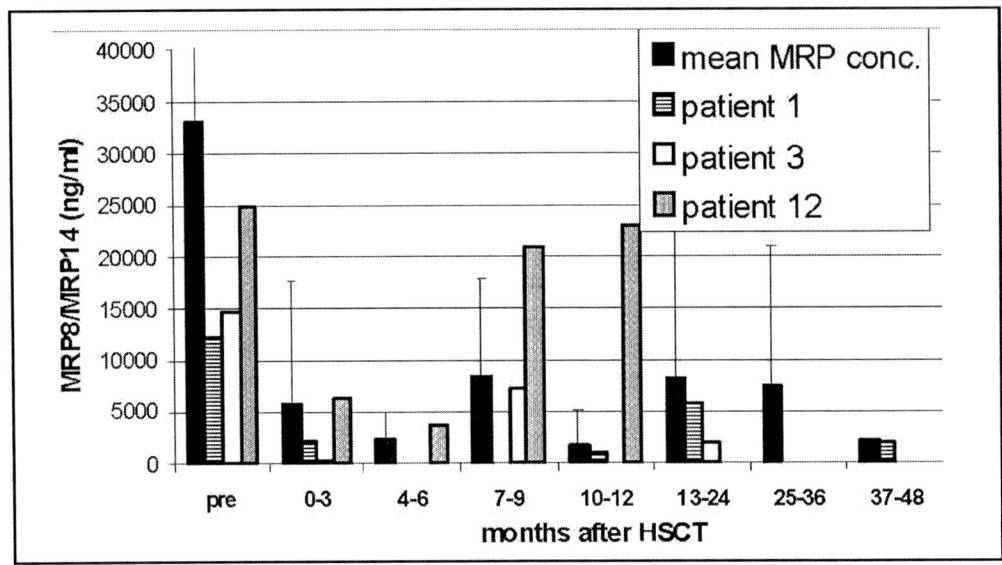

Figure 9. Myeloid related protein (MRP)8/MRP14 serum-concentration decreases after stem cell transplantation. Mean MRP8/MRP14 serum concentrations divided over different intervals prior and after HSCT are shown as solid black bars in a 48 month follow up. MRP8/MRP14 concentrations decrease dramatically after HSCT and show a direct correlation with disease activity represented by mean child health assessment questionnaire (CHAQ). Individual values MRP from 3 patients with transient or persistent relapses are shown as striped bars (patient 1), open bars (patient 3) and shaded bars (patient 12).

positive correlation with the CHAQ (r= 0.80)and ESR (r= 0.45). During transient relapse there is an increase in MRP.

Summary

In conclusion, HSCT in this severely ill patient group induces a very significant and drug free remission of the disease in the majority of patients, but carries a significant risk of developing fatal MAS. Factors that may predispose an individual to MAS, such as viral infections, must be identified and less profound T-cell depletion, control of systemic disease prior to transplant and a slow tapering of steroids after HSCT is advised. One of the most difficult aspects is to carefully weigh the risks of the prolonged immunosuppression of "conventional" treatment against those of the short but intense immunosuppression of HSCT. This treatment related mortality illustrates that the selection of patients with this non-lethal disorder must be restricted to those with severe, refractory disease before the stage of severe permanent joint destruction.

Acknowledgements

Patient data were obtained from 8 pediatric university centers in Europe. I wish to thank the colleagues from Utrecht, Leiden (the Netherlands), Bruxelles (Belgium), Halle, Jena (Germany), Paris (France), London and Newcastle upon Tyne (United Kingdom) for their kind cooperation. The clinical studies in the Netherlands were supported in part by a grant from the Dutch League against Rheumatism. Serum analysis for MRP8/14 was performed by Dr. J Roth and M Frosch, University of Münster, Germany.

References

1. Cassidy JE, Petty RE. Juvenile rheumatoid arthritis. In: Cassidy JE, Petty RE, eds. Textbook of Paediatric Rheumatology. 3rd ed. Philadelphia: WB Saunders Company, 1995:133-223.
2. Wallace CA, Levinson JE. Juvenile rheumatoid arthritis: Outcome and treatment for the 1990s. Rheum Dis Clin North Am 1991; 17:891-905.
3. Gäre BA, Fasth A. The natural history of juvenile chronic arthritis: A population based cohort study. I. Onset and disease process. J Rheumatol 1995; 22:295-307.
4. Gäre BA, Fasth A. The natural history of juvenile chronic arthritis: A population based cohort study. II. Outcome. J Rheumatol 1995; 22:308-19.
5. Lovell DJ, Giannini EH, Reiff PHA et al. Etanercept in children with polyarticular juvenile rheumatoid arthritis. The Pediatric Rheumatology Collaborative Study Group. N Eng J Med 2000; 342:763-69.
6. Burt K, Slavin S, Burns WH et al. Induction of tolerance in autoimmune diseases by hematopoietic stem cell transplantation: Getting close to a cure? Blood 2002; 99:768-84.
7. Joske DJL. Autologous bone marrow transplantation for rheumatoid arthritis. Lancet 1997; 350:337-8.
8. Tyndall A, Black C, Finke J et al. Treatment of systemic sclerosis with autologous haemopoietic stem cell transplantation. Lancet 1997; 349:254
9. Traynor A, Schroeder J, Rosa R et al. Treatment of severe lupus erythematosus with high-dose intense chemotherapy and hematopoietic stem cell transplantation: A phase I study. Lancet 2000; 356:701-7.
10. Wulffraat NM, van Royen A, Bierings M. Autologous haemopoietic stem cell transplantation in four patients with refractory juvenile chronic arthritis. Lancet 1999; 353:550-3
11. Brewer EJ Jr, Bass J, Baum J et al. Current proposed revision of JRA Criteria. JRA Criteria Subcommittee of the Diagnostic and Therapeutic Criteria Committee of the American Rheumatism Section of The Arthritis Foundation. Arthritis Rheum 1977; 20(2):195-9.
12. European League Against Rheumatism: EULAR bulletin N04: Nomenclature and Classification of Arthritis in Children. National Zeitung AG, 1977.
13. Fink CW. Proposal for the development of classification criteria for idiopathic arthritides of childhood. J Rheumatol 1995; 22(8):1566-9.
14. Petty RE, Southwood TR, Baum J et al. Revision of the proposed classification criteria for juvenile idiopathic arthritis: Durban, 1997. J Rheumatol 1998; 25(10):1991-4.
15. Gewanter HL, Roghmann KJ, Baum J. The prevalence of juvenile arthritis. Arthritis Rheum 1983; 26(5):599-603.
16. Baum J, Gutowska G. Death in juvenile rheumatoid arthritis. Arthritis Rheum Suppl 1977; 20:253.

17. Ansell BM, Wood PHN. Prognosis in juvenile chronic arthritis. Clin Rheum Dis 1976; 2:397.
18. Coulton CJ, Zborowsky E, Lipton J et al. Assessment of the reliability and validity of the arthritis impact measurement scales for children with juvenile arthritis. Arthritis Rheum 1987; 30(7):819-24.
19. Singh G, Athreya BH, Fries JF et al. Measurement of health status in children with juvenile rheumatoid arthritis. Arthritis Rheum 1994; 37(12):1761-9.
20. Lovell DJ. Newer functional outcome measurements in juvenile rheumatoid arthritis: A progress report. J Rheumatol 1992; 33(suppl):28-31.
21. Tucker LB, De Nardo BA, Abetz LN et al. The Childhood Arthritis Health profile (CAHP): Validity and reliability of the conditioning specific scales. Arthritis and Rheum 1995; 38:S183.
22. Duffy CM, Arsenault L, Duffy KN et al. The Juvenile Arthritis Quality of Life Questionnaire—development of a new responsive index for juvenile rheumatoid arthritis and juvenile spondyloarthritides. J Rheumatol 1997; 24(4):738-46.
23. Wright FV, Law M, Crombie V et al. Development of a self-report functional status index for juvenile rheumatoid arthritis. J Rheumatol 1994; 21(3):536-44.
24. Giannini EH, Ruperto N, Ravelli A et al. Preliminary definition of improvement in juvenile arthritis. Arthritis Rheum 1997; 40:1202-9.
25. Ruperto N, Ravelli A, Falcini F et al. Performance of the preliminary definition of improvement in juvenile chronic arthritis patients treated with methotrexate. Ann Rheum Dis 1998, 57:38-41.
26. Fuchs HA, Pincus T. Reduced joint counts in controlled clinical trials in rheumatoid arthritis. Arthritis Rheum 1994; 37:470-5.
27. Len C, Ferraz MB, Goldenberg J et al. Paediatric Escola Paulista de Medicina Range of Motion Scale: A reduced joint count scale for general use in juvenile rheumatoid arthritis. J Rheumatol 1999, 26:909-13.
28. Slaper-Cortenbach ICM, Wijngaarden-duBois MJGJ, de Vries-van Rossen A et al. The depletion of T cells from haematopoietic stem cell transplants. Rheumatology 1999, 38: 751-54.
29. Vlieger AM, Brinkman D, Quartier P et al. Infection Associated Macrophage Activating Syndrome in 3 patients receiving Autologous Stem Cell Transplantation (ASCT) For Refractory JCA. Proceedings of the 26th annual European Blood and Marrow Transfusion (EBMT) meeting 2000, Innsbruck. Bone Marrow Transplant; 25(suppl1, abstr).
30. Quartier P, Prieur AM, Fischer A. Disseminated toxoplasmosis following autologous bone marrow transplantation for systemic onset juvenile chronic arthritis. Lancet 1999; 353:1885-86.
31. Mouy R, Stephan JL, Pillet P et al. Efficacy of cyclopsorin A in the treatment of macrophage activation syndrome in juvenile arthritis: Report of five cases. J Pediatr 1996; 129:750-54.
32. Wulffraat NM. Autologous stem cell transplantation (ASCT) in refractory polyarticular and systemic JIA. Proceedings of the European League Against Rheumatism (EULAR), Glasgow. Ann Rheum Dis 1999 (abstr).
32a. Ramanan AV, Baildam EM. Macrophage activation syndrome is hemphagocytic lymphohistiocytosis—need for the right terminology. J Rheumatol 2002; 29(5):1105.
33. Goransdotter EK, Fadeel B, Nilsson-Ardnor S et al. Spectrum of perforin gene mutations in familial hemophagocytic lymphohistiocytosis. Am J Hum Genet 2001; 68:590-7.
33a. Arico M, Danesino C, Pende D et al. Pathogenesis of haemophagocytic lymphohistiocytosis. Br J Haematol 2001; 114(4):761-9.
34. Stepp SE, Mathew PA, Bennett M et al. Perforin: More than just an effector molecule. Immunol Today 2000; 21:254-5.
35. Stepp SE, Dufourcq-Lagelouse R, Le Deist F et al. Perforin gene defects in familial hemophagocytic lymphohistiocytosis. Science 1999; 286:1957-59.
36. Ohadi M, Lalloz MR, Sham P et al. Localization of a gene for familial hemophagocytic lymphohistiocytosis at chromosome 9q21.3-22 by homozygosity mapping. Am J Hum Genet 1999; 64:165-71.
37. Normand NJ, Elkon KB, Karen B. Onel. lower expression of perforin in cd8+ t cells of patients with systemic onset juvenile rheumatoid arthritis. Abstr in Arthritis and Rheum vol 43 (no 9, supplement).
38. Wulffraat NM, Rijkers GT, Brooimans R et al. Reduced perforin expression in systemic juvenile idiopathic arthritis is restored by autologous stem cell transplantation (Auto-SCT). Rheumatology 2003; 42:375-79.
39. Rammes A, Roth J, Goebeler M et al. Myeloid-related protein (MRP) 8 and MRP14, calcium-binding proteins of the S100 family, are secreted by activated monocytes via a novel, tubulin-dependent pathway. J Biol Chem 1997; 272(14):9496-9502.
40. Frosch M, Strey A, Vogl T et al. Myeloid-related proteins 8 and 14 are specifically secreted during interaction of phagocytes and activated endothelium and are useful markers for monitoring disease activity in pauciarticular-onset juvenile rheumatoid arthritis. Arthritis Rheum 2000; 43(3):628-37.
41. Wulffraat NM, Haas PJ, Frosch M et al. Increased myeloid related protein 8 and 14 secretion reflects phagocyte activation and correlates with disease activity in juvenile idiopathic arthritis treated with autologous stem-cell transplantation. Ann Rheum Dis 2003; 62(3)236-241.

CHAPTER 45

Immunology of Scleroderma

Carol M. Artlett

Introduction

Systemic sclerosis, or scleroderma (SSc), is an autoimmune disease, which manifests clinically by progressive cutaneous and visceral fibrosis. SSc is a complex, heterogeneous disease with clinical forms ranging from limited skin involvement (lcSSc) to diffuse disease (dcSSc) causing extensive cutaneous sclerosis and severe internal organ alterations. Age, race and genetic factors have all been found to influence the development and course of SSc,[1] but the strongest genetic factor is gender. The incidence of SSc in females is in great excess compared to its incidence in males, ranging from 3:1 to 8:1.[1,2] The etiology of SSc remains unknown but its occurrence has been associated with numerous risk factors including various environmental agents, genetic influences, viruses and more recently with microchimerism.[3,4] Environmental agents associated with SSc or closely related syndromes include organic solvents such as vinyl chloride and trichloroethylene, adulterated rapeseed oil, certain tryptophan products, bleomycin and silica.[5,6] Genetic factors have also been associated with the onset of SSc.[5] Vinyl chloride-induced disease has an HLA-DR5 association in those individuals who progress to SSc[7] and specific HLA alleles have been associated with autoantibody subgroups expressed in SSc patients.[8] Viruses have also been implicated in the pathogenesis of SSc, and recent interest has been focused on retroviruses and cytomegalovirus.[10,11] Sequence homologies have been identified between several retroviruses and topoisomerase I, the target antigen of the SSc-specific scleroderma-70 (Scl-70) autoantibody[12] and sera from SSc patients containing anti- Scl-70 autoantibodies were able to recognize the retrovirus responsible for equine infectious anemia.[13] In spite of the initiating exposure in SSc, cellular and humoral immunological changes are early events and precede vasculopathy and fibrosis. Although aggressive treatments may suppress SSc disease activity in some cases, there are few, if any, complete cures. Since these conditions arise as a direct result of dysregulation of the immune system, modification of immune stem cells may be important in the control of SSc.

Cellular Immunology in Systemic Sclerosis

Immune activation is an early event in SSc, however, it is not known whether it is the initiating event in SSc or is a secondary disease process. T cells are important in the development of tissue damage and dominate the inflammatory infiltrates in the lesions early in disease. T cells regulate many of the functions of fibroblasts and endothelial cells by the production of soluble mediators or by cytotoxic effects, and provide specificity to the immune response and show evidence of selection. There is accumulated evidence on the SSc immune response to suggest that it is Th2 mediated.

T Cells in Peripheral Blood

The T cell subsets in the blood have been investigated extensively and the results are frequently found to be contradictory. These contradictions may result from differences in T cell patterns in early and late stages of disease and/or with the clinical subsets. The clinical expression of SSc is heterogeneous and has different stages of expression and activation during the disease course. Variations between peripheral blood T cell studies from patients with SSc have been outlined in Table 1.

A higher CD4:CD8 ratio was found to be associated with a shorter disease duration and more extensive skin involvement.[24] In vivo T cell responses to exogenous antigens appear to be normal in SSc, and patients have normal responses to immunization, cutaneous delayed-type hypersensitivity, and antibody production.[26]

T Cells in Lesions

In early SSc lesions, lymphocytes infiltrate the skin and are scattered throughout the subcutaneous tissue and dermis, or are localized around small vessels, nerves and skin appendages, whereas later in the course of disease, fibroblasts and collagen are more prominent.[27-29] Immunohistochemical study of skin biopsies from patients with SSc of recent onset have identified increased numbers of perivascular CD3⁺ T lymphocytes, which are mostly CD4⁺ helper cells expressing CD45RO⁺ memory phenotype and HLA class II. The T cell infiltrate precedes the findings of small vessel vasculopathy and alterations of interstitial tissues including fibroblast activation and proliferation.[30] The degree of cellular infiltration correlates with the extent and progression of skin thickening. Fibroblasts displaying increased production of various collagens are located in close proximity to lymphocytes.[31] In vitro studies have shown that extracellular matrix components includ-

Table 1. Peripheral blood studies in patients with SSc

Cell Subset	Alteration: Reference
Absolute lymphocyte counts	Normal: 14 Decreased: 15, 16
Activated T cells	Increased: 14, 15, 17-19 Decreased: 20, 21
Percentage of CD4+ T cells	Increased: 22 Normal: 17, 20, 21, 23 Decreased: 23
Ratio of CD4:CD8 cells	Increased: 14, 15, 17, 19-22, 24 Normal: 23, 24
Absolute numbers and percentages of CD8+ cells	Increased: 23 Normal: 25 Decreased: 14

ing collagens type I, III, and VI, fibronectin, decorin and glycosaminoglycans are increased in SSc lesions.[32-36]

Patients with active lung disease have increased numbers and percentages of T cells and macrophages in the interstitium[37-39] and in bronchoalveolar (BAL) fluids.[40-42] In contrast to the T cell repertoire in the skin, patients with alveolitis have more CD8+ T cells than CD4+ T cells in the BAL fluids.

T Cell Receptor (TCR) Gene Expression

Evidence of antigen driven TCR cell expansion in patients with SSc was determined by the analysis of the TCR gene families. White et al showed an increase in the expression of Vdelta 1 (Vδ1) gene on CD3+ T cells and gamma/delta (γ/δ) T cells in lungs.[43] A restricted diversity of the TCR junctional regions were identified in Vδ1+ γ/δ T cells isolated from BAL and peripheral blood from patients with SSc compared to controls.[44] Absolute numbers and percentages of Vδ1+ T cells were found to be increased in peripheral blood and in skin biopsies.[25] Furthermore, these cells were activated and expressed HLA-DR and very late activation antigen (alpha) (CD49d).[25] In studies from the active lesions from patients with SSc with less than 18 months disease duration, Sakkas et al found that the Vbeta 13 (Vβ13), Vβ14 and Vβ21 TCR gene segments were more frequent.[45] These results suggests expansion, possibly by antigen-driven selection, of the T cells in SSc patients.

Soluble Mediators of the Immune System

The functions of fibroblasts and vascular cells are affected by the soluble mediators, which are secreted by cells of the immune system. Patients with SSc have increased levels of IL-2, IL-4, IL-6, IL-8, transforming growth factor (TGF)-β, tumor necrosis factor (TNF)-α and platelet derived growth factor in the sera and tissues. These mediators have been implicated in the pathogenesis of SSc and not only affect fibroblasts, endothelial cells and vascular smooth muscle cells, but may also in turn be produced by some of these cells.

Along with the low numbers of CD8+ T cells, patients with SSc also have low numbers of natural killer (NK) cells and natural killer cell activity is reduced.[46] These patients also have a reduced lymphokine-activate killer (LAK) cell function as determined by [51]CR release[20,22,47-49] Furthermore, patients with SSc have increased soluble levels of the activation marker CD30, which is associated with Th2-type immune responses,[50,51] and these levels correlate with skin score and erythrocyte sedimentation rate. Immunohistochemical analysis has shown CD30 expression by large numbers of CD4+ T cells in the active lesions from patients with SSc, which produce IL-4.[51] In addition, a large proportion of these CD4+ T cells express the Vδ1+ subset and are activated (expressed HLA-DR). The majority of these cells can be found in the perivascular areas.[25]

The skin of patients with SSc shows evidence of an active immune response associated with collagen overproduction. There is marked fibrosis with tightly packed collagen fibers and scattered infiltrates of mononuclear cells clustered around vessels, sweat glands and nerves[27,52] with a coordinate increase in expression of Type I and Type III collagen genes in SSc fibroblasts.[34] Furthermore, not all fibroblasts over produce collagen, and there is a distinct fibroblast subset, which produce the high levels of collagen.[53]

Interleukin-1 (IL-1)

IL-1 was originally thought to be produced by the monocyte/macrophage lineage of cells only, however, it has been found to be produced by a variety of other cells including keratinocytes, endothelial cells, and mesangial cells. It is involved in the immune regulation of fibroblast proliferation and collagen deposition.[54-56] In SSc, the production of IL-1α and IL-1β is expressed on fibroblasts[57] and has been found to be either increased in sera[58,59] or normal in sera from SSc patients.[48,60] IL-1 levels correlate with disease duration of SSc, and levels are increased in patients with less than 5 years disease duration compared to patients who have a longer disease course.[61] In vitro studies have shown that SSc fibroblasts have a higher density of IL-1 receptor (R) on the cell surface with a corresponding increase in IL-1R mRNA.[62] Furthermore, dermal fibroblasts were found to express higher levels of the intracellular isoform of IL-1R antagonist compared to normal fibroblasts.[63]

Interleukin-2 (IL-2)

IL-2 is involved in antigen specific-proliferation of T cells and acts on other cells of the immune system including thymocytes, B-lymphocytes, macrophages, NK cells and lymphokine activation/killer cells. Spontaneous production of IL-1 has been reported in some SSc patients,[61] which enhances the production of IL-2 by CD4+ T cells.[64,65] Furthermore, patients with SSc have increased serum levels of IL-2[66,67] and there is a correlation between serum soluble IL-2 levels and the extent of skin involvement[68] and mortality.[67] Serum levels of soluble IL-2R were found to be elevated in patients with SSc,[66] and also correlated with disease activity and extent of skin thickening.[60,69] Lymphocytes from SSc patients were found to have an enhanced expression of the high-affinity IL-2R.[70] The proportion of IL-2R expressing lymphocytes is elevated in patients with SSc.[14] Clearly, IL-2 plays an important role in the progression of SSc. In a patient with SSc treated for renal cell carcinoma with IL-2, exacerbation of SSc and increased activation of the immune system occurred.[71]

Interleukin-4 (IL-4)

IL-4 is a Th2 cytokine that is expressed by T cells, mast cells[72] and fibroblasts.[73] IL-4 has also been found to upregulate the expression of tenascin, a large extracellular matrix molecule found in wound healing, in patients with SSc and healthy skin fibro-

Table 2. HLA-DRB1 risk factors and SSc Disease Subsets

DRB1* Allele	Disease Association	Geography	Reference
DRB1*01	DSSc	USA	109, 110
	SSc	USA	111
	LSSc	USA	112
	LSSc	Europe	113, 114
	SSc	Europe	115
	LSSc	Japan	116
DRB1*15	SSc	Japan	117, 118
	DSSc	Japan	119
	LSSc	Japan	120
DRB1*17	LSSc	USA	111, 121
	SSc	Europe	122, 123
	DSSc	Europe	113
	SSc (male)	Europe	114
DRB1*11	SSc	Canada	124
	SSc	USA	111
	DSSc	USA	112
	DSSc (male)	USA	125
	SSc	Mexico	126
	LSSc	Europe	113
	SSc	Europe	114, 115
	DSSc	Europe	127
	DSSc (male)	Europe	125
	DSSc	Australia	128
	SSc	Australia	129
DRB1*08	SSc	Japan	117

DSSc= Diffuse systemic sclerosis; LSSc= Limited skin involved systemic sclerosis

blasts[74] and stimulates alpha 2(I) collagen synthesis by fibroblasts.[72,73,75] CD8+ T cells isolated from BAL fluid from patients with SSc who have pulmonary fibrosis, express increased levels of IL-4.[76] Furthermore, an alternative spliced form of IL-4 (IL-4δ2) transcript was found to be increased in mononuclear cells from patients with SSc.[77] IL-4 levels significantly correlate with skin involvement and are inversely associated with disease duration.[78] Other studies have not confirmed the increased levels of IL-4 in patients with SSc.[66]

Interleukin-6 (IL-6)

IL-6 is a pleiotropic inflammatory cytokine involved in B cell activation[79] and plays a role in acute phase response of T cells in tissue trauma.[80] IL-1 and TNF-α can induce IL-6 production from a variety of cells, including fibroblasts,[81,82] and can act synergistically with fibroblasts to elicit this response.[83,84] IL-6 is involved in the inflammation process of other rheumatic diseases[85] and was found to be elevated in the sera from patients with SSc.[48,86] In vitro studies have shown that fibroblasts from SSc lesions have elevated IL-6 mRNA compared to normal fibroblasts.[87] However, IL-6 expression was found to be lower on endothelial cells and fibroblasts in early disease of less than 1 year compared to later disease of greater than 1 year.[86] Mononuclear cells isolated from patients with SSc and stimulated by collagen were found to have an augmented secretion of IL-6.[88]

Interleukin-8 (IL-8)

IL-8 is an angiogenic cytokine[89] and is chemotactic for neutrophils, T lymphocytes and endothelial cells.[90] An upregulation in the expression of IL-8 was demonstrated in skin endothelium from patients with SSc,[86] however, this increase in the expression of IL-8 on endothelial cells correlated with a disease duration of less than one year.[86] Elevated levels of IL-8 have been identified in serum[91] and in BAL fluids from SSc patients with pulmonary fibrosis.[92]

Other Cytokines

Other cytokines have also been found to be elevated in SSc sera or tissue biopsies. These include increased serum levels of IL-10,[93] IL-12,[94] IL-13[49] and IL-17.[95] IL-17 was found to be overproduced by T cells from the peripheral blood and fibrotic lesions of SSc patients.[95] Transforming growth factor (TGF)-β induces a persistent increase in mRNA levels of type I and type III collagen and fibronectin[96] and is able to stimulate collagen production in normal fibroblasts.[97] Elevated levels of TGF-β mRNA was identified in the skin of patients with SSc[98] before the onset of fibrosis[99] and exposure to TGF-β increased the adhesion capacity of SSc fibroblasts to collagens I, IV, VI, fibronectin and laminin.[100] Tumor necrosis factor (TNF)-α has been found to be an important cytokine in the early stages of SSc and correlates with prominent inflammation and endothelial cell damage.[101]

Adhesion Molecules

Fibroblasts express intercellular adhesion molecule (ICAM)-1 in response to TNF-α. Serum levels of ICAM-1 are higher in patients with dcSSc than in patients with lcSSc.[102] Furthermore, in vitro studies of SSc fibroblasts demonstrated that ICAM-1 expression in response to cytokine stimulation (IL-1β, TNF-α, IFN-γ) was increased when compared to normal fibroblast cultures.[103] ICAM-1 and HLA class II antigens can be detected by in situ hybridization on most endothelial cells and fibroblasts, especially those located in areas of infiltrating lymphocytes.[104]

Immunogenetics

Immunogenetic studies support a more specific defect in SSc and many patients have been found to produce antibodies with well-defined target epitopes. HLA class II genes (DRB1*01, DRB1*17, DRB1*11) have been implicated as strong risk factors in the pathogenesis of SSc[105] and that these alleles are associated with particular subsets. Class III risk factors have centered on complement 4 null alleles, however, these alleles are in tight linkage disequilibrium with class II alleles, particularly DRB1*17, which is a significant risk factor for SSc.[106-108] Although DRB1*01, DRB1*17 and DRB1*11 are risk factors with SSc across several continents, it is the subsets of the disease that particularly show variation in the level of risk with HLA alleles (Table 2).

Frequently, the autoantibody pattern has been found to be associated with particular HLA risk factors. Studies suggest that there is a primary association between Scl-70 (topoisomerase I antibody) and HLA class II.[9,115,116,130-132] Risk factors have been identified with DQB1* alleles[130,132,133] and DRB1*11[7,134] in Caucasians, DRB1*11 in Blacks,[116] DRB1*15, DQB1*0601 and DQB1*0301 in Japanese[116,131] and DRB1*16 in American Indians.[116] Furthermore, a recent study has shown DPB1*1301 to be a very strong risk factor for Scl-70.[135] Anti-centromere antibody (ACA) is associated with the HLA class II DRB1*01 and

Table 3. Comparison of systemic sclerosis and graft-versus-host disease

Clinical Feature	Systemic Sclerosis	Graft-Versus-Host Disease
Cutaneous fibrosis	Yes	Yes
Skin pigmentary changes	Yes	Yes
Raynaud's phenomenon	Yes	Rare
Microvascular alterations	Yes	Yes
Lung fibrosis	Yes	Yes
Myocardial fibrosis	Yes	Yes
Joint & muscle involvement	Yes	Yes
Esophageal disease	Yes	Yes
Hepatic disease	Rare	Yes
Lymphadenopathy, splenomegaly	Rare	Yes
Bowel disease	Yes	Rare
Kidney fibrosis	Yes	No
Glomerulonephritis	Rare	Yes
Pleural disease	Yes	No
Calcinosis	Yes	No

DRB1*11[110,115,136] or DQB1* alleles.[133,137] DRB1*01 has been found to be strong risk factor for ACA in Japanese.[116]

Microchimerism and Systemic Sclerosis

Recent research has centered on the involvement of microchimeric cells in the pathogenesis of SSc.[3,4] The hypothesis that microchimeric fetal cells left from previous pregnancies may be involved in the pathogenesis of SSc by mediating a chronic graft-versus-host-like disease was first proposed by Scott Pereira.[138,139] Subsequently, other investigators have proposed the involvement of fetal cells in the pathogenesis of SSc.[140] However, the discovery of the presence of male fetal cells in a normal woman 27 years after the birth of her infant[141] made the hypothesis of microchimeric cell involvement in the pathogenesis of SSc tangible.[142]

SSc has many clinical features similar to those of chronic graft-versus-host disease (GVHD)[2,136-139,142-149] and, therefore, it has been postulated that SSc may be a form of chronic GVHD.[136-139] Skin, lung, and esophageal involvement are prominent features of both chronic GVHD and SSc.[143,144,150] Furthermore, immune activation and lymphocytic infiltration in affected tissues,[66,151,152] upregulation of inflammatory cytokines,[28,153,154] and fibrosis in the dermis and visceral organs[155,156] characterize both diseases. Early SSc lesions appear to be similar histologically to early chronic GVHD in rats and mice. Mononuclear cells localized deep within the dermis are associated topographically with fibroblasts, which have increased collagen production with loss of dermal appendages.[157] The clinical similarities and differences between SSc and chronic GVHD have been outlined in Table 3.

The activation of the immune system appears to be an early event in both SSc and chronic GVHD, that T cells are central to the development of tissue damage, and dominate the inflammatory infiltrates. Furthermore, a recent analysis of antinuclear antibodies in chronic GVHD identified the SSc-associated antibodies, Scl-70 and Pm-Scl, in 32% of individuals with chronic GVHD who presented with clinical symptoms similar to those of patients with dcSSc.[158] Furthermore, anti-mitochondrial antibodies are also found in patients with SSc and in 81% of patients with chronic GVHD.[159]

To provide additional support to this hypothesis, it was important to demonstrate the presence of microchimeric cells not only in peripheral blood but also in affected SSc tissues. Microchimeric cells have been found in the active lesions from women with SSc who have had male fetuses.[136] These results indicated that nonautologous cells were present in SSc lesions and correlated with a history of having male offspring. Microchimeric cells have also has been identified in peripheral blood CD3⁺ T lymphocytes.[136,160] Furthermore, microchimeric maternal cells have been detected in the active lesions of other autoimmune diseases, the juvenile idiopathic inflammatory myopathies.[161-163]

The first animal model investigating the potential role that microchimeric cells may be mediating fibrosis was developed by Christner et al.[164] The numbers of microchimeric cells in the peripheral blood of the mice increased an average of 48-fold following injections with vinyl chloride. Histological examination of the skin showed inflammation, with abundant fibroblasts and mononuclear infiltrate in the dermis. These results suggest that an environmental agent may activate microchimeric cells, which causes them to divide and multiply. The correlation between the increase in microchimeric cells and resulting dermal inflammation and fibrosis similar to GVHD suggest that activated microchimeric cells may be a necessary factor in the pathogenesis of SSc.

Humoral Immunology in Systemic Sclerosis

Serologic alterations and specific antinuclear antibodies (ANA) were among the first immune abnormalities noted in patients with SSc. Recent evidence has confirmed that sera from the majority of SSc patients contain one of three mutually exclusive, SSc-associated autoantibodies: Scl-70/topoisomerase I antibody, anti-centromere antibodies or anti-RNA polymerase III antibody.[7,165] Each autoantibody is associated with the presence of specific clinical features[134,166-168] and certain human leukocyte antigen (HLA) alleles.[113,128,130] Thus, while these SSc-associated autoantibodies do not appear to be directly involved in disease pathogenesis, they are associated with disease-specific pathologic phenomena. ANAs have also been detected in the relatives and spouses of patients with SSc, and some have antigens specific for scleroderma.[130] Antibodies most commonly identified in SSc have been outlined in Table 4.

Scl-70/Topoisomerase I Antibodies

Scl-70 is a nuclear antibody that targets the topoisomerase I protein. Topoisomerase I catalyzes the relaxation of supercoiled DNA by breaking one of the strands at a specific sequence, unwinding the helix and repairing the break.[180,181] Scl-70 has been associated with dcSSc with pulmonary interstitial fibrosis and peripheral vascular disease.[134,166] Scl-70 has a slightly higher prevalence in African-Americans than in Caucasians, and is significantly increased in Japanese.[182,183] The differences observed in the frequency of Scl-70 in SSc are most likely due to HLA risk factors and ethnic variability.

Anti-Centromeric Antibodies (ACA)

The centromere comprised of the kinetochore, which is a trilaminar disc-shaped structure, serves as the attachment site for the spindle microtubules. The microtubules facilitate the alignment and separation of the chromosome during mitosis.[184,185]

Table 4. Autoantibodies in scleroderma

Antigen	Immunofluorescence Pattern	Identity	Disease and Frequency	References
Scl-70 (topoisomerase I)	Nuclear (diffuse fine speckles)	100 kD protein that degrades to 70 kD	30% of dcSSc 21% chronic GVHD	110,158
Kinetochore (centromere)	Centromere	18-400 kD proteins at the inner and out kinetochore plates	70 to 80% of CREST	110
RNA Polymerase I & III	Nucleolar (speckled)	13 proteins 12-210 kD	4% of dcSSc with internal organ involvement 22% of female Caucasian SSc	169-172
RNA Polymerase II	Nucleolar	220 kD CTD heptapeptide	SSc/SLE overlap	173
Fibrillarin	Nucleolar (clumpy)	34 kD protein component of U3 RNP	8% of men with SSc 48% SSc	169,174
Fibrillin 1	Nucleolar	30 kD protein	High prevalence in American Indians, Japanese and Blacks	175
Mitochondrial M2	Cytoplasmic (rod like)	70, 50 and 45 kD proteins	25% CREST 95% primary biliary cirrhosis High prevalence in SLE 81% chronic GVHD	176-178
PM-Scl	Nucleolar (homogeneous)	11 proteins 20 to 110 kD	17% SSc 3% SSc/myositis overlap 11% chronic GVHD	158,169

CREST= calcinosis, Raynaud's phenomenon, esophageal dysfunction, sclerodactyly and telangiectasia; CTD= carboxyl-terminal domain; dcSSc= diffuse cutaneous systemic sclerosis; GVHD= graft versus host disease; kD= kilodalton; RNP= ribonucleoprotein; SLE= systemic lupus erythematosus; SSc= systemic sclerosis. Adapted from Duvas A, 1996.[179]

The prevalence of ACA in SSc varies but has been found to be strongly associated with the lcSSc.[134] ACA has also been identified in normal individuals[186] and patients with Raynaud's,[166,186] SLE,[186] primary biliary cirrhosis,[187] and morphea.[187] ACA comprises at least 6 centromere polypeptides (CENP): CENP-A to CENP-F. More than 90% of ACA-positive sera from SSc patients reacts with CENP-A, -B, and -C[188] and studies have demonstrated that these antibodies are capable of disrupting mitosis.[189] CENP-A is a protein that is very similar to histone H3 of the H3/H4 nucleosomal core and is approximately 18 kD in size. It copurifies with H3/H4, but it has a centromere specific binding domain and is therefore distinct from H3 and H4.[190,191] CENP-B is a DNA binding protein of 80kD and can be identified in 100% of all ACA-positive sera. It is distributed throughout the centromeric alpha-satellite heterochromatin below the kinetochore.[192] CENP-C is a 140 kD protein that is a component of the kinetochore plate, which is essential for normal centromere function.[193]

RNA Polymerase (RNAP) Antibodies

Anti-RNAP I and III antibodies are associated with female Caucasian patients with dcSSc, particularly with a high prevalence of internal organ involvement and poor prognosis.[169,172,194,195] Anti-RNAP I antibodies have also been identified in 78% of patients with rheumatoid arthritis.[196] Antibodies to RNAP III have been detected in 45% of sera from SSc patients with dcSSc and only 6% of patients with lcSSc, and in some instances occur more frequently than antibodies to Scl-70.[195] Antibodies to RNAP II recognize a CTD heptapeptide repeat that contains a high content of charged residues.[173] Antibodies to RNAP II occur in SSc but are not disease specific and can be identified in mixed connective tissue disease and SLE.[170,197] Anti-RNAP II antibodies are often accompanied by anti-Ku and anti-nRNP antibodies.[197]

Other Autoantibodies Identified in Patients with SSc

Antibodies directed against other proteins have been reported in SSc. Pm-Scl is found in approximately 17% of patients with SSc and identifies a group of SSc patients with a high prevalence of myositis and renal involvement.[169] Anti-fibrillin 1 antibodies were detected in the majority of Japanese, African American and Choctaw Indian patients, however, there are ethnic differences in antigenic epitope specificity. The sera from African American patients recognized the N-terminal end of fibrillin 1, whereas in Japanese and Choctaw Indian patients the sera recognized 2 or 3 epitopes. In the few Caucasian patients who had anti-fibrillin 1 antibodies, the EGF-cb repeat and proline-rich C regions were recognized.[175] Anti-fibrillarin antibodies occur in patients who develop severe fatal primary arterial hypertension.[174,198] Anti-fibrillin antibodies have been identified in a large proportion of patients with other autoimmune diseases including rheumatoid arthritis, SLE and mixed connective tissue diseases.[199] Anti-mitochondrial antibodies are found in a high proportion of patients with other connective tissue diseases, however, reactivity to the mitochondrial peptides appears to be disease specific.[176] Antibodies to nonhistone nuclear proteins or nucleolar antigens have been described and have been termed Ku, Ro (SS-A), La (SS-B) Sm, nRNP, and Jo-1 have been found in subgroups of patients with SSc.[200]

References

1. Mayes MD. Scleroderma epidemiology. Rheum Dis Clin NA 1996; 22:751-764.
2. Silman AJ, Jannini S, Symmons D et al. An epidemiologic study of scleroderma in the West Midlands. Br J Rheumatol 1988; 27:286-290.
3. Artlett CM, Smith JB, Jimenez SA. Identification of fetal DNA and cells in skin lesions from women with systemic sclerosis. N Eng J Med 1998; 338:1186-1191.
4. Nelson JL, Furst DE, Maloney S et al. Microchimerism and HLA-compatible relationships of pregnancy in scleroderma. Lancet 1998; 351:559-562.
5. Haustein U-F, Herrmann K. Environmental scleroderma. Clin Dermatol 1994; 12:467-473.
6. Silman AJ, Hochberg MC. Occupational and environmental influences on scleroderma. Rheum Dis Clin NA 1996; 22:737-749.
7. Black CM, Welsh KI. Genetics of scleroderma. Clin Dermatol 1994; 12:337-347.
8. Black CM, Welsh KI, Walker AE et al. Genetic susceptibility to scleroderma-like syndrome induced by vinyl chloride. Lancet 1983; I:53-55.
9. Fanning GC, Welsh KI, Bunn C et al. HLA associations in three mutually exclusive autoantibody subgroups in UK systemic sclerosis patients. Br J Rheumatol 1998; 37:201-207.
10. Jimenez SA, Diaz A, Khalili K. Retroviruses and the pathogenesis of systemic sclerosis. Int Rev Immunol 1995; 12:159-175.
11. Pandey JP, LeRoy EC. Human cytomegalovirus and the vasculopathies of autoimmune diseases (especially scleroderma), allograft rejection, and coronary restenosis. Arthritis Rheum 1998; 41:10-15.
12. Maul GG, Jimenez SA, Riggs E et al. Determination of an epitope of the diffuse systemic sclerosis marker antigen DNA topoisomerase I: Sequence similarity with retroviral p30^{gag} protein suggests a possible cause for autoimmunity in systemic sclerosis. Proc Natl Acad Sci USA 1989; 86:8492-8496.
13. Priel E, Showalter SD, Roberts M et al. Topoisomerase I activity associated with human immunodeficiency virus (HIV) particles and equine infectious anemia virus core. EMBO J 1990; 9:4167-4172.
14. Degiannis D, Seibold JR, Czarnecki M et al. Soluble and cellular markers of immune activation in patients with systemic sclerosis. Clin Immunol Immunopathol 1990; 56:259-270.
15. Frieri M, Angadi C, Paolano A et al. Altered T cell subpopulations and lymphocytes expressing natural killer cell phenotypes in patients with progressive systemic sclerosis. J Allergy Clin Immunol 1991; 87:773-779.
16. Inoshita T, Whiteside TL, Rodnan GP et al. Abnormalities of T lymphocyte subsets in patients with progressive systemic sclerosis (PSS, scleroderma). J Lab Clin Med 1981; 97:264-277.
17. Freundlich B, Jimenez SA. Phenotype of peripheral blood lymphocytes in patients with progressive systemic sclerosis: Activated T lymphocytes and the effect of D-penicillamine therapy. Clin Exp Immunol 1987; 69:375-384.
18. Fiocco U, Rosada M, Cozzi L et al. Early phenotypic variation of circulating helper memory T cells in scleroderma: Correlation with disease activity. Ann Rheum Dis 1993; 52:272-277.
19. Gustafsson R, Totterman TH, Klareskog L et al. Increase in activated T cells and reduction in suppressor inducer T cells in systemic sclerosis. Ann Rheum Dis 1990; 49:40-45.
20. Kantor TV, Whiteside TL, Friberg D et al. Lymphokine-activated killer cell and natural killer cell activities in patients with systemic sclerosis. Arthritis Rheum 1992; 35:694-699.
21. Umehara H, Kumagai S, Ishida H et al. Enhanced production of interleukin-2 in patients with progressive systemic sclerosis. Hyperactivity of CD4-positive T cells? Arthritis Rheum 1988; 31:401-407.
22. Whiteside TL, Kumagai Y, Roumm AD et al. Suppressor cell function and T lymphocyte subpopulations in peripheral blood of patients with progressive systemic sclerosis. Arthritis Rheum 1983; 26:841-847.
23. Kahan A, Picard F, Menkes CJ et al. Abnormalities of T lymphocyte subsets in systemic sclerosis demonstrated with anti-CD29 monoclonal antibodies. Ann Rheum Dis 1991; 50:354-358.
24. Keystone EC, Lau C, Gladman D et al. Immunoregulatory T cell subpopulations in patients with scleroderma using monoclonal antibodies. Clin Exp Immunol 1982; 48:443-448.
25. Giacomelli R, Matucci-Cerinic M, Cipriani P et al. Circulating Vdelta1+ T cells are activated and accumulate in the skin of systemic sclerosis patients. Arthritis Rheum 1998; 41:327-334.
26. Lupoli S, Amlot P, Black C. Normal immune responses in systemic sclerosis. J Rheumatol 1990; 17:323-327.
27. Kraling BM, Maul GG, Jimenez SA. Mononuclear cellular infiltrates in clinically involved skin from patients with systemic sclerosis of recent onset predominantly consist of monocytes/macrophages. Pathobiology 1995; 63:48-56.
28. Prescott RJ, Freemont AJ, Jones CJ et al. Sequential dermal microvascular and perivascular changes in the development of scleroderma. J Pathol 1992; 166:255-263.
29. Roumm AD, Whiteside TL, Medsger TA et al. Lymphocytes in the skin of patients with progressive systemic sclerosis. Quantification, subtyping, and clinical correlations. Arthritis Rheum 1984; 27:645-653.
30. Fagundus DM, LeRoy EC. Cytokines and systemic sclerosis. Clin Dermatol 1994; 12:407-417.
31. Kahari V, Sandberg M, Kalimo H et al. Identification of fibroblasts responsible for increased collagen production in localized scleroderma by in situ hybridization. J Invest Dermatol 1988; 90:664-670.
32. Peltonen J, Kahari L, Uitto J et al. Increased expression of Type VI collagen genes in systemic sclerosis. Arthritis Rheum 1990; 33:1829-1835.
33. Rudnicka L, Varga J, Christiano AM et al. Elevated expression of type VI collagen in the skin of patients with systemic sclerosis. Regulation by transforming growth factor beta. J Clin Invest 1994; 93:1709-1715.
34. Jimenez SA, Feldman G, Bashey RI et al. Coordinate increase in expression of Type I and Type III collagen genes in progressive systemic sclerosis fibroblasts. Biochem J 1986; 237:837-843.
35. Kuroda K, Shinkai H. Decorin and glycosaminoglycan synthesis in skin fibroblasts from patients with systemic sclerosis. Arch Dermatol Res 2001; 289:481-485.
36. Cooper SM, Keyser AJ, Beaulieu AD et al. Increase in fibronectin in the deep dermis of involved skin in progressive systemic sclerosis. Arthritis Rheum 1979; 22:983-987.
37. Rossi GA, Bitterman PB, Rennard SI et al. Evidence for chronic inflammation as a component of the interstitial lung disease associated with progressive systemic sclerosis. Am Rev Respir Dis 1985; 131:612-617.
38. Wells AU, Lorimer S, Majumdar S et al. Fibrosing alveolitis in systemic sclerosis: Increase in memory T-cells in the interstitium. Eur Respir J 2001; 8:266-271.
39. Smith EA. Connective tissue metabolism including cytokines in scleroderma. Curr Opin Rheumatol 1992; 4:869-877.
40. Silver RM, Miller KS, Kinsella MB et al. Evaluation and management of scleroderma lung disease using bronchoalveolar lavage. Am J Med 1990; 88:470-476.
41. Gustafsson R, Fredens K, Nettelbladt O et al. Eosinophil activation in systemic sclerosis. Arthritis Rheum 1991; 34:414-422.

42. Yurovsky VV, Wigley FM, Wise RA et al. Skewing of the CD8+ T-cell repertoire in the lungs of patients with systemic sclerosis. Hum Immunol 1996; 48:84-97.
43. White B, Yurovsky VV. Oligoclonal expansion of V delta 1+ gamma/delta T cells in systemic sclerosis patients. Ann NY Acad Sci 1995; 756:382-391.
44. Yurovsky VV, Sutton PA, Schulze DH et al. Expansion of selected V delta 1+ gamma delta T cells in systemic sclerosis patients. J Immunol 1994; 153:881-891.
45. Sakkas LI, Xu B, Artlett CM et al. Oligoclonal T cell expansion in the skin of patients with systemic sclerosis. J Immunol (In Press)
46. Majewski S, Blaszczyk M, Wasik M et al. Natural killer cell activity of peripheral blood mononuclear cells from patients with various forms of systemic scleroderma. Br J Dermatol 1987; 116:1-8.
47. Kantor TV, Friberg D, Medsger TA et al. Cytokine production and serum levels in systemic sclerosis. Clin Immunol Immunopathol 1992; 65:278-285.
48. Needleman BW, Wigley FM, Stair RW. Interleukin-1, interleukin-2, interleukin-4, interleukin-6, tumor necrosis factor alpha, and interferon-gamma levels in sera from patients with scleroderma. Arthritis Rheum 1992; 35:67-72.
49. Hasegawa M, Fujimoto M, Kikuchi K et al. Elevated serum levels of interleukin 4 (IL-4), IL-10, and IL-13 in patients with systemic sclerosis. J Rheumatol 1997; 24:328-332.
50. Giacomelli R, Cipriani R, Lattanzio R et al. Circulating levels of soluble CD30 are increased in patients with systemic sclerosis (SSc) and correlate with serological and clinical features of the disease. Clin Exp Immunol 1997; 108:42-46.
51. Mavilia C, Scaletti C, Romagnani P et al. Type 2 helper T-cell predominance and high CD30 expression in systemic sclerosis. Am J Pathol 1997; 151:1751-1758.
52. Torres JE, Sanchez JL. Histopathologic differentiation between localized and systemic scleroderma. Am J Dermatopathol 1998; 20:242-245.
53. White Needleman B, Ordonez JV, Taramelli D et al. In vitro identification of a subpopulation of fibroblasts that produces high levels of collagen in scleroderma patients. Arthritis Rheum 1990; 33:842-852.
54. Wahl SM, Wahl LM, McCarthy JB. Lymphocyte-mediated activation of fibroblast proliferation and collagen production. J Immunol 1978; 121:942-946.
55. Mizel SB, Ben-Zvi A. Studies on the role of lymphocyte-activating factor (interleukin 1) in antigen-induced lymph node lymphocyte proliferation. Cell Immunol 1980; 54:382-389.
56. Kahari V-M, Heino J, Vuorio E. Interleukin-1 increases collagen production and mRNA levels in cultured skin fibroblasts. Biochim Biophys Acta 1987; 929:142-147.
57. Kawaguchi Y. IL-1 alpha gene expression and protein production by fibroblasts from patients with systemic sclerosis. Clin Exp Immunol 1994; 97:445-450.
58. Umehara H, Kumagai S, Murakami M et al. Enhanced production of interleukin-1 and tumor necrosis factor alpha by cultured peripheral blood monocytes from patients with scleroderma. Arthritis Rheum 1990; 33:893-897.
59. Kahaleh MB. Soluble immunologic products in scleroderma sera. Clin Immunol Immunopathol 1991; 58:139-144.
60. Patrick MR, Kirkham BW, Graham M et al. Circulating interleukin-1 beta and soluble interleukin 2 receptor: Evaluation as markers of disease activity in scleroderma. J Rheumatol 1995; 22:654-658.
61. Alcocer-Varela J, Martinez-Cordero E, Alarcon-Segovia D. Spontaneous production of, and defective response to, interleukin-1 by peripheral blood mononuclear cells from patients with scleroderma. Clin Exp Immunol 1985; 59:666-672.
62. Kawaguchi Y, Harigai M, Hara M et al. Increased interleukin-1 receptor, type I, at messenger RNA and protein level in skin fibroblasts from patients with systemic sclerosis. Biochem Biophys Res Commun 1992; 184:1504-1510.
63. Higgins GC, Wu Y, Postlethwaite AE. Intracellular IL-1 receptor antagonist is elevated in human dermal fibroblasts that over express intracellular precursor IL-1 alpha. J Immunol 1999; 163:3969-3975.
64. Mizel SB. Interleukin 1 and T cell activation. Immunol Rev 1982; 63:51-72.
65. Hawrylko E, Spertus A, Mele CA et al. Increased interleukin-2 production in response to human type I collagen stimulation in patients with systemic sclerosis. Arthritis Rheum 1991; 34:580-587.
66. Famularo G, Procopio A, Giacomelli R et al. Soluble interleukin-2 receptor, interleukin-2 and interleukin-4 in sera and supernatants from patients with progressive systemic sclerosis. Clin Exp Immunol 1990; 81:368-372.
67. Degiannis D, Seibold JR, Czarnecki M et al. Soluble interleukin-2 receptors in patients with systemic sclerosis. Clinical and laboratory correlations. Arthritis Rheum 1990; 33:375-380.
68. Kahaleh MB, LeRoy EC. Interleukin-2 in scleroderma: Correlation of serum level with extent of skin involvement and disease duration. Ann Intern Med 1989; 110:446-450.
69. Steen VD, Engel EE, Charley MR et al. Soluble serum interleukin 2 receptors in patients with systemic sclerosis. J Rheumatol 1996; 23:646-649.
70. Kahaleh MB, Yin T. Enhanced expression of high affinity interleukin-2 receptors in scleroderma: Possible role for IL-6. Clin Immunol Immunopathol 1992; 62:97-102.
71. Puett DW, Fuchs HA. Rapid exacerbation of scleroderma in a patient treated with interleukin 2 and lymphokine activated killer cells for renal cell carcinoma. J Rheumatol 1994; 21:752-753.
72. Fertin C, Nicolas JF, Gillery P et al. Interleukin-4 stimulates collagen synthesis by normal and scleroderma fibroblasts in dermal equivalents. Cell Mol Biol 1991; 37:823-829.
73. Salmon-Ehr V, Serpier H, Nawrocki B et al. Expression of interleukin-4 in scleroderma skin specimens and scleroderma fibroblast cultures. Potential role in fibrosis. Arch Dermatol 1996; 132:802-806.
74. Makhluf HA, Stepniakowska J, Hoffman S et al. IL-4 upregulates tenascin synthesis in scleroderma and healthy skin fibroblasts. J Invest Dermatol 1996; 107:856-859.
75. Lee KS, Ro YJ, Ryoo YW et al. Regulation of interleukin-4 on collagen gene expression by systemic sclerosis fibroblasts in culture. J Dermatol Sci 1996; 12:110-117.
76. Atamas SP, Yurovsky VV, Wise R et al. Production of type 2 cytokines by CD8+ lung cells is associated with greater decline in pulmonary function in patients with systemic sclerosis. Arthritis Rheum 1999; 42:1168-1178.
77. Sakkas LI, Tourtellotte C, Berney S et al. Increased levels of alternatively spliced interleukin 4 (IL-4delta2) transcripts in peripheral blood mononuclear cells from patients with systemic sclerosis. Clin Diag Lab Immunol 1999; 6:660-664.
78. Linder PS, Frieri M. Analysis and clinical correlation of serum interleukin-4 levels in progressive systemic sclerosis. J Allergy Clin Immunol 2001; 87S:160.
79. Muraguchi A, Hirano T, Tang B et al. The essential role of B cell stimulatory factor 2 (BSF-2/IL-6) for the terminal differentiation of B cells. J Exp Med 1988; 167:332-344.
80. Lotz M, Jirik F, Kabouridis P et al. B cell stimulating factor 2/interleukin 6 is a costimulant for human thymocytes and T lymphocytes. J Exp Med 1988; 167:1253-1258.
81. Elias JA, Kotloff R. Mononuclear cell-fibroblast interactions in the human lung. Chest 1991; 99:73S-79S.
82. Takemura H, Susuki H, Fujisawa H et al. Enhanced interleukin 6 production by cultured fibroblasts from patients with systemic sclerosis in response to platelet derived growth factor. J Rheumatol 1998; 25:1534-1539.
83. Harigai M, Hara M, Kitani A et al. Interleukin 1 and tumor necrosis factor-alpha synergistically increase the production of interleukin 6 in human synovial fibroblast. J Clin Lab Immunol 1991; 34:107-113.
84. Elias JA, Lentz V. IL-1 and tumor necrosis factor synergistically stimulate fibroblast IL-6 production and stabllize IL-6 messenger RNA. J Immunol 1990; 145:161-166.
85. Swaak AJG, van Rooyen A, Nieuwenhuis E et al. Interleukin 6 (IL-6) in synovial fluid and serum of patients with rheumatic disease. Scand J Rheumatol 1988; 17:469-474.
86. Koch AE, Kronfeld-Harrington LB, Szekanecz Z et al. In situ expression of cytokines and cellular adhesion molecules in the skin of patients with systemic sclerosis. Pathobiology 1993; 61:239-246.
87. Feghali CA, Bost KL, Boulware DW et al. Control of IL-6 expression and response in fibroblasts from patients with systemic sclerosis. Autoimmunity 1994; 17:309-318.

88. Gurram M, Pahwa S, Frieri M. Augmented interleukin-6 secretion in collagen-stimulated peripheral blood mononuclear cells from patients with systemic sclerosis. Ann Allergy 1994; 73:493-496.
89. Koch AE, Polverini PJ, Kunkel SL et al. Interleukin-8 as a macrophage-derived mediator of angiogenesis. Science 1992; 258:1798-1801.
90. Larsen CG, Anderson AO, Appella E et al. The neutrophil-activating protein (NAP-1) is also chemotactic for T lymphocytes. Science 1989; 243:1464-1466.
91. Reitamo S, Remitz A, Varga J et al. Demonstration of interleukin 8 and autoantibodies to interleukin 8 in the serum of patients with systemic sclerosis and related disorders. Arch Dermatol 1993; 129:189-193.
92. Southcott AM, Jones KP, Li D et al. Interleukin-8. Differential expression in lone fibrosing alveolitis and systemic sclerosis. Am J Respir Crit Care 1995; 151:1604-1612.
93. Kucharz EJ, Brzezinska-Wcislo L, Kotulska A et al. Elevated serum level of interleukin-10 in patients with systemic sclerosis. Clin Rheumatol 1997; 16:638-639.
94. Sato S, Hanakawa H, Hasegawa M et al. Levels of interleukin 12, a cytokine of type 1 helper T cells, are elevated in sera from patients with systemic sclerosis. J Rheumatol 2000; 27:2838-2842.
95. Kurasawa K, Hirose K, Sano H et al. Increased interleukin-17 production in patients with systemic sclerosis. Arthritis Rheum 2000; 43:2455-2463.
96. Varga J, Rosenbloom J, Jimenez SA. Transforming growth factor-β (TGFβ) causes a persistent increase in steady state amounts of types I and III collagen and fibronectin mRNAs in normal human dermal fibroblasts. Biochem J 1987; 247:597-604.
97. Varga J, Jimenez SA. Stimulation of normal human fibroblast collagen production and processing by transforming growth factor-beta. Biochem Biophys Res Commun 1986; 138:974-980.
98. Gruschwitz M, Muller PU, Sepp N et al. Transcription and expression of transforming growth factor type beta in the skin of progressive systemic sclerosis: A mediator of fibrosis? J Invest Dermatol 1990; 94:197-203.
99. Higley H, Persichitte K, Chu S et al. Immunocytochemical localization and serologic detection of transforming growth factor beta. Association with type I procollagen and inflammatory cell markers in diffuse and limited systemic sclerosis, morphea, and Raynaud's phenomenon. Arthritis Rheum 1994; 37:278-288.
100. Majewski S, Hunzelmann N, Schirren CG et al. Increased adhesion of fibroblasts from patients with scleroderma to extracellular matrix components: In vitro modulation by INF-gamma but not by TGF-beta. J Invest Dermatol 1992; 98:86-91.
101. Gruschwitz MS, Albrecht M, Vieth G et al. In situ expression and serum levels of tumor necrosis factor-alpha receptors in patients with early stages of systemic sclerosis. J Rheumatol 1997; 24:1936-1943.
102. Ihn H, Sato S, Fukimoto M et al. Circulating intercellular adhesion molecule-1 in the sera of patients with systemic sclerosis: Enhancement by inflammatory cytokines. Br J Rheumatol 1997; 36:1270-1275.
103. Cho MM, Jimenez SA, Johnson BA et al. In vitro cytokine modulation of intercellular adhesion molecule-1 expression on systemic sclerosis dermal fibroblasts. Pathobiology 1994; 62:73-81.
104. Gruschwitz MS, Vieth G. Up-regulation of class II major histocompatibility complex and intercellular adhesion molecule I expression on scleroderma fibroblasts and endothelial cells by interferon-gamma and tumor necrosis factor alpha in the early disease stage. Arthritis Rheum 1997; 40:540-550.
105. Briggs DC, Black CM, Welsh KI. Genetic factors in scleroderma. Rheum Dis Clin NA. 1990; 16:31-51.
106. Briggs DC, Welsh K, Pereira RS et al. A strong association between null alleles at the C4A locus in the major histocompatibility complex and systemic sclerosis. Arthritis Rheum 1986; 29:1274-1277.
107. Rittner G, Schwanitz G, Baur MP et al. Family studies in scleroderma (systemic sclerosis) demonstrating an HLA-linked increased chromosomal breakage rate in cultured lymphocytes. Hum Genet 1988; 81:64-70.
108. Laurent MR, Welsh KI. Genetic markers in rheumatological diseases. J Immunogenet 1983; 10:275-291.
109. Lynch CJ, Singh G, Whiteside TL et al. Histocompatibility antigens in progressive systemic sclerosis (PSS; Scleroderma). J Clin Immunol 1982; 2:314-318.
110. Whiteside TL, Medsger TA, Rodnan GP. HLA-DR antigens in progressive systemic sclerosis (scleroderma). J Rheumatol 1983; 10:128-131.
111. Livingstone JZ, Scott TE, Wigley FM et al. Systemic sclerosis (scleroderma): Clinical, genetic and serologic subsets. J Rheumatol 1987; 14:512-518.
112. Steen VD, Powell DL, Medsger TA. Clinical correlations and prognosis based on serum autoantibodies in patients with systemic sclerosis. Arthritis Rheum 1988; 31:196-203.
113. Black CM, Welsh KI, Maddison PJ et al. HLA antigens, autoantibodies and clinical subsets in scleroderma. Br J Rheumatol 1984; 23:267-271.
114. Luderschmidt C, Scholz S, Mehlhaff E et al. Association of progressive systemic scleroderma to several HLA-B and HLA-DR alleles. Arch Dermatol 1987; 123:1188-1191.
115. Genth E, Mierau R, Genetzky P et al. Immunogenetic associations of scleroderma-related antinuclear antibodies. Arthritis Rheum 1990; 33:657-665.
116. Kuwana M, Kaburaki J, Arnett FC et al. Influence of ethnic background on clinical and serologic features in patients with systemic sclerosis and anti-DNA topoisomerase I antibody. Arthritis Rheum 1999; 42:465-474.
117. Jazwinska EC, Olive C, Dunckley H et al. HLA-DRw15 is increased in frequency in Japanese scleroderma. Dis Markers 1990; 8:323-326.
118. Takeuchi F, Nakano K, Yamada H et al. Association of HLA-DR with progressive systemic sclerosis in Japanese. J Rheumatol 1994; 21:857-863.
119. Satoh M, Akizuki M, Kuwana M et al. Genetic and immunological differences between Japanese patients with diffuse scleroderma and limited scleroderma. J Rheumatol 1994; 21:111-114.
120. Kuwana M, Okano Y, Kaburaki J et al. HLA class II genes associated with anticentromere antibody in Japanese patients with systemic sclerosis (scleroderma). Ann Rheum Dis 1995; 54:983-987.
121. Germain BF, Espinoza LR, Bergen LL et al. Increased prevalence of DRw3 in the CREST syndrome. Arthritis Rheum 1981; 24:857-859.
122. Kallenberg CGM, van der Voort-Beelen JM, D'Amaro J et al. Increased frequency of B8/DR3 in scleroderma and association of the haplotype with impaired cellular immune response. Clin Exp Immunol 1981; 43:478-485.
123. Ercilla MG, Arriaga F, Gratacos MR et al. HLA antigens and scleroderma. Arch Dermatol Res 1981; 271:381-385.
124. Gladman DD, Keystone EC, Baron M et al. Increased frequency of HLA-DR5 in scleroderma. Arthritis Rheum 1981; 24:854-856.
125. Lambert NC, Distler O, Muller-Ladner U et al. HLA-DQA1*0501 is associated with diffuse systemic sclerosis in Caucasian men. Arthritis Rheum 2000; 43:2005-2010.
126. Vargas-Alarcon G, Granados J, Ibanez de Kasep G et al. Association of HLA-DR5 (DR11) with systemic sclerosis (scleroderma) in Mexican patients. Clin Exp Rheumatol 1995; 13:11-16.
127. Vlachoyiannopoulos PG, Dafni UG, Pakas I et al. Systemic scleroderma in Greece: Low mortality and strong linkage with DRB1*1104 allele. Ann Rheum Dis 2000; 59:359-367.
128. Barnett AJ, Tait BD, Barnett MA et al. T lymphocyte subset abnormalities and HLA antigens in scleroderma (systemic sclerosis). Clin Exp Immunol 1989; 76:24-29.
129. Dunckley H, Jazwinska EC, Gatenby PA et al. DNA-DR typing shows HLA-DRw11 RFLPs are increased in frequency in both progressive systemic sclerosis and CREST variants of scleroderma. Tissue Antigens 1989; 33:418-420.
130. Reveille JD, Durban E, MacCleod-St.Clair MJ et al. Association of amino acid sequences in the HLA-DQB1 first domain with the antitopoisomerase I autoantibody response in scleroderma (progressive systemic sclerosis). J Clin Invest 1992; 90:1-8.
131. Kuwana M, Kaburaki J, Okano Y et al. The HLA-DR and DQ genes control the autoimmune response to DNA topoisomerase I in systemic sclerosis. J Clin Invest 1993; 92:1296-1301.
132. Whyte J, Artlett CM, Harvey G et al. HLA-DQB1 associations with anti-topoisomerase-1 antibodies in patients with systemic sclerosis and their first degree relatives. United Kingdom Systemic Sclerosis Group. J Autoimmunity 1994; 7:509-520.

133. McHugh NJ, Whyte J, Artlett CM et al. Anti-centromere antibodies (ACA) in systemic sclerosis patients and their relatives: A serological and HLA study. Clin Exp Immunol 1994; 96:267-274.
134. Rands AL, Whyte J, Cox B et al. MHC class II associations with autoantibody and T cell immune responses to the scleroderma autoantigen topoisomerase I. J Autoimmunity 2000; 15:451-458.
135. Gilchrist FC, Bunn C, Foley PJ et al. Class II HLA associations with autoantibodies in scleroderma: A highly significant role for HLA-DP. Genes Immunity 2001; 2:76-81.
136. Steen VD, Ziegler GL, Rodnan GP et al. Clinical and laboratory associations of anticentromere antibody in patients with progressive systemic sclerosis. Arthritis Rheum 1984; 27:125-131.
137. Reveille JD, Owerbach D, Goldstein R et al. Association of polar amino acids at position 26 of the HLA-DQB1 first domain with anticentromere antibody response in systemic sclerosis (scleroderma). J Clin Invest 1992; 89:1208-1213.
138. Silman AJ, Black CM. Increased incidence of spontaneous abortion and infertility with scleroderma before disease onset: A controlled study. Ann Rheum Dis 1988; 47:441-444.
139. Black CM, Stevens WM. Scleroderma. Rheum Dis Clin NA 1989; 15:193-212.
140. Mullinax, F. Chimerism and autoimmunity. Proceedings of the Fourth ASEAN Congress of Rheumatology 1993; 30-40.
141. Bianchi DW, Zickwolf GK, Weil GJ et al. Male fetal progenitor cells persist in maternal blood for as long as 27 years postpartum. Proc Natl Acad Sci USA 1996; 93:705-708.
142. Nelson JL. Maternal-fetal immunology and autoimmune disease: Is some autoimmune disease auto-alloimmune or allo-autoimmune? Arthritis Rheum 1996; 39:191-194.
143. Lawley TJ, Peck GL, Moutsopoulos HM et al. Scleroderma, Sjögren-like syndrome and chronic graft-versus-host disease. Ann Intern Med 1977; 87:707-709.
144. Graham-Brown RAC, Sarkani I. Scleroderma-like changes due to chronic-graft-versus host disease. Clin Exp Dermatol 1983; 8:531-538.
145. Herzog P, Clements PJ, Roberts NK et al. Progressive systemic sclerosis-like syndrome after bone marrow transplantation. J Rheumatol 1980; 7:56-64.
146. Valenta LJ. Familial scleroderma in a kindred with high incidence of autoimmune disease: Correlation with HLA-A1/B8 haplotype. Arch Dermatol 1987; 123:1438-1440.
147. Bos GMJ, Majoor GD, Slaaf DW et al. Similarity of scleroderma-like skin lesions in allogeneic and syngeneic bone marrow transplantation models. Transplant Proc 1989; 21:3262-3263.
148. Chosidow O, Bagot M, Vernant J-P et al. Sclerodermatous chronic graft-versus-host disease. J Am Acad Dermatol 1992; 26:49-55.
149. Artlett CM, Welsh KI, Black CM et al. Fetal-maternal HLA compatibility confers susceptibility to systemic sclerosis. Immunogenetics 1997; 47:17-22.
150. Fleischmajer R, Perlish JS, Reeves JRT. Cellular infiltrates in scleroderma skin. Arthritis Rheum 1977; 20:975-984.
151. Lampert IA, Janossy G, Suitters AJ et al. Immunological analysis of the skin in graft versus host disease. Clin Exp Immunol 1982; 50:123-131.
152. Jimenez SA. Cellular immune dysfunction and the pathogenesis of scleroderma. Semin Arthritis Rheum 1983; 13 (Suppl. 1):104-113.
153. Postlethwaite AE. Connective tissue metabolism including cytokines in scleroderma. Curr Opin Rheumatol 1993; 5:766-772.
154. Janin-Mercier A, Devergie A, van Cauwenberge D et al. Immunohistologic and ultrastructural study of the sclerotic skin in chronic graft-versus-host disease in man. Am J Pathol 1984; 115:296-306.
155. Jimenez SA, Hitraya E, Varga J. Pathogenesis of scleroderma: Collagen. Rheum Dis Clin NA 1996; 22:647-674.
156. White B, Korn JH, Piela-Smith TH. Preferential adherence of human gamma delta, CD8+, and memory T cells to fibroblasts. J Immunol 1994; 152:4912-4918.
157. LeRoy EC, Black CM, Fleischmajer R et al. Scleroderma (systemic sclerosis): Classification, subsets and pathogenesis. J Rheumatol 1988; 15:202-205.
158. Bell SA, Faust H, Mittermuller J et al. Specificity of antinuclear antibodies in scleroderma-like chronic graft-versus-host disease: Clinical correlation and histocompatibility locus antigen association. Br J Dermatol 1996; 134:848-854.
159. Siegert W, Stemerowicz R, Hopf U. Antimitochondrial antibodies in patients with chronic graft-versus-host disease. Bone Marrow Transplant 1992; 10:221-227.
160. Evans PC, Lambert N, Maloney S et al. Long-term fetal microchimerism in peripheral blood mononuclear cell subsets in healthy women and women with scleroderma. Blood 1999; 93:2033-2037.
161. Artlett CM, Ramos R, Jimenez SA et al. Chimeric cells of maternal origin in juvenile idiopathic inflammatory myopathies. Lancet 2000; 356:2155-2156.
162. Reed AM, Picornell YJ, Harwood A et al. Chimerism in children with juvenile dermatomyositis. Lancet 2000; 356:2156-2157.
163. Tanaka A, Lindor K, Gish R et al. Fetal microchimerism alone does not contribute to the induction of primary biliary cirrhosis. Hepatology 1999; 30:833-838.
164. Christner PJ, Artlett CM, Conway RF et al. Increased numbers of microchimeric cells of fetal origin and dermal fibrosis in mice following injection of vinyl chloride. Arthritis Rheum 2000; 43:2598-2605.
165. Harvey GR, McHugh NJ. Serologic abnormalities in systemic sclerosis. Curr Opin Rheumatol 1999; 11:495-502.
166. Weiner ES, Earnshaw WC, Senecal J-L et al. Clinical associations of anticentromere antibodies and antibodies to topoisomerase I: A study of 355 patients. Arthritis Rheum 1988; 31:378-385.
167. Fritzler MJ, Kinsella TD. The CREST syndrome: A distinct serologic entity with anticentromere antibodies. Am J Med 1980; 69:520-526.
168. McCarty GA, Rice JR, Bembe ML et al. Anticentromere antibody. Clinical correlations and association with favorable prognosis in patients with scleroderma variants. Arthritis Rheum 1983; 26:1-7.
169. Reimer G, Steen VD, Penning CA et al. Correlates between autoantibodies to nucleolar antigens and clinical features in patients with systemic sclerosis (scleroderma). Arthritis Rheum 1988; 31:525-532.
170. Hirakata M, Okano Y, Pati U et al. Identification of autoantibodies to RNA polymerase II: Occurrence in systemic sclerosis and association with autoantibodies to RNA polymerases I and III. J Clin Invest 1993; 91:2665-2672.
171. Harvey GR, Butts S, Rands AL et al. Clinical and serological associations with anti-RNA polymerase antibodies in systemic sclerosis. Clin Exp Immunol 1999; 117:395-402.
172. Chang M, Wang RJ, Yangco DT et al. Analysis of autoantibodies against RNA polymerases using immunoaffinity-purified RNA polymerase I, II and III antigen in an enzyme-linked immunosorbent assay. Clin Immunol Immunopathol 1998; 89:71-78.
173. Hirakata M, Kanungo J, Suwa A et al. Autoimmunity to RNA polymerase II is focused at the carboxyl terminal domain of the large subunit. Arthritis Rheum 1996; 39:1886-1891.
174. Okano Y, Steen VD, Medsger TA. Autoantibody to U3 nucleolar ribonucleoprotein (fibrillarin) in patients with systemic sclerosis. Arthritis Rheum 1992; 35:95-100.
175. Tan FK, Arnett FC, Reveille JD et al. Autoantibodies to fibrillin 1 in systemic sclerosis: Ethnic differences in a recognition and lack of correlation with specific clinical features and HLA alleles. Arthritis Rheum 2000; 43:2464-2471.
176. Mouritsen S, Demant E, Permin H et al. High prevalence of anti-mitochondrial antibodies among patients with some well-defined connective tissue diseases. Clin Exp Immunol 1986; 66:68-76.
177. Black CM, Briggs DC, Welsh KI. The immunogenetic background of scleroderma - an overview. Clin Exp Dermatol 1992; 17:73-78.
178. Chou MJ, Lai MY, Lee SL et al. Reactivity of anti-mitochondrial antibodies in primary biliary cirrhosis and systemic sclerosis. J Formos Med Assoc 1992; 91:1075-1080.
179. Douvas A. Pathogenesis: Serologic correlates. Clements PJ, Furst DE, eds. Systemic Sclerosis. Baltimore: Williams & Wilkins, 1996:175
180. Champoux JJ, Dulbecco R. An activity from mammalian cells that untwists superhelical DNA - a possible swivel for DNA replication (polyoma-ethidium bromide-mouse-embryo cell-dye binding assay). Proc Natl Acad Sci USA 1972; 69:143-146.

181. Liu LF, Miller KG. Eukaryotic DNA topoisomerases: Two forms of type I DNA topoisomerases from HeLa cell nuclei. Proc Natl Acad Sci USA 1981; 78:3487-3491.
182. Reveille JD, Durban E, Goldstein R et al. Racial differences in the frequencies of scleroderma-related autoantibodies. Arthritis Rheum 1992; 35:216-218.
183. Kuwana M, Okano Y, Kaburaki J et al. Racial differences in the distribution of systemic sclerosis-related serum antinuclear antibodies. Arthritis Rheum 1994; 37:902-906.
184. Rieder CL. The formation, structure, and composition of the mammalian kinetochore and kinetochore fiber. Int Rev Cytol 1982; 79:1-58.
185. Pluta AF, Cooke CA, Earnshaw WC. Structure of the human centromere at metaphase. Trends Biochem Sci 1990; 15:181-185.
186. Rothfield N, Whitaker D, Bordwell B et al. Detection of anticentromere antibodies using cloned autoantigen CENP-B. Arthritis Rheum 1987; 30:1416-1419.
187. Powell FC, Winkelmann RK, Venencie-Lemarchand F et al. The anticentromere antibody: Disease specificity and clinical significance. Mayo Clin Proc 1984; 59:700-706.
188. Earnshaw W, Bordwell B, Marino C et al. Three human chromosomal autoantigens are recognized by sera from patients with anti-centromere antibodies. J Clin Invest 1986; 77:426-430.
189. Bernat RL, Borisy GG, Rothfield NF et al. Injection of anticentromere antibodies in interphase disrupts events required for chromosome movement in mitosis. J Cell Biol 1990; 111:1519-1533.
190. Palmer DK, O'Day K, Trong HL et al. Purification of the centromerespecific protein CENP-A and demonstration that it is a distinctive histone. Proc Natl Acad Sci USA 1991; 88:3724-3738.
191. Billings PB, Martinez A, Haselby JA et al. Protein blot assays specific for the discrimination of the centromere autoantigen, CENP-A, from human cells. Electrophoresis 1993; 14:909-916.
192. Cooke CA, Bernat RL, Earnshaw WC. CENP-B: A major human centromere protein located beneath the kinetochore. J Cell Biol 1990; 110:1475-1488.
193. Saitoh H, Tomkiel J, Cooke CA et al. CENP-C, an autoantigen in scleroderma, is a component of the human inner kinetochore plate. Cell 1992; 70:115-125.
194. Kuwana M, Kaburaki J, Okano Y et al. Clinical and prognostic associations based on serum antinuclear antibodies in Japanese patients with systemic sclerosis. Arthritis Rheum 1994; 37:75-83.
195. Okano Y, Steen VD, Medsger TA. Antibody reactive with RNA polymerase III in systemic sclerosis. Ann Intern Med 1993; 119:1005-1013.
196. Stetler DA, Rose KM, Wenger ME et al. Antibodies to distinct polypeptides of RNA polymerase I in sera from patients with rheumatic autoimmune disease. Proc Natl Acad Sci USA 1982; 79:7499-7503.
197. Satoh M, Ajmani AK, Ogasawara T et al. Autoantibodies to RNA polymerase II are common in systemic lupus erythematosus and overlap syndrome: Specific recognition of the phosphorylated (IIO) form by a subset of human sera. J Clin Invest 1994; 94:1981-1989.
198. Sacks DG, Okano Y, Steen VD et al. Isolated pulmonary hypertension in systemic sclerosis with diffuse cutaneous involvement: Association with serum anti-U3RNP antibody. J Rheumatol 1996; 23:639-642.
199. Kasturi KN, Hatakeyama A, Spiera H et al. Antifibrillarin autoantibodies present in systemic sclerosis and other connective tissue diseases interact with similar epitopes. J Exp Med 1995; 181:1027-1036
200. Isern RA, Yaneva M, Weiner E et al. Autoantibodies in patients with primary pulmonary hypertension: Association with anti-Ku. Am J Med 1992; 93:307-312.

CHAPTER 46

Hematopoietic Stem Cell Transplantation for Systemic Sclerosis

Andrew M. Yeager, Diane BuchBarker, Thomas A. Medsger, Jr. and Albert D. Donnenberg

Overview

Systemic sclerosis is a chronic disorder of connective tissue characterized by inflammation and fibrosis of, and degenerative changes in, the blood vessels, skin, synovium, skeletal muscle, and certain internal organs, especially the gastrointestinal tract, lung, heart and kidney.[1] Although there is an early and often clinically unappreciated inflammatory component, the hallmark of the disease is skin thickening (scleroderma) caused by excessive accumulation of connective tissue. Subintimal proliferative vascular changes are prominent and lead to Raynaud's phenomenon and obliterative arteriolar and capillary lesions.

The incidence of systemic sclerosis is reported to be between four and 12 patients per million per year, and the disease affects females from three to eight times more often than males. The course of systemic sclerosis is extremely variable. Early in the illness, it is difficult to judge prognosis with respect to either survival or disability. Many patients with diffuse cutaneous involvement experience progressive sclerosis of the fingers and hands, leading to deforming flexion contractures. In others, joint mobility is not impaired despite progressive skin changes. The most reliable early predictors of subsequent severe diffuse skin involvement are the appearance of cutaneous thickening before the onset of Raynaud's phenomenon, rapid progression of scleroderma in the more proximal parts of the extremities, and palpable tendon friction rubs. Later in the course of systemic sclerosis, spontaneous improvement in skin thickening may occur; therefore, the prefix "progressive" is not uniformly applicable and should be abandoned.[2]

The five-year survival after diagnosis of systemic sclerosis ranges from 34% to more than 75%.[3,4] Improved survival in more recent surveys may relate to advances in supportive care, especially in the management of renal crisis.[5] Diffuse cutaneous involvement, male gender, older age and pulmonary, cardiac or renal involvement adversely affect outcome.[6-8] The presence of serum autoantibodies such as anti-topoisomerase I (anti-Scl-70) or anti-RNA polymerase I or III are associated with serious visceral disease.[9] An analysis from the University of Pittsburgh Scleroderma Databank indicates that patients whose total skin thickness score (modified Rodnan score) (Table 1)[10] increases more than 20 points per year (66 patients) are at especially high risk for organ dysfunction and have a five-year survival that is inferior to that of patients with less than 20 points per year (34 patients) (64% versus 83%, respectively; p= 0.062).

Several factors contribute to the challenge of evaluating the effectiveness of treatment for systemic sclerosis,[11,12] including: the need for at least 18 to 24 months of therapy to reverse fibrosis; the observation of spontaneous improvement among many patients late in the course of the disease; the limited availability of valid outcome measures; the influence of psychological determinants on symptoms; the need for classification of patients into subsets; and the need for multicenter trials to study adequate numbers of patients. For these reasons, a multidisciplinary group of experts have developed guidelines for clinical trials of potential disease-modifying therapies in systemic sclerosis.[13]

Conventional Therapeutic Approaches

Although there is no consensus on the selection or effectiveness of drug therapy in systemic sclerosis, the use of these agents is most clearly justified in patients with early, rapidly progressive, life-threatening and/or disabling diffuse disease. Corticosteroid administration remains the mainstay of conventional therapy but is generally restricted to patients with symptomatic serositis or myositis, and the latter may require higher doses of corticosteroids. However, high-dose corticosteroid therapy may precipitate acute renal failure.[14] Treatment of systemic sclerosis with D-penicillamine, a compound that interferes with the intermolecular cross-linking of collagen and that also is immunosuppressive, has been encouraging, with reported reduction in skin thickness, reduced frequency of subsequent visceral involvement (especially renal) and increased survival.[15-17]

A variety of other chemotherapeutic or immunosuppressive treatments has been evaluated in patients with systemic sclerosis, but no single drug or combination of drugs has been shown to be of consistent value. Azathioprine[18] or methotrexate,[19,20] but not chlorambucil,[21] have been reported to have beneficial effects. Improvements in cutaneous manifestations after 5-fluorouracil were outweighed by unacceptable gastrointestinal toxicity.[22] Colchicine, which inhibits the accumulation of collagen by blocking the conversion of procollagen to collagen, has not consistently led to objective reduction in skin thickness or other

Table 1. Modified Rodnan Skin Score: calculated by summing skin scored from all evaluated anatomic areas

Skin Thickness Rated by Clinical Palpation:
- 0 = normal skin thickness
- 1 = mild skin thickness
- 2 = moderate skin thickness
- 3 = severe skin thickness (inability to pinch skin into a fold)

Number of Anatomic Areas Evaluated (N= 17)

Area	Side	SCORE
Face		0-3
Anterior Chest		0-3
Abdomen		0-3
Fingers	Rt.	0-3
	Lt.	0-3
Dorsum of Hands	Rt.	0-3
	Lt.	0-3
Forearms	Rt.	0-3
	Lt.	0-3
Upper Arms	Rt.	0-3
	Lt.	0-3
Thighs	Rt.	0-3
	Lt.	0-3
Lower Legs	Rt.	0-3
	Lt.	0-3
Dorsum of Feet	Rt.	0-3
	Lt.	0-3
TOTAL		**0-51**

Lt= left; Rt= right

From: Clements PJ, Lachenbruch PA, Seibold JR et al. Skin thickness score in systemic sclerosis: an assessment of interobserver variability in 3 independent studies. J Rheumatol 1993; 20:1892-6.

manifestations of systemic sclerosis.[23,24] Uncontrolled trials of cyclosporine[25-27] or interferon-gamma[28-30] have suggested favorable effects on skin thickening. Several case reports have used antithymocyte globulin, to which responses were prompt[31,32] and associated with nearly complete elimination of CD4+ and CD8+ T lymphocytes from the peripheral blood.[32] Other therapeutic modalities have included plasmapheresis, the benefits of which are both unclear and difficult to evaluate because of confounding concurrent immunosuppressive and corticosteroid treatments.[33-36] Extracorporeal photopheresis was found to be effective when compared with D-penicillamine,[37] but the results of this study have been challenged on methodologic grounds.[38,39]

Rationale for High-Dose Immunosuppression and Autologous Stem Cell Transplantation in Systemic Sclerosis

As with other autoimmune disorders, the rationale for autologous HSCT in systemic sclerosis is based on the ability to provide effective, intensive in vivo immunosuppression to eradicate autoreactive immune cells and their precursors, and to restore normal lymphohematopoiesis without re-introduction of host autoimmune cells in the infused autologous hematopoietic stem cell product. The goal of the in vivo pre-transplant conditioning regimen is maximal lymphoablation with minimal nonhematologic toxicities, which differs substantially from the goal of tumor eradication in autologous HSCT for neoplastic diseases, in which myelosuppression or even myeloablation is a goal or an expected, accepted toxicity of the preparative regimen. The agents and modalities used in the preparative regimens for autologous HSCT in systemic sclerosis and other autoimmune diseases should focus on in vivo immunoablation, rather than relying on the preparative regimens that have been developed for autologous HSCT in oncologic diseases. For example, total body irradiation (TBI) is immunosuppressive, myelosuppressive (or even myeloablative) and antineoplastic, and has been used extensively as a component of pre-transplant conditioning for various cancers. However, the short-and long-term extramedullary toxicities of TBI and the availability of other agents with equal or better immunosuppressive effects may not justify its routine use in patients undergoing autologous HSCT for autoimmune disorders. This is underscored by the recent report of fatal regimen-related pulmonary toxicity in two patients with systemic sclerosis who received fractionated TBI (800 cGy over four fractions) as part of the pre-transplant conditioning.[40] In contrast, preparative regimens that incorporate agents such as fludarabine, anti-thymocyte globulin and cyclophosphamide may provide the required immunoablation without significant myelotoxicity or non-hematologic toxicities in the context of autologous HSCT for systemic sclerosis and other autoimmune diseases.

The effectiveness of a chemoimmunotherapeutic preparative regimen in inducing rapid lymphoablation for autologous HSCT in systemic sclerosis is illustrated in Figure 1, which shows the decrease in circulating T cells in a patient who received intravenous fludarabine (25 mg/m^2/day for 5 days), rabbit anti-thymocyte globulin (Thymoglobulin®; 2.5 mg/kg/day for 3 days) and cyclophosphamide (in the example given, 200 mg/m^2/day for 5 days) before infusion of CD34 selected, CD3 depleted autologous peripheral blood progenitor cells. The level of peripheral T cells decreased from 0.9×10^6 cells/L to 0.6×10^6 cells/L during the pre-transplant period (a 3.1-log reduction). The half-life of these T cells was 0.28 days, compared with a T-cell half-life of approximately 2.5 days in recipients of conventional fully myeloablative regimens (e.g., busulfan, cyclophosphamide and etoposide).

The second requirement for successful autologous HSCT in autoimmune diseases is eradication or drastic reduction of immune cells, specifically autoreactive T cells, from the autologous hematopoietic stem cell product. Data in patients with neoplastic diseases show high levels of T cells in cyclophosphamide- and G-CSF-mobilized autologous peripheral blood progenitor cells (PBPCs), and this is likely to be the case in patients with systemic sclerosis as well. However, the T cells in these products are often hyporesponsive, as shown by decreased in vitro proliferation in response to antigens or mitogens. We hypothesize that the T cell expansion that occurs in response to cyclophosphamide-induced T cell loss or injury comes from a naïve compartment, with functional hyporesponsiveness despite expression of markers normally associated with activated memory T cells (e.g., HLA-DR, bright CD38).[41] Techniques for clinical-scale positive selection of CD34+ cells and negative selection of T cells from PBPCs are available and feasible. Combined CD34 selection and T-cell depletion with immunomagnetic methods (e.g., Isolex device; Baxter Oncology) may provide as much as a 4.5- to 5-log reduction of CD3+ cells without a clinically significant loss of CD34+ cells from autologous PBPC products.

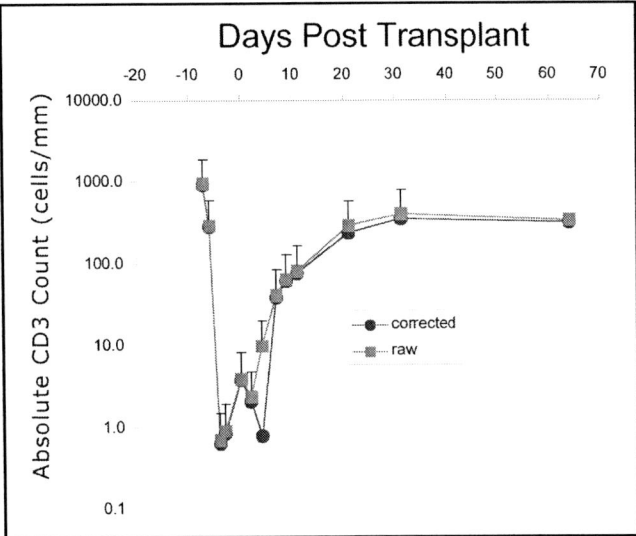

Figure 1. Kinetics of loss and post-transplant reconstitution of T cells in a patient undergoing autologous transplantation with CD34-selected, CD3-depleted peripheral blood progenitor cells for systemic sclerosis. The half-life of T cells during the preparative regimen (cyclophosphamide, fludarabine and rabbit anti-thymocyte globulin) was 0.28 day, and the reduction in T cells was 3.1 log. The T cells recovered rapidly after auto HSCT, reaching about one third of pretransplant levels by day +31. Raw CD3$^+$ cell levels reflect all T cells enumerated by the single-platform flow cytometry assay, and corrected CD3$^+$ cell levels exclude T cells that are measured in the single-platform assay but which are nonviable, as determined by binding of annexin V or by expression of caspase 3 activity.

Results of Autologous Stem Cell Transplantation in Systemic Sclerosis

At a consensus meeting in 1997, Clements and Furst proposed these criteria for entry of patients with systemic sclerosis onto clinical trials of autologous HSCT: diffuse cutaneous involvement with modified Rodnan total skin score greater than 20, disease duration less than three years and evidence of mild pulmonary, cardiac and/or renal involvement.[42] These criteria also recommended excluding from study patients age 60 years or greater and those with severe organ impairment. As indicated above, patients whose skin thickness scores increase by more than 20 points per year represent a high-risk group. By modifying the Clements and Furst criteria to include only those with rapid skin progression, one can restrict eligibility for autologous HSCT to a subset of systemic sclerosis patients who are at greatest risk for serious, life-threatening or disabling complications and who have inferior survival with conventional treatments.

In 1997, Tyndall et al reported the first autologous HSCT in an adult with systemic sclerosis, using a pre-transplant conditioning regimen of high-dose cyclophosphamide (200 mg/kg). By 6 months after autologous HSCT, this 27-year-old woman had decreased skin tightness, reduction in skin thickness score, decreased ischemic changes in distal digits and stabilization of pulmonary function.[43] In 1999, Martini et al reported the first autologous HSCT in a pediatric patient with systemic sclerosis, using a preparative regimen of high-dose cyclophosphamide plus alemtuzumab (IgG anti-CD52; Campath). This 12-year-old girl had significant improvement in skin score, pulmonary symptoms and linear growth, despite persistence of high titers of both antinuclear antibody and of anti-DNA topoisomerase I after autologous HSCT.[44]

Since those initial case reports, three relatively large series have described outcomes of autologous HSCT for systemic sclerosis (Table 2). In a heterogenous treatment group report from the European Group for Blood and Marrow Transplantation (EBMT) that included patients treated with TBI in Seattle and non-TBI patients in Europe, 41 patients (37 with diffuse disease, four with limited disease) were enrolled on a study of autologous peripheral blood progenitor cells (PBPC) or bone marrow transplantation.[45] Thirty-four of the patients were female, and seven were male; the median age was 41 years (range, 11.5-58.1 years). Mobilization of PBPCs consisted of cyclophosphamide and granulocyte colony-stimulating factor (G-CSF) in 29 patients, cyclophosphamide alone in one patient, and G-CSF alone in ten patients. One patient had autologous marrow collection without mobilization. Four patients died after mobilization chemotherapy but before undergoing pre-transplant conditioning: one from progression of systemic sclerosis, two from cardiac dysfunction and one from ischemic cardiac disease. Of the 37 patients who proceeded to autologous HSCT, 35 received CD34-selected or unmanipulated PBPCs, and two received unmanipulated bone marrow. The types of conditioning regimens were heterogeneous: 19 patients received cyclophosphamide only (150-200 mg/kg); nine received cyclophosphamide, anti-thymocyte globulin and TBI (800 cGy); four received cyclophosphamide and anti-thymocyte globulin; two each received either cyclophosphamide plus alemtuzumab or other chemotherapy; and one received cyclophosphamide plus total lymphoid irradiation. Seven patients died after autologous HSCT. Four of these deaths occurred early (11 to 79 days after transplant) and were related to the procedure (two interstitial pneumonitis, one central nervous system hemorrhage, one diffuse pulmonary alveolar hemorrhage). One of the patients who died with interstitial pneumonitis had received TBI as part of the pre-transplant conditioning regimen. The treatment-related mortality was 17% overall (including the three post-mobilization deaths and four post-transplant deaths) and 10.8% for the autologous HSCT procedure only. Of the 37 autologous HSCT recipients, 25 had positive clinical responses, four had no responses, one had a partial response after mobilization chemotherapy, and seven (19%), including six patients who had had initial responses, had disease progression. The median time for disease progression after autologous HSCT was 69 days (range, 49-255 days). Three late deaths (217 to 527 days after transplant) were related to disease progression. Of 29 evaluable patients, 20 (69%) had at least 25% improvement in skin scores, and two (7%) had deterioration in these scores. Although there was no significant change in pre-existing pulmonary function or pulmonary hypertension, pulmonary manifestations of systemic sclerosis did not progress. One patient with pre-existing renal disease had further deterioration after autologous HSCT, but there were no new occurrences of renal disease in the other patients. The median duration of followup was 12 months (range, 3–55 months), and the overall survival at one year after autologous HSCT was 73% (95% confidence interval, 58% to 88%).[45]

In a multicenter prospective phase I/II study in France, 11 patients with refractory systemic sclerosis were enrolled on a study of autologous PBPC mobilization with cyclophosphamide and granulocyte colony-stimulating factor (G-CSF), collection and CD34$^+$ cell selection of the PBPC product, and autologous HSCT.[46] Six of the patients were female, and five were male; the median age was 43 years (range, 17-62 years). The CD34 selec-

Table 2. Autologous hematopoietic stem cell transplantation for systemic sclerosis

No. Patients Enrolled/ No. Transplanted	Age (Range)	Gender M/F	Mobilization Regimen (No. of Patients)	HSC Source PB/BM	Preparative Regimen (No. of Patients)	TRM% (No. of Patients)	Outcome of HSCT (No. of Patients)	Survival (95% CI)/ Median Followup (Range)	References
41/37	41 (11.5-58.1)	7/34	CY + G-CSF (29) CY (1) G-CSF (10)	36/1	CY 150-200* (19) CY 120 + ATG + TBI 800** (9) CY 200 + ATG (4) CY 200 + ALEM (2) CY 200 + TLI (1) Other (2)	Overall 17% (7) HSCT 10.8% (4)	Resp (19) Resp then relapse (6) NR (4) Prog (1) Died w/disease (3)	73% (58-88%)/ 12 mo (3-55)	Binks et al, 2001[45]
12/11	43 (17-62)	5/6	CY + G-CSF (11)	10/1	CY 200 (9) CY + ALG (1) MEL 140*** + ALG	9.1% (1)	Resp (3) Resp then relapse (5) NR (2)	63.6% (ND)/ 18 mo (1-26)	Farge et al, 2002[46]
19/19	40 (23-61)	3/16	G-CSF	19/0	CY 120 + ATG + fx TBI 800 (19)	15.8% (3)	Resp (12) NR (1) Prog (3)	78.9% (60.6 -97.3%)/ 15 mo (8-45)	McSweeney et al, 2002[40]

* total cyclophosphamide dose in mg/kg; ** dose of total body irradiation in cGy; *** total melphalan dose in mg/m².

Abbreviations for Table 2: ALEM= alemtuzumab; ALG= anti-lymphocyte globulin; ATG= anti-thymocyte globulin; BM= bone marrow; CI= confidence interval; CY= cyclophosphamide; fx= fractionated; G-CSF= granulocyte colony-stimulating factor; HSC= hematopoietic stem cell; HSCT= hematopoietic stem cell transplantation; MEL= melphalan; ND= not determined; NR= no response; PB= peripheral blood progenitor cells; Prog= progression; Resp= response; TBI= total body irradiation; TLI= total lymphoid irradiation; TRM= treatment-related mortality

tion procedure resulted in a three-log reduction in the content of T cells in the PBPCs. Nine patients received a preparative regimen of cyclophosphamide (200 mg/kg) alone, followed by transplantation with CD34 selected autologous PBPCs. One patient received melphalan (140 mg/m^2) plus anti-lymphocyte globulin, followed by infusion of unselected autologous PBPCs because of a low yield of CD34$^+$ cells in the apheresis product. One patient received cyclophosphamide plus anti-lymphocyte globulin, followed by infusion of autologous bone marrow cells because of failure of mobilization of PBPCs. All patients had hematologic reconstitution, but one patient with underlying cardiac dysfunction died with pneumonitis and multiorgan failure four weeks after autologous HSCT, for a procedure-related mortality of 9.1%. Eight of 11 patients had major (six patients) or partial (two patients) initial responses after autologous HSCT, and three patients had no responses. However, five of the eight initial responders (three major, two partial) had relapses of systemic sclerosis within 12 months (range, 9-12) after autologous HSCT. At a median of 18 months after autologous HSCT (range, 1-26), seven patients were alive, including five with sustained initial or secondary (salvage) major responses, one with no initial response, and one with progression of disease after an initial partial response. The four deaths occurred as a result of transplant-related complications (as noted above), progressive systemic sclerosis without any initial response to autologous HSCT, or progression of disease after either major or minor response to autologous HSCT.[46]

A North American (Seattle) multicenter study[40] previously reported in part with the EBMT experience[45] evaluated autologous transplantation of G-CSF-mobilized, CD34 selected PBPCs and a preparative regimen of TBI (800 cGy in four fractions), equine anti-thymocyte globulin and cyclophosphamide (120 mg/kg) in 19 patients with systemic sclerosis who met the proposed consensus criteria for autologous HSCT.[42] Sixteen of the patients were female, and three were male; the median age was 40 years (range, 23-61 years). Four patients had apparent self-limited exacerbations of skin inflammation during G-CSF administration. Three of the ten patients who were seropositive for cytomegalovirus (CMV) developed CMV antigenemia that required and responded to treatment with ganciclovir. Three patients (15.8%) died with transplant-related complications. Two patients who had pulmonary manifestations of systemic sclerosis before autologous HSCT died with pulmonary toxicity and respiratory failure at 58 days and 79 days, respectively, after transplant. After these deaths, the protocol was amended to include shielding of the lungs during TBI, an approach successfully used for the last 11 patients enrolled on this trial. A third patient died 64 days after autologous HSCT with Epstein-Barr virus (EBV)-associated post-transplant lymphoproliferative disease (PTLD). This patient also had recurrent acyclovir-resistant herpes simplex virus infections after autologous HSCT and, in contrast to the other patients, had no evidence of T cell recovery at one month after transplant. Thirteen of the 15 evaluable patients (86.7%) at three months after transplant and 12 of 12 evaluable patients at one year after transplant had objective disease responses, as measured by at least 25% improvement in modified Rodnan skin score or by at least 0.4 point improvement in the modified Health Assessment Questionnaire Disability Index (mHAQ-DI).[47] Substantial and significant improvement in modified Rodnan skin score occurred within the first three months after autologous HSCT and continued for two to three years afterwards. Four patients had progressive or nonresponsive disease (two pulmonary, two renal), and one of these patients died 123 days after autologous HSCT. In contrast to the studies from the EBMT and French multicenter groups, no patients in this series who had initial responses developed subsequent relapses of systemic sclerosis. At a median of 15 months after transplant (range, 8-45 months), 15 of 19 patients were alive, with a projected two-year survival of 78.9% (95% confidence interval, 60.6%-97.3%). Among nine evaluable patients with detectable anti-Scl-70 antibodies before autologous HSCT, only one became seronegative after transplant.[40]

Conclusions

Autologous HSCT can lead to objective responses in from 50% to more than 60% of patients with systemic sclerosis. The observed relapses of disease in some patients who had initial major responses suggest the need for more intensive immunosuppression during the preparative regimen, post-transplant maintenance immunotherapy, or both. The absence of reported relapses in patients with systemic sclerosis who received aggressive pre-transplant conditioning with cyclophosphamide, anti-thymocyte globulin and TBI,[40] compared with the higher relapse rate observed after cyclophosphamide alone,[46] supports the value of multiagent preparative regimens to optimize in vivo depletion of residual immunocompetent cells. However, the unique sensitivity of scleroderma to TBI related toxicity indicates that TBI should be eliminated from future phase II and III trials. The use of cyclophosphamide to mobilize autologous PBPCs may also provide cytoreduction of lymphocytes. Maintenance immunosuppression after autologous HSCT, analogous to post-transplant therapies in neoplastic diseases, is an intriguing and potentially important question that can only be addressed in a multicenter study and that is problematic because of the lack of proven effectiveness of any specific agent in the treatment of systemic sclerosis.

The procedure-related mortality (approximately 10% to 15%) reported after autologous HSCT in patients with systemic sclerosis is of concern and is substantially higher than that observed after autologous HSCT for neoplastic diseases (1-2% or less). Underlying organ dysfunction, especially cardiac and perhaps pulmonary or renal, may contribute to this unacceptably high mortality. Using stringent criteria for patient selection,[42] performing careful pre-transplant assessment of cardiopulmonary and renal status and minimizing non-hematologic, non-immunologic toxicities of the preparative regimens should help decrease the procedure-related mortality.

Data on autoimmune T and B cell responses and immune reconstitution after autologous HSCT for systemic sclerosis are still limited. For example, in the case report of Martini et al[44] and the series reported by McSweeney et al,[40] sustained clinical improvement occurred despite persistence of autoimmune antibodies (e.g., anti-Scl-70, anti-nuclear antibody). Thoughtful, in-depth mechanistic studies are essential to understanding the immunoregulatory effects of autologous HSCT on systemic sclerosis and other autoimmune disorders and should be incorporated as correlative companion studies to the phase II and phase III clinical trials of autologous HSCT in these diseases.

References

1. Medsger TA Jr. Systemic sclerosis (scleroderma): Clinical aspects. In: Koopman WJ, ed. Arthritis and Allied Conditions. 13th ed. Williams & Wilkins, 1997:1433-1464.
2. Black C, Dieppe PK, Huskisson T et al. Regressive systemic sclerosis. Ann Rheum Dis 1986; 45:384-8.

3. Medsger TA Jr, Masi AT. Epidemiology of progressive systemic sclerosis. Clin Rheum Dis 1979; 5:15-25.
4. Lee P, Langevitz P, Alderdice CA et al. Mortality in systemic sclerosis (scleroderma). QJ Med 1992; 82:139-48.
5. Steen VD, Costantino JP, Shapiro AP et al. Outcome of renal crisis in systemic sclerosis; relation to availability to angiotensin converting enzyme (ACE) inhibitors. Ann Intern Med 1990; 113:352-7.
6. Altman RD, Medsger TA Jr, Bloch DA et al. Predictors of survival in systemic sclerosis (scleroderma). Arthritis Rheum 1991; 34:403-13.
7. Silman AJ. Scleroderma and survival. Ann Rheum Dis 1991; 50:267-9.
8. Spooner MS, LeRoy EC. The changing face of severe scleroderma in five patients. Clin Exp Rheumatol 1990; 8:101-5.
9. Giordano M, Valentini G, Migliaresi S et al. Different antibody patterns and different prognoses in patients with scleroderma with various extent of skin sclerosis. J Rheumatol 1986; 13:911-6.
10. Clements P, Lachenbruch P, Seibold J et al. Inter-and intraobserver variability of total skin thickness score (modified Rodnan TSS) in systemic sclerosis. J Rheumatol 1995; 22:1281-5.
11. Medsger TA Jr. Progressive systemic sclerosis. Clin Rheum Dis 1983; 9:655-70.
12. Seibold JR, Furst D, Medsger TA Jr. Why everything (or nothing) seems to work in the treatment of scleroderma. J Rheumatol 1992; 19:673-6.
13. White B, Bauer EA, Goldsmith LA et al. Guidelines for clinical trials in systemic sclerosis (scleroderma): I. Disease-modifying intervention. Arthritis Rheum 1995; 38:351-60.
14. Steen VD, Medsger TA Jr. Case-control study of corticosteroids and other drugs that either precipitate or protect from the development of scleroderma renal crisis. Arthritis Rheum 1999; 42:1613-9.
15. Steen VD, Medsger TA Jr, Rodnan GP. D-penicillamine therapy in progressive systemic sclerosis (scleroderma). Ann Intern Med 1982; 97:652-8.
16. Jimenez SA, Andrews RP, Myers AR. Treatment of rapidly progressive scleroderma (PSS) with D-penicillamine: A prospective study. In: Black CM, Myers AR, eds. Systemic Sclerosis (Scleroderma): Current Topics in Rheumatology. New York: Gower Medical Publishing, 1985:387-93.
17. Sattar MA, Guindi TR, Sugathan TN. Penicillamine in systemic sclerosis: A reappraisal. Clin Rheumatol 1990; 9:517-22.
18. Jansen GT, Baraza DF, Ballard JL et al. Generalized scleroderma. Treatment with an immunosuppressive agent. Arch Dermatol 1968; 97:690-8.
19. van den Hoogen FMJ, Boerbooms AMT, van den Putte LBA et al. Low dose methotrexate treatment in systemic sclerosis. J Rheumatol 1991; 18:1763-4.
20. Bode BY, Yocum DE, Gall EP et al. Methotrexate (MTX) in scleroderma: Experience in 10 patients. Arthritis Rheum 1990; 33:566.
21. Furst DE, Clements PJ, Hillis S et al. Immunosuppression with chlorambucil vs. placebo for scleroderma. Results of a three-year, parallel, randomized, double-blind study. Arthritis Rheum 1989; 32:584-93.
22. Casas J, Saway PA, Villareal I et al. 5-fluorouracil in the treatment of scleroderma: A randomized, double-blind, placebo controlled international collaborative study. Ann Rheum Dis 1990; 49:926-8.
23. Alarcon-Segovia D. Progressive systemic sclerosis: Management. Part IV. Colchicine. Clin Rheum Dis 1979; 5:294-302.
24. Guttadauria M, Diamond H, Kaplan D. Colchicine in the treatment of scleroderma. J Rheumatol 1977; 4:272-5.
25. Zachariae H, Halkier-Sorensen L, Heickendorff L et al. Cyclosporin A treatment of systemic sclerosis. Br J Dermatol 1990; 122:677-81.
26. Vayssairat M, Baudot N, Boitard C et al. Cyclosporin therapy for severe systemic sclerosis associated with anti-Scl-70 autoantibody. J Am Acad Dermatol 1990; 22:695-6.
27. Clements PJ, Lachenbruch PA, Sterz M et al. Cyclosporine in systemic sclerosis: Results of a forty-eight-week open safety study in ten patients. Arthritis Rheum 1993; 36:75-83.
28. Kahan A, Amor B, Menkes CJ et al. Recombinant interferon-gamma in the treatment of systemic sclerosis. Am J Med 1989; 87:237-47.
29. Freundlich B, Jimenez SA, Steen VD et al. Treatment of systemic sclerosis with recombinant interferon-gamma. A phase I/II clinical trial. Arthritis Rheum 1992; 35:1134-42.
30. Hein R, Behr J, Hundgen M et al. Treatment of systemic sclerosis with gamma-interferon. Br J Dermatol 1992; 126:496-501.
31. Tarkowski A, Lindgren I. Beneficial effects of antithymocyte globulin in severe cases of progressive systemic sclerosis. Transplantation Proc 1994; 26:3197-9.
32. Goronzy JJ, Weyand CM. Long-term immunomodulation effects of T-lymphocyte depletion in patients with systemic sclerosis. Arthritis Rheum 1990; 33:511-9.
33. Dau PC, Kahalen MD, Sagebiel RW. Plasmapheresis and immunosuppressive drug therapy in scleroderma. Arthritis Rheum 1981; 24:1128-36.
34. Schmidt C, Schooneman F, Siebert P et al. Traitement de la sclerodermie generalisee par echanges plasmatiques. Ann Med Interne (Paris) 1988; 139(1):20-2.
35. Capodicasa G, DeSanto NG, Galione A et al. Clinical effectiveness of apheresis in the treatment of progressive systemic sclerosis. Int J Artific Organs 1983; 6:81-6.
36. Guillevin L, Amoura Z, Merviel PH et al. Treatment of progressive systemic sclerosis by plasma exchange: Long-term results in 40 patients. Int J Artific Organs 1990; 13: 125-8.
37. Rook AH, Freundlich B, Jegasothy BV et al. Treatment of systemic sclerosis with extracorporeal photochemotherapy: Results of a multicenter trial. Arch Dermatol 1992; 128:337-46.
38. Trentham D. Photochemotherapy in systemic sclerosis: the stage is set. Arch Dermatol 1992; 128:389-90.
39. Fries JF, Seibold JR, Medsger TA Jr. Photopheresis for scleroderma? No! J Rheumatol 1992; 19:1011-3.
40. McSweeney PA, Nash RA, Sullivan KM et al. High-dose immunosuppressive therapy for severe systemic sclerosis: Initial outcomes. Blood 2002; 100:1602-10.
41. Donnenberg AD, Margolick JB, Donnenberg VS. Lymphopoiesis, apoptosis and immune amnesia. Ann NY Acad Sci 1996; 770:213-26.
42. Clements PJ, Furst DE. Choosing appropriate patients with systemic sclerosis for treatment by autologous stem cell transplantation. J Rheumatol 1997; 48(suppl):85-8.
43. Tyndall A, Black C, Finke J et al. Treatment of systemic sclerosis with autologous haemopoietic stem cell transplantation. Lancet 1997; 349:254.
44. Martini A, Maccario R, Ravelli A et al. Marked and sustained improvement two years after autologous stem cell transplantation in a girl with systemic sclerosis. Arthritis Rheum 1999; 42:807-11.
45. Binks M, Passweg JR, Furst D et al. A Phase I/II trial of autologous stem cell transplantation in systemic sclerosis: Procedure related mortality and impact on skin disease. Ann Rheum Dis 2001; 60:577-84.
46. Farge D, Marolleau JP, Zohar S et al. Autologous bone marrow transplantation in the treatment of refractory systemic sclerosis: Early results from a French multicentre phase I-II study. Br J Haematol 2002; 119:726-39.
47. Clements PJ, Wong WK, Hurwitz EL et al. The Disability Index of the Health Assessment Questionnaire is a predictor and correlate of outcome in the high-dose versus low-dose penicillamine in systemic sclerosis trial. Arthritis Rheum 2001; 44:653-61.

CHAPTER 47

High-Dose Immunosuppressive Chemotherapy with Autologous Stem Cell Support for Chronic Autoimmune Thrombocytopenia

Richard D. Huhn, Patrick F. Fogarty, Ryotaro Nakamura and Cynthia E. Dunbar

Introduction

Autoimmune idiopathic thrombocytopenia (AITP) is a disorder of low platelet counts in which antibodies directed against platelet and megakaryocyte surface proteins cause platelet destruction in reticuloendothelial organs and inhibit platelet production.[1,2] A compensatory increase in marrow megakaryocytes is an important diagnostic hallmark but is often not observed. Platelet destruction outpaces platelet production, leading to varying degrees of thrombocytopenia. Acute AITP is more common in children than adults and usually has a short self-limited course following an acute viral infection.[3,4] Chronic AITP is more common in adults and is frequently refractory to conventional therapies.[5,6] Chronic AITP is typically idiopathic (although it may be associated with a variety of other autoimmune disorders or lymphoproliferative diseases).[7-12]

The immunopathology of chronic AITP is complex and is not well understood. In the majority of cases, anti-platelet autoantibodies are directed against glycoproteins GPIIb/IIIa (the fibrinogen receptor) or GPIb/IX (the von Willebrand factor receptor).[13-15] Many additional surface protein targets have also been identified.[16] T-lymphocytes recognizing the specific surface antigens are responsible for initiation and propagation of the autoimmune reactions, providing helper functions for generation of autoantibodies by mature B-lymphocytes.[17,18] The initial immunizing events are not known—it has been suggested that mimicry of self antigens or exposure of cryptic epitopes due to the actions of infectious agents could lead to autoimmunization in AITP.[19] According to credible hypotheses, epitope spreading may amplify the cellular and humoral antiplatelet reactivity, as is postulated in other autoimmune disorders.[20]

AITP in adults usually presents with spontaneous microvascular bleeding in the skin or mucosal surfaces. The most common presenting symptoms and signs are petechiae, epistaxis and unexplained bruising.[21] More serious bleeding such as central nervous system or gastrointestinal hemorrhages do not usually occur unless there has been significant thrombocytopenia for long periods of time or there have been compound pathological insults related to primary autoimmune disease or therapy (e.g., endothelial cell injuries related to chemotherapy of concomitant malignancies).

Conventional Therapies for Adult AITP

Mild thrombocytopenia without hemorrhagic complications does not require urgent treatment.[3] For those patients with significant thrombocytopenia (platelet count <30,000/µL) the usual first line of treatment is a short course of prednisone or other immunosuppressive corticosteroid, to which approximately 60% of new cases of adult AITP will have an initially favorable response. Unfortunately, most such responses are transient and thrombocytopenia often recurs rapidly upon tapering the steroid dosage. Splenectomy is frequently effective in restoring platelet counts but must be considered carefully in light of the potential surgical and infectious complications.[22,23] Alternatively, therapeutic trials of high-dose intravenous immune globulin or anti-Rh(D) may be considered before splenectomy or if thrombocytopenia recurs following splenectomy.[24] Response to infusion of immunoglobulin or anti-Rh(D) may predict responsiveness to splenectomy.[25] Conversely, the response to anti-Rh(D) is generally lost following splenectomy.[25a] Some experts recommend doing scintigraphic or surgical searches for accessory spleen tissue if thrombocytopenia recurs following splenectomy, especially if Howell-Jolly bodies that appeared after the initial splenectomy disappear in conjunction with the recurrence of thrombocytopenia.[26]

For patients for whom the above measures above are ineffectual, options for subsequent therapies are selected empirically.[3,6] Many case reports and small non-randomized case series describe a wide variety of immunsuppressive therapies. In one case series, high dose dexamethasone infusion resulted in clinically significant responses in 10 patients with chronic AITP.[27] However, subsequent reports on high-dose corticosteroids have been inconclusive.[28,29]

Recently, possible associations between some autoimmune diseases and infection with *Helicobacter pylori* have been described.[30] There have been a few case series reports in which patients with chronic AITP were successfully treated with antibiotics effective against *H. pylori*. Although one report indicated a very high rate of durable responses, another case series found no responses at all (ref. 31 and James Bussel, personal communication).

A wide variety of immunomodulatory therapies have been tried with limited success for treatment of chronic refractory AITP. Examples include cyclosporin A, mycophenolate mofetil, dapsone, vitamin C, interferons and plasmapheresis with or without protein A column immunoabsorption. Recently, successful amelioration of chronic AITP, among various autoimmune diseases, has been effected with directed anti-B lymphocyte therapy using rituximab (anti-CD20 monoclonal antibody).[31a] Several oral cytotoxic agents that non-specifically suppress immunity through their toxicity against activated or proliferating lymphocytes, including cyclophosphamide, azathioprine, etoposide and the vinca alkaloids, have been reported to effect low rates of remission of chronic autoimmune diseases.[3] For AITP, cyclophosphamide has been the most extensively studied of these drugs. In a small proportion of cases, low-dose oral cytotoxic drugs may induce durable remissions of thrombocytopenia.[32] However, chronic exposure to low-dose cyclophosphamide or other cytotoxic drugs may induce neoplastic changes, especially of hematopoietic tissues.[33] In a study reported by Reiner et al, larger doses of cyclophosphamide, administered parenterally to 20 patients with refractory AITP in cyclical pulses of 1.0 to 1.5 gm/M^2, were effective in obtaining complete responses in 13 patients and partial responses in an additional 4 patients.[34] Five of the complete responders relapsed (two were rescued with repeat courses of pulse cyclophosphamide) and two of the partial responders relapsed. Cytotoxic drugs in moderately high doses are highly myelosuppressive with attendant risks including bacterial and opportunistic infections and exacerbation of transfusion requirements.

Intensified immunosuppression with combination chemotherapy, especially cyclophosphamide-based regimens (as have been effective for lymphoproliferative diseases) have been used for refractory patients. Figueroa et al treated 10 refractory patients (having failed an average of 6.8 previous therapies) with three to eight cycles of regimens consisting of cyclophosphamide and prednisone combined with vinca alkaloids, procabazine and/or etoposide.[35] Six patients had complete responses (platelet counts >180,000/μL) of which 4 were considered durable, and two patients had partial responses (platelet counts >50,000/μL). In a follow-up report on these and two additional patients, there were described a total of 5 durable complete responses, two complete responses with subsequent relapse (one of whom achieved complete response from a later alternative therapy), two partial responses (one with relapse) and three non-responses.[36] Two of the non-responding patients, and the patient who had obtained a partial response followed by relapse, died of hemorrhagic events. The patient who had a complete response followed by relapse subsequently obtained partial response with prednisone and danazol but died of a fungal infection. The other patient with partial remission died of transfusion-associated hepatitis C and one of the complete responding patients died of a stroke (probably coincidental). Very little else has been published on the long-term morbidity and mortality of chronic refractory ITP. One study of longitudinal outcomes relative to response also indicated that patients who failed to achieve and maintain platelet counts >30,000/μL had significantly increased rates of hospitalization and high risks of bleeding and infection.[37] These experiences reinforce the urgency of attempting to obtain durable responses for patients with chronic AITP, as those patients who remain chronically thrombocytopenic are apparently at great risk of serious hemorrhagic or infectious complications.

High-Dose Chemotherapy with Autologous Stem Cell Support for Autoimmune Cytolytic Diseases

Background

Proposals to treat autoimmune diseases by high dose chemotherapy and hematopoietic stem cell began to be propounded and tested in the 1990's. Support for this approach was initially based on observations that: (1) patients with malignancies (e.g., lymphomas) and concomitant autoimmune diseases who underwent autologous transplantation sometimes obtained significant clinical improvements of the associated autoimmune syndromes, and (2) occasionally it was observed that bone marrow transplant recipients of hematopoietic stem cell allografts from donors with autoimmune diseases eventually developed autoimmune disorders themselves.[38,39] Presumably, high-dose chemotherapy would ablate autoreactive T cells with the immune recovery following transplantation allowing 're-education' with tolerance to autoantigens. Thus, it was subsequently suggested that lymphocyte depletion of the autografts should be performed to eliminate or reduce the burden of autoaggressive lymphocytes that could re-establish autoimmunity after transplantation, so as to facilitate immune re-education without autoreactivity.[39,40]

In 1997, Lim et al described the treatment of two patients with AITP by high-dose cyclosphosphamide with unmanipulated autologous peripheral blood stem cells.[41] Both patients had previously failed corticosteroids, intravenous immunoglobulin and splenectomy. Their myelo/immunosuppression regimen consisted of cyclophosphamide 50 mg/kg daily for 4 consecutive days, a regimen commonly used in allogeneic transplantation. Complications were minimal—neutropenia resolved on days 11 and 13, respectively, and neither patient developed culture-proven infections. Although both patients required platelet transfusions in the peri-transplant period, complete remissions of AITP (platelet counts >150,000/μL) were achieved within 5 weeks of treatment. As of the publication date, complete remissions were maintained for more than 11 and 7 months, respectively, but both patients subsequently relapsed.[42] Concurrently, however, Skoda et al reported their experience in treating a chronic AITP patient with high-dose cyclophosphamide and lymphocyte-depleted autologous blood stem cells.[43] The method of lymphocyte depletion of the graft was a combination of positive selection using a Ceprate A column and negative selection with a cocktail of anti-T and -B cell specific antibodies. The dose of lymphocytes administered in the graft was 1.9 x 10^5/kg. This patient did not obtain improvement of his platelet count. Marmont et al subsequently reported a similar case of failure of autotransplantation with a lymphocyte-depleted stem cell graft for AITP after conditioning with thiotepa and cyclophosphamide.[42] These results obviously ran counter to the theoretical expectation that lymphocyte depletion of the autograft should be more likely to induce remission of autoimmune disease. Of course, a sample size of one or two patients is inadequate to draw reliable conclusions from any therapeutic trial. In addition to the studies reported above, several anecdotal case reports of auto- and allotransplantation for autoimmune hemocytolytic diseases (e.g., Evans syndrome and autoimmune hemolytic anemia) have presented mixed results with respect to safety and therapeutic outcomes.[44-46]

Table 1. Responses to a phase I trial of high-dose cyclophosphamide with hematopoietic support

Complete response	Self-sustained platelet counts >100,000/µL independent of transfusions and all other therapies for at least 6 weeks.
Partial response	(a) Self-sustained platelet counts >50,000/µL without hemostatic complications for 6 weeks, or (b) Reduced bleeding complications and transfusion requirements related to clear and sustained increases in platelet counts.

On the presumption that the risks of adverse consequences of prolonged myelosuppression are unacceptable for patients with non-malignant diseases, it has been widely believed that hematopoietic stem cell support would be needed for safe administration of intensive therapy such as high-dose cyclophosphamide. Brodsky et al challenged this assumption and investigated high-dose cyclophosphamide without stem cell support for the treatment of 8 patients with various refractory severe autoimmune diseases, including 3 patients with cytolytic syndromes (one autoimmune hemolytic anemia, one Evans syndrome and one AITP).[47] Two of these patients (autoimmune hemolytic anemia and AITP) did not obtain remissions and eventually died of complications of their autoimmune diseases. The patient with Evans syndrome did obtain a partial remission but continued to require oral steroids. It is important to note that the time to neutrophil recovery among the eight patients in the study was much more prolonged than would be expected with stem cell rescue (median 17 days, range 11 to 22 days). During that time, patients would be at high risk of neutropenic complications. Not surprisingly, 6 of the 8 patients required empiric antibiotics for febrile neutropenia, although no details of infectious pathogens or complications were reported.

Systematic Phase I/II Trial of Lymphocyte-Depleted Grafts—National Heart Lung & Blood Institute (NHLBI) Trial 97H-0154

The limited and contradictory data regarding feasibility, safety, and potential therapeutic benefits of high-dose cytotoxic chemotherapy with blood stem cell support ('autotransplantation') with or without lymphocyte depletion, highlight the need for systematic clinical investigations of this approach to the treatment of chronic autoimmune thrombocytopenia. Moreover, AITP represents an excellent clinical model for an autoimmune disease to be treated by immunoablative therapy. In contrast to the myriad symptoms, signs and outcome parameters of other autoimmune syndromes, the platelet count is a single quantitative parameter to follow in AITP

To begin to address the questions of feasibility and safety and to pave the way for further systematic study, the National Heart Lung and Blood Institute conducted a phase I trial of high-dose cyclophosphamide with hematopoietic support using lymphocyte-depleted autologous blood stem cells for treatment of patients with chronic refractory AITP. Preliminary evidence of efficacy was a lesser objective of the study. Responses were prospectively defined in Table 1.

The treatment schema is shown in Figure 1. Fourteen patients, aged 17 to 52 years old with disease durations of 6 months to 40 years, were evaluable for safety and therapeutic response as of November, 2001 (the study continues to enroll patients at the time of this writing). All patients had failed to obtain durable responses or had relapsed after standard treatments (corticosteroids, splenectomy and intravenous immune globulin) and had unsuccessfully undergone one or more second-line therapies. All patients had significant hemostatic complications jeopardizing life, health or daily activities. Bone marrow examinations documented normal or increased numbers of megakaryocytes; absence of morphologic evidence of marrow failure, dysplasia and neoplastic or infiltrative diseases; and absence of detectable cytogenetic abnormalities. Five patients had concurrent autoimmune hemolytic anemia (Evans syndrome).

Blood Stem Cell Mobilization, Collection and Processing

Blood stem cells were mobilized with G-CSF 10 µg/kg/day by intravenous infusion on days ⁻12 to ⁻7 (day of stem cell infusion is Day 0) in the evening. On the day following the fifth dose of G-CSF, large-bore triple-lumen apheresis catheters were inserted in femoral veins (with platelet transfusion support) and leukapheresis was performed, processing 15-20 liter blood volume per session. Leukapheresis products were enriched in CD34⁺ cells by immunomagnetic selection using the Isolex™ 300i device (with concomitant depletion of lymphocytes). If the final product did not contain significantly greater than 2x10⁶ CD34⁺ cells per kg (patient weight), an additional dose of G-CSF was administered and the apheresis procedure was repeated on the next day. Three patients failed to obtain sufficient collections after two sessions and underwent repeat mobilization courses with G-CSF after a rest period of approximately 2 weeks. Mobilization of stem/progenitor cells was generally brisk in comparison to previously-treated cancer patients undergoing mobilization for transplantation, reaching a mean peak peripheral blood concentration of 109±73 CD34⁺cells/µL. The stem cell grafts after CD34 enrichment were generous, containing a mean of 4.5±2.0 x10⁶ CD34⁺ cells/kg. The overall mean lymphocyte content of the grafts after CD34⁺ cell selection was 0.9±1.2x10⁵ CD3⁺ cells/kg.

Patients generally tolerated the mobilization and collection procedures well. Side effects of G-CSF were minor, consisting of musculoskeletal discomfort that was transient and treatable with oral analgesics. There was no discernable effect of the G-CSF on platelet counts (such effect would have been difficult to identify in the setting of ongoing autoimmune thrombocytopenia and apheresis procedures). Apheresis catheters were inserted with platelet transfusion support without significant difficulty. Two patients developed femoral hematomas that were managed by administration of local pressure and platelet transfusion. There were no episodes of significant citrate toxicity or other adverse events requiring discontinuation of the apheresis procedures.

High-Dose Cyclosphosphamide and Post-Stem Cell Hematopoietic Recovery

Cyclophosphamide (four doses of 50 mg/kg IV given consecutively on days ⁻5 to ⁻2 (with MESNA for uroprotection) was tolerated well. One patient experienced hemorrhagic cystitis that resolved in 4 days with platelet transfusion support. The stem

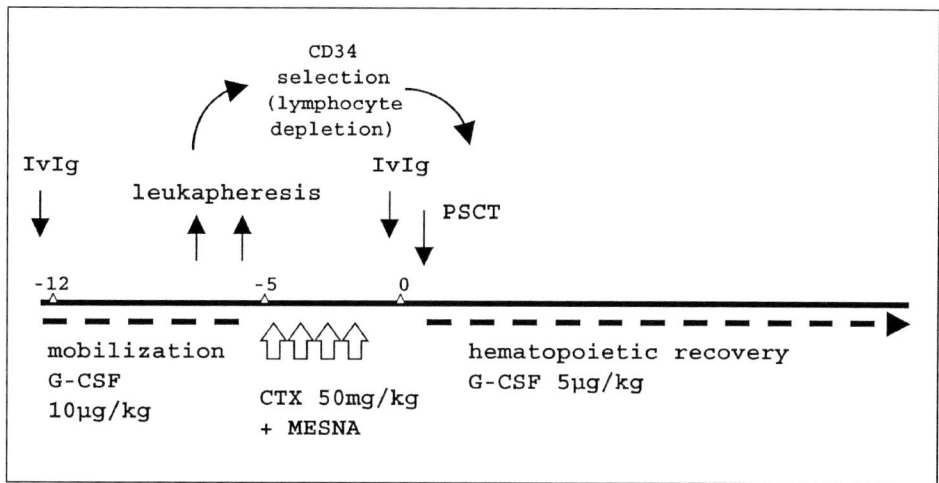

Figure 1. Treatment schema. See text for details.

cell graft was thawed and infused in one step on day 0. G-CSF 5 μg/kg/day was given to accelerate hematopoietic recovery. The mean time to neutrophil recovery post-stem cell infusion was 9.0±0.6 days. All patients experienced neutropenic fever that was immediately responsive to empiric antibiotics.

Platelet Transfusions

Patients received prophylactic platelet transfusion support if required to maintain platelet counts >10,000/μL and otherwise (a) to prepare for line insertion or (b) to treat bleeding. In addition, intravenous immune globulin 1 g/kg was infused on day ⁻12 and day 0 to help reduce concurrent platelet destruction during apheresis (platelet consumption) and during the post-transplant period of myelosuppression. During the phase of stem cell mobilization and collection, the median platelet usage was 2 transfusions (3 patients required only one platelet transfusion to prepare for line insertion). Three patients experienced minor hemostatic problems (vaginal bleeding and epistaxis) during the period of myelosuppression; all were managed satisfactorily by platelet transfusion and supportive care. Overall, patients required a median of 9 platelet transfusions during this period. Corrected count increments from platelet transfusions in these patients were not substantially smaller than would be expected in patients with thrombocytopenia due to non-cytolytic diseases (i.e., marrow failure).[48]

Therapeutic Responses

Six patients obtained complete responses and 2 patients obtained partial responses, as defined by the prospective criteria listed in Table 1. In most cases, responses were immediately apparent but, in two cases, complete responses occurred over a period of several weeks following transplant. Interestingly, both of those patients eventually responded to therapies they had previously failed (in one case, prednisone reversed a decline in the platelet count; in the other, a gradual but stable elevation of platelet count occurred while on danazol). All responses are durable as of the time of this writing. A platelet count response curve of one representative responding patient is shown in Figure 2. The longest follow-up duration is 45 months.

Immunological Parameters

Humoral and cellular immunologic correlates were sought to attempt to identify pathophysiologic parameters that could predict responses or relapses and to possibly better explain disease mechanisms underlying chronic autoimmune platelet destruction. Serum anti-platelet glycoprotein Ib and IIIa antibody titers were measured by a modification of the microbead adaptation of the monoclonal antibody to immobilized platelet antigen (MAIPA) assay pre-transplant and at various times post-transplant.[49,50] Unfortunately, while a few patients had substantial decreases of anti-glycoprotein titers after transplant, most did not, and there was no apparent relationship between change of anti-glycoprotein titer and clinical response to transplant. Additionally, immunophenotyping of circulating lymphocytes was performed pre- and post-transplant to attempt to estimate and compare the 'level' of immune activation (i.e., expression of CD25 and HLA-DR on $CD3^+$, $CD4^+$ and $CD8^+$ cells) and its response to transplant. Again, no clear relationship was discernable in the small sample of patients studied in this trial.

Context

An important principle of clinical research is reinforced by examining the history of therapeutic approaches to AITP: in order to detect outcome trends, it is necessary to study substantial numbers of relatively homogeneous subjects treated by uniform methods. The few anecdotal reports and small case series regarding high-dose therapy with hematopoietic stem cell support for autoimmune hemocytolytic diseases gave little indication that immunoablation could effectively ameliorate the autoimmune processes. The results of the NHLBI trial suggest that approximately 50% of patients with chronic AITP may obtain satisfactory outcomes from high-dose therapy. A larger trial (or extension of the ongoing trial) will be necessary to solidify this conclusion and to examine the issue of whether high-dose therapy without stem cell support may be a comparable approach. A number of other investigators have shown that immunoablative therapy with blood stem cell support can effectively halt or reverse the pathology of a variety of autoimmune diseases, as described in detail in other chapters of this text. AITP appears to be an appropriate target for this therapeutic approach. Furthermore, the platelet count presents a convenient quantitative response parameter for further study to refine the methods and to study the underlying mechanisms.

The role of lymphocyte depletion of the autograft in promoting therapeutic responses has yet to be determined. It has been previously demonstrated that some patients had early relapses of

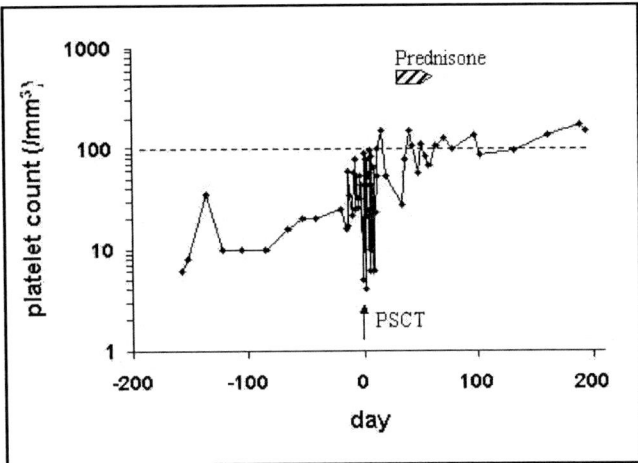

Figure 2. Platelet response of one representative patient following high-dose cyclophosphamide with lymphocyte-depleted blood stem cell support (PSCT). Note that a short tapering course of prednisone was required to 'rescue' a threatened relapse and thereafter the complete response was durable without therapy. As of this writing, the response has persisted for >45 mos. The dashed line represents the platelet count prospectively designated for complete response.

concomitant autoimmune syndromes after receiving unmanipulated hematopoietic cell grafts for a variety of malignant disorders.[40] T cell expansions present in these patients before transplantation may regenerate by ten months post-transplantation, suggesting that the expansions regenerate from mature lymphocytes present in the autografts.[51] Thus, elimination of potentially autoaggressive T cell clones may be possible only by extensive T cell depletion. It will be important to perform larger prospective randomized trials comparing outcomes of high-dose therapy with lymphocyte-depleted vs. unmodified autograft support or no autograft support using otherwise comparable subjects and methods between the arms of such a trial.

Stem cell mobilization and collection procedures were not associated with serious adverse events in the anecdotal reports or the NHLBI trial. Furthermore, blood stem cell support promotes accelerated neutrophil recovery after high-dose cyclophosphamide therapy, and thus appears to make the immunoablative therapy safer and more tolerable. Although it is yet too early to recommend high-dose therapy for non-refractory patients, if the risks of transplant-related morbidity can be shown to be no greater than the morbidity from splenectomy, such a trial should be considered.

In responding patients, achievement of self-sustained safe platelet counts was quite delayed in a few cases. Theoretically, ablation of activated autoreactive T lymphocytes might not be expected to result in immediate cessation of platelet destruction because the anti-platelet antibodies and the plasma cells that make them could persist for many weeks or months post-transplant. Therefore, some patients may require short term support with other immunosuppressive agents after transplant before achieving sustained remission.

Open Questions and Important Future Investigations

As mentioned above, titers of anti-platelet antibodies have not been generally useful in predicting or understanding therapeutic responses in AITP. At present, there are no predictive immune parameters known. Some patients with AITP have been shown to have T-cells that proliferate in response to glycoprotein IIb/IIIa and to provide helper function for anti-IIb/IIIa autoantibodies in vitro.[52] Undoubtedly, T lymphocytes directed against additional platelet antigens will also be shown to play roles in the pathology of AITP in some patients.

Recently developed molecular techniques using polymorphisms in the beta-variable region of the T cell receptor DNA could be used to study T cell repertoires in patients with AITP. Limited TCR diversity has been demonstrated in patients with various diseases involving autoimmunity (e.g., rheumatoid arthritis, renal allograft rejection, graft-vs.-host disease and alloimmune thrombocytopenia) and immune dysregulation (e.g., paroxysmal nocturnal hemoglobinuria and aplastic anemia).[53-60] The T cell repertoire tended to normalize in aplastic anemia patients who achieved unmaintained remission after immunosuppressive treatment with cyclosporine, suggesting that the establishment and propogation of T cell clones contribute to the pathogenesis of that disease.[60]

The availability of precise quantitative assays for anti-platelet T-lymphocytes would enable mechanistically significant studies of important immune parameters. Such assays could be based on the so-called tetramer assay in which T-lymphocytes having receptors for specific antigens can be detected by flow cytometry after they bind peptides that secondarily bind fluorochrome-linked avidin-biotin complexes.[61] Alternatively, flow cytometric detection of intracellular cytokine production by T lymphocytes exposed to autoantigens (the so-called Fastimmune™ assay) could be used in a similar manner.[62,63] It would also be of significant interest to monitor the depletion of autoreactive lymphocytes from blood stem cell grafts using such assays and to attempt to examine associations between relative depletion and clinical responses.

Finally, the potential benefit of graft-vs.-autoimmunity effect that could be provided by allotransplantation may deserve further investigation. Marmont and colleagues treated a patient with Evans syndrome by allogeneic reduced intensity transplantation and graded incremental donor lymphocyte infusions.[64] They were able to achieve donor chimerism associated with grade II graft-vs.-host disease and complete clinical and biological remission of autoimmune cytolysis in this patient. Further case series and/or formal trials of this approach with longer follow-up will be needed in order to better assess whether such a mechanism may contribute to remission of AITP to justify the risks of allogeneic transplantation.

References

1. George JN, El-Harake MA, Aster RH. Thrombocytopenia due to enhanced platelet destruction by immunologic mechanisms. In: Beutler E, Lichtman MA, Coller BS et al, eds. Williams Hematology. New York: McGraw Hill; 1995:1314-1355.
2. George JN, Raskob GE. Idiopathic thrombocytopenic purpura: Diagnosis and management. Am J Med Sci 1998; 316:87-93.
3. George JN, Woolf SH, Raskob GE et al. Idiopathic thrombocytopenic purpura: A practice guideline developed by explicit methods for the American Society of Hematology. Blood 1996; 88:3-40.
4. Ritchey AK, Keller FG. Hematologic manifestations of childhood illness. In: Hoffman R, Benz EJ Jr, Shattil SJ et al, eds. Hematology—Basic Principles and Practice. New York: Churchill Livingstone; 2000:2391-2410.
5. McMillan R. Therapy for adults with refractory chronic immune thrombocytopenic purpura. Ann Intern Med 1997; 126:307-314.
6. George JN, Kojouri K, Perdue JJ et al Management of patients with chronic, refractory idiopathic thrombocytopenic purpura. Semin Hematol 2000; 37:290-298.

7. McDonagh JE, Isenberg DA. Development of additional autoimmune diseases in a population of patients with systemic lupus erythematosus. Ann Rheum Dis 2000; 59(3):230-2.
8. Kirshner JJ, Zamkoff KW, Gottlieb AJ. Idiopathic thrombocytopenic purpura and Hodgkin's disease: Report of two cases and a review of the literature. Am J Med Sci 1980; 280:21-8.
9. Bradley SJ, Hudson GV, Linch DC. Idiopathic thrombocytopenic purpura in Hodgkin's disease: A report of eight cases. Clin Oncol (R Coll Radiol) 1993; 5:355-7.
10. Gupta V, Hegde UM, Parameswaran R et al. Multiple myeloma and immune thrombocytopenia. Clin Lab Haematol 2000; 22(4):239-42.
11. Diehl LF, Ketchum LH. Autoimmune disease and chronic lymphocytic leukemia: Autoimmune hemolytic anemia, pure red cell aplasia, and autoimmune thrombocytopenia. Semin Oncol 1998; 25:80-97.
12. Lim SH, Ifthikharuddin JJ. Autoimmune thrombocytopenic purpura complicating lymphoproliferative disorders. Leuk Lymphoma 1994; 15:61-64.
13. Berchtold P, Wenger M. Autoantibodies against platelet glycoproteins in autoimmune thrombocytopenic purpura: Their clinical significance and response to treatment. Blood 1993; 81:1246-1250.
14. Kuwana M, Kaburaki J, Kitasato H et al. Immunodominant epitopes on glycoprotein IIb-IIIa recognized by autoreactive T cells in patients with immune thrombocytopenic purpura. Blood 2001; 98(1):130-9.
15. McMillan R. Autoantibodies and autoantigens in chronic immune thrombocytopenic purpura. Semin Hematol 2000; 37(3):239-48.
16. Beardsley DS, Ertem M. Platelet autoantibodies in immune thrombocytopenia. Transfus Sci 1998; 19:237-244.
17. Kuwana M, Kaburaki J, Ikeda Y. Autoreactive T cells to platelet GPIIb-IIIa in immune thrombocytopenic purpura: Role in production of anti-platelet autoantibody. J Clin Invest 1998; 102:1393-1402.
18. Semple JW, Lazarus AH, Freedman J. The cellular immunology associated with autoimmune thrombocytopenic purpura: An update. Transfus Sci 1998; 19:245-251.
19. Wucherpfennig KW. Mechanisms for the induction of autoimmunity by infectious agents. J Clin Invest 2001; 108(8):1097-104.
20. Tuohy VK, Kinkel RP. Epitope spreading: A mechanism for progression of autoimmune disease. Arch Immunol Ther Exp (Warsz) 2000; 48(5):347-51.
21. George JN, Rizvi MA. Thrombocytopenia. In: Beutler E, Lichtman MA, Coller BS et al, eds. Williams Hematology. New York: McGraw-Hill; 2001:1495-1539.
22. Bell WR Jr. Long-term outcome of splenectomy for idiopathic thrombocytopenic purpura. Semin Hematol 2000; 37(1Suppl1):22-5.
23. Lozano-Salazar RR, Herrera MF, Vargas-Vorackova F et al. Laparoscopic versus open splenectomy for immune thrombocytopenic purpura. Am J Surg 1998; 176:366-369.
24. Bussel JB. Overview of idiopathic thrombocytopenic purpura: New approach to refractory patients. Semin Oncol 2000; 27(6Suppl12):91-8.
25. Choi CW, Kim BS, Seo JH et al. Response to high-dose intravenous immune globulin as a valuable factor predicting the effect of splenectomy in chronic idiopathic thrombocytopenic purpura patients. Am J Hematol 2001; 66(3):197-202.
25a. McCrae KR, Bussel JB, Mannucci PM et al. Platelets: an update on diagnosis and management of thrombocytopenic disorders. Hematology (Am Soc Hematol Educ Program) 2001; 282-305.
26. Facon T, Caulier MT, Fenaux P et al. Accessory spleen in recurrent chronic immune thrombocytopenic purpura. Am J Hematol 1992; 41:184-189.
27. Andersen JC. Response of resistant idiopathic thrombocytopenic purpura to pulsed high-dose dexamethasone therapy. N Engl J Med 1994; 330:1560-1564.
28. Caulier MT, Rose C, Roussel MT et al. Pulsed high-dose dexamethasone in refractory chronic idiopathic thrombocytopenic purpura: A report on 10 cases. Br J Haematol 1995; 91:477-479.
29. Stasi R, Brunetti M, Pagano A et al. Pulsed intravenous high-dose dexamethasone in adults with chronic idiopathic thrombocytopenic purpura. Blood Cells Mol Dis 2000; 26(6):582-6.
30. Sorrentino D, Ferraccioli GF, De Vita S et al. Helicobacter pylori infection and autoimmune processes: An emerging field of study. Ital J Gastroenterol Hepatol 1998; 30Suppl3:S310-2.
31. Emilia G, Longo G, Luppi M et al. Helicobacter pylori eradication can induce platelet recovery in idiopathic thrombocytopenic purpura. Blood 2001; 97(3):812-4.
31a. Silverman GJ, Weisman S. Rituximab therapy and autoimmune disorders. Prospects for anti-B cell therapy. Arthritis Rheum 2003; 48:1484-1492.
32. Verlin M, Laros RK Jr, Penner JA. Treatment of refractory thrombocytopenic purpura with cyclophosphamide. Am J Hematol 1976; 1:97-104.
33. Smith AG, Prentice AG, Lucie NP et al. Acute myelogenous leukaemia following cytotoxic therapy: Five cases and a review. Q J Med 1982; 51:227-40.
34. Reiner A, Gernsheimer T, Slichter SJ. Pulse cyclophosphamide therapy for refractory autoimmune thrombocytopenic purpura. Blood 1995; 85:351-358.
35. Figueroa M, Gehlsen J, Hammond D et al. Combination chemotherapy in refractory immune throbocytopenic purpura. N Engl J Med 1993; 328:1226-1229.
36. McMillan R. Long-term outcomes after treatment for refractory immmune thrombocytopenic purpura. N Engl J Med 2001; 344:1402-1403.
37. Portielje JE, Westendorp RG, Kluin-Nelemans HC et al. Morbidity and mortality in adults with idiopathic thrombocytopenic purpura. Blood 2001; 97(9):2549-54.
38. Tyndall A, Gratwohl A. Blood and marrow stem cell transplants in auto-immune disease: A consensus report written on behalf of the European League against Rheumatism (EULAR) and the European Group for Blood and Marrow Transplantation (EBMT). Br J Rheumatol 1997; 36:390-392.
39. Marmont AM. Immunoablation followed or not by hematopoietic stem cells as an intense therapy for severe autoimmune diseases. New perspectives, new problems. Haematologica 2001; 86(4):337-45.
40. Euler HH, Marmont AM, Bacigalupo A et al. Early recurrence or persistence of autoimmune diseases after unmanipulated autologous stem cell transplantation. Blood 1996; 88:3621-3625.
41. Lim SH, Kell J, Al-Sabah A et al. Peripheral blood stem-cell transplantation for refractory autoimmune thrombocytopenic purpura. Lancet 1997; 349:475.
42. Marmont AM, Van Lint MT, Occhini D et al. Failure of autologous stem cell transplantation in refractory thrombocytopenic purpura. Bone Marrow Transplant 1998; 22:827-828.
43. Skoda RC, Tichelli A, Tyndall A et al. Autologous peripheral blood stem cell transplantation in a patient with chronic autoimmune thrombocytopenia. Br J Haematol 1997; 99:56-57.
44. Jindra P, Koza V, Fiser J et al. Autologous CD34+ cells transplantation after FAMP treatment in a patient with CLL and persisting AIHA: Complete remission of lymphoma with control of autoimmune complications. Bone Marrow Transplant 1999; 24:215-217.
45. Oyama Y, Papadopoulos EB, Miranda M et al. Allogeneic stem cell transplantation for Evans syndrome. Bone Marrow Transplant 2001; 28:903-905.
46. De Stefano P, Zecca M, Giorgiani G et al. Resolution of immune haemolytic anemia with allogeneic marrow transplantation after an unsuccessful autograft. Br J Haematol 1999; 106:1063-1064.
47. Brodsky RA, Petri M, Smith BD et al. Immunoablative high-dose cyclophosphamide without stem-cell rescue for refractory, severe autoimmune disease. Ann Intern Med 1998; 129:1031-1035.
48. Murphy S. Preservation and clinical use of platelets. In: Beutler E, Lichtman MA, Coller BS et al, eds. Williams Hematology. New York: McGraw Hill; 2001:1905-1916.
49. McMillan R. Clinical role of antiplatelet antibody assays. Seminars in Thrombosis and Hemostasis 1995; 21:37-45.
50. Brighton TA, Evans S, Castaldi PA et al. Prospective evaluation of the clinical usefulness of an antigen-specific assay (MAIPA) in idiopathic thrombocytopenic purpura and other immune thrombocytopenia. Blood 1996; 88:194-201.

51. Protheroe AS, Pickard C, Johnson PW et al. Persistence of clonal T-cell expansions following high-dose chemotherapy and autologous peripheral blood progenitor cell rescue. Br J Haematol 2000; 111:766-773.
52. Kuwana M, Kaburaki J, Ikeda Y. Autoreactive T cells to platelet GPIIb-IIIa in immune thrombocytopenic purpura. Role in production of anti-platelet autoantibody. J Clin Invest 1998; 102:1393-402.
53. Lim A, Toubert A, Pannetier C et al. Spread of clonal T-cell expansions in rheumatoid arthritis patients. Hum Immunol 1996; 48:77-83.
54. Even J, Lim A, Puisieux I et al. T-cell repertoires in healthy and diseased human tissues analysed by T-cell receptor beta-chain CDR3 size determination: Evidence for oligoclonal expansions in tumours and inflammatory diseases. Res Immunol 1995; 146:65-80.
55. Gagne K, Brouard S, Giral M et al. Highly altered V beta repertoire of T cells infiltrating long-term rejected kidney allografts. J Immunol 2000; 164:1553-1563.
56. Dietrich PY, Caignard A, Lim A et al. In vivo T-cell clonal amplification at time of acute graft-versus-host disease. Blood 1994; 84:2815-2820.
57. Wang L, Tadokoro K, Tokunaga K et al. Restricted use of T-cell receptor V beta genes in posttransfusion graft-versus-host disease. Transfusion 1997; 37:1184-1191.
58. Yassai M, McFarland JG, Newton-Nash D et al. T cell receptor and alloimmune thrombocytopenias: A model for autoimmune diseases? Ann Med Interne (Paris) 1992;143:365-370.
59. Karadimitris A, Manavalan JS, Thaler HT et al. Abnormal T-cell repertoire is consistent with immune process underlying the pathogenesis of paroxysmal nocturnal hemoglobinuria. Blood 2000; 96:2613-2620.
60. Zeng W, Nakao S, Takamatsu H et al. Characterization of T-cell repertoire of the bone marrow in immune-mediated aplastic anemia: Evidence for the involvement of antigen-driven T-cell response in cyclosporine-dependent aplastic anemia. Blood 1999; 93:3008-3016.
61. Altman JD, Moss PA, Goulder PJ et al. Phenotypic analysis of antigen-specific T lymphocytes. Science 1996; 274:94-6.
62. Maecker HT, Maino VC, Picker LJ. Immunofluorescence analysis of T-cell responses in health and disease. J Clin Immunol 2000; 20(6):391-9.
63. Maino VC, Picker LJ. Identification of functional subsets by flow cytometry: Intracellular detection of cytokine expression. Cytometry 1998; 34:207-15.
64. Marmont AM, Gualandi F, van Lint MT et al. Refractory Evans' syndrome treated with allogeneic SCT followed by DLI. Demonstration of a graft-vs.-autoimmunity effect. Bone Marrow Transplant 2003; 31:399-402.

CHAPTER 48

High-Dose Chemotherapy with Haematopoietic Stem Cell Transplantation in Primary Systemic Vasculitis, Behçet's Disease and Sjögren's Syndrome

Christoph Fiehn and Manfred Hensel

Introduction

The potential role of intensive immunosuppression and hematopoietic stem cell transplantation in the treatment of severe autoimmune diseases has been evaluated for several years. A cure seems to be possible in some cases. "Resetting" of the immune system (rather than deletion of the auto-aggressive T-cell clones) seems to be the mechanism of action. Limiting severe or fatal organ damage and reducing high cumulative doses of conventionally administered immunosuppressive drugs are the major goals of high-dose chemotherapy (HDCT). Data presented at an international meeting in Basel, Switzerland, in October 2000[1] and the data which are described in this book show high response rates in various autoimmune disease categories.

Primary systemic vasculitis (PSV) is a heterogeneous group of diseases which involve vessels of different sizes, manifest with a broad range of clinical signs and are caused by immunopathologic events, which are thought to be of autoimmune origin. A summary of the most frequent forms of PSV is given in Table 1. Patients with primary systemic vasculitis (PSV) seem to be ideal candidates for HDCT because complete remissions can be achieved with standard immunosuppressive therapy and the disease may have a fatal outcome if not sufficiently controlled by therapy. Severe organ manifestations are less frequent in Behçet's disease and Sjögren's syndrome. However, particular vascular system involvement in Behçet's syndrome[2-4] and manifestations of Sjögren's syndrome in the lung or brain[5-7] can be life threatening in some of the patients.

The possibility to achieve complete remission with standard therapy was an essential requirement for successful dose-escalation and bone marrow, or stem cell transplantation in hematologic malignancies 20 years ago. In analogy, PSV should be a good candidate disease to achieve a remission by a dose-intensified induction therapy and hematopoietic stem cell transplantation (HSCT). Appropriate patient selection is a major concern in HSCT because of possible procedure related mortality.[8] The patients must be identified in who disease severity and disease-related mortality overweighs the toxicity of the high-dose chemotherapy.

Beginning in the 1950s, the prognosis of PSV dramatically improved with the introduction of corticosteroids, increasing the 5-year survival rate from 13% without treatment to 48-57%.[9] Patients with renal disease had poor prognosis.[9-13] Fauci et al[14] demonstrated that cyclophosphamide further improved prognosis of vasculitis refractory to corticosteroids. In 2002, there still exists a subgroup of patients with PSV that: (1) cannot be cured by standard therapy; (2) need long-term immunosuppression; and (3) eventually die due to progression of the disease or cumulative therapy-related toxicity. Therefore, in studies which analyze long term outcome of Wegener's granulomatosis (WG) patients, death rates from 14 up to 56% over the longest observation time were reported.[15-17] This death rate has to be contributed to both the disease activity and the side effects of therapy. In polyarteriitis nodosa (PAN), if cyclophosphamide is given in combination with corticosteroids at the time of diagnosis, an advantage in terms of survival has been clearly demonstrated, except in one retrospective study.[18] In a recently published metaanalysis of four prospective trials including 278 patients with PAN, microscopic polyangiitis, and Churg-Strauss-syndrome, significantly prolonged survival with cyclophosphamide treatment has been shown for patients with very active disease, measured by two prognostic scoring systems (FFS and BVAS. See Table 2).[19] Patients were treated with the combination of corticosteroids and cyclophosphamide. The authors concluded that if the intensity of initial treatment is consistent with disease severity (measured by their scoring system), survival can be prolonged.

Furthermore, in some disease groups such as Wegener's granulomatosis there are high relapse rates.[17,20] These recurrent episodes of disease activity lead to accumulation of organ damage and increase the risk of both morbidity and mortality. In a large follow-up study with 155 patients with WG, 70% of the patients developed renal involvement and, in 16 patients (10.3%), this led to terminal renal failure.[17] Death due to vasculitis or its complications are not primarily related to the type of vasculitis, but rather the extent of organ involvement, e.g., kidney, GI tract or heart. This is true not only for PSV, but for Behçet's disease and Sjögren's syndrome as well.

In Behçet's disease, in particular, arterial vascular involvement can be life threatening.[2,3,21-23] About 4% of the patients with Behçet's disease have arterial involvement, with vasculitis of the

Stem Cell Therapy for Autoimmune Disease, edited by Richard K. Burt and Alberto M. Marmont. ©2004 Landes Bioscience/Eurekah.com.

Table 1. Definition and characteristics of the most frequent types of systemic vasculitis (according to the Chapel Hill Classification 1992)

Vasculitis of large vessels	
Giant cell arteriitis (temporal arteriitis, cranial arteriitis)	Granulomatous vasculitis which occurs primarily in the elderly and involves most frequently the extracranial branches of the carotid artery.
Takayasu's arteriitis	Vasculitis which manifests primarily in the aorta and its major branches and occurs most commonly in females under 40 years of age.
Vasculitis of medium vessels	
Polyarteriitis nodosa (PAN)	Vasculitis with small and medium sized arterial inflammation involving the skin, kidney, peripheral nerves, muscle and gut.
Vasculitis of small vessels	
Wegener's granulomatosis	Multisystem disease with necrosis, granuloma formation, vasculitis of the upper and lower respiratory tracts, frequently glomerulonephritis and variable degrees of small and occasionally medium-sized vessel vasculitis. Association to anti-neutrophile cytoplasmatic antibodies with a cytoplasmatic fluorescence pattern (cANCA) and proteinase-3 specifity.
Microscopic polyangiitis	Necrotizing vasculitis of small vessels with frequent occurrence of glomerulonephritis and pulmonary capillaritis.
Churg-Strauss-syndrome (allergic angiitis and granulomatosis)	Granulomatous eosinophil-rich vasculitis which frequently involves the skin, peripheral nerves and lungs and which is associated with peripheral blood eosinophilia and asthma.
Cryoglobulinemic vasculitis	Small vessel vasculitis with cryoglobulin-deposits which frequently involves skin and kidney. Association to hepatitis C infection.

Table 2. Simplified overview of the clinical scoring systems for systemic vasculitis. Number of possible items/organ manifestations in different organ systems which are included in the calculation of the score.

Organ System Involved	Activity (-FFS)	Activity (BVAS)	Extension (DEI)	Damage (VDI)
ENT	-	0-6	0-1	0-6
Lung	-	0-5	0-1	0-7
Kidney	0-2	0-7	0-1	0-3
Skin	-	0-6	0-1	0-2
Mucosa	-	0-2	-	0-1
Eye	-	0-5	0-1	0-7
Joints/muscles	-	0-2	0-1	0-5
CNS	0-1	0-4	0-1	0-6
PNS		0-3	0-1	0-1
Cardiovascular	0-1	0-7	0-1	0-7/9
Gastrointestinal	0-1	0-2	0-1	0-4
Systemic manifestations	-	0-5	0-1	-
Side effect of therapy	-	-	-	0-5
Maximum score	**5**	**63**	**21**	**64**

ENT= ear, nose and throat; CNS= central nervous system; PNS= peripheral nervous system; FFS= Five factor scale;[19] BVAS= Birmingham vasculitis activity score;[30] VDI= Vasculitis activity index;[32] DEI = Disease extension index.[31]

pulmonary artery which results in pulmonary arterial aneurysms being the most frequent manifestation.[2,4] In an overview of 24 cases with this type of vasculitis, 50% of the patients died after a mean of 9.5 month.[4]

In Sjögren's syndrome, much less data are available about frequency and mortality of severe organ manifestations of the disease. In particular, involvement of the central nervous system, which was reported to occur in 28% of patients with Sjögren's syndrome,[6] and lung manifestations represent a potential risk for organ damage and mortality. However, the risk of progressive organ failure due to Sjögren's syndrome is not clear yet and in spite of several reports of fatal outcome due to myelinolysis[5] and lung disease,[7,24] the majority of the patients might respond well to immunosuppressive treatment.[7] Therefore, the careful selection of patients who need HSCT is critical and particularly difficult in Sjögren's syndrome.

The first step of treatment either in PSV, Behçet's disease or in severe organ involvement of Sjögren's syndrome should be effective and result in rapid remission induction, in order to limit organ damage. Induction therapy is followed by mild, long-term maintenance therapy for prevention or relapse. This maintenance therapy may lead to chronic and severe drug toxicity. The National Institutes of Health experience with Wegener's granulomatosis reported the contribution of treatment toxicity to permanent damage in over 50% of their patients.[25] In a large, international randomized trial in patients with ANCA-positive systemic vasculitis, severe or life-threatening adverse effects have been observed in 26% of patients (maintenance therapy with oral cyclophosphamide or azathioprine).[26]

For the described long-term side effects of maintenance therapy, mostly high cumulative doses of cyclophosphamide and high mean doses of corticosteroids are responsible. Patients that received prolonged daily oral cyclophosphamide are at high risk for bladder cancer, which was estimated to be 5% at 10 years and 16% at 15 years, depending on the cumulative dose.[27] A Swedish study found a 11-fold increase in bladder cancer rates in patients receiving oral cyclophosphamide for more than one year and an increase in dermatological malignancy related to azathioprine and steroid exposure.[28] In a German study, 155 patients with Wegener's granulomatosis, which were mostly treated with long term cyclophosphamide, 11 developed a myelodysplastic syndrome after a median dose of 112 g cyclophosphamide, and one patient developed cancer of the bladder.[17]

Patient Selection

The first step in treatment of PSV should be standard remission induction with cyclophosphamide and corticosteroids. After remission has been achieved, patients with worse prognosis have to be identified with the help of clinical scoring systems. While several clinical scoring systems for vasculitis exist (Table 2), initial disease activity may be assessed by the five-factor scale (FFS), developed by Guillevin et al.[29] This scoring system allows the identification of subgroups of patients with panarteriitis nodosa or Churg-Strauss-Syndrome, who have worse prognosis (Table 2). In this analysis, patients with two or more risk factors (proteinuria >1g/d, creatinine >140 μmol/l, cardiomyopathy, gastrointestinal or CNS-involvement), and treated with a combination of corticosteroids and cyclophosphamide, had a 46% 1-year mortality, in contrast to 11% in those with a score of 0. Based on the data of a meta-analysis that included 278 patients, the FFS allows comparison of initial severity of the different types of vasculitis.[19]

The Birmingham vasculitis activity score (BVAS) (Table 3)[30] is another clinical scoring system to assess disease activity. This numerical score depends on the specific organ involvement and the severity of that involvement. For patients with Wegener's granulomatosis, the disease extent index can be used.[31] These clinical tools can help to identify subgroups of patients with severe disease and worse prognosis, in which more intense immunosuppressive therapy including HSCT may be required.

These scales may be used to select patients with PSV for controlled randomized trials. Patients with a FFS (>2),[19] or an initial high BVAS (>20), should be considered for such trials including HSCT instead of long-term conventional maintenance therapy. This might be the case in particular if the patients respond to standard therapy but fail to reach a lasting remission, or a long term therapy with cyclophosphamide with a high risk of cumulative toxicity can be predicted because of a high frequency of flares.

At the moment, based on these considerations, our strategy of patient selection for HSCT in PSV at our institution includes the following guidelines: patients with high disease activity and involvement of internal organs, as measured by increased FFS and BVAS, who either reach only a partial response to standard therapy or develop recurrent flares with subsequent progressive organ damage, or are at high risk of organ damage as a consequence of long-term or recurrent immunosuppressive therapy.

Comparable to high dose trials in tumor patients or with other autoimmune diseases such as systemic sclerosis, the major problem of patient selection in systemic vasculitis is early identification, before severe organ damage has developed. These patients can be identified with the help of scoring systems like the vasculitis damage index (VDI) (Table 4). VDI measures cumulative and irreversible organ damage by 64 symptoms in 11 organ systems.[13,32,33] Patients with a VDI > 5 should be excluded, as well as patients with severe heart or lung failure. Furthermore, we propose to use exclusion criteria similar to those used in the ASTIS-trial, a multi-center, randomized trial comparing HSCT and standard therapy for patients with severe systemic sclerosis.[1]

Exclusion Criteria:
- VDI >5
- Left ventricular ejection fraction <45%
- Uncontrolled cardiac arrhythmias
- Uncontrolled hypertension
- Pulmonary diffusion capacity <40% predicted
- Respiratory failure - PaO2 <8 kPa (60 mmHg)

In contrast, kidney failure must not exclude HSCT. In our institution and in other hematologic centers where high dose therapies of multiple myeloma are performed, considerable experience has accumulated in patients with AL-amyloidosis or multiple myeloma, which have terminal renal failure due to light-chain deposition and who were treated with a protocol based on a cyclophosphamide mobilization and a melphalan conditioning-regimen for HSCT, which is very similar to the protocol which we use for patients with severe autoimmune diseases.[34] This protocol, with dose adjustments depending on the kidney function, has shown to be safe and effective and might be applied easily to patients with PSV and kidney involvement.

Table 3. Birmingham Vasculitis Activities Score (BVAS). Check box only if abnormality is newly present or worsening within the previous four weeks and ascribable to vasculitis.

SYSTEM	Score	SYSTEM	Score
Systemic maximum total 3		**Mucosalmembranes/eyes maximum total 6**	
None	0	None	0
Malaise	1	Mouth ulcer	1
Myalgia	1	Genital ulcer	1
Arthralgias/arthritis	1	Conjunctivitis	1
Fever <38.5	1	Episcleritis	2
Fever >38.5	2	Uveitis	6
Weight loss 1-2 kg in 1 month	2	Retinal exudates	6
Weight loss >2 kg within last month	3	Retinal hemorrhage	6
Cutaneous maximum total 6		**Cardiovascular maximum total 6**	
None	0	None	0
Infarct	2	Bruits	2
Purpura	2	New loss of pulse	4
Other skin vasculitis	2	Aortic incompetence	4
Ulcer	4	Pericarditis	4
Gangrene	6	New myocardial infarct	6
Multiple digit gangrene	6	CHF/cardiomyopathy	6
Chest maximum total 6		**Nervous system maximum total 9**	
None	0	None	0
Dyspnea or wheezing	2	Confusion/dementia	3
Nodules or fibrosis	2	Seizure	9
Pleural effusions or pleurisy	4	Stroke	9
Infiltrate	4	Cord lesion	9
Hemoptysis/hemorrhage	4	Peripheral neuropathy	9
Massive hemoptysis	6	Motor mononeuritis multiplex	9
ENT maximal total 6		**Renal maximum total 12**	
None	0	None	0
Nasal discharge/obstruction	2	Diastolic >90mmHg	4
Sinusitis	2	Proteinuria >1+ or >0.2 g/24 hours	4
Epistaxis	4	Hematuria >1+ or >10 RBC/ml x	8
Crusting	4	Creatinine >125-249 µmol/l	8
Aural discharge	4	Creatinine >250-499 µmol/l	10
Otitis media	4	Creatinine >500 µmol/l	12
New deafness	6	Rise in creatinine >10%	12
Hoarseness/laryngitis	2		
Subglottic involvement	6		
Abdominal maximum total 9			
None	0		
Abdominal pain	3		
Bloody diarrhea	6		
Gall bladder perforation	9		
Gut infarction	9		
Pancreatitis	9		

Maximum Score = 63. Ref: Q J Med 1994; 87: 671-678. APPENDIX III

Experience with High-Dose Chemotherapy and HSCT in Primary Systemic Vasculitis, Behçet's Disease and Sjögren's Syndrome

The international experience with HSCT in autoimmune diseases was presented at an international meeting in October 2000 in Basel, Switzerland and is described in the present book. At the Basel meeting, data on 390 patients from all over the world were reported. This combined international data base included 260 patients from the EBMT/EULAR Basel European/Asian database, 87 from North America (55 from the IBMTR), 39 from Australia and 4 others.[1] Experience with PSV remains anecdotal at present. Within this database, 9 cases of systemic vasculitis

Table 4. Vasculitis Damage Index (VDI)

System/Item	+ / -	System/Item	+ / -
I. Musculoskeletal Significant atrophy or weakness Deforming or erosive arthritis Avascular necrosis Osteoporosis- fractures or vertebral collapse Osteomyelitis		**VII. Gastrointestinal** Gut infarction Mesenteric insufficiency or pancreatitis Chronic peritonitis Esophageal stricture or upper GI tract surgery	
II. Skin Alopecia Skin ulcerations Oral ulceration		**VIII. Peripheral Vascular** Absence of peripheral pulse in 1 limb Second episode of absent pulse in 1 limb Absent peripheral pulse in >2 limbs Major vessel stenosis Claudification of the extermities Complicated venous thrombosus Minor tissue loss Major tissue loss Second episode of major tissue loss	
III. Ear, Nose, and Throat Hearing loss Nasal blocking or chronic discharge Chronic sinusitis or bone destruction Subglottic stenosis without surgery Subglottic stenosis with surgery		**IX. Ocular** Cataract Retinal change Optic atrophy Visual impairment/diplopia Blindness in 1 eye Blindness in 2nd eye Orbital wall destruction	
IV. Pulmonary Pulmonary hypertension Pulmonary fibrosis/cavity Pleural fibrosis Pulmonary infarction Chronic asthma Chronic breathlessness Impaired pulmonary function tests		**X. Neuropsychiatric** Cognitive impairment Major psychosis Seizures Cerebrovascular accident Cranial nerve lesion Peripheral neuropathy Transverse myelitis	
V. Cardiovascular Angina/coronary artery bypass Myocardial infarct Second myocardial infarct Cardiomyopathy Vascular disease Pericarditis Hypertension		**XI. Other damage** Premature gonadal failure Marrow failure Diabetes mellitus Chronic chemical cystitis Maligancy Other	
VI. Renal GFR <50% Proteinuria >0.5 g/24 hours End stage renal failure			

From Exley AR et al. Arthritis and Rheum 1997; 40(2):371-379.

including Behçet's syndrome were reported. In 3 cases of Wegener's granulomatosis, initially complete remissions were observed. Two of them relapsed at 2 and 3 years, respectively. In two of three patients with cryoglobulinemia, with variable vasculitic features, complete responses were reported (Table 5). Three patients with Behçet's syndrome were treated by HSCT, two of them in our institution. In one case there was a relapse following autologous HSCT, and this patient then received an allogeneic HSCT from her HLA-identical brother.[34a] For Sjögren's syndrome, only one patient was reported. However, there are several reports of HSCT

Table 5. Patients with PSV, Behçet's disease and Sjögren's syndrome which were treated with HDCT. Data from the Basel database (1) and from personal communication.

Disease	No. of Patients	Response to HDCT
Wegener's Granulomatosis	3	3 CR, in 2 relapse after 2.3 and 3 yrs
Cryoglobulinemia	3	2 CR
Behçet's disease	3	1 CR, 1 PR, 1 relapse after 2 months
Undifferentiated Vasculitis	1	minimal response
Polyarteritis Nodosa	1	CR

CR= complete response; PR= partial response

Heidelberg Experience

In the high dose protocol developed at our institution in Heidelberg, Germany, so far 10 patients, 5 of them with PSV, were included (Table 6). Two patients with Behçet's disease and pulmonary involvement,[38] one patient with unclassified vasculitis, and one patient with PAN have been transplanted. All patients had been refractory to conventional immunosuppressive therapy including oral and pulse cyclophosphamide. With a follow-up of 53 and 15 months, respectively, one patient with Behçet's disease, and one patient with PAN (the latter transplanted at the University of Ulm, Germany) are still in complete remission without additional therapy. The other patient with Behçet's disease experienced a partial remission 41 months after HSCT without further pulmonary complications. He requires only low-dose maintenance therapy with corticosteroids to control mild disease activity. The patient with unclassified vasculitis has been transplanted 32 months ago and had minimal improvement. One patient with highly active and intensively treated ANCA-positive systemic vasculitis accompanied by glomerulonephritis markedly improved after mobilization therapy with cyclophosphamide (4 g/m^2). He requires only low-dose corticosteroids 40 months after mobilization. Because of this dramatic improvement, he refused further dose-escalation.

in patients with Sjögren's syndrome and consecutive lymphoma. Unfortunately, in these reports, mainly no response or only short term response of the autoimmune disease, in contrast to the lymphoma, was described.[35,36] However, one paper reports a lasting remission of nonerosive polyarthritis in Sjögren's syndrome after HSCT for lymphoma.[37]

Our study protocol consists of two cycles of mobilization therapy with cyclophosphamide (2 g/m^2 and 4 g/m^2) and G-CSF, followed by stem cell harvest and CD34 selection. Conditioning

Table 6. Patients with systemic vasculitis, who were treated according to the high dose protocol from Heidelberg/Germany

Diagnosis	Age at SCT	Reason for Dose Escalation	CD34$^+$ Dose Collected After 2g/m^2 Cy + G-CSF	CD34$^+$ Dose Collected After 4g/m2 Cy + G-CSF Before Enrichment	Response to Mobilization	Response to Melphalan and SCT
Behçet	32	Pulmonary bleeding Progress under Cy	2.9 x 10^6/kg BW	7.1 x 10^6/kg BW	Partial Remission, Withdrawal of CS	Complete remission since 53 months
Behçet	49	Pulmonary bleeding Relapse under Cy	2.5 x 10^6/kg BW	7.8 x 10^6/kg BW	Stable disease	Partial remission, low-dose CS since 41 months
Unclassified Vasculitis	41	Requirement of high doses of CS and long time Cy	13.1 x 10^6/kg BW	23.7 x 10^6/kg BW	Minimal response	Minimal response low-dose CS since 16 months
Polyarteritis nodosa (transplanted in Ulm/Germany; Dr.A.Breitbart)	41	Recurrent relapses under intensive conventional immunosuppr. (Cy, CS)	<2.5 x 10^6/kg BW	>5 x 10^6/kg BW	Minimal response	Complete remission since 15 months
cANCA pos. Vasculitis	54	No disease control with long-term conventional immunosuppr. (Cy, CS)	<1 x 10^6/kg BW	2.7 x 10^6/kg BW	Partial Remission, withdrawal of oral Cy, low-dose CS since 37 month	Not done

BW= body weight; CS= corticosteroids; Cy= Cyclophosphamide; SCT= Stem cell transplantation

consists of melphalan 200 mg/m^2. Cyclophosphamide is the most effective immunosuppressive drug for remission induction in PSV. Furthermore, Brodsky et al 1998[39] reported that dose escalation of cyclophosphamide (with doses up to 200 mg/m^2) without stem cell rescue might be sufficient for long-term remission in autoimmune diseases in some cases. Mobilization with cyclophosphamide leads to successful stem cell harvest in most patients.[40] This regimen has been used in several phase I and II-studies and currently in the multicenter ASTIS trial.

The conditioning regimen with melphalan is used because (1) it is myeloablative without total body irradiation; (2) it has been used successfully in hematologic malignancies for 20 years; (3) it has low toxicity as compared to other conditioning regimens; and (4) it is safely used in patients with renal failure due to AL-amyloidosis or multiple myeloma,[34] a fact which might be important for the use in PSV with severe kidney involvement. The toxicity of the melphalan conditioning regimen has been evaluated intensively as part of multiple myeloma transplantation programs. It has been demonstrated that long-term administration of standard-dose alkylating agents, rather than autotransplant-supported high dose therapy with melphalan, was associated with myelodysplastic syndrome or acute leukemia.[41]

Future Perspectives and Conclusion

While data for Sjögren's syndrome hardly exists, evidence is accumulating, that HSCT might be considered in selected cases of PSV and vasculitic manifestations of Behçet's disease. The extent to which different forms of PSV need different approaches to treatment is unclear, but it is possible that Wegener's granulomatosis, microscopic polyangiitis, Churg-Strauss-syndrome and PAN respond in a similar way to dose-escalated therapy. The severity of vasculitis rather than distinction between subgroups, should determine the decision about eligibility for HSCT. Current experience with HSCT in PSV is anecdotal, but the described cases have good outcome and are promising. Based on the described selection strategy, more pilot data in PSV should be collected. There are many open questions: Should we transplant earlier in the course of the disease? Which selection criteria are the best to identify high-risk patients that require such aggressive therapies (FFS, BVAS, etc)? Which conditioning regimen should be used? For the next years, controlled trials for patients with PSV should be considered to answer some of those questions.

References

1. Tyndall A, Passweg J, Gratwohl A. Haemopoietic stem cell transplantation in the treatment of severe autoimmune diseases 2000. Ann Rheum Dis 2001; 60(7):702-707.
2. Koc Y, Gullu I, Akpek G et al. Vascular involvement in Behçet's disease. J Rheumatol 1992; 19(3):402-410.
3. Basak M, Gul S, Yazgan Y et al. A case of rapidly progressive pulmonary aneurysm as a rare complication of Behçet's syndrome—a case report. Angiology 1998; 49(5):403-408.
4. Hamuryudan V, Yurdakul S, Moral F et al. Pulmonary arterial aneurysms in Behçet's syndrome: A report of 24 cases. Br J Rheumatol 1994; 33(1):48-51.
5. Yoon KH, Fong KY, Koh DR et al. Central pontine myelinolysis—a rare manifestation of CNS Sjogren's syndrome. Lupus 2000; 9(6):471-473.
6. Lafitte C, Amoura Z, Cacoub P et al. Neurological complications of primary Sjogren's syndrome. J Neurol 2001; 248(7):577-584.
7. Davidson BK, Kelly CA, Griffiths ID. Ten year follow up of pulmonary function in patients with primary Sjogren's syndrome. Ann Rheum Dis 2000; 59(9):709-712.
8. Bacon PA, Carruthers DM. New therapeutic aspects: Haematopoietic stem cell transplantation. Best Prac Res Clin Rheum 2001; 15(2):299-313.
9. Frohnert PP, Sheps SG. Long-term follow-up study of periarteritis nodosa. Am J Med 1967; 43(1):8-14.
10. Bradley JD, Brandt KD, Katz BP. Infectious complications of cyclophosphamide treatment for vasculitis. Arthritis Rheum 1989; 32(1):45-53.
11. Cacoub P, Le Thi Huong, Guillevin P et al. Causes of death in systemic vasculitis of polyarteriitis nodosa. Analysis of a series of 165 patients. Ann Med Interne Paris 1988; 139:381-390.
12. Cohen RD, Conn DL, Ilstrup DM. Clinical features, prognosis, and response to treatment in polyarteritis. Mayo Clin Proc 1980; 55:146-155.
13. Exley AR, Carruthers DM, Luqmani RA et al. Damage occurs early in systemic vasculitis and is an index of outcome. QJM 1997; 90(6):391-399.
14. Fauci AS, Katz P, Haynes BF et al. Cyclophosphamide therapy of severe systemic necrotizing vasculitis. N Engl J Med 1979; 301(5):235-238.
15. Matteson EL, Gold KN, Bloch DA et al. Long-term survival of patients with Wegener's granulomatosis from the American College of Rheumatology Wegener's Granulomatosis Classification Criteria Cohort. Am J Med 1996; 101(2):129-134.
16. Savage CO, Winearls CG, Evans DJ et al. Microscopic polyarteritis: Presentation, pathology and prognosis. Q J Med 1985; 56(220):467-483.
17. Reinhold-Keller E, Beuge N, Latza U et al. An interdisciplinary approach to the care of patients with Wegener's granulomatosis: Long-term outcome in 155 patients. Arthritis Rheum 2000; 43(5):1021-1032.
18. Leib ES, Restivo C, Paulus HE. Immunosuppressive and corticosteroid therapy of polyarteritis nodosa. Am J Med 1979; 67(6):941-947.
19. Gayraud M, Guillevin L, le Toumelin P et al. Long-term followup of polyarteritis nodosa, microscopic polyangiitis, and Churg-Strauss syndrome: Analysis of four prospective trials including 278 patients. Arthritis Rheum 2001; 44(3):666-675.
20. Gordon M, Luqmani RA, Adu D et al. Relapses in patients with a systemic vasculitis. Q J Med 1993; 86(12):779-789.
21. Yazici H, Basaran G, Hamuryudan V et al. The ten-year mortality in Behçet's syndrome. Br J Rheumatol 1996; 35(2):139-141.
22. Roguin A, Edoute Y, Milo S et al. A fatal case of Behçet's disease associated with multiple cardiovascular lesions. Int J Cardiol 1997; 59(3):267-273.
23. Numan F, Islak C, Berkmen T et al. Behçet disease: Pulmonary arterial involvement in 15 cases. Radiology 1994; 192(2):465-468.
24. Tavoni A, Vitali C, Cirigliano G et al. Shrinking lung in primary Sjogren's syndrome. Arthritis Rheum 1999; 42(10):2249-2250.
25. Hoffman GS, Kerr GS, Leavitt RY et al. Wegener granulomatosis: An analysis of 158 patients. Ann Intern Med 1992; 116(6):488-498.
26. Jayne D, Gaskin G. Randomised trial of cyclophosphamide versus azathioprine during remission in ANCA-associated vasculitis (CYCAZAREM). J Am Soc Nephrol 1999; 10:105A.
27. Talar-Williams C, Hijazi YM, Walther MM et al. Cyclophosphamide-induced cystitis and bladder cancer in patients with Wegener granulomatosis. Ann Intern Med 1996; 124(5):477-484.
28. Westman KW, Bygren PG, Olsson H et al. Relapse rate, renal survival, and cancer morbidity in patients with Wegener's granulomatosis or microscopic polyangiitis with renal involvement. J Am Soc Nephrol 1998; 9(5):842-852.
29. Guillevin L, Lhote F, Gayraud M et al. Prognostic factors in polyarteritis nodosa and Churg-Strauss syndrome. A prospective study in 342 patients. Medicine Baltimore 1996; 75(1):17-28.
30. Luqmani RA, Bacon PA, Moots RJ et al. Birmingham Vasculitis Activity Score (BVAS) in systemic necrotizing vasculitis. QJM 1994; 87(11):671-678.
31. Reinhold-Keller E, Kekow J, Schnabel A et al. Influence of disease manifestation and antineutrophil cytoplasmic antibody titer on the response to pulse cyclophosphamide therapy in patients with Wegener's granulomatosis. Arthritis Rheum 1994; 37(6):919-924.

32. Exley AR, Bacon PA, Luqmani RA et al. Examination of disease severity in systemic vasculitis from the novel perspective of damage using the vasculitis damage index (VDI). Br J Rheumatol 1998; 37(1):57-63.
33. Exley AR, Bacon PA, Luqmani RA et al. Development and initial validation of the Vasculitis Damage Index for the standardized clinical assessment of damage in the systemic vasculitides. Arthritis Rheum 1997; 40(2):371-380.
34. Badros A, Barlogie B, Siegel E et al. Results of autologous stem cell transplant in multiple myeloma patients with renal failure. Br J Haematol 2001; 114(4):822-829.
34a. Marmont AM, Gualandi F, van Lint MT et al. Autologous hematopoietic stem cell transplant for refractory Behcet's disease. Bone Marrow Transplant 2003; 31(Suppl1):15(abstr).
35. Rosler W, Manger B, Repp R et al. Autologous PBPCT in a patient with lymphoma and Sjogren's syndrome: Complete remission of lymphoma without control of the autoimmune disease. Bone Marrow Transplant 1998; 22(2):211-213.
36. Ferraccioli G, Damato R, De Vita S et al. Haematopoietic stem cell transplantation (HSCT) in a patient with Sjogren's syndrome and lung malt lymphoma cured lymphoma not the autoimmune disease. Ann Rheum Dis 2001; 60(2):174-176.
37. Jondeau K, Job-Deslandre C, Bouscary D et al Dreyfus F. Remission of nonerosive polyarthritis associated with Sjogren's syndrome after autologous hematopoietic stem cell transplantation for lymphoma. J Rheumatol 1997; 24(12):2466-2468.
38. Hensel M, Breitbart A, Ho AD. Autologous hematopoietic stem-cell transplantation for Behcet's disease with pulmonary involvement. N Engl J Med 2001; 344(1):69.
39. Brodsky RA, Petri M, Smith BD et al. Immunoablative high-dose cyclophosphamide without stem-cell rescue for refractory, severe autoimmune disease. Ann Intern Med 1998; 129(12):1031-1035.
40. Burt RK, Fassas A, Snowden J et al. Collection of hematopoietic stem cells from patients with autoimmune diseases. Bone Marrow Transplant 2001; 28(1):1-12.
41. Govindarajan R, Jagannath S, Flick JT et al. Preceding standard therapy is the likely cause of MDS after autotransplants for multiple myeloma. Br J Haematol 1996; 95(2):349-353.

CHAPTER 49

Hematopoietic Stem Cell Transplantation in the Treatment of Chronic Inflammatory Demyelinating Polyradiculoneuropathy

George Hutton, Yu Oyama, Richard K. Burt and Uday Popat

Introduction

Chronic inflammatory demyelinating polyradiculoneuropathy (CIDP) is an acquired immune mediated neuropathy. The characteristic clinical picture is one of slowly progressive weakness and sensory loss. Muscle stretch reflexes are depressed or absent. The diagnosis is supported by findings of elevated CSF protein, demyelinating electrodiagnostic features and an abnormal nerve biopsy. There are several lines of evidence that suggest CIDP is an autoimmune disorder involving autoreactive antibodies and T cells. Beneficial treatments include corticosteroids, intravenous immunoglobulin and plasma exchange. Patients with a progressive course who are refractory to such traditional therapy may be candidates for a more aggressive treatment approach involving high dose immune suppression followed by hematopoietic stem cell transplantation (HSCT).

CIDP–Pathophysiology

Although CIDP is assumed to be an autoimmune disease, the pathogenesis of the disorder is not fully understood. The clinical and laboratory similarities of CIDP to Guillain-Barré syndrome, and response of both disorders to plasma exchange and intravenous immunoglobulin suggest an immune pathogenesis. Such observations and the association of CIDP with other autoimmune diseases,[18] have led to the general hypothesis that CIDP, like Guillain-Barré syndrome, is caused by an autoimmune reaction to myelin or Schwann cell antigens, possibly triggered by one or several infective agents.[15]

There are several animal models of the acquired demyelinating polyradiculoneuropathies, referred to as experimental allergic neuritis (EAN). Most of the animal models correspond to the acute human form, Guillain-Barré syndrome. Experimental autoimmune neuritis sometimes develops into a chronic relapsing form in rats and especially in rabbits.[19,20] These models of chronic EAN were induced with whole myelin. These models faithfully reproduced the chronic inflammatory changes in the endoneurium and onion bulb formation seen in CIDP. Chronic EAN in the rabbit also showed electrophysiologic evidence of demyelination.[20]

Acute experimental autoimmune neuritis can be induced by myelin protein 2 (P2)[21,22] which does not have an extracellular domain, and by myelin protein 0 (P0) and peripheral myelin protein 22 (PMP22), both of which have extracellular domains.[23,24] Chronic experimental autoimmune neuritis can be induced more readily in rabbits by immunization with either galactocerebroside[25] or large amounts of myelin.[20] The search for autoantibodies directed against myelin antigens has been relatively successful in Guillain-Barré syndrome, identifying antibodies to several myelin proteins and glycolipids (Tables 1 and 2).[26] Recent studies have demonstrated various antibodies in the sera of patients with CIDP (Tables 1 and 2), but no single antibody has been consistently demonstrated.[27-32]

A study involving 40 CIDP patients found antibodies to glycolipids and glycoconjugates infrequently, and not significantly more commonly than in control sera.[27] Only 15% of CIDP patients showed antibodies to P0, not differing significantly from controls.[27] Similar low percentages were found in a series investigating antibody responses to P0 and P2 myelin proteins in CIDP patients.[28] High titers of IgM antibodies to sulphated glucuronyl paragloboside (SGPG), a glycosphingolipid component of peripheral myelin, were detected in 6/30 (20%) CIDP patients.[31] Antibodies to PMP22 or its extracellular domain were detected in 7 of 17 (41%) patients with CIDP.[32] However, in another study, antibodies to PMP22 were found in 3 of 6 CIDP patients, but also in the sera of patients with Charcot Marie Tooth disease types 1 and 2,[33] so further study is needed.

A recent study provides evidence that antibodies to P0 glycoprotein are present in the serum of a more respectable minority of patients with CIDP.[29] In this study, 6/21 (28%) CIDP patients had serum anti P0 IgG antibodies, and four of these caused conduction block and demyelination following intraneural injection in rats.[29] Thus, antibodies against P0 glycoprotein have the most convincing support as a putative autoimmune target in CIDP. This glycoprotein is the major molecule responsible for compaction of the myelin sheath. Antibody targeting of this glycoprotein with its large extracellular domain may result in myelin disintegration (Table 3 and 4).

Antibodies alone are not a sufficient explanation for the production of demyelination because they would not penetrate the blood nerve barrier unless it were first rendered leaky. It is

Table 1. Antibodies to peripheral nerves and myelin proteins in CIDP and GBS

Antigen	Description/Role in EAN	CIDP	GBS	Normal
Peripheral nerve myelin		2-22%	18-29%	0%
Myelin basic protein	Protein component of central and peripheral myelin	1/1 case report		
Bovine P2 myelin protein	Minor protein component of peripheral myelin (5%), may be important in myelin assembly. P2 protein can induce EAN in rats. Can inhibit P2-induced EAN by administering nasal or oral P2 protein (tolerance).	0-34%	6-40%	0-15%
Human P0 myelin protein	Protein zero: the major structural protein of peripheral myelin; about 50% of myelin protein. May be important in myelin compaction. Has an extracellular domain. Serum anti-P0 IgG antibodies from CIDP patients caused conduction block in rats after intraneural injection.	0-29%	11-19%	0-6%
Human PMP-22	Peripheral myelin protein 22. Has an extracellular domain. Recombinant, homologous PMP-22 induces EAN in the Lewis rat, a model of GBS.	35%	52%	4%

CIDP= chronic inflammatory demyelinating polyneuropathy; GBS= Guillain Barre syndrome; P0= myelin protein zero; P2= myelin protein 2; PMP= peripheral myelin protein 22. Studies have used different testing methods (ELISA, RIA, CFT, indirect immunoflourescence) and involved different numbers of subjects, accounting for the variability in antibody positivity. Refer to the text for full details. References 19-24, 26-33.

Table 2. Antibodies to glycolipids in CIDP and GBS

Antigen	Description	CIDP	GBS	Normal
Galactocerebroside	Essential galactolipid for the proper compaction of myelin	0-9%	0-12%	0%
Ganglioside GM1	Gangliosides are acidic glycolipids composed of a lipid moiety linked to sphingosine that is conjugated to a complex oligosaccharide containing at least one sialic acid residue (GM1 has one sialic residue). Particularly concentrated in neural plasma membranes.	0-25%	17-39%	0-12%
Asialo-GM1	As above, but without the sialic acid group.	25% (1/4)	75% (3/4)	0%
LM1/SPG	Most abundant ganglioside in human peripheral nerve myelin	3-67%	8-43%	0-20%
Sulfatide	The major acidic glycosphingolipid in peripheral nerve myelin. Its concentration is 100-fold higher than any of the gangliosides.	0-87%	9-65%	0-15%
SGPG	Same structure as LM1/SPG, except that it has a terminal sulfated glucuronic acid instead of sialic acid.	0-40%	13-29%	0-10%
SLPG	Similar to the above but with a lactosylaminyl group	7-30%	2-25%	0-4%

CIDP= chronic inflammatory demyelinating polyneuropathy; GBS= Guillain Barre syndrome; LM1= sialosyl neolacto tetraosylceramide, a major component of peripheral nerve ganglioside; SGPG= sulphated glucuronyl paragloboside; SLPG= sialosyl lactosaminyl paragloboside; SPG= sialosyl paragloboside, a peripheral nerve myelin glycoshingolipid. Studies have used different testing methods (ELISA, RIA, TLC) and involved different numbers of subjects, accounting for the variability in antibody positivity. Refer to the text for full details. References 25-27, 29, 31.

Table 3. Features of chronic inflammatory demyelinating polyradiculoneuropathy

Usually presents as an isolated disorder but may occur with a variety of concurrent illnesses
Symmetric proximal and distal weakness
Distal sensory loss to all modalities
Depressed or absent muscle stretch reflexes
Elevated protein in the cerebrospinal fluid without pleocytosis
Electrodiagnostic features of a demyelinating neuropathy
Nerve biopsy shows demyelination, thinly myelinated fibers, and onion-bulbs in 60-70% of cases
Treatment involves corticosteroids, IVIG, plasmapheresis, or other immunosuppressive treatments

---

likely that a T cell response is also involved. CIDP patients had increased percentages of activated circulating T lymphocytes[34] and raised serum concentrations of both IL-2 and IL-2 receptors.[35,36] In a single study, primary T cell proliferated responses to P2 and P0 proteins or peptides were recorded in almost half of the patients studied.[37] Tumor necrosis factor alpha (TNF-alpha) is a cytokine secreted by both macrophages and T cells which has toxic effects on myelin, Schwann cells and endothelial cells. Elevated serum TNF-alpha levels were found in 25% of CIDP patients.[38] TNF-alpha levels correlated with clinical severity and electrophysiologic abnormalities, and decreased after immunotherapy.[38]

CIDP–Pathology

Pathologic studies of CIDP come from relatively few autopsy studies of patients with unexplained demyelinating polyneuropathies as well as from sural nerve biopsies. Most cases of remitting CIDP to have been examined at autopsy have shown evidence of a multifocal, predominantly proximal demyelinating disorder affecting chiefly spinal nerve roots, spinal nerves, major plexuses and/or proximal nerve trunks, with lesions extending throughout the peripheral nervous system (reviewed in ref.15). In several cases, affected nerves have been enlarged. Axonal changes including axon loss, wallerian degeneration and clusters of regenerating axons are also frequently seen on nerve biopsy. Nerve biopsies that have included new lesions have shown phagocytosis by macrophages of the myelin portion of the Schwann cell plasma membrane. In cases with significant cell infiltration, the cells are predominantly macrophages and cytotoxic $CD8^+$ lymphocytes, although $CD4^+$ cells may also be present.[14] There is also evidence that T cells may actively cross the blood nerve barrier as mirrored by increased levels of circulating levels of E-selectin (endothelial leukocyte adhesion molecule).[16,17]

CIDP–History, Epidemiology and Clinical Features

The true incidence of CIDP is not known, but it may represent as many as 10%-30% of previously undiagnosed cases of neuropathy at tertiary care referral centers.[1,2] Recent prevalence studies have shown rates of 1 to 2 per 100,000 people.[3,4] Although diagnostic criteria for the condition have been established, not all cases fulfill these criteria, and the condition is often misdiagnosed, especially in the early stages.

Probable cases of CIDP were reported in the early decades of this century under a number of different names (for a review see ref. 5). In 1958, Austin described two patients with a chronic corticosteroid responsive polyneuropathy.[6] In 1975, Dyck and colleagues published a seminal report of 53 patients with a condition they called "chronic inflammatory polyradiculoneuropathy".[7] Since then, multiple large series have been published expanding the description of the clinical and electrophysiologic features, laboratory studies and treatment responsiveness of the condition.[8-10]

Although CIDP may occur at any age, including childhood, it is much more common in the adult population, with a peak incidence in the 40-60 year age group.[9,11] The characteristic clinical picture is one of slowly progressive weakness and sensory loss (Table 3). CIDP usually presents as an isolated disorder but may occur with concurrent illnesses, such as HIV infection, monoclonal gammopathy, connective tissue disease and inflammatory bowel disease.[2] The major symptom is symmetric proximal and distal weakness. The weakness may be profound, but is usually not accompanied by muscle atrophy or fasciculations. In contrast to Guillain-Barré syndrome (acute inflammatory demyelinating polyradiculoneuropathy or AIDP), the respiratory muscles in CIDP are usually not affected. Most patients with CIDP have prominent sensory complaints, such as numbness or paresthesias. Muscle stretch reflexes are depressed or absent. Infrequently, the cranial nerves or central nervous system may be affected.[10,12]

The disease course can be slowly and steadily progressive, stepwise progressive or relapsing (with or without treatment). A key issue in distinguishing CIDP from Guillain-Barré syndrome is the duration of progression. By definition, Guillain-Barré syndrome reaches its nadir in four weeks. Most criteria for CIDP require progression beyond two months.[11,13] Occasional patients may be encountered in whom it is difficult to determine whether CIDP or Guillain-Barré syndrome is the most appropriate diagnosis. Other varieties of neuropathy that need be excluded include inherited demyelinating neuropathies and neuropathies related to metabolic derangements, paraneoplastic syndromes, monoclonal gammopathies, HIV infection or Lyme disease.[5]

CIDP–Diagnosis

There have been several proposed diagnostic criteria for CIDP including those of Barohn et al[11] and those of the Ad Hoc Subcommittee of the American Academy of Neurology AIDS Task Force published in 1991.[13] (Table 4). Such criteria are best suited for research purposes. If they were strictly interpreted, many patients who have the natural history and responses of CIDP would not qualify as definite cases. The latter criteria include clinical, electrodiagnostic, cerebrospinal fluid studies and pathologic features, including mandatory, supportive and exclusion criteria. Based on the number of criteria satisfied, the disease is categorized as definite, probable or possible CIDP.

The diagnosis of CIDP is supported by a profile of laboratory findings. As indicated above, the three most important laboratory studies in CIDP are cerebrospinal fluid examination, electrodiagnostic studies and nerve biopsy. An elevated cerebrospinal fluid protein is found in 95% of CIDP patients and is the most sensitive but least specific of the laboratory studies for CIDP.[11] As in patients with Guillain-Barré syndrome, spinal fluid pleocytosis is distinctly uncommon, and more than 10-20 cells/mm^3 in the spinal fluid should always suggest another condition. This has been referred to as albuminocytologic dissociation.

Table 4. Definite or probable CIDP according to the criteria of the Ad Hoc Subcommittee of the American Academy of Neurology AIDS Task Force (Neurology 1991; 41:617-618)

I Clinical
A. Mandatory
    1. Progressive or relapsing motor and sensory, rarely only motor or sensory, dysfunction of more than one limb of a peripheral nerve nature, developing over at least 2 months
    2. Hypo- or areflexia. This will usually involve all four limbs
B. Supportive
    1. Large fiber sensory loss predominates over small fiber sensory loss
C. Exclusion
    1. Mutilation of hands or feet, retinitis pigmentosa, ichthyosis, appropriate history of drug or toxic exposure known to cause a similar perpheral neuropathy or family history of a genetically based peripheral neuropathy
    2. Sensory level
    3. Unequivocal sphincter disturbance

II Electrodiagnostic (Physiologic) studies
A. Mandatory
    Must have 3 of the 4 following:
    1. Reduction in conduction velocity (CV) in two or more motor nerves
        a. <80% of lower limit of normal (LLN) if amplitude >80% of LLN.
        b. <70% of LLN if amplitude <80% of LLN
    2. Partial conduction block or abnormal temporal dispersion in one or more motor nerves: either peroneal nerve between ankle and below fibular head, median nerve between wrist and elbow, or ulnar nerve between wrist and below elbow. Criteria suggestive of partial conduction block: <15% change in duration between proximal and distal sites and >20% drop in peak (-p) area or peak to peak (p-p) amplitude between proximal and distal sites. Criteria for abnormal temporal dispersion and possible conduction block: >15% change in duration between proximal and distal sites and >20% drop in -p area or p-p amplitude between proximal and distal sites. These criteria are only suggestive of partial conduction block as they are derived from studies of normal individuals. Additional studies, such as stimulation across short segments or recording of individual motor unit potentials, are required for confirmation
    3. Prolonged distal latency in two or more nerves.
        a. >125% of upper limit of normal (ULN) if amplitude >80% of LLN
        b. >150% of ULN if amplitude <80% of LLN
    4. Absent F waves or prolonged minimum F wave latencies (10-15 trials) in two or more motor nerves:
        a. >120% of ULN if amplitude >80% of LLN
        b. >150% of ULN if amplitude <80% of LLN
B. Supportive
    1. Reduction in sensory CV <80% of LLN
    2. Absent H reflex

III Pathologic features
A. Mandatory
    Nerve biopsy showing unequivocal evidence of demyelination and remyelination
    1. Demyelination by either electron microscopy (>5 fibers) or teased fiber studies(>12% of 50 teased fibers, minimum of 4 internodes each, demonstrating demyelination/remyelination)
B. Supportive
    1. Subperineurial or endoneurial edema
    2. Mononuclear cell infiltration
    3. Onion-bulb formation
    4. Prominent variation in the degree of demyelination between fascicles
C. Exclusion
    1. Vasculitis,neurofilamentous swollen axon, amyloid deposits, intracytoplasmic inclusions in Schwann cells or macrophages indicating adrenoleukodystrophy, metachromatic leukodystrophy, globoid cell leukodystrophy, or other evidence of specific pathology

Cerebrospinal fluid studies
A. Mandatory
    1. Cell count <10/mm^3 if HIV seronegative or <50/mm^3 if HIV seropositive.
    2. Negative VDRL
B. Supportive
    1. Elevated protein

Definite CIDP= Clinical A and C, Physiologic A, Pathologic A and C, and CSF A; Probable CIDP= Clinical A and C, Physiologic A, and CSF A; Possible CIDP= Clinical A and C, and Physiologic A

Table 5. Disease pattern, course and prognosis

Report	#*	Disease Course	Prognosis and Mortality
Dyck[7]	53	Monophasic 15%, relapsing 34%, progressive 49%	2 (3.8%) complete recovery; 36 (67.9%) mild to moderate neurologic impairment; 6 (11.3%) confinement to wheelchair or to bed; 9 (17%) died (Average follow-up of 7.5 years). Among 9 deaths, 6 were due to CIDP and 3 were due to other reasons.
McCombe[48]	92	Monophasic 34.8%, relapsing 65.2%	59 (73%) improved with mild or less residual symptoms; 14 (17%) moderate, disability; 3 (3%) severe disability or non-ambulatory; 10 (11%) died. Among 10 deaths, 5 (5.4%) were due to CIDP and 5 were due to intercurrent illness.
Barohn[11]	60	Monophasic 53.3%, relapsing (with average frequency of relapse 0.7±0.4 per year) 46.6%	18 (30%) complete remission; 6 (10%) partial remission; 3 (5%) did not respond to therapy; 2 (3.3%) died of complications of CIDP (over the 10-year follow-up).
Gorson[8]	60	45 patients with idiopathic CIDP and 15 patients with MGUS. Monophasic 29%, relapsing 31%, progressive 40%.	9% free of symptoms; 62% minimal impairment but independent with activities of daily living; 25% moderate disability; 4% bedbound, chairbound or required constant supervision (at median follow-up of 3 years). 7% (3 out of 45 patients) subsequently died, 2 as a complication of CIDP or therapy (1 pulmonary embolus and the other pneumonia during cyclophosphamide therapy).
Bouchard[9]	100	Monophasic 41%, relapsing 14%, gradually progressive 45%, (excluded patients monoclonal gammopathy, HIV infection and paraneoplastic disorders).	24 (28.9%) without handicap; 11 (13.3%) with mild handicap; 23 (27.7%) with moderate handicap; 6 (7.2%) with severe handicap; 5 (6%) bedridden; and 14 (17%) died (over average 6 year follow-up). Among 14 deaths, 9 from progression of neurologic deficit, 4 from unknown cause and 1 from pulmonary embolism.
Sghirlanzoni[64]	60	Monophasic 48%, relapsing 52%	8 (13%) complete remission off therapy; 7 (12%) remission but on therapy; 17 (28%) unchanged; 7 (13%) worsened; 3 (5%) died (after mean follow up 4.4 years).

* Number of patients in the report, EPS (electrophysiologic studies)

Specific electrodiagnostic features are included in the research criteria for diagnosis of CIDP, but are beyond the scope of this review.[11,13] Such electrodiagnostic features of acquired demyelination include some combination of nonuniform slowing of nerve conduction velocities, partial conduction block or abnormal temporal dispersion, prolonged distal latencies and absence or prolongation of F waves. Only about 70% of patients with CIDP ever meet strict demyelinating criteria.[2,11]

Sural nerve biopsy is undertaken in many patients with a clinical syndrome suggestive of CIDP. The classic histopathological feature of demyelination accompanied by thinly myelinated fibers with Schwann cell proliferation (onion bulbs) is found in only 60%-70% of nerve biopsies.[9,11] In many cases, the demyelinating features are subtle, with scattered demyelinated internodal segments best observed by nerve fiber teasing. Other features may include endoneurial and subperineurial edema and inflammatory cell infiltration. In only a minority of cases, however, is the number of inflammatory cells greater than that found in other, noninflammatory neuropathies and normal controls.[9,14] CIDP is multifocal in nature, and a sural nerve biopsy may not be representative of what is taking place at other sites, particularly at the level of the nerve roots. Despite its inclusion in the research criteria, a sural nerve biopsy may not be necessary in every patient suspected of having CIDP.[2]

CIDP–Therapy

Randomized controlled trials have confirmed that corticosteroids,[39] plasma exchange[40-42] and intravenous immunoglobulin (IVIG)[41-44] are beneficial in CIDP. Plasma exchange and IVIG were shown to be equally effective in a controlled trial.[42] A recent crossover trial comparing IVIG and oral prednisolone showed that both treatments reduced impairment and disability in CIDP, with no significant difference between the two treatments.[45] Prednisone is considered by some to be first line treatment, though a recent study demonstrated the efficacy of IVIG as initial treatment of CIDP.[46] Medications that have been used in CIDP patients refractory to prednisone, plasma exchange and IVIG are similar to those used for other immune mediated neurological diseases. Such treatments include azathioprine,[47] cyclophosphamide,[48,49] cyclosporine,[50,51] total lymphoid irradiation[52] and interferon alpha and beta,[53,54] all of which have been used anecdotally with varying degrees of success.

Table 6. MRC muscle strength grading

5	Normal strength
5-	Barely detectable weakness
4S	Same as 4 but stronger than reference muscle
4	Muscle is weak but moves the joint against a combination of gravity and some resistance.
4W	Same as 4 but weaker than reference muscle
3+	The muscle is capable of transient resistance but collapses abruptly. This degree of weakness is difficult to put into words, but it is a muscle which is able to move the joint against gravity and an additional small amount of resistance. It is not to be used for muscle capable of sustained resistance throughout the whole range of movement.
3	Muscle cannot move against resistance but moves the joint fully against gravity. With the exception of knee extensors, the joint must be moved through the full mechanical range against gravity. If a patient has contractures that limit movement of the joint, the mechanical range will obviously be to the point at which the contractures cause a significant resistance to the movement.
3-	Muscle moves the joint against gravity but not through the full extent of the mechanical range of the joint.
2	Muscle moves the joint when gravity is eliminated.
1	A flicker of movement is seen or felt in the muscle.
0	No movement

Prognosis of CIDP and Rationale of HSCT

When considering an aggressive form of therapy such as hematopoietic stem cell transplantation, it is necessary to consider the risk-benefit ratio very carefully. Therefore, a discussion of the prognosis of CIDP is necessary. Six retrospective reports are available and are summarized below and in Table 5. In the first report of 53 patients (average follow up of 7.5 years), patients' clinical course of monophasic, relapsing, and progressive were 15% (8 patients), 34% (18 patients), and 49% (26 patients), respectively. Two out of 53 (3.8%) patients completely recovered neurologically, 32 out of 53 (60.4%) patients had mild to moderate neurologic impairment but were able to walk and to work, 4 out of 53 (7.5%) were ambulatory but not working, 6 out of 53 (11.3%) were confined to a wheelchair or to bed, and 9 out of 53 (17%) died. Among 9 deaths, 6 were due to CIDP (average duration from onset of illness was 7.3 years) and 3 were due to other reasons (average duration from onset of illness was 5.5 years).[7]

In the second report of 92 patients, 32 (34.8%) patients had a monophasic course and 60 (65.2%) had a relapsing course. Forty-nine (65%) out of 76 patients treated with prednisone responded. Twelve patients did not respond to prednisone and received immunosuppressive drugs. Four out of 7 responded to azathioprine and 4 out of 5 patients responded to cyclophosphamide. Eight out of 12 patients responded to plasma exchange and 2 needed maintenance exchanges. Fifty-nine (73%) patients improved with mild or less residual symptoms, 14 (17%) patients improved with moderate disability, 3 (3%) patients developed severe disability requiring assistance or were nonambulatory, 10 (11%) patients died, and 6 patients were lost to follow up. Among deaths, 5 (5.4%) were from CIDP (between 1 year to 41 years after onset of disease) and another 5 were from inter-current illness.[48]

Table 7. Rankin's disability score

0	asymptomatic
1	nondisabling symptoms that do not interfere with daily activities
2	slight disablity, unable to carry out all activities but still able to look after themselves
3	moderate disability requiring assistance with some activities but able to walk without assistance
4	moderately severe disability, unable to walk without assistance, and unable to attend to bodily needs without assistance
5	severe disability totally dependent requiring constant nursing care.

Improvement is defined as at least 1 point decrease in the Rankin scale. Remission is defined as achieving a level of 0 or 1 for at least 6 months off all immune modulating therapy.

In the third report of 60 patients, 32 (53.3%) patients had a monophasic course, and 28 (46.6 %) patients had a relapsing course with average frequency of relapse of 0.7±0.4 per year. Eighteen (30%) patients had a complete remission and 6 (10%) had a partial remission. Three (5%) patients did not respond to therapy and 2 (3.3%) patients died of complications of the CIDP over the 10 year follow up. Neither age, duration of illness, motor nerve conduction velocity, compound motor action potential amplitude, CSF protein level, nor nerve biopsy were predictive of the clinical outcome. Only the average muscle score (AMS) at the time of diagnosis showed correlation with the AMS at the time of maximal improvement, indicating that a patient with greater weakness at the initiation of treatment was also more likely to have residual weakness at the time of plateau.[11]

In the fourth report of 60 patients, 45 had idiopathic CIDP and 15 had CIDP with MGUS (monoclonal gammopathy of undetermined significance). Among idiopathic CIDP patients, monophasic, relapsing (one or more relapse) and progressive course were present in 13 (29%), 14 (31%) and 18 (40%) patients, respectively. At median follow up of 3 years, 9% were free of symptoms, 62% had minimal impairment but were independent with activities of daily living, 25% had moderate disability and 4% were bedbound, chairbound or required constant supervision. Seven percent (3 out of 45 patients) of patients subsequently died, two as a complication of CIDP or therapy (one pulmonary embolus and the other pneumonia during cyclophosphamide therapy). In the 15 patients with MGUS, 10 patients had a progressive course and 5 had a relapsing course. At median follow up of 3.1 years, 10 (67%) patients had minimal impairment (Rankin score 1 or 2) and 5 (33%) had moderate disability (Rankin score 3).[8]

In the fifth report of 100 patients (excluding patients with monoclonal gammopathy, HIV infection and paraneoplastic disorders), monophasic, relapsing and gradually progressive course were present in 41 (41%), 14 (14%) and 45 (45%), respectively. Among the 100 patients, 83 were evaluable over an average follow up of 6 years. Patients without handicap, with mild handicap, with moderate handicap, with severe handicap, bedridden, and death were 24 (28.9%), 11(13.3%), 23 (27.7%), 6 (7.2%), 5 (6%), and 14 (17%), respectively. Among 14 deaths, 9 were as a result of progression of neurologic deficit, 4 were of unknown cause and 1 was from pulmonary embolism. Poor prognostic factors included CNS manifestations, presence of four limb weak-

Table 8. Barthel Index[67]

	Activity Score
Feeding 0 = unable 5 = needs help cutting, spreading butter, etc., or requires modified diet 10 = independent	0,5,10
Bathing 0 = dependent 5 = independent (or in shower)	0,5
Grooming 0 = needs help with personal care 5 = independent face/hair/teeth/shaving (implements provided)	0,5
Dressing 0 = dependent 5 = needs help but can do about half unaided 10 = independent (including buttons, zips, laces, etc.)	0,5,10
Bowels 0 = incontinent (or needs to be given enemas) 5 = occasional accident 10 = continent	0,5,10
Bladder 0 = incontinent, or catheterized and unable to manage alone 5 = occasional accident 10 = continent	0,5,10
Toilet Use 0 = dependent 5 = needs some help, but can do something alone 10 = independent (on and off, dressing, wiping)	0,5,10
Transfers (bed to chair and back) 0 = unable, no sitting balance 5 = major help (one or two people, physical), can sit 10 = minor help (verbal or physical) 15 = independent	0,5,10,15
Mobility (on level surfaces) 0 = immobile or <50 yards 5 = wheelchair independent, including corners, >50 yards 10 = walks with help of one person (verbal or physical) >50 yards 15 = independent (but may use any aid; for example, stick) >50 yards	0,5,10,15
Stairs 0 = unable 5 = needs help (verbal, physical, carrying aid) 10 = independent	0,5,10

TOTAL (0 - 100)
Patient Name: _____ Rater: _____ Date:__/__/__ __:__

ness at onset, increased CSF protein, severe demyelinating features on electrophysiologic testing, axonal loss and active demyelination on biopsy.[9]

In the sixth report, 18 (30%) out of 60 patients had MGUS. At the end of follow up (mean follow up period of 4.4 years), 36 out of 60 patients (60%) had improved. CIDP without MGUS (pure CIDP) improved in 29 out of 42 (69%) and CIDP with MGUS (with MGUS) improved in 7 out of 18 (39%). Seven patients worsened (3 pure CIDP, 4 with MGUS); and 17 (28%) did not change (10 pure CIDP, 7 with MGUS). Complete remission was achieved in 8 (13%) patients and 7 (12%) were in remission but still receiving therapy. Three patients died. Favorable prognostic factors were female gender, young age at onset (45 years old or less), relapsing remitting course and absence of axonal damage on neurophysiologic study. Poor prognostic features were older age of onset (more than 45 years old) and presence of axonal damage on neurophysiologic study.[64]

CIDP follows a relapsing or progressive course. First line therapy for remission is generally considered to be IVIG, prednisone or plasma exchange. When these fail, subsequent regimens such as cyclophosphamide, methotrexate, or azathioprine are empiric, often leading to prolonged immunosuppression. Because of disease exacerbations during medication tapering, some patients may require long term immunosuppressive treatment. Such treatment subjects the patients to the many long term side effects of these agents. It is this group of patients who need high doses of immunosuppressive agents to maintain remission, for whom a new, more aggressive treatment approach may be warranted. Waiting too long until the patient has failed multiple immunosuppressive medications with persistent and insidious progressive disability may be too late for intervention with aggressive immune-based therapies directed towards peripheral demyelination. These patients have probably developed chronic disability due to axonal atrophy.[55] In a study of patients receiving monthly pulse IV cyclophosphamide, patients with a disease duration of less than 14 months and patients who had some response to prior immunosuppression improved with monthly pulse cyclophosphamide.[56] These same considerations of early disease that is still immune responsive may be important criteria to respond to HSCT.

Case Reports

Four patients with CIDP who were refractory to conventional treatment were treated with high-dose cyclophosphamide (200 mg/kg over 4 days) without stem cell transplant.[65] The patients ranged in age from 39 to 61 and had partial but incomplete response to previous immunotherapy, including prednisone, plasmapheresis, IVIG, azathioprine, fludarabine, and oral and intravenous pulse cyclophosphamide. All patients had improvements in strength and functional status, with follow up ranging from 11 to 40 months. Summated compound motor action potential (CMAP) amplitudes improved in 3 out of 4. All patients were able to discontinue all other immunomodulatory medications.

A group in The Netherlands performed myeloablative high dose chemotherapy and autologous HSCT in a patient with CIDP.[66] A 38 year old male developed numbness and tingling in his fingertips in 1988 that progressed to sensory loss in the arms and legs and muscle weakness. In 1990, weakness progressed to MRC (Medical Research Council) grade 4 in all four extremities. Sensory loss was in a glove and stocking distribution. Electrophysiologic studies revealed decreased nerve conduction velocities and prolonged distal latencies and F wave responses. Sural nerve biopsy was consistent with CIDP. Prednisone improved muscle strength to normal, but subsequent attempts of tapering to less than 20 mg per day were unsuccessful due to recurrence of weakness. Addition of azathioprine or methotrexate did not allow prednisone tapering. IVIG therapy was needed repetitively to maintain response. Autologous HSCT was performed 8 years after disease onset. Peripheral blood stem cells were mobilized by cyclophosphamide (4 g/m^2) and granulocyte colony stimulating factor (5 mcg/kg) and were $CD34^+$ selected. The conditioning regimen was BEAM (carmustine 300 mg/m^2, etoposide 800 mg/m^2, cytarabine 800 mg/m^2, melphalan 140 mg/m^2). Two years after HSCT, neurophysiologic studies have improved, and performance status is normal with only residual fingertip numbness on 5 mg of prednisone per day.

Suggested Outcome Measures

HSCT outcome may be followed by MRC muscle strength grading (Table 6), quantitative isometric strength, mean motor nerve conduction velocity, the modified Rankin scale (Table 7)[57] and Barthel Index (Table 8).[67]

Conclusion

CIDP has a mortality between 3.3 to 17%, but morbidity is substantial as demonstrated in retrospective data, and subsequent loss in quality of life. Patients who fail standard therapies may be considered candidates for carefully controlled trials of HSCT.

References

1. Dyck PJ, Oviatt KF, Lambert EH. Intensive evaluation of referred unclassified neuropathies yields improved diagnosis. Ann Neurol 1981; 10:222-26.
2. Barohn RJ, Saperstein DS. Guillain-Barré syndrome and chronic inflammatory demyelinating polyneuropathy. Semin Neurol 1998; 18:49-61.
3. Lunn MPT, Manji H, Choudhary PP et al. Chronic inflammatory demyelinating polyradiculoneuropathy: A prevalence study in south east England. J Neurol Neurosurg Psychiatry 1999; 66:269-271.
4. McLeod JG, Pollard JD, Macaskill P et al. Prevalence of chronic inflammatory demyelinating polyneuropathy in the New South Wales, Australia. Ann Neurol 1999; 46:910-913.
5. Kissel JT, Mendell JR. Chronic inflammatory demyelinating polyradiculoneuropathy. In: Mendell JR, Kissel JT, Cornblath DR, eds. Diagnosis and Management of Peripheral Nerve Disorders. New York: Oxford University Press, 2001:173-191.
6. Austin JH. Recurrent polyneuropathies and their corticosteroid treatment. Brain 1958; 81:157-94.
7. Dyck PJ, Lais AC, Ohta M et al. Chronic inflammatory polyradiculoneuropathy. Mayo Clin Proc 1975; 50:621-37.
8. Gorson KC, Allam G, Ropper AH. Chronic inflammatory demyelinating polyneuropathy: Clinical features and response to treatment in 67 consecutive patients with and without a monoclonal gammopathy. Neurology 1997; 48:321-28.
9. Bouchard C, Lacroix C, Plante V et al. Clinicopathologic findings and prognosis of chronic inflammatory demyelinating polyneuropathy. Neurology 1999; 52:498-503.
10. Maisonobe T, Chassande B, Verin M et al. Chronic dysimmune demyelinating polyneuropathy: A clinical and electrophysiological study of 93 patients. J Neurol Neurosurg Psychiatry 1996; 61:36-42.
11. Barohn RJ, Kissel JT, Warmolts JR et al. Chronic inflammatory demyelinating polyradiculoneuropathy. Clinical characteristics, course, and recommendations for diagnostic criteria. Arch Neurol 1989; 46:878-84.
12. Mendell JR, Kolkin S, Kissel JT et al. Evidence for central nervous system demyelination in chronic inflammatory demyelinating polyradiculoneuropathy. Neurology 1987; 37:1291-94.

13. Ad Hoc Subcommittee of the American Academy of Neurology AIDS task force. Research criteria for diagnosis of chronic inflammatory demyelinating polyradiculoneuropathy (CIDP). Neurology 1991; 41:617-618.
14. Krendel DA, Parks HP, Anthony DC et al. Sural nerve biopsy in chronic inflammatory demyelinating polyradiculoneuropathy. Muscle Nerve 1989; 12:257-64.
15. Dyck PJ, Prineas J, Pollard J. Chronic inflammatory demyelinating polyradiculoneuropathy. In: Dyck PJ, Thomas PK, Griffin JW et al, eds. Peripheral Neuropathy. Philadelphia: W.B. Saunders Company, 1993:1498-1517.
16. Oka N, Akiguchi I, Kawasaki T et al. Elevated serum levels of endothelial leukocyte adhesion molecules in Guillain-Barré syndrome and chronic inflammatory demyelinating polyneuropathy. Ann Neurol 1994; 35:621-624.
17. Hartung H-P, Reiners K, Michels M et al. Serum levels of soluble E-selectin (ELAM-1) in immune-mediated neuropathies. Neurology 1994; 44:1153-58.
18. Hughes RAC. Guillain-Barré Syndrome. Heidelberg: Springer-Verlag, 1990.
19. Adam AM, Atkinson PF, Hall SM et al. Chronic experimental allergic neuritis in Lewis rat. Neuropathol Appl Neurobiol 1989; 15:249-64.
20. Harvey GK, Pollard JD, Schindhelm et al. Chronic experimental allergic neuritis. An electrophysiological and histological study in the rabbit. J Neurol Sci 1987; 81:215-26.
21. Rostami AM, Ventura E, Kimura H et al. Induction of severe experimental allergic neuritis with a synthetic peptide corresponding to the 53-78 amino acid sequence of the myelin P2 protein. Neurology 1988; 38:375.
22. Hahn AF, Feasby TE, Wilkie L et al. P2-peptide induced experimental allergic neuritis: A model to study axonal degeneration. Acta Neuropathol 1991; 82:60-5.
23. Milner P, Lovelidge CA, Taylor WA et al. P0 myelin produces experimental allergic neuritis in Lewis rats. J Neurol Sci 1987; 79:275-85.
24. Gabriel CM, Hughes RAC, Moore SE et al. Induction of experimental allergic neuritis with peripheral myelin protein-22. Brain 1998; 121:1895-1902.
25. Saida T, Siada K, Dorfman SJ. Experimental allergic neuritis induced by sensitization with galactocerebroside. Science 1979; 204:1103.
26. Hartung H-P, Toyka KV, Griffin JW. Guillain-Barré syndrome and chronic inflammatory demyelinating polyradiculoneuropathy. In: Antel JA, Birnbaum G, Hartung H-P, eds. Clinical Neuroimmunology. Malden: Blackwell Science, 1998:294-306.
27. Melendez-Vasquez C, Redford J, Choudhary PP et al. Immunological investigation of chronic inflammatory demyelinating polyradiculoneuropathy. J Neuroimmunol 1997; 73:124-34.
28. Khalili-Shirazi A, Atkinson P, Gregson N et al. Antibody responses to P0 and P2 myelin proteins in Guillain-Barré syndrome and chronic inflammatory demyelinating polyradiculoneuropathy. J Neuroimmunol 1993; 46:245-52.
29. Yan WX, Archelos JJ, Hartung H-P et al. P0 protein is a target antigen in chronic inflammatory demyelinating polyradiculoneuropathy. Arch Neurol 2001; 50:286-92.
30. Hughes RAC. Chronic inflammatory demyelinating polyradiculoneuropathy. Arch Neurol 2001; 50:281-82.
31. Yuki N, Tagawa Y, Handa S. Autoantibodies to peripheral nerve glycosphingolipids SPG, SLPG, and SGPG in Guillain-Barré syndrome and chronic inflammatory demyelinating polyradiculoneuropathy. J Neuroimmunol 1996; 70:1-6.
32. Gabriel CM, Gregson NA, Hughes RAC. Anti-PMP22 antibodies in patients with inflammatory neuropathy. J Neuroimmunol 2000; 104:139-46.
33. Ritz MF, Lechner-Scott J, Scott RJ et al. Characterization of autoantibodies to peripheral myelin protein 22 in patients with hereditary and acquired neuropathies. J Neuroimmunol 2000; 104:155-63.
34. Van den Berg LH, Mollee I, Wokke JH et al. Increased frequencies of HPRT mutant T lymphocytes in patients with Guillain-Barré syndrome and chronic inflammatory demyelinating polyradiculoneuropathy: Further evidence for a role of T cells in the etiopathogenesis of peripheral demyelinating diseases. J Neuroimmunol 1994; 58:37-42.
35. Hartung H-P, Reiners K, Schmidt B et al. Serum interleukin-2 concentrations in Guillain-Barré syndrome and chronic inflammatory demyelinating polyradiculoneuropathy: Comparison with other neurological diseases of presumed immunopathogenesis. Ann Neurol 1991; 30:48-53.
36. Hartung H-P, Hughes RAC, Taylor WA et al. T cell activation in Guillain-Barré syndrome and in MS: Elevated levels of IL-2 receptors. Neurology 1990; 40:215-18.
37. Khalili-Shirazi A, Hughes RAC, Brostoff SW et al. T cell responses to myelin proteins in Guillain-Barré syndrome. J Neurol Sci 1992; 111:200-3.
38. Misawa S, Kuwabara S, Mori M et al. Serum levels of tumor necrosis factor-alpha in chronic inflammatory demyelinating polyneuropathy. Neurology 2001; 56:666-69.
39. Dyck PJ, O'Brien PC, Oviatt KF et al. Prednisone improves chronic inflammatory demyelinating polyradiculoneuropathy more than no treatment. Ann Neurol 1982; 11:136-41.
40. Dyck PJ, Daube J, O'Brien P et al. Plasma exchange in chronic inflammatory demyelinating polyradiculoneuropathy. N Engl J Med 1986; 314:461-65.
41. Hahn AF, Bolton CF, Pillay N et al. Plasma-exchange therapy in chronic inflammatory demyelinating polyneuropathy. A double-blind, sham controlled, cross-over study. Brain 1996; 119:1055-1066.
42. Dyck PJ, Litchy WJ, Kratz KM et al. A plasma exchange versus immune globulin infusion trial in chronic inflammatory demyelinating polyradiculoneuropathy. Ann Neurol 1994; 36:838-45.
43. Vermeulen M, van Doorn PA, Brand A et al. Intravenous imunnoglobulin treatment in patients with chronic inflammatory demyelinating polyneuropathy: A double blind, placebo controlled study. J Neurol Neurosurg Psychiatry 1993; 56:36-39.
44. Van Doorn PA, Brand A, Strengers PF et al. High dose intravenous immunoglobulin treatment in chronic inflammatory demyelinating polyneuropathy: A double-blind, placebo-controlled crossover study. Neurology 1990; 40:209-12.
45. Hughes R, Bensa S, Willison H et al. Randomized controlled trial of intravenous immunoglobulin versus oral prednisolone in chronic inflammatory demyelinating polyradiculoneuropathy. Ann Neurol 2001; 50:195-201.
46. Mendell JR, Barohn RJ, Freimer ML et al. Randomized controlled trial of IVIg in untreated chronic inflammatory demyelinating polyneuropathy. Neurology 2001; 56:445-49.
47. Pentland B, Adams GG, Mawdsley C. Chronic idiopathic polyneuropathy treated with azathioprine. J Neurol Neurosurg Psychiatry 1982; 45:866-69.
48. McCombe PA, Pollard JD, McLeod JG. Chronic inflammatory demyelinating polyradiculoneuropathy. A clinical and electrophysiological study of 92 cases. Brain 1987; 110:1617-30.
49. Prineas JW, McLeod JG. Chronic relapsing polyneuritis. J Neurol Sci 1976; 27:427-58.
50. Hodgkinson SJ, Pollard JD, McLeon JG. Cyclosporin A in the treatment of chronic demyelinating polyradiculoneuropathy. J Neurol Neurosurg Psychiatry 1990; 53:327-30.
51. Barnett MH, Pollard JD, Davies L et al. Cyclosporin A in resistant chronic inflammatory demyelinating polyradiculoneuropathy. Muscle Nerve 1998; 21:454-60.
52. Rosenberg NL, Lacy JR, Kennaugh RC et al. Treatment of refractory chronic demyelinating polyneuropathy with lymphoid irradiation. Muscle Nerve 1985; 8:223-32.
53. Gorson KC, Allam G, Simovic D et al. Improvement following interferon-alpha 2A in chronic inflammatory demyelinating polyneuropathy. Neurology 1997; 48:777-80.
54. Choudhary PP, Thompson N, Hughes RA. Improvement following interferon beta in chronic inflammatory demyelinating polyradiculoneuropathy. J Neurol 1995; 242:252-53.
55. Oda M, Satake M, Nakamura N et al. Axonal and perikaryal involvement in chronic inflammatory demyelinating polyneuropathy. J Neurol Neurosurg Psychiatry 1999; 66(6):727-33.
56. Good JL, Chehrenama M, Mayer RF et al. Pulse cyclophosphamide therapy in chronic inflammatory demyelinating polyneuropathy. Neurology 1998; 51(6):1735-8.
57. Van Swieten JC, Koudstaal PJ, Visser MC et al. Interobserver agreement for assessment of handicap in stroke. 1988; 19:604-7.

58. Burt RK, Traynor AE, Oyama Y et al. High-dose immune suppression and autologous hematopoietic stem cell transplantation in refractory Crohn's Disease. Blood Oct 10 2002. [epub ahead of print].
59. Traynor AE, Barr WG, Rosa RM et al. Hematopoietic stem cell transplantation for severe and refractory lupus. Analysis after five years and fifteen patients. Arthritis Rheum 2002; 46(11):2917-23.
60. Burt RK, Traynor AE, Pope R et al. Treatment of autoimmune disease by intense immunosuppressive conditioning and autologous hematopoietic stem cell transplantation. Blood 1998; 92(10):3505-14.
61. Burt RK, Slavin S, Burns WH et al. Induction of tolerance in autoimmune diseases by hematopoietic stem cell transplantation: Getting closer to a cure? Blood 2002; 99(3):768-84.
62. Binks M, Passweg JR, Furst D et al. Phase I/II trial of autologous stem cell transplantation in systemic sclerosis: Procedure related mortality and impact on skin disease. Ann Rheum Dis 2001; 60(6):577-84.
63. Traynor AE, Schroeder J, Rosa RM et al. Treatment of severe systemic lupus erythematosus with high-dose chemotherapy and haemopoietic stem cell transplantation: A phase I study. Lancet 2000; 356:701-7.
64. Sghirlanzoni A, Solari A, Ciano C et al. Chronic inflammatory demyelinating polyradiculoneuropathy: Long-term course and treatment of 60 patients. Neurol Sci 2000; 21:31-37.
65. Brannagan TH, Pradhan A, Heiman-Patterson T et al. High-dose cyclophosphamide without stem-cell rescue for refractory CIDP. Neurology 2002; 58:1856-58.
66. Vermeulen M, Van Oers MH. Successful autologous stem cell transplantation in a patient with chronic inflammatory demyelinating polyneuropathy. J Neurol Neurosurg Psychiatry 2002; 72:127-28.
67. www.neuro.mcg.edu/mcgstrok/Indices/Barthel Ind.htm

CHAPTER 50

Hematopoietic Stem Cell Therapy for Patients with Refractory Myasthenia Gravis

Richard K. Burt

Introduction

Myasthenia gravis (MG), which means severe muscle weakness, is due to antibody mediated loss of motor end plate acetylcholine (ACh) receptors.[1,2] In normal muscle, acetylcholine released from a nerve ending binds to the acetylcholine receptor (AChR) on the post synaptic motor end plate of muscle inducing a depolarization potential.[2,3] If enough AChR depolarizations occur, the firing threshold is exceeded and an action potential or contraction of the muscle follows. AchR antibody reduce the number of ACh receptors causing loss of synaptic folds on the motor end plate and impaired neuromuscular transmission. Ex vivo transfer of ACh receptor antibody induces disease across a wide range of species.[4,5] While in utero, transplacental transfer from mother to fetus causes neonatal MG.[6-8] The clinical manifestations are muscular fatigue and weakness.[9,10] Treatment of MG includes acetylcholine esterase inhibitors and immune suppressive or immune modulating medications.[11-15] For patients with severe disease, HSCT may be considered. In fact, the first animal hematopoietic stem cell transplantation (HSCT) for an autoimmune disease was reported in 1983 for experimental autoimmune myasthenia gravis (EAMG).[16] Syngeneic HSCT cured EAMG related muscle weakness.

Experimental Autoimmune Myasthenia Gravis

EAMG can be induced in a variety of species including mice, rats, and nonhuman primates by immunization with Torpedo californica (pacific electric ray) acetylcholine receptors in complete Freund's adjuvant.[17] Immunization activates MHC class II AChR epitope restricted CD4+ cells that in turn stimulate B cells to produce pathogenic anti-AChR antibodies. Within a species, EAMG susceptibility is age and gender-associated, being easier to induce in females.[18,19] The AChR is a multi-unit membrane protein.[20] The main immunogenic region (MIR) for induction of MG is a 9-mer peptide sequence of the AChR alpha unit (α67-76), although epitope or determinate spreading occurs.[20-22] Passive transfer of serum antibodies from an animal with EAMG transfers disease.[3] Similarly, human serum from MG patients transfers weakness and electromyographic (EMG) decremental changes to mice.[2] EAMG appears to be a reasonably good model for MG. However, thymic lymphofollicular hyperplasia (LFH) seen in most humans with MG is not present in EAMG.[23] Unlike EAMG, the initiating event(s) for MG is unknown.

Rats with EAMG had a rapid and sustained disappearance of anti-AChR antibodies after treatment with transplant doses of cyclophosphamide (200 mg/kg).[16] The addition of total body irradiation (600 cGy) to cyclophosphamide, and bone marrow reconstitution from syngeneic rats with active EAMG, eliminated not only anti-AChR antibodies but also anamnestic responses upon AChR rechallenge.[16] If EAMG is an accurate model for MG, then intense immune suppressive or immune ablative therapy with autologous HSCT may provide a durable remisssion or cure of EAMG.

Myasthenia Gravis

MG may be neonatal, congenital, or autoimmune.[9,10] Neonatal MG arises from transplacental transfer of ACh receptor antibodies from a mother causing autoimmune MG in the fetus. Neonatal MG resolves with post delivery clearance of maternal antibodies. Congenital MG results from a genetic defect or mutations in the AChR. Patients with congenital MG do not have AChR antibodies.

The diagnosis of MG is made by clinical manifestations, improvement to the anticholinesterase edrophonium chloride (Tensilon), and EMG. MG is characterized by weakness, often fluctuating, being worsened by exercise. Fatigue and weakness may occur in ocular, facial, bulbar, and/or limb muscles.[9,10] Ocular ptosis, ophthalmoparesis, dysarthria, and dysphagia are common. In severe cases, respiratory muscles are affected. Electrophysiologic studies reveal loss of amplitude with repetitive nerve stimulation. Approximately 85% of patients with MG have anti-AChR antibodies. These patients may have other genetic or autoimmune-mediated problems with the nerve-muscle synapse. Achieving action potential threshold depends on clustering of the AChR at the motor endplate. A neuronal protein, agrin, activates muscle-specific kinase (MuSK) to cluster AChRs via rapsyn, a muscle cytoplasmic synapse protein (Fig. 1).[24-26] Some MG patients without anti-AchR antibodies have antibodies to MuSK. Therefore, disruption of AChR clustering by either antibodies to AChR or MuSK results in the same clinical manifestations and EMG findings. MG must also be differentiated from other myasthenic syndromes such as Eaton Lambert syndrome, which is a malignancy-associated disorder with antibodies against PQ-type voltage-gated calcium channels.[27] Eaton Lambert syndrome can be distinguished from MG by EMG.

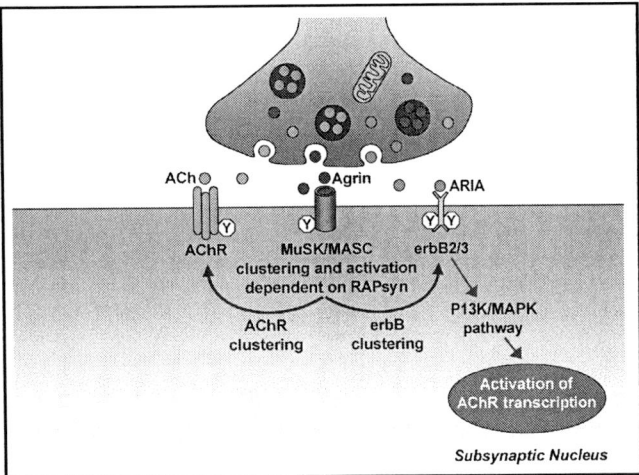

Figure 1. Neuromuscular junction. Agrin secreted by the motor nerve binds muscle specific kinase (MUSK) via myotube-associated specificity component (MASC) resulting in Rapsyn dependent clustering of Acetylcholine receptors (AChRs) and erbB2/3 receptors. Acetylcholine receptor inducing activity (ARIA) also released from the motor nerve activates P13 kinase and MAP kinase via erbB2/3 leading to increased AChR transcription. Acetylcholine (ACh) released by the motor neuron binds to the AChR causing muscle contraction.

Table 1. Osserman classification system of myasthenia gravis

Grade 0:	Asymptomatic
Grade I:	Ocular signs and symptoms only
Grade IIA:	Mild generalized weakness
Grade IIB:	Moderate generalized weakness
Grade III:	Acute fulminating, bulbar, and generalized with respiratory failure
Grade IV:	Late severe
Grade V:	With muscle atrophy

Table 2. Survival according to Osserman class

Osserman Class	2 Year Survival	4 Year Survival
I	100%	98%
IIA	95%	90%
IIB	90%	80%
III	70%	60%

From: Christensen PB et al. Neurosurgery and Psychiatry. Mortality and Survival in Myasthenia Gravis. J Neurol 1998; 64(1):78-83.

MG affects approximately 25,000 Americans.[28,29] Like most autoimmune diseases, it is associated with particular HLA genotypes, has a female predominance with about two-thirds of affected patients being female.[30,31] MG generally occurs in women in their second to fourth decade of life. Age of onset for men is generally later, in the fifth decade. The prevalence of MG has been increasing possibly due to better treatments and prolonged survival of patients with MG.[29] About 10% of MG patients have severe generalized disease. Approximately 30-50% of patients with thymoma have MG, while roughly 10-15% of MG patients have a thymoma which are epithelial cell (not lymphoid) tumors of the thymus.[32,33] Thymic lymphofollicular hyperplasia is common in MG patients without thymomas. Pure red cell aplasia is also associated with thymomas and MG.[34] While disease severity varies, spontaneous remission of MG is rare.

Disease severity is rated and treated according to the Osserman scoring system[35] (Table 1). Ocular only symptoms (Grade I) require only anti-cholinesterases such as physostigmine or neostigmine bromide. Immune suppressive drugs such as corticosteroids, azathrioprine, cyclophosphamide, cyclosporin, or mycophenolate are added for mild and moderate disease (Grade IIA and IIB). In addition to immune suppression and anti-cholinesterases, Grade III and IV disease is treated with plasmapheresis or intravenous immunoglobulin. Thymectomy is performed on all patients with thymomas and patients without thymomas who have moderate or severe disease.[11,36] After thymectomy, about 50% of patients enter a medication free remission which may take 6 or more months.[11]

Myasthenic crisis is defined as respiratory failure requiring mechanical ventilation.[37-39] It is a potentially lethal complication and is precipitated by infections, emotional stress, and medications. Myasthenic crises are complicated by lethal cardiac arrhythmias, usually asystole. This may be related to anti-cholinesterase related cholinergic-induced asystole, however, most asystolic events do not display signs of cholinergic overdose such as bradycardia, miosis or hypersalivation. Early intervention with cardiac pacing should be considered. Since infection is a crises trigger, patients undergoing HSCT should be treated aggressively with antibiotics even before onset of fever in order to prevent infections. A variety of medications including Penicillamine, beta-adrenergic blockers, and aminoglycosides aggravate MG symptoms.[38] Several MG treatment modalities, such as initiation of corticosteroids and anesthesia for thymectomy, are also triggers for MG crises. HSCT immune ablative regimens may also be crises triggers unless carefully designed.

MG has developed in a small number of patients after hematopoietic stem cell transplantation.[40-45] These cases occurred after allogeneic (not autologous) HSCT, arose from donor cells, were anti-AChR antibody positive, and were almost always associated with clinical graft versus host disease (GVHD). GVHD is a disease of immune dysregulation associated with lymph node and thymic involution, absence of immune responses to vaccines, and an inverted CD4/CD8 ratio. It is perhaps not surprising that GVHD may rarely lead to MG, another disease of immune dysregulation. In contrast, MG has not been reported after an autologous HSCT and virtually never occurs in allogeneic transplants uncomplicated by GVHD.

Hematopoietic Stem Cell Transplantation of Myasthenia Gravis

Eligibility for HSCT would be determined by selecting patients failing standard therapy and at high risk for mortality. Survival correlates with Osserman class (Table 2). In the first 4 years, approximately 20% per year of class IIB and 40% per year of class III myasthenia gravis patients expire.[46-48] Therefore, candidates should be either Osserman IIB, III, or IV. Class V, atrophy, is excluded because it may be unlikely that atrophy will reverse.

Table 3. Possible eligibility criteria for HSCT of Myasthenia Gravis

1. Established diagnosis of myasthenia gravis defined as clinical evidence of muscle weakness and fatigability and supported by a positive response to edrophonium chloride, and abnormal repetitive nerve stimulation or single-fiber EMG
2. Ages 15-55 years
3. Positive anti-AChR antibody or positive anti muscle specific kinase (MuSK) antibody
4. Failure of thymectomy
5. Failure of anticholinesterase therapy, corticosteroids, and at least two of the following: azathrioprine, cyclosporin, cellcept, cyclophosphamide, plasma exchange or IVIG. Failure is defined as at least 6 months after thymectomy and at least 3 months of above drug therapy and an Osserman score of IIB, III, or IV

And at least one of the following:
1. History of myasthenia crises (requiring mechanical ventilation)
2. Hospitalized for myasthenia gravis within the last 18 months
3. Inability to maintain nutrition due to muscle weakness
4. A Karnofsky performance status of 70% or less (may or may not be able to care for self but unable to carry on normal activity or unable to do active work)

Table 4. Quantitative Myasthenia Gravis Score

Test Items Weakness	None (Score 0)	Mild (Score 1)	Moderate (Score 2)	Severe (Score 3)
Double vision on lateral gaze (right or left) (circle one)	61 seconds	11-60 seconds	1-10 seconds	Spontaneous
Ptosis (upward spontaneous gaze)	61 seconds	11-60 seconds	1-10 seconds	Spontaneous
Facial muscles	Normal lid	Complete with some resistance	Complete without resistance	Incomplete
Swallowing 4 oz water (1/2 cup)	Normal	Minimal cough or throat clear	Severe cough, choking, nasal regurgitation	Cannot swallow—not attempted
Speech following counting aloud from 0-50 (onset of dysarthria)	None at #50	Dysarthria at #30-49	Dysarthria at #10-29	Dysarthia at #9
Right arm outstretched (90° sitting)	240 seconds	90-239 seconds	10-89 seconds	0-9 seconds
Left arm outstretched (90° sitting)	240 seconds	90-239 seconds	10-89 seconds	0-9 seconds
Vital capacity % predicted	>80%	65-79%	50-64%	<50%
Right hand grip (kg) Male Female	>45 >30	15-44 10-29	5-14 5-9	0-4 0-4
Left hand grip (kg) Male Female	>35 >25	15-34 10-24	5-14 5-9	0-4 0-4
Head lifted (45° supine)	120 seconds	30-119 seconds	1-29 seconds	0
Right leg outstretched (45° supine)	100 seconds	31-99 seconds	1-30 seconds	0
Left leg outstretched (45° supine)	100 seconds	31-99 seconds	1-30 seconds	0

Total QMG score ranges from 0-39

The criteria for HSCT of MG at Northwestern University is shown in Table 3. Since the Osserman criteria of moderate or severe weakness are somewhat arbitrary, further criteria that may be objective were added. These include a history of myasthenia crises (requiring mechanical ventilation), or hospitalized for myasthenia gravis within the last 18 months, or inability to maintain nutrition due to muscle weakness, or unable to carry on normal activity or unable to do active work.

The immune suppressive conditioning regimen is cyclophosphamide and ATG. This is based on the low mortality of this regimen in diseases such as systemic lupus erythematosus and the use of both agents as treatments for MG without reports of associated myasthenic crises.[48-55] Medications known to precipitate myasthenia crises are to be avoided unless absolutely medically necessary. Medications that are relative contraindications include: beta blockers, aminoglycosides, D-penicillamine, quinolones, benzodiazepines (valium, etc.), neuromuscular blockers (pancuronium, etc.), verapamil, anesthetics, dilantin, or overuse of anticholinesterases.[56]

Post HSCT outcome measures will be survival, anticholinesterase medication usage, immune suppressive medication usage, Osserman score, EMG, and the Quantitative Myasthenia Gravis Score (QMGS). A complete response is defined as freedom of symptoms and no disease-specific medications, partial response as freedom of symptoms on disease-specific medications, and clinical improvement as a decrease in the Osserman scale by at least 1 class. The QMGS is shown in Table 4. Since to be candidates all patients will have a thymectomy, HSCT offers a unique opportunity to study immune reconstitution in athymic adults. Evaluation of recent thymic emigrants by TREC (refer to Chapter 25) could help clarify existence of extra-thymic sites of lymphocyte maturation. Anti-AChR antibodies are also an easy immunologic marker to follow.[57] Pre-transplant lymphocytes should be cryopreserved and held in reserve in case immune reconstitution does not occur in patients without a thymus.

Summary

While EAMG was one of the first animal autoimmune diseases successfully treated with HSCT, a patient with myasthenia gravis has yet to undergo HSCT. This will probably change relatively shortly, as a US FDA approved phase I trial of HSCT for MG has recently opened at Northwestern University.

References

1. Patrick J, Lindstrom J. Autoimmune response to acetylcholine receptor. Science 1973; 180:871-872.
2. Lindstrom JM. Acetylcholine receptors and myasthenia. [Review] Muscle Nerve 2000; 23(4):453-77.
3. Plomp JJ, Van Kempen GT, De Baets et al. Acetylcholine release in myasthenia gravis:Regulation at single end-plate level. Ann Neurol 1995; 37(5):627-36.
4. Buschman E, van Oers N, Katz M et al. Experimental myasthenia gravis induced in mice by passive transfer of human myasthenic immunoglobulin. Evidence for an ameliorating effect by alpha-fetoprotein. J Neuroimmunol 1987; 13(3):315-30.
5. Tzartos S, Hochschwender S, Vasquez P et al. Passive transfer of experimental autoimmune myasthenia gravis by monoclonal antibodies to the main immunogenic region of the acetylcholine receptor. J Neuroimmunol 1987; 15(2):185-94.
6. Batocchi AP, Majolini L, Evoli A et al. Course and treatment of myasthenia gravis during pregnancy. [Review] Neurology 1999; 52(3):447-52.
7. Gardnerova M, Eymard B, Morel E et al. The fetal/adult acetylcholine receptor antibody ratio in mothers with myasthenia gravis as a marker for transfer of the disease to the newborn. Neurology 1997; 48(1):50-4.
8. Keesey K, Lindstrom J, Cokely H et al. Antiacetylcholine receptor antibody in neonatal myasthenia gravis. N Engl J Med 1977; 296:55.
9. Vincent A, Drachman DB. Myasthenia gravis. [Review] Adv Neurol 2002; 88:159-88.
10. Drachman DB. Myasthenia gravis. N Engl J Med 1994; 330:1797-1810.
11. Busch C, Machens A, Pichlmeier U et al. Long-term outcome and quality of life after thymectomy for myasthenia gravis. [Review] Ann Surg 1996; 224(2):225-32.
12. Marano E, Pagano G, Persico G et al. Thymectomy for myasthenia gravis: Predictive factors and long term evolution. A retrospective study on 46 patients. Acta Neurologica 1993; 15(4):277-88.
13. Venuta F, Rendina EA, De Giacomo T et al. Thymectomy for myasthenia gravis: A 27-year experience. Eur J Cardiothorac Surg 1999; 15(5):621-4. Discussion 624-5.
14. Walker MB. Treatment of myasthenia gravis with physostigmine. Lancet 1934; 1:1200-1201
15. Spring PJ, Spies JM. Myasthenia gravis: Options and timing of immunomodulatory treatment. [Review] BioDrugs 2001; 15(3):173-83.
16. Pestronk A, Drachman DB, Teoh R et al. Combined short-term immunotherapy for experimental autoimmune myasthenia gravis. Ann Neurol 1983; 14(2):235-41.
17. Christadoss P, Poussin M, Deng C. Animal models of myasthenia gravis. [erratum appears in Clin Immunol May 2000; 95(2):170.] Clin Immunol 2000; 94(2):75-87.
18. Hoedemaekers AC, Verschuuren JJ, Spaans F et al. Age-related susceptibility to experimental autoimmune myasthenia gravis: Immunological and electrophysiological aspects. Muscle Nerve 1997; 20(9):1091-101.
19. Hoedemaekers A, Graus Y, van Breda Vriesman P et al. Age- and sex-related resistance to chronic experimental autoimmune myasthenia gravis (EAMG) in Brown Norway rats. Clin Exper Immunol 1997; 107(1):189-97.
20. Vincent A, Willcox N, Hill M et al. Determinant spreading and immune responses to acetylcholine receptors in myasthenia gravis. [Review] Immunol Rev 1998; 164:157-68.
21. Tzartos SJ, Kokla A, Walgrave SL et al. Localization of the main immunogenic region of human muscle acetylcholine receptor to residues 67-76 of the [alpha] subunit. Proc Natl Acad Sci USA 1988; 85:2899-2903.
22. Tzartos SJ et al. The main immunogenic region of the acetylcholine receptor. Structure and role in myasthenia gravis. Autoimmunity 1991; 8:257-270.
23. Meinl E, Klinkert WE, Wekerle H. The thymus in myasthenia gravis. Changes typical for the human disease are absent in experimental autoimmune myasthenia gravis of the Lewis rat. Am J Pathol 1991; 139(5):995-1008.
24. Willmann R, Fuhrer C. Neuromuscular synaptogenesis: Clustering of acetylcholine receptors revisited. [Review] Cell Mol Life Sci 2002; 59(8):1296-316.
25. Liyanage Y, Hoch W, Beeson D et al. The agrin/muscle-specific kinase pathway: New targets for autoimmune and genetic disorders at the neuromuscular junction. [Review] Muscle Nerve 2002; 25(1):4-16.
26. Lin W, Burgess RW, Dominguez B et al. Distinct roles of nerve and muscle in postsynaptic differentiation of the neuromuscular synapse. Nature 2001; 410(6832):1057-64.
27. Pascuzzi RM. Myasthenia gravis and Lambert-Eaton syndrome. [Review] Therapeutic Apheresis 2002; 6(1):57-68.
28. Kalb B, Matell G, Pirskanen R et al. Epidemiology of myasthenia gravis: A population-based study in Stockholm, Sweden. Neuroepidemiology 2002; 21(5):221-5.
29. Phillips 2nd LH, Torner JC. Epidemiologic evidence for a changing natural history of myasthenia gravis. Neurology 1996; 47(5):1233-8.
30. Poulas K, Tzartos SJ. The gender gap in autoimmune disease. [Letter] Lancet 2001; 357(9251):234.
31. Poulas K, Tsibri E, Papanastasiou D et al. Equal male and female incidence of myasthenia gravis. Neurology 2000; 54(5):1202-3.

32. Lovelace RE, Younger DS. Myasthenia gravis with thymoma. Neurology 1997; 48(suppl5):S76-S81.
33. Levy Y, Afek A, Sherer Y et al. Malignant thymoma associated with autoimmune diseases: A retrospective study and review of the literature. [Review] Semin Arthritis Rheum 1998; 28(2):73-9.
34. Mizobuchi S, Yamashiro T, Nonami Y et al. Pure red cell aplasia and myasthenia gravis with thymoma: A case report and review of the literature. [Review] Jpn J Clin Oncol 1998; 28(11):696-701.
35. Osserman KE. Myasthenia Gravis. New York: Grune and Stratton: 1958:79-86
36. Lanska DJ. Indications for thymectomy in myasthenia gravis. Neurology 1990; 40:1828-1829.
37. Berrouschot J, Baumann I, Kalischewski P et al. Therapy of myasthenic crisis. Crit Care Med 1997; 25(7):1228-35.
38. Kaeser HE. Drug-induced myasthenic syndromes. [Review] Acta Neurol Scand Supplementum 1984; 100:39-47.
39. Thomas CE et al. Myasthenic crises: Clinical features, mortality, complications, and risk factors for prolonged intubation. Neurology 48:1253-1260.
40. Mackey JR, Desai S, Larratt L et al. Myasthenia gravis in association with allogeneic bone marrow transplantation: Clinical observations, therapeutic implications and review of literature. Bone Marrow Transplant 1997; 19(9):939-42.
41. Christadoss P, Poussin M, Deng C. Animal models of myasthenia gravis. [erratum appears in Clin Immunol 2000; 95(2):170]. [Review] Clin Immunol 2000; 94(2):75-87.
42. Baron F, Sadzot B, Wang F et al. Myasthenia gravis without chronic GVHD after allogeneic bone marrow transplantation. Bone Marrow Transplant 1998; 22(2):197-200.
43. Nelson KR, McQuillen MP. Neurologic complications of graft-versus-host disease. [Review] Neurol Clin 1988; 6(2):389-403.
44. Bolger GB, Sullivan KM, Spence AM et al. Myasthenia gravis after allogeneic bone marrow transplantation: Relationship to chronic graft-versus-host disease. Neurology 1986; 36(8):1087-91.
45. Smith CI, Aarli JA, Biberfeld P et al. Myasthenia gravis after bone-marrow transplantation. Evidence for a donor origin. N Engl J Med 1983; 309(25):1565-8.
46. Christensen PB, Jensen TS, Tsiropoulos I et al. Mortality and survival in myasthenia gravis. J Neurol Neurosurg Psychiatry 1998; 64(1):78-83.
47. Christensen PB, Jensen TS, Tsiropoulos I et al. Mortality and survival in myasthenia gravis: A Danish population based study. J Neurol Neurosurg Psychiatry 1998; 64(1):78-83.
48. Cosi V, Romani A, Lombardi M et al. Prognosis of myasthenia gravis: A retrospective study of 380 patients. J Neurol 1997; 244(9):548-55.
49. Pirofsky B. The effect of anti-thymocyte antiserum in progressive myasthenia gravis. Ann NY Acad Sci 1981; 377:779-85.
50. Pirofsky B, Reid RH, Bardana Jr EJ et al. Myasthenia gravis treated with purified antithymocyte antiserum. Neurology 1979; 29(1):112-6.
51. Pirofsky B, Reid RH, Bardana Jr EJ et al. Antithymocyte antiserum therapy in myasthenia gravis. Postgrad Med J 1976; 52(5Suppl):112-7.
52. Barnes AD. The use of antithymocyte globulin in myasthenia gravis. Postgrad Med J 1976; 52(5 Suppl):110-1.
53. Leovey A, Szobor A, Szegedi G et al. Myasthenia gravis: ALG treatment of seriously ill patients. Eur Neurol 1975; 13(5):422-32.
54. de Feo LG, Schottlender J, Martelli NA et al. Use of intravenous pulsed cyclophosphamide in severe, generalized myasthenia gravis. [Clinical Trial. Journal Article. Randomized Controlled Trial] Muscle Nerve J 2002; 26(1):31-6.
55. Niakan E, Harati Y, Rolak LA. Immunosuppressive drug therapy in myasthenia gravis. Arch Neurology 1986; 43(2):155-6.
56. Barrons RW. Drug-induced neuromuscular blockade and myasthenia gravis. [Review] Pharmacotherapy 1997; 17(6):1220-32.
57. Lindstrom JM, Seybold ME, Lennon VA et al. Antibody to acetylcholine receptor in myasthenia gravis: Prevalence, clinical correlates, and diagnostic value. Neurology 1976; 26:1054-1059.

CHAPTER 51

Hematopoietic Stem Cell Transplantation in Patients with Autoimmune Bullous Skin Disorders

Joan Guitart and Richard K. Burt

Introduction

There is an array of primary skin disorders that have been considered autoimmune diseases in their pathogenesis. Most of them are proven to have pathogenic autoantibodies that react against specific antigens on the cell surface of the keratinocytes or basement membrane components. Patients with autoimmune blistering diseases are commonly treated with systemic corticosteroids and immunosuppressant agents. Although most patients can be controlled with standard medications, a minority of patients have a severe clinical course and are resistant to conventional treatment. Most of these patients succumb from the complications of treatment, especially long-term corticosteroids. Recently, "immunoablation" with high dose chemotherapy followed by autologous hematopoietic transplantation has demonstrated that it can induce durable remissions for severe autoimmune diseases.[1] For autoimmune bullous skin diseases, case reports of high-dose cyclophosphamide without stem cell support also suggest promise as a treatment option.[2] In patients with extensive denuded skin lesions who have open skin lesions, a fear of serious infection during the neutropenic period after sub-myeloablative chemotherapy is a concern. Autologous hematopoietic stem cells can minimize the duration of the cyclophosphamide-induced neutropenic period.

Autoimmune Blistering Skin Disorders

Autoimmune bullous skin disorders include a number of blistering disorders characterized by the presence of autoantibodies against epidermal or basement membrane epitopes. The target autoantigens are generally desmosomal, hemidesmosomal, and basement membrane proteins. In some cases, the antibody can be detected on serum by indirect immunofluorescence or ELISA methods, but on many occasions, direct immunofluorescence of the biopsied lesion is the only method available to detect the pathogenic antibodies.

Autoimmune blistering disorders can be divided into three groups:

1) Conditions with a more indolent course and good prognosis
2) Severe cases with an aggressive and refractory course
3) Conditions associated with a secondary triggering condition.

Generally, bullous skin diseases with a good prognosis include bullous pemphigoid, parapemphigoid gestationalis (herpes gestationis), dermatitis herpetiformes, and linear IgA bullous disease. In general, these diseases carry good prognosis with conventional treatment. In the pre-steroid era, these diseases had a poor prognosis, but corticosteroids have diminished mortality and morbidity dramatically. These conditions will not be discussed further since, due to favorable prognosis, stem cell transplantation is not an appropriate therapeutic option.

Bullous skin diseases due to other conditions include paraneoplastic pemphigus and drug-induced bullous conditions. Paraneoplastic pemphigus is primarily associated with B cell lymphoproliferative diseases, although other malignancies, especially thymoma and certain sarcomas, have also been reported. Furthermore, many cases associated with Castelman's disease have been reported. Autoantibodies are IgG1 subclass and recognize a number of intra-and extracellular antigen domains including desmoplakin I & II, desmoglein and bullous pemphigoid-I antigen. Drug-induced pemphigus is caused by certain drugs such as penicillamine, captopril, enalapril, penicillins, cephalosporins, and rifampin. Vancomycin is commonly associated with drug induced linear IgA dermatosis. Both paraneoplastic and drug-induced pemphigus will be excluded from further discussion since treatment for paraneoplastic pemphigus is directed against the underlying malignancy, and drug-induced pemphigus recovers in 15-50% of cases with withdrawal of the offending drug.

Potential life-threatening autoimmune bullous skin diseases include pemphigus vulgaris, pemphigus foliaceus, cicatricial pemphigoid, and epidermolysis bullosa acquisita.

Features of Life-Threatening Bullous Skin Diseases

Pemphigus Vulgaris (PV)

PV is the most common form of pemphigus and affects approximately 1 to 5 people per 100,000 and can occur at any age, but most commonly develops in the fourth to sixth decades of life.[3-5] Lesions typically consist of blisters that coalesce and rupture, resulting in large denuded areas of skin. Healing occurs without scarring, but pigmentation changes are common. Manual pressure to the skin may elicit the separation of epidermis (Nikolsky's sign).

About 50-70% of cases present with oral mucosal lesions with or without skin lesions. Some patients achieve remission after variable periods, some require long-term immunosuppression, while others are refractory to numerous treatment options. Morbidity and mortality are proportional to the extent of skin involvement, the number of failed treatments, and dose of systemic steroids. Death is usually from infections, predominately *staphylococcus aureus*,[6] that gain access from breakdown of skin/mucosal barriers. Other causes of death include electrolyte disturbances, dehydration, and malnutrition. Overall mortality is approximately 10% over 10 years.

Patients have circulating IgG 1 and 4 autoantibodies that react to desmoglein 3 (Dsg-3).[7-9] Most studies suggest a correlation between disease activity and antibody titer. Transplacental transfer of maternal pemphigus vulgaris antibodies may cause transient blisters in neonates. Purified pemphigus vulgaris IgG injected into neonatal mice causes intraepidermal acantholysis.

Pemphigus vulgaris, like all diseases in the pemphigus group, is pathologically characterized by intraepidermal suprabasal acantholytic blisters with detachment of adjacent epidermal keratinocytes. The diagnosis of pemphigus is confirmed by direct immunofluorescence, which shows IgG deposited on the surface of keratinocytes in and around lesions. Anti-desmoglein antibodies can also be detected using ELISA methods in 80-90% of patients.

Pemphigus Foliaceus (PF)

PF is similar to PV[10] but does not have mucosal blisters, which helps to distinguish it from PV. Unlike PV, the clinical presentation can resemble a papulosquamous condition, like psoriasis or seborrheic dermatitis rather than a blistering disease. A rare variant of pemphigus folliaceous, called fogo sevageum, is endemic in the Amazon basin and it has been linked to an arthropod vector. Subcorneal vesicles favor the diagnosis of PF, while suprabasilar vesicles favor the diagnosis of PV. Mortality for PF is not as high as for PV, but some PF patients also become refractory to treatment and can evolve into a clinical presentation similar to pemphigus vulgaris. The cause of death is generally infection. The antibody in pemphigus foliaceus is directed against desmoglein 1 (Dsg-1).[11] Similar to PV, titers of circulating autoantibodies correlate with extent and activity of the disease.

Cicatricial Pemphigoid (CP)

Cicatricial pemphigoid has recurring blisters on mucous membranes or on skin near orifices (mouth, oropharynx, nasopharynx, esophagus, genitals, and conjunctiva).[12] The loss of function due to scarring and adhesions often necessitates surgical interventions. Genital areas, conjunctivae, larynx, pharynx and esophagus involvement causes significant morbidity. Brunsting and Perry described a variant of CP characterized by deep skin bullous lesions involving the head and neck that heal with scars.[13]

The major autoantigens are the hemidesmosome-associated proteins BP180 & BP230 (same as bullous pemphigoid) and the anchoring filament component laminin 5.[14] The autoantibodies are IgG or IgA but cannot always be detected in patients serum even when the patient has active disease. Cicatricial pemphigoid can be a very devastating disease. Although during the early stages the disease shows primarily inflammation of mucosal surfaces, eventually the chronic damage of mucosal epithelia can result in blindness and significant scarring of the buccal, nasal, genital and anal regions.

Epidermolysis Bullosa Acquisita (EBA)

EBA is a rare disorder with bullous distribution that is more acral (hands and feet and at joint flexion/extension points) or mucosal and is often associated with trauma (mechanobullous disease). EBA resolves with extensive scar formation with small inclusion cysts or milia. Loss of hair and nails are potential complications. The target antigen is in the anchoring fibrils below the lamina lucida type VII collagen.[15] The autoantibodies belong to the IgG type and are pathogenic. EBA can be recalcitrant and lead to scarring, alopecia, disability, and significant morbidity and mortality.

Treatment of Life-Threatening Bullous Skin Diseases

Introduction of corticosteroids has greatly reduced mortality, but significant morbidity still remains. Many steroid regimens with or without immunosuppressive drugs have been tried. Patients with only oral lesions can be treated with topical steroids and oral hygiene. Patients with widespread disease need systemic corticosteroids. In general, the dosage recommended is prednisolone 1.0-1.5 mg/kg/day in combination with topical or intralesional steroids. Dosage will be adjusted based on clinical response, and the titer of circulating pemphigus antibody can be of help as an adjunctive tool for dosage adjustment. If disease is not controlled with this regimen, various dosage recommendations exist, such as higher daily dose or monthly pulse intravenous steroids. As an adjunctive treatment for steroids or as steroid-sparing agents, various immunosuppressive drugs are suggested including azathioprine, cyclophosphamide, gold, cyclosporin, methotrexate and mycophenolate mofetil. Plasmapheresis has been tried and showed variable responses. Intravenous immunoglobulin may also be a safe alternative modality for steroid-sparing effect in some recalcitrant cases.[16]

Complications due to scarring and adhesion require various local mechanical or surgical procedures. Recently, a group at Johns Hopkins treated cases of autoimmune skin diseases with high dose cyclophosphamide without stem cell transplant.[2] Although complicated by a relatively long period of neutropenia, the results seemed promising in patients with severe and resistant bullous skin disease.

Hematopoietic Stem Cell Transplantation (HSCT)

Potential HSCT candidates should have an established diagnosis of an autoimmune skin disorder that preferably includes pemphigus vulgaris, although some cases with severe recalcitrant pemphigus foliaceus as well as carefully selected patients with cicatricial pemphigoid or epidermolysis bullosa acquisita may also be considered. At Northwestern University, candidates must have failed standard therapies such as prednisone 0.5 mg/kg/day for more than 3 months, and at least two other immunosuppressive agents such as: cyclophosphamide, azathioprine, mycophenolate mofetil, gold, tetracycline (or minocycline), cyclosporin, methotrexate, or plasmapheresis. Failure is defined as the inability to wean steroids to less than 0.5 mg/kg/day, and involvement of more than 10% of skin body surface area, involvement of one or more mucosal lesions, or recurrent infections requiring more than two hospitalizations.

Due to long term immunosuppression and breakdown of cutaneous/mucosal surfaces, these patients are at high risk of sepsis. Aggressive anti-microbial prophylaxis is necessary especially for

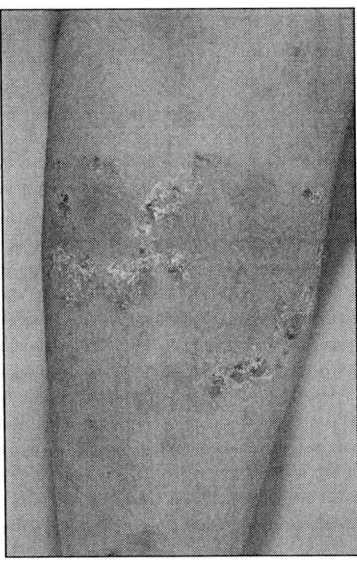

Figure 1. Pemphigus foliaceus.

gram positive organisms. A history of vancomycin-resistant *Staphlococcus aureus* should be obtained and skin lesions cultured. The conditioning regimen used at Northwestern is cyclophosphamide 200 mg/kg and rabbit ATG (6.0 mg/kg). This regimen was selected due to its efficacy in other autoimmune diseases and lack of mucositis. One patient has recently undergone autologous HSCT at Northwestern University (manuscript in progress). This patient's pre-HSCT refractory cutaneous pemphigus foliaceus skin lesions are shown in Figure 1.

References

1. Burt RK, Slavin S, Burns WH et al. Induction of tolerance in autoimmune disease by hematopoietic stem cell transplantation; Getting closer to a cure? Blood 2002; 99(3):870-887.
2. Brodsky RA, Petri M, Smith BD et al. Immunoablative high-dose cyclophosphamide without stem-cell rescue for refractory, severe autoimmune disease. Ann Intern Med 1998; 129(12):1031-5.
3. Pisanti S, Sharav Y, Kaufman E et al. Pemphigus vulgaris: Incidence in Jews of different ethnic groups, according to age, sex, and initial lesion. Oral Surg Oral Med Oral Pathol 1974; 38(3):382-7.
4. Hietanen J, Salo OP. Pemphigus: An epidemiological study of patients treated in Finnish hospitals between 1969 and 1978. J Acta Derm Venereol 1982; 62(6):491-6.
5. Simon DG, Krutchkoff D, Kaslow RA et al. Pemphigus in Hartford County, Connecticut, from 1972 to 1977. Arch Dermatol 1980; 116(9):1035-7.
6. Rosenberg FR, Sanders S, Nelson CT. Pemphigus: A 20-year review of 107 patients treated with corticosteroids. Arch Dermatol 1976; 112(7):962-70.
7. Amagai M, Nishikawa T, Nousari HC et al. Antibodies against desmoglein 3 (pemphigus vulgaris antigen) are present in sera from patients with paraneoplastic pemphigus and cause acantholysis in vivo in neonatal mice. J Clin Invest 1998; 102(4):775-82.
8. Amagai M, Hashimoto T, Shimizu N et al. Absorption of pathogenic autoantibodies by the extracellular domain of pemphigus vulgaris antigen (Dsg3) produced by baculovirus. J Clin Invest 1994; 94(1):59-67.
9. Amagai M, Karpati S, Prussick R et al. Autoantibodies against the amino-terminal cadherin-like binding domain of pemphigus vulgaris antigen are pathogenic. J Clin Invest 1992; 90(3):919-26.
10. Castro RM, Roscoe JT, Sampaio SA. Brazilian pemphigus foliaceus. [Review] Clin Dermatol 1983; 1(2):22-41.
11. Stanley JR, Klaus-Kovtun V, Sampaio SA. Antigenic specificity of fogo selvagem autoantibodies is similar to North American pemphigus foliaceus and distinct from pemphigus vulgaris autoantibodies. J Invest Dermatol 1986; 87(2):197-201.
12. Foster CS. Cicatricial pemphigoid. Trans Am Ophthalmol Soc 1986; 84:527-663.
13. Brunsting LA, Perry HO. Benign pemphigoid? A report of seven cases with scarring herpetiform plaques about the head and neck. Arch Dermatol 1956; 68:128-31
14. Bernard P, Prost C, Durepaire N et al. The major cicatricial pemphigoid antigen is a 180-kD protein that shows immunologic cross-reactivities with the bullous pemphigoid antigen. J Invest Dermatol 1992; 99(2):174-9.
15. Woodley DT, Briggaman RA, O'Keefe EJ et al. Identification of the skin basement-membrane autoantigen in epidermolysis bullosa acquisita. N Engl J Med 1984; 310(16):1007-13.
16. Sami N, Qureshi A, Ruocco E et al. Corticosteroid-sparing effect of intravenous immunoglobulin therapy in patients with pemphigus vulgaris. Arch Dermatol 2002; 138(9):1158-62.

CHAPTER 52

Autologous Hematopoietic Stem Cell Transplantation for Idiopathic Inflammatory Myositis

Yu Oyama, Walter G. Barr and Richard K. Burt

Introduction

The term idiopathic inflammatory myopathies (IIM) refers to a group of disorders of unknown cause in which immune-mediated inflammation results in muscle injury and complaints of weakness. IIM consist of six distinct subtypes including: type I- primary idiopathic polymyositis (PM), type II- primary idiopathic dermatomyositis (DM), type III- dermatomyositis or polymyositis associated with malignancy, type IV- juvenile dermatomyositis or polymyositis (JDM), type V- myositis associated with another connective tissue disease, and type VI- inclusion body myositis (IBM).[1] IIM is believed to be triggered by environmental factors in genetically susceptible individuals. Response to immunosuppressive therapies, frequent co-existing autoimmune diseases, existence of autoantibodies in patients' serum, and experimental animal models all suggest an autoimmune pathogenesis. Patients develop proximal muscle weakness with or without tenderness of involved muscle. Laboratory tests reveal elevated serum muscle enzymes, myopathic changes by electromyography (EMG) and biopsy evidence of mononuclear cell infiltration with lymphocytes and plasma cells. Treatment includes corticosteroids, immunosuppressive drugs such as azathioprine, methotrexate, cyclophosphamide, cyclosporin and IVIG. While the overall prognosis for patients with IIM has improved in the last 20 years, there remain subsets of patients who continue to have active disease despite conventional therapy, for whom effects from long-term corticosteroids or immunosuppressive therapies are of concern, and for whom hematopoietic stem cell transplantation (HSCT) may be considered.

Pathophysiology

The case for an immune basis for IIM includes the widespread presence of a heterogeneous group of autoantibodies. Autoantibodies found specifically in patients with IIM are referred to as myositis-specific autoantibodies (MSA), whereas autoantibodies found in IIM as well as other disorders are termed myositis associated autoantibodies (MAA). Either MSA or MAA are found in 50-60% of patients with IIM, whereas MSA are found in 35-40% of patients with IIM.[2] There are three MSA antibody classes: (1) autoantibodies to signal recognition particle (anti-SRP); (2) autoantibodies to a 220kDa nuclear protein which binds a nuclear-transcription helicase (anti-Mi-2); and (3) autoantibody to aminoacyl-transfer RNA (tRNA) synthetases. The most common anti-aminoacyl-tRNA synthetase antibody is anti-histidyl-tRNA synthetase (anti-Jo-1). Other less frequent anti-synthetases are anti-alanyl-tRNA synthetase (PL-12), anti-threonyl-tRNA synthetase (PL-7), anti-isoleucyl-tRNA synthetase (anti-OJ), anti-asparaginyl-tRNA synthetase (anti-KS), and anti-glycyl-tRNA synthetase (anti-EJ). An association with certain clinical features have been reported for particular autoantibodies[3] (Table 1). Whether these autoantibodies are themselves pathogenic or simply epiphenomenona remains unknown.

Myositis may be associated with other autoimmune disorders such as Hashimoto's thyroiditis, Graves disease, myasthenia gravis, type I diabetes mellitus, primary biliary cirrhosis, primary vitiligo, and various systemic collagen vascular diseases. In some patients with scleroderma, systemic lupus erythematosus, mixed connective tissue diseases and Sjögren's syndrome, myositis can be a prominent feature.

The role of genetic factors in the development of myositis is suggested by evidence of an increased incidence of myositis in certain families, an increased frequency in certain racial groups and a strong association of certain HLA genes. Specific MHC genes that appear to increase the risk of myositis include HLA-DR3 (DRB1*0301), HLA DR6, HLA-DRw52, and HLA-DQA1*0501.[3,4]

Certain types of IIM may be marked by seasonal variation in occurrence and flareups suggesting environmental triggers. Anti-Jo-1 antibody positive disease frequently occurs between February and July. Anti-SRP antibody positive disease often develops between September and February.[5]

Both HIV and HTLV-1 can cause myositis as either isolated clinical phenomenon or concurrently with other manifestations of AIDS.[6] Other viruses such as picornavirus, coxsackievirus, echovirus, adenovirus, Epstein-Barr virus, hepatitis B virus, cytomegalovirus, influenza virus, mumps, rubella and varicella-zoster virus have also been associated with myositis. Since pathologic examination of muscle in IIM has not revealed viral antigens or genomes,[7] a viral "hit and run" type of injury has been hypothesized. Other microorganisms such as bacteria, fungi and parasites may cause myositis, although a causal relationship with IIM is unproven. D-penicillamine may cause a myositis that is indistinguishable from PM, whereas most other drugs such as zidovudine,

Stem Cell Therapy for Autoimmune Disease, edited by Richard K. Burt and Alberto M. Marmont. ©2004 Landes Bioscience/Eurekah.com.

Table 1. IIM autoantibodies

Auto-Antibody	Anti-aminoacyl-transfer RNA (tRNA) synthetase antibodies	Anti-SRP	Anti-Mi-2
Symptoms and Clinical Manifestation	Interstitial lung disease (ILD), fever, arthritis, Raynaud's phenomenon, mechanic's hands, sometimes overlap with SLE or RA	Rapid onset, Often develops in autumn, rash palpitations, cardiac disease	Classic DM rashes with "V" sign skin
IIM Subtype	Anti-Jo-1: PM>DM, others DM>PM	PM	DM
Prognosis			
Steroid Response	Moderate	Poor	Good
Response to Taper	Flare	Flare	Good

Anti-Mi-2= an antibody that binds a nuclear-transcription helicase; Anti-SRP= anti-signal recognition particle antibody; DM= dermatomyositis; anti-Jo-1= anti-histidyl-tRNA synthetase; IIM= idiopathic inflammatory myositis; PM= polymyositis; RA= rheumatoid arthritis; SLE= systemic lupus erythematosus. Table modified from Love et al.[3]

3-hydroxy-3- methylglutaryl-coenzyme A (HMG-CoA) reductase inhibitors, and chloroquine cause muscle damage pathologically distinguishable from IIM.[8]

In both PM and IBM, muscle fibers are surrounded and invaded by oligoclonal CD8+ mononuclear cells, implicating processes driven by defined antigens. However, a specific antigenic target has yet to be identified. In IBM, inflammatory infiltration abates over time and amyloid deposits develop. Despite similar pathological mechanisms, IBM does not respond well to immune-based therapies. A predominantly perivascular infiltrate of lymphocytes and complement activation is characteristic of DM. The inflammatory cells in the perimysial and perivascular regions are mostly B cells and CD4+ cells suggesting involvement of a humoral mediated mechanism activated by antigen driven CD4+ cells. Although abundant indirect evidence exists, target antigens in PM/DM/IBM and triggering events of complement activation in DM remain undefined.[9,10] An autoimmune pathogenesis for IIM is indirectly supported by myosin-induced autoimmune polymyositis in rats.[11]

Epidemiology, Clinical Features and Treatment

Estimated incidence of IIM is 0.5 to 8.4 cases per million. Blacks are more affected than whites and except for IBM, females are more affected than males. Peak incidence in children and adults is 10-15 years and 45-60 years, respectively.

The cardinal feature of IIM is symmetric proximal muscle weakness. Some patients also develop muscle pain or tenderness. Both polymyositis (PM) and dermatomyositis (DM) share clinical features except for a cutaneous rash in dermatomyositis. PM/DM start with the insidious onset of weakness of the limb girdles. Some patients develop pharyngeal, laryngeal and esophageal muscle involvement that could cause dysfunction such as aphonia and aspiration. Ocular muscles are usually spared, but other facial muscles or distal muscles can be affected. As with other collagen-vascular diseases, Raynaud's phenomenon, arthritis, malaise, fatigue, anorexia, fever, or weight loss can occur. In DM, symmetric purplish skin changes on eyelids (heliotrope sign) and symmetric reddish scaly rash over dorsal aspect of interphalangeal joints, elbows, toes, knees or medial malleoli (Gottron's sign), often precede muscle weakness. As a systemic autoimmune disease, organ involvement can develop especially in the lungs and heart. Pulmonary involvement has been reported to be as high as 45% and is recognized to produce substantial morbidity in patients with PM/DM, leading to interstitial lung disease that can progress to diffuse lung damage or pulmonary hypertension.[12] Ventilatory insufficiency due to respiratory muscle involvement and aspiration pneumonia due to laryngeal involvement are additional causes of pulmonary morbidity.[13] Cardiac involvement has been reported to be as high as 70%, including silent subclinical injury to overt myocarditis leading to heart failure.[14-16] Pathological features include fibrosis of the conduction system, generalized myocarditis and pericardial effusions. Anti-Ro antibody is associated with neonatal heart block.[17] Coexistent glomerulonephritis may rarely occur.[18] At the time of diagnosis, evaluation for esophageal involvement, and ventilatory insufficiency should be carried out to avoid life-threatening pulmonary complications.[19]

Muscle derived enzymes such as creatine phosphokinase (CPK), aldolase and aspartate aminotransferase (AST) are usually elevated. Electrophysiological studies with electromyography (EMG) typically show increased insertional activity, fibrillations and sharp positive waves, spontaneous, bizarre high-frequency discharges, and polyphasic motor unit potentials of low amplitude and short duration. Occasionally, electrophysiologic studies may be normal.[20] Muscle biopsy may also be normal,[20] perhaps representing a sampling error. Magnetic resonance imaging (MRI) with combination of T1 and the fat-suppressed T2 (STIR) sequences is useful in defining the extent of involvement and targeting biopsy sites.[21] EMG is routinely ordered as a unilateral study in order to target the most involved muscle for biopsy in the contralateral side. This avoids the problem of needle induced inflammatory artifact on biopsy.

The initial treatment for IIM is corticosteroids 1-2 mg/kg per day in single or divided doses. In severe cases, pulse intravenous methylprednisolone can be used. It may take 4-8 weeks for significant improvement. If disease flares while tapering steroids or fails to respond, weekly oral (5-15 mg) methotrexate,[22] or weekly parenteral (25-50 mg) methotrexate, or daily azathioprine (100-150 mg)[23] is added. A combination of methotrexate and azathioprine may be helpful in patients refractory to both drugs alone.[24] Other agents reported to be effective are cyclophosphamide,[25] cyclosporin,[26] tacrolimus, mycophenolate mofetil,[27] etanercept,[28] IVIG,[29] rituximab[30] and CAMPATH-1H.[31] Hydroxychloroquine is effective for DM-associated skin rash.

IBM, which makes up 15-28% of inflammatory myositis, is a recently recognized category. The main population is older males and the disease onset is slower than PM/DM. Muscle weakness can be asymmetric, focal and distal. Quadriceps and forearm flexor muscles are often affected. Falling due to weakness of knee extensors and swallowing dysfunction are typical symptoms. As opposed to other forms of myositis, CPK and other muscle enzymes often demonstrate only minor elevations or are normal. IBM is sometimes associated with other connective tissue diseases and may also be associated with malignancy. Myositis specific autoantibodies (MSA) are detected in some patients. The electromyography (EMG) may demonstrate myopathic, neuropathic or mixed patterns. The diagnosis is made by typical, although not specific, pathologic features including intracellular lined vacuoles, intranuclear and cytoplasmic inclusion bodies, and deposits of amyloidogenic proteins in lined vacuoles in addition to the more typical features of myositis. There are two forms of IBM including sporadic and hereditary (familial) forms. The hereditary form usually does not have signs of inflammation on biopsy, whereas the sporadic form does. The natural history of this disease is not well described and the exact morbidity and mortality is unknown. Progressive dysphagia that could lead to aspiration pneumonia and progressive weakness that could lead to nonambulatory status would increase the risk of mortality.[32-35] Therapy to IBM is, in general, disappointing. Immunosuppressive therapy is far less effective compared with other inflammatory myopathies. Nonetheless, prednisone, methotrexate and azathioprine may be effective in some patients as part of a strategy to slow the progression of weakness. Placebo controlled trials of IVIG and β-interferon did not show benefit.

Juvenile dermatomyositis (JDM) presents with skin rash, muscle weakness, myalgias and later, patients may develop vasculitis, joint contractures, calcinosis and oral and skin ulcers. Elevated ESR, CPK, LDH, ANA, myogenic pattern of EMG and histologic findings similar to adult PM/DM are seen. Patients generally respond to corticosteroids and immunosuppressive therapy better than adults, but there are some patients who need long-term corticosteroids or other immunosuppressive therapies due to chronic continuous disease activity. Severe disability or even deaths have been reported.[36,37]

Various malignancies are associated with inflammatory myopathy (especially DM), notably, lung, breast, gynecologic cancers especially ovarian,[38] lymphoma, angiosarcoma, stomach, prostate,[39-42] colon,[43] nasopharyngeal, hepatocellular, uterine, renal cell, and bladder.[44] Malignancy can develop before, simultaneously, or after the diagnosis of DM. It is more common in elderly males, but may occur in younger patients including pediatric patients. The mortality of this patient population is 74.6% in 2 years.[45]

Various other collagen diseases can develop findings consistent with PM/DM, such as scleroderma, SLE, mixed connective tissue diseases (MCTD), and less commonly, rheumatoid arthritis, vasculitides, Adult-onset Still's disease, Wegener's granulomatosis, and Sjögren's syndrome. The prognosis and response to therapy are not well documented in this group of patients.[46]

Prognosis

Reports regarding the prognosis of IIM[13,15,20,39,47-53] are summarized in Table 2. Poor prognostic and adverse features are old age, pulmonary involvement, cardiac involvement, presence of dysphagia, nonwhite race, failure to induce remission with corticosteroids, leukocytosis (>10,000) at diagnosis, fever (>38°C) and diagnosis of coexistent malignancy. Data on survival vary. There has been a general improvement in overall outcomes in recent decades, perhaps related to earlier diagnosis, improvement in supportive care and more judicious use of immunosuppressive drugs. Sultan et al[48] report a 5 and 10 year survival of 95 and 83.8%, while Marie et al[47] report 2, 5, 15 year survivals of 82, 77 and 61%, respectively. Five-year survival ranges from 95% to 64%. Patients with pneumonitis and dysphasia have 2 year survival of 30 to 56%.[49] Patients with pulmonary involvement have 27% mortality over 14 years compared to 9% without pulmonary involvement.[13] The presence of substantial ongoing morbidity and mortality related to IIM and its treatment demands new and more effective therapeutic options.

Case Reports of Autologous HSCT

Two cases of autologous transplant for polymyositis have been reported from Europe. A 28 year-old female with Jo-1 positive polymyositis, severe lung involvement {interstitial infiltrate, decreased total lung capacity (TLC) (52%) and carbon monoxide transfer (DLCO) (38%)}, and elevated transaminases, failed high-dose steroids, azathioprine, weekly methotrexate, and pulse cyclophosphamide. Peripheral blood stem cells (PBCS) were mobilized by cyclophosphamide, etoposide, and granulocyte-colony stimulating factor (G-CSF). The graft was manipulated by both CD34 positive and CD4 and CD8 negative selection. Myeloablative conditioning regimen consisted of busulfan (16 mg/kg), cyclophosphamide (120 mg/kg) and anti-thymocyte globulin (ATG) (90 mg/kg). The immediate post-transplantation course was complicated with adult respiratory distress syndrome (ARDS) requiring mechanical ventilation and fever that was attributed to ATG. Hospital discharge was on day 13. At 15 months after the transplant, Karnofsky performance score, muscle strength, dyspnea, pulmonary functions, and interstitial changes on CT scan all showed improvement in spite of discontinuation of all immune suppressive medications.[54]

A 38 year-old Jo-1 positive polymyositis female with restrictive lung disease failed prednisone, azathioprine, pulse cyclophosphamide, IVIG, cyclosporin and plasmapheresis. PBSC were mobilized with cyclophosphamide (2 g/m^2) and G-CSF. HSCT conditioning consisted of a reduced-intensity nonmyeloablative regimen of cyclophosphamide 6 g/m^2. While the regimen was well tolerated and induced early improvement, disease recurred on post HSCT day 21.[55]

Outcome Measures

As found in the CIDP chapter of this book, MRC muscle strength grading, isometric muscle strength, electrophysiological, Barthel index, SF-36 and HAQ may be used for outcome measurements.

Eligibility for HSCT

Potential HSCT candidates should have failed standard therapies and have poor prognosis risk disease defined by at least some of the following criteria: (1) objective persistent muscle weakness (MRC <5); (2) persistent elevation of muscle derived enzymes

Table 2. Prognosis in polymyositis/dermatomyositis

Report (Reference)	Prognosis	Adverse Features
Rose[51]	30% overall morality over 12 years	Onset of age >40 years old
Medsger[49]	Overall survival at 2, 5 and 7 years is 72%, 64% and 53%	Pneumonitis (the most common etiology of pneumonitis was aspiration), dysphasia, adult onset (≥50 years old), and nonwhite race
DeVere[53]	28% mortality over 26 year-follow up	Malignancy
Bohan[20]	13.7% mortality over 16 year-follow up	Malignancies, old age, fibrosis and degeneration on muscle biopsy
Carpenter[50]	53% to 75% survival at 8 years among adult patients	Severe muscle weakness and dysphasia
Hochberg[52]		Nonwhite female
Benbassat[39]	30 out of 92 patients (32.6%) died after follow-up from 1956 to 1980	Failure to induce remission, leukocytosis (>10,000), older age, fever (>38°C)
Maugars[15]	Survivals at 2, 5, and 9 years are 73.9%, 66.7%, and 55.4%	Old age at onset, pulmonary fibrosis, low DLCo, cancer, lack of myalgia, lack of weakness reversibility
Marie[13,47]	82%, 77% and 61% survival at 2, 5, and 15 years. Patients with pulmonary involvement had 27.3% mortality over 14 years whereas patients without pulmonary involvement had 9.1%	Pulmonary involvement
Sultan[48]	Survival at 5 and 10 years were 95% and 83.8%	

DLCo= diffusion capacity

(CPK, LDH, aldolase); (3) persistent systemic inflammatory signs such as fever, anorexia, weight loss, malaise; (4) worsening pulmonary function tests; (5) abnormal EKG; and/or (6) presence of joint contracture, calcinosis, vasculitis, or skin ulcers. Because IBM poorly responds to immunosuppressive therapy, it is debatable whether patients with IBM should be included as potential autologous HSCT candidates. Protocols designed to investigate the involvement of allogeneic stem cells (such as cord blood stem cells) in myocyte regeneration may be more appropriate for IBM (see Chapters 10 and 11). Malignancy, unless otherwise indicated as treatment for the cancer, is probably a contraindication for HSCT. Patients deemed cured 5 years after the last treatment of cancer, may still be considered HSCT candidates.

Conclusion

IIM has significant morbidity and mortality that persists in spite of our currently available immunosuppressive therapies. Long-term data in the era of newer immunosuppressive drugs are not available. Patients who do not respond to conventional dose immunosuppressive drugs or need long-term immunosuppressive therapy, may benefit from immunoablative therapy with HSCT. This therapy may also form the basis to clarify further the role of the immune system in IIM.

References

1. Bohan A, Peter JB. Polymyosits and dermatomyositis (first of two parts). N Eng J Med 1975; 292:344-347.
2. Brouwer R, Hengstman GJD, Vree Egberts W et al. Autoantibody profiles in the sera of European patients with myositis. Ann Rheum Dis 2001; 60:116-123.
3. Love LA, Leff RL, Fraser DD et al. A new approach to the classification of idiopathic inflammatory myopathy: Myositis-specific autoantibodies define useful homogenous patient groups. Medicine (Baltimore) 1991; 70:360-374.
4. Hirsh TJ, Enlow RW, Bias WB et al. HLA D related (DR) antigens in various kinds of myositis. Hum Immunol 1981; 3:181-186.
5. Leff RL, Burgess SH, Miller FW et al. Distinct seasonal patterns in the onset of adult idiopathic inflammatory myopathy in patients with anti-Jo-1 and anti-signal recognition particle autoantibodies. Arthritis Rheum 1991; 34:1391-1396.
6. Itescu S. Rheumatic aspects of acquired immunodeficiency syndrome. Curr Opin Rheumatol 1996; 8:346-353.
7. Leff RL, Love LA, Miller FW et al. Viruses in idiopathic inflammatory myopathies: Absence of candidate viral genomes in muscle. Lancet 1992; 339:1192-1195.
8. Pascuzzi RM. Drugs and toxins associated with myopathies. Curr Opin Rheumatol 1998; 10:511-520.
9. Wortmann RL. Inflammatory diseases of muscle and other myopathies. Chapter 86. Ruddy: Kelley's Textbook of Rheumatology. 6th ed. W B Saunders Company.
10. Dalakas MC. Progress in inflammatory myopathies: Good but not good enough. J Neurol Neurosurg Psychiatry 2001; 70:569-573.

11. Kojima T, Tanuma N, Aikawa Y et al. Myosin-induced autoimmune polymyositis in the rat. J Neurol Sci 1997; 151:141-148.
12. Takizawa H, Shiga J, Moroi Y et al. Interstitial lung disease in dermatomyositis: Clinicopathologic study. J Rheumatol 1987; 14:102-107.
13. Marie I, Hatron PY, Hachualla E et al. Pulmonary involvement in polymyositis and in dermatomyositis. J Rheumatol 1998; 25:1336-1343.
14. Askari AO, Huetther TL. Cardiac abnormalities in polymyositis/dermatomyositis. Semin Arthritis Rheum 1982; 12:208-219.
15. Maugars YM, Berthelot JMM, Abbas AA et al. Long-term prognosis of 69 patients with dermatomyositis or polymyositis. Clin Exp Rheumatol 1996; 14:263-274.
16. Denbow CE, Lie JT, Tancredi RG et al. Cardicac involvement in polymyositis. A clinicopathologic study of 20 autopsied patients. Arthritis Rheum 1979; 22:1088-1092.
17. Behan WMH, Aichison M, Behan PO. Pathogenesis of heart block in a fatal case of dermatomyositis. Br Heart J 1986; 56:479-482.
18. Valenzuela OF, Reisner IW, Porush JG. Idiopathic polymyositis and glomerulonephritis. J Nephrol 2001; 14:120-124.
19. Marie I, Hatron PY, Levesque H et al. Influence of age on characteristics of polymyositis and dermatomyositis in adults. Medicine 1999; 78:139-147.
20. Bohan A, Peter JB, Bowman RL et al. A computer-assisted analysis of 153 patients with polymyositis and dermatomyositis. Medicine (Baltimore) 1977; 56:255-286.
21. Adams EM, Chow CK, Premkumar A et al. The idiopathic inflammatory myopathies: Spectrum of MRI imaging findings. Radiographics 1995; 15:563-574.
22. Metzger AL, Bohan A, Goldberg LS et al. Polymyositis and dermatomyositis: Combined methotrexate and corticosteroid therapy. Ann Int Med 1974; 81:182-189.
23. Bunch T. Prednisone and azathioprine for polymyositis. Long-term follow up. Arthritis Rheum 1981; 24:45-48.
24. Villalba L, Hicks JE, Adams EM et al. Treatment of refractory myositis: A randomized crossover study of two new cytotoxic regimens. Arthritis Rheum 1998; 41:392-399.
25. Schnabel A, Reuter M, Gross WL. Intravenous pulse cyclophosphamide in the treatment of interstitial lung disease due to collagen vascular diseases. Arthritis Rheum 1998; 41:1215-1220.
26. Qushmaq KA, Chalmers A, Esdaile JM. Cyclosporin A in the treatment of refractory adult polymyositis/dermatomyositis: Population based experience in 6 patients and literature review. J Rheumatol 2000; 27:2855-2859.
27. Schneider C, Gold R, Schafers M et al. Mycophenolate mofetil in the therapy of polymyositis associated with a polyautoimmune syndrome. Muscle Nerve 2002; 25:286-288.
28. Phillips K, Husni ME, Karlson EW et al. Experience with etanercept in an academic medical center: Are infection rates increased? Arthritis Rheum 2002; 47:17-21.
29. Cherin P, Pelletier S, Teixeira A et al. Results and long-term followup of intravenous immunoglobulin infusion in chronic, refractory polymyositis. Arthritis Rheum 2002; 46:467-474.
30. Levine TD. A pilot study of Rituximab therapy for refractory dermatomyositis. Arthritis Rheum 2002; 46(suppl):S488(abstract).
31. Reiff AO. Campath 1h administration as immunoablative therapy for a patient with treatment refractory polymyositis. Arthritis Rheum 2002; 46(supple):S314(abstract).
32. Leonard H, Calabrese DO, Chou SM. Inclusion body myositis. Rheum Dis Clin North Am 1994; 20:955-972.
33. Lindberg C, Persson LI, Brorkander J et al. Inclusion body myositis: Clinical, morphological, physical and laboratory findings in 18 cases. Acta Neurol Scand 1994; 89:123-131.
34. Peng A, Koffman BM, Malley JD et al. Disease progression in sporadic inclusion myositis: Observation in 78 patients. Neurology 2000; 55:296-298.
35. Lotz BP, Engel AG, Nishino H et al. Inclusion body myositis, observation in 40 patients. Brain 1989; 112:727-747.
36. Spencer CH, Hanson V, Singsen BH et al. Course of treated juvenile dermatomyositis. J Pediatrics 1984; 105:399-408.
37. Peloro TM, Miller F, Hahn TF et al. Juvenile dermatomyositis: A retrospective review of 30-year experience. J Am Acad Dermaol 2001; 45:28-34.
38. Cherin P, Piette JC, Herson S et al. Dermatomyositis and ovarian cancer: A report of 7 cases and literature review. J Rheumatol 1993; 20:1897-1899.
39. Benabassat J, Gefel D, Larholt K et al. Prognostic factors in polymyositis/dermatomyositis A computer-assisted analysis of ninety-two cases. Arthritis Rheum 1985; 28:249-255.
40. Airio A, Pukkala E, Isomaki H. Elevated cancer incidence in patients with dermatomyositis: A population based study. J Rheumatol 1995; 22:1300-1303.
41. Bonnetblanc JM, Bernard P, Fayol J. Dermatomyositis and malignancy: A multicenter cooperative study. Dermatologica 1990; 180:212-216.
42. Callen JP. Myositis and malignancy. Current Opin Rheumatol 1994; 6:590-594.
43. Callen JP. Relationship of cancer to inflammatory muscle diseases. Dermatomyositis, polymyositis, and inclusion body myositis. Rheum Dis Clin North Am 1994; 20:943-953.
44. Chen YJ, Wu CY, Shen JL. Predicting factors of malignancy in dermatomyositis and polymyositis: A case-control study. Br J Dermatol 2001; 144:825-831.
45. Bassett-Seguin N, Roujeau JC, Gherardi R et al. Prognostic factors and predictive signs of malignancy in adult dermatomyositis. Arch Dermatol 1990; 126:633-637.
46. Tymms K, Webb J. Dermatopolymyositis and other connective tissue diseases: A review of 105 cases. 1985; 12:1140-1148.
47. Marie I, Hachulla E, Hatron PY et al. Polymyositis and dermatomyositis: Short term and long term outcome, and predictive factors of prognosis. J Rhematol 2001; 28:2230-2237.
48. Sultan SM, Ioannou Y, Moss K et al. Outcome in patients with idiopathic inflammatory myositis: Morbidity and mortality. Rheumatology 2002; 41:22-26.
49. Medsger Jr TA, Robinson H, Masi AT. Factors affecting survivorship in polymyositis: A life-table study of 124 patients. Arthritis Rheum 1971; 14:249-258.
50. Carpenter JR, Bunch TW, Engel AG et al. Survival in polymyositis: Corticosteroid and risk factors. J Rheumatol 1977; 4:207-214.
51. Rose AL, Walton JN. Polymyositis: A survey of 89 patients with particular reference to treatment and prognosis. Brain 1966; 89:747-768.
52. Hochberg MC, Lopez-Acuna D, Gittelsohn AM. Mortality from polymyositis and dermatomyositis in the United States, 1968-1978. Arthritis Rheum 1983; 26:1465-1471.
53. DeVere R, Bradley WG. Polymyositis: Its presentation, morbidity and mortality. Brain 1975; 98:637-666.
54. Baron F, Ribbens C, Kaye O et al. Effective treatment of Jo 1-associated polymyositis with T-cell depleted autologous peripheral blood stem cell transplantation. Br J Haematology 2000; 110:339-342.
55. Bingham S, Griffith B, McGonagle D et al. Autologous stem cell transplantation for rapidly progressive Jo-1-positive polymyositis with long-term follow-up. Br J Haematology 2001; 113:839-842.

CHAPTER 53

Hematopoietic Stem Cell Transplantation as Treatment for Type 1 Diabetes

Júlio C. Voltarelli, Richard K. Burt, Norma Kenyon, Dixon B. Kaufman and Elizabeth C. Squiers

Introduction

Type 1 diabetes mellitus is an autoimmune disease associated with B cell derived antibodies and T cell proliferative responses to a variety of islet cell peptides. Near normalization of blood sugar levels as monitored by glycosylated hemoglobin (HgbA1C) diminishes diabetic secondary complications. Insulin, while prolonging life, is not a cure and intensive insulin therapy to control blood sugar is not always practical, especially in lower economic classes or developing countries. Even in developed countries, serious sequela and mortality still occur.[1-5] In America, diabetes remains the most common cause of blindness and renal failure, and the sixth leading cause of death. Hematopoietic stem cell transplantation in early onset diabetes, and in established diabetes when combined with either pancreas or islet cell transplantation, offers hope of curing diabetes by reintroduction of islet cell tolerance.

Type 1 Diabetes Mellitus

Etiology

Diabetes mellitus which means "to run through a siphon" refers to the polyuria/glucosuria and wasting/emaciation caused by hyperglycemia. The prevalence of type 1 diabetes is 0.2% in most countries.[6] Forty percent of affected type 1 diabetics develop disease before 14 years of age, 70% before 34 years of age and 30% after 35 years of age.[7] Like most suspected autoimmune diseases, diabetes is associated with particular HLA genes[8] and while HLA predisposes to susceptibility, like other autoimmune diseases, it is not sufficient to cause disease.

Patients with type 1 diabetes have elevated CD4/CD8 ratios in the peripheral blood, and T cells proliferate to islet cell antigens such as glutamic acid decarboxylase (GAD) antigens.[9] Islet cell specific antibodies such as anti-GAD65 may be identified in the blood.[10] In siblings of patients with type 1 diabetes, preclinical disease may be detected by both the presence of circulating islet cell antibodies[11] and poor tolerance of an oral glucose challenge.

The nonobese diabetic (NOD) mouse is an animal model of diabetes that develops a preclinical insulinitis prior to onset of clinical diabetes. Symptoms may be adoptively transferred by T cells from a diabetic to nondiabetic but disease prone mouse.[12] A human autoimmune etiology is supported by adoptive transfer of type 1 diabetes following an HLA matched sibling hematopoietic stem cell transplantation performed for leukemia.[13,14]

Infectious agents have been suggested to be involved in disease initiation.[15-18] In animal models, viruses such as coxsackie, rubella, and CMV, as well as chemicals such as strepozocin, can induce diabetes. Potential mechanisms for virus-induced diabetes include direct beta cell destruction, molecular mimicry of viral and islet cell epitopes, and bystander activation against self epitopes in viral infected islet cells. Like most human autoimmune diseases, it appears that the combination of HLA genotype, other nonHLA genes, hormonal milieu, and environmental exposures may be necessary to induce disease.

Complications of Diabetes

Acute complications of Type 1 diabetes include DKA and hyperosmolar coma. Long term complications are largely due to the effects of diabetes on blood vessels including: retinopathy, nephropathy, cerebrovascular disease, peripheral vascular disease, and autonomic and peripheral neuropathy. Vascular complications are due to hypertension and accelerated atherosclerosis due to hyperlipidemia and advanced glycosylated end products (AGEs) which leads to free radical injury of endothelium.[19,20] AGEs result from hyperglycemic post-translational glycosylation of intra and extracellular proteins. Age-adjusted incidence of myocardial infarction is 4-6 times higher than in nondiabetics.[21] Diabetics with myocardial infarcts have higher post infarct-related morbidity and mortality.[22,23] Cerebrovascular accidents (CVA) including infarcts, hemorrhage, and aneurysms are more common in diabetics.[24-28] Peripheral vascular disease with claudication, ulcers, and gangrene, and extremity amputations occur in 9-11% of diabetics. For diabetics, minor trauma, often unnoticed due to diabetic associated neuropathy, may be the precipitating event leading to amputation.

Progressive renal disease is a common cause of mortality in Type 1 diabetics[29] and is associated with hypertension that further increases vascular complications. Finally, diabetics are prone to an increased incidence of infections including urinary tract infections, pneumonia (mycobacterium tuberculosis and bacterial), wound infections, osteomyelitis especially with diabetic foot ulcers, and mucocutaneous (e.g., vaginal) and disseminated fungal infections.

Treatment of Type 1 Diabetes

Insulin Therapy

Prior to discovery and therapeutic use of the islet cell hormone insulin, all patients with type 1 diabetes died, usually within 6-12 months. Insulin is, however, not a cure. It forestalls, but does not prevent, the morbidity, mortality, and suffering associated with diabetes. The Diabetes Control and Complications Trials (DCCT) demonstrated that tight control of blood sugar reduced the incidence of secondary complications but at the expense of increased hypoglycemia reactions.[30] This has led to the evaluation of curative interventions, which normalize blood sugars without use of exogenous insulin.

Diabetes is treated with either conventional insulin therapy or intensive insulin therapy (IIT). The goal of IIT is tight control of blood sugar. The risk of secondary complications (retinopathy, neuropathy, cardiovascular disease, nephropathy, extremity amputation, etc) from type 1 diabetes has changed over time and data on long term survival (i.e., 10, 20, and 25 years) using IIT is not yet available. For every 1% increase in HbA1c above normal (HgbA1c <6.5%), mortality increases 11%.[31-34] With conventional insulin therapy, HgbA1c often remains between 9-12%. Since mortality in diabetes correlates with long term control of blood sugar, the current therapy for decreasing or delaying complications of diabetes is tight control of blood sugars using intensive insulin therapy.[31-34] Intensive insulin therapy requires meticulous monitoring of blood sugar (4-10 times a day), frequent insulin injections (more than 3 times per day or an insulin pump), close control of diet, and is generally limited to motivated persons with regular access to health care. As quoted from the literature "Achieving optimal blood glucose control, without an unacceptable rate of hypoglycaemia or unacceptable restrictions on lifestyle, is not simple with presently available insulin preparations and monitoring tools. Accordingly, the appropriate use of insulin to obtain good metabolic control requires the continued and informed expertise of both patient and advising professional, but also attention from both to self-motivation in order to make the desired lifestyle changes possible".[35]

Diabetics on insulin often suffer from negative psychological and social implications.[36] IIT is complicated by a higher incidence of hypoglycemic reactions (approximately 3 times higher than conventional insulin treatment) that may result in seizures and/or death.[37] The incremental cost per year of life gained by intensive insulin therapy has been estimated at $28,661 US dollars.[38] In America, only 20-30% of type 1 diabetics are on IIT,[39] approximately 10% on an insulin pump, and 20% on multiple injections. Therefore, in clinical practice, therapy associated with higher mortality, i.e., conventional insulin therapy, remains the standard of care for approximately 70% of type 1 diabetics in America. The access to medical care, education, and motivation for IIT is influenced by socioeconomic status. This results in a disproportionate percentage of conventional insulin therapy in lower socioeconomic groups. Finally, patients in developing countries, in which most of the world's population reside, due to limited medical resources, are even less likely to initiate or remain on IIT.

On the other hand, with the proper conditioning regimen design, HSCT for new onset diabetes should have a mortality of less than 0.5 to 1%. The goal of HSCT for new onset type 1 diabetes is to prevent any further post transplant medical intervention, treatments, or need for insulin therapy.

Pancreatic and Islet Cell Transplants

Pancreas or islet cell transplantation is the only treatment of type 1 diabetes that reestablishes an insulin independent normoglycemic state. The down side of this therapy is a limited supply of donors, as well as treatment related mortality from infectious and neoplastic complications secondary to life-long immune suppression.

Immune Suppression for Early Onset Disease

Immune-mediated islet cell destruction is not complete until sometime after onset of diabetic ketoacidosis (DKA). Measurements of C-peptide, a marker for endogenous insulin, indicates persistence of islet cells with low normal C-peptide levels for an interval of up to 1 year after DKA onset. This led, beginning in the 1980s, to immune suppressive trials for new onset DKA.[40-49] Several trials including French, Canadian-European, Australian, and American indicated that cyclosporine and/or azathioprine preserved insulin secretion and/or increased the duration of insulin independence. The best results seemed to occur for patients starting immune suppressive treatment with 8 weeks of DKA onset. A recent study of anti-CD3 antibody infused for 14 days in patients within 6 weeks of diagnosis demonstrated improved endogenous insulin production, reduced glycosylated hemoglobin levels, and reduced exogenous insulin requirements for up to one year after immune suppressive treatment.[50] These studies indicate that immune suppressive therapy needs to be initiated early in the course of type 1 diabetes to be effective. In patients with malignancies and coincidental long standing diabetes, allogeneic HSCT from normal donors has not ameliorated insulin requirements.[51]

Hematopoietic Stem Cell Therapy of New Onset Diabetes

Pre-Clinical Model of Stem Cell Transplantion for Diabetes

Hematopoietic stem cell transplantation from a nondiabetic strain into a diabetic prone NOD mouse will prevent diabetes and reverse prediabetic insulinitis.[52] After onset of irreversible diabetes, a combined HSC/pancreas (or islet cell) transplant from a diabetic resistant strain is necessary for cure.[53] The stage of disease, therefore, determines the HSC transplant approach.

Clinical Application of Hematopoietic Stem Cell Transplantation for Diabetes: Conditioning Regimen

Cyclophosphamide or cyclophosphamide with ATG is one of the most common conditioning regimens used for HSCT of autoimmune diseases.[54] These regimens have demonstrated minimal regimen related toxicity especially in diseases such as rheumatoid arthritis (refer to Chapter 43). Due to younger age at onset, type 1 diabetics would have an even lower anticipated mortality. Therefore, cyclophosphamide and ATG may be chosen as a conditioning regimen for diabetes due to its known safety and efficacy in other autoimmune diseases. There is no increase in risk of late malignancies from a cyclophosphamide/ATG conditioning regimen. Transplant doses of cyclophosphamide (200 mg/kg) do not cause infertility in females less than age 26, while approximately 1/3 over age 26 maintain ovarian function.[55] Infertility may, therefore, be avoided by limiting HSCT to patients under 26 years of age.

Patient Selection

Diabetes begins years before onset of symptoms and evidence of residual islet cell function may persist for one year after disease onset.[56-64] An aggressive approach such as HSCT would probably have to be initiated after onset of clinical diabetes, rather than in high risk patients with preclinical disease. As mentioned earlier, patients with new onset diabetes, if treated with nonablative immune suppression within 6 weeks of diagnosis, experience reduced deterioration in endogenous insulin production. A more aggressive attempt at immune ablation with HSC support may further preserve beta-cell function and mitigate insulin requirements. Based on these concepts, a phase I trial of autologous HSC transplantation will soon be opened in Brazil. Eligibility criteria are: (1) Age less than 26 years old; (2) enrolled within 6 weeks of diagnosis; and (3) financial, social, or economic inability to comply with IIT or maintain close and frequent access to medical facilities required for IIT.

Study Endpoints

The primary endpoint to be considered in this study would be insulin requirements (Units/kg/day) and glycosylated hemoglobin (HgbA1c). Secondary parameters would include fasting C-peptide levels, anti-GAD65 antibody titer, and CD4 and CD8 T cell GAD peptide (15 mer) specific responses. Fifteen mer peptides, overlapping by 11 residues, which cover the entire sequence of GAD (about 130 peptides) could be used to stimulate the peripheral blood collected before and at set times after HSCT. The frequency of GAD-specific responses before and after HSCT at the level of single peptide epitopes, in both CD4 and CD8 T cells, could be assessed and analyzed including the cytokines produced to determine how they change between pre and post transplant samples.

Treatment of Chronic Diabetes: Combined Hematopoietic Stem Cell and Pancreas/Islet Cell Transplantation

A major complication of pancreas or islet transplantation is allograft rejection. Other major toxicities arise from life-long immune suppression including opportunistic infections, enhanced cardiovascular risk and bone demineralization as well as Epstein Barr virus post transplant lymphoproliferative disease (PTLD). These problems could be eliminated if donor specific tolerance could be induced. Clinical tolerance may be defined as failure to reject an organ or tissue despite no immune suppressive or modulating medications with intact normal 3rd party immune responses. One approach to achieve tolerance is to combine organ or islet cell transplantation with hematopoietic stem cells from the same donor who provided the organ. In animal models, mixed hematopoietic chimerism, that is, hematopoiesis arising from both the host and organ donor, induces donor specific tolerance.[65]

Hematologic chimerism induced tolerance was first recognized in the 1940s and 1950s when it was discovered that in utero sharing of cells between dizygotic twins, or in utero innoculation with donor hematopoietic cells, resulted in post natal donor specific organ tolerance.[66-68] Subsequently, it was discovered that induction of tolerance by sharing of hematopoietic cells is not unique to the fetal or in utero environment. Mixed chimerism induces tolerance even across MHC barriers in adult animals who receive combined hematopoietic and organ transplant from the same donor.[69-72] Hematopoietic transplant induced donor specific tolerance is, in fact, so potent that it can induce organ specific tolerance across species.[73-76] Hematopoietic chimerism induced tolerance also occurs in nonhuman primates.[77,78] Diabetes has been cured in murine models by allogeneic marrow and islet cell grafts.[79] In macaques monkeys, streptozotocin induced diabetes may be reversed by combined infusion of islet cells and stem cells from MHC-mismatched donors despite lack of immunosuppression.[78]

The mechanisms of hematologic chimerism induced tolerance remain incompletely understood. Central or thymic deletion of donor reactive cells is considered an important mechanism.[70-83] Donor hematopoietic cells migrate to the thymus and induce deletion of donor reactive host cells during their maturation. Similarly, thymic host hematopoietic cells induce deletion of host reactive donor cells. The end result is bi-directional host/donor tolerance with otherwise intact immune competence. Peripheral mechanisms of tolerance may also be involved such as suppressor, regulatory, or "veto" mechanisms.[84-88] A veto cell is any cell that causes anergy, death, or inactivation of a T cell upon its recognition by that T cell.

Patients who have undergone an allogeneic bone marrow transplant, and then later received a kidney transplant for an unrelated condition, have not required immune suppression to prevent rejection of the kidney confirming the ability of hematopoietic chimerism to induce organ tolerance in humans.[89,90] However, clinical trials of bone marrow or peripheral blood stem cell infusions combined with organ transplant have generally been disappointing in terms of tolerance induction.[91] Unlike the animal models and the anecdotal human case reports of organ transplant after confirmed donor hematopoietic engraftment, these trials have not employed myeloablative conditioning regimens and either did not monitor or failed to obtain significant chimerism (generally less than 1% donor hematopoiesis). Future clinical trials will require more intense conditioning regimens, as well as infusion of megadose donor CD34+ HSCs, in an attempt to increase donor chimerism prior to or concurrent with organ allograft transplantation.

Approximately 1000 pancreas transplants are performed in the USA each year and about 85% involve simultaneous pancreas kidney transplants. The current contraindications for pancreatic or islet cell transplantation in type 1 diabetes are patients who have normal renal function and do not exhibit a brittle course or hypoglycemic unawareness. Due to donor risks from partial pancreatectomy, living-related pancreas or living-related islet cell transplants are currently not feasible, and an unrelated cadaver organ and stem cell source would be required. Hematopoeitic stem cells have been successfully collected from cadaveric vertebral bodies and iliac crests at the time of organ procurement.

ABO matching is mandatory for organ transplantation because ABO antigens are presented on vascular endothelial cells and a mismatch would cause antibody mediated acute organ rejection. Most organ transplants are, however, HLA mismatched. Since the donor and recipient pair would be unrelated and HLA mismatched, T cell depletion and/or progenitor cell CD34+ enrichment could be considered to prevent graft versus host disease. To ensure engraftment, an intense but nonmyeloablative immune suppressive conditioning regimen would be needed. Nonmyeloablative conditioning regimens have been used in unrelated hematopoietic transplants for malignancies. An example of a sucessful regimen is 100 mg CAMPATH-1H, 150 mg/m^2 fludarabine, and 140 mg/m^2 melphalan.[92]

Summary

A HSC transplantation trial for newly diagnosed type 1 diabetes will soon begin in Brazil. In America, where access to medical care and intensive insulin treatment has limited severe diabetic complications to 5-10% of diabetics, a trial of combined HSC and panreas/kidney transplantation is being considered for pancreas transplant candidates. While not discussed in this chapter, stem cells whether embryonic, mesenchymal, or hematopoietic are also being evaluated in animal models to regenerate islet cells. Clinical trials utilizing various combinations of stem cell therapies for diabetes are on the horizon.

References

1. Muller WA. Diabetes mellitus—Long term survival. J Insur Med 1998; 30:17-27.
2. Wibell L, Nystrom L, Ostman J et al. Increased mortality in diabetes during the first 10 years of the disease. A population-based study (DISS) in Swedish adults 15-34 years old at diagnosis. J Intern Med 2001; 249(3):263-70.
3. Edge JA, Ford-Adams ME, Dunger DB. Causes of death in children with insulin dependent diabetes 1990-96. Arch Dis Child 1999; 81(4):318-23.
4. Laing SP, Swerdlow AJ, Slater SD et al. The British Diabetic Association Cohort Study, II: Cause-specific mortality in patients with insulin-treated diabetes mellitus. Diabet Med 1999; 16(6):466-71.
5. Nathan DM. Long-term complications of diabetes mellitus. N Engl J Med 1993; 328(23):1676-85.
6. Green A, Gale EA, Patterson CC. Incidence of childhood-onset insulin-dependent diabetes mellitus:The EURODIABACE Study. Lancet 1992; 339(8798):905-9.
7. DPT-1 Study group:Diabetes in America. 2nd ed. Bethesda: National Institutes of Health, 1995.
8. Wassmuth R, Lernmark A. The genetics of susceptibility to diabetes. Clin Immunol Immunopathol 1989; 53:358-3999.
9. Honeyman MC, Stone N, de Aizpurua H et al. High T cell responses to the glutamic acid decarboxylase (GAD) isoform 67 reflect a hyperimmune state that precedes the onset of insulin-dependent diabetes. J Autoimmun 1997; 10(2):165-73.
10. Baekkeskov S, Aanstoot HJ, Christgau S et al. Identification of the 64K autoantigen in insulin-dependent diabetes as the GABA-synthesizing enzyme glutamic acid decarboxylase. Nature 1990; 347(6295):151-6. 347(6289) 25; 782.
11. Verge CF, Gianani R, Kawasaki E et al. Prediction of type I diabetes in first-degree relatives using a combination of insulin, GAD, and ICA512bdc/IA-2 autoantibodies. Diabetes 1996; 45(7):926-33.
12. Wicker LS, Miller BJ, Mullen Y. Transfer of autoimmune diabetes mellitus with splenocytes from nonobese diabetic (NOD) mice. Diabetes 1986; 35(8):855-60.
13. Lampeter EF, Homberg M, Quabeck K et al. Transfer of insulin-dependent diabetes between HLA-identical siblings by bone marrow transplantation. Lancet 1993; 15341(8855):1243-4.
14. Vialettes B, Maraninchi D, San Marco MP et al. Autoimmune polyendocrine failure-type 1 (insulin-dependent) diabetes mellitus and hypothyroidism—after allogeneic bone marrow transplantation in a patient with lymphoblastic leukaemia. Diabetologia 1993; 36(6):541-6.
15. Benoist C, Mathis D. Autoimmune diabetes. Retrovirus as trigger, precipitator or marker? Nature 1997; 388(6645):833-4.
16. Vague P, Vialettes B, Prince MA et al. Coxsackie B viruses and autoimmune diabetes. N Engl J Med 1981; 305(19):1157-8.
17. Karam JH, Grodsky GM, Forsham PH. Coxsackie viruses and diabetes. Lancet 1971; 2(7735):1209.
18. Anonymous. Coxsackie viruses and diabetes. Lancet 1971; 2(7728):804.
19. Schleicher ED, Wagner E, Nerlich AG. Increased accumulation of the glycoxidation product N(epsilon)-(carboxymethyl)lysine in human tissues in diabetes and aging. J Clin Invest 1997; 99(3):457-68.
20. Mullarkey CJ, Edelstein D, Brownlee M. Free radical generation by early glycation products: A mechanism for accelerated atherogenesis in diabetes. Biochem Biophys Res Commun Dec 31 1990; 173(3):932-9.
21. Fontbonne A, Thibult N, Eschwege E. Body fat distribution and coronary heart disease mortality in subjects with impaired glucose tolerance or diabetes mellitus: The Paris Prospective Study, 15-year follow-up. Diabetologia 1992; 35(5):464-8.
22. Herlitz J, Malmberg K, Karlson BW et al. Mortality and morbidity during a five-year follow-up of diabetics with myocardial infarction. Acta Med Scand 1988; 224(1):31-8.
23. Abbott RD, Donahue RP, Kannel WB et al. The impact of diabetes on survival following myocardial infarction in men vs women. The Framingham Study JAMA 1988; 260(23):3456-60.
24. Tunbridge WMG. Factors contributing to deaths of diabetics under 50 years of age. Lancet 1981; 1:569-572.
25. Kessler I. Mortality experience of diabetic patients. Am J Med 1971; 51:715-724.
26. Garcia M, McNamara P, Gordon T et al. Cardiovascular complications in diabetics. Adv Metab Disord 1973; 2Suppl 2:493-9.
27. Kannel WB, McGee DL. Diabetes and cardiovascular disease. The Framingham study. JAMA 1979; 241(19):2035-8.
28. Abbott RD, Donahue RP, Kannel WB et al. The impact of diabetes on survival following myocardial infarction in men vs women. The Framingham Study. JAMA 1988; 260(23):3456-60.
29. Borch-Johnsen K, Andersen PK, Deckert T. The effect of proteinuria on relative mortality in type 1 (insulin-dependent) diabetes mellitus. Diabetologia 1985; 28(8):590-6.
30. The Diabetes Control and Complications Trial (DCCT) Research Group: The effects of intensive treatment of diabetes on the development and progression of long term complications in IDDM. N Engl J Med 1993; 329:977-986.
31. The Diabetes Control and Complications Trial Research Group: The effect of intensive treatment of diabetes on the development and progression of long-term complications in insulin-dependent diabetes mellitus. N Engl J Med 1993; 329:977–986.
32. The Diabetes Control and Complications Trial Research Group: The absence of a glycemic threshold for the development of long-term complications: The perspective of the Diabetes Control and Complications Trial. Diabetes 45:1289–1298.
33. The Diabetes Control and Complications Trial Research Group: The relationship of glycemic exposure (HbA1c) to the risk of development and progression of retinopathy in the Diabetes Control and Complications Trial. Diabetes 1995; 44:968–983.
34. Lawson ML, Gerstein HC, Tsui E et al. Effect of intensive therapy on early macrovascular disease in young individuals with type 1 diabetes. A systematic review and meta-analysis. Diabetes Care 1999; 22 Suppl 2:B35-9.
35. Hypoglycemia in the Diabetes Control and Complications Trial. The Diabetes Control and Complications Trial Research Group. Diabetes 1997; 46(2):271-86.
36. Azar ST, Kanaan N. Intensive insulin therapy compared with conventional insulin therapy does not reduce depressive symptoms in parents of children with type 1 diabetes. Diabetes Care 1999; 22(8):1372-3.
37. Rosenn BM, Miodovnik M, Holcberg G et al. Hypoglycemia: The price of intensive insulin therapy for pregnant women with insulin-dependent diabetes mellitus. Obstet Gynecol 1995; 85(3):417-22.
38. Lifetime benefits and costs of intensive therapy as practiced in the diabetes control and complications trial. The Diabetes Control and Complications Trial Research Group. JAMA 1996; 276(17):1409-15.
39. Home PD. Intensive insulin therapy in clinical practice. Diabetologia Jul 1997; 40 Suppl 2:S83-7.
40. Bougneres PF, Carel JC, Castano L et al. Factors associated with early remission of type I diabetes in children treated with cyclosporine. N Engl J Med 1988; 318(11):663-70.
41. Bougneres PF, Landais P, Boisson C et al. Limited duration of remission of insulin dependency in children with recent overt type I diabetes treated with low-dose cyclosporin. Diabetes 1990; 39(10):1264-72.

42. Assan R, Feutren G, Sirmai J et al. Plasma C-peptide levels and clinical remissions in recent-onset type I diabetic patients treated with cyclosporin A and insulin. Diabetes 1990; 39(7):768-74.
43. Harrison LC, Colman PG, Dean B et al. Increase in remission rate in newly diagnosed type I diabetic subjects treated with azathioprine. Diabetes 1985; 34(12):1306-8.
44. Feutren G, Papoz L, Assan R et al. Cyclosporin increases the rate and length of remissions in insulin-dependent diabetes of recent onset. Results of a multicentre double-blind trial. Lancet 1986; 2(8499):119-24.
45. The Canadian European Randomized Trial Group Cyclosporin-induced remission of IDDM after early intervention. Diabetes 1988; 37:1574-82.
46. Skyler JS, Rabinovitch A. Cyclosporine in recent onset type I diabetes mellitus. Effects on islet beta cell function. Miami Cyclosporine Diabetes Study Group. J Diabetes Complications 1992; 6(2):77-88.
47. Chase HP, Butler-Simon N, Garg SK et al. Cyclosporine A for the treatment of new-onset insulin-dependent diabetes mellitus. Pediatrics 1990; 85(3):241-5.
48. Cook JJ, Hudson I, Harrison LC et al. Double-blind controlled trial of azathioprine in children with newly diagnosed type I diabetes. Diabetes 1989; 38(6):779-83.
49. Silverstein J, Maclaren N, Riley W et al. Immunosuppression with azathioprine and prednisone in recent-onset insulin-dependent diabetes mellitus. N Engl J Med 1988; 319(10):599-604.
50. Herold KC, Hagopian W, Auger JA et al. Anti-CD3 monoclonal antibody in new-onset type 1 diabetes mellitus. N Engl J Med 2002; 346(22):1692-8.
51. Nelson JL, Torrez R, Louie FM et al. Preexisting autoimmune disease in patients with long-term survival after allogeneic bone marrow transplantation. J Rheumatol 1997; 48:23-9(suppl).
52. LaFace DM, Peck AB. Reciprocal allogeneic bone marrow transplantation between NOD mice and diabetes-nonsusceptible mice associated with transfer and prevention of autoimmune diabetes. Diabetes 1989; 38(7):894-901.
53. Yasumizu R, Sugiura K, Iwai H et al. Treatment of type 1 diabetes mellitus in nonobese diabetic mice by transplantation of allogeneic bone marrow and pancreatic tissue. Proc Nat Acad Sci USA 1987; 84(18):6555-7.
54. Burt RK, Fassas A, Snowden J et al. Collection of hematopoietic stem cells from patients with autoimmune diseases. Bone Marrow Transplant 2001; 28(1):1-12.
55. Sanders JE, Hawley J, Levy W et al. Pregnancies following high-dose cyclophosphamide with or without high-dose busulfan or total-body irradiation and bone marrow transplantation. Blood 1996; 87(7):3045-52.
56. Srikanta S, Ganda OP, Eisenbarth GS et al. Islet-cell antibodies and beta-cell function in monozygotic triplets and twins initially discordant for Type I diabetes mellitus. N Engl J Med 1983; 308(6):322-5.
57. Block MB, Rosenfield RL, Mako ME et al. Sequential changes in beta-cell function in insulin-treated diabetic patients assessed by C-peptide immunoreactivity. N Engl J Med 1973; 288(22):1144-8.
58. Glatthaar C, Beaven DW, Donald RA et al. Residual pancreatic function in insulin dependent diabetics. Aus N Z J Med Feb 1982; 12(1):43-7.
59. Rubenstein AH, Kuzuya H, Horwitz DL. Clinical significance of circulating C-peptide in diabetes mellitus and hypoglycemic disorders. Arch Intern Med 1977; 137(5):625-32.
60. Wallensteen M, Dahlquist G, Persson B et al. Factors influencing the magnitude, duration, and rate of fall of B-cell function in type 1 (insulin-dependent) diabetic children followed for two years from their clinical diagnosis. Diabetologia Sep 1988; 31(9):664-9.
61. Marner B, Agner T, Binder C et al. Increased reduction in fasting C-peptide is associated with islet cell antibodies in type 1 (insulin-dependent) diabetic patients. Diabetologia 1985; 28(12):875-80.
62. Agner T, Damm P, Binder C. Remission in IDDM: Prospective study of basal C-peptide and insulin dose in 268 consecutive patients. Diabetes Care 1987; 10(2):164-9.
63. Effect of intensive therapy on residual beta-cell function in patients with type 1 diabetes in the diabetes control and complications trial. A randomized, controlled trial. The Diabetes Control and Complications Trial Research Group. Ann Intern Med 1998; 128(7):517-23.
64. Madsbad S. Prevalence of residual B cell function and its metabolic consequences in Type 1 (insulin-dependent) diabetes. Diabetologia 1983; 24(3):141-7.
65. Charlton B, Auchincloss Jr H, Fathman CG. Mechanisms of transplantation tolerance. Ann Rev Immunol 1994; 12:707-34.
66. Owen RD. Immunogeneic consequences of vascular anastomosis between bovine twins. Science 1945; 102:400-401
67. Anderson D, Billingham RE, Lampkin GH et al. Heredity 1951; 5(172):379-397.
68. Billingham RE, Brent L, Medawar P. Actively acquired tolerance of foreign cells. Nature 1953; 172:603-606.
69. Ildstad ST, Wren SM, Oh E et al. Mixed allogeneic reconstitution (A+B——A) to induce donor-specific transplantation tolerance. Permanent acceptance of a simultaneous donor skin graft. Transplantation 1991; 51(6):1262-7.
70. Murase N, Starzl TE, Tanabe M et al. Variable chimerism, graft-versus-host disease, and tolerance after different kinds of cell and whole organ transplantation from Lewis to brown Norway rats. Transplantation 1995; 60(2):158-71.
71. Schwarze ML, Menard MT, Fuchimoto Y et al. Mixed hematopoietic chimerism induces long-term tolerance to cardiac allografts in miniature swine. Ann Thorac Surg 2000; 70:131.
72. de Vries-van der Zwan A, van der Pol MA, Besseling AC et al. Haematopoietic stem cells can induce specific skin graft acceptance across full MHC barriers. Bone Marrow Transplant 1998; 22(1):91-8.
73. Saito T, Kuang JQ, Bittira B et al. Xenotransplant cardiac chimera: Immune tolerance of adult stem cells. Ann Thorac Surg 2002; 74(1):19-24(discussion:24).
74. Abe M, Qi J, Sykes M et al. Mixed chimerism induces donor-specific T-cell tolerance across a highly disparate xenogeneic barrier. Blood 2002; 99(10):3823-9.
75. Buhler L, Awwad M, Treter S et al. Pig hematopoietic cell chimerism in baboons conditioned with a nonmyeloablative regimen and CD154 blockade. Transplantation 2002; 73(1):12-22.
76. Neipp M, Exner BG, Ildstad ST. A nonlethal conditioning approach to achieve engraftment of xenogeneic rat bone marrow in mice and to induce donor-specific tolerance. Transplantation 1998; 66(8):969-75.
77. Kawai T, Sogawa H, Koulmanda M et al. Long-term islet allograft function in the absence of chronic immunosuppression: A case report of a nonhuman primate previously made tolerant to a renal allograft from the same donor. Transplantation 2001; 72(2):351-4.
78. Gaur LK, Kennedy E, Nitta Y et al Induction of donor-specific tolerance to islet allografts in nonhuman primates. Ann New York Acad Sci 2002; 958:194-8.
79. Wu T, Levay-Young B, Heuss N et al. Inducing tolerance to MHC-matched allogeneic islet grafts in diabetic NOD mice by simultaneous islet and bone marrow transplantation under nonirradiative and nonmyeloablative conditioning therapy. Transplantation 2002; 74(1):22-7.
80. Kappler JW, Roehm N, Marrack P. T cell tolerance by clonal elimination in the thymus. Cell 1987; 49(2):273-80.
81. Wood PJ, Streilein JW. Immunogenetic basis of acquired transplantation tolerance. Transplantation 1984; 37(3):223-6.
82. Cutler AJ, Bell EB. Neonatally tolerant rats actively eliminate donor-specific lymphocytes despite persistent chimerism. Eur J Immunol 1996; 26(2):320-8.
83. Khan A, Tomita Y, Sykes M. Thymic dependence of loss of tolerance in mixed allogeneic bone marrow chimeras after depletion of donor antigen. Peripheral mechanisms do not contribute to maintenance of tolerance. Transplantation 1996; 62(3):380-7.
84. Tutschka PJ, Ki PF, Beschorner WE et al. Suppressor cells in transplantation tolerance. II. Maturation of suppressor cells in the bone marrow chimera. Transplantation 1981; 32(4):321-5.
85. Roser BJ. Cellular mechanisms in neonatal and adult tolerance. Immunol Rev 1989; 107:179-202.

86. Qin SX, Cobbold S, Benjamin R. Induction of classical transplantation tolerance in the adult. J Exp Med 1989; 169(3):779-94.
87. Qin S, Cobbold SP, Pope H et al. "Infectious" transplantation tolerance. Science 1993; 259(5097):974-7.
88. Martin PJ. Prevention of allogeneic marrow graft rejection by donor T cells that do not recognize recipient alloantigens: Potential role of a veto mechanism. Blood 1996; 88(3):962-9.
89. Butcher JA, Hariharan S, Adams MB et al. Renal transplantation for end-stage renal disease following bone marrow transplantation: A report of six cases, with and without immunosuppression. Clin Transplant 1999; 13:330.
90. Millan MT, Shizuru JA, Hoffmann P et al. Mixed chimerism and immunosuppressive drug withdrawal after HLA-mismatched kidney and hematopoietic progenitor transplantation. Transplantation 2002; 73(9):1386-91.
91. Ciancio G, Burke GW, Garcia-Morales R et al. Effect of living-related donor bone marrow infusion on chimerism and in vitro immunoregulatory activity in kidney transplant recipients. Transplantation 2002; 74(4):488-96.
92. Chakraverty R, Peggs K, Chopra R et al. Limiting transplantation-related mortality following unrelated donor stem cell transplantation by using a nonmyeloablative conditioning regimen. Blood 2002; 99(3):1071-8.

CHAPTER 54

Autologous Hematopoietic Stem Cell Transplantation for Crohn's Disease

Robert M. Craig and Richard K. Burt

Introduction

Autologous hematopoietic stem cell transplantation (HSCT) is being explored as therapy for autoimmune diseases including rheumatoid arthritis, multiple sclerosis, and systemic lupus erythematosus. Recently, our medical center performed the first four HSCTs in the world for patients with severe Crohn's disease (CD), and the salutary response to therapy was dramatic. This chapter will review the immunology of the intestinal immune system and pathophysiology of CD. The natural history of CD and identification of high risk and refractory disease is described to counterpose the morbidity and mortality of CD with HSCT. CD therapy is discussed to show that standard therapies are symptomatic, directed at the manifestations of CD, rather than attempting to be curative and directed at the underlying cause. Finally, patients who have undergone bone marrow transplantations for other reasons, who incidentally had inflammatory bowel disease, and our first two patients are described.

Immunology of the Intestinal Immune System

The gastrointestinal tract may be viewed as a unique immune organ which is involved in establishing a mucosal barrier to infectious agents, determining tolerance or immunity to a variety of dietary foods and colonizing bacteria, having a unique lymphocyte trafficking and homing network, and possibly functioning as an extra-thymic site for lymphocyte maturation. The gut associated lymphoid tissues (GALT) are composed of Peyer's patches (Fig. 1), mesenteric lymph nodes, lamina propria lymphocytes (LPL), intraepithelial lymphocytes (IEL), as well as epithelial cells (enterocytes).[1]

Lymphoid follicles are present in the intestinal tract from the duodenum to the rectum. Submucosal aggregates of follicles, termed Peyer's patches, are located throughout the small intestine in humans as well as the large bowel of some species such as rabbits.[2] Although the number of Peyer's patches decline with age, unlike the thymus, they do not begin to involute until puberty and are present throughout life. Peyer's patches are separated from the gut lumen by a single layer of epithelial cells termed follicle associated epithelium (FAE)[2] (Fig. 2). The intestinal epithelial cells, although permeable to low molecular weight nutrients, are designed to exclude larger molecular weight molecules as well as pathologic organisms and toxins. FAE, which is located over Peyer's patches, is interrupted by specialized epithelial cells with short villous projects that are termed M cells for microvilli (Fig. 2).[2] These M cells allow for the transport of antigens from the gut lumen to immune cells in the underlying Peyer's patch. M cells can be separated from epithelial cells by light microscopy using immunohistochemistry for vimentin, a cytoskeletal protein present within M cells but absent from epithelium. M cells can also be easily distinguished on electron microscopy by the shortened height of their microvilli.

M cells are the first step in initiating a mucosal immune response or tolerance. Lumenal antigens are taken up by M cells and transported to intraepithelial lymphocytes and/or macrophages that have been invaginated or surrounded by the cytoplasm of M-cells. The origin of M-cells remains controversial and may be differentiated from crypt stem cells or represent a modified epithelial cell since ex vivo co-culture of lymphocytes from Peyer's patches induces conversion of enterocytes into antigen transporting M cells.[3,4] M cells do not present MHC class II antigens or co-stimulatory molecules. After uptake and transport by M-cells, antigens are processed and represented within Peyer's patches by follicular dendritic and other antigen presenting cells.

IEL are usually CD8$^+$ T cells. IEL may express either the alpha/beta (α/β) or gamma delta (γ/δ) T cell receptor (TCR), the ratio of which varies by species. Approximately 10% of human IEL express the γ/δ TCR. Besides playing a role in differentiating enterocytes into M-cells, γ/δ IEL are thought to play a role in suppressing immune responses to gut antigens.[6] γ/δ T cells may arise from different lymphoid sites than α/β T cells.[7,8]

α/β lymphocytes are positively and negatively selected within the thymus to deplete or functionally silence autoreactive repertoires while simultaneously providing a diverse TCR repertoire against infections. γ/δ T cells can originate from gut in humans and mice, and the gut may even account for most γ/δ T cell production.[7-10]

LPL are located between the epithelial basement membrane and muscularis mucosa.[2] In contrast to IEL, most LPL are CD4$^+$ with a CD4/CD8 ratio of approximately 2:1. LPL may express either the γ/δ TCR or more commonly α/β TCR. While LPL function remains poorly understood, most are considered to be helper memory α/β T lymphocytes involved in B cell secretory IgA production. IgA is the predominate immunoglobulin pro-

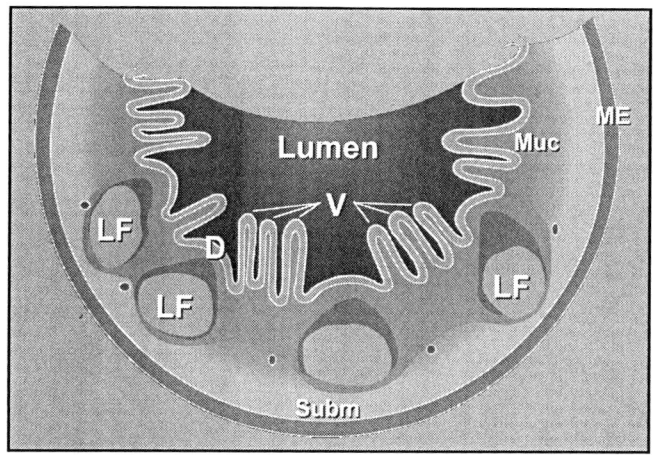

Figure 1. Peyer's patch in small intestine. D= dome epithelium, LF= lymphoid follicle, ME= Muscularis Externa, Muc= mucosa, V= villi; Subm= submucous.

duced by gut B cells and is secreted into the lumen. Intraluminal IgA agglutinates infectious agents preventing their adherence to and penetration through enterocytes.

Immune cells exit from Peyer's patches or enter from the peripheral blood via neighboring high endothelial venules (HEV). Trafficking of immune cells into and out of Peyer's patches is regulated by multiple adhesion molecules. Blocking of L-selectin prevents trafficking of naïve CD4+ T cells into lymph nodes but not Peyer's patches.[11] Naïve CD4+ cells may be prevented from trafficking into Peyer's patches by blocking both L-selectin and alpha4 beta7 integrins.[11,12] Variances in tissue HEV may allow for differential control of lymphocyte recruitment and may also target specific T cell subsets to a restricted recirculation between lymphoid structures and unique peripheral organs.

The gastrointestinal mucosa is one of the largest interfaces of the body with a single epithelial layer separating the interior from the exterior environment. Therefore, it would be phylogenetically reasonable for extra-thymic T cell lymphopoiesis to occur within gut lymphoid follicles.[13-17] It has been demonstrated that antigen fed orally may induce tolerance.[18] High levels of oral antigen may cause deletion of antigen specific repertoires, while low concentrations of oral antigen may induce T cell regulatory or suppressor cytokine profiles.[19] Peyer's patches and/or mesenteric lymph nodes may be necessary for oral tolerance, since germ free mice which have a significant shortage of T cells within Peyer's patches do not develop oral tolerance upon antigen feeding.[20]

There are several phenotypes of dendritic cells within Peyer's patches that may play an important role in gut immunity. If a lumenal antigen is presented as an innocuous food protein, dendritic cells generate anti-inflammatory IL-10, IL-4 and TGF-β resulting in oral tolerance.[21] If the gut antigen is presented along with a danger signal from a microbial pathogen, dentritic cells within Peyer's patches may release pro-inflammatory IL-12 and INF-γ, which may result in immunity towards both the pathogen as well as normally innocuous foods or even self-antigens, either of which could precipitate inflammatory bowel disease.

Pathophysiology of Crohn's Disease

Crohn's disease is an immune-mediated disease. An autoimmune etiology remains controversial and unproven. No intestinal self antigen (initiating or spread epitope) has been identified. On the other hand, several animal gene knockout models suggest

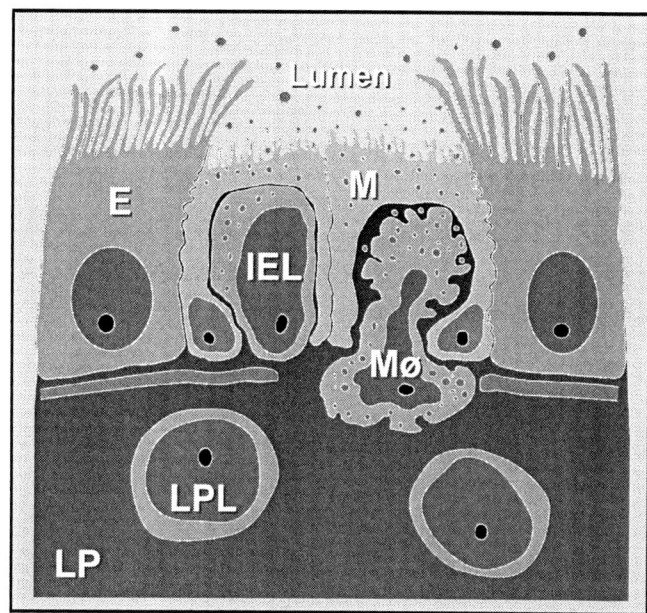

Figure 2. Follicle associated epithelium. Single cell gut epithelium. E= enterocyte; IEL= intraepithelial lymphocyte; LPL= lamina propria lymphocytes; M= M cell; MO= macrophage.

that inflammatory bowel disease may be a result of immune dsyregulation between Th1 and Th2 cytokines. Deficiency of multiple Th2 cytokines (such as IL-10, IL-2, TGF-β) may cause colitis. Interleukin-10 deficient mice develop acute and chronic colitis.[22] IL-2 deficient,[23] double mutant IL-2 and IL-4 deficient, and transforming growth factor beta deficient[24] mice develop colitis. When raised in a germ free environment, these gene knockout mice remain disease free. Animal models demonstrate the need for both cytokine imbalance and gut bacterial flora as disease triggers. Infusion of the T cell subset CD4+ CD45 RBhigh into SCID mice causes colitis.[25] When raised in a germ free environment, CD4+ CD45 RBhigh SCID mice do not develop disease.[26] Infusion of the normal broad T cell repertoire or treatment with IL-10 or anti-Th1 antibodies into a CD4+CDRBhigh SCID mouse induces remission of colitis.[25]

Histologically, Crohn's disease manifests as a Th1 delayed type hypersensitivity of the gut wall with granuloma formation.[27] Failure to induce disease by immunization with intestinal auto-epitopes and murine models of cytokine gene knockout dysregulation suggests that Crohn's disease is a Th1 immune-mediated imbalance, perhaps caused by a delayed type hypersensitivity against intestinal flora. High dose immune ablative therapy and HSCT may reset the immune balance resulting in regeneration of a normal phenotype and prolonged disease remission.

Further support for intestinal flora as a cause of disease is the presence of anti-Saccharomyces cerevisiae antibody (ASCA) in patients with Crohn's disease. Saccharomyces cerevisiae is Baker's yeast. Yeast exert a strong adjuvant effect upon dendritic cells resulting in IL-12 production and priming of T cell responses, and are a commonly used assay for demonstrating intact Th1 delayed type hypersensitivity. ASCA may be used as a diagnostic marker to help differentiate Crohn's disease from ulcerative colitis.[28,29] Proliferation assays of peripheral blood lymphocytes incubated with Saccharomyces cerevisiae and pulsed with tritiated thymidine demonstrate a three fold increase in proliferative response from Crohn's patients compared to normal controls, including healthy normal bakers.[30] Pre and post transplant peripheral blood ASCA antibody, T cell proliferative responses, and T cell lines to intact Saccharomyces cerevisiae and/or extracts from Saccharomyces cerevisiae may be generated and analyzed to determine the role of gut flora in causing Crohn's disease.

Adoptive transfer of enteric bacterial antigen activated CD4+ T cells from C3H/HeJBir mice into scid/scid recipients causes colitis and provides additional evidence that enteric flora causes Crohn's.[31] The precursor frequency of IL-2 producing Th1 skewed (INF-γ producing) CD4+ T cells to enteric bacteria is significantly higher in C3H/HeJBir mice with colitis compared to normal controls.[31] Pre and post transplant peripheral blood CD4+ T cell precursor frequency and cytokine profile following stimulation with enteric bacteria may be monitored before and after HSCT.

An imbalance in cytokine regulation to gut bacteria is the presumed etiology of Crohn's. Since tissue can be easily obtained from the gastrointestinal track via colonoscopy, immune deviation may be further analyzed by comparison of pre and post HSCT colon biopsy samples. T cell (interferon gamma, IL-10, IL-4, IL-2) and monocyte (TNF, IL-12) cytokine imbalance could be analyzed by PCR analysis and in situ hybridization from normal and involved tissues. Clones of cells could be generated and analyzed for T regulatory or suppressor cell phenotype (refer to Chapter 20) and compared before and after HSCT.

The occurrence of a Th1 cytokine imbalance in Crohn's is probably multi-factorial involving multiple environmental stimuli interacting with genetic predisposition resulting in a skewed but sometimes reversible cytokine imbalance. It is anticipated, but unproven, that HSCT will restore the normal cytokine balance. NOD2 is a Crohn's associated gene contributing to, but not sufficient to cause, disease.[32] NOD (nucleotide-binding oligomerization domain) proteins are involved in programmed cell death and innate immune responses.[33] NOD2 is restricted to monocytes and activates NF-kappaB in response to bacterial products. Premature truncation of the NOD2 gene confers disease susceptibility in a subset of Crohn's patients.[34] Besides cytokine imbalance, and T cell immune responses to Saccharomyces cerevisiae and enteric bacteria, NOD2 gene polymorphisms which probably contribute to the immune deviation seen in Crohns, could be analyzed in patients undergoing HSCT.

Crohn's Disease—Clinical Features and Natural History

CD usually presents with diarrhea, abdominal pain and/or weight loss. The illness characteristically waxes and wanes, and is often not diagnosed for many months.[35] Other presenting symptoms or signs include fever, vomiting, perianal disease, fistulae, rectal bleeding or an abdominal mass. The abdominal pain is characteristically in the right lower quadrant, overlying the most frequently involved segment, the terminal ileum and cecum. CD can affect either the small intestine or colon, or both. More severe varieties can affect the mouth, esophagus, stomach and pancreatobiliary systems. The diarrhea can be large volume with less frequent passages in small intestinal Crohn's, or small volume with frequent passages in colitis. Weight loss can be secondary to anorexia or food avoidance due to the disease, or to malabsorption from decreased functioning bowel or bile salt malabsorption.

Although mycobacterial, ischemic, lymphomatous or fungal disease can mimic CD in its initial presentation, the characteristic radiographic, endoscopic and histologic findings, and the chronic clinical course allow for a high degree of assurance in diagnosis. Sometimes it is difficult to differentiate CD from chronic ulcerative colitis (CUC), and there are some patients who present with overlap features. The diagnosis is generally clear on the basis of clinical, radiographic, endoscopic and histologic criteria.[36]

The clinical course is variable and may be persistent, progressive, or may improve or go into a remission. The clinical triad of abdominal pain, diarrhea and weight loss attends most patients upon presentation. Often the patient also has fever. Typically the patient has waxing and waning of his illness throughout his life, with medical therapy adjusted depending upon clinical presentations. The fluctuating nature of the illness makes analysis of clinical trials suspect, unless the trials are randomized. Various scoring systems have been used to quantify response to therapy in these treatments, and to follow the natural course of the illness. The most popular scoring system is the Crohn's Disease Activity Index (CDAI) (Table 1).[37] Another instrument, the Inflammatory Bowel Disease Questionnaire (IBDQ), which addresses quality of life issues (Table 2), is also used for following patients, especially in efficacy studies.[38]

Patients with CD may have involvement of the colon, the small intestine or both, although about half have involvement of both the distal ileum and proximal colon.[39] The prognosis might be better in those with disease confined to the colon, especially if there is only segmental involvement.[40] Although those with colonic Crohn's tend to have more systemic symptoms, including fever and arthralgias, the need for surgery is less than in those with small intestinal involvement. The mortality rate directly related to CD in one large series was 6%,[41] and in another with Crohn's colitis 3%.[42] However, it is difficult to relate mortality directly to the sites of involvement. A study from Minnesota showed a statistically significant increased mortality above expected in patients with CD (p= 0.007), although the numerical increase was slight.[43]

CD is a morbid illness, fraught with the potential of many complications. The complications associated with the intestinal involvement (local complications) include fistulae to other segments of bowel, skin, bladder, or vagina; abscess formation; anorectal disease with ulceration, fissuring, fistulae or perianal abscess; intestinal strictures with obstruction; intestinal perforation; hemorrhage; toxic megacolon; and colon carcinoma. Systemic complications include cutaneous lesions (pyoderma gangrenosa, erythema nodosa, erythema multiforme, dermatitis), hepatobiliary lesions (chronic active hepatitis, fatty liver, sclerosing cholangitis, pericholangitis); ocular (iritis); and arthritis. Often, the complications cause severe disability, markedly diminishing one's quality of life, and some of the complications are life threatening (hemorrhage, abscess, intestinal obstruction, sclerosing cholangitis).

Surgery is performed in CD only for complications and not for cure. The usual indications are intestinal hemorrhage, obstruction, perforation, fistulas, or abscess, that do not respond to medical management. Severe, intractable abdominal pain is an occasional indication. Recurrence is the rule following surgery, and approximately half of the patients will require additional surgery.[44-46] The patient is subjected to increasing risk of malabsorption and diarrhea with each surgical resection.

Over the past decade, there has been an increasing awareness that CD predisposes to malignant transformation, particularly in the colon, although not to the same extent as ulcerative colitis.[47,48] In one group of patients with extensive Crohn's colitis, the authors found an incidence of colonic carcinoma or dysplasia com-

Table 1. Crohn's Disease Activity Index (CDAI)

Variable	Quantity	Multiple	Total
Number of liquid or soft stools per day		2	
Abdominal pain (0= none, 1= mild, 2= moderate, 3= severe)		5	
General well being (0= well, 1= slightly under par, 2= poor, 3= very poor, 4= terrible)		7	
Number of complications: arthralgias, iritis, erythema nodosum, pyoderma gangrenosa, aphthous ulcerations, anal fissure, anal fistula, anal abscess, fever >37 degrees past week, intestinal obstruction		20	
Opiates for diarrhea (no= 0, yes= 1)		30	
Abdominal mass (no= 0, questionable= 2, yes= 5)		0	
Deviation from normal hematocrit (N= 42 for female, 47 for male)		6	
% deviation from standard weight		1	
TOTAL CDAI			

From ref 37: CDAI <150 = remission; >450 = severely ill

parable to that found in CUC.[45] Small intestinal cancer has also been reported in Crohn's disease, usually in excluded segments of bowel.[49] At this point it is not clear how prevalent Crohn's colitis is complicated by cancer, and how this impacts on overall Crohn's survival.

It is clear from the above discussion that there is a definite mortality that can be attributed to CD. The mortality rate directly attributed to CD is probably in excess of 10% in the sicker patients considered for bone marrow transplantation. This compares favorably with the <5% mortality rate from an autologous HSCT. In addition, HSCT may eliminate the severe morbidity from CD outlined above.

Crohn's Disease—Therapy

Conventional therapy uses the most benign drugs first, adding drugs with more potential side effects later. Sulfasalazine or mesalamine products are first line therapy. 5-aminosalicylate (5-ASA) is the active pharmacologic ingredient.[50,51] If this anti-inflammatory agent is not effective, metronidazole or other antibiotics can be tried. Metronidazole is particularly effective in patients with anorectal fistulas.[52] Failure to promote improvement or a remission requires the institution of immunosuppressive agents or corticosteroids. Whereas corticosteroids are extremely effective in producing a remission in most patients, they have the side effects of osteoporosis, obesity, cutaneous changes, decreased resistance to infection, diabetes mellitus, and aseptic necrosis of the femoral head.[50] Budesonide is a corticosteroid with a large first-pass metabolism in the liver and consequently less systemic effects from therapy. Studies have shown its equal efficacy to prednisone in Crohn's ileitis and ileocolitis.[53]

Immunosuppressive agents such as azathioprine and 6-mercaptopurine are mostly effective as corticosteroid-sparing agents, although they have some activity in promoting resolution of the inflammatory process in their own right.[54-56] A few patients remain extremely ill or have a complication that does not resolve with these agents. They have been candidates for newer therapy, such as the use of cyclosporin A, anti-TNF alpha (infliximab), interleukin 10, mycophenolate mofetil or methotrexate; or the surgical resection of diseased bowel.[57-62] However, surgical ablation of CD does not cure the illness, which usually recurs in another segment of bowel.

All of the above therapy is directed at the manifestations of CD, rather than its cause. 5-ASA products and corticosteroids serve as non-specific anti-inflammatory agents. Immunosuppressive agents oppose the actions of immunocytes. Antibiotic actions are not clear, but probably decrease the exposure to culpable antigens. Cytokines or cytokine inhibitors work on one of the expressions of CD inflammation. The HSCT approach is unique in that it is acting more proximally in the disease process, in the hopes of producing a long term remission.

Rationale for Stem Cell Transplantation in Crohn's Disease

Regardless of the molecular events that confer tolerance or its converse immunity, there appear to be both susceptibility factors, presumably genetic, and trigger mechanisms involved.[63] A model for understanding the process is a genetic predisposition that is dormant until exposed to the appropriate environmental and hormonal triggers. Theoretically, stem cell transplantation following immune ablation allows the elimination or marked decrease of auto destructive T-lymphocyte repertoires and the reintroduction of tolerance from the stem cell compartment. Although the susceptibility for a relapse of CD is not eliminated using autologous HSCs, the patient may remain in permanent remission so long as he/she is not exposed to the appropriate trigger(s) in the future. Using allogeneic HSCs may also elimi-

Table 2. Inflammatory Bowel Disease Questionnaire (IBDQ)

	Score 1-7		Score 1-7
(B) 1) How frequent have your bowel movements been during the last 2 weeks?		(B) 17) Overall, in the last 2 weeks, how much of a problem have you had passing large amounts of gas?	
(S) 2) How often has the feeling of fatigue or of being tired and worn out been a problem for you during the last 2 weeks?		(S) 18) Overall, in the last 2 weeks, how much of a problem have you had maintaining, or getting to, the weight you would like to be at?	
(E) 3) How often during the last 2 weeks have you felt frustrated, impatient, or restless?		(E) 19) Many patients with bowel problems often have worries and anxieties related to their illness. These include worries about getting cancer, worries about never feeling better, and worries about having a relapse. In general, in the last 2 weeks, how often have you felt worried or anxious?	
(SF) 4) How often in the last 2 weeks have you been unable to attend school or work because of your bowel problem?		(B) 20) How much of the time in the last 2 weeks have you been troubled by a feeling of abdominal bloating?	
(B) 5) How much of the time in the last 2 weeks have your bowel movements been loose?		(E) 21) How often in the last 2 weeks have you felt relaxed and free of tension?	
(S) 6) How much energy have you had in the last 2 weeks?		(B) 22) How much of the time in the last 2 weeks have you had a problem with rectal bleeding?	
(E) 7) How often in the last 2 weeks did you feel worried about the possibility of needing to have surgery because of your bowel problem?		(E) 23) How much of the time in the last 2 weeks have you felt embarrased as a result of your bowel problem?	
(SF) 8) How often in the last 2 weeks did you have to delay or cancel a social engagement because of your bowel problem?		(B) 24) How much of the time in the last 2 weeks have you been troubled by a feeling of having to go to the bathroom even though your bowels are empty?	
(B) 9) How often in the last 2 weeks have you been troubled by cramps in your abdomen?		(E) 25) How much of the time in the last 2 weeks have you felt tearful, or upset?	
(S) 10) How often in the last 2 weeks have you felt generally unwell?		(B) 26) How much of the time in the last 2 weeks have you had trouble by accidental soiling of your underpants?	
(E) 11) How often in the last 2 weeks have you been troubled because of fear of not finding a wash-room?		(E) 27) How much of the time in the last 2 weeks have you felt angry as a result of your bowel problem?	
(SF) 12) How much difficulty have you had, as a result of your bowel problems, doing leisure or sport activities you would like to have done during the last 2 weeks?		(SF) 28) To what extend has your bowel problem limited sexual activity during the last 2 weeks?	
(B) 13) How often in the last 2 weeks have you been troubled by pain in the abdomen?		(B) 29) How mush time in the last 2 weeks have you been troubled by feeling sick to your stomach?	

continued on next page

Table 2. Continued

	Score 1-7		Score 1-7
(S) 14) How often in the last 2 weeks have you had problems getting a good night's sleep, or been troubled by waking during the night?		(E) 30) How much of the time in the last 2 weeks have you felt irritable?	
(E) 15) How often in the last 2 weeks have you felt depressed or discouraged?		(E) 31) How often in the last 2 weeks have you felt lack of understanding from others?	
(SF) 16) How often in the last 2 weeks have you had to avoid attending events where there was no washroom close at hand?		(E) 32) How satisfied, happy, or pleased have you been with your personal life during the past 2 weeks?	

From ref 38

The IBDQ is 32 questions grouped in 4 categories: bowel symptoms (B), systemic symptoms (S), emotional function (E), and social function (SF). Response are presented as a 7 point scale. As an example for question 1, the response may be:

1. Bowel movements as or more frequent than they have ever been.
2. Extremely frequent.
3. Very frequent.
4. Moderate increase in frequency of bowel movements.
5. Some increase in frequency of bowel movements.
6. Slight increase in frequency of bowel movements.
7. Normal, no increase in frequency of bowel movements.

nate a genetic predisposition to disease. Allogeneic HSCT protocols are currently being evaluated in phase I safety trials for rheumatoid arthritis and scleroderma and, if found safe and effective, allogeneic HSCT may be used in future transplant trials for patients with CD. Attempting immune ablation and regeneration from either autologous or allogeneic hematopoietic stem cells, i.e., a HSCT, for cases of severe inflammatory bowel disease allows a unique opportunity to simultaneously biopsy an easily accessible but poorly understood immune compartment and disease affected organ.

Anecdotal Case Reports of Patients with CD Undergoing HSCT for Other Reasons

Six patients with CD have been reported who underwent allogeneic bone marrow transplantation for other reasons.[64] One was in CD remission at the time of the transplantation and remained in remission for 15 years in spite of discontinuation of immunosuppression. Three of the 5 with active CD at the time of transplantation went into remission for 6-10 years and remained in remission at the time of the report. The fourth had significant fistulous disease in the year following the transplantation, requiring ileal resection. The fifth died from sepsis three months following the transplantation, and the activity of the CD was not fully evaluated. Another report of a patient with an allogeneic bone marrow transplantation described improvement in CD following the transplantation, at least for the short-term.[65] Another patient was described in some detail, who had CD prior to an autologous bone marrow transplantation for non-Hodgkin's lymphoma, and who went into remission from both his lymphoma and CD for 7 years and remained in remission at the time of the report.[66] Finally, another patient with incidental Crohn's colitis who underwent an autologous bone marrow transplantation was reported in an abstract to enter a symptomatic remission, although there remained colonic inflammation on colonoscopy.[67]

Although the evidence supporting HSCT in patients with CD in these reports is encouraging, there have been dissenters who have suggested that the data are not compelling enough to put patients through this risk.[68,69] The authors point out that most of the described patients had undergone the more rigorous allogeneic rather than autologous transplantation. The authors also reiterate the mortality risk that is posed by any HSCT and its attendant bone marrow suppression. In regards to CUC, most practitioners would elect to have a patient with severe unresponsive CUC undergo a proctocolectomy, which is curative, rather than consider HSCT. However, if there is an institutional record of HSCT being performed safely in CD, in the future, patients with severe CUC may be offered the option of either proctocolectomy or HSCT.

HSCT for CD

Although Northwestern University Medical Center has had Institutional Review Board and U.S. FDA approval for autologous stem cell transplantation in severe Crohn's Disease since 1998, suitable patients were not identified until 2001 when four successful autologous peripheral stem cell transplantations were performed.[70] The first two patients initial course will be reported here. HSCs were mobilized with cyclophosphamide 2.0 g/m^2 and G-CSF. The transplant regimen is shown in Table 3. The two patients had been ill almost continuously for 6 years in one (age 16) and 12 years for the other (age 22). See Tables 4 and 5 for their clinical data and course.

One had so much abdominal pain that she required high doses of opiates. Each had a Crohn's disease activity index (CDAI) greater that 250 (severe disease). Each had failed therapy with 5-ASA, corticosteroids, antibiotics, azathioprine or methotrexate, and infliximab. One had undergone resection of part of her ileum and colon. The stem cell administration was uneventful in each, and each is in remission in terms of the CDAI (<100 in each; remission <150),off all medications, including narcotics.

Conclusion

It should be clear from this discussion that there is a large body of data establishing the immune pathophysiology of CD. The natural history of CD can be severely morbid in some pa-

Table 3. Hematopoietic stem cell transplant regimen for Crohn's disease

Day	-6	-5	-4	-3	-2	-1	0
Mesna		X	X	X	X		
Cyclophosphamide		50 mg/kg	50 mg/kg	50 mg/kg	50 mg/kg		
Equine Anti-thymocyte globulin	Skin Test		30 mg/kg	30 mg/kg	30 mg/kg		
Methylprednisolone			1 gm	1 gm	1 gm		
CD34$^+$ infusion							X

Table 4. 1st Stem cell transplant for Crohn's disease

Test/Symptom	Pre-Transplant	2 Months Post Transplant	6 Months Post Transplant
Diarrhea	Average 25 times a day, sometimes bloody	none	none
Entry immune suppressive medications	Prednisone 22 mg/day 6-mercaptopurine	Prednisone 15 mg/day	Prednisone 7.5 mg/day*
Opiates (dilaudid)	3.0 mg/hour	2.3 mg/hour	1.1 mg/hour*
Hemoglobin (mg/dl) (normal 12-15 mg/dl)	9.7	10.3	11.9
CRP (normal <2.5)	N/A	11	4.5
Sedimentation Rate (normal <20)	N/A	49	42
Albumin g/dl (normal 3.5—5.0 g/dl)	3.3	3.7	4.1
CDAI	305	116	84
Small bowel x-ray	Terminal ileitis	Mild terminal ileum inflammation	normal
Colonoscopy	Severe colitis	Mild colitis	Slight ulceration of ileocecal area

* Evaluation to be performed at 1 year post transplantation is pending, but by 10 months post transplant the patient has been tapered off all corticosteroids and opiates. CDAI= Crohn's disease activity index (CDAI 250-400 severe disease, <100 remission); CRP= C reactive protein; N/A= not available.

Table 5. 2nd Crohn's disease stem cell transplant

Test/Symptom	Pre-Transplant	2 Months Post Transplant	6 Months Post Transplant
Diarrhea	Average 10-12 times a day	none	none
Entry immune suppressive medications	Methotrexate, 6-mercaptopurine, 5-aminosalicylic acid	none	none
Hemoglobin (mg/dl) (normal 12-15 mg/dl)	12	12	12
CRP (normal <2.5)	10.8	1.1	1.4
Sedimentation Rate (normal <20)	15	10	10
Albumin (g/dl) (normal 3.5—5.0 g/dl)	2.8	3.0	3.2
CDAI	271	62	<50
Small bowel x-ray	Multiple skip areas of inflammation		normal
Colonoscopy	Severe colitis and terminal ileitis		Slight ulceration of colon

CDAI= Crohn's disease activity index (CDAI 250-400 = severe disease, CDAI <100 = remission); CRP= C reactive protein

tients, and the most severe patients probably carry a mortality >10%. A trial of HSCT can be considered appropriate for these very sick patients, as the mortality rate for autologous bone marrow transplantation should be less than 5%. So far, the results from incidental bone marrow transplantations in patients with CD and our first four patients with HSCTs are encouraging. Long-term follow up of these patients will be required to determine the ultimate impact of this unique treatment.

References

1. Mowat AM, Viney JL. The anatomical basis of intestinal immunity. Immunol Rev 1997; 156:145-66.
2. Gebert A, Rothkotter HJ, Pabst R. M cells in Peyer's patches of the intestine. Int Rev Cytol 1996; 167:91-159.
3. Kerneis S, Bogdanova A, Kraehenbuhl JP et al. Conversion by Peyer's patch lymphocytes of human enterocytes into M cells that transport bacteria. Science 1997; 277(5328):949-52.
4. Borghesi C, Taussig MJ, Nicoletti C. Rapid appearance of M cells after microbial challenge is restricted at the periphery of the follicle-associated epithelium of Peyer's patch. Lab Invest 1999; 79(11):1393-401.
5. Croitoru K, Ernst PB. Leukocytes in the intestinal epithelium: An unusual immunological compartment revisited. Reg Immunol 1992; 4(2):63-9.
6. Hanninen A, Harrison LC. Gamma delta T cells as mediators of mucosal tolerance: The autoimmune diabetes model. Immunol Rev 2000; 173:109-19.
7. Rocha B, Vassalli P, Guy-Grand D. Thymic and extrathymic origins of gut intraepithelial lymphocyte populations in mice. J Exp Med 1994; 180(2):681-6.
8. Rocha B, Guy-Grand D, Vassalli P. Extrathymic T cell differentiation. Current Opin Immunol 1995; 7(2):235-42.
9. McVay LD, Carding SR. Generation of human gamma delta T-cell repertoires. Crit Rev Immunol 1999; 19(5-6):431-60.
10. Saito H, Kanamori Y, Takemori T et al. Generation of intestinal T cells from progenitors residing in gut cryptopatches. Science 1998; 280(5361):275-8.
11. Bradley LM, Malo ME, Fong S et al. Blockade of both L-selectin and alpha4 integrins abrogates naive CD4 cell trafficking and responses in gut-associated lymphoid organs. Int Immunol 1998; 10(7):961-8.
12. Williams MB, Butcher EC. Homing of naive and memory T lymphocyte subsets to Peyer's patches, lymph nodes, and spleen. J Immunol 1997; 159(4):1746-52.
13. Farstad IN, Halstensen TS, Fausa O et al. Do human Peyer's patches contribute to the intestinal intraepithelial gamma/delta T-cell population? Scand J Immunol 1993; 38(5):451-8.
14. Lefrancois L, Puddington L. Extrathymic intestinal T-cell development: Virtual reality? Immunol Today 1997; 16(1):16-21.
15. Bandeira A, Itohara S, Bonneville M et al. Extrathymic origin of intestinal intraepithelial lymphocytes bearing T-cell antigen receptor gamma delta. Proc Nat Acad Sci USA 1991; 88(1):43-7.
16. Hedrick SM, Sharp LL. T-cell fate. Immunol Rev 1998; 165:95-110.
17. McVay LD, Carding SR. Generation of human gamma delta T-cell repertoires. Crit Rev Immunol 1999; 19(5-6):431-60.
18. Garside P, Mowat AM. Oral tolerance. Semin Immunol 2001; 13(3):177-85.
19. Chen Y, Inobe J, Marks R et al. Peripheral deletion of antigen-reactive T cells in oral tolerance. Nature 1995; 376(6536):177-80.
20. Maeda Y, Noda S, Tanaka K et al. The failure of oral tolerance induction is functionally coupled to the absence of T cells in Peyer's patches under germfree conditions. Immunobiology 2001; 204(4):442-57.
21. Alpan O, Rudomen G, Matzinger P. The role of dendritic cells, B cells, and M cells in gut-oriented immune responses. J Immunol 2001; 166(8):4843-52.
22. Kuhn R, Lohler J, Rennick D et al. Interleukin-10-deficient mice develop chronic enterocolitis. Cell 1993; 75(2):263-74.
23. Ehrhardt RO, Ludviksson B. Induction of colitis in IL2-deficient-mice: The role of thymic and peripheral dysregulation in the generation of autoreactive T cells. Res Immunol 1997; 148(8-9):582-8.
24. Kulkarni AB, Ward JM, Yaswen L et al. Transforming growth factor-beta 1 null mice. An animal model for inflammatory disorders. Am J Pathol 1995; 146(1):264-75.
25. Morrissey PJ, Charrier K. Induction of wasting disease in SCID mice by the transfer of normal CD4+/CD45RBhi T cells and the regulation of this autoreactivity by CD4+/CD45RBlo T cells. Res Immunol 1994; 145(5):357-62.
26. Hudcovic T, Stepankova R, Cebra J et al. The role of microflora in the development of intestinal inflammation: Acute and chronic colitis induced by dextran sulfate in germ-free and conventionally reared immunocompetent and immunodeficient mice. Folia Microbiologica 2001; 46(6):565-72.
27. Ramzan NN, Leighton JA, Heigh RI et al. Clinical significance of granuloma in Crohn's disease. Inflammatory Bowel Diseases 2002; 8(3):168-73.
28. Joossens S, Reinisch W, Vermeire S et al. The value of serologic markers in indeterminate colitis: A prospective follow-up study. Gastroenterology 2002; 122(5):1242-7.
29. Sutton CL, Yang H, Li Z et al. Targan SR. Braun J. Familial expression of anti-Saccharomyces cerevisiae mannan antibodies in affected and unaffected relatives of patients with Crohn's disease. Gut 2000; 46(1):58-63.
30. Young CA, Sonnenberg A, Burns EA. Lymphocyte proliferation response to baker's yeast in Crohn's disease. Digestion 1994; 55(1):40-3.
31. Cong Y, Brandwein SL, McCabe RP et al. CD4+ T cells reactive to enteric bacterial antigens in spontaneously colitic C3H/HeJBir mice: Increased T helper cell type 1 response and ability to transfer disease. J Exp Med 1998; 187(6):855-64.
32. Cuthbert AP, Fisher SA, Mirza MM et al. The contribution of NOD2 gene mutations to the risk and site of disease in inflammatory bowel disease. Gastroenterology 2002; 122(4):867-74.
33. Inohara N, Nunez G. The NOD: A signaling module that regulates apoptosis and host defense against pathogens. Oncogene 2001; 20(44):6473-81.
34. Ogura Y, Bonen DK, Inohara N et al. A frameshift mutation in NOD2 associated with susceptibility to Crohn's disease. Nature 2001; 411(6837):603-6.
35. Farmer RG, Hawk WA, Turnbull RB. Clinical patterns in Crohn's disease: A statistical study of 615 cases. Gastroenterology 1975; 68:627-39.
36. Ogorek CP, Caroline DF, Fisher RS. Presentation, evaluation and natural history of inflammatory bowel disease. Chapter 19. In: Inflammatory bowel disease. MacDermott RP, Stenson WF, eds. New York, Elsevier 1992:355-86.
37. Best WR, Becktel JM, Singleton JW et al. Development of a Crohn's disease activity index. National Cooperative Crohn's Disease Study. Gastroenterology 1976; 70:843-50.
38. Irvine EJ, Feagan B, Rochon J et al. Quality of life: A valid and reliable measure of therapeutic efficacy in the treatment of inflammatory bowel disease. Gastroenterology 1994; 106:287-96.
39. Guyatt G, Mitchell A, Irvine EJ et al. A new measure of health status for clinical trials in inflammatory bowel disease. Gastroenterology 1989; 96:804-10.
40. Farmer RG, Hawk WA, Turnbull RB. Clinical patterns in Crohn's disease: A statistical study of 615 cases. Gastroenterology 1975; 68:627-37.
41. Farmer RG, Whelan G, Fazio VW. Long-term follow-up of patients with Crohn's disease. Relationship of the clinical pattern and prognosis. Gastroenterology 1985; 88:1818-27.
42. Lapidus A, Bernell O, Hellers G et al. Clinical course of colorectal Crohn's disease: A 35-year follow-up study of 507 patients. Gastroenterology 1998;114:1151-60.
43. Loftus EV, Silverstein MD, Sandborn WJ et al. Crohn's disease in Olmsted county, Minnesota, 1940-1993: Incidence, prevalence, and survival. Gastroenterology 1998;114:1161-68.
44. Shivanando S, Hordijk ML, Pena AS et al. Risk of recurrence and reoperation in a defined population. Gut 1989; 30:990-95.

45. Sachar DB, Subramani K, Mauer K et al. Patterns of postoperative recurrence in fistulizing and stenotic Crohn's disease. J Clin Gastroenterol 1996; 22:114-16.
46. Lautenbach E, Berlin JA, Lichtenstein GR. Risk factors for early postoperative recurrence of Crohn's disease. Gastroenterology 1998;115:259-67.
47. Savoca PE, Ballantyne GH, Cahoe CE. Gastrointestinal malignancies in Crohn's disease: A 20 year experience. Dis Colon Rectum 1990; 33:7-11.
48. Friedman S, Rubin PH, Bodian C et al. Screening and surveillance colonoscopy in chronic Crohn's colitis. Gastroenterology 2001; 120:820-26.
49. Senay E, Sachar DB, Keohane M et al. Small bowel carcinoma in Crohn's disease. Cancer 1989; 63:360-63
50. Summers RW, Switz DM, Sessions JT Jr et al. National Cooperative Crohn's Disease Study: Results of drug treatment. Gastroenterology 1979; 77:847-69.
51. Rao SS, Cann PA, Holdsworth CD. Clinical experience of the tolerance of mesalazine and olsalazine in patients intolerant of sulfasalazine. Scand J Gastroenterol 1987; 22:332-36.
52. Bernstein LH, Frank MS, Brant LJ et al. Healing of perineal Crohn's disease with metronidazole. Gastroenterology 1980; 79:357-65
53. Bar-Meir S, Chowers Y, Lavy A et al. Budesonide versus prednisone in the treatment of active Crohn's disease. Gastroenterology 1998; 115:835-40.
54. Willoughby JM, Becket J, Kumar PJ et al. Controlled trial of azathioprine in Crohn's disease. Lancet 1971; 2:944-47.
55. Present DH, Korelitz BI, Wisch N et al. Treatment of Crohn's disease with 6-mercaptopurine. A long-term, randomized, double-blind study. N Engl J Med 1980; 302:981-87.
56. Markowitz J, Grancher K, Kohn N et al. A multicenter trial of 6-mercaptopurine and prednisone in children with newly diagnosed Crohn's disease. Gastroenterology 2000; 119:895-902.
57. Brynshov J, Freund L, Rasmussen SN et al. A placebo-controlled, double-blind, randomized trial of cyclosporine therapy in active chronic Crohn's disease. N Engl J Med 1989; 321:845-50.
58. Kozarek RA, Patterson DJ, Gelfand MD et al. Methotrexate induces clinical and histologic remission in patients with refractory inflammatory bowel disease. Ann Intern Med 1989; 110:353-56.
59. Targan S, Hanauer SB, van Deventer SJH et al. A short-term study of chimeric monoclonal antibody cA2 to tumor necrosis factor a for Crohn's disease. N Engl J Med 1997; 337:1029-35.
60. Present DP, Rutgeerts P, Targan S et al. Infliximab for the treatment of fistulas in patients with Crohn's disease. N Engl J Med 1999; 340:1398-1405.
61. Fedorak RN, Gangl A, Elson CO et al. Recombinant human interleukin 10 in the treatment of patients with mild to moderately active Crohn's disease. Gastroenterology 2000; 119:1473-82.
62. Neurath MF, Wanitschke R, Peters M et al. Randomized trial of mycophenolate mofetil versus azathioprine for treatment of chronic active Crohn's disease. Gut 1999; 44:625-28.
63. Tysk C, Lindberg E, Jarnerot G et al. Ulcerative colitis and Crohn's disease in unselected population of monozygotic and dizygotic twins: A study of heredibility and the influence of smoking. Gut 1988; 29:990-96.
64. Lopez-Cubera SO, Sullivan KM, McDonald GB. Course of Crohn's disease after allogeneic bone marrow transplantation. Gastroenterology 1998; 114:433-40.
65. Drakos PE, Nagler A, Or R. Case of Crohn's disease in bone marrow transplantation. Am J Hematol 1993; 43:157-58.
66. Kashyap A, Foreman SJ. Autologous bone marrow transplantation for non-Hodgkin's lymphoma resulting in long term remission of coincidental Crohn's disease. Br J Haematol 1998; 103:651-52.
67. Castro J, Benich HI, Smith HL. Prolonged clinical remission in patients with inflammatory bowel disease after high dose chemotherapy and autologous bone marrow stem cell transplantation. Blood 1996; 88:133A.
68. Hawkey CJ, Snowden JA, Lobo A et al. Stem cell transplantation for inflammatory bowel disease: Practical and ethical issues. Gut 2000; 46:869-72.
69. James SP. Allogeneic bone marrow transplantation in Crohn's disease. Gastroenterology 1998; 114:596-98.
70. Burt RK, Traynor A, Oyama Y et al. High dose immune suppression and autologous hematopoietic stem cell transplantation in refractory Crohn's Disease. Blood 2003; 101(5):2064-66.

CHAPTER 55

Bronchial Asthma and Idiopathic Pulmonary Fibrosis as Potential Targets for Hematopoietic Stem Cell Transplantation

Júlio C. Voltarelli, Eduardo A. Donadi, José A. B. Martinez, Elcio O. Vianna and Willy Sarti

Introduction

In this chapter we examined existing evidence supporting the application of hematopoietic stem cell transplantation (HSCT) for the treatment of severe and refractory cases of bronchial asthma or idiopathic pulmonary fibrosis (IPF). Both diseases share chronic inflammatory features mediated by bone marrow derived T cells, but there may be a contribution of local tissue factors to their pathogenesis. Very limited animal and clinical studies support the use of high dose immunosuppression and stem cell rescue in these diseases. Selected cases of near fatal asthma or progressive idiopathic pulmonary fibrosis may benefit from a clinical trial of less toxic HSCT, as seen in some forms of severe autoimmune diseases.

HSCT has been used for more than three decades to treat hematological and nonhematological disorders, including, more recently, severe autoimmune diseases.[1] The rationale for the use of HSCT in autoimmune disorders is manifold, as revised by Klassen.[2] Firstly, preclinical studies using HSCT in animal models of autoimmunity have suggested efficacy in preventing, reversing and even curing autoimmune disorders. Secondly, T lymphocytes have a central role in the pathogenesis of autoimmune diseases, and allogeneic hematopoietic cells from donors presenting with autoimmune disorders can transfer the same disease to the recipient. A corollary of these observations is that autoreactive progenitor cells may be replaced by normal hematopoietic stem cell repopulation. Thirdly, several lines of evidence have shown improvement or cure of a concomitant autoimmune disorder in patients submitted to HSCT for the treatment of hematological and nonhematological disorders. Fourthly, reconstitution of a new immune system may follow bone marrow or peripheral blood stem cell transplantation. Fifthly, increased therapeutic responses in autoimmune disorders are seen with increasing doses of immunosuppressive agents, and HSCT can be used both as a rescuing agent after chemotherapy and as a source of lympho-hematopoietic progenitor cells. Sixthly, immune mediated cytopenias may be treated with HSCT. Finally, stem cell therapy may allow repair of damaged nonhematopoietic tissue (refer to Chapters 10 and 11).

The relative success of HSCT in the treatment of systemic autoimmune diseases led to the hypothesis that the procedure could benefit patients with other forms of chronic inflammatory diseases, including lung diseases. This chapter reviews the possible application of HSCT for the treatment of bronchial asthma and idiopathic pulmonary fibrosis.

Hematopoietic Stem Cell Transplantation for Bronchial Asthma

Bronchial asthma is not considered an autoimmune disorder, but a chronic inflammatory disorder.[3] On the other hand, many autoimmune disorders do present with chronic inflammatory features,[4] and bone marrow derived cells, particularly T cells, are implicated in the pathogenesis of autoimmune diseases and bronchial asthma.[1,5,6] Although T cells have been implicated in the pathogenesis of both allergic bronchial asthma and autoimmune diseases, and although some susceptibility genomic regions are shared by these diseases,[7] the cytokine pattern observed in each condition has been shown to be different. Thus, in several autoimmune diseases including rheumatoid arthritis,[8] type 1 diabetes,[9] coeliac disease,[10] and others, T lymphocytes are usually polarized to the Th1 response, whereas in allergic bronchial asthma T lymphocytes are polarized to the Th2 response.[11] According to the Th1/Th2 paradigm, a negative association between diseases caused by Th1 or Th2 responses should be expected; however, the incidence of bronchial asthma in patients presenting with some autoimmune diseases of the Th1 phenotype has not been shown to be decreased.[12,13] The coexistence of Th1 and Th2-mediated diseases further support a link between these disorders.

Evidence supporting the use of HSCT for bronchial asthma includes: (1) inflammatory pathogenesis of asthma mediated predominantly by bone marrow derived cells (Table 1); (2) experimental and clinical studies showing transfer or cure of atopy and asthma by HSCT; (3) beneficial effect of immunosuppressive agents in severe asthma; (4) existence of a subset of patients with high probability of death from asthma (near fatal asthma) who would benefit from an aggressive immunomodulatory approach. On the other hand, there are some drawbacks of this proposal, including the strong influence of local factors on asthma pathophysiology, which may worsen pulmonary complications in asth-

Table 1. Inflammatory factors associated with the pathogenesis of bronchial asthma

Increased mRNA for Th2 cytokines (IL-3, IL-4, IL-5, IL-13) and IgE receptors
Decreased mRNA for Th1-inducing factors (T-bet, IL-12, IFN-γ)
Increased number of dendritic cells in bronchial mucosa
T cell mediated bronchial hyperresponsiveness
T cell mediated recruitment and activation of eosinophils, granulocytes, monocytes and lymphocytes

matic patients submitted to HSCT for hematological disorders (Table 2).

Pathogenesis of Bronchial Asthma

Approximately 90% of individuals under the age of 30 years who have asthma also have atopy,[14] a term which has been coined to allergic individuals who present a positive immediate skin test to common environmental allergens, and increased total serum IgE or specific IgE levels to these allergens.[15] The major difference between atopic and nonatopic individuals is mainly related to the type of immune response produced after the exposure to ubiquitous aeroallergens, mainly house dust mite, cockroach, fungal, pollen, and animal dander allergens. Nonatopic individuals produce IgG1 and IgG4 antibodies against aeroallergens, exhibiting an immunologic response mediated by polarized Th1 cells, and present no airway-related symptoms to a later exposure to these allergens. In contrast, atopic asthmatic individuals produce high amounts of IgE against aeroallergens, exhibit an immunologic response mediated by polarized Th2 cells, and react with bronchoconstriction to a later exposure to allergens.[16]

Genetic and environmental factors interact to induce immunological and inflammatory changes that culminate in the production of airway obstruction, bronchial hyperresponsiveness and chronic airway inflammation, the shared features among asthmatic patients. Multiple genes and multiple genome regions have been associated with susceptibility to bronchial asthma, including several cytokines (IL-3, IL-4, IL-5, IL-9, IL-13) and β2-adrenergic receptor genes, which are located in chromosome 5q; the major histocompatibility complex (MHC) genes, on chromosome 6q; the high-affinity receptor for carboxy terminal constant region of IgE molecule (FcεRI) gene,[17,18] and the Th1-specific T-box transcription factor (T bet) gene,[19] on chromosome 11; interferon gamma (IFN-γ), nitric oxide synthase-1 and mast cell growth factor genes, on chromosome 12q; nuclear transcription factor (NF-κB) gene, on chromosome 14q; and others.[17,18] These studies indicate that major gene components associated with bronchial asthma are responsible for the coding of proteins highly implicated in the development of diseases characterized by Th2 cell polarization, and most of these genes are located outside the MHC. The exposure of genetically susceptible individuals to environmental allergens may induce a presymptomatic state of sensitization, characterized by a Th2-mediated inflammatory response to allergens, which in turn may induce chronic airway inflammation and bronchial hyperresponsiveness. This condition, referred as asymptomatic asthmatic state or intermediate phenotype, is not sufficient to trigger bronchial asthma, but may provide adequate conditions for progression to asthma.[20]

Bronchial hyperresponsiveness is defined as an increased ability of the airway to narrow its caliber after exposure to nonspecific stimuli, including bronchoconstrictor pharmacologic agonists, such as histamine, acetylcholine, methacoline, and many other stimuli. After nonspecific stimuli provocation, patients presenting with bronchial hyperresponsiveness exhibit a 20% fall in the forced expiratory volume in the first second (FEV_1). Usually, the magnitude of airway hyperresponsiveness correlates with the severity of asthma and with variations of the peak expiratory flow rate. An improvement in FEV_1 may be observed after the inhalation of bronchodilators. The development of bronchial hyperresponsiveness in asthmatics has been associated with persistent airway inflammation, mainly caused by the activation of inflammatory cells such as mast cells, eosinophils, neutrophils and lymphocytes. Although the mechanisms responsible for airway hyperresponsiveness are not completely understood, the consequences of the persistent inflammation include airway wall thickening, loss of airway epithelium, airway edema, and altered airway smooth muscle function.[21]

The study of bronchoalveolar lavage fluid and bronchial biopsy obtained from allergic asthmatic patients shows a large number of lymphocytes, predominantly of the Th2 phenotype, expressing mRNA for IL-3, IL-4, IL-5, IL-13, granulocyte-monocyte colony stimulating factor (GM-CSF),[22-26] and mRNA for GATA-3, a transcription factor confined to Th2 cells.[27] Most of the mediators arisen from chronically activated airway CD4+ T cells may recruit eosinophils, monocytes, granulocytes and other lymphocytes.[11] These findings, together with those related to the

Table 2. Pros and cons of HSCT in bronchial asthma

PROS	References
Role of BM derived T cells in atopy, bronchial hyperresponsiveness and inflammation	5,6,19,60
Therapeutic response of asthma to immunosuppressive agents	79-82
Transfer of atopy/hyperresponsiveness by clinical and experimental HSCT	60,62
Existence of a subset of patients with near fatal asthma	68,69
CONS	**References**
Experimental and clinical lung transplantation suggest role of local factors in asthma	54-56
Some patients with asthma or atopy have pulmonary complications after HSCT	64,65
Unknown effects of HSCT on asthma patients transplanted for hematological diseases in large series	IBMTR
Most patients with severe asthma improve with optimal conservative treatment	77

increased expression of genes associated with the Th2 cytokine profile, has led to the proposal that T lymphocyte is the major orchestrating cell of the asthmatic response.[28]

The induction of a Th2 immunologic response needs 3 known signals. The first is the recognition of allergen peptide by the CD4+ T cell receptor (TCR) presented in association with an MHC class II molecule by a professional antigen presenting cell (APC).[29] Dendritic cells are considered to be the major APC in airway mucosa of patients with asthma,[30] reside primarily within the airway epithelium in close contact with inhaled allergen, and the interaction of dendritic cells of atopic patients with allergen favors local activation of polarized Th2 cells.[31] Although no strong associations between MHC class II and TCR genes have been reported in patients with asthma,[32] the number of dendritic cells expressing the α-chain of the high-affinity IgE receptor (FcεRI-α is increased in relation to nonasthmatic individuals.[33] The second signal is generated after the interaction of the costimulatory CD28 molecule on CD4 cell and CD80 (B7-1) and CD86 (B7-2) molecules on APC.[34] The blockage of CD28/B7 signaling by anti-CD80 or anti-CD86 antibodies inhibits the allergen-induced secretion of IL-5 and IL-13 by bronchial tissue of atopic asthmatics after ex vivo allergen stimulation.[35] In addition, interactions of CD28 with CD80 and CD86 molecules are reported to induce Th2 cell activation, and interaction of CD28 with CD86 is also associated with the development of antigen-induced airway hyperresponsiveness.[36] The conditions in which antigen priming occurs provide the third signal to the polarization towards the Th2 phenotype, a phenomenon which has been basically related to the influence of the local mediator milieu. There is no doubt that antigen priming in the presence of IL-12 promotes the diferentiation of naive CD4+ T cells to polarize towards Th1 cells, which in turn produce IL-2 and IFN-γ. On the other hand, IL-4 is critical for T cells to polarize towards Th2 cells, which produces IL-4, IL-5, IL-9 and IL-13.[29] Although mast cells and eosinophils produce IL-4, mouse models indicate that the major endogenous source of this cytokine comes from T cells.[37] Moreover, IL-4, prostaglandin E_2 (PGE_2) and nitric oxide produced by APCs may induce Th2 polarization by inhibiting the production of IL-12.[38,39] The balance between Th1 and Th2 cells may be influenced by IFN-γ, i.e., the development of the Th2 phenotype is induced in the absence of IFN-γ whereas Th1 cells are induced in the presence of IFN-γ.[11,40] The early polarization by IL-12 has been considered to be a crucial component associated with the third signal. At high IL-12 concentration, T cells polarize to the Th1 phenotype, whereas, at low concentration of IL-12, T cells polarize to the Th2 phenotype. IL-18 additionally acts as a Th1 polarization factor by inducing IFN-γ production by prepolarized Th1 cells.[40] Besides the role of IL-12 in T cell polarization, a Th1-specific T-box transcription factor (T-bet) has been recently described as having the unique ability to redirect fully polarized Th2 cells into Th1 cells by transactivating the IFN-γ gene in Th1 cells.[41] Mice presenting with a target deletion of T-bet spontaneously develop many physiological and inflammatory features of asthma.[19]

Several lines of evidence point to the role of deficiency of IL-12 in the development of bronchial asthma.[29] Decreased mRNA expression for IL-12,[42] and increased IL-4 and IL-13 mRNA expression have been reported in bronchial biopsies of patients with asthma.[23-25] The decreased IL-12 production by peripheral monocytes is associated with a decreased production of IFN-γ in patients presenting with bronchial asthma.[43] Airway hyperresponsiveness and eosinophil-mediated inflammation are abolished when recombinant IL-12 is administered together with allergens in a mouse model of bronchial asthma.[44] Based on the assumption that dendritic cells, which are the major producers of IL-12, may develop and mature in different conditions, Kalinski et al[45] proposed that dendritic cells may be functionally different, i.e., they may carry an additional signal 3 contributing towards the commitment of naive T cells into Th1 or Th2 subsets. Dendritic cells of type 1 would induce the development of Th1 cells, whereas dendritic cells of type 2 would induce the development of Th2 cells.[46] The same group of investigators have questioned whether the preferential polarization of APCs from atopic individuals might be due to an intrinsic aberrancy of APC or the result of imposed extrinsic factors.[29] A corollary of these studies is that dendritic cells and T cells are key elements involved in the initiation and maintenance of the activation of Th2 cells.

The major consequence of the activation of Th2 cells in allergic bronchial asthma is the B cell gene switch to produce IgE in response to ubiquitous environmental allergens. The initial interaction of B and Th2 cells depends on the recognition of an allergen peptide in association with an MHC class II molecule on the B cell membrane, followed by the CD28/B7 costimulation, as occurs for other APCs. Two further signals are necessary to induce the switching of antibody synthesis from IgM to IgE, the first is generated after the binding of CD40 molecule present on B cell surface to the CD40 ligand (CD154), expressed on Th2 cells.[47] The interaction of these molecules leads to activation of the nuclear transducer NFκB, one of the transcription factors required for the production of IgE.[48] The activated Th2 cells secrete IL-4 and IL-13, the most important inducers of the production of IgE,[47] and then, provide the second signal to the switching to IgE. The binding of IL-4 or its homolog IL-13 to specific receptors (IL4R and IL-13R, respectively) on B cell membrane leads to the activation of another transcription factor, the signal transducer and activator of transcription 6 (STAT-6).[48] Once formed, IgE antibodies briefly circulate in the blood before binding to the FcεRI-α on mast cells and basophils. The cross-linking of allergens to IgEs on the membrane of mast cells and basophils triggers the release of preformed mediators, including histamine, tryptase and chymase, and newly-formed membrane-derived lipid mediators, including leukotrienes (LTA_4, LTB_4, LTC_4, LTD_4 and LTE_4), prostanoids (PGD_2, thromboxane A_2), and platelet-activating factor.[49] Mast cell-derived cysteinyl-leukotrienes (LTC_4, LTD_4 and LTE_4) are considered to be the major mediators of the early airflow obstruction, also known as early asthmatic response, which occurs immediately after allergen exposure and subsides within one hour. Approximately half of asthmatic patients also present a four to six-hour late airflow obstruction, also known as late asthmatic response. The late response develops as a consequence of the production of cytokines and chemokines by mast cells, macrophages, epithelial cells, lymphocytes and eosinophils. Besides producing cysteinyl leukotrienes, eosinophils also contribute to the late asthmatic response by releasing granule-associated proteins such as major basic protein, eosinophil-derived neurotoxin, peroxidase and cationic proteins.[11,16] In concert or individually, the myriad of preformed and newly formed mediators released after allergen exposure contributes to the development of chronic airway inflammation, airway hyperresponsiveness and bronchoconstriction, pathologic features associated with the production of respiratory crises manifest by dyspnea and wheezing.

Bone Marrow Derived Cells and Local Factors in the Pathogenesis of Asthma

In order to evaluate the possible role of HSCT in bronchial asthma, one should know whether pathogenic features of the disease are exclusively dependent on cells that arise from bone marrow or whether these pathogenic features depends on the presence of lung factors which activate bone marrow-derived cells, or both mechanisms. If one considers that bronchial asthma is directly and exclusively dependent on bone marrow derived immune cells, autologous HSCT would be a reasonable alternative to treat severe chronic asthma. On the other hand, if we consider that bronchial asthma is an organ specific disease, then if allogenic HSC can differentiate into pulmonary parenchyma, an allogenic HSCT may be an option. Otherwise a lung transplant with or without HSC would be required to "cure" asthma.

Several lines of evidence support the central role of T cells in the development of atopic clinical features of atopic manifestations, bronchial hyperresponsiveness, and persistent chronic inflammation, although other bone marrow-derived cell types may also contribute to the pathogenetic events of allergic bronchial asthma. Th2 memory cells predominate in bronchoalveolar lavage fluid and bronchial biopsies, being the major source of IL-4, IL-13 and IL-5.[22-26] Th2 cytokines recruit eosinophils and are associated with IgE synthesis.[5,47] Nude mice or T cell-depleted mice do not develop eosinophilic infiltration of airways following allergen challenge.[6] Mice presenting with a target deletion of the Th1 transcription factor (T-bet) present increased amounts of IL-4 and IL-13 in bronchoalveolar lavage fluid compared to mice without this deletion. The adoptive transfer of CD4+ cells from T-bet deficient mice into a histocompatible severe combined immunodeficiency (SCID) mouse induced an increased production of IL-4 in broncoalveolar lavage fluid of the SCID mouse.[19] In addition, the treatment of bronchial asthma with glucocorticoids which reduce peripheral number of T cells is associated with reduction of T cell numbers in bronchial mucosa and with clinical improvement.[50] Immunomodulatory drugs such as cyclosporin A, mycophenolate mofetil, rapamycin, and dexamethasone inhibit allergen-driven T cell proliferation and IL-5 production,[51] and the treatment of severe asthmatic patients with cyclosporin A is associated with improved lung function.[52] Finally, in human asthma, besides all the reported descriptions stressing the role of Th2 cells and of Th2-induced mediators in the development of asthma, a reduced expression of the Th1 transcription factor, T-bet, is observed in T cells recovered from airways of patients presenting with allergic bronchial asthma.[19] The latter findings are suggestive that intrinsic T cell defects may be present in patients with atopic bronchial asthma, and these defects may be reverted with HSCT, as have been reported for many immunodeficiency and immune-mediated disorders.[53]

In support of the concept that local mechanisms are responsible for the development of asthma, lessons from clinical and experimental lung transplantation have emphasized the idea that bronchial asthma may be a localized lung disorder. In humans, Corris and Dark[54] reported that two severely asthmatic atopic patients who received normal lungs did not develop asthma up to three years after transplantation. In addition, these patients did not have diurnal variation in peak expiratory flow rate in the first six months, and did not present bronchial hyperresponsiveness to nebulized histamine. Conversely, two other nonasthmatic recipients who were transplanted with lungs from donors presenting with mild bronchial asthma developed diurnal variation in peak flow expiratory rate and airflow obstruction in the first week after transplantation, which persisted despite the treatment with cyclosporin, azathioprine and prednisolone. Posttransplant transbronchial biopsy findings showed pathological evidence compatible with bronchial asthma, which were worst than those observed in pretransplant biopsy.

Denburg et al[55] studying in vivo and in vitro allergic inflammation reported that airway-derived factors may provide favorable conditions to up or down-regulate differentiation of bone marrow or blood progenitor cells, particularly of eosinophil/basophil progenitors. The number of these cells is constitutively increased in blood and bone marrow of patients presenting with atopy and asthma, and further increased during seasonal allergen exposure or after allergen challenge. In contrast, the number of eosinophil/basophil progenitors decreased in response to treatment with inhaled corticosteroid. In addition, in vivo studies in dogs showed similar results with respect to the number of eosinophil/basophil progenitors. The authors proposed that cytokines released by inflamed airway structural cells may provide a communication between the allergic response observed in the lungs and the bone marrow-peripheral blood axis. Taken together, these results suggest that some mechanisms responsible for asthma would be localized in the lungs despite the possible presence of some systemic components.

Experimental lung transplantation has also shown interesting results. When a single lung lobe of a donor dog is immunized by the local instillation of the neoantigen keyhole-limpet hemocyanin and implanted into a nonsensitized dog, the production of anti-hemocyanin antibody is observed only in the implanted sensitized lobe. Anti-hemocyanin antibody is detected in the circulation as long as 320 days after the single lobe transplantation. According to these results, the authors concluded that immune cells in lung tissue can function independently from cells of the systemic circulation, and that donor lung immune cells continued to produce antibodies for a long time after transplantation.[56]

The issues discussed above permit us to conclude that both local and systemic factors may operate in bronchial asthma. Based on the concept that T cells recruit other relevant allergic inflammatory cells,[5,6,11] produce inflammatory changes seen in allergic airways,[11,22-26,28] may transfer bronchial hyperresponsiveness,[6] and may have particular intrinsic defects in atopy,[19] it is reasonable to propose HSCT as an alternative or even as a curative treatment for bronchial asthma patients, although many questions remains to be answered, especially those related to asthma as a localized lung disease. The observations that a transplanted asthmatic lung into a normal subject continue to produce asthmatic symptoms raises an important question of whether dendritic cells, which may survive longer than other cells in the transplanted lung, might also be as important as T cells. The issue of distinct dendritic cell subsets (DC1/DC2), and inactivation of DC with monoclonal antibodies pretransplantation and whether HSC can differentiate into lung parenchyma are interesting subjects to be further explored. The systematic evaluation of the outcome of moderate and severe asthma after HSCT for other diseases would also be informative.

Experimental and Clinical Studies Associating Asthma and HSCT

The mechanisms which induce, maintain and aggravate bronchial asthma have been extensively studied with the aim of improving or preventing asthmatic attacks; however, the cure of the disease has not yet been accomplished. The observation that an inverse correlation has been observed between the infection rate and the prevalence of asthma has led to the proposal that an early-life induction of a Th1-mediated response might decrease the prevalence of allergy,[57,58] and consequently of asthma. According to this "hygiene hypothesis", infection with *Mycobacterium tuberculosis*, measles and hepatitis A viruses, and living in a rural environment or with older siblings are important factors associated with decreased prevalence of allergy. This hypothesis seems to be logical, since the population of newborn T cells are preferentially polarized to Th2 cells, and an early exposure to infectious agents might induce a Th1 polarization.[59] If one could shift the polarization from Th2 to Th1 cells, in theory, one would expect a possible cure for allergy. Besides changing the T cell polarization, one would also achieve the modification of the atopic profile, the reversal of persistent airway inflammation, and finally the reversal of airway hyperresponsiveness, to finally reach the cure for allergy and asthma. Nowadays, one possible way to modify all these mainstays of bronchial asthma pathogenesis is with HSCT; however, there are few studies available regarding the effects of cell transplantation on these mechanisms to further decide whether HSCT is a valid alternative for the treatment of bronchial asthma.

Experimental studies indicate that the transference of progenitor cells may modify a preexisting allergic condition. DeSanctis et al[60] investigated the role of genetic factors on bronchial hyperresponsiveness studying two strains of mice, one presenting with an asthma-like bronchial hyperresponsiveness to methacholine and the other lacking this hyperresponsiveness. The genetically determined bronchial hyperresponsiveness observed in C57BL/6XA "asthma-like" mouse is transferred to the hyporesponsive C57BL/6 "nonasthma-like" mouse by a T-cell enriched cell preparation. In addition, the previous treatment with anti-T cell monoclonal antibody prevents the transference of methacholine hyperresponsiveness to the recipient mouse. Besides stressing the role of genetic factors associated with bronchial hyperresponsiveness, the authors emphasize the pivotal role of T cells on bronchial hyperresponsiveness, even in the absence of a defined antigen challenge or previous tissue inflammation. This animal model indicates that genetically determined hyperresponsiveness may be overcome by cell transference, which may repopulate lung tissue and modify bronchial responsiveness even in the absence of identified environmental influences.

Clinical studies have shown that atopic manifestations can be transferred by HSCT. A five year old boy without a previous history of atopy developed severe atopic dermatitis and food allergy after treatment with HSCT obtained from his HLA-identical sister.[61] Because donor and recipient were HLA identical, it is possible that they also shared other atopy-related genes, which might have facilitated the transference of atopic manifestations. There is only one study systematically evaluating the transference of allergic manifestations after an HSCT performed to treat hematological cancers. Agosti et al[62] studied 12 donor-recipient pairs before and after HSCT, with respect to atopic symptoms, immediate skin test reactivity for aeroallergens, and presence of specific IgE antibodies. Considering the 11 recipients who survived for more than one year after transplantation, 8 patients became positive to allergens for which pretransplantation skin test had been negative in recipient and positive in donors. Seven recipients either acquired or had an exacerbation of allergic rhinitis, 2 recipients without a previous history of asthma developed bronchial asthma one year after transplantation, and in another recipient a preexisting asthma was aggravated. In addition, a long term transfer of donor derived mite-specific IgE antibody was demonstrated in the serum of 2 recipients. These findings are indicative of the transference of skin sensitization and cell sensitization from donor to recipient; however, some points should be stressed. Recipients who had previous negative skin tests and received hematopoietic stem cells from skin test positive donors turned positive to some aeroallergens. This posttransplant sensitization may be due either to natural exposure to allergens as well as to events frequently observed after HSCT favoring further polarization to Th2 cells, including down-regulation of Th1 cells induced by acute graft-versus-host disease (GVHD), immunosuppression and cytomegalovirus infection. Additionally, we should take into account that all donors were siblings, enhancing the chance that both donors and recipients may share some atopy genes. Overall, these findings indicate that allergen-specific IgE hypersensitivity can be passively transferred by HSCT from donor to recipient by allergen-specific memory B cells, corroborating the classical findings of Landsteiner and Chase who, in 1942, reported the transference of delayed hypersensitivity by blood cells.[63]

Unfortunately, there are no systematic studies of allergic reactivity or clinical outcome of asthmatic or atopic patients transplanted with nonatopic donors. In addition, no information is available with respect to the outcome of an underlying severe bronchial asthma in patients subjected to HSCT for the treatment of a superimposed disease. Data from the International Bone Marrow Transplant Registry (IBMTR) show that only about 1% (130/11337) of leukemic patients who received HSCT from an HLA identical sibling reported asthma as a preexisting condition. However, no information is available regarding the outcome of bronchial asthma after transplantation, except that the survival rate of asthmatic patients is closely similar to nonasthmatic ones. In only 27 patients, bronchial asthma was reported as a posttransplant complication. In addition, no cases of HSCT have been reported for the specific treatment of bronchial asthma (The data presented here were obtained from the Statistical Center of the IBMTR/ABMTR. The analysis has not been reviewed or approved by the Advisory Committees of the IBMTR/ABMTR). On the other hand, obstructive lung disease has been associated with a previous history of asthma in allogeneic HSCT performed to treat hematological disorders,[64] and a history of atopy was highly predictive of idiopathic pulmonary complications in autologous HSCT.[65]

Overall, these studies have shown that atopic manifestations,[61,62] bronchial hyperresponsiveness,[60] immediate skin-test sensitization and cell sensitization[62] can be adoptively transferred from donor to recipient by hematopoietic progenitor cells. As a corollary, the transference of nonatopic progenitor cells or the transference of manipulated atopic progenitor cells may treat bronchial asthma by repopulating bone marrow microenvironment, lymphoid organs and tissues. Notwithstanding, many questions still have to be addressed and many problems have to be solved or circumvented to achieve those objectives (Table 2). Among these questions, we would emphasize: Who should be the candidates for HSCT? What type of HSCT should be indicated: syngeneic, related allogeneic, unrelated allogeneic, myeloablative, nonmyeloablative or subablative, autologous or even cord blood cells? May complications of HSCT overcome possible benefits? What are future directions to reach these objectives?

Who Would Be the Candidates for HSCT?

The prevalence of asthma has risen drastically in the last two decades, with a worldwide impact in health care systems. The increasing prevalence rate throughout the world since the 1960s is probably twofold to threefold.[66] In regards to mortality, there is also a great concern related to the apparent increased asthma related death rate as observed in developed countries, where the mortality rate presently stands at between 1.0 and 1.5 per 100,000 of the general population.[67]

The majority of patients with bronchial asthma can be classified as having mild to moderate asthma and are adequately treated with topical steroids, dissodium cromoglycate and bronchodilator drugs; however, some patients develop severe or even fatal asthma. Many predisposing factors are described in association with fatal asthma, which may be dependent on patient, doctor and family characteristics, and on features of asthma itself.[68] An association of several of these features has been considered to determine the severity of asthma that, in the most severe cases, has been named near fatal asthma.

Due to the increasing asthma mortality worldwide, the circumstances surrounding death from asthma have been a major subject for investigative and public health care purposes. It would be helpful for clinicians to identify patients who are at risk of death. Some studies have focused on the evaluation of patients with a history of near fatal attacks. In a recent survey performed in Australia, the clinical characteristics of patients who had near fatal attacks were similar to those of patients who died.[69] Near fatal asthma can be defined as attacks of asthma requiring treatment with mechanical ventilation or resulting in unconsciousness and severe respiratory failure along with retention of carbon dioxide.

Turner et al[70] studied 19 patients admitted with near fatal asthma in order to evaluate historical and physiologic features associated with the risk of development of life-threatening asthma.[70] Their findings confirmed the results obtained in other studies which identified previous admissions to intensive care units and prior need of mechanical ventilation as strong predictors for near fatal asthma.[71,72] Although inhaled corticosteroids were prescribed to 89% of these patients, compliance with this therapy was observed only in 29% of the cases. In addition, the authors suggested a gender difference in perceiving dyspnea episodes, i.e., more male patients than females would only experience dyspnea when the degree of airway obstruction is already severe. Indeed, one study found that asthmatic males may be at a greater risk of sudden asphyxic death (see below) due to a precipitous decline in lung function.[73] On the other hand, Kikuchi et al[74] showed a blunted perception of severe dyspnea among patients with near fatal asthma, but they did not report a gender difference in this regard. These observations have important implications for developing strategies to prevent death from acute asthma. Physicians should know that reliance on patient's own assessment of his or her condition, without an objective determination of airway narrowing, carries a risk of undertreatment, which may lead to death.

In a cohort of 2242 subjects admitted for asthma, 85 deaths occurred within three years of discharge. The failure to prescribe inhaled steroids on discharge was associated with an increased risk of subsequent death.[75] More data on long-term survival of patients with near-fatal asthma and on the efficacy of therapeutic interventions targeted to this high-risk population are still required. Most deaths follow a period of unstable and deteriorating control of asthma and pathological examination of the lungs shows

Table 3. Predisposing factors for fatal asthma attacks

Prior history of mechanical ventilation
Previous admission to intensive care units
Failure to use inhaled steroids
Blunted perception of airway obstruction
Exposure to aeroallergen
History of sudden onset asthma attack
Psychosocial problems

an intense inflammatory response with widespread mucous plugging.[76] A small proportion of patients die from asthma attacks that progress from minimal symptoms to respiratory arrest within one to two hours, a condition which has been named "sudden asphyxic asthma".[73] In several retrospective surveys, this subset accounts for 10 to 25% of deaths from asthma which occurred within three hours after the onset of an attack. Pathological examination of the lungs showed little inflammation and it is assumed that the patients died from extreme airway narrowing.

Even knowing the predisposing factors for fatal attacks (Table 3), we have also to raise the troublesome issue of what to do with such patients once they are identified. Obviously, one must endeavor to prevent a recurrence, but how? In most cases, the treatment of aggravating comorbidities, and the prescription of systemic steroids and/or topical steroids have been enough to treat such patients. Otherwise, there is a small group of patients that do not improve and are kept at death risk despite treatment with high dose of systemic steroids,[77,78] immunosuppressive agents such as methotrexate, cyclophosphamide or cyclosporin,[79,80] or even immuno-modulators such as intravenous human gamma-globulin.[81,82] Taken together, those facts indicate that patients at death risk; i.e., patients with near fatal asthma, should be considered as candidates for more aggressive therapies including HSCT.

What Type of HSCT Would Be the Most Beneficial for Bronchial Asthma?

Syngeneic HSCT provides the most safe conditions to treat nonmalignant disorders, since rejection and graft versus host disease (GVHD) almost never occur, and the survival rate is usually high. HLA-identical siblings share major histocompatibility genes; however, many minor histocompatibility genes are not shared, and this fact is associated with important alloimmune reactions such as graft rejection, acute and chronic GVHD and graft-versus-tumor or graft-versus-autoimmunity effect.[1,83,84] With respect to the best HSCT donor for bronchial asthma, a syngeneic, an autologous, and maybe a related HLA-identical sibling donor also possess many atopy-related genes. Indeed, it has been reported that HSCT from HLA-identical sibling donors can transfer IgE response, allergic rhinitis and bronchial asthma to recipients,[61,62] leading to the acquisition of different atopic diseases or exacerbation of previous ones. In addition, the concordance rate for the development of bronchial asthma among monozygotic twins is high, and the frequency of bronchial hyperresponsiveness without asthma in siblings of asthmatic patients is also high.[85,86] Animals that spontaneously develop autoimmune diseases are not cured by a syngeneic HSCT, and the disease may be transferred to a normal strain of mouse by HSCT from the autoimmune-prone donor.[87] Curing a spontaneous onset autoimmune-like disease requires allogeneic HSCT from a nonautoimmune-prone donor.[88,89] These situations regarding autoimmune diseases would be expected after syngeneic or related allogeneic HSCT, since donor and recipients presented the same or similar allergic gene profile. Therefore, with respect to bronchial asthma, it is possible that a HSCT performed with a syngenic, autologous or sibling donor may transfer progenitor cells committed to develop or worsen allergic symptoms including asthma.

In theory, an HSCT performed between an HLA identical but unrelated and nonatopic donor should be an adequate donor, since these individuals do not present atopy-associated genes. Thus, normal progenitor cells might repopulate bone marrow, and bone marrow-derived cells including normal T and dendritic cells may repopulate tissues, including the lungs. The observation that lung transplantation from an asthmatic patient to a nonasthmatic patient is followed by asthma symptoms[54] suggest that resident airway cells or airway-derived mediators may be responsible for the maintenance of asthma. Those cells could be eliminated by transplanted allogeneic cells in a graft-versus-allergy or graft-versus-atopy effect.

HSCT using cord blood cells has been used for treatment of several disorders and has many advantages in part due to the naivity of alloreactive T cells of the graft.[90] However, as a consequence of several trophic factors derived mainly from the placenta, the population of T cells recovered from newborn infants are skewed towards the Th2 polarization.[59] Therefore, it is possible that HSCT performed with unmanipulated cord blood cells may worsen a preinstalled Th2-mediated disease. Notwithstanding, manipulation of cord blood cells in order to stimulate or upregulate Th1-mediated response might be of benefit. Cord blood cells from an infant born from nonatopic parents, and manipulated with IL-12 would be a an interesting perspective.

Autologous HSCT has been successfully performed in many autoimmune disorders with or without T cell manipulation.[1] Autologous HSCT is associated with very low mortality (<5%) and it has demonstrated long term control of disease activity for many autoimmune diseases. Mobilization of peripheral blood stem cells from the bone marrow of autoimmune patients has been accomplished by the administration of the single agent granulocyte colony-stimulating factor (G-CSF) or by combining G-CSF plus cyclophosphamide. The major complications of these mobilization regimens, including infections and disease flares, seem to be specific for the underlying autoimmune disease and dependent on the extent of organ involvement.[1] Part of those complications are due to inflammatory and Th2 skewness caused by G-CSF. In addition, ex vivo lymphocyte depletion from the graft has been used in most centers, while in others unmanipulated grafts associated with in vivo depletion with anti-T cell antibodies has been employed.[1] Regarding the possible use of autologous HSCT in bronchial asthma, one would also expect relapses of bronchial asthma during G-CSF mobilization to increase the amount of peripheral blood stem cells, as observed for autoimmune diseases.

In spite of theoretical considerations, fully ablative allogeneic HSCT, especially from nonrelated donors, would be too risky to be used in a starting clinical trial for bronchial asthma. The risk of complications could be reduced using low-intensity conditioning regimens (mini-allo or subablative transplants). Likewise, cord blood transplantation, even HLA-identical, does not provide optimal engraftment for adult patients in most cases. Thus, based on the evidence presented above, the best choice of HSCT for a pilot trial in bronchial asthma would probably be an autologous or mini-allo transplant with in vivo or in vitro T cell depletion.

Conclusions: Risks and Benefits of HSCT for Bronchial Asthma

Bronchial asthma and autoimmune diseases share several pathogenetic features, and the rationale for the use of HSCT for these diseases is analogous. Among the similarities we may include the pivotal role of T lymphocytes, the sharing of some susceptibility genes, the persistent inflammatory features, the improvement after immunosuppressive treatment, the coexistence of these diseases in the same patient, evidence from animal models indicating adoptive transference of the disease features by progenitor cells, and, finally, the transfer of autoimmune or atopic diseases by donor cells into recipients undergoing HSCT for malignant disorders. However, there are some dissimilarities between asthma and autoimmune diseases: different patterns of T helper cell polarization (Th1/Th2), the existence of a possible intrinsic defect of T cell polarization in atopic bronchial asthma, which has not been described for autoimmune disorders, and very importantly, the presence of a localized persistent inflammation in bronchial asthma. Although this localized persistent inflammation seems to be a major obstacle to HSCT for severe bronchial asthma, evidence from lung transplantation and in some systemic autoimmune disorders, including multiple sclerosis and rheumatoid arthritis in which the major inflammatory alterations are also observed in particular tissues, i.e., myelin and synovium, HSCT appears beneficial. One of the most compelling rationale for the use of HSCT in autoimmune diseases, which is the well-described cure of coincidental autoimmunity seen after HSCT for hematological and nonhematological disorders, has not been systematically evaluated in recipients of HSCT who present with bronchial asthma as a comorbidity. It would be of great interest to evaluate the outcome of bronchial asthma, particularly among recipients treated with HSCT obtained from nonrelated HLA-identical donors. Although much progress has been obtained from the fast growing clinical experience with HSCT in autoimmune diseases, little is known about the possible use of HSCT in bronchial asthma. Despite these caveats, and based on the available information, we believe that HSCT may be a plausible indication for severe bronchial asthma refractory to optimal medical treatment. The input from experts in respiratory medicine, allergy, clinical immunology, and HSCT would be very helpful to discuss this kind of treatment as an option for bronchial asthma.

If one suggests a transplantation treatment, whether bone marrow or lung transplantation, it would be necessary to consider the risk/benefits of both procedures. The risk is related to the chance of death by asthma that is high among those selected patients with near fatal asthma. On the other hand, there is a risk related to the transplantation procedure itself. If one consider the magnitude of the body invasion and/or the grade of trauma of lung transplantation, it appears that the bone marrow transplantation would be a better choice. However, as stated before, there are important clues on both sides and maybe only the clinical experience will decide if a bone marrow or a lung transplantation is the best choice to treat near fatal asthma.

HSCT for Idiopathic Pulmonary Fibrosis

Idiopathic pulmonary fibrosis (IPF) is an interstitial lung disease of unknown cause, most of the time associated with a poor prognosis.[91] IPF shares a substantial number of clinical, roentgenographic and physiologic features with other idiopathic interstitial pneumonias and many studies imprecisely included different histopathological patterns under such denomination. However, it is now clear that the clinical label IPF should be reserved for patients with a specific form of fibrosing interstitial pneumonia denominated usual interstitial pneumonia (UIP).[92] The cause, or causes, of IPF are presently unknown, but cigarette smoking has been identified as a potential risk factor for the disease. Several viruses have also been implicated in the pathogenesis of IFP, but no clear evidence points to a viral etiology. A familial form of the disease provides evidence for an important role of genetic mechanisms, however, the involved factors are still obscure. The IPF annual incidence has been estimated at 10 and 7 cases per 100,000 for males and females, respectively, on the basis of US registry data, while population-based studies put the prevalence between 3 and 20 cases per 100,000.[93] The disease generally strikes subjects in the fifth and sixth decade of life.

Patients with IPF typically complain of exertional dyspnea and nonproductive cough. As the disorder progresses, the patients become greatly impaired and show a very poor quality of life, mainly due to disabling dyspnea.[94] Typical chest roentgenographic findings in IPF are bilateral reticular opacities, most prominent in the periphery and lower lobes of the lungs. Progressive fibrosis leads to dilatation of the distal air spaces, which is seen as peripheral honeycombing. A definitive diagnosis of IPF requires a compatible clinical history, the exclusion of other known causes of interstitial lung disease such as environmental exposures or collagen vascular diseases, and a surgical lung biopsy showing UIP.[91] Median survival of newly diagnosed subjects with IPF is about three to four years, although a less aggressive evolution of the disease may be observed sometimes.[91,95,96] Therapeutics of IPF has been based in the use of corticosteroids and immunosuppressive agents such as cyclophosphamide and azathioprine.[97] Unfortunately, less than 30% of the patients show objective evidence of improvement or better survival with this approach. IPF patients typically are good candidates for lung transplantation and usually show remarkable improvements in quality of life just after the organ transplant. However, there is a shortage of lungs in satisfactory condition to be transplanted, and a significant number of patients die while on the waiting list. In addition, most of lung transplant recipients will develop bronchiolitis obliterans, and the mean survival for this group of patients after the surgery approaches that of the disease itself.[98] It is a consensus among clinicians that novel therapeutic options for IPF need to be developed.[99]

Pathogenesis of Idiopathic Pulmonary Fibrosis

Although our knowledge about the mechanisms involved in the development of pulmonary fibrosis have substantially increased along the past decades, the pathogenesis of IPF is still unknown. A substantial amount of information has been obtained studying animal models of pulmonary fibrosis. Most of the time, chemicals like bleomycin and paraquat have been employed as fibrotic agents. In some of the experiments, the histological changes resemble those observed in IPF but, strictly speaking, they are really models of drug-induced lung disease and not of the idiopathic human illness.

The usual view about the pathogenesis of IPF states that an unidentified pulmonary insult initiates a chronic inflammatory process leading to persistent injury and fibrosis. Important cellular elements in this process would be neutrophils, eosinophils, lymphocytes and alveolar macrophages.[100] However, this view has recently been challenged and there are suggestions that the disease could be primarily an epithelial-fibroblastic disorder.[101]

Although the pathogenetic phenomena leading to the development of irreversible pulmonary scar lesions in IPF are not completely understood, some findings may support a trial of hematopoietic stem cell transplantation (HSCT) for this disease (Table 4):

Table 4. Evidences supporting the use of HSCT for idiopathic pulmonary fibrosis

Low survival (3-4 y) of newly diagnosed IPF even with chronic immunosuppression
High incidence of *bronchiolitis obliterans* in lung transplant recipients
Autoimmune features in the blood and in the lung of IFP patients
Lung fibrosis mediated by BM-derived cells and cytokines
Improvement of IPF native lung following intensive immunosuppression and single lung transplantation

1. Polyclonal B-lymphocyte stimulation resulting in increased peripheral blood immunoglobulin concentrations has been described in IPF. Nonorgan-specific circulating autoantibodies have been found in approximately 40% of the patients.[102] These include antinuclear antibodies reacting to DNA topoisomerase II or cytokeratin-8. In addition, immune complexes are present in blood, lung lavage fluid and lung tissue. It has been suggested that the presence of autoantibodies are unlikely to be a primary cause of tissue damage, although they may augment the inflammatory process. It is possible that they represent a nonspecific consequence of lung inflammation and injury. More specific immunohistochemical data have demonstrated the presence of circulating IgG autoantibodies to pulmonary epithelial cells in IPF patients, thus implicating a putative autoantigen as endogenous and specific to the lung. The question is still open. Do autoantibodies in IPF represent evidence of ongoing injury, or could they be responsible for initiating the damaging process? Furthermore, could they amplify an already established cycle of inflammation/fibrogenesis? In this scenario, could IPF be considered a 'local' autoimmune process?

2. There are several factors that are able to modify wound healing and the final degree of pulmonary fibrosis including the type of inflammatory response. It is believed that a Th2 type response predominates in the pulmonary interstitium of IPF patients.[91,102] There are mast cells, eosinophils and increased amounts of interleukin-4 and interleukin-13. The finding of reduced expression of the Th1-derived cytokine interferon-g, which may activate cell-mediated mechanisms for removal of cellular antigens and restoration of normal tissue, led to a recent clinical trial of interferon-g-1b in IPF with some reported benefit.[103]

3. Several growth factors appear to be involved in the development of pulmonary fibrosis, invariably regulating other cell functions, as well as cell proliferation.[104] They may originate from a variety of sources including immune cells. For instance, TGF-b1 is known to cause severe pulmonary fibrosis when overexpressed in animal models, and a significant overexpression of the mediator is found in IPF lung tissue. An important source of TGF-b1 is alveolar macrophages, a cellular type originated mainly from blood monocytes.

4. There are case reports of single lung transplantation for IPF where intensive immunosuppression, including the use of cyclosporin, was associated with roentgenographic and functional improvement of the remaining native lung.[105,106] These data suggest that IPF may be responsive to aggressive regimens of immunosuppressive therapy.

Risks and Benefits of HSCT in Idiopathic Pulmonary Fibrosis

HSCT would give IPF patients a new "immunological beginning". The pathogenetical mechanisms involved in its genesis could then be reset, interrupting or changing into normal the aberrant patterns of immunological and inflammatory responses. This phenomenon was observed, at least in part, in patients transplanted for autoimmune diseases. In addition, if IPF really is initiated by an environmental insult occurring at very early stages of the disorder, new flares of the disease could also be avoided, provided the patients do not meet the initiating agents again.

When conceiving HSCT for IPF, some difficulties may be anticipated. What would be the best candidates and the right moment to perform the proceeding? Ideally, HSCT would be indicated for patients with disease in the early stages, because advanced lung scarring is most probably irreversible. Some patients, however, show slow progressing forms of the disease, and there are no effective tools to predict the evolution of the condition in every case. IPF frequently hits subjects after the fifth decade, ages with the highest risk for complications related to chemotherapy. HSCT would probably be offered to IPF patients less than 60 or 65 years old, showing objective evidences of a progressive disorder, but without advanced lung scarring. Another possible problem of HSCT for IPF patients is the use of chemotherapeutic agents with potential pulmonary toxicity during the phases of stem cell mobilization and conditioning. The use of such drugs in patients with fibrotic lung disease could more easily lead to pulmonary side effects. Finally, the present hypothesis about IPF suggests that the still unidentified stimulus can produce repeated episodes of acute lung injury, justifying the findings of patchy distribution of lesions in lung biopsies with UIP.[91] In the case of continuous environmental exposure, the disease may relapse after HSCT and progress latter.

In conclusion, IPF is a disabling disease associated with poor prognosis. No drug therapy has clearly been demonstrated to benefit patients with IPF until the present time. Although its pathogenesis is not completely known, there are some immunological findings supporting a trial of HSCT in this condition. The planning of a clinical trial of HSCT for IPF should consider particular problems in this type of patients, including their old age. This fact and others discussed above for asthma would recommend the choice of an autologous HSCT (with in vivo or in vitro T cell depletion) or an allogeneic transplantation with subablative conditioning (mini-allo transplants) for the initial clinical trials in idiopathic pulmonary fibrosis.

HSCT for Bronchial Asthma and IPF: General Conclusions and Future Directions

Bronchial asthma and idiopathic pulmonary fibrosis share many chronic inflammatory features observed in systemic autoimmune diseases which may predict for successful HSCT in severe and refractory "autoimmune" lung diseases. However, there are fewer clinical and experimental studies in asthma and IPF than in autoimmune diseases showing direct benefit of HSCT in the clinical course of the disease. Indeed, in limited series, the presence of atopy or asthma predicted pulmonary complications after HSCT. We could argue that studies in patients with bronchial asthma transplanted for coincidental hematological diseases, are required before we could propose a pilot trial of HSCT for asthma or IPF. We could also argue that lung transplantation would be more likely to correct local inflammatory abnormalities operating in asthma and IFP than a HSCT. However, lung transplantation has a complexity of organ procurement and of the surgical procedure much higher than HSCT and it is associated with significant morbidity and mortality.

A large proportion of patients with idiopathic pulmonary fibrosis and a much smaller proportion of asthmatics can be identified with severe prognostic factors that could justify adoption of aggressive alternative therapies. The use of less toxic forms of HSCT, such as autologous transplantation or subablative transplantation associated with in vivo or in vitro T cell depletion, would justify initiation of a pilot clinical trial in those severe forms of lung diseases. Besides all the evidence discussed above, allogeneic approaches of HSCT could take advantage of the tissue-repair properties of hematopoietic stem cells.[107]

References

1. Burt K, Slavin S, Burns WH et al. Induction of tolerance in autoimmune diseases by hematopoietic stem cell transplantation: Getting close to a cure? Blood 2002; 99:768-784.
2. Klassen LW. Bone marrow transplantation or hematopoietic stem cell transplantation. In: Koopman WJ, ed. Arthritis and allied conditions. 14th ed. Philadelphia: Lippincott Williams &Wilkins, 2001:921-30.
3. Lemanske RF Jr. Inflammatory events in asthma: An expanding equation. J Allergy Clin Immunol 2000; 105:S633-36.
4. Robinson DR. In: Klippel JD & Dieppe PA, eds. Inflammation, 2nd ed. London: Mosby, 1998:(Section 1):7.1-7.10.
5. Kay AB. T cells as orchestrators of the asthmatic response. Ciba Found Symp 1997; 206:56-70.
6. Gelfand EW, Irvin CG. T lymphocytes: Setting the tone of the airways. Nature Med 1997; 3(4):382-83.
7. Cookson W. The alliance of genes and environment in asthma and allergy. Nature 1999; 402(Suppl 6760):B5-11.
8. Miosse P. Cytokines in rheumatoid arthritis: Is it all TNF-alpha? Cell Mol Biol 2001; 47:675-78.
9. Yoon JW, Jun HS. Cellular and molecular pathogenic mechanisms of insulin-dependent diabetes mellitus. Ann NY Acad Sci 2001; 928:200-11.
10. Salvati VM, MacDonald TT, Bajaj-Elliott M et al. Interleukin 18 and associated markers of T helper cell type 1 activity in coeliac disease. Gut 2002; 50:186-90.
11. Busse WW, Lemanske RF Jr. Asthma. N Engl J Med 2001; 350-62.
12. Kero J, Gissler M, Hemminki E et al. Could Th1 and Th2 diseases coexist? Evaluation of asthma incidence in children with coeliac disease, type 1 diabetes, or rheumatoid arthritis: A register study. J Allergy Clin Immunol 2001; 108:781-83.
13. Stene LC, Nafstad P. Relation between occurrence of type 1 diabetes and asthma. Lancet 2001; 357:607-8.
14. Peat JK, Li J. Reversing the trend: Reducing the prevalence of asthma. J Allergy Clin Immunol 1999; 103:1-10.
15. Burrows B, Martinez MD, Halonen M et al. Association of asthma with serum IgE levels and skin-test reactivity to allergens. N Engl J Med 1989; 320:271-77.
16. Kay AB. Allergy and allergic diseases—First of two parts. N Engl J Med 2001; 344:30-38.
17. The Collaborative Study on the Genetics of Asthma (CSGA). A genome-wide search for asthma susceptibility loci in ethnically diverse populations. Nat Genet 1997; 15:389-392.
18. Meyers DA, Bleeckers ER. Genetics of allergic disease. In: Middleton Jr E, Reed C, Ellis EF, et al eds. Allergy: Principles and Practice. 5th ed. St Louis: Mosby, 1998:40-45.

19. Finotto S, Neurath MF, Glickman JN et al. Development of spontaneous airway changes consistent with human asthma in mice lacking T-bet. Science 2002; 295:336-38.
20. Laprise C, Boulet LP. Asymptomatic airway hyperresponsiveness: A three year follow-up. Am J Resp Crit Care Med 1997; 156:403-9.
21. O'Byrne PM. Airway hyperresponsiveness. In: Middleton E Jr, Reed C, Ellis EF et al eds. Allergy: Principles and Practice. 5th ed. St Louis: Mosby, 1998:859-66.
22. Robinson DS, Hamid Q, Ying S et al. Predominant Th2-like bronchoalveolar T-lymphocyte population in atopic asthma. N Engl J Med 1992; 326:298-304.
23. Ying S, Durham SR, Corrigan CJ et al. Phenotype of cells expressing mRNA for Th2-type (interleukin 4 and interleukin 5) and Th1-type (interleukin 2 and interferon gamma) cytokines in bronchoalveolar lavage and biopsies from atopic asthmatic and normal control subjects. Am J Resp Cell Mol Biol 1995; 12:477-87.
24. Humbert M, Durham SR, Ying S et al. IL-4 and IL-5 mRNA and protein in bronchial biopsies from patients with atopic and nonatopic asthma: Evidence against "intrinsic" atopic asthma being a distinctive immunopathologic entity. Am J Resp Crit Care Med 1996; 154:1497-1504.
25. Humbert M, Durham SR, Kimmit P et al. Elevated expression of messenger ribonucleic acid encoding IL-13 in the bronchial mucosa of atopic and nonatopic subjects with asthma. J Allergy Clin Immunol 1997; 99:657-65.
26. Ying S, Humbert M, Barkans J et al. Expression of IL-4 and IL-5 mRNA and protein product by CD4+ and CD8+ T cells, eosinophils, and mast cells in bronchial biopsies obtained from atopic and nonatopic (intrinsic) asthmatics. J Immunol 1997; 158:3539-44.
27. Nakamura Y, Ghaffar O, Olivenstein R et al. Gene expression of the GATA-3 transcription factor is increased in atopic asthma. J Allergy Clin Immunol 1999; 103:215-22.
28. Kay AB. T cells as orchestrators of the asthmatic response. Ciba Found Symp 1997; 206:56-67.
29. Kapsenberg ML, Hilkens CMU, van der Pouw Kraan TCMT et al. Atopic allergy: A failure of antigen-presenting cells to properly polarize helper cells? Am J Resp Crit Care Med 2000; 162:S76-80.
30. Moller GM, Overbeek SE, van Helden-Meeuwsen CG et al. Increased numbers of dendritic cells in the bronchial mucosa of atopic asthmatic patients: Downregulation by inhaled corticosteroids. Clin Exp Allergy 1996; 26:517-24.
31. Bellini A, Vittori E, Marini M et al. Intraepithelial dendritic cells and selective activation of Th2-like lymphocytes in patients with atopic asthma. Chest 1993; 103:997-1005.
32. Holloway JW, Beghe B, Holgate ST. The genetic basis of atopic asthma. Clin Exp Allergy 1999; 29:1023-32.
33. Tunon De-Lara JM, Redington AE, Bradding P et al. Dendritic cells in normal and asthmatic airways: Expression of the alpha subunit of the high affinity immunoglobulin E receptor (Fc epsilon RI-alpha). Clin Exp Allergy 1996; 26:648-55.
34. Lenschow DJ, Walunas TL, Bluestone JA. CD28/B7 system of T cell costimulation. Annu Rev Immunol 1996; 14:233-58.
35. Jaffar ZH, Stanciu L, Pandit A et al. Essential role for both CD80 and CD86 costimulation, but not CD40 interactions, in allergen-induced Th2 cytokine production from asthmatic bronchial tissue: Role for alpha beta, but not gamma delta, T cells. J Immunol 1999; 163:6283-91.
36. Tsuyuki S, Tsuyuki J, Einsle K et al. Costimulation through B7-2 (CD86) is required for the induction of a lung mucosal T helper cell 2 (Th2) immune response and altered airway responsiveness. J Exp Med 1997; 185:1671-79.
37. Launois P, Maillard I, Pingel K et al. IL-4 rapidly produced by V beta 4 alpha CD4+ T cells instructs Th2 development and susceptibility to Leishmania major in BALB/c mice. Immunity 1997; 6:541-49.
38. van den Pouw Kraan TCMT, Boeije LC, Smeenk RJ et al. Prostaglandin E2 is a potent inhibitor of human IL-12 production. J Exp Med 1995; 181:775-79.
39. Huang FP, Niedbala W, Wei XQ et al. Nitric oxide regulates Th1 development through the inhibition of IL-12 synthesis by macrophages. Eur J Immunol 28:4062-70.
40. Robinson D, Shibuya K, Mui A et al. IGIF does not drive Th1 development but synergizes with IL-12 for IFN-γ production and activated IRAK and NFκB. Immunity 1997; 7:571-81.
41. Szabo SJ, Kim ST, Costa GL et al. A novel transcription factor, T-bet, directs Th1 lineage commitment. Cell 2000; 100(6):655-69.
42. Ying S, Durham SR, Corrigan CJ et al. Phenotype of cells expressing mRNA for Th2-type (IL-4 and IL-5) and Th1-type (IL-2 and interferon gamma) cytokines in bronchoalveolar lavage and bronchial biopsies from atopic asthmatics and normal control subjects. Am J Resp Cell Mol Biol 1995; 12:477-87.
43. van der Pouw kraan TC, Boeije LCM, de Groot R et al. Reduced production of IL-12 and IL-12-dependent IFN-g release by allergic asthma patients. J Immunol 1997; 158:5560-65.
44. Wills-Karp M, Ewart SL. The genetics of allergen-induced airway hyperresponsiveness in mice. Am J Respir Crit Care Med 1997; 156(4):S89-96.
45. Kalinski P, Hilkens CMU, Wierenga EA et al. T cell priming by type-1 and type-2 polarized dendritic cells: The concept of a third signal? Immunol Today 1999; 20(12):561-67.
46. Kapsenberg ML, Kalinski P. The concept of type 1 and type 2 antigen-presenting cells. Immunol Letters 1999; 69:5-6.
47. Yssel H, Abbal C, Pene J et al. The role of IgE in asthma. Clin Exp Allergy 1998; 28(Suppl 5):104-18.
48. Corry DB, Kheradmand F. Induction and regulation of the IgE response. Nature 1999; 402(Suppl):B18-B23.
49. Church MK, Levi-Schaffer F. The human mast cell. J Allergy Clin Immunol 1997; 99:155-60.
50. Bentley AM, Hamid Q, Robinson DS et al. Prednisolone treatment in asthma. Reduction in the number of eosinophils, T-cells, tryptase-only positive mast cells, and modulation of IL-4, IL-5 and interferon gamma cytokine gene expression within the bronchial mucosa. Am J Respir Crit Care Med 1996; 153:551-56.
51. Powell N, Till S, Bungre J et al. The immunomodulatory drugs cyclosporine A, mycophenolate mofetil, and sirolimus (rapamycin) inhibit allergen-induced proliferation and IL-5 production by PBMC from atopic asthmatic patients. J Allergy Clin Immunol 2001; 108:915-17.
52. Alexander AG, Barnes NC, Kay AB. Trial of cyclosporin in corticosteroid-dependent chronic severe asthma. Lancet 1992; 339:324-28.
53. Sullivan KM, Parkman R, Walters MC. Bone marrow transplantation for nonmalignant disease. Hematology (Am Soc Hematol Educ Program) 2000; 1:319-38.
54. Corris PA, Dark JH. Aetiology of asthma: Lessons from lung transplantation. Lancet 1993; 341:1369-71.
55. Denburg JA, Inman MD, Leber B et al. The role of bone marrow in allergy and asthma. Allergy 1996; 51:141-48.
56. Bice DE, Williams AJ, Muggenburg BA. Long-term antibody production in canine lung allografts: Implication in pulmonary immunity and asthma. Am J Respir Cell Mol Biol 1996; 14(4):341-47.
57. Strachan DP. Hay fever, hygiene, and household size. B M J 1989; 299:1259-60.
58. Mattes J, Karmaus W. The use of antibiotics in the first year of life and development of asthma: Which comes first? Clin Exp Allergy 1999; 29:729-32.
59. Prescott SL, Macaubas C, Holt BJ et al. Transplacental proming of the human immune system to environmental allergens: Universal skewing of initial T cell responses towards the Th2 cytokine profile. J Immunol 1998; 160:4730-37.
60. De Sanctis GT, Itoh A, Green FH et al. T-lymphocytes regulate genetically determined airway hyperresponsiveness in mice. Nat Med 1997; 3(4):460-62.
61. Bellou A, Kanny G, Fremont S et al. Transfer of atopy following bone marrow transplantation. Ann Allergy Asthma Immunol 1997; 78:513-16.
62. Agosti JM, Sprenger JD, Lum LG et al. Transfer of allergen-specific IgE-mediated hypersensitivity with allogeneic bone marrow transplantation. N Engl J Med 1988; 319:1623-28.
63. Landsteiner K, Chase MW. Experiments on transfer of cutaneous sensitivity to simple compounds. Proc Soc Exp Biol Med 1942; 49:688-90.

64. Socié G, Mary JI, Esperou H et al. Health and functional status of adult recipients 1 year after allogeneic haematopoietic stem cell transplantation. Br J Haematol 2001; 113:194-201.
65. Frankovich J, Donaldson S, Lee Y et al. High-dose therapy and autologous hematopoietic cell transplantaion in children with primary refractory and relapsed Hodgkin's disease: Atopy predicts idiopathic diffuse lung injury syndromes. Biol Blood Marrow Transplant 2001; 7:49-57.
66. Von Mutius E. The rising trends in asthma and allergic disease. Clin Exp Allergy 1998; 28(Suppl 5):45-49.
67. Campbell MJ, Holgate ST, Johston SL. Trends in asthma mortality. BMJ 1997; 315:1012-18.
68. Strunk CR, Milgrom H, Iklé DN et al. Risk factors for asthma. In: Sheffer AL, ed. Fatal Asthma. New York: Marcel Decker Inc, 1998:31-44.
69. Campbell DA, McLennan G, Coates JR et al. A comparison of asthma deaths and near fatal asthma attacks in South Australia. Eur Respir J 1994; 7:490-97.
70. Turner MA, Noertjojo K, Vedal S et al. Risk factors for near fatal asthma. A case control study in hospitalized patients with asthma. Am J Respir Crit Care Med 1998; 157:1804-9.
71. Richards GN, Kolbe J, Fenwick J et al. Demographic characteristics of patients with severe life threatening asthma: Comparison with asthma deaths. Thorax 1993; 48:1105-9.
72. Rea HH, Scragg R, Jackson R et al. A case-control study of deaths from asthma. Thorax 1986; 41:833-39.
73. Wasserfallen JB, Schaller MD, Feihl F et al. Sudden asphyxic asthma: A distinct entity? Am Rev Respir Dis 1990; 142:108-11.
74. Kikuchi Y, Okabe S, Tamura G et al. Chemosensitivity and perception of dyspnea in patients with a history of near fatal asthma. N Engl J Med 1994; 330:1329-34.
75. Guite HF, Dundas R, Burney PGJ. Risk factors for death from asthma, chronic obstructive pulmonary disease and cardiovascular disease after a hospital admission for asthma. Thorax 1999; 54:301-7.
76. Surs S, Crotty TB, Kephart GM et al. Sudden-onset fatal asthma: A distinct entity with few eosinophils and relatively more neutrophils in the airway submucosa? Am Rev Respir Dis 1993; 148:713-19.
77. Carmichel J, Paterson JC, Diaz P et al. Corticosteroids resistance in chronic asthma. BMJ 1981; 282:1419-22.
78. Kamada AK, Leung DYM, Gleason Mc et al. High dose systemic glucocorticoid therapy in treatment of asthma: A case of resistance and patterns of response. J Allergy Clin Immunology 1992; 90:685-87.
79. Alexander AG, Barnes NC, Kay AB. Trial of cyclosporin in corticosteroid dependent chronic severe asthma. Lancet 1992; 339:324-28.
80. Alexander AG, Barnes NC, Kay AB et al. Can clinical response to cyclosporin in chronic severe asthma be predicted by an in vitro T lymphocyte proliferation assay? Eur Respir J 1996; 9:1421-26.
81. Gelfand EW, Landwher LP, Esterl B et al. Intravenous immunoglobulin: An alternative on steroid dependent allergic diseases. Clin Exp Immunol 1996; 104:61-66.
82. Spahn JO, Leung DYM, Chaçn MTS et al. Mechanisms of glucocorticoid reduction in asthmatics treated with intravenous immunoglobulin. J Allergy Clin Immunol 1999; 103:421-26.
83. Letendre L, Hoagland HC, Moore SB et al. Mayo Clinic experience with allogeneic and syngeneic bone marrow transplantation, 1982 through 1990. Mayo Clin Proc 1992; 67:109-16.
84. Gahrton G, Svensson H, Bjorkstrand B et al. Syngeneic transplantation in multiple myeloma – a case-matched comparison with autologous and allogeneic transplantation. European Group for Blood and Marrow Transplantation. Bone Marrow Transplant 1999; 24(7):741-45.
85. Dold S, Wjst M, von Mutius E et al. Genetic risk for asthma, allergic rhinitis and atopic dermatitis. Arch Dis Child 1992; 67:1018-22.
86. Holgate ST. Genetic and environmental interactions in allergy and asthma. J Allergy Clin immunol 1999; 104:1136-46.
87. LaFace DM, Peck AB. Reciprocal allogeneic bone marrow transplantation between NOD mice and diabetes-nonsusceptible mice associated with transfer and prevention of autoimmune diabetes, Diabetes 1989; 38:894-901.
88. Kushida T, Inaba M, Takeuchi K et al. Treatment of intractable autoimmune diseases in MRL/lpr mice using a new strategy for allogeneic bone marrow transplantation. Blood 2000; 51:862-68.
89. Himeno K, Good RA. Marrow transplantation from tolerant donors to treat and prevent autoimmune disease in BXSB mice. Proc Natl Acad Sci USA 1988; 85:2235-39.
90. Rocha V, Wagner Jr JE, Sobocinski KA et al. Graft-versus-host disease in children who have received a cord-blood or bone marrow transplant from an HLA-identical sibling. Eurocord and International Bone Marrow Transplant Registry Working Committee on Alternative Donor and Stem Cell Sources. N Engl J Med 2001; 342:1846-54.
91. Gross TJ, Hunninghake GW. Idiopathic pulmonary fibrosis. N Engl J Med 2001; 345:517-25.
92. Katzenstein ALA, Myers JL. Idiopathic pulmonary fibrosis. Clinical relevance of pathologic classification. Am J Respir Crit Care Med 1998; 157:1301-15.
93. Coultas DB, Zumwalt RE, Black WC et al. The epidemiology of interstitial lung diseases. Am J Respir Crit Care Med 1994; 150:967-73.
94. Martinez TY, Pereira CA, dos Santos ML et al. Evaluation of the short-form 36-item questionnaire to measure health-related quality of life in patients with idiopathic pulmonary fibrosis. Chest 2000; 117:1627-32.
95. King Jr TE, Schwarz MI, Brown K et al. Idiopathic pulmonary fibrosis. Relationship between histopathologic features and mortality. Am J Respir Crit Care Med 2001; 164:1025-32.
96. Xaubet A, Agusti C, Luburich P et al. Is it necessary to treat all patients with idiopathic pulmonary fibrosis? Sarcoidosis Vasc Diffuse Lung Dis 2001; 18:289-95.
97. American Thoracic Society & European Respiratory Society. Idiopathic pulmonary fibrosis. Diagnosis and treatment: A international consensus statement. Am J Respir Crit Care Med 2000; 161:646-64.
98. Hosenpud JD, Bennett LE, Keck BM et al. Effect of diagnosis on survival benefit of lung transplantation for end stage lung disease. Lancet 1998; 351:24-27.
99. Mason RJ, Schwarz MI, Hunninghake GW et al. Pharmacological therapy for idiopathic pulmonary fibrosis: Past, present and future. Am J Respir Crit Care Med 1999; 160:1771-77.
100. Chan ED, Worthen GS, Augustin A et al. Inflammation in the pathogenesis of interstitial lung disease. In: Schwarz MI, King Jr TE, eds. Interstitial lung disease. 3rd ed. Hamilton: BC Becker, 1998:135-64.
101. Selman M, King TE Jr, Pardo A. Idiopathic pulmonary fibrosis: Prevailing and evolving hypotheses about its pathogenesis and implications for therapy. Ann Intern Med 2001; 134:136-51.
102. Singh S, du Bois R. Autoantibodies in cryptogenic fibrosing alveolitis. Respir Res 2001; 2:61-63.
103. Ziesche R, Hofbauer E, Wittmann K et al. A preliminary study of long-term treatment with interferon Gamma-1b and low dose prednisolone in patients with idiopathic pulmonary fibrosis. N Engl J Med 1999; 341:1264–69.
104. Allen JT, Spiteri MA. Growth factors in idiopathic pulmonary fibrosis: Relative roles. Respir Res 2002; 3:13-22.
105. Lok SS, Smith E, Dorah HM et al. Idiopathic pulmonary fibrosis and cyclosporine: A lesson from single lung transplantation. Chest 1998; 114:1478-81.
106. Owuens JP, van der Berg JW, van der Bij W et al. Long term survival despite early loss of graft function after single lung transplantation for pulmonary fibrosis. J Heart Lung Transplant 2002; 21:395-401.
107. Korbling M, Katz RL, Ruifrok AC et al. Hepatocytes and epithelial cells of donor origin in recipients of peripheral-blood stem cells. N Engl J Med 2002; 346:770-72.

CHAPTER 56

Autologous Stem Cell Transplantation in Relapsing Polychondritis

Falk Hiepe, Andreas Thiel, Oliver Rosen, Gero Massenkeil, Gerd-Rüdiger Burmester, Andreas Radbruch and Renate Arnold

Introduction

Relapsing polychondritis is a rare multisystem autoimmune disorder of unknown etiology that was first described by Jaksch-Wartenhorst in 1923.[1] It is an episodic and progressive inflammatory disease of the cartilaginous structures, including the elastic cartilage of the ear and nose, the hyaline cartilage of peripheral joints, the fibrocartilage at axial sites, and cartilaginous structures of the tracheobronchial tree. Inflammation of other proteoglycan-rich structures such as the eyes, heart, blood vessels, inner ear, and kidneys may also occur. Since relapsing polychondritis was first described, more than 300 cases of the disease have been reported. The diagnosis is based on the clinical findings and may be confirmed histologically by biopsy of the affected ear. The disease is treated by corticosteroids and immunsuppressive agents. Here, we report on a woman suffering from refractory relapsing polychondritis whom we successfully treated by autologous stem cell transplantation (ASCT), resulting in a long-term remission which has lasted for 55 months now. We, therefore, conclude that ASCT may be of therapeutic benefit to patients with relapsing polychondritis who failed to respond to conventional therapy. Eligibility criteria for ASCT in relapsing polychondritis are discussed.

Etiopathogenesis

Although the etiology and pathogenesis of relapsing polychondritis are still unknown, several findings suggest that both humoral and cell-mediated immune responses are involved. Cartilage contains large quantities of type II collagen. Serum autoantibodies to native collagen II as well as to collagen IX and XI were found during acute attacks.[2-5] Granular deposits of immunoglobulins and complement were detected at the chondrofibrous junction of affected cartilage by immunofluorescence.[6,7] A predominance of HLA-DR positive cells and significant quantities of CD4+ T lymphocytes in cellular infiltrates were found by immunohistology. Susceptibility to relapsing polychondritis is significantly associated with HLA-DR4, and the extent of organ involvement is negatively associated with HLA-DR6.[8,9] Experimental immunization of rats with type II collagen can induce auricular chondritis and arthritis. As in humans, the murine lesions were characterized by severe chondritis, positive immunofluorescence reactions to IgG and C3, and circulating IgG that reacted with native type II collagen.[10,11]

Clinical Features

Clinical features are the key to diagnosis of relapsing polychondritis. The disease is generally characterized by a sudden, flagrant onset. More than 20% of patients originally present with fever, which also frequently accompanies acute flares. Fatigue and weight loss are also common. Acute uni- or bilateral auricular chondritis is a cardinal feature that occurs in up to 85% of patients during the course of the disease.[12-14] It leads to reddening or violaceous discoloration, warmth, swelling and pain of the affected auricular cartilage (helix, antihelix, tragus, and external auditory canal), yet spares the lobulus. This is a key distinguishing feature important for the differential diagnosis. After repeated episodes of inflammation, the ears ultimately become soft and deformed, resulting in the typical "cauliflower" ear deformity. Involvement of the external auditory canal leads to conductive deafness. Approximately 50% of patients develop nasal chondritis during the course of the disease, resulting in saddle nose deformity.[12-14] Respiratory manifestations are present in 25% of patients at initial presentation. Roughly half of them ultimately develop severe symptoms of glottis or tracheobronchial tree involvement,[12-16] including tenderness over the anterior cervical trachea, larynx and thyroid cartilage, hoarseness, persistent cough, choking spells, and wheezing and dyspnea on exertion.

Seronegative, non-deforming and non-erosive polyarthritis or oligoarthritis may be present at onset. In this case, the arthritis affects both small or large joints and is often episodic and migratory.

Roughly 20 to 30% of patients develop cardiovascular manifestations, including aortic and mitral regurgitation, conduction abnormalities and aneurysms of the aorta or other arteries. Vasculitis may also occur.

Associated Diseases

Although relapsing polychondritis is a separate clinical entity, up to 30% of patients also suffer from another inflammatory disease (Table 1). The most common associated disease is vasculitis, which can affect vessels of all sizes. Five to 14% of all patients with relapsing polychondritis have biopsy-confirmed leukocytoclastic vasculitis, ranging from isolated cutaneous leukocytoclastic vasculitis to aortitis. Although vasculitis most commonly leads to dilatation of the aortic ring and ascending aorta, aneurysm of the descending thoracic or abdominal aorta can also occur. Aortic rupture has also been reported. Dermato-

Table 1. Diseases associated with relapsing polychondritis

Rheumatic Diseases

Rheumatoid arthritis
Juvenile chronic arthritis
Psoriatic arthritis
Reactive arthritis
Ankylosing spondylitis
Systemic lupus erythematosus
Sjögren's syndrome
Systemic sclerosis
Wegener's granulomatosis
Panarteritis nodosa
Microscopic polyangiitis
Churg-Strauss syndrome
Takayasu's arteritis
Temporal arteritis
Behçet's syndrome (MAGIC syndrome)

Hematological Diseases

Myelodysplastic syndromes
Lymphoma
Pernicious anemia
Acute leukemia
Mixed cryglobulinemia

Gastrointestinal Diseases

Crohn's disease
Ulcerative colitis
Primary biliary cirrhosis

Skin Diseases

Psoriasis
Cutaneous leukocytoclastic vasculitis
Atopic dermatitis
Vitiligo
Lichen planus

Endocrinological Diseases

Thyroid autoimmune diseases
Diabetes mellitus
Thymoma

Other Diseases

Myasthenia gravis
Familial Mediterranean fever
Glomerulonephritis

---

logical and renal manifestations, neuropathies, audiovestibular abnormalities, and episcleritis are most likely due to microscopic angiitis. Vasculitis seems to worsen the prognosis of relapsing polychondritis.[17] Relapsing polychondritis is infrequently accompanied by defined vasculitides like Wegener's granulomatosis and polyarteritis nodosa.[18] The co-occurrence of relapsing polychondritis and Behçet's disease has also been reported; this constellation was named "MAGIC syndrome" (mouth and genital ulcers with inflamed cartilage syndrome).[19] Roughly 25% of patients with relapsing polychondritis have associated autoimmune diseases such as systemic lupus erythematosus, rheumatoid arthritis, mixed connective tissue disease, Sjögren's syndrome, and hypothyroidism.

The coexistence of myelodysplasia has been observed in at least 40 cases. According to recent estimates, 30% of patients with relapsing polychondritis have an associated myelodysplastic syndrome and, conversely, 0.6% of patients with myelodysplasia have relapsing polychondritis. Myelodysplasia is more common in males than in females. The patient is more likely to die of complications of myelodysplasia than of relapsing polychondritis.[20,21]

Laboratory Studies

Unspecific markers of inflammation, including an elevated erythrocyte sedimentation rate (ESR), anemia, leukocytosis, thrombocytosis, and hypergammaglobulinemia may be detected during acute flares.[22,23] The possibility of macrocytic anemia should be considered, as this infrequently occurs due to an associated early myelodysplastic syndrome.[21] Anti-collagen type II antibodies are found during acute episodes of relapsing polychondritis, and their concentration in serum seems to correlate with the severity of the disease.[2,3] However, since the specificity of these antibodies is low and their frequency ranges from only 20 to 50%, their usefulness as a diagnostic marker is limited.[4,5] The prevalence of antinuclear antibodies (ANA) observed in relapsing polychondritis is low. The significant ANA titers detected in a patient with relapsing polychondritis is more likely due to the presence of an associated disorder, such as SLE, MCTD, Sjögren's syndrome or acquired myelodysplasia.[24] Likewise, other serological tests for rheumatoid factor, antineutrophil cytoplasmic antibodies (ANCA), etc., are helpful only in the diagnosis and management of accompanying rheumatologic disorders.[25,26] Serum creatinine and first morning urine should be routinely tested to screen for glomerulonephritis.

Diagnosis

The diagnosis of relapsing polychondritis is clinical, i.e., based on the observed symptoms and signs. McAdam et al[12] propose the use of clinical diagnostic criteria, at least three of which must be present to establish the diagnosis (Table 2). Biopsy of the affected ear may confirm the diagnosis of relapsing polychondritis. All patients should be evaluated for laryngotracheal disease because of the potential for serious respiratory complications. This is done by means of pulmonary function tests and computed tomography.[15] Echocardiography may be indicated to evaluate large vessel involvement. The renal status should always be investigated because of the possibility of concomitant glomerulonephritis.[27-29]

Treatment

Nonsteroidal anti-inflammatory drugs can control mild episodes of inflammation in a few cases. Low-dose colchicine may also be helpful.[13] In most cases, however, corticosteroids are needed to suppress disease activity. They reduce the frequency, duration

Table 2. Diagnostic criteria for relapsing polychondritis. A diagnosis is certain when three or more of the criteria are met.[12]

- Bilateral auricular chondritis
- Nasal chondritis
- Non-erosive, seronegative inflammatory polyarthritis
- Ocular inflammation (conjunctivitis, keratitis, scleritis, and/or episcleritis, uveitis)
- Respiratory tract chondritis (laryngeal and/or tracheal cartilage)
- Cochlear and/or vestibular dysfunction (neurosensory hearing loss, tinnitus and/or vertigo)
- Cartilage biopsy confirmation of a compatible histological picture

and severity of flares, but do not stop the progression of relapsing polychondritis in severe cases.[12] Prednisolone (1 mg/kg) is indicated in patient with serious manifestations. Many immunosuppressive drugs have been used in attempts to allow tapering of the steroid dose, to achieve a lower maintenance dose, or to treat refractory cases. These include dapsone,[30,31] penicillamine,[32] azathioprine[12,33] cyclophosphamide,[12,29,33] cyclosporine,[34] methotrexate,[35] and monoclonal anti-CD4 antibody.[36,37]

Prognosis

Most patients with relapsing polychondritis experience intermittent or fluctuating inflammatory manifestations. Relapsing polychondritis is generally a progressive disease.[14] The majority of patients develop some degree of disability such as bilateral deafness, impaired vision, phonation difficulties and cardiorespiratory problems during the later stages. Complications of long-term corticosteroid use cause additional morbidity.[13,23]

Overall survival rates were reported to be 74% at 5 years and 55% at 10 years after diagnosis. The need for corticosteroid therapy did not influence survival.[14] The outcome was far less dismal in another study, which documented a survival rate of 94% and an average disease duration of 8 years.[13] The most common causes of death are infection, systemic vasculitis and malignancy. Only 10% of the deaths could be attributed to chondritis-related airway involvement. Other causes of death include respiratory failure from airway collapse or obstruction and complications of heart disease. Anemia at diagnosis was a marker for decreased survival in the entire group. For patients less than 51 years old, saddle-nose deformity and systemic vasculitis were the worst prognostic signs. In older patients, anemia was the only predictor of outcome.[14]

Autologous Stem Cell Transplantation—Case Report

Case Description

A 41-year-old female was admitted to our hospital with relapsing polychondritis, which was first diagnosed in 1985. Manifestations of the disease included severe arthritis, costosternal pain, vasculitis, scleritis, saddle nose and tracheal involvement; the patient had also suffered a life-threatening episode of pyoderma gangrenosum. Despite continuous and intensive conventional therapy for several years, no remission was achieved. The patient was at risk of developing a tracheo-esophageal fistula upon further disease progression. The previous treatment regimens had included intravenous immunoglobulins, high-dose methylprednisolone, methotrexate, monoclonal anti-CD4 antibody and intravenous cyclophosphamide (cumulative dose 6.0 g) with concomitant steroid therapy. At the time of admission, the patient was receiving a daily dose of 30 mg methylprednisolone, and her Karnofsky score was 60%.

Stem Cell Mobilization and Collection

Stem cells were mobilized using 2 g/m^2 of cyclophosphamide. Five days later, G-CSF (10 µg/kg/day) was administered until CD34$^+$ cells were harvested. Leukapheresis (Cobe Spectra; Cobe BCT, Lakewood, CO, USA) was performed when the leukocyte count reached 4.0 x 10^9/l. Only one leukapheresis was required to collect the minimum number of cells required for transplantation, i.e., 4 x 10^6 CD34$^+$ cells/kg of patient weight. The number of CD34$^+$ cells collected was 7.7 x 10^6 CD34$^+$ cells/kg of patient weight. Severe arthralgia developed one day after the first application of G-CSF. Febrile episodes of unknown origin also occurred. The symptoms disappeared in response to high-dose steroids, which were gradually tapered.

Transplant Engineering

CD34$^-$ cells in the stem cell transplant were depleted by selecting CD34$^+$ cells via high-gradient magnetic cell sorting using a CliniMacs™ device (Miltenyi Biotec GmbH, Bergisch Gladbach, Germany).[38-40] After CD34$^+$ cell enrichment, the transplant contained 7.3 x 10^6 CD34$^+$ cells/kg. Ex vivo purging reduced the number of contaminating mononuclear cells to 0.37 x 10^3 CD3$^+$ cells/kg of patient weight. CD3$^+$ cells from the CD34$^-$ fraction were therefore added to the purified CD34$^+$ cells to yield a minimum of 1.0 x 10^4 CD3$^+$ cells/kg of patient weight. The CD34$^+$ cell suspensions were cryopreserved in 5 vol% dimethyl sulfoxide until transplantation.

Preparatory Regimen and Autologous Stem Cell Transplantation

The preparatory regimen consisted of 200 mg cyclophosphamide/kg of patient weight (divided days -5 to -2) and antithymocyte globulin (ATG; rabbit; obtained from Fresenius, Bad Homburg, Germany) 90 mg/kg (3 x 30 mg/kg over three days) of patient weight (days -4 to -2).[41] During ATG treatment, 500 mg methylprednisone was administered twice a day. The time between stem cell mobilization with cyclophosphamide and ASCT was 30 days. 25 hours after conditioning, the patient received the transplant containing a total of 10.5 x 10^9 cells/l and 7.3 x 10^6 CD34$^+$ cells/kg of patient weight in a volume of 42 ml. Systemic inflammatory response syndrome (SIRS) occurred during the first ATG infusion in the immunoablative phase. Supportive care was provided according to standard protocols for bone marrow transplantation, including isolation of the patient and infection prophylaxis. Immunoglobulin substitution therapy (10 g every other week) was administered for prevention of hypoimmunoglobinemia, and was ended in 6 months. Nevertheless, the patient developed signs of sepsis and interstitial pneumonia five and seven weeks after ASCT. Disseminated intravascular coagulation also occurred. CMV reactivation was excluded in either case. The infections were controlled using broad-spectrum antibiotics, including amphotericin B and ganciclovir. The coagulopathy was treated with heparin and fresh-frozen plasma. Dopamine was temporarily needed to control arterial hypotension and to maintain renal perfusion. The patient was dismissed from the hospital 59 days after ASCT.

Clinical Outcome

Now, after 55 months of follow-up, the patient is still in complete remission from the disease, as defined by the disappearance of all clinical symptoms and signs of polychondritis. Corticosteroid therapy was gradually reduced from 50 mg/day methylprednisolone before ASCT to 5 mg/d prednisolone 12 months after ASCT. In the absence of clinical signs of the disease, prednisolone was further tapered and withdrawn 36 months after ASCT. Forty-five months after ASCT, the patient suddenly complained of impaired vision in the right eye. The ophthalmologic examination revealed chorioretinitis. The serologic data confirmed our suspicion of a reactivation of Toxoplasma infection. The visual disturbance resolved completely under antibiotics. Nevertheless, this incident suggests that the patient's immune system—which had apparently returned to normal after ASCT—might be weakened and susceptible to opportunistic infections.

Hematological and Immunological Reconstitution After ASCT

Reconstitution of granulocytes and platelets occurred rapidly, i.e., within 2 weeks. The absolute number of nucleated cells reached $1.0 \times 10^9/l$ on day +12 after ASCT. The platelet count was $20 \times 10^9/l$ on day +8. The number of platelet transfusions given during bone marrow aplasia was 5 units. Twelve units of red blood cells were administered. The absolute number of $CD8^+$ cells was low during the first 7 months, but had increased fivefold within 1 year after ASCT relative to the patient's status at admission. For up to 2 months after ASCT, the absolute $CD4^+$ cell counts remained below the limit of detection. Almost all $CD4^+$ cells detected during the second phase of reconstitution (2 to 5 months after ASCT) were $CD4^+$, $CD45RO^+$, and $CD45RA^-$ memory/effector cells. The activation marker HLA-DR was expressed on up to 50% of these cells. This transient appearance of activated memory/effector cells was in concurrence with the findings in viral and bacterial infections, such as interstitial pneumonia, as well as localized infections in the perianal region and urinary tract. Naive $CD4^+$, $CD45RA^+$, and $CD45RO^-$ cells were nearly undetectable until 5 months after ASCT. The $CD4^+$ cell counts had returned to pretransplantation levels 6 to 7 months after ASCT. Notably, the absolute number of $CD4^+$ cells was already low before ASCT, which had already been observed since the patient was treated with monoclonal anti-CD4 antibody. The absolute $CD4^+$ cell count did not rise above the pretransplantation level until 12 months after ASCT, and then gradually continued to rise to the maximum counts detected at the last follow-up 48 months after ASCT. It is noteworthy that the absolute numbers of naive $CD4^+$, $CD45RA^+$, and $CD45RO^-$ cells have been higher than those of the $CD4^+$, $CD45RO^+$ and $CD45RA^-$ memory/effector cells since month 12 after ASCT (Fig. 1). The complete reconstitution of B cells was apparent 12 months after ASCT.

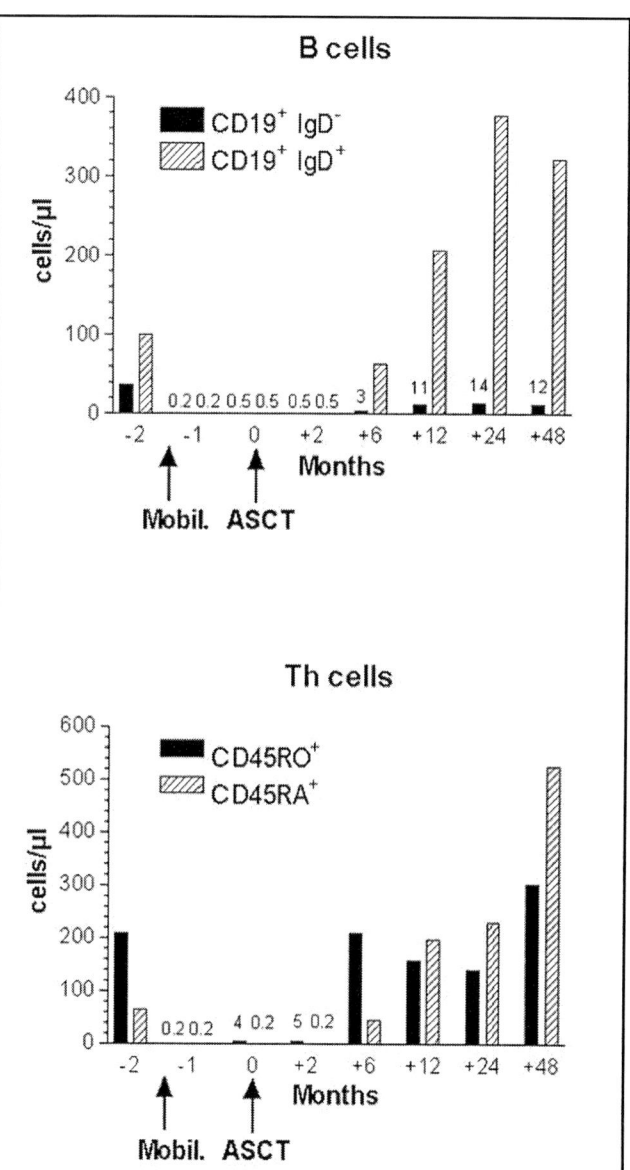

Figure 1. B and T-helper (Th) cell reconstitution after autologous stem cell transplantation (ASCT) in a patient with refractory relapsing polychondritis

There has been a predominance of naive $CD19^+/IgD^+$ B cells (Fig. 1).

Conclusion

Autologous stem cell transplantation (ASCT) may be considered as an effective therapeutic alternative in patients with relapsing polychondritis who failed conventional treatment attempts with corticosteroids and immunosuppressive agents.

Recommendations for Selecting Patients for ASCT

There still are no clear-cut optimal ASCT eligibility criteria for patients with relapsing polychondritis. Since relapsing polychondritis is a rare disease, little data exists that might facilitate the identification of patients with a poor prognosis. In our opinion, patients showing disease progression despite treatment with 20 mg/day of corticosteroids and at least two different im-

Table 3. Proposed inclusion and exclusion criteria for ASCT

Inclusion Criteria
- Age younger than 65 years
- Relapsing polychondritis as defined according to the diagnostic criteria by McAdam et al[12]
- Failure of standard treatment: glucocorticoids and at least two immunosuppressive drugs
- Beginning organ manifestations which can lead to life-threatening conditions (airway involvement, vasculitis)
- Adequate function of all majors organs to tolerate ASCT

Probable Exclusions
- Infections
- Severe organ dysfunction
    - Aortic aneurysm
    - Heart
        - Uncontrolled malignant arrhythmia
        - Clinical evidence of congestive heart failure (New York class III and IV)
    - Lung
        - Airways collapse or severe obstruction
        - Severe pulmonary dysfunction with DLCO corrected for hemoglobin <45%
    - Kidney
        - Serum creatinine >2 mg/dl or glomerular filtration rate <40 ml/min

---

munosuppressive drugs can be considered as candidates for stem cell therapy. Moreover, the optimal candidate should show beginning signs of organ manifestations associated with a risk of developing life-threatening complications after the failure of conventional therapy. This especially includes vasculitis with the risk of serious cardiovascular complications and/or airway involvement that can result in a fatal collapse of the tracheobronchial system. On the other hand, progressive organ involvement may be a contraindication for ASCT because these patients have only limited tolerance to cytoreduction therapy (Table 3). The degree of damage to the patient's tracheobronchial system, cardiovascular system and kidneys should therefore be evaluated prior to the decision to perform stem cell therapy. The diagnostic tests should include a pulmonary function test, high-resolution computed tomography, bronchoscopy, electrocardiography, echocardiography, renal status, and unspecific markers of inflammation.

References

1. Jaksch-Wartenhorst R. Polychondropathia. Wiener Z Inn Med 1923; 6: 93-100.
2. Apostoloff E, Schlaak B, Mielke F et al. Über rezidivierende Polychondritis mit Nachweis von Kollagen II-Autoantikörpern-Kasuistischer Beitrag. Dt. Gesundh. -Wesen 1983; 38:81-84.
3. Foidart JM, Abe S, Martin GR et al. Antibodies to type II collagen in relapsing polychondritis. N Engl J Med 1978; 299:1203-1207.
4. Terato K, Shimozuru Y, Katayama K et al. Specificity of antibodies to type II collagen in rheumatoid arthritis. Arthritis Rheum 1990; 33:1493-1500.
5. Yang CL, Brinckmann J, Rui HF et al. Autoantibodies to cartilage collagens in relapsing polychondritis. Arch Dermatol Res 1993; 285:245-249.
6. Valenzuela R, Cooperrider PA, Gogate P et al. Relapsing polychondritis. Immunomicroscopic findings in cartilage of ear biopsy specimens. Hum Pathol 1980; 11:19-22.
7. Dolan DL, Lemmon GB Jr, Teitelbaum SL. Relapsing polychondritis. Analytical literature review and studies on pathogenesis. Am J Med 1966; 41:285-299.
8. Zeuner M, Straub RH, Rauh G et al. Relapsing polychondritis: Clinical and immunogenetic analysis of 62 patients. J Rheumatol 1997; 24:96-101.
9. Lang B, Rothenfusser A, Lanchbury JS et al. Susceptibility to relapsing polychondritis is associated with HLA-DR4. Arthritis Rheum 1993; 36:660-664.
10. McCune WJ, Schiller AL, Dynesius-Trentham RA et al. Type II collagen-induced auricular chondritis. Arthritis Rheum 1982; 25:266-273.
11. Cremer MA, Pitcock JA, Stuart JM et al. Auricular chondritis in rats. An experimental model of relapsing polychondritis induced with type II collagen. J Exp Med 1981; 154:535-540.
12. McAdam LP, O'Hanlan MA, Bluestone R et al. Relapsing polychondritis: Prospective study of 23 patients and a review of the literature. Medicine (Baltimore) 1976; 55:193-215.
13. Trentham DE, Le CH. Relapsing polychondritis. Ann Intern Med 1998; 129:114-122.
14. Michet CJ Jr, McKenna CH, Luthra HS et al. Relapsing polychondritis. Survival and predictive role of early disease manifestations. Ann Intern Med 1986; 104:74-78.
15. Tillie-Leblond I, Wallaert B, Leblond D et al. Respiratory involvement in relapsing polychondritis. Clinical, functional, endoscopic, and radiographic evaluations. Medicine (Baltimore) 1998; 77:168-176.
16. Lee-Chiong TL Jr. Pulmonary manifestations of ankylosing spondylitis and relapsing polychondritis. Clin Chest Med 1998; 19:747-57, ix.
17. Michet CJ. Vasculitis and relapsing polychondritis. Rheum Dis Clin North Am 1990; 16:441-444.
18. Handrock K, Gross WL. Relapsing polychondritis as a secondary phenomenon of primary systemic vasculitis. Ann Rheum Dis 1993; 52:895-897.
19. Firestein GS, Gruber HE, Weisman MH et al. Mouth and genital ulcers with inflamed cartilage: MAGIC syndrome. Five patients with features of relapsing polychondritis and Behcet's disease. Am J Med 1985; 79:65-72.
20. Hebbar M, Brouillard M, Wattel E et al. Association of myelodysplastic syndrome and relapsing polychondritis: Further evidence. Leukemia 1995; 9:731-733.
21. Myers B, Gould J, Dolan G. Relapsing polychondritis and myelodysplasia: A report of two cases and review of the current literature. Clin Lab Haematol 2000; 22:45-48.
22. Letko E, Zafirakis P, Baltatzis S et al. Relapsing polychondritis: A clinical review. Semin Arthritis Rheum 2002; 31(6):384-95.
23. Molina JF, Espinoza LR. Relapsing polychondritis. Baillieres Best Pract Res Clin Rheumatol 2000; 14:97-109.

24. Piette JC, el Rassi R, Amoura Z. Antinuclear antibodies in relapsing polychondritis. Ann Rheum Dis 1999; 58:656-657.
25. Isaak BL, Liesegang TJ, Michet Jr CJ. Ocular and systemic findings in relapsing polychondritis. Ophthalmology 1986; 93:681-689.
26. Papo T, Piette JC, Le Thi HD et al. Antineutrophil cytoplasmic antibodies in polychondritis. Ann Rheum Dis 1993; 52:384-385.
27. Chang-Miller A, Okamura M, Torres VE et al. Renal involvement in relapsing polychondritis. Medicine (Baltimore) 1987; 66:202-217.
28. Daniel L, Granel B, Dussol B et al. Recurrent glomerulonephritis in relapsing polychondritis. Nephron 2001; 87:190-191.
29. Ruhlen JL, Huston KA, Wood WG. Relapsing polychondritis with glomerulonephritis. Improvement with prednisone and cyclophosphamide. JAMA 1981; 245:847-848.
30. BarrancoVP, Minor DB, Soloman H. Treatment of relapsing polychondritis with dapsone. Arch Dermatol 1976; 112:1286-1288.
31. Ridgway HB, Hansotia PL, Schorr WF. Relapsing polychondritis: Unusual neurological findings and therapeutic efficacy of dapsone. Arch Dermatol 1979; 115:43-45.
32. Crockford MP, Kerr IH. Relapsing polychondritis. Clin Radiol 1988; 39:386-390.
33. Hoang-Xaun T, Foster CS, Rice BA. Scleritis in relapsing polychondritis. Response to therapy. Ophthalmology 1990; 97:892-898.
34. Svenson KL, Holmdahl R, Klareskog L et al. Cyclosporin A treatment in a case of relapsing polychondritis. Scand J Rheumatol 1984; 13:329-333.
35. Park J, Gowin KM, Schumacher HR Jr. Steroid sparing effect of methotrexate in relapsing polychondritis. J Rheumatol 1996; 23:937-938.
36. Choy EH, Chikanza IC, Kingsley GH et al. Chimaeric nti-CD4 monoclonal antibody for relapsing polychondritis. Lancet 1991; 338:450.
37. van der Lubbe PA, Miltenburg AM, Breedveld FC. Anti-CD4 monoclonal antibody for relapsing polychondritis. Lancet 1991; 337:1349.
38. Miltenyi S, Muller W, Weichel W et al. High gradient magnetic cell separation with MACS. Cytometry 1990; 11:231-238.
39. Kato K, Radbruch A. Isolation and characterization of CD34+ hematopoietic stem cells from human peripheral blood by high-gradient magnetic cell sorting. Cytometry 1993; 14:384-392.
40. McNiece I, Briddell R, Stoney G et al. Large-scale isolation of CD34+ cells using the Amgen cell selection device results in high levels of purity and recovery. J Hematother 1997; 6:5-11.
41. Storb R, Etzioni R, Anasetti C et al. Cyclophosphamide combined with antithymocyte globulin in preparation for allogeneic marrow transplants in patients with aplastic anemia. Blood 1994; 84:941-949.

CHAPTER 57

Allogeneic Hematopoietic Stem Cell Transplantation for Autoimmune Diseases

Shimon Slavin, Alberto Marmont and Richard K. Burt

Introduction

Autoimmune diseases result from self-reactive T-lymphocytes and autoantibodies, produced most likely in cooperation with T-cell dependent B-cells. Until recently, non-specific suppression of self-reactive lymphocytes or the inflammatory process mediated by the ongoing anti-self reactivity represented the main goal of therapy, but in the large majority of cases, neither cure nor remission can be obtained. Patients with severe, life threatening, manifestations of autoimmune diseases such as multiple sclerosis (MS), systemic lupus erythematosus (SLE) and rheumatoid arthritis (RA) may require long-term maintenance immunosuppressive treatment similar to organ allograft recipients, with all the anticipated side effects related to chronic immunosuppression on the one hand and side effects directly related to the immunosuppressive drugs (e.g., corticosteroids, cytotoxic agents and cyclosporin A to mention just a few) on the other. Unfortunately, none of the approaches available to date can offer effective and safe regulation of anti-self reactivity. Clearly, reinduction of unresponsiveness towards self antigens remains the yet unaccomplished final goal.

Recent data from animal models with induced or spontaneous autoimmune diseases,[1-6] as well as clinical observations in patients with autoimmune disease accompanying a primary malignancy,[7,8] suggest that reinduction of unresponsiveness to self antigens and alloantigens appears be a realistic goal. However, whereas induction of unresponsiveness to neo-antigens (primary response) may be relatively easy to accomplish, even when the ligands presented to the immune systems are strong alloantigens, re-induction of unresponsiveness in primed recipients with memory cells (secondary response) is much harder to accomplish, suggesting in analogy, that reinduction of unresponsiveness towards self-antigens may also be difficult to accomplish in patients with ongoing autoimmune diseases, since memory cells against self-antigens may be much more difficult to eliminate.

We will briefly review the basic concepts for allogeneic hematopoietic stem cell transplantation (HSCT) that makes us believe that it may be performed safely in patients with life threatening autoimmune diseases. Once proven to be feasible, safe and effective, allogeneic stem cell therapy may become an important modality for the treatment of otherwise incurable autoimmune diseases.

Rationale for Allogeneic HSCT as Treatment of Life Threatening Autoimmune Diseases

Allogeneic HSCT had been demonstrated in the 1970s and 1980s to cure animal autoimmune disorders. Case reports of patients with coincidental autoimmune diseases that were cured in patients undergoing allogeneic HSCT for malignancies or aplastic anemia were also reported around the same time. However, allogeneic HSCT were not developed as a treatment for autoimmune diseases because it remained a toxic therapeutic modality predominantly due to graft versus host disease (GVHD). In the early 1990s, autologous HSCT was demonstrated to cure some animal autoimmune diseases. Since autologous HSCT are not encumbered by the complication of GVHD, clinical trials using autologous transplants were subsequently initiated on patients with severe and refractory autoimmune disorders. It is thought that the conditioning prior to the transplantation procedure normally involves myeloablative treatment that could result in elimination of host-type immunohematopoietic cells, T cells included. The autologous stem cell rescue would result in regeneration of immunocompetent T cells tolerant to self-antigens from the pool of uncommitted stem cells administered. Theoretically, if the autograft is T-cell-depleted, all the newly regenerating T cells are likely to become tolerant to self-antigens, since high affinity self-reactive T cells are likely to undergo apoptosis in status nascendi in the thymus.

While autologous HSCT appears beneficial in ameliorating and perhaps even curing or inducing long-term remissions in some patients, others are relapsing after autologous HSCT. Different categories of animal disease require different sources of stem cells for cure. Environmentally induced autoimmune diseases (refer to Chapter 29) may be cured with syngeneic or autologous HSCT; spontaneous onset autoimmune diseases, while ameliorated by syngeneic HSCT, require allogeneic stem cells from a non-autoimmune prone strain for cure (Chapter 30). Finally, viral induced autoimmune disease, in which a latent viral infection persists, may be exacerbated by autologous HSCT but cured by an allogeneic HSCT from a virus resistant donor strain.[9] Since multiple genetic loci are involved in the development of human autoimmune diseases, it is likely that the more disease-associated genes present, the more likely an allogeneic source of stem cells would be needed for long term disease remission. Allogeneic HSC, even if HLA matched, would provide unique non-MHC autoimmune resistant genes (refer to Chapter 22).

In addition to replacement of host stem cells which may be genetically resistant to development of a particular autoimmune disorder, an unselected allogeneic graft may provide additional potential immunotherapy against self-reactive lymphocytes that may survive the chemoradiotherapy conditioning regimen. This has been termed a graft versus autoimmune (GVA) effect, theoretically mediated by donor lymphocytes. It is therefore important to understand the principles of immunotherapy of leukemia, including elimination of malignant lymphocytes by donor T-cells through a process termed graft versus leukemia (GVL), as a model for the feasibility of GVA by alloreactive donor T cells.[10] We and others have established the feasibility of inducing remission of a relapsed malignancy following allogeneic HSCT by donor lymphocyte infusion (DLI).[11,12] The data suggests that the major therapeutic component of allogeneic HSCT can be attributed to immunocompetent donor T lymphocytes recognizing and eliminating tumor cells of host origin. Closely similar effects have been observed following allogeneic HSCT in coincident autoimmune diseases.[8,13] In addition, a parallism between the insurgence of GVHD and the achievement of complete remission was reported after allogeneic SCT for refractory Evans syndrome.[14] In a similar case, the reinduction of complete clinical and immune remission was achieved following a series of DLI.[15,16]

Therefore, we postulate that allogeneic cell-mediated immunotherapy (i.e., GVA) inducible with DLI, may be used to displace autoreactive immune cells in patients with life-threatening autoimmune diseases. Hence, the use of allogeneic HSCT for effective treatment of autoimmune disorders can be envisioned by either changing the genetic susceptibility of the stem cells that generate the immune compartment and/or an immunotherapeutic GVA effect of donor lymphocytes against autoreactive host T cells. However, methods must be developed to prevent or minimize GVHD before allogeneic HSCT can become an accepted therapy for autoimmune diseases.

Currently, acute and chronic GVHD can be neither safely prevented nor adequately treated unless the graft is lymphocyte depleted. Since donor T and NK cells facilitate graft acceptance, lymphocyte depletion, which appears to be the only method for consistent prevention of GVHD, requires high myeloablative doses of chemotherapy or chemoradiotherapy to prevent graft rejection. However, successful transplantation of T cell depleted stem cells can be accomplished even with no supportive donor lymphocytes by using mega-doses of donor stem cells (>10^7 donor CD34$^+$ cells/kg). This overcomes the high incidence of rejection of purified stem cells, even in recipients of less than optimal conditioning.[17,18] In addition, complete replacement of host with donor hematopoietic cells, including eradication of immune cells of host origin, can be accomplished by applying incremental doses of DLI, even following HSCT with T cell depletion, with no early GVHD.[19,20]

Principles of Regimen for Allogeneic HSCT of Autoimmunity

Low dose intensity conditioning regimens, also known as non-myeloablative stem cell transplantation (NST), has replaced classical myeloablative allogeneic conditioning for the treatment of acute and chronic leukemia, non-Hodgkin's and Hodgkin's lymphoma, multiple myeloma as well as a large variety of non-malignant diseases with indication for allogeneic HSCT, including Fanconi's anemia, Gaucher's disease, Blackfan Diamond syndrome and β thalassemia.[21] Successful eradication of host stem cells and all cells derived from stem cells can be accomplished by NST, most commonly focusing on immunosuppression with fludarabine in combination with either busulfan, cytoxan, melphlan or low dose total body irradiation (TBI)[21-25] suggesting that donor T-cells present in the graft can eliminate all host type immunohematopoietic cells while avoiding the need for high dose chemotherapy or radiation therapy. In principle, such an approach could also be applied for patients with life-threatening autoimmune disease, rather than conventional myeloablative conditioning as was already documented as a proof of principle.[8] The use of NST may be useful for elimination of self-reactive lymphocytes of host origin, while replacing host with donor stem cells with no short or long-term conditioning regimen-related morbidity. Availability of safer and more effective NST procedures for patients with autoimmune diseases is likely to increase the use of such a biologic therapy mediated by immunocompetent lymphocytes for treatment of life-threatening autoimmune diseases.

Elimination and down-regulation of self-reactive T lymphocytes with DLI following non-myeloablative conditioning, by establishing mixed chimeras with bilateral transplantation tolerance of graft vs host and host vs graft responses, also holds a great clinical potential. Anti-host responses which may result in GVHD may be down-regulated by induction of anergy, mostly likely through a mechanism of "veto" induced by hematopoietic cells of host origin following establishment of mixed chimerism, thus resulting in repopulation of host cells with immunocompetent donor cells, yet without resulting in GVHD, as previously described in mice.[26-28] This is a theoretical suggestion since the details for successful induction of intentional mixed chimerism in man have not yet been established. However, this therapeutic option is supported by experimental animal data in the literature of mixed chimerism.[17,18] Unfortunately, the latter approach is still experimental and further progress is required for consistent successful induction of bilateral transplantation tolerance based on the bilateral veto capacity of host versus graft and graft versus host responses, using unmodified or T cell depleted allogeneic stem cells as has been shown in experimental animals.[29-33]

We predict that once tolerance is established, if autoimmune disease persists in mixed chimeras, donor derived T lymphocytes may be subsequently infused to increase the proportion and efficacy of displacement of host hematopoietic stem cells, thus converting the hematopoiesis to 100% donor type and inducing disease remission.

Suggested Scheme of Protocols for NST for Treatment of Autoimmune Diseases

As can be seen in Figure 1, based on available data in patients treated with NST for malignant or non-malignant diseases, it seems reasonable to design protocols that focus on immunosuppression rather than myeloablation. The most effective immunosuppressive agents available today include fludarabine, CAMPATH-1H (humanized anti-CD52) expressed on human lymphocytes, particularly T cells, polyclonal anti-T lymphocyte globulin (ATG) and cytoxan or other alkylating agents including busulfan, melphalan or low dose (200 cGy) total body irradiation (TBI) or total lymphoid irradiation (TLI). Since low dose radiation may subject recipients to the risk of late secondary malignancies, at least theoretically, we prefer regimens that do not contain TBI for conditioning of recipients of stem cell allografts. In general, protocols shown to be effective in patients with hematologic malignancies are likely to be effective also for the treat-

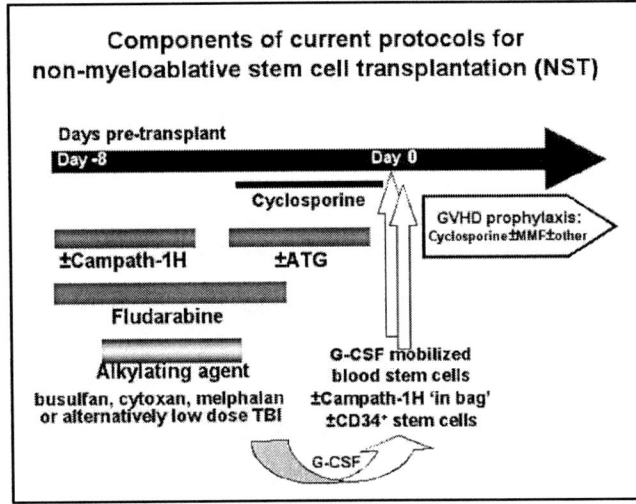

Figure 1. Outlines of protocol options for the use of non-myeloablative stem cell transplantation (NST) for the treatment of autoimmune diseases, focusing on lymphoablative rather than myeloablative conditioning, based on the cumulative experience in patients treated with reduced intensity conditioning for the treatment of malinant and non-malignant disorders.

ment of life threatening autoimmune diseases. Hence, for achieving engraftment, a combination of fludarabine and one of the alkylating agents with post-transplant immunosuppression with cyclosporine A (CSA), or tacrolimus (FK506), or either CSA or FK506 and mycophenolate mofetil (MMF). CSA may be given preventively to improve immunosuppression for ensuring prevention of rejection of the allograft. ATG appears to be effective for prevention of rejection, as well as for partial control of early GVHD since ATG remaining in the circulation may block T cells present in the graft. CAMPATH-1H is a humanized anti-lymphocyte antibody and therefore remains in the circulation for no less than 4 weeks, thus effectively reducing the risk of GVHD, but at the cost of increased risk of infections. Administration of CAMPATH-1H in the stem cell transfusion bag (CAMPATH-1H 'in the bag') appears to be useful for control of GVHD, however, it may be associated with increased risk of graft rejection.[20,34] In combination with fludarabine, an alkylating agents can be most effective even at doses that are well tolerated such as cytoxan 300 mg/m^2 for 3 days; busulfan 2-4 mg/kg for 2 days or melphalan 100 mg/m^2 and possibly less. TBI 200 cGy instead of alkylating agent is also very effective and well tolerated even by patients with poor performance status. Post grafting, CSA and/or other immunosuppressive agents must be carefully and slowly tapered off, because many of the patients are likely to develop late acute GVHD following discontinuation of anti-GVHD prophylaxis. The role of anti-lymphocyte antibodies remains unknown, because durable engraftment was observed by several transplant teams even without using ATG or CAMPATH-1H. Whereas CAMPATH-1H appears to be most effective for prevention of GVHD at the cost of more profound immunosuppression, the role of ATG in prevention of GVHD remains to be determined.

Disease Specific Conditioning Regimens

The regimen should be designed in terms of organ toxicity and risk of GVHD according to the severity of the autoimmune

Table 1. Scleroderma allogeneic NST regimen

Day	-6	-5	-4	-3	-2	-1	0
Cyclophosphamide (50 mg/kg/d)		X	X	X	X		
CAMPATH-1H (20 mg/d)	X	X	X	X	X		
Unmanipulated PBSC infused							X

disease. Therefore, no single regimen or approach would be appropriate for all autoimmune diseases.

Scleroderma

Scleroderma with diffuse skin involvement and/or visceral involvement has a 5 year mortality of 60% (12% mortality per year). This mortality is higher than chronic myelogenous leukemia (CML) or low grade lymphomas for which allogeneic HSCT is an accepted treatment. For a carefully designed regimen, the high mortality of scleroderma justifies the risk of allogeneic HSCT from an HLA matched sibling using unmanipulated bone marrow or peripheral blood stem cells. High risk patients eligible for this therapy may be defined as: An established diagnosis of scleroderma and two or more of the following high risk features:

a) Diffuse cutaneous scleroderma with involvement proximal to the elbow or knee and a Rodnan score (refer to Chapter 46) of >14.[85]

b) DLCO <80% of predicted or decrease in lung function (TLC, DLCO or FEV1) of 10% or more over 12 months.

c) Active alveolitis on bronchoalveolar lavage.

d) Pulmonary fibrosis or alveolitis on CT scan or CXR.

e) Elevated ESR ≥25 mm/hour.

f) Proteinuria (greater than trace on dipstick).

g) Urine blood on dipstick or sediment.

h) Abnormal EKG (non-specific ST-T wave abonormalities, low QRS voltage, or ventricular hypertrophy).

An example of an allogeneic NST protocol that is active at both Northwestern University (Chicago) and Hadassah University (Jerusalem) for scleroderma is shown in Table 1.

This regimen is very immune suppressive but not myeloablative. CAMPATH-1H due to its long half-life and effect on T cells and NK cells has been reported to decrease the incidence and severity of GVHD in allogeneic HSCT of malignancies and was therefore included in the regimen. Due to the higher morbidity of scleroderma, an unmanipulated stem cell product is given in order to theoretically preserve some lymphocyte mediated GVA effects. On the other hand, due to a high disease related mortality, the risk from unmanipulated PBSC of GVHD, albeit reduced by CAMPATH-1H, could be deemed acceptable. Scleroderma involves the lungs and heart and may compromise cardiovascular reserve to stress (infection or fever). We, therefore, exclude scleroderma patients with pulmonary artery hypertension >40 mm Hg from eligibility.

Rheumatoid Arthritis

In case reports of allogeneic HSCT in patients with coincidental rheumatoid arthritis (RA), almost all patients had sustained

Table 2. Rheumatoid arthritis allogeneic NST regimen

Day	-6	-5	-4	-3	-2	-1	0
Fludarabine (25 mg/m²/d)	X	X	X	X	X		
Cyclophosphamide (50 mg/kg/d)			X	X	X		
CAMPATH-1H 4 mg/d	X	X	X	X	X		
CD34⁺ selected stem cells							X

remissions without evidence of RA after HSCT. A subset of RA patients with more than 20 involved joints or significant limitations on their health assessment questionnaire have a 5 year mortality of 40%. For these patients, HSCT from an HLA matched sibling could be considered. High risk patients eligible for this therapy may be defined as: An established clinical diagnosis of rheumatoid arthritis by American College of Rheumatology criteria and failure to respond to at least 3 months of either methotrexate or leflunomide in combination with a TNF inhibitor. Failure being defined as at least 12 swollen joints and 20 involved joints or inability to answer at least 70% of HAQ questions with "no difficulty".

An example of an allogeneic NST protocol that is active at both Northwestern University (Chicago) and Hadassah University (Jerusalem) for RA is shown in Table 2.

Since RA is generally considered less morbid a disease than scleroderma, the graft is T cell depleted in order to avoid GVHD. The regimen, although non-myeloablative, is immune intense to improve CD34⁺ selected cell engraftment. High doses of CD34⁺ cells (>10^7/kg) will also be infused to facilitate engraftment. The scleroderma protocol uses donor T cells to obtain full chimerism. In contrast, the RA regimen with a non-myeloablative regimen and T cell depleted is attempting to obtain mixed chimerism (i.e., both recipient and donor hematopoeisis). If engraftment fails, autologous hematopoiesis will reconstitute and the rheumatoid symptoms will at least temporarily improve. DLI are not included in this initial phase I study. If RA symptoms persist despite mixed chimerism, small doses and gradual increments of DLI may be given to induce remission in future versions of the protocol. Low dose post transplant cyclosporine and cellcept will be used to help prevent graft rejection. The RA regimen being intensely immune suppressive and CD34⁺ selected will require aggressive post transplant infection prophylaxis for at least 6-12 months.

Future Directions

Experiments carried out in animal models of human diseases suggest that disease manifestations and immunological processes leading to the development of autoimmunity may be reversed by cyoreductive therapy followed by either syngeneic, autologous or preferably allogeneic stem cell transplantation. The available data suggest that some patients may benefit from autologous HSCT, preferably T-cell-depleted to prevent re-infusion of auto-reactive cells, others may require allogeneic HSCT for complete eradication of disease manifestation by establishment of a new genetic susceptibility, and by eradication of recipient autoimmune reactive cells.

In considering allogeneic HSCT for the treatment of autoimmunity, cumulative data from a cohort of patients treated successfully with NST suggests that a lower intensity of non-myeloablative conditions may be sufficient to provide a window of opportunity for induction of host vs. graft transplantation tolerance through either mixed or full donor chimerism. Clearly, safer forms of transplantation are mandatory for considering the use of allogeneic stem cell transplantation for the treatment of autoimmune diseases, especially diseases not associated with immediate life-threatening complications. Whereas patients with advanced autoimmune diseases may qualify for a relatively hazardous treatment modality, such patients, for example with MS or RA, to mention just a few, may already have irreversible manifestations of disease. On the other hand, it may not be justified to risk patients at an early stage of their disease, although such individuals are unquestionably the ideal candidates for reversal of autoimmunity if treated before manifesting irreversible signs of disease. For patients with a matched donor available, until the risks of acute and chronic GVHD will be better controlled, NST must be restricted only for patients with severe yet reversible and life-threatening autoimmune diseases.

Future allogeneic NST protocols may focus more on selective conditioning regimen depletion of self-reactive and graft-reactive recipient T and B cell clones, rather than non-specific lymphoablation, and/or transient post transplant blocking of costimulatory signals, providing 'signal 1' without 'signal 2', which is likely to help induce self tolerance. Protocols using cord blood stem cells transplants for severe autoimmune diseases such as systemic lupus erythematosus are also in development at Northwestern University. Whereas the use of autologous HSCT is already acceptable for the treatment of patients with life-threatening autoimmune diseases[35,36] it remains to be seen if allogeneic HSCT, particularly the better tolerated newer NST regimen, will provide a safer and effective tool in fighting, controlling or hopefully curing autoimmune diseases.[37]

Acknowledgments

We wish to thank Baxter International Corporation; the Gabrielle Rich Leukemia Research Foundation; the Danny Cunniff Leukemia Research Laboratory; The Cancer Treatment Foundation;The Szydlowsky Foundation; Joanne & David Morrison; Ronne & Donald Hess; Pepy & Jerry Silverstein for their continuous support of our ongoing basic and clinical research in cell therapy.

References

1. Slavin S. Successful treatment of autoimmune disease in (NZB/NZW)F1 female mice by using fractionated total lymphoid irradiation. Proc Natl Acad Sci USA 1979; 76:5274-76.
2. Rossini AA, Slavin S, Woda BA et al. Total lymphoid irradiation prevents diabetes mellitus in the bio-breeding/Worcester (BB/W) rat. Diabetes 1984; 33:543-47.
3. Karussis DM, Slavin S, Ben-Nun A et al. Chronic-relapsing experimental autoimmune encephalomyelitis (CR-EAE): Treatment and indution of tolerance, with high dose cyclophosphamide followed by syngeneic bone marrow transplantation. J Neuroimmunol 1992; 39:201-10.
4. Knaan-Shanzer S, Houben P, Kinwel Bohre EPM et al. Remission induction of adjuvant arthritis in rats by total body irradiation and autologous bone marrow transplantation. Bone Marrow Transplant 1991; 8:333-38.
5. Ikehara S, Inaba M, Yasumizu R et al. Autoimmune diseases as stem cell disorders. J Exp Med 1994; 173(1):141-55.
6. Karussis DM, Slavin S, Lehmann D et al. Prevention of experimental autoimmune encephalomyelitis and induction of tolerance with acute immunosuppression followed by syngeneic bone marrow transplantation. J Immunol 1992; 148:1693-98.

7. Marmont, AM. New horizons in the treatment of autoimmune diseases: Immunoablation and stem cell transplantation. Ann Rev Med 2000; 51:115-34.
8. Slavin S, Nagler A, Varadi G et al. Graft vs autoimmunity following allogeneic non-myeloablative blood stem cell transplantation in a patient with chronic myelogenous leukemia and severe systemic psoriasis and psoriatic polyarthritis. Exp Hematol 2000; 28(7):853-7.
9. Burt, RK, Padilla, J, Dal Canto MC et al. Viral hyperinfection of the central nervous system and high mortality following hematopoietic stem cell transplantation for treatment of Theiler's murine encephaleomyelitis virus (TMEV)– induced demyelineating disease. Blood 1999; 94(8):2915-22.
10. Slavin S, Weiss L, Morecki S et al. Eradication of murine leukemia with histoincompatible marrow grafts in mice conditioned with total lymphoid irradiation (TLI). Cancer Immunol Immunother 1981; 11:155-58.
11. Slavin S, Naparstek E, Nagler A et al. Allogeneic cell therapy for relapsed leukemia following bone marrow transplantation with donor peripheral blood lymphocytes. Exp Hematol 1995; 23:1553-62.
12. Slavin S, Naparstek E, Nagler A et al. Allogeneic cell therapy with donor peripheral blood cells and recombinant human interleukin-2 to treat leukemia relapse post allogeneic bone marrow transplantation. Blood 1996; 87:2195-2204.
13. Hinterberger W, Hinterberger-Fisher M, Marmont AM. Clinically demonstrable anti-autoimmunity mediated by allogeneic immune cells favourably affects outcome after stem cell transplantation in human autoimmune diseases. Bone Marrow Transplant 2002; 30(11)753-9.
14. Oyama Y, Papadopoulos EB, Miranda M et al. Allogeneic stem cell transplantation for Evans syndrome. Bone Marrow Transplant 2001; 28:903-5.
15. Marmont AM, Gualandi F, van Lint MT et al. Refractory Evans syndrome treated with allogeneic SCT followed by DLI. Demonstration of a Graft-versus-Autoimmunity effect. Bone Marrow Transplant 2003; 31(5):399-402.
16. Marmont AM. Historical perspectives, rationale and future directions for hematopoietic stem cell transplantation for severe autoimmune diseases. Haematologica 2002; 87(8):24-7.
17. Bachar-Lustig E, Rachamim N, Li HW et al. Megadose of T cell-depleted bone marrow overcomes MHC barriers in sublethally irradiated mice. Nat Med 1995; 1:1268.
18. Reisner Y, Martelli MF. Bone marrow transplantation across HLA barriers by increasing the number of transplanted cells. Immunol Today 1995; 16:437.
19. Naparstek E, Nagler A, Or R et al. Allogeneic cell mediated immunotherapy using donor lymphocytes for prevention of relapse in patients treated with allogeneic BMT for hematological malignancies. Clin Transplants 1996; 281-90.
20. Naparstek E, Or R, Nagler A et al. T-cell-depleted allogeneic bone marrow transplantation for acute leukaemia using Campath-1 antibodies and post-transplant administration of donor's peripheral blood lymphocytes for prevention of relapse. Brit J Haematol 1995; 89:506-15.
21. Slavin S, Nagler A, Naparstek E et al. Non-myeloablative stem cell transplantation and cell therapy as an alternative to conventional bone marrow transplantation with lethal cytoreduction for the treatment of malignant and non-malignant hematologic diseases. Blood 1998; 91(3): 756-63.
22. Giralt S, Estey E, Albitar M et al. Engraftment of allogeneic hematopoietic progenitor cells with purine analog-containing chemotherapy: Harnessing graft-versus-leukemia without myeloablative therapy. Blood 1997; 89:4531-36.
23. Khouri IF, Keating M, Korbling M et al. Transplant-lite: Induction of graft-versus-malignancy using fludarabine-based nonablative chemotherapy and allogeneic blood progenitor-cell transplantation as treatment for lymphoid malignancies. J Clin Oncol 1998; 16:2817-24.
24. Sweeney PA, Wagner JL, Maloney DG et al. Outpatient PBSC allografts using immunosuppression with low-dose TBI before, and cyclosporine (CSP) and mycophenolate mofetil (MMF) after transplant. Blood 1998; 92(1):519a (abstr. 2133).
25. Carella AM, Champlin R, Slavin S et al. "Mini-allografts": ongoing trials in humans. Bone Marrow Transplant 2000; 25(4):345-50.
26. Weiss L, Slavin S. Prevention and treatment of graft vs host disease by down-regulation of anti-host reactivity with veto cells of host origin. Bone Marrow Transplantation 1999; 23:1139-43.
27. Prigozhina T, Gurevitch O, Zhu J et al. Permanent and specific transplantation tolerance induced by a non-myeloablative treatment to a wide variety of allogeneic tissues. Transplantation 1997; 63;(10):1394-99.
28. Prigozhina T, Gurevitch O, Slavin S. Non-myeloblative conditioning to induce bilateral tolerance after allogeneic bone marrow transplantation in mice. Experimental Hematology, 1999; 27:1503-10.
29. Slavin S, Strober S, Fuks Z et al. Long-term survival of skin allogafts in mice treated with fractionated total lymphoid irradiation. Science 1976; 193:1252-54.
30. Slavin S, Strober S, Fuks Z et al. Induction of specific tissue transplantation tolerance using fractionated total lymphoid irradiation in adult mice: Long-term survival of allogeneic bone marrow and skin grafts. J Exp Med 1977; 146:34-48.
31. Ildstad ST, Sachs DH. Reconstitution with syngeneic plus allogeneic or xenogeneic bone marrow leads to specific acceptance of allografts or xenografts. Nature 1984; 307:168.
32. Slavin S. Total lymphoid irradiation (TLI). Immunol Today 1987; 8:88-92.
33. Sharabi Y, Abraham VS, Sykes M et al. Mixed allogeneic chimeras prepared by a non-myeloablative regimen: Requirement for chimerism to maintain tolerance. Bone Marrow Transplant 1992; 9:191.
34. Naparstek E, Delukina M, Or R et al. Engraftment of marrow allografts treated with Campath-1 monoclonal antibodies. Experimental Hematol 1999; 27:1210-18.
35. Gratwohl A, Passweg J, Gerber I et al. Stem cell transplantation for autoimmune diseases. Best Practice & Research Clin Haematol 2001; 14: 755-76.
36. Moore J, Brooks P. Stem cell transplantation for autoimmune diseases. In: Harrison WB, Dijkmans BAC, eds. Combination treatment in autoimmune diseases. Berlin-Heidelberg-New York: Springer, 2002: 193-213.41.
37. Burt RK, Traynor AE, Craig R et al. The promise of hematopoietic stem cell transplantation for autoimmune diseases. Bone Marrow Transplant 2003; 31:521-524.

INDEX

A

α-7-integrin 65
α-melanoyte-stimulating hormone (α-MSH) 135
α-myosin heavy chain 40, 41
α1-adrenergic receptor 42, 43, 45
Acetylcholine (ACh) 45, 107, 160, 239-241, 429, 430
Acetylcholine receptor (AChR) 107, 160, 239-241, 429-432
Acetylcholine receptor inducing activity (ARIA) 430
Acid fibroblast growth factor (aFGF) 18
Acute lymphocytic leukemia (ALL) 32, 33, 257
Acute myeloid leukemia (AML) 32, 102, 226, 228, 257, 352, 369
Acute T cell leukemia lymphoma (ATLL) 101, 102
Acyclovir 270, 271, 305, 382, 402
Adenovirus (AV) 28, 185, 186, 267, 268, 382, 437
Adipocytes 24, 26, 27, 29, 35-37, 39, 40, 79
Adipose derived adult stem cells (ADAS cells) 24, 26-29
Adrenergic receptor 42, 43, 45, 186
Adult stem cell 2, 3, 5, 15, 20, 24, 26, 27, 29, 35, 37, 59-61, 68, 69
Agrin 429, 430
Aldehyde dehydrogenase (ALDH) 31-33, 233
Alemtuzumab 266, 401
Allogeneic HSCT 5, 8, 117, 212-214, 217, 225-228, 253, 259, 269, 280, 311, 348, 350, 352, 369, 375, 415, 443, 453, 461, 463, 474-477
Alopecia 328, 329, 331-333, 336, 360, 362, 415, 435
Ambulation index (AI) 291, 293, 306
Amyloid precursor protein (APP) 284
Anakinra 358, 362, 364, 365
Angiogenesis 47, 49-53, 65
Angiotensin converting enzyme inhibitor (ACEI) 259
Anti-centromere antibody (ACA) 390-392
Anti-coagulation 256, 350, 353
Anti-convulsants 175
Anti-dsDNA antibodies 166, 168, 169, 350
Anti-hemocyanin antibody 460
Anti-phospholipid 107
Anti-Rh(D) 404
Anti-thymocyte globulin (ATG) 214, 215, 226, 234, 241, 242, 254, 256-259, 263, 266, 269, 270, 304, 305, 307, 308, 312, 314-318, 344, 349-355, 367, 369, 371-373, 380, 382, 399-402, 432, 436, 439, 443, 454, 470, 475, 476
Antiarrhythmics 175
Antibiotic 77, 174, 175, 186, 234, 270, 305, 306, 348, 372, 382, 404, 406, 407, 430, 451, 453, 471
Antigen presentation 108, 113, 145, 157, 158
Antigen presenting cell (APC) 113-117, 133, 135, 136, 139, 140, 142, 148-151, 156, 157, 177, 180, 182-186, 188, 190, 194, 195, 207, 245, 448, 459
Antigenic mimicry 106
Antihypertensives 175, 347
Antimalarial drugs 359
Antinuclear antibodies (ANA) 145, 161, 174, 176, 177, 180, 228, 237, 238, 265, 321, 329, 330, 339, 400, 347, 350, 352, 379, 391, 402, 439, 464, 469
Antiphospholipid syndrome 108, 223, 247, 329, 347, 348, 350, 353
Antipsychotics 175
Aplastic anemia (AA) 32, 226, 227, 232-235, 237, 241, 242, 256, 257, 349, 368, 408, 474
APO-1 115, 124, 126
Apoptosis 47, 56, 75, 76, 106, 108, 109, 114, 115, 119-128, 135, 169, 184, 189, 195, 197, 206, 209, 253, 317, 323, 349, 474
Aspergillus fumigatus 264
Atgam® 266
Atrial natriuretic peptide (ANP) 39, 45, 46
Auranofin 359
Autoimmune idiopathic thrombocytopenia (AITP) 226, 404-408
Autoimmune lymphoproliferative syndrome (ALPS) 31, 109, 115, 127, 128, 133
Autoimmune polyendocrinopathy candidiasis ectoderm 223
Autoimmune polyglandular endocrinopathy with candidiasis and ectodermal dysplasia (APECED) 109, 223
Autoimmune regulator (AIRE) 143, 223
Autoimmune thrombocytopenia 406
Autoimmune thrombocytopenic purpura 225, 226, 228, 349
Autoimmunity 9, 31, 33, 106-112, 114-116, 119, 127, 128, 135, 136, 139, 143-145, 147, 148, 151, 157, 158, 164-166, 168, 169, 173, 174, 176-180, 182-190, 215, 223-227, 232, 240, 243, 259, 280, 320, 325, 348, 350, 353, 355, 405, 408, 457, 462, 463, 475, 477
Autologous HSCT 8
Autologous stem cell transplantation (ASCT) 232, 242, 243, 264, 266, 267, 303-309, 318, 344, 347, 371, 373, 374, 385, 399, 400, 453, 468, 470-472
Autologous Stem Cell Transplantation International Rheumatoid Arthritis (ASTIRA) trial 367, 374
Axon 11, 14, 63, 284, 286, 287, 317, 421, 422
Axonal loss 284-288, 297, 299, 426
Azathioprine 265, 295, 303, 332, 335, 339, 343, 344, 355, 398, 405, 413, 423, 424, 426, 435, 437, 439, 443, 451, 453, 460, 464, 470

B

β-adrenergic receptor 42, 43
β-gal gene 82
β-galactosidase 26, 67
β-myosin heavy chain (B-MHC) 40, 65
β1 integrin family 75
β1-adrenergic receptor 186
β2-adrenergic receptor 42, 458
B cell 32, 74, 78, 79, 108, 109, 113, 114, 116, 117, 119, 135, 139, 141, 142, 166, 168, 169, 177, 179, 180, 182, 187, 189, 194-196, 199-202, 206, 208-214, 216, 217, 223, 228, 242, 245-247, 266, 278, 279, 320, 321, 344, 347, 354, 373, 374, 390, 402, 405, 429, 434, 438, 442, 448, 449, 459, 461, 471, 474, 477
B cell activation factor from the TNF family (BAFF) 169
B cell receptor (BCR) 108, 113, 114, 116, 200, 206, 209, 210
B lymphocyte 113, 119, 166, 169, 194, 206, 209, 321, 389, 404, 405, 464
B lymphocyte antigen receptor 113
B7-1 114, 459
    see also CD80
B7-2 114, 459
    see also CD86
Barthel index 291, 425, 426, 439
Basic fibroblast growth factor (bFGF) 25, 28, 49, 51, 80
Bcl-2 21, 62, 75, 119, 120, 123, 133, 169
Bcl-2 homologue antagonist/killer (Bak) 120
Bcl-2 inhibitor death agonist (Bid) 120
Bcl-2 interacting mediator (Bim) 120
BCNU 100, 257, 304, 312, 314, 315, 317, 349
BCNU, etoposide, cytosine arabinoside and melphalan (BEAM) 256-258, 305, 307, 308, 312, 314-318, 349, 351, 369, 426
Beta cell 7-9, 226, 442, 444
Beta-IFN-1a 298, 299
Beta-IFN-1b (Betaseron) 298, 299
BILAG Index 331
Birkbeck granules 140
Birmingham Vasculitis Activities Score (BVAS) 411-414, 417
Blastocyst 1, 2, 5-7, 15, 16, 59, 60, 79
Blindness 415, 435, 442
Blood-brain barrier (BBB) 11, 63, 278, 295, 296, 298
BlyS 169
Bone marrow derived stem cells (BMSC) 20, 59-69
Bone marrow transplantation (BMT) 1, 3, 24, 31-33, 76, 77, 82, 98, 107, 115, 134, 136, 150, 177, 198, 212-214, 224-228, 233, 234, 237-243, 245-248, 250, 269, 279, 280, 308, 310-313, 315, 352, 368, 369, 400, 448, 451, 453, 455, 463, 470
Bone morphogenic protein (BMP) 13, 27, 28
Bone morphogenetic protein-2 (BMP-2) 27, 28
Borrelia burgdorferi 185, 186, 188, 190
Brain 3, 11-15, 24, 25, 27, 35, 36, 39, 45, 46, 63, 64, 68, 80, 89, 147, 182, 185, 188, 239, 240, 242, 263, 277-280, 286, 287, 292, 294-299, 303, 306-308, 315, 317, 331, 333
    atrophy 278, 286, 294, 299, 308, 315, 317
Brain natriuretic peptide (BNP) 39, 45, 46
Bronchial asthma 457-463, 465
Budesonide 451
Busalfan/cyclophosphamide (BuCy) 367, 369, 373
Bystander activation 183, 184, 187-190, 311, 442

C

C-kit 19-21, 35, 55, 61, 62, 64-66, 68
C-MetR 65
C-peptide 443, 444
C. glabrata 272
C. krusei 272
CAMPATH 228, 258, 259, 260, 266, 318, 374, 375, 400, 444, 475-477
Canale-Smith syndrome 166
Candida albicans 139, 146, 264, 272
Cardiac
    hypertrophy 42, 45
    myosin 186, 189
    stem cell 39, 81
Cardiomyocyte 15, 29, 36, 39-45, 47, 53, 60, 65, 81
Cardiomyogenic (CMG) cell 39-47
Cardiomyopathy 66, 186, 336, 413-415
Carmustine 258, 314, 426
Cartilage 24, 27, 29, 35, 36, 57, 60, 65, 89, 263, 468-470
Caspase 120, 122-128, 197, 400
CD1 115, 169, 194
CD4⁺CD25⁺ Tr cell 148, 151, 152
CD10 26, 28
CD11 26, 101
CD11/CD18 101
CD13 26, 28, 96, 141
CD18 99, 101, 321
CD22 116, 169
CD30 116, 120, 389
CD31 50, 51, 216, 217
CD33 93, 96, 141
CD34 19-21, 26, 28, 31-33, 36, 49-53, 62, 64-66, 74-78, 92, 93, 95-101, 104, 105, 136, 141, 142, 145-147, 208, 212, 214, 215, 232, 233, 242, 255, 256, 258, 259, 266-270, 280, 304, 305, 312, 314-318, 348-350, 352, 355, 367, 369-375, 380, 399, 400, 402, 406, 416, 426, 439, 444, 454, 470, 475, 477
CD38 74, 75, 78, 95, 97, 141, 399
CD44 26, 28, 64, 88
CD45 26, 64-67, 78, 79, 207, 215, 280, 323, 324, 373, 449
CD56 26, 209, 211, 215, 384
CD69 149, 166, 168, 179, 197, 198, 344
CD79 209
CD80 114, 139, 145, 459
    see also B7-1
CD86 114, 139, 141, 143, 145, 166, 459
    see also B7-2
CD95 115, 119-128, 198
CD95L 119-124, 126-128, 198
CD95L/CD95 119-121, 123
CD123 141
CD154 116, 198, 226, 459
CD205 140
Celiac disease 160, 237
CFLIP 122
Chagas disease 185, 186, 188, 189
Child health assessment questionnaire (CHAQ) 259, 379-381, 383, 385, 386
Chondrocytes 24, 27, 29, 35-37, 79
Chronic inflammatory demyelinating polyneuropathy (CIDP) 33, 235, 419-424, 426, 439
Chronic myeloid leukemia (CML) 5, 32, 226, 257, 349, 476
Churg-Strauss-syndrome 411-413, 417, 469
Cicatricial pemphigoid (CP) 434, 435
Ciliary neurotrophic factor (CNTF) 13

Class switching recombination (CSR) 209, 210
CLIP 157, 158
Clonal selection 132
Clostridium difficile 266
Cobblestone area-forming cell (CAFC) 8, 74, 93, 96
Colchicine 398, 469
Cold agglutinin disease (CAD) 223, 235
Collagen 26-28, 47, 79, 87-89, 108, 109, 116, 158, 239, 263, 368, 388-391, 398, 435, 437-439, 464, 468, 469
Collagen-induced arthritis (CIA) 108, 158, 239, 368
Complement 11, 26, 68, 69, 94, 99, 101, 107-109, 114, 136, 155, 164, 168, 169, 176, 177, 183, 194, 200, 201, 206, 233, 263, 264, 278, 279, 320, 321, 328, 332, 333, 335, 339-342, 345, 390, 438, 468
  deficiency 111, 164
Complementarity determining region (CDR) 194, 200
Complete Freund's adjuvant 107, 116, 145, 189, 243, 429
Conditioning regimen 32, 33, 73, 92, 136, 212, 214, 215, 217, 225, 227, 228, 232-235, 240-243, 253, 254, 256, 258-260, 262, 265, 266, 268-271, 280, 304-307, 309, 310, 312, 315-317, 349, 352, 354, 367, 374, 375, 380, 381, 399, 400, 417, 426, 432, 436, 439, 443, 444, 463, 475-477
Congenic dissection 166, 167, 169
Connective tissue growth factor (CTGF) 89
Cord blood (CB) 8, 28, 49, 73-75, 77-79, 92, 93, 99-101, 141, 207, 211, 212, 214, 225, 227, 348, 440, 461, 463, 477
Cornea 68, 89, 185, 187, 189, 190
Cortical thymic epithelial cells (CTEC) 158
Corticosteroid 9, 232, 254, 259, 260, 263-265, 271, 272, 297, 299, 316, 317, 320, 330, 332, 343, 347, 348, 350, 358, 359, 365, 367, 370, 371, 378, 384, 398, 399, 404-406, 411, 413, 416, 419, 421, 423, 430, 431, 434, 435, 437-439, 451, 453, 454, 460, 462, 464, 468-471, 474
Costimulation 116, 133, 134, 459
Coxsackie virus (CV) 185-189
Creatinine clearance 331, 353, 354
Crohn's disease 145, 147, 159, 160, 165, 183, 224, 226, 228, 237, 238, 254, 257, 318, 448-451, 453, 454, 469
Crohn's disease activity index (CDAI) 450, 451, 453, 454
Cryptic antigen 183-185, 189
Crytpococcus neoformans 264
CXCR4 20, 75, 93, 198
Cyclic adenosine monophosphate response element modulator (CREM) 323, 325
Cyclic adenosine monophosphate response element-binding (CREB) 63, 325
Cyclin-dependent kinase 2 (Cdk2) 120
Cyclophosphamide (CY) 32, 33, 215, 226, 227, 232-235, 238, 239, 241, 242, 250, 253-260, 265, 266, 280, 303-312, 314-318, 332, 335, 339, 343, 344, 347-355, 367-375, 380, 399-402, 405, 406, 408, 411, 413, 416, 417, 423, 424, 426, 429-432, 434-437, 439, 443, 453, 454, 462-464, 470, 476, 477
Cyclosporin (CSA) 134, 173, 174, 177, 213, 233, 234, 241, 260, 263-265, 304, 339, 343, 344, 355, 364, 369, 371-373, 378, 380, 383, 399, 404, 408, 423, 430, 431, 435, 437, 439, 443, 451, 460, 462, 465, 470, 474, 476, 477
Cytokine 6, 8, 15, 24, 26, 28, 36, 39, 47, 49, 55-57, 74, 75, 78, 81, 88, 96-98, 101, 108, 109, 113-117, 119, 121-123, 128, 133, 135, 139, 141, 143-145, 148-152, 155, 165, 169, 174, 183, 184, 186, 189, 194, 196-199, 208, 209, 211, 212, 214-217, 232, 233, 253, 254, 265, 278, 279, 284, 295, 304, 309, 317, 339, 345, 348, 362, 364, 374, 389-391, 408, 421, 444, 449-451, 457-460, 464
Cytokine synthesis inhibitory factor (CSIF) 135
Cytokine toxicity 174

Cytomegalovirus (CMV) 80, 94, 151, 185, 187-189, 263-271, 305, 307, 312, 315, 316, 378, 388, 402, 437, 442, 461, 471
Cytosine arabinoside 79, 258, 349
Cytotoxic T lymphocyte-associated molecule-4 (CTLA-4) 115, 136, 144, 152

D

D-penicillamine 175, 358, 359, 398, 399, 432, 437
Danger theory 132
DC-SIGN 143, 147
Death domain (DD) 121-128, 333, 336
Death effector domain (DED) 125-127
Death inducing signaling complex (DISC) 126, 127
Dectin-1 143, 147
Deep venous thrombosis 353
Delayed type hypersensitivity (DTH) 188, 189, 210, 309, 311, 449
Dermatomyositis 264, 329, 385, 437-440
Diabetes control and complications trials (DCCT) 443
Diabetes models 189
Diffuse proliferative lupus nephritis (DPLN) 340-342, 344
Disability 259, 277, 281, 284-288, 290-294, 296-299, 303, 307, 311, 312, 314-318, 332, 335, 347, 358, 360, 367, 368, 372, 373, 383, 398, 402, 423, 424, 426, 435, 439, 450, 470
Discoid lupus 328
Disease activity score (DAS) 358, 372
Disease modifying anti-rheumatic drug (DMARD) 358, 359, 364, 365, 367-369, 372, 373, 375, 378
Disease progression 189, 279, 281, 286, 288, 290, 298, 299, 308, 309, 315, 361, 374, 400, 470, 471
Drug-induced lupus 173-180, 342

E

Elastase 52, 132
Electromyography (EMG) 429, 431, 432, 437-439
Elf-1 321, 323, 325
Embryonal carcinoma (EC) 5-7, 39, 147
Embryonic stem cell 1, 2, 3, 5, 14, 15, 20, 27, 35, 36, 59, 60, 68, 70, 72, 79, 81, 86-88, 90, 95
Encephalitis 188, 238, 245, 309
End stage renal failure (ESRF) 256, 415
Endothelial progenitor cells (EPC) 49-51, 81
Enhancer binding (E) proteins 120
Enterocytes 66, 208, 448, 449
Enterovirus 185, 186, 267, 268
Env 93, 94
Enzyme-linked immunospot (ELISPOT) assay 197
Eosinophilia-myalgia syndrome 173
Epidermolysis bullosa acquisita (EBA) 434, 435
Epitope spread 107, 109, 116, 183, 184, 188, 189, 309, 311, 404
Epstein-Barr virus (EBV) 107, 151, 185, 186, 258, 263, 267-269, 278, 316, 378, 382, 402, 437, 444
Erythematosus 33, 106, 114, 157, 159, 161, 164, 173, 175, 176, 180
Erythropoietin (Epo) 76
Escherichia coli 264, 364
Escola Paulista de Medicina range of motion (EPM-ROM) 379, 380, 382, 384
Etanercept 108, 228, 264, 265, 358, 362-365, 371, 372, 439
Etoposide 257, 258, 264, 304, 312, 314, 349, 370, 399, 405, 426, 439

European consensus lupus activity measure (ECLAM) 334
Evoked potentials (EP) 290, 295, 306
Expanded disability status scale (EDSS) 281, 290, 291, 303
　　see also extended disability status scale
Experimental allergic myasthenia gravis (EAMG) 224, 239-241, 429, 432
　　see also experimental autoimmune myasthenia gravis
Experimental allergic neuritis (EAN) 419, 420
Experimental autoimmune encephalomyelitis (EAE) 108, 109, 128, 152, 167, 188, 215, 223, 225, 226, 237-243, 245, 246, 254, 278-280, 285-288, 309-312, 316, 348
Experimental autoimmune myasthenia gravis (EAMG) 239, 429
　　see also experimental allergic myasthenia gravis
Experimental autoimmune thyroiditis 108
Experimental autoimmune uveoretinitis 108
Extended disability status scale (EDSS) 281, 285-287, 290-292, 297-299, 303-307, 312, 314-318
　　see also expanded disability status scale
Extracellular matrix (ECM) 47, 87-90, 95, 388, 389
　　scaffold 89, 90

F

Fabry disease 99, 101
Facial erythema 328
FADD-like IL-1b-converting enzyme (FLICE) 122, 126
Familial hemophagocytic lymphohistiocytosis (FHL) 115, 382-384
Fanconi anemia (FA) 99, 101
Fas 31, 32, 62, 75, 109, 115, 121-128, 133, 136, 166, 168, 169, 180, 197, 198, 232, 233, 321, 349
Fas ligand 32, 75, 115, 128, 133, 197, 198, 321
Fas-associated death domain (FADD) 121, 122, 124-127
Fastimmune™ assay 408
Fcγ receptor 107, 165
Felty's syndrome 33
Ferritin 264, 384
Fibronectin 19, 28, 36, 79, 80, 87-89, 95, 389, 390
Flk-1 6
Flt-1 6
Fluconazole 266, 270-272
Fludarabine 242, 258, 259, 354, 373, 375, 399, 400, 426, 444, 475-477
Fluid attenuated inversion recovery (FLAIR) 295, 296
Focal proliferative lupus nephritis (FPLN) 340, 341
Follicle associated epithelium (FAE) 448, 449
Fragmin 256, 353
Freund's adjuvant 368
Fuchs swelling index (FSI) 379, 383

G

GABA 12
Gadolinium 227, 286, 287, 295, 297, 303, 312, 314, 315, 317
Gag gene 93, 95, 98
Ganciclovir 270, 271, 305, 402, 471
Gastrointestinal tract 66, 135, 263, 267, 336, 398, 448
Gaucher disease 101
Gd 295-299, 303-308
Gene
　　expression 14, 16, 41, 45, 55, 56, 66, 87-89, 92, 98, 101, 122, 132, 139, 146, 149, 166, 168, 199, 200, 212, 325, 389
　　marking 92, 98, 100, 101
　　transfer 92-101
Genital ridge 2, 79

Glatiramer acetate 277, 288, 294, 298, 299, 303
Gld mice 127
Glial cells 5, 11-13, 284, 285
Glial fibrillary acidic protein (GFAP) 28, 64
Glial growth factor-2 (GGF2) 13
Glomerulonephritis 109, 164-169, 176, 225, 238, 340, 342, 349, 350, 391, 412, 416, 438, 469
Glucocerebrosidase (GC) 99-101
Glucocorticoid 120, 264, 265, 295, 460, 472
Glucocorticoid receptors (GR) 120
Glutamic acid decarboxylase (GAD) 116, 187, 189, 442, 444
Glutamic acid decarboxylase-65 (GAD65) 116, 187, 442, 444
Glycosaminoglycans (GAGs) 87, 88, 389
Glycosylated hemoglobin 442-444
Gold 173, 174, 176, 177, 226, 255, 328, 336, 358, 359, 370, 435
Goodpasture's syndrome 107
Gp130 6, 7
Graft versus autoimmune (GVA) 224, 225, 227, 228, 240, 242, 253, 259, 355, 368, 375, 408, 462, 475, 476
　　effect 225, 227, 228, 242, 253, 375, 475, 476
Graft versus host disease (GVHD) 62, 73, 77, 107, 110, 150, 151, 173, 177, 180, 212-214, 217, 226-228, 241, 242, 245, 250, 259, 260, 262, 269, 317, 368, 369, 375, 391, 392, 408, 430, 444, 461, 462, 474-477
Graft versus leukemia (GVL) 225, 240, 253, 259, 475
Granulocyte-colony stimulating factor (G-CSF) 28, 35, 57, 64, 75-78, 92, 96, 101, 214, 232, 254, 255, 260, 312, 314-316, 348, 350, 354, 367, 369, 370, 372, 374, 380, 399-402, 406, 407, 416, 439, 453, 463, 470
Graves disease 107, 159, 160, 437
Green fluorescent protein (GFP) 57, 63, 64, 66, 68, 69, 78, 95, 280
Guinea pig myelin basic protein (GMBP) 243
Gut associated lymphoid tissue (GALT) 135, 209, 448

H

Haemophilus influenzae 188, 263
Hair follicle 57, 67, 68
Hashimoto's
　　disease 107
　　thyroiditis 110, 159, 437
Heart 3, 36, 39, 41, 42, 45-47, 53, 59, 60, 65, 68, 81, 89, 145, 185, 186, 189, 227, 263, 331, 378, 398, 406, 411, 413, 438, 468, 470, 472, 476
Helicobacter pylori 404
Hematopoiesis 1, 3, 6, 20, 31-33, 56, 57, 74-76, 78, 232-234, 256, 318, 375, 399, 444, 475, 477
Hematopoietic stem cell (HSC) 3, 5, 6, 8, 9, 15, 18, 24, 28, 31-33, 35, 36, 39, 49, 53, 56, 60-62, 64-69, 73-75, 81, 89, 92-101, 117, 128, 145, 146, 150, 152, 195, 200, 206, 208, 220-228, 232, 233, 235, 237, 241, 245, 247, 253, 254, 256, 257, 259, 267, 271, 275, 279, 280, 308, 312-318, 320, 347, 348, 351, 358, 365, 375, 378, 379, 399, 401, 405-407, 411, 419, 424, 429, 430, 434, 435, 437, 442-445, 448, 451, 453, 454, 457, 461, 464, 465, 474, 475
Hematopoietic stem cell transplantation (HSCT) 5, 7, 8, 28, 92, 93, 117, 150, 195, 206, 212-217, 223-228, 232, 239, 253-257, 259, 260, 262, 263, 265-272, 277, 279-281, 309, 311-318, 320, 345, 347-355, 358, 365, 367-376, 378-386, 399-402, 411, 413-417, 419, 424, 426, 429-432, 435-437, 439, 440, 442-444, 448-451, 453, 455, 457, 458, 460-465, 474-477
Hemolytic anemia 173, 175, 176, 225, 235, 328, 330, 335, 405, 406
Hemorrhagic cystitis 256, 265, 304, 307, 315, 343, 406
Hepatic stem cell 5, 20, 81

Hepatitis
　A virus 461
　B surface antigen 200, 269
　B virus (HBV) 94, 270, 437
　C (HCV) 21, 62, 94, 109, 185, 186, 269, 270, 368, 405, 412
Hepatoblast 21, 61
Hepatocyte growth factor (HGF) 18, 81, 88, 89
Herpes simplex virus (HSV) 267-270, 402
Herpes simplex virus-1 (HSV-1) 185, 187-190
Herpes zoster 264, 317, 343
Herpetic stromal keratitis (HSK) 187, 189, 190
HgbA1c 442-444
High proliferative potential colony forming cell (HPP-CFC) 57, 74, 75, 93
High-dose immunosuppression 232, 352, 370, 399, 457
HIV-1 94, 97, 98, 101
HIV-II 94
HOX 75
HTLV-I 94
HTLV-II 94
Human herpes virus-6 (HHV-6) 185, 186, 270
Human leukocyte antigen (HLA) 8, 9, 26, 28, 31, 32, 77, 93, 95, 106, 109, 141, 149, 150, 155-161, 183, 186, 190, 197, 198, 210, 212-215, 227, 228, 233, 235, 238, 239, 243, 256, 259, 277-279, 320, 347, 352, 373, 375, 388-391, 399, 407, 415, 430, 437, 442, 444, 461-463, 468, 471, 475-477
Humoral immunity 108, 148, 149, 196, 199, 200, 263
Hyaluronic acid 88
Hydralazine 173-177, 342
Hydroxychloroquine 330, 358, 364, 371, 372, 439

I

Idiopathic inflammatory myositis (IIM) 264, 391, 437-440
Idiopathic pulmonary fibrosis (IPF) 457, 463-465
Idiotypes 107, 108, 110, 111
IgA 106, 135, 160, 199-201, 209, 212, 340, 434, 435, 448, 449
IgD 199, 201, 213, 216, 471
IgE 108, 142, 199-201, 325, 458-462
IgG 166, 168, 177-179, 185, 186, 199, 200, 206, 209, 212, 235, 269, 278, 295, 329, 330, 339, 340, 373, 400, 419, 420, 435, 464, 468
IgM 161, 166, 168, 177, 185, 187, 200, 201, 209, 212, 223, 235, 329, 330, 340, 419, 459
Immunoglobulin (Ig) 206, 209, 210, 212, 214, 223, 265, 267, 269, 279, 295, 320, 321, 330, 340, 341, 344, 363, 404, 405, 419, 423, 430, 435, 448, 464, 468, 470
Immunoglobulin (IgSF) superfamily 156, 169
In vitro fertilization 2, 6, 7
Inducible cyclic adenosine monophosphate early repressor (ICER) 325
Inflammatory bowel disease 128, 159-161, 421, 448-450, 452, 453
Inflammatory bowel disease questionnaire (IBDQ) 450, 452, 453
Inflammatory demyelination 279, 284, 285, 287, 317
Infliximab 108, 228, 265, 358, 362-365, 372, 374, 451, 453
Inhibitor of DNA binding (Id) protein 120, 141
Innate immunity 113, 114, 143, 182
Insulin 7, 18-20, 25-27, 107, 108, 116, 132, 187, 189, 223, 225, 237, 246, 348, 442-445
Insulin-dependent diabetes mellitus (IDDM) 107, 108, 223, 348
Insulin-like growth factor II (IGF-II) 18
Interferon beta 1a 294, 306
Interferon-β 277, 281
Interleukin-1 (IL-1) 49, 55, 75, 76, 108, 121, 122, 126, 145, 174, 211, 362, 364, 389, 390

Interleukin-3 (IL-3) 55, 56, 75, 76, 96, 99-101, 141, 145, 458
Interleukin-6 (IL-6) 28, 55, 56, 75, 76, 78, 96-101, 108, 144, 145, 209-212, 321, 373, 389, 390
Interleukin-7 (IL-7) 28, 121, 208, 209, 214, 217
Interleukin-8 (IL-8) 28, 390
Interleukin-10 (IL-10) 108, 115-117, 135, 136, 142-145, 148-152, 165, 209, 214, 390, 449, 450
Interleukin-11 (IL-11) 28, 55, 56, 96-99
Interleukin-13 (IL-13) 390, 458-460
Interleukin-17 (IL-17) 390
Internal ribosome entry site (IRES) 94, 95, 98
Intracellular adhesion molecule-1 (ICAM-1) 26, 134, 390
Intracellular cytokine staining (ICS) assay 197, 198
Intraepithelial lymphocytes (IEL) 208, 448, 449
Intrathymic T progenitor (ITTP) 206
Intravenous immunoglobulin (IVIG) 264, 271, 279, 344, 405, 419, 421, 423, 426, 430, 431, 435, 437, 439, 470
Invariant chain (Ii) 157, 158, 177
Islet cell transplantation 442-444
Isoniazid 173, 175, 342
Isoproterenol 26, 43-46

J

Juvenile chronic arthritis (JCA) 224, 241, 242, 263, 264, 267, 378-380, 382, 469
Juvenile dermatomyositis (JDM) 437, 439
Juvenile idiopathic arthritis (JIA) 159, 161, 254, 259, 378-380, 382, 384, 385

K

Kaleidoscope of autoimmunity 106, 110
Keratinocytes 57, 67, 68, 75, 87, 88, 109, 145, 169, 389, 434, 435
Kidney 8, 9, 49, 59, 64, 67, 68, 134, 150, 206, 241, 329, 341, 344, 349, 353, 391, 398, 411-413, 417, 444, 472

L

Lamina propria lymphocytes (LPL) 448, 449
Laminin 7, 79, 87-89, 390, 435
Langerhans cells (LC) 114, 139-141, 146
Langerin 140
LE cell 328, 329
Leflunomide 358, 360-362, 365, 371, 372, 374, 477
Legionella pneumoniae 382
Leptin 26
Leukemia inhibitory factor (LIF) 6, 7, 28, 79, 80, 88, 89, 96-99
Leukocyte adhesion deficiency (LAD) 99, 101
Leukotriene 265, 459
Limiting dilution assay (LDA) 196, 197
Linkage of disequilibrium (LOD) score 164
Linkage studies 165, 166, 169
Lipid rafts 134, 136, 324
Listeria monocytogenes 264
Liver 3, 5, 18-21, 24, 28, 35, 49, 59-64, 66, 68, 69, 73-75, 79, 81, 82, 110, 115-117, 120, 122, 127, 128, 141, 142, 147, 150, 157, 206, 209, 233, 246, 248, 259, 264, 312, 315, 316, 336, 384, 450, 451
　stem cells 19, 59, 61, 74, 82
Long term culture initiating cell (LTCIC) assay 74
Lpr mice 31, 127, 128, 166, 169, 225, 238, 246, 248-250
Lung 35, 57, 59, 60, 64-66, 68, 79, 145, 227, 254, 256, 259, 263, 315, 316, 331, 336, 349, 353, 372, 373, 389, 391, 398, 406, 411-413, 438, 439, 457, 458, 460-465, 472, 476
　transplantation 458, 460, 463-465

Lupus activity index (LAI) 330, 332-335
Lupus nephritis 107, 166, 245, 247, 248, 339-345, 355
Lyme arthritis 185, 186, 188, 190
Lymphocyte function-associated antigen-1 (LFA-1) 116, 134, 190, 321, 325
Lymphoma 3, 8, 101, 227, 263, 349, 369, 374, 416, 439, 453, 469, 475
Lymphopoiesis 100, 101, 141, 146, 210, 212, 217, 449

M

6-mercaptopurine 451, 454
M-cadherin 64
Macrocytic anemia 469
Macrophage activation syndrome (MAS) 259, 263, 264, 382-386
Magnetic resonance imaging (MRI) 227, 277-279, 281, 284-288, 290, 294-299, 303-308, 312, 314-318, 438
Major histocompatibility complex (MHC) 28, 40, 41, 64, 82, 108, 109, 113-115, 119, 120, 135, 136, 139, 141, 143-145, 147, 149, 155-161, 166, 167, 176-180, 182-184, 194, 195, 197, 198, 206, 207, 216, 225, 238, 245, 246, 248, 250, 277, 279, 321, 429, 437, 444, 448, 458, 459, 475
   class I 82, 115, 136, 156-158, 161, 206
   class II 28, 108, 113, 135, 139, 141, 144, 145, 147, 149, 156-159, 176, 177, 183, 207, 429, 448, 459
Mapping strategies 168
Marrow stromal cells 24, 26-28, 35, 39, 40, 49, 64
MEF2-A 41, 42
MEF2-D 41, 42
Megakaryocyte growth and development factor (MGDF) 75-78
Melphalan 241, 257-259, 304, 312, 314, 349, 401, 402, 413, 416, 417, 426, 444, 475, 476
Membranous lupus nephritis 341, 343
Memory T cell 116, 169, 182, 183, 185, 190, 195-197, 210, 212, 214, 227, 241, 281, 353, 399
Mesangial proliferation 340, 343
Mesenchymal stem cell (MSC) 27, 35-37, 39, 40, 47, 61, 64-67, 69, 79, 81, 250
Metamoirosis 60-66, 69
Methicillin-resistant *S. aureus* (MRSA) 269-271
Methotrexate 242, 255, 263, 265, 303, 332, 335, 358-365, 367, 369-372, 374, 375, 378, 384, 398, 426, 435, 437, 439, 451, 453, 454, 462, 470, 477
Metronidazole 451
MEX-SLEDAI 332
Microarray analyses 168
Microchimerism 62, 109, 110, 375, 388, 391
Microsatellite markers 164, 166, 167
Microscopic polyangiitis 411, 412, 417, 469
Minimum record of disability (MRD) 290
Minocycline 173-175, 435
Mitoxantrone 277, 279, 303, 308, 318
Mixed chimerism 8, 32, 115, 225, 226, 259, 375, 444, 475, 477
Mixed lymphocyte reaction/response (MLR) 28, 132, 135
MoAb 228
Modified Rodnan skin score 399, 402
Molecular mimicry 108, 109, 183-190, 223, 278, 311, 442
Monoclonal antibody 79, 96-98, 108, 201, 226, 228, 242, 265, 298, 405, 407, 460, 461
Monoclonal antibody to immobilized platelet antigen (MAIPA) 407
Mosaic of autoimmunity 106, 107, 110, 223
Mouse embryonic fibroblast (MEF) 6, 7, 79, 88, 89
MRC muscle strength grading 426, 439
MRNA Analysis 198, 202

Multi-drug resistant (MDR-1) gene 55, 96, 100
Multiple myeloma 8, 99, 413, 417, 475
Multiple sclerosis (MS) 9, 109, 114, 116, 145, 147, 152, 159, 160, 165, 182, 183, 185-188, 215, 216, 217, 224, 227, 238-243, 253-259, 262, 267, 269, 270, 277-281, 284-288, 290-292, 294-299, 303, 305, 307-309, 311, 312, 314-318, 378, 463, 474, 477
Multipotent adult progenitor cells (MAPCs) 35-37, 62, 64, 67, 79, 80, 89
Murine cytomegalovirus (MCMV) 189
Murine embryonal stem cell virus (MESV) 98
Muscarinic receptor 45, 46
Myasthenia gravis (MG) 33, 107, 114, 159, 160, 224, 226, 237, 239, 241, 257, 265, 330, 429-432, 437, 469
Mycobacterium tuberculosis 147, 151, 241, 264, 271, 368, 442, 461
Mycophenolate mofetil (MMF) 260, 339, 344, 350, 355, 404, 435, 439, 451, 460, 476
Mycoplasma arthritidis superantigen 109
Myelin 14, 35, 109, 111, 152, 160, 183, 185, 186, 188, 189, 217, 243, 278-281, 284-287, 303, 308, 309, 311, 419-421, 463
Myelin associated glycoprotein (MAG) 14, 285
Myelin basic protein (MBP) 14, 109, 152, 160, 183, 185, 186, 189, 243, 279, 280, 295, 309, 420
Myelin oligodendroglia glycoprotein (MOG) 189, 279, 280
Myelodysplastic syndrome (MDS) 32, 228, 234, 413, 417, 469
Myeloid related protein (MRP) 384-386
Myeloid-lymphoid initiating cell (ML-IC) assay 74
Myf-5 64, 65
Myocarditis 145, 147, 185, 186, 188, 189, 331, 378, 438
MyoD1 28
Myosin light chain (MLC) 41
Myositis-specific autoantibodies (MSA) 437, 439

N

N-CAM (CD56) 26
Narcolepsy 159, 160
Neisseria brasiliensis 264
Neisseria gonorrhea 264
Neisseria meningitides 264
Nerve growth factor receptor (NGFR) 102, 109
Neural stem cells
   see neuronal stem cells
Neuregulin-1 (Nrg-1) 13
Neurodegeneration 286-288
Neuromuscular junction 430
Neuronal stem cells (NSCs) 3, 5, 11-16, 27, 28, 35, 36, 39, 59, 61, 63, 68, 80, 259, 317, 318
Neurons 5, 7, 11-15, 35, 39, 59, 60, 63, 64, 68, 69, 80, 92, 135, 185, 188, 278, 284-288, 298, 308, 317, 318, 429, 430
NK T cell 115, 135
Nocardia asteroides 264
Non-Hodgkin's lymphoma 228, 369, 453
Nonobese diabetic (NOD) 223-226, 238, 245-247, 442, 443, 450
   mice 189, 238, 245, 246
Nonobese diabetic severe combined immunodeficiency (NOD-SCID mice) 93, 95
Nonsteroidal anti-inflammatory drugs (NSAIDs) 175, 180, 358, 365, 378, 380, 469
Nuclear factor-κB (NF-κB) 124, 125, 207, 323, 325, 450, 458

O

OCT-4 89
Oligodendrocyte 9, 11-15, 39, 80, 185, 186, 188, 189, 259, 278, 280, 284, 285, 317, 318
Osserman scoring system 430, 431
Osteoblast 24, 27, 29, 35-37, 39, 40, 49, 79, 225
Osteopontin 27, 89, 169
Oval cell 18-21, 61, 62, 81, 82
Oval cell marker-6 (OV-6) 19, 21

P

Paced Auditory Serial Addition Test—3-second version (PASAT-3) 293, 294
Pancreas/islet cell transplantation 442-444
Paraneoplastic pemphigus 235, 434
Paraproteinemic 223
Parkinson's disease 2, 7, 8, 287
Pathogen-associated molecular patterns (PAMPS) 132
Pattern recognition receptors (PRR) 132, 143
Pax-7 64
Pemphigus 33, 107, 159, 160, 223, 235, 254, 258, 259, 434-436
Pemphigus foliaceus (PF) 159, 160, 434-436
Pemphigus vulgaris (PV) 159, 160, 223, 235, 246, 254, 259, 434, 435
Perforin 115, 149, 197, 198, 384, 385
Peripheral blood progenitor cells (PBPC) 73, 76-78, 304, 380, 399-402
Peripheral tolerance 114-117, 132-136, 143, 145, 146, 177, 182, 185, 188
Pernicious anemia 107, 469
Peyer's patch 135, 139, 146, 209, 448, 449
Phenylephrine 43-46
Phospholipase A_2 150
Photosensitivity 328, 330, 335
Plasmacytoid monocytes 141, 146
Plasmapheresis 265, 342, 344, 399, 404, 421, 426, 430, 435, 439
Plasticity 3, 9, 11, 13-15, 24, 36, 57, 59-61, 63-66, 68, 69, 81, 139, 228
Platelet transfusion 235, 405-407, 471
Platelet-activating factor 459
Platelet-derived growth factor (PDGF) 14, 52, 78, 79, 88, 141
Pneumocystis carinii 266, 268, 272
Pol gene 93
Poly (ADP-ribose) polymerase (PARP) 165, 321
Polyarteritis nodosa (PAN) 258, 411, 412, 416, 417, 469
Polymerase chain reaction (PCR) 24, 40, 43, 46, 62, 78, 98, 99, 109, 122, 164, 168, 178, 180, 186, 195, 198-200, 202, 208, 270, 271, 450
Polymyositis (PM) 107, 255, 258, 259, 264, 329, 437-440
Porcine microvascular endothelial cells (PMEC) 19
Post transplant lymphoproliferative disease/disorder (PTLD) 258, 267-269, 314, 316, 318, 402, 444
Pre-ligand assembling domain (PLAD) 123
Primary biliary cirrhosis 110, 223, 392, 437, 469
Primary systemic vasculitis (PSV) 411, 413, 414, 416, 417
Primary vitiligo 437
Procabazine 405
Procainamide 173-180, 342
Procainamide-hydroxylamine (PAHA) 177-180
Propranolol 43-46
Prostaglandin E_2 (PGE$_2$) 144, 459
Prostanoids (PGD$_2$) 459
Protein kinase A (PKA) 321, 323-325
Protein kinase C (PKC) 207, 211, 323, 324
Protein kinase R (PKR) 324
Proteinuria 108, 145, 169, 177, 248, 250, 328-333, 336, 339-345, 349, 350, 353, 354, 413-415, 476
Proteolipid protein (PLP) 14, 152, 183, 185, 188, 239, 279, 280, 285, 309-311
Pseudomonas aeruginosa 264
Psoriasis 160, 226, 237, 257, 265, 435, 469
Pulmonary artery hypertension (PAH) 255, 259, 260, 476
Pulmonary embolus 342, 353, 423, 424
Purkinje cells 64, 68

Q

Quantitative Myasthenia Gravis Score (QMGS) 431, 432

R

Radiation 1, 57, 61, 212, 226, 241, 248, 250, 256, 259, 265, 317, 318, 349, 475
RAM-1 95
Rapamycin 134, 136, 460
Raynaud's phenomenon 328, 329, 331, 391, 392, 398, 438
Rebif 298, 299
Receptor for hyaluronate mediated motility (RHAMM) 88
Reiter's syndrome 161
Relapsing polychondritis 468-472
Renal
  biopsy 339-343
  failure 9, 107, 145, 248, 256, 264, 340, 342, 343, 353, 398, 411, 413, 415, 417, 442
Reovirus 185, 187
Responder index for lupus erythematosus (RIFLE) 335
Retinoic acid-related orphan receptor (RORγ) 120
Retronectin™ 95, 98
Retroviral vector 80, 93-95, 97, 98, 101, 102
Retrovirus 12, 66, 92-97, 99, 101, 146, 186, 388
Reverse transcriptase 43, 46, 62, 93
Rheumatoid arthritis (RA) 33, 107, 108, 114, 117, 128, 145, 147, 149, 157-160, 165, 186, 207, 215, 223-228, 237, 238, 240-243, 246, 248, 253-259, 263, 265, 270, 280, 318, 328, 358-365, 367-375, 378-382, 392, 393, 408, 438, 439, 443, 448, 453, 457, 463, 469, 474, 476, 477
Rheumatoid factor (RF) 161, 176, 226, 259, 358, 359, 370, 372, 373, 378, 379, 469
Rituximab 228, 354, 374, 405, 439
RNA polymerase 391, 392, 398
RNA Polymerase (RNAP) antibodies 392
Rubella virus 185, 187

S

Schistosomes 142
Scl-70 388, 390-392, 398, 402
Scleroderma 173, 176, 215, 224, 228, 250, 254-256, 258-260, 328, 329, 378, 388, 391, 392, 398, 402, 437, 439, 453, 476, 477
Scripps neurological rating scale (SNRS) 291, 293, 303, 306
Sebaceous gland 67, 68
SELENA-SLEDAI 332
Semliki forest virus (SFV) 188, 189
Severe aplastic anemia (SAA) 32, 33, 226, 227, 232-235, 242, 368, 369

Severe autoimmune disease (SAD) 5, 32, 223, 226-228, 234, 256, 262, 263, 266-270, 347, 352, 378, 406, 411, 413, 434, 457, 477
Severe combined immunodeficient (SCID) 74, 78, 93, 95, 212, 214, 223, 225, 348, 449, 450, 460
Silencer of death domain (SODD) 125, 127
Silencing 94, 98
Single nucleotide polymorphism (SNP) 165, 168, 278
Skeletal muscle 15, 24, 26, 28, 39-41, 46, 57, 60, 63-65, 80, 81, 89, 398
Skin 5, 24, 35, 36, 57, 59, 67, 68, 81, 89, 92, 107, 110, 114, 132, 140, 145-147, 149, 151, 160, 161, 174, 177, 186, 242, 246, 248, 254, 256, 262, 263, 268, 269, 271, 317, 328-331, 336, 388-391, 398-400, 402, 404, 412, 414, 415, 434-436, 438-440, 450, 454, 458, 461, 469, 476
SLICC/ACR damage index 335, 337
Smooth-muscle-like cell (SMC) 47
Somatic hypermutation (SHM) 200, 209, 210, 212
Somatostatin 132
Spinal cord atrophy 286, 299
Spleen focus-forming provirus integration B (Spi-B) 141, 146
Src family kinase 134
Stage-specific embryonal antigens (SSEA) 6, 7
    SSEA-1 7
    SSEA-3 7
    SSEA-4 7
Staphlococcus aureus 263, 264, 270, 436
Steel factor (SF) 55-57
Stem cell factor (SCF) 26, 28, 35, 64, 75-78, 81, 96-101, 141, 208, 254
Stem cell transplantation 5, 28, 74, 92, 93, 117, 136, 150, 195, 206, 213, 214, 223, 224, 226, 232, 234, 235, 237, 242, 243, 245, 253, 257, 262, 264, 266, 267, 269, 272, 277, 303, 309, 313, 314, 318, 320, 339, 344, 345, 347, 349, 352, 358, 365, 367, 369, 371, 373-376, 378, 379, 385, 386, 399-401, 411, 416, 419, 424, 429, 430, 434, 435, 437, 442, 443, 448, 451, 453, 457, 464, 468, 470, 471, 474, 475, 477
Stenotrophomonas maltophilia 264
Steroids 108, 120, 235, 264, 303, 305-307, 312, 335, 339, 342-344, 369-372, 378, 380, 383, 386, 404, 406, 413, 434, 435, 438, 439, 462, 470
Still's disease 258, 259, 439
Streptococcus pneumoniae 263, 264
STRO-1 26
Stromal cell-derived factor-1 α (SDF1α) 20
Stromal keratitis 185, 187-189
Sulfasalazine 173, 175, 358, 359, 361, 362, 364, 365, 371, 451
Superantigen 108, 109, 123, 183
Suppressor cell 133, 135, 136, 152, 309, 450
Susceptibility loci 165-168, 182, 278, 321
Syngeneic HSCT 311, 367, 371, 429, 462, 474
Systemic inflammatory response syndrome (SIRS) 470
Systemic lupus erythematosus (SLE) 31, 33, 106-109, 114, 157, 159, 161, 164-169, 173-175, 177, 180, 215-217, 223-225, 227, 228, 235, 237, 238, 240, 242, 247, 248, 253-258, 263-266, 269, 270, 272, 318, 320, 321, 323-325, 328-331, 332-337, 339-342, 344, 347-352, 354, 355, 364, 378, 385, 392, 393, 432, 437-439, 448, 469, 474, 477
Systemic lupus erythematosus disease activity index (SLEDAI) 330, 332-335, 337, 352, 353
Systemic sclerosis (SSc) 107, 110, 227, 228, 242, 253, 255, 256, 258, 259, 263, 265, 388-393, 398-402, 413, 469

T

T cell receptor (TCR) 113, 114, 116, 119-123, 126-128, 132-136, 141, 143, 144, 148-152, 156, 160, 161, 165, 167, 169, 177-180, 182-185, 189, 190, 194, 195, 197-200, 206-213, 215-217, 226, 227, 233, 253, 278, 279, 309, 320, 321, 323-325, 371, 389, 408, 448, 459
T cell receptor excision circle (TREC) 178-180, 195, 197, 207, 208, 210, 212, 213, 216, 217, 227, 373, 432
T cell receptor spectratyping 198
Taqman® 199
TCR rearrangement excision circles 178, 195
Telomerase 89
Tetracycline 80, 175, 435
Th3 cells 115, 135, 152
Theiler's murine encephalomyelitis virus (TMEV) 109, 183, 187, 188, 239, 309, 311, 313
Thrombocytopenia 107, 174, 177, 235, 247, 248, 255, 328, 330, 333, 335, 342, 404-408
Thrombopoietin (TPO) 57, 77, 78, 96, 254
Thromboxane A_2 459
Thy-1.1 19
Thymocyte 36, 108, 119, 120, 143, 145, 158, 177-180, 195, 196, 206-208, 214, 226, 233-235, 254, 256-258, 263, 266, 280, 312, 314, 316, 344, 349, 352, 367, 371, 380, 389, 399-402, 439, 454, 470
Thymus 108, 113-115, 119, 120, 132, 133, 135, 136, 143, 146, 147, 151, 152, 158, 177-180, 182, 185, 195, 206-210, 212, 214-217, 227, 250, 355, 430, 432, 444, 448, 474
Thymus epithelial cells (TEC) 120, 179
Tolerance 5, 8, 9, 92, 106, 108, 109, 113-117, 119, 120, 132-136, 139, 143-148, 150-152, 168, 169, 174, 177, 178, 180, 182, 185, 187, 188, 214, 223, 227, 234, 242, 243, 246-248, 253, 277, 279, 280, 324, 347, 353, 360, 371, 373-375, 405, 420, 442, 444, 448, 449, 451, 472, 475, 477
Toll receptors 132
Toll-like receptor (TLR) 143, 144, 146, 147
Topoisomerase I 388, 390-392, 398, 400
Topoisomerase II 464
Total body irradiation (TBI) 134, 212, 214, 235, 238-243, 253, 256-259, 263, 265, 309-312, 314-318, 349, 351, 352, 368, 369, 371-373, 380, 381, 383, 399-402, 417, 429, 475, 476
Toxic oil syndrome 173, 174
Toxoplasma gondii 264
Tr1 cells 135, 145, 148-152
TRA-1-60 6, 7
TRA-1-81 6, 7
Transcription 6 459
Transcription factor 6, 11, 13, 27, 41, 42, 63, 66, 75, 81, 89, 95, 98, 102, 141, 143, 146, 207, 321, 323, 325, 458-460
Transforming growth factor (TGF) 209, 389, 390, 449, 464
Transforming growth factor-α (TGF-α) 13, 18
Transforming growth factor-β (TGF-β) 18, 52, 88, 96, 115-117, 135, 136, 148-152, 390
Transit amplifying (TA) cells 11, 67, 68
Transitional cell cancer 343
Transporters associated with antigen processing (TAP) 156, 157
Triple negative (TN) cell 206
Trypanosoma cruzi 185, 186, 188, 269, 270
Trypsin2 132
Tuberculosis 147, 151, 241, 264, 265, 269, 271, 364, 368, 442, 461

Tumor necrosis factor (TNF) 108, 109, 117, 119-123, 125-128, 133, 144, 155, 169, 186, 198, 211, 216, 228, 232, 265, 362-364, 367, 371, 372, 374, 375, 378, 380, 384, 389, 390, 421, 450, 451, 477
    family 169
Tumor necrosis factor α (TNF-α) 28, 49, 124, 127, 141, 144-147, 167, 211, 212, 215, 216, 265, 362, 363, 367, 378, 384, 390, 421, 451
Tumor necrosis factor receptor (TNFR) 119-121, 123, 126-128, 265, 362
Tumor necrosis factor receptor (TNFR)-associated death domain (TRADD) 124, 125, 127
Tumor necrosis factor-related apoptosis-inducing ligand (TRAIL) 123, 124, 128
Type I collagen 24, 25, 27, 87, 89

U

Ulcerative colitis 160, 237, 449, 450, 469
Umbilical cord blood (UCB) 8, 28, 74, 92

V

Valganciclovir 270
Vancomycin-resistant enterococci (VRE) 269, 270
Variable, joining and diversity (VJD) genes 182, 210
Varicella zoster virus (VZV) 268-271, 380, 382
Vascular endothelial growth factor (VEGF) 6, 49, 51, 52, 88, 89
Vasculitis damage index (VDI) 412, 413, 415
Vasculogenesis 47, 49, 50
Vector producing cells (VPC) 93-95
Vesicular stomatitis virus glycoprotein (VSV-G) 95, 98
Vinca alkaloids 99, 405
Visual evoked potentials (VEPs) 295, 306
VLA2 28
VLA4 28, 93, 95
VLA5 28, 95

W

Wegener's disease 228
Wegener's granulomatosis (WG) 259, 411-413, 415, 417, 439, 469

X

Xenobiotic 173, 174, 176, 177, 179, 238

Y

Y-chromosome 61-63, 65-69

Chapter 2, Figure 2A, page 8. Undifferentiated and differentiated human ES cells. Undifferentiated human embryonic stem (ES) cells maintained on irradiated mouse fibroblast feeder cells. Colony of cells with uniform morphology and large nuclei characteristic of human ES cells, original magnification 200X.